D1684658

ENCYCLOPEDIA OF PHYSICS

CHIEF EDITOR

S. FLÜGGE

VOLUME XXV/2b

LIGHT AND MATTER Ib

EDITOR

L. GENZEL

WITH 34 FIGURES

SPRINGER-VERLAG
BERLIN · HEIDELBERG · NEW YORK
1974

HANDBUCH DER PHYSIK

HERAUSGEGEBEN VON

S. FLÜGGE

BAND XXV/2b

LICHT UND MATERIE Ib

BANDHERAUSGEBER

L. GENZEL

MIT 34 FIGUREN

SPRINGER-VERLAG
BERLIN · HEIDELBERG · NEW YORK
1974

Professor Dr. Siegfried Flügge

Physikalisches Institut der Universität Freiburg i. Br.

Professor Dr. Ludwig Genzel

Max-Planck-Institut für Festkörperforschung, Stuttgart

ISBN 3-540-06638-1 Springer-Verlag Berlin Heidelberg New York
ISBN 0-387-06638-1 Springer-Verlag New York Heidelberg Berlin

Das Werk ist urheberrechtlich geschützt. Die dadurch begründeten Rechte, insbesondere die der Übersetzung, des Nachdruckes, der Entnahme von Abbildungen, der Funksendung, der Wiedergabe auf photomechanischem oder ähnlichem Wege und der Speicherung in Datenverarbeitungsanlagen bleiben, auch bei nur auszugsweiser Verwertung, vorbehalten. Bei Vervielfältigungen für gewerbliche Zwecke ist gemäß § 54 UrhG eine Vergütung an den Verlag zu zahlen, deren Höhe mit dem Verlag zu vereinbaren ist. © by Springer-Verlag Berlin Heidelberg 1974. Library of Congress Catalog Card Number A 56-2942. Printed in Germany. Monophotosatz, Offsetdruck und Bindearbeiten: Universitätsdruckerei H. Stürtz AG, Würzburg.

Die Wiedergabe von Gebrauchsnamen, Handelsnamen, Warenbezeichnungen usw. in diesem Werk berechtigt auch ohne besondere Kennzeichnung nicht zu der Annahme, daß solche Namen im Sinne der Warenzeichen- und Markenschutz-Gesetzgebung als frei zu betrachten wären und daher von jedermann benutzt werden dürften.

This work is dedicated
to the memory of my esteemed father

Max S. Birman
(1901–1970)

J.L.B.

Preface

The present article was written in order to provide research workers with a comprehensive and thorough treatment of the application of Group Theory to a branch of Solid State (Condensed Matter) Optics namely the theory of infra-red and Raman Lattice Processes. The same group theory methods are applicable in other branches of Solid State Physics, with appropriate modifications, so that a thorough knowledge can be valuable in analyses of electronic properties of crystals, magnetic properties, and of other elementary excitations.

The reader is presumed to possess a modest knowledge of elementary group theory and lattice dynamics such as can be obtained from standard introductory textbooks [*1–4*] but the present work is self-contained and ample references to more elementary literature are given in case the reader wishes to refresh understanding of some matters.

The reader's attention is now directed to the introductory survey in Sects. 1 and 2 which give an overview of the contents of the article. A more intrepid reader may also wish to begin by reading the concluding statements in Part P (pp. 460–465) in order to gain an additional perspective on the article.

A bird's eye view is that Sects. 1–65 deal with the structure, irreducible representations, reduction coefficients and Clebsch-Gordan coefficients of crystal space groups. Sects. 66–110 develop the intertwined themes of classical lattice dynamics and crystal symmetry including corepresentation theory. Sects. 111–124 present the quantum theory of lattice dynamics and of infra-red and Raman processes, and demonstrate the general utility of group theory analysis. Finally Sects. 125–154 give specific and detailed application of the previous general theory to optical infra-red and Raman lattice processes in insulators with diamond and rocksalt structure: space groups O_h^7 and O_h^5. Illustrations are drawn from perfect and imperfect crystals of both types.

Joseph L. Birman

Contents

Theory of Crystal Space Groups and Infra-Red and Raman Lattice Processes of Insulating Crystals

By Dr. Joseph L. Birman, The Henry Semat Professor of Physics, The City College of the City University of New York, New York, NY 10031 (USA)*. (With 34 Figures)

A. Scope and plan of the article 1

1. General survey 1
2. Plan of the article: An overview 3

B. The crystal space group 6

3. Crystal symmetry – Introduction 6
4. The translation subgroup of a crystal 9
   α) Translation operators $\{\varepsilon|\boldsymbol{R}_L\}$ 9
   β) The translation group $\mathfrak{T}$ 11
   γ) Structure of $\mathfrak{T}$ 11
   δ) Born-Karman boundary conditions 12
   ε) A property of the $\{\varepsilon|\boldsymbol{t}\}$ 13
5. Rotational symmetry elements: The crystal point group 13
   α) Rotational operators $\{\varphi|0\}$ 14
   β) The point group $\mathfrak{P}$ 14
6. General symmetry element in a crystal: Space group $\mathfrak{G}$ 15
   α) The operator $\{\varphi|\boldsymbol{t}(\varphi)\}$ 15
   β) Group property of the set $\{\varphi|\boldsymbol{t}(\varphi)\}$ 16
   γ) Compatibility of rotation and translation 16
   δ) The operator $\{\varphi|\boldsymbol{t}\}$ in non-Cartesian axes 17
   ε) Order of the space group $\mathfrak{G}$ 18
   ζ) Normality of translation subgroup $\mathfrak{T}$ 18
   η) Factor group 19
   θ) Site symmetry 19
7. The space group $\mathfrak{G}$ as a central extension of $\mathfrak{T}$ by $\mathfrak{P}$ 19
8. Symmorphic space groups 22
9. Non-symmorphic space groups 22
10. Some subgroups of a space group 23

C. Irreducible representations and vector spaces for finite groups 25

11. Introduction 25
12. Transformation operators on functions 26
13. Group of transformation operators on functions 27
14. Functions and representations 28
15. Irreducible representations and spaces 29
16. Idempotent transformation operators 31

---

* Formerly Professor of Physics, New York University, New York, NY 10003 (USA).

17. Direct products . . . 32
α) Direct products of representations . . . 32
β) Reduction coefficients . . . 33
γ) Irreducible representations of direct product groups . . . 33
18. Clebsch-Gordan coefficients . . . 34

D. Irreducible representations of the crystal translation group $\mathfrak{T}$ . . . 39
19. Introduction . . . 39
20. Irreducible representations of $\mathfrak{T}$ . . . 40
21. The reciprocal lattice . . . 40
22. Irreducible representations of $\mathfrak{T}=\mathfrak{T}_1 \otimes \mathfrak{T}_2 \otimes \mathfrak{T}_3$ . . . 41
23. Wave vector: First Brillouin zone . . . 42
24. Completeness and orthonormality for $D^{(\mathbf{k})}$ . . . 44
25. Irreducible vector spaces for $\mathfrak{T}$: Bloch vectors . . . 46
26. Direct products in $\mathfrak{T}$ . . . 47

E. Irreducible representations and vector spaces of space groups . . . 47
27. Introduction . . . 47
28. Irreducible representation $D^{(*\mathbf{k})(m)}$ of $\mathfrak{G}$ . . . 48
29. Representation of $\mathfrak{T}$ subduced by $D^{(*\mathbf{k})(m)}$ of $\mathfrak{G}$ . . . 49
30. Transformation of Bloch vectors by rotation operators . . . 50
31. Conjugate representations of $\mathfrak{T}$ . . . 51
32. Characterization of the subduced representation . . . 52
33. Block structure of $D^{(*\mathbf{k})(m)}$ of $\mathfrak{G}$ . . . 53
34. Group of the canonical $\mathbf{k}$: $\mathfrak{G}(\mathbf{k})$ . . . 55
35. Irreducibility of the acceptable representations $D^{(\mathbf{k}_1)(m)}$ of $\mathfrak{G}(\mathbf{k}_1)$ . . . 56
36. $D^{(*\mathbf{k})(m)}$ of $\mathfrak{G}$ induced from $D^{(\mathbf{k}_1)(m)}$ of $\mathfrak{G}(\mathbf{k}_1)$ . . . 57
37. Characters of $D^{(*\mathbf{k})(m)}$ of $\mathfrak{G}$; induced characters . . . 60
38. Allowable irreducible $D^{(\mathbf{k})(m)}$: General star with $\mathfrak{G}(\mathbf{k})=\mathfrak{T}$ . . . 61
39. Allowable irreducible $D^{(\mathbf{k})(m)}$: Special star: Little group technique . . . 62
40. Non-allowable irreducible $D^{(\mathbf{k})(\mu)}$: Little group technique . . . 64
41. Allowable irreducible $D^{(\mathbf{k})(m)}$ as ray representations . . . 66
42. Ray representations of $\mathfrak{P}(\mathbf{k})$: The covering group $\mathfrak{P}^\star(\mathbf{k})$ . . . 68
43. Gauge transformations of ray representations . . . 70
44. Relationship between little group and ray representation methods . . . 71
45. Full $D^{(*\mathbf{k})(m)}$ for symmorphic groups: Illustration . . . 74
46. Full $D^{(*\mathbf{k})(m)}$ for non-symmorphic groups . . . 76
47. Complete set of all $D^{(*\mathbf{k})(m)}$ for a space group . . . 77
48. Verification of completeness of $D^{(*\mathbf{k})(m)}$ . . . 77
49. Verification of orthonormality relations for $D^{(*\mathbf{k})(m)}$ . . . 79
50. Induction of $D^{(\mathbf{k})(m)}$ from sub-space groups . . . 81
51. Compatibility relations for $D^{(*\mathbf{k})(m)}$ and subduction . . . 85

F. Reduction coefficients for space groups: Full group methods . . . 86
52. Introduction . . . 86
53. Direct product $D^{(*\mathbf{k})(m)} \otimes D^{(*\mathbf{k}')(m')}$ . . . 87
54. Symmetrized powers $[D^{(*\mathbf{k})(m)}]_{(p)}$ . . . 88
α) Ordinary Kronecker powers . . . 88
β) Symmetrized Kronecker powers . . . 89

55. Definition of reduction coefficients . . . 91
56. Wave vector selection rules . . . 92
α) Star reduction coefficients for the ordinary product . . . 92
β) Star reduction coefficients for the symmetrized product . . . 94
57. Determination of reduction coefficients: Method of linear algebraic equations . . 95
58. Determination of reduction coefficients: Method of the reduction group . . . 97
59. Determination of reduction coefficients: Use of basis functions . . . 99
60. Theory of Clebsch-Gordan coefficients for space groups . . . 101

G. Reduction coefficients for space groups: Subgroup methods . . . 105

61. Introduction . . . 105
62. Complete subgroup character system . . . 105
63. Subgroup reduction coefficients . . . 107
64. Comparison of full group and subgroup methods . . . 109
65. Reduction coefficients: A little group technique . . . 112

H. Space group theory and classical lattice dynamics . . . 113

66. Introduction . . . 113
67. Equations of motion in the harmonic approximation . . . 114
68. Translation symmetry and particle displacements . . . 117
69. Translation symmetry and force matrix . . . 118
70. General symmetry and particle displacements . . . 119
71. General symmetry and force matrix . . . 123
72. Solution of the equations of motion: Eigenvectors $[\boldsymbol{e}_j]$ . . . 125
73. Real normal coordinates q_j . . . 129
74. Crystal symmetry and the eigenvectors $[\boldsymbol{e}_j]$ of $[\boldsymbol{D}]$ . . . 131
75. Necessary degeneracy of the eigenvectors $[\boldsymbol{e}_j]$ . . . 132
76. Crystal symmetry and the transformation of normal coordinates q_j . . . 133
77. Fourier transformations . . . 136
78. Fourier transformed displacements and force matrix: The dynamical matrix $[D(\boldsymbol{k})]$ 136
79. Eigenvectors of the dynamical matrix $[\boldsymbol{D}(\boldsymbol{k})]$ . . . 138
80. Complex normal coordinates . . . 141
81. Crystal symmetry and the dynamical matrix $[\boldsymbol{D}(\boldsymbol{k})]$ and its eigenvectors . . . 142
82. Eigenvectors of $[\boldsymbol{D}(\boldsymbol{k})]$ as bases for representation $D^{(\boldsymbol{k})(e)}$ of $\mathfrak{G}(\boldsymbol{k})$ . . . 146
83. Eigenvectors of $[\boldsymbol{D}(\boldsymbol{k})]$ as bases for representations $D^{(\boldsymbol{k})(j)}$ of $\mathfrak{G}(\boldsymbol{k})$ . . . 150
84. Equivalence of $D^{(\boldsymbol{k})(e)}$ and $D^{(\boldsymbol{k})(j)}$ . . . 151
85. Necessary degeneracy under $\mathfrak{G}(\boldsymbol{k})$ and the eigenvectors of $[\boldsymbol{D}(\boldsymbol{k})]$ . . . 152
86. Complex normal coordinates $Q\begin{pmatrix}\boldsymbol{k}\\ j_\mu\end{pmatrix}$ as bases for the representation $D^{(\boldsymbol{k})(j)}$ of $\mathfrak{G}(\boldsymbol{k})$ 155

I. Space-time symmetry and classical lattice dynamics . . . 159

87. Introduction . . . 159
88. The antilinear, antiunitary transformation operator K and time reversal . . . 159
89. The complete space-time symmetry group $\mathscr{G}$ . . . 164
90. Eigenvectors $\boldsymbol{e}\left(\left|\begin{matrix}\boldsymbol{k}\\ j_\lambda\end{matrix}\right.\right)$ and normal coordinates $Q\begin{pmatrix}\boldsymbol{k}\\ j_\lambda\end{pmatrix}$ as bases for representation of $\mathscr{G}$ 165
91. Necessary degeneracy under the full space-time crystal symmetry group $\mathscr{G}$ . . . 166
92. Test for reality of $D^{(^*\boldsymbol{k})(j)}$ of $\mathfrak{G}$ . . . 168
93. Simplification of the reality test of $D^{(^*\boldsymbol{k})(m)}$ . . . 173

94. Classification of $D^{(\star k)(m)}$ according to reality by use of a new test 176
95. Physically irreducible representations of $\mathfrak{G}$ as corepresentations of $\mathscr{G}$ 180
96. Structure of corepresentations of $\mathscr{G}$: The costar, co $\star \boldsymbol{k}$ 184
97. Corepresentations of $\mathscr{G}$: Class III costar 187
98. Corepresentations of $\mathscr{G}$: Class II costar and general theory 188
99. Corepresentations of $\mathscr{G}$: Class I costar 195
100. Acceptable irreducible corepresentations of $\mathscr{G}(\boldsymbol{k})$ as irreducible ray corepresentations 196
101. Complex normal coordinates as bases for irreducible corepresentations of $\mathscr{G}$ . . . 199
102. Eigenvectors of $D(\boldsymbol{k})$ as bases for irreducible corepresentations of $\mathscr{G}$ 202
103. Determination of actual normal mode symmetry in a crystal 202
104. Determination of eigenvectors $\boldsymbol{e}\left(\left|{\boldsymbol{k} \atop j}\right.\right)$ by symmetry: Factorization of the dynamical matrix 203

J. Applications of results on symmetry adapted eigenvectors in classical lattice dynamics 209

105. Introduction 209
106. Tensor calculus for lattice dynamics 210
α) Effect of unitary elements 210
β) Effect of antiunitary elements 216
107. Critical points 220
α) Representation theory for the "symmetry set" 220
β) Determination of potential critical points by point symmetry 228
108. Compatibility or connectivity theory for representations 230
109. Construction crystal invariants 231
α) The crystal Hamiltonian: Harmonic and anharmonic 235
β) Force constant coupling parameters 238
γ) Anharmonic terms in the potential 239
110. Construction of crystal covariants: Electric moment and polarizability 244

K. Space-time symmetry and quantum lattice dynamics 250

111. Introduction 250
112. The many-body electron-ion Hamiltonian 252
113. Born-Oppenheimer adiabatic approximation 253
114. Normal coordinates and quantization 259
115. Lattice eigenfunctions in harmonic adiabatic approximation 261
116. Symmetry of harmonic lattice eigenfunctions: Introduction 263
117. Transformations of products of Hermite polynomials: Symmetrized Kronecker product 264
118. Transformation of the lattice eigenfunction: Summary and generalities 269

L. Interaction of radiation and matter: Infra-red absorption and Raman scattering by phonons 271

119. Introduction 271
120. Infra-red absorption by phonons 272
α) Semi-classical radiation theory for ions and electrons 272
β) Transition rate 274
γ) Analysis of the transition matrix element for infra-red lattice absorption 276
δ) Symmetry of the matrix element for infra-red absorption 277
ε) One phonon and multiphonon processes 279
121. Raman scattering by phonons: Generalized Placzek theory 282
α) Hamiltonian 283
β) Transition rate for scattering 285

γ) Simplification of the scattering matrix elements for an insulator 287
δ) Symmetry of the Raman scattering matrix element 289
ε) One phonon and multiphonon processes 290
ζ) The $\boldsymbol{A} \cdot \boldsymbol{A}$ term in scattering Hamiltonian 293

122. A mutual exclusion selection rule for certain two phonon overtones in infra-red and Raman processes in crystals with an inversion 295

123. Polarization effects in infra-red and Raman lattice processes 298
α) Raman tensor and Clebsch-Gordan coefficients 299
β) Polarization effects in Raman scattering: The Raman tensor in a cubic crystal with inversion 301
γ) Polarization effects due to macroscopic electric field 305
i) Cubic crystals with inversion symmetry 305
ii) Cubic crystals without inversion 308
δ) Polarization effects and two phonon bound states 310
ε) Polarization in infra-red absorption due to anisotropy 312

124. Aspects of modern quantum theories of lattice Raman scattering and infra-red absorption 313
α) Many-body polarizability theory of Raman scattering 315
β) Many-body theory of infra-red absorption 322
γ) Group theory and the thermal phonon Green functions 324
δ) Microscopic theory of Raman scattering: Bloch picture 328
ε) Microscopic theory of Raman scattering: Exciton picture 333
ζ) Microscopic theory of Raman scattering: Polariton picture 336
η) Resonance Raman scattering and symmetry breaking 339

M. Group theory of diamond and rocksalt space groups 342

125. Introduction 342
126. Geometry of the rocksalt and diamond space groups 342
127. Irreducible representations in rocksalt 345
128. Some wave vector selection rules in rocksalt 352
129. Reduction of $^\star\boldsymbol{X}^{(4-)} \otimes {}^\star\boldsymbol{X}^{(5-)}$ in rocksalt 352
130. Reduction of $^\star\boldsymbol{L}^{(3-)} \otimes {}^\star\boldsymbol{L}^{(3+)}$ in rocksalt 358
131. Additional reduction coefficients in rocksalt 360
132. Irreducible representations $D^{(\boldsymbol{\Gamma})(m)}$, $D^{(^\star\boldsymbol{X})(m)}$, $D^{(^\star\boldsymbol{L})(m)}$ in diamond 362
133. Reduction coefficients 367
α) Products of $D^{(\boldsymbol{\Gamma})(m)}$, $D^{(^\star\boldsymbol{X})(m)}$, $D^{(^\star\boldsymbol{L})(m)}$ in diamond 367
β) Additional reduction coefficients in diamond 367
134. Clebsch-Gordan coefficients in diamond structure for $D^{(^\star\boldsymbol{X})(m)} \otimes D^{(^\star\boldsymbol{X})(m')}$ 367
135. Test of effect of time reversal symmetry in diamond and rocksalt structure 371
136. Connectivity and labelling of irreducible representations in diamond and rocksalt structures: Consequences for selection rules 373

N. Phonon symmetry, infra-red absorption and Raman scattering in diamond and rocksalt space groups 377

137. Introduction 377
138. Phonon symmetry in rocksalt and diamond 378
139. Compatibility and phonon symmetry in diamond and rocksalt 384
140. Critical points for phonons in diamond structure: Germanium, silicon and diamond 386
Diamond structure: Point $\boldsymbol{\Gamma}$ 388
Diamond structure: Point $\boldsymbol{X}_1$ 388
Point $\boldsymbol{L}_1$ 394
Point $\boldsymbol{W}_1$ 395

Line Σ . . . 395
Line Q . . . 396
141. Two phonon density of states and critical points in diamond structure . . . 397
142. Interpretation of lattice Raman and infra-red spectra in crystals of the diamond structure . . . 398
Diamond . . . 404
Silicon . . . 407
Germanium . . . 412
C, Si, Ge . . . 415
143. Symmetry set of critical points in rocksalt structure . . . 416
144. Two phonon density of states and critical points in rocksalt—NaCl . . . 420
145. Interpretation of lattice Raman and infra-red spectra in some rocksalt structure crystals . . . 422
NaCl . . . 422
NaF . . . 426
Other alkali halides . . . 428
146. Polarization effects in two phonon Raman scattering in rocksalt and diamond structures . . . 431
α) Rocksalt . . . 431
β) Diamond . . . 432

O. Some aspects of the optical properties of crystals with broken symmetry: Point imperfections and external stresses . . . 436
147. Introduction . . . 436
148. Symmetry group of the imperfect crystal with a point defect . . . 437
149. "Band" phonons in imperfect diamond and rocksalt crystals . . . 438
150. Local phonons in imperfect diamond and rocksalt crystals . . . 441
151. Dynamical aspects of perturbed crystal vibrations . . . 443
152. Infra-red absorption in the perturbed system . . . 448
153. Raman scattering in the perturbed system . . . 453
154. Symmetry breaking and induced lattice absorption and scattering . . . 455
α) Symmetry breaking . . . 455
β) Morphic Effects . . . 456

P. Respice, adspice, prospice . . . 460

Q. Acknowledgements . . . 465

Appendix A: Complete tables of reduction coefficients-selection rules for rocksalt structure O_h^5 (Tables A.1 to A.11) . . . 467
Appendix B: Complete tables of reduction coefficients-selection rules for the diamond space group O_h^7 (Tables B.1 to B.10) . . . 479
Appendix C: Illustration of ray representation method: Point X in diamond (Table C.1) . . . 494
Appendix D: Tables for the zincblende structure: $F\bar{4}3m$; T_d^2 (Tables D.1 to D.10) . . . 497

References . . . 507

Index of key equations . . . 510

Index of tables . . . 518

Index of figures . . . 521

Sachverzeichnis (Deutsch-Englisch) . . . 523

Subject Index (English-German) . . . 531

Principal symbols and notation

A. *Description of the crystal*

$\boldsymbol{a}_1, \boldsymbol{a}_2, \boldsymbol{a}_3$	Basis vectors of the crystal lattice
$\boldsymbol{R}_L$	Bravais lattice vector
l_1, l_2, l_3	Integer components of $\boldsymbol{R}_L$: $\boldsymbol{R}_L = l_1\,\boldsymbol{a}_1 + l_2\,\boldsymbol{a}_2 + l_3\,\boldsymbol{a}_3$
$\boldsymbol{r}$	General vector of crystal space
x_1, x_2, x_3	Components of $\boldsymbol{r}$
$\boldsymbol{b}_1, \boldsymbol{b}_2, \boldsymbol{b}_3$	Basis vectors of the reciprocal lattice
$\boldsymbol{B}_H$	Reciprocal lattice vector
h_1, h_2, h_3	Integer components of $\boldsymbol{B}_H$: $\boldsymbol{B}_H = h_1\,\boldsymbol{b}_1 + h_2\,\boldsymbol{b}_2 + h_3\,\boldsymbol{b}_3$
$\boldsymbol{k}_i$ $(i=1, 2, 3)$	Wave vector parallel to $\boldsymbol{b}_i$ $(i=1, 2, 3)$: $\boldsymbol{k}_i = \left(\frac{2\pi p_i}{N_i}\right)\boldsymbol{b}_i$
$\boldsymbol{k} = \boldsymbol{k}_1 + \boldsymbol{k}_2 + \boldsymbol{k}_3$	General wave vector
$\boldsymbol{r}^{\,l}_{\kappa} \equiv \boldsymbol{r}\begin{pmatrix} l \\ \kappa \end{pmatrix}$	Location of the κ-th atom of the l-th elementary cell
$\boldsymbol{r}_\kappa$	Location of the κ-th atom of the elementary cell chosen as origin
$r_\alpha\begin{pmatrix} l \\ \kappa \end{pmatrix}$	α-th component of $\boldsymbol{r}\begin{pmatrix} l \\ \kappa \end{pmatrix}$

B. *Symmetry elements*

$\boldsymbol{\varphi}$	Matrix of an orthogonal transformation
$\tilde{\boldsymbol{\varphi}}$	Transpose of $\boldsymbol{\varphi}$
$\boldsymbol{\varphi}^{-1}$	Inverse of $\boldsymbol{\varphi}$
ε	Unit matrix of three dimensional space
φ	Angle of the rotation defined by matrix $\boldsymbol{\varphi}$
$\{\varepsilon \mid \boldsymbol{R}_L\}$	Translation by vector $\boldsymbol{R}_L$
$\{\boldsymbol{\varphi} \mid \boldsymbol{0}\}$	Pure rotation defined by $\boldsymbol{\varphi}$
$\{\boldsymbol{\varphi} \mid \boldsymbol{t}(\boldsymbol{\varphi})\}$	General symmetry element of the crystal consisting of rotation defined by matrix $\boldsymbol{\varphi}$ and translation by vector $\boldsymbol{t}(\boldsymbol{\varphi})$
$\boldsymbol{\tau}(\boldsymbol{\varphi})$	Smallest vector associated with rotation $\boldsymbol{\varphi}$: $\boldsymbol{t}(\boldsymbol{\varphi}) = \boldsymbol{\tau}(\boldsymbol{\varphi}) + \boldsymbol{R}_L$
$\{\varepsilon \mid \boldsymbol{R}_L\}^\sigma$	Translation conjugate of $\{\varepsilon \mid \boldsymbol{R}_L\}$ by $\{\boldsymbol{\varphi}_\sigma \mid \boldsymbol{\tau}_\sigma\}$: $\{\varepsilon \mid \boldsymbol{R}_L\}^\sigma = \{\boldsymbol{\varphi}_\sigma \mid \boldsymbol{\tau}_\sigma\}^{-1} \cdot \{\varepsilon \mid \boldsymbol{R}_L\} \cdot \{\boldsymbol{\varphi}_\sigma \mid \boldsymbol{\tau}_\sigma\}$
$\boldsymbol{R}_{L_{\sigma\rho}} = (\boldsymbol{\varphi}_\sigma, \boldsymbol{\varphi}_\rho)$	Factor group element: $\boldsymbol{R}_{L_{\varphi\rho}} = (\boldsymbol{\varphi}_\sigma, \boldsymbol{\varphi}_\rho) = \{\varepsilon \mid \boldsymbol{\varphi}_\sigma \cdot \boldsymbol{\tau}_\rho + \boldsymbol{\tau}_\sigma - \boldsymbol{\tau}_{\sigma\rho}\}$
K	Time reversal operator

C. *Groups and representations*

$\mathfrak{T}_1$	Group of translations $\{\boldsymbol{\varepsilon} \mid l\,\boldsymbol{a}_1\}$
$\mathfrak{T}$	Translation group
$\mathfrak{T}(\boldsymbol{k})$	Translation group of wave vector $\boldsymbol{k}$
$\mathfrak{P}$	Point group
$\mathfrak{P}(\boldsymbol{r})$	Point group at site $\boldsymbol{r}$
$\mathfrak{G}$	Space group
$\mathfrak{G}_{\{\boldsymbol{\varphi} \mid \boldsymbol{t}(\boldsymbol{\varphi})\}} = \mathfrak{G}_S$	Alternative notation for space group elements
$\mathfrak{G}_{P(R)}$	Group of transformation operators
$\mathfrak{G}_{D(R)}$	Group of matrices $D(R)$
$\mathscr{G}$	Space time symmetry group
$\star\boldsymbol{k}$	Star of wave vector $\boldsymbol{k}$
$\star\mathfrak{P}(\boldsymbol{k})$	Covering group of wave vector $\boldsymbol{k}$
$\mathfrak{G}(\boldsymbol{k})$	Space group of canonical wave vector $\boldsymbol{k}$
$\Pi(\boldsymbol{k})$	Point group of $\boldsymbol{k} = \mathfrak{G}(k)/\mathfrak{T}(k)$
$\mathfrak{C}(\boldsymbol{k})$	k-th class of $\mathfrak{G}(\boldsymbol{k})$
$C(\boldsymbol{k})$	Element of $\mathfrak{C}(\boldsymbol{k})$
$c(\boldsymbol{k})$	Order of $c(\boldsymbol{k})$
$P_{mn}^{(l)}$	Idempotent transformation operator
$D^{(l)}(R)$	Matrix corresponding to R in the l-th representation of the space group
$\chi^{(l)}(R)$	Character or trace of $D^{(l)}(R)$
$D^{(\boldsymbol{k})(m)}$	m-th irreducible representation of $\mathfrak{G}(\boldsymbol{k})$
$D^{(\star\boldsymbol{k})(m)}$	m-th irreducible representation of $\mathfrak{G}$ corresponding to $\star\boldsymbol{k}$
$\Sigma_l = \{\psi_\alpha\}$	Set of wave functions $\psi_1, \psi_2, \ldots, \psi_l$
$\psi_\lambda^{(\boldsymbol{k}_1)(m)}$	λ-th Bloch function with wave vector $\boldsymbol{k}$, transforming like the representation $D^{(\boldsymbol{k}_1)(m)}$ of $\mathfrak{G}(\boldsymbol{k}_1)$
$\dot{D}^{(\boldsymbol{k})(m)}(R)$	Matrix identical to $D^{(\boldsymbol{k})(m)}(R)$ if $R \in \mathfrak{G}(\boldsymbol{k})$, 0 otherwise
$\dot{\chi}^{(\boldsymbol{k})(m)}(R)$	Character of $\dot{D}^{(\boldsymbol{k})(m)}(R)$
$r^{(\boldsymbol{k})}(\lambda, \mu)$	Factor system of projective representation
$\mathfrak{K}^{(\star\boldsymbol{k})(m)}$	Kernel of the representation $D^{(\star\boldsymbol{k})(m)}$
$\mathfrak{R}$	Reduction group
$(\star\boldsymbol{k}\, m\, \star\boldsymbol{k}'\, m' \mid \star\boldsymbol{k}''\, m'')$	Reduction coefficients
$(\{\boldsymbol{k}_\sigma \dotplus \boldsymbol{k}'_{\sigma'}\}\, m\, m' \mid \boldsymbol{k}''\, m'')$	Subgroup reduction coefficient
$D^{(\star\boldsymbol{k})(m)} \otimes D^{(\star\boldsymbol{k}')(m')}$	Direct product of representations
$[D^{(\star\boldsymbol{k})(m)}]_{(p)}$	p-th symmetrised power of $D^{(\star\boldsymbol{k})(m)}$

D. *Lattice dynamics*

T	Kinetic energy
φ	Potential energy of the crystal
V	$\varphi - \varphi_0$ variation in potential energy
$\mathscr{L}$	Lagrangian
N	Number of unit cells in a Born-Von Karman approximation
r	Number of atoms per unit cell

D. *Lattice dynamics (continued)*

$\boldsymbol{u}\begin{pmatrix} l \\ \kappa \end{pmatrix}$	Displacement of the κ-th atom of the l-th cell
$\Phi_{\alpha\beta}\begin{pmatrix} l & l' \\ \kappa & \kappa' \end{pmatrix}$	Force matrices defined by: $\left.\left(\dfrac{\partial^2 \varphi}{\partial u_\alpha\begin{pmatrix} l \\ \kappa \end{pmatrix} \partial u_\beta\begin{pmatrix} l' \\ \kappa' \end{pmatrix}}\right)\right\vert_{u_\alpha\binom{l}{\kappa}=0;\; u_\beta\binom{l'}{\kappa'}=0}$
$[\boldsymbol{\Phi}]$	Force matrix with elements $\Phi_{\alpha\beta}\begin{pmatrix} l & l' \\ \kappa & \kappa' \end{pmatrix}$
$[\boldsymbol{u}]$	Column matrix of elements $u_\alpha\begin{pmatrix} l \\ \kappa \end{pmatrix}$
$[\boldsymbol{M}]$	Diagonal mass matrix $[M]_{\substack{\kappa\kappa' \\ l l' \\ \alpha\beta}} = M_\kappa\, \delta_{\kappa\kappa'}\, \delta_{ll'}\, \delta_{\alpha\beta}$
$\boldsymbol{w}\begin{pmatrix} l \\ \kappa \end{pmatrix}$	Weighted displacement $=\sqrt{M_\kappa}\,\boldsymbol{u}\begin{pmatrix} l \\ \kappa \end{pmatrix}$
$\zeta\begin{pmatrix} l \\ \kappa \end{pmatrix}$	Time independent displacement amplitude $\boldsymbol{w}\begin{pmatrix} l \\ \kappa \end{pmatrix} = e^{-i\omega t}\,\zeta\begin{pmatrix} l \\ \kappa \end{pmatrix}$
$[\boldsymbol{D}]$	Matrix related to $[\varphi]$ by: $D_{\alpha\beta}\begin{pmatrix} l & l' \\ \kappa & \kappa' \end{pmatrix} = \dfrac{1}{\sqrt{M_\kappa M_{\kappa'}}}\,\Phi_{\alpha\beta}\begin{pmatrix} l & l' \\ \kappa & \kappa' \end{pmatrix}$
$[\boldsymbol{e}_j]$	Eigenvector of $[D]$ associated with eigenvalue ω_j^2
$\boldsymbol{e}_\alpha\left(\begin{matrix} l \\ \kappa \end{matrix}\middle\vert j\right)$	Component of $[\boldsymbol{e}_j]$
q_j	Normal coordinate along eigenvector $[\boldsymbol{e}_j]$
$\boldsymbol{e}\left(\middle\vert\begin{matrix} \boldsymbol{k} \\ j \end{matrix}\right)$	Eigenvector of the dynamical matrix
$\boldsymbol{e}_\alpha\left(\kappa\middle\vert\begin{matrix} \boldsymbol{k} \\ j \end{matrix}\right)$	Component of $\boldsymbol{e}\left(\middle\vert\begin{matrix} \boldsymbol{k} \\ j \end{matrix}\right)$
$\boldsymbol{e}_\alpha\left(\kappa\middle\vert\boldsymbol{k}\right)$	Vector of j dimensional space with coordinate $\boldsymbol{e}_\alpha\left(\middle\vert\begin{matrix} \boldsymbol{k} \\ j \end{matrix}\right)$
$Q\begin{pmatrix} \boldsymbol{k} \\ j \end{pmatrix}$	Complex normal coordinate
$\boldsymbol{\varepsilon}\left(\middle\vert\begin{matrix} \boldsymbol{k} \\ j \end{matrix}\middle\vert t\right)$	Time dependent eigenvector at wave vector $\boldsymbol{k}$

Theory of Crystal Space Groups and Infra-Red and Raman Lattice Processes of Insulating Crystals

By

Joseph L. Birman

With 34 Figures

A. Scope and plan of the article

1. General survey. This article was written with the objective of developing, and illustrating, the principles and practice of the application of group theory methods to the analysis of the infra-red and Raman lattice optical processes in insulating crystals. The group theory methods have proven to be very powerful tools in the interpretation and prediction of optical processes. It is also our objective to make these methods as accessible, and transparent, and hence as widely used as possible.

To achieve these objectives we need to have a clear idea of the strategy which will be pursued here. First we shall briefly develop the general theory of the structure of the crystal space groups. We then need to develop the consequences of the existence of the space group symmetry. These consequences flow in large measure from the assertion of necessary degeneracy which is that the physical states of the system are bases for the irreducible representations of the symmetry group. Hence we need to develop the theory of the irreducible representations of the symmetry groups, and of the functions which are bases for them. Owing to the close connection between states of the system, and representations, we devote a great deal of effort to the elaboration of this theory of the irreducible representations of space groups and we present a variety of methods which have recently been used to obtain them. An immediate natural extension of this analysis is to obtain the selection rules for transitions between states. This requires the reduction of the direct product of two irreducible representations into irreducible components.

The first part of the work is then complete, when we have achieved a description of the symmetry of the physical states, and when we have obtained the selection rules for transitions.

Next we turn to the physical side, and discuss the classical and quantum theories of lattice vibrations. In particular we develop the theory in such a way that the consequences of the symmetry can be obtained. Because the equations of motion are invariant under the symmetry operations of the crystal space group and also under the time reversal operation, we must enlarge the symmetry group to include antiunitary operations. Since this enlargement may have profound consequences for the characterization of the states of the system we give a very thorough discussion of the effect of incorporating anti-unitary symmetry: physical states are now required to be bases of irreducible corepresentations.

We now return again to physics and develop the theory of infra-red and Raman lattice optical properties of crystals. This theory treats the interaction between matter and radiation in a quantum mechanical framework. A key result which we obtain, relates the transition rate for absorption and scattering to the values of certain matrix elements. Exactly at this point symmetry plays a decisive role in determining whether the matrix element does or does not vanish. The mathematical selection rules find their physical application in determining the amplitude for transition processes. It is just here that the whole analysis can be applied in the interpretation, or prediction, of optical spectra. The group theory methods, taken in conjunction with other dynamical work is illustrated by determining the phonon energy and symmetry, and by analyzing the optical lattice spectra of crystals of diamond structure such as diamond, silicon, germanium, and of rocksalt structure such as sodium chloride. Examples are taken from perfect crystals and crystals with point imperfections.

To recapitulate: symmetry plays a central role in the classification of the eigenstates of the crystal, considered as a many-body system of ions plus electrons. The case of interest here relates to those elementary excitations representing lattice vibrations or phonons. Transitions between eigenstates are produced by perturbing fields, and a transition between a specified pair of states is allowed if the matrix element is non-zero. The symmetry of initial state, final state, and perturbing field determines whether or not the matrix element vanishes. More specifically, the group theory methods permit us to analyse the question: for a particular process involving specified changes in the lattice vibrational state of the crystal, and the concomitant change in the radiation field, can infra-red absorption or Raman scattering occur?

The work of this article is then, the development of three major, interrelated subjects:

The theory of crystal space groups:
The classical and quantum theory of lattice dynamics;
The theory of infra-red and Raman lattice optical properties,

with applications given to analysis of

The phonon spectra and
The optical properties of crystals with diamond and rocksalt structure.

Although these subjects are interrelated the development of each subject in this article is intended to be reasonably complete so that the article is self-contained. Introductory material of an intermediate level is presented for review and reference and notational consistency. As with most living branches of physics in which there is intense activity various aspects of the theory and applications have developed to a different degree, and this is reflected in various places in the text. By carrying through various parts of the theoretical analysis in detail it is our intention to illuminate difficult, and interesting, matters of general validity which may make it easier to carry out the analysis of other space groups as the need arises. The illustrative applications, on diamond and rocksalt structure were also chosen both for methodological and paedagogical reasons as well as owing to the great importance of these two crystal structures in current research, and the likelihood of their continued importance in the future.

In writing this article it was assumed that the reader possesses some working background in group theory as could be obtained from standard references [*1–3*], but is not expert. In particular, easily available material will be briefly referred to but not elaborated. Likewise it is assumed that the reader is familiar with the standard material on lattice dynamics, given in the well known texts in the field [*4*].

As a contribution to this Encyclopedia, this article is intended to fill a place not already occupied, and also to make contact with some related articles. The discussion of the representation and corepresentation theory of the space groups gives the method of analysis of those symmetry groups of crystals which were already described in the article on crystallography [*5*], and which is needed to derive modern physical consequences such as selection rules for optical and other transition processes. The discussion of the effect of symmetry in lattice dynamics elaborates the treatment given in [*4*]. The development of the theory of interaction of radiation with matter makes contact with discussions of the many-body theory of phonons and optical processes given in [*20*] and [*22*], but the contact focusses here upon certain aspects of the theory amenable to symmetry analysis such as: the structure of the anharmonic phonon Hamiltonian, the question of mode-mixing and effect of symmetry upon the structure of the thermal phonon Green function. Owing to deepening theoretical analysis and advances in experimental measurements, great progress has been made since the earlier reports on the interpretation of infra-red [*38*] and Raman [*39*] spectra of crystals. Particularly noteworthy in addition to the deepening analysis related to symmetry is the increased understanding of the dynamical, microscopic basis of the phenomena involved in the radiation matter interaction.

Finally we remark on the style of this article. We have deliberately refrained from adopting a condensed treatment of the theory presented here. This is because our aim here in this monograph is to be throughout explicit and concrete even at the expense of brevity and elegance. We have not hesitated to give detailed explanation, and derivation in all detail, where we felt it to be needed. The criteria used in deciding whether a particular topic should be developed in great detail is whether or not it is readily available in a clearly written form and also our experience in communicating the subject to colleagues. We preferred to err on the side of caution giving too much, rather than compressing to the point of incomprehensibility or obscurity. If the presentation of some topics seems to some readers too explicit and detailed, we hope this word of explanation about style will suffice to maintain their general interest.

2. Plan of the article: An overview. Owing to the length of this article it may be helpful to provide an overview to guide the reader. Although the entire work is interconnected, the major subject divisions given in Sect. 1 are independent to a degree and it is possible to read each major subject division more or less separately, referring back as needed, to earlier matter.

The article is structured into sections, numbered 1, ..., 154. The sections are grouped into parts, denoted A, ..., Q. Equations are numbered serially (decimal notation) in each section. It should be noted that the first section of each part of this article is an introduction to that part. The introduction serves several purposes. It is intended to point out some major aspects of the development to be given, and to provide orientation and continuity and some explicit interconnection with other sections or parts. Generally, several parts taken together comprise the development of each major subject.

The development of space group theory is covered by Parts B, C, D, E, F, G and I. In Part B we give a review of the structure of the crystal space groups $\mathfrak{G}$ as a symmetry group of three dimensional crystal space. The mathematical structure of the crystal space group is emphasized. We do not give a comprehensive enumeration of the 230 space groups, since that is already given, with illustrations in another article. In Part C we review some standard material on the representation theory of finite groups. Although this material is well known, we require it as background for our development of the representation theory of the space groups. In Part D the representation theory of the translation group $\mathfrak{T}$ is given. The irreducible representations of the crystal translation group play a central role in the theory and it is important to give their analysis properly, and to introduce correctly the notion of the first Brillouin zone. Then in Part E a detailed development is given of the construction and properties of irreducible representations and vector spaces of the crystal space group $\mathfrak{G}$. This work is central to the characterization of eigenfunctions and eigenvalues in terms of their symmetry classification. The discussion in Parts F and G relates to the determination of the reduction coefficients for space groups. These reduction coefficients are the basic ingredients in the determination of selection rules. Mathematically they are the elements in the Clebsch-Gordon series for the reduction of the direct product of two space group irreducible representations.

At this point we interrupt the development of group theory and turn to discuss the physics of classical lattice vibrations. After obtaining the equations of motion in harmonic approximation in Part H we apply the group theory analysis to demonstrate the connection: the eigenvectors span a linear vector space for representations of the covering or symmetry group $\mathfrak{G}$. Then in Part I we give a development of the theory of the effect of antiunitary symmetry. Because of the relative unfamiliarity of this work we go into some detail on the analyses of the space-time symmetry group $\mathscr{G}$ which comprises: the ordinary unitary spatial symmetry, plus the antiunitary symmetry including the time reversal. It is worthwhile emphasizing that here we have to do with a mathematical operator-symmetry group deriving directly from the dynamical equations, rather than having a naive geometrical meaning. The consequences of antiunitary symmetry include the new connection with lattice dynamics: the physical eigenvectors span a linear vector space for corepresentations of the space time symmetry group $\mathscr{G}$. (In the mathematics literature corepresentations are also known as semilinear representations.)

The development of classical and quantum theory of lattice dynamics is given in Parts H, J, K, L. First the classical theory of lattice vibrations especially in harmonic approximation is reviewed in Part H: the development is closely tied to the use of symmetry in determining the eigenvectors. Most important from the viewpoint of applications is the assertion in Sect. 85 of the "Lemma of Necessary Degeneracy", which enables us to connect the physical theory with the symmetry theory. This assertion that the permitted degeneracy of the eigenvalues and eigenvectors in a physical system is a consequence of the underlying symmetry is made over and again in physics, and provides key insight in many different kinds of situation. Here it appears simply and naturally in an easily grasped illustration in classical lattice dynamics. In fact its validity goes beyond the harmonic approximation as a general assertion. In Part J various aspects of classical lattice dynamics are examined and systemetized on the basis of representation theory. This includes invariant theory, a tensor calculus, anharmonic theory and the discussion of how to determine actual normal mode eigenvectors by symmetry, and thus factorize the dynamical matrix. The development of quantum lattice dynamics in Part K follows the traditional Born-Oppenheimer adiabatic treatment. But, extending the usual treatment, we develop here the symmetry theory of the eigenfunctions. The transformation of lattice eigenfunctions under symmetry provides a useful characterization of the ground and excited eigenstates of the lattice system of coupled harmonic oscillators. Also it provides an interesting interconnection between the symmetry theory based upon the space group $\mathfrak{G}$, or space-time group $\mathcal{G}$, and the theory of the permutation symmetry of a system of identical partices. The latter is essentially a group of type $SU(n)$. Unfortunately the development of corepresentation theory has not yet advanced to the point where these results can all be easily put into that more general framework. In summary then, the objective of this theory is the characterization of each classical, or quantum, lattice eigenstate in terms of its transformation under symmetry; more particularly in terms of the irreducible representation or irreducible linear vector space to which it belongs.

The major subject which is treated in Part L is a central theme of the work: the theory of the interaction of radiation with matter. We develop this theory with particular emphasis on the infra-red and Raman optical lattice processes. The development is given first in terms of the first quantized approach based upon conventional perturbation theory. Following this path one is immediately led to the analysis of optical processes in terms of the transition matrix element for the infra-red or Raman process. In this analysis the central role of symmetry is played by the Wigner-Eckart theorem governing the vanishing or non-vanishing of the matrix element. All the ingredients necessary to analyse this question are now in our hands; the symmetry of initial and final crystal lattice states and of the transition operator. Together with the reduction coefficients the symmetry analysis can then be completed, and selection rules can be assembled. This work provides a direct, concrete, and easily visualized and used application of symmetry. The application of selection rules in interpretation of lattice spectra is, besides, one of the most successful chapters of the work. Following this we briefly develop aspects of the modern many-body theories of infra-red and Raman lattice optical properties, in terms of the thermal Green function or response function theory. This is done to make contact with some current work and to demonstrate albeit briefly the use of symmetry also in this branch of the theory. Finally a short introduction is given to the most rapidly developing of all aspects of modern work namely the microscopic theory of these lattice optic effects. Because of study of effects near "resonance" the Raman scattering processes are coming to achieve a very high degree of sophistication both in experimental results and theoretical interpretation. Likewise highly precise infra-red lattice optic studies reveal a wealth of detail now being interpreted. It is intended that our presentation shall be useful in providing orientation in this developing field.

The final major subject of application of the general theory to diamond and rocksalt is given in Parts M, N, and O. As explained in the previous section we have the dual objective here of illustrating the theory on some important and relevent systems, and also providing a clear enough dicussion so the reader can learn how to apply the theory to any other systems of his own choice. Also given in Part O is an introduction to the theory of the imperfect crystal with point defects, emphasizing the symmetry-related aspects of the optical properties. Again, the theory is applied in diamond and rocksalt structures.

Finally a word on references to other literature. I have found it manifestly impossible to give comprehensive and detailed references to all the many scientists who worked on problems related, even directly, to the subject matter of this article. I have attempted to give a reasonable selection of citations to help the reader find his own way in the literature. A conscientious reader can, by following a chain of references, easily immerse himself in the present and past work on a given topic. I have tried to be helpful by providing the initial leads, or references to proceedings of international conferences and the like. I have been selective in my own use of literature and I often cite work which I have found useful, in the hope that the reader will, also. But I have not deliberately intended to slight any worker nor to distort any historical process or priority.

B. The crystal space group

3. Crystal symmetry – Introduction. In this part of the article the subject of crystal symmetry in configuration space will be presented [*5–7*, *9*, *42*].

We begin descriptively. The crystal is a three dimensional regular array of constituents: neutral atoms, charged ions, molecular complexes, etc. A symmetry operation or symmetry transformation of a crystal is a rearrangement, interchange, or permutation of the constituents which leaves unchanged, or invariant, the relative positions of the constituents and so produces an equivalent crystal in regard to its orientation. Such a symmetry operation in configuration space can correspond to a "real", or physically performable transformation such as a rigid rotation into an equivalent direction of the entire crystal about some axis in configuration space. It can, on the contrary, correspond to an "imagined" transformation such as inversion of all atom positions through a fixed point. When such a transformation has occurred, another crystal, which we call a replica of the original, has been produced. The term replica, which we occasionally use is intended to convey a relationship between initial object and transformed object which is weaker than identity, but stronger than similarity. The replica can be superimposed upon the original crystal, it also has the same orientation as the original with respect to fixed external axes. The transformation in question is mathematically: a congruent mapping of the crystal onto itself. The congruent mapping of the crystal onto itself produces a replica. The converse is *not*-true: all congruent mappings, or rearrangements are *not* crystallographic symmetry operations.

The basic symmetry of a crystal is the translational periodicity which characterizes the locations of constituents. Then, the translational symmetry implies that we may choose as an origin of coordinates any site containing an atom of one particular sort: all such atoms are equivalent. Likewise all such choices of origin must lead to identical physical results. Evidently, if a crystal contains several types of atoms or ions (e.g. NaCl, KCl, etc.) then the equivalent atoms or ions form a subset. The atoms of the subset are in the relationship to one another of equivalence under translation symmetry operations. Cases also arise of crystals composed of one chemical species e.g. C, Si, Ge, for which chemical identity does not imply structural equivalence. As a result of the crystalline arrangement, chemically identical constituents such as two Ge atoms may belong to distinct translational subsets. Only the atoms within any one subset may be translationally equivalent. One speaks, in the latter case, of the crystal having a basis. In case of a monatomic crystal for which all atoms are translationally equivalent the crystal lattice is then a Bravais lattice [5–7, 9].

As a result of the particular translational symmetry which a crystal may possess, it may happen that rotational symmetry may exist also. In this case the burden of compatibility is so to speak entirely upon the rotational symmetry element: it must be compatible with the existing translation symmetry.

Again, as a result of the detailed arrangement, composite symmetry operations consisting of rotations combined with fractional translations may exist in a crystal. These glide planes, or screw axes produce essentially new transformations. Of course, every permitted rotational symmetry operation may be combined with a full lattice translation to produce a composite symmetry operation.

The totality of translational, rotational, and composite symmetry operations comprise the crystal space group $\mathfrak{G}$. We now proceed to discuss the mathematical structure and theory of these groups.

The mathematical description of the types of symmetry transformations which arise in the space groups is that they are linear, real, inhomogeneous, discrete (special affine) transformations of three dimensional Euclidean space. An affine transformation can be understood as a transformation of a point in three dimensional space into another point in three dimensional space (active interpretation). Alternatively an affine transformation can be interpreted as a relabelling of the coordinates of a fixed point as a result of a change in the coordinate system used to describe the point (passive interpretation). In either interpretation, the transformation is an "association" of a point whose coordinates in a rectilinear coordinate system are (x_1, x_2, x_3), with a point whose coordinates are (x'_1, x'_2, x'_3). This association, or correspondence, of two points, or two triads of numbers, e.g. the components, is the transformation. The most general type of transformation which arises in crystals is given by:

$$\begin{aligned} x_1' &= \varphi_{11} x_1 + \varphi_{12} x_2 + \varphi_{13} x_3 + t_1(\varphi_{ij}),\\ x_2' &= \varphi_{21} x_1 + \varphi_{22} x_2 + \varphi_{23} x_3 + t_2(\varphi_{ij}),\\ x_3' &= \varphi_{31} x_1 + \varphi_{32} x_2 + \varphi_{33} x_3 + t_3(\varphi_{ij}), \end{aligned} \tag{3.1}$$

where φ_{ij} are real constants and the t_k are real and may depend upon the set of φ_{ij}. Of all the general affine transformations (3.1) relating the triads (x_1, x_2, x_3) to (x_1', x_2', x_3'), we shall select those which: a) leave the square of the distance, or metric, between any two points invariant, and b) are consistent with the lattice hypothesis.

Let the coordinates of two points, before the transformation, be (x_1, x_2, x_3), and (y_1, y_2, y_3) on cartesian coordinate axes. The square of the Euclidean distance between these points is

$$d^2 = (x_1 - y_1)^2 + (x_2 - y_2)^2 + (x_3 - y_3)^2. \tag{3.2}$$

Under the affine transformation (3.1), the triad (y_1, y_2, y_3) is also transformed as:

$$\begin{aligned} y_1' &= \varphi_{11} y_1 + \varphi_{12} y_2 + \varphi_{13} y_3 + t_1(\varphi_{ij}),\\ y_2' &= \varphi_{21} y_1 + \varphi_{22} y_2 + \varphi_{23} y_2 + t_2(\varphi_{ij}),\\ y_3' &= \varphi_{31} y_1 + \varphi_{32} y_2 + \varphi_{33} y_3 + t_3(\varphi_{ij}), \end{aligned} \tag{3.3}$$

where φ_{ij} and t_k are identical in (3.1) and (3.3). Then the invariance of the square of the Euclidean distance under the transformation is expressed by

$$d'^2 = (x_1' - y_1')^2 + (x_2' - y_2')^2 + (x_3' - y_3')^2 = d^2. \tag{3.4}$$

Evidently (3.4) puts a restriction upon the φ_{ij} which can occur. This is the restriction that the homogeneous part of the transformation be orthogonal when all $t_k = 0$. (We shall examine the orthogonal restriction a little more closely later: it takes a different form in cartesian or non-cartesian axes.)

The lattice hypothesis (b above) places restrictions upon certain of the translations $t_k(\varphi_{ij})$. Of greatest importance is the restriction which arises in the case of pure translation symmetry. For a pure translation, only the diagonal elements of φ do not vanish and then $\varphi_{jj} = 1$ so (3.1) becomes for pure translations

$$x_j' = x_j + t_j(\varphi_{ij} = \delta_{ij}). \tag{3.5}$$

The restriction on t_j is then

$$t_j(\varphi_{ij} = \delta_{ij}) \quad \text{is integer.} \tag{3.6}$$

The lattice hypothesis asserts that the crystal lattice can be replicated by integer pure translations only [6].

Other restrictions arise upon permissable φ_{ij} and $t_k(\varphi_{ij})$ so that all symmetry elements shall be mutually consistent. For example, a well-known consequence of this requirement of compatibility is that the only permitted crystal rotations in three dimensions are through angles of $2\pi/n$ where $n=1, 2, 3, 4, 6$; $n=5$ is excluded [6]; see (6.19) below.

4. The translation subgroup of a crystal. Let us introduce three linearly independent vectors in crystal space: $\boldsymbol{a}_1, \boldsymbol{a}_2\ \boldsymbol{a}_3$. These may be chosen in a variety of ways, depending upon convention, particular convenience, etc. We shall assume that in any case of interest to us, the choice has been made *a priori* following international convention [6]. The magnitudes of these vectors are $|\boldsymbol{a}_j| = a_j$. The vectors have then the dimension of length. Assume an origin of coordinates 0 has been selected. Usually the convention dictating choice of origin in a crystal, and the choice of $\boldsymbol{a}_j$ will be made simultaneously, and on the same ground.

At this point let us assume further that 0, the origin, coincides with the site at which an atom or ion resides. Then we take a vector from 0 to an arbitrary point in the crystal to be

$$\boldsymbol{r} = x_1\,\boldsymbol{a}_1 + x_2\,\boldsymbol{a}_2 + x_3\,\boldsymbol{a}_3, \tag{4.1}$$

where x_j are arbitrary real numbers (dimensionless). Thus the triad of real numbers (x_1, x_2, x_3) are the components of $\boldsymbol{r}$ in the coordinate system defined by $\boldsymbol{a}_1, \boldsymbol{a}_2, \boldsymbol{a}_3$. Consider a vector

$$\boldsymbol{R}_L \equiv l_1\,\boldsymbol{a}_1 + l_2\,\boldsymbol{a}_2 + l_3\,\boldsymbol{a}_3, \tag{4.2}$$

where the l_j are real integers. Such a vector is a lattice vector in crystal space. The set of all lattice vectors in crystal space is obtained when the l_j are permitted to assume all real integer values.

We now consider an elementary lattice translation. Let it be defined by the transformation by which vector $\boldsymbol{r}$ is transformed into vector $\boldsymbol{r}'$ where

$$\boldsymbol{r}' = x_1'\,\boldsymbol{a}_1 + x_2'\,\boldsymbol{a}_2 + x_3'\,\boldsymbol{a}_3 \tag{4.3}$$

and

$$x_1' = x_1 + l_1, \quad x_2' = x_2 + l_2, \quad x_3' = x_3 + l_3, \tag{4.4}$$

or

$$\boldsymbol{r}' = \boldsymbol{r} + \boldsymbol{R}_L. \tag{4.5}$$

According to the lattice hypothesis, the only pure translational symmetry (congruent) transformation which a crystal may possess is translation under a lattice vector $\boldsymbol{R}_L$. Hence, (4.5) can also be interpreted as associating together two equivalent points, $\boldsymbol{r}$ and $\boldsymbol{r}'$. That is, physical properties of the crystal defined at $\boldsymbol{r}$ and $\boldsymbol{r}'$ must be identical.

α) *Translation operators* $\{\varepsilon|\boldsymbol{R}_L\}$. It proves convenient to define an operator $\{\varepsilon|\boldsymbol{R}_L\}$, consisting of two symbols, to represent the transformation (4.5) and later, the more general transformation (3.1). In this section, the first symbol will be unnecessary, but is brought along for consistency and to keep in mind the sequel. The operator is defined by

$$\{\varepsilon|\boldsymbol{R}_L\} \cdot \boldsymbol{r} = \boldsymbol{r}' = \boldsymbol{r} + \boldsymbol{R}_L. \tag{4.6}$$

From (4.6) it follows that the inverse, or reciprocal, of $\{\varepsilon|\boldsymbol{R}_L\}$ can be defined as

$$\{\varepsilon|\boldsymbol{R}_L\}^{-1} \equiv \{\varepsilon|-\boldsymbol{R}_L\}, \tag{4.7}$$

where

$$\{\varepsilon|\boldsymbol{R}_L\}^{-1} \cdot \boldsymbol{r} = \boldsymbol{r} - \boldsymbol{R}_L. \tag{4.8}$$

Thus

$$\{\varepsilon|\boldsymbol{R}_L\}^{-1} \cdot \boldsymbol{r}' = \boldsymbol{r}' - \boldsymbol{R}_L$$

and if we substitute for $\boldsymbol{r}'$ from (4.6) we have

$$\{\varepsilon|\boldsymbol{R}_L\}^{-1} \cdot \{\varepsilon|\boldsymbol{R}_L\} \cdot \boldsymbol{r} = \{\varepsilon|\boldsymbol{R}_L\} \cdot \{\varepsilon|\boldsymbol{R}_L\}^{-1} \cdot \boldsymbol{r} = \boldsymbol{r}. \tag{4.9}$$

Hence if we define the identity operator (no translation) as $\{\varepsilon|0\}$, where

$$\{\varepsilon|0\} \cdot \boldsymbol{r} = \boldsymbol{r} \tag{4.10}$$

then

$$\{\varepsilon|\boldsymbol{R}_L\} \cdot \{\varepsilon|\boldsymbol{R}_L\}^{-1} = \{\varepsilon|\boldsymbol{R}_L\}^{-1} \cdot \{\varepsilon|\boldsymbol{R}_L\} = \{\varepsilon|0\}. \tag{4.11}$$

Evidently we can generalize to the rule by which two translation operations may be performed in succession. Thus if

$$\{\varepsilon|\boldsymbol{R}_L\} \cdot \boldsymbol{r} = \boldsymbol{r}' = \boldsymbol{r} + \boldsymbol{R}_L.$$

Then

$$\{\varepsilon|\boldsymbol{R}_L\} \cdot \boldsymbol{r}' = \boldsymbol{r}'' = \boldsymbol{r}' + \boldsymbol{R}_L = \boldsymbol{r} + 2\boldsymbol{R}_L \tag{4.12}$$

or

$$\{\varepsilon|\boldsymbol{R}_L\} \cdot \{\varepsilon|\boldsymbol{R}_L\} = \{\varepsilon|2\boldsymbol{R}_L\} = \{\varepsilon|\boldsymbol{R}_L\}^2. \tag{4.13}$$

Hence if $\{\varepsilon|\boldsymbol{R}_L\}$ is taken as in (4.6) and we define

$$\boldsymbol{R}_M = m_1\,\boldsymbol{a}_1 + m_2\,\boldsymbol{a}_2 + m_3\,\boldsymbol{a}_3, \tag{4.14}$$

where m_j are integers, and

$$\boldsymbol{R}_N = n_1\,\boldsymbol{a}_1 + n_2\,\boldsymbol{a}_2 + n_3\,\boldsymbol{a}_3, \tag{4.15}$$

where n_j are integers, so that $\boldsymbol{R}_M$ and $\boldsymbol{R}_N$ are lattice vectors, then we can define operators as:

$$\{\varepsilon|\boldsymbol{R}_M\} \cdot \boldsymbol{r} = \boldsymbol{r} + \boldsymbol{R}_M, \tag{4.16}$$

$$\{\varepsilon|\boldsymbol{R}_N\} \cdot \boldsymbol{r} = \boldsymbol{r} + \boldsymbol{R}_N. \tag{4.17}$$

Evidently the general rule of composition (successive application) of these operators is

$$\{\varepsilon|\boldsymbol{R}_M\} \cdot \{\varepsilon|\boldsymbol{R}_L\} \cdot \boldsymbol{r} = \{\varepsilon|\boldsymbol{R}_M\} \cdot (\boldsymbol{r} + \boldsymbol{R}_L) = \boldsymbol{r} + \boldsymbol{R}_L + \boldsymbol{R}_M \tag{4.18}$$

and

$$\{\varepsilon|\boldsymbol{R}_M\} \cdot \{\varepsilon|\boldsymbol{R}_L\} = \{\varepsilon|\boldsymbol{R}_M + \boldsymbol{R}_L\} = \{\varepsilon|\boldsymbol{R}_L + \boldsymbol{R}_M\}. \tag{4.19}$$

If we take

$$\boldsymbol{R}_{M+L} \equiv \boldsymbol{R}_{L+M} = \boldsymbol{R}_L + \boldsymbol{R}_M, \tag{4.20}$$

where

$$\boldsymbol{R}_{M+L} = (m_1 + l_1)\,\boldsymbol{a}_1 + (m_2 + l_2)\,\boldsymbol{a}_2 + (m_3 + l_3)\,\boldsymbol{a}_3 \tag{4.21}$$

then evidently $\boldsymbol{R}_{M+L}$ is a lattice vector as are $\boldsymbol{R}_M$ and $\boldsymbol{R}_L$. Thus (4.18) gives

$$\{\varepsilon|\boldsymbol{R}_M\} \cdot \{\varepsilon|\boldsymbol{R}_L\} = \{\varepsilon|\boldsymbol{R}_{L+M}\} = \{\varepsilon|\boldsymbol{R}_L\} \cdot \{\varepsilon|\boldsymbol{R}_M\}. \tag{4.22}$$

Finally from (4.18) and (4.22) we easily verify that

$$\{\varepsilon|\boldsymbol{R}_N\}\cdot\{\varepsilon|\boldsymbol{R}_{M+L}\}=\{\varepsilon|\boldsymbol{R}_{N+M}\}\cdot\{\varepsilon|\boldsymbol{R}_L\}, \tag{4.23}$$

where

$$\boldsymbol{R}_{N+M}=(n_1+m_1)\,\boldsymbol{a}_1+(n_2+m_2)\,\boldsymbol{a}_2+(n_3+m_3)\,\boldsymbol{a}_3. \tag{4.24}$$

Incidentally, because of the simple additive properties of translations we can reduce (4.23) as

$$\{\varepsilon|\boldsymbol{R}_N\}\cdot\{\varepsilon|\boldsymbol{R}_{M+L}\}=\{\varepsilon|\boldsymbol{R}_{N+M+L}\}. \tag{4.25}$$

β) *The translation group* $\mathfrak{T}$. At this point let us summarize. The operators $\{\varepsilon|\boldsymbol{R}_L\}$ are defined by (4.6). The operator is defined by means of its effect upon a vector $\boldsymbol{r}$ in the crystal space. To each lattice vector $\boldsymbol{R}_L$ we may define an associated translation operator $\{\varepsilon|\boldsymbol{R}_L\}$. Since a lattice vector is any vector of type (4.2) with $-\infty<l_j<+\infty$, we see that the product of two lattice translation operators is a lattice translation operator. Among the collection of translation operators (4.6) there is one unique one, which is the identity operator. Each translation operator can be paired with another one (its inverse), so that the product of the two operators is the identity operator: see (4.7). Finally, the product (successive application) of translation operators has the mathematical property of associativity: see (4.23).

Thus, the collection of all translation operators $\{\varepsilon|\boldsymbol{R}_L\}$ form a group [*1–3*]. We denote this group as $\mathfrak{T}$, the translation operator group of the crystal. Certain properties of this group, which are immediate consequences of the additivity of lattice vectors, are of considerable consequence for our later work.

Consider the three vectors:

$$\boldsymbol{R}_1\equiv\boldsymbol{a}_1;\qquad \boldsymbol{R}_2=\boldsymbol{a}_2;\qquad \boldsymbol{R}_3=\boldsymbol{a}_3 \tag{4.26}$$

the corresponding operators are

$$\{\varepsilon|\boldsymbol{R}_1\};\qquad \{\varepsilon|\boldsymbol{R}_2\};\qquad \{\varepsilon|\boldsymbol{R}_3\}. \tag{4.27}$$

Now observe

$$\{\varepsilon|l_1\,\boldsymbol{a}_1\}=\{\varepsilon|\boldsymbol{R}_1\}^{l_1}, \tag{4.28}$$

$$\{\varepsilon|l_2\,\boldsymbol{a}_2\}=\{\varepsilon|\boldsymbol{R}_2\}^{l_2}, \tag{4.29}$$

$$\{\varepsilon|l_3\,\boldsymbol{a}_3\}=\{\varepsilon|\boldsymbol{R}_3\}^{l_3}. \tag{4.30}$$

Hence

$$\{\varepsilon|\boldsymbol{R}_L\}=\{\varepsilon|l_1\,\boldsymbol{a}_1+l_2\,\boldsymbol{a}_2+l_3\,\boldsymbol{a}_3\} \tag{4.31}$$

$$=\{\varepsilon|\boldsymbol{a}_1\}^{l_1}\cdot\{\varepsilon|\boldsymbol{a}_2\}^{l_2}\cdot\{\varepsilon|\boldsymbol{a}_3\}^{l_3}. \tag{4.32}$$

γ) *Structure of* $\mathfrak{T}$. Hence it follows that the group $\mathfrak{T}$ is a direct product group. If we denote by $\mathfrak{T}_1$ the group consisting of all translation operators based on $\{\varepsilon|\boldsymbol{R}_1\}$,

$$\{\varepsilon|l_1\,\boldsymbol{a}_1\}=\{\varepsilon|\boldsymbol{a}_1\}^{l_1}=\{\varepsilon|\boldsymbol{R}_1\}^{l_1} \tag{4.33}$$

for all l_1, and similarly for $\mathfrak{T}_2$ and $\mathfrak{T}_3$, we observe

$$\mathfrak{T}=\mathfrak{T}_1\otimes\mathfrak{T}_2\otimes\mathfrak{T}_3. \tag{4.34}$$

Each of the groups $\mathfrak{T}_1$, $\mathfrak{T}_2$, $\mathfrak{T}_3$ is a translation group. Each is separately generated by one element, respectively $\{\varepsilon|\boldsymbol{R}_1\}$, $\{\varepsilon|\boldsymbol{R}_2\}$, $\{\varepsilon|\boldsymbol{R}_3\}$, and the powers of that element.

Evidently each of the groups $\mathfrak{T}_1, \mathfrak{T}_2, \mathfrak{T}_3$ is an abelian group on one generator. Since

$$\{\varepsilon|\boldsymbol{R}_1\}\cdot\{\varepsilon|\boldsymbol{R}_2\}=\{\varepsilon|\boldsymbol{R}_2\}\cdot\{\varepsilon|\boldsymbol{R}_1\} \quad \text{etc.} \tag{4.35}$$

all generators commute; hence $\mathfrak{T}$ is itself an abelian direct product group.

Clearly, since the l_j range over all real integers, the number of elements, or order, of each of $\mathfrak{T}_1, \mathfrak{T}_2, \mathfrak{T}_3$ is infinite, and $\mathfrak{T}$ is of "triply infinite" order. From a physical viewpoint it is inconvenient to deal with groups of infinite order. Consequently it usual, at this point to introduce the Born-Karman periodic boundary conditions.

δ) *Born-Karman boundary conditions* [8].[1] Consider the group $\mathfrak{T}_1$ and the translations which it represents. Suppose that there exists a large positive real integer N_1 such that the points in crystal space defined by $\boldsymbol{r}-N_1\,\boldsymbol{a}_1$ and $\boldsymbol{r}+N_1\,a_1$ are identical. Such an identity could hold rigorously only in case the crystal in the $\boldsymbol{a}_1$ direction were a chain. However, if N_1 is large enough, this assertion can be understood as meaning that the crystal is divided into large blocks, or fundamental regions (Grundgebiete). These large blocks repeat so as to produce the entire crystal. As there are $2N_1$ steps, each of length $\boldsymbol{a}_1$, between $\boldsymbol{r}-N_1\,\boldsymbol{a}_1$ and $\boldsymbol{r}+N_1\,\boldsymbol{a}_1$ we may write the Born-Karman cyclic condition as

$$\boldsymbol{r}-N_1\,\boldsymbol{a}_1=\boldsymbol{r}+N_1\,\boldsymbol{a}_1 \tag{4.36}$$

or

$$\boldsymbol{r}=\boldsymbol{r}+2N_1\,\boldsymbol{a}_1. \tag{4.37}$$

Consequently we may take (4.37) into operator form as

$$\{\varepsilon|0\}\cdot\boldsymbol{r}=\boldsymbol{r}=\{\varepsilon|2N_1\,\boldsymbol{a}_1\}\cdot\boldsymbol{r}=\{\varepsilon|\boldsymbol{a}_1\}^{2N_1}\cdot\boldsymbol{r}. \tag{4.38}$$

Similarly let N_2, N_3 be large positive real integers such that for any $\boldsymbol{r}$

$$\{\varepsilon|0\}\cdot\boldsymbol{r}=\boldsymbol{r}=\{\varepsilon|2N_2\,\boldsymbol{a}_2\}\cdot\boldsymbol{r}=\{\varepsilon|\boldsymbol{a}_2\}^{2N_2}\cdot\boldsymbol{r}, \tag{4.39}$$

$$\{\varepsilon|0\}\cdot\boldsymbol{r}=\boldsymbol{r}=\{\varepsilon|2N_3\,\boldsymbol{a}_3\}\cdot\boldsymbol{r}=\{\varepsilon|\boldsymbol{a}_3\}^{2N_3}\cdot\boldsymbol{r}. \tag{4.40}$$

Hence

$$\{\varepsilon|\boldsymbol{a}_1\}^{2N_1}=\{\varepsilon|\boldsymbol{a}_2\}^{2N_2}=\{\varepsilon|\boldsymbol{a}_3\}^{2N_3}=\{\varepsilon|0\}. \tag{4.41}$$

Since $\{\varepsilon|0\}$ is the identity element of $\mathfrak{T}_1$ or $\mathfrak{T}_2$, or $\mathfrak{T}_3$ or $\mathfrak{T}$ we see that the Born-Karman condition in effect converts various groups of infinite order into groups of finite (if large) order. An equivalent way of stating this result is in terms of the permissable values which the integers l_j can take. Thus, a lattice vector in crystal space, consistent with the Born-Karman cyclic boundary conditions can now be defined as $\boldsymbol{R}_L$ where

$$\boldsymbol{R}_L=l_1\,\boldsymbol{a}_1+l_2\,\boldsymbol{a}_2+l_3\,\boldsymbol{a}_3 \qquad -N_j<l_j\leqq N_j \tag{4.42}$$

and

$$\boldsymbol{R}_L+2N_1\,\boldsymbol{a}_1=\boldsymbol{R}_L+2N_2\,\boldsymbol{a}_2=\boldsymbol{R}_L+2N_3\,\boldsymbol{a}_3=\boldsymbol{R}_L. \tag{4.43}$$

Evidently $\mathfrak{T}_1$ is now to be considered an abelian group on one generator $\{\varepsilon|\boldsymbol{a}_1\}$, of order $2N_1$, and likewise for $\mathfrak{T}_2$ and $\mathfrak{T}_3$. The full translation group $\mathfrak{T}$ is to be

[1] M. Born and T. von Karman: Physik. Z. **14**, 15 (1913).

considered an abelian group on three generators of order N where

$$N \equiv (2N_1)\cdot(2N_2)\cdot(2N_3) = 8N_1 N_2 N_3. \tag{4.44}$$

Equivalently $\mathfrak{T}$ is a direct product of smaller abelian groups, as indicated in (4.34).

ε) *A property of the* $\{\varepsilon|\boldsymbol{t}\}$. Let us now observe that the operators $\{\varepsilon|\boldsymbol{t}\}$ are not entirely defined when applied to differences of vectors $\boldsymbol{r}$, using the definitions (4.6). Observe

$$\{\varepsilon|\boldsymbol{t}\}\cdot\boldsymbol{r} = \boldsymbol{r}+\boldsymbol{t}, \tag{4.45}$$

$$\{\varepsilon|\boldsymbol{t}\}\cdot\boldsymbol{r}' = \boldsymbol{r}'+\boldsymbol{t}, \tag{4.46}$$

so

$$\{\varepsilon|\boldsymbol{t}\}\cdot\boldsymbol{r}-\{\varepsilon|\boldsymbol{t}\}\cdot\boldsymbol{r}' = \boldsymbol{r}-\boldsymbol{r}'. \tag{4.47}$$

But if $\boldsymbol{\rho}\equiv\boldsymbol{r}-\boldsymbol{r}'$

$$\{\varepsilon|\boldsymbol{t}\}\cdot\boldsymbol{\rho} = \boldsymbol{\rho}+\boldsymbol{t}, \tag{4.48}$$

so

$$\{\varepsilon|\boldsymbol{t}\}\cdot(\boldsymbol{r}-\boldsymbol{r}') = (\boldsymbol{r}-\boldsymbol{r}')+\boldsymbol{t}. \tag{4.49}$$

Hence

$$\{\varepsilon|\boldsymbol{t}\}\cdot\boldsymbol{r}-\{\varepsilon|\boldsymbol{t}\}\cdot\boldsymbol{r}' \neq \{\varepsilon|\boldsymbol{t}\}\cdot(\boldsymbol{r}-\boldsymbol{r}'). \tag{4.50}$$

We must then distinguish between the operation (4.47), and (4.49). In this case the distinction is clearly provided by the physics of the transformation. If our intention is to shift rigidly the individual points $\boldsymbol{r}$ and $\boldsymbol{r}'$ by an equal amount $\boldsymbol{t}$, clearly preserving, or leaving invariant, the distance $\boldsymbol{r}-\boldsymbol{r}'$, we use (4.47). If our intention is to rigidly shift the vector $\boldsymbol{\rho}\equiv(\boldsymbol{r}-\boldsymbol{r}')$ we must use (4.49). There should be no cause for confusion or error.

5. Rotational symmetry elements: The crystal point group. Consider the case of a simple cubic crystal. Then we may take the vectors $\boldsymbol{a}_1, \boldsymbol{a}_2, \boldsymbol{a}_3$ of (4.1) to be an orthogonal triad, of equal magnitude, so $\boldsymbol{a}_i\cdot\boldsymbol{a}_j = a^2\delta_{ij}$ and $|a_i| = a$. Then using a cartesian matrix designation for $\boldsymbol{\varphi}$, we take the elements of the matrix $\boldsymbol{\varphi}$ as $(\boldsymbol{\varphi})_{ij} = \varphi_{ij}$. The requirement that $\boldsymbol{\varphi}$ be orthogonal as in (3.4) then restricts the transposed matrix, which we call $\tilde{\boldsymbol{\varphi}}$, with elements $(\tilde{\boldsymbol{\varphi}})_{ij} = \tilde{\varphi}_{ij}$:

$$\boldsymbol{\varphi}\cdot\tilde{\boldsymbol{\varphi}} = \tilde{\boldsymbol{\varphi}}\cdot\boldsymbol{\varphi} = \boldsymbol{\varepsilon}, \tag{5.1}$$

where $\boldsymbol{\varepsilon}$ is the 3×3 unit matrix:

$$(\boldsymbol{\varepsilon})_{ij} = \delta_{ij}. \tag{5.2}$$

Taking the determinant of (5.1) we have

$$\det\boldsymbol{\varphi}\cdot\tilde{\boldsymbol{\varphi}} = (\det\boldsymbol{\varphi})^2 = 1 \tag{5.3}$$

or

$$\det\boldsymbol{\varphi} = \pm 1. \tag{5.4}$$

For a proper rotation

$$\det\boldsymbol{\varphi} = +1 \quad \text{(proper)}. \tag{5.5}$$

For an improper rotation (or reflection)

$$\det\boldsymbol{\varphi} = -1 \quad \text{(improper)}. \tag{5.6}$$

The general term rotation includes both proper and improper cases; these will be distinguished as needed.

It is well known [1] that a given real unitary, or orthogonal, matrix $\boldsymbol{\varphi}$ can be transformed by an orthogonal transformation to the form

$$\boldsymbol{\varphi}=\pm\begin{pmatrix}\cos\varphi & -\sin\varphi & 0\\ \sin\varphi & \cos\varphi & 0\\ 0 & 0 & 1\end{pmatrix}. \tag{5.7}$$

As will be seen below, the consistency requirement between rotation $\boldsymbol{\varphi}$ and lattice translation requires

$$\varphi=2\pi/n; \qquad n=1,2,3,4,6. \tag{5.8}$$

Note that the transformation which brings one $\boldsymbol{\varphi}$ of a set to the form (5.7) will not, in general bring any other $\boldsymbol{\varphi}'$ of even the same set to that form.

α) *Rotational operators* $\{\boldsymbol{\varphi}|0\}$. The operator $\{\boldsymbol{\varphi}|0\}$ can be defined as

$$\{\boldsymbol{\varphi}|0\}\cdot\boldsymbol{r}\equiv\boldsymbol{\varphi}\cdot\boldsymbol{r}=\boldsymbol{r}', \tag{5.9}$$

where

$$\boldsymbol{\varphi}\cdot r=x_1'\,\boldsymbol{a}_1+x_2'\,\boldsymbol{a}_2+x_3'\,\boldsymbol{a}_3 \tag{5.10}$$

and x_1', x_2', x_3' are given as in (3.1) but with $t_j(\varphi)=0$:

$$\begin{aligned} x_1'&=\varphi_{11}\,x_1+\varphi_{12}\,x_2+\varphi_{13}\,x_3,\\ x_2'&=\varphi_{21}\,x_1+\varphi_{22}\,x_2+\varphi_{23}\,x_3,\\ x_3'&=\varphi_{31}\,x_1+\varphi_{32}\,x_2+\varphi_{33}\,x_3. \end{aligned} \tag{3.1}$$

The inverse rotation $\boldsymbol{\varphi}^{-1}$ has the property that $\boldsymbol{\varphi}^{-1}=\tilde{\boldsymbol{\varphi}}$ for an orthogonal matrix, and $\boldsymbol{\varphi}^{-1}\cdot\boldsymbol{\varphi}=\boldsymbol{\varepsilon}$, as in (5.1), so in terms of the operator symbol

$$\{\boldsymbol{\varphi}|0\}^{-1}=\{\boldsymbol{\varphi}^{-1}|0\} \tag{5.11}$$

and

$$\{\boldsymbol{\varphi}|0\}^{-1}\cdot\{\boldsymbol{\varphi}|0\}=\{\boldsymbol{\varphi}^{-1}\cdot\boldsymbol{\varphi}|0\}=\{\boldsymbol{\varepsilon}|0\}, \tag{5.12}$$

where $\{\boldsymbol{\varepsilon}|0\}$ is the identity (4.10).

A particular crystal will in general possess a variety of matrices $\boldsymbol{\varphi}_\sigma$ which appear in (3.1), and which we denote

$$\boldsymbol{\varepsilon}, \boldsymbol{\varphi}_2, \boldsymbol{\varphi}_3, \ldots, \boldsymbol{\varphi}_p. \tag{5.13}$$

Each of the $\boldsymbol{\varphi}_\sigma$ will be real unitary, and each satisfies (5.1). The product of two rotations will be

$$\boldsymbol{\varphi}_\sigma\cdot\boldsymbol{\varphi}_\tau=\boldsymbol{\varphi}_{\sigma\tau}. \tag{5.14}$$

It is in general one of the set (5.13).

β) *The point group* $\mathfrak{P}$. Take the set (5.13) to represent the complete collection of all distinct rotations $\boldsymbol{\varphi}_\sigma$ which appear as the homogeneous parts of the general transformations (3.1). The set of rotations is a group called the point group $\mathfrak{P}$ of the crystal. Using the operator symbols, the set of p operators

$$\{\boldsymbol{\varepsilon}|0\}, \{\boldsymbol{\varphi}_2|0\}, \ldots, \{\boldsymbol{\varphi}_p|0\} \tag{5.15}$$

is 𝔓. Evidently the multiplication of these objects is

$$\{\varphi_\sigma|0\}\cdot\{\varphi_\tau|0\}=\{\varphi_{\sigma\tau}|0\}, \tag{5.16}$$

where $\{\varphi_{\sigma\tau}|0\}$ is one of the set (5.15).

Note that it is not claimed that any point in the crystal possesses symmetry 𝔓. To repeat, 𝔓 is simply the collection of rotational parts of the general transformations (3.1). The set of operators (5.15) satisfies the group postulates: among the set is the identity $\{\varepsilon|0\}$, the product of two operators is in the set (by 5.16) each operator has an inverse, and the product is associative: hence 𝔓 is a mathematical group.

6. General symmetry element in a crystal: Space group 𝔊

α) *The operator* $\{\varphi|\boldsymbol{t}(\varphi)\}$. Now introduce the operator $\{\varphi|\boldsymbol{t}(\varphi)\}$ to represent the general symmetry transformation (3.1):

$$\{\varphi|\boldsymbol{t}(\varphi)\}\cdot\boldsymbol{r}=\varphi\cdot\boldsymbol{r}+\boldsymbol{t}(\varphi)=\boldsymbol{r}', \tag{6.1}$$

where the expression (6.1) is given in components in (3.1). Then we may enumerate all the operations in the set, or totality, of type (3.1). The enumeration is performed by grouping together all those operations which have the same rotational part

$$\begin{aligned}
&\{\varepsilon|0\},\ \{\varepsilon|\boldsymbol{a}_1\},\ \{\varepsilon|\boldsymbol{a}_2\},\ \{\varepsilon|\boldsymbol{a}_3\},\ \dots,\ \{\varepsilon|\boldsymbol{R}_L\},\ \dots;\\
&\{\varphi_2|\boldsymbol{\tau}(\varphi_2)\},\ \{\varphi_2|\boldsymbol{\tau}(\varphi_2)+\boldsymbol{a}_1\},\ \{\varphi_2|\boldsymbol{\tau}(\varphi_2)+\boldsymbol{a}_2\},\ \{\varphi_2|\boldsymbol{\tau}(\varphi_2)+\boldsymbol{a}_3\},\ \dots,\\
&\qquad\qquad\qquad\qquad\{\varphi_2|\boldsymbol{\tau}(\varphi_2)+\boldsymbol{R}_L\},\ \dots;\\
&\dots,\ \dots,\ \dots,\\
&\{\varphi_p|\boldsymbol{\tau}(\varphi_p)\},\ \{\varphi_p|\boldsymbol{\tau}(\varphi_p)+\boldsymbol{a}_1\},\ \{\varphi_p|\boldsymbol{\tau}(\varphi_p)+\boldsymbol{a}_2\},\ \{\varphi_p|\boldsymbol{\tau}(\varphi_p)+\boldsymbol{a}_3\},\ \dots,\\
&\qquad\qquad\qquad\qquad\{\varphi_p|\boldsymbol{\tau}(\varphi_p)+\boldsymbol{R}_L\},\ \dots.
\end{aligned} \tag{6.2}$$

Note that the enumeration (6.1) has been carried out in such a fashion that $\boldsymbol{\tau}(\varphi_\sigma)$ is defined as

$$\boldsymbol{\tau}(\varphi_\sigma)=\boldsymbol{t}(\varphi_\sigma)-\boldsymbol{R}_L\equiv\boldsymbol{\tau}_\sigma, \tag{6.3}$$

i.e. $\boldsymbol{\tau}(\varphi_\sigma)$ is the smallest translation which can arise combined with φ_σ, modulo a crystal lattice vector $\boldsymbol{R}_L$. Then there are two possibilities:

$$\boldsymbol{\tau}(\varphi) \text{ is the null vector;} \tag{6.4}$$

or

$$\boldsymbol{\tau}(\varphi) \text{ is a translation vector, but not a lattice translation vector.} \tag{6.5}$$

That is for any space group operation either the translational part $\boldsymbol{t}(\varphi)$ is a lattice vector $\boldsymbol{R}_L(\varphi)$, or it is a sum of a lattice vector (possibly null) and a translation vector $\boldsymbol{\tau}(\varphi)$, called a fractional translation. The enumeration of the possible consistent combinations of φ_σ and $\boldsymbol{\tau}(\varphi_\sigma)$ is the task of crystallography, and leads to the result that 230 three dimensional space groups can occur [6, 7]. We shall assume this task of enumeration completed, and the results available to us as needed. We shall rather focus on the mathematical structure of the space groups, and those properties which we need for our later analysis. Note: *vide infra*; the choice of $\boldsymbol{\tau}(\varphi)$ may not be unique since several equivalent fractional vectors may be smallest—an arbitrary choice is then made.

A space group $\mathfrak{G}$ is that collection of transformations of type (3.1) which replicates the crystal via a congruent mapping. Equivalently, a space group $\mathfrak{G}$ is that collection of transformation operators (one operator corresponds to each transformation) of type (6.1) which transforms equivalent points $\boldsymbol{r}$ and $\boldsymbol{r}'$ in configuration space into each other.

β) Group property of the set $\{\varphi|\boldsymbol{t}(\varphi)\}$. To demonstrate that the set (6.1) is a mathematical group we first establish the result of two successive transformations. Thus if

$$\{\varphi_2|\boldsymbol{t}(\varphi_2)\}\cdot\boldsymbol{r}=\varphi_2\cdot\boldsymbol{r}+\boldsymbol{t}(\varphi_2)=\boldsymbol{r}' \tag{6.6}$$

and

$$\{\varphi_3|\boldsymbol{t}(\varphi_3)\}\cdot\boldsymbol{r}'=\varphi_3\cdot\boldsymbol{r}'+\boldsymbol{t}(\varphi_3)=\varphi_3\cdot\varphi_2\cdot\boldsymbol{r}+\varphi_3\,\boldsymbol{t}(\varphi_2)+\boldsymbol{t}(\varphi_3)=\boldsymbol{r}'' \tag{6.7}$$

then

$$\{\varphi_3|\boldsymbol{t}(\varphi_3)\}\cdot\{\varphi_2|\boldsymbol{t}(\varphi_2)\}=\{\varphi_3\cdot\varphi_2|\varphi_3\cdot\boldsymbol{t}(\varphi_2)+\boldsymbol{t}(\varphi_3)\} \tag{6.8}$$

by direct observation. Now the product operation $\{\varphi_3\cdot\varphi_2|\varphi_3\cdot\boldsymbol{t}(\varphi_2)+\boldsymbol{t}(\varphi_3)\}$ is clearly a symmetry transformation, with rotational part: $\varphi_3\cdot\varphi_2$, and translational part: $\varphi_3\cdot\boldsymbol{t}(\varphi_2)+\boldsymbol{t}(\varphi_3)$. If $\varphi_3\cdot\varphi_2=\varphi_4$, then $\boldsymbol{t}(\varphi_4)=\varphi_3\cdot\boldsymbol{t}(\varphi_2)+\boldsymbol{t}(\varphi_3)$.

Using the product law (6.8) we can immediately establish the reciprocal operator of any $\{\varphi_\sigma|\boldsymbol{t}(\varphi_\sigma)\}$. The inverse rotation to φ_σ is φ_σ^{-1}. Then

$$\{\varphi_\sigma|\boldsymbol{t}(\varphi_\sigma)\}^{-1}=\{\varphi_\sigma^{-1}|-\varphi_\sigma^{-1}\cdot\boldsymbol{t}(\varphi_\sigma)\}. \tag{6.9}$$

Eq. (6.9) can easily be verified:

$$\begin{aligned}\{\varphi_\sigma|\boldsymbol{t}(\varphi_\sigma)\}\cdot\{\varphi_\sigma^{-1}|-\varphi_\sigma^{-1}\cdot\boldsymbol{t}(\varphi_\sigma)\}&=\{\varphi_\sigma\cdot\varphi_\sigma^{-1}|-\varphi_\sigma\cdot\varphi_\sigma^{-1}\cdot\boldsymbol{t}(\varphi_\sigma)+\boldsymbol{t}(\varphi_\sigma)\}\\&=\{\varepsilon|0\}.\end{aligned} \tag{6.10}$$

Note that (6.9) is an operator in the set (6.1) since its rotational part is the inverse of an allowed rotation and is in the set (5.13), while its translational part is a translation: $-\boldsymbol{t}(\varphi_\sigma)$ rotated by φ_σ^{-1}.

Finally we may verify that the multiplication is associative:

$$\begin{aligned}\{\varphi_\sigma|\boldsymbol{t}(\varphi_\sigma)\}\cdot\{\varphi_\tau|\boldsymbol{t}(\varphi_\tau)\}\cdot\{\varphi_\mu|\boldsymbol{t}(\varphi_\mu)\}&=(\{\varphi_\sigma|\boldsymbol{t}(\varphi_\sigma)\}\cdot\{\varphi_\tau|\boldsymbol{t}(\varphi_\tau)\})\cdot\{\varphi_\mu|\boldsymbol{t}(\varphi_\mu)\}\\&=\{\varphi_\sigma|\boldsymbol{t}(\varphi_\sigma)\}\cdot(\{\varphi_\tau|\boldsymbol{t}(\varphi_\tau)\}\cdot\{\varphi_\mu|\boldsymbol{t}(\varphi_\mu)\}).\end{aligned} \tag{6.11}$$

Summarizing: the set of operators given in (6.1): $\{\varphi_\sigma|\boldsymbol{t}(\varphi_\sigma)\}$ are a mathematical group [*1*]. We have verified the properties of closure under binary composition (multiplication), the existence of an identity element, the existence of a reciprocal for each element, and associativity of the binary composition. Further, owing to the Born-Karman condition imposed on $\mathfrak{T}$, the group $\mathfrak{G}$ is finite: that is, it consists of a finite number of operators only.

γ) Compatibility of rotation and translation. From (6.9) it follows that if a translation $\boldsymbol{t}(\varphi_2)$ is present as the translational part of a composite symmetry element in a space group $\mathfrak{G}$, then its rotated counterparts such as $\varphi_3\cdot\boldsymbol{t}(\varphi_2)$ are present. This permits an investigation and systematization of compatible rotation and translation operators. The complete examination of these compatibility requirements is available elsewhere [5–7]. Hence we merely note one facet of this subject. If the operation $\{\varphi_2|t(\varphi_2)\}$ is a pure lattice translation $\{\varepsilon|R_L\}$:

$$\{\varphi_\sigma|t(\varphi_\sigma)\}\cdot\{\varepsilon|R_L\}\cdot\{\varphi_\sigma^{-1}|-\varphi_\sigma^{-1}\cdot t(\varphi_\sigma)\}=\{\varepsilon|\varphi_\sigma\cdot R_L\} \tag{6.12}$$

is also a translation. But a rotation of a lattice vector must again produce a lattice vector, and hence $\varphi_\sigma\cdot R_L$ must be a lattice vector:

$$\varphi_\sigma\cdot R_L=R_{L'}. \tag{6.13}$$

Taking components

$$\sum_j(\varphi_\sigma)_{ij}\,l_j=l'_i, \tag{6.14}$$

where l'_j is integer and l_j is also an integer. It follows that in some coordinate system

$$(\varphi_\sigma)_{ij}\ \text{is integer.} \tag{6.15}$$

In particular the trace of the matrix φ_σ is

$$\operatorname{tr}\varphi_\sigma=\sum_j(\varphi_\sigma)_{jj}=\text{integer}. \tag{6.16}$$

But there is some coordinate system in which the matrix φ_σ appears as (5.7)

$$\varphi_\sigma=\pm\begin{pmatrix}\cos\varphi & -\sin\varphi & 0\\ \sin\varphi & \cos\varphi & 0\\ 0 & 0 & 1\end{pmatrix}. \tag{5.7}$$

It is known that the trace of a matrix is invariant to rotations in coordinate system so

$$\operatorname{tr}\varphi_\sigma=\pm(2\cos\varphi+1). \tag{6.17}$$

Hence

$$\pm(2\cos+1)=\text{integer} \tag{6.18}$$

for which the solutions are

$$\varphi=2\pi/n;\qquad n=1,2,3,4,6. \tag{6.19}$$

Thus (6.19) is the well known condition of compatibility of rotations and translations in a crystal, in three dimensions.

δ) *The operator* $\{\varphi|t\}$ *in non-Cartesian axes* [6]. In space groups for which the natural axis triad a_1, a_2, a_3 is non-orthogonal, it is generally most convenient to use dyadic operators for the rotational symmetry elements [6]. Since the dyadic operator effects a linear transformation between r and r' it can be expressed in the following form

$$\varphi=\sum_{ij}\varphi_{ij}\,a_i\,b_j, \tag{6.20}$$

where b_1, b_2, b_3 are the reciprocal set to the a_1, a_2, a_3:

$$a_i\cdot b_j=\delta_{ij}. \tag{6.21}$$

Evidently when we form

$$r'=\varphi\cdot r \tag{6.22}$$

we have produced a linear homogeneous transformation in which the components (x_1, x_2, x_3) of $\boldsymbol{r}$ are transformed to the components (x'_1, x'_2, x'_3) of $\boldsymbol{r}'$. The requirement that the rotational transformation preserve distances gives rise to the unitary restriction on $\boldsymbol{\varphi}$. If $\boldsymbol{\varphi}_c$ is the conjugate dyadic to $\boldsymbol{\varphi}$ then

$$\boldsymbol{\varphi}_c \equiv \sum_{ij} \varphi_{ij}\, \boldsymbol{b}_j\, \boldsymbol{a}_i. \tag{6.23}$$

The orthogonal restriction is then

$$\boldsymbol{\varphi}_c = \boldsymbol{\varphi}^{-1}, \tag{6.24}$$

where

$$\boldsymbol{\varphi} \cdot \boldsymbol{\varphi}^{-1} = \boldsymbol{\varphi}^{-1} \cdot \boldsymbol{\varphi} = \varepsilon \tag{6.25}$$

and ε is the unit dyadic:

$$\varepsilon = \sum_i \boldsymbol{a}_i\, \boldsymbol{b}_i. \tag{6.26}$$

Clearly

$$\boldsymbol{r} = \varepsilon \cdot \boldsymbol{r} \tag{6.27}$$

for all $\boldsymbol{r}$. Dyadic operators may be manipulated in ways essentially similar to the Cartesian matrices which we shall use in this article.

An equivalent representation of the dyadic operator $\boldsymbol{\varphi}$ can be given in terms of the axis of rotation, specified by the unit vector $\boldsymbol{u}$ parallel to that axis, and the angle of rotation about that axis:

$$\boldsymbol{\varphi} = \pm\big(\boldsymbol{u}\,\boldsymbol{u} + (\varepsilon - \boldsymbol{u}\,\boldsymbol{u}) \cos\varphi + \varepsilon \times \boldsymbol{u} \sin\varphi\big). \tag{6.28}$$

The explicit representation of $\boldsymbol{\varphi}$ as dyadic or matrix is incidental to the principal purpose of this article, so this topic will not be exhaustively developed [*6*].

The two space groups which will be analysed in detail are diamond: $Fd\,3m$, O_h^7 and rocksalt: $Fm\,3m$, O_h^5. Both of these cubic cases admit a natural orthogonal triad, and hence for them a Cartesian matrix representation of $\boldsymbol{\varphi}$ is simplest.

ε) *Order of the space group* $\mathfrak{G}$. From the definition of the space group operators in configuration space, it is immediately observed that the space group $\mathfrak{G}$ is a group of order $g_p N$ where g_p is the order of the point group $\mathfrak{P}$, and N is the order of the translation group $\mathfrak{T}$, as given in (4.44). In symbols

$$O(\mathfrak{G}) = O(\mathfrak{P})\, O(\mathfrak{T}) = g_p N. \tag{6.29}$$

ζ) *Normality of translation subgroup* $\mathfrak{T}$. It is clear that the set of space group operators in $\mathfrak{T}$:

$$\{\varepsilon | \boldsymbol{R}_L\} \tag{6.30}$$

is a subgroup of the full space group $\mathfrak{G}$. Consider the conjugate of an operator (6.30) by any general space group operator in $\mathfrak{G}$. That is,

$$\{\boldsymbol{\varphi} | \boldsymbol{t}(\boldsymbol{\varphi})\}^{-1} \cdot \{\varepsilon | \boldsymbol{R}_L\} \cdot \{\boldsymbol{\varphi} | \boldsymbol{t}(\boldsymbol{\varphi})\} = \{\varepsilon | \boldsymbol{\varphi}^{-1}\, \boldsymbol{R}_L\}. \tag{6.31}$$

By the lattice hypothesis, the operator (6.31) which is a pure translation must be a lattice translation

$$\{\varepsilon | \boldsymbol{\varphi}^{-1} \cdot \boldsymbol{R}_L\} = \{\varepsilon | \boldsymbol{R}_{L'}\}. \tag{6.32}$$

Thus the subgroup $\mathfrak{T}$ is closed under conjugation i.e. it is normal, or invariant [*1*].

η) *Factor group.* The decomposition of $\mathfrak{G}$ with respect to $\mathfrak{T}$ is then given as

$$\mathfrak{G}=\mathfrak{T}+\{\varphi_2|\tau_2\}\,\mathfrak{T}+\cdots+\{\varphi_\sigma|\tau_\sigma\}\,\mathfrak{T}+\cdots+\{\varphi_p|\tau_p\}\,\mathfrak{T}, \tag{6.33}$$

where τ_σ may be a null vector. For given space group $\mathfrak{G}$, the coset representatives in (6.33) are not unique (unless all $\tau_\sigma=0$). This is evident since any lattice translation operator may be added to the representative $\{\varphi_\sigma|\tau_\sigma\}$ without changing the coset. The factor group

$$\mathfrak{G}/\mathfrak{T}=\mathfrak{F} \tag{6.34}$$

is isomorphic to the point group $\mathfrak{P}$ defined by setting all translations equal to zero in the set of coset representatives:

$$\mathfrak{P}=\{\varepsilon|0\},\ \{\varphi_2|0\},\ \ldots,\ \{\varphi_p|0\} \tag{6.35}$$

as in (5.15).

Of course, the elements in $\mathfrak{F}$ are cosets or sets of operators; the elements in $\mathfrak{P}$ are individual operators. The isomorphism $\mathfrak{F}=\mathfrak{P}$ refers to the abstract groups.

θ) *Site symmetry.* In general no point in the crystal will have symmetry $\mathfrak{P}$, even if $\mathfrak{G}/\mathfrak{T}=\mathfrak{P}$. Only in case the space group is one for which all $\tau_\sigma=0$ i.e. is symmorphic, will a point have this symmetry.

The symmetry group of a point $\boldsymbol{r}$ in the crystal is defined as the set of operations which replicate the entire crystal, *and* which have that point fixed. Such a group may be variously indicated as $\mathfrak{P}_{\text{site}}(\boldsymbol{r})$ or $\mathfrak{P}(\boldsymbol{r})$ in what follows. This group is called the site group. Site groups for all space groups can be easily obtained from the international tables [9]; in some cases specific enumerations have been given. Another common notation for the site group is $\mathfrak{P}\binom{l}{\kappa}$ where $\boldsymbol{r}\equiv\boldsymbol{r}\binom{l}{\kappa}\equiv\binom{l}{\kappa}$ is the specific site in the crystal lattice. This notation is used in Sect. 148 (see Eqs. (148.2)).

7. The space group $\mathfrak{G}$ as a central extension of $\mathfrak{T}$ by $\mathfrak{P}$ [*10*]*. A group $\mathfrak{G}$ which has a normal subgroup $\mathfrak{T}$, with factor group $\mathfrak{G}/\mathfrak{T}=\mathfrak{P}$ is an extension of $\mathfrak{T}$ by $\mathfrak{P}$. To fully specify the structure of the group $\mathfrak{G}$ it is necessary to give: 1) the normal subgroup $\mathfrak{T}$; 2) the factor group $\mathfrak{P}$; 3) the automorphisms of $\mathfrak{T}$ which correspond to each element in $\mathfrak{P}$; 4) the factor set. We next show how to accommodate the space groups into mathematical extension theory.

The elements of $\mathfrak{T}$ are $\{\varepsilon|\boldsymbol{R}_L\}$. The elements of $\mathfrak{P}$ are the $\{\varphi_\sigma|0\}$. Every element of $\mathfrak{P}$ corresponds to a coset in $\mathfrak{G}$. In particular, the element $\{\varphi_\sigma|0\}$ of $\mathfrak{P}$ corresponds to the coset $\{\varphi_\sigma|\tau_\sigma\}\,\mathfrak{T}$ of $\mathfrak{G}$, whose representative is $\{\varphi_\sigma|\tau_\sigma\}$. Then

$$\{\varphi_\sigma|\tau_\sigma\}\,\mathfrak{T}\rightarrow\{\varphi_\sigma|0\}. \tag{7.1}$$

To each element $\{\varphi_\sigma|0\}$ of $\mathfrak{P}$ we associate a mapping of $\mathfrak{T}$ onto $\mathfrak{T}$ (automorphism) by taking

$$\{\varepsilon|\boldsymbol{R}_L\}\rightleftarrows\{\varphi_\sigma|\tau_\sigma\}^{-1}\,\{\varepsilon|\boldsymbol{R}_L\}\,\{\varphi_\sigma|\tau_\sigma\}=\{\varepsilon|\boldsymbol{R}_L\}^\sigma \tag{7.2}$$

or

$$\{\varepsilon|\boldsymbol{R}_L\}^\sigma\equiv\{\varepsilon|\varphi_\sigma^{-1}\cdot\boldsymbol{R}_L\}. \tag{7.3}$$

Consider the decomposition of $\mathfrak{G}$ with respect to cosets of $\mathfrak{T}$, which is given as

$$\mathfrak{G}=\mathfrak{T}+\{\varphi_2|\tau_2\}\,\mathfrak{T}+\cdots+\{\varphi_\sigma|\tau_\sigma\}\,\mathfrak{T}+\cdots+\{\varphi_\rho|\tau_\rho\}\,\mathfrak{T}. \tag{7.4}$$

According to (6.8) the product of two coset representatives is

$$\{\varphi_\sigma|\tau_\sigma\}\cdot\{\varphi_\rho|\tau_\rho\}=\{\varphi_\sigma\cdot\varphi_\rho|\varphi_\sigma\cdot\tau_\rho+\tau_\sigma\}. \tag{7.5}$$

Call $\varphi_\sigma\cdot\varphi_\rho\equiv\varphi_{\sigma\rho}$. Now in (7.4) the coset representative with this rotational part is $\{\varphi_{\sigma\rho}|\tau_{\sigma\rho}\}$. Hence we rewrite (7.5) as

$$\{\varphi_\sigma|\tau_\sigma\}\cdot\{\varphi_\rho|\tau_\rho\}=\{\varepsilon|\boldsymbol{R}_{\sigma\rho}\}\cdot\{\varphi_{\sigma\rho}|\tau_{\sigma\rho}\}, \tag{7.6}$$

where

$$\boldsymbol{R}_{\sigma\rho}\equiv\varphi_\sigma\cdot\tau_\rho+\tau_\sigma-\tau_{\sigma\rho}. \tag{7.7}$$

The vector $\boldsymbol{R}_{\sigma\rho}$ is a lattice translation vector, and $\{\varepsilon|\boldsymbol{R}_{\sigma\rho}\}$ is an element in $\mathfrak{T}$. Let us define a special symbol $(\varphi_\sigma, \varphi_\rho)$ by

$$(\varphi_\sigma,\varphi_\rho)\equiv\{\varepsilon|\boldsymbol{R}_{\sigma\rho}\}. \tag{7.8}$$

Hence we observe that the product of two elements in $\mathfrak{P}$ corresponds to the product of two coset representatives in $\mathfrak{G}$, modulo an element in $\mathfrak{T}$. Thus

$$\text{in } \mathfrak{P}:\quad \{\varphi_\sigma|0\}\cdot\{\varphi_\rho|0\}=\{\varphi_{\sigma\rho}|0\} \tag{7.9}$$

and

$$\text{in } \mathfrak{G}:\quad \{\varphi_\sigma|\tau_\sigma\}\cdot\{\varphi_\rho|\tau_\rho\}=(\varphi_\sigma,\varphi_\rho)\,\{\varphi_{\sigma\rho}|\tau_{\sigma\rho}\}. \tag{7.10}$$

The product law of $\mathfrak{P}$ is preserved in $\mathfrak{G}$, modulo an element in $\mathfrak{T}$. The set of p^2 elements

$$(\varphi_\sigma,\varphi_\rho)\equiv\{\varepsilon|\boldsymbol{R}_{\sigma\rho}\},\qquad \sigma,\rho=1\ldots p \tag{7.11}$$

is called the *factor set* in the extension.

The automorphisms $\{\varepsilon|\boldsymbol{R}_L\}^\sigma$ must satisfy certain conditions for consistency with the product (7.10). Thus the product of two automorphisms is given as the automorphism of the products or explicitly:

$$\begin{aligned}(\{\varepsilon|\boldsymbol{R}_L\}^\sigma)^\rho&=\{\varepsilon|\varphi_\rho^{-1}\cdot\varphi_\sigma^{-1}\cdot\boldsymbol{R}_L\}=\{\varepsilon|(\varphi_\sigma\cdot\varphi_\rho)^{-1}\cdot\boldsymbol{R}_L\}\\&=\{\varepsilon|\varphi_{\sigma\rho}^{-1}\cdot\boldsymbol{R}_L\}=\{\varepsilon|\boldsymbol{R}_L\}^{\sigma\rho}.\end{aligned} \tag{7.12}$$

In obtaining (7.12) we use the definition (7.2), the product rule (7.10), (7.11), and the property of $\mathfrak{T}$ as an Abelian, in fact cyclic, group so that elements in the factor set commute:

$$(\varphi_\sigma,\varphi_\rho)\,(\varphi_\mu,\varphi_\lambda)=(\varphi_\mu,\varphi_\lambda)\,(\varphi_\sigma,\varphi_\rho) \tag{7.13}$$

since these are individually members of $\mathfrak{T}$. Owing to (7.12) the extensions of $\mathfrak{T}$ by $\mathfrak{P}$ are called central.

In $\mathfrak{G}$ the elements obey an associative multiplication, so that

$$\{\varphi_\sigma|\tau_\sigma\}\cdot(\{\varphi_\rho|\tau_\rho\}\cdot\{\varphi_\pi|\tau_\pi\})=(\{\varphi_\sigma|\tau_\sigma\}\cdot\{\varphi_\rho|\tau_\rho\})\cdot\{\varphi_\pi|\tau_\pi\}. \tag{7.14}$$

Using (7.5), (7.8) in (7.14) we find on the left side:

$$\begin{aligned}\{\varphi_\sigma|\tau_\sigma\}\cdot(\varphi_\rho,\varphi_\pi)\cdot\{\varphi_{\rho\pi}|\tau_{\rho\pi}\}&=\{\varphi_\sigma|\tau_\sigma\}\cdot(\varphi_\rho,\varphi_\pi)\cdot\{\varphi_\sigma|\tau_\sigma\}^{-1}\cdot\{\varphi_\sigma|\tau_\sigma\}\cdot\{\varphi_{\rho\pi}|\tau_{\rho\pi}\}\\&=(\varphi_\rho,\varphi_\pi)^{\sigma^{-1}}\cdot(\varphi_\sigma,\varphi_{\rho\pi})\cdot\{\varphi_{\sigma\rho\pi}|\tau_{\sigma\rho\pi}\}\end{aligned} \tag{7.15}$$

and on the right side

$$(\varphi_\sigma, \varphi_\rho) \cdot \{\varphi_{\sigma\rho} | \tau_{\sigma\rho}\} \cdot \{\varphi_\pi | \tau_\pi\} = (\varphi_\sigma, \varphi_\rho) \cdot (\varphi_{\sigma\rho}, \varphi_\pi) \cdot \{\varphi_{\sigma\rho\pi} | \tau_{\sigma\rho\pi}\}. \tag{7.16}$$

Then we must have

$$(\varphi_\rho, \varphi_\pi)^{\sigma^{-1}} \cdot (\varphi_\sigma, \varphi_{\rho\pi}) = (\varphi_\sigma, \varphi_\rho) \cdot (\varphi_{\sigma\rho}, \varphi_\pi). \tag{7.17}$$

Working these terms out we find for the left side

$$\begin{aligned} \{\varepsilon | \varphi_\sigma \cdot \boldsymbol{R}_{\rho\pi}\} \cdot \{\varepsilon | \boldsymbol{R}_{\sigma\rho\pi}\} &= \{\varepsilon | \varphi_\sigma \cdot (\varphi_\rho \cdot \tau_\pi + \tau_\rho - \tau_{\rho\pi}) + \varphi_\sigma \cdot \tau_{\rho\pi} + \tau_\sigma - \tau_{\sigma\rho\pi}\} \\ &= \{\varepsilon | \varphi_\sigma \varphi_\rho \cdot \tau_\pi + \varphi_\sigma \cdot \tau_\rho - \varphi_\sigma \tau_{\rho\pi} + \varphi_\sigma \cdot \tau_{\rho\pi} + \tau_\sigma - \tau_{\sigma\rho\pi}\} \end{aligned} \tag{7.18}$$

and for the right

$$\begin{aligned} &\{\varepsilon | \varphi_\sigma \cdot \tau_\rho + \tau_\sigma - \tau_{\sigma\rho}\} \cdot \{\varepsilon | \varphi_{\sigma\rho} \cdot \tau_\pi + \tau_{\sigma\rho} - \tau_{\sigma\rho\pi}\} \\ &\quad = \{\varepsilon | \varphi_\sigma \cdot \tau_\rho + \tau_\sigma - \tau_{\sigma\rho} + \varphi_{\sigma\rho} \cdot \tau_\pi + \tau_{\sigma\rho} - \tau_{\sigma\rho\pi}\}. \end{aligned} \tag{7.19}$$

Thus, comparing term by term it is verified that (7.17) is obeyed by the elements of the factor set.

The final step is to recast the product law (6.8) for space group operators in the appropriate form for connection with extension theory. The general space group operator $\{\varphi_\sigma | \boldsymbol{t}(\varphi_\sigma)\}$ can be written

$$\{\varphi_\sigma | \boldsymbol{t}(\varphi_\sigma)\} = \{\varepsilon | \boldsymbol{R}_s\} \cdot \{\varphi_\sigma | \tau_\sigma\}, \tag{7.20}$$

where $\boldsymbol{R}_s$ is a lattice vector, and $\{\varphi_\sigma | \tau_\sigma\}$ a coset representative. Then taking $\boldsymbol{R}_p$ as a lattice vector we have

$$\begin{aligned} &\{\varphi_\sigma | \boldsymbol{t}(\varphi_\sigma)\} \cdot \{\varphi_\pi | \boldsymbol{t}(\varphi_\pi)\} \\ &\quad = \{\varepsilon | \boldsymbol{R}_s\} \cdot \{\varphi_\sigma | \tau_\sigma\} \cdot \{\varepsilon | \boldsymbol{R}_p\} \cdot \{\varphi_\pi | \tau_\pi\} \\ &\quad = \{\varepsilon | \boldsymbol{R}_s\} \cdot \{\varphi_\sigma | \tau_\sigma\} \cdot \{\varepsilon | \boldsymbol{R}_p\} \cdot \{\varphi_\sigma | \tau_\sigma\}^{-1} \cdot \{\varphi_\sigma | \tau_\sigma\} \cdot \{\varphi_\pi | \tau_\pi\} \\ &\quad = \{\varepsilon | \boldsymbol{R}_s\} \cdot \{\varepsilon | \boldsymbol{R}_p\}^{\sigma^{-1}} \cdot \{\varphi_\sigma | \tau_\sigma\} \cdot \{\varphi_\pi | \tau_\pi\} \\ &\quad = \{\varepsilon | \boldsymbol{R}_s\} \cdot \{\varepsilon | \boldsymbol{R}_p\}^{\sigma^{-1}} \cdot (\varphi_\sigma, \varphi_\pi) \cdot \{\varphi_{\sigma\pi} | \tau_{\sigma\pi}\}. \end{aligned} \tag{7.21}$$

Summarizing we observe that every element $\{\varphi_\sigma | \boldsymbol{t}(\varphi_\sigma)\}$ in $\mathfrak{G}$ can be written as the ordered product of an element from $\mathfrak{T}$ times a coset representative which corresponds to an element in $\mathfrak{P}$:

$$\{\varphi_\sigma | \boldsymbol{t}(\varphi_\sigma)\} = \{\varepsilon | \boldsymbol{R}_s\} \cdot \{\varphi_\sigma | \tau_\sigma\}. \tag{7.22}$$

The product rule of two such elements is

$$\begin{aligned} &\{\varepsilon | \boldsymbol{R}_s\} \cdot \{\varphi_\sigma | \tau_\sigma\} \cdot \{\varepsilon | \boldsymbol{R}_p\} \cdot \{\varphi_\pi | \tau_\pi\} \\ &\quad = \{\varepsilon | \boldsymbol{R}_s\} \cdot \{\varepsilon | \boldsymbol{R}_p\}^{\sigma^{-1}} \cdot (\varphi_\sigma, \varphi_\pi) \cdot \{\varphi_{\sigma\pi} | \tau_{\sigma\pi}\}, \end{aligned} \tag{7.23}$$

where the automorphisms of $\mathfrak{T}$ are given as

$$\{\varepsilon | \boldsymbol{R}_p\}^{\sigma^{-1}} \equiv \{\varepsilon | \varphi_\sigma \cdot \boldsymbol{R}_p\} \tag{7.24}$$

and the factor set as

$$(\varphi_\sigma, \varphi_\pi) = \{\varepsilon | \boldsymbol{R}_{\sigma\pi}\} = \{\varepsilon | \varphi_\sigma \cdot \tau_\pi + \tau_\sigma - \tau_{\sigma\pi}\}. \tag{7.25}$$

The automorphisms satisfy (7.12), and the factor set (7.17). Consequently the defining properties of the space group $\mathfrak{G}$ satisfy the conditions of Schreier's theorem. The space group $\mathfrak{G}$ is a central extension of $\mathfrak{T}$ by $\mathfrak{P}$ [*10*].

The mathematical problem of determining all space groups $\mathfrak{G}$ is then identical to that of finding all translation groups $\mathfrak{T}$, crystal point groups $\mathfrak{P}$, and central extensions. This problem has been solved completely, long ago by SCHÖNFLEISS and FEDEROV. The result and complete enumeration of the 230 crystallographic space groups are given in the article by H. JAGODZINSKI in this Encyclopedia, Vol. VII/1 [*5*]. We take these results as needed. We use both the Schönfleiss notation, and the short Hermann-Mauguin symbols [*9*].

It is useful in discussing space group representation theory by the method of ray representations as in Sects. 41–44 to understand the space groups as extensions. Then this present section should be referred to while reading Sects. 41–44.

8. Symmorphic space groups. Of the 230 space groups 73 are symmorphic. That is, all coset representatives are pure rotations $\{\varphi_\sigma|\mathbf{0}\}$; $\tau_\sigma=0$ all σ. Evidently the case of symmorphic space groups is one for which the extension of $\mathfrak{T}$ by $\mathfrak{P}$ is a split extension [*10*]. That is, $\mathfrak{G}$ is the semi-direct product of $\mathfrak{T}$ by $\mathfrak{P}$. Equivalently put, for a symmorphic space group the factor set obeys the rule

$$(\varphi_\sigma, \varphi_\rho) \equiv \{\varepsilon|0\} \qquad \text{for all } \sigma, \rho. \tag{8.1}$$

It follows that the coset decomposition of $\mathfrak{G}$ with respect to $\mathfrak{T}$ is particularly simple in this case

$$\mathfrak{G}/\mathfrak{T} = \mathfrak{T} + \{\varphi_2|0\}\,\mathfrak{T} + \cdots + \{\varphi_{g_p}|0\}\,\mathfrak{T}. \tag{8.2}$$

Evidently the set of coset representatives themselves

$$\{\varepsilon|0\},\ \{\varphi_2|0\}, \ldots, \{\varphi_{g_p}|0\} \equiv \mathfrak{P} \tag{8.3}$$

is closed under multiplication and so forms a subgroup. Then this group, which is $\mathfrak{P}$, the crystal point group, corresponds to the symmetry of some actual, physical point in the crystal. That is, $\mathfrak{P}$ is the site group of the origin, $\mathfrak{P}(0)$, taken as the point about which all rotational operations are to be performed.

The space group of the rocksalt structure which is one of the groups whose analysis will be presented is a symmorphic space group.* It is O_h^5, $Fm3m$. In the coset enumeration (8.2) of this group $\mathfrak{T}$ the translation group, is the non-primitive group F (face centered cubic), and $\mathfrak{P}$ the point group is O_h, the full cubic group. Hence g_p is 48, there are 48 cosets in (8.2).

9. Non-symmorphic space groups. In the remaining 157 non-symmorphic space groups, some or all the $\tau_\sigma \neq 0$. These groups are the general central extension of $\mathfrak{T}$ by $\mathfrak{P}$.

The second explicit case which we shall study [1] in this article is that of the diamond space group O_h^7: $F\,d\,3m$. This is a fairly typical non-symmorphic case. The

* This will be fully discussed in Sects. 126 et seq.

[1] This will be discussed in Sects. 126 et seq.

translation group is again F (face centered cubic), the point group of the crystal $\mathfrak{P}$ is O_h. Thus O_h^7 is a (non-split) extension of F by $\mathfrak{P}$. It happens that a simplifying feature exists here however, which may prove useful in later exploitations of the representation theory. In the enumeration of the 48 cosets in $\mathfrak{G}/F$, it is found that

$$\mathfrak{G}/F=\mathfrak{T}+\{\varphi_2|0\}\,\mathfrak{T}+\cdots+\{\varphi_{24}|0\}\,\mathfrak{T}+\{\boldsymbol{i}|\boldsymbol{\tau}\}\,\mathfrak{T}+\cdots+\{\varphi_{48}|\boldsymbol{\tau}\}\,\mathfrak{T}. \tag{9.1}$$

There are 24 coset representatives which have no fractional translation associated: $\boldsymbol{\tau}_\sigma=0$, $\sigma=1,\ldots,24$ and $\boldsymbol{\tau}_\sigma=\boldsymbol{\tau}$, $\sigma=25,\ldots,48$ so the fractional $\boldsymbol{\tau}$ associated with the remaining 24 coset representatives is the same. Actually the subgroup of $\mathfrak{G}$ consisting of

$$\mathfrak{S}=\mathfrak{T}+\{\varphi_2|0\}+\cdots+\{\varphi_{24}|0\}\,\mathfrak{T} \tag{9.2}$$

is a symmorphic space group T_d^2: $F\bar{4}3m$ in its own right. Consequently $\mathscr{S}\equiv T_d^2$ is a subgroup of index 2 and hence normal; the factor group $\mathfrak{G}/\mathfrak{S}$ is isomorphic to the parity group. Then it follows that

$$\mathfrak{G}/\mathfrak{S}=\mathfrak{S}+\{\boldsymbol{i}|\boldsymbol{\tau}\}\,\mathfrak{S}, \tag{9.3}$$

where

$$\mathfrak{G}=O_h^7;\quad \mathfrak{S}=T_d^2. \tag{9.4}$$

Every coset representative in (9.1) is then either

$$\{\varphi_\sigma|0\},\qquad \sigma=1,\ldots,24 \tag{9.5}$$

or

$$\{\boldsymbol{i}|\boldsymbol{\tau}\}\cdot\{\varphi_\sigma|0\},\qquad \sigma=1,\ldots,24 \tag{9.6}$$

where (9.5) applies to the first 24 cosets enumerated in (9.1), and (9.6) to the next 24 cosets. This simplifies the task of writing the factor set in this case. Thus for

$$\{\varphi_\sigma|0\}\cdot\{\varphi_\rho|0\}=\{\varphi_{\sigma\rho}|0\} \tag{9.7}$$

the factor set is

$$(\varphi_\sigma,\varphi_\rho)=\{\varepsilon|0\}. \tag{9.8}$$

From (9.6)

$$\{\varphi_\sigma|0\}\cdot\{\boldsymbol{i}|\boldsymbol{\tau}\}=(\varphi_\sigma,\boldsymbol{i})\cdot\{\boldsymbol{i}\,\varphi_\sigma|\boldsymbol{\tau}\} \tag{9.9}$$

and

$$(\varphi_\sigma,\boldsymbol{i})\equiv\{\varepsilon|\varphi_\sigma\cdot\boldsymbol{\tau}-\boldsymbol{\tau}\}=\{\varepsilon|\varphi_\sigma\cdot\boldsymbol{\tau}+\boldsymbol{i}\cdot\boldsymbol{\tau}\}. \tag{9.10}$$

Finally

$$\{\boldsymbol{i}\,\varphi_\sigma|\boldsymbol{\tau}\}\cdot\{\boldsymbol{i}\,\varphi_\rho|\boldsymbol{\tau}\}=(\boldsymbol{i}\,\varphi_\sigma,\boldsymbol{i}\,\varphi_\rho)\cdot\{\varphi_\sigma\cdot\varphi_\rho|0\}, \tag{9.11}$$

$$(\boldsymbol{i}\,\varphi_\sigma,\boldsymbol{i}\,\varphi_\rho)\equiv\{\varepsilon|\boldsymbol{i}\,\varphi_\sigma\cdot\boldsymbol{\tau}+\boldsymbol{\tau}\}. \tag{9.12}$$

Evidently the specification of the entire factor set requires evaluation of only the 24 translations ($\varphi_\sigma\cdot\boldsymbol{\tau}+\boldsymbol{i}\cdot\boldsymbol{\tau}$), since these are the basic objects which arise in (9.10) and (9.12). Also it will be evident that the translations (9.12) will be merely taken from a small set: the null translation, and the basic primitive vector set of the (face centered cubic) structure: see Sect. 126.

10. Some subgroups of a space group. From the structure of a space group it is evident that a space group possesses a variety of subgroups. In the later work in this article we shall have occasion to make use of various of these subgroups.

All space groups possess the translation group $\mathfrak{T}$ as normal space group. Any subgroup $\mathfrak{T}_s$ of $\mathfrak{T}$ will be a subgroup of $\mathfrak{G}$ as well. Let the subgroup $\mathfrak{T}_s$ of $\mathfrak{T}$ be normal in $\mathfrak{G}$, it is evidently normal in $\mathfrak{T}$, as $\mathfrak{T}$ is Abelian. Then as well as $\mathfrak{G}/\mathfrak{T}$ (which is the factor group $\mathfrak{P}$), the factor group $\mathfrak{G}/\mathfrak{T}_s$ can be defined. But, if

$$(\mathfrak{T}/\mathfrak{T}_s) = \mathfrak{T}_s + \{\varepsilon|\boldsymbol{R}_M\}\,\mathfrak{T}_s + \cdots. \tag{10.1}$$

Then

$$\mathfrak{G}/\mathfrak{T}_s = (\mathfrak{G}/\mathfrak{T}) \cdot (\mathfrak{T}/\mathfrak{T}_s). \tag{10.2}$$

In the collection of coset representatives in $\mathfrak{G}/\mathfrak{T}_s$ we will find all objects of form

$$\{\varepsilon|\boldsymbol{R}_M\} \cdot \{\varphi_\sigma|\tau_\sigma\} \tag{10.3}$$

that is combinations (products) of a representative from $(\mathfrak{T}/\mathfrak{T}_s)$ with one from $\mathfrak{P}$. All such must occur. If the index of $\mathfrak{T}_s$ in $\mathfrak{T}$ is s, then the order of the factor group $\mathfrak{G}/\mathfrak{T}_s$ is $(g_p \cdot s)$ where, as before, g_p is the order of $\mathfrak{P}$. We shall have occasion later to use explicitly such decompositions as (10.2).

For the symmorphic space groups, the point group $\mathfrak{P}$ is an actual subgroup which is useful at times to deal with in the representation theory.

Often an useful decomposition of a space group $\mathfrak{G}$ is in terms of subgroups which are themselves space groups. (For example as was shown in Sect. 9, the diamond space group O_h^7, which is a typical non-symmorphic space group, admits as a subgroup of index 2 the zincblende space group T_d^2.) Suppose that a space group $\mathfrak{G}$ admits a subgroup $\mathfrak{G}_a$ which is also a space group. $\mathfrak{G}_a$ may or may not possess the full translation group $\mathfrak{T}$ as a subgroup; for generality suppose that it does not. Let $\mathfrak{T}_a$ be the normal subgroup of translations contained in $\mathfrak{G}_a$. Then let the elements in $\mathfrak{T}_a$ be

$$\mathfrak{T}_a = \{\varepsilon|\boldsymbol{t}_{a1}\}, \quad \{\varepsilon|\boldsymbol{t}_{a2}\}, \quad \{\varepsilon|\boldsymbol{t}_{a3}\}, \ldots \tag{10.4}$$

those in $\mathfrak{G}_a$ may be written as:

$$\mathfrak{G}_a = \mathfrak{T}_a, \{\varphi_\alpha|\boldsymbol{t}(\varphi_\alpha)\}\,\mathfrak{T}_a, \ldots, \{\varphi_p|\boldsymbol{t}(\varphi_p)\}\,\mathfrak{T}_\alpha. \tag{10.5}$$

Since $\mathfrak{T}_a$ is normal in $\mathfrak{G}_a$ we may define the factor group $\mathfrak{P}_a$ of the (sub) space group as

$$\mathfrak{G}_a/\mathfrak{T}_a = \mathfrak{P}_a. \tag{10.6}$$

Evidently $\mathfrak{P}_a$ will be a subgroup of $\mathfrak{P} = \mathfrak{G}/\mathfrak{T}$. Now since $\mathfrak{G}_a$ is a subgroup of $\mathfrak{G}$, we may decompose $\mathfrak{G}$ into cosets relative to $\mathfrak{G}_a$:

$$\mathfrak{G} = \mathfrak{G}_a, \{\varepsilon|\boldsymbol{t}_{a1}\}\,\mathfrak{G}_a, \ldots, \{\varphi_\beta|\boldsymbol{t}(\varphi_\beta)\}\,\mathfrak{G}_a, \ldots, \{\varphi_\sigma|\boldsymbol{t}(\varphi_\sigma)\}\,\mathfrak{G}_a. \tag{10.7}$$

In (10.7) coset representatives include as usual those elements in $\mathfrak{G}$ but not in $\mathfrak{G}_a$, and there is the usual arbitrariness in choosing a translational part of an operator like $\{\varphi_\beta|\boldsymbol{t}(\varphi_\beta)\}$. Since $\mathfrak{G}_a$ is a group in its own right it possesses a full complement of classes, possible subgroups, etc.

Consider an element $\{\varphi_\beta|\boldsymbol{t}(\varphi_\beta)\}$, in $\mathfrak{G}$ but not in $\mathfrak{G}_a$. If we form the set $\mathfrak{G}_b$ by taking

$$\{\varphi_\beta|\boldsymbol{t}(\varphi_\beta)\}^{-1} \cdot \mathfrak{G}_a \cdot \{\varphi_\beta|\boldsymbol{t}(\varphi_\beta)\} = \mathfrak{G}_b \tag{10.8}$$

then $\mathfrak{G}_b$ is a space group, whose elements can be obtained easily from those of $\mathfrak{G}_a$. In fact $\mathfrak{G}_b$ is also a subgroup of $\mathfrak{G}$, and $\mathfrak{G}_a$ and $\mathfrak{G}_b$ are conjugate subgroups of $\mathfrak{G}$. If p is the smallest integer for which

$$\{\varphi_\beta | \boldsymbol{t}(\varphi_\beta)\}^p = \{\varepsilon | \boldsymbol{R}_L\} \tag{10.9}$$

then we may form a series of conjugate subgroups based on $\{\varphi_\beta | \boldsymbol{t}(\varphi_\beta)\}$ and its powers. Thus

$$\{\varphi_\beta | \boldsymbol{t}(\varphi_\beta)\}^{-1} \cdot \mathfrak{G}_b \cdot \{\varphi_\beta | \boldsymbol{t}(\varphi_\beta)\} = \mathfrak{G}_c \tag{10.10}$$

is another such space group. For each such space group there will be a related coset decomposition (10.7). It is often useful to present the coset decomposition of $\mathfrak{G}$ in these related fashions, via conjugate subgroups. Clearly conjugate subgroups are isomorphic.

C. Irreducible representations and vector spaces for finite groups

11. Introduction. This part of the article concerns itself with the general theory of irreducible representations and irreducible vector spaces of finite groups. It is assumed the reader is generally familiar with the standard elementary material [*1–3*] of representation theory of finite groups; and this will be briefly reviewed for notational and orientational purposes, in Sects. 12–18, which comprise this part.

To connect functions and representations it is necessary to introduce transformation operators on the vector space of functions. The operator group will be homomorphic to the set of coordinate operators comprising the crystal space group.

Let us now briefly look ahead to parts D and E, which follow. Every space group $\mathfrak{G}$ contains a normal subgroup of translations $\mathfrak{T}$. Since $\mathfrak{T}$ is an Abelian group (in fact the direct product of three cyclic groups) its irreducible representations and the irreducible linear vector space are one dimensional. The wave vector $\boldsymbol{k}$, and Bloch vector[1] $\psi^{(\boldsymbol{k})}$, characterize the irreducible representation $D^{(\boldsymbol{k})}$. The set of all admissible $\boldsymbol{k}$ span the first Brillouin zone for the crystal, and characterize all irreducible representations $D^{(\boldsymbol{k})}$ of $\mathfrak{T}$.

For each $\boldsymbol{k}$ a set of operators from $\mathfrak{G}$ is defined which transforms the Bloch vector $\psi^{(\boldsymbol{k})}$ to a function with an equivalent wave vector $\boldsymbol{k}$. This set of operators is the space group of wave vector $\boldsymbol{k}$, denoted $\mathfrak{G}(\boldsymbol{k})$. It is a subgroup of $\mathfrak{G}$. The irreducible representations of $\mathfrak{G}(\boldsymbol{k})$ are determined. Two methods can be used for this purpose. The method of ray, or projective, or multiplier representations, which makes use of the structure of $\mathfrak{G}(\boldsymbol{k})$ as an extension, will be discussed. The little group method will also be presented. Of all the irreducible representations of $\mathfrak{G}(\boldsymbol{k})$, only certain ones are admissable for our purposes. These allowable irre-

[1] F. Bloch: Z. Physik **52**, 555 (1928).

ducible representations $D^{(k)(m)}$ will be determined, and the corresponding vector space, also.

The final step is to decompose $\mathfrak{G}$ into cosets with respect to $\mathfrak{G}(\boldsymbol{k})$. An irreducible representation of $\mathfrak{G}(\boldsymbol{k})$ is then used to induce an irreducible representation of $\mathfrak{G}$. The irreducible representation of $\mathfrak{G}$ is characterized by a star of wave vectors $\star\boldsymbol{k}$, and the index m from the allowable $D^{(k)(m)}$. This procedure also determines the structure of the irreducible vector space on which $D^{(\star k)(m)}$ is based.

All the work outlined above is based on the assumption that the transformation operators are linear unitary operators such as is encountered for purely spatial symmetry transformations. To account for time reversal effects, it is necessary to introduce antiunitary, antilinear operators. These produce semi-linear representations or the "corepresentations" so named by WIGNER. The structure of these will be indicated, and discussed in Part I.

All the above material will apply to any system which admits crystal space group symmetry $\mathfrak{G}$. The case of electrons, or generally spinor particles, is more complicated owing to the need to enlarge the transformation operator group to include transformations of spinor indices which arise when the basic function vector space is spanned by Bloch vectors which are not scalar functions but which possess spinor indices. The material presented here is basic to that generalization.

The reader who finds him/her-self unfamiliar with the material in Sects. 12–18 is urged to refresh his/her understanding by consulting the standard texts [1–3] prior to continuing with the work of this article.

12. Transformation operators on functions. In order to proceed with our analysis it is now necessary to enlarge the scope of discussion to include an operator which transforms a function. The definition of $\{\varphi|\boldsymbol{t}(\varphi)\}$ is, from (6.1)

$$\{\varphi|\boldsymbol{t}(\varphi)\}\cdot\boldsymbol{r}=\boldsymbol{r}'=\varphi\cdot\boldsymbol{r}+\boldsymbol{t}(\varphi). \tag{12.1}$$

Let $\psi(\boldsymbol{r})$ be a scalar function of the position variable in configuration space, $\boldsymbol{r}$. By this one means that at every point $\boldsymbol{r}$ in configuration space, the value of the function ψ is determined, or known.

For purposes of compactness of exposition we shall, in the next few paragraphs, abbreviate the symmetry operators $\{\varphi|\boldsymbol{t}(\varphi)\}$ by the single symbol $S, Q, \ldots$. Thus $\{\varphi|\boldsymbol{t}(\varphi)\}=S$, etc. We shall now define an operator P_S which operates upon, or transforms a function ψ into a function $P_S\psi$. This is to be understood as signifying that given the values of ψ at all points $\boldsymbol{r}$ in configuration space, the values of the function $P_S\psi$ are determined. Let S stand for the symmetry operation which transforms point $\boldsymbol{r}$ into point $\boldsymbol{r}'$:

$$S\cdot\boldsymbol{r}=\boldsymbol{r}', \tag{12.2}$$

where S is of course a space group symmetry transformation like $\{\varphi|\boldsymbol{t}(\varphi)\}$. Then we define

$$P_S\psi(\boldsymbol{r}')=\psi(\boldsymbol{r}). \tag{12.3}$$

That is, the function $P_S\psi$ has the same value at the point $\boldsymbol{r}'$ as the function ψ has at the point $\boldsymbol{r}$. The definition of the operation P_S in no way implies a symmetry for ψ but simply establishes a rule for the analysis. Eq. (12.3) can be manipulated in several equivalent forms:

$$P_S\,\psi(S\,\boldsymbol{r})=\psi(\boldsymbol{r}) \tag{12.4}$$

or

$$P_S\,\psi(\boldsymbol{r})=\psi(S^{-1}\,\boldsymbol{r}) \tag{12.5}$$

In all cases (12.2), (12.3), (12.4) are to be interpreted as relations among functions. The operators P_S are always linear-unitary operators unless otherwise stated.

13. Group of transformation operators on functions. We are given the group of transformations in the crystal configuration or coordinate space $\mathfrak{G}$. When required, we may identify this group as $\mathfrak{G}_{\{\varphi|\boldsymbol{t}(\varphi)\}}$ indicating by the subscript the nature of the operators comprising the group. For compactness, as before we may take $\mathfrak{G}_S$ as the symbol denoting the group.

The group law in $\mathfrak{G}_S$ may be epitomized by observing that if P and R are in $\mathfrak{G}_S$ then

$$P\cdot R \quad \text{is in} \quad \mathfrak{G}_S. \tag{13.1}$$

Now to every $P, R, S, \ldots$ in $\mathfrak{G}_S$ a transformation operator P_R can be associated by (12.2)–(12.5). In this fashion a set of operators is obtained: $P_R, P_S, \ldots$. To demonstrate that this set forms a group, observe that if

$$\boldsymbol{r}'=S\cdot\boldsymbol{r};\qquad \boldsymbol{r}''=R\cdot\boldsymbol{r}'=(RS)\cdot\boldsymbol{r} \tag{13.2}$$

then

$$\begin{aligned}(P_S\,\psi)\,(\boldsymbol{r}')&=\psi(\boldsymbol{r})\\ \big(P_R(P_S\,\psi)\big)\,(\boldsymbol{r}'')&=(P_S\,\psi)\,(\boldsymbol{r}')=\psi(\boldsymbol{r}).\end{aligned} \tag{13.3}$$

Hence, the function $\big(P_R(P_S\,\psi)\big)$ has the value at $\boldsymbol{r}''$ that the function ψ has at $\boldsymbol{r}$. But consider the operator P_{RS} applied to a function:

$$(P_{RS}\,\psi)\,(\boldsymbol{r}'')=\psi(\boldsymbol{r}). \tag{13.4}$$

Then

$$P_R\,P_S=P_{RS} \tag{13.5}$$

that is we may identify the product $P_R\,P_S$ with the single operator P_{RS}. It follows that the set

$$P_E, P_R, P_S, \ldots \quad \text{forms group } \mathfrak{G}_{P_R} \tag{13.6}$$

and that

$$\mathfrak{G}_R \text{ and } \mathfrak{G}_{P_R} \text{ are homomorphic}. \tag{13.7}$$

According to the development just given by which to each operator R there is defined one and only one operator P_R by (12.3) the groups $\mathfrak{G}_R$ and $\mathfrak{G}_{P_R}$ are actually isomorphic.

It needs to be pointed out explicitly here, however, that one may proceed in an actual physical problem in the reverse order. That is, one may first define the transformation operator P_R via transformations on functions, and then show that the set P_R forms a group $\mathfrak{G}_{P_R}$. The corresponding configuration space transformations may be obtained from the P_R. Then $\mathfrak{G}_{P_R}$ may be homomorphic not isomorphic onto $\mathfrak{G}_R$. When we discuss symmetry of the lattice hamiltonian this approach will be taken in part. It is however most important when considering particles with spin.

14. Functions and representations. Let us now assume that we are given a closed linearly independent set of l functions for some physical problem or situation:

$$\psi_1, \psi_2, \ldots, \psi_l \equiv \{\psi_a\}. \tag{14.1}$$

By closed we mean that for the problem at hand, any arbitrary function φ_α can be expressed as a linear combination of the set $\{\psi_a\}$:

$$\varphi_\alpha = \sum_{a=1}^{l} c_{a\alpha} \psi_a. \tag{14.2}$$

Observe that to emphasize the functional relationship, we suppress $\boldsymbol{r}$, the configuration space label of the point. The set $\{\psi_a\}$ can be considered as a set of vectors spanning a Hilbert space.

Now let us apply an operator of type P_S to one of the functions

$$P_S \psi_a(S\boldsymbol{r}) = \psi_a(\boldsymbol{r}). \tag{14.3}$$

But the set $\{\psi_a\}$ is closed by hypothesis. Hence $P_S \psi_a$ must be dependent, and expressible as a linear combination of the set $\{\psi_a\}$:

$$P_S \psi_a = \sum_{n=1}^{l} D(S)_{na} \psi_n. \tag{14.4}$$

Again observe that (14.4) is a relation between functions. If we now keep P_S fixed and allow the specific function ψ_a to be in turn each of the members of the set $\{\psi_a\}$ we have

$$P_S \psi_a = \sum_{n=1}^{l} D(S)_{na} \psi_n, \qquad a = 1, \ldots, l. \tag{14.5}$$

In this fashion we obtain the elements of an $l \times l$ matrix $D(S)$, from (14.5). If we consider the operator P_Q we obtain a matrix $D(Q)$. Allowing the operator P_S to be, in turn, each possible transformation operator in $\mathfrak{G}_{P_S}$ we obtain a set of $l \times l$ matrices

$$D(E),\ D(P), \ldots, D(Q), \ldots. \tag{14.6}$$

We easily find, then that from (14.4) and (13.5)

$$\begin{aligned} P_R P_S \psi_a &= P_R \big(\sum_m D(S)_{ma} \psi_m\big) = \sum_m D(S)_{ma} P_R \psi_m = \sum_m D(S)_{ma} \sum_n D(R)_{nm} \psi_n \\ &= \sum_n \big(\sum_m D(R)_{nm} D(S)_{ma}\big) \psi_n = \sum_n D(RS)_{na} \psi_n. \end{aligned} \tag{14.7}$$

But

$$P_{RS} \psi_a = \sum_n D(RS)_{na} \psi_n. \tag{14.8}$$

Hence the set of matrices combine under matrix multiplication as

$$D(R)\, D(S) = D(RS). \tag{14.9}$$

The set of matrices (14.6) then form a matrix group. When it is necessary for us to distinguish this matrix group in a special fashion we will write $\mathfrak{G}_{D(R)}$.

The group $\mathfrak{G}_{D(R)}$ is the homomorphic image of $\mathfrak{G}_{P_R}$ and $\mathfrak{G}_R$. The matrix group $\mathfrak{G}_{D(R)}$ based upon the closed function set $\{\psi_m\}$ is a representation of $\mathfrak{G}_R$ and $\mathfrak{G}_{P_R}$. The set $\{\psi_a\}$ spans a certain linear vector, Hilbert, space Σ, and the individual functions ψ_m in the set may be considered as base vectors spanning the vector space. The situation is analogous to the fashion in which the rectilinear base vectors $\boldsymbol{a}_1, \boldsymbol{a}_2, \boldsymbol{a}_3$ span the three dimensional Euclidean space in which our crystal is embedded. The set $\{\psi_a\}$ is also often denoted as an invariant space of functions, under the group action. As a rule we shall not use the term complete set for (14.1) since that term may imply some analytic properties not necessary in the present contexts.

15. Irreducible representations and spaces. Let the invariant function space of Sect. 14 be of dimension l. If there is a subset of functions

$$\psi_1, \psi_2, \dots, \psi_m, \qquad m<l \tag{15.1}$$

which is also invariant then the invariant function space splits into two invariant subspaces. In our work if this splitting occurs, we may assume these are disjoint invariant subspaces.

The set of matrices D of the representation group $\mathfrak{G}_{D(R)}$ can be then fully reduced or decomposed. When fully reduced, each matrix in the representation takes the form of a direct sum of two matrices:

$$D(\{\varphi|\boldsymbol{t}(\varphi)\}) = \begin{pmatrix} \bar{D}(\{\varphi|\boldsymbol{t}(\varphi)\}) & 0 \\ 0 & H(\{\varphi|\boldsymbol{t}(\varphi)\}) \end{pmatrix} \tag{15.2}$$

for all $\{\varphi|\tau(\varphi)\}$ in $\mathfrak{G}$. In (15.2), $\bar{D}$ and H are matrices of dimension $m\times m$ and $(l-m)\times(l-m)$ respectively.

If there are several invariant, disjoint spaces then the decomposition (full reduction) (15.2) is generalized to a direct sum of several square matrices.

Correspondingly one can write

$$P_{\{\varphi|\boldsymbol{t}(\varphi)\}}\psi_a = \sum_{n=1}^{m} \bar{D}(\{\varphi|\boldsymbol{t}(\varphi)\})_{na}\psi_n, \qquad a=1,\dots,m \tag{15.3}$$

and

$$P_{\{\varphi|\boldsymbol{t}(\varphi)\}}\psi_a = \sum_{m+1}^{l} H(\{\varphi|\boldsymbol{t}(\varphi)\})_{na}\psi_n, \qquad a=m+1,\dots,l \tag{15.4}$$

for all $\{\varphi|\boldsymbol{t}(\varphi)\}$ in $\mathfrak{G}$. Again (15.3) and (15.4) can be generalized if there are several invariant, disjoint, spaces.

An irreducible representation is a matrix system D or $\mathfrak{G}_{D(R)}$ which cannot be reduced i.e. cannot be transformed into the form (15.2) for all matrices simultaneously. If a matrix representation D of $\mathfrak{G}$ is given, it may be tested for irreducibility by the Schur lemma [*1–3*]. Thus given the set $D(\{\varphi|\boldsymbol{t}(\varphi)\})$, then if there is an M:

$$MD(\{\varphi|\boldsymbol{t}(\varphi)\}) = D(\{\varphi|\boldsymbol{t}(\varphi)\})M \tag{15.5}$$

and the *only* matrix M satisfying this equation is $M = m D(\{\varepsilon|\boldsymbol{0}\})$ where m is a constant, then D is irreducible.

The irreducible representations are conventionally denoted by affixing a superscript to the matrix symbol as $D^{(l)}$. We adopt this convention.

If a set of irreducible representations, $D^{(l)}$, $l=1, \ldots, r$ is given for a finite group $\mathfrak{G}$, one may use several (essentially equivalent) criteria to decide whether one possesses all the irreducible representations of $\mathfrak{G}$. In case of the space groups, not all the criteria can be conveniently applied. Thus, if r is the number of classes of $\mathfrak{G}$, then there are r distinct irreducible representations of $\mathfrak{G}$. Further if the trace or character of $D^{(l)}\{\varphi|\boldsymbol{t}(\varphi)\}$ is denoted

$$\chi^{(l)}(\{\varphi|\boldsymbol{t}(\varphi)\}) \equiv \sum_n D^{(l)}(\{\varphi|\boldsymbol{t}(\varphi)\})_{nn} \tag{15.6}$$

then

$$\sum_{l=1}^{r} |\chi^{(l)}(\{\varepsilon|0\})|^2 = g_p N, \tag{15.7}$$

where $\{\varepsilon|0\}$ is the group identity and $g_p N$ is the order of the group $\mathfrak{G}$. Let a typical element in class $\mathfrak{C}_k$ be denoted C_k; a typical element in the inverse class $\mathfrak{C}_k^{-1}$ is denoted C_k^{-1}; the order of $\mathfrak{C}_k$ is denoted c_k then

$$\sum_{k=1}^{r} c_k \chi^{(l)}(C_k) \chi^{(l')}(C_k)^* = g_p N \delta_{ll'}, \tag{15.8}$$

Finally

$$\sum_{l=1}^{r} c_k \chi^{(l)}(C_k) \chi^{(l)}(C_j) = g_p N \delta_{j,k^{-1}}. \tag{15.9}$$

The results (15.7)–(15.9) are taken over from the well-known results for finite groups. Note of course that these formulae apply to the entire group: the sums are over *all* irreducible representations l, or all classes k, of the group $\mathfrak{G}$. Formulae such as (15.7)–(15.9) would apply as well to a subgroup of $\mathfrak{G}$, but only if it were treated in and of itself as an entity i.e. as a group, with its own classes, etc.

Finally, we recall Maschke's theorem: for a finite group every representation is either irreducible or fully reducible (decomposable).

In the following we often denote the set of functions which span the irreducible linear vector space which is a basis for irreducible representation $D^{(l)}$ of $\mathfrak{G}$ by

$$\Sigma^{(l)} \equiv \{\psi_1^{(l)}, \ldots, \psi_m^{(l)}, \ldots, \psi_l^{(l)}\} \tag{15.10}$$

where l is the dimension of $D^{(l)}$.

In what follows in this article, unless explicitly assumed to the contrary it will be assumed that be basis set of functions (15.10) are an orthonormal set. Thus a suitable scalar product exists such that

$$(\psi_m^{(l)}, \psi_{m'}^{(l')}) = \delta_{mm'} \delta_{ll'} \tag{15.11}$$

where $D^{(l)}$, $D^{(l')}$ $(l' \neq l)$ are inequivalent irreducible representations. Usually

$$(\psi_m^{(l)}, \psi_{m'}^{(l')}) = \int d^3 \boldsymbol{r}\, \psi_m^{(l)}(\boldsymbol{r})^* \psi_{m'}^{(l')}(\boldsymbol{r}). \tag{15.12}$$

Since (15.11) is assumed to hold we may take the irreducible representations $D^{(l)}$ to be unitary:

$$D^{(l)^+} = D^{(l)^{-1}} \tag{15.13}$$

or

$$\widetilde{D^{(l)^*}} = D^{(l)^{-1}}. \tag{15.14}$$

16. Idempotent transformation operators. To obtain the compete set of irreducible representations of the group $\mathfrak{G}_{P_R}$ we may proceed in a systematic fashion using the group algebra of $\mathfrak{G}$ over the complex field [3, 11]. In this fashion we can obtain a reduction or decomposition of this algebra into a direct sum of simple two-sided ideals. When this is accomplished there will be $g_p N$ operators $P_{mn}^{(l)}$ with the property

$$P_{\mu\nu}^{(l)} \cdot P_{\sigma\tau}^{(l')} = P_{\mu\tau}^{(l)} \, \delta_{ll'} \, \delta_{\nu\sigma}. \tag{16.1}$$

The property (16.1) is the key property of the reduced algebra. When one has completely decomposed the group algebra the operators $P_{\mu\nu}^{(l)}$ are completely determined as linear combinations of the basic elements $P_{\{\varphi|\boldsymbol{t}(\varphi)\}}$ of $\mathfrak{G}_{P_R}$. It then follows that

$$P_{\{\varphi|\boldsymbol{t}(\varphi)\}} \, P_{\mu\nu}^{(l)} = \sum_{\mu'} D^{(l)}(\{\varphi|\boldsymbol{t}(\varphi)\})_{\mu\mu'} \, P_{\mu'\nu}^{(l)} \tag{16.2}$$

so that the set of operators $P_{\mu'\nu}^{(l)}$ can be used to obtain the matrices $D^{(l)}$ of the irreducible representation. The set of algebraic objects $P_{\mu\nu}^{(l)}$ have a series of related properties which insure that they are correct. It is to be emphasized that the complete determination of the $P_{\mu\nu}^{(l)}$ is strictly an algebraic problem which is in principle and in practice soluble, given the structure of $\mathfrak{G}$.

The more conventional [12] if more difficult, procedure is the reverse. One then first obtains all the matrix irreducible representation $D^{(l)}$ of $\mathfrak{G}$. Then one constructs the objects $P_{mn}^{(l)}$:

$$P_{mn}^{(l)} = \frac{l}{N g_p} \sum_{\{\varphi|\boldsymbol{t}(\varphi)\}} D^{(l)}(\{\varphi|\boldsymbol{t}(\varphi)\})_{mn}^* \, P_{\{\varphi|\boldsymbol{t}(\varphi)\}}. \tag{16.3}$$

The complete set of these operators express the decomposition of the algebra into the direct sum of simple two sided ideals.

In either event the operators $P_{\mu\nu}^{(l)}$ serve to determine functions $\psi_{\mu\nu}^{(l)}$ which are properly symmetrized: i.e. are the correct linear combinations. Let $\Psi(\boldsymbol{r})$ be an arbitrary function. Then

$$P_{\mu\nu}^{(l)} \, \Psi \equiv \psi_{\mu\nu}^{(l)} \tag{16.4}$$

is a function belonging to the μ row, ν column of a linear vector space. The space is closed, irreducible, and is a basis for irreducible representation $D^{(l)}$ of $\mathfrak{G}$. It may be noted that to employ the full benefit of the operators $P_{\mu\nu}^{(l)}$ it is necessary to characterize the functions which they project in terms of both row and column indices, in order to correspond to the indices of the operators of the ideal. With μ (or ν) held fixed, the set of l functions ν(or μ)$=1, \ldots, l$ are partners in the irreducible vector space.

A weaker set of projection operators can be constructed than those of (16.3) from the characters of the irreducible representations of the group. The operator

$$P^{(l)} \equiv \frac{l}{N g_p} \sum_{\mathfrak{G}} \chi^{(l)}(\{\varphi|\boldsymbol{t}(\varphi)\})^* P_{\{\varphi|\boldsymbol{t}(\varphi)\}} \tag{16.5}$$

has the property

$$P^{(l)} \Psi \equiv \psi^{(l)}. \tag{16.6}$$

That is $P^{(l)}$ produces a function belonging to the l-th irreducible representation, irrespective of row. The function $\psi^{(l)}$ will in general be decomposable into a sum of functions which are bases for different rows of the irreducible representation $D^{(l)}$. The "weak" projection operators are often all that can be conveniently constructed owing to the availability of the characters and not the full matrices. In principle however, both strong and weak projection operators can be obtained with sufficient labor using the straightforward algebraic techniques discussed in [3, 11].

17. Direct products

α) *Direct products of representations.* Let the vector space $\Sigma^{(l)}$ as defined in (15.10) be a basis for the representation of $D^{(l)}$ of $\mathfrak{G}$, and let the set of l' functions $\{\psi_{a'}\}$ which span the linear vector space $\Sigma^{(l')}$ be a basis for the representation $D^{(l')}$ of $\mathfrak{G}$. Then the set of $l \cdot l'$ functions $\{\psi_a \cdot \psi_{a'}\}$, $a=1, \ldots, l$, $a'=1, \ldots, l'$, span a linear vector space $\Sigma^{(l \otimes l')}$ and are a basis for the representation $D^{(l)} \otimes D^{(l')}$ of $\mathfrak{G}$. The vector space $\Sigma^{(l \otimes l')}$ is the direct product vector space and the representation $D^{(l)} \otimes D^{(l')}$ is the direct product representation. It is often convenient, but not necessary, to take $D^{(l)}$ and $D^{(l')}$ to be irreducible. We shall assume this is the case in what immediately follows so that the direct product is taken between irreducible representations.

The matrix elements of the direct product matrices are

$$(D^{(l)} \otimes D^{(l')})_{\alpha\gamma\beta\delta} = D^{(l)}_{\alpha\beta} \cdot D^{(l')}_{\gamma\delta}, \tag{17.1}$$

where the dot on the right hand side of (17.1) is ordinary multiplication of the two complex numbers $D^{(l)}_{\alpha\beta} \cdot D^{(l')}_{\gamma\delta}$. It is understood that each matrix in (17.1) is labelled by the identical group element:

$$\begin{aligned}(D^{(l)} \otimes D^{(l')})(\{\varphi|\boldsymbol{t}(\varphi)\}) &= D^{(l \otimes l')}(\{\varphi|\boldsymbol{t}(\varphi)\})\\ &= D^{(l)}(\{\varphi|\boldsymbol{t}(\varphi)\}) \otimes D^{(l')}(\{\varphi|\boldsymbol{t}(\varphi)\}).\end{aligned} \tag{17.2}$$

In (17.2) we have given several alternate and equivalent ways of writing the direct product matrix.

The direct product representation (17.2) is a representation of $\mathfrak{G}$ [1]. Then $\mathfrak{G}$ is homomorphic onto $\mathfrak{G}_{D^{(l \otimes l')}}(\{\varphi|\boldsymbol{t}(\varphi)\})$ so that every element $\{\varphi|\boldsymbol{t}(\varphi)\}$ is represented by a matrix in (17.2).

Every representation of a finite group is decomposable (Maschke's theorem), hence (17.2) can be decomposed into a sum of irreducible representations. Using (15.6) for definition of the character or trace, we obtain by taking trace of (17.2):

$$\begin{aligned}\chi^{(l \otimes l')}(\{\varphi|\boldsymbol{t}(\varphi)\}) &= \chi^{(l)}(\{\varphi|\boldsymbol{t}(\varphi)\})\, \chi^{(l')}(\{\varphi|\boldsymbol{t}(\varphi)\})\\ &= \mathrm{Tr}\left[D^{(l)} \otimes D^{(l')}(\{\varphi|\boldsymbol{t}(\varphi)\})\right].\end{aligned} \tag{17.3}$$

β) Reduction coefficients.

Reduction coefficients for the direct product of irreducible representations of $\mathfrak{G}$ can now be defined. If the integer $(l\,l'|m) \geqq 0$ denotes the number of times the irreducible representation $D^{(m)}$ appears in the reduction of $D^{(l \otimes l')}$ into irreducible constitutents, then

$$D^{(l \otimes l')} = \sum_m \oplus (l\,l'|m)\, D^{(m)}. \tag{17.4}$$

Also for the vector spaces

$$\Sigma^{(l \otimes l')} = \sum_m \oplus (l\,l'|m)\, \Sigma^{(m)}. \tag{17.5}$$

In (17.4), (17.5) the sum is to be understood as the direct sum, and is symbolized by $\oplus$:

$$D^{(l \otimes l')} = (l\,l'|1)\, D^{(1)} \oplus (l\,l'|2)\, D^{(2)} \oplus \cdots. \tag{17.6}$$

The series (17.6), or (17.4) is also known as the Clebsch-Gordan Series, and the integers $(l\,l'|m)$ are often called "coefficients of the Clebsch-Gordan Series". We prefer to use the term "reduction coefficients" to avoid confusion (see below, Sect. 18), with Clebsch-Gordan Coefficients.

Taking the trace of each matrix in (17.4) we have

$$\chi^{(l \otimes l')}(\{\varphi|\boldsymbol{t}(\varphi)\}) = \sum_m (l\,l'|m)\, \chi^{(m)}(\{\varphi|\boldsymbol{t}(\varphi)\}) \tag{17.7}$$

in (17.7) the sum is over all irreducible representations of $\mathfrak{G}$, and the sum is now an ordinary arithmetic sum. Equivalently the reduction coefficients can be determined by use of the orthonormality relations so that

$$(l\,l'|m) = \frac{1}{g_p N} \sum_{\mathfrak{G}} \chi^{(l \otimes l')}(\{\varphi|\boldsymbol{t}(\varphi)\}) \cdot \chi^{(m)}(\{\varphi|\boldsymbol{t}(\varphi)\})^*, \tag{17.8}$$

where the sum is taken over all elements in the group $\mathfrak{G}$. A slightly more compact form for (17.8) is

$$(l\,l'|m) = \frac{1}{g_p N} \sum_k c_k\, \chi^{(l \otimes l')}(C_k)\, \chi^{(m)}(C_k)^*, \tag{17.9}$$

where the sum is ordinary arithmetic sum and the quantities in (17.9) are c_k, the order of class k, C_k, a typical element in class $\mathfrak{C}_k$, and the various characters previously defined.

The character reduction coefficients are of direct use in physical applications. From them, selection rules are obtained for the allowed optical processes, scattering processes, and others. One principal objective of our work is the determination of the reduction coefficients for space groups. As will be seen later, formulae (17.4) and (17.7) can be used to determine the complete set of coefficients $(l\,l'|m)$.

γ) Irreducible representations of direct product groups [*10*]. A useful result in the analysis of the translation subgroup $\mathfrak{T}$ concerns irreducible representations of an inner direct product group. A group $\mathfrak{H}$ is an inner direct product group of two groups $\mathfrak{A}$ and $\mathfrak{B}$,

$$\mathfrak{H} = \mathfrak{A} \otimes \mathfrak{B} \tag{17.10}$$

if every element of $\mathfrak{H}$ can be uniquely written as a "product" of one element from $\mathfrak{A}$ times one element from $\mathfrak{B}$:

$$H_k = A_i \cdot B_j = B_j \cdot A_i \tag{17.11}$$

and all elements from $\mathfrak{A}$ and $\mathfrak{B}$ commute. If $\Gamma^{(a)}$ is an irreducible representation of $\mathfrak{A}$, and $\Gamma^{(b)}$ is an irreducible representation of $\mathfrak{B}$, then all irreducible representations $\Gamma^{(h)}$ of $\mathfrak{H}$ can be obtained as

$$\Gamma^{(h)} = \Gamma^{(a)} \otimes \Gamma^{(b)} = \Gamma^{(b)} \otimes \Gamma^{(a)}, \tag{17.12}$$

where a and b are permitted to run through all possible indices (respectively irreducible representations) of $\mathfrak{A}$ and $\mathfrak{B}$.

As will soon be apparent, the analysis of irreducible representations of semidirect product groups, or groups which are extensions of one group by another is considerably more complex and cannot be put into such compact and simple form as (17.12). In fact, that analysis will be the burden of the detailed work of the next several sections.

18. Clebsch-Gordan coefficients*. In this section the work begun in Sect. 17 will be completed. There we introduced the direct product of two representations, and in (17.4), the reduction coefficient. The reduction coefficient gives the number of times each irreducible representation is contained in the direct product.

A more complete analysis is obtained when the direct product representation (17.1), (17.2) is similarity transformed by a unitary matrix and thereby brought into fully reduced, or block diagonal, form. The matrix elements of the unitary matrix which simulataneously transforms all matrices into reduced form are denoted Clebsch-Gordan coefficients. These matrix elements also have another very closely related and important significance: they are the elements of the matrix which transforms the direct product space (left hand side of (17.5)) into irreducible spaces (right hand side of (17.5)). Or in other words these matrix elements enable one to determine the "correct linear combinations" of the product functions (each such function is a product of one function from each of the factors) which are bases for an irreducible representation of the product space. The reduction coefficient is a weaker piece of information as soon will become clear.

α) *Definition of Clebsch-Gordan coefficient.* Now recall (15.10) and the discussion in Sect. 17. Let

$$\Sigma^{(l)} \equiv \{\psi_1^{(l)}, \ldots, \psi_l^{(l)}\} \to D^{(l)} \quad \text{of } \mathfrak{G}, \tag{18.1}$$

$$\Sigma^{(l')} \equiv \{\psi_1^{(l')}, \ldots, \psi_{l'}^{(l')}\} \to D^{(l')} \quad \text{of } \mathfrak{G}, \tag{18.2}$$

and

$$\Sigma^{(l \otimes l')} \equiv \{\psi_1^{(l)} \psi_1^{(l')}, \ldots, \psi_m^{(l)} \psi_{m'}^{(l')}, \ldots, \psi_l^{(l)} \psi_{l'}^{(l')}\} \to D^{(l \otimes l')} \quad \text{of } \mathfrak{G} \tag{18.3}$$

where the arrow indicates that the function is a basis of the indicated representation. Then assume $(l\,l'|l'')=1$. We wish to determine the unique set of l'' functions which are bases for $D^{(l'')}$ and which are some specified linear combinations of

* A considerable portion of this section was written by Dr. RHODA BERENSON.

products $\psi_\mu^{(l)}\psi_{\mu'}^{(l')}$. If $(l\,l'|l'')>1$ there will be $(l\,l'|l'')$ correct linear combinations of the basic set (18.3), each such linear combination will give a linearly independent function $\psi_\mu^{(l)}\psi_{\mu'}^{(l')}$ where $\gamma=1,\ldots,(l\,l'|l'')$, transforming as a basis for row μ'' of $D^{(l'')}$. This is the case where multiplicity exists, and sometimes γ is denoted a multiplicity index.

The coefficients $\begin{pmatrix} l & l' & l'' & \gamma \\ \mu & \mu' & \mu'' & \end{pmatrix}$, of these correct linear combinations are called Clebsch-Gordan or vector coupling coefficients and are defined by

$$\psi_{\mu''}^{(l'')\gamma}=\sum_{\mu\,\mu'}\begin{pmatrix} l & l' & l'' & \gamma \\ \mu & \mu' & \mu'' & \end{pmatrix}\psi_\mu^{(l)}\psi_{\mu'}^{(l')}. \tag{18.4}$$

For $(l\,l'|l'')=1$ these coefficients can be uniquely determined up to a phase, i.e. if we multiply each coefficient (all μ, μ' and μ'') by the same phase we get the same space $\Sigma^{(l'')}$ multiplied by a phase.

If $(l\,l'|l'')>1$, say $(l\,l'|l'')=2$, then

$$\psi_{\mu''}^{(l'')\gamma_1}=\sum_{\mu\,\mu'}\begin{pmatrix} l & l' & l'' & \gamma_1 \\ \mu & \mu' & \mu'' & \end{pmatrix}\psi_\mu^{(l)}\psi_{\mu'}^{(l')} \tag{18.5}$$

and

$$\psi_{\mu''}^{(l'')\gamma_2}=\sum_{\mu\,\mu'}\begin{pmatrix} l & l' & l'' & \gamma_2 \\ \mu & \mu' & \mu'' & \end{pmatrix}\psi_\mu^{(l)}\psi_{\mu'}^{(l')}. \tag{18.6}$$

But any linear combination of $\psi_{\mu''}^{(l'')\gamma_1}$ and $\psi_{\mu''}^{(l'')\gamma_2}$ also transforms as a basis for $D^{(l'')}$ and, therefore, any linear combination of the coefficients $\begin{pmatrix} l & l' & l'' & \gamma_1 \\ \mu & \mu' & \mu'' & \end{pmatrix}$ and $\begin{pmatrix} l & l' & l'' & \gamma_2 \\ \mu & \mu' & \mu'' & \end{pmatrix}$ would also give correct linear combinations of the products $\psi_\mu^{(l)}\psi_{\mu'}^{(l')}$. Hence for $(l\,l'|l'')>1$ the choice of Clebsch-Gordan coefficients is not unique. Any linear combination of these coefficients is equally good.

Alternatively, the products $\psi_\mu^{(l)}\psi_{\mu'}^{(l')}$ can be written as a linear combination of the $\psi_{\mu''}^{(l'')\gamma}$

$$\psi_\mu^{(l)}\psi_{\mu'}^{(l')}=\sum_{\gamma\,l''\,\mu''}\begin{pmatrix} l'' & \gamma & l & l' \\ \mu'' & & \mu & \mu' \end{pmatrix}\psi_{\mu''}^{(l'')}. \tag{18.7}$$

Assuming all of the irreducible representations involved are unitary, then

$$\begin{pmatrix} l'' & \gamma & l & l' \\ \mu'' & & \mu & \mu' \end{pmatrix}=\begin{pmatrix} l & l' & l'' & \gamma \\ \mu & \mu' & \mu'' & \end{pmatrix}^* \tag{18.8}$$

and

$$\sum_{l''\,\mu''\,\gamma}\begin{pmatrix} l & l' & l'' & \gamma \\ \mu & \mu' & \mu'' & \end{pmatrix}\begin{pmatrix} l'' & \gamma & l & l' \\ \mu'' & & \bar\mu & \bar\mu' \end{pmatrix}=\sum_{l''\,\mu''\,\gamma}\begin{pmatrix} l & l' & l'' & \gamma \\ \mu & \mu' & \mu'' & \end{pmatrix}\begin{pmatrix} l & l' & l'' & \gamma \\ \bar\mu & \bar\mu' & \mu'' & \end{pmatrix}^*=\delta_{\mu\bar\mu}\,\delta_{\mu'\bar\mu'} \tag{18.9}$$

and

$$\sum_{\mu\,\mu'}\begin{pmatrix} l'' & \gamma & l & l' \\ \mu'' & & \mu & \mu' \end{pmatrix}\begin{pmatrix} l & l' & \bar l'' & \bar\gamma \\ \mu & \mu' & \bar\mu'' & \end{pmatrix}=\sum_{\mu\,\mu'}\begin{pmatrix} l & l' & l'' & \gamma \\ \mu & \mu' & \mu'' & \end{pmatrix}^*\begin{pmatrix} l & l' & \bar l'' & \bar\gamma \\ \mu & \mu' & \bar\mu'' & \end{pmatrix}=\delta_{l''\bar l''}\,\delta_{\mu''\bar\mu''}\,\delta_{\gamma\bar\gamma}. \tag{18.10}$$

It should be remarked explicitly that in order to satisfy the condition of unitarity (18.8) for the case of multiplicity, with $\gamma>1$ it is necessary to assume that the distinct bases $\psi_{\mu''}^{(l'')\gamma}$ have all been made orthonormal. That is, that a suitable scalar (inner) product can be defined so that

$$(\psi_{\mu''}^{(l'')\gamma}, \psi_{\mu''}^{(l'')\gamma'}) = \int d^3\boldsymbol{r}\, \psi_{\mu''}^{(l'')\gamma *}\, \psi_{\mu''}^{(l'')\gamma'} = \delta_{\gamma\gamma'}.$$

This can always be achieved by using the Gram-Schmidt method to prepare orthonormal functions from a linearly independent set.

Now we prove that a matrix constructed from the Clebsch-Gordan elements will reduce the direct product matrix (17.1), (17.2). Recall the definition of an operator P_R which transforms a function as in (14.3), (14.4). If we operate on Eq. (18.4) with P_R we have

$$P_R\, \psi_{\mu''}^{(l'')\gamma} = \sum_{\mu\mu'} \left(\begin{matrix} l & l' \\ \mu & \mu' \end{matrix}\middle|\begin{matrix} l'' & \gamma \\ \mu'' & \end{matrix}\right) \sum_{\nu\nu'} D^{(l)}(R)_{\nu\mu}\, D^{(l')}(R)_{\nu'\mu'}\, \psi_\nu^{(l)}\, \psi_{\nu'}^{(l')}. \tag{18.11}$$

But

$$P_R\, \psi_{\mu''}^{(l'')\gamma} = \sum_{\bar{\mu}''} D^{(l'')}(R)_{\bar{\mu}''\mu''}\, \psi_{\bar{\mu}''}^{(l'')} \tag{18.12}$$

or

$$\begin{aligned}\sum_{\bar{\mu}''} D^{(l'')}(R)_{\bar{\mu}''\mu''} \sum_{\bar{\nu}\bar{\nu}'} \left(\begin{matrix} l & l' \\ \bar{\nu} & \bar{\nu}' \end{matrix}\middle|\begin{matrix} l'' & \gamma \\ \bar{\mu}'' & \end{matrix}\right) \psi_{\bar{\nu}}^{(l)}\, \psi_{\bar{\nu}'}^{(l')} \\ = \sum_{\mu\mu'} \sum_{\nu\nu'} \left(\begin{matrix} l & l' \\ \mu & \mu' \end{matrix}\middle|\begin{matrix} l'' & \gamma \\ \mu'' & \end{matrix}\right) D^{(l)}(R)_{\nu\mu}\, D^{(l')}(R)_{\nu'\mu'}\, \psi_\nu^{(l)}\, \psi_{\nu'}^{(l')}.\end{aligned} \tag{18.13}$$

Since the products $\psi_\nu^{(l)}\, \psi_{\nu'}^{(l')}$ are linearly independent,

$$\sum_{\bar{\mu}''} \left(\begin{matrix} l & l' \\ \nu & \nu' \end{matrix}\middle|\begin{matrix} l'' & \gamma \\ \bar{\mu}'' & \end{matrix}\right) D^{(l'')}(R)_{\bar{\mu}''\mu''} = \sum_{\mu\mu'} \left(\begin{matrix} l & l' \\ \mu & \mu' \end{matrix}\middle|\begin{matrix} l'' & \gamma \\ \mu'' & \end{matrix}\right) D^{(l)}(R)_{\nu\mu}\, D^{(l')}(R)_{\nu'\mu'}. \tag{18.14}$$

Eq. (18.14) can be written in a more useful form if we multiply both sides by $\left(\begin{matrix} l'' & \gamma \\ \mu'' & \end{matrix}\middle|\begin{matrix} l & l' \\ \bar{\nu} & \bar{\nu}' \end{matrix}\right)$, sum on $l''\,\mu''\,\gamma$ and use Eq. (18.9). Then we have

$$\begin{aligned}\sum_{l''\mu''\gamma} \sum_{\bar{\mu}''} \left(\begin{matrix} l & l' \\ \nu & \nu' \end{matrix}\middle|\begin{matrix} l'' & \gamma \\ \bar{\mu}'' & \end{matrix}\right) D^{(l'')}(R)_{\bar{\mu}''\mu''} \left(\begin{matrix} l'' & \gamma \\ \mu'' & \end{matrix}\middle|\begin{matrix} l & l' \\ \bar{\nu} & \bar{\nu}' \end{matrix}\right) \\ = \sum_{\mu\mu'} \delta_{\bar{\nu}\mu}\, \delta_{\bar{\nu}'\mu'}\, D^{(l)}(R)_{\nu\mu}\, D^{(l')}(R)_{\nu'\mu'} = D^{(l)}(R)_{\nu\bar{\nu}}\, D^{(l')}(R)_{\nu'\bar{\nu}'}.\end{aligned} \tag{18.15}$$

Hence, the Clebsch-Gordan coefficients are elements of the unitary matrix which brings the direct product $D^{(l)} \otimes D^{(l')}$ into reduced form.

This may be seen more clearly if we consider the unitary matrix $U^{(l\otimes l')}$ which brings $D^{(l)} \otimes D^{(l')}$ into the fully reduced matrix Δ. Then

$$U^{-1}\, D^{(l)}(R) \otimes D^{(l')}(R)\, U = \Delta(R)$$

or

$$D^{(l)}(R) \otimes D^{(l')}(R) = U\, \Delta(R)\, \tilde{U}^*. \tag{18.16}$$

The elements of Δ are usually written as $\Delta_{l''\nu'',\bar{l}''\bar{\nu}''}$ where

$$\Delta_{l''\nu'',\bar{l}''\bar{\nu}''} = \delta_{l''\bar{l}''}\, D_{\nu''\bar{\nu}''}^{(l'')}. \tag{18.17}$$

If $(l\,l'|l'')>1$, it is convenient to arrange Δ (without any loss of generality) so that all $D^{(l'')}$ appear in consecutive blocks. Then we can write

$$\Delta_{l''\gamma\nu'',\bar{l}''\bar{\gamma}\bar{\nu}''}=\delta_{l''\bar{l}''}\,\delta_{\gamma\bar{\gamma}}\,D^{(l'')}_{\nu''\bar{\nu}''} \tag{18.18}$$

for $\gamma=1\dots(l\,l'|l'')$.

Eq. (18.16) can be rewritten as

$$\sum_{l''\gamma\nu''}\sum_{\bar{l}''\bar{\gamma}\bar{\nu}''} U^{(l\otimes l')}_{\nu\nu',l''\gamma\nu''}\,\Delta(R)_{l''\gamma\nu'',\bar{l}''\bar{\gamma}\nu''}\,U^{(l\otimes l')^{-1}}_{\bar{l}''\bar{\gamma}\bar{\nu}'',\bar{\nu}\bar{\nu}'}=\big(D^{(l)}(R)\otimes D^{(l')}(R)\big)_{\nu\nu',\bar{\nu}\bar{\nu}'}$$

or

$$\sum_{l''\gamma}\sum_{\nu''\bar{\nu}''} U^{(l\otimes l')}_{\nu\nu',l''\gamma\nu''}\,D^{(l'')}(R)_{\nu''\bar{\nu}''}\,U^{(l\otimes l')*}_{\bar{\nu}\bar{\nu}',l''\gamma\bar{\nu}''}=D^{(l)}(R)_{\nu\bar{\nu}}\,D^{(l')}(R)_{\nu'\bar{\nu}'}. \tag{18.19}$$

Comparing Eqs. (18.19) and (18.15) we see that

$$U^{(l\otimes l')}_{\nu\nu',l''\gamma\nu''}=\begin{pmatrix} l & l' & l'' & \gamma \\ \nu & \nu' & \nu'' & \end{pmatrix} \tag{18.20}$$

and that the Clebsch-Gordan coefficients are the elements of the unitary matrix that reduces $D^{(l)}\otimes D^{(l')}$. It may be useful to point out that the matrix (18.20) is labelled asymmetrically, since (as is seen from (18.16)) the rows of U carry indices appropriate to the direct product matrix while the columns carry indices relevant to the reduced matrix i.e. labels of a single irreducible representation plus the multiplicity label. In fully block reduced form Δ appears as

$$\Delta(R)=\begin{pmatrix} D^{(1)}(R)\dots & \dots & \dots & \dots \\ \vdots & \ddots & \dots & \vdots \\ \vdots & \vdots & D^{(l'')}(R)\dots & \vdots \\ \vdots & \vdots & \vdots \ddots & \vdots \\ \vdots & \vdots & \vdots & \dots D^{(r)}(R) \end{pmatrix}. \tag{18.21}$$

β) *Calculation of Clebsch-Gordan coefficients.* An alternate form of Eq. (18.15) which enables us to calculate the Clebsch-Gordan coefficients most readily is obtained by multiplying both sides of (18.17) by $D^{(\bar{l}'')}(R)^*_{\nu''\bar{\nu}''}$ summing on R and using orthogonality of irreducible representations. Then

$$\begin{aligned}\sum_{l''\gamma}\sum_{\bar{\mu}''\mu''}\begin{pmatrix} l & l' & l'' & \gamma \\ \nu & \nu' & \bar{\mu}'' & \end{pmatrix}\begin{pmatrix} l'' & \gamma & l & l' \\ \mu'' & & \bar{\nu} & \bar{\nu}' \end{pmatrix}\delta_{l''\bar{l}'}\,\delta_{\bar{\mu}''\nu''}\,\delta_{\mu''\bar{\nu}''}\frac{g}{l''}\\ =\sum_R D^{(\bar{l}'')}(R)^*_{\nu''\bar{\nu}''}\,D^{(l)}(R)_{\nu\bar{\nu}}\,D^{(l')}(R)_{\nu'\bar{\nu}'}\end{aligned} \tag{18.22}$$

or

$$\sum_\gamma\begin{pmatrix} l & l' & l'' & \gamma \\ \nu & \nu' & \nu'' & \end{pmatrix}\begin{pmatrix} l & l' & l'' & \gamma \\ \bar{\nu} & \bar{\nu}' & \bar{\nu}'' & \end{pmatrix}^*=\frac{l''}{g}\sum_R D^{(l'')}(R)^*_{\nu''\bar{\nu}''}\,D^{(l)}(R)_{\nu\bar{\nu}}\,D^{(l')}(R)_{\nu'\bar{\nu}'}. \tag{18.23}$$

First consider the case of $(l\,l'|l'')=1$. Then

$$\begin{pmatrix} l & l' & l'' \\ \nu & \nu' & \nu'' \end{pmatrix}\begin{pmatrix} l & l' & l'' \\ \bar{\nu} & \bar{\nu}' & \bar{\nu}'' \end{pmatrix}^*=\frac{l''}{g}\sum_R D^{(l'')}(R)^*_{\nu''\bar{\nu}''}\,D^{(l)}(R)_{\nu\bar{\nu}}\,D^{(l')}(R)_{\nu'\bar{\nu}'}. \tag{18.24}$$

If in Eq. (18.24) we set $\nu=\bar{\nu}$, $\nu'=\bar{\nu}'$, $\nu''=\bar{\nu}''$, we can determine which coefficients are non-zero. Having found a non-zero coefficient, say for $\bar{\nu}=\nu_0$, $\bar{\nu}'=\nu'_0$ and $\bar{\nu}''=\nu''_0$, we can choose its phase as real, fix $\bar{\nu}=\nu_0$, $\bar{\nu}'=\nu'_0$ and $\bar{\nu}''=\nu''_0$ and then let ν, ν' and ν'' range over all allowable values. This procedure yields the entire Clebsch-Gordan matrix for $(l\,l'|l'')=1$.

For $(l\,l'|l'')>1$ a systematic procedure to obtain all $(l\,l'|l'')$ sets of coefficients has been given by KOSTER[1]. Notice that what appears in the right hand side of Eq. (18.23) is

$$\sum_{\gamma}\begin{pmatrix} l & l' & l'' & \gamma \\ \nu & \nu' & \nu'' & \end{pmatrix}\begin{pmatrix} l & l' & l'' & \gamma \\ \bar{\nu} & \bar{\nu}' & \bar{\nu}'' & \end{pmatrix}^*$$

which for fixed $\bar{\nu}$, $\bar{\nu}'$, $\bar{\nu}''$ is a linear combination of Clebsch-Gordan coefficients. But since any linear combination of coefficients is equally good, we can separately calculate

$$\begin{pmatrix} l & l & l'' & \gamma_1 \\ \nu & \nu' & \nu'' & \end{pmatrix}, \ldots, \begin{pmatrix} l & l' & l'' & \gamma_{(l\,l'|l'')} \\ \nu & \nu' & \nu'' & \end{pmatrix}$$

as long as we assure that the matrices obtained for γ and γ' are orthogonal. Hence, we proceed as in the case for $(l\,l'|l'')=1$ by finding a non-zero coefficient, fixing $\bar{\nu}=\nu_0$, $\bar{\nu}'=\nu'_0$, $\bar{\nu}''=\nu''_0$, and letting ν, ν' and ν'' range over all values. This yields the Clebsch-Gordan matrix for $\gamma=1$. Now select ν_0, ν'_0, ν''_0 which are different from the first set, yielding a different non-zero coefficient. Let $\bar{\nu}=\nu_0$, $\bar{\nu}'=\nu'_0$ and $\bar{\nu}''=\nu''_0$ and let ν, ν', ν'' range over all allowed values, yielding a second set of coefficients, which can be made orthogonal to the first set. We continue in this manner until all $(l\,l'|l'')$ sets of coefficients are obtained.

γ) *Transformation of Clebsch-Gordan coefficients.* It will be of later interest to demonstrate how the matrix of Clebsch-Gordan coefficients transforms when the representations $D^{(l)}$, $D^{(l')}$ and $D^{(l'')}$ undergo similarity transformations. Let A be a unitary matrix; and

$$\bar{D}^{(l)}=AD^{(l)}A^{-1};\qquad \bar{D}^{(l')}=A'D^{(l')}A'^{-1};\qquad \bar{D}^{(l'')}=A''D^{(l'')}A''^{-1}. \tag{18.25}$$

Then the transformed basis functions are

$$\varphi_{\mu}^{(\bar{l})}=\sum_{\lambda}A_{\lambda\mu}^{-1}\psi_{\lambda}^{(l)};\qquad \varphi_{\mu'}^{(\bar{l}')}=\sum_{\lambda'}A'^{-1}_{\lambda'\mu'}\psi_{\lambda'}^{(l')};\qquad \varphi_{\mu''}^{(\bar{l}'')}=\sum_{\lambda''}A''^{-1}_{\lambda''\mu''}\psi_{\lambda''}^{(l'')}. \tag{18.26}$$

If for simplicity we write the Clebsch-Gordan matrix for the original basis functions $\{\psi\}$ as $U^{(l\otimes l')}$ and for the transformed basis $\{\varphi\}$ as $V^{(\bar{l}\otimes\bar{l}')}$, then we have

$$\begin{aligned}
\varphi_{\mu''}^{(\bar{l}'')}&=\sum_{\lambda''}A''^{-1}_{\lambda''\mu''}\psi_{\mu''}^{(l'')}\\
&=\sum_{\lambda''}A''^{-1}_{\lambda''\mu''}\sum_{\lambda\lambda'}U^{(l\otimes l')}_{\lambda\lambda',l''\lambda''}\psi_{\lambda}^{(l)}\psi_{\lambda'}^{(l')}\\
&=\sum_{\lambda\lambda'\lambda''}A''^{-1}_{\lambda''\mu''}U^{(l\otimes l')}_{\lambda\lambda',l''\lambda''}\sum_{\mu\mu'}A_{\mu\lambda}A'_{\mu'\lambda'}\varphi_{\mu}^{(\bar{l})}\varphi_{\mu'}^{(\bar{l}')}.
\end{aligned} \tag{18.27}$$

But

$$\varphi_{\mu''}^{(\bar{l}'')}=\sum_{\mu\mu'}V^{(\bar{l}\otimes\bar{l}')}_{\mu\mu',\bar{l}''\mu''}\varphi_{\mu}^{(\bar{l})}\varphi_{\mu'}^{(\bar{l}')} \tag{18.28}$$

[1] G.F. KOSTER: Phys. Rev. **109**, 227 (1958).

so that

$$V^{(l \otimes l')}_{\mu\mu', l''\mu''} = \sum_{\lambda\lambda'\lambda''} (A \otimes A')_{\mu\mu', \lambda\lambda'} U_{\lambda\lambda', l''\lambda''} A''^{-1}_{\lambda''\mu''} \tag{18.29}$$

or

$$V = (A \otimes A') U A''^{-1}. \tag{18.30}$$

δ) *Method of projection operators.* An alternate method for obtaining Clebsch-Gordan coefficients is to use projection operators. The projection operator $P^{(l'')}_{\mu''\nu''}$, defined as

$$P^{(l'')}_{\mu''\nu''} = \frac{l''}{g} \sum_R D^{(l'')}(R)^*_{\mu''\nu''} P_R \tag{18.31}$$

has the following properties:

$$P^{(l'')}_{\mu''\mu''} \psi^{(l'')}_{\mu''} = \psi^{(l'')}_{\mu''}, \tag{18.32}$$

$$P^{(l'')}_{\mu''\nu''} \psi^{(l'')}_{\nu''} = \psi^{(l'')}_{\mu''} \tag{18.33}$$

$$P^{(l'')}_{\mu''\nu''} F = \psi^{(l'')}_{\mu''}, \tag{18.34}$$

where F is an arbitrary linear combination of basis functions,

$$F = \sum_l \sum_\mu \psi^{(l)}_\mu.$$

We can use projection operators to obtain the basis functions for $D^{(l)}$ and $D^{(l')}$, form all products $\psi^{(l)}_\mu \psi^{(l')}_{\mu'}$ and then use Eq. (18.34) to project those linear combinations of products which transform as $\psi^{(l'')}_{\mu''}$. Hence, projection operators can be used to obtain Clebsch-Gordan coefficients. Since this procedure requires first calculating the basis functions it is easier to use Eq. (18.23) which only requires the representations. However, if basis functions are available, the projection operator provides a good check on the coefficients computed using Eqs. (18.23) since from Eq. (18.32) we must have

$$P^{(l'')}_{\mu''\mu''} \sum_{\mu\mu'} \left(\begin{matrix} l & l' \\ \mu & \mu' \end{matrix} \middle| \begin{matrix} l'' & \gamma \\ \mu'' & \end{matrix} \right) \psi^{(l)}_\mu \psi^{(l')}_{\mu'} = \sum_{\mu\mu'} \left(\begin{matrix} l & l' \\ \mu & \mu' \end{matrix} \middle| \begin{matrix} l'' & \gamma \\ \mu'' & \end{matrix} \right) \psi^{(l)}_\mu \psi^{(l')}_{\mu'}. \tag{18.35}$$

D. Irreducible representations of the crystal translation group $\mathfrak{T}$

19. Introduction. In the next few Sects. 20–25 the irreducible representations of the crystal translation group $\mathfrak{T}$ will be analysed. The essential concepts of wave vector $\boldsymbol{k}$, Bloch vector $\psi^{(\boldsymbol{k})}$, Brillouin zone, and completeness and orthonormality relations for the irreducible representations will be discussed, as well as the direct product of irreducible representations of $\mathfrak{T}$. Since $\mathfrak{T}$ is an Abelian group, actually the direct product of three simpler Abelian groups, the mathematical theory is very simple. It is necessary however in order to discuss the space group representations that this material be in useful form.

20. Irreducible representations of $\mathfrak{T}$. In this section we consider the irreducible representations of the one-dimensional translation group $\mathfrak{T}_1$. From (4.33) it follows that $\mathfrak{T}_1$ is an Abelian group on the generator $\{\varepsilon|\boldsymbol{a}_1\}$.

An Abelian group has as many classes $\mathfrak{C}_k$ as elements, and each class is comprised of a single element. Thus, in $\mathfrak{T}_1$, every element $\{\varepsilon|l_1\,\boldsymbol{a}_1\}$ is the only member of its own class: $\mathfrak{C}_{l_1}$. Thus an Abelian group has as many one dimensional irreducible representations as it has elements. Let us denote an irreducible representation of $\mathfrak{T}_1$ by $D^{(k_1)}$. Then since, from (4.41)

$$\{\varepsilon|\boldsymbol{a}_1\}^{2N_1}=\{\varepsilon|\boldsymbol{0}\} \tag{20.1}$$

we must have

$$D^{(k_1)}(\{\varepsilon|\boldsymbol{a}_1\}^{2N_1})=D^{(k_1)}(\{\varepsilon|\boldsymbol{0}\})=1. \tag{20.2}$$

But

$$D^{(k_1)}(\{\varepsilon|\boldsymbol{a}_1\}^{2N_1})=[D^{(k_1)}(\{\varepsilon|\boldsymbol{a}_1\})]^{2N_1}=1. \tag{20.3}$$

Hence $D^{(k_1)}(\{\varepsilon|\boldsymbol{a}_1\})$ must be one of the $2N_1$ th roots of unity:

$$D^{(k_1)}(\{\varepsilon|\boldsymbol{a}_1\})=\exp-(2\pi i p_1/2N_1), \tag{20.4}$$

where

$$p_1=0, 1, 2, \ldots, (2N_1-1). \tag{20.5}$$

It is more convenient to choose for p_1:

$$p_1=-N_1, (-N_1+1), \ldots, 0, \ldots, (N_1-1). \tag{20.6}$$

The set of $2N_1$ choices for p_1 produces the set of $2N_1$ permissible roots of unity in (20.4). An equivalent way of stating (20.6) is that p_1 can take on the integer values (20.6) *modulo* $(2N_1)$. That is any permitted value of p_1 is equivalent to p_1+2N_1:

$$p_1 \doteqdot p_1+2N_1. \tag{20.7}$$

The equivalence of p_1 and p_1+2N_1 means that these two choices produce the identical root of unity in (20.4), and hence cannot be distinguished. To simplify our notation still further, we introduce vector terminology, based on the construction of the reciprocal lattice.

21. The reciprocal lattice. Given the triad of vectors $\boldsymbol{a}_1, \boldsymbol{a}_2, \boldsymbol{a}_3$ in configuration space (Sect. 4) we can define [6] a reciprocal triad $\boldsymbol{b}_1, \boldsymbol{b}_2, \boldsymbol{b}_3$. These have the property

$$\boldsymbol{a}_i\cdot\boldsymbol{b}_j=\delta_{ij} \tag{21.1}$$

and may be obtained by

$$\boldsymbol{b}_1=\frac{\boldsymbol{a}_2\times\boldsymbol{a}_3}{|\boldsymbol{a}_1\cdot\boldsymbol{a}_2\times\boldsymbol{a}_3|} \tag{21.2}$$

and cyclic permutations. The reciprocal vectors $\boldsymbol{b}_j$ span a space which is called reciprocal space. Evidently, in the reciprocal space, vectors of form

$$\boldsymbol{\eta}=y_1\,\boldsymbol{b}_1+y_2\,\boldsymbol{b}_2+y_3\,\boldsymbol{b}_3 \tag{21.3}$$

may be defined where the y_j are numbers on the range $\infty<y_j<-\infty$. A vector such as (21.3) has the dimension of an inverse length since evidently the $\boldsymbol{b}_j$ have that dimension. If the components y_j range continuously over their permitted range then vectors $\boldsymbol{\eta}$ fill reciprocal space.

Let us define a lattice vector in reciprocal space as

$$\boldsymbol{B}_H=h_1\,\boldsymbol{b}_1+h_2\,\boldsymbol{b}_2+h_3\,\boldsymbol{b}_3, \tag{21.4}$$

where h_j are integers. The set of all such reciprocal lattice vectors, or *rel vectors*, define a net in the reciprocal lattice. We may simply consider that net or lattice to be defined by the assembly of points at the termini of the rel vectors. In our work the reciprocal lattice is to be taken as simply a collection of points; these points will be seen to be related to the irreducible representations of the translation group $\mathfrak{T}$. We do not make use of the theory of weighted reciprocal (or Fourier) space developed by EWALD[1].

A vector parallel to $\boldsymbol{b}_1$ may be defined by

$$\boldsymbol{k}_1\equiv\left(\frac{2\pi p_1}{2N_1}\right)\boldsymbol{b}_1 \tag{21.5}$$

and

$$p_1=-N_1,\,-N_1+1,\,\ldots,\,N_1-1, \tag{21.6}$$

where $2N_1$ is given in (4.41) and (20.1) as the edge of the Born-Karman cell in the $\boldsymbol{b}_1$ direction. The set of such vectors with components which are rationalfractions evidently form a dense, but discontinuous array. These vectors have the property

$$-\pi\,\boldsymbol{b}_1\leqq k_1<\pi\,\boldsymbol{b}_1. \tag{21.7}$$

Observe the inequality sign on the left hand side of (21.7), which comes from the requirement that the $(2N_1)$ roots of unity which are used in (20.4) shall all be independent.

Notice that the vectors are defined with (20.4)–(20.7) in mind and are restricted to a region about the origin in the reciprocal lattice. Now to make the vector (21.4) consonant with (20.4) we identify $\boldsymbol{k}_1=\pi\,\boldsymbol{b}_1$ and $\boldsymbol{k}_1=-\pi\,\boldsymbol{b}_1$ as being the same vector.

Thus we can rewrite (20.4) using (20.5) as:

$$D^{(\boldsymbol{k}_1)}(\{\varepsilon|\boldsymbol{a}_1\})=\exp-i\,\boldsymbol{k}_1\cdot\boldsymbol{a}_1. \tag{21.8}$$

This is an extremely useful form in which to have the one dimensional matrix irreducible representation of $\mathfrak{T}_1$.

22. Irreducible representations of $\mathfrak{T}=\mathfrak{T}_1\otimes\mathfrak{T}_2\otimes\mathfrak{T}_3$. Evidently we can repeat for the group $\mathfrak{T}_2$ the identical argument just used for $\mathfrak{T}_1$. Thus, call

$$\boldsymbol{k}_2\equiv\left(\frac{2\pi p_2}{2N_2}\right)\boldsymbol{b}_2 \tag{22.1}$$

---

[1] P. P. EWALD: Z. Physik **2**, 232 (1920).

with

$$p_2 = -N_2, -N_2+1, \ldots, N_2-1 \tag{22.2}$$

and hence

$$-\pi \boldsymbol{b}_2 \leqq \boldsymbol{k}_2 < \pi \boldsymbol{b}_2. \tag{22.3}$$

Then if we denote an irreducible representation of $\mathfrak{T}_2$ by $D^{(\boldsymbol{k}_2)}$, we obtain

$$D^{(\boldsymbol{k}_2)}(\{\varepsilon|\boldsymbol{a}_2\}) = \exp - i\, \boldsymbol{k}_2 \cdot \boldsymbol{a}_2. \tag{22.4}$$

Finally, repeating the argument for $\mathfrak{T}_3$, we call

$$\boldsymbol{k}_3 \equiv \left(\frac{2\pi p_3}{2N_3}\right) \boldsymbol{b}_3 \tag{22.5}$$

with

$$p_3 = -N_3, -N_3+1, \ldots, N_3-1 \tag{22.6}$$

and hence

$$-\pi \boldsymbol{b}_3 \leqq \boldsymbol{k}_3 < \pi \boldsymbol{b}_3. \tag{22.7}$$

Then denoting an irreducible representation of $\mathfrak{T}_3$ by $D^{(\boldsymbol{k}_3)}$, we obtain

$$D^{(\boldsymbol{k}_3)}(\{\varepsilon|\boldsymbol{a}_3\}) = \exp - i\boldsymbol{k}_3 \cdot \boldsymbol{a}_3. \tag{22.8}$$

Now, the group $\mathfrak{T}$ is a direct product group so that the discussion of Eqs. (17.10)–(17.12) applies. That is let us denote the irreducible representations of $\mathfrak{T}$ by $D^{(\boldsymbol{k})}$ then

$$D^{(\boldsymbol{k})} = D^{(\boldsymbol{k}_1)} \otimes D^{(\boldsymbol{k}_2)} \otimes D^{(\boldsymbol{k}_3)}. \tag{22.9}$$

Hence using (22.9) we can find the matrix representing any arbitrary element in $\mathfrak{T}$, in each of the permitted irreducible representations of $\mathfrak{T}$. Observe that there are $(2N_1)(2N_2)(2N_3)$ such irreducible representations, that is choices of p_1, p_2, p_3. This is the number of elements, and classes, in $\mathfrak{T}$. Hence (22.9) does in fact give all irreducible representations of $\mathfrak{T}$.

23. Wave vector: First Brillouin zone. Consider now an element in $\mathfrak{T}$

$$\{\varepsilon|\boldsymbol{R}_L\} = \{\varepsilon|l_1 \boldsymbol{a}_1 + l_2 \boldsymbol{a}_2 + l_3 \boldsymbol{a}_3\}. \tag{23.1}$$

It is represented, in $D^{(\boldsymbol{k})}$, by

$$D^{(\boldsymbol{k})}(\{\varepsilon|\boldsymbol{R}_L\}) = \exp - i\, \boldsymbol{k} \cdot \boldsymbol{R}_L, \tag{23.2}$$

where we define the total wave vector $\boldsymbol{k}$, by

$$\boldsymbol{k} \equiv \boldsymbol{k}_1 + \boldsymbol{k}_2 + \boldsymbol{k}_3 = 2\pi\left(\frac{p_1}{2N_1}\right)\boldsymbol{b}_1 + 2\pi\left(\frac{p_2}{2N_2}\right)\boldsymbol{b}_2 + 2\pi\left(\frac{p_3}{2N_3}\right)\boldsymbol{b}_3 \tag{23.3}$$

with p_1, p_2, p_3 chosen as indicated in (21.5), (22.2), and (22.3). Now explicitly

$$\boldsymbol{k} \cdot \boldsymbol{R}_L = l_1\, \boldsymbol{k}_1 \cdot \boldsymbol{a}_1 + l_2\, \boldsymbol{k}_2 \cdot \boldsymbol{a}_2 + l_3\, \boldsymbol{k}_3 \cdot \boldsymbol{a}_3, \tag{23.4}$$

where l_j are integers, or

$$\boldsymbol{k} \cdot \boldsymbol{R}_L = 2\pi\left(\frac{l_1 p_1}{2N_1} + \frac{l_2 p_2}{2N_2} + \frac{l_3 p_3}{2N_3}\right). \tag{23.5}$$

Consider two wave vectors $\boldsymbol{k}$, and $\boldsymbol{k}'$ where

$$\boldsymbol{k}' = \boldsymbol{k} + 2\pi \boldsymbol{B}_h. \tag{23.6}$$

It is easy to see that if $D^{(\boldsymbol{k})}$ is given by (23.2), then

$$\begin{aligned} D^{(\boldsymbol{k}')}(\{\varepsilon|\boldsymbol{R}_L\}) &= \exp - i\,\boldsymbol{k}' \cdot \boldsymbol{R}_L \\ &= \exp - i\,(\boldsymbol{k} \cdot \boldsymbol{R}_L + 2\pi \boldsymbol{B}_h \cdot \boldsymbol{R}_L). \end{aligned} \tag{23.7}$$

But

$$2\pi \boldsymbol{B}_h \cdot \boldsymbol{R}_L = 2\pi (h_1 l_1 + h_2 l_2 + h_3 l_3), \tag{23.8}$$

where all h_j and l_j are integers. Hence

$$D^{(\boldsymbol{k}')}(\{\varepsilon|\boldsymbol{R}_L\}) = \exp - i\,\boldsymbol{k}' \cdot \boldsymbol{R}_L = D^{(\boldsymbol{k})}(\{\varepsilon|\boldsymbol{R}_L\}). \tag{23.9}$$

Thus if $\boldsymbol{k}$ and $\boldsymbol{k}'$ are two wave vectors related by (23.6), they define the identical irreducible representation of $\mathfrak{T}$. Also note the equivalence of this argument to that given in (20.7).

Let us recapitulate. The set of all vectors $\boldsymbol{k}$ in the reciprocal lattice, where $\boldsymbol{k} = \boldsymbol{k}_1 + \boldsymbol{k}_2 + \boldsymbol{k}_3$ and

$$-\pi \boldsymbol{b}_1 \leqq \boldsymbol{k}_1 < \pi \boldsymbol{b}_1, \qquad -\pi \boldsymbol{b}_2 \leqq \boldsymbol{k}_2 < \pi \boldsymbol{b}_2, \qquad -\pi \boldsymbol{b}_3 \leqq \boldsymbol{k}_3 < \pi \boldsymbol{b}_3, \tag{23.10}$$

gives rise to the set of all irreducible representations $D^{(\boldsymbol{k})}$ of $\mathfrak{T}$. This set of all independent $\boldsymbol{k}$ vectors in the reciprocal lattice is contained within and upon the surface of a cellular polyhedron surrounding the origin in the reciprocal lattice. The cellular polyhedron so defined is called the first Brillouin zone of the structure.[1] It possesses the full rotational symmetry of the point group of the space group. Thus the first Brillouin zone contains all the $\boldsymbol{k}$ vectors which satisfy the inequalities (23.10). However if two vectors $\boldsymbol{k}$ and $\boldsymbol{k}'$ satisfy (23.6) then only one of them should be included in the zone.

An alternative definition of the first Brillouin zone of a structure can be given, so as to be more consonant with conventional treatments of mathematical crystallography. According to SCHÖNFLEISS, the fundamental cell of a lattice can be defined by a geometrical construction. From the origin of the lattice construct vectors to the first, second, ... neighbours of the origin. Erect planes perpendicularly bisecting these vectors. The locus of all the points on the surface, and in the interior of these planes, define a region in space known as the Schönfleiss fundamental cell. This cell has the point group symmetry of the lattice. Also when the origin of such a cell is shifted successively to the termini of lattice vectors in the lattice, it is found that the cells pack tightly, filling space without interstices. The Schönfleiss cell is the solution in three dimensions to a problem of HILBERT on the possible inequivalent ways of packing three dimensional space with polyhedra. Some years after SCHÖNFLEISS, the cell polyhedron and its properties just mentioned, were rediscovered by WIGNER and SEITZ.[2] The cell has often been called the Wigner-Seitz polyhedron.

[1] G.F. KOSTER, in: Solid State Physics (eds. F. SEITZ and D. TURNBULL), Vol. 5. New York: Academic Press 1957.

[2] E. WIGNER and F. SEITZ: Phys. Rev. **43**, 804 (1933).

Note that to construct the Brillouin zone or polyhedron by the method of SCHÖNFLEISS one must first modify the reciprocal lattice based on the reciprocal lattice vectors (21.3). Consider the space lattice defined by vectors

$$2\pi \boldsymbol{B}_H = 2\pi h_1 \boldsymbol{b}_1 + 2\pi h_2 \boldsymbol{b}_2 + 2\pi h_3 \boldsymbol{b}_3. \tag{23.11}$$

The modification is trivial amounting to multiplication of all reciprocal lattice vectors by 2π. Following EWALD, call the lattice defined by the points at the termini of vectors $2\pi \boldsymbol{B}_H$ the Fourier lattice. Then the first Brillouin zone is obtained by the Schönfleiss construction in the Fourier lattice.

The $(2N_1)\cdot(2N_2)\cdot(2N_3)$ points $\boldsymbol{k}$ contained in the first Brillouin zone define the complete set of irreducible representations $D^{(\boldsymbol{k})}$ of the group $\mathfrak{T}$. Conversely the first Brillouin zone can be defined as the set of $(2N_1)(2N_2)(2N_3)$ points $\boldsymbol{k}$ such that they define a complete set of irreducible representations $D^{(\boldsymbol{k})}$ of $\mathfrak{T}$. Using the latter definition[3] we are not constrained to having the points $\boldsymbol{k}$ of the first Brillouin zone delineate a closed polyhedron. The only restriction on the $\boldsymbol{k}$ is that of all the family of vectors $\boldsymbol{k}_1, \boldsymbol{k}_1', \ldots$ satisfying (23.6) we must use only one. So, for example in the conventional definition of the Brillouin zone only one of the pair of points on opposite faces of the zone are to be taken as belonging to the zone. The other is redundant; which of the pair is selected is arbitrary.

24. Completeness and orthonormality for $D^{(\boldsymbol{k})}$. Now we verify the completeness and orthonormality relations (15.7)–(15.9) using the irreducible representations $D^{(\boldsymbol{k})}$ based on Bloch vectors $\psi^{(\boldsymbol{k})}$. Note that as written (15.7)–(15.9) apply to the entire group $\mathfrak{G}$. When used for $\mathfrak{T}$, we take $g_p = 1$.

As all irreducible representations are one-dimensional, the sum (15.7) is

$$\sum_{l=1}^{r} \cdot 1 = \sum_{\mathfrak{T}} \cdot 1 = (2N_1)(2N_2)(2N_3), \tag{24.1}$$

where the last steps follow since in an Abelian group the number of classes satisfies

$$r = (2N_1)\cdot(2N_2)\cdot(2N_3) \tag{24.2}$$

and equals the number of group elements.

Since $D^{(\boldsymbol{k})}$ is one-dimensional, the matrix and its character are identical: $D^{(\boldsymbol{k})} = \chi^{(\boldsymbol{k})}$. Then from (15.8) we have

$$\begin{aligned} &\sum_{\{\varepsilon|\boldsymbol{R}_L\}} D^{(\boldsymbol{k})}(\{\varepsilon|\boldsymbol{R}_L\})\, D^{(\boldsymbol{k}')}(\{\varepsilon|\boldsymbol{R}_L\})^* \\ &\quad = \sum_{l_1}\sum_{l_2}\sum_{l_3} \exp - i(\boldsymbol{k}\cdot\boldsymbol{R}_L - \boldsymbol{k}'\cdot\boldsymbol{R}_L) \\ &\quad = \sum_{l_1}\sum_{l_2}\sum_{l_3} \exp - 2\pi i \left(\frac{l_1(p_1-p_1')}{2N_1} + \frac{l_2(p_2-p_2')}{2N_2} + \frac{l_3(p_3-p_3')}{2N_3}\right). \end{aligned} \tag{24.3}$$

In (24.3), (p_1, p_2, p_3) and (p_1', p_2', p_3') are to be chosen as indicated in (21.6), (22.2), (22.6), and in the summation over (l_1, l_2, l_3) the values consistent with the Born-

[3] L. BOUCKAERT, R. SMOLUCHOWSKI, and E. WIGNER: Phys. Rev. **50**, 58 (1936).

Karman conditions are to be used so that, as in (4.42) $-N_j < l_j \leqq N_j$. The summations are easily carried out. We illustrate the sum on l_j but to do this it is most convenient to change the summation range to $0 \leqq l_j < 2N_1$. Then

$$\sum_{l_1=0}^{(2N_1-1)} \exp -(2\pi i\, y_1\, l_1) = \frac{(1-e^{-2\pi i y_1 (2N_1)})}{(1-e^{-2\pi i y_1})} \tag{24.4}$$

with

$$y_1 \equiv (p_1 - p_1')/2N_1. \tag{24.5}$$

The change in range of the summation does not affect the result. The function on the right hand side of (24.4) will be vanishingly small except when y_1 equals an integer:

$$y_1 = h_1 \tag{24.6}$$

when it equals $(2N_1)$. The triple sum (24.3) is then the product of three such factors, and we have

$$\sum_{\{\varepsilon|\boldsymbol{R}_L\}} D^{(\boldsymbol{k})}(\{\varepsilon|\boldsymbol{R}_L\}) \cdot D^{(\boldsymbol{k}')}(\{\varepsilon|\boldsymbol{R}_L\})^* = (2N_1)(2N_2)(2N_3)\, \Delta(\boldsymbol{k}-\boldsymbol{k}'). \tag{24.7}$$

The function $\Delta(\boldsymbol{\eta})$ has the property [13]

$$\Delta(\boldsymbol{\eta}) = \begin{cases} 1 & \text{for } \boldsymbol{\eta} = 2\pi \boldsymbol{B}_H \\ 0 & \text{otherwise.} \end{cases} \tag{24.8}$$

In (24.8) $\boldsymbol{B}_H$ is a reciprocal lattice vector. But since we have specifically restricted the permitted k vectors to lie within or upon the surface of the first Brillouin zone, $\boldsymbol{B}_H = 0$, as two wave vectors which differ by a reciprocal lattice vector give equivalent irreducible representations and only one must be taken. Evidently (24.1) is properly a delta function in the limit $N_1 \to \infty$, $N_2 \to \infty$, $N_3 \to \infty$, but with N_1, N_2, N_3 sufficiently large, we assume that any error is small, thus we verify (15.8).

Now let us verify (15.9) on these representation matrices $D^{(\boldsymbol{k})}$. That equation takes the form:

$$\begin{aligned} &\sum_{\boldsymbol{k}} D^{(\boldsymbol{k})}(\{\varepsilon|\boldsymbol{R}_L\})\, D^{(\boldsymbol{k})}(\{\varepsilon|\boldsymbol{R}_M\}) \\ &= \sum_{p_1}\sum_{p_2}\sum_{p_3} \exp -2\pi i \left(\frac{p_1(l_1+m_1)}{2N_1} + \frac{p_2(l_2+m_2)}{2N_2} + \frac{p_3(l_3+m_3)}{2N_3} \right). \end{aligned} \tag{24.9}$$

Now we carry out the summation over permitted values of (p_1, p_2, p_3) as given in (21.6), (22.2), (22.6). From the result of (24.3) and (24.7) we can immediately write the result:

$$\sum_{\boldsymbol{k}} D^{(\boldsymbol{k})}(\{\varepsilon|\boldsymbol{R}_L\})\, D^{(\boldsymbol{k})}(\{\varepsilon|\boldsymbol{R}_M\}) = (2N_1)(2N_2)(2N_3)\, \Delta(\boldsymbol{R}_L + \boldsymbol{R}_M), \tag{24.10}$$

where

$$\Delta(\boldsymbol{r}) = \begin{cases} 1 & \text{for } \boldsymbol{r} = 0 \\ 0 & \text{otherwise.} \end{cases} \tag{24.11}$$

The delta function (24.11) is taken incrystal or configuration space, and so requires

$$\boldsymbol{R}_L = -\boldsymbol{R}_M \tag{24.12}$$

for a non-vanishing result (24.10). However (24.12) is just the requirement that, in $\mathfrak{T}$, considered as a group, $\{\varepsilon|\boldsymbol{R}_L\}$ and $\{\varepsilon|\boldsymbol{R}_M\}$ shall be reciprocal classes. This follows since every element $\{\varepsilon|\boldsymbol{R}_L\}$ is the only element in its own class in $\mathfrak{T}$, so the reciprocal class to $\{\varepsilon|\boldsymbol{R}_L\}$ comprises the single element $\{\varepsilon|\boldsymbol{R}_L\}^{-1}=\{\varepsilon|-\boldsymbol{R}_L\}$. Thus we verify (15.9) for the $D^{(\boldsymbol{k})}$.

25. Irreducible vector spaces for $\mathfrak{T}$: Bloch vectors. We now turn to the irreducible linear vector spaces $\Sigma^{(\boldsymbol{k})}$ which are bases for the irreducible representations $D^{(\boldsymbol{k})}$ of $\mathfrak{T}$. The vectors which define this space are functions which generally are called Bloch functions. Since each irreducible representation $D^{(\boldsymbol{k})}$ of $\mathfrak{T}$ is one-dimensional, a single Bloch function or Bloch vector defines each corresponding irreducible linear vector space $\Sigma^{(\boldsymbol{k})}$. Let us call the Bloch function $\psi^{(\boldsymbol{k})}$. Then from (14.5) we have

$$P_{\{\varepsilon|\boldsymbol{R}_L\}}\psi^{(\boldsymbol{k})}=D^{(\boldsymbol{k})}(\{\varepsilon|\boldsymbol{R}_L\})\,\psi^{(\boldsymbol{k})}=\exp-i\,\boldsymbol{k}\cdot\boldsymbol{R}_L\,\psi^{(\boldsymbol{k})}. \tag{25.1}$$

Equivalently from (12.5)

$$\begin{aligned} P_{\{\varepsilon|\boldsymbol{R}_L\}}\psi^{(\boldsymbol{k})}(\boldsymbol{r}) &= \psi^{(\boldsymbol{k})}(\{\varepsilon|\boldsymbol{R}_L\}^{-1}\cdot\boldsymbol{r}) \\ &= \psi^{(\boldsymbol{k})}(\{\varepsilon|-\boldsymbol{R}_L\}\cdot\boldsymbol{r}) \\ &= \psi^{(\boldsymbol{k})}(\boldsymbol{r}-\boldsymbol{R}_L). \end{aligned} \tag{25.2}$$

Hence for a Bloch function $\psi^{(\boldsymbol{k})}$

$$\psi^{(\boldsymbol{k})}(\boldsymbol{r}-\boldsymbol{R}_L)=\exp-i\,\boldsymbol{k}\cdot\boldsymbol{R}_L\,\psi^{(\boldsymbol{k})}(\boldsymbol{r}). \tag{25.3}$$

Since each Bloch function spans an irreducible linear vector space for group $\mathfrak{T}$, each of the $(2N_1)\cdot(2N_2)\cdot(2N_3)$ functions is linearly independent of any other. This is a very important property which we shall make use of shortly. Observe that at this stage of our argument we have postulated the existence of such a Bloch vector on general group theoretic grounds. We have not yet related such a vector to a particular differential equation or partial differential equation describing the dynamics of any physical system.

To obtain a Bloch function $\psi^{(\boldsymbol{k})}$ from an arbitrary function Ψ we may use the projection operator (16.3) for the group $\mathfrak{T}$. We know and have available all the ingredients needed to form and apply this operator. The irreducible representations are in this case all one-dimensional, consequently we substitute the one-dimensional matrix or complex number $D^{(\boldsymbol{k})*}$ and then obtain for the operator $P^{(\boldsymbol{k})}$

$$P^{(\boldsymbol{k})}=\frac{1}{N}\sum_{\{\varepsilon|\boldsymbol{R}_L\}}\exp(-i\boldsymbol{k}\cdot\boldsymbol{R}_L)\,P_{\{\varepsilon|\boldsymbol{R}_L\}} \tag{25.4}$$

we may easily verify that

$$P^{(\boldsymbol{k})}\Psi=\psi^{(\boldsymbol{k})} \tag{25.5}$$

and that for the $\psi^{(\boldsymbol{k})}$ defined by (25.5)

$$P_{\{\varepsilon|\boldsymbol{R}_L\}}\psi^{(\boldsymbol{k})}=D^{(\boldsymbol{k})}(\{\varepsilon|\boldsymbol{R}_L\})\,\psi^{(\boldsymbol{k})} \tag{25.6}$$

as required.

It may be helpful to emphasize that the operator $P_{\{\varepsilon|\boldsymbol{R}_L\}}$ can be considered as a testing operator, in the sense that it tests a Bloch vector $\psi^{(\boldsymbol{k})}$ and determines its wave vector $\boldsymbol{k}$ from the matrix $D^{(\boldsymbol{k})}(\{\varepsilon|\boldsymbol{R}_L\})=\exp-i\,\boldsymbol{k}\cdot\boldsymbol{R}_L$. Conversely a function which is multiplied by $D^{(\boldsymbol{k})}(\{\varepsilon|\boldsymbol{R}_L\})$ when tested by $P_{\{\varepsilon|\boldsymbol{R}_L\}}$ is a Bloch vector at wave vector $\boldsymbol{k}$.

26. Direct products in $\mathfrak{T}$. Let us now illustrate the direct product theorems (17.4)–(17.6). Thus if $\psi^{(\boldsymbol{k})}$ is a Bloch function which spans an invariant, and irreducible space $\Sigma^{(\boldsymbol{k})}$, and $\psi^{(\boldsymbol{k}')}$ is a Bloch function which spans space $\Sigma^{(\boldsymbol{k}')}$ then the product function

$$\psi^{(\boldsymbol{k})}\cdot\psi^{(\boldsymbol{k}')} \tag{26.1}$$

spans the space $\Sigma^{(\boldsymbol{k}\otimes\boldsymbol{k}')}$. But from

$$P_{\{\varepsilon|\boldsymbol{R}_L\}}\psi^{(\boldsymbol{k})}\cdot\psi^{(\boldsymbol{k}')}=D^{(\boldsymbol{k})}(\{\varepsilon|\boldsymbol{R}_L\})\,D^{(\boldsymbol{k}')}(\{\varepsilon|\boldsymbol{R}_L\})\,\psi^{(\boldsymbol{k})}\cdot\psi^{(\boldsymbol{k}')} \tag{26.2}$$

$$=\exp-[i(\boldsymbol{k}+\boldsymbol{k}')\cdot\boldsymbol{R}_L]\,\psi^{(\boldsymbol{k})}\cdot\psi^{(\boldsymbol{k}')} \tag{26.3}$$

$$=D^{(\boldsymbol{k}\oplus\boldsymbol{k}')}(\{\varepsilon|\boldsymbol{R}_L\})\,\psi^{(\boldsymbol{k})}\cdot\psi^{(\boldsymbol{k}')} \tag{26.4}$$

we see that $\Sigma^{(\boldsymbol{k}\otimes\boldsymbol{k}')}$ is also an irreducible space. That is, the wave vector $\boldsymbol{k}+\boldsymbol{k}'$ is either already a wave vector in the first Brillouin zone, or equivalent (modulo $2\pi\boldsymbol{B}_H$) to a wave vector in the first Brillouin zone.

The reduction coefficients for direct products in $\mathfrak{T}$ are particularly simple, only one such coefficient differs from zero corresponding to the permitted value $\boldsymbol{k}+\boldsymbol{k}'$.

$$D^{(\boldsymbol{k}\otimes\boldsymbol{k}')}=D^{(\boldsymbol{k}+\boldsymbol{k}')}$$

$$\Sigma^{(\boldsymbol{k}\otimes\boldsymbol{k}')}=\Sigma^{(\boldsymbol{k}+\boldsymbol{k}')}$$

as follows from (17.4) and (17.5).

E. Irreducible representations and vector spaces of space groups

27. Introduction. This part of the article, comprising Sect. 27–51, is one of the most important in the entire article. Here the general theory of the irreducible representations of space groups will be presented. Both symmorphic and non-symmorphic cases will be treated. The material presented here is applicable to any many-body system with space group symmetry and, conversely, any such system which admits as an invariance group of transformations the space group $\mathfrak{G}$ will have properties in accordance with the irreducible representations of $\mathfrak{G}$.

The development here will proceed by first assuming that we possess the irreducible representations of $\mathfrak{G}$ and establishing a variety of properties possessed by this matrix group. Essential in this regard is the property of $\mathfrak{T}$ as a normal subgroup of $\mathfrak{G}$, which has a consequence that we may always use the Bloch vectors $\psi^{(\boldsymbol{k})}$, which are bases for irreducible representations of $\mathfrak{T}$, as the building blocks of the irreducible representations of $\mathfrak{G}$. The irreducible representations of $\mathfrak{G}$ are based upon a set of s Bloch vectors: $\{\psi^{(k)}, \psi^{(k_2)}, \ldots, \psi^{(k_s)}\}$. The s wave vectors $\{\boldsymbol{k}, \boldsymbol{k}_2, \ldots, \boldsymbol{k}_s\}$ of this set form the star $\star\boldsymbol{k}$. Two cases arise depending on whether $s=p$, or $s<p$, where p is the index of $\mathfrak{T}$ in $\mathfrak{G}$; i.e. the order of the factor group $\mathfrak{G}/\mathfrak{T}$. The former case is called the "general" star; these representations are simple to deal with.

The latter case is the "special" star and the novel features of space group irreducible representations arise in the consideration of it. Consideration of the special star leads to the definition of the space group of the wave vector $\boldsymbol{k}$, $\mathfrak{G}(\boldsymbol{k})$ and to the determination of the irreducible representations of $\mathfrak{G}(\boldsymbol{k})$. The allowable irreducible representations of $\mathfrak{G}(\boldsymbol{k})$ are used to induce the irreducible representations of $\mathfrak{G}$. When the irreducible representations of $\mathfrak{G}$ have been obtained the orthonormality and completeness relations will be verified upon them.

The mathematical aspects of the work due to CLIFFORD,[1] which are related to the structure of $\mathfrak{G}$, or $\mathfrak{G}(\boldsymbol{k})$, as extensions of $\mathfrak{T}$ by $\mathfrak{P}$, or $\mathfrak{P}(\boldsymbol{k})$, will be reviewed in Sect. 36. The reader most interested in the more abstract part of the theory could read this Section first, then return to Sects. 28–35. In our experience the somewhat more inductive procedure adopted here is preferred by theoretical and experimental physicists to whom this article is addressed.[2]

28. Irreducible representation $D^{(\star k)(m)}$ of $\mathfrak{G}$. It is necessary for us to introduce some notation. We use the symbol

$$D^{(\star k)(m)}(\{\varphi \mid \boldsymbol{t}(\varphi)\}) \tag{28.1}$$

for the matrix representing the element $\{\varphi \mid \boldsymbol{t}(\varphi)\}$ in the irreducible representations of $\mathfrak{G}$ labelled by the symbols $(\star\boldsymbol{k})(m)$. The latter symbols will be fully explained below. The matrices (28.1) in any irreducible representation (fixed $(\star\boldsymbol{k})(m)$) must multiply as do the group elements, so that e.g.

$$\begin{aligned} D^{(\star k)(m)}(\{\varphi_l \mid \boldsymbol{t}(\varphi_l)\}) \cdot D^{(\star k)(m)}(\{\varphi_n \mid \boldsymbol{t}(\varphi_n)\}) \\ = D^{(\star k)(m)}(\{\varphi_l\,\varphi_n \mid \varphi_l \cdot \boldsymbol{t}(\varphi_n) + \boldsymbol{t}(\varphi_l)\}). \end{aligned} \tag{28.2}$$

Let us assume that the representation is of dimension

$$\dim D^{(\star k)(m)} = (s \cdot l_m). \tag{28.3}$$

Thus there are $(s \cdot l_m)$ vectors in the irreducible linear vector space $\Sigma^{(\star k)(m)}$ which is a basis for $D^{(\star k)(m)}$.

At this point we emphasize the importance of keeping clearly in mind in the following discussion the following mathematical objects:

[1] For the mathematical aspects of the work we follow A.H. CLIFFORD: Ann. of Math. **38**, 533 (1937). Also see [*12*].

[2] Other references in the physics literature are: F. SEITZ: Ann. of Math. **37**, 17 (1936). – L. BOUCKAERT, R. SMOLUCHOWSKI, and E. WIGNER: Phys. Rev. **50**, 58 (1936). – C. HERRING: J. Franklin Inst. **233**, 525 (1942). – G.F. KOSTER, in: Solid State Physics (eds. F. SEITZ and D. TURNBULL), Vol. 5. New York: Academic Press 1957. – R.J. ELLIOTT: Phys. Rev. **96**, 130 (1954).

a) The space group $\mathfrak{G}$ consisting of the coordinate transformation operations $\{\varphi|t(\varphi)\}$, and the isomorphic space group, also $\mathfrak{G}$, consisting of the function operators $P_{\{\varphi|t(\varphi)\}}$;

b) The matrix group $D^{(\star k)(m)}$, which constitutes the matrix representation (in this case irreducible) of the group $\mathfrak{G}$;

c) The underlying vector space of functions, which when operated upon by the operators $P_{\{\varphi|t(\varphi)\}}$, generates the representation $D^{(\star k)(m)}$.

To proceed, we consider the space $\Sigma^{(\star k)(m)}$ to be symbolically represented by a row vector comprising the $(s \cdot l_m)$ basic vectors $\boldsymbol{\eta}_\alpha^{(m)}$:

$$\Sigma^{(\star k)(m)} = \{\boldsymbol{\eta}_1^{(m)}, \boldsymbol{\eta}_2^{(m)}, \ldots, \boldsymbol{\eta}_{s \cdot l_m}^{(m)}\}. \tag{28.4}$$

The representation $D^{(\star k)(m)}$ is produced by this space, as

$$P_{\{\varphi|t(\varphi)\}} \eta_\alpha^{(m)} = \sum_{\beta=1}^{(s \cdot l_m)} D^{(\star k)(m)}(\{\varphi|t(\varphi)\})_{\beta\alpha} \eta_\beta^{(m)}, \qquad \alpha = 1 \ldots (s \cdot l_m). \tag{28.5}$$

In Eq. (28.5) $\eta_\alpha^{(m)}$ can be any of the $(s \cdot l_m)$ basic vectors in $\Sigma^{(\star k)(m)}$. Thus we could write

$$P_{\{\varphi|t(\varphi)\}} \Sigma^{(\star k)(m)} = \Sigma^{(\star k)(m)} D^{(\star k)(m)}(\{\varphi|t(\varphi)\}) \tag{28.6}$$

for the entire space and representation.

29. Representation of $\mathfrak{T}$ subduced by $D^{(\star k)(m)}$ of $\mathfrak{G}$. As $D^{(\star k)(m)}(\{\varphi|t(\varphi)\})$ is a representation of $\mathfrak{G}$, so

$$D^{(\star k)(m)}(\{\varepsilon|\boldsymbol{R}_L\}) \tag{29.1}$$

is a representation of $\mathfrak{T}$. This representation is said to be subduced: it is obtained by restricting the represented elements $\{\varphi|t(\varphi)\}$ to the subgroup $\mathfrak{T}$. As a representation of $\mathfrak{T}$, $D^{(\star k)(m)}$ is by Maschke's theorem either irreducible or decomposable into a direct sum of irreducible representations of $\mathfrak{T}$. Let us take (29.1) to be in fully reduced form for every translation in $\mathfrak{T}$:

$$D^{(\star k)(m)}(\{\varepsilon|\boldsymbol{R}_L\}) = \begin{pmatrix} \boldsymbol{\Pi}_1 \exp -i\,\boldsymbol{k}_1 \cdot \boldsymbol{R}_L & 0 & 0 \ldots 0 \\ 0 & \boldsymbol{\Pi}_2 \exp -i\,\boldsymbol{k}_2 \cdot \boldsymbol{R}_L & \ldots \\ 0 & \ldots & \boldsymbol{\Pi}_s \exp -i\,\boldsymbol{k}_s\, \boldsymbol{R}_L \end{pmatrix}. \tag{29.2}$$

In (29.2) the $\boldsymbol{\Pi}_\alpha$ are $l_\alpha \times l_\alpha$ unit matrices.

$$\boldsymbol{\Pi}_1 \equiv \begin{pmatrix} 1 & 0 & \ldots & 0 \\ 0 & 1 & & \\ \vdots & & 1 & \vdots \\ \vdots & & & \ddots \\ 0 & & 0 \ldots & 1 \end{pmatrix} \quad (l_1 \times l_1), \ldots \text{ etc.} \tag{29.3}$$

Thus we have assumed that in $D^{(\star k)(m)}$, the irreducible representation $D^{(k_1)}$ of $\mathfrak{T}$ appears l_1 times, $D^{(k_2)}$ of $\mathfrak{T}$ appears l_2 times, ..., $D^{(k_s)}$ appears l_s times. The assumption that $D^{(\star k)(m)}(\{\varepsilon|\boldsymbol{R}_L\})$ is of form (29.2) imposes no essential restriction on the analysis. The quantities $\exp -i\,\boldsymbol{k}_\alpha \cdot \boldsymbol{R}_L$ are the inequivalent irreducible representations (one dimensional) matrices of $\mathfrak{T}$. Since these are inequivalent, the $\boldsymbol{k}_\alpha$ are distinct and inequivalent.

$$\boldsymbol{k}_1 \neq \boldsymbol{k}_2 + 2\pi\,\boldsymbol{B}_H \quad \text{etc.} \tag{29.4}$$

Then we may identify the first l_1 functions in $\Sigma^{(\star\boldsymbol{k})\,(m)}$ as Bloch functions, or Bloch vectors which are characterized by wave vector $\boldsymbol{k}_1$:

$$\{\eta_1^{(m)}, \ldots, \eta_{l_1}^{(m)}\} \equiv \{\psi_1^{(\boldsymbol{k}_1)\,(m)}, \ldots, \psi_{l_1}^{(\boldsymbol{k}_1)\,(m)}\}. \tag{29.5}$$

In (29.5) we introduce the symbol

$$\psi_{\lambda_1}^{(\boldsymbol{k}_1)\,(m)}, \qquad \lambda_1 = 1, \ldots, l_1 \tag{29.6}$$

to represent the Bloch function with wave vector $\boldsymbol{k}_1$; the additional index m will be explained below, and the index λ_1 numbers the l_1 possible functions. In a similar fashion, the next l_2 functions can be grouped into a set

$$\psi_{\lambda_2}^{(\boldsymbol{k}_2)\,(m)}, \qquad \lambda_2 = 1, \ldots, l_2 \tag{29.7}$$

etc., so the general case is

$$\psi_{\lambda_\alpha}^{(\boldsymbol{k}_\alpha)\,(m)}, \qquad \lambda_\alpha = 1, \ldots, l_\alpha, \ \alpha = 1, \ldots, s. \tag{29.8}$$

These results may be concisely summarized as

$$D^{(\star\boldsymbol{k})\,(m)} \downarrow l_1\, D^{(\boldsymbol{k}_1)} \oplus l_2\, D^{(\boldsymbol{k}_2)} \oplus \cdots \oplus l_s\, D^{(\boldsymbol{k}_s)} \tag{29.9}$$

and

$$\Sigma^{(\star\boldsymbol{k})\,(m)} \downarrow l_1\, \Sigma^{(\boldsymbol{k}_1)} \oplus l_2\, \Sigma^{(\boldsymbol{k}_2)} \oplus \cdots \oplus l_s\, D^{(\boldsymbol{k}_s)}. \tag{29.10}$$

It is understood that in Eqs. (29.9) and (29.10) the left side refers to the irreducible representation of $\mathfrak{G}$, the right side to the decomposition of the subduced representation into irreducible components in $\mathfrak{T}$.

Finally, from (29.2) we obtain for the trace

$$\operatorname{Tr} D^{(\star\boldsymbol{k})\,(m)}(\{\varepsilon|\boldsymbol{R}_L\}) = \sum_{\alpha=1}^{(s)} l_\alpha \exp -i\,\boldsymbol{k}_\alpha \cdot \boldsymbol{R}_L. \tag{29.11}$$

30. Transformation of Bloch vectors by rotation operators. Before proceeding with the characterization of the representation $D^{(\star\boldsymbol{k})\,(m)}$ we require an intermediate result. The basic building blocks of the representations are Bloch vectors $\psi^{(\boldsymbol{k})}$, and we need to consider transformed functions. Thus, let $\psi^{(\boldsymbol{k})}$ be a Bloch vector with wave vector $\boldsymbol{k}$ so, as in (25.6)

$$P_{\{\varepsilon|\boldsymbol{R}_L\}}\, \psi^{(\boldsymbol{k})} = D^{(\boldsymbol{k})}(\{\varepsilon|\boldsymbol{R}_L\})\, \psi^{(\boldsymbol{k})}. \tag{30.1}$$

Consider the function

$$P_{\{\varphi_\lambda|\boldsymbol{\tau}_\lambda\}}\, \psi^{(\boldsymbol{k})}. \tag{30.2}$$

From

$$\{\varepsilon|\boldsymbol{R}_L\} \cdot \{\varphi_\lambda|\boldsymbol{\tau}_\lambda\} = \{\varphi_\lambda|\boldsymbol{\tau}_\lambda\} \cdot \{\varepsilon|\varphi_\lambda^{-1} \cdot \boldsymbol{R}_L\}. \tag{30.3}$$

We obtain for (30.2):

$$P_{\{\varepsilon|\boldsymbol{R}_L\}}\cdot(P_{\{\boldsymbol{\varphi}_\lambda|\boldsymbol{\tau}_\lambda\}}\psi^{(\boldsymbol{k})})=P_{\{\boldsymbol{\varphi}_\lambda|\boldsymbol{\tau}_\lambda\}}\cdot P_{\{\varepsilon|\boldsymbol{\varphi}_\lambda^{-1}\cdot\boldsymbol{R}_L\}}\psi^{(\boldsymbol{k})}$$
$$=D^{(\boldsymbol{k})}(\{\varepsilon|\varphi_\lambda^{-1}\cdot\boldsymbol{R}_L\})\,P_{\{\boldsymbol{\varphi}_\lambda|\boldsymbol{\tau}_\lambda\}}\psi^{(\boldsymbol{k})}. \tag{30.4}$$

But

$$D^{(\boldsymbol{k})}(\varepsilon|\varphi_\lambda^{-1}\cdot\boldsymbol{R}_L\})=\exp-i\,\boldsymbol{k}\cdot(\varphi_\lambda^{-1}\cdot\boldsymbol{R}_L)$$
$$=\exp-i(\varphi_\lambda\cdot\boldsymbol{k})\cdot\boldsymbol{R}_L \tag{30.5}$$
$$=D^{(\boldsymbol{k}_\lambda)}(\{\varepsilon|\boldsymbol{R}_L\}),$$

where

$$\boldsymbol{k}_\lambda\equiv\varphi_\lambda\cdot\boldsymbol{k} \tag{30.6}$$

is some wave vector. Two cases may be distinguished:

$$\boldsymbol{k}_\lambda=\varphi_\lambda\cdot\boldsymbol{k}=\boldsymbol{k}+2\pi\,\boldsymbol{B}_H, \tag{30.7}$$

$$\boldsymbol{k}_\lambda=\varphi_\lambda\cdot\boldsymbol{k}\neq\boldsymbol{k}+2\pi\,\boldsymbol{B}_H. \tag{30.8}$$

In case (30.7) $D^{(\boldsymbol{k}_\lambda)}$ is equivalent to $D^{(\boldsymbol{k})}$, in case (30.8) they are inequivalent. Combining (30.4) and (30.5) we observe

$$P_{\{\varepsilon|\boldsymbol{R}_L\}}\cdot(P_{\{\boldsymbol{\varphi}_\lambda|\boldsymbol{\tau}_\lambda\}}\psi^{(\boldsymbol{k})})=D^{(\boldsymbol{k}_\lambda)}(\{\varepsilon|\boldsymbol{R}_L\})\cdot(P_{\{\boldsymbol{\varphi}_\lambda|\boldsymbol{\tau}_\lambda\}}\psi^{(\boldsymbol{k})}). \tag{30.9}$$

Consequently if we consider the operator $P_{\{\varepsilon|\boldsymbol{R}_L\}}$ to be a testing operator, as in (25.6), we determine that the function

$$P_{\{\boldsymbol{\varphi}_\lambda|\boldsymbol{\tau}_\lambda\}}\psi^{(\boldsymbol{k})} \tag{30.10}$$

is a function with wave vector $\boldsymbol{k}_\lambda$.

Evidently, for any operator in the coset $\{\varphi_\lambda|\tau_\lambda\}\,\mathfrak{T}$ the transformed function will behave similarly. Thus

$$P_{\{\boldsymbol{\varphi}_\lambda|\boldsymbol{t}(\boldsymbol{\varphi}_\lambda)\}}\psi^{(\boldsymbol{k})}\quad\text{is in }\Sigma^{(\boldsymbol{k}_\lambda)} \tag{30.11}$$

and

$$P_{\{\boldsymbol{\varphi}_\lambda|\boldsymbol{t}(\boldsymbol{\varphi}_\lambda)\}}\Sigma^{(\boldsymbol{k})}=\Sigma^{(\boldsymbol{k}_\lambda)}. \tag{30.12}$$

31. Conjugate representations of $\mathfrak{T}$. The results of Sect. 30 permit us to introduce an important idea. In Sect. 10 the definition of conjugate subgroups of a group $\mathfrak{G}$ was given. We apply this to the subgroup $\mathfrak{T}$ of $\mathfrak{G}$. Consider the fixed element $\{\varphi_\lambda|\boldsymbol{t}(\varphi_\lambda)\}$ in $\mathfrak{G}$ as producing a mapping of $\mathfrak{T}$ onto itself by

$$\{\varphi_\lambda|\boldsymbol{t}(\varphi_\lambda)\}^{-1}\,\mathfrak{T}\,\{\varphi_\lambda|\boldsymbol{t}(\varphi_\lambda)\}\to\mathfrak{T}. \tag{31.1}$$

Clearly the mapping is an inner automorphism (conjugation) of $\mathfrak{G}$ defined by the rule

$$\{\varphi_\lambda|\boldsymbol{t}(\varphi_\lambda)\}^{-1}\cdot\{\varepsilon|\boldsymbol{R}_L\}\cdot\{\varphi_\lambda|\boldsymbol{t}(\varphi_\lambda)\}=\{\varepsilon|\varphi_\lambda^{-1}\cdot\boldsymbol{R}_L\}. \tag{31.2}$$

Corresponding elements of the isomorphic subgroup are related via (31.2).

Now in the irreducible representation $D^{(\boldsymbol{k})}$ of $\mathfrak{T}$ consider the matrices for $\{\varepsilon|\boldsymbol{R}_L\}$ and $\{\varepsilon|\varphi_\lambda^{-1}\cdot\boldsymbol{R}_L\}$. These are

$$D^{(\boldsymbol{k})}(\{\varepsilon|\boldsymbol{R}_L\})=\exp-i\,\boldsymbol{k}\cdot R_L \tag{31.3}$$

and

$$D^{(k)}(\{\varepsilon|\varphi_\lambda^{-1}\cdot R_L\})=\exp -i\,\boldsymbol{k}\cdot(\varphi_\lambda^{-1}\cdot R_L)=D^{(k_\lambda)}(\{\varepsilon|\boldsymbol{R}_L\}) \tag{31.4}$$

by (30.5). We say

$$D^{(k)} \text{ and } D^{(k_\lambda)} \text{ are conjugate} \tag{31.5}$$

representations of $\mathfrak{T}$. Observe that $\{\varepsilon|\boldsymbol{R}_L\}$ is represented by the same matrix based on the space $\Sigma^{(k_\lambda)}$ of (30.12), as $\{\varepsilon|\varphi_\lambda^{-1}\cdot\boldsymbol{R}_L\}$ which is the conjugate element as in (31.2), based on the space $\Sigma^{(k)}$.

Clearly as $D^{(k)}$ is irreducible, so $D^{(k_\lambda)}$ is irreducible; which follows immediately from Schur's lemma. Observe that from the viewpoint of the subgroup $\mathfrak{T}$ above, the mapping of elements

$$\{\varepsilon|\boldsymbol{R}_L\}\leftrightarrows\{\varepsilon|\varphi_\lambda^{-1}\cdot\boldsymbol{R}_L\} \tag{31.6}$$

is an outer automorphism. From the viewpoint of $\mathfrak{G}$, (31.6) indicates that the mapping is inner. Recall the mathematical distinction between inner and outer automorphism relates to whether the element to which mapping corresponds, here: $\{\varphi_\lambda|\boldsymbol{t}(\varphi_\lambda)\}$ is (inner) or is not (outer) within the set comprising the group [*10*].

32. Characterization of the subduced representation. Return to (29.2) and the representation space $\Sigma^{(*k)(m)}$. Consider the representation subspace (29.5), consisting of the first l_1 functions each of which is a basis for $D^{(k_1)}$ of $\mathfrak{T}$. Two possibilities exist: either these l_1 functions completely span $\Sigma^{(*k)(m)}$ or they do not. If

$$\Sigma^{(*k)(m)}\to\Sigma^{(k_1)}\oplus\cdots\oplus\Sigma^{(k_1)} \quad (l_1 \text{ times}) \tag{32.1}$$

then the set of wave vectors $\boldsymbol{k}_\alpha$ consists only of the single wave vector $\boldsymbol{k}_1$. The l_1 functions (29.5) may be distinct, but all have the same wave vector $\boldsymbol{k}_1$.

If $\Sigma^{(*k)(m)}$ is not completely spanned by the first l_1 functions, then these l_1 functions are not an invariant, or closed, subspace under all the operators in $\mathfrak{G}$. Then there must be an operator $P_{\{\varphi_\lambda|\tau_\lambda\}}$ in $\mathfrak{G}$ such that

$$P_{\{\varphi_\lambda|\tau_\lambda\}}\,\Sigma^{(k_1)}\neq\Sigma^{(k_1)}. \tag{32.2}$$

By (30.12)

$$P_{\{\varphi_\lambda|\tau_\lambda\}}\,\Sigma^{(k_1)}=\Sigma^{(k_{1\lambda})}, \tag{32.3}$$

where

$$\boldsymbol{k}_{1\lambda}=\varphi_\lambda\cdot\boldsymbol{k}_1=\boldsymbol{k}_\lambda. \tag{32.4}$$

Then under $P_{\{\varphi_\lambda|\tau_\lambda\}}$ each of the l_1 functions (29.5) is sent into a function with wave vector $\boldsymbol{k}_{1\lambda}$, producing exactly l_1 functions in the space

$$\Sigma^{(k_{1\lambda})}\oplus\Sigma^{(k_{1\lambda})}\oplus\cdots\oplus\Sigma^{(k_{1\lambda})} \quad (l_1 \text{ times}).$$

These l_1 functions are linearly independent, and linearly independent of the initial l_1 functions.

We now have $2l_1$ functions. If these do not completely span the invariant space $\Sigma^{(*k)(m)}$ then there must be an operator $P_{\{\varphi_\mu|\tau_\mu\}}$ such that

$$P_{\{\varphi_\mu|\tau_\mu\}}\,\Sigma^{(k_1)}\neq\Sigma^{(k_1)}\neq\Sigma^{(k_{1\lambda})}. \tag{32.5}$$

Clearly

$$P_{\{\varphi_\mu|\tau_\mu\}}\,\Sigma^{(k_1)}=\Sigma^{(k_{1\mu})} \tag{32.6}$$

with

$$k_{1\mu}=\varphi_\mu\cdot k_1. \tag{32.7}$$

In this fashion we proceed until $\Sigma^{(\star k)(m)}$ is exhausted, and we find

$$\Sigma^{(\star k)(m)}=l_1\,\Sigma^{(k_1)}\oplus l_1\,\Sigma^{(k_{1\lambda})}\oplus\cdots\oplus l_1\,\Sigma^{(k_{1\mu})}\oplus\cdots\oplus l_1\,\Sigma^{(k_{1s})}. \tag{32.8}$$

It is to be emphasized that $\Sigma^{(\star k)(m)}$ must be completely exhausted after a finite number of steps. At each step in the procedure we confront the alternatives that either the invariant space has been completely decomposed into the preceding components or not. If not, then the reduction continues.

Each resulting space (such as in (32.5) and (32.6)) contains l_1 linearly independent functions, and the entire decomposition is disjoint, because the k_α are inequivalent. Comparing (32.8) and (29.2), (29.9)–(29.11) we immediately can identify the set $\{l_\alpha\}$ and the set $\{k_\alpha\}$ of (29.2). Thus

$$l_1=l_2=\cdots=l_s\equiv l_m \tag{32.9}$$

so that the dimension of each subspace occurring in $\Sigma^{(\star k)(m)}$ must be equal. Each such subspace corresponds to a distinct irreducible representation of $\mathfrak{T}$ taken l_m times. The set of wave vectors k_α is not arbitrary but is entirely specified when one member is given. Let k be the canonical wave vector of the set, then define the entire set of wave vectors as the star of k:

$$\star k\equiv\{k_1=k,\, k_2=\varphi_2\cdot k,\,\ldots,\, k_s\equiv\varphi_s\cdot k\}. \tag{32.10}$$

Summarizing: the entire irreducible representation space $\Sigma^{(\star k)(m)}$ of a single irreducible representation of $\mathfrak{G}$ can be decomposed into s disjoint subspaces

$$\Sigma^{(\star k)(m)}=\Sigma^{(k)(m)}\oplus\Sigma^{(k_2)(m)}\oplus\cdots\oplus\Sigma^{(k_s)(m)}, \tag{32.11}$$

where each such subspace goes with one irreducible representation of $\mathfrak{T}$. Each subspace $\Sigma^{(k_\mu)(m)}$ is of the same dimension, and produces its irreducible representation $D^{(k_\mu)}$ of $\mathfrak{T}$ the same number of times l_m for all μ. The representations $D^{(k_\mu)}$ $\mu=1,\ldots,s$ are inequivalent irreducible representations of $\mathfrak{T}$.

33. Block structure of $D^{(\star k)(m)}$ of $\mathfrak{G}$. From (32.10) and (32.11) it is observed that a particular ordering in the irreducible vector space $\Sigma^{(\star k)(m)}$ has been imposed. Thus, the first l_m functions have wave vector $k_1=k$ the next l_m have $k_2,\ldots$ etc. as in (32.10). Hence it follows that each matrix in $D^{(\star k)(m)}$ can be blocked off into sub-blocks of $(l_m\times l_m)$ dimensional matrices corresponding to the ordering of the wave vectors in (32.10). The $l_m\times l_m$ matrix in the upper left corner will be denoted $D^{(k_1)(m)}$, and each $(l_m\times l_m)$ matrix block will be denoted similarly: $D^{(k_\lambda)(m)}{}_{\mu\lambda}$ signifying the matrix connecting spaces $\Sigma^{(k_\mu)(m)}$ and $\Sigma^{(k_\lambda)(m)}$. Then we have

$$D^{(\star k)(m)}=\begin{pmatrix} D^{(k_1)(m)}{}_{11}\cdots D^{(k_s)(m)}{}_{1s}\\ \vdots \\ D^{(k_1)(m)}{}_{s1}\cdots D^{(k_s)(m)}{}_{ss}\end{pmatrix} \tag{33.1}$$

for every matrix in the representation. In particular for any element in $\mathfrak{T}$ (33.1) is diagonal so

$$D^{(\star k)(m)}(\{\varepsilon|R_L\})_{\mu\nu} = (D^{(\star k_\mu)(m)}(\{\varepsilon|R_L\}))_{\mu\mu}\,\delta_{\mu\nu} = \Pi_m(\exp - i\,k_\mu\cdot R_L)\,\delta_{\mu\nu}, \tag{33.2}$$

where Π_m is the $(l_m \times l_m)$ unit matrix. Now consider the identity

$$\{\varepsilon|R_L\}\cdot\{\varphi_\lambda|t(\varphi_\lambda)\} = \{\varphi_\lambda|t(\varphi_\lambda)\}\cdot\{\varepsilon|\varphi_\lambda^{-1}\cdot R_L\}. \tag{33.3}$$

For the representation (33.1) we have

$$D^{(\star k)(m)}(\{\varepsilon|R_L\})\,D^{(\star k)(m)}(\{\varphi_\lambda|t(\varphi_\lambda)\}) = D^{(\star k)(m)}(\{\varphi_\lambda|t(\varphi_\lambda)\})\,D^{(\star k)(m)}(\{\varepsilon|\varphi_\lambda^{-1}\cdot R_L\}). \tag{33.4}$$

Take $\mu\nu$ block matrix element of (33.4) and use (33.2):

$$D^{(k_\mu)(m)}(\{\varepsilon|R_L\})_{\mu\mu}\,D^{(k_\nu)(m)}(\{\varphi_\lambda|t(\varphi_\lambda)\})_{\mu\nu} = D^{(k_\nu)(m)}(\{\varphi_\lambda|t(\varphi_\lambda)\})_{\mu\nu}\,D^{(k_\nu)(m)}(\{\varepsilon|\varphi_\lambda^{-1}\cdot R_L\})_{\nu\nu}. \tag{33.5}$$

For fixed, but arbitrary $\{\varphi_\lambda|t(\varphi_\lambda)\}$, this equation holds for all $\{\varepsilon|R_L\}$ in $\mathfrak{T}$. All matrices in (33.5) are l_m dimensional.

Now by (33.2) we have for (33.5):

$$(\exp - i\,k_\mu\cdot R_L)\cdot D^{(k_\nu)(m)}(\{\varphi_\lambda|t(\varphi_\lambda)\})_{\mu\nu} = D^{(k_\nu)(m)}(\{\varphi_\lambda|t(\varphi_\lambda)\})_{\mu\nu}\exp - i\,k_\nu\cdot(\varphi_\lambda^{-1}\cdot R_L). \tag{33.6}$$

From (33.6) the matrix

$$D^{(k_\mu)(m)}(\{\varphi_\lambda|t(\varphi_\lambda)\})_{\mu\nu} = 0 \tag{33.7}$$

unless

$$k_\mu = \varphi_\lambda\cdot k_\nu + 2\pi\,B_H \tag{33.8}$$

i.e. unless the conjugate irreducible representation $D^{(k_\mu)(m)}$ and $D^{(k_{\nu\lambda})(m)}$ of $\mathfrak{T}$ are equivalent. But for each μ only one k_ν can satisfy (33.8) since the wave vectors in the star (32.10) are by definition inequivalent. Also, there must be one such k_ν as can be seen from the definition of the k_ν in terms of the canonical wave vector k, and the group property of the φ_λ.

It is clear that (33.7) indicates that in each row and column in the block decomposition (33.1), only one $(l_m \times l_m)$ matrix block differs from zero. Further the block form of the entire representation shows that the now diagonal matrices in $D^{(\star k)(m)}$ permute the entire l_1 dimensional subspaces among one another.

Summarizing, the block diagonal form of $D^{(\star k)(m)}$ corresponding to the order of enumerating the subspaces (32.11) is

$$D^{(\star k)(m)}(\{\varepsilon|R_L\}) = \begin{pmatrix} D^{(k_1)(m)}(\{\varepsilon|R_L\})_{11} & 0\ \dots\ 0 \\ 0 & \ddots \\ 0 & D^{(k_s)(m)}(\{\varepsilon|R_L\})_{ss} \end{pmatrix} \tag{33.9}$$

or, more explicitly:

$$D^{(\star k)}(\{\varepsilon|\boldsymbol{R}_L\})=\begin{pmatrix} \boldsymbol{\Pi}_m \exp -i\,\boldsymbol{k}_1\cdot\boldsymbol{R}_L & 0 & \cdots & 0 \\ 0 & \boldsymbol{\Pi}_m \exp -i\,\boldsymbol{k}_\nu\cdot\boldsymbol{R}_L & & \\ \vdots & & \ddots & \\ & & & \boldsymbol{\Pi}_m \exp -i\,\boldsymbol{k}_s\cdot\boldsymbol{R}_L \end{pmatrix} \tag{33.10}$$

where $\boldsymbol{\Pi}_m$ is the unit $l_m\times l_m$ matrix. In addition, in this representation:

$$D^{(\star k)(m)}(\{\varphi_\lambda|\,\boldsymbol{t}(\varphi_\lambda)\})=\begin{pmatrix} 0 & \cdots & 0 & 0 & \cdots \\ 0 & & \vdots & \vdots & \\ \vdots & & D^{(k_\nu)(m)}(\{\varphi_\lambda|\boldsymbol{t}(\varphi_\lambda)\})_{\mu\nu} & & \\ & & & \vdots & \\ D^{(k_\lambda)(m)}(\{\varphi_\lambda|\boldsymbol{t}(\varphi_\lambda)\})_{\lambda 1} & & & 0 & \cdots \\ \vdots & & & \vdots & \\ 0 & & & & \ddots \end{pmatrix} \tag{33.11}$$

for general operators in $\mathfrak{G}$. The representation space for $D^{(\star k)(m)}$ is $\Sigma^{(\star k)(m)}$ cf. (32.11) composed of the s disjoint spaces.

34. Group of the canonical $\boldsymbol{k}$: $\mathfrak{G}(\boldsymbol{k})$. As we envisage the entire irreducible representation $D^{(\star k)(m)}$ we observe that certain of the matrices (33.10) and (33.11) have non-zero elements (block matrices) in the $\mu=1$, $\nu=1$ position. For such matrices the subspace $\Sigma^{(k_1)(m)}$ is invariant. That is, for these matrices, the represented operator $P_{\{\varphi_\lambda|\boldsymbol{t}(\varphi_\lambda)\}}$ is diagonal in the subspace $\Sigma^{(k_1)(m)}$. Adapting our notation from (33.1), we have

$$P_{\{\varphi_\lambda|\boldsymbol{t}(\varphi_\lambda)\}}\,\Sigma^{(k_1)(m)}=\Sigma^{(k_1)(m)}\,D^{(k_1)(m)}(\{\varphi_\lambda|\boldsymbol{t}(\varphi_\lambda)\})_{11}. \tag{34.1}$$

We will first study this set of operators.

Clearly for all operators in $\mathfrak{T}$ (34.1) applies. It may be that there are no other operators in $\mathfrak{G}$ with this property.

However, from (30.12) we know that

$$P_{\{\varphi_\lambda|\boldsymbol{\tau}_\lambda\}}\,\Sigma^{(k_1)}=\Sigma^{(k_{1\lambda})}, \tag{34.2}$$

where

$$\boldsymbol{k}_{1\lambda}=\varphi_\lambda\cdot\boldsymbol{k}_1. \tag{34.3}$$

In (34.2) and (34.3), $\Sigma^{(k_1)}$ refers to a single Bloch vector $\psi^{(k_1)}$. Then if, as in (30.7)

$$\varphi_\lambda\cdot\boldsymbol{k}_1=\boldsymbol{k}_1+2\pi\boldsymbol{B}_H \tag{34.4}$$

$\boldsymbol{k}_1$ and $\boldsymbol{k}_{1\lambda}$ define equivalent irreducible representation of $\mathfrak{T}$. If (34.4) applies then the entire l_m dimensional space consisting of the l_m linearly independent families (29.5) each one of which has wave vector $\boldsymbol{k}_1$ will be invariant under this operator $P_{\{\varphi_\lambda|\boldsymbol{\tau}_\lambda\}}$. If the rotational part φ_λ if the operator $\{\varphi_\lambda|\boldsymbol{\tau}_\lambda\}$ obeys (34.4) then also for any power of this operator the same applies.

Let $\{\varphi_\mu|\tau_\mu\}$ be another operator such that

$$\varphi_\mu\cdot \boldsymbol{k}_1=\boldsymbol{k}_1+2\pi \boldsymbol{B}'_H, \tag{34.5}$$

where $\boldsymbol{B}'_H$ is a reciprocal lattice vector, perhaps distinct from $\boldsymbol{B}_H$ of (34.4). Then the powers of this operator also have the same property.

Finally, all products $\{\varphi_\lambda|\tau_\lambda\}\cdot\{\varphi_\mu|\tau_\mu\}$ and powers or products of these operators also have the same property of sending $\boldsymbol{k}_1$ into an equivalent wave vector.

In this fashion we enumerate a set of operators $P_{\{\varphi_{l_\lambda}|\tau_{l_\lambda}\}}$ such that when applied to a single Bloch vector each operator sends that Bloch vector into an equivalent Bloch vector

$$P\{\varphi_{l_\lambda}|\tau_{l_\lambda}\}\,\Sigma^{(\boldsymbol{k}_1)}=\Sigma^{(\boldsymbol{k}_1)}. \tag{34.6}$$

The collection of all these operators evidently form a group called $\mathfrak{G}(\boldsymbol{k})$, the space group of the wave vector $\boldsymbol{k}$. $\mathfrak{G}(\boldsymbol{k})$ will be a subgroup of $\mathfrak{G}$. It may be decomposed into cosets with respect tp $P_{\mathfrak{T}}$, where $P_{\mathfrak{T}}$ is the function operator group isomorphic to $\mathfrak{T}$:

$$\mathfrak{G}(\boldsymbol{k})=P_{\mathfrak{T}}+\cdots+P_{\{\varphi_{l_\lambda}|\tau(\varphi_{l_\lambda})\}}P_{\mathfrak{T}}+\cdots+P_{\{\varphi_{l_\kappa}|\tau(\varphi_{l_\kappa})\}}P_{\mathfrak{T}}. \tag{34.7}$$

Clearly for every operator in (34.7) the l_m dimensional space $\Sigma^{(\boldsymbol{k}_1)(m)}$ is closed.

The set of matrices $D^{(\boldsymbol{k}_1)(m)}(\{\varphi_{l_\lambda}|\tau_{l_\lambda}\})_{11}$ which are l_m dimensional matrix components of $D^{(\star\boldsymbol{k})(m)}$ as in (33.1), are then perforce a representation of the group $\mathfrak{G}(\boldsymbol{k})$ of (34.7). It is essential to recognize, however, that this representation is not general, but is based upon the invariant space $\Sigma^{(\boldsymbol{k}_1)(m)}$ so that in this representation:

$$D^{(\boldsymbol{k}_1)(m)}(\{\varepsilon|\boldsymbol{R}_L\})_{11}=\boldsymbol{\Pi}_m\exp -i\,\boldsymbol{k}_1\cdot\boldsymbol{R}_L. \tag{34.8}$$

Hence we can obtain the group $\mathfrak{G}(\boldsymbol{k})$ as we have done, using quite general arguments. However, of all the representations of $\mathfrak{G}(\boldsymbol{k})$ only those for which (34.8) applies may be used in $D^{(\star\boldsymbol{k})(m)}$. The representations of $\mathfrak{G}(\boldsymbol{k})$ obeying (34.8) are called *acceptable*, or *allowable*.

It may prove useful on occasion to define the factor group or point group of the wave vector

$$\mathfrak{P}(\boldsymbol{k})\equiv\mathfrak{G}(\boldsymbol{k})/\mathfrak{T} \tag{34.9}$$

and we reserve this notation as needed.

35. Irreducibility of the acceptable representations $D^{(\boldsymbol{k}_1)(m)}$ of $\mathfrak{G}(\boldsymbol{k}_1)$. Now we establish that the acceptable representation $D^{(\boldsymbol{k}_1)(m)}$ of $\mathfrak{G}(\boldsymbol{k}_1)$ must be an irreducible representation of $\mathfrak{G}(\boldsymbol{k}_1)$. By hypothesis $D^{(\star\boldsymbol{k})(m)}$ is an irreducible representation of $\mathfrak{G}$, and is based upon the irreducible space $\Sigma^{(\star\boldsymbol{k})(m)}$ of (32.11).

The space $\Sigma^{(\boldsymbol{k}_1)(m)}$ is invariant and a basis for $D^{(\boldsymbol{k}_1)(m)}$. Suppose $\Sigma^{(\boldsymbol{k}_1)(m)}$ was itself decomposable into invariant subspaces and hence $D^{(\boldsymbol{k}_1)(m)}$ decomposable into corresponding irreducible components, with respect to $\mathfrak{G}(\boldsymbol{k}_1)$:

$$\Sigma^{(\boldsymbol{k}_1)(m)}=\Sigma^{(\boldsymbol{k}_1)(m')}\oplus\Sigma^{(\boldsymbol{k}_1)(m'')} \tag{35.1}$$

$$D^{(\boldsymbol{k}_1)(m)}=D^{(\boldsymbol{k}_1)(m')}\oplus D^{(\boldsymbol{k}_1)(m'')}. \tag{35.2}$$

Let the dimension of $\Sigma^{(\boldsymbol{k}_1)(m')}$ be $l_{m'}$, and that of $\Sigma^{(\boldsymbol{k}_1)(m'')}$ be $l_{m''}$ so $l_{m'}+l_{m''}=l_m$.

Using the $l_{m'}$ functions of $\Sigma^{(k_1)(m')}$ we construct the sequence of spaces as in (32.2), (32.5) and obtain

$$\Sigma^{(k_\nu)(m')} \equiv P_{\{\varphi_\nu|\tau_\nu\}} \Sigma^{(k_1)(m')}, \tag{35.3}$$

$$\Sigma^{(k_s)(m')} \equiv P_{\{\varphi_s|\tau_s\}} \Sigma^{(k_1)(m')}. \tag{35.4}$$

Hence we obtain the space which is the union of these

$$\Sigma^{(\star k)(m')} = \Sigma^{(k_1)(m')} \oplus \Sigma^{(k_\nu)(m')} \oplus \cdots \oplus \Sigma^{(k_s)(m')} \tag{35.5}$$

and a second space

$$\Sigma^{(\star k)(m'')} = \Sigma^{(k_1)(m'')} \oplus \Sigma^{(k_\nu)(m'')} \oplus \cdots \oplus \Sigma^{(k_s)(m'')} \tag{35.6}$$

so

$$\Sigma^{(\star k)(m)} = \Sigma^{(\star k)(m')} \oplus \Sigma^{(\star k)(m'')}. \tag{35.7}$$

Just following the argument in Sect. 32 in reverse order we see that $\Sigma^{(\star k)(m')}$ must be an invariant space under all the operations in $\mathfrak{G}$, and it is a basis for the representation $D^{(\star k)(m')}$. Similarly for $\Sigma^{(\star k)(m'')}$ and $D^{(\star k)(m'')}$.

Then if $\Sigma^{(\star k)(m)}$ is decomposable as in (35.7), so will be $D^{(\star k)(m)}$:

$$D^{(\star k)(m)} = D^{(\star k)(m')} \oplus D^{(\star k)(m'')}. \tag{35.8}$$

But this is contrary to hypotheses that $D^{(\star k)(m)}$ is irreducible. Hence $\Sigma^{(\star k)(m)}$ cannot decompose as in (35.7). But then $\Sigma^{(k_1)(m)}$ must be an irreducible invariant space, which proves that $D^{(k_1)(m)}$ of $\mathfrak{G}(k_1)$ must be irreducible as well as acceptable.

The two requirements of acceptability (34.8), and irreducibility completely determine the $D^{(k_1)(m)}$ of $\mathfrak{G}(k_1)$ which arise. Before discussing the methods used to determine $D^{(k_1)(m)}$, it will be demonstrated that the entire irreducible representation $D^{(\star k)(m)}$ of $\mathfrak{G}$ can be constructed from $D^{(k_1)(m)}$ of $\mathfrak{G}(k_1)$.

36. $D^{(\star k)(m)}$ of $\mathfrak{G}$ induced from $D^{(k_1)(m)}$ of $\mathfrak{G}(k_1)$. In this section it will be demonstrated that the irreducible representation $D^{(\star k)(m)}$ of the full space group $\mathfrak{G}$ is a representation induced from the allowable irreducible representation $D^{(k_1)(m)}$ of $\mathfrak{G}(k_1)$ to $\mathfrak{G}$. Thus complete knowledge of the $D^{(k_1)(m)}$ suffices to determine the $D^{(\star k)(m)}$.

As in (34.7) a decomposition of $\mathfrak{G}(k)$ with respect to $\mathfrak{T}$ can be given as

$$\mathfrak{G}(k) = \mathfrak{T} + \{\varphi_{l_2}|\tau_{l_2}\}\mathfrak{T} + \cdots + \{\varphi_{l_\lambda}|\tau_{l_\lambda}\}\mathfrak{T} + \cdots + \{\varphi_{l_k}|\tau_{l_k}\}\mathfrak{T}. \tag{36.1}$$

Observe that we reserve the index l to always refer to one of the l_k coset representatives in $\mathfrak{G}(k)$. The group $\mathfrak{G}(k)$ is evidently a subgroup of $\mathfrak{G}$, the full space group. Hence a coset decomposition of $\mathfrak{G}$ with respect to $\mathfrak{G}(k)$ may be given as

$$\mathfrak{G} = \mathfrak{G}(k) + \cdots + \{\varphi_\sigma|\tau_\sigma\}\mathfrak{G}(k) + \cdots + \{\varphi_s|\tau_s\}\mathfrak{G}(k). \tag{36.2}$$

We always reserve the index σ to refer to one of the s coset representatives in (36.2): clearly $\{\varphi_\sigma|\tau_\sigma\}$ is in $\mathfrak{G}$ but *not* in $\mathfrak{G}(k)$. This point is obvious, but most important in what follows.

Now, let us assume that we possess a complete set of r allowable irreducible representations of the space group $\mathfrak{G}(\boldsymbol{k})$. These will be designated

$$D^{(\boldsymbol{k})(m)}(\{\varphi_{l_\lambda}|\tau_{l_\lambda}+\boldsymbol{R}_L\}), \qquad m=1,\ldots,r. \tag{36.3}$$

Observe that the matrices $D^{(\boldsymbol{k})(m)}$ are only defined for elements in $\mathfrak{G}(\boldsymbol{k})$. In (36.3) the index m is reserved for one of the r allowable irreducible representations of $\mathfrak{G}(\boldsymbol{k})$, and we take the matrices $D^{(\boldsymbol{k})(m)}$ to be l_m dimensional. Now we introduce a notational device [14]. Let the "dotted matrix" $\dot{D}^{(\boldsymbol{k})(m)}(\{\varphi_p|\boldsymbol{t}_p\})$ be defined by

$$\dot{D}^{(\boldsymbol{k})(m)}(X)=\begin{cases}0 & \text{if } X \text{ is not in } \mathfrak{G}(\boldsymbol{k})\\ D^{(\boldsymbol{k})(m)} & \text{if } X \text{ is in } \mathfrak{G}(\boldsymbol{k}).\end{cases} \tag{36.4}$$

In (36.4) X is a space group element, and 0 is the l_m dimensional null matrix.

As before, write the matrix $D^{(\star\boldsymbol{k})(m)}(\{\varphi_p|\tau_p\})$ in block matrix form e.g. $D^{(\star\boldsymbol{k})(m)}(\{\varphi_p|\tau_p\})_{\sigma\tau}$ where the indices σ and τ refer respectively to all those rows labelled by the same wave vector $\boldsymbol{k}_\sigma$ and to all those columns labelled by the same wave vector $\boldsymbol{k}_\tau$, where $\boldsymbol{k}_\sigma=\varphi_\sigma\cdot\boldsymbol{k}$ and $\boldsymbol{k}_\tau=\varphi_\tau\cdot\boldsymbol{k}$. By comparison with (33.1) $D^{(\star\boldsymbol{k})(m)}(\{\varphi_p|\boldsymbol{t}_p\})_{\sigma\tau}$ is the $(l_m\times l_m)$ block matrix which describes how, under the operator transformation $\{\varphi_p|\boldsymbol{t}_p\}$, the l_m functions $\psi_\alpha^{(\boldsymbol{k}_\tau)(m)}$, $\alpha=1,\ldots,l_m$ are transformed into the l_m functions $\psi_\beta^{(\boldsymbol{k}_\sigma)(m)}$, $\beta=1,\ldots,l_m$.

Now, starting from the set which belongs to $\boldsymbol{k}$, (29.5):

$$\psi_1^{(\boldsymbol{k})(m)},\ldots,\psi_{l_m}^{(\boldsymbol{k})(m)}, \tag{36.5}$$

where (36.5) is the space $\Sigma^{(\boldsymbol{k})(m)}$, let us assume that any member of the space $\Sigma^{(\boldsymbol{k}_\sigma)(m)}$ can be obtained by

$$\psi_\alpha^{(\boldsymbol{k}_\sigma)(m)}=\{\varphi_\sigma|\tau_\sigma\}\,\psi_\alpha^{(\boldsymbol{k})(m)}, \qquad \alpha=1,\ldots,l_m, \tag{36.6}$$

where (36.6) prescribes that each Bloch vector in $\Sigma^{(\boldsymbol{k}_\sigma)(m)}$ can be achieved by transforming the corresponding Bloch vector in $\Sigma^{(\boldsymbol{k})(m)}$ by the same coset representative $\{\varphi_\sigma|\tau_\sigma\}$. There is thus a one-one relation between individual Bloch vectors in $\Sigma^{(\boldsymbol{k})(m)}$ and individual Bloch vectors in $\Sigma^{(\boldsymbol{k}_\sigma)(m)}$. Further, there is no loss in generality if we assume that the choice of basis vectors in each of the spaces $\Sigma^{(\boldsymbol{k}_\sigma)(m)}$, $\sigma=1,\ldots,s$ can be taken as in (36.6). Recall the definition of indices σ and τ from (36.2).

With the choice (36.6) for the basis vectors we can immediately write down the block matrices in the representation $D^{(\star\boldsymbol{k})(m)}$ for the coset representatives in (36.2), and we obtain:

$$D^{(\star\boldsymbol{k})(m)}(\{\varphi_\sigma|\tau_\sigma\})_{\sigma 1}=\boldsymbol{\Pi}_m \tag{36.7}$$

and

$$D^{(\star\boldsymbol{k})(m)}(\{\varphi_\sigma|\tau_\sigma\}^{-1})_{1\sigma}=\boldsymbol{\Pi}_m, \tag{36.8}$$

where $\boldsymbol{\Pi}_m$ is the l_m dimensional unit matrix, and the index 1 refers to $\boldsymbol{k}_1\equiv\boldsymbol{k}$. The matrices (36.7) and (36.8) are the only non-vanishing blocks in the first column and row of the matrices for the coset representatives.

Now consider the space group element

$$\{\varphi_\sigma|\tau_\sigma\}\,\{\varphi_p|\tau_p\}\,\{\varphi_\tau|\tau_\tau\}^{-1}, \tag{36.9}$$

where as always, σ and τ refer to coset representatives in (36.3). Form the block matrix element $\sigma\tau$ in the matrix representing the element (36.9) in the space group irreducible representation. Thus

$$\begin{aligned} D^{(\star k)(m)}&(\{\varphi_\sigma|\tau_\sigma\}\{\varphi_p|\tau_p\}\{\varphi_\tau|\tau_\tau\}^{-1})_{\sigma\tau} \\ &=\sum_\mu\sum_\nu D^{(\star k)(m)}(\{\varphi_\sigma|\tau_\sigma\})_{\sigma\mu} D^{(\star k)(m)}(\{\varphi_p|\tau_p\})_{\mu\nu} D^{(\star k)(m)}(\{\varphi_\tau|\tau_\tau\}^{-1})_{\nu\tau} \\ &=D^{(\star k)(m)}(\{\varphi_p|\tau_p\})_{11}. \end{aligned} \tag{36.10}$$

In obtaining (36.10) we used (36.7) and (36.8). Now however, note that the $\sigma=1$, $\tau=1$ block matrix always refers to the situation where the operator $P_{\{\varphi_p|\tau_p\}}$ sends the space $\Sigma^{(k_1)(m)}$ onto itself i.e. the operator is in $\mathfrak{G}(k)$. Consequently, if $\{\varphi_p|\tau_p\}$ is not in $\mathfrak{G}(k)$, the $\sigma=1$, $\tau=1$ block $D^{(\star k)(m)}(\{\varphi_p|\tau_p\})_{11}$ is certainly the l_m dimensional null matrix. Hence

$$D^{(\star k)(m)}(\{\varphi_p|\tau_p\})_{11}=\begin{cases}0 & \text{if } \{\varphi_p|\tau_p\} \text{ is not in } \mathfrak{G}(k)\\ D^{(k)(m)}(\{\varphi_p|\tau_p\}) & \text{if } \{\varphi_p|\tau_p\} \text{ is in } \mathfrak{G}(k)\end{cases} \tag{36.11}$$

where $D^{(k)(m)}$ is irreducible.

Thus if $\{\varphi_p|\tau_p\}$ is an arbitrary element in $\mathfrak{G}$ and $\{\varphi_\sigma|\tau_\sigma\}$ and $\{\varphi_\tau|\tau_\tau\}$ are among the coset representatives in (36.2) then let $\{\varphi_p|\tau_p\}\cdot\{\varphi_\tau|\tau_\tau\}$ be in the coset $\{\varphi_\sigma|\tau_\sigma\}\,\mathfrak{G}(k)$. Then

$$\{\varphi_p|\tau_p\}\cdot\{\varphi_\tau|\tau_\tau\}=\{\varphi_\sigma|\tau_\sigma\}\cdot\{\varphi_{l_\lambda}|\tau_{l_\lambda}\}\cdot\{\varepsilon|R_L\}, \tag{36.12}$$

where $\{\varepsilon|R_L\}$ is some element in $\mathfrak{T}$. Also

$$\{\varphi_\sigma|\tau_\sigma\}^{-1}\cdot\{\varphi_p|\tau_p\}\cdot\{\varphi_\tau|\tau_\tau\}=\{\varphi_{l_\lambda}|\tau_{l_\lambda}\}\cdot\{\varepsilon|R_L\}, \tag{36.13}$$

where the right hand side of (36.13) is an element in $\mathfrak{G}(k)$. Then using (36.4) and (36.11) we can immediately write down an expression for the block matrices $D^{(\star k)(m)}(\{\varphi_p|\tau_p\})_{\sigma\tau}$. We have

$$D^{(\star k)(m)}(\{\varphi_p|\tau_p\})_{\sigma\tau}=\dot{D}^{(k)(m)}(\{\varphi_\sigma|\tau_\sigma\}^{-1}\cdot\{\varphi_p|\tau_p\}\cdot\{\varphi_\tau|\tau_\tau\}). \tag{36.14}$$

Eq. (36.14) solves the problem for all space groups $\mathfrak{G}$, symmorphic or non-symmorphic. It gives the block matrices comprising the full matrix, in terms of the known $D^{(k)(m)}$ for each element in $\mathfrak{G}(k)$, and $D^{(\star k)(m)}$ is then simply constructed from the blocks (36.14) by putting the blocks in appropriate locations in the big matrix.

It is necessary to emphasize an important matter at this point in the analyses. In each irreducible representation $D^{(\star k)(m)}$ of $\mathfrak{G}$ it is necessary to select one wave vector $\boldsymbol{k}$ which refers to the $\sigma=1$, $\tau=1$ block matrix and to define this wave vector $\boldsymbol{k}$ as the canonical wave vector $\boldsymbol{k}$ of the star (32.10). It is a completely arbitrary matter which of the wave vectors in $\star k$ is selected as canonical for a given star. However, once a vector $\boldsymbol{k}$ is picked it defines a particular subspace $\Sigma^{(k)(m)}$ and a $\mathfrak{G}(k)$ with specific elements, and coset representatives. It is then not permissable to change the choice of $\boldsymbol{k}$ for that star. While arbitrary, it is important to keep $\boldsymbol{k}$ fixed in the course of the work. If a different $\boldsymbol{k}$ in the same star is selected as the canonical wave vector, it will merely produce a similarity transformation in the representation $D^{(\star k)(m)}$ sending it into an equivalent one. As long as we deal with representations we must concern ourselves with the specific form of the matrix $D^{(\star k)(m)}$ and the block submatrices (33.1), and we do not leave free a similarity transformation.

The process epitomized in (36.14), by which the acceptable irreducible matrices $D^{(\boldsymbol{k})(m)}$ of the subgroup $\mathfrak{G}(\boldsymbol{k})$ are utilized to construct the irreducible representation $D^{(*\boldsymbol{k})(m)}$ of $\mathfrak{G}$ is called *inducing* the representation of the group from that of its subgroup. It has been known in the mathematical literature for a long time going back to the work of FROEBENIUS and SCHUR. The most compact recent statements of this process are available in standard references [*11*], but the development given above in Sects. 26–36 is closest to that of CLIFFORD.

Finally we consider that there are occasions when we require the actual complete matrix element of the matrix $D^{(*\boldsymbol{k})(m)}$. Thus let $\psi_\alpha^{(\boldsymbol{k}_\tau)(m)}$ be a member of the space

$$\Sigma^{(\boldsymbol{k}_\tau)(m)} \equiv \{\psi_1^{(\boldsymbol{k}_\tau)(m)}, \ldots, \psi_{l_m}^{(\boldsymbol{k}_\tau)(m)}\}. \tag{36.15}$$

Let

$$P_{\{\varphi_p | \boldsymbol{t}(\varphi_p)\}} \equiv P_{\{\varphi_p\}} \tag{36.16}$$

be an element in $\mathfrak{G}$. Then

$$P_{\{\varphi_p\}} \psi_b^{(\boldsymbol{k}_\tau)(m)} = \sum_{\sigma=1}^{s} \sum_{\beta=1}^{l_m} D^{(*\boldsymbol{k})(m)}(\{\varphi_p\})_{(\sigma a)(\tau b)} \psi_a^{(\boldsymbol{k}_\sigma)(m)}, \tag{36.17}$$

where

$$D^{(*\boldsymbol{k})(m)}(\{\varphi_p\})_{(\sigma a)(\tau b)}$$

is the ab element of the block matrix on the right hand side of (36.14):

$$\begin{aligned} D^{(*\boldsymbol{k})(m)}(\{\varphi_p\})_{(\sigma a)(\tau b)} &\equiv \left(D^{(*\boldsymbol{k})(m)}(\{\varphi_p\})_{\sigma\tau}\right)_{ab} \\ &= \left(\dot{D}^{(\boldsymbol{k})(m)}(\{\varphi_\sigma\}^{-1} \cdot \{\varphi_p\} \cdot \{\varphi_\tau\})\right)_{ab}. \end{aligned} \tag{36.18}$$

Of course if $\{\varphi_p\}$ is in $\mathfrak{G}(\boldsymbol{k}_\tau)$ then the only non-zero block matrix will be with $\sigma = \tau$ and the space $\Sigma^{(\boldsymbol{k}_\tau)(m)}$ is sent onto itself. We will adopt the notation (36.17), (36.18) when we require explicit statement of the result of operation upon one member of the space $\Sigma^{(*\boldsymbol{k})(m)}$ by a group element in $\mathfrak{G}$. Note the abbreviation (36.16)–(36.18) which will occasionally be used – but only when redefined at each use, to avoid notational confusion. This abbreviates $\{\varphi_\sigma | \boldsymbol{t}(\varphi_\sigma)\}$ by $\{\varphi_\sigma\}$.

37. Characters of $D^{(*\boldsymbol{k})(m)}$ of $\mathfrak{G}$; induced characters. Continuing with the situation of Sect. 36 particularly regarding notation, we denote the character of the matrix (36.3) as

$$\chi^{(\boldsymbol{k})(m)}(\{\varphi_{l_\lambda} | \boldsymbol{t}_{l_\lambda}\}) \equiv \operatorname{trace} D^{(\boldsymbol{k})(m)}(\{\varphi_{l_\lambda} | \boldsymbol{t}_{l_\lambda}\}). \tag{37.1}$$

Again, the notational device (36.4) allows us to define dotted characters $\dot{\chi}^{(\boldsymbol{k})(m)}$ for all the elements in the full group $\mathfrak{G}$:

$$\dot{\chi}^{(\boldsymbol{k})(m)}(X) = \begin{cases} 0 & \text{if } X \text{ is not in } \mathfrak{G}(\boldsymbol{k}) \\ \chi^{(\boldsymbol{k})(m)}(X) & \text{if } X \text{ is in } \mathfrak{G}(\boldsymbol{k}). \end{cases} \tag{37.2}$$

From (36.14), (37.1), and (37.2) we obtain the trace of any element in $\mathfrak{G}$ in the irreducible representation $D^{(*\boldsymbol{k})(m)}$. It is

$$\chi^{(*\boldsymbol{k})(m)}(\{\varphi_p | \tau_p\}) = \sum_{\sigma=1}^{s} \dot{\chi}^{(\boldsymbol{k})(m)}(\{\varphi_\sigma | \tau_\sigma\}^{-1} \cdot \{\varphi_p | \tau_p\} \cdot \{\varphi_\sigma | \tau_\sigma\}). \tag{37.3}$$

Observe that the argument of each dotted character,

$$\{\varphi_\sigma|\tau_\sigma\}^{-1}\cdot\{\varphi_p|\tau_p\}\cdot\{\varphi_\sigma|\tau_\sigma\} \tag{37.4}$$

is either in $\mathfrak{G}(\boldsymbol{k})$, or else the dotted character vanishes. Now further consider the case where $\{\varphi_p|\tau_p\}$ is an element in $\mathfrak{G}(\boldsymbol{k})$ i.e. consider

$$\chi^{(\star k)(m)}(\{\varphi_{l_\lambda}|\tau_{l_\lambda}\})=\sum_{\sigma=1}^{s}\dot{\chi}^{(k)(m)}(\{\varphi_\sigma|\tau_\sigma\}^{-1}\cdot\{\varphi_{l_\lambda}|\tau_{l_\lambda}\}\cdot\{\varphi_\sigma|\tau_\sigma\}). \tag{37.5}$$

Now, in $\mathfrak{G}(\boldsymbol{k})$ the element $\{\varphi_{l_\lambda}|\tau_{l_\lambda}\}$ is in some class. The class of $\{\varphi_{l_\lambda}|\tau_{l_\lambda}\}$in $\mathfrak{G}(\boldsymbol{k})$ is the maximal set of elements in $\mathfrak{G}(\boldsymbol{k})$ conjugate to $\{\varphi_{l_\lambda}|\tau_{l_\lambda}\}$, where the conjugation is performed with respect to elements in $\mathfrak{G}(\boldsymbol{k})$. However, the conjugation indicated in (37.5) is performed with respect to coset representatives $\{\varphi_\sigma|\tau_\sigma\}$ which are specifically *not* in $\mathfrak{G}(\boldsymbol{k})$. Hence the conjugate element $\{\varphi_\sigma|\tau_\sigma\}^{-1}\cdot\{\varphi_{l_\lambda}|\tau_{l_\lambda}\}\cdot\{\varphi_\sigma|\tau_\sigma\}$ can be in $\mathfrak{G}(\boldsymbol{k})$, but may well be in a different class in $\mathfrak{G}(\boldsymbol{k})$ than $\{\varphi_{l_\lambda}|\tau_{l_\lambda}\}$. Thus the individual non vanishing terms in (37.5) may or may not be equal; that is

$$\dot{\chi}^{(k)(m)}(\{\varphi_{l_\lambda}|\tau_{l_\lambda}\})=\dot{\chi}^{(k)(m)}(\{\varphi_\sigma|\tau_\sigma\}^{-1}\cdot\{\varphi_{l_\lambda}|\tau_{l_\lambda}\}\cdot\{\varphi_\sigma|\tau_\sigma\})$$

if the arguments (elements) are indeed in the same class in $\mathfrak{G}(\boldsymbol{k})$, or if, in the particular representation considered two different classes happen to have the same character. Each case must be examined individually: while it is possible to give some general rules, these seem too unwieldy for practical interest.

As in the case of the matrices $D^{(\star k)(m)}$ it is worth emphasizing that the choice of one canonical wave vector for a star is arbitrary but fixed during the entire analysis. Particularly observe that to evaluate (37.5) the indices σ of the coset representatives must be specified with regard to $\mathfrak{G}(\boldsymbol{k})$, requiring the explicit choice of one $\boldsymbol{k}$ in $\star\boldsymbol{k}$.

Evidently several tasks lie ahead of us still. It must be shown how to obtain all the acceptable irreducible representations $D^{(k)(m)}$ of each $\mathfrak{G}(\boldsymbol{k})$, in practical fashion. It must be shown then how to obtain all the irreducible representations of $\mathfrak{G}$. Finally, using the orthonormality and completeness relations for the group characters it must be verified that indeed all irreducible representations of $\mathfrak{G}$ have been determined. These tasks occupy us in the following sections.

38. Allowable irreducible $D^{(k)(m)}$: General star with $\mathfrak{G}(k)=\mathfrak{T}$. In determining the allowable irreducible representations of $\mathfrak{G}(\boldsymbol{k})$, two major cases can be distinguished; depending on the structure, and dimension, of the star of the wave vector: $\star\boldsymbol{k}$ (32.10). In the first case $\star\boldsymbol{k}$ contains g_p independent inequivalent wave vectors:

$$\star\boldsymbol{k}=\{\boldsymbol{k}_1=\boldsymbol{k},\boldsymbol{k}_2=\varphi_2\cdot\boldsymbol{k}_1,\ldots,\boldsymbol{k}_{g_p}=\varphi_{g_p}\cdot\boldsymbol{k}\}. \tag{38.1}$$

That is, every distinct rotation in $\mathfrak{P}=\mathfrak{G}/\mathfrak{T}$ produces a distinct wave vector when applied to the canoncial wave vector $\boldsymbol{k}$. This case is one for which the star $\star\boldsymbol{k}$ has as many arms as there are rotations in $\mathfrak{P}$ so $s=g_p$ and it is called the "general star". The second case in which $s<g_p$ is called the special star: it will be dealt with in the later sections.

For the case of the general star (38.1), the space group $\mathfrak{G}(\boldsymbol{k})=\mathfrak{T}$ the translation group. In this case the allowable irreducible representation $D^{(\boldsymbol{k})(m)}$ has dimension 1

$$D^{(\boldsymbol{k})(m)}(\{\varepsilon|0\})=1. \tag{38.2}$$

Further it is clear that in this case the entire subspace $\Sigma^{(\boldsymbol{k})(m)}$ consists of the single Bloch vector $\psi^{(\boldsymbol{k})}$ so

$$D^{(\boldsymbol{k})(m)}(\{\varepsilon|\boldsymbol{R}_L\})=\exp -i\,\boldsymbol{k}\cdot\boldsymbol{R}_L. \tag{38.3}$$

The irreducible space $\Sigma^{(\star\boldsymbol{k})(m)}$ (32.11) is

$$\Sigma^{(\star\boldsymbol{k})(m)}=\Sigma^{(\boldsymbol{k}_1)(m)}\oplus\cdots\oplus\Sigma^{(\boldsymbol{k}_{g_p})(m)} \tag{38.4}$$

or

$$\Sigma^{(\star\boldsymbol{k})(m)}=(\psi^{(\boldsymbol{k})},\psi^{(\boldsymbol{k}_2)},\ldots,\psi^{(\boldsymbol{k}_{g_p})}). \tag{38.5}$$

The irreducible representation $D^{(\star\boldsymbol{k})(m)}$ is g_p dimensional. Explicit forms for $D^{(\star\boldsymbol{k})(m)}$ and the character system will be given below. Note also that in this case the index m is unnecessary since $\mathfrak{T}$ has only one dimensional irreducible representations; we retain it for notational consistency.

39. Allowable irreducible $D^{(\boldsymbol{k})(m)}$: Special star: Little group technique. Now we consider the case where $s<g_p$ so that the star $\star\boldsymbol{k}$ contains fewer arms than there are distinct rotations in $\mathfrak{P}$. Evidently in this case $\mathfrak{G}(\boldsymbol{k})$ is not trivial, but is a space group as in (36.1) with additional coset representatives besides the identity. To determine the allowable irreducible $D^{(\boldsymbol{k})(m)}$ several useful techniques have been brought forward, which we shall discuss seriatim. In the present section we present the little group technique which seems to have been first applied in the context of space group theory by HERRING.[1]

The allowable irreducible representations which we seek are all characterized by the fact that for every element in $\mathfrak{T}$

$$D^{(\boldsymbol{k})(m)}(\{\varepsilon|\boldsymbol{R}_L\})=\exp -i\,\boldsymbol{k}\cdot\boldsymbol{R}_L\,\boldsymbol{\Pi}_m, \tag{39.1}$$

where $\boldsymbol{\Pi}_m$ is the $l_m\times l_m$ unit matrix. Suppose there is an element in $\mathfrak{T}$, $\{\varepsilon|\boldsymbol{R}_L(\boldsymbol{k})\}$, with the property

$$\boldsymbol{k}\cdot\boldsymbol{R}_L(\boldsymbol{k})=2\pi\kappa, \tag{39.2}$$

where κ is any integer. Then clearly for this element

$$D^{(\boldsymbol{k})(m)}(\{\varepsilon|\boldsymbol{R}_L(\boldsymbol{k})\})=\boldsymbol{\Pi}_m=D^{(\boldsymbol{k})(m)}(\{\varepsilon|\boldsymbol{0}\}). \tag{39.3}$$

The element $\{\varepsilon|\boldsymbol{R}_L(\boldsymbol{k})\}$ is a pure translation to which the vector index $\boldsymbol{k}$ has been adduced to indicate that $\boldsymbol{R}_L(\boldsymbol{k})$ statisfies (39.2). The set of all elements $\{\varepsilon|\boldsymbol{R}_L(\boldsymbol{k})\}$ evidently form a group which is a subgroup of $\mathfrak{T}$. This subgroup may be trivial, consisting only of $\{\varepsilon|\boldsymbol{0}\}$ but often it is not. Let us assume it is not trivial and denote it by $\mathfrak{T}(\boldsymbol{k})$, where

$$\mathfrak{T}(\boldsymbol{k}) \text{ is the set } \{\varepsilon|\boldsymbol{R}_L(\boldsymbol{k})\}. \tag{39.4}$$

[1] C. HERRING: J. Franklin Inst. **233**, 525 (1942).

Then in regard to the allowable irreducible representation $D^{(k)(m)}$ all the matrices representing elements in $\mathfrak{T}(\boldsymbol{k})$ are equal to the identity (39.3). The set of matrices $D^{(k)(m)}(\{\varepsilon|\boldsymbol{R}_L(\boldsymbol{k})\})$ are the center of the representation.

Define the coset representatives in the abelian factor group $\mathfrak{T}/\mathfrak{T}(\boldsymbol{k})$ by

$$\mathfrak{T}=\mathfrak{T}(\boldsymbol{k})+\{\varepsilon|\boldsymbol{R}_{L_2}(\bar{\boldsymbol{k}})\}\,\mathfrak{T}(\boldsymbol{k})+\cdots+\{\varepsilon|\boldsymbol{R}_{L_t}(\bar{\boldsymbol{k}})\}\,\mathfrak{T}(\boldsymbol{k}), \tag{39.5}$$

where

$$\boldsymbol{R}_{L_\alpha}(\bar{\boldsymbol{k}}), \qquad \alpha=1,\ldots,t \tag{39.6}$$

is a lattice vector, *not* in $\mathfrak{T}(\boldsymbol{k})$, and we take $\mathfrak{T}/\mathfrak{T}(\boldsymbol{k})$ to be a group of order t. We use $\bar{\boldsymbol{k}}$ (barred $\boldsymbol{k}$) to indicate that (39.6) is not in $\mathfrak{T}(\boldsymbol{k})$. Then we can decompose $\mathfrak{G}(\boldsymbol{k})$ into cosets with respect to $\mathfrak{T}(\boldsymbol{k})$. To avoid notational clumsiness we will express this decomposition in terms of the coordinate transformation group operators $\{\varphi|\tau\}$ directly instead of the function operators of (34.7), but the reversion is easily made from one set to the other. Then

$$\begin{aligned}\mathfrak{G}(\boldsymbol{k})=&\,\mathfrak{T}(\boldsymbol{k})+\{\varepsilon|\boldsymbol{R}_{L_2}(\bar{\boldsymbol{k}})\}\,\mathfrak{T}(\boldsymbol{k})+\cdots+\{\varepsilon|\boldsymbol{R}_{L_t}(\bar{\boldsymbol{k}})\}\,\mathfrak{T}(\boldsymbol{k})\\&+\{\varphi_{l_2}|\tau_{l_2}\}\,\mathfrak{T}(\boldsymbol{k})+\{\varphi_{l_2}|\tau_{l_2}+\boldsymbol{R}_{L_2}(\bar{\boldsymbol{k}})\}\,\mathfrak{T}(\boldsymbol{k})\\&+\cdots+\{\varphi_{l_2}|\tau_{l_2}+\boldsymbol{R}_{L_t}(\bar{\boldsymbol{k}})\}\,\mathfrak{T}(\boldsymbol{k})\\&+\cdots+\{\varphi_{l_k}|\tau_{l_k}+\boldsymbol{R}_{L_t}(\bar{\boldsymbol{k}})\}\,\mathfrak{T}(\boldsymbol{k}).\end{aligned} \tag{39.7}$$

The factor group $\mathfrak{G}(\boldsymbol{k})/\mathfrak{T}(\boldsymbol{k})$ consists of the $(l_k\cdot t)$ cosets

$$\{\varphi_{l_\alpha}|\tau_{l_\alpha}+\boldsymbol{R}_{L_\beta}(\bar{\boldsymbol{k}})\}\,\mathfrak{T}(\boldsymbol{k}), \qquad l_\alpha=1,\ldots,l_k,\ \beta=1,\ldots,t. \tag{39.8}$$

An acceptable irreducible representation of $\mathfrak{G}(\boldsymbol{k})/\mathfrak{T}(\boldsymbol{k})$ will be an acceptable irreducible representation of $\mathfrak{G}(\boldsymbol{k})$. The only difference between them is in regard to those elements in $\mathfrak{T}(\boldsymbol{k})$ which are in the center or kernel of the representation group and hence are equivalent to the identity $\{\varepsilon|0\}$ of $\mathfrak{G}(\boldsymbol{k})$.

In the little group technique one considers the group homomorphic to the set (39.8)

$$\Pi(\boldsymbol{k})\equiv\mathfrak{G}(\boldsymbol{k})/\mathfrak{T}(\boldsymbol{k}) \tag{39.9}$$

as an abstract group with $(l_k\cdot t)$ elements, and determines all of its irreducible representations by any standard method. Of the set of all irreducible representations only the allowable ones are selected for use. To obtain the character system for the irreducible representation of $\Pi(\boldsymbol{k})$ as an abstract group it is required that one first obtain the group multiplication table.

Now observe that the collection of coset representatives

$$\{\varphi_{l_\alpha}|\tau_{l_\alpha}+\boldsymbol{R}_{L_\beta}(\bar{\boldsymbol{k}}\}, \qquad \alpha=1,\ldots,l_k,\ \beta=1,\ldots,t \tag{39.10}$$

is not closed under multiplication. To establish closure, we need to keep in mind that the matrix representing any element in $\mathfrak{T}(\boldsymbol{k})$ satisfies (39.3) in an allowable representation and so behaves as the abstract identity element $\{\varepsilon|0\}$. Hence we define the multiplication of the $(l_k\cdot t)$ objects (39.10) in such a fashion that closure is achieved, by:

$$\begin{aligned}\{\varphi_{l_\alpha}|\tau_{l_\alpha}+\boldsymbol{R}_{L_\beta}(\bar{\boldsymbol{k}})\}\cdot\{\varphi_{l_\gamma}|\tau_{l_\gamma}+\boldsymbol{R}_{L_\delta}(\bar{\boldsymbol{k}})\}&=\{\varphi_{l_\alpha}\cdot\varphi_{l_\gamma}|\tau_{l_{\alpha\gamma}}+\boldsymbol{R}_{L_\varepsilon}(\bar{\boldsymbol{k}})\}\\&=\{\varepsilon|\boldsymbol{R}_{L_\varepsilon}(\bar{\boldsymbol{k}})\}\cdot\{\varphi_{l_{\alpha\gamma}}|\tau_{l_{\alpha\gamma}}\}\end{aligned} \tag{39.11}$$

with β, δ, $\varepsilon=1,\ldots,t$ and with

$$\tau_{l_{\alpha\gamma}}+\boldsymbol{R}_{L_\varepsilon}(\bar{\boldsymbol{k}})\equiv\varphi_{l_\alpha}\cdot\{\tau_{l_\gamma}+\boldsymbol{R}_{L_\delta}(\bar{\boldsymbol{k}})\}+\tau_{l_\alpha}+\boldsymbol{R}_{L_\beta}(\bar{\boldsymbol{k}})+\boldsymbol{R}_L(\boldsymbol{k}). \tag{39.12}$$

In (39.12), $\boldsymbol{R}_L(\boldsymbol{k})$ is any lattice vector in $\mathfrak{T}(\boldsymbol{k})$. In all practical cases in which this method has been applied, the lattice vector $\boldsymbol{R}_{L_\varepsilon}(\bar{\boldsymbol{k}})$ is uniquely defined. Assuming that it is, the multiplication law (39.11), (39.12) uniquely determines the product coset which arises upon multiplication.

Under the rule (39.11), the $l_k t$ coset representatives are closed and isomorphic to an abstract group $\boldsymbol{\Pi}(\boldsymbol{k})$. The multiplication table, class structure, and class multiplication coefficients can be determined for this group in the usual fashion, and then also the characters of the group elements. If the class multiplication coefficients for classes $\mathfrak{C}_i^{(\boldsymbol{k})}$, $\mathfrak{C}_j^{(\boldsymbol{k})}$, $\mathfrak{C}_l^{(\boldsymbol{k})}$ in $\Pi(\boldsymbol{k})$ are defined as

$$\mathfrak{C}_i^{(\boldsymbol{k})}\,\mathfrak{C}_j^{(\boldsymbol{k})}=\sum(ij\,|\,l)^{(\boldsymbol{k})}\,\mathfrak{C}_l^{(\boldsymbol{k})}. \tag{39.13}$$

Then the characters in the irreducible representations satisfy

$$c_i^{(\boldsymbol{k})}\,\chi^{(\boldsymbol{k})(\mu)}(\mathfrak{C}_i^{(\boldsymbol{k})})\cdot c_j^{(\boldsymbol{k})}\,\chi^{(\boldsymbol{k})(\mu)}(\mathfrak{C}_j^{(\boldsymbol{k})})=l_\mu\sum_l(ij\,|\,l)^{(\boldsymbol{k})}\,\chi^{(\boldsymbol{k})(\mu)}(\mathfrak{C}_l^{(\boldsymbol{k})})\cdot c_l^{(\boldsymbol{k})}, \tag{39.14}$$

where l_μ is the dimension of the irreducible representation $\chi^{(\boldsymbol{k})(\mu)}$ and $c_i^{(\boldsymbol{k})}$, is the order of the class $\mathfrak{C}_i^{(\boldsymbol{k})}$, ... in $\Pi(\boldsymbol{k})$. The characters of allowable irreducible representations $\chi^{(\boldsymbol{k})(m)}$ of $\Pi(\boldsymbol{k})$ and $\mathfrak{G}(\boldsymbol{k})$ must satisfy

$$\chi^{(\boldsymbol{k})(m)}(\{\varepsilon\,|\,\boldsymbol{R}_L(\bar{\boldsymbol{k}})\})=l_m\exp-i\,\boldsymbol{k}\cdot\boldsymbol{R}_L(\bar{\boldsymbol{k}}),\qquad\text{for } m \text{ allowable}. \tag{39.15}$$

Note that in (39.15) we use the lattice translation $\boldsymbol{R}_L(\bar{\boldsymbol{k}})$ since for any $\boldsymbol{R}_L(\boldsymbol{k})$ in $\mathfrak{T}(\boldsymbol{k})$ the result is trivial owing to (39.2). Since $\boldsymbol{k}$ is known, (39.15) is an essential restriction on the index of the irreducible representation and permits unequivocal identification of the allowable ones which are always indexed with $\mu=m$, in the following.

Finally, the characters of the irreducible representations of $\Pi(\boldsymbol{k})$ satisfy the completeness rule

$$\sum_j c_j^{(\boldsymbol{k})}\,\chi^{(\boldsymbol{k})(\mu)}(\mathfrak{C}_j)\,\chi^{(\boldsymbol{k})(\mu)}(\mathfrak{C}_j)^*=(l_k t), \tag{39.16}$$

where $(l_k t)$ is the order of $\Pi(\boldsymbol{k})$. It is very important to observe that (39.16) applies as well to the allowable as to the non-allowable $\chi^{(\boldsymbol{k})(\mu)}$ and further that all objects in (39.16) refer to the group $\Pi(\boldsymbol{k})$, in which classes, etc. are defined by conjugation of elements in $\boldsymbol{\Pi}(\boldsymbol{k})$, by elements in $\Pi(\boldsymbol{k})$. Recall again that $\Pi(\boldsymbol{k})\equiv\mathfrak{G}(\boldsymbol{k})/\mathfrak{T}(\boldsymbol{k})$.

40. Non-allowable irreducible $D^{(\boldsymbol{k})(\mu)}$: Little group technique. The usefulness of the non-allowable irreducible representations of $\Pi(\boldsymbol{k})=\mathfrak{G}(\boldsymbol{k})/\mathfrak{T}(\boldsymbol{k})$ was pointed out by ELLIOTT and LOUDON, in their analysis of some selection rules in diamond structure.[1] We return later to their analyses of selection rules, but in the present context we shall focus in the significance of $D^{(\boldsymbol{k})(\mu)}$ of $\Pi(\boldsymbol{k})$ for μ a non-allowable index.

Evidently *all* irreducible representations $D^{(\boldsymbol{k})(\mu)}$ are homomorphic images of the abstract group $\Pi(\boldsymbol{k})$ which consists of the set of coset representatives (39.10),

[1] R.J. ELLIOTT and R. LOUDON: J. Phys. Chem. Solids 15, 146 (1960).

closed under group multiplication by (39.11) and (39.12). Now suppose we have an allowable irreducible representation $D^{(\boldsymbol{k})(m)}$ of $\Pi(\boldsymbol{k})$, and consider the matrix representing element $\{\varepsilon|\boldsymbol{R}_L(\bar{\boldsymbol{k}})\}$:

$$D^{(\boldsymbol{k})(m)}(\{\varepsilon|\boldsymbol{R}_L(\bar{\boldsymbol{k}})\})=\exp-i\,\boldsymbol{k}\cdot\boldsymbol{R}_L(\bar{\boldsymbol{k}})\,\boldsymbol{\Pi}_m. \tag{40.1}$$

Take the direct product of this matrix with itself to find:

$$D^{(\boldsymbol{k})(m)}(\{\varepsilon|\boldsymbol{R}_L(\bar{\boldsymbol{k}})\})\otimes D^{(\boldsymbol{k})(m)}(\{\varepsilon|\boldsymbol{R}_L(\bar{\boldsymbol{k}})\})=\exp-2i\,\boldsymbol{k}\cdot\boldsymbol{R}_L(\bar{\boldsymbol{k}})\,\Pi_m\otimes\Pi_m. \tag{40.2}$$

Clearly (40.2) is a $(2l_m\times 2l_m)$ square matrix which, insofar as translation behaves as a matrix system going with wave vector

$$\boldsymbol{k}'=2\boldsymbol{k}. \tag{40.3}$$

Then, the entire direct product representation of $P(\boldsymbol{k})$

$$D^{(\boldsymbol{k})(m)}\otimes D^{(\boldsymbol{k})(m)} \tag{40.4}$$

must be connected with wave vector $\boldsymbol{k}'$ of (40.3).

On the other hand any representation of an abstract group is either irreducible or decomposable, by Maschke's theorem. Hence, (40.4), as a representation of $\Pi(\boldsymbol{k})$ is decomposable into the direct sum of irreducible components, at $\boldsymbol{k}$:

$$D^{(\boldsymbol{k})(m)}\otimes D^{(\boldsymbol{k})(m)}=\sum_{\mu}(m\,m|\mu)\,D^{(\boldsymbol{k})(\mu)}, \tag{40.5}$$

where in (40.5) μ admits all values. By the previous argument the representation (40.5) goes with wave vector $\boldsymbol{k}$. Consequently we identify some "non-allowable" irreducible representations $D^{(\boldsymbol{k})(\mu)}$, $\mu\neq m$ of $\Pi(\boldsymbol{k})$ as actually representations related to the wave vector $2\boldsymbol{k}$. Consequently direct product representations like (40.4) are *simultaneously* representations at $\boldsymbol{k}$ (which may be non-allowable), and representations at $2\boldsymbol{k}$, which may be irreducible.

Again, starting with the $D^{(\boldsymbol{k})(m)}$ known to be allowable and belonging to $\Pi(\boldsymbol{k})$ we may take repeated inner Kronecker products

$$D^{(\boldsymbol{k})(m)}\otimes\cdots\otimes D^{(\boldsymbol{k})(m)}=[D^{(\boldsymbol{k})(m)}]_p,\qquad p\leq t \tag{40.6}$$

finding in each case again a representation of the abstract group of operators $\Pi(\boldsymbol{k})$ which actually is related to wave vector $p\,\boldsymbol{k}$.

Consequently we conclude that all the non-allowable irreducible representations $D^{(\boldsymbol{k})(\mu)}$ $\mu\neq m$ are related to a family of wave vectors based on $\boldsymbol{k}$

$$\boldsymbol{k}, 2\boldsymbol{k}, \dots, t\,\boldsymbol{k}. \tag{40.7}$$

By inspecting the character table for the non-allowable $D^{(\boldsymbol{k})(\mu)}$ and comparing

$$D^{(\boldsymbol{k})(\mu)}\ \mu\neq m \qquad \text{of } \Pi(\boldsymbol{k}) \tag{40.8}$$

with

$$D^{(p\boldsymbol{k})(m)} \qquad \text{of } \Pi(p\,\boldsymbol{k}),\ p=1,\dots,t \tag{40.9}$$

where t is an integer equal to the index of $\mathfrak{T}(\boldsymbol{k})$ in $\mathfrak{T}$, we may discover the relationship between $D^{(\boldsymbol{k})(\mu)}$ $\mu\neq m$ and the irreducible representations of $\Pi(p\,\boldsymbol{k})$.

It is worth recapitulating the argument of the last two sections. The abstract group $\Pi(\boldsymbol{k})=\mathfrak{G}(\boldsymbol{k})/\mathfrak{T}(\boldsymbol{k})$ was constructed in order to determine allowable irreducible representation $D^{(\boldsymbol{k})(m)}$ going with canonical wave vector $\boldsymbol{k}$ to be used in inducing the irreducible $D^{(\star\boldsymbol{k})(m)}$ of $\mathfrak{G}$. However, when $\Pi(\boldsymbol{k})$ is treated as an abstract group, allowable $D^{(\boldsymbol{k})(m)}$ as well as non-allowable $D^{(\boldsymbol{k})(\mu)}$ arise. The latter, irreducible representations of the abstract group of operators, $\Pi(\boldsymbol{k})$, have been identified as belonging to the family of wave vectors (40.7), for which $D^{(\boldsymbol{k})(\mu)}$ is either irreducible or reducible. In the former case

$$D^{(\boldsymbol{k})(\mu)} \doteqdot D^{(p\boldsymbol{k})(m)} \tag{40.10}$$

in the latter

$$D^{(\boldsymbol{k})(\mu)} \doteqdot \sum_{m} \oplus (\mu|m)\, D^{(p\boldsymbol{k})(m)}. \tag{40.11}$$

In either event the essential point is that once the group $\Pi(\boldsymbol{k})$ is treated as an abstract group i.e. a collection of cosets (39.10), closed by (39.11) and (39.12) the irreducible representation of $\Pi(\boldsymbol{k})$ which arise must only *a posteriori* be sorted into allowable or non-allowable; it is not possible in general to directly only obtain allowable ones, except in cases where the entire family (40.7) collapses; such as $\boldsymbol{k}=(0,0,0)$.

41. Allowable irreducible $D^{(\boldsymbol{k})(m)}$ as ray representations. The little group method discussed in Sects. 39 and 40 requires construction of a new group $\mathfrak{G}(\boldsymbol{k})/\mathfrak{T}(\boldsymbol{k})=\Pi(\boldsymbol{k})$, and the determination of all of its irreducible representations so that the allowable subset may be found. The ray representation method to be discussed in the next several sections, in principal requires the construction of projective, or ray representations of *existing* groups only, in particular of the 32 crystal point groups. In practice the ray representation method involves some additional labor to determine the allowable $D^{(\boldsymbol{k})(m)}$ so no practical advantage results in most cases from either method. However, the ray representation method is an important one in principle in regard to the insight it gives into the structure of the representation $D^{(\boldsymbol{k})(m)}$ and the relationship of this representation to the underlying crystal point group $\mathfrak{P}$. This method was first applied in space group theory by DÖRING[1] [*14*, *15*].

In (36.1) the structure of $\mathfrak{G}(\boldsymbol{k})$ was given. Evidently $\mathfrak{G}(\boldsymbol{k})$ is an extension of $\mathfrak{T}$. $\mathfrak{T}$ is normal, and we call the factor group

$$\mathfrak{G}(\boldsymbol{k})/\mathfrak{T}=\mathfrak{P}(\boldsymbol{k}). \tag{41.1}$$

The factor group $\mathfrak{P}(\boldsymbol{k})$ is isomorphic to some crystal point group $\mathfrak{P}$. For the product of two coset representatives in $\mathfrak{G}(\boldsymbol{k})$ we have

$$\{\varphi_{l_\lambda}|\tau_{l_\lambda}\}\cdot\{\varphi_{l_\mu}|\tau_{l_\mu}\}=\{\varepsilon|\boldsymbol{R}_{L\lambda,\mu}\}\cdot\{\varphi_{l_{\lambda\mu}}|\tau_{l_{\lambda\mu}}\}, \tag{41.2}$$

where

$$\varphi_{l_{\lambda\mu}}=\varphi_{l_\lambda}\cdot\varphi_{l_\mu}, \tag{41.3}$$

$$\boldsymbol{R}_{L\lambda,\mu}\equiv\varphi_{l_\lambda}\cdot\tau_{l_\mu}-\tau_{l_{\lambda\mu}}+\tau_{l_\lambda}. \tag{41.4}$$

---

[1] W. DÖRING: Z. Naturforsch. **14**a, 343 (1959). — A. KITZ: phys. stat. solidi **8**, 813 (1965). — See A.C. HURLEY: Phil. Trans. Roy. Soc. London, Ser. A **260**, 1 (1966) for other references.

Here $\boldsymbol{R}_{L\lambda,\mu}$ is a lattice vector, and $\tau_{l_{\lambda\mu}}$ is the fractional associated with the operators $\varphi_{l_{\lambda\mu}}$. Owing to the associative property (6.7) of multiplication for the space group elements the lattice vectors $\boldsymbol{R}_{L\lambda,\mu}$ are related. From

$$
\begin{aligned}
(\{\varphi_{l_\lambda}|\tau_{l_\lambda}\}\cdot\{\varphi_{l_\mu}|\tau_{l_\mu}\})\cdot\{\varphi_{l_\nu}|\tau_{l_\nu}\} &= \{\varepsilon|\boldsymbol{R}_{L\lambda,\mu}\}\cdot\{\varphi_{l_{\lambda\mu}}|\tau_{l_{\lambda\mu}}\}\cdot\{\varphi_{l_\nu}|\tau_{l_\nu}\}\\
&= \{\varepsilon|\boldsymbol{R}_{L\lambda,\mu}\}\cdot\{\varepsilon|\boldsymbol{R}_{L\lambda\mu,\nu}\}\cdot\{\varphi_{l_{\lambda\mu\nu}}|\tau_{l_{\lambda\mu\nu}}\}\\
&= \{\varepsilon|\boldsymbol{R}_{L\lambda,\mu}+\boldsymbol{R}_{L\lambda\mu,\nu}\}\cdot\{\varphi_{l_{\lambda\mu\nu}}|\tau_{l_{\lambda\mu\nu}}\}\\
&= \{\varphi_{l_\lambda}|\tau_{l_\lambda}\}\cdot(\{\varphi_{l_\mu}|\tau_{l_\mu}\}\cdot\{\varphi_{l_\nu}|\tau_{l_\nu}\}) \qquad (41.5)\\
&= \{\varphi_{l_\lambda}|\tau_{l_\lambda}\}\cdot\{\varepsilon|\boldsymbol{R}_{L\mu,\nu}\}\cdot\{\varphi_{l_{\mu\nu}}|\tau_{l_{\mu\nu}}\}\\
&= \{\varepsilon|\varphi_{l_\lambda}\cdot\boldsymbol{R}_{L\mu,\nu}\}\cdot\{\varphi_{l_\lambda}|\tau_{l_\lambda}\}\cdot\{\varphi_{l_{\mu\nu}}|\tau_{l_{\mu\nu}}\}\\
&= \{\varepsilon|\varphi_{l_\lambda}\cdot\boldsymbol{R}_{L\mu,\nu}+\boldsymbol{R}_{L\lambda,\mu\nu}\}\cdot\{\varphi_{l_{\lambda\mu\nu}}|\tau_{l_{\lambda\mu\nu}}\}
\end{aligned}
$$

we have by direct comparison

$$\{\varepsilon|\boldsymbol{R}_{L\lambda,\mu}+\boldsymbol{R}_{L\lambda\mu,\nu}\}=\{\varepsilon|\varphi_{l_\lambda}\cdot\boldsymbol{R}_{L\mu,\nu}+\boldsymbol{R}_{L\lambda,\mu\nu}\}. \qquad (41.6)$$

The definition of $\boldsymbol{R}_{L\lambda\mu,\nu}$ and $\boldsymbol{R}_{L\lambda,\mu\nu}$ follows that in (41.4). Now in the allowable irreducible representation $D^{(k)(m)}$ of $\mathfrak{G}(\boldsymbol{k})$ we have the product (41.2) as

$$
\begin{aligned}
D^{(k)(m)}(\{\varphi_{l_\lambda}|\tau_{l_\lambda}\})\cdot D^{(k)(m)}(\{\varphi_{l_\mu}|\tau_{l_\mu}\}) &= D^{(k)(m)}(\{\varepsilon|\boldsymbol{R}_{L\lambda,\mu}\})\cdot D^{(k)(m)}(\{\varphi_{l_{\lambda\mu}}|\tau_{l_{\lambda\mu}}\})\\
&= \exp -i\boldsymbol{k}\cdot\boldsymbol{R}_{L\lambda,\mu}\boldsymbol{\Pi}_m\cdot D^{(k)(m)}(\{\varphi_{l_{\lambda\mu}}|\tau_{l_{\lambda\mu}}\}).
\end{aligned} \qquad (41.7)
$$

Define the quantity

$$r^{(k)}(\lambda,\mu)\equiv\exp -i\,\boldsymbol{k}\cdot\boldsymbol{R}_{L\lambda,\mu}. \qquad (41.8)$$

Then (41.7) may be written

$$D^{(k)(m)}(\{\varphi_{l_\lambda}|\tau_{l_\lambda}\})\cdot D^{(k)(m)}(\{\varphi_{l_\mu}|\tau_{l_\lambda}\})=r^{(k)}(\lambda,\mu)\boldsymbol{\Pi}_m D^{(k)(m)}(\{\varphi_{l_{\lambda\mu}}|\tau_{\lambda\mu}\}). \qquad (41.9)$$

Now evidently for every pair of coset representatives $\{\varphi_{l_\lambda}|\tau_{l_\lambda}\}$, $\{\varphi_{l_\mu}|\tau_{l_\mu}\}$ in $\mathfrak{G}(\boldsymbol{k})$ there is defined a quantity $r^{(k)}(\lambda,\mu)$:

$$r^{(k)}(\lambda,\mu)=\exp -i\,\boldsymbol{k}\cdot\boldsymbol{R}_{L\lambda,\mu},\qquad \lambda,\mu=1,\ldots,l_k. \qquad (41.10)$$

Consequently there are $(l_k)^2$ objects $r^{(k)}(\lambda,\mu)$. These are called the factor system.

Consider now the point group $\mathfrak{P}$, isomorphic to $\mathfrak{P}(\boldsymbol{k})$ whose elements are the l_k rotations

$$\mathfrak{P}=\varepsilon,\varphi_{l_2},\ldots,\varphi_{l_\lambda},\ldots,\varphi_{l_k}\cong\mathfrak{P}(\boldsymbol{k}). \qquad (41.11)$$

These multiply as

$$\varphi_{l_\lambda}\cdot\varphi_{l_\mu}=\varphi_{l_{\lambda\mu}}. \qquad (41.12)$$

Now assume we had an irreducible representation $D^{(j)}$ of $\mathfrak{P}$. Then in this representation of $\mathfrak{P}$

$$D^{(j)}(\varphi_{l_\lambda})\cdot D^{(j)}(\varphi_{l_\mu})=D^{(j)}(\varphi_{l_{\lambda\mu}}). \qquad (41.13)$$

Comparing (41.9) and (41.13) observe that the matrices $D^{(k)(m)}$ for the l_k coset representatives $\{\varphi_{l_\lambda}|\tau_{l_\lambda}\}$ in the allowable irreducible representations of $\mathfrak{G}(\boldsymbol{k})$, "almost" multiply together as the matrices in an irreducible representation $D^{(j)}$ of the group $\mathfrak{P}$, of order l_k. The "discrepancy" consists of the numerical factors $r^{(\boldsymbol{k})}(\lambda, \mu)$, present in the product (41.9). Otherwise stated, the set of matrices $D^{(\boldsymbol{k})(m)}$ almost are an ordinary or vector representation of the group $\mathfrak{P}$.

A set of matrices

$$D^{(\boldsymbol{k})(m)} \tag{41.14}$$

which are images of the elements of a group $\mathfrak{P}(\boldsymbol{k})$

$$D^{(\boldsymbol{k})(m)}(\{\varphi_{l_\lambda}|\tau_{l_\lambda}\}) \rightleftarrows \varphi_{l_\lambda} \tag{41.15}$$

such that if group multiplication in $P(\boldsymbol{k})$ is

$$\varphi_{l_\lambda} \cdot \varphi_{l_\mu} = \varphi_{l_{\lambda\mu}} \tag{41.16}$$

and matrix multiplication obeys

$$D^{(\boldsymbol{k})(m)}(\{\varphi_{l_\lambda}|\tau_{l_\lambda}\}) \cdot D^{(\boldsymbol{k})(m)}(\{\varphi_{l_\mu}|\tau_{l_\mu}\}) = r^{(\boldsymbol{k})}(\lambda, \mu)\, D^{(\boldsymbol{k})(m)}(\{\varphi_{l_{\lambda\mu}}|\tau_{l_{\lambda\mu}}\}), \tag{41.17}$$

where $|r^{(\boldsymbol{k})}(\lambda, \mu)| = 1$, and

$$r^{(\boldsymbol{k})}(\lambda, \mu)\, r^{(\boldsymbol{k})}(\lambda\mu, \nu) = r^{(\boldsymbol{k})}(\lambda, \mu\nu)\, r^{(\boldsymbol{k})}(\mu, \nu) \tag{41.18}$$

define a ray or projective, or multiplier representation of the abstract group $\mathfrak{P}(\boldsymbol{k})$ with factor set $r^{(\boldsymbol{k})}(\lambda, \mu)$ [*14*].

It remains to verify that (41.18) holds for the factor set $r^{(\boldsymbol{k})}(\lambda, \mu)$ defined by (41.8) and (41.10). From (41.18) we have

$$r^{(\boldsymbol{k})}(\lambda, \mu)\, r^{(\boldsymbol{k})}(\lambda\mu, \nu) = \exp - i\boldsymbol{k} \cdot (\boldsymbol{R}_{L_{\lambda\mu}} + \boldsymbol{R}_{L_{\lambda\mu,\nu}}) \tag{41.19}$$

and, using (41.6)

$$= \exp - i\boldsymbol{k} \cdot (\varphi_{l_\lambda} \cdot \boldsymbol{R}_{L_{\mu\nu}} + \boldsymbol{R}_{L_{\lambda,\mu\nu}}). \tag{41.20}$$

But since from (34.7)

$$\boldsymbol{k} \cdot \varphi_{l_\lambda} = \varphi_{l_\lambda}^{-1} \cdot \boldsymbol{k} = \boldsymbol{k} + 2\pi \boldsymbol{B}_H \tag{41.21}$$

we may rewrite (41.20) as

$$= \exp - i\boldsymbol{k} \cdot (\boldsymbol{R}_{L_{\mu,\nu}} + \boldsymbol{R}_{L_{\lambda,\mu\nu}}) \tag{41.22}$$

and then recognize (41.22) as

$$= r^{(\boldsymbol{k})}(\lambda, \mu\nu)\, r^{(\boldsymbol{k})}(\mu, \nu) \tag{41.23}$$

which verified (41.18). The set of l_k^2 quantities $r^{(\boldsymbol{k})}(\lambda, \mu)$ is then suitable for the factor system of an irreducible ray representation of the crystallographic point group $\mathfrak{P}(\boldsymbol{k})$.

42. Ray representations of $\mathfrak{P}(\boldsymbol{k})$: The covering group $\mathfrak{P}^\star(\boldsymbol{k})$ [*14*]. Our problem is now to determine those irreducible ray representations, or projective representations, of the crystallographic point group $\mathfrak{P}(\boldsymbol{k})$ which have the correct factor system $r^{(\boldsymbol{k})}(\lambda, \mu)$. In order to make this determination it is necessary first to obtain and examine all the irreducible ray representations of $\mathfrak{P}(\boldsymbol{k})$. It may then be that some of these are a priori suitable; the more likely situation which one encounters is that a gauge transformation is required to the correct factor set. It is necessary then to develop the theory of ray, or projective, representations of a group $\mathfrak{P}(\boldsymbol{k})$. For consistency of notation we will always call the elements in $\mathfrak{P}(\boldsymbol{k})$ by the symbols defined in (36.1) and (41.11). However, the theory to be developed is general and can apply to the ray representations of any finite group $\mathfrak{P}$. Gauge transformations are discussed in Sect. 43.

Following SCHUR, let us assume that we had a group $\mathfrak{P}^\star(\boldsymbol{k})$ called the representation group, or covering group of $\mathfrak{P}$. The properties of $\mathfrak{P}^\star(\boldsymbol{k})$ will be given below. Let $\mathfrak{P}^\star(\boldsymbol{k})$ be an extension of an Abelian group A by $\mathfrak{P}(\boldsymbol{k})$, such that A is a normal subgroup of $\mathfrak{P}^\star(\boldsymbol{k})$, A is contained in the center of $\mathfrak{P}^\star(\boldsymbol{k})$ and thus

$$\mathfrak{P}^\star(\boldsymbol{k})/A \cong \mathfrak{P}. \tag{42.1}$$

Let the group A be of order m, so every element of A is of order $\leqq m$. Call the elements in A:

$$A: \quad E_1\, A_2 \ldots A_m \tag{42.2}$$

then for all elements in $\mathfrak{P}^\star$

$$A_k\, \varphi_{l_\lambda} = \varphi_{l_\lambda}\, A_k, \tag{42.3}$$

Where A_k is any element in A, and φ_{l_λ} every element in $\mathfrak{P}^\star$, by the definition of the center of a group. Suppose we had such a group $\mathfrak{P}^\star$, and we made a coset decomposition of $\mathfrak{P}^\star$ with respect to A:

$$\mathfrak{P}^\star = A + \Phi_{l_2} A + \cdots + \Phi_{l_k} A, \tag{42.4}$$

where Φ_{l_λ} is a coset representative in $\mathfrak{P}^\star/A$. The coset representatives in (42.4) are chosen to correspond to the elements of $\mathfrak{P}(\boldsymbol{k})$. For the product of two coset representatives we have

$$\Phi_{l_\lambda}\, \Phi_{l_\mu} = \Phi_{l_{\lambda\mu}}\, A(\lambda, \mu) = A(\lambda, \mu)\, \Phi_{l_{\lambda\mu}}, \tag{42.5}$$

where $A(\lambda, \mu) \equiv A_{\lambda\mu}$ is some element in A, and we use (42.3).

Now let ${}^\star D^{(j)}$ be an ordinary irreducible representation of $\mathfrak{P}^\star(\boldsymbol{k})$, then from (42.5)

$$\begin{aligned} {}^\star D^{(j)}(\Phi_{l_\lambda}) \cdot {}^\star D^{(j)}(\Phi_{l_\mu}) &= {}^\star D^{(j)}(\Phi_{\lambda\mu})\, {}^\star D^{(j)}(A_{\lambda\mu}) \\ &= {}^\star D^{(j)}(A_{\lambda\mu}) \cdot {}^\star D^{(j)}(\Phi_{\lambda\mu}). \end{aligned} \tag{42.6}$$

But since $A_{\lambda\mu}$ is in the center, then for *fixed* $A_{\lambda\mu}$ we have from (42.3) for *any* Φ_{l_ν}:

$${}^\star D^{(j)}(A_{\lambda\mu}) \cdot {}^\star D^{(j)}(\Phi_{l_\nu}) = {}^\star D^{(j)}(\Phi_{l_\nu}) \cdot {}^\star D^{(j)}(A_{\lambda\mu}). \tag{42.7}$$

But ${}^\star D^{(j)}$ is irreducible, hence by SCHUR's lemma

$${}^\star D^{(j)}(A_{\lambda\mu}) = r(\lambda, \mu)\, \Pi_j, \tag{42.8}$$

where Π_j is the l_j dimensional unit matrix and $r(\lambda, \mu)$ is a constant. Further, since the group A is Abelian, then $|r(\lambda, \mu)| = 1$. Then, if we did have any irreducible representations of the group $\mathfrak{P}^\star$, it has the property

$${}^\star D^{(j)}(\Phi_{l_\lambda}) \cdot {}^\star D^{(j)}(\Phi_{l_\mu}) = r(\lambda, \mu)\, {}^\star D^{(j)}(\Phi_{l_{\lambda\mu}}). \tag{42.9}$$

This is exactly what is required for the set of matrices ${}^\star D^{(j)}$ to be considered as a projective representation of the group $\mathfrak{P}(\boldsymbol{k})$ since the $r(\lambda, \mu)$ are a set which obeys the requirement (41.18) owing to the associativity of multiplication in $\mathfrak{P}^\star(\boldsymbol{k})$. Of course, in order for (42.9) not to be trivial for $\mathfrak{P}(\boldsymbol{k})$ we need $r(\lambda, \mu) \neq 1$.

This, in turn requires that the group $\mathfrak{P}^\star$ should not be a semidirect product of A by P, but a more general extension. That is, if $\mathfrak{P}^\star$ was a semi direct product then in (42.5) the product of two coset representatives would again be a coset representative, *not* mod. an element in A. In case of a split extension all $A_{\lambda\mu}=E$ in (42.5), and the $r(\lambda,\mu)=1$.

SCHUR's basic theorem is: *All* the projective irreducible representations of $\mathfrak{P}(\boldsymbol{k})$ are irreducible representations of $\mathfrak{P}^\star(\boldsymbol{k})$ subduced on $\mathfrak{P}(\boldsymbol{k})$. Owing to the associative law (41.18) which can now be applied to (42.5) to constrain the order of the generators A_μ of A the number of distinct inequivalent extensions of A by $\mathfrak{P}(\boldsymbol{k})$ which give rise to distinct projective representations of $\mathfrak{P}(\boldsymbol{k})$ is limited. The extension $\mathfrak{P}^\star(\boldsymbol{k})$ of minimal order producing all the distinct projective irreducible representations of $\mathfrak{P}(\boldsymbol{k})$ is the representation group, or covering group of $\mathfrak{P}(\boldsymbol{k})$. Any greater extension of $\mathfrak{P}(\boldsymbol{k})$, to other such groups will, it can be shown, simply produce projective representations of $\mathfrak{P}(\boldsymbol{k})$ which are not distinct. The determination of all representation groups $\mathfrak{P}^\star(\boldsymbol{k})$ for all crystallographic groups was completed by SCHUR. Tables of the irreducible character systems for all such groups $\mathfrak{P}^\star(\boldsymbol{k})$ are available.[1] Hence we take the task of determining all the inequivalent irreducible representations of $\mathfrak{P}^\star(\boldsymbol{k})$ as completed, as also the task of subducing the projective representations from $\mathfrak{P}^\star(\boldsymbol{k})$ to $\mathfrak{P}(\boldsymbol{k})$.

43. Gauge transformations of ray representations [*14*]. When the projective representations of $\mathfrak{P}(\boldsymbol{k})$ are determined by homomorphism from $\mathfrak{P}^\star(\boldsymbol{k})$, a particular factor system for the projective representation is thereby specified: namely that one of (42.7). Consider a projective representation of $\mathfrak{P}(\boldsymbol{k})$, ${}^\star D^{(j)}$ such that

$$ {}^\star D^{(j)}(\Phi_{l_\lambda})\,{}^\star D^{(j)}(\Phi_{l_\mu})=r(\lambda,\mu)\,{}^\star D^{(j)}(\Phi_{l_{\lambda\mu}}) \tag{43.1}$$

and another projective representation ${}^\star\bar{D}^{(j)}$ related to the first by

$$ {}^\star\bar{D}^{(j)}(\Phi_{l_\lambda})=c_\lambda\,{}^\star D^{(j)}(\Phi_{l_\lambda}),\qquad \lambda=1,\ldots,l_k, \tag{43.2}$$

where c_λ is a constant of modulus unity. Evidently another projective representation of $\mathfrak{P}(\boldsymbol{k})$ arises when the barred matrices are used, with factor system $\bar{r}(\lambda,\mu)$

$$ {}^\star\bar{D}^{(j)}(\Phi_{l_\lambda})\,{}^\star\bar{D}^{(j)}(\Phi_{l_\mu})=\bar{r}(\lambda,\mu)\,{}^\star\bar{D}^{(j)}(\Phi_{l_{\lambda\mu}}), \tag{43.3}$$

where

$$ \bar{r}(\lambda,\mu)=r(\lambda,\mu)\cdot\frac{c_\lambda c_\mu}{c_{\lambda\mu}}. \tag{43.4}$$

Two projective representations related by (43.2) are said to be related by gauge transformation, the factor systems are related by (43.4). The projective representations ${}^\star D^{(j)}$ and ${}^\star\bar{D}^{(j)}$ are also said to be associated. Projective representations of a group $\mathfrak{P}(\boldsymbol{k})$ which are not associated, are of distinct type.

A particularly necessary gange transformation from the projective representations which arise by subduction from ${}^\star D^{(j)}$ of the representation group $\mathfrak{P}^\star(\boldsymbol{k})$ is required in order to ensure that the projective representation obtained does indeed correspond to the allowable $D^{(\boldsymbol{k})(m)}$ of $\mathfrak{P}(\boldsymbol{k})$ and hence the allowable $D^{(\boldsymbol{k})(m)}$ of $\mathfrak{G}(\boldsymbol{k})$.

[1] HURLEY: *op. cit.*, Sect. 41, footnote 1.

That is one task needed is to find the constants c_λ such that

$$D^{(k)(m)}(\{\varphi_{l_\lambda}|\tau_{l_\lambda}\})=c_\lambda {}^\star D^{(j)}(\Phi_{l_\lambda}). \tag{43.5}$$

More specifically, we require a gauge transformation from the gauge (42.7) to the gauge $r^{(k)}(\lambda,\mu)$ of (41.10):

$$r^{(k)}(\lambda,\mu)=\frac{c_\lambda c_\mu}{c_{\lambda\mu}}\, r(\lambda,\mu)=\exp -i\,\boldsymbol{k}\cdot\boldsymbol{R}_{L_{\lambda,\mu}}. \tag{43.6}$$

The choice of c_λ required to effect (43.6) is certainly not unique but this is of no significance to our work. We shall assume that a set $\{c_\lambda\}$ satisfying (43.6) has been determined so that we have found the allowable $D^{(k)(m)}$ of $\mathfrak{G}(\boldsymbol{k})$. In practice we shall most easily obtain the characters $\chi^{(k)(m)}$. The full matrices may also be found with some additional labor, and tabulation of these have appeared.[1]

44. Relationship between little group and ray representation methods. In order to discuss the relationship between little group and ray representation methods, it is necessary first to contrast the little group $\Pi(\boldsymbol{k})$ of (39.9):

$$\Pi(\boldsymbol{k})\equiv\mathfrak{G}(\boldsymbol{k})/\mathfrak{T}(\boldsymbol{k}) \tag{44.1}$$

and the representation group $\mathfrak{P}^\star(\boldsymbol{k})$ obtained by extension of the point group

$$\mathfrak{P}(\boldsymbol{k})\equiv\mathfrak{G}(\boldsymbol{k})/\mathfrak{T}. \tag{44.2}$$

The simplest and most direct comparison of the two groups is via the multiplication of those coset representatives corresponding to the elements in the underlying point group $\mathfrak{P}$.

First consider $\Pi(\boldsymbol{k})$, and use (39.11). In $\Pi(\boldsymbol{k})$ we have

$$\{\varphi_{l_\alpha}|\tau_{l_\alpha}\}\cdot\{\varphi_{l_\beta}|\tau_{l_\beta}\}=\{\varepsilon|\boldsymbol{R}_{L_\varepsilon}(\bar{\boldsymbol{k}})\}\cdot\{\varphi_{l_\alpha}\cdot\varphi_{l_\beta}|\tau_{l_{\alpha\beta}}\}, \tag{44.3}$$

where

$$\boldsymbol{R}_{L_\varepsilon}(\bar{\boldsymbol{k}})=\varphi_{l_\alpha}\cdot\tau_{l_\beta}+\tau_{l_\alpha}-\tau_{l_{\alpha\beta}} \tag{44.4}$$

is a lattice vector not in $\mathfrak{T}(\boldsymbol{k})$. The element $\{\varepsilon|\boldsymbol{R}_{L_\varepsilon}(\bar{\boldsymbol{k}})\}$ as an abstract element does not commute with other coset representatives in $\mathfrak{G}(\boldsymbol{k})/\mathfrak{T}(\boldsymbol{k})$. Hence $\{\varepsilon|\boldsymbol{R}_{L_\varepsilon}(\bar{\boldsymbol{k}})\}$ is not in the center of $\Pi(\boldsymbol{k})$.

However, as an allowable irreducible representation $D^{(k)(m)}$ we have from (39.15):

$$D^{(k)(m)}(\{\varepsilon|\boldsymbol{R}_{L_\varepsilon}(\bar{\boldsymbol{k}})\})=\exp -i\,\boldsymbol{k}\cdot\boldsymbol{R}_{L_\varepsilon}(\bar{\boldsymbol{k}})\,\Pi_m. \tag{44.5}$$

The elements $\{\varepsilon|\boldsymbol{R}_{L_\alpha}(\bar{\boldsymbol{k}})\}$ are coset representatives in $\mathfrak{T}/\mathfrak{T}(\boldsymbol{k})$, which is an abelian group of order t actually generated by one generating element of order t. Then $\exp -i\,\boldsymbol{k}\cdot\boldsymbol{R}_{L_\varepsilon}(\bar{\boldsymbol{k}})$ is a t-th root of unity.

While $\{\varepsilon|\boldsymbol{R}_{L_\varepsilon}(\bar{\boldsymbol{k}})\}$ does not commute with the other elements in $\Pi(\boldsymbol{k})$ (aside from the pure translations) we have

$$\{\varepsilon|\boldsymbol{R}_{L_\varepsilon}(\bar{\boldsymbol{k}})\}\cdot\{\varphi_{l_{\alpha\beta}}|\tau_{\alpha\beta}\}=\{\varphi_{l_{\alpha\beta}}|\tau_{\alpha\beta}\}\cdot\{\varepsilon|\varphi_{l_{\alpha\beta}}^{-1}\cdot\boldsymbol{R}_{L_\varepsilon}(\bar{\boldsymbol{k}})\}. \tag{44.6}$$

[1] HURLEY: *op. cit.*, Sect. 41, footnote 1.

Eq. (44.6) holds for the abstract elements. In the allowable irreducible representation $D^{(\boldsymbol{k})(m)}$ we must have

$$\begin{aligned} D^{(\boldsymbol{k})(m)}(\{\varepsilon|\varphi_{l_{\alpha\beta}}^{-1}\cdot \boldsymbol{R}_{L_\varepsilon}(\bar{\boldsymbol{k}})\}) &= \exp -i\boldsymbol{k}\cdot(\varphi_{l_{\alpha\beta}}^{-1}\cdot \boldsymbol{R}_{L_\varepsilon}(\bar{\boldsymbol{k}}))\boldsymbol{\Pi}_m \\ &= \exp -i(\varphi_{l_{\alpha\beta}}\cdot \boldsymbol{k})\cdot \boldsymbol{R}_{L_\varepsilon}(\bar{\boldsymbol{k}})\boldsymbol{\Pi}_m . \end{aligned} \tag{44.7}$$

But from the definition of $\mathfrak{G}(\boldsymbol{k})$:

$$\varphi_{l_{\alpha\beta}}\cdot \boldsymbol{k} = \boldsymbol{k} + 2\pi \boldsymbol{B}_H \tag{44.8}$$

hence

$$D^{(\boldsymbol{k})(m)}(\{\varepsilon|\varphi_{l_{\alpha\beta}}^{-1}\cdot \boldsymbol{R}_{L_\varepsilon}(\bar{\boldsymbol{k}})\}) = \exp -i\boldsymbol{k}\cdot \boldsymbol{R}_{L_\varepsilon}(\bar{\boldsymbol{k}}) = D^{(\boldsymbol{k})(m)}(\{\varepsilon|\boldsymbol{R}_{L_\varepsilon}(\bar{\boldsymbol{k}})\}). \tag{44.9}$$

Hence in $D^{(\boldsymbol{k})(m)}$ we have from (44.3), (44.6), (44.9)

$$\begin{aligned} & D^{(\boldsymbol{k})(m)}(\{\varphi_{l_\alpha}|\tau_{l_\alpha}\})\cdot D^{(\boldsymbol{k})(m)}(\{\varphi_{l_\beta}|\tau_{l_\beta}\}) \\ &\quad = D^{(\boldsymbol{k})(m)}(\{\varepsilon|\boldsymbol{R}_{L_\varepsilon}(\bar{\boldsymbol{k}})\})\cdot D^{(\boldsymbol{k})(m)}(\{\varphi_{l_{\alpha\beta}}|\tau_{\alpha\beta}\}) \\ &\quad = D^{(\boldsymbol{k})(m)}(\{\varphi_{l_{\alpha\beta}}|\tau_{\alpha\beta}\})\cdot D^{(\boldsymbol{k})(m)}(\{\varepsilon|\boldsymbol{R}_{L_\varepsilon}(\bar{\boldsymbol{k}})\}). \end{aligned} \tag{44.10}$$

Consequently, in $D^{(\boldsymbol{k})(m)}$ the matrices $D^{(\boldsymbol{k})(m)}(\{\varepsilon|\boldsymbol{R}_{L_\alpha}(\bar{\boldsymbol{k}})\})$ are in the center of the matrix group, and thus by SCHUR's lemma, since $D^{(\boldsymbol{k})(m)}$ is an irreducible representation of $\Pi(\boldsymbol{k})$, these matrices are constant (scalar times the identity matrix). The constant is of course given in (44.5).

Consequently, we see that the little group, allowable irreducible matrices $D^{(\boldsymbol{k})(m)}$ form a ray representation of the abstract group $\mathfrak{P}$, or $\mathfrak{P}(\boldsymbol{k})$, with factor system

$$r^{(\boldsymbol{k})}(\alpha, \beta) = \exp -i\boldsymbol{k}\cdot \boldsymbol{R}_{L_\varepsilon}(\bar{\boldsymbol{k}}). \tag{44.11}$$

Now compare (44.3) and (41.2). Clearly

$$\boldsymbol{R}_{L_\varepsilon}(\bar{\boldsymbol{k}}) \quad \text{in (44.3)}$$

equals

$$\boldsymbol{R}_{L_{\alpha,\beta}} \quad \text{in (41.2)}.$$

Apart from the different nomenclature in the two cases, the translations are identical. Hence the factor system (44.11) indeed corresponds to

$$r^{(\boldsymbol{k})}(\alpha, \beta) = \exp -i\boldsymbol{k}\cdot \boldsymbol{R}_{L_{\alpha,\beta}}. \tag{44.12}$$

Comparing with (43.6) it is seen that (44.12) is indeed the canonical gauge required of the ray representation of $\mathfrak{P}$. Hence the allowable irreducible representation $D^{(\boldsymbol{k})(m)}$ of $\Pi(\boldsymbol{k})$ is a ray representation of $\mathfrak{P}$ with the canonical gauge required to correspond to wave vector $\boldsymbol{k}$.

The group $\Pi(\boldsymbol{k}) = \mathfrak{G}(\boldsymbol{k})/\mathfrak{T}(\boldsymbol{k})$ may then be recognized as an extension of the Abelian group $\mathfrak{T}/\mathfrak{T}(\boldsymbol{k})$ by $\mathfrak{P}(\boldsymbol{k})$. The elements in $\mathfrak{T}/\mathfrak{T}(\boldsymbol{k})$ are not in the center of the extended *abstract* group, although in the matrix representation $D^{(\boldsymbol{k})(m)}$, the corresponding matrices are in the center of the *matrix* group. Actually if $\Pi(\boldsymbol{k})$ is understood as an extension of $\mathfrak{T}/\mathfrak{T}(\boldsymbol{k})$ by $\mathfrak{P}(\boldsymbol{k})$ then it is realized that the elements

$$\{\varepsilon|\boldsymbol{R}_{L_\alpha}(\bar{\boldsymbol{k}})\}, \qquad \alpha = 1, \ldots, t \tag{44.13}$$

are a cyclic normal subgroup of $\Pi(\boldsymbol{k})$, of order t. The irreducible representations of $\mathfrak{T}/\mathfrak{T}(k)$ are then the t-th roots of unity. Hence, if the generator of $\mathfrak{T}/\mathfrak{T}(\boldsymbol{k})$ is the element

$$\{\varepsilon|\boldsymbol{R}_{L_\alpha}(\bar{\boldsymbol{k}})\} \tag{44.14}$$

then the irreducible representations of $\mathfrak{T}/\mathfrak{T}(\boldsymbol{k})$ are all of form

$$D(\{\varepsilon|\boldsymbol{R}_{L_\alpha}(\bar{\boldsymbol{k}})\}) = \exp - 2\pi i\, p/t, \qquad p=0, \ldots, (t-1). \tag{44.15}$$

For $\boldsymbol{k}$ prescribed, as a rule only one value of p will be compatible with

$$\exp - i\,\boldsymbol{k}\cdot\boldsymbol{R}_{L_\alpha}(\bar{\boldsymbol{k}}) = \exp - 2\pi i\, p/t. \tag{44.16}$$

All the irreducible representations (44.15) of $\mathfrak{T}/\mathfrak{T}(\boldsymbol{k})$ may be used to induce irreducible representations of $\Pi(\boldsymbol{k})$, following exactly the procedure of Sects. 36 and 37. The allowable $D^{(\boldsymbol{k})(m)}$ of $\Pi(\boldsymbol{k})$ evidently correspond to the representations induced from that one irreducible representation of $\mathfrak{T}/\mathfrak{T}(\boldsymbol{k})$ with the proper value of p given in (44.15). Then it is also clear why the non-allowable $D^{(\boldsymbol{k})(\mu)}$ correspond to the family of wave vectors (40.6). The $D^{(\boldsymbol{k})(\mu)}$ are induced from the other values of p than that one given in (44.15). This proof complements the direct proof given in Sect. 40.

It will be recalled that according to SCHUR the *smallest* order extension $\mathfrak{P}^\star(\boldsymbol{k})$ of an Abelian group A by $\mathfrak{P}$ is the representation group if the complete set of irreducible, non-associated, inequivalent ray representations of $\mathfrak{P}$ is obtained from all the vector representations $^\star D$ of $\mathfrak{P}^\star(k)$ by subduction. The group $\Pi(\boldsymbol{k})$ is, in this terminology, called a "sufficiently extended group". Its irreducible vector representations include by subduction all the irreducible ray representations of $\mathfrak{P}(\boldsymbol{k})$ but these occur more than once, with different but associated factor sets. This is clear since a change from the canonical wave vector $\boldsymbol{k}$ of the representation, to one of the other in the family simply corresponds to a gauge transformation by unimodular factors like $\exp - i(p-p')\,\boldsymbol{k}\cdot\boldsymbol{R}_{L_\varepsilon}(\bar{\boldsymbol{k}})$.

Summarizing, the group $\mathfrak{P}^\star(\boldsymbol{k})$ is the representation group whose irreducible vector representation $^\star D$ subduce all the irreducible non-associated ray representations of $\mathfrak{P}$. These subduced representations generally require a gauge transformation in order to be put as the allowable irreducible $D^{(\boldsymbol{k})(m)}$. The allowable irreducible representations $D^{(\boldsymbol{k})(m)}$ of $\Pi(\boldsymbol{k})$ are immediately ray representations of $\mathfrak{P}$ with the correct factor system.

The advantage of working with the little group $\Pi(\boldsymbol{k})$ is that one does not require an added gauge transformation to obtain $D^{(\boldsymbol{k})(m)}$, and also that the non-allowable $D^{(\boldsymbol{k})(\mu)}$ can be useful. Disadvantage is that $\Pi(\boldsymbol{k})$ is generally a large group whose character system needs to be worked out usually *a priori* in each case, and this is often quite troublesome.

The advantage of the ray representation method in working with $\mathfrak{P}^\star(\boldsymbol{k})$ is that all the basic ray representations ever needed can be obtained once and for all for the groups $\mathfrak{P}$. In fact, now that these are tabulated, additional calculation is not required. Disadvantage is the need to make a gauge transformation to obtain the desired factor system $r^{(\boldsymbol{k})}(\lambda, \mu)$ from the factor system which comes with the group $\mathfrak{P}^\star(\boldsymbol{k})$.

From a theoretical viewpoint it would appear that the ray representation method is generally preferable in regard to conciseness as well as generality. The task of finding suitable factor sets, and making the relevant gauge transformation to this factor set, seems a small price to pay for the degree of generality; see Sect. 126 and Appendix D for application to diamond structure.

45. Full $D^{(\star k)(m)}$ for symmorphic groups: Illustration. The symmorphic space group is a split extension of $\mathfrak{T}$ by $\mathfrak{P}$; or equivalently a semidirect product of $\mathfrak{T}$ by $\mathfrak{P}$. In this case, the coset representatives in the decomposition of $\mathfrak{G}$ into cosets with respect to $\mathfrak{T}$ are closed without any artificial or arbitrary assumptions on multiplication. Hence we have the point group $\mathfrak{P}$ contained in $\mathfrak{G}$ as an actual subgroup. The coset decomposition of $\mathfrak{G}$, with respect to $\mathfrak{T}$ in terms of coset representatives $\{\varphi|\mathbf{0}\}$, which are the pure rotations in $\mathfrak{P}$ is

$$\mathfrak{G}/\mathfrak{T}=\mathfrak{T}+\{\varphi_2|0\}\mathfrak{T}+\cdots+\{\varphi_{g_p}|0\}\mathfrak{T}. \tag{45.1}$$

α) *General star.* For a general star, (38.1) has g_p distinct wave vectors and the irreducible vector space $\Sigma^{(\star k)(m)}$ consists of g_p linearly independent Bloch functions

$$\{\psi^{(k)},\psi^{(k_2)},\ldots,\psi^{(k_p)}\}\equiv\Sigma^{(\star k)(m)}. \tag{45.2}$$

As each vector space (Bloch function) is one dimensional the irreducible representation of the $\mathfrak{G}(\boldsymbol{k})$ is simply the irreducible representation of $\mathfrak{T}$, i.e.

$$D^{(k)(m)}(\{\varepsilon|\boldsymbol{R}_L\})=\exp-i\,\boldsymbol{k}\cdot\boldsymbol{R}_L \tag{45.3}$$

for the canonical wave vector. The coset decomposition $\mathfrak{G}(\boldsymbol{k})/\mathfrak{T}$ is then identical with $\mathfrak{G}/\mathfrak{T}$, i.e. (36.1) and (45.1) are identical. Now from (36.14) we can write down explicitly the full matrices $D^{(\star k)(m)}$.

For the elements in $\mathfrak{T}$ we have

$$D^{(\star k)(m)}(\{\varepsilon|\boldsymbol{R}_L\})=\begin{pmatrix}\exp-i\,\boldsymbol{k}\cdot\boldsymbol{R}_L & 0 & \\ 0 & \exp-i\,\boldsymbol{k}_2\cdot\boldsymbol{R}_L & \ddots \\ 0 & 0\ \ldots & \exp-i\,\boldsymbol{k}_{g_p}\cdot\boldsymbol{R}_L\end{pmatrix}. \tag{45.4}$$

For the coset representative elements $\{\varphi_p|0\}$ we have from (36.14) the bloc matrices:

$$D^{(\star k)(m)}(\{\varphi_p|0\})_{\sigma\tau}=\begin{cases}1 & \text{if } \varphi_\sigma^{-1}\cdot\varphi_p\cdot\varphi_\tau=\varepsilon\\ 0 & \text{otherwise}.\end{cases} \tag{45.5}$$

For the full matrix the general form is that of a permutation matrix:

$$D^{(\star k)(m)}(\{\varphi_p|0\})=\begin{pmatrix}0 & \ldots & 1 & \ldots & 0\\ \vdots & & & & \vdots\\ . & \ldots & 0 & \ldots & 1\\ 1 & & 0 & & \vdots\\ \vdots & & & 0 & \\ . & & \ldots & & 0\end{pmatrix}. \tag{45.6}$$

For an arbitrary element $\{\varphi_p|\boldsymbol{R}_L\}$ a simple matrix multiplication gives from (45.4) and (45.6)

$$D^{(\star k)(m)}(\{\varphi_p|\boldsymbol{R}_L\}) = D^{(\star k)(m)}(\{\varepsilon|\boldsymbol{R}_L\}) \cdot D^{(\star k)(m)}(\{\varphi_p|0\})$$

$$= \begin{pmatrix} 0 & \dots & \exp -i\boldsymbol{k}\cdot\boldsymbol{R}_L & \dots & 0 \\ \vdots & \dots & 0 & \dots & \exp -i\boldsymbol{k}_\tau\cdot\boldsymbol{R}_L \\ \exp -i\boldsymbol{k}_\sigma\cdot\boldsymbol{R}_L & & \dots & & 0 \\ \vdots & & & & \vdots \end{pmatrix}. \tag{45.7}$$

β) *Special star.* The special star of a symmorphic space group has a canonical wave vector $\boldsymbol{k}$ such that $\mathfrak{G}(\boldsymbol{k})$ is itself a non-trivial symmorphic space group:

$$\mathfrak{G}(\boldsymbol{k}) = \mathfrak{T} + \{\varphi_{l_2}|0\}\,\mathfrak{T} + \cdots + \{\varphi_{l_k}|0\}\,\mathfrak{T}. \tag{45.8}$$

Since the point group $\mathfrak{P}(\boldsymbol{k}) = \mathfrak{G}(\boldsymbol{k})/\mathfrak{T}(\boldsymbol{k})$ is a subgroup of $\mathfrak{G}(\boldsymbol{k})$, the ray representations are vector representations. That is $\mathfrak{P}^\star(\boldsymbol{k}) \cong \mathfrak{P}(\boldsymbol{k})$: the representation group is isomorphic to an ordinary crystallographic point group. This important point follows owing to the fact that the product of two coset representatives in (45.8) is again a coset representative, so that factor set $r^{(\boldsymbol{k})}(\lambda, \mu) = 1$. Then the irreducible representations of $\mathfrak{P}(\boldsymbol{k})$ are ordinary irreducible representations of point groups, which are well known. For example, for group O_h, or O the irreducible representations can have dimension $l_m = 1, 2, 3$. Again, of course, (36.14) solves the problem in its entirety.

Now consider the situation where $\star\boldsymbol{k}$ has s distinct wave vectors in it, and as before the order of $\mathfrak{P}$ is l_k. The coset representatives are labelled in accord with our conventions of (36.1) and (36.2). Further, to avoid cumbersome illustrations consider the particular case where

$$\{\varphi_l|0\} \quad \text{is in } \mathfrak{G}(\boldsymbol{k}), \tag{45.9a}$$

$$\{\varphi_2^{-1}\,\varphi_l\,\varphi_\sigma|0\} \quad \text{is in } \mathfrak{G}(\boldsymbol{k}), \tag{45.9b}$$

and

$$\{\varphi_\tau^{-1}\,\varphi_l\,\varphi_\tau\} \quad \text{is in } \mathfrak{G}(\boldsymbol{k}). \tag{45.9c}$$

Under the stated circumstances the irreducible matrix is:

$$D^{(\star k)(m)}(\{\varphi_l|0\}) = \begin{pmatrix} D^{(k)(m)}(\{\varphi_l|0\}) & 0 & & \dots & \dots \\ 0 & 0 & \dots & D^{(k)(m)}(\{\varphi_2^{-1}\,\varphi_l\,\varphi_\sigma|0\}) & \dots \\ \vdots & 0 & & & \\ \vdots & & 0 & & \\ \vdots & & & \ddots & \\ & & & D^{(k)(m)}(\{\varphi_\tau^{-1}\,\varphi_l\,\varphi_\tau|0\}) & \dots \\ & & & & \ddots \end{pmatrix} \tag{45.10}$$

As before for the general star, the matrix representing an arbitrary element $\{\varphi_p|\boldsymbol{R}_L\}$ is obtained by multiplying the diagonal matrix representing the translation by (45.10).

In this fashion, every irreducible representation of the symmorphic group can be obtained when the star, and canonical wave vector $\boldsymbol{k}$, of that star, are specified. The irreducible vector space $\Sigma^{(k)(m)}$ for $D^{(k)(m)}$, and $\Sigma^{(\star k)(m)}$ for $D^{(\star k)(m)}$ are correspondingly determined. When $D^{(k)(m)}$ is multidimensional with $l_m=2$, or 3, the vector space $\Sigma^{(k)(m)}$ is, correspondingly. It may be useful to refer back to the original discussion in Sects. 32 and 33 to refresh this point.

46. Full $D^{(\star k)(m)}$ for non-symmorphic group. For the general star the non-symmorphic group irreducible representations $D^{(\star k)(m)}$ are of just the same form as those of the symmorphic group (45.3), (45.4), except that coset representatives are now elements $\{\varphi_\alpha|\tau_\alpha\}$, with $\tau_\alpha \neq 0$, for at least some α.

In the case of the special star the method of ray-representations is the easiest to utilize. Consider that we have obtained the allowable $D^{(k)(m)}$ by gauge transformation from the $\star D$ of $\mathfrak{P}^\star(k)$. Then for each coset representative $\{\varphi_{l_\lambda}|\tau_{l_\lambda}\}$ in $\mathfrak{G}(k)$ we have determined a matrix $D^{(k)(m)}(\{\varphi_{l_\lambda}|\tau_{l_\lambda}\})$. Again we must use the key induction formula (36.14) to obtain the full matrix irreducible representation $D^{(\star k)(m)}$ of $\mathfrak{G}$. Again consider the particular case where

$$\{\varphi_l|\tau_l\} \quad \text{is in } \mathfrak{G}(k), \tag{46.1a}$$

$$\{\varphi_{l_2}|\tau_2\}^{-1}\cdot\{\varphi_l|\tau\}\cdot\{\varphi_\sigma|\tau_\sigma\} \quad \text{is in } \mathfrak{G}(k), \tag{46.1b}$$

$$\{\varphi_\tau|\tau_\tau\}^{-1}\cdot\{\varphi_l|\tau\}\cdot\{\varphi_\tau|\tau_\tau\} \quad \text{is in } \mathfrak{G}(k). \tag{46.1c}$$

The elements (46.1 b) and (46.1 c) can be written

$$\{\varphi_{l_2}|\tau_2\}^{-1}\cdot\{\varphi_l|\tau_l\}\cdot\{\varphi_\sigma|\tau_\sigma\}=\{\varepsilon|R_{L_{\bar{2}l\sigma}}\}\cdot\{\varphi_{l_2}^{-1}\cdot\varphi_l\cdot\varphi_\sigma|\tau_{\bar{2}l\sigma}\}, \tag{46.2}$$

where $R_{L_{\bar{2}l\sigma}}$ is a lattice vector, and

$$\{\varphi_\tau|\tau_\tau\}^{-1}\cdot\{\varphi_l|\tau_l\}\cdot\{\varphi_\tau|\tau_\tau\}=\{\varepsilon|R_{L_{\bar{\tau}l\tau}}\}\cdot\{\varphi_\tau^{-1}\cdot\varphi_l\cdot\varphi_\tau|\tau_{\bar{\tau}l\tau}\}. \tag{46.3}$$

In (46.2) and (46.3) the elements on the extreme right side are coset representatives in $\mathfrak{G}(k)$, premultiplied by pure translations. Then, in the acceptable $D^{(k)(m)}$ we have

$$\begin{aligned} &D^{(k)(m)}(\{\varphi_{l_2}|\tau_2\}^{-1}\cdot\{\varphi_l|\tau_l\}\cdot\{\varphi_\sigma|\tau_\sigma\}) \\ &\quad=\exp -i\,k\cdot R_{L_{\bar{2}l\sigma}}\,D^{(k)(m)}(\{\varphi_{l_2}^{-1}\cdot\varphi_l\cdot\varphi_\sigma|\tau_{\bar{2}l\sigma}\}) \end{aligned} \tag{46.4}$$

and similarly for the matrix representing the element (46.3). Then the structure of the full matrix $D^{(\star k)(m)}(\{\varphi_l|\tau_l\})$ is similar to that shown in (45.10), except that the coset representatives all have fractional translations associated, and more importantly the elements such as (46.1 b), (46.1 c) may have non-zero translations associated, producing non-trivial phase factors, such as are illustrated in (46.4). It is worth emphasizing that the ray representation matrices are only applicable to the coset representatives which are isomorphic to the abstract elements in $\mathfrak{P}$. Thus, phase factors due to the translations in (46.2) and (46.3) will arise and necessarily must be associated for in the matrices.

On the other hand, if the little group method is used to obtain the $D^{(k)(m)}$ then not only are the matrices $D^{(k)(m)}$ specified for coset representatives $\{\varphi_{l_\lambda}|\tau_\lambda\}$ but also for elements such as $\{\varphi_{l_\lambda}|\tau_\lambda+R_{L_\alpha}(k)\}$ which comprise the elements in $\mathfrak{G}(k)/\mathfrak{T}(k)$. Again, this detail produces no problem in principle, but one must be certain in constructing the matrices to include all needed factors: in particular this mean checking all products such as (46.1 b) and (46.1 c).

47. Complete set of all $D^{(\star k)(m)}$ for a space group. With the previous sections analysis at our disposal, we are now in a position to specify the procedure by which all the irreducible representations $D^{(\star k)(m)}$ of the space group $\mathfrak{G}$ can be obtained.

One must construct the first Brillouin zone as analysed in Sect. 23. The first Brillouin Zone is the locus of all wave vectors $\boldsymbol{k}$ which define the complete set of inequivalent irreducible representations of the translation group $\mathfrak{T}$. For each wave vector $\boldsymbol{k}$ in the first Brillouin zone we determine the $\star\boldsymbol{k}$, by applying each of the g_p distinct rotations φ_μ of $\mathfrak{P}$ to $\boldsymbol{k}$. Of the set

$$\varphi_\mu \cdot \boldsymbol{k} \equiv \boldsymbol{k}_\mu, \qquad \mu = 1, \ldots, g_p \tag{47.1}$$

only $s \leqq g_p$ are inequivalent i.e. satisfy

$$\boldsymbol{k}_\sigma \equiv \varphi_\sigma \cdot \boldsymbol{k} \neq \boldsymbol{k} + 2\pi \boldsymbol{B}_H. \tag{47.2}$$

The set of inequivalent $\boldsymbol{k}$ form $\star\boldsymbol{k}$:

$$\star\boldsymbol{k} = \{\boldsymbol{k}, \boldsymbol{k}_2, \ldots, \boldsymbol{k}_s\}. \tag{47.3}$$

In this manner the N wave vectors in the zone are grouped into distinct and disjoint stars. Each star is characterized by giving its canonical wave vector $\boldsymbol{k}$. Once the canonical $\boldsymbol{k}$ is specified, for a given star, it is fixed.

Consider the star: $\star\boldsymbol{k}$. The group $\mathfrak{G}(\boldsymbol{k})$ is determined for the canonical wave vector $\boldsymbol{k}$ of this star. Then the complete set of allowable inequivalent irreducible representations $D^{(\boldsymbol{k})(m)}$ of $\mathfrak{G}(\boldsymbol{k})$ is determined by the ray representation, or little group methods. From each distinct allowable irreducible representation $D^{(\boldsymbol{k})(m)}$ of $\mathfrak{G}(\boldsymbol{k})$ a distinct irreducible representation $D^{(\star\boldsymbol{k})(m)}$ of the full space group $\mathfrak{G}$ is obtained by the method of induction epitomized in (36.14).

The set of all

$$D^{(\star\boldsymbol{k})(m)} \text{ for } \begin{cases} \text{all allowable } m, \text{ and} \\ \text{all } \star\boldsymbol{k} \end{cases} \tag{47.4}$$

are the complete set of inequivalent irreducible representations of the space group $\mathfrak{G}$.

48. Verification of completeness of $D^{(\star k)(m)}$. To demonstrate completeness of the set $D^{(\star\boldsymbol{k})(m)}$ of (47.4), we shall use the character relation (15.7) transcribed to the space group case. The character of any element is, from (37.3):

$$\chi^{(\star\boldsymbol{k})(m)}(\{\varphi_p|\tau_p\}) = \sum_{\sigma=1}^{s} \dot{\chi}^{(\boldsymbol{k})(m)}(\{\varphi_\sigma|\tau_\sigma\}^{-1} \cdot \{\varphi_p|\tau_p\} \cdot \{\varphi_\sigma|\tau_\sigma\}), \tag{48.1}$$

where the index $\sigma \neq 1$ belongs to a coset representative in $\mathfrak{G}$, but not in $\mathfrak{G}(\boldsymbol{k})$, while $\sigma = 1$ corresponds to the identity.

The proof of completeness modifies a similar proof first given by Döring and Zehler.[1] It is required to show

$$\sum_{\star\boldsymbol{k}} \sum_{m} |\chi^{(\star\boldsymbol{k})(m)}(\{\varepsilon|0\})|^2 = g_p N, \tag{48.2}$$

[1] W. Döring and V. Zehler: Ann. Physik 13, 214 (1953).

where g_p is the order of $\mathfrak{G}/\mathfrak{T}$ (i.e. the number of distinct rotational elements in $\mathfrak{G}$), N is the number of translations in $\mathfrak{T}$, the sum on the left hand side is over all stars in the zone: $\star\boldsymbol{k}$ and over all allowable irreducible representations of $\mathfrak{G}(\boldsymbol{k})$. The restriction to "allowable" m in (48.2) is awkward, and can be relieved. For example, if one used the little group method of obtaining allowable irreducible representations of $\mathfrak{G}(\boldsymbol{k})$ one works with the group $\mathfrak{G}(\boldsymbol{k})/\mathfrak{T}(\boldsymbol{k})$ where $\mathfrak{T}(\boldsymbol{k})$ consists of all symmetry elements in $\mathfrak{T}$ whose matrix in the representation is identical to the matrix for the identity. As was discussed in Sects. 39 and 40, $\mathfrak{G}(\boldsymbol{k})/\mathfrak{T}(\boldsymbol{k})$ is a group of order $(t \cdot l_k)$. It thus follows that if we consider all irreducible representations $D^{(\boldsymbol{k})(m)}$ of the group $\mathfrak{G}(\boldsymbol{k})/\mathfrak{T}(\boldsymbol{k})$; the corresponding characters satisfy

$$\sum_{\text{all } m} |\chi^{(\boldsymbol{k})(m)}(\{\varepsilon|0\})|^2 = \sum_{\text{all } m} |l_m|^2 = l_k t, \tag{48.3}$$

where l_m is the dimension of the allowable irreducible representation. Of the t integers: $0, 1, \ldots, (t-1)$ only one such integer $t'=1$, corresponds to an allowable irreducible representation, as we discussed in Sects. 38, 39, and 44. Then

$$\sum_{\text{allowable } m} |\chi^{(\boldsymbol{k})(m)}(\{\varepsilon|0\})|^2 = \sum_{\text{allowable } m} |l_m|^2 = l_k. \tag{48.4}$$

The identical result is obtained from the ray representation method, and can be understood as the completeness relationship for the ray representations of $\mathfrak{P}(\boldsymbol{k})$ with the particular acceptable factor set (43.6) obtained by gauge transformation, as discussed in Sect. 43. It will be recalled that the group $\mathfrak{P}(\boldsymbol{k})$ has l_k cosets, or $\mathfrak{P}^\star(\boldsymbol{k})$ has l_k elements. For any star, with s arms

$$\chi^{(\star\boldsymbol{k})(m)}(\{\varepsilon|0\}) = s \cdot l_m. \tag{48.5}$$

Thus

$$|\chi^{(\star\boldsymbol{k})(m)}(\{\varepsilon|0\})|^2 = s^2\, l_m^2. \tag{48.6}$$

But as there are s wave vectors in the star, each wave vector may be taken to contribute $(s \cdot l_m^2)$ to the sum (48.2). We then change the sum over stars in (48.2) to a sum over all wave vectors in the zone as:

$$\sum_{\star\boldsymbol{k}} \sum_{m} |\chi^{(\star\boldsymbol{k})(m)}(\{\varepsilon|0\})|^2 = \sum_{\star\boldsymbol{k}} \sum_{m} s \cdot (l_m)^2 = \sum_{\boldsymbol{k}} s \cdot l_k. \tag{48.7}$$

However, l_k is the order of $\mathfrak{P}(\boldsymbol{k})$ and $s \cdot l_k$ is

$$s \cdot l_k = g_p, \tag{48.8}$$

where g_p is the order of $\mathfrak{P}$, the entire crystal point group. Using (48.8) the summand can be taken outside of the sum sign, and then with

$$\sum_{\boldsymbol{k}} = N, \tag{48.9}$$

where N is the number of distinct inequivalent wave vectors in the first Brillouin zone, which equals the number of translations in $\mathfrak{T}$, we verify (48.2). Consequently we see that the set (48.4) are indeed a complete set of inequivalent irreducible representations. Hence the procedure reviewed in Sect. 47 gives all the $D^{(\star\boldsymbol{k})(m)}$ of $\mathfrak{G}$.

49. Verification of orthonormality relations for $D^{(*k)(m)}$. We now shall verify the orthonormality relations, which of course must be obeyed, since the representations are known to be irreducible. We need first to extend the sum in (48.1) from the coset representatives $\{\varphi_\sigma|\tau_\sigma\}$ defined by (36.2), to all coset representatives. Observe that a necessary and sufficient condition for the element

$$\{\varphi_{l_\lambda}|\tau_{l_\lambda}\}^{-1}\cdot\{\varphi_p|\tau_p\}\cdot\{\varphi_{l_\lambda}|\tau_{l_\lambda}\} \tag{49.1}$$

to be in $\mathfrak{G}(\boldsymbol{k})$, if $\{\varphi_{l_\lambda}|\tau_{l_\lambda}\}$ is in $\mathfrak{G}(\boldsymbol{k})$, is that $\{\varphi_p|\tau_p\}$ be in $\mathfrak{G}(\boldsymbol{k})$. Consequently, permitting the sum in (48.1) to range over all coset representatives merely introduces a factor $(1/l_k)$ where l_k is the order of $\mathfrak{G}(k)/\mathfrak{T}$, since any arbitrary coset representative can be written as

$$\{\varphi_q|\tau_q\}=\{\varphi_\sigma|\tau_\sigma\}\cdot\{\varphi_{l_\lambda}|\tau_{l_\lambda}\} \tag{49.2}$$

by the uniquences of coset decomposition of $\mathfrak{G}$ with respect to $\mathfrak{G}(\boldsymbol{k})$. Thus we can write instead of (48.1):

$$\chi^{(*k)(m)}(\{\varphi_p|\tau_p\})=(1/l_k)\sum_{q=1}^{g_p}\dot{\chi}^{(k)(m)}(\{\varphi_q|\tau_q\}^{-1}\cdot\{\varphi_p|\tau_p\}\cdot\{\varphi_q|\tau_q\}) \tag{49.3}$$

which is an unrestricted sum over all coset representatives $\{\varphi_q|\tau_q\}$ in $\mathfrak{G}$.

Let a typical element in $\mathfrak{G}$ be taken as

$$\{\varphi_p|\tau_p+\boldsymbol{R}_L\}. \tag{49.4}$$

Then the orthonormality relation to be verified is

$$\sum_{\boldsymbol{R}_L}\sum_p\chi^{(*k)(m)}(\{\varphi_p|\tau_p+\boldsymbol{R}_L\})\,\chi^{(*k')(m')}(\{\varphi_p|\tau_p+\boldsymbol{R}_L\})^*=g_p N\,\delta(\boldsymbol{k}-\boldsymbol{k}')\,\delta_{mm'} \tag{49.5}$$

where the sum over all elements in $\mathfrak{G}$.

The dotted character of the general element (49.4) is simply related to the dotted character of the coset representative element alone. Thus

$$\dot{\chi}^{(k)(m)}(\{\varphi_p|\tau_p+\boldsymbol{R}_L\})=(\exp-i\,\boldsymbol{k}\cdot\boldsymbol{R}_L)\,\dot{\chi}^{(k)(m)}(\{\varphi_p|\tau_p\}). \tag{49.6}$$

Then the character of this element in the entire representation is:

$$\begin{aligned}\chi^{(*k)(m)}(\{\varphi_p|\tau_p+\boldsymbol{R}_L\})&=(1/l_k)\sum_{q=1}^{g_p}\dot{\chi}^{(k)(m)}(\{\varphi_q|\tau_q\}^{-1}\cdot\{\varphi_p|\tau_p+\boldsymbol{R}_L\}\cdot\{\varphi_q|\tau_q\})\\&=(1/l_k)\sum_q\dot{\chi}^{(k)(m)}(\{\varepsilon|\varphi_q^{-1}\cdot\boldsymbol{R}_L\}\cdot\{\varphi_q^{-1}\cdot\varphi_p\cdot\varphi_q|\boldsymbol{t}_{\bar{q}pq}\}).\end{aligned} \tag{49.7}$$

In (49.7) we separate out the translation $\varphi_q^{-1}\cdot\boldsymbol{R}_L$ because it bears the lattice index L, and

$$\boldsymbol{t}_{\bar{q}pq}\equiv\varphi_q^{-1}\cdot\varphi_p\,\tau_q+\varphi_q^{-1}\cdot\tau_p-\varphi_q^{-1}\cdot\tau_q. \tag{49.8}$$

Now we must substitute (49.7) into the left hand side of (49.5). We find for the left side

$$\begin{aligned}&\sum_{\boldsymbol{R}_L}\sum_p\sum_q\sum_{q'}(1/l_k)^2\,\dot{\chi}^{(k)(m)}(\{\varepsilon|\varphi_q^{-1}\cdot\boldsymbol{R}_L\}\cdot\{\varphi_q^{-1}\cdot\varphi_p\cdot\varphi_q|\boldsymbol{t}_{\bar{q}pq}\})\\&\quad\times\dot{\chi}^{(k')(m)}(\{\varepsilon|\varphi_{q'}^{-1}\cdot\boldsymbol{R}_L\}\cdot\{\varphi_{q'}^{-1}\cdot\varphi_p\cdot\varphi_{q'}|\boldsymbol{t}_{\bar{q}'pq'}\})^*.\end{aligned} \tag{49.9}$$

Using (49.6) we simplify the last two factors in (49.9) and obtain for their product

$$\exp - i(\boldsymbol{k}_q - \boldsymbol{k}'_{q'}) \cdot \boldsymbol{R}_L \, \dot{\chi}^{(\boldsymbol{k})(m)}(\{\varphi_q^{-1} \cdot \varphi_p \cdot \varphi_q | \boldsymbol{t}_{\bar{q}pq}\}) \, \dot{\chi}^{(\boldsymbol{k}')(m)}(\{\varphi_{q'}^{-1} \cdot \varphi_p \cdot \varphi_{q'} | \boldsymbol{t}_{\bar{q}'pq'}\}), \quad (49.10)$$

where

$$(\boldsymbol{k}_q - \boldsymbol{k}_{q'}) \cdot \boldsymbol{R}_L = \boldsymbol{k} \cdot (\varphi_q^{-1} \cdot \boldsymbol{R}_L - \varphi_{q'}^{-1} \cdot \boldsymbol{R}_L) \quad (49.11)$$

as consistent with our convention. Using (49.10) the sum over all $\boldsymbol{R}_L$ can be carried out, and we obtain:

$$\sum_L \exp - i(\boldsymbol{k}_q - \boldsymbol{k}'_{q'}) \cdot \boldsymbol{R}_L = N \, \delta(\boldsymbol{k}_q - \boldsymbol{k}'_{q'}), \quad (49.12)$$

where

$$\delta(\boldsymbol{k}_q - \boldsymbol{k}'_{q'}) = \begin{cases} 1 & \text{if } \varphi_q \boldsymbol{k} = \varphi_{q'} \cdot \boldsymbol{k}' + 2\pi \boldsymbol{B}_H \\ 0 & \text{otherwise.} \end{cases} \quad (49.13)$$

But $\boldsymbol{k}$ and $\boldsymbol{k}'$ are *both* canonical wave vectors of their stars, so we derive two conditions from (49.13):

$$\boldsymbol{k} = \boldsymbol{k}' \quad (49.14)$$

and

$$\varphi_q^{-1} \cdot \varphi_{q'} = \varphi_{l_\lambda}, \quad (49.15)$$

where φ_{l_λ} is in $\mathfrak{G}(\boldsymbol{k})$. The two conditions (49.14) and (49.15) permit the remaining sums to be carried out, since now, in the product (49.10) both dotted characters refer to the same group $\mathfrak{G}(\boldsymbol{k})$, and the arguments of the two characters may now be related. Consequently (49.9) becomes

$$\begin{aligned} &\sum_p \sum_q \sum_\lambda (1/l_k)^2 \, \dot{\chi}^{(\boldsymbol{k})(m)}(\{\varphi_q^{-1} \cdot \varphi_p \cdot \varphi_q | \boldsymbol{t}_{\bar{q}pq}\}) \\ &\quad \times \dot{\chi}^{(\boldsymbol{k})(m)}(\{\varphi_{l_\lambda}^{-1} \cdot \varphi_q^{-1} \cdot \varphi_p \varphi_q \cdot \varphi_{l_\lambda} | \boldsymbol{t}_{\lambda\bar{q}pq\lambda}\})^* \end{aligned} \quad (49.16)$$

where the sum index q' is replaced by λ.

But since φ_{l_λ} is one of the l_κ rotations in $\mathfrak{G}(\boldsymbol{k})$, there are only l_k values of q' such that for given q, (49.15) can be satisfied. Thus there are only l_k terms in the sum on λ. But by the argument given in (49.1) for an element $\{\varphi_{l_\lambda} | \tau_{l_\lambda}\}$ in $\mathfrak{G}(\boldsymbol{k})$,

$$\dot{\chi}^{(\boldsymbol{k})(m')}(\{_{l_\lambda}^{-1} \cdot \varphi_q^{-1} \cdot \varphi_p \cdot \varphi_q \cdot \varphi_{l_\lambda} | \boldsymbol{t}_{\lambda\bar{q}pq\lambda}\})^* = \dot{\chi}^{(\boldsymbol{k})(m')}(\{\varphi_q^{-1} \cdot \varphi_p \cdot \varphi_q | \boldsymbol{t}_{\bar{q}pq}\})^* \quad (49.17)$$

which is independent of λ. Hence the sum over λ simply cancels one factor of l_k. We are then left with

$$(N/l_k) \sum_p \sum_q \dot{\chi}^{(\boldsymbol{k})(m)}(\{\varphi_q^{-1} \cdot \varphi_p \cdot \varphi_q | \boldsymbol{t}_{\bar{q}pq}\}) \, \dot{\chi}^{(\boldsymbol{k})(m')}(\{\varphi_q^{-1} \cdot \varphi_p \cdot \varphi_q | \boldsymbol{t}_{\bar{q}pq}\})^*. \quad (49.18)$$

In (49.18) the sums on p and q are over all coset representatives in $\mathfrak{G}$, the irreducible representation $\boldsymbol{k}$ is fixed.

For fixed q, the sum over p gives every element in $\mathfrak{G}(\boldsymbol{k})/\mathfrak{T}$ exactly once, so we have

$$\sum_p \dot{\chi}^{(\boldsymbol{k})(m)}(\varphi_q^{-1} \cdot \varphi_p \cdot \varphi_q | \boldsymbol{t}_{\bar{q}pq}\}) \, \dot{\chi}^{(\boldsymbol{k})(m')}(\{\varphi_q^{-1} \varphi_p \varphi_q | \boldsymbol{t}_{\bar{q}pq}\})^* = l_k \, \delta_{mm'}. \quad (49.19)$$

The remaining sum gives

$$\sum_q = g_\rho \tag{49.20}$$

which is the order of $\mathfrak{P}$.

Assembling the results (49.13), (49.20), (49.21), we verify (49.5), which was to be proved.

50. Induction of $D^{(k)(m)}$ from sub-space groups. In this section we briefly review an alternate method for obtaining the character system $\chi^{(k)(m)}$ of the allowable irreducible representations $D^{(k)(m)}$ of $\mathfrak{G}(k)$. This alternate method is based upon the general theory of inducing a representation of a group from representations of a subgroup. The theory of Sects. 36 and 37 for the construction of the entire space group irreducible representation $D^{(*k)(m)}$ by induction from $D^{(k)(m)}$ was also based upon the same general procedure. The particular application to be discussed here was first proposed by ZAK[1,2] in the context of space group theory.

ZAK's application is based upon constructing a series of normal subgroups of the space group $\mathfrak{G}(k)$, terminating in $\mathfrak{T}$ with the property that the factor group of the group with respect to its normal subgroup is of order 2 or 3 only. That is, consider that the group $\mathfrak{G}(k)$ has a normal subgroup $\mathfrak{N}_1(k)$, such that

$$\mathfrak{G}(k) = \mathfrak{N}_1(k) + \cdots + x_{1\,n_1}\mathfrak{N}_1(k), \qquad n_1 \leqq 3, \tag{50.1}$$

where $x_{1\,n_1} \equiv \{\varphi_{l_\lambda} | \tau_\lambda\}$ is some coset representative in $\mathfrak{G}(k)$. Let $\mathfrak{N}_1(k)$ have a normal subgroup $\mathfrak{N}_2(k)$ such that

$$\mathfrak{N}_1(k) = \mathfrak{N}_2(k) + \cdots + x_{2\,n_2}\mathfrak{N}_2(k), \qquad n_2 \leqq 3. \tag{50.2}$$

The process continues until the final normal subgroup $\mathfrak{T}$ is reached:

$$\mathfrak{N}_{i-1} = \mathfrak{T} + \cdots + x_{i\,n_i}\mathfrak{T}, \qquad n_i \leqq 3. \tag{50.3}$$

Now we focus on the process of inducing an irreducible representation from the normal subgroup $\mathfrak{N}_\beta(k)$ to the next group $\mathfrak{N}_\alpha(k)$ where

$$\mathfrak{N}_\alpha(k) = \mathfrak{N}_\beta(k) + \cdots + x_{\beta\,n_\beta}\mathfrak{N}_\beta(k). \tag{50.4}$$

Let ${}^\beta\chi^{(j)}$ be the character system in the j-th irreducible representation of the group $\mathfrak{N}_\beta(k)$. Hence to every element N_β in $\mathfrak{N}_\beta(k)$, we have a character assigned: ${}^\beta\chi^{(j)}(N_\beta)$. The character system ${}^\alpha\chi$ of the representation ${}^\alpha D$ induced from ${}^\beta\chi^{(j)}$ can be obtained in a fashion exactly analogous to (37.3). Thus the character system ${}^\alpha\chi$ is given as

$$ {}^\alpha\chi(x_{\beta\,n_\beta} N_\beta) = \sum_{\beta'=1}^{n_\beta} {}^\beta\dot{\chi}^{(j)}(x^{-1}_{\beta'\,n_{\beta'}} \cdot x_{\beta\,n_\beta} N_\beta \cdot x_{\beta'\,n_{\beta'}}) \tag{50.5}$$

with

$$\begin{aligned} {}^\beta\dot{\chi}^{(j)}(y) &= 0 && \text{if } y \text{ is not in } \mathfrak{N}_\beta(k) \\ &= {}^\beta\chi^{(j)}(y) && \text{if } y \text{ is in } \mathfrak{N}_\beta(k). \end{aligned} \tag{50.6}$$

Note that $x_{\beta\,n_\beta}$ is a coset representative in the coset decomposition (50.4), thus it is an element in $\mathfrak{N}_\alpha(k)$ but not in $\mathfrak{N}_\beta(k)$, except for $n_\beta = 1$ in which case $x_{\beta\,n_1} = E$,

[1] J. ZAK: J. Math. Phys. **1**, 165 (1960).

[2] L.J. KLAUDER, JR., and J.G. GAY: J. Math. Phys. **9**, 1488 (1968).

the identity element. Note that for an element $x_{\beta n_\beta} N_\beta$ which is *not in* $\mathfrak{N}_\beta$, (i.e. if $n_\beta \neq 1$), ${}^\alpha\chi=0$, while for an element N_β, the right hand side of (50.5) is a sum of terms corresponding to conjugate irreducible representations of $\mathfrak{N}_\beta(\boldsymbol{k})$.

To be specific, let $n_\beta=2$ then from (50.5)

$$ {}^\alpha\chi(N_\beta)={}^\beta\chi^{(j)}(N_\beta)+{}^\beta\chi^{(j)}(x^{-1} N_\beta x). \tag{50.7}$$

If ${}^\beta D^{(j)}$ is the irreducible representation of $\mathfrak{N}_\beta$ in which the element N_β has character ${}^\beta\chi^{(j)}(N_\beta)$, then the conjugate irreducible representation ${}^\beta\bar{D}^{(j)}$ is defined as

$$ {}^\beta\bar{D}^{(j)}(N_\beta)\equiv{}^\beta D^{(j)}(x^{-1} N_\beta x). \tag{50.8}$$

Hence

$$ {}^\beta\bar{\chi}^{(j)}(N_\beta)\equiv{}^\beta\chi^{(j)}(x^{-1} N_\beta x) \tag{50.9}$$

In (50.7)–(50.9) the coset representative is denoted as x, omitting subscripts. For an element $x N_\beta$ not in $\mathfrak{N}_\beta$ we have

$$ {}^\alpha\chi(x N_\beta)=0. \tag{50.10}$$

In the case at hand, the induced representation ${}^\alpha D$ must be carefully examined for irreducibility, by contrast to the induction of $D^{(\star k)(m)}$ in (37.3); there $D^{(\star k)(m)}$ was irreducible as demonstrated in Sects. 30–37. Two cases must now be distinguished (we continue with the case $n_\beta=2$):

$$ \text{a)}\quad {}^\beta D^{(j)} \text{ inequivalent to } {}^\beta\bar{D}^{(j)}, \tag{50.11}$$

$$ \text{b)}\quad {}^\beta D^{(j)} \text{ equivalent to } {}^\beta\bar{D}^{(j)}. \tag{50.12}$$

In case a) (50.11) ${}^\alpha D$ is irreducible, the proof follows that of the earlier sections owing to the inequivalence of the representation subduced upon $\mathfrak{N}_\beta$ by ${}^\alpha D$ of $\mathfrak{N}_\alpha$.

In case b) with ${}^\beta D^{(j)}$ equivalent to ${}^\beta\bar{D}^{(j)}$ we can easily demonstrate that the induced ${}^\alpha D$ is a reducible representation of $\mathfrak{N}_\alpha$. First we explicitly write the character system for this representation.

$$ {}^\alpha\chi(N_\beta)=2\,{}^\beta\chi^{(j)}(N_\beta) \quad \text{for } N_\beta \text{ in } \mathfrak{N}_\beta, \tag{50.13}$$

$$ {}^\alpha\chi(x)=0 \quad \text{for } x \text{ not in } \mathfrak{N}_\beta. \tag{50.14}$$

Form the sum over all elements N_α in $\mathfrak{N}_\alpha$:

$$ \sum_{\mathfrak{N}_\alpha}|{}^\alpha\chi(N_\alpha)|^2=\sum_{\mathfrak{N}_\beta}|{}^\alpha\chi(N_\beta)|^2=4\sum_{\mathfrak{N}_\beta}|{}^\beta\chi^{(j)}(N_\beta)|^2=4g_\beta, \tag{50.15}$$

where g_β is the order of $\mathfrak{N}_\beta$ and we used (50.13) and (50.14). But the order of $\mathfrak{N}_\alpha$ is $g_\alpha=2g_\beta$. Hence, if we denote an irreducible representation of $\mathfrak{N}_\alpha$ by ${}^\alpha D^{(j)}$, with character system ${}^\alpha\chi^{(j)}$, we would have for it

$$ \sum_{\mathfrak{N}_\alpha}|{}^\alpha\chi^{(j)}(N_\alpha)|^2=g_\alpha=2g_\beta. \tag{50.16}$$

Comparing (50.16) and (50.15) we immediately see that ${}^\alpha D$ is reducible in $\mathfrak{N}_\alpha$. The representation ${}^\alpha D$ can be expressed as a sum of irreducible constituents as

$$ {}^\alpha\chi(N_\alpha)=\sum_j (j| \quad {}^\alpha\chi^{(j)}(N_\alpha), \tag{50.17}$$

where $(j|$ are positive integers or zero. Substitute (50.17) into (50.15) to obtain

$$\sum_{\mathfrak{N}_\alpha}\sum_{j}\sum_{j'}(j|\,(j'|^*\,{}^{\alpha}\chi^{(j)}(N_\alpha)\,{}^{\alpha}\chi^{(j')}(N_\alpha)^* = \sum_j |(j||^2\, g_\alpha = 4g_\beta = 2g_\alpha, \tag{50.18}$$

$$\sum_j |(j||^2 = 2. \tag{50.19}$$

Then the only possible solution is for $(j|=1$ for two distinct irreducible representations of $\mathfrak{N}_\alpha$. Or

$$(j| = (\bar{j}| = 1 \tag{50.20}$$

and

$$^{\alpha}\chi = {}^{\alpha}\chi^{(j)} + {}^{\alpha}\chi^{(\bar{j})}. \tag{50.21}$$

Evidently the dimension of $^{\alpha}D$ is $2l_j$ where l_j is the dimension of $^{\beta}D^{(j)}$. But the Froebenius reciprocity theorem shows that as representations of $\mathfrak{N}_\beta$ we identify the subduced characters as

$$^{\alpha}\chi^{(j)}(N_\beta) = {}^{\beta}\chi^{(j)}(N_\beta), \tag{50.22}$$

$$^{\alpha}\chi^{(\bar{j})}(N_\beta) = {}^{\beta}\chi^{(j)}(N_\beta) \tag{50.23}$$

so

$$\dim {}^{\alpha}\chi^{(j)} = \dim {}^{\alpha}\chi^{(\bar{j})} = l_j. \tag{50.24}$$

Finally then, we obtain from (50.7), (50.22), and (50.23)

$$^{\alpha}\chi^{(j)}(x N_\beta) = -{}^{\alpha}\chi^{(\bar{j})}(x N_\beta). \tag{50.25}$$

It does not seem possible to give an explicit, and general statement for the actual characters $^{\alpha}\chi^{(j)}(x N_\beta)$ of the elements $x N_\beta$ in the j-th irreducible representation of the group $\mathfrak{N}_\alpha$. In the special case where x commutes with all elements in $\mathfrak{N}_\alpha$ (and hence $\mathfrak{N}_\beta$):

$$x N_\beta = N_\beta x \tag{50.26}$$

such a statement can be given. This is of course the case where the group $\mathfrak{N}_\alpha$ is a direct product of $\mathfrak{N}_\beta$ and the factor group $\mathfrak{N}_\alpha/\mathfrak{N}_\beta$. Also if in the matrix system $^{\alpha}D^{(j)}$ of the irreducible representation, it happens that $^{\alpha}D^{(j)}(x)$ is in the center, then:

$$^{\alpha}D^{(j)}(x)\,{}^{\beta}D^{(j)}(N_\beta) = {}^{\beta}D^{(j)}(N_\beta)\,{}^{\alpha}D^{(j)}(x). \tag{50.27}$$

But in $\mathfrak{N}_\beta$, $^{\beta}D^{(j)}$ is irreducible, hence by SCHUR lemma

$$^{\alpha}D^{(j)}(x) = (\text{const})\,{}^{\beta}D^{(j)}(E). \tag{50.28}$$

To determine the constant we form

$$^{\alpha}D^{(j)}(x)^2 = {}^{\alpha}D^{(j)}(x^2) = (\text{const})^2\,{}^{\beta}D^{(j)}(E). \tag{50.29}$$

Since x^2 is an element in $\mathfrak{N}_\beta$

$$^{\alpha}D^{(j)}(x^2) = {}^{\beta}D^{(j)}(x^2). \tag{50.30}$$

Take the trace of (50.29) and (50.30):

$$^{\beta}\chi^{(j)}(x^2) = (\text{const})^2 \cdot l_j \tag{50.31}$$

or

$$\text{const} = [{}^{\beta}\chi^{(j)}(x^2)/l_j]^{\frac{1}{2}}. \tag{50.32}$$

Then

$$^{\alpha}\chi^{(j)}(x) = [{}^{\beta}\chi^{(j)}(x^2)/l_j]^{\frac{1}{2}}\, l_j. \tag{50.35}$$

In general for any element in $x\,\mathfrak{N}_\beta$, in this case:

$$^{\alpha}\chi^{(j)}(x\,N_\beta) = \pm [{}^{\beta}\chi^{(j)}(x^2)/l_j]^{\frac{1}{2}}\, {}^{\beta}\chi^{(j)}(N_\beta). \tag{50.34}$$

The symbols in (50.34) have their usual meaning.

In an analogous manner we may analyse the case where $N_\beta = 3$. The coset decomposition (50.4) is then

$$\mathfrak{N}_\alpha(\boldsymbol{k}) = \mathfrak{N}_\beta(\boldsymbol{k}) + x\,\mathfrak{N}_\beta(\boldsymbol{k}) + x^{-1}\,\mathfrak{N}_\beta(\boldsymbol{k}). \tag{50.35}$$

Again the character system of the induced representation is given by (50.5). The cases which must be distinquished are, as in (50.11) and (50.12):

$$\text{a)}\quad {}^{\beta}D^{(j)} \quad \text{inequivalent to } {}^{\beta}\bar{D}^{(j)},\ {}^{\beta}\bar{\bar{D}}^{(j)}, \tag{50.36}$$

$$\text{b)}\quad {}^{\beta}D^{(j)} \quad \text{equivalent to } {}^{\beta}\bar{D}^{(j)},\ {}^{\beta}\bar{\bar{D}}^{(j)}. \tag{50.37}$$

Observe that the conjugate representations to ${}^{\beta}D^{(j)}$ are in this case (compare (50.8))

$$^{\beta}\bar{D}^{(j)}(N_\beta) \equiv {}^{\beta}D^{(j)}(x^{-1}\,N_\beta\,x) \tag{50.38}$$

and

$$^{\beta}\bar{\bar{D}}^{(j)}(N_\beta) \equiv {}^{\beta}D^{(j)}(x\,N_\beta\,x^{-1}). \tag{50.39}$$

In case a) the induced character system

$$^{\alpha}\chi^{(j)}(N_\beta) = {}^{\beta}\chi^{(j)}(N_\beta) + {}^{\beta}\bar{\chi}^{(j)}(N_\beta) + {}^{\beta}\bar{\bar{\chi}}^{(j)}(N_\beta), \qquad {}^{\alpha}\chi^{(j)}(x\,N_\beta) = 0 \tag{50.40}$$

is an irreducible character system, corresponding to an irreducible representation ${}^{\alpha}D^{(j)}$ of $\mathfrak{N}_\alpha$. In case b) the induced representation ${}^{\alpha}\chi$ can be reduced into the sum of three irreducible representations of equal dimension l_j, equal to the dimension of ${}^{\beta}\chi^{(j)}$. The character system of the irreducible representation is then obtained by the same argument which led to (50.34): the matrices ${}^{\alpha}D^{(j)}(x)$, and ${}^{\alpha}D^{(j)}(x^{-1})$ are, in this case, in the center of the matrix group ${}^{\alpha}D^{(j)}$ and hence these matrices are by SCHUR lemma constants. As the element x^3 is in $\mathfrak{N}_\beta$, its character: ${}^{\beta}\chi^{(j)}(x^3)$, is known from the reduction of $\mathfrak{N}_\beta$. Following an argument analogous to (50.27)–(50.34) we have the entire character system of the irreducible components as

$$^{\alpha}\chi^{(j)}(N_\beta) = {}^{\beta}\chi^{(j)}(N_\beta), \tag{50.41}$$

$$^{\alpha}\chi^{(j)}(x\,N_\beta) = \exp(2\pi i\, p/3)\,[{}^{\beta}\chi^{(j)}(x^3)/l_j]^{\frac{1}{3}}\, {}^{\beta}\chi^{(j)}(N_\beta), \qquad p = 1, 2, 3, \tag{50.42}$$

$$^{\alpha}\chi^{(j)}(x^{-1}\,N_\beta) = \exp(2\pi i\, p/3)\,[{}^{\beta}\chi^{(j)}(x^{-3})/l_j]^{\frac{1}{3}}\, {}^{\beta}\chi^{(j)}(N_\beta), \qquad p = 1, 2, 3. \tag{50.43}$$

It is clear that ZAK's method of inducing on subgroups of index 2 or 3 makes essential use of the property of the matrix representing the coset representative (x, and x^2) being in the center of the matrix representation and hence a constant, whose value is determined by the arguments of (50.27)–(50.34). It is therefore restricted to certain non-symmorphic groups. It might be observed that ZAK's method makes essential use of the property that space groups $\mathfrak{G}$ are solvable (i.e. the indices of successive normal subgroups in the chain (50.1)–(50.3) are prime and terminate in 1). This same property was emphasized by SEITZ in his initial work on reduction of space groups.

Clearly the process of inducing from subgroups may be continued until we obtain the set of allowable irreducible characters $\chi^{(k)(m)}$ of $\mathfrak{G}(\boldsymbol{k})$. At that point the complete character system $\chi^{(\star k)(m)}$ is obtained as in (37.3) by inducing from $\mathfrak{G}(\boldsymbol{k})$ to $\mathfrak{G}$. It is also clear that the identical character system $\chi^{(k)(m)}$ must be obtained either by using ZAK's method or by using the little group or ray representation methods. A direct comparison of the Zak procedure with the others can be made if one utilizes the structure of the crystallographic point groups, and particularly their composition series, as they are all solvable. Their comparison would be somewhat laborious and so we do not give it here.

Summarizing we observe that three methods for obtaining the allowable irreducible representations $\chi^{(k)(m)}$ have been presented. Convenience alone dictates whether or not to use the ray representation, little group method, or when applicable, the induction method of ZAK. This part of the problem may now be assumed concluded with the assumption that construction of all the $D^{(\star k)(m)}$ is complete.

51. Compatibility relations for $D^{(\star k)(m)}$ and subduction. One is often interested in the connection between an irreducible representation $D^{(\star k)(m)}$ and an irreducible representation $D^{(\star k')(m')}$ corresponding to an adjacent star.[1] That is, let the canonical wave vectors $\boldsymbol{k}$ and $\boldsymbol{k}'$ of the stars $\star\boldsymbol{k}$, and $\star\boldsymbol{k}'$ be close to one another in the zone:

$$\boldsymbol{k}'=\boldsymbol{k}+\boldsymbol{\kappa} \tag{51.1}$$

where $\boldsymbol{\kappa}$ is a small vector. It suffices to discuss the allowable $D^{(k)(m)}$, and $D^{(k')(m')}$ and their connection, owing to the property of the induced representations being entirely determined by those of the small representations. The properties of the allowable $D^{(k)(m)}$ and $D^{(k')(m')}$ are entirely determined by their respective groups $\mathfrak{G}(\boldsymbol{k})$ and $\mathfrak{G}(\boldsymbol{k}')$. We shall concern ourselves with the situation where we pass from a point $\boldsymbol{k}$ of high symmetry to an adjacent point $\boldsymbol{k}'$ such that:

$$\text{a)}\quad \mathfrak{G}(\boldsymbol{k}) \text{ isomorphic to } \mathfrak{G}(\boldsymbol{k}') \tag{51.2}$$

or else:

$$\text{b)}\quad \mathfrak{G}(\boldsymbol{k}') \text{ a subgroup of } \mathfrak{G}(\boldsymbol{k}). \tag{51.3}$$

In case a) with $\mathfrak{G}(\boldsymbol{k})$ isomorphic to $\mathfrak{G}(\boldsymbol{k}')$, the same set of allowable irreducible $D^{(k)(m)}$ and $D^{(k')(m')}$ arise. As the two groups are isomorphic, the set $D^{(k)(m)}$ can be put into one-one correspondence to the set $D^{(k')(m')}$. Further we assume that as $\boldsymbol{\kappa}\to 0$

$$D^{(k')(m')}\to D^{(k)(m)} \tag{51.4}$$

then we can identify the indices m and m'. The assumption that $D^{(k')(m')}$ goes over smoothly into $D^{(k)(m)}$ as $\boldsymbol{\kappa}\to 0$ for case a) seems not to require further elaboration.

For case b), with $\mathfrak{G}(\boldsymbol{k}')$ a subgroup of $\mathfrak{G}(\boldsymbol{k})$ we understand that symmetry elements in $\mathfrak{G}(\boldsymbol{k})$ have been "lost" in going to $\mathfrak{G}(\boldsymbol{k}')$. This can arise when $\boldsymbol{k}$ is a

[1] L. BOUCKAERT, R. SMOLUCHOWSKI, and E. WIGNER: Phys. Rev. **50**, 58 (1936).

point of high symmetry while $\boldsymbol{k}'$ is on a neighboring line, or is a point lower in symmetry in the neighborhood of $\boldsymbol{k}$. We may consider the representation $D^{(\boldsymbol{k})(m)}$ of $\mathfrak{G}(\boldsymbol{k})$ restricted to $\mathfrak{G}(\boldsymbol{k}')$, as *subducing* [*11–12*] a representation on $\mathfrak{G}(\boldsymbol{k}')$:

$$D^{(\boldsymbol{k})(m)} \downarrow \text{ on } \mathfrak{G}(\boldsymbol{k}') \tag{51.5}$$

As a representation of $\mathfrak{G}(\boldsymbol{k}')$, $D^{(\boldsymbol{k})(m)}$ will in general be reducible into irreducible constituents:

$$D^{(\boldsymbol{k})(m)} \downarrow = \sum_{m'} (\boldsymbol{k}\, m \downarrow | m')\, D^{(\boldsymbol{k}')(m')}. \tag{51.6}$$

In (51.6) the reduction coefficients $(\boldsymbol{k}\, m \downarrow | m')$ are positive integers.

The examination of compatibility concerns itself with the study of the relations between allowable $D^{(\boldsymbol{k})(m)}$ at neighboring points such as $\boldsymbol{k}$ and $\boldsymbol{k}'$ of (51.1). Compatible representations are related via (51.6) when $\mathfrak{G}(\boldsymbol{k}')$ is a subgroup of $\mathfrak{G}(\boldsymbol{k})$, and when the coefficient $(\boldsymbol{k}\, m \downarrow | m')$ of the compatible pair is non-zero.

We return to compatibility considerations in Part N of this article, when we analyse specific cases.

F. Reduction coefficients for space groups: Full group methods

52. Introduction. In this part of the article, Part F, the mathematical problem of obtaining the reduction coefficients for the space group will be discussed, using what has come to be called full-group methods.[1, 2] In the next Part, G, sub-group methods[3–6] will be presented.

The problem to be analyzed in both Parts F and G is to be considered as strictly a mathematical one of determining the irreducible representations contained in a reducible representation which is the direct product of two irreducible representations. The mathematical problem is then precisely that discussed in general in Sect. 17 for finite groups. By the rubric full group method, one simply means using at all stages the irreducible representations $D^{(\star\boldsymbol{k})(m)}$ of the full group $\mathfrak{G}$, and correspondingly carrying out all calculations in principle by utilizing *all* the symmetry elements of $\mathfrak{G}$.

So far, the full group methods have been successfully utilized in obtaining complete reduction coefficients for a number of distinct space groups, symmorphic and non-symmorphic. The full group methods have also been applied to analysis of cases of considerably greater complexity than the sub-group methods to be

[1] J.L. BIRMAN: Phys. Rev. **127**, 1093 (1962); **131**, 1489 (1963).
[2] J. ZAK: J. Math. Phys. **3**, 1278 (1962).
[3] R.J. ELLIOTT and R. LOUDON: J. Phys. Chem. Solids **15**, 146 (1960).
[4] M. LAX and J.J. HOPFIELD: Phys. Rev. **124**, 115 (1961).
[5] J. ZAK: Phys. Rev. **151**, 464 (1966).
[6] H. POULET: J. de Physique (Paris) **26**, 684 (1965).

presented in the next part. It behooves the serious student of this subject to understand the full group methods completely in order to put the entire subject of space groups on an identical footing with the usual finite groups: use of the sub-group methods, with due precautions to avoid error, can then be undertaken in special cases.

In the next Sects. 53–60, we discuss the structure of the direct product representations produced by taking ordinary and symmetrized direct (Kronecker) products of the space group irreducible representations $D^{(\star k)(m)}$. Then the key concept of wave-vector, or star, selection rules will be presented. Using these, the complete character reduction coefficients can be obtained and the reduction completed.

53. Direct product $D^{(\star k)(m)} \otimes D^{(\star k')(m')}$. As analysed in Sects. 29 and 32, the representation space $\Sigma^{(\star k)(m)}$ which is a basis for $D^{(\star k)(m)}$ is a space of dimension $(s \cdot l_m)$. It is spanned by Bloch vectors $\psi_\alpha^{(k_\sigma)(m)}$:

$$\{\psi_1^{(k)(m)}, \ldots, \psi_{l_m}^{(k_s)(m)}\} = \Sigma^{(\star k)(m)} \tag{53.1}$$

and the space $\Sigma^{(\star k)(m)}$ generates a representation $D^{(\star k)(m)}$

$$P_{\{\varphi|t(\varphi)\}} \Sigma^{(\star k)(m)} = D^{(\star k)(m)}(\{\varphi|t(\varphi)\}) \Sigma^{(\star k)(m)}. \tag{53.2}$$

The space $\Sigma^{(\star k')(m')}$ which is a basis for $D^{(\star k')(m')}$ is of dimension $(s' \cdot l_{m'})$. It is spanned by Bloch vectors $\psi_{\alpha'}^{(k'_\sigma)(m')}$. Likewise the space

$$(\psi_1^{(k')(m')}, \ldots, \psi_{l_{m'}}^{(k'_s)(m')}) = \Sigma^{(\star k')(m')} \tag{53.3}$$

generates $D^{(\star k')(m')}$

$$P_{\{\varphi|t(\varphi)\}} \Sigma^{(\star k')(m')} = D^{(\star k')(m')}(\{\varphi|t(\varphi)\}) \Sigma^{(\star k')(m')}. \tag{53.4}$$

The set of $(s \cdot l_m)(s' \cdot l_{m'})$ functions, each one of which is a product of a function $\psi_\alpha^{(k)(m)}$ and a function $\psi_{\alpha'}^{(k')(m')}$ is:

$$\begin{aligned}\{\psi_1^{(k)(m)} \cdot \psi_1^{(k')(m')}, \ldots, \psi_1^{(k)(m)} \cdot \psi_{l_{m'}}^{(k')(m')}, \ldots, \psi_{l_m}^{(k_s)(m)} \cdot \psi_{l_{m'}}^{(k'_{s'})(m')}\} \\ \equiv \Sigma^{(\star k \otimes \star k')(m \otimes m')}\end{aligned} \tag{53.5}$$

and is the direct product space. It generates the direct product representation

$$P_{\{\varphi|t(\varphi)\}} \Sigma^{(\star k \otimes \star k')(m \otimes m')} = D^{(\star k \otimes \star k')(m \otimes m')}(\{\varphi|t(\varphi)\}) \Sigma^{(\star k \otimes \star k')(m \otimes m')}. \tag{53.6}$$

As in (17.3), the character system of the direct product representation $D^{(\star k \otimes \star k')(m \otimes m')}$ is easily found as the ordinary arithmetic product of the characters of the factors:

$$\chi^{(\star k \otimes \star k')(m \otimes m')} = \chi^{(\star k)(m)} \cdot \chi^{(\star k')(m')} \tag{53.7}$$

or substituting a specific space group element as argument of (53.7):

$$\chi^{(\star k \otimes \star k')(m \otimes m')}(\{\varphi|t(\varphi)\}) = \chi^{(\star k)(m)}(\{\varphi|t(\varphi)\}) \cdot \chi^{(\star k')(m')}(\{\varphi|t(\varphi)\}). \tag{53.8}$$

The expression (53.8) could be written out in terms of various products of dotted characters referred to $\mathfrak{G}(k)$, and $\mathfrak{G}(k')$ by substituting into (53.8) from (37.3), or from (49.3). The resulting formulae would then be far more complicated in appearance than (53.8) involving a double sum over coset representatives and the resulting sums of products of dotted characters. For our purposes in this section we shall not do this but merely consider that the product (53.8) involves a single individual character which is obtained from (37.3) and (49.3). The latter expressions fully involve the entire theory of space group representations including the theory of induced representations. In Sect. 62 explicit use of dotted characters will be made.

In the direct product representation $D^{(\star k)(m)} \otimes D^{(\star k')(m')}$ the matrices representing elements in $\mathfrak{T}$ assume a particularly simple, diagonal form. This is plain from (33.10) for the individual matrices in each of the factors, representing elements in $\mathfrak{T}$. Thus the matrix is

$$D^{(\star k \otimes \star k')(m \times m')}(\{\varepsilon | \boldsymbol{R}_L\})$$
$$= \begin{pmatrix} \boldsymbol{\Pi}_{mm'} \exp i(\boldsymbol{k}_1 + \boldsymbol{k}'_1) \cdot \boldsymbol{R}_L & \dots & 0 & \dots & 0 \\ & & \vdots & & \\ 0 & \dots & \boldsymbol{\Pi}_{mm'} \exp i(\boldsymbol{k}_1 + \boldsymbol{k}'_{s'}) \cdot \boldsymbol{R}_L & & \vdots \\ & \dots & & & \dots \boldsymbol{\Pi}_{mm'} \exp i(\boldsymbol{k}_s + \boldsymbol{k}'_{s'}) \cdot \boldsymbol{R}_L \end{pmatrix} \tag{53.9}$$

where $\boldsymbol{\Pi}_{mm'}$ is the $l_m l_{m'} \times l_m l_{m'}$ unit matrix. The values of $\boldsymbol{k}_\sigma + \boldsymbol{k}'_{\sigma'}$ which appear down the diagonal in (53.9) evidently include all possible pairs of terms with one wave vector from each of the stars $\star\boldsymbol{k}$ and $\star\boldsymbol{k}'$.

For a general space group element the direct product matrix does not assume as simple a form as (53.9) but can be obtained if needed from (33.11). In any event (53.8) prescribes the characters as completely as we need them for the work of reducing products.

54. Symmetrized powers $[D^{(\star k)(m)}]_{(p)}$

α) *Ordinary Kronecker powers.* The Kronecker powers of a representation $D^{(\star k)(m)}$ can be defined in exactly the same fashion as the Kronecker product of Sect. 53 [*16*, *17*]. We shall examine some of these powers explicitly, to provide a basis for our later work. This subject will be discussed again in Sect. 117.

First consider the Kronecker square of a representation $D^{(\star k)(m)}$. The irreducible space $\Sigma^{(\star k)(m)}$ is as in (53.1), spanned by $(s \cdot l_m)$ Bloch functions $\psi_\alpha^{(k_\sigma)(m)}$, $\sigma = 1, \dots, s$; $\alpha = 1, \dots, l_m$. Consider a second, linearly independent vector space $\Sigma'^{(\star k)(m)}$ spanned by the $(s \cdot l_m)$ Bloch functions $\psi'^{(k_\sigma)(m)}_\alpha$

$$\Sigma'^{(\star k)(m)} = \{\psi'^{(k)(m)}_1, \dots, \psi'^{(k_s)(m)}_{l_m}\} \tag{54.1}$$

which also is a basis for the same irreducible representation $D^{(\star k)(m)}$ of $\mathfrak{G}$. It is to be emphasized that the $(s \cdot l_m)$ functions $\psi_\alpha^{(k_\sigma)(m)}$, and also the $(s \cdot l_m)$ $\psi'^{(k_\sigma)(m)}_\alpha$ are linearly independent although both sets produce spaces (53.1) and (54.1) which give rise to the same representation $D^{(\star k)(m)}$. Now consider the $(s \cdot l_m)^2$ functions formed by taking all possible products of one function from $\Sigma^{(\star k)(m)}$ and one function from $\Sigma'^{(\star k)(m)}$:

$$\psi_1^{(k)(m)} \cdot \psi'^{(k)(m)}_1, \dots, \psi_1^{(k)(m)} \cdot \psi'^{(k)(m)}_{l_m}, \dots, \psi_{l_m}^{(k_s)(m)} \psi'^{(k_s)(m)}_{l_m}. \tag{54.2}$$

It is evident that this case is in principle no different then any other direct product of two representations. Since however the representation by which the product space (54.2) transforms is the direct product of $D^{(\star k)(m)}$ with itself, this representation is called the Kronecker square of $D^{(\star k)(m)}$. It is of dimension $(s \cdot l_m)^2$. It is denoted

$$D^{(\star k)(m)} \otimes D^{(\star k)(m)} = [D^{(\star k)(m)}]_2. \tag{54.3}$$

Note that in (54.3) the subscript indicating the power is *not* written with parenthesis, because this is the *ordinary*, not symmetrized Kronecker square.

Clearly the process may be continued. We can form the $(s \cdot l_m)^3$ functions by taking each factor in the product of three factors from amongst the $s \cdot l_m$ terms in each of three linearly independent vector spaces: $\Sigma^{(\star k)(m)}$, $\Sigma'^{(\star k)(m)}$, $\Sigma''^{(\star k)(m)}$, spanned by functions

$$\psi_\alpha^{(k_\sigma)(m)} \qquad \sigma = 1, \ldots, s;\ \alpha = 1, \ldots, l_m, \tag{54.4}$$

$$\psi_\alpha'^{(k_\sigma)(m)} \qquad \sigma = 1, \ldots, s;\ \alpha = 1, \ldots, l_m, \tag{54.5}$$

$$\psi_\alpha''^{(k_\sigma)(m)} \qquad \sigma = 1, \ldots, s;\ \alpha = 1, \ldots, l_m \tag{54.6}$$

respectively. Each of these spaces is separately a basis for the same irreducible representation $D^{(\star k)(m)}$. Then the set of $(s \cdot l_m)^3$ functions

$$\{\psi_1^{(k)(m)} \cdot \psi_1'^{(k)(m)} \cdot \psi_1''^{(k)(m)}, \ldots, \psi_{l_m}^{(k_s)(m)} \cdot \psi_{l_m}'^{(k_s)(m)} \cdot \psi_{l_m}''^{(k_s)(m)}\} \tag{54.7}$$

is a basis for the ordinary Kronecker cube of $D^{(\star k)(m)}$ denoted:

$$D^{(\star k)(m)} \otimes D^{(\star k)(m)} \otimes D^{(\star k)(m)} = [D^{(\star k)(m)}]_3. \tag{54.8}$$

A symbol such as

$$[D^{(\star k)(m)}]_p \tag{54.9}$$

denotes the Kronecker p-th power of $D^{(\star k)(m)}$. It is clear, too, that the character system of the Kronecker p-th power representation is easily obtained from the character system of the representation $D^{(\star k)(m)}$. From (54.9):

$$\mathrm{Tr}\,[D^{(\star k)(m)}]_p = [\chi^{(\star k)(m)}]_p = (\chi^{(\star k)(m)})^p \tag{54.10}$$

for any element in $\mathfrak{G}$.

β) Symmetrized Kronecker powers. We turn next to a representation different in principle from the ordinary Kronecker powers just discussed. It is the symmetrized Kronecker power of a representation $D^{(\star k)(m)}$. We return to the space $\Sigma^{(\star k)(m)}$ of (53.1). Consider the set of independent product functions composed by taking all possible pairs of basis functions from the same set (53.1). That is we form the space

$$\begin{aligned}\{&\psi_1^{(k)(m)} \cdot \psi_1^{(k)(m)}, \psi_1^{(k)(m)} \cdot \psi_2^{(k)(m)}, \ldots, \psi_1^{(k)(m)} \cdot \psi_{l_m}^{(k)(m)}\\ &\psi_2^{(k)(m)} \cdot \psi_2^{(k)(m)}, \ldots, \psi_2^{(k)(m)} \cdot \psi_{l_m}^{(k)(m)}, \ldots, \psi_{l_m}^{(k_s)(m)} \cdot \psi_{l_m}^{(k_s)(m)}\}\\ &\equiv [\Sigma^{(\star k \otimes \star k)(m \otimes m)}]_{(2)}.\end{aligned} \tag{54.11}$$

The general function in this space is of form

$$\psi_\alpha^{(k_\sigma)(m)} \cdot \psi_\beta^{(k_\tau)(m)}, \qquad \beta \geqq \alpha;\ \tau \geqq \sigma. \tag{54.12}$$

The restrictions $\beta \geqq \alpha$ and $\tau \geqq \sigma$ eliminates the possibility of counting the same product twice. In the space (54.11), there are $\frac{1}{2}(s\,l_m)(s\,l_m+1)$ independent functions. We may call the space (54.12) the space generated by the symmetrized product of $\Sigma^{(\star k)(m)}$ with itself. The representation of $\mathfrak{G}$ produced by this space is obtained as:

$$P_{\{\varphi|t(\varphi)\}}[\Sigma^{(\star k \otimes \star k)(m\otimes m)}]_{(2)} = [D^{(\star k)(m)}(\{\varphi|t(\varphi)\})]_{(2)}[\Sigma^{(\star k\otimes \star k)(m\otimes m)}]_{(2)}. \tag{54.13}$$

The representation

$$[D^{(\star k)(m)}]_{(2)} \tag{54.14}$$

is called the symmetrized Kronecker square of $D^{(\star k)(m)}$; note particularly the conventional notation with the subscript in parenthesis used to denote this object; and contrast it with (54.3). The characters in this representation may be obtained[1] from the basic set of characters $\chi^{(\star k)(m)}$ by the rule:

$$\mathrm{Tr}[D^{(\star k)(m)}]_{(2)} \equiv [\chi^{(\star k)(m)}]_{(2)}, \tag{54.15}$$

$$[\chi^{(\star k)(m)}(\varphi|t(\varphi)\})]_{(2)} = \tfrac{1}{2}\{(\chi^{(\star k)(m)}(\{\varphi|t(\varphi)\}))^2 + \chi^{(\star k)(m)}(\{\varphi|t(\varphi)\}^2)\}. \tag{54.16}$$

In (54.16) the element given as argument in the second term is simply the square of the element $\{\varphi|t(\varphi)\}$.

In a similar fashion we may obtain[2] the symmetrized Kronecker cube of a representation $D^{(\star k)(m)}$ based upon the vector space $\Sigma^{(\star k)(m)}$. We may form the symmetrized cube vector space

$$[\Sigma^{(\star k\otimes \star k\otimes \star k)(m\otimes m\otimes m)}]_{(3)} \tag{54.17}$$

composed of all the independent products of three factors each one a Bloch vector taken from the same space $\Sigma^{(\star k)(m)}$. The form of the functions comprising the vector space (4.25) is

$$\psi_\alpha^{(k_\rho)(m)}\,\psi_\beta^{(k_\sigma)(m)}\,\psi_\sigma^{(k_\tau)(m)}, \qquad \rho,\sigma,\tau=1,\dots,s;\ \alpha\leqq\beta\leqq\gamma;\ \rho\leqq\sigma\leqq\tau. \tag{54.18}$$

There are $(\frac{1}{6})(s\,l_m)(s\,l_m+1)(s\,l_m+2)$ such functions (54.18). The vector space (54.17), with Bloch vectors (54.18), is a basis for the representation

$$[D^{(\star k)(m)}]_{(3)} \tag{54.19}$$

which is called the symmetrized Kronecker cube. Again, it can be shown that

$$\mathrm{Tr}[D^{(\star k)(m)}]_{(3)} = [\chi^{(\star k)(m)}]_{(3)}, \tag{54.20}$$

where

$$\begin{aligned}[\chi^{(\star k)(m)}(\{\varphi|t(\varphi)\})]_{(3)} = \tfrac{1}{6}\{&(\chi^{(\star k)(m)}(\{\varphi|t(\varphi)\}))^3\\ &+3\chi^{(\star k)(m)}(\{\varphi|t(\varphi)\})\cdot\chi^{(\star k)(m)}(\{\varphi|t(\varphi)\}^2)\\ &+2\chi^{(\star k)(m)}(\{\varphi|t(\varphi)\}^3)\}.\end{aligned} \tag{54.21}$$

Formulae for the higher symmetrized powers of a representation can also be given corresponding to the expressions (54.16) and (54.21).

In Sect. 117 of this article we give a derivation of the characters of the symmetrized Kronecker power of a representation which is adapted specifically to the

[1] See Sect. 117 *infra*.

[2] See Sect. 117 *infra*.

lattice vibration problem. We shall have no need, in our work, for the antisymmetrized Kronecker power of a representation, which is also discussed in standard literature [*2*, *16*, *17*].

55. Definition of reduction coefficients. According to MASCHKE'S theorem [*18*], a representation of a finite group over the field of complex numbers is either irreducible or decomposable into a direct sum of irreducible representations. This can be applied to the product representations discussed in Sects. 53 and 54. Our need is to reduce the direct product representation $D^{(\star\boldsymbol{k}\otimes\star\boldsymbol{k}')(m\otimes m')}$ into its irreducible components. We define the full reduction coefficients $(\star\boldsymbol{k}\, m\, \star\boldsymbol{k}'\, m' \,|\, \star\boldsymbol{k}''\, m'')$ by the basic equation as in (17.4)

$$D^{(\star\boldsymbol{k}\otimes\star\boldsymbol{k}')(m\otimes m')} = \sum_{\boldsymbol{k}''}\sum_{m''} \oplus\, (\star\boldsymbol{k}\, m\, \star\boldsymbol{k}'\, m' \,|\, \star\boldsymbol{k}''\, m'')\, D^{(\star\boldsymbol{k}'')(m'')}. \tag{55.1}$$

The sum in (55.1) is over all stars $\star\boldsymbol{k}''$, and over all allowable irreducible representations m'' which may arise for given $\boldsymbol{k}''$. The coefficients

$$(\star\boldsymbol{k}\, m\, \star\boldsymbol{k}'\, m' \,|\, \star\boldsymbol{k}''\, m'') \tag{55.2}$$

are integers and simply denote which irreducible representations $D^{(\star\boldsymbol{k}'')(m'')}$ occur and how many times, in the reduction.

An alternate definition of the reduction coefficients can be given as in (17.7) using the character systems $\chi^{(\star\boldsymbol{k})(m)}$. Then (55.1) becomes

$$\chi^{(\star\boldsymbol{k}\otimes\star\boldsymbol{k}')(m\otimes m')} = \sum_{\star\boldsymbol{k}''}\sum_{m''} (\star\boldsymbol{k}\, m\, \star\boldsymbol{k}'\, m' \,|\, \star\boldsymbol{k}''\, m'')\, \chi^{(\star\boldsymbol{k}'')(m'')} \tag{55.3}$$

or, using the rule for the character of a direct product

$$\chi^{(\star\boldsymbol{k})(m)}\, \chi^{(\star\boldsymbol{k}')(m')} = \sum_{\star\boldsymbol{k}''}\sum_{m''} (\star\boldsymbol{k}\, m\, \star\boldsymbol{k}'\, m' \,|\, \star\boldsymbol{k}''\, m'')\, \chi^{(\star\boldsymbol{k}'')(m'')}. \tag{55.4}$$

It is understood in (55.1), (55.3) and (55.4) that one particular group element $P_{\{\varphi|\boldsymbol{t}(\varphi)\}}$ is being represented (or in (55.3) and (55.4), one particular class), so the argument is omitted here for brevity.

To determine the character reduction coefficients, we may utilize the orthonormality rule (49.5), in the conventional manner, as in (17.8)–(17.9). Then for these coefficients we have the character reduction coefficients which may be determined by:

$$\begin{aligned}(\star\boldsymbol{k}\, m\, \star\boldsymbol{k}'\, m' \,|\, \star\boldsymbol{k}''\, m'')\\ = \frac{1}{g_p N} \sum_{\{\varphi|\boldsymbol{t}(\varphi)\}} \chi^{(\star\boldsymbol{k})(m)}(\{\varphi|\boldsymbol{t}(\varphi)\}) \cdot \chi^{(\star\boldsymbol{k}')(m')}(\{\varphi|\boldsymbol{t}(\varphi)\})\, \chi^{(\star\boldsymbol{k}'')(m'')}(\{\varphi|\boldsymbol{t}(\varphi)\})^*.\end{aligned} \tag{55.5}$$

In (55.5) the sum is over *all* group elements in $\mathfrak{G}$. In spite of the apparent complexity of carrying out a sum over all the group elements such as is involved in (55.5), it does prove possible to evaluate the reduction coefficients by use of (55.5). This will be discussed in the next few sections.

In analogy to the definition of the ordinary reduction coefficients $(\star\boldsymbol{k}\, m\, \star\boldsymbol{k}'\, m' \,|\, \star\boldsymbol{k}''\, m'')$ by which we can reduce the product of any two space group irreducible representations, we can define reduction coefficients for the ordinary and symmetrized powers of an irreducible representation. Thus

$$[\chi^{(\star k)(m)}]_p = \sum_{\star k''} \sum_{m''} ([\star k\, m]_p | \star k''\, m'')\, \chi^{(\star k'')(m'')}. \tag{55.6}$$

This defines the reduction coefficient $([\star k\, m]_p | \star k''\, m'')$, for the ordinary Kronecker power of $D^{(\star k)(m)}$. Clearly this coefficient can be determined by an equation like (55.5):

$$([\star k\, m]_p | \star k''\, m'') = \frac{1}{g_p N} \sum_{\{\varphi | t(\varphi)\}} [\chi^{(\star k)(m)}(\{\varphi | t(\varphi)\})]_p \cdot \chi^{(\star k'')(m'')}(\{\varphi | t\})^*. \tag{55.7}$$

In (55.6) and (55.7) p can be any integer. Similarly reduction coefficients for the symmetrized powers are defined by

$$[\chi^{(\star k)(m)}]_{(p)} = \sum_{\star k''} \sum_{m''} ([\star k\, m]_{(p)} | k''\, m'')\, \chi^{(\star k'')(m'')} \tag{55.8}$$

and

$$([\star k\, m]_{(p)} | k''\, m'') = \frac{1}{g_p N} \sum_{\{\varphi | t(\varphi)\}} [\chi^{(\star k)(m)}(\{\varphi | t(\varphi)\})]_{(p)} \cdot \chi^{(\star k'')(m'')}(\{\varphi | t\}).^* \tag{55.9}$$

56α. Wave vector selection rules

α) *Star reduction coefficients for the ordinary product.* We shall proceed to obtain the complete reduction coefficients in two steps. In Eqs. (55.4), (55.7), (55.9) we sum over all stars $\star k''$. But, only a limited number of stars $\star k''$ will actually arise.

To be specific consider the reduction coefficients (55.4) $(\star k\, m\, \star k'\, m' | \star k''\, m'')$. These arise in the reduction of the ordinary direct product of two distinct space group irreducible representations. Consider the space (53.5). A typical function in (53.5) is

$$\psi_\alpha^{(k_\sigma)(m)} \cdot \psi_\beta^{(k'_\tau)(m')}, \qquad \sigma = 1, \dots, s;\ \tau = 1, \dots, s';\ \alpha = 1, \dots, l_m;\ \beta = 1, \dots, l_{m'}. \tag{56.1}$$

The wave vector by which the product (56.1) transforms is

$$k_\sigma + k'_\tau \equiv k''_\nu, \tag{56.2}$$

where k''_ν is one of the arms of $\star k''$. It is clear that in the direct product space (53.5), product functions occur with resultant wave vectors corresponding to all $(s \cdot s')$ possible distinct ways of adding each wave vector in $\star k$ to each wave vector in $\star k'$.

These $(s \cdot s')$ resultant vectors must be expressible in terms of the sum of entire stars, owing to the decomposibility of the representation $D^{(\star k \otimes \star k')(m \otimes m')}$ and the corresponding space (53.5), into entire irreducible representations. Each such $D^{(\star k'')(m'')}$ which can arise is characterized by one entire star. For the process of determining the complete set of stars $\star k''$ which arise when all possible resultants (56a.2) are formed we introduce a direct product notation for the stars alone:

$$\star k \otimes \star k' = \sum_{\star k''} \oplus (\star k\, \star k' | \star k'')\, \star k''. \tag{56.3}$$

The coefficients $({}^\star\boldsymbol{k}\,{}^\star\boldsymbol{k}'|{}^\star\boldsymbol{k}'')$ are wave vector reduction coefficients; they are integers. In terms of these coefficients, we can write the statement that the $(s\cdot s')$ vectors are expressible as the sum of entire stars as:

$$s\cdot s'=\sum_{{}^\star\boldsymbol{k}''}({}^\star\boldsymbol{k}\,{}^\star\boldsymbol{k}'|{}^\star\boldsymbol{k}'')\,s''. \tag{56.4}$$

Eq. (56.4) can also be understood as a dimensionality conservation requirement for the sum is over the stars ${}^\star\boldsymbol{k}''$. Since addition of wave vectors is commutative

$${}^\star\boldsymbol{k}\otimes{}^\star\boldsymbol{k}'={}^\star\boldsymbol{k}'\otimes{}^\star\boldsymbol{k} \tag{56.5}$$

and

$$({}^\star\boldsymbol{k}\,{}^\star\boldsymbol{k}'|{}^\star\boldsymbol{k}'')=({}^\star\boldsymbol{k}'\,{}^\star\boldsymbol{k}|{}^\star\boldsymbol{k}''). \tag{56.6}$$

The coefficients $({}^\star\boldsymbol{k}\,{}^\star\boldsymbol{k}'|{}^\star\boldsymbol{k}'')$ give the total dimensionality of all irreducible representations of the space group, based on ${}^\star\boldsymbol{k}''$, which arise in the reduction of the direct product of irreducible representations based on ${}^\star\boldsymbol{k}$ and ${}^\star\boldsymbol{k}'$ (irrespective of m and m') as long as $l_m=l_{m'}=1$.

Often we can obtain the wave vector reduction coefficients by inspection. This is particularly true for stars of very high symmetry. For other cases a particular systematic enumeration of the terms in (56.2) proves useful. Thus we can construct a rectangular table whose rows are labelled by the arms in ${}^\star\boldsymbol{k}$, and whose columns are labelled by the arms in ${}^\star\boldsymbol{k}'$. Call the arms in the star of $\boldsymbol{k}$:

$${}^\star\boldsymbol{k}=(\boldsymbol{k},\varphi_2\cdot\boldsymbol{k},\ldots,\varphi_s\cdot\boldsymbol{k}) \tag{56.7}$$

and in the star of $\boldsymbol{k}'$:

$${}^\star\boldsymbol{k}'=(\boldsymbol{k}',\varphi_2'\cdot\boldsymbol{k}',\ldots,\varphi_{s'}'\cdot\boldsymbol{k}'). \tag{56.8}$$

At the intersection of the σ-th row and τ-th column in the table, place the element (wave vector)

$$\varphi_\sigma\cdot(\boldsymbol{k}+\varphi_\tau'\cdot\boldsymbol{k}')=\varphi_\sigma\cdot\boldsymbol{k}+\varphi_\sigma\cdot\varphi_\tau'\cdot\boldsymbol{k}'. \tag{56.9}$$

This is a wave vector in ${}^\star(\boldsymbol{k}+\varphi_\tau'\cdot\boldsymbol{k}')$. By this construction we create a table in which, in every row, one of the addends is the same (e.g. $\varphi_\sigma\cdot\boldsymbol{k}$); this is not so of the columns. However every column in the table contains wave vectors in the same star. In the case illustrated in (56.9), the wave vectors in the τ-th column are all members of ${}^\star(\boldsymbol{k}+\varphi_\tau'\cdot\boldsymbol{k}')$. In particular cases it may happen that a particular star ${}^\star\boldsymbol{k}$ contains more than (s) arms, i.e. $s''>s$. But then it will merely be necessary to take together several columns in the table in order to assemble all the arms in ${}^\star\boldsymbol{k}''$. Conversely when $s''<s$, a given column may contain the same star several times. Again this causes no difficulties. In either case, recognizing which columns should be taken together is simple, once the table has been constructed.

It may be remarked that by finding the wave vector reduction coefficients $({}^\star\boldsymbol{k}\,{}^\star\boldsymbol{k}'|{}^\star\boldsymbol{k}'')$ we are in essence carrying out the reduction of those diagonal matrices in the direct product representation which represent the elements in $P_{\mathfrak{T}}$. The matrices (53.9) are already diagonal but our procedure consists in rearranging the diagonal elements so that all vectors belonging to one star are assembled.

When $l_m>1$ and $l_{m'}>1$, only a slight modification is needed. Again, inspecting the vector space (53.5) we see that $l_m>1$ and/or $l_{m'}>1$ merely means that the elements of given star: ${}^\star\boldsymbol{k}$ or ${}^\star\boldsymbol{k}'$ occur l_m times on the left hand side of (56.2).

In this case the multiplicity is easily accounted for. We have

$$(l_m {}^\star\boldsymbol{k}) \otimes (l_{m'} {}^\star\boldsymbol{k}') = \sum_{{}^\star\boldsymbol{k}''} \oplus (l_m l_{m'}) ({}^\star\boldsymbol{k}\, {}^\star\boldsymbol{k}' | {}^\star\boldsymbol{k}'')\, {}^\star\boldsymbol{k}'' \tag{56.10}$$

or in terms of dimensions

$$(l_m s) \cdot (l_{m'} s') = \sum (l_m l_{m'}) ({}^\star\boldsymbol{k}\, {}^\star\boldsymbol{k}' | {}^\star\boldsymbol{k}'')\, s''. \tag{56.11}$$

It may also be useful to observe that the case in which ${}^\star\boldsymbol{k}$ and ${}^\star\boldsymbol{k}'$ are identical presents nothing new. We have here to deal with the ordinary Kronecker power of a representation with itself. We may obtain the corresponding wave vector reduction coefficient by analogy to (56.10)

$$(l_m {}^\star\boldsymbol{k}) \otimes (l_m {}^\star\boldsymbol{k}) = \sum_{{}^\star\boldsymbol{k}''} \oplus\, l_m^2 ([{}^\star\boldsymbol{k}]_2 | {}^\star\boldsymbol{k}'')\, {}^\star\boldsymbol{k}'' \tag{56.12}$$

or we could write

$$(l_m {}^\star\boldsymbol{k}) \otimes (l_m {}^\star\boldsymbol{k}) = [l_m {}^\star\boldsymbol{k}]_2 . \tag{56.13}$$

We should then obtain the identical coefficients as in (56.12), but we would now call them $([l_m {}^\star\boldsymbol{k}]_2 | {}^\star\boldsymbol{k}'')$. Then we would write

$$[l_m {}^\star\boldsymbol{k}]_2 = \sum_{{}^\star\boldsymbol{k}''} \oplus ([l_m {}^\star\boldsymbol{k}]_2 | {}^\star\boldsymbol{k}'')\, {}^\star\boldsymbol{k}'' . \tag{56.14}$$

β) *Star reduction coefficients for the symmetrized product.* Now let us turn to the wave vector reduction coefficients for the symmetrized Kronecker powers. In this case, it is necessary to include the dimensionality (value of l_m) in the argument, *a priori*: That is if the vector space $\Sigma^{(\boldsymbol{k})(m)}$ is of dimension $l_m > 1$ then, in the enumeration of the wave vectors $\boldsymbol{k}$, each wave vector in the star must appear l_m times, or equivalently stated, each star appears l_m times. Consider first the case $l_m = 1$. Now the set of wave vectors which arise in the symmetrized Kronecker square space can be written:

$$\boldsymbol{k}_\sigma + \boldsymbol{k}_\tau , \qquad \sigma \leqq \tau = 1, \ldots, s. \tag{56.15}$$

Let us rewrite (56.15) to conform with the table discussed in (56.7)–(56.9). Thus we get

$$\boldsymbol{k}_\sigma + \boldsymbol{k}_\tau \equiv \boldsymbol{\varphi}_\sigma \cdot (\boldsymbol{k} + \boldsymbol{\varphi}_\sigma^{-1} \cdot \boldsymbol{\varphi}_\tau \cdot \boldsymbol{k}), \qquad \sigma \leqq \tau = 1, \ldots, s. \tag{56.16}$$

Now clearly when $\sigma = \tau$ in (56.16), the elements of ${}^\star(2\boldsymbol{k})$ are obtained, where $2\boldsymbol{k} \equiv \boldsymbol{k} + \boldsymbol{k}$. Thus we write

$${}^\star\boldsymbol{k} \otimes {}^\star\boldsymbol{k} = {}^\star(2\boldsymbol{k}) \oplus \{\boldsymbol{k}_\sigma + \boldsymbol{k}_\tau\}, \qquad \sigma \neq \tau . \tag{56.17}$$

However, in the set $\{\boldsymbol{k}_\sigma + \boldsymbol{k}_\tau\}$ defined by (56.17) a given combination $\boldsymbol{k}_\sigma + \boldsymbol{k}_\tau$ will occur twice, owing to the unrestricted inequality in (56b.3). On the contrary, in (56.15) the inequality requires $\sigma < \tau$. Hence

$$\begin{aligned} &\{\boldsymbol{k}_\sigma + \boldsymbol{k}_\tau\} && \text{with } \sigma \neq \tau \\ &= 2\{\boldsymbol{k}_\sigma + \boldsymbol{k}_\tau\} && \text{with } \sigma < \tau . \end{aligned} \tag{56.18}$$

Writing

$$[{}^\star\boldsymbol{k}]_{(2)} \tag{56.19}$$

for the symmetrized square of the star $\star k$, which is the set (56.15), we evidently have

$$2[\star k]_{(2)} = \star k \otimes \star k \oplus \star(2k) \tag{56.20}$$

or

$$[\star k]_{(2)} = \tfrac{1}{2}[\star k \otimes \star k \oplus \star(2k)]. \tag{56.21}$$

Now, if we repeat the argument, for the general case with $l_m > 1$, we simply perform the appropriate grouping of wave vectors in the table, and obtain the result

$$[l_m \star k]_{(2)} = \tfrac{1}{2}[l_m^2 \star k \otimes \star k \oplus l_m \star(2k)] \tag{56.22}$$

of which (56.21) is evidently a special case with $l_m = 1$. The reduction coefficient appropriate to equation (56.22) can be written

$$([l_m \star k]_{(2)} | \star k''). \tag{56.23}$$

In a similar fashion, we may obtain the symmetrized cube of a star as

$$[l_m \star k]_{(3)} = \tfrac{1}{6}\{l_m^3 \star k \otimes \star k \otimes \star k \oplus 3 l_m^2 \star k \otimes \star(2k) \oplus 2 l_m \star(3k)\} \tag{56.24}$$

and define the associated reduction coefficient as

$$([l_m \star k]_{(3)} | \star k''). \tag{56.25}$$

The importance of the wave vector reduction coefficients or wave vector selection rules is that they delineate the only stars which arise in taking the direct product. That is, in reducing product $D^{(\star k)(m)} \otimes D^{(\star k')(m')}$ we need only seek terms $D^{(\star k'')(m'')}$ for which the associated, or corresponding, wave vector selection rule is non-vanishing. A necessary, but not sufficient condition for $D^{(\star k'')(m'')}$ to appear is that

$$(\star k \star k' | \star k'') \neq 0 \tag{56.26}$$

or the appropriate modification for ordinary and symmetrized powers of a specified representation. Of course, the appearance of a given $\star k''$ is necessary, but not sufficient for the appearance of a particular m'' in the reduction.

The wave vector reduction coefficients, or star reduction coefficients bear an analogy to the reduction coefficients in the Clebsch-Gordan series. Actually, the series (56.3), is the analogue for the discrete, finite, groups of the Clebsch-Gordan series for the continuous rotation group.

57. Determination of reduction coefficients: Method of linear algebraic equations. The complete reduction coefficients for products of space group irreducible representations are defined in (55.1), and we may now return to this equation to complete their determination. Eqs. (55.7) and (55.9) are of course analogous, and identical methods can be used, in their cases, too.

Now consider that in (55.1) the two factors $D^{(\star k)(m)}$ and $D^{(\star k')(m')}$ are specified. The corresponding star reduction coefficients (56.3) are assumed determined. Then for each $\star k''$ such that

$$(\star k \star k' | \star k'') \neq 0 \tag{57.1}$$

representations $D^{(\star k'')(m'')}$ do occur. Suppose that for this star there are $r_{k''}$ allowable irreducible representations, each of dimensionality $l_{k''m''}$. Then if, as before, s'' is the number of arms in $\star \boldsymbol{k}''$, the total dimensionality of the irreducible representations based on $\star \boldsymbol{k}''$ which will arise is:

$$(\star \boldsymbol{k} \star \boldsymbol{k}' | \star \boldsymbol{k}'') \cdot s''. \tag{57.2}$$

Then the conservation of dimensionality requires that the coefficients $(\star \boldsymbol{k}\, m \star \boldsymbol{k}'\, m' | \star \boldsymbol{k}''\, m'')$ satisfy

$$\sum_{m''=1}^{r_{k''}} (\star \boldsymbol{k}\, m \star \boldsymbol{k}'\, m' | \star \boldsymbol{k}''\, m'')\, l_{k''m''} = s'' (\star \boldsymbol{k} \star \boldsymbol{k}' | \star \boldsymbol{k}''). \tag{57.3}$$

Now if the number of distinct (i.e. non-equivalent) stars $\star \boldsymbol{k}''$ which occur in $(\star \boldsymbol{k} \star \boldsymbol{k}' | \star \boldsymbol{k}'')$ is ξ'' then the number of distinct coefficients to be determined is

$$\sum_{k''=1}^{\xi''} r_{k''}. \tag{57.4}$$

Hence, the Eq. (55.4) are

$$\chi^{(\star k)(m)}(\{\varphi_p | \boldsymbol{t}_p\})\, \chi^{(\star k')(m)}(\{\varphi_p | \boldsymbol{t}_p\}) = \sum_{\star k''\, m''} (\star \boldsymbol{k}\, m \star \boldsymbol{k}'\, m' | \star \boldsymbol{k}''\, m'')\, \chi^{(\star k'')(m'')}(\{\varphi_p | \boldsymbol{t}_p\}). \tag{57.5}$$

The Eq. (57.5) are now a set of linear algebraic equations in which only the reduction coefficients are unknown. That is, for each space group element in $\mathfrak{G}$ all the characters in (57.5) are known. While there are $g_p N$ space group elements, there are only a very much smaller number of unknown reduction coefficients (57.4).

The solution of (57.5) now depends upon choosing as many distinct space group elements, one from each class in $\mathfrak{G}$, as there are unknown reduction coefficients in (57.4). The space group elements may include coset representatives $\{\varphi_p | \tau_p\}$ as well as general space group elements $\{\varphi_p | \tau_p + \boldsymbol{R}_L\}$. It is easily seen that when Eq. (57.3) is used, to delimit the number of possible reduction coefficients at a given, permitted $\boldsymbol{k}''$, the task of solving the algebraic linear equations (57.5) is greatly simplified. Finally, one utilizes the property of the reduction coefficients as positive integers including zero to eliminate spurious possibilities for the solution of (57.5).

This method of obtaining reduction coefficients, based directly upon the fundamental definition of the reduction coefficients (55.1) or (55.3) is of complete generality and can always be used for any space group, symmorphic or non-symmorphic. It could be called the method of linear algebraic equations, and can as well be applied when any, or all, the stars in the reduction are of high symmetry, i.e. $\mathfrak{G}(\boldsymbol{k})$ is of high order, or are of low symmetry. The general theorem on the uniqueness of the reduction of a representation into irreducible components establishes that a solution of (57.5) and (57.3) is the unique, and correct one. As a check on the numerical work, one may test the Eq. (57.5) by using a different space group element than those employed in the initial process of reduction.

In Sects. 129, 130 this method will be applied to obtain reduction coefficients in rocksalt space group and specific illustrations will be given, to demonstrate the practicality of this method.

58. Determination of reduction coefficients: Method of the reduction group. The method of linear algebraic equations presented in Sect. 57 is, so to speak, the direct method of determining reduction coefficients. The converse method is based on the sum over the group as exemplified in (17.8), but now transcribed to the full space group as in (55.5). To sum over the entire group is only feasible when some representations of particularly simple structure are involved so that one can take advantage of their structure by working with certain factor groups. In particular, for stars of high symmetry, the matrix group $D^{(\star k)(m)}$ is a homomorphic image of the abstract group $\mathfrak{G}$ such that the kernel of the homomorphism $\mathfrak{K}$ is a group of high order (i.e. of order N) and the index of $\mathfrak{K}$ in $D^{(\star k)(m)}$ is a fairly small integer (i.e. of order g_p). Explicit discussion should make this point clear.

Call the collection of all matrices in $D^{(\star k)(m)}$ which are in the kernel

$$\mathfrak{K}^{(\star k)(m)} \equiv \{D^{(\star k)(m)}(Y)\}, \quad \text{where } D^{(\star k)(m)}(Y) = D^{(\star k)(m)}(\{\varepsilon|0\}). \tag{58.1}$$

This set of elements $\{Y\}$ evidently forms a normal subgroup of $\mathfrak{G}$. Then we can obtain the factor group of $\mathfrak{G}$ with respect to this subgroup. Call the subgroup

$$\mathfrak{N}_{\{Y\}} \equiv \{Y\} \tag{58.2}$$

which is the collection of abstract elements in $\mathfrak{G}$ such that in the given irreducible representation (matrix group) $D^{(\star k)(m)}$, each element Y corresponds to a matrix:

$$D^{(\star k)(m)}(Y) = D^{(\star k)(m)}(\{\varepsilon|0\}). \tag{58.3}$$

It will generally turn out that the elements in $\mathfrak{N}_{\{Y\}}$ are pure translations although only in the case $\star k = \Gamma$ will $\mathfrak{N}_{\{Y\}} = \mathfrak{T}$. Thus usually we can consider a decomposition of the translation group $\mathfrak{T}$ with respect to $\mathfrak{N}_{\{Y\}}$ which will produce a number of coset representatives in the factor series:

$$\mathfrak{T}/\mathfrak{N}_{\{Y\}} = \{\varepsilon|0\}, \{\varepsilon|t_2\}, \ldots, \{\varepsilon|t_{n_j}\}. \tag{58.4}$$

Note that the argument made here is quite similar to the simpler argument used in Sect. 39 when the little group technique was presented. In that case the group $\Pi(k)$ arose in (39.9), as the factor group of $\mathfrak{G}(k)/\mathfrak{T}(k)$ where $\mathfrak{T}(k)$ was defined as the set of abstract elements in $\mathfrak{T}$, such that in the representation, the matrices for elements in $\mathfrak{T}(k)$ were the same as those of the identity $\{\varepsilon|0\}$. Thus the matrices for the elements in $\mathfrak{T}(k)$ are in the center of the matrix group which is the acceptable irreducible representation of $\mathfrak{G}(k)$ i.e. in the center of $D^{(k)(m)}$. In the present case, it is the entire group $\mathfrak{G}$ whose irreducible representations $D^{(\star k)(m)}$ define $\mathfrak{N}_{\{Y\}}$ and hence the series (58.4).

Hence we form the factor group

$$\mathfrak{G}/\mathfrak{N}_{\{Y\}} \tag{58.5}$$

it will consist of a number of cosets. Each such independent coset in (58.5) is represented by a matrix in $D^{(\star k)(m)}$. The set of these matrices is closed: i.e. forms a matrix group. Now consider the coset representatives

$$\{\varepsilon|0\}, \ldots, \{\varepsilon|t_{n_j}\}, \quad \{\varphi_2|0\}, \ldots, \{\varphi_{g_p}|t_{n_j}\}. \tag{58.6}$$

The set of $g_p n_j$ coset representatives defines a set of "independent" group elements in $\mathfrak{G}$, insofar as the irreducible representation $D^{(\star k)(m)}$ of $\mathfrak{G}$ is concerned.

We call the collection of elements (58.6) $\mathfrak{R}(k)$ for "reduction group of k". Now, evidently $D^{(\star k)(m)}$ is a representation which is only a homomorphic image of $\mathfrak{G}$ but often an isomorphic image of $\mathfrak{R}(k)$. Thus, the set of matrices $D^{(\star k)(m)}(Z)$ where Z is an element in $\mathfrak{R}(k)$, forms a matrix group, closed under matrix multiplication. Further the set $D^{(\star k)(m)}(Z)$ is a matrix representation of the set of abstract elements $\mathfrak{R}(k)$ defined by (58.6). Then we can simplify the relation

$$\sum_{Z\in\mathfrak{G}} |\chi^{(\star k)(m)}(Z)|^2 = g_p N \tag{58.7}$$

which is a sum over all elements in the space group $\mathfrak{G}$, to

$$\sum_{Z\in\mathfrak{R}(k)} |\chi^{(\star k)(m)}(Z)|^2 = g_p n_j \tag{58.8}$$

The Eq. (58.8) is a completeness relation which arises from the property of the $D^{(\star k)(m)}$ as an irreducible representation of $\mathfrak{G}$.

Now in just the same way we may define the kernal of the representation $D^{(\star k')(m')}$ and call it $\mathfrak{K}'^{(\star k')(m')}$. We may also define the group $\mathfrak{R}(k')$ in the same fashion. The translations in $\mathfrak{R}(k')$ may or may not coincide with those in $\mathfrak{R}(k)$. Finally we may obtain a group $\mathfrak{R}(k'')$ which applies to the irreducible representation $D^{(\star k'')(m'')}$. Then we have three separate groups of matrices which are individually representations of the collections of abstract elements $\mathfrak{R}(k)$, $\mathfrak{R}(k')$, $\mathfrak{R}(k'')$. These matrix groups correspond to the essential (non-redundant) information in the complete matrix groups $D^{(\star k)(m)}$, $D^{(\star k')(m')}$, and $D^{(\star k'')(m'')}$ respectively. The first two of these are the factors in the product to be reduced and the third is the representation whose occurrance (m'') in the product is to be tested, when we know that $\star k''$ does appear, from the relevant wave vector reduction coefficient.

Now from the three groups $\mathfrak{R}(k)$, $\mathfrak{R}(k')$, $\mathfrak{R}(k'')$ we form a single group $\mathfrak{R}$ as follows. We augment each of these three collections of abstract elements by adding elements from its own center. In this way we find the least (smallest dimension) common group $\mathfrak{R}$. If for example the kernal of $D^{(\star k'')(m'')}$ is $\mathfrak{K}''^{(\star k'')(m'')}$ and the elements in $\mathfrak{K}''^{(\star k'')(m'')}$ are Y'' so that $\mathfrak{T}/\mathfrak{N}_{\{Y''\}}$ consists of n_j'' coset representatives and $n_{j''} > n_{j'}$ and $n_{j''} > n_j$ then we may have a case where we can define a common group $\mathfrak{R}$ as the collection of abstract elements

$$\mathfrak{G}/\mathfrak{N}_{\{Y''\}} = (\mathfrak{G}/\mathfrak{N}_{\{Y\}})\,(\mathfrak{N}_{(Y)}/\mathfrak{N}_{\{Y''\}}) \tag{58.9}$$

and then also

$$\mathfrak{G}/\mathfrak{N}_{\{Y''\}} = (\mathfrak{G}/\mathfrak{N}_{\{Y'\}})\,(\mathfrak{N}_{\{Y'\}}/\mathfrak{N}_{\{Y''\}}). \tag{58.10}$$

It may happen that the elements in $\mathfrak{N}_{\{Y\}}/\mathfrak{N}_{\{Y''\}}$ are in the center of $D^{(\star k)(m)}$ and those in $\mathfrak{N}_{\{Y'\}}/\mathfrak{N}_{\{Y''\}}$ are in the center of $D^{(\star k')(m')}$. In this case we may proceed immediately with the reduction.

With the group $\mathfrak{R}$, we possess a common reduction group. To be precise, this is a group of abstract elements for each of which we possess a matrix representation of the two factors $D^{(\star k)(m)}$ and $D^{(\star k')(m')}$ and of the product $D^{(\star k'')(m'')}$: As a result of their being irreducible we have the character relations

$$\sum_{Z\in\mathfrak{R}} |\chi^{(\star k)(m)}(Z)|^2 = g_p n_{j''} = \text{order of } \mathfrak{R} \tag{58.11}$$

for each of the representations as well as

$$\sum_{Z\in\mathfrak{R}} \chi^{(\star k)(m)}(Z)\,\chi^{(\star k')(m')}(Z)^* = \delta_{kk'}\,\delta_{mm'}\,g_p\,n_{j''}. \tag{58.12}$$

The orthonormality relation (58.12) must be understood to be restricted, in the present context, to any pair of the three stars $\star k$, $\star k'$, $\star k''$ and the associated m, m', m''; the element Z is any element in $\mathfrak{R}$. Now, with this orthonormality relation (58.12) we may complete the reduction using the equivalent of (55.5). Thus we have

$$(\star k\, m\, \star k'\, m' \mid \star k''\, m'') = \frac{1}{g_p\, n_{j''}} \sum_{Z\in\mathfrak{R}} \chi^{(\star k)(m)}(Z)\,\chi^{(\star k')(m')}(Z)\,\chi^{(\star k'')(m'')}(Z)^*. \tag{58.13}$$

The Eq. (58.13) completes the reduction. By the device of defining the reduction group $\mathfrak{R}$, common to the three matrix groups involved in the reduction to be carried out, we actually put the complete sum (55.5) into a form such that it can be carried out rigorously. This method of using the reduction group has the advantage of directly determining the reduction coefficients of interest, without any trial or error.

The modification in the procedure which is needed if the elements in $\mathfrak{N}_{\{Y\}}/\mathfrak{N}_{\{Y''\}}$ and $\mathfrak{N}_{\{Y'\}}/\mathfrak{N}_{\{Y''\}}$ are not in the centers of their respective representations are, in principle, simple. The three basic groups $\mathfrak{G}/\mathfrak{N}_{\{Y\}}$, $\mathfrak{G}/\mathfrak{N}_{\{Y'\}}$, and $\mathfrak{G}/\mathfrak{N}_{\{Y''\}}$ are each augmented with additional elements until a common reduction group $\mathfrak{R}$ is in fact achieved. Again, in this case (58.13) can be used and the reduction achieved.

In summary, the reduction group method is a rigorous device by which one takes advantage of the homorphism of $D^{(\star k)(m)}$ into the full space group $\mathfrak{G}$, and likewise for the other representations involved in the reduction, in order to be able to carry out the sum (55.6) for the determination of the reduction coefficients. When it can be applied conveniently, this method has the advantage of permitting one to make orthonormality crosschecks upon the coefficients. As in the simpler case of the little group method for finding irreducible representations of $\mathfrak{G}(k)$ by using the irreducible representations of $\Pi(k)$, the utility of the reduction group procedure seems restricted to cases of stars of high symmetry or equivalently, canonical wave vectors k of high symmetry.

The reduction group method is illustrated in Sect. 132, wherein reduction coefficients are obtained for products of representations of high symmetry in the diamond space group.

59. Determination of reduction coefficients: Use of basis functions. A partial determination of reduction coefficients can be achieved for those representations belonging to *one* fixed star in the framework of the full-group method by utilizing a complete set of basis functions for that star. That is, now assume that for the given product, whose reduction is required, we have determined all the wave vector or star reduction coefficients. Now with these, we turn to the reduction of the corresponding spaces. It is clear that in terms of the full group reduction coefficients $(\star k\, m\, \star k'\, m' \mid \star k''\, m'')$ the decomposition of the product space is

$$\Sigma^{(\star k)(m)} \otimes \Sigma^{(\star k')(m')} = \sum_{\star k''\, m''} \oplus (\star k\, m\, \star k'\, m' \mid \star k''\, m'')\, \Sigma^{(\star k'')(m'')}. \tag{59.1}$$

Now using the dimensionality arguments epitomized in the wave vector selection rules (57.3) we may consider the total dimensionality of all representations belonging to $\star k''$ as known; these arise from the product being reduced:

$$D^{(\star k)(m)} \otimes D^{(\star k')(m')}.$$

Let us assume that we possess the complete set of functions (59.1). From the complete set we select all those product pairs such that

$$k_{\sigma} + k'_{\sigma'} = k''_{\sigma''}, \tag{59.2}$$

where $k''_{\sigma''}$ is any one of the wave vectors in $\star k''$. In this manner we formally assemble the set of functions which span all the spaces $\{\Sigma^{(\star k'')(m'')}\}$ for the given $\star k''$ and all m'' which arise.

If we now apply to this set of eigenfunctions all the rotational symmetry operations (coset representatives in $\mathfrak{G}/\mathfrak{T}$) we may then obtain a complete reducible representation *at this star*. That is, let us call the space spanned by the set so chosen from $\Sigma^{(\star k'')}$:

$$\Sigma^{(\star k'')} \equiv \{\psi_{\alpha}^{(k_{\sigma})(m)} \psi_{\alpha'}^{(k'_{\sigma'})(m')}\}, \qquad k_{\sigma} + k'_{\sigma'} = k''_{\sigma''};\ \alpha = 1, \ldots, l_m;\ \alpha' = 1, \ldots, l_{m'}. \tag{59.3}$$

Then

$$P_{\{\varphi|t(\varphi)\}} \Sigma^{(\star k'')} = D(\{\varphi|t(\varphi)\}) \Sigma^{(\star k'')}. \tag{59.4}$$

But in general (if more than one m'' arises in the sum (59.4)) the matrix $D(\{\varphi|t(\varphi)\})$ is reducible. To carry out the reduction we should take the trace as

$$\operatorname{tr} D(\{\varphi|t(\varphi)\}) = \chi(\{\varphi|t(\varphi)\}). \tag{59.5}$$

Then with the reducible character system (59.5) available and also the character system for all the allowable irreducible representations $\chi^{(\star k)(m'')}$ at the fixed star $\star k''$ the reduction of the reducible character system (59.5) can be carried out to determine the coefficients $(\star k\, m\, \star k'\, m' | \star k''\, m'')$. Also in this case, methods which can be used are the method of linear algebraic equations (Sect. 57) and the reduction group method (Sect. 58).

An entirely equivalent procedure to determine which m'' arise can be formulated in terms of the idempotent projection operators relating to the irreducible representations $D^{(\star k'')(m'')}$. With the characters $\chi^{(\star k'')(m'')}(\{\varphi_p|t_p\})$ available, the character projection operator, or the weak operator $P^{(l)}$ of (16.5)

$$P^{(\star k'')(m'')} = \frac{1}{N g_p} \sum_{\mathfrak{G}} \chi^{(\star k'')(m'')}(\{\varphi_p|t_p\})^* P_{\{\varphi_p|t_p\}} \tag{59.6}$$

can be found. When this operator for given m'' is applied to the functions (59.3) the result will be either null, if m'' is not contained in (59.3), or the correct function. It is to be noted that the set of functions so found belongs to the entire space group irreducible representation $D^{(\star k)(m)}$, and not to a particular row.

The use of basis functions and projection operators is evidently desirable when one needs explicit enumeration of the functions spanning $\Sigma^{(\star k'')(m'')}$. As a rule, the partiality or incompleteness of the reduction achieved in this manner makes it useful either in specific cases, or as a check on other results. This method also is cumbersome insofar as one must determine partial character systems separately, and repeatedly for each allowed component $\star\boldsymbol{k}''$ of those which may arise in the reduction. Then each such is separately reduced. The other methods, by contrast, produce all the reduction coefficients at the same time rather than piecemeal. It will be noted later that the same remark will be made regarding certain subgroup methods.

60. Theory of Clebsch-Gordan coefficients for space groups*. As was discussed in Sect. 18, the work of finding reduction coefficients is only a partial solution to the problem of reducing the direct product of two irreducible representations. The final step is to obtain Clebsch-Gordan coefficients.

Despite the fact that crystal space groups have been known for more than 80 years [7] the task of finding reduction coefficients and Clebsch-Gordan coefficients for these groups has only been relatively recently pursued. In this section we shall present some of this recent work. The reader is urged to consult ongoing literature since it is to be expected that work on the theory, calculation, and applications of the Clebsch-Gordan coefficients for space groups will be moving rapidly ahead in the near future. In Sect. 134 the work of this section will be applied to a calculation of certain of the coefficients for the diamond structure.

In this section we shall follow the discussion given by ITZKAN [1] and BERENSON [1]. Similar discussion was presented by LITVIN and ZAK [2], SAULEVICH, SVIRIDOV, and SMIRNOV [3], SAKATA [4], and CORNWELL.[3]

α) *Definition of space group Clebsch-Gordan coefficients.* Clebsch-Gordan coefficients for space groups are defined just as in (18.4), or (18.15), for any finite group. We require the actual form of the representation matrices which we write following (36.18). In particular for element $D^{(\star\boldsymbol{k})(l)}_{\sigma a \tau b}$:

$$D^{(\star\boldsymbol{k})(l)}(\{\varphi|\boldsymbol{t}\})_{\sigma a \tau b} = \dot{D}^{(\boldsymbol{k})(l)}(\{\varphi_\sigma|\tau_\sigma\}^{-1}\cdot\{\varphi|\boldsymbol{t}\}\cdot\{\varphi_\tau|\tau_\tau\})_{ab} \tag{60.1}$$

where $D^{(\boldsymbol{k})(l)}$ is an irreducible representation of $\mathfrak{G}(\boldsymbol{k})$ and $\{\varphi_\sigma|\tau_\sigma\}$ and $\{\varphi_\tau|\tau_\tau\}$ are coset representatives in the decomposition of $\mathfrak{G}$ with respect to $\mathfrak{G}(\boldsymbol{k})$; recall the definition of the dotted matrix in (36.4). For the direct product we write:

$$D^{(\star\boldsymbol{k})(l)} \otimes D^{(\star\boldsymbol{k}')(l')} = \sum_{(\boldsymbol{k}''l'')} \oplus (\star\boldsymbol{k}\, l\, \star\boldsymbol{k}'\, l' | \star\boldsymbol{k}''\, l'')\, D^{(\star\boldsymbol{k}'')(l'')}. \tag{60.2}$$

The Clebsch-Gordan coefficients are defined by

$$\psi^{(\boldsymbol{k}''_{\sigma''})(l'')}_{a''} = \sum_{\sigma a} \sum_{\sigma' a'} \left(\begin{matrix} \boldsymbol{k} & l & \boldsymbol{k}' & l' \\ \sigma & a & \sigma' & a' \end{matrix}\middle| \begin{matrix} \boldsymbol{k}'' & l'' & \gamma \\ \sigma'' & a'' & \end{matrix}\right) \psi^{(\boldsymbol{k}_\sigma)(l)}_a\, \psi^{(\boldsymbol{k}'_{\sigma'})l'}_{a'}. \tag{60.3}$$

---

* A considerable portion of this section was written by Dr. RHODA BERENSON. See Sects. 18 and 134.

[1] I. ITZKAN: The Clebsch-Gordan Coefficients for the Crystallographic Space Groups. Ph. D. thesis, Physics Department, New York University, June 1969. – R. BERENSON: Theory of Crystal Clebsch-Gordan Coefficients. Ph. D. Thesis, New York University, February 1974. – R. BERENSON, I. ITZKAN and J.L. BIRMAN: J. Math. Phys. (to be published). Available from University Microfilms Inc., Ann Harbor, Mich.

[2] D.B. LITVIN and J. ZAK: J. Math. Phys. **9**, 212 (1969).

[3] B.K. NOVOSADOV, L.K. SAULEVICH, D.T. SVIRIDOV and Y.F. SMIRNOV: Soviet Phys. Doklady **14**, 50 (1969). – L.K. SAULEVICH, D.T. SVIRDOV, and Y.F. SMIRNOV: Soviet Phys. Cryst. **15**, 351, 355 (1970). – J.F. CORNWELL: phys. stat. solidi **37**, 225 (1970).

[4] I. SAKATA: J. Math. Phys. (in press) (1973).

It is convenient to write

$$U^{\gamma}_{\sigma a \sigma' a', \sigma'' a''} \equiv U^{(\boldsymbol{k} l \otimes \boldsymbol{k}' l')}_{\sigma a \sigma' a', \boldsymbol{k}'' l'' \gamma \sigma'' a''} \equiv \left(\begin{matrix} \boldsymbol{k} & l & \boldsymbol{k}' & l' \\ \sigma & a & \sigma' & a' \end{matrix}\middle| \begin{matrix} \boldsymbol{k}'' & l'' & \gamma \\ \sigma'' & a'' & \end{matrix}\right).$$

Following the discussion for Clebsch-Gordan coefficients of finite groups given in Sect. 18, we obtain the following equation which is analogous to Eq. (18.23).

$$\begin{aligned} &\sum_{\gamma} U^{\gamma}_{\sigma a \sigma' a', \sigma'' a''} U^{\gamma *}_{\bar{\sigma} \bar{a} \bar{\sigma}' \bar{a}', \bar{\sigma}'' \bar{a}''} \\ &= \frac{l''}{g} \sum_{x} D^{(\star \boldsymbol{k})(l)}(\{\varphi_x | \boldsymbol{t}_x\})_{\sigma a \bar{\sigma} \bar{a}} D^{(\star \boldsymbol{k}')(l')}(\{\varphi_x | \boldsymbol{t}_x\})_{\sigma' a' \bar{\sigma}' \bar{a}'} D^{(\star \boldsymbol{k}'')(l'')}(\{\varphi_x | \boldsymbol{t}_x\})^{*}_{\sigma'' a'' \bar{\sigma}'' \bar{a}''} \end{aligned} \tag{60.4}$$

where l'' is the dimension of $D^{(\star \boldsymbol{k}'')(l'')}$ and g is the order of the space group.

From (6.33) we can decompose the space group into cosets with respect to the translation invariant subgroup $\mathfrak{T}$ as:

$$\mathfrak{G} = \mathfrak{T} + \{\varphi_2 | \tau_2\} \mathfrak{T} + \cdots + \{\varphi_h | \tau_h\} \mathfrak{T}. \tag{60.5}$$

Hence, any element $\{\varphi | \boldsymbol{t}\}$ can be written as a product of a coset representative $\{\varphi | \tau\}$ and a pure translation $\{\varepsilon | \boldsymbol{R}_L\}$ so

$$\begin{aligned} D^{(\star \boldsymbol{k})(l)}(\{\varphi_x | \boldsymbol{t}_x\})_{\sigma a \sigma' a'} &= \sum_{\beta b} D^{(\star \boldsymbol{k})(l)}(\{\varphi_x | \tau_x\})_{\sigma a \beta b} D^{(\star \boldsymbol{k})(l)}(\{\varepsilon | \boldsymbol{R}_L\})_{\beta b \bar{\sigma} \bar{a}} \\ &= D^{(\star \boldsymbol{k})(l)}(\{\varphi_x | \tau_x\})_{\sigma a \beta b} \delta_{\beta \bar{\sigma}} \delta_{b \bar{a}} e^{-i \boldsymbol{k}_{\bar{\sigma}} \cdot \boldsymbol{R}_L} \\ &= e^{-i \boldsymbol{k}_{\bar{\sigma}} \cdot \boldsymbol{R}_L} D^{(\star \boldsymbol{k})(l)}(\{\varphi_x | \tau_x\})_{\sigma a \bar{\sigma} \bar{a}}. \end{aligned} \tag{60.6}$$

Eq. (60.4) can, therefore, be rewritten as

$$\begin{aligned} \sum U^{\gamma}_{\sigma a \sigma' a', \sigma'' a''} U^{\gamma *}_{\bar{\sigma} \bar{a} \bar{\sigma}' \bar{a}', \bar{\sigma}'' \bar{a}''} &= \frac{l''}{g} \sum_{\boldsymbol{R}_L} e^{-i(\boldsymbol{k}_{\bar{\sigma}} + \boldsymbol{k}'_{\bar{\sigma}'} - \boldsymbol{k}''_{\bar{\sigma}''}) \cdot \boldsymbol{R}_L} \\ &\times \sum_{X} D^{(\star \boldsymbol{k})(l)}(\{\varphi_x | \tau_x\})_{\sigma a \bar{\sigma} \bar{a}} D^{(\star \boldsymbol{k}')(l')}(\{\varphi_x | \tau_x\})_{\sigma' a' \bar{\sigma}' \bar{a}'} D^{(\star \boldsymbol{k}'')(a'')}(\{\varphi_x | \tau_x\})^{*}_{\sigma'' a'' \bar{\sigma}'' \bar{a}''}. \end{aligned} \tag{60.7}$$

Now recall (49.12), (49.13) so

$$\sum_{\boldsymbol{R}_L} \exp - i \boldsymbol{k} \cdot \boldsymbol{R}_L = N \Delta(\boldsymbol{k}) \tag{60.8 a}$$

where

$$\Delta(\boldsymbol{k}) = \begin{cases} 1 & \text{if } \boldsymbol{k} = 2\pi \boldsymbol{B}_H \\ 0 & \text{otherwise} \end{cases} \tag{60.8 b}$$

and N is the order of $\mathfrak{T}$ which also equals g/h. Therefore, the Clebsch-Gordan coefficients are zero unless

$$\boldsymbol{k}_{\bar{\sigma}} + \boldsymbol{k}'_{\bar{\sigma}'} - \boldsymbol{k}''_{\sigma''} = 2\pi \boldsymbol{B}_H. \tag{60.8 c}$$

If (60.8 c) is fullfilled then

$$\begin{aligned} &\sum_{\gamma} U^{\gamma}_{\sigma a \sigma' a', \sigma'' a''} U^{\gamma *}_{\bar{\sigma} \bar{a} \bar{\sigma}' \bar{a}', \bar{\sigma}'' \bar{a}''} \\ &= \frac{l''}{h} \sum_{x} D^{(\star \boldsymbol{k})(l)}(\{\varphi_x | \tau_x\})_{\sigma a \bar{\sigma} \bar{a}} D^{(\star \boldsymbol{k}')(l')}(\{\varphi_x | \tau_x\})_{\sigma' a' \bar{\sigma}' \bar{a}'} \\ &\quad \times D^{(\star \boldsymbol{k}'')(l'')}(\{\varphi_x | \tau_x\})^{*}_{\sigma'' a'' \bar{\sigma}'' \bar{a}''}. \end{aligned} \tag{60.9}$$

β) *Calculation of the* (111) *block of coefficients.* Since the selection of the canonical vectors of the various stars: $\boldsymbol{k}_1=\boldsymbol{k}$, $\boldsymbol{k}_1'=\boldsymbol{k}'$ and $\boldsymbol{k}_1''=\boldsymbol{k}''$ is arbitrary, it is convenient to choose $\boldsymbol{k}_1+\boldsymbol{k}_1'-\boldsymbol{k}_1''=2\pi\,\boldsymbol{B}_H$ so that for $\sigma=\bar{\sigma}=\sigma'=\bar{\sigma}'=\sigma''=\bar{\sigma}''=1$. Eq. (60.9) becomes

$$\begin{aligned} U_{1a1a',1a''}U^*_{1\bar{a}1\bar{a}',1\bar{a}''} &= \frac{l''}{h}\sum_x D^{(\star\boldsymbol{k})(l)}(\{\varphi_x|\tau_x\})_{1a1\bar{a}}D^{(\star\boldsymbol{k}')(l')}(\{\varphi_x|\tau_x\})_{1a'1\bar{a}'}\\ &\quad\times D^{(\star\boldsymbol{k}'')(l'')*}(\{\varphi_x|\tau_x\})_{1a''1\bar{a}''}\\ &=\frac{l''}{h}\sum_x \dot{D}^{(\boldsymbol{k})(l)}(\{\varphi_x|\tau_x\})_{a\bar{a}}\dot{D}^{(\boldsymbol{k}')(l')}(\{\varphi_x|\tau_x\})_{a'\bar{a}'}\\ &\quad\times\dot{D}^{(\boldsymbol{k}'')(l'')*}(\{\varphi_x|\tau_x\})_{a''\bar{a}''}. \end{aligned}\tag{60.10}$$

Where x will be summed over those elements which simultaneously belong to $\mathfrak{G}(\boldsymbol{k})$, $\mathfrak{G}(\boldsymbol{k}')$ and $\mathfrak{G}(\boldsymbol{k}'')$. Actually the dot on the matrix includes this restriction.

For simplicity of notation Eq. (60.10) has been written for the case of no multiplicity. The general case including multiplicity can also be analysed similarly[5] but we will not give details here. The Clebsch-Gordan coefficients defined in Eq. (60.10) will be called the (111) block of coefficients. They are determined exactly as for the case given in Sect. 18, Eqs. (18.15) *et seq.* that is, we select a particular $a=\bar{a}=a_0$, $a'=\bar{a}'=a_0'$ and $a''=\bar{a}''=a_0''$ so that the right hand side of Eq. (60.10) is non-zero. We then fix the phase of $U_{1a_0 1a_0',1a_0''}$ as real, keep $\bar{a}=a_0$, $\bar{a}'=a_0'$, and $\bar{a}''=a_0''$ and let a, a' and a'' take on all allowable values. If

$$(\boldsymbol{k}\,l\,\boldsymbol{k}'\,l'|\boldsymbol{k}''\,l'')>1$$

we repeat this procedure until all (111) blocks of coefficients are determined.

γ) *Calculation of the* $(\sigma\sigma'\sigma'')$ *block of coefficients.* To evaluate the Clebsch-Gordan coefficients for the $(\sigma\sigma'\sigma'')$ block, we proceed as follows: Select a non-zero element $U_{\sigma\bar{a}\sigma'\bar{a}',\sigma''\bar{a}''}$ and call it U'. Then we have

$$\begin{aligned} U'^*\,U_{\sigma a\sigma' a'\sigma'' a''} &=\frac{l''}{L}\sum_X D^{(\star\boldsymbol{k})(l)}(\{\varphi_x|\tau_x\})_{\sigma a\sigma\bar{a}}D^{(\star\boldsymbol{k}')(l')}(\{\varphi_x|\tau_x\})_{\sigma' a'\sigma'\bar{a}'}\\ &\quad\times D^{(\star\boldsymbol{k}'')(l'')*}(\{\varphi_x|\tau_x\})_{\sigma'' a''\sigma''\bar{a}''}\end{aligned}\tag{60.11}$$

where X is summed over those elements that simultaneously belong to $\mathfrak{G}(\boldsymbol{k}_\sigma)$, $\mathfrak{G}(\boldsymbol{k}'_{\sigma'})$ and $\mathfrak{G}(\boldsymbol{k}''_{\sigma''})$:

$$\varphi_x\boldsymbol{k}_\sigma\doteq\boldsymbol{k}_\sigma;\qquad \varphi_x\boldsymbol{k}'_{\sigma'}\doteq\boldsymbol{k}'_{\sigma'};\qquad \varphi_x\boldsymbol{k}''_{\sigma''}\doteq\boldsymbol{k}''_{\sigma''}\tag{60.12}$$

where $\doteq$ indicates equality modulo $2\pi\,\boldsymbol{B}_H$.

Now consider the element φ_Σ which rotates the (111) block into to $(\sigma\sigma'\sigma'')$ block so that

$$\varphi_\Sigma\boldsymbol{k}\doteq\boldsymbol{k}_\sigma;\qquad \varphi_\Sigma\boldsymbol{k}'\doteq\boldsymbol{k}'_{\sigma'};\qquad \varphi_\Sigma\boldsymbol{k}''\doteq\boldsymbol{k}''_{\sigma''}.\tag{60.13}$$

Then

$$\varphi_\Sigma^{-1}\varphi_x\varphi_\Sigma\boldsymbol{k}\doteq\boldsymbol{k};\qquad \varphi_\Sigma^{-1}\varphi_x\varphi_\Sigma\boldsymbol{k}'\doteq\boldsymbol{k}';\qquad \varphi_\Sigma^{-1}\varphi_x\varphi_\Sigma\boldsymbol{k}''\doteq\boldsymbol{k}''.\tag{60.14}$$

[5] I. ITZKAN: Ph. D. Thesis. New York University 1969. — R. BERENSON: Ph. D. Thesis. New York University 1974. See footnote 1.

Therefore, $\varphi_\Sigma^{-1}\varphi_x\varphi_\Sigma$ belongs simultaneously to $\mathfrak{G}(\boldsymbol{k})$, $\mathfrak{G}(\boldsymbol{k}')$ and $\mathfrak{G}(\boldsymbol{k}'')$. Let $\varphi_y=\varphi_\Sigma^{-1}\varphi_x\varphi_\Sigma$ or $\varphi_x=\varphi_\Sigma\varphi_y\varphi_\Sigma^{-1}$. Then

$$\begin{aligned}D^{(\star\boldsymbol{k})(l)}&(\{\varphi_x|\tau_x\})_{\sigma a\sigma\bar{a}}\\&=D^{(\star\boldsymbol{k})(l)}(\{\varphi_\Sigma|\tau_\Sigma\}\{\varphi_y|\tau_y\}\{\varphi_\Sigma|\tau_\Sigma\}^{-1})_{\sigma a\sigma\bar{a}}\\&=\dot{D}^{(\boldsymbol{k})(l)}(\{\varphi_\sigma|\tau_\sigma\}^{-1}\{\varphi_\Sigma|\tau_\Sigma\}\{\varphi_y|\tau_y\}\{\varphi_\Sigma|\tau_\Sigma\}^{-1}\{\varphi_\sigma|\tau_\sigma\})_{a\bar{a}}.\end{aligned}\tag{60.15}$$

But since $\varphi_\Sigma\boldsymbol{k}\doteq\boldsymbol{k}_\sigma$, $\{\varphi_\Sigma|\tau_\Sigma\}$ must be in the σ-th coset in the decomposition of $\mathfrak{G}$ with respect to $\mathfrak{G}(\boldsymbol{k})$ so we can write $\{\varphi_\Sigma|\tau_\Sigma\}=\{\varphi_\sigma|\tau_\sigma\}\{\varphi_k|\boldsymbol{t}_k\}$ where $\{\varphi_k|\boldsymbol{t}_k\}\in\mathfrak{G}(\boldsymbol{k})$. Eq. (60.15) then can be written

$$\begin{aligned}D^{(\star\boldsymbol{k})(l)}&(\{\varphi_x|\tau_x\})_{\sigma a\sigma\bar{a}}\\&=D^{(\boldsymbol{k})(l)}(\{\varphi_k|\boldsymbol{t}_k\}\{\varphi_y|\tau_y\}\{\varphi_k|\boldsymbol{t}_k\}^{-1})_{a\bar{a}}\\&=[D^{(\boldsymbol{k})(l)}(\{\varphi_k|\boldsymbol{t}_k\})\,D^{(\boldsymbol{k})(l)}(\{\varphi_y|\tau_y\})\,D^{(\boldsymbol{k})(l)}(\{\varphi_k|\boldsymbol{t}_k\}^{-1})]_{a\bar{a}}.\end{aligned}\tag{60.16}$$

Now for space group representations

$$D^{(\boldsymbol{k})(l)}(\{\varphi_k|\boldsymbol{t}_k\}^{-1})=\omega_{kk^{-1}}\big(D^{(\boldsymbol{k})(l)}(\{\varphi_k|\boldsymbol{t}_k\})\big)^{-1}=D^{(\boldsymbol{k})(l)}(\{\varphi_k|\boldsymbol{t}_k\})^{-1}.\tag{60.17}$$

Therefore, since the ray factor $\omega_{k,k^{-1}}=1$ for space groups we can write

$$D^{(\star\boldsymbol{k})(l)}(\{\varphi_x|\tau_x\})_{\sigma a\sigma\bar{a}}=\bar{D}^{(\boldsymbol{k})(l)}(\{\varphi_y|\tau_y\})_{a\bar{a}}.\tag{60.18}$$

$\bar{D}^{(\boldsymbol{k})(l)}$ is a similarity transformed representation of $D^{(\boldsymbol{k})(l)}$:

$$\bar{D}^{(\boldsymbol{k})(l)}=D^{(\boldsymbol{k})(l)}(\{\varphi_k|\boldsymbol{t}_k\})\,D^{(\boldsymbol{k})(l)}\,D^{(\boldsymbol{k})(l)}(\{\varphi_k|\boldsymbol{t}_k\})^{-1}.$$

Using the same procedure for $D^{(\star\boldsymbol{k}')(l')}(\{\varphi_x|\tau_x\})$ and $D^{(\star\boldsymbol{k}'')(l'')}(\{\varphi_x|\tau_x\})$ Eq. (60.11) becomes

$$\begin{aligned}U'^{*}\,U_{\sigma a\sigma' a',\sigma'' a''}=\frac{l''}{g}\sum_{Y}&\bar{D}^{(\boldsymbol{k})(l)}(\{\varphi_y|\tau_y\})_{a\bar{a}}\bar{D}^{(\boldsymbol{k}')(l')}(\{\varphi_y|\tau_y\})_{a'\bar{a}'}\\&\times\bar{D}^{(\boldsymbol{k}'')(l'')*}(\{\varphi_y|\tau_y\})_{a''\bar{a}''}.\end{aligned}\tag{60.19}$$

But this is just the equation for the coefficients for the (111) block for similarity transformed representations $\bar{D}^{(\boldsymbol{k})(l)}$, $\bar{D}^{(\boldsymbol{k}')(l')}$ and $\bar{D}^{(\boldsymbol{k}'')(l'')}$. These are then related to the coefficients for the (111) block for $D^{(\boldsymbol{k})(l)}$, $D^{(\boldsymbol{k}')(l')}$ and $D^{(\boldsymbol{k}'')(l'')}$ by Eq. (18.29) of Chap. 18.

$$\begin{aligned}U_{\sigma a\sigma' a'\sigma'' a''}=\sum_{bb'b''}&[D^{(\boldsymbol{k})(l)}(\{\varphi_k|\boldsymbol{t}_k\})\otimes D^{(\boldsymbol{k}')(l')}(\{\varphi_{k'}|\boldsymbol{t}_{k'}\})]_{aa',bb'}\\&\times U_{1b1b',1b''}D^{(\boldsymbol{k}'')(l'')}(\{\varphi_{k''}|\boldsymbol{t}_{k''}\})^{-1}_{b'',a''}\end{aligned}\tag{60.20}$$

or

$$U^{\sigma\sigma'\sigma''}=D^{(\boldsymbol{k})(l)}(\{\varphi_k|\boldsymbol{t}_k\})\otimes D^{(\boldsymbol{k}')(l)}(\{\varphi_{k'}|\boldsymbol{t}_{k'}\})\,U'''\,D^{(\boldsymbol{k}'')(l'')}(\{\varphi_{k''}|\boldsymbol{t}_{k''}\})^{-1}.\tag{60.21}$$

$\{\varphi_k|\boldsymbol{t}_k\}$, $\{\varphi_{k'}|\boldsymbol{t}_{k'}\}$ and $\{\varphi_{k''}|\boldsymbol{t}_{k''}\}$ must satisfy

$$\begin{aligned}\{\varphi_\Sigma|\tau_\Sigma\}&=\{\varphi_\sigma|\tau_\sigma\}\{\varphi_k|\boldsymbol{t}_k\}\\\{\varphi_\Sigma|\tau_\Sigma\}&=\{\varphi_{\sigma'}|\tau_{\sigma'}\}\{\varphi_{k'}|\boldsymbol{t}_{k'}\}\\\{\varphi_\Sigma|\tau_\Sigma\}&=\{\varphi_{\sigma''}|\tau_{\sigma''}\}\{\varphi_{k''}|\boldsymbol{t}_{k''}\}.\end{aligned}\tag{60.22}$$

To summarize, the complete set of space group Clebsch-Gordan coefficients is determined by using Eq. (60.10) for the (111) block and then using Eq. (60.21) to find the other blocks where Eqs. (60.13) and (60.22) determine $\{\varphi_k|\boldsymbol{t}_k\}$, $\{\varphi_{k'}|\boldsymbol{t}_{k'}\}$ and $\{\varphi_{k''}|\boldsymbol{t}_{k''}\}$. Note that we do not require the entire space group representation $D^{(\star \boldsymbol{k})(l)}$ but only the representations of $\mathfrak{G}(\boldsymbol{k})$, $\mathfrak{G}(\boldsymbol{k}')$ and $\mathfrak{G}(\boldsymbol{k}'')$, in this method. Consequently this technique for obtaining the Clebsch-Gordan coefficients is close to the sub-group methods described below in Sects. 61–65. The reader may wish to return to the material in this section after reading Sect. 65.

G. Reduction coefficients for space groups: Subgroup methods

61. Introduction. Now we follow some conventional usage in discussing alternate methods for obtaining reduction coefficients for the space groups. These have been called subgroup methods.

The basic distinction between full-group and sub-group methods concerns their focal points of interest in regard to the manipulations needed to determine the reduction coefficients. The full-group methods discussed in Sects. 52–60 deal with the entire space group $\mathfrak{G}$, and its irreducible representations $D^{(\star \boldsymbol{k})(m)}$, and uses these to determine the relevant reduction coefficients. The subgroup methods focus as much as possible on the single subgroup $\mathfrak{G}(\boldsymbol{k})$ and its allowable irreducible representations $D^{(\boldsymbol{k})(m)}$. Since the $D^{(\star \boldsymbol{k})(m)}$ is constructed by induction from $D^{(\boldsymbol{k})(m)}$, it must be possible to obtain an equivalent set of results either way if either approach is carried through to a successful conclusion.

In the following sections we need to examine the details of carrying out subgroup type calculations and then contrast these with the full-group procedure.

62. Complete subgroup character system. In the subgroup method we focus upon a *particular* resultant star: $\star \boldsymbol{k}''$ in which we are interested, in the reduction and further, upon a particular wave vector in $\star \boldsymbol{k}''$. Let this be the canonical wave vector $\boldsymbol{k}''$ which specifies $\star \boldsymbol{k}''$. The question is asked: given the appropriate pieces of the two factors $D^{(\star \boldsymbol{k})(m)}$ and $D^{(\star \boldsymbol{k}')(m')}$, which m'' will arise, for specified $\boldsymbol{k}''$, when the reduction is effected? To analyze this problem requires that the entire discussion of the previous sections be recast into a form in which particular subgroups of $\mathfrak{G}$, and particular block matrices and characters and particular vector subspaces play a dominant rôle.

Thus consider the representation $D^{(\star \boldsymbol{k})(m)}$. Since we shall only be concerned with characters, in obtaining selection rules, we only need consider the diagonal bloc matrices in this representation. Observe that for a given, arbitrary element $\{\varphi_p|\boldsymbol{t}_p\}$ in $\mathfrak{G}$, the diagonal bloc matrices are the dotted matrices:

$$D^{(\star \boldsymbol{k})(m)}(\{\varphi_p|\boldsymbol{t}_p\})_{\sigma\sigma} = \dot{D}^{(\boldsymbol{k})(m)}(\{\varphi_\sigma|\boldsymbol{t}_\sigma\}^{-1}\cdot\{\varphi_p|\boldsymbol{t}_p\}\cdot\{\varphi_\sigma|\boldsymbol{t}_\sigma\}). \tag{62.1}$$

Thus $\dot{D}^{(\boldsymbol{k})(m)}(\{\varphi_p|\boldsymbol{t}_p\})$ "belongs" to the rows and columns whose corresponding Bloch vectors have wave vectors $\boldsymbol{k}$, while $\dot{D}^{(\boldsymbol{k})(m)}(\{\varphi_\sigma|\boldsymbol{t}_\sigma\}^{-1}\cdot\{\varphi_p|\boldsymbol{t}_p\}\cdot\{\varphi_\sigma|\boldsymbol{t}_\sigma\})$ "belongs" to the rows and columns whose corresponding Bloch vectors have wave vectors $\boldsymbol{k}_\sigma$. The same applies, for the full representations $D^{(\star\boldsymbol{k}')(m')}$ and $D^{(\star\boldsymbol{k}'')(m'')}$. Likewise, we may take any single term (character) in (37.3) as belonging to a particular wave vector. In this way, we may look at a given irreducible representation $D^{(\star\boldsymbol{k})(m)}$ in the varied forms assumed when this representation is "projected" into the subspaces with different wave vectors which belong to $\star\boldsymbol{k}$.

Now consider the groups $\mathfrak{G}(\boldsymbol{k}')$ and $\mathfrak{G}(\boldsymbol{k}'')$ which are relevant to the canonical wave vectors $\boldsymbol{k}'$ and $\boldsymbol{k}''$ of the stars $\star\boldsymbol{k}'$ and $\star\boldsymbol{k}''$ respectively. We write

$$\mathfrak{G}(\boldsymbol{k}')=\mathfrak{T}+\cdots+\{\varphi_{l_{\lambda'}}|\tau_{l_{\lambda'}}\}\,\mathfrak{T}+\cdots+\{\varphi_{l_{k'}}|\tau_{l_{k'}}\}\,\mathfrak{T} \tag{62.2}$$

and decompose $\mathfrak{G}$ into cosets relative to $\mathfrak{G}(\boldsymbol{k}')$ as:

$$\mathfrak{G}=\mathfrak{G}(\boldsymbol{k}')+\cdots+\{\varphi_{\sigma'}|\tau_{\sigma'}\}\,\mathfrak{G}(\boldsymbol{k}')+\cdots+\{\varphi_{s'}|\tau_{s'}\}\,\mathfrak{G}(\boldsymbol{k}'). \tag{62.3}$$

Also

$$\mathfrak{G}(\boldsymbol{k}'')=\mathfrak{T}+\cdots+\{\varphi_{l_{\lambda''}}|\tau_{l_{\lambda''}}\}\,\mathfrak{T}+\cdots+\{\varphi_{l_{k''}}|\tau_{l_{k''}}\}\,\mathfrak{T} \tag{62.4}$$

and

$$\mathfrak{G}=\mathfrak{G}(\boldsymbol{k}'')+\cdots+\{\varphi_{\sigma''}|\tau_{\sigma''}\}\,\mathfrak{G}(\boldsymbol{k}'')+\cdots+\{\varphi_{s''}|\tau_{s''}\}\,\mathfrak{G}(\boldsymbol{k}''). \tag{62.5}$$

Observe the notational conventions adopted in (62.2)–(62.5) and compare (36.1), (36.2). Let $\boldsymbol{k}_\sigma$ and $\boldsymbol{k}'_{\sigma'}$ be any two wave vectors from $\star\boldsymbol{k}$ and $\star\boldsymbol{k}'$ respectively, such that

$$\boldsymbol{k}_\sigma+\boldsymbol{k}'_{\sigma'}=\boldsymbol{k}''(\mathrm{mod}\;2\pi\boldsymbol{B}_H) \tag{62.6}$$

(in general $\boldsymbol{k}_\sigma$ and $\boldsymbol{k}'_{\sigma'}$ will not be the canonical wave vectors of their stars if $\boldsymbol{k}''$ is). Next, consider the totality of independent vectors $\boldsymbol{k}_\sigma$ and $\boldsymbol{k}'_{\sigma'}$ which can combine to produce the fixed $\boldsymbol{k}''$. These may be obtained by applying to the given $\boldsymbol{k}_\sigma$ in (62.1) all the rotational elements in $\mathfrak{G}(\boldsymbol{k}'')$ to obtain

$$\varphi_{l_{\lambda''}}\cdot\boldsymbol{k}_\sigma,\qquad \lambda''=1,\ldots,k''. \tag{62.7}$$

However for each element $\{\varphi_{l_{\lambda''}}|\boldsymbol{t}_{l_{\lambda''}}\}$ common to $\mathfrak{G}(\boldsymbol{k}'')$ and $\mathfrak{G}(\boldsymbol{k}_\sigma)$

$$\varphi_{l_{\lambda''}}\cdot\boldsymbol{k}_\sigma=\boldsymbol{k}_\sigma(\mathrm{mod}\;2\pi\boldsymbol{B}_H) \tag{62.8}$$

and we do not obtain a distinct $\boldsymbol{k}_\sigma$. Clearly, as λ'' ranges from $1,\ldots,k''$ in (62.7) we obtain $(l_{k''}/n_{k''k_\sigma})$ distinct vectors. For each such distinct wave vector in $\star\boldsymbol{k}$ there must be a corresponding distinct wave vector in $\star\boldsymbol{k}'$ since if (62.8) applies, then

$$\varphi_{l_{\lambda''}}\cdot\boldsymbol{k}'_{\sigma'}=\boldsymbol{k}'_{\sigma'}(\mathrm{mod}\;2\pi\boldsymbol{B}_H) \tag{62.9}$$

and $\{\varphi_{l_{\lambda''}}|\boldsymbol{t}_{l_{\lambda''}}\}$ is also in $\mathfrak{G}(\boldsymbol{k}'_{\sigma'})$. The integer $n_{k''k_\sigma}$ is the order of the group which is the intersection of the factor groups $\mathfrak{G}(\boldsymbol{k}'')/\mathfrak{T}$ and $\mathfrak{G}(\boldsymbol{k}_\sigma)/\mathfrak{T}$, or, the number of rotational elements in common to both factor groups.

Now return to the expression (53.8) for the character of an element $\{\varphi_p|\boldsymbol{t}_p\}$ in $\mathfrak{G}$ in the direct product representation:

$$\chi^{(\star\boldsymbol{k}\otimes\star\boldsymbol{k}')(m\otimes m')}(\{\varphi_p|\boldsymbol{t}_p\})=\chi^{(\star\boldsymbol{k})(m)}(\{\varphi_p|\boldsymbol{t}_p\})\cdot\chi^{(\star\boldsymbol{k}')(m')}(\{\varphi_p|\boldsymbol{t}_p\}). \tag{62.10}$$

The translation $\boldsymbol{t}_p$ in (62.10) may be general: lattice plus fractional. Then (62.10) may be expressed in terms of the dotted characters:

$$\begin{aligned}\chi^{(\star \boldsymbol{k})(m)}(\{\varphi_p|\boldsymbol{t}_p\})\cdot\chi^{(\star \boldsymbol{k}')(m')}(\{\varphi_p|\boldsymbol{t}_p\}) \\ =\sum_{\sigma=1}^{s}\sum_{\sigma'=1}^{s'}\dot{\chi}^{(\boldsymbol{k})(m)}(\{\varphi_\sigma|\boldsymbol{t}_\sigma\}^{-1}\cdot\{\varphi_p|\boldsymbol{t}_p\}\cdot\{\varphi_\sigma|\boldsymbol{t}_\sigma\}) \\ \times\dot{\chi}^{(\boldsymbol{k}')(m')}(\{\varphi_{\sigma'}|\boldsymbol{t}_{\sigma'}\}^{-1}\cdot\{\varphi_p|\boldsymbol{t}_p\}\cdot\{\varphi_{\sigma'}|\boldsymbol{t}_{\sigma'}\}).\end{aligned} \tag{62.11}$$

We now consider the subset of all products selected from (62.11) such that the sum of the wave vectors of the factors satisfies (62.5). One manner of writing this object is

$$\sum_{\sigma}\sum_{\sigma'}\dot{\chi}^{(\boldsymbol{k})(m)}(\{\varphi_p^{\sigma}\})\,\dot{\chi}^{(\boldsymbol{k}')(m')}(\{\varphi_p^{\sigma'}\})\,\delta(\boldsymbol{k}''-\boldsymbol{k}_\sigma-\boldsymbol{k}'_{\sigma'}), \tag{62.12}$$

where

$$\{\varphi_p^{\sigma}\}\equiv\{\varphi_\sigma|\boldsymbol{t}_\sigma\}^{-1}\cdot\{\varphi_p|\boldsymbol{t}_p\}\cdot\{\varphi_\sigma|\boldsymbol{t}_\sigma\} \tag{62.13}$$

and similarly for σ', and

$$\delta(\boldsymbol{k}''-\boldsymbol{k}_\sigma-\boldsymbol{k}'_{\sigma'})=\begin{cases}1 & \text{if } \boldsymbol{k}_\sigma+\boldsymbol{k}'_{\sigma'}=\boldsymbol{k}''\,(\text{mod }2\pi\,\boldsymbol{B}_H)\\ 0 & \text{otherwise}.\end{cases} \tag{62.14}$$

A better way of writing (62.12) for our purposes is

$$\left(\frac{1}{n_{k''k_\sigma}}\right)\sum_{\lambda''=1}^{k''}\dot{\chi}^{(\boldsymbol{k})(m)}(\{\varphi_p^{l_{\lambda''}\sigma}\})\,\dot{\chi}^{(\boldsymbol{k}')(m')}(\{\varphi_p^{l_{\lambda''}\sigma'}\}), \tag{62.15}$$

where $(n_{k''k_\sigma})$ is the redundancy factor previously defined, and

$$\{\varphi_p^{l_{\lambda''}\sigma}\}\equiv(\{\varphi_{l_{\lambda''}}|\boldsymbol{t}_{l_{\lambda''}}\}\cdot\{\varphi_\sigma|\boldsymbol{t}_\sigma\})^{-1}\cdot\{\varphi_p|\boldsymbol{t}_p\}\cdot(\{\varphi_{l_{\lambda''}}|\boldsymbol{t}_{l_{\lambda''}}\}\cdot\{\varphi_\sigma|\boldsymbol{t}_\sigma\}) \tag{62.16}$$

is the conjugate indicated in (62.16). The sum in (62.15) goes over all rotational elements in $\mathfrak{G}(\boldsymbol{k}'')/\mathfrak{T}$.

The sum (62.15) represents a character system which is "complete" in the sense of the subgroup analysis. All possible contributions to the character of all representations m'' based on the specified wave vector $\boldsymbol{k}''$ which can be obtained by taking direct products of appropriate parts of representations $D^{(\star\boldsymbol{k})(m)}$ and $D^{(\star\boldsymbol{k}')(m)}$ are contained in (62.15).

63. Subgroup reduction coefficients. Since the character system (62.15) presents a complete character system for a direct sum of allowable irreducible representations of $\mathfrak{G}(\boldsymbol{k}'')$, it can be reduced, following Maschke's theorem. To differentiate this case from the former cases in which we reduced direct products of entire representations, we define a special symbol: the subgroup reduction coefficient $(\{\boldsymbol{k}_\sigma+\boldsymbol{k}'_{\sigma'}\}\,m m'|\boldsymbol{k}''\,m'')$. The definition is, from (62.15):

$$\begin{aligned}\frac{1}{(n_{k''k_\sigma})}\sum_{\lambda''=1}^{k''}\dot{\chi}^{(\boldsymbol{k})(m)}(\{\varphi_p^{l_{\lambda''}\sigma}\})\,\dot{\chi}^{(\boldsymbol{k}')(m')}(\{\varphi_p^{l_{\lambda''}\sigma'}\}) \\ =\sum_{m''}(\{\boldsymbol{k}_\sigma+\boldsymbol{k}'_{\sigma'}\}\,m m'|\boldsymbol{k}''\,m'')\,\dot{\chi}^{(\boldsymbol{k}'')(m'')}(\{\varphi_p|\boldsymbol{t}_p\}).\end{aligned} \tag{63.1}$$

The symbol $(\{\boldsymbol{k}_\sigma+\boldsymbol{k}'_{\sigma'}\}\, m\, m' | \boldsymbol{k}''\, m'')$ is intended to convey the idea that we combine all possible $\boldsymbol{k}_\sigma$ and $\boldsymbol{k}'_{\sigma'}$ to obtain the given $\boldsymbol{k}''$. As in the full group method, Eq. (63.1) can be solved directly by treating a suitable set of linear inhomogeneous equations (one for each element) and finding their solution. Alternatively we may use the orthonormality and completeness of the character system for the group $\mathfrak{G}(\boldsymbol{k}'')$, or $\mathfrak{G}(\boldsymbol{k}'')/\mathfrak{T}(\boldsymbol{k}'')$, to obtain:

$$(\{\boldsymbol{k}_\sigma+\boldsymbol{k}'_{\sigma'}\}\, m\, m' | \boldsymbol{k}''\, m'')$$
$$=\left(\frac{1}{g(k'')}\right)\left(\frac{1}{n_{k''\,k_\sigma}}\right)\sum_{p}\sum_{\lambda''=1}^{k''}\dot{\chi}^{(\boldsymbol{k})(m)}(\{\varphi_p^{l_{\lambda''}\sigma}\})\,\dot{\chi}^{(k')(m')}(\{\varphi_p^{l_{\lambda''}\sigma'}\})\,\dot{\chi}^{(\boldsymbol{k}'')(m'')}(\{\varphi_p|\boldsymbol{t}_p\})^*. \tag{63.2}$$

The sum in (63.2) over p may be taken over all group elements in $\mathfrak{G}$, then the dotted characters provide all needed restrictions, and $g(k'')$ is the order of either $\mathfrak{G}(\boldsymbol{k}'')$ or $\mathfrak{G}(\boldsymbol{k}'')/\mathfrak{T}(\boldsymbol{k}'')$, depending on whether or not all translations are included in the sum.

Observe the peculiar structure of the arguments (symmetry elements) of the various characters in (63.2). It is important to note that (63.2) is an equation which basically derives from matrices all of which represent the identical single abstract symmetry element $\{\varphi_p|\boldsymbol{t}_p\}$. Those are full group matrices, and their characters are as in (37.3).

Now we ask the following question. Choose a different wave vector in $\star\boldsymbol{k}''$ upon which to base the analysis; what is the relationship between the subgroup reduction coefficient so obtained, and (63.2)? We consider the chosen wave vector $\boldsymbol{k}''_{\sigma''}\equiv\varphi_{\sigma''}\cdot\boldsymbol{k}''$ which is evidently in $\star\boldsymbol{k}''$. Then if $\boldsymbol{k}_\sigma$ and $\boldsymbol{k}'_{\sigma'}$ are a pair of wave vectors which satisfy (62.6), then

$$\varphi_{\sigma''}\cdot(\boldsymbol{k}_\sigma+\boldsymbol{k}'_{\sigma'})=\varphi_{\sigma''}\cdot\boldsymbol{k}'' \tag{63.3}$$

or

$$\boldsymbol{k}_{\sigma''\sigma}+\boldsymbol{k}'_{\sigma''\sigma'}=\boldsymbol{k}''_{\sigma''}. \tag{63.4}$$

Further, if $\{\varphi_{l_{\lambda''}}|\boldsymbol{t}_{l_{\lambda''}}\}$ is in $\mathfrak{G}(\boldsymbol{k}'')$, then from

$$\varphi_{l_{\lambda''}}\cdot\boldsymbol{k}''=\boldsymbol{k}''\,(\mathrm{mod}\;2\pi\,\boldsymbol{B}_H) \tag{63.5}$$

it follows

$$\varphi_{\sigma''}\cdot\varphi_{l_{\lambda''}}\cdot\varphi_{\sigma''}^{-1}\cdot\varphi_{\sigma''}\cdot\boldsymbol{k}''=\varphi_{\sigma''}\cdot\boldsymbol{k}''\,(\mathrm{mod}\;2\pi\,\boldsymbol{B}_H) \tag{63.6}$$

or

$$\{\varphi_{\sigma''}|\boldsymbol{t}_{\sigma''}\}\cdot\{\varphi_{l_{\lambda''}}|\boldsymbol{t}_{l_{\lambda''}}\}\cdot\{\varphi_{\sigma''}|\boldsymbol{t}_{\sigma''}\}^{-1}\equiv\{\varphi_{l_{\lambda''}}^{\bar{\sigma}''}\} \tag{63.7}$$

is in $\mathfrak{G}(\boldsymbol{k}''_{\sigma''})$.

Thus we reexpress (63.1) so that it refers now to the new problem as follows

$$\left(\frac{1}{n_{k''_{\sigma''}\,k_{\sigma''\sigma}}}\right)\sum_{l''_\lambda}\dot{\chi}^{(\boldsymbol{k})(m)}(Z^{-1}\cdot\{\varphi_p|\boldsymbol{t}_p\}\cdot Z)\cdot\dot{\chi}^{(\boldsymbol{k}')(m')}(Y^{-1}\cdot\{\varphi_p|\boldsymbol{t}_p\}\cdot Y)$$
$$=\sum_{m''}(\{\boldsymbol{k}_{\sigma''\sigma}+\boldsymbol{k}'_{\sigma''\sigma'}\}\, m\, m'|\boldsymbol{k}''_{\sigma''}\, m'')\,\dot{\chi}^{(\boldsymbol{k}'')(m'')}(\{\varphi_p^{\bar{\sigma}''}\}), \tag{63.8}$$

where

$$Z\equiv\{\varphi_{l_{\lambda''}}^{\bar{\sigma}''}\}\cdot\{\varphi_{\sigma''}|\boldsymbol{t}_{\sigma''}\}\cdot\{\varphi_\sigma|\boldsymbol{t}_\sigma\}$$

and

$$Y\equiv\{\varphi_{l_{\lambda''}}^{\bar{\sigma}''}\}\cdot\{\varphi_{\sigma''}|\boldsymbol{t}_{\sigma''}\}\cdot\{\varphi_{\sigma'}|\boldsymbol{t}_{\sigma'}\}$$

and $\{\varphi_{l_{\lambda''}}^{\bar{\sigma}''}\}$ is defined in (63.7).

To prove the independence of the subgroup reduction coefficients of which wave vector in $\star\boldsymbol{k}''$ is selected for analysis, it is necessary to show that for any σ''

$$(\{\boldsymbol{k}_\sigma+\boldsymbol{k}'_{\sigma'}\}\, m\, m' \,|\, \boldsymbol{k}''\, m'')=(\{\boldsymbol{k}_{\sigma''\sigma}+\boldsymbol{k}'_{\sigma''\sigma'}\}\, m\, m' \,|\, \boldsymbol{k}''_{\sigma''}\, m''). \tag{63.9}$$

Comparing (63.1) and (63.8) we observe

$$n_{k''\,k_\sigma}=n_{k''_{\sigma''}\,k_{\sigma''\sigma}} \tag{63.10}$$

since the order of the intersection of $\mathfrak{G}(\boldsymbol{k}'')/\mathfrak{T}$ and $\mathfrak{G}(\boldsymbol{k}_\sigma)/\mathfrak{T}$ is identical to the order of the intersection of $\mathfrak{G}(\boldsymbol{k}''_{\sigma''})/\mathfrak{T}$ and $\mathfrak{G}(\boldsymbol{k}_{\sigma''\sigma})/\mathfrak{T}$, (these subgroups are conjugate by pairs). Then working out the products of elements in (63.8) we obtain:

$$\begin{aligned}\left(\frac{1}{n_{k''\,k_\sigma}}\right) \sum_{l_{\lambda''}} \dot{\chi}^{(\boldsymbol{k})(m)}\left(\left\{\{\boldsymbol{\varphi}_p^{\bar{\sigma}''}\}^{l_{\lambda''}\sigma}\right\}\right) \dot{\chi}^{(\boldsymbol{k}')(m')}\left(\left\{\{\boldsymbol{\varphi}_p^{\bar{\sigma}''}\}^{l_{\lambda''}\sigma'}\right\}\right)&\\ =\sum_{m''} (\{\boldsymbol{k}_{\sigma''\sigma}+\boldsymbol{k}'_{\sigma''\sigma'}\}\, m m' \,|\, \boldsymbol{k}''_{\sigma''}\, m'')\, \dot{\chi}^{(k'')(m'')}(\{\boldsymbol{\varphi}_p^{\bar{\sigma}''}\}),&\end{aligned} \tag{63.11}$$

where

$$\left\{\{\boldsymbol{\varphi}_p^{\bar{\sigma}''}\}^{l_{\lambda''}\sigma}\right\} \equiv [\{\boldsymbol{\varphi}_{l_{\lambda''}}|\boldsymbol{t}_{l_{\lambda''}}\}\cdot\{\boldsymbol{\varphi}_\sigma|\boldsymbol{t}_\sigma\}]^{-1}\cdot\{\boldsymbol{\varphi}_p^{\bar{\sigma}''}\}\cdot[\{\boldsymbol{\varphi}_{l_{\lambda''}}|\boldsymbol{t}_{l_{\lambda''}}\}\cdot\{\boldsymbol{\varphi}_\sigma|\boldsymbol{t}_\sigma\}].$$

But on comparing (63.11) and (63.1) we recall that the arbitrary element chosen in (63.1) was $\{\boldsymbol{\varphi}_p|\boldsymbol{t}_p\}$. We could just as well have chosen as the arbitrary element $\{\boldsymbol{\varphi}_p^{\sigma''}\}$ instead, and thus (63.1) and (63.11) are identical.

Consequently we have established the result that the identical subgroup reduction coefficients arise, irrespective of which arm of the final star is chosen. It is of great importance for this analysis to know how the same irreducible representation $D^{(\star\boldsymbol{k})(m)}$ appears in the various subgroups $\mathfrak{G}(\boldsymbol{k}), \ldots, \mathfrak{G}(\boldsymbol{k}_\sigma), \ldots, \mathfrak{G}(\boldsymbol{k}_s)$.

We conclude this section by remarking that our demonstration that (63.8) applies, is based on direct comparison of the equations which define the subgroup reduction coefficients (63.1) and (63.11). We do not make any assumptions regarding the equality of individual characters such as $\dot{\chi}^{(\boldsymbol{k})(m)}(\{\boldsymbol{\varphi}_{l_\lambda}|\boldsymbol{t}_{l_\lambda}\})$ and $\dot{\chi}^{(\boldsymbol{k})(m)}(\{\boldsymbol{\varphi}_\sigma|\boldsymbol{t}_\sigma\}^{-1}\cdot\{\boldsymbol{\varphi}_{l_\lambda}|\boldsymbol{t}_{l_\lambda}\}\cdot\{\boldsymbol{\varphi}_\sigma|\boldsymbol{t}_\sigma\})$, since as described in the discussion immediately following (37.5) these individual objects may or may not be equal.

64. Comparison of full group and subgroup methods. In carrying out a full group method determination of reduction coefficients $(\star\boldsymbol{k}\, m\, \star\boldsymbol{k}'\, m' \,|\, \star\boldsymbol{k}''\, m'')$ we must initially construct full group character tables. In this manner we obtain a character $\chi^{(\star\boldsymbol{k})(m)}(\{\boldsymbol{\varphi}_p|\boldsymbol{t}_p\})$ for every element in the space group $\mathfrak{G}$ in every irreducible representation. It is then a matter of proceeding as with any other finite group, to reduce the direct product of two irreducible representations into irreducible components of the full space group. The full group reduction coefficients may be determined directly from (55.4) or from (55.5). In the former case, Eq. (55.2) are unknown coefficients. In the latter case the sum over all group elements may be performed via introduction of the "reduction group". The reduction group technique is a rigorous device, taking advantage of the homomorphism onto $\mathfrak{G}$, of each of the matrix groups, $D^{(\star\boldsymbol{k})(m)}$, $D^{(\star\boldsymbol{k}')(m')}$, and $D^{(\star\boldsymbol{k}'')(m'')}$. Using this method, the sum over all group elements in $\mathfrak{G}$ is carried out. In carrying out the reductions in (55.4) and (55.5), no ambiguity can arise since all sums are perfectly well defined, as are all objects (characters, group elements) appearing in them, in respect to the full group $\mathfrak{G}$.

In the full group method all states of physical systems are labelled by the complete space group irreducible representation which they generate, i.e., the states are considered, with all their partners, in the complete irreducible linear vector space $\Sigma^{(\star \boldsymbol{k})(m)}$.

The use of full group reduction coefficients means that we are dealing with complete irreducible representations of $\mathfrak{G}$. In particular it means that when the reduction is effected, representations going with more than one star $\star\boldsymbol{k}''$ may and generally will arise.

The task of carrying out a full group procedure involves construction of the basic set of full group character tables using (37.3) or (49.3). This presupposes that reduction of each $\mathfrak{G}(\boldsymbol{k})$, for the canonical $\boldsymbol{k}$ in each star, has been accomplished. When the full tables have been assembled and the reduction accomplished we possess all the irreducible representations which result from taking the direct product $D^{(\star \boldsymbol{k})(m)} \otimes D^{(\star \boldsymbol{k}')(m')}$ and decomposing. In general, there will be complete representations going with several distinct stars in the final result.

The disadvantages of the full group method are that it is necessary to set up complete character tables *a priori*, and that it is necessary to do the complete reduction to get a set of coefficients. The advantages are the possibilities of checking the results (the full set of coefficients) using standard completeness and orthonormality rules, discussed in Sects. 48 and 49.

By contrast, to carry out a subgroup procedure it is necessary to construct the relevant objects in the sum (62.15). These are the same dotted characters which would appear in the full group table, but in principle (62.15) is concerned with fewer of them. The only elements whose characters are non-zero and which appear in (62.15) are the conjugates: $\{\varphi_p^{l_{\lambda''}\sigma}\}$, (see (62.16)), which are in $\mathfrak{G}(\boldsymbol{k})$. This is certainly a smaller set than $\mathfrak{G}$. So, too, in (63.2), a sum over fewer elements is required.

In the subgroup method, attention is focussed on that particular $\boldsymbol{k}''$ desired, and only those relevant factors and coefficients are dealt with; it follows that fewer group elements are dealt with in the sum (63.2) than in (55.5). In this sense the subgroup methods appear somewhat more economical than full group methods. Naturally, when the subgroup methods are used in their totality to obtain all the resultant irreducible representations $D^{(\star \boldsymbol{k}'')(m'')}$ which arise from taking direct product of all pairs: $D^{(\boldsymbol{k}_\sigma)(m)}$ and $D^{(\boldsymbol{k}'_{\sigma'})(m')}$ the same work must be performed and the same results obtained as in using full group methods.

A problem more serious with subgroup methods than full group methods is immediately apparent when the structure of (63.2) and (55.5) is contrasted. It is vital, in the subgroup methods, that the correct pieces of the entire representations be used e.g. the individual terms in the sum (63.2). Further, because of the complicated structure of the representations it is not a trivial matter to identify the correct pieces to be used in obtaining the "complete" character system exhibited in (63.2) in order to carry out the reduction. To put the matter equivalently, an irreducible representation $D^{(\star \boldsymbol{k})(m)}$ of $\mathfrak{G}$ (33.1) has as its $\sigma=1$, $\tau=1$ bloc $(D^{(\star \boldsymbol{k})(m)})_{11} = \dot{D}^{(\boldsymbol{k})(m)}$ corresponding to the allowable irreducible representations of $\mathfrak{G}(\boldsymbol{k})$, where $\boldsymbol{k}$ is the canonical wave vector specifying $\star\boldsymbol{k}$. When the same representation is viewed from the vantage of the wave vector $\boldsymbol{k}_\sigma$ another arm of the same star, the appearance of the representation is complicated because of the necessity of transforming the symmetry elements which are the arguments of the matrices such as (36.14). In principal there is no difficulty in carrying out these subgroup method calculations, but in practice the matter of using the correct elements, and forms of matrices is important.

Another way of contrasting full and subgroup methods is in terms of the relevant vector spaces. The full group method works with entire inequivalent irreducible vector spaces and their products, i.e.:

$$\Sigma^{(\star\boldsymbol{k})(m)} \otimes \Sigma^{(\star\boldsymbol{k}')(m')}. \tag{64.1}$$

The entire product space (64.1) is decomposed into a direct sum of entire irreducible vector spaces

$$\Sigma^{(\star\boldsymbol{k})(m)} \otimes \Sigma^{(\star\boldsymbol{k}')(m')} = \sum_{\star\boldsymbol{k}''\, m''} \oplus (\star\boldsymbol{k}\, m\, \star\boldsymbol{k}'\, m' | \star\boldsymbol{k}''\, m'')\, \Sigma^{(\star\boldsymbol{k}'')(m'')}. \tag{64.2}$$

In the subgroup method only those subspaces are considered, whose direct product produces the desired subspace. In terms of our previous notation

$$\begin{aligned} &\left(\frac{1}{n_{k''k_\sigma}}\right) \sum_{\lambda''=1}^{k''} \left(\Sigma^{(\varphi_{\lambda''}\cdot \boldsymbol{k}_\sigma)(m)} \otimes \Sigma^{(\varphi_{\lambda''}\cdot \boldsymbol{k}'_{\sigma'})}\right) \\ &\qquad = \sum_{m''} (\{\boldsymbol{k}_\sigma + \boldsymbol{k}'_{\sigma'}\}\, m\, m' | \boldsymbol{k}''\, m'')\, \Sigma^{(\boldsymbol{k}'')(m'')}. \end{aligned} \tag{64.3}$$

Now as a practical matter suppose in the subgroup method one erroneously omits certain products (left side of (64.3)) which would give rise to an entire representation (subspace) $\Sigma^{(\boldsymbol{k}'')(\mu'')}$. This error may pass undetected since the only completeness check available in the subgroup method is that all direct product representations introduced on the left of (64.3) shall appear on the right of (64.3). Of course if errors were never made in calculations, this would not be a matter of concern. It is intrinsically not possible to make this error of complete omission of parts of vector spaces in the full group method. This is because at each stage the completeness and orthonormality relation of the entire irreducible representation of the same group $\mathfrak{G}$, the space group, help us and provide obvious checks on the construction of the character table and on the reduction, just as with any well known finite group.

Finally we may compare the full group and subgroup methods by stating the obvious connection between the totality of subgroup reduction coefficients and the full group reduction coefficients. Clearly, for fixed $\star\boldsymbol{k}''\, m''$, the full group reduction coefficient $(\star\boldsymbol{k}\, m\, \star\boldsymbol{k}'\, m' | \star\boldsymbol{k}''\, m'')$ gives the number of times the entire representation $D^{(\star\boldsymbol{k}'')(m'')}$ occurs in the reduction. But if the entire representation occurs then certainly the subgroup representation $D^{(\boldsymbol{k}'')(m'')}$ of $\mathfrak{G}(\boldsymbol{k}'')$ occurs equally often. Hence the subgroup coefficient which couples all possible $D^{(\boldsymbol{k}_\sigma)(m)}$ and $D^{(\boldsymbol{k}'_{\sigma'})(m')}$ to give for fixed $\boldsymbol{k}''$ all resultant $D^{(\boldsymbol{k}'')(m'')}$ must equal the corresponding full group coefficient. Thus, interpreting each coefficient appropriately (as subgroup or full group reduction coefficient) we get

$$(\star\boldsymbol{k}\, m\, \star\boldsymbol{k}'\, m' | \star\boldsymbol{k}''\, m'') = (\{\boldsymbol{k}_\sigma + \boldsymbol{k}'_{\sigma'}\}\, m\, m' | k''\, m'').$$

65. Reduction coefficients: A little group technique. In Sect. 40 it was pointed out that the non-allowable irreducible $D^{(\boldsymbol{k})(\mu)}$ of the little group $\Pi(\boldsymbol{k})$ are representations associated with the set of wave vectors (40.7)

$$\boldsymbol{k},\, 2\boldsymbol{k},\, \ldots,\, t\,\boldsymbol{k}. \tag{65.1}$$

This observation was first made in the context of reduction coefficients obtained for products of space group representations by ELLIOTT and LOUDON.[1] In this section we shall follow these authors by indicating the use of this observation to reduce direct products of irreducible representations within the family (65.1). The other type of application, proposed by ELLIOTT and LOUDON was to the analysis of direct products of representations from a series of different families. It is now clear that the latter problem should be analysed by the more general subgroup methods of the previous sections, or by full group methods.

The complete set of irreducible representations $D^{(\boldsymbol{k})(m)}$ and $D^{(\boldsymbol{k})(\mu)}$ of $\Pi(\boldsymbol{k})$ are assumed known, following the discussion of Sect. 40. Assume that we wish to analyse the representations contained in the product space $D^{(\boldsymbol{k})(m)} \otimes D^{(\boldsymbol{k})(m')}$. Clearly this product will contain representations going with wave vector $2\boldsymbol{k}$. That is the vector spaces

$$\Sigma^{(\boldsymbol{k})(m)} \equiv \{\psi_1^{(\boldsymbol{k})}, \ldots, \psi_{l_m}^{(\boldsymbol{k})}\} \tag{65.2}$$

and

$$\Sigma^{(\boldsymbol{k})(m')} \equiv \{\psi_1'^{(\boldsymbol{k})}, \ldots, \psi_{l_{m'}}'^{(\boldsymbol{k})}\} \tag{65.3}$$

combine to produce the vector space

$$\Sigma^{(\boldsymbol{k}\otimes\boldsymbol{k})(m\otimes m')} \equiv \{\psi_1^{(\boldsymbol{k})} \cdot \psi_1'^{(\boldsymbol{k})}, \ldots, \psi_{l_m}^{(\boldsymbol{k})} \cdot \psi_{l_{m'}}'^{(\boldsymbol{k})}\} \tag{65.4}$$

which evidently contains functions in $\Sigma^{(2\,\boldsymbol{k})}$. Then the representation

$$D^{(\boldsymbol{k})(m)} \otimes D^{(\boldsymbol{k})(m')} \tag{65.5}$$

produces a character system

$$\chi^{(\boldsymbol{k})(m)} \otimes \chi^{(\boldsymbol{k})(m')} \tag{65.6}$$

which can be reduced.

Now the representations of the little group $\Pi(\boldsymbol{k})$ which have been determined are a complete set of inequivalent irreducible representations of $\Pi(\boldsymbol{k})$, of which some belong to $2\boldsymbol{k}, \ldots$. The representations $D^{(\boldsymbol{k})(\mu)}$ of $\Pi(\boldsymbol{k})$ which belong to $2\boldsymbol{k}$ can be easily identified from the complete character system, since if

$$\chi^{(\boldsymbol{k})(\mu)}(\{\varepsilon|\boldsymbol{R}_L(\bar{\boldsymbol{k}})\}) = \exp -i(2\boldsymbol{k})\cdot\boldsymbol{R}_L(\bar{\boldsymbol{k}}) \tag{65.7}$$

then the entire representation $D^{(\boldsymbol{k})(\mu)}$ must be identified as going with $2\boldsymbol{k}$. Similarly by inspecting the other characters for this translation, the representation belonging to the other members of the family (65.1) can be identified. Hence for each μ in $\chi^{(\boldsymbol{k})(\mu)}$ we can make an association

$$\mu_1 \approx 2\boldsymbol{k};\ \ldots;\ \mu_p \approx t\,\boldsymbol{k}. \tag{65.8}$$

Then the reduction of (62.6) can be carried out as follows. Define a reduction coefficient for (62.6) by

$$\chi^{(\boldsymbol{k})(m)} \otimes \chi^{(\boldsymbol{k})(m')} = \sum_{\mu \approx 2\,\boldsymbol{k}} (m\,m'|\mu)\,\chi^{(\boldsymbol{k})(\mu)}, \tag{65.9}$$

[1] R.J. ELLIOTT and R. LOUDON: J. Phys. Chem. Solids **15**, 146 (1960).

where the sum is over those μ corresponding to wave vector $2\boldsymbol{k}$ i.e. just the non-allowable μ at wave vector $\boldsymbol{k}$. Then the coefficient can be obtained by

$$(m\,m'|\mu)=\frac{1}{g_\Pi}\sum_{\Pi(\boldsymbol{k})}\chi^{(\boldsymbol{k})(m)}(\{\varphi|\boldsymbol{t}(\varphi)\})\,\chi^{(\boldsymbol{k})(m')}(\{\varphi|\boldsymbol{t}(\varphi)\})\,\chi^{(\boldsymbol{k})(\mu)}(\{\varphi|\boldsymbol{t}(\varphi)\})^*. \quad (65.10)$$

The sum in (65.10) is over all elements in $\Pi(\boldsymbol{k})$, and μ corresponds to $2\boldsymbol{k}$, for this example. The order of $\Pi(\boldsymbol{k})$ is denoted g_Π. Since, in this method it is assumed that we possess the entire character table, the calculation (65.10) is simple.

Having determined the coefficients $(m\,m'|\mu)$ it remains to associate them with the allowable irreducible representations of the little group $\Pi(2\boldsymbol{k})$. Let use be explicit on this point. As long as all representations $D^{(\boldsymbol{k})(m)}$ and $D^{(\boldsymbol{k})(\mu)}$ refer to the identical group $\Pi(\boldsymbol{k})$, the character systems $\chi^{(\boldsymbol{k})(m)}$, $\chi^{(\boldsymbol{k})(\mu)}$ can be constructed and used in a straightforward fashion to reduce direct products. The reduction coefficients $(m\,m'|\mu)$ are then perfectly well defined in the context of the group $\Pi(\boldsymbol{k})$. However, an association of $D^{(\boldsymbol{k})(\mu)}$, and hence $(m\,m'|\mu)$ with an allowable irreducible representation $D^{(p\boldsymbol{k})(m)}$ at $p\,\boldsymbol{k}$ is only possible if $D^{(\boldsymbol{k})(\mu)}$ is an irreducible representation $D^{(p\boldsymbol{k})(m'')}$ of $\Pi(p\,\boldsymbol{k})$. In this case, the reduction coefficients (65.10) are the subgroup reduction coefficients and evidently give the number of times the irreducible allowable $D^{(2\boldsymbol{k})(m'')}$ is contained in (65.5).

If $D^{(\boldsymbol{k})(\mu)}$ is not irreducible, as a representation of $\Pi(p\,\boldsymbol{k})$, or if it is desired to obtain reduction coefficients for products of representations belonging to different families, by use of subgroup methods, then the more general procedures of the previous sections are required.

The illustration given above explicitly for the case of a product producing $2\boldsymbol{k}$, must also apply when reducing $[D^{(\boldsymbol{k})(m)}]_3$ etc. Again, the more general subgroup methods of the previous sections should be considered except in the special case of irreducibility of the $D^{(\boldsymbol{k})(\mu)}$.

H. Space group theory and classical lattice dynamics

66. Introduction. The present part of the article concerns itself with applications of space group theory to the classical lattice dynamics [*4*] of a crystal.[1-6] A major effect of this complete space group symmetry is to simplify the secular problem for determination of normal mode eigenfrequencies and eigenvectors in the harmonic approximation. The secular equation is seen to factorize according to the irreducible representations of the space group $\mathfrak{G}$ which is involved. The

[1] E. P. Wigner: Nachr. Akad. Wiss. Göttingen, Math-Physik. Kl. 133 (1930).
[2] S. Yanagawa: Progr. Theoret. Phys. (Kyoto) **10**, 83 (1953).
[3] I. V. V. Raghavacharyulu: Can. J. Phys. **39**, 1704 (1961).
[4] H. W. Streitwolf: phys. stat. solidi **5**, 383 (1964).
[5] S. H. Chen: Ph. D. Thesis (1964) cited in [*47*], p. 138.
[6] A. A. Maradudin and S. H. Vosko: Rev. Mod. Phys. **40**, 1 (1968).

factorization due to the pure translational symmetry results in the introduction of normal coordinates depending on the wave vector $\boldsymbol{k}$ of the irreducible representation. The effect of complete symmetry is to enable further characterization of the individual eigenstates according to allowable $D^{(\boldsymbol{k})(m)}$ of $\mathfrak{G}(\boldsymbol{k})$ i.e. according to actual row of the irreducible representation $D^{(\star\boldsymbol{k})(m)}$ of $\mathfrak{G}$.

Once the characterization of the eigenvectors has been achieved according to the irreducible representation of $\mathfrak{G}$, it is immediately possible to apply these results to determination of functions of the eigenvectors. Such functions include lattice invariants and covariants which can be constructed from the eigenvectors, such as: anharmonic crystal potential energy terms and higher order electric dipole moment of the crystal. These functions can be simplified using the results of the group theory analysis. In each degree of the function in terms of eigenvectors the essential, minimum number of non-vanishing independent coupling coefficients can be identified. This is the maximum simplification achievable by group theory.

Another important classical function of the eigenvectors which can be determined partially by the use of the group theory is the Green function which can be used in the analysis of both perfect and imperfect crystal properties [*19–22*].

In the analyses of the symmetry properties of the classical lattice dynamic eigenvalue problem we shall encounter the time reversal transformation operator. It will then be necessary to enlarge the symmetry group of the physical system, from the group of transformation operators isomorphic to the space group $\mathfrak{G}$, to include the time reversal operator. Changes in the previous analysis will be produced. These may include a doubling of the necessary degeneracy owing to the added operator symmetry and also changes in the reduction coefficients, and hence in selection rules. The corepresentation theory required here will be presented, in the next Part I.

Finally, it needs to be emphasized that much of the analysis of this Part H, can be freed from the restrictive assumption that the harmonic approximation is valid. These generalities will be discussed to the extent presently possible in the context of general lattice dynamics and many body theory.

67. Equations of motion in the harmonic approximation. Let us now consider a crystal in which the atoms, when at rest, are located at the positions

$$\boldsymbol{r}\begin{matrix} l \\ \kappa \end{matrix} = \boldsymbol{R}_L + \boldsymbol{r}_\kappa, \tag{67.1}$$

where, as before (refer to Sect. 4 for definitions):

$$\boldsymbol{R}_L = l_1\,\boldsymbol{a}_1 + l_2\,\boldsymbol{a}_2 + l_3\,\boldsymbol{a}_3, \qquad l_i \text{ integers} \tag{67.2}$$

is a lattice vector and

$$\boldsymbol{r}_\kappa = \kappa_1\,\boldsymbol{a}_1 + \kappa_2\,\boldsymbol{a}_2 + \kappa_3\,\boldsymbol{a}_3, \qquad 0 \leqq \kappa_i < 1 \tag{67.3}$$

are basis vectors. The $\boldsymbol{R}_L$ may be considered to specify the cell index in which an atom is located, while the $\boldsymbol{r}_\kappa$ gives the location of a particular basis particle within the cell. Another notation which is often used for (67.1) is

$$r_\alpha \begin{pmatrix} l \\ \kappa \end{pmatrix} \tag{67.4}$$

in which the indices l and κ have the identical meaning as in (67.1) and $\alpha=1, 2, 3$ represents a Cartesian label; indices (1, 2, 3) may equivalently be denoted (x, y, z).

Let the displacements from rest of the particle initially at $\boldsymbol{r}\begin{pmatrix} l \\ \kappa \end{pmatrix}$ be denoted $\boldsymbol{u}\begin{pmatrix} l \\ \kappa \end{pmatrix}$ so that the instantaneous position of each particle can be taken as

$$\boldsymbol{r}\begin{pmatrix} l \\ \kappa \end{pmatrix} + \boldsymbol{u}\begin{pmatrix} l \\ \kappa \end{pmatrix} \equiv \boldsymbol{\rho}\begin{pmatrix} l \\ \kappa \end{pmatrix}. \tag{67.5}$$

In (67.5), the $\boldsymbol{u}\begin{pmatrix} l \\ \kappa \end{pmatrix}$ are to be considered as the time-varying, independent, displacements which are the basic dynamical variables of the problem.

The potential energy of the crystal may be taken as a general function of the instantaneous positions $\boldsymbol{\rho}\begin{pmatrix} l \\ \kappa \end{pmatrix}$ of the particles. We call this function Φ. In the conventional classical treatment, we assume the approximation in which the magnitudes of the displacements $\left|\boldsymbol{u}\begin{pmatrix} l \\ \kappa \end{pmatrix}\right|$ are small, and the inital, rest positions correspond to stable equilibrium. Then the first two non-vanishing terms in the expansion of Φ with respect to deviations from equilibrium are

$$\Phi = \Phi\left(\left\{\boldsymbol{r}\begin{pmatrix} l \\ \kappa \end{pmatrix}\right\}\right) + \tfrac{1}{2}\sum_{l\kappa\alpha}\sum_{l'\kappa'\beta} u_\alpha\begin{pmatrix} l \\ \kappa \end{pmatrix} \Phi_{\alpha\beta}\begin{pmatrix} l & l' \\ \kappa & \kappa' \end{pmatrix} u_\beta\begin{pmatrix} l' \\ \kappa' \end{pmatrix}, \tag{67.6}$$

where

$$\Phi_{\alpha\beta}\begin{pmatrix} l & l' \\ \kappa & \kappa' \end{pmatrix} = \left(\frac{\partial^2 \Phi}{\partial u_\alpha\begin{pmatrix} l \\ \kappa \end{pmatrix} \partial u_\beta\begin{pmatrix} l' \\ \kappa' \end{pmatrix}}\right)\Bigg|_{u_\alpha\left(\begin{smallmatrix} l \\ \kappa \end{smallmatrix}\right)=0;\; u_\beta\left(\begin{smallmatrix} l' \\ \kappa' \end{smallmatrix}\right)=0} . \tag{67.7}$$

The first term in (67.6) represents the reference potential energy of the lattice i.e. the value of Φ when the set of instantaneous particle positions $\left\{\boldsymbol{\rho}\begin{pmatrix} l \\ \kappa \end{pmatrix}\right\}$ corresponds to the set of rest or reference positions: $\left\{\boldsymbol{r}\begin{pmatrix} l \\ \kappa \end{pmatrix}\right\}$; it is thus a constant. The generalized force constants of the lattice dynamical problem, in the harmonic approximation are defined by (67.7). Evidently the series (67.6) may be continued beyond the first two terms to obtain anharmonic terms. We shall later give certain results based on consideration of the anharmonic terms, but for the moment we neglect them. Since the reference configuration is assumed to be one of stable equilibrium the quantity

$$V \equiv \Phi - \Phi\left\{\boldsymbol{r}\begin{pmatrix} l \\ \kappa \end{pmatrix}\right\} = \tfrac{1}{2}\sum_{\substack{l\kappa\alpha \\ l'\kappa'\beta}} u_\alpha\begin{pmatrix} l \\ \kappa \end{pmatrix} \Phi_{\alpha\beta}\begin{pmatrix} l & l' \\ \kappa & \kappa' \end{pmatrix} u_\beta\begin{pmatrix} l' \\ \kappa' \end{pmatrix} \tag{67.8}$$

must be positive if there is at least one non-vanishing $u_\alpha\begin{pmatrix} l \\ \kappa \end{pmatrix}$, or zero if all the $u_\alpha\begin{pmatrix} l \\ \kappa \end{pmatrix}$ identically vanish. Thus (67.8) is a positive semi-definite quadratic form:

$$V \geqq 0. \tag{67.9}$$

Since the order in which the differentiation in (67.7) is carried out is evidently immaterial, the force constants (67.7) satisfy the relation

$$\Phi_{\alpha\beta}\begin{pmatrix} l & l' \\ \kappa & \kappa' \end{pmatrix} = \Phi_{\beta\alpha}\begin{pmatrix} l' & l \\ \kappa' & \kappa \end{pmatrix}. \tag{67.10}$$

Let us define the $(3Nr)$ rowed column matrix whose elements are the components of the individual displacement vectors $u_\alpha\begin{pmatrix} l \\ \kappa \end{pmatrix}$. Here, N is the number of unit cells in the crystal subject to Born-Karman boundary conditions, r is the number of ions (atoms) in a unit cell i.e. the number of basis particles so that $3rN$ is the total number of mechanical degrees of freedom of the vibrating crystal. Call this column matrix $[\boldsymbol{u}]$:

$$[\boldsymbol{u}] = \begin{pmatrix} u_1\begin{pmatrix} 1 \\ 1 \end{pmatrix} \\ \vdots \\ u_3\begin{pmatrix} N \\ r \end{pmatrix} \end{pmatrix}. \tag{67.11}$$

Its transpose is the row matrix $\widetilde{[u]}$

$$\widetilde{[\boldsymbol{u}]} = \left(u_1\begin{pmatrix} 1 \\ 1 \end{pmatrix}, \dots, u_3\begin{pmatrix} N \\ r \end{pmatrix}\right). \tag{67.12}$$

If we call the $(3rN)\times(3rN)$ matrix whose elements are the $\Phi_{\alpha\beta}\begin{pmatrix} l & l' \\ \kappa & \kappa' \end{pmatrix}$ defined in (67.10) as $[\boldsymbol{\Phi}]$, then we may write for (67.8)

$$V = \tfrac{1}{2}\widetilde{[\boldsymbol{u}]}\,[\boldsymbol{\Phi}]\,[\boldsymbol{u}]. \tag{67.13}$$

This condensed notation is sometimes useful.

The cartesian components of the displacements $u_\alpha\begin{pmatrix} l \\ \kappa \end{pmatrix}$ are independent so that the kinetic energy of the vibrating lattice is

$$T = \tfrac{1}{2}\sum_{l\kappa\alpha} M_\kappa\left(\dot{u}_\alpha\begin{pmatrix} l \\ \kappa \end{pmatrix}\right)^2, \tag{67.14}$$

where M_κ is the mass of the κ-th basis particle. The Lagrangian is

$$\mathcal{L} = T - V = \tfrac{1}{2}\sum_{l\kappa\alpha} M_\kappa\left(\dot{u}_\alpha\begin{pmatrix} l \\ \kappa \end{pmatrix}\right)^2 - \tfrac{1}{2}\sum_{\substack{l\kappa\alpha \\ l'\kappa'\beta}} u_\alpha\begin{pmatrix} l \\ \kappa \end{pmatrix}\Phi_{\alpha\beta}\begin{pmatrix} l & l' \\ \kappa & \kappa' \end{pmatrix} u_\beta\begin{pmatrix} l' \\ \kappa' \end{pmatrix} \tag{67.15}$$

or in terms of the notation (67.13)

$$\mathcal{L} = \tfrac{1}{2}\widetilde{[\dot{\boldsymbol{u}}]}\,[\boldsymbol{M}]\,[\dot{\boldsymbol{u}}] - \tfrac{1}{2}\widetilde{[\boldsymbol{u}]}\,[\boldsymbol{\Phi}]\,[\boldsymbol{u}] \tag{67.16}$$

where $[\dot{\boldsymbol{u}}]$ is the matrix (67.11) whose elements are $\dot{u}_\alpha\begin{pmatrix} l \\ \kappa \end{pmatrix}$ and $[M]$ is the diagonal mass matrix:

$$[\boldsymbol{M}]_{\substack{l\,l' \\ \kappa\,\kappa' \\ \alpha\,\beta}} = M_\kappa\,\delta_{l\,l'}\,\delta_{\kappa\,\kappa'}\,\delta_{\alpha\,\beta}. \tag{67.17}$$

In (67.14)–(67.16) the dot refers to total time derivative as customary. Taking the $u_\alpha\begin{pmatrix} l \\ \kappa \end{pmatrix}$ as the generalized coordinate variables, and the $\dot{u}_\alpha\begin{pmatrix} l \\ \kappa \end{pmatrix}$ as generalized velocity variables the equation of motion can be found from

$$\frac{d}{dt}\frac{\partial \mathcal{L}}{\partial \dot{u}_\alpha\begin{pmatrix} l \\ \kappa \end{pmatrix}} - \frac{\partial \mathcal{L}}{\partial u_\alpha\begin{pmatrix} l \\ \kappa \end{pmatrix}} = 0. \tag{67.18}$$

We then find as the equations of motion of the dynamical variables $u_\alpha\begin{pmatrix} l \\ \kappa \end{pmatrix}$:

$$M_\kappa \ddot{u}_\alpha\begin{pmatrix} l \\ \kappa \end{pmatrix} + \sum_{l'\kappa'\beta} \Phi_{\alpha\beta}\begin{pmatrix} l & l' \\ \kappa & \kappa' \end{pmatrix} u_\beta\begin{pmatrix} l' \\ \kappa' \end{pmatrix} = 0. \tag{67.19}$$

Evidently Eq. (67.5) indicates that the time varying, independent displacements $\boldsymbol{u}\begin{pmatrix} l \\ \kappa \end{pmatrix}$ which are the basic dynamical variable of the lattice vibration problem may be considered as a vector field $\boldsymbol{u}(\boldsymbol{r})$ whose values are only defined at the lattice sites $\begin{pmatrix} l \\ \kappa \end{pmatrix}$ i.e. for the rest positions of the particles. Likewise, through Eq. (67.5), an individual displacement vector is associated with each undisturbed lattice site $\boldsymbol{r}\begin{pmatrix} l \\ \kappa \end{pmatrix}$, and the equations of motion are then given in terms of this basic set of variables as in (67.19).

68. Translation symmetry and particle displacements. We consider separately the effect of translational and rotational symmetry upon the basic displacement variables $\boldsymbol{u}\begin{pmatrix} l \\ \kappa \end{pmatrix}$, and upon the force matrix (67.7). In this section we only consider the effect of translational symmetry $\{\varepsilon|\boldsymbol{R}_L\}$ in the lattice.

The transformation operator corresponding to $\{\varepsilon|\boldsymbol{R}_L\}$ is $P_{\{\varepsilon|\boldsymbol{R}_L\}}$. Apply this operator $P_{\{\varepsilon|\boldsymbol{R}_L\}}$ first to the vector field $\boldsymbol{u}(\boldsymbol{r})$. Clearly

$$(P_{\{\varepsilon|\boldsymbol{R}_L\}}\boldsymbol{u})(\boldsymbol{r}) = \boldsymbol{u}(\boldsymbol{r} - \boldsymbol{R}_L) \tag{68.1}$$

that is, the effect of a rigid translation by a lattice vector $\boldsymbol{R}_L$ upon a displacement vector $\boldsymbol{u}$, is to associate the displacement with the point which is translated by the lattice vector $\boldsymbol{R}_L$. In this, "active" interpretation of the symmetry operation we consider that we shift the displacement field while keeping the lattice points fixed in space. Thus "new", or shifted, displacements are associated with "old" or undisplaced sites. Evaluating (68.1) at $\boldsymbol{r}\begin{matrix} l \\ \kappa \end{matrix} = \boldsymbol{R}_L + \boldsymbol{r}_\kappa$ one finds

$$(P_{\{\varepsilon|\boldsymbol{R}_L\}}\boldsymbol{u})\left(\boldsymbol{r}\begin{matrix} l \\ \kappa \end{matrix}\right) = \boldsymbol{u}\begin{pmatrix} l-\bar{l} \\ \kappa \end{pmatrix}. \tag{68.2}$$

The label $\boldsymbol{r}$ in (68.2) is evidently superfluous, consequently we shall usually write instead of (68.2)

$$(P_{\{\varepsilon|\boldsymbol{R}_{\bar{L}}\}}\,\boldsymbol{u})\begin{pmatrix} l \\ \kappa \end{pmatrix}=\boldsymbol{u}\begin{pmatrix} l-\bar{l} \\ \kappa \end{pmatrix}. \tag{68.3}$$

69. Translation symmetry and force matrix. In terms of the transformed displacements (68.3) the expression for the potential energy (67.8) becomes

$$V=\Phi-\Phi\left(\left\{r\begin{pmatrix} l \\ \kappa \end{pmatrix}\right\}\right)=\tfrac{1}{2}\sum_{\substack{l\kappa\alpha \\ l'\kappa'\beta}} u_\alpha\begin{pmatrix} l-\bar{l} \\ \kappa \end{pmatrix}\Phi_{\alpha\beta}\begin{pmatrix} l-\bar{l} & l'-\bar{l} \\ \kappa & \kappa' \end{pmatrix}u_\beta\begin{pmatrix} l'-\bar{l} \\ \kappa \end{pmatrix} \tag{69.1}$$

where, from (67.7)

$$\Phi_{\alpha\beta}\begin{pmatrix} l-\bar{l} & l'-\bar{l} \\ \kappa & \kappa' \end{pmatrix}\equiv\left[\frac{\partial^2\Phi}{\partial u_\alpha\begin{pmatrix} l-\bar{l} \\ \kappa \end{pmatrix}\partial u_\beta\begin{pmatrix} l'-\bar{l} \\ \kappa' \end{pmatrix}}\right]_{(0)}. \tag{69.2}$$

Clearly since the displacement merely shifts all atoms into equivalent positions we have for the force constants:

$$\Phi_{\alpha\beta}\begin{pmatrix} l-\bar{l} & l'-\bar{l} \\ \kappa & \kappa' \end{pmatrix}=\Phi_{\alpha\beta}\begin{pmatrix} l & l' \\ \kappa & \kappa' \end{pmatrix}. \tag{69.3}$$

Because of the symmetry of the lattice the force constant tensor is a second rank tensor invariant field, unchanged by a lattice symmetry operation.

Thus letting $\bar{l}=l'$

$$\Phi_{\alpha\beta}\begin{pmatrix} l-l' & 0 \\ \kappa & \kappa' \end{pmatrix}=\Phi_{\alpha\beta}\begin{pmatrix} l & l' \\ \kappa & \kappa' \end{pmatrix} \tag{69.4}$$

and the force constants $\Phi_{\alpha\beta}\begin{pmatrix} l & l' \\ \kappa & \kappa' \end{pmatrix}$ can only depend upon the difference between the lattice vectors $\boldsymbol{R}_L$ and $\boldsymbol{R}_{L'}$, and not upon the actual *individual* cell indices.

We may obtain an equivalent statement of this result from the passive interpretation of the effect of a symmetry transformation. That is, let the displacement field $\boldsymbol{u}(\boldsymbol{r})$ be considered fixed but the lattice points themselves shifted by the lattice vector $\boldsymbol{R}_{\bar{L}}$. Then the lattice point which was located at $\boldsymbol{R}_{\bar{L}}$ is now located as $\boldsymbol{R}_{\bar{L}}$, while that one which was at $\boldsymbol{R}_{L'}$ is now at $\boldsymbol{R}_{L'}-\boldsymbol{R}_{\bar{L}}$, with the displacement field $\boldsymbol{u}(\boldsymbol{r})$ considered fixed. Since the increment added to the crystal potential energy V must depend only upon the individual displacements, we again come to the result that the addition of a common lattice vector to all the cell indices in the individual force coefficient leaves it invariant (69.4). At times it will be found useful to emphasize the dependence of $\Phi_{\alpha\beta}\begin{pmatrix} l & l' \\ \kappa & \kappa' \end{pmatrix}$ upon the difference $l-l'$, only. We shall then write

$$\Phi_{\alpha\beta}\begin{pmatrix} l & l' \\ \kappa & \kappa' \end{pmatrix}=\Phi_{\alpha\beta}\begin{pmatrix} \lambda \\ \kappa\kappa' \end{pmatrix},$$

where

$$\boldsymbol{R}_\lambda=\boldsymbol{R}_L-\boldsymbol{R}_{L'} \tag{69.5}$$

and the asymmetric notation is intended to emphasize that only one cell index is important namely a *relative* index. In other cases, formal inclusion of all sets of indices may be useful and they will then be displayed.

70. General symmetry and particle displacements. Consider the effect of a general space symmetry transformation operator $P_{\{\boldsymbol{\varphi}|\boldsymbol{t}\}}$ applied to the displacement field $\boldsymbol{u}(\boldsymbol{r})$. Under this general symmetry transformation the vector displacement field $\boldsymbol{u}$ is transformed into the vector displacement field $\boldsymbol{u}_{\{\boldsymbol{\varphi}\}}$ where:

$$\boldsymbol{u}_{\{\boldsymbol{\varphi}\}} \equiv P_{\{\boldsymbol{\varphi}|\boldsymbol{t}\}}\, \boldsymbol{u}. \tag{70.1}$$

Since $\boldsymbol{u}$ is a vector we must have

$$(P_{\{\boldsymbol{\varphi}|\boldsymbol{t}\}}\, \boldsymbol{u})_\alpha = (\boldsymbol{u}_{\{\boldsymbol{\varphi}\}})_\alpha = \sum_\beta \varphi_{\alpha\beta} (\boldsymbol{u})_\beta. \tag{70.2}$$

In (70.2)

$$\varphi_{\alpha\beta} \equiv (\boldsymbol{\varphi})_{\alpha\beta} \tag{70.3}$$

is the $\alpha\beta$ element of the rotation matrix $\boldsymbol{\varphi}$; the latter is the rotational part of the general space symmetry operation $\{\boldsymbol{\varphi}|\boldsymbol{t}\}$. At times in what follows, it may prove useful to use the convention that

$$\varphi_{\alpha\beta} \equiv D^{(\boldsymbol{r})}(\boldsymbol{\varphi})_{\alpha\beta}. \tag{70.4}$$

In (70.4) we write $D^{(\boldsymbol{r})}$ for the matrix by which an ordinary polar vector transforms. Thus as in Sect. 5, we may write

$$\boldsymbol{r}' = \boldsymbol{r}_{\{\boldsymbol{\varphi}\}} = \boldsymbol{\varphi} \cdot \boldsymbol{r} = D^{(\boldsymbol{r})}(\boldsymbol{\varphi})\, \boldsymbol{r}, \tag{70.5}$$

where (70.5) should now be compared with (5.9). Coming back to (70.2), the components of $\boldsymbol{u}$ transform into one another under rotation just as the components of any polar vector e.g. the position vector. Eq. (70.2) then asserts the linear homogeneous relationship between the components of the vector function $\boldsymbol{u}$, and those of the vector function $\boldsymbol{u}_{\{\boldsymbol{\varphi}\}}$. But the transformation $P_{\{\boldsymbol{\varphi}|\boldsymbol{t}\}}$ is also a point-point mapping. That is, in addition to rotating the components of the vector field $\boldsymbol{u}$, the transformation also associates the rotated components with other field points. Then the complete rule for the effect of the transformation operator applied to $\boldsymbol{u}$ is:

$$(P_{\{\boldsymbol{\varphi}|\boldsymbol{t}\}}\, \boldsymbol{u})_\alpha(\boldsymbol{r}) = \sum_\beta \varphi_{\alpha\beta}\, u_\beta(\{\boldsymbol{\varphi}|\boldsymbol{t}\}^{-1} \cdot \boldsymbol{r}). \tag{70.6}$$

For the crystal lattice, $\boldsymbol{u}$ is only defined at lattice sites so we may write (70.6) for $\boldsymbol{r} = \binom{l}{\kappa}$

$$(\boldsymbol{u}_{\{\boldsymbol{\varphi}\}})_\alpha \binom{l}{\kappa} = \sum_\beta \varphi_{\alpha\beta}\, u_\beta \binom{l_{\bar{\varphi}}}{\kappa_{\bar{\varphi}}}. \tag{70.7}$$

In (70.7) we adopt the convention

$$\{\boldsymbol{\varphi}|\boldsymbol{t}\}^{-1} \cdot \boldsymbol{r}\binom{l}{\kappa} \equiv \boldsymbol{\varphi}^{-1} \cdot \boldsymbol{r}\binom{l}{\kappa} - \boldsymbol{\varphi}^{-1} \cdot \boldsymbol{t} \tag{70.8}$$

or

$$\{\varphi|t\}^{-1}\cdot\binom{l}{\kappa}\equiv\binom{l_{\bar\varphi}}{\kappa_{\bar\varphi}}. \tag{70.9}$$

On the other hand, if we define

$$\binom{l_\varphi}{\kappa_\varphi}\equiv\{\varphi|t\}\cdot r\binom{l}{\kappa} \tag{70.10}$$

then we may write (70.7) in the form inverse to the way it stands, as:

$$(u_{\{\varphi\}})_\alpha\binom{l_\varphi}{\kappa_\varphi}=\sum_\beta \varphi_{\alpha\beta}\,u_\beta\binom{l}{\kappa}. \tag{70.11}$$

Owing to the property that φ is an orthogonal transformation, we have

$$\sum_\lambda \varphi_{\lambda\mu}\,\varphi_{\lambda\nu}=\sum_\sigma \varphi_{\mu\sigma}\,\varphi_{\nu\sigma}=\delta_{\mu\nu}. \tag{70.12}$$

The use of (70.12) permits us to invert either (70.7) or (70.11). From (70.11) we have

$$\sum_\alpha \varphi_{\alpha\gamma}\sum \varphi_{\alpha\beta}\,u_\beta\binom{l}{\kappa}=\sum_\alpha \varphi_{\alpha\gamma}(\boldsymbol{u}_{\{\varphi\}})_\alpha\binom{l_\varphi}{\kappa_\varphi} \tag{70.13}$$

or, using (70.12):

$$u_\gamma\binom{l}{\kappa}=\sum_\alpha \varphi_{\alpha\gamma}(\boldsymbol{u}_{\{\varphi\}})_\alpha\binom{l_\varphi}{\kappa_\varphi}, \tag{70.14}$$

$$u_\gamma\binom{l}{\kappa}=\sum_\alpha (\varphi^{-1})_{\alpha\gamma}(u_{\{\varphi\}})_\alpha\binom{l_\varphi}{\kappa_\varphi}. \tag{70.15}$$

From (70.7) we obtain

$$u_\delta\binom{l_{\bar\varphi}}{\kappa_{\bar\varphi}}=\sum_\alpha \varphi_{\alpha\delta}(\boldsymbol{u}_{\{\varphi\}})_\alpha\binom{l}{\kappa}. \tag{70.16}$$

The expressions (70.14)–(70.16) could also be written in terms of $D^{(r)}$, using (70.4).

At the expense of a little redundancy, it may be useful to recapitulate at this point. The coordinate transformation $\{\varphi|t\}$ can be taken as

$$\{\varphi|t\}\equiv\{\varphi|\tau(\varphi)+\boldsymbol{R}_M\} \tag{70.17}$$

with $\boldsymbol{R}_M$ a lattice vector, $\tau(\varphi)$ a fractional. Then under this coordinate transformation the point $r\binom{l}{\kappa}\equiv\boldsymbol{R}_L+\boldsymbol{r}_\kappa$ is transformed into

$$\{\varphi|t\}\cdot r\binom{l}{\kappa}=\varphi\cdot r\binom{l}{\kappa}+t \tag{70.18}$$

or

$$\binom{l}{\kappa}\to\binom{l_\varphi}{\kappa_\varphi}, \tag{70.19}$$

where

$$\binom{l_\varphi}{\kappa_\varphi}\equiv\begin{pmatrix}\varphi\cdot\boldsymbol{R}_L+\boldsymbol{R}_M\\ \varphi\cdot\boldsymbol{r}_\kappa+\tau(\varphi)\end{pmatrix}. \tag{70.20}$$

It is very important to recognize that the abbreviated symbol on the left hand side of (70.20) stands for the entire object on the right hand side. The right hand side merely makes explicit what is contained on the left side and in particular exhibits the "pure" lattice part: $\boldsymbol{\varphi}\cdot\boldsymbol{R}_L+\boldsymbol{R}_M$, and the "fractional" part: $\boldsymbol{\varphi}\cdot\boldsymbol{r}_\kappa+\boldsymbol{\tau}(\boldsymbol{\varphi})$ of the coordinate of the transformed point, subject to remarks of the next paragraph. Similarly for the inverse of (70.17)

$$\{\boldsymbol{\varphi}|\boldsymbol{t}\}^{-1}=\{\boldsymbol{\varphi}^{-1}|-\boldsymbol{\varphi}^{-1}\cdot\boldsymbol{\tau}(\boldsymbol{\varphi})-\boldsymbol{\varphi}^{-1}\cdot\boldsymbol{R}_M\}=\{\boldsymbol{\varphi}^{-1}|-\boldsymbol{\varphi}^{-1}\cdot\boldsymbol{t}\} \tag{70.21}$$

we obtain the transformation

$$\{\boldsymbol{\varphi}|\boldsymbol{t}\}^{-1}\cdot r\begin{pmatrix} l\\ \kappa\end{pmatrix}=\boldsymbol{\varphi}^{-1}\cdot r\begin{pmatrix} l\\ \kappa\end{pmatrix}-\boldsymbol{\varphi}^{-1}\cdot\boldsymbol{t} \tag{70.22}$$

or

$$\begin{pmatrix} l\\ \kappa\end{pmatrix}\rightarrow\begin{pmatrix} l_{\bar\varphi}\\ \kappa_{\bar\varphi}\end{pmatrix}, \tag{70.23}$$

where

$$\begin{pmatrix} l_{\bar\varphi}\\ \kappa_{\bar\varphi}\end{pmatrix}\equiv\begin{pmatrix} \boldsymbol{\varphi}^{-1}\cdot\boldsymbol{R}_L-\boldsymbol{\varphi}^{-1}\cdot\boldsymbol{R}_M\\ \boldsymbol{\varphi}^{-1}\cdot\boldsymbol{r}_\kappa-\boldsymbol{\varphi}^{-1}\cdot\boldsymbol{\tau}(\boldsymbol{\varphi})\end{pmatrix} \tag{70.24}$$

and again the lattice and the fractional parts are exhibited explicitly in (70.24).

In both (70.20) and (70.24) the lattice translation and fractional translation are exhibited explicitly. It is noteworthy, however, that an additional, implicit lattice translation vector may be contained in (70.20) and (70.24) in the nominally fractional parts

$$\boldsymbol{r}_{\kappa_\varphi}\equiv\boldsymbol{\varphi}\cdot\boldsymbol{r}_k+\boldsymbol{\tau}(\boldsymbol{\varphi}), \tag{70.25}$$

$$\boldsymbol{r}_{\kappa_{\bar\varphi}}\equiv\boldsymbol{\varphi}^{-1}\cdot\boldsymbol{r}_k-\boldsymbol{\varphi}^{-1}\cdot\boldsymbol{\tau}(\boldsymbol{\varphi}) \tag{70.26}$$

respectively. Either or both of these transformations, even with $\boldsymbol{\tau}=0$ may take the atom of κ-th sort into an identical atom of κ-th sort, in another unit cell, thus actually shifting the initial position by an added lattice vector not explicitly recognized in (70.20) and (70.24). Certainly

$$\boldsymbol{\varphi}\cdot\boldsymbol{R}_L+\boldsymbol{R}_M \tag{70.27}$$

and

$$\boldsymbol{\varphi}^{-1}\cdot\boldsymbol{R}_L-\boldsymbol{\varphi}^{-1}\cdot\boldsymbol{R}_M \tag{70.28}$$

are lattice vectors. However, the notation (70.20) and (70.24), while very useful in general does suffer the drawback of obscuring the possibility that e.g. in (70.25) and (70.26)

$$\boldsymbol{r}_{\kappa_\varphi}=\boldsymbol{R}_N(\kappa_\varphi,\kappa')+\boldsymbol{r}_{k'}, \tag{70.29}$$

$$\boldsymbol{r}_{\kappa_{\bar\varphi}}=\boldsymbol{R}_N(\kappa_{\bar\varphi},\kappa'')+\boldsymbol{r}_{k''}, \tag{70.30}$$

where $\boldsymbol{R}_N(\kappa_\varphi,\kappa')$, and $\boldsymbol{R}_N(\kappa_{\bar\varphi},\kappa'')$ are lattice vectors. It will be necessary for us to return to the possibilities (70.29) and (70.30) in our later discussion of Sects. 80 et seq.

It should also be understood, that because $\{\boldsymbol{\varphi}|\boldsymbol{t}\}$ is a symmetry operation, the physical property at $\begin{pmatrix} l\\ \kappa\end{pmatrix}$ is identical to that at $\begin{pmatrix} l_\varphi\\ \kappa_\varphi\end{pmatrix}$. That is the same atom or ion resides at these two places, the values of physical quantities are the same at these two positions.

Finally, it is useful to verify the unitarity of the transformation operators $P_{\{\varphi|t\}}$. A state of the crystal with instantaneous displacements is specified when the $3Nr$ dimensional vector $[\boldsymbol{u}]$ of (67.11) is given. The norm or length squared of this vector may be defined as

$$[\boldsymbol{u}]^2 = \widetilde{[\boldsymbol{u}]} \cdot [\boldsymbol{u}] \equiv (\boldsymbol{u}, \boldsymbol{u}) = \sum_{\kappa l \alpha} \left(u_\alpha \begin{pmatrix} l \\ \kappa \end{pmatrix} \right)^2. \tag{70.31}$$

Clearly

$$[\boldsymbol{u}]^2 \geqq 0, \tag{70.32}$$

where the equality only applies for the null displacement. Consider the state of the crystal specified by the set of displacements $[P_{\{\varphi|t\}}\, \boldsymbol{u}]$, whose components are given in (70.2). The norm of this vector is

$$\begin{aligned}(P_{\{\varphi|t\}}\, \boldsymbol{u}, P_{\{\varphi|t\}}\, \boldsymbol{u}) &= (\boldsymbol{u}_\varphi, \boldsymbol{u}_\varphi) = \sum_{\kappa l \alpha} \left((P_{\{\varphi|t\}}\, \boldsymbol{u})_\alpha \begin{pmatrix} l \\ \kappa \end{pmatrix} \right)^2 \\ &= \sum_{\kappa l \alpha} \left(\sum_\beta \varphi_{\alpha\beta}\, u_\beta \begin{pmatrix} l_{\bar{\varphi}} \\ \kappa_{\bar{\varphi}} \end{pmatrix} \right) \cdot \left(\sum_{\bar{\beta}} \varphi_{\alpha\bar{\beta}}\, u_{\bar{\beta}} \begin{pmatrix} l_{\bar{\varphi}} \\ \kappa_{\bar{\varphi}} \end{pmatrix} \right) = \sum_{\kappa l \beta} \left(u_\beta \begin{pmatrix} l_{\bar{\varphi}} \\ \kappa_{\bar{\varphi}} \end{pmatrix} \right)^2,\end{aligned} \tag{70.33}$$

where we have used (70.12). For given $\{\varphi|t\}$ we may change the sum indices in (70.33)

$$\sum_{\kappa l} \rightarrow \sum_{\kappa_{\bar{\varphi}} l_{\bar{\varphi}}} \tag{70.34}$$

as the sum goes over each atom in the crystal once and only once. Hence using notation of (70.2):

$$(\boldsymbol{u}, \boldsymbol{u}) = (\boldsymbol{u}_\varphi, \boldsymbol{u}_\varphi). \tag{70.35}$$

This is evidently true for all $P_{\{\varphi|t\}}$ in $\mathfrak{G}_P$, the group of transformation operators isomorphic to the space group $\mathfrak{G}$. Hence the operators $P_{\{\varphi|t\}}$ defined by (70.7) are unitary.

The displacement field $[\boldsymbol{u}]$ is real. Now consider the operator K which corresponds to taking the complex conjugate:

$$K\, u_\alpha \begin{pmatrix} l \\ \kappa \end{pmatrix} \equiv u_\alpha \begin{pmatrix} l \\ \kappa \end{pmatrix}^*. \tag{70.36}$$

Owing to the reality of the physical displacements

$$K\, u_\alpha \begin{pmatrix} l \\ \kappa \end{pmatrix} = u_\alpha \begin{pmatrix} l \\ \kappa \end{pmatrix}. \tag{70.37}$$

From (70.37) and (70.7)

$$K\, P_{\{\varphi|t\}}\, u_\alpha \begin{pmatrix} l \\ \kappa \end{pmatrix} = P_{\{\varphi|t\}}\, K\, u_\alpha \begin{pmatrix} l \\ \kappa \end{pmatrix} = P_{\{\varphi|t\}}\, u_\alpha \begin{pmatrix} l \\ \kappa \end{pmatrix}. \tag{70.38}$$

The last step follows (70.37). Finally it is evident that

$$(K\, \boldsymbol{u}, K\, \boldsymbol{u}) = (\boldsymbol{u}, \boldsymbol{u}) \tag{70.39}$$

owing to the reality of the displacements.

71. General symmetry and force matrix. The physical content of (70.7) or (70.11) is that the transformation $P_{\{\varphi|t\}}$ associates with the lattice point $\begin{pmatrix} l \\ \kappa \end{pmatrix}$ the rotated displacement $\boldsymbol{\varphi}\cdot\boldsymbol{u}$, which was formerly associated with the equivalent lattice point $\begin{pmatrix} l_{\bar{\varphi}} \\ \kappa_{\bar{\varphi}} \end{pmatrix}$. Let us consider the $\boldsymbol{u}_{\{\boldsymbol{\varphi}\}}\begin{pmatrix} l_\varphi \\ \kappa_\varphi \end{pmatrix}$ of (70.11) as the basic dynamical variables of the problem of the vibrating lattice. Then in terms of these variables, the expression (67.8) for V becomes

$$V=\tfrac{1}{2}\sum_{\substack{l_\varphi \kappa_\varphi \alpha' \\ l'_\varphi \kappa'_\varphi \beta'}} (u_{\{\boldsymbol{\varphi}\}})_{\alpha'}\begin{pmatrix} l_\varphi \\ \kappa_\varphi \end{pmatrix} \Phi'_{\alpha'\beta'}\begin{pmatrix} l_\varphi & l'_\varphi \\ \kappa_\varphi & \kappa'_\varphi \end{pmatrix} (u_{\{\boldsymbol{\varphi}\}})_{\beta'}\begin{pmatrix} l'_\varphi \\ \kappa'_\varphi \end{pmatrix}, \tag{71.1}$$

where, in the new variables,

$$\Phi'_{\alpha'\beta'}\begin{pmatrix} l_\varphi & l'_\varphi \\ \kappa_\varphi & \kappa'_\varphi \end{pmatrix} \equiv \left[\frac{\partial^2 \Phi}{\partial (u_{\{\varphi\}})_{\alpha'}\begin{pmatrix} l_\varphi \\ \kappa_\varphi \end{pmatrix} \partial (u_{\{\varphi\}})_{\beta'}\begin{pmatrix} l'_\varphi \\ \kappa'_\varphi \end{pmatrix}}\right]_{(0)}. \tag{71.2}$$

Now for fixed $\{\varphi\}$ the sum over $l_\varphi \kappa_\varphi$ and $l'_\varphi \kappa'_\varphi$ in (71.1) may just as well be considered as sum on $l\kappa$ and $l'\kappa'$. Let us consider the reduction of (71.1) for comparison with (67.8). First we observe that by the chain rule.

$$\begin{aligned} \frac{\partial}{\partial (u_{\{\varphi\}})_{\alpha'}\begin{pmatrix} l_\varphi \\ \kappa_\varphi \end{pmatrix}} &= \sum_{l,\kappa,\alpha} \frac{\partial u_\alpha\begin{pmatrix} l \\ \kappa \end{pmatrix}}{\partial (u_{\{\varphi\}})_{\alpha'}\begin{pmatrix} l_\varphi \\ \kappa_\varphi \end{pmatrix}} \cdot \frac{\partial}{\partial u_\alpha\begin{pmatrix} l \\ \kappa \end{pmatrix}} \\ &= \sum_{\alpha=1}^{3} \varphi_{\alpha\alpha'} \frac{\partial}{\partial u_\alpha\begin{pmatrix} l \\ \kappa \end{pmatrix}}. \end{aligned} \tag{71.3}$$

In (71.3) each component of a transformed displacement (vector) is considered to depend upon all $3Nr$ components of all the original displacements. Then, substituting (71.3) into (71.2) and using (67.7) we have

$$\Phi'_{\alpha'\beta'}\begin{pmatrix} l_\varphi & l'_\varphi \\ \kappa_\varphi & \kappa'_\varphi \end{pmatrix} = \sum_\alpha \sum_\beta \varphi_{\alpha\alpha'} \varphi_{\beta\beta'} \Phi_{\alpha\beta}\begin{pmatrix} l & l' \\ \kappa & \kappa' \end{pmatrix}, \tag{71.4}$$

or

$$\Phi'_{\alpha'\beta'}\begin{pmatrix} l_\varphi & l'_\varphi \\ \kappa_\varphi & \kappa'_\varphi \end{pmatrix} = \sum_{\alpha\beta} (\varphi^{-1})_{\alpha'\alpha} \Phi_{\alpha\beta}\begin{pmatrix} l & l' \\ \kappa & \kappa' \end{pmatrix} \varphi_{\beta\beta'}. \tag{71.5}$$

Now we put (70.11) into (71.1), and carry out the sum over $l\kappa$ and $l'\kappa'$ to obtain

$$V=\tfrac{1}{2}\sum_{\substack{l\kappa\alpha' \\ l'\kappa'\beta}} \sum_{\alpha\beta} \varphi_{\alpha\alpha'} u_\alpha\begin{pmatrix} l \\ \kappa \end{pmatrix} \Phi'_{\alpha'\beta'}\begin{pmatrix} l_\varphi & l'_\varphi \\ \kappa_\varphi & \kappa'_\varphi \end{pmatrix} \varphi_{\beta\beta'} u_\beta\begin{pmatrix} l' \\ \kappa' \end{pmatrix}. \tag{71.6}$$

Then regrouping

$$V=\tfrac{1}{2}\sum_{\substack{l\kappa\alpha \\ l'\kappa'\beta}} u_\alpha\begin{pmatrix} l \\ \kappa \end{pmatrix} \left\{\sum_{\alpha'\beta'} \varphi_{\alpha\alpha'} \Phi'_{\alpha'\beta'}\begin{pmatrix} l_\varphi & l'_\varphi \\ \kappa_\varphi & \kappa'_\varphi \end{pmatrix} \varphi_{\beta\beta'}\right\} u_\beta\begin{pmatrix} l' \\ \kappa' \end{pmatrix}. \tag{71.7}$$

But, if we compare (71.7) and (67.8) we see that since V is the identical, invariant, object,

$$\Phi_{\alpha\beta}\begin{pmatrix} l & l' \\ \kappa & \kappa' \end{pmatrix} = \sum_{\alpha'\beta'} \varphi_{\alpha\alpha'} \Phi'_{\alpha'\beta'} \begin{pmatrix} l_\varphi & l'_\varphi \\ \kappa_\varphi & \kappa'_\varphi \end{pmatrix} \varphi_{\beta\beta'}. \tag{71.8}$$

Or, if we take the inverse transformation of (71.8) we obtain by using (70.12)

$$\Phi'_{\alpha'\beta'}\begin{pmatrix} l_\varphi & l'_\varphi \\ \kappa_\varphi & \kappa'_\varphi \end{pmatrix} = \sum_{\alpha\beta} \varphi_{\alpha\alpha'} \Phi_{\alpha\beta} \begin{pmatrix} l & l' \\ \kappa & \kappa' \end{pmatrix} \varphi_{\beta\beta'} \tag{71.9}$$

$$= \sum_{\alpha\beta} (\varphi^{-1})_{\alpha'\alpha} \Phi_{\alpha\beta} \begin{pmatrix} l & l' \\ \kappa & \kappa' \end{pmatrix} \varphi_{\beta\beta'}. \tag{71.10}$$

From (71.1)–(71.10) it is clear that the primed force matrix is appropriate to the transformed displacement field and thus, that we may formally consider the primed force matrix as resulting from transformation of the unprimed matrix. Perhaps the simplest way this may be seen is via the matrix formalism (67.13). Thus, let us call the column matrix whose elements are the $(u_{\{\varphi\}})_\alpha \begin{pmatrix} l_\varphi \\ \kappa_\varphi \end{pmatrix}$ as $[\boldsymbol{u}_{\{\varphi\}}]$. Hence we may proceed as follows, from

$$V = \tfrac{1}{2}\,\widetilde{[\boldsymbol{u}]}\,[\boldsymbol{\Phi}]\,[\boldsymbol{u}], \tag{71.11}$$

$$V = \tfrac{1}{2}\,\widetilde{[\boldsymbol{u}]}\,P^{-1}_{\{\varphi|\boldsymbol{t}\}}\,P_{\{\varphi|\boldsymbol{t}\}}\,[\boldsymbol{\Phi}]\,P^{-1}_{\{\varphi|\boldsymbol{t}\}}\,P_{\{\varphi|\boldsymbol{t}\}}\,[\boldsymbol{u}]. \tag{71.12}$$

Now

$$P_{\{\varphi|\boldsymbol{t}\}}\,[\boldsymbol{u}] = [\boldsymbol{u}_{\{\varphi\}}] \tag{71.13}$$

and

$$\widetilde{[\boldsymbol{u}]}\,P^{-1}_{\{\varphi|\boldsymbol{t}\}} = \widetilde{P_{\{\varphi|\boldsymbol{t}\}}\,[\boldsymbol{u}]} \tag{71.14}$$

$$= \widetilde{[\boldsymbol{u}_{\{\varphi\}}]}. \tag{71.15}$$

Then

$$V = \tfrac{1}{2}\,\widetilde{[\boldsymbol{u}_{\{\varphi\}}]}\,[\boldsymbol{\Phi}']\,[\boldsymbol{u}_{\{\varphi\}}], \tag{71.16}$$

where

$$[\boldsymbol{\Phi}'] \equiv P_{\{\varphi|\boldsymbol{t}\}}\,[\boldsymbol{\Phi}]\,P^{-1}_{\{\varphi|\boldsymbol{t}\}} \tag{71.17}$$

and denoting the components of $[\boldsymbol{\Phi}']$ as before: $\Phi'_{\alpha'\beta'}\begin{pmatrix} l_\varphi & l'_\varphi \\ \kappa_\varphi & \kappa_\varphi \end{pmatrix}$, then (71.16) and (71.1) are identical. But (71.17) can be interpreted to signify that the matrix $[\boldsymbol{\Phi}]$ transforms under crystal symmetry operations as a second-rank tensor field. The same argument which led to (69.4) and (69.5) also implies for $[\boldsymbol{\Phi}']$:

$$\Phi'_{\alpha'\beta'}\begin{pmatrix} l_\varphi & l'_\varphi \\ \kappa_\varphi & \kappa'_\varphi \end{pmatrix} = \Phi'_{\alpha'\beta'}\begin{pmatrix} \lambda_\varphi & 0 \\ \kappa_\varphi & \kappa'_\varphi \end{pmatrix}, \tag{71.18}$$

where

$$\boldsymbol{R}_{\lambda_\varphi} \equiv \boldsymbol{R}_{L_\varphi} - \boldsymbol{R}_{L'_\varphi}. \tag{71.19}$$

We now need to make explicit use of the property of symmetry transformations producing a replica of the initial system. That is, a point symmetry transformation $P_{\{\varphi|\boldsymbol{t}\}}$ connects two points $\boldsymbol{r}$ and $\boldsymbol{r}_{\{\varphi\}}$ at which the crystal possesses identical physical properties. If we consider the tensor field quantities $[\boldsymbol{\Phi}]$ and $[\boldsymbol{\Phi}']$ to represent point physical properties then symmetry requires that

$$\Phi'_{\alpha'\beta'}\begin{pmatrix} l_\varphi & l'_\varphi \\ \kappa_\varphi & \kappa_\varphi \end{pmatrix} = \Phi_{\alpha'\beta'}\begin{pmatrix} l_\varphi & l'_\varphi \\ \kappa_\varphi & \kappa_\varphi \end{pmatrix} \tag{71.20}$$

or

$$\Phi'_{\alpha'\beta'}\begin{pmatrix} l & l' \\ \kappa & \kappa' \end{pmatrix} = \Phi_{\alpha'\beta'}\begin{pmatrix} l & l' \\ \kappa & \kappa' \end{pmatrix}. \tag{71.21}$$

It is essential that we understand (71.20) and (71.21) to mean that the symmetry transformation is one which preserves the functional form of the (tensor) matrix field $[\boldsymbol{\Phi}]$. In these equations we are equating the same component (indices $\alpha\beta$) of primed and unprimed (transformed, and initial) matrices evaluated at the *same* field point. The field point can be indifferently taken as $\begin{pmatrix} l & l' \\ \kappa & \kappa' \end{pmatrix}$ or the transformed set $\begin{pmatrix} l_\varphi & l'_\varphi \\ \kappa_\varphi & \kappa'_\varphi \end{pmatrix}$. In contrast, (71.4) and (71.5) represent the relation between the force matrices at two *different* sets of points related by symmetry transformation $\{\varphi|\boldsymbol{t}\}$. Eqs. (71.20) and (71.21) are the statement of form invariance of the physical tensor field $[\boldsymbol{\Phi}]$. Omitting the coordinate label, (71.20) and (71.21) can be written

$$\Phi'_{\alpha\beta} = \Phi_{\alpha\beta} \tag{71.22}$$

or

$$P_{\{\varphi|\tau(\varphi)\}}[\boldsymbol{\Phi}]\,P^{-1}_{\{\varphi|\tau(\varphi)\}} \equiv [\boldsymbol{\Phi}'] = [\boldsymbol{\Phi}]. \tag{71.23}$$

If we combine (71.9) and (71.20) we have

$$\Phi'_{\alpha'\beta'}\begin{pmatrix} l_\varphi & l'_\varphi \\ \kappa_\varphi & \kappa'_\varphi \end{pmatrix} = \sum_\alpha\sum_\beta \varphi_{\alpha\alpha'}\,\Phi_{\alpha\beta}\begin{pmatrix} l & l' \\ \kappa & \kappa' \end{pmatrix}\varphi_{\beta\beta'} = \Phi_{\alpha'\beta'}\begin{pmatrix} l_\varphi & l'_\varphi \\ \kappa_\varphi & \kappa'_\varphi \end{pmatrix}. \tag{71.24}$$

From (71.8) and (71.21) we have the inverse of (71.24). The physical interpretation of (71.24) is that it connects the value of the same field, the physical force matrix field $[\boldsymbol{\Phi}]$, at two sets of equivalent points which are connected by a space symmetry operation. Using (71.24) one could determine the minimum set of non-vanishing force matrix elements (coupling parameters) for any given space group by permitting the symmetry operations $\{\varphi|\tau(\varphi)\}$ to run through all elements in $\mathfrak{G}$, or at least those elements which give independent results. This has been discussed in another article, in this Encyclopedia, by Liebfried [*4*] and Lax [*58*].

From the definition of the potential energy (67.6) it follows that the force matrix $[\boldsymbol{\Phi}]$ is real. This can be expressed, for our present purposes by writing the transformation of $[\boldsymbol{\Phi}]$ by the operator K, as in (70.36) *et seq.*

$$K[\boldsymbol{\Phi}]\,K^{-1} = \Phi^* = \Phi. \tag{71.25}$$

72. Solution of the equations of motion: Eigenvectors $[e_j]$. In the next few sections it will be demonstrated that owing to the space group symmetry in configuration space, the solutions of the equations of motion (67.19), fall into irreducible vector spaces. Each such irreducible vector space is a basis for an $(s\cdot l_m)$ dimensional irreducible representation of $\mathfrak{G}$; and each such irreducible vector space is associated with a particular eigenfrequency of (67.19).

To solve (67.19) we proceed in conventional fashion [13]. Introduce mass weighted displacement coordinates $\boldsymbol{w}$, by

$$w_\alpha \begin{pmatrix} l \\ \kappa \end{pmatrix} = \sqrt{M_\kappa}\, u_\alpha \begin{pmatrix} l \\ \kappa \end{pmatrix}. \tag{72.1}$$

The time dependence is removed by taking

$$w_\alpha \begin{pmatrix} l \\ \kappa \end{pmatrix} = \xi_\alpha \begin{pmatrix} l \\ \kappa \end{pmatrix} \mathrm{e}^{-i\omega t}. \tag{72.2}$$

In (72.2) the $\xi_\alpha \begin{pmatrix} l \\ \kappa \end{pmatrix}$ are time independent amplitudes. If we define

$$D_{\alpha\beta} \begin{pmatrix} l & l' \\ \kappa & \kappa' \end{pmatrix} \equiv \frac{1}{\sqrt{M_\kappa M_\kappa}}\, \Phi_{\alpha\beta} \begin{pmatrix} l & l' \\ \kappa & \kappa' \end{pmatrix} \tag{72.3}$$

then the equations of motion (67.19) become

$$-\omega^2 \xi_\alpha \begin{pmatrix} l \\ \kappa \end{pmatrix} + \sum_{l'\kappa'\beta} D_{\alpha\beta} \begin{pmatrix} l & l' \\ \kappa & \kappa' \end{pmatrix} \xi_\beta \begin{pmatrix} l' \\ \kappa' \end{pmatrix} = 0. \tag{72.4}$$

To achieve a condensation of form for the Eqs. (72.4), we follow the procedure (67.11)–(67.18). Thus defining a column matrix $[\boldsymbol{\xi}]$ whose components are $\xi_\alpha \begin{pmatrix} l \\ \kappa \end{pmatrix}$ and a matrix $[\boldsymbol{D}]$ from the components $D_{\alpha\beta} \begin{pmatrix} l & l' \\ \kappa & \kappa' \end{pmatrix}$ we can rewrite (72.4) as

$$[\boldsymbol{D}]\,[\boldsymbol{\xi}] = [\omega^2 \boldsymbol{I}]\,[\boldsymbol{\xi}], \tag{72.5}$$

where $[\boldsymbol{I}]$ is a unit matrix and ω^2 are the eigenfrequencies of the lattice problem, to be determined. The solubility condition for (72.5) is that the secular determinant vanish, or:

$$\left\| D_{\alpha\beta} \begin{pmatrix} l & l' \\ \kappa & \kappa' \end{pmatrix} - \omega^2 \delta_{\alpha\beta}\, \delta_{ll'}\, \delta_{\kappa\kappa'} \right\| = 0 \tag{72.6}$$

when written in this way the Eq. (72.6) possesses $3Nr$ roots $\omega_j^2, j=1, \ldots, 3Nr$.

Consider a particular eigenvector which is a $3Nr$ rowed column matrix belonging to the eigenvalue ω_j^2 of (72.5). We denote the eigenvector $[\boldsymbol{e}_j]$ and then (72.5) becomes

$$[\boldsymbol{D}]\,[\boldsymbol{e}_j] = \omega_j^2\,[\boldsymbol{e}_j]. \tag{72.7}$$

It is convenient to call the components of $[\boldsymbol{e}_j]$ as $e_\alpha \left(\begin{matrix} l \\ \kappa \end{matrix}\middle| j\right)$, so that, when written out, (72.7) becomes, for this eigenvalue (fixed j):

$$-\omega_j^2\, e_\alpha \left(\begin{matrix} l \\ \kappa \end{matrix}\middle| j\right) + \sum_{l'\kappa'\beta} D_{\alpha\beta} \begin{pmatrix} l & l' \\ \kappa & \kappa' \end{pmatrix} e_\beta \left(\begin{matrix} l' \\ \kappa' \end{matrix}\middle| j\right) = 0. \tag{72.8}$$

Clearly, the component $e_\alpha \left(\begin{matrix} l \\ \kappa \end{matrix}\middle| j\right)$ of the matrix $[\boldsymbol{e}_j]$ represents the α-th component of the time independent amplitude of the mass weighted displacement of the atom which was initially at $r\begin{pmatrix} l \\ \kappa \end{pmatrix}$, when the j-th normal mode of frequency ω_j is excited in the lattice.

From (67.8)–(67.10), (71.25) and (72.3), it follows that the matrix $[D]$ is positive semidefinite, real and symmetric

$$[D] \geqq 0, \tag{72.9}$$

$$[D] = [D]^*, \tag{72.10}$$

$$\widetilde{[D]} = [D]. \tag{72.11}$$

For such a matrix, the eigenvalues are real, non-negative:

$$\omega_j^2 \geqq 0. \tag{72.12}$$

The eigenvectors of such a matrix as $[D]$ are then a complete set, which, irrespective of degeneracy, may be taken as real, and orthonormalized in the real sense:

$$\sum_{\kappa l \alpha} e_\alpha \left(\begin{matrix} l \\ \kappa \end{matrix}\middle| j\right) e_\alpha \left(\begin{matrix} l \\ \kappa \end{matrix}\middle| j'\right) = \delta_{jj'} \tag{72.13}$$

and

$$\sum_{j} e_\alpha \left(\begin{matrix} l \\ \kappa \end{matrix}\middle| j\right) e_\beta \left(\begin{matrix} l' \\ \kappa' \end{matrix}\middle| j\right) = \delta_{\alpha\beta}\, \delta_{ll'}\, \delta_{\kappa\kappa'}. \tag{72.14}$$

To allow for degeneracy we need to introduce another index. That is, if the linearly independent eigenvectors

$$[e_{j_1}], \ldots, [e_{j_\rho}], \ldots, [e_{j_{l_j}}] \tag{72.15}$$

all satisfy (72.7)

$$[D]\,[e_{j_\rho}] = \omega_j^2\, [e_{j_\rho}] \tag{72.16}$$

then the orthonormality rules (72.13) and (72.14) can be taken as

$$\sum_{\kappa l \alpha} e_\alpha \left(\begin{matrix} l \\ \kappa \end{matrix}\middle| j_\rho\right) e_\alpha \left(\begin{matrix} l \\ \kappa \end{matrix}\middle| j'_{\rho'}\right) = \delta_{jj'}\, \delta_{\rho\rho'}, \tag{72.17}$$

$$\sum_{j} \sum_{\rho} e_\alpha \left(\begin{matrix} l \\ \kappa \end{matrix}\middle| j_\rho\right) e_\beta \left(\begin{matrix} l' \\ \kappa' \end{matrix}\middle| j_\rho\right) = \delta_{\alpha\beta}\, \delta_{ll'}\, \delta_{\kappa\kappa'}. \tag{72.18}$$

To obtain the orthonormality rules (72.17) and (72.18) in case of degeneracy we have made an arbitrary choice of the linear combinations of degenerate eigenvectors. This is always possible by using the Schmidt-Gram orthogonalization procedure.

Equivalently, we can consider a particular $[e_{j_\rho}]$ as a $3Nr$ rowed column vector, each component of which requires 3 indices for specification: $\kappa\, l\, \alpha$. Then (72.17) can be understood as an orthogonality in the sense of a scalar product[1]

$$[e_{j_\rho}] \cdot [e_{j'_{\rho'}}] \equiv (e_{j_\rho}, e_{j'_{\rho'}}) = \delta_{jj'}\, \delta_{\rho\rho'}. \tag{72.19}$$

A row vector $[e_{\alpha l \kappa}]$ can be similarly defined consisting of $3Nr$ columns; each column is specified by the index j_ρ. Then (72.18) could define another scalar

[1] Single dot scalar product is sum on $(\kappa\, l\, \alpha)$.

product by [2]

$$[\boldsymbol{e}_{\alpha l \kappa}]:[\boldsymbol{e}_{\beta l' \kappa'}] \equiv (\boldsymbol{e}_{\alpha l \kappa}, \boldsymbol{e}_{\beta l' \kappa'}) = \delta_{\alpha\beta}\,\delta_{ll'}\,\delta_{\kappa\kappa'}. \tag{72.20}$$

The totality of column eigenvectors

$$[\boldsymbol{e}_{j_\rho}], \qquad j=1,\ldots;\ \rho=1,\ldots,l_j \tag{72.21}$$

can be assembled into a $3Nr \times 3Nr$ matrix. Similarly the totality of row eigenvectors can be so assembled into a $3Nr \times 3Nr$ matrix:

$$[\boldsymbol{e}_{\alpha l \kappa}], \qquad \alpha=1,\ldots,3;\ l=1,\ldots,N;\ \kappa=1,\ldots,r. \tag{72.22}$$

The matrix so formed – either from the rows or columns, is the same one. Call it $\boldsymbol{E}$. Then the rows of $\boldsymbol{E}$ are labelled by three indices $\kappa\, l\, \alpha$, the columns by the index j_ρ. Written out, in this convention, $\boldsymbol{E}$ is a $3Nr \times 3Nr$ matrix with components as follows

$$\boldsymbol{E} = \begin{pmatrix} e_1\begin{pmatrix}0\\1\end{pmatrix}\!\Big|1\Big) & e_1\begin{pmatrix}0\\1\end{pmatrix}\!\Big|2\Big) & \ldots e_1\begin{pmatrix}0\\1\end{pmatrix}\!\Big|3Nr\Big) \\ e_2\begin{pmatrix}0\\1\end{pmatrix}\!\Big|1\Big) & e_2\begin{pmatrix}0\\1\end{pmatrix}\!\Big|2\Big) & \ldots e_2\begin{pmatrix}0\\1\end{pmatrix}\!\Big|3Nr\Big) \\ \vdots & \vdots & \\ e_3\begin{pmatrix}N-1\\r\end{pmatrix}\!\Big|1\Big) & e_3\begin{pmatrix}N-1\\r\end{pmatrix}\!\Big|2\Big) & \ldots e_3\begin{pmatrix}N-1\\r\end{pmatrix}\!\Big|3Nr\Big) \end{pmatrix}. \tag{72.23}$$

In writing (72.23) we adopt the convention that the range of l is $0,\ldots,N-1$ purely for reasons of typographical convenience. The orthonormality rules (72.19) and (72.10) then demonstrate that the columns of $\boldsymbol{E}$, and the rows of $\boldsymbol{E}$ are real-orthogonal vectors. Consequently $\boldsymbol{E}$ is a real unitary or orthogonal matrix:

$$\boldsymbol{E}^{-1} = \tilde{\boldsymbol{E}}. \tag{72.24}$$

Then using (72.24) and (72.16) we obtain

$$\boldsymbol{E}^{-1}[\boldsymbol{D}]\cdot E = \boldsymbol{\Delta}, \tag{72.25}$$

where $\boldsymbol{\Delta}$ is a diagonal matrix of the eigenvalues

$$\boldsymbol{\Delta} = \begin{pmatrix} \omega_1^2 & 0 & 0 & & & \\ 0 & \omega_1^2 & & & & \\ & & \ddots & & & \\ 0 & & \omega_1^2 & & & \\ 0 & & & \omega_2^2 & & \\ & & & & \ddots & \\ & & & & \omega_2^2 & \\ & & & & & \ddots \end{pmatrix}. \tag{72.26}$$

Each eigenvalue of $[\boldsymbol{D}]$ appears in (72.26) as often as its degree of degeneracy. Note that the specific form (72.26) does follow from the orthonormality rules (72.17) and (72.18).

The property of completeness of the set (72.15) requires that, if $[\boldsymbol{\varepsilon}_{j_\nu}]$ is an arbitrary eigenvector of $[\boldsymbol{D}]$ with eigenvalue ω_j^2:

$$[\boldsymbol{D}][\boldsymbol{\varepsilon}_{j_\nu}] = \omega_j^2 [\boldsymbol{\varepsilon}_{j_\nu}] \tag{72.27}$$

---

[2] Double dot scalar product is sum on (j_ρ).

then $[\varepsilon_{j_v}]$ can be expressed as a linear combination of the members of the set (72.15):

$$[\varepsilon_{j_v}] = \sum_{\rho=1}^{l_j} (\rho|v)\,[e_{j_\rho}]. \tag{72.28}$$

In components, (72.28) becomes

$$\varepsilon_\alpha\begin{pmatrix} l \\ \kappa \end{pmatrix}\!\!\left|\, j_v \right) = \sum_\rho (\rho|v)\, e_\alpha\begin{pmatrix} l \\ \kappa \end{pmatrix}\!\!\left|\, j_\rho \right). \tag{72.29}$$

The coefficients $(\rho|v)$ determine the numbers of times the entire eigenvector $[e_{j_\rho}]$ occurs in $[\varepsilon_{j_v}]$. Using (72.17), $(\rho|v)$ can be determined as

$$(\rho|v) = \sum_{\kappa l \alpha} e_\alpha\begin{pmatrix} l \\ \kappa \end{pmatrix}\!\!\left|\, j_\rho \right)\, \varepsilon_\alpha\begin{pmatrix} l \\ \kappa \end{pmatrix}\!\!\left|\, j_v \right) = (e_{j_\rho}, \varepsilon_{j_v}). \tag{72.30}$$

Anticipating the later result to be proved below, the set (72.15) span an irreducible linear vector space $\Sigma^{(\star \boldsymbol{k})(m)}$ of $\mathfrak{G}$. The indices j, ρ of the present work must then be related to the indices $\boldsymbol{k}$, m, which characterize the irreducible representation in a definite fashion to be exploited below.

73. Real normal coördinates q_j. To introduce a set of normal coordinates, the mass weighted displacements (72.1) are written in terms of the complete set of eigenvectors, allowing for degeneracy. That is we take

$$w_\alpha\begin{pmatrix} l \\ \kappa \end{pmatrix} = \sum_j \sum_\rho e_\alpha\begin{pmatrix} l \\ \kappa \end{pmatrix}\!\!\left|\, j_\rho \right)\, q_{j_\rho}. \tag{73.1}$$

The q_{j_ρ} are real, time dependent objects, which from (73.1) can be interpreted as the coefficient of the eigenvector $[e_{j_\rho}]$ in the vector $\boldsymbol{w}$. Using (72.19) we invert (73.1)

$$q_{j_\rho} = \sum_{\kappa l \alpha} e_\alpha\begin{pmatrix} l \\ \kappa \end{pmatrix}\!\!\left|\, j_\rho \right)\, w_\alpha\begin{pmatrix} l \\ \kappa \end{pmatrix}, \tag{73.2}$$

$$q_{j_\rho} = [e_{j_\rho}] \cdot \boldsymbol{w}. \tag{73.3}$$

Thus q_{j_ρ} is the projection of $\boldsymbol{w}$ upon $[e_{j_\rho}]$. We also can write (73.1) as

$$\boldsymbol{w} = \sum_j \sum_\rho [e_{j_\rho}]\, q_{j_\rho}. \tag{73.4}$$

Taking the time derivative of (73.4)

$$\dot{\boldsymbol{w}} = \sum_j \sum_\rho [e_{j_\rho}]\, \dot{q}_{j_\rho}. \tag{73.5}$$

The kinetic energy (67.14) is, using the scalar product (72.17),

$$T = \tfrac{1}{2}[\dot{\boldsymbol{w}}] \cdot [\dot{\boldsymbol{w}}] = \tfrac{1}{2} \sum_{j\rho} \sum_{j'\rho'} [e_{j_\rho}] \cdot [e_{j'_{\rho'}}]\, \dot{q}_{j_\rho}\, \dot{q}_{j'_{\rho'}} = \tfrac{1}{2} \sum_{j\rho} \dot{q}_{j_\rho}^2, \tag{73.6}$$

where we used (72.19). For the potential energy (67.8) or (67.13), we obtain

$$V=\tfrac{1}{2}[\boldsymbol{u}][\boldsymbol{\Phi}][\boldsymbol{u}] \qquad (73.7)$$
$$=\tfrac{1}{2}[\boldsymbol{w}]\cdot[\boldsymbol{D}]\cdot[\boldsymbol{w}]$$

where $[\boldsymbol{D}]$ is defined in (72.3). Substituting (73.4) into (72.16) we obtain

$$[\boldsymbol{D}][\boldsymbol{w}]=\sum_j\sum_\rho \omega_j^2 [\boldsymbol{e}_{j_\rho}]\, q_{j_\rho}. \qquad (73.8)$$

Hence

$$V=\tfrac{1}{2}\sum_{j\rho}\sum_{j'\rho'} \omega_j^2 [\boldsymbol{e}_{j'_{\rho'}}]\cdot[\boldsymbol{e}_{j_\rho}]\, q_{j'_{\rho'}}\, q_{j_\rho} \qquad (73.9)$$

$$=\tfrac{1}{2}\sum_{j\rho} \omega_j^2\, q_{j_\rho}^2. \qquad (73.10)$$

Consequently the classical Hamiltonian is

$$\mathscr{H}(q_{j_\rho}, \dot{q}_{j_\rho})=\tfrac{1}{2}\sum_{j\rho}\{\dot{q}_{j_\rho}^2+\omega_j^2\, q_{j_\rho}^2\} \qquad (73.11)$$

and the equations of motion are

$$\dot{p}_{j_\rho}=\ddot{q}_{j_\rho}=-\frac{\partial\mathscr{H}}{\partial q_{j_\rho}}=-\omega_j^2\, q_{j_\rho}. \qquad (73.12)$$

Hence

$$\ddot{q}_{j_\rho}+\omega_j^2\, q_{j_\rho}=0, \qquad j=1,\ldots,; \ \rho=1,\ldots,l_j. \qquad (73.13)$$

The ability to obtain the separable Eq. (73.13) clearly justifies calling the q_{j_ρ} normal coordinates. In a “normal mode” of vibration the coordinate $q_{j_\rho}(t)$ is excited and the total time dependent coordinate

$$q_{j_\rho}(t)=q_{j_\rho}^{s,c}(0)\begin{cases}\sin\omega_j t\\ \cos\omega_j t.\end{cases} \qquad (73.14)$$

When the crystal is in the classical state of one normal mode excited, the displacements are from (73.1)

$$u_\alpha\begin{pmatrix}l\\ \kappa\end{pmatrix}=\left(\frac{1}{\sqrt{M_\kappa}}\right) e_\alpha\left(\begin{matrix}l\\ \kappa\end{matrix}\middle| j_\rho\right) q_{j_\rho}^{s,c}(0)\begin{cases}\sin\omega_j t\\ \cos\omega_j t.\end{cases} \qquad (73.15)$$

In this case all atoms are oscillating at frequency ω_j; the relative amplitudes of atom displacements is governed by the eigenvector components of $[\boldsymbol{e}_{j_\rho}]$. More often one cannot excite a single normal mode, but, in principle, consider arbitrary displacements at frequency ω_j. The most general situation could then be described as a superposition of the degenerate normal modes at ω_j:

$$u_\alpha\begin{pmatrix}l\\ \kappa\end{pmatrix}=\left(\frac{l}{\sqrt{M_k}}\right)\sum_\rho e_\alpha\left(\begin{matrix}l\\ \kappa\end{matrix}\middle| j_\rho\right) q_{j_\rho}^{s,c}(0)\begin{cases}\sin\omega_j t\\ \cos\omega_j t\end{cases} \qquad (73.16)$$

The $q_{j_\rho}^{s,c}(0)$ are constants which may be prescribed by initial conditions.

Having chosen the eigenvectors $[\boldsymbol{e}_{j_\rho}]$ real, $u_\alpha\begin{pmatrix}l\\ \kappa\end{pmatrix}$ will be real if the $q_{j_\rho}^{s,c}(0)$ are taken real. We shall assume this to be the case in what follows. For generality it may be necessary to take the phase as $(\omega_j t+\delta_j)$ in (73.14)–(73.16), but we generally choose $\delta_j=0$.

74. Crystal symmetry and the eigenvectors $[e_j]$ of $[D]$. The basic property of completeness of the eigenvectors $[e_{j_\rho}]$ of the dynamical matrix for particular frequency ω_j, as in (72.28) permits the application of the group theory analysis. Coming back to (72.7), or (72.16), we transform the equation by means of a transformation operator $P_{\{\varphi|t\}}$:

$$P_{\{\varphi|t\}}[D]P_{\{\varphi|t\}}^{-1}\cdot P_{\{\varphi|t\}}[e_{j_\rho}]=\omega_j^2 P_{\{\varphi|t\}}[e_{j_\rho}]. \tag{74.1}$$

However from (71.17), and (71.23)

$$P_{\{\varphi|t\}}[D]P_{\{\varphi|t\}}^{-1}=[D']=[D]. \tag{74.2}$$

Hence

$$[D]P_{\{\varphi|t\}}[e_{j_\rho}]=\omega_j^2 P_{\{\varphi|t\}}[e_{j_\rho}]. \tag{74.3}$$

This is equivalent to

$$[D][e_{\{\varphi\}j_\rho}]=\omega_j^2[e_{\{\varphi\}j_\rho}] \tag{74.4}$$

with

$$[e_{\{\varphi\}j_\rho}]\equiv P_{\{\varphi|t\}}[e_{j_\rho}]. \tag{74.5}$$

Now there are two approaches we will take. First observe that owing to the completeness of the set $[e_{j_\rho}]$ of (72.15) the transformed eigenvector $[e_{\{\varphi\}j_\rho}]$, which is an eigenvector at eigenfrequency ω_j^2 can be written as a linear combination of the basic set:

$$[e_{\{\varphi\}j_\rho}]=\sum_{\rho'=1}^{l_j} D^{(j)}(\{\varphi|t\})_{\rho'\rho}[e_{j_{\rho'}}]. \tag{74.6}$$

The index j here in (74.6) is reserved to refer to the eigenstates at frequency ω_j. It thus follows that the l_j eigenvectors of (72.15) are a basis for a representation $D^{(j)}$ of the group of operators $\mathfrak{G}_P$. The development can follow that of Sects. 14–15.

On the other hand, a component of $[e_{j_\rho}]$ is $e_\alpha\left(\begin{smallmatrix}l\\ \kappa\end{smallmatrix}\middle| j_\rho\right)$. It has the physical significance of being the α-th cartesian component of displacement of atom at $\left(\begin{smallmatrix}l\\ \kappa\end{smallmatrix}\right)$ at frequency ω_j. The $[e_{j_\rho}]$ is then a physical vector (first rank tensor) field and has appropriate transformation properties. Thus

$$(e_{\{\varphi\}})_\alpha\left(\begin{matrix}l\\ \kappa\end{matrix}\middle| j_\rho\right)=\sum_\beta \varphi_{\alpha\beta}e_\beta\left(\begin{matrix}l_{\bar\varphi}\\ \kappa_{\bar\varphi}\end{matrix}\middle| j_\rho\right). \tag{74.7}$$

Observe that (74.6) relates entire eigenvectors – the transformed, as well as the initial ones, while (74.7) gives the relation between components. The two Eqs. (74.6) and (74.7) can be combined if we take a component of (74.6) and use (74.7), to give

$$\begin{aligned}(e_{\{\varphi\}})_\alpha\left(\begin{matrix}l\\ \kappa\end{matrix}\middle| j_\rho\right)&=\sum_\beta \varphi_{\alpha\beta}e_\beta\left(\begin{matrix}l_{\bar\varphi}\\ \kappa_{\bar\varphi}\end{matrix}\middle| j_\rho\right)\\ &=\sum_{\rho'=1}^{l_j} D^{(j)}(\{\varphi|t\})_{\rho'\rho}\,e_\alpha\left(\begin{matrix}l\\ \kappa\end{matrix}\middle| j_{\rho'}\right).\end{aligned} \tag{74.8}$$

Note in (74.8) that the value of eigenvectors at different field points enters.

The elements of $D^{(j)}(\{\varphi|t\})_{\rho'\rho}$ may be obtained from (74.6) by use of the orthonormality relations (72.19), as:

$$\sum_{\kappa l\alpha}(e_{\{\varphi\}})_\alpha\begin{pmatrix}l\\\kappa\end{pmatrix}j_\rho\Big)\, e_\alpha\begin{pmatrix}l\\\kappa\end{pmatrix}j_{\rho'}\Big)=\sum_{\kappa l\alpha\beta}e_\alpha\begin{pmatrix}l\\\kappa\end{pmatrix}j_{\rho'}\Big)\,\varphi_{\alpha\beta}\,e_\beta\begin{pmatrix}l_{\bar\varphi}\\\kappa_{\bar\varphi}\end{pmatrix}j_\rho\Big) \equiv D^{(j)}(\{\varphi|t\})_{\rho'\rho} \tag{74.9}$$

or

$$D^{(j)}(\{\varphi|t\})_{\rho'\rho}=(e_{\{\varphi\}\,j_\rho},e_{j_{\rho'}}). \tag{74.10}$$

The eigenvector $[e_{\{\varphi\}\,j_\rho}]$ may be interpreted as a rotated or transformed eigenvector corresponding to $[e_{j_\rho}]$. Since the set of rotated eigenvectors for fixed $\{\varphi|t\}$

$$[e_{\{\varphi\}\,j_1},\ldots,e_{\{\varphi\}\,j_\rho},\ldots,e_{\{\varphi\}\,j_{l_j}}] \tag{74.11}$$

is complete, the orthonormality relations of (72.19) and (72.20) apply as well for the set (74.11) so that both types of scalar product apply:

$$[e_{\{\varphi\}\,j_\rho}]\cdot[e_{\{\varphi\}\,j'_{\rho'}}]=\delta_{jj'}\,\delta_{\rho\rho'}, \tag{74.12}$$

$$[e_{\{\varphi\}\,\alpha l\kappa}]:[e_{\{\varphi\}\,\beta l'\kappa'}]=\delta_{\alpha\beta}\,\delta_{ll'}\,\delta_{\kappa\kappa'}. \tag{74.13}$$

Recall the definition of the two scalar products: single dot in (72.17) and (72.19); double dot in (72.18) and (72.20).

75. Necessary degeneracy of the eigenvectors $[e_j]$. The representation $D^{(j)}$ of $\mathfrak{G}$ is defined by (74.6) or equivalently by (74.10). It is a physical representation corresponding to the degenerate eigenvalue ω_j^2 of $[D]$, and based on the set of eigenvectors (72.15), which can be taken as spanning a linear vector space $\Sigma^{(j)}$. By MASCHKE's theorem $D^{(j)}$ is either irreducible or decomposable (see Sect. 15).

Now we introduce the physical assertion of necessary degeneracy. From (70.36), (72.10), we observe that the space

$$\Sigma^{(j)}=\{[e_{j_1}],\ldots,[e_{j_\rho}],\ldots,[e_{j_{l_j}}]\} \tag{75.1}$$

is identical to the space

$$K\Sigma^{(j)}=\Sigma^{(j)*}=\{[e_{j_1}]^*,\ldots,[e_{j_{l_j}}]^*\}=\Sigma^{(j)} \tag{75.2}$$

owing to the reality of the eigenvectors

$$[e_{j_\rho}]^*=[e_{j_\rho}]. \tag{75.3}$$

Consequently, the representation $D^{(j)}$ based upon $\Sigma^{(j)}$ is real:

$$D^{(j)}=D^{(j)*}. \tag{75.4}$$

Hence we assert: If $D^{(j)}$ is the representation by which the linearly independent eigenvectors $[e_{j_\rho}]$ transform, for fixed frequency ω_j^2 then:

$$D^{(j)}\text{ is a real, irreducible, representation of }\mathfrak{G}. \tag{75.5}$$

This real irreducible representation has also been called a "physically" irreducible representation. Then (75.5) is the assertion of necessary degeneracy. The cases of accidental degeneracy are situations where $D^{(j)}$ is reducible and we shall encounter them in the sequel.

When an irreducible representation $D^{(j')}$ of $\mathfrak{G}$ is not real, it cannot correspond to a physical eigenvalue. In that case the physically irreducible representation is evidently

$$D^{(j)} = D^{(j')} \oplus D^{(j')*}. \tag{75.6}$$

This case will be discussed separately in the sequel also.

For emphasis, it should be noted that (75.5) and (75.6) assert that the set of all degenerate physical eigenstates specified by the degenerate eigenvectors of the dynamical matrix must be a basis for a real, irreducible representation of $\mathfrak{G}$. However, the converse is manifestly not true. That is, every irreducible representation of $\mathfrak{G}$ does *not* correspond to a physical eigenstate. One reason for the falsity of the converse is that not every irreducible representation $D^{(\star\boldsymbol{k})(m)}$ of $\mathfrak{G}$ is real. This matter of the reality of $D^{(\star\boldsymbol{k})(m)}$ will be discussed below in more detail in Sect. 92. A second reason will be developed later, also, when it is demonstrated in Sect. 82 that the physical irreducible representations $D^{(\boldsymbol{k})(m)}$ which arise, as symmetry species of the normal modes, are only: those induced (in the special sense discussed below) from the vector, or cartesian displacement representation.

The assertion of necessary degeneracy permits a connection to be made between the purely mathematical analyses of Parts B to F of the irreducible representations of the space group $\mathfrak{G}$, and the physics of the symmetry property of eigenvectors of the dynamical matrix $[\boldsymbol{D}]$. Then the indices j_ρ which specify the eigenvectors in (75.1) can be correlated with the indices $(\star\boldsymbol{k})(m)$ of $D^{(\star\boldsymbol{k})(m)}$:

$$\Sigma^{(j)} \leftrightarrow \Sigma^{(\star\boldsymbol{k})(m)} \tag{75.7}$$

and the individual basis vectors

$$[\boldsymbol{e}_{j_\rho}] \leftrightarrow [\boldsymbol{e}_{\boldsymbol{k}_\sigma;\, m;\, \lambda}] \equiv \psi_\lambda^{(\boldsymbol{k}_\sigma)(m)}. \tag{75.8}$$

That is, a connection can be made between the eigenvectors $[\boldsymbol{e}_{j_\rho}]$ and Bloch vectors

$$\psi_\lambda^{(\boldsymbol{k}_\sigma)(m)}, \qquad \sigma = 1, \ldots, s;\ \lambda = 1, \ldots, l_m. \tag{75.9}$$

Evidently the $[\boldsymbol{e}_{j_\rho}]$ can be so chosen, as the correct linear combinations to achieve (75.8), (75.9). This program will be carried through with the introduction of the complex normal coordinates, starting in Sect. 80.

76. Crystal symmetry and the transformation of normal coordinates q_j. Coming back to the discussion in Sect. 73, the normal coordinate q_{j_ρ} is the projection of $\boldsymbol{w}$ upon the eigenvector $[\boldsymbol{e}_{j_\rho}]$ as in (73.1) and (73.4), so

$$w_\alpha\begin{pmatrix} l \\ \kappa \end{pmatrix} = \sum_j \sum_\rho e_\alpha\left(\begin{matrix} l \\ \kappa \end{matrix}\middle| j_\rho\right) q_{j_\rho}. \tag{76.1}$$

Consider the rotated displacement field $[\boldsymbol{w}_{\{\varphi\}}]$. This rotated physical displacement field may be expressed in terms of the unrotated eigenvectors $[\boldsymbol{e}_{j_\rho}]$ as

$$\boldsymbol{w}_{\{\varphi\}}=\sum_j \sum_\rho [\boldsymbol{e}_{j_\rho}]\, q_{\{\varphi\}\, j_\rho}. \tag{76.2}$$

In (76.2) the new, or rotated components $q_{\{\varphi\}\, j_\rho}$ of $\boldsymbol{w}_{\{\varphi\}}$ are defined as the projection of the rotated displacement field upon $[\boldsymbol{e}_{j_\rho}]$.

Alternatively we may express the rotated displacements $[\boldsymbol{w}_{\{\varphi\}}]$ upon rotated eigenvectors $[\boldsymbol{e}_{\{\varphi\}\, j_\rho}]$. It is easily seen that this case is given as

$$\boldsymbol{w}_{\{\varphi\}}=\sum_j \sum_\rho [\boldsymbol{e}_{\{\varphi\}\, j_\rho}]\, q_{j_\rho}, \tag{76.3}$$

where the q_{j_ρ} are the same in (76.3) and (76.1). Comparing (70.7), (74.7), and (76.1) we have

$$\begin{aligned}
(\boldsymbol{w}_{\{\varphi\}})_\alpha \begin{pmatrix} l \\ \kappa \end{pmatrix} &= \sum_\beta \varphi_{\alpha\beta}\, w_\beta \begin{pmatrix} l_{\bar\varphi} \\ \kappa_{\bar\varphi} \end{pmatrix} \\
&= \sum_\beta \sum_j \sum_\rho \varphi_{\alpha\beta}\, e_\beta \left(\begin{matrix} l_{\bar\varphi} \\ \kappa_{\bar\varphi} \end{matrix} \middle| j_\rho \right) q_{j_\rho} \\
&= \sum_j \sum_\rho (\boldsymbol{e}_{\{\varphi\}})_\alpha \left(\begin{matrix} l \\ \kappa \end{matrix} \middle| j_\rho \right) q_{j_\rho}.
\end{aligned} \tag{76.4}$$

Eq. (76.3) expresses the result that components of a rotated vector upon rotated axes are the same as those of the unrotated vector upon initial axes, consistent with the usual result in the familiar vector analysis of three dimensions. In our case this is the q_{j_ρ}, which are the components of the rotated vector.

To determine the $q_{\{\varphi\}\, j_\rho}$ of (76.2), we proceed in a fashion utilizing (74.6). From (76.2) and (72.19):

$$\begin{aligned}
q_{\{\varphi\}\, j_\rho} &= [\boldsymbol{e}_{j_\rho}]\cdot[\boldsymbol{w}_{\{\varphi\}}] \\
&= \sum_{j'\rho'} [\boldsymbol{e}_{j_\rho}]\cdot[\boldsymbol{e}_{\{\varphi\}\, j'_{\rho'}}]\, q_{j'_{\rho'}} \\
&= \sum_{j'\rho'} [\boldsymbol{e}_{j_\rho}]\cdot \sum_{\rho''} D^{(j')}(\{\varphi|\boldsymbol{t}\})_{\rho''\rho'} [\boldsymbol{e}_{j'_{\rho''}}]\, q_{j'_{\rho'}} \\
&= \sum_{j'\rho'\rho''} D^{(j')}(\{\varphi|\boldsymbol{t}\})_{\rho''\rho'}\, \delta_{jj'}\, \delta_{\rho\rho''}\, q_{j'\rho'} \\
&= \sum_{\rho'} D^{(j)}(\{\varphi|\boldsymbol{t}\})_{\rho\rho'}\, q_{j\rho'}.
\end{aligned} \tag{76.5}$$

Observe that use has been made of (74.6) in going from the second to the third line and this is essential to achieve (76.5).

It is important to recognize that from (76.5) we can obtain a rule by which the normal coordinates q_{j_ρ} "transform". Since these objects are abstract dynamical variables which are not simply related to the physical displacements, it is necessary to have a somewhat artificial rule in order to define their transformation properties under transformation operators $P_{\{\varphi|\boldsymbol{t}\}}$: the latter are defined via the transformation properties of functions (scalar, vector, tensor functions) of coordinates in configuration space. The q_{j_ρ} are themselves not properly functions of configuration space coordinates. This point can be a cause of confusion and it is important to distinguish transformation laws based upon physical objects (like displacements) from those based upon a defined rule.

We may then write

$$P_{\{\varphi|t\}}\, \boldsymbol{w} \equiv \boldsymbol{w}_{\{\varphi\}} = \sum_j \sum_\rho [\boldsymbol{e}_{j_\rho}]\, q_{\{\varphi\} j_\rho} = \sum_j \sum_\rho [\boldsymbol{e}_{\{\varphi\} j_\rho}]\, q_{j_\rho} \tag{76.6}$$

and

$$P_{\{\varphi|t\}}\, q_{j_\rho} \equiv q_{\{\varphi\} j_\rho} = \sum_{\rho'} D^{(j)}(\{\varphi|t\})_{\rho\rho'}\, q_{j_{\rho'}}. \tag{76.7}$$

Again observe that $D^{(j)}$ is a real physically irreducible representation of $\mathfrak{G}$, by (75.5).

It is useful to emphasize that the use of (74.6) has the effect of incorporating the assertion of necessary degeneracy into the analysis. This comes about by utilizing the completeness of the set of eigenvectors $[\boldsymbol{e}_{j_\rho}]$ at fixed ω_j^2. Otherwise a general expansion of rotated normal coordinates in terms of the complete set of $[\boldsymbol{e}_{j_\rho}]$ for all frequencies would produce:

$$\begin{aligned} q_{\{\varphi\} j_\rho} &= \sum_{\kappa l \alpha} e_\alpha \begin{pmatrix} l \\ \kappa \end{pmatrix} j_\rho \Big) (w_{\{\varphi\}})_\alpha \begin{pmatrix} l \\ \kappa \end{pmatrix} \\ &= \sum_{\kappa l \alpha\beta} e_\alpha \begin{pmatrix} l \\ \kappa \end{pmatrix} j_\rho \Big)\, \varphi_{\alpha\beta}\, w_\beta \begin{pmatrix} l_{\bar\varphi} \\ \kappa_{\bar\varphi} \end{pmatrix} \\ &= \sum_{\kappa l \alpha\beta} \sum_{j'\rho'} e_\alpha \begin{pmatrix} l \\ \kappa \end{pmatrix} j_\rho \Big)\, \varphi_{\alpha\beta}\, e_\beta \begin{pmatrix} l_{\bar\varphi} \\ \kappa_{\bar\varphi} \end{pmatrix} j'_{\rho'} \Big)\, q_{j'_{\rho'}}. \end{aligned} \tag{76.8}$$

This expression must be identical to (76.5) although it is not obvious. In fact taking (76.8) and comparing with (76.5) we obtain

$$\begin{aligned} &\sum_{\kappa l \alpha\beta} \sum_{j'\rho'} e_\alpha \begin{pmatrix} l \\ \kappa \end{pmatrix} j_\rho \Big)\, \varphi_{\alpha\beta}\, e_\beta \begin{pmatrix} l_{\bar\varphi} \\ \kappa_{\bar\varphi} \end{pmatrix} j'_{\rho'} \Big)\, q_{j'_{\rho'}} \\ &\quad = \sum_{j'\rho'} \left\{ \sum_{\kappa l \alpha\beta} e_\alpha \begin{pmatrix} l \\ \kappa \end{pmatrix} j_\rho \Big)\, \varphi_{\alpha\beta}\, e_\beta \begin{pmatrix} l_{\bar\varphi} \\ \kappa_{\bar\varphi} \end{pmatrix} j'_{\rho'} \Big) \right\} q_{j'_{\rho'}} \\ &\quad = \sum_{j'\rho'} \{ D^{(j)}(\{\varphi|t\})_{\rho\rho'}\, \delta_{jj'} \}\, q_{j'_{\rho'}} \end{aligned} \tag{76.9}$$

so comparing with (74.9):

$$\sum_{\kappa l \alpha\beta} e_\alpha \begin{pmatrix} l \\ \kappa \end{pmatrix} j_\rho \Big)\, \varphi_{\alpha\beta}\, e_\beta \begin{pmatrix} l_{\bar\varphi} \\ \kappa_{\bar\varphi} \end{pmatrix} j'_{\rho'} \Big) = D^{(j)}(\{\varphi|t\})_{\rho\rho'}\, \delta_{jj'}. \tag{76.10}$$

Evidently (76.10) expresses the orthogonality of eigenvectors belonging to different manifolds (different ω_j^2 values) as well as the definition (74.9) of the transformation coefficients within one manifold. Upon using the definition of the inverse rotation, it is apparent that (76.10) is entirely equivalent to (74.9), which has thus been obtained in two different, but related ways. In words (74.9), and (76.10) simply give the matrix elements of the matrix $D^{(j)}$ by which the eigenvectors, or normal coordinates, transform in terms of linear combinations of components of transformed and original eigenvectors.

77. Fourier transformations. The work of Sects. 68–76 has resulted in the demonstration that the set of all normal coordinates q_{j_ρ} contains within itself bases for the entire irreducible representation $D^{(\star k)(m)}$ of $\mathfrak{G}$. The assertion of necessary degeneracy was given in (75.5) and plays an essential rôle in this. The next steps in the work are to determine the correct linear combinations of the q_{j_ρ} which are bases for the irreducible representation $D^{(k)(m)}$ of $\mathfrak{G}(\boldsymbol{k})$. In the next few sections we shall construct the basic Bloch vectors which reduce $\mathfrak{T}$. The relationship of these to the Fourier transformed displacements, to the dynamical matrix, and to the eigenvectors of the dynamical matrix will be presented. The complex normal coordinates which arise in this connection will be introduced and their transformation properties under the operators in $\mathfrak{G}(\boldsymbol{k})$ will be presented.

Let us recall (24.7) and (24.10) which we now rewrite as

$$\sum_{\boldsymbol{R}_L} \left(\frac{\exp i\,\boldsymbol{R}_L\cdot\boldsymbol{k}}{\sqrt{N}}\right)\left(\frac{\exp i\,\boldsymbol{R}_L\cdot\boldsymbol{k}'}{\sqrt{N}}\right)^* = \delta_{\boldsymbol{k},\boldsymbol{k}'} \tag{77.1}$$

and

$$\sum_{\boldsymbol{k}} \left(\frac{\exp i\,\boldsymbol{k}\cdot\boldsymbol{R}_L}{\sqrt{N}}\right)\left(\frac{\exp i\,\boldsymbol{k}\cdot\boldsymbol{R}_{L'}}{\sqrt{N}}\right)^* = \delta_{LL'}. \tag{77.2}$$

These relations were earlier obtained from the row and column orthonormality relations for the irreducible representations of the translation group $\mathfrak{T}$. They may equally be considered as a set of completeness relations for the sets of functions

$$\frac{1}{\sqrt{N}} \exp i\,\boldsymbol{k}\cdot\boldsymbol{R}_L \tag{77.3}$$

which are defined for all $\boldsymbol{k}$ in the zone (fixed $\boldsymbol{R}_L$), or for all lattice vectors $\boldsymbol{R}_L$ (fixed $\boldsymbol{k}$). As before, $N=8\,N_1\,N_2\,N_3$ is the order of $\mathfrak{T}$.

The plane waves (77.3) are the basic objects in the Fourier transformation theory of the physical objects in question.

78. Fourier transformed displacements and force matrix: The dynamical matrix $[D(\boldsymbol{k})]$. Now take as the expansion of the Cartesian components of elementary displacements

$$u_\alpha\begin{pmatrix} l \\ \kappa \end{pmatrix} = \frac{1}{\sqrt{M_\kappa}}\,\frac{1}{\sqrt{N}} \sum_{\boldsymbol{k}} w_\alpha(\kappa\,|\,\boldsymbol{k})\, e^{i\boldsymbol{k}\cdot\boldsymbol{R}_L}, \tag{78.1}$$

where κ refers to one of the basis atoms in the cell, and $\boldsymbol{k}$ is a permitted wave vector. Clearly $w_\alpha(\kappa\,|\,\boldsymbol{k})$ are complex variables and we may take

$$w_\alpha^*(\kappa\,|\,\boldsymbol{k}) = w_\alpha(\kappa\,|\,-\boldsymbol{k}) \tag{78.2}$$

so that the $u_\alpha\begin{pmatrix} l \\ \kappa \end{pmatrix}$ are real. Consider the $w_\alpha(\kappa\,|\,\boldsymbol{k})$ as the new independent dynamical variables.

In order to obtain the equations of motion of these variables we first express the potential and kinetic energies in terms of the $w_\alpha(\kappa|\boldsymbol{k})$. Substituting (78.1) into (67.8) we find

$$V=\frac{1}{2}\sum_{\substack{l\kappa\alpha\\ l'\kappa'\beta}}\sum_{\boldsymbol{k},\boldsymbol{k}'}\Phi_{\alpha\beta}\begin{pmatrix} l & l' \\ \kappa & \kappa' \end{pmatrix}\frac{1}{N\sqrt{M_\kappa M_{\kappa'}}}\, w_\alpha(\kappa|\boldsymbol{k})\, w_\beta(\kappa'|\boldsymbol{k}')\, e^{i(\boldsymbol{k}\cdot\boldsymbol{R}_L+\boldsymbol{k}'\cdot\boldsymbol{R}_{L'})}. \tag{78.3}$$

Now using (77.1)–(77.2) and the property (69.7) of the force matrix we find for V:

$$V=\tfrac{1}{2}\sum_{\substack{\kappa\alpha\\ \kappa'\beta}}\sum_{\boldsymbol{k}} w_\alpha(\kappa|-\boldsymbol{k})\, D_{\alpha\beta}\begin{pmatrix} \boldsymbol{k} \\ \kappa\,\kappa' \end{pmatrix} w_\beta(\kappa'|\boldsymbol{k}), \tag{78.4}$$

where

$$D_{\alpha\beta}\begin{pmatrix} \boldsymbol{k} \\ \kappa\,\kappa' \end{pmatrix}\equiv\sum_{\lambda}\Phi_{\alpha\beta}\begin{pmatrix} \lambda \\ \kappa\,\kappa' \end{pmatrix}\frac{\exp -i\,\boldsymbol{k}\cdot\boldsymbol{R}_\lambda}{\sqrt{M_\kappa M_{\kappa'}}}. \tag{78.5}$$

The matrix $[\boldsymbol{D}(\boldsymbol{k})]$, whose matrix elements are given in (78.5) is the dynamical matrix of the crystal. $[\boldsymbol{D}(\boldsymbol{k})]$ is the Fourier transform of the force constant matrix $[\boldsymbol{\Phi}]$. The dynamical matrix which is to be considered complex, is Hermitian

$$D_{\alpha\beta}\begin{pmatrix} \boldsymbol{k} \\ \kappa\,\kappa' \end{pmatrix}=D_{\beta\alpha}\begin{pmatrix} \boldsymbol{k} \\ \kappa'\,\kappa \end{pmatrix}^*$$

or

$$[\boldsymbol{D}(\boldsymbol{k})]=[\boldsymbol{D}(\boldsymbol{k})]^+=\widetilde{[\boldsymbol{D}(\boldsymbol{k})]}^* \tag{78.6}$$

owing to the fact that the force matrix is symmetric

$$\Phi_{\alpha\beta}\begin{pmatrix} l & l' \\ \kappa & \kappa' \end{pmatrix}=\Phi_{\beta\alpha}\begin{pmatrix} l' & l \\ \kappa' & \kappa \end{pmatrix} \tag{78.7}$$

so

$$\Phi_{\alpha\beta}\begin{pmatrix} l-l' & 0 \\ \kappa & \kappa \end{pmatrix}=\Phi_{\beta\alpha}\begin{pmatrix} l'-l & 0 \\ \kappa' & \kappa \end{pmatrix} \tag{78.8}$$

(recall (69.4)). Then (78.4) becomes

$$V=\tfrac{1}{2}\sum_{\substack{\kappa\alpha\\ \kappa'\beta}}\sum_{\boldsymbol{k}} w_\alpha^*(\kappa|\boldsymbol{k})\, D_{\alpha\beta}\begin{pmatrix} \boldsymbol{k} \\ \kappa\,\kappa' \end{pmatrix} w_\beta(\kappa'|\boldsymbol{k}). \tag{78.9}$$

Similarly for the kinetic energy we find after substituting (78.1) into (67.14); and simplifying

$$T=\tfrac{1}{2}\sum_{\kappa\alpha}\sum_{\boldsymbol{k}}\dot{w}_\alpha(\kappa|\boldsymbol{k})^*\,\dot{w}_\alpha(\kappa|\boldsymbol{k}). \tag{78.10}$$

In terms of these complex variables we have as Lagrangian

$$\mathscr{L}=T-V=\tfrac{1}{2}\sum\dot{w}_\alpha(\kappa|\boldsymbol{k})^*\,\dot{w}_\alpha(\kappa|\boldsymbol{k})-\tfrac{1}{2}\sum\sum w_\alpha^*(\kappa|\boldsymbol{k})\, D_{\alpha\beta}\begin{pmatrix} \boldsymbol{k} \\ \kappa\,\kappa' \end{pmatrix} w_\beta(\kappa'|\boldsymbol{k}). \tag{78.11}$$

Clearly if we call $w_\alpha(\kappa|\boldsymbol{k})^*$ a coordinate, then its conjugate momentum is

$$\pi_\alpha(\kappa|\boldsymbol{k}) = \frac{\partial \mathscr{L}}{\partial \dot{w}_\alpha(\kappa|\boldsymbol{k})^*} = \dot{w}_\alpha(\kappa|\boldsymbol{k}). \tag{78.12}$$

From Hamilton's equations we have [13]

$$\dot{\pi}_\alpha(\kappa|\boldsymbol{k}) = \ddot{w}_\alpha(\kappa|\boldsymbol{k}) = -\sum_{\kappa'\beta} D_{\alpha\beta}\begin{pmatrix} \boldsymbol{k} \\ \kappa\,\kappa' \end{pmatrix} w_\beta(\kappa'|\boldsymbol{k}) \tag{78.13}$$

or

$$\ddot{w}_\alpha(\kappa|\boldsymbol{k}) + \sum_{\kappa'\beta} D_{\alpha\beta}\begin{pmatrix} \boldsymbol{k} \\ \kappa\,\kappa' \end{pmatrix} w_\beta(\kappa'|\boldsymbol{k}) = 0. \tag{78.14}$$

Note that we used (78.2) in order to eliminate the factor of $(\frac{1}{2})$, and also note that the coordinate $w_\alpha(\kappa|\boldsymbol{k})^*$ and the momentum $\dot{w}_\alpha(\kappa|\boldsymbol{k})$ are conjugate variables. Owing to the complex exponential in (78.1), this difficulty of dealing with complex displacement field variables $\boldsymbol{w}(|\boldsymbol{k})$ arises. A useful fashion of looking at the variable or object $w_\alpha(\kappa|\boldsymbol{k})$ is that it is the $\alpha\,\kappa$ component of the vector field object $\boldsymbol{w}(|\boldsymbol{k})$. Clearly there are $3r$ components

$$w_\alpha(\kappa|\boldsymbol{k}), \qquad \alpha = 1, \ldots, 3; \; \kappa = 1, \ldots, r \tag{78.15}$$

so

$$\boldsymbol{w}(|\boldsymbol{k}) \tag{78.16}$$

for fixed $\boldsymbol{k}$ is an object in $3r$ dimensional space. Now we assume a harmonic time dependence writing

$$w_\alpha(\kappa|\boldsymbol{k}) = \xi_\alpha(\kappa|\boldsymbol{k}) \exp i\,\omega(\boldsymbol{k}|j)\,t. \tag{78.17}$$

Substituting into (78.14) we obtain an eigenvalue problem

$$-\omega^2(\boldsymbol{k}|j)\,\xi_\alpha(\kappa|\boldsymbol{k}) + \sum_{\kappa'\beta} D_{\alpha\beta}\begin{pmatrix} \boldsymbol{k} \\ \kappa\,\kappa' \end{pmatrix} \xi_\beta(\kappa'|\boldsymbol{k}) = 0. \tag{78.18}$$

In the next section we solve (78.18) by finding eigenvectors and eigenvalues.

79. Eigenvectors of the dynamical matrix $[\boldsymbol{D}(\boldsymbol{k})]$. The $3r$ eigenvectors of the dynamical matrix Eq. (78.18) are taken as

$$\boldsymbol{e}\left(\left|\begin{matrix} \boldsymbol{k} \\ j \end{matrix}\right.\right), \qquad j = 1, \ldots, 3r \tag{79.1}$$

with each such eigenvector (79.1) having $3r$ components, for fixed j

$$e_\alpha\left(\kappa\left|\begin{matrix} \boldsymbol{k} \\ j \end{matrix}\right.\right), \qquad \alpha = 1, \ldots, 3; \; \kappa = 1, \ldots, r. \tag{79.2}$$

The eigenvalues of $[\boldsymbol{D}(\boldsymbol{k})]$ which correspond are labelled

$$\omega^2(\boldsymbol{k}|j), \qquad j = 1, \ldots, 3r. \tag{79.3}$$

Then, for fixed $\boldsymbol{k}$ and j, these satisfy:

$$\sum_{\kappa'\beta} D_{\alpha\beta}\begin{pmatrix}\boldsymbol{k}\\ \kappa\,\kappa'\end{pmatrix} e_\beta\left(\kappa'\left|\begin{matrix}\boldsymbol{k}\\ j\end{matrix}\right.\right)=\omega^2(\boldsymbol{k}|j)\, e_\alpha\left(\kappa\left|\begin{matrix}\boldsymbol{k}\\ j\end{matrix}\right.\right). \tag{79.4}$$

If there is no degeneracy, then the eigenvectors (79.2) can be chosen to satisfy a complex (Hermitian) orthonormality rule [13]:

$$\sum_{\kappa\alpha} e_\alpha^*\left(\kappa\left|\begin{matrix}\boldsymbol{k}\\ j\end{matrix}\right.\right) e_\alpha\left(\kappa\left|\begin{matrix}\boldsymbol{k}\\ j'\end{matrix}\right.\right)=\delta_{jj'}, \tag{79.5}$$

$$\sum_{j} e_\beta^*\left(\kappa'\left|\begin{matrix}\boldsymbol{k}\\ j\end{matrix}\right.\right) e_\alpha\left(\kappa\left|\begin{matrix}\boldsymbol{k}\\ j\end{matrix}\right.\right)=\delta_{\alpha\beta}\,\delta_{\kappa\kappa'}. \tag{79.6}$$

These follow from the hermiticity of the dynamical matrix $[\boldsymbol{D}(\boldsymbol{k})]$.

In case of degeneracy the eigenvectors of $[\boldsymbol{D}(\boldsymbol{k})]$ must be indexed with an added index to label the degenerate eigenvectors. We write, for an l_m-fold degenerate eigenvalue

$$[\boldsymbol{D}(\boldsymbol{k})]\cdot \boldsymbol{e}\left(\left|\begin{matrix}\boldsymbol{k}\\ j_\lambda\end{matrix}\right.\right)=\omega^2(\boldsymbol{k}|j)\,\boldsymbol{e}\left(\left|\begin{matrix}\boldsymbol{k}\\ j_\lambda\end{matrix}\right.\right),\qquad \lambda=1,\dots,l_m. \tag{79.7}$$

As before, in (72.17) and (72.18), in case of degenerate eigenvalue it is possible to produce a set of eigenvectors which are orthonormal with respect to the hermitian scalar product. The two scalar products (79.5) and (79.6) may then be rewritten, using the definition of single dot, and double dot scalar product as in (72.17)–(72.20).

$$\boldsymbol{e}\left(\left|\begin{matrix}\boldsymbol{k}\\ j_\lambda\end{matrix}\right.\right)^*\cdot \boldsymbol{e}\left(\left|\begin{matrix}\boldsymbol{k}\\ j'_{\lambda'}\end{matrix}\right.\right)=\delta_{jj'}\,\delta_{\lambda\lambda'} \tag{79.8}$$

and

$$\boldsymbol{e}_\alpha\left(\kappa\left|\boldsymbol{k}\right.\right):\boldsymbol{e}_\beta\left(\kappa'\left|\boldsymbol{k}\right.\right)^*=\delta_{\alpha\beta}\,\delta_{\kappa\kappa'} \tag{79.9}$$

or in components

$$\sum_{\kappa\alpha} e_\alpha^*\left(\kappa\left|\begin{matrix}\boldsymbol{k}\\ j_\lambda\end{matrix}\right.\right) e_\alpha\left(\kappa\left|\begin{matrix}\boldsymbol{k}\\ j'_{\lambda'}\end{matrix}\right.\right)=\delta_{jj'}\,\delta_{\lambda\lambda'}, \tag{79.10}$$

$$\sum_{j}\sum_{\lambda} e_\beta^*\left(\kappa'\left|\begin{matrix}\boldsymbol{k}\\ j_\lambda\end{matrix}\right.\right) e_\alpha\left(\kappa\left|\begin{matrix}\boldsymbol{k}\\ j'_{\lambda'}\end{matrix}\right.\right)=\delta_{\alpha\beta}\,\delta_{\kappa\kappa'}. \tag{79.11}$$

Owing to the hermiticity of $[\boldsymbol{D}(\boldsymbol{k})]$ as in (78.6), the eigenvalues (79.3) satisfy the condition

$$\omega^2(\boldsymbol{k}|j)\qquad\text{real.} \tag{79.12}$$

Owing to the positive semi-definiteness of the force matrix $[\boldsymbol{\Phi}]$, the dynamical matrix $[\boldsymbol{D}(\boldsymbol{k})]$ is also positive semi-definite and we have [13]

$$\omega^2(\boldsymbol{k}|j)>0\qquad\text{for stability.} \tag{79.13}$$

As in the earlier case (72.23)–(72.26) the eigenvectors can be composed into a matrix which we call here $\boldsymbol{E}(\boldsymbol{k})$. In this case it is a $3r\times 3r$ dimensional matrix, by contrast to $\boldsymbol{E}$ in (72.23) which was $3Nr \times 3Nr$ dimensional. Then

$$\boldsymbol{E}(\boldsymbol{k}) = \begin{pmatrix} e_1\left(1\middle|\begin{matrix}\boldsymbol{k}\\1\end{matrix}\right) & \dots & e_1\left(1\middle|\begin{matrix}\boldsymbol{k}\\3r\end{matrix}\right) \\ \vdots & & \\ e_3\left(r\middle|\begin{matrix}\boldsymbol{k}\\1\end{matrix}\right) & \dots & e_3\left(r\middle|\begin{matrix}\boldsymbol{k}\\3r\end{matrix}\right) \end{pmatrix}. \tag{79.14}$$

The matrix $\boldsymbol{E}(\boldsymbol{k})$ is unitary

$$\boldsymbol{E}(\boldsymbol{k})^{-1} = \boldsymbol{E}(\boldsymbol{k})^{+} = \widetilde{\boldsymbol{E}(\boldsymbol{k})}^{*} \tag{79.15}$$

owing to (79.8) and (79.9).

Then, it follows from (79.7) and (79.15) that

$$\boldsymbol{E}(\boldsymbol{k})^{-1}\,[\boldsymbol{D}(\boldsymbol{k})]\,\boldsymbol{E}(\boldsymbol{k}) = \boldsymbol{\Delta}(\boldsymbol{k}), \tag{79.16}$$

where $\boldsymbol{\Delta}(\boldsymbol{k})$ is the diagonal matrix of real eigenvalues at wave vector $\boldsymbol{k}$:

$$\boldsymbol{\Delta}(\boldsymbol{k}) = \begin{pmatrix} \omega^2(\boldsymbol{k}|1) & 0 & 0 \\ 0 & \omega^2(\boldsymbol{k}|1) & \\ 0 & \ddots & 0 \\ 0 & 0 & \omega^2(\boldsymbol{k}|j_m) \end{pmatrix}. \tag{79.17}$$

In (79.17) each eigenvalue is repeated as often as the degree of its degeneracy; an eigenvalue may appear an integral multiple (c_j) of its degeneracy (l_m) in case of *accidental degeneracy*.

Contrasting (79.4) and (72.8) we find that introducing the Fourier transformation has converted a problem of a secular equation of dimension $3Nr \times 3Nr$ into N distinct secular equations each $3r \times 3r$. Each of these equations such as (79.4) is specified by a fixed $\boldsymbol{k}$, one of the N distinct choices in the Brillouin zone. That is, from the single matrix $[\boldsymbol{D}]$ of (72.8), N Hermitian $3r \times 3r$ matrices $[\boldsymbol{D}(\boldsymbol{k})]$ have been constructed.

The complex eigenvectors $\boldsymbol{e}\left(\middle|\begin{matrix}\boldsymbol{k}\\j\end{matrix}\right)$ of (79.4) are related to the real $[\boldsymbol{e}_j]$ of (72.8) by

$$e_\alpha\left(\kappa\middle|\begin{matrix}\boldsymbol{k}\\j\end{matrix}\right) = \frac{1}{\sqrt{N}}\sum_L \mathrm{e}^{-i\boldsymbol{k}\cdot\boldsymbol{R}_L}\, e_\alpha\left(\begin{matrix}l\\\kappa\end{matrix}\middle|j\right). \tag{79.18}$$

The inverse of (79.18) is

$$e_\alpha\left(\begin{matrix}l\\\kappa\end{matrix}\middle|j\right) = \frac{1}{\sqrt{N}}\sum_{\boldsymbol{k}} \mathrm{e}^{+i\boldsymbol{k}\cdot\boldsymbol{R}_L}\, e_\alpha\left(\kappa\middle|\begin{matrix}\boldsymbol{k}\\j\end{matrix}\right). \tag{79.19}$$

It could be observed that (79.18) may be rewritten in the form of a projection operator applied to a basic displacement. Thus defining

$$P_{\{\varepsilon|-\boldsymbol{R}_L\}}\, e_\alpha\left(\begin{matrix}0\\\kappa\end{matrix}\middle|j\right) = e_\alpha\left(\begin{matrix}l\\\kappa\end{matrix}\middle|j\right) \tag{79.20}$$

we can write for (79.18)

$$e_\alpha\left(\kappa\middle|\begin{matrix}\boldsymbol{k}\\j\end{matrix}\right) = P^{(\boldsymbol{k})}\, e_\alpha\left(\begin{matrix}0\\\kappa\end{matrix}\middle|j\right), \tag{79.21}$$

where $P^{(\boldsymbol{k})}$ is the idempotent projection operator of Eq. (25.4) which produces Bloch function of wave vector $\boldsymbol{k}$. Notice also that some care is required in associating the degeneracy indices j on both sides of (79.18).

80. Complex normal coordinates. Now we introduce a set of complex normal coordinates $Q\begin{pmatrix}\boldsymbol{k}\\ j\end{pmatrix}$ by writing for the amplitudes $w_\alpha(\kappa|\boldsymbol{k})$ of (78.1)

$$w_\alpha(\kappa|\boldsymbol{k})=\sum_{j=1}^{3r} e_\alpha\left(\kappa\middle|\begin{matrix}\boldsymbol{k}\\ j\end{matrix}\right) Q\begin{pmatrix}\boldsymbol{k}\\ j\end{pmatrix}. \tag{80.1}$$

Using (79.8) we can invert (80.1)

$$Q\begin{pmatrix}\boldsymbol{k}\\ j\end{pmatrix}=\sum_{\kappa\alpha} e_\alpha^*\left(\kappa\middle|\begin{matrix}\boldsymbol{k}\\ j\end{matrix}\right) w_\alpha(\kappa|\boldsymbol{k}). \tag{80.2}$$

Substituting (80.1) into (78.9), and using (79.4) we obtain for the potential energy in the harmonic approximation

$$V=\tfrac{1}{2}\sum_{\boldsymbol{k}}\sum_{j} \omega^2(\boldsymbol{k}|j)\, Q\begin{pmatrix}\boldsymbol{k}\\ j\end{pmatrix}^* Q\begin{pmatrix}\boldsymbol{k}\\ j\end{pmatrix}. \tag{80.3}$$

The time derivative of (80.1) is

$$\dot{w}_\alpha(\kappa|\boldsymbol{k})=\sum_{j=1}^{3r} e_\alpha\left(\kappa\middle|\begin{matrix}\boldsymbol{k}\\ j\end{matrix}\right) \dot{Q}\begin{pmatrix}\boldsymbol{k}\\ j\end{pmatrix}. \tag{80.4}$$

Substituting (80.4) into (78.10) the kinetic energy is given as

$$T=\tfrac{1}{2}\sum_{\boldsymbol{k}}\sum_{j} \dot{Q}\begin{pmatrix}\boldsymbol{k}\\ j\end{pmatrix}^* \dot{Q}\begin{pmatrix}\boldsymbol{k}\\ j\end{pmatrix}. \tag{80.5}$$

Following (78.9)–(78.14) the equations of motion of the complex normal coordinates are derivable from the Hamiltonian

$$\mathscr{H}=T+V=\tfrac{1}{2}\sum_{\boldsymbol{k}}\sum_{j}\left\{\dot{Q}\begin{pmatrix}\boldsymbol{k}\\ j\end{pmatrix}^* \dot{Q}\begin{pmatrix}\boldsymbol{k}\\ j\end{pmatrix}+\omega^2(\boldsymbol{k}|j)\, Q\begin{pmatrix}\boldsymbol{k}\\ j\end{pmatrix}^* Q\begin{pmatrix}\boldsymbol{k}\\ j\end{pmatrix}\right\} \tag{80.6}$$

and are

$$\ddot{Q}\begin{pmatrix}\boldsymbol{k}\\ j\end{pmatrix}+\omega^2(\boldsymbol{k}|j)\, Q\begin{pmatrix}\boldsymbol{k}\\ j\end{pmatrix}=0 \tag{80.7}$$

thus justifying the $Q\begin{pmatrix}\boldsymbol{k}\\ j\end{pmatrix}$ as normal coordinates. The solutions of (80.7) are

$$Q\begin{pmatrix}\boldsymbol{k}\\ j\end{pmatrix}=Q_0\begin{pmatrix}\boldsymbol{k}\\ j\end{pmatrix}\exp\left[\pm i\,\omega(\boldsymbol{k}|j)\,t\right] \tag{80.8}$$

where $Q_0\begin{pmatrix}\boldsymbol{k}\\ j\end{pmatrix}$ is time independent.

Again observe that the energy (80.6) is not simply a sum of squares owing to the $Q\begin{pmatrix}\boldsymbol{k}\\ j\end{pmatrix}$ being complex. In (80.3), (80.5), (80.6) the sum is over all wave vectors $\boldsymbol{k}$ in the Brillouin zone and over all j at each $\boldsymbol{k}$.

The physical displacements $u_\alpha\begin{pmatrix} l \\ \kappa \end{pmatrix}$ are given as

$$u_\alpha\begin{pmatrix} l \\ \kappa \end{pmatrix}=\frac{1}{\sqrt{M_\kappa N}}\sum_j\sum_{\boldsymbol{k}} e^{i\boldsymbol{k}\cdot\boldsymbol{R}_L}\, e_\alpha\left(\kappa\left|\begin{matrix}\boldsymbol{k}\\ j\end{matrix}\right.\right) Q\begin{pmatrix}\boldsymbol{k}\\ j\end{pmatrix}. \tag{80.9}$$

The inverse of (80.9) is

$$Q\begin{pmatrix}\boldsymbol{k}\\ j\end{pmatrix}=\frac{1}{\sqrt{N}}\sum_L\sum_{\kappa\alpha} e^{-i\boldsymbol{k}\cdot\boldsymbol{R}_L}\, e_\alpha^*\left(\kappa\left|\begin{matrix}\boldsymbol{k}\\ j\end{matrix}\right.\right) u_\alpha\begin{pmatrix} l \\ \kappa \end{pmatrix}\sqrt{M_\kappa}. \tag{80.10}$$

From (80.9) and (80.10) we can see that the complex normal coordinate $Q\begin{pmatrix}\boldsymbol{k}\\ j\end{pmatrix}$ is the component of the mass weighted cartesian displacement, projected upon the vector $e^{-i\boldsymbol{k}\cdot\boldsymbol{R}_L}\, e_\alpha^*\left(\kappa\left|\begin{matrix}\boldsymbol{k}\\ j\end{matrix}\right.\right)$. The latter vector is a tensor product of the eigenvector and of a plane wave. Thus the conceptual situation here mirrors that of the real normal coordinates as discussed in Sect. 73.

81. Crystal symmetry and the dynamical matrix $[D(\boldsymbol{k})]$ and its eigenvectors. First we demonstrate that the $3r$ eigenvectors

$$\boldsymbol{e}\left(\left|\begin{matrix}\boldsymbol{k}\\ j\end{matrix}\right.\right),\qquad j=1,\ldots,3r$$

are Bloch vectors at wave vector $\boldsymbol{k}$. From (79.18)

$$\begin{aligned} e_\alpha\left(\kappa\left|\begin{matrix}\boldsymbol{k}\\ j\end{matrix}\right.\right)&=\frac{1}{\sqrt{N}}\sum_{\boldsymbol{R}_L} \mathrm{e}^{-i\boldsymbol{k}\cdot\boldsymbol{R}_L}\, e_\alpha\left(\begin{matrix} l\\ \kappa\end{matrix}\middle| j\right)\\ &=\frac{1}{\sqrt{N}}\sum_{\{\varepsilon|-\boldsymbol{R}_L\}} D^{(\boldsymbol{k})}(\{\varepsilon|-\boldsymbol{R}_L\})^*\, P_{\{\varepsilon|-\boldsymbol{R}_L\}}\, e_\alpha\left(\begin{matrix} 0\\ \kappa\end{matrix}\middle| j\right)\\ &=P^{(\boldsymbol{k})}\, e_\alpha\left(\begin{matrix} 0\\ \kappa\end{matrix}\middle| j\right). \end{aligned} \tag{81.1}$$

We identified the operator $P^{(\boldsymbol{k})}$ with that of (25.4) by comparing the summand in (81.1) and identifying $e^{-i\boldsymbol{k}\cdot\boldsymbol{R}_L}$ with $D^{(\boldsymbol{k})}(\{\varepsilon|-\boldsymbol{R}_L\})^*$ following (23.2). Rewriting (81.1) in a vector notation

$$\boldsymbol{e}\left(\left|\begin{matrix}\boldsymbol{k}\\ j\end{matrix}\right.\right)=P^{(\boldsymbol{k})}\,\boldsymbol{e}\left(0\middle| j\right) \tag{81.2}$$

where $\boldsymbol{e}\left(0\middle| j\right)$ is a vector whose $\alpha\kappa$ component is $e_\alpha\left(\begin{matrix} 0\\ \kappa\end{matrix}\middle| j\right)$, and which represents the amplitude of displacement of the atoms in cell $l=0$, in the j-th mode. We must understand the operation of $P_{\{\varepsilon|-\boldsymbol{R}_L\}}$ upon the displacement in the usual sense

$$P_{\{\varepsilon|-\boldsymbol{R}_L\}}\, e_\alpha\left(\begin{matrix} 0\\ \kappa\end{matrix}\middle| j\right)=e_\alpha\left(\begin{matrix} l\\ \kappa\end{matrix}\middle| j\right). \tag{81.3}$$

In this manner the displacement in the 0-th cell ($l=0$) may be decomposed into components with various Bloch vectors. It then follows from the properties of the operator $P^{(\boldsymbol{k})}$ that as in (25.6)

$$P_{\{\varepsilon|\boldsymbol{R}_{L'}\}}\, P^{(\boldsymbol{k})} = D^{(\boldsymbol{k})}(\{\varepsilon|\boldsymbol{R}_{L'}\})\, P^{(\boldsymbol{k})} \tag{81.4}$$

so

$$P_{\{\varepsilon|\boldsymbol{R}_{L'}\}}\, e_\alpha\left(\kappa\middle|\begin{matrix}\boldsymbol{k}\\ j\end{matrix}\right) = D^{(\boldsymbol{k})}(\{\varepsilon|\boldsymbol{R}_{L'}\})\, e_\alpha\left(\kappa\middle|\begin{matrix}\boldsymbol{k}\\ j\end{matrix}\right). \tag{81.5}$$

Direct application of the test operator $P_{\{\varepsilon|\boldsymbol{R}_{L'}\}}$ to (79.18) gives the same result:

$$\begin{aligned} P_{\{\varepsilon|\boldsymbol{R}_{L'}\}}\, e_\alpha\left(\kappa\middle|\begin{matrix}\boldsymbol{k}\\ j\end{matrix}\right) &= \frac{1}{\sqrt{N}}\sum_L \mathrm{e}^{-i\boldsymbol{k}\cdot\boldsymbol{R}_L}\, e_{\{\varepsilon|\boldsymbol{R}_{L'}\}\alpha}\left(\begin{matrix}l\\ \kappa\end{matrix}\middle| j\right) \\ &= \frac{1}{\sqrt{N}}\sum_L \mathrm{e}^{-i\boldsymbol{k}\cdot\boldsymbol{R}_L}\, e_\alpha\left(\begin{matrix}l-l'\\ \kappa\end{matrix}\middle| j\right) \\ &= \mathrm{e}^{-i\boldsymbol{k}\cdot\boldsymbol{R}_{L'}}\sum_L \mathrm{e}^{-i\boldsymbol{k}\cdot(\boldsymbol{R}_L-\boldsymbol{R}_{L'})}\, e_\alpha\left(\begin{matrix}l-l'\\ \kappa\end{matrix}\middle| j\right) \\ &= D^{(\boldsymbol{k})}(\{\varepsilon|\boldsymbol{R}_{L'}\})\, \mathrm{e}_\alpha\left(\kappa\middle|\begin{matrix}\boldsymbol{k}\\ j\end{matrix}\right). \end{aligned} \tag{81.6}$$

To obtain (81.5) we replace the sum over L by an equivalent one over $(L-L')$, which is permitted owing to the Born-Karman condition.

Since $e_\alpha\left(\kappa\middle|\begin{matrix}\boldsymbol{k}\\ j\end{matrix}\right)$ is a Bloch vector at wave vector $\boldsymbol{k}$, the work of Sects. 34–44 can be taken over in toto, with some additions. The basic objects is (81.1); then for fixed $\boldsymbol{k}$ and j, the eigenvector (81.1) has components labelled by $\alpha\,\kappa$. Now consider the coordinate transformation $\{\varphi|\boldsymbol{t}\}\equiv\{\varphi|\boldsymbol{\tau}(\varphi)\}$ of (70.24) with $\boldsymbol{R}_M=0$ to which there corresponds the transformation operator $P_{\{\varphi|\boldsymbol{t}\}}$. Consider

$$(P_{\{\varphi|\boldsymbol{t}\}}\, \boldsymbol{e})\left(\middle|\begin{matrix}\boldsymbol{k}\\ j\end{matrix}\right) \equiv \boldsymbol{e}_{\{\varphi\}}\left(\middle|\begin{matrix}\boldsymbol{k}\\ j\end{matrix}\right). \tag{81.7}$$

From Sect. 30, this is a Bloch vector at wave vector $\boldsymbol{\varphi}\cdot\boldsymbol{k}$. In terms of the operators this follows directly as:

$$P_{\{\varepsilon|\boldsymbol{R}_{L'}\}}\, P_{\{\varphi|\boldsymbol{t}\}} = P_{\{\varphi|\boldsymbol{t}\}}\, P_{\{\varepsilon|\varphi^{-1}\cdot\boldsymbol{R}_{L'}\}}. \tag{81.8}$$

Hence applying left and right hand side of (81.8) to (81.1) we have

$$\begin{aligned} P_{\{\varepsilon|\boldsymbol{R}_{L'}\}}\, \boldsymbol{e}_{\{\varphi\}}\left(\middle|\begin{matrix}\boldsymbol{k}\\ j\end{matrix}\right) &= D^{(\boldsymbol{k})}(\{\varepsilon|\varphi^{-1}\cdot\boldsymbol{R}_{L'}\})\, \boldsymbol{e}_{\{\varphi\}}\left(\middle|\begin{matrix}\boldsymbol{k}\\ j\end{matrix}\right) \\ &= D^{(\varphi\cdot\boldsymbol{k})}(\{\varepsilon|\boldsymbol{R}_{L'}\})\, \boldsymbol{e}_{\{\varphi\}}\left(\middle|\begin{matrix}\boldsymbol{k}\\ j\end{matrix}\right). \end{aligned} \tag{81.9}$$

To obtain (81.9) directly by the explicit determination of components, is somewhat more tedius. Note that (81.9) demonstrates that (81.7) is a Bloch vector at wave vector $\varphi \cdot \boldsymbol{k}$. Let us first consider the object

$$\begin{aligned} P_{\{\varphi|\boldsymbol{t}\}} P^{(\boldsymbol{k})} &= \frac{1}{\sqrt{N}} \sum_{\{\varepsilon|-\boldsymbol{R}_L\}} D^{(\boldsymbol{k})}(\{\varepsilon|-\boldsymbol{R}_L\})^* P_{\{\varphi|\boldsymbol{t}\}} P_{\{\varepsilon|-\boldsymbol{R}_L\}} \\ &= \frac{1}{\sqrt{N}} \sum_{\{\varepsilon|-\boldsymbol{R}_L\}} D^{(\boldsymbol{k})}(\{\varepsilon|-\boldsymbol{R}_L\})^* P_{\{\varepsilon|-\varphi\cdot\boldsymbol{R}_L\}} P_{\{\varphi|\boldsymbol{\tau}(\varphi)\}}. \end{aligned} \tag{81.10}$$

Now let

$$\varphi \cdot \boldsymbol{R}_L \equiv \boldsymbol{R}_{L_\varphi} \quad \text{so} \quad \boldsymbol{R}_L = \varphi^{-1} \cdot \boldsymbol{R}_{L_\varphi} \tag{81.11}$$

then

$$\begin{aligned} D^{(\boldsymbol{k})}(\{\varepsilon|-\boldsymbol{R}_L\})^* &= D^{(\boldsymbol{k})}(\{\varepsilon|-\varphi^{-1}\cdot\boldsymbol{R}_{L_\varphi}\})^* \\ &= D^{(\varphi\cdot\boldsymbol{k})}(\{\varepsilon|-\boldsymbol{R}_{L_\varphi}\})^*. \end{aligned} \tag{81.12}$$

Changing the sum over $-\boldsymbol{R}_L$ to one over $-\boldsymbol{R}_{L_\varphi}$ we get

$$\frac{1}{\sqrt{N}} \sum_{\{\varepsilon|-\boldsymbol{R}_{L_\varphi}\}} D^{(\varphi\cdot\boldsymbol{k})}(\varepsilon|-\boldsymbol{R}_{L_\varphi}\})^* P_{\{\varepsilon|-\boldsymbol{R}_{L_\varphi}\}} P_{\{\varphi|\boldsymbol{\tau}(\varphi)\}}. \tag{81.13}$$

However, in (81.13) the index $\boldsymbol{R}_{L_\varphi}$ is a dummy summation index which may be replaced, e.g. by $\boldsymbol{R}_N$; while the transformation operator on the extreme right is a constant operator in regard to summation, so we obtain

$$\frac{1}{\sqrt{N}} \sum_{\{\varepsilon|-\boldsymbol{R}_N\}} D^{(\varphi\cdot\boldsymbol{k})}(\{\varepsilon|-\boldsymbol{R}_N\})^* P_{\{\varepsilon|-\boldsymbol{R}_N\}} P_{\{\varphi|\boldsymbol{\tau}(\varphi)\}} = P^{(\varphi\cdot\boldsymbol{k})} P_{\{\varphi|\boldsymbol{\tau}(\varphi)\}}. \tag{81.14}$$

So

$$P_{\{\varphi|\boldsymbol{\tau}(\varphi)\}} P^{(\boldsymbol{k})} = P^{(\varphi\cdot\boldsymbol{k})} P_{\{\varphi|\boldsymbol{\tau}(\varphi)\}}. \tag{81.15}$$

To obtain explicit expressions which may be compared to (79.12) we can apply the operator (81.15) to $[e_j]$ to obtain

$$\begin{aligned} P^{(\varphi\cdot\boldsymbol{k})} P_{\{\varphi|\boldsymbol{\tau}(\varphi)\}} e_\alpha \left(\begin{matrix}0\\ \kappa\end{matrix}\middle| j\right) &= P^{(\varphi\cdot\boldsymbol{k})} \sum_\beta \varphi_{\alpha\beta} e_\beta \left(\begin{matrix}0\\ \kappa_{\bar{\varphi}}\end{matrix}\middle| j\right) \\ &= P^{(\varphi\cdot\boldsymbol{k})} \sum_\beta \varphi_{\alpha\beta} e_\beta \left(\begin{matrix}0\\ \varphi^{-1}\boldsymbol{r}_\kappa - \varphi^{-1}\cdot\boldsymbol{\tau}(\varphi)\end{matrix}\middle| j\right). \end{aligned} \tag{81.16}$$

We also used (70.24) to obtain (81.16). Continuing with the operator $P^{(\varphi\cdot\boldsymbol{k})}$ and, for convenience going back to notation with l denoting a lattice vector, or cell index, (81.16) becomes

$$\frac{1}{\sqrt{N}} \sum_{\boldsymbol{R}_L} e^{-i(\varphi\cdot\boldsymbol{k})\cdot\boldsymbol{R}_L} \sum_\beta \varphi_{\alpha\beta} e_\beta \left(\begin{matrix}l\\ \varphi^{-1}\cdot\boldsymbol{r}_\kappa - \varphi^{-1}\cdot\boldsymbol{\tau}(\varphi)\end{matrix}\middle| j\right). \tag{81.17}$$

Summarizing, we obtain the effect of a (rotational plus fractional) operator upon an eigenvector:

$$P_{\{\varphi|\boldsymbol{\tau}(\varphi)\}} e_\alpha \left(\kappa\middle|\begin{matrix}\boldsymbol{k}\\ j\end{matrix}\right) = \sum_\beta \varphi_{\alpha\beta} e_\beta \left(\varphi^{-1}\cdot\boldsymbol{r}_\kappa - \varphi^{-1}\cdot\boldsymbol{\tau}(\varphi)\middle|\begin{matrix}\varphi\cdot\boldsymbol{k}\\ j\end{matrix}\right). \tag{81.18}$$

This demonstrates simultaneously the property of eigenvectors $e\left(\left|\begin{matrix}k\\ j\end{matrix}\right.\right)$, or $e\left(\left|\begin{matrix}\varphi\cdot k\\ j\end{matrix}\right.\right)$ a first rank tensor, or vector, field and also that they are Bloch vectors at the prescribed wave vector; respectively $\boldsymbol{k}$ and $\varphi\cdot\boldsymbol{k}$.

As pointed out by WIGNER [1] (p. 106) in a related but different case, it is essential to make use of the property of the $P_{\{\varphi|t\}}$ as operators upon functions. Failing this, we may make serious errors by having the operators act upon numbers e.g. the value of an eigenvector like $e\left(\left|\begin{matrix}k\\ j\end{matrix}\right.\right)$ at a "point", or coordinate, $l\kappa\alpha$, etc. The latter will usually give incorrect results.

Having examined the transformation properties of the eigenvectors, we now turn to study of the transformation of the entire dynamical problem (79.7). Consequently, we have to apply the operator $P_{\{\varphi|\tau(\varphi)\}}$ to this equation as

$$P_{\{\varphi|\tau(\varphi)\}}\cdot[D(k)]\cdot P^{-1}_{\{\varphi|\tau(\varphi)\}}\cdot P_{\{\varphi|\tau(\varphi)\}}\, e\left(\left|\begin{matrix}k\\ j_\lambda\end{matrix}\right.\right)=\omega^2(k|j)\,P_{\{\varphi|\tau(\varphi)\}}\, e\left(\left|\begin{matrix}k\\ j_\lambda\end{matrix}\right.\right). \tag{81.19}$$

Now consider the transformed dynamical matrix

$$[D'(k)]\equiv P_{\{\varphi|\tau(\varphi)\}}\,[D(k)]\,P^{-1}_{\{\varphi|\tau(\varphi)\}}. \tag{81.20}$$

The most straightforward way to proceed is via (78.5) using a projection operator, as we did in transforming the displacement eigenvectors. Observe that (78.5) can be rewritten as in (81.2), utilizing the projection operator $P^{(k)}$:

$$\begin{aligned}D_{\alpha\beta}\left(\begin{matrix}k\\ \kappa\,\kappa'\end{matrix}\right)&=\sum_\lambda\frac{\exp-i\,k\cdot R_\lambda}{\sqrt{M_\kappa M_{\kappa'}}}\,\Phi_{\alpha\beta}\left(\begin{matrix}\lambda\\ \kappa\,\kappa'\end{matrix}\right)\\ &=\frac{1}{\sqrt{M_\kappa M_{\kappa'}}}\sum_{\{\varepsilon|-R_\lambda\}}D^{(k)}(\{\varepsilon|-R_\lambda\})^*\,P_{\{\varepsilon|-R_\lambda\}}\,\Phi_{\alpha\beta}\left(\begin{matrix}0\\ \kappa\,\kappa'\end{matrix}\right)\end{aligned} \tag{81.21}$$

where as in (81.3)

$$P_{\{\varepsilon|-R_\lambda\}}\,\Phi_{\alpha\beta}\left(\begin{matrix}0\\ \kappa\,\kappa'\end{matrix}\right)=\Phi_{\alpha\beta}\left(\begin{matrix}\lambda\\ \kappa\,\kappa'\end{matrix}\right). \tag{81.22}$$

So

$$D_{\alpha\beta}\left(\begin{matrix}k\\ \kappa\,\kappa'\end{matrix}\right)=\frac{1}{\sqrt{M_\kappa M_{\kappa'}}}\,P^{(k)}\,\Phi_{\alpha\beta}\left(\begin{matrix}0\\ \kappa\,\kappa'\end{matrix}\right) \tag{81.23}$$

or in operator form

$$[D(k)]=P^{(k)}\,M^{-\frac{1}{2}}[\Phi(0)]\,M^{-\frac{1}{2}} \tag{81.24}$$

where $[M]^{-\frac{1}{2}}$ is the square root of the diagonal mass matrix defined in Eq. (67.17). Then (81.23) is to be understood as the $\alpha\beta;\kappa\kappa'$ component of (81.24). Now with (81.24) we may evaluate (81.20). Clearly from (81.15) in (81.24):

$$\begin{aligned}P_{\{\varphi|\tau(\varphi)\}}\cdot[D(k)]\,P^{-1}_{\{\varphi|\tau(\varphi)\}}&=P_{\{\varphi|\tau(\varphi)\}}\,(P^{(k)}\,M^{-\frac{1}{2}}[\Phi(0)]\,M^{-\frac{1}{2}})\,P^{-1}_{\{\varphi|\tau(\varphi)\}}\\ &=P^{(\varphi\cdot k)}\,P_{\{\varphi|\tau(\varphi)\}}\,M^{-\frac{1}{2}}[\Phi(0)]\,M^{-\frac{1}{2}}\,P^{-1}_{\{\varphi|\tau(\varphi)\}}.\end{aligned} \tag{81.25}$$

But $\{\varphi|\tau(\varphi)\}$ is a symmetry operation of the dynamical matrix so that we may invoke (71.23) to obtain from (81.25)

$$P_{\{\varphi|\tau(\varphi)\}} \cdot [\boldsymbol{D}(\boldsymbol{k})] \cdot P^{-1}_{\{\varphi|\tau(\varphi)\}} = P^{(\varphi \cdot \boldsymbol{k})} [M]^{-\frac{1}{2}} [\boldsymbol{\Phi}(0)] [M]^{-\frac{1}{2}} = [\boldsymbol{D}(\varphi \cdot \boldsymbol{k})]. \tag{81.26}$$

Evidently the matrix elements of $[\boldsymbol{D}(\varphi \cdot \boldsymbol{k})]$ are, from (81.24) and (81.23):

$$D_{\alpha\beta}\begin{pmatrix} \varphi \cdot \boldsymbol{k} \\ \kappa\, \kappa' \end{pmatrix} = \sum_{\lambda} \frac{\exp i(\varphi \cdot \boldsymbol{k}) \cdot \boldsymbol{R}_\lambda}{\sqrt{M_\kappa M_{\kappa'}}} \boldsymbol{\Phi}_{\alpha\beta}\begin{pmatrix} \lambda \\ \kappa\, \kappa' \end{pmatrix}. \tag{81.27}$$

Using (81.26) in (81.19) we have

$$[\boldsymbol{D}(\varphi \cdot \boldsymbol{k})] \cdot P_{\{\varphi|\tau(\varphi)\}}\, \boldsymbol{e}\left(\left|\begin{matrix} \boldsymbol{k} \\ j_\lambda \end{matrix}\right.\right) = \omega^2(\boldsymbol{k}|j) \cdot P_{\{\varphi|\tau(\varphi)\}}\, \boldsymbol{e}\left(\left|\begin{matrix} \boldsymbol{k} \\ j_\lambda \end{matrix}\right.\right). \tag{81.28}$$

But, the dynamical equation for eigenvectors and eigenfrequencies at wave vector $\varphi \cdot \boldsymbol{k}$ is

$$[\boldsymbol{D}(\varphi \cdot \boldsymbol{k})] \cdot \boldsymbol{e}\left(\left|\begin{matrix} \varphi \cdot \boldsymbol{k} \\ j_\lambda \end{matrix}\right.\right) = \omega^2(\varphi \cdot \boldsymbol{k}|j_\lambda)\, \boldsymbol{e}\left(\left|\begin{matrix} \varphi \cdot \boldsymbol{k} \\ j_\lambda \end{matrix}\right.\right). \tag{81.29}$$

Invoking (81.28) it is clear that

$$P_{\{\varphi|\tau(\varphi)\}}\, \boldsymbol{e}\left(\left|\begin{matrix} \boldsymbol{k} \\ j_\lambda \end{matrix}\right.\right) \quad \text{is an eigenvector at wave vector } \varphi \cdot \boldsymbol{k} \tag{81.30}$$

and that

$$\omega^2(\boldsymbol{k}|j) = \omega^2(\varphi \cdot \boldsymbol{k}|j_\lambda). \tag{81.31}$$

Thus it follows that for each operator in $\mathfrak{G}$, and particularly for the coset representatives $\{\varphi|\tau(\varphi)\}$

$$\omega^2(\boldsymbol{k}|j_\lambda) = \omega^2(\varphi_2 \cdot \boldsymbol{k}|j_\lambda) = \cdots = \omega^2(\varphi_{g_p} \cdot \boldsymbol{k}|j_\lambda), \qquad \lambda = 1, \ldots, l_j. \tag{81.32}$$

Then the effect of the rotation operator is to transform the dynamical matrix problem to an equivalent one; in the equivalent one the eigenvalues are related as in (81.32). This is then an aspect of the necessary degeneracy of the lattice dynamic problem caused by space group symmetry $\mathfrak{G}$.

Finally, note that for each operator $P_{\{\varphi_{l_\lambda}|\tau_{l_\lambda}\}}$ in $\mathfrak{G}(\boldsymbol{k})$, we have from (81.26)

$$P_{\{\varphi_{l_\lambda}|\tau_{l_\lambda}\}} \cdot [\boldsymbol{D}(\boldsymbol{k})] \cdot P^{-1}_{\{\varphi_{l_\lambda}|\tau_{l_\lambda}\}} = [\boldsymbol{D}(\boldsymbol{k})]. \tag{81.33}$$

Thus $[\boldsymbol{D}(\boldsymbol{k})]$ is invariant under the operations in $\mathfrak{G}(\boldsymbol{k})$. Since it is an invariant operator, its eigenvectors can be arranged to be bases for irreducible representations of $\mathfrak{G}(\boldsymbol{k})$.

82. Eigenvectors of $[D(k)]$ as bases for representation $D^{(k)(e)}$ of $\mathfrak{G}(k)$. Because the $\boldsymbol{e}\left(\left|\begin{matrix} \boldsymbol{k} \\ j \end{matrix}\right.\right)$ of (81.1) are Bloch vectors, we recognize that they can be utilized as bases for irreducible representations of $\mathfrak{G}$. Exactly which representations will occur is a matter for further analysis. We restrict attention now to the $3r$ eigenvectors (81.1) at fixed $\boldsymbol{k}$; which transform like a physical vector field.

Consider the set of operators $P_{\{\varphi_{l_\lambda}|\tau_{l_\lambda}\}}P_{\mathfrak{T}}$ which comprise $\mathfrak{G}(\boldsymbol{k})$, as in (36.1). For each operator in the set (34.4) holds, so

$$\varphi_{l_\lambda}\cdot\boldsymbol{k}=\boldsymbol{k}+2\pi\boldsymbol{B}_H. \tag{82.1}$$

Consequently from (81.17)

$$\begin{aligned}P_{\{\varphi_{l_\lambda}|\tau_{l_\lambda}\}}e_\alpha\left(\kappa\middle|\begin{matrix}\boldsymbol{k}\\ j\end{matrix}\right)&=\sum_\beta(\varphi_{l_\lambda})_{\alpha\beta}e_\beta\left(\kappa_{\bar\varphi_{l_\lambda}}\middle|\begin{matrix}\boldsymbol{k}\\ j\end{matrix}\right)\\&=\sum_\beta(\varphi_{l_\lambda})_{\alpha\beta}e_\beta\left(\varphi_{l_\lambda}^{-1}\cdot\kappa-\varphi_{l_\lambda}^{-1}\cdot\tau_{l_\lambda}\middle|\begin{matrix}\boldsymbol{k}\\ j\end{matrix}\right).\end{aligned} \tag{82.2}$$

Note that adding $2\pi\boldsymbol{B}_H$ to $\boldsymbol{k}$ does not alter the operator $P^{(\boldsymbol{k})}$.

Before proceeding with (82.2) observe that for all $\boldsymbol{e}\left(\middle|\begin{matrix}\boldsymbol{k}\\ j\end{matrix}\right)$ a particular canonical or fixed choice of α,κ has been made once and for all. That is, the Cartesian axes corresponding to $\alpha=1,2,3$ are fixed. Also, the particular basis atoms $\kappa=1,\ldots,r$ are fixed, one for each Bravais sublattice. For example these may be chosen all to be in cell $l=0$. Now recall (70.29) and (70.30). Under the transformation $\{\varphi_{l_\lambda}|\tau_{l_\lambda}\}$ of (82.2) the canonical basis atom is transformed as

$$\boldsymbol{r}_\kappa\to\boldsymbol{r}_{\kappa\bar\varphi_{l_\lambda}}=\boldsymbol{R}_N(\kappa_{\bar\varphi_{l_\lambda}},\kappa'')+\boldsymbol{r}_{\kappa''} \tag{82.3}$$

by (70.30), where again $\boldsymbol{r}_{\kappa''}$ denotes a canonical basis atom. It is symmetrically equivalent to $\boldsymbol{r}_\kappa$ under the transformation $\{\varphi_{l_\lambda}|\tau_{l_\lambda}\}$ but may, or may not belong to the same Bravais sublattice as $\boldsymbol{r}_\kappa$. That is, whether or not $\boldsymbol{R}_N=0$, $\boldsymbol{r}_{\kappa''}$ may or may not equal $\boldsymbol{r}_\kappa$. There are evidently four cases to distinguish.

$$\boldsymbol{r}_\kappa=\boldsymbol{r}_{\kappa''};\quad \boldsymbol{R}_N(\kappa_{\bar\varphi_{l_\lambda}},\kappa'')=0, \tag{82.4}$$

$$\boldsymbol{r}_\kappa=\boldsymbol{r}_{\kappa''};\quad \boldsymbol{R}_N(\kappa_{\bar\varphi_{l_\lambda}},\kappa'')\neq0, \tag{82.5}$$

$$\boldsymbol{r}_\kappa\neq\boldsymbol{r}_{\kappa''};\quad \boldsymbol{R}_N(\kappa_{\bar\varphi_{l_\lambda}},\kappa'')=0, \tag{82.6}$$

$$\boldsymbol{r}_\kappa\neq\boldsymbol{r}_{\kappa''};\quad \boldsymbol{R}_N(\kappa_{\bar\varphi_{l_\lambda}},\kappa'')\neq0, \tag{82.7}$$

with

$$\boldsymbol{R}_N(\kappa_{\bar\varphi_{l_\lambda}},\kappa'')\equiv\varphi_{l_\lambda}^{-1}\cdot\boldsymbol{r}_\kappa-\varphi_{l_\lambda}^{-1}\cdot\tau_{l_\lambda}-\boldsymbol{r}_{\kappa''}, \tag{82.8}$$

where $\boldsymbol{r}_{\kappa''}$ is again, a canonical basis atom vector, and $\boldsymbol{R}_N$ is a lattice vector.

Thus the closure of the set of eigenvectors

$$\boldsymbol{e}\left(\middle|\begin{matrix}\boldsymbol{k}\\ j\end{matrix}\right),\qquad j=1,\ldots,3r \tag{82.9}$$

with components

$$\boldsymbol{e}_\alpha\left(\kappa\middle|\begin{matrix}\boldsymbol{k}\\ j\end{matrix}\right),\qquad \alpha=1,\ldots,3;\ \kappa=1,\ldots,r\text{ (canonical)} \tag{82.10}$$

requires that on the right hand side of (82.2) the components

$$e_\beta\left(\kappa_{\bar\varphi}\middle|\begin{matrix}\boldsymbol{k}\\ j\end{matrix}\right) \tag{82.11}$$

shall also be expressed in terms of the canonical variables $\boldsymbol{r}_\kappa$. Hence, invoking (82.3) and (82.8) in (82.11) we have

$$\begin{aligned} e_\beta\left(\kappa_{\bar{\varphi}_{l_\lambda}}\middle|\begin{matrix}\boldsymbol{k}\\ j\end{matrix}\right) &= e_\beta\left(\boldsymbol{R}_N+\boldsymbol{r}_{\kappa''}\middle|\begin{matrix}\boldsymbol{k}\\ j\end{matrix}\right) \\ &= P_{\{\varepsilon|-\boldsymbol{R}_N\}}\, e_\beta\left(\boldsymbol{r}_{\kappa''}\middle|\begin{matrix}\boldsymbol{k}\\ j\end{matrix}\right) \\ &= D^{(\boldsymbol{k})}(\{\varepsilon|-\boldsymbol{R}_N\})\, e_\beta\left(\boldsymbol{r}_{\kappa''}\middle|\begin{matrix}\boldsymbol{k}\\ j\end{matrix}\right) \end{aligned} \tag{82.12}$$

with $\boldsymbol{R}_N$ defined via (82.3). Thus for (82.2)

$$P_{\{\varphi_{l_\lambda}|\tau_{l_\lambda}\}}\, e_\alpha\left(\kappa\middle|\begin{matrix}\boldsymbol{k}\\ j\end{matrix}\right) = \sum_\beta (\varphi_{l_\lambda})_{\alpha\beta} D^{(\boldsymbol{k})}(\{\varepsilon|-\boldsymbol{R}_N\})\, e_\beta\left(\kappa''\middle|\begin{matrix}\boldsymbol{k}\\ j\end{matrix}\right). \tag{82.13}$$

Now in (82.3) κ and κ'' are both among the canonical set (82.10). Hence (82.13) is a general relationship between the canonical components of the same vector $\boldsymbol{e}\left(\middle|\begin{matrix}\boldsymbol{k}\\ j\end{matrix}\right)$ fixed j. Then, if for fixed j we consider the $3r$ components (79.2) each such component is labelled by two indices: α and κ:

$$e_\alpha\left(\kappa\middle|\begin{matrix}\boldsymbol{k}\\ j\end{matrix}\right), \qquad \alpha=1,\dots,3;\ \kappa=1,\dots,r. \tag{82.14}$$

The relation (82.13) can then be written in matrix form

$$P_{\{\varphi_{l_\lambda}|\tau_{l_\lambda}\}}\, e_\alpha\left(\kappa\middle|\begin{matrix}\boldsymbol{k}\\ j\end{matrix}\right) = \sum_\beta\sum_{\kappa''} D^{(\boldsymbol{k})(e)}(\{\varphi_{l_\lambda}|\tau_{l_\lambda}\})_{\alpha\kappa,\beta\kappa''}\, e_\beta\left(\kappa''\middle|\begin{matrix}\boldsymbol{k}\\ j\end{matrix}\right), \tag{82.15}$$
$$\alpha=1,\dots,3;\ \kappa=1,\dots,r.$$

For fixed $\boldsymbol{k}$ and j the matrix $D^{(\boldsymbol{k})(e)}$ is $3r$ dimensional, having $3r$ rows and columns. We can easily obtain the components of $D^{(\boldsymbol{k})(e)}$. From (82.13)

$$D^{(\boldsymbol{k})(e)}(\{\varphi_{l_\lambda}|\tau_{l_\lambda}\})_{\alpha\kappa,\beta\kappa''} = (\varphi_{l_\lambda})_{\alpha\beta} D^{(\boldsymbol{k})}(\{\varepsilon|-\boldsymbol{R}_N(\kappa_{\bar{\varphi}_{l_\lambda}},\kappa'')\})\,\delta_{\kappa_{\bar{\varphi}_{l_\lambda}},\kappa''}, \tag{82.16}$$

where

$$\delta_{\kappa_{\bar{\varphi}_{l_\lambda}},\kappa''} = \begin{cases} 1 & \text{for that pair related by (82.3)} \\ 0 & \text{for any other pair.} \end{cases}$$

Now it is important to recognize that the matrix $D^{(\boldsymbol{k})(e)}$ is independent of j. That is, the definition of $D^{(\boldsymbol{k})(e)}$ which is given in (82.15) can be written in terms of the vectors $\boldsymbol{e}_\alpha\left(\kappa\middle|\boldsymbol{k}\right)$:

$$P_{\{\varphi_{l_\lambda}|\tau_{l_\lambda}\}}\, \boldsymbol{e}_\alpha\left(\kappa\middle|\boldsymbol{k}\right) = \sum_\beta\sum_{\kappa''} D^{(\boldsymbol{k})(e)}(\{\varphi_{l_\lambda}|\tau_{l_\lambda}\})_{\alpha\kappa,\beta\kappa''}\, \boldsymbol{e}_\beta\left(\kappa''\middle|\boldsymbol{k}\right). \tag{82.17}$$

Recall here the discussion in Sect. 79, particularly (79.9) and (79.14). Thus denoting the $3r$ row vectors

$$\boldsymbol{e}_\alpha\left(\kappa\middle|\boldsymbol{k}\right), \qquad \alpha=1,\dots,3;\ \kappa=1,\dots,r \tag{82.18}$$

as in the array (79.14), then (82.17) can be understood as a relation between entire row vectors which holds for each j.

It should be no surprise that a pair of indices is needed for row, and for column. In the representation (82.13), the matrix $D^{(\boldsymbol{k})(e)}$ is seen to be a direct product of the representation $D^{(r)}$ by which a polar vector transforms, with a permutation representation, modulo a set of phase factors like $D^{(\boldsymbol{k})}(\{\boldsymbol{\varepsilon}|-\boldsymbol{R}_N\})$. That is, $D^{(\text{perm})}$ permutes the canonical basis set $\kappa=1,\dots,r$ with added phase included if needed in the matrix element depending on which of (82.4)–(82.7) applies. Hence

$$D^{(\boldsymbol{k})(e)}=D^{(r)}\otimes D^{(\text{perm})}. \tag{82.19}$$

The matrix elements of $D^{(r)}$ bear labels $\alpha\,\beta$ such as $D^{(r)}_{\alpha\beta}$, those of $D^{(\text{perm})}$ bear labels κ, κ'' such that the elements are $D^{(\text{perm})}_{\kappa''\kappa}$. Clearly the elements of each of those could be obtained from (82.16), but we only require an understanding that such construction is possible and not the specific form involved so

$$(D^{(\boldsymbol{k})(e)})_{\alpha\kappa,\,\beta\kappa''}=D^{(r)}_{\alpha\beta}\cdot D^{(\text{perm})}_{\kappa''\kappa}. \tag{82.20}$$

Of greater importance is the trace of $D^{(\boldsymbol{k})(e)}$. This is easily determined from (82.13). We find

$$\begin{aligned}\operatorname{Tr} D^{(\boldsymbol{k})(e)}(\{\varphi_{l_\lambda}|\tau_{l_\lambda}\})&\equiv\chi^{(\boldsymbol{k})(e)}(\{\varphi_{l_\lambda}|\tau_{l_\lambda}\})\\&=\sum_{\kappa''}(\pm(1+2\cos\varphi_{l_\lambda}))\,D^{(\boldsymbol{k})}(\{\boldsymbol{\varepsilon}|-\boldsymbol{R}_N(\kappa_{\bar\varphi_{l_\lambda}},\kappa'')\})\,\delta_{\kappa_{\bar\varphi_{l_\lambda}},\kappa''}\end{aligned} \tag{82.21}$$

In (82.21) $\delta_{\kappa\kappa''}=1$ if $\kappa=\kappa''$ as in (82.4) and (82.5), while $\delta_{\kappa\kappa''}=0$ if $\kappa\neq\kappa''$, as in (82.6) and (82.7). Also

$$\pm(1+2\cos\varphi_{l_\lambda})\equiv\sum_\alpha(\varphi_{l_\lambda})_{\alpha\alpha} \tag{82.22}$$

is the trace of the proper or improper rotation matrix $\boldsymbol{\varphi}$, and the sum on the right hand side is over all canonical basis atoms κ''. Since $D^{(\boldsymbol{k})(e)}$ is a representation of $\mathfrak{G}(\boldsymbol{k})$ it can be decomposed into a direct sum of allowable physically irreducible representations. We return to this below in Sect. 103.

Finally it is to be noted that the representation $D^{(\boldsymbol{k})(e)}$ is unitary. This follows from the equation

$$\boldsymbol{e}_\alpha\left(\kappa\middle|\boldsymbol{k}\right):\,\boldsymbol{e}_\beta\left(\kappa'\middle|\boldsymbol{k}\right)=P_{\{\varphi_{l_\lambda}|\tau_{l_\lambda}\}}\,\boldsymbol{e}_\alpha\left(\kappa\middle|\boldsymbol{k}\right):\;P_{\{\varphi_{l_\lambda}|\tau_{l_\lambda}\}}\,\boldsymbol{e}_\beta\left(\kappa'\middle|\boldsymbol{k}\right) \tag{82.23}$$

which expresses the unitarity of the operator P with respect to the scalar product (:), defined in (79.9), (79.11). Then

$$D^{(\boldsymbol{k})(e)}(\{\varphi_{l_\lambda}|\tau_{l_\lambda}\})^{-1}=D^{(\boldsymbol{k})(e)}(\{\varphi_{l_\lambda}|\tau_{l_\lambda}\})^{+} \tag{82.24}$$

the inverse is equal to the Hermitian adjoint.

83. Eigenvectors of $[D(\boldsymbol{k})]$ as bases for representations $D^{(\boldsymbol{k})(j)}$ of $\mathfrak{G}(\boldsymbol{k})$. Now return to (82.17) and (81.28). Thus for any element $\{\varphi_{l_\lambda}|\tau_\lambda\}$ in $\mathfrak{G}(\boldsymbol{k})$, the group leaving the dynamical matrix invariant:

$$[D(\varphi_{l_\lambda}\cdot \boldsymbol{k})]=[D(\boldsymbol{k})] \tag{83.1}$$

and (81.28) becomes:

$$[D(\boldsymbol{k})]\cdot P_{\{\varphi_{l_\lambda}|\tau_{l_\lambda}\}}\,\boldsymbol{e}\left(\left|\begin{matrix}\boldsymbol{k}\\ j_\mu\end{matrix}\right.\right)=\omega^2(\boldsymbol{k}|j)\,P_{\{\varphi_{l_\lambda}|\tau_{l_\lambda}\}}\,\boldsymbol{e}\left(\left|\begin{matrix}\boldsymbol{k}\\ j_\mu\end{matrix}\right.\right). \tag{83.2}$$

Hence

$$P_{\{\varphi_{l_\lambda}|\tau_{l_\lambda}\}}\,\boldsymbol{e}\left(\left|\begin{matrix}\boldsymbol{k}\\ j_\mu\end{matrix}\right.\right) \tag{83.3}$$

is an eigenvector at wave vector $\boldsymbol{k}$. But the set of degenerate eigenvectors corresponding to one eigenvalue

$$\boldsymbol{e}\left(\left|\begin{matrix}\boldsymbol{k}\\ j_1\end{matrix}\right.\right),\ldots,\boldsymbol{e}\left(\left|\begin{matrix}\boldsymbol{k}\\ j_\mu\end{matrix}\right.\right),\ldots,\boldsymbol{e}\left(\left|\begin{matrix}\boldsymbol{k}\\ j_{l_j}\end{matrix}\right.\right) \tag{83.4}$$

is complete in regard to spatial symmetry. Hence the eigenvector (83.3) can be expressed as a linear combination of the set (83.4), or

$$P_{\{\varphi_{l_\lambda}|\tau_{l_\lambda}\}}\,\boldsymbol{e}\left(\left|\begin{matrix}\boldsymbol{k}\\ j_\mu\end{matrix}\right.\right)=\sum_{\nu=1}^{l_j} D^{(\boldsymbol{k})(j)}(\{\varphi_{l_\lambda}|\tau_{l_\lambda}\})_{\nu\mu}\,\boldsymbol{e}\left(\left|\begin{matrix}\boldsymbol{k}\\ j_\nu\end{matrix}\right.\right),\qquad \mu=1,\ldots,l_j. \tag{83.5}$$

Notice that this is a relationship between entire eigenvectors. As usual, the completeness of the set or space (83.4) permits us to define a matrix $D^{(\boldsymbol{k})(j)}$, in terms of which the representation produced by the space (83.4) is given (refer to Sect. 102). Then using (83.5), an l_j dimensional matrix is defined, producing a representation of $\mathfrak{G}(\boldsymbol{k})$.

To obtain matrix elements of $D^{(\boldsymbol{k})(j)}$, we consider the component $\alpha\kappa$ of (83.5). Then we find

$$P_{\{\varphi_{l_\lambda}|\tau_{l_\lambda}\}}\,e_\alpha\left(\kappa\left|\begin{matrix}\boldsymbol{k}\\ j_\mu\end{matrix}\right.\right)=\sum_{\nu=1}^{l_j} D^{(\boldsymbol{k})(j)}(\{\varphi_{l_\lambda}|\tau_{l_\lambda}\})_{\nu\mu}\,e_\alpha\left(\kappa\left|\begin{matrix}\boldsymbol{k}\\ j_\nu\end{matrix}\right.\right). \tag{83.6}$$

Now use (82.15) and (82.13) on the left hand side of (83.6) to obtain

$$\begin{aligned}\sum_\beta\sum_{\kappa''} D^{(\boldsymbol{k})(e)}(\{\varphi_{l_\lambda}|\tau_{l_\lambda}\})_{\alpha\kappa,\beta\kappa''}\,e_\beta\left(\kappa''\left|\begin{matrix}\boldsymbol{k}\\ j_\mu\end{matrix}\right.\right)\\ =\sum_\beta(\varphi_{l_\lambda})_{\alpha\beta}\,D^{(\boldsymbol{k})}(\{\varepsilon|-\boldsymbol{R}_N(\kappa_{\bar\varphi_{l_\lambda}},\kappa'')\})\,e_\beta\left(\kappa''\left|\begin{matrix}\boldsymbol{k}\\ j_\mu\end{matrix}\right.\right)\\ =\sum_{\nu=1}^{l_j} D^{(\boldsymbol{k})(j)}(\{\varphi_{l_\lambda}|\tau_{l_\lambda}\})_{\nu\mu}\,e_\alpha\left(\kappa\left|\begin{matrix}\boldsymbol{k}\\ j_\nu\end{matrix}\right.\right).\end{aligned} \tag{83.7}$$

Then using the orthogonality rule (79.8), we find.

$$\begin{aligned}D^{(\boldsymbol{k})(j)}(\{\varphi_{l_\lambda}|\tau_{l_\lambda}\})_{\nu\mu}&=\sum_{\alpha\kappa}\sum_{\beta\kappa''} e_\alpha^*\left(\kappa\left|\begin{matrix}\boldsymbol{k}\\ j_\nu\end{matrix}\right.\right) D^{(\boldsymbol{k})(e)}(\{\varphi_{l_\lambda}|\tau_{l_\lambda}\})_{\alpha\kappa,\beta\kappa''}\,e_\beta\left(\kappa''\left|\begin{matrix}\boldsymbol{k}\\ j_\mu\end{matrix}\right.\right)\\ &=\sum_{\alpha\kappa}\sum_{\beta\kappa''} e_\alpha^*\left(\kappa\left|\begin{matrix}\boldsymbol{k}\\ j_\nu\end{matrix}\right.\right)(\varphi_{l_\lambda})_{\alpha\beta}\,D^{(\boldsymbol{k})}(\{\varepsilon|-\boldsymbol{R}_N(\kappa_{\bar\varphi_{l_\lambda}},\kappa'')\})\,\delta_{\kappa_{\bar\varphi_{l_\lambda}},\kappa''}\,e_\beta\left(\kappa''\left|\begin{matrix}\boldsymbol{k}\\ j_\mu\end{matrix}\right.\right).\end{aligned} \tag{83.8}$$

This is an explicit expression for the elements of the matrix $D^{(\boldsymbol{k})(j)}$, in terms of those elements of $D^{(\boldsymbol{k})(e)}$. We return to (83.8) in Sect. 85, where we will discuss necessary degeneracy and the symmetry group $\mathfrak{G}(\boldsymbol{k})$ of $[\boldsymbol{D}(\boldsymbol{k})]$.

84. Equivalence of $D^{(\boldsymbol{k})(e)}$ and $D^{(\boldsymbol{k})(j)}$. Now observe that we are dealing with numbers which can be viewed as the $3r$ components of the $(3r)$ vectors $\boldsymbol{e}\left(\left|\begin{smallmatrix}\boldsymbol{k}\\ j\end{smallmatrix}\right.\right)$, or equivalently as the $3r$ components of the $3r$ vectors $\boldsymbol{e}_\alpha\left(\kappa\left|\boldsymbol{k}\right.\right)$. These appear in the analysis in two distinct, but related fashions. Thus, the component $e_\alpha\left(\kappa\left|\begin{smallmatrix}\boldsymbol{k}\\ j\end{smallmatrix}\right.\right)$ can be viewed as the $\alpha\kappa$ component of $\boldsymbol{e}\left(\left|\begin{smallmatrix}\boldsymbol{k}\\ j\end{smallmatrix}\right.\right)$, or as the j component of $\boldsymbol{e}_\alpha\left(\kappa\left|\boldsymbol{k}\right.\right)$. The set of $3r$ vectors

$$\boldsymbol{e}\left(\left|\begin{matrix}\boldsymbol{k}\\ j\end{matrix}\right.\right),\qquad j=1,\ldots,3r \tag{84.1}$$

is then linearly dependent upon the set of $3r$ vectors.

$$\boldsymbol{e}_\alpha\left(\kappa\left|\boldsymbol{k}\right.\right),\qquad \alpha=1,\ldots,3;\ \kappa=1,\ldots,r. \tag{84.2}$$

Since both sets (84.1) and (84.2) are orthonormal by (79.8)–(79.11), they can be related by a unitary transformation

$$\boldsymbol{e}_\alpha\left(\kappa\left|\boldsymbol{k}\right.\right)=[\boldsymbol{U}]\,\boldsymbol{e}\left(\left|\begin{matrix}\boldsymbol{k}\\ j\end{matrix}\right.\right). \tag{84.3}$$

The elements of the matrix $[\boldsymbol{U}]$ must carry asymetric indices corresponding to (84.3):

$$[\boldsymbol{U}]_{\alpha\kappa,j}. \tag{84.4}$$

The unitarity is then expressed by

$$[\boldsymbol{U}]^{-1}=[\boldsymbol{U}]^{+}. \tag{84.5}$$

Owing to the unitarity of $[\boldsymbol{U}]$ the inverse of (84.3) can be written as

$$\boldsymbol{e}\left(\left|\begin{matrix}\boldsymbol{k}\\ j\end{matrix}\right.\right)=[\boldsymbol{U}]^{+}\,\boldsymbol{e}_\alpha\left(\kappa\left|\boldsymbol{k}\right.\right). \tag{84.6}$$

Clearly $[\boldsymbol{U}]$ is a permutation matrix, with a single non-zero element, equal to 1, in each row and column. But from (82.17), (83.5), (84.3), (84.6) we must have that the representations $D^{(\boldsymbol{k})(e)}$ and $D^{(\boldsymbol{k})(j)}$ are equivalent, since the bases for the two representations are related by a unitary transformation $[\boldsymbol{U}]$. So that

$$D^{(\boldsymbol{k})(j)}=[\boldsymbol{U}]^{-1}\,D^{(\boldsymbol{k})(e)}[\boldsymbol{U}]. \tag{84.7}$$

Then evidently for the traces

$$\operatorname{Tr} D^{(\boldsymbol{k})(j)}=\operatorname{Tr} D^{(\boldsymbol{k})(e)}. \tag{84.8}$$

We can actually compute (84.8) using (82.21), but we use this later.

According to MASCHKE's theorem Sect. 15 the representation $D^{(\boldsymbol{k})(e)}$, and $D^{(\boldsymbol{k})(j)}$, is either irreducible or decomposable. Now let us emphasize that $D^{(\boldsymbol{k})(j)}$ is the representation by which the *totality* of all actual eigenvectors, of all $3r$ branches, transform. Similarly $D^{(\boldsymbol{k})(e)}$ is a $3r$ dimensional representation referring to the transformation of the totality of $(3r)$ eigenvectors. In the latter case, however the basic transformation property simply utilizes the vector field property of the displacement eigenvectors. But, it is the eigenvectors $\boldsymbol{e}\left(\left|\begin{matrix}\boldsymbol{k}\\ j\end{matrix}\right.\right)$ which are related to physical eigenvalues $\omega^2(\boldsymbol{k}|j)$. We turn to this in the next section in which we shall also find the matrix $[\boldsymbol{U}]$.

85. Necessary degeneracy under $\mathfrak{G}(\boldsymbol{k})$ and the eigenvectors of $[\boldsymbol{D}(\boldsymbol{k})]$. In this section we shall assemble some of the previous results and examine the relationship between eigenvectors of the dynamical matrix $[\boldsymbol{D}(\boldsymbol{k})]$ and the necessary degeneracy connected with symmetry group $\mathfrak{G}$.

The dynamical equation (79.7) is

$$[\boldsymbol{D}(\boldsymbol{k})]\cdot\left(\boldsymbol{e}\left|\begin{matrix}\boldsymbol{k}\\ j_\mu\end{matrix}\right.\right)=\omega^2(\boldsymbol{k}|j)\,\boldsymbol{e}\left(\left|\begin{matrix}\boldsymbol{k}\\ j_\mu\end{matrix}\right.\right),\qquad \mu=1,\dots,l_j. \tag{85.1}$$

Now transform this equation using the transformation operator $P_{\{\varphi_{l_\lambda}|\tau_{l_\lambda}\}}$ in $\mathfrak{G}(\boldsymbol{k})$:

$$\begin{aligned}&P_{\{\varphi_{l_\lambda}|\tau_{l_\lambda}\}}\cdot[\boldsymbol{D}(\boldsymbol{k})]\cdot P^{-1}_{\{\varphi_{l_\lambda}|\tau_{l_\lambda}\}}\cdot P_{\{\varphi_{l_\lambda}|\tau_{l_\lambda}\}}\,\boldsymbol{e}\left(\left|\begin{matrix}\boldsymbol{k}\\ j_\mu\end{matrix}\right.\right)\\ &\qquad=\omega^2(\boldsymbol{k}|j)\,P_{\{\varphi_{l_\lambda}|\tau_{l_\lambda}\}}\,\boldsymbol{e}\left(\left|\begin{matrix}\boldsymbol{k}\\ j_\mu\end{matrix}\right.\right).\end{aligned} \tag{85.2}$$

From (81.33) this is

$$[\boldsymbol{D}(\boldsymbol{k})]\cdot P_{\{\varphi_{l_\lambda}|\tau_{l_\lambda}\}}\,\boldsymbol{e}\left(\left|\begin{matrix}\boldsymbol{k}\\ j_\mu\end{matrix}\right.\right)=\omega^2(\boldsymbol{k}|j)\,P_{\{\varphi_{l_\lambda}|\tau_{l_\lambda}\}}\,\boldsymbol{e}\left(\left|\begin{matrix}\boldsymbol{k}\\ j_\mu\end{matrix}\right.\right). \tag{85.3}$$

Taking matrix elements of (85.3) we have

$$\sum_{\beta\kappa'} D\begin{pmatrix}\boldsymbol{k}\\ \kappa\,\kappa'\end{pmatrix}_{\alpha\beta}\left(P_{\{\varphi_{l_\lambda}|\tau_{l_\lambda}\}}\,e_\beta\left(\kappa'\left|\begin{matrix}\boldsymbol{k}\\ j_\mu\end{matrix}\right.\right)\right)=\omega^2(\boldsymbol{k}|j)\,P_{\{\varphi_{l_\lambda}|\tau_{l_\lambda}\}}\,e_\alpha\left(\kappa\left|\begin{matrix}\boldsymbol{k}\\ j_\mu\end{matrix}\right.\right) \tag{85.4}$$

using (82.17) we obtain

$$\begin{aligned}&\sum_{\beta\kappa'}\sum_{\gamma\kappa''} D\begin{pmatrix}\boldsymbol{k}\\ \kappa\,\kappa'\end{pmatrix}_{\alpha\beta} D^{(\boldsymbol{k})(e)}(\{\varphi_{l_\lambda}|\tau_{l_\lambda}\})_{\beta\kappa';\gamma\kappa''}\,e_\gamma\left(\kappa''\left|\begin{matrix}\boldsymbol{k}\\ j_\mu\end{matrix}\right.\right)\\ &\qquad=\sum_{\bar\beta\bar\kappa'} D^{(\boldsymbol{k})(e)}(\{\varphi_{l_\lambda}|\tau_{l_\lambda}\})_{\alpha\kappa,\bar\beta\bar\kappa'}\,e_{\bar\beta}\left(\bar\kappa'\left|\begin{matrix}\boldsymbol{k}\\ j_\mu\end{matrix}\right.\right)\omega^2(\boldsymbol{k}|j).\end{aligned} \tag{85.5}$$

From (79.4) and (85.1)

$$\omega^2(\boldsymbol{k}|j)\,e_{\bar\beta}\left(\bar\kappa'\left|\begin{matrix}\boldsymbol{k}\\ j_\mu\end{matrix}\right.\right)=\sum_{\bar\gamma\bar\kappa''} D\begin{pmatrix}\boldsymbol{k}\\ \bar\kappa',\bar\kappa''\end{pmatrix}_{\bar\beta\bar\gamma} e_{\bar\gamma}\left(\bar\kappa''\left|\begin{matrix}\boldsymbol{k}\\ j_\mu\end{matrix}\right.\right). \tag{85.6}$$

Then substituting (85.6) into (85.5) we have

$$\begin{aligned}\sum_{\beta\kappa'}\sum_{\gamma\kappa''} D\begin{pmatrix}\boldsymbol{k}\\ \kappa\,\kappa'\end{pmatrix}_{\alpha\beta} D^{(\boldsymbol{k})(e)}(\{\varphi_{l_\lambda}|\tau_{l_\lambda}\})_{\beta\kappa';\gamma\kappa''}\, e_\gamma\left(\kappa''\middle|\begin{matrix}\boldsymbol{k}\\ j_\mu\end{matrix}\right)\\ =\sum_{\bar\beta\bar\kappa'}\sum_{\bar\gamma\bar\kappa''} D^{(\boldsymbol{k})(e)}(\{\varphi_{l_\lambda}|\tau_{l_\lambda}\})_{\alpha\kappa,\bar\beta\bar\kappa'}\, D\begin{pmatrix}\boldsymbol{k}\\ \bar\kappa'\,\bar\kappa''\end{pmatrix}_{\bar\beta\bar\gamma} e_{\bar\gamma}\left(\bar\kappa''\middle|\begin{matrix}\boldsymbol{k}\\ j_\mu\end{matrix}\right).\end{aligned}\tag{85.7}$$

Now to proceed we need to avail ourselves of the hermitian property of the dynamical matrix: $[\boldsymbol{D}(\boldsymbol{k})]=[\boldsymbol{D}(\boldsymbol{k})]^+$ or in components

$$D\begin{pmatrix}\boldsymbol{k}\\ \kappa\,\kappa'\end{pmatrix}_{\alpha\beta}=D\begin{pmatrix}\boldsymbol{k}\\ \kappa'\,\kappa\end{pmatrix}^*_{\beta\alpha}.\tag{85.8}$$

Then for (85.7) we obtain

$$\begin{aligned}\sum_{\beta\kappa'}\sum_{\gamma\kappa''} D^{(\boldsymbol{k})(e)}(\{\varphi_{l_\lambda}|\tau_{l_\lambda}\})_{\beta\kappa',\gamma\kappa''}\, D\begin{pmatrix}\boldsymbol{k}\\ \kappa'\,\kappa\end{pmatrix}^*_{\beta\alpha} e_\gamma\left(\kappa''\middle|\begin{matrix}\boldsymbol{k}\\ j_\mu\end{matrix}\right)\\ =\sum_{\bar\beta\bar\kappa'}\sum_{\bar\gamma\bar\kappa''} D\begin{pmatrix}\boldsymbol{k}\\ \bar\kappa''\,\bar\kappa'\end{pmatrix}^*_{\bar\gamma\bar\beta} D^{(\boldsymbol{k})(e)}(\{\varphi_{l_\lambda}|\tau_{l_\lambda}\})_{\alpha\kappa,\bar\beta\bar\kappa'}\, e_{\bar\gamma}\left(\bar\kappa''\middle|\begin{matrix}\boldsymbol{k}\\ j_\mu\end{matrix}\right)\end{aligned}\tag{85.9}$$

or collecting

$$\begin{aligned}\sum_{\gamma\kappa''}(\boldsymbol{D}(\boldsymbol{k})\cdot D^{(\boldsymbol{k})(e)}(\{\varphi_{l_\lambda}|\tau_{l_\lambda}\}))_{\gamma\kappa'',\alpha\kappa}\, e_\gamma\left(\kappa''\middle|\begin{matrix}\boldsymbol{k}\\ j_\mu\end{matrix}\right)\\ =\sum_{\bar\gamma\bar\kappa''}(D^{(\boldsymbol{k})(e)}(\{\varphi_{l_\lambda}|\tau_{l_\lambda}\})\cdot\boldsymbol{D}(\boldsymbol{k}))_{\bar\gamma\bar\kappa'',\alpha\kappa}\, e_{\bar\gamma}\left(\bar\kappa''\middle|\begin{matrix}\boldsymbol{k}\\ j_\mu\end{matrix}\right).\end{aligned}\tag{85.10}$$

Equating the coefficient of the same component $e_\gamma\left(\kappa''\middle|\begin{matrix}\boldsymbol{k}\\ j_\mu\end{matrix}\right)$ on both sides of (85.10) we recognize that (85.10) derives from the matrix equation

$$D^{(\boldsymbol{k})(e)}(\{\varphi_{l_\lambda}|\tau_{l_\lambda}\})\cdot D(\boldsymbol{k})=D(\boldsymbol{k})\cdot D^{(\boldsymbol{k})(e)}(\{\varphi_{l_\lambda}|\tau_{l_\lambda}\}).\tag{85.11}$$

Consequently we demonstrated that the dynamical matrix commutes with each of the matrices of the representation $D^{(\boldsymbol{k})(e)}$ of $\mathfrak{G}(\boldsymbol{k})$. The form of $D(\boldsymbol{k})$ is known from (78.5); it is *not* a scalar, i.e. a number times the unit matrix. Since a non-constant matrix $[\boldsymbol{D}(\boldsymbol{k})]$ commutes with all matrices of a representation $D^{(\boldsymbol{k})(e)}$ of $\mathfrak{G}(\boldsymbol{k})$, SCHUR's lemma informs us that $D^{(\boldsymbol{k})(e)}$ must be reducible. To reduce $D^{(\boldsymbol{k})(e)}$ to a direct sum of irreducible components, we shall first bring $[\boldsymbol{D}(\boldsymbol{k})]$ to diagonal form.

From (79.14)–(79.17) the unitary matrix of eigenvectors $[\boldsymbol{E}(\boldsymbol{k})]$ brings $[\boldsymbol{D}(\boldsymbol{k})]$ to diagonal form $[\boldsymbol{\Delta}(\boldsymbol{k})]$, so

$$[\boldsymbol{E}(\boldsymbol{k})]^{-1}[\boldsymbol{D}(\boldsymbol{k})]\cdot[\boldsymbol{E}(\boldsymbol{k})]=[\boldsymbol{\Delta}(\boldsymbol{k})]\tag{85.12}$$

since $[\boldsymbol{\Delta}(\boldsymbol{k})]$ is real. For the moment we suppress the argument (operator) of $D^{(\boldsymbol{k})(e)}$ and then transform (85.11) with $[\boldsymbol{E}(\boldsymbol{k})]$, obtaining:

$$[\boldsymbol{E}(\boldsymbol{k})]^{-1}\cdot[\boldsymbol{D}^{(\boldsymbol{k})(e)}][\boldsymbol{E}(\boldsymbol{k})]\cdot\boldsymbol{\Delta}(\boldsymbol{k})=\boldsymbol{\Delta}(\boldsymbol{k})\cdot[\boldsymbol{E}(\boldsymbol{k})]^{-1}\cdot[\boldsymbol{D}^{(\boldsymbol{k})(e)}]\cdot[\boldsymbol{E}(\boldsymbol{k})].\tag{85.13}$$

Define the transform of $D^{(k)(e)}$ by the barred matrix:

$$\overline{D^{(k)(e)}} \equiv [E(k)]^{-1} \cdot D^{(k)(e)} \cdot [E(k)] = \widetilde{[E(k)]^{*}} \cdot D^{(k)(e)} \cdot [E(k)] \tag{85.14}$$

so from (85.13)

$$\overline{D^{(k)(e)}} \cdot \Delta(k) = \Delta(k) \cdot \overline{D^{(k)(e)}}. \tag{85.15}$$

Rewrite $[\Delta(k)]$ from (79.17) by collecting the degenerate eigenvalues

$$[\Delta(k)] = \begin{pmatrix} \omega^2(k|1)\,\Pi_1 & 0 & 0 & \dots & 0 \\ 0 & \omega^2(k|2)\,\Pi_2 & & & \vdots \\ & & \ddots & & \\ & & & \omega^2(k|j)\,\Pi_j & \\ 0 & \dots & & & \omega^2(k|n)\,\Pi_{l_n} \end{pmatrix}. \tag{85.16}$$

The Π_j are l_j dimensional unit matrices. Now from (85.14) and (85.16) we conclude that since each of the $\overline{D^{(k)(e)}}$ commutes with the same bloc diagonal matrix (85.16), hence each of the $\overline{D^{(k)(e)}}$ for each $\{\varphi_{l_\lambda}|\tau_{l_\lambda}\}$ partitions corresponding to (85.16) as:

$$\overline{D^{(k)(e)}} = \begin{pmatrix} D^{(k)(1)} & 0 & 0 & \dots & \\ 0 & D^{(k)(2)} & 0 & & \\ 0 & 0 & D^{(k)(j)} & & \\ & & & \ddots & D^{(k)(n)} \end{pmatrix}. \tag{85.17}$$

Each of the blocks in (85.17) is a non-zero matrix, with $D^{(k)(1)}\ l_1 \times l_1, \dots, D^{(k)(n)}\ l_n \times l_{n'}$, each one corresponding to eigenvalue $\omega^2(k|1), \dots, \omega^2(k|n)$ respectively.

Taking components of (85.14) we obtain a connection with (83.8). Recall that the components of $[E(k)]$ as indicated in (79.14) are labelled by row indices $\alpha\kappa(=1, \dots, 3r)$ and column indices $j_\nu(=1, \dots, 3r)$, so $\widetilde{[E(k)]^{*}}$ has the inverse labelling. Then taking the $(j_\mu j'_\nu)$ matrix element of $\overline{D^{(k)(e)}}$ we obtain for (85.14):

$$(\overline{D^{(k)(e)}})_{j_\mu j'_\nu} = \sum_{\beta\kappa'} \sum_{\alpha\kappa} e^{*}_{\alpha}\left(\kappa \middle| \begin{matrix} k \\ j'_\nu \end{matrix}\right) (D^{(k)(e)})_{\alpha\kappa,\, \beta\kappa'}\, e_\beta \left(\kappa' \middle| \begin{matrix} k \\ j_\mu \end{matrix}\right). \tag{85.18}$$

Comparing (85.18), (85.17), and (83.8) we can write

$$(\overline{D^{(k)(e)}})_{j_\mu, j'_\nu} = (D^{(k)(j)})_{\mu\nu}\, \delta_{jj'}. \tag{85.19}$$

Consequently the transformation $[U]$ which transforms $D^{(k)(e)}$ into a direct sum is $E(k) = [U]$.

Now we invoke the assertion of necessary degeneracy under the spatial symmetry operators $\mathfrak{G}(k)$. We assert that each representation $D^{(k)(j)}$ in the partitioned form which corresponds to a distinct $\omega^2(k|j)$ is an allowable physically irreducible representation of $\mathfrak{G}(k)$. It should be noted that a particular allowable irreducible representation $D^{(k)(m)}$ of $\mathfrak{G}(k)$ may appear more than once in (85.17). Which of the allowable $D^{(k)(m)}$ do in fact appear is to be determined by methods discussed below.

Observe that the theory does not assume that a given $D^{(k)(m)}$ occurs only once in (85.17). However in order to apply (85.18) it is necessary to have solved the eigenvalue problem, and to have the eigenvectors $e_\alpha\left(\kappa\left|\begin{matrix}\boldsymbol{k}\\ j_\mu\end{matrix}\right.\right)$. Then keeping in mind the delta in (85.19) we would take the l_j distinct eigenvectors for a given branch $\omega^2(\boldsymbol{k}|j)$ in order to evaluate (85.18). For another branch of the same symmetry those distinct eigenvectors would be used to obtain (85.18).

To obtain the full space group consequences of necessary degeneracy under the spatial operators in $\mathfrak{G}$, the full crystal space group, it is necessary to induce from the $D^{(\boldsymbol{k})(m)}$ which arise in solving the lattice dynamic problem, to the set of corresponding full group representations $D^{(^*\boldsymbol{k})(m)}$.

The entire necessary degeneracy of this problem is that related to the full space-time symmetry group $\mathcal{G}$ which will be analysed below.

86. Complex normal coordinates $Q\begin{pmatrix}\boldsymbol{k}\\ j_\mu\end{pmatrix}$ as bases for the representation $D^{(\boldsymbol{k})(j)}$ of $\mathfrak{G}(\boldsymbol{k})$. As in Sect. 76 it is desirable to have a rule by which to discuss the transformation of the complex normal coordinates $Q\begin{pmatrix}\boldsymbol{k}\\ j_\mu\end{pmatrix}$. The $Q\begin{pmatrix}\boldsymbol{k}\\ j_\mu\end{pmatrix}$ are dynamical variables for the lattice dynamic problem, as seen in (80.7) and (80.8) so it is a matter of some importance. We can proceed as in Sect. 76 by relating the transformation of the $Q\begin{pmatrix}\boldsymbol{k}\\ j_\mu\end{pmatrix}$ to that of physical displacements.

To allow for degeneracy, rewrite (80.9) as

$$u_\alpha\begin{pmatrix}l\\ \kappa\end{pmatrix}=\frac{1}{\sqrt{M_\kappa N}}\sum_j\sum_\nu\sum_{\boldsymbol{k}}\exp(i\,\boldsymbol{k}\cdot\boldsymbol{R}_L)\,e_\alpha\left(\kappa\left|\begin{matrix}\boldsymbol{k}\\ j_\nu\end{matrix}\right.\right)Q\begin{pmatrix}\boldsymbol{k}\\ j_\nu\end{pmatrix} \tag{86.1}$$

and the inverse follows from the orthonormality rules:

$$Q\begin{pmatrix}\boldsymbol{k}\\ j_\nu\end{pmatrix}=\frac{1}{\sqrt{N}}\sum_L\sum_{\kappa\alpha}\exp(-i\,\boldsymbol{k}\cdot\boldsymbol{R}_L)\,e_\alpha^*\left(\kappa\left|\begin{matrix}\boldsymbol{k}\\ j_\nu\end{matrix}\right.\right)u_\alpha\begin{pmatrix}l\\ \kappa\end{pmatrix}\sqrt{M_\kappa}\,. \tag{86.2}$$

It is helpful to define a vector $\boldsymbol{v}\left(\left|\begin{matrix}\boldsymbol{k}\\ j_\nu\end{matrix}\right.\right)$ whose $\alpha\,l\,\kappa$ component is

$$v_\alpha\left(\begin{matrix}l\\ \kappa\end{matrix}\left|\begin{matrix}\boldsymbol{k}\\ j_\nu\end{matrix}\right.\right)\equiv\frac{1}{\sqrt{M_\kappa N}}\exp(i\,\boldsymbol{k}\cdot\boldsymbol{R}_L)\,e_\alpha\left(\kappa\left|\begin{matrix}\boldsymbol{k}\\ j_\nu\end{matrix}\right.\right). \tag{86.3}$$

Then for (86.1)

$$\boldsymbol{u}(\;)=\sum_{j\nu\boldsymbol{k}}\boldsymbol{v}\left(\left|\begin{matrix}\boldsymbol{k}\\ j_\nu\end{matrix}\right.\right)Q\begin{pmatrix}\boldsymbol{k}\\ j_\nu\end{pmatrix} \tag{86.4}$$

and for (86.2)

$$Q\begin{pmatrix}\boldsymbol{k}\\ j_\nu\end{pmatrix}=\sum_{l\kappa\alpha}v_\alpha^*\left(\begin{matrix}l\\ \kappa\end{matrix}\left|\begin{matrix}\boldsymbol{k}\\ j_\nu\end{matrix}\right.\right)u_\alpha\begin{pmatrix}l\\ \kappa\end{pmatrix}M_\kappa \tag{86.5}$$

$$=\boldsymbol{v}^*\left(\left|\begin{matrix}\boldsymbol{k}\\ j_\nu\end{matrix}\right.\right)\cdot\boldsymbol{u}\,M_\kappa=\left(\left(v\left|\begin{matrix}\boldsymbol{k}\\ j_\nu\end{matrix}\right.\right),\boldsymbol{u}\,M_\kappa\right) \tag{86.6}$$

using the scalar product (79.8).

Let us first demonstrate that $Q\begin{pmatrix}\boldsymbol{k}\\ j_\nu\end{pmatrix}$ is a Bloch vector at wave vector $\boldsymbol{k}$. This is most easily seen from (86.2) by examining the translated physical displacement field

$$P_{\{\varepsilon|\boldsymbol{R}_M\}}\,\boldsymbol{u}\equiv\boldsymbol{u}_{\{\boldsymbol{R}_M\}}. \tag{86.7}$$

This may be written

$$\boldsymbol{u}_{\{\boldsymbol{R}_M\}}=\sum_{j\nu\boldsymbol{k}}\boldsymbol{v}_{\{\boldsymbol{R}_M\}}\left(\left|\begin{matrix}\boldsymbol{k}\\ j_\nu\end{matrix}\right.\right)Q\begin{pmatrix}\boldsymbol{k}\\ j_\nu\end{pmatrix} \tag{86.8}$$

or

$$\boldsymbol{u}_{\{\boldsymbol{R}_M\}}=\sum_{j\nu\boldsymbol{k}}\boldsymbol{v}\left(\left|\begin{matrix}\boldsymbol{k}\\ j_\nu\end{matrix}\right.\right)Q_{\{\boldsymbol{R}_M\}}\begin{pmatrix}\boldsymbol{k}\\ j_\nu\end{pmatrix} \tag{86.9}$$

where (86.9) defines the transformed $Q\begin{pmatrix}\boldsymbol{k}\\ j_\nu\end{pmatrix}$. But

$$\begin{aligned}(\boldsymbol{v}_{\{\boldsymbol{R}_M\}})_\alpha\left(\begin{matrix}l\\ \kappa\end{matrix}\left|\begin{matrix}\boldsymbol{k}\\ j_\nu\end{matrix}\right.\right)&=v_\alpha\left(\begin{matrix}l-m\\ \kappa\end{matrix}\left|\begin{matrix}\boldsymbol{k}\\ j_\nu\end{matrix}\right.\right)\\
&=\exp(-i\,\boldsymbol{k}\cdot\boldsymbol{R}_M)\,v_\alpha\left(\begin{matrix}l\\ \kappa\end{matrix}\left|\begin{matrix}\boldsymbol{k}\\ j_\nu\end{matrix}\right.\right)\\
&=D^{(\boldsymbol{k})}(\{\varepsilon|\boldsymbol{R}_M\})\,v_\alpha\left(\begin{matrix}l\\ \kappa\end{matrix}\left|\begin{matrix}\boldsymbol{k}\\ j_\nu\end{matrix}\right.\right).\end{aligned} \tag{86.10}$$

Hence

$$\boldsymbol{u}_{\{\boldsymbol{R}_M\}}=D^{(\boldsymbol{k})}(\{\varepsilon|\boldsymbol{R}_M\})\sum_{j\nu\boldsymbol{k}}\boldsymbol{v}\left(\left|\begin{matrix}\boldsymbol{k}\\ j_\nu\end{matrix}\right.\right)Q\begin{pmatrix}\boldsymbol{k}\\ j_\nu\end{pmatrix}, \tag{86.11}$$

when

$$P_{\{\varepsilon|\boldsymbol{R}_M\}}Q\begin{pmatrix}\boldsymbol{k}\\ j_\nu\end{pmatrix}\equiv Q_{\{\boldsymbol{R}_M\}}\begin{pmatrix}\boldsymbol{k}\\ j_\nu\end{pmatrix}=D^{(\boldsymbol{k})}(\{\varepsilon|\boldsymbol{R}_M\})\,Q\begin{pmatrix}\boldsymbol{k}\\ j_\nu\end{pmatrix}. \tag{86.12}$$

Then $Q\begin{pmatrix}\boldsymbol{k}\\ j_\nu\end{pmatrix}$ can be taken as a Bloch vector at wave vector $\boldsymbol{k}$.

Consider the effect of a transformation operator $P_{\{\boldsymbol{\varphi}|\boldsymbol{t}\}}$ which is not in $\mathfrak{G}(\boldsymbol{k})$. By the argument given in Sect. 30, it immediately follows that

$$P_{\{\boldsymbol{\varphi}|\boldsymbol{t}\}}Q\begin{pmatrix}\boldsymbol{k}\\ j_\nu\end{pmatrix} \tag{86.13}$$

is a Bloch vector at wave vector $\boldsymbol{\varphi}\cdot\boldsymbol{k}$:

$$P_{\{\boldsymbol{\varphi}|\boldsymbol{t}\}}Q\begin{pmatrix}\boldsymbol{k}\\ j_\nu\end{pmatrix}=Q\begin{pmatrix}\boldsymbol{\varphi}\cdot\boldsymbol{k}\\ j_\nu\end{pmatrix}. \tag{86.14}$$

We may choose it to belong to the same row as $Q\begin{pmatrix}\boldsymbol{k}\\ j_\nu\end{pmatrix}$.

To consider the effect of a transformation operator $P_{\{\boldsymbol{\varphi}_{l_\lambda}|\boldsymbol{\tau}_{l_\lambda}\}}$ which is in $\mathfrak{G}(\boldsymbol{k})$, it is necessary to return to the definitions (86.1)–(86.6). Then the rotated physical displacement field is

$$\boldsymbol{u}_{\{\varphi_{l_\lambda}\}} \equiv P_{\{\varphi_{l_\lambda}|\tau_{l_\lambda}\}}\,\boldsymbol{u}. \tag{86.15}$$

We may express (86.15) as

$$\boldsymbol{u}_{\{\varphi_{l_\lambda}\}} = \sum_{j\nu\boldsymbol{k}} \boldsymbol{v}_{\{\varphi_{l_\lambda}\}}\left(\left|\begin{matrix}\boldsymbol{k}\\ j_\nu\end{matrix}\right.\right) Q\begin{pmatrix}\boldsymbol{k}\\ j_\nu\end{pmatrix} \tag{86.16}$$

or

$$\boldsymbol{u}_{\{\varphi_{l_\lambda}\}} = \sum_{j\nu\boldsymbol{k}} v\left(\left|\begin{matrix}\boldsymbol{k}\\ j_\nu\end{matrix}\right.\right) Q_{\{\varphi_{l_\lambda}\}}\begin{pmatrix}\boldsymbol{k}\\ j_\nu\end{pmatrix}. \tag{86.17}$$

Then inverting (86.17) we obtain from (86.6)

$$Q_{\{\varphi_{l_\lambda}\}}\begin{pmatrix}\boldsymbol{k}\\ j_\nu\end{pmatrix} = \boldsymbol{v}^*\left(\left|\begin{matrix}\boldsymbol{k}\\ j_\nu\end{matrix}\right.\right)\cdot \boldsymbol{u}_{\{\varphi_{l_\lambda}\}} M_\kappa \tag{86.18}$$

$$\begin{aligned} &= \sum_{l\kappa\alpha} v_\alpha^*\left(\begin{matrix}l\\ \kappa\end{matrix}\left|\begin{matrix}\boldsymbol{k}\\ j_\nu\end{matrix}\right.\right)(\boldsymbol{u}_{\{\varphi_{l_\lambda}\}})_\alpha\begin{pmatrix}l\\ \kappa\end{pmatrix} M_\kappa \\ &= \sum_{l\kappa\alpha}\sum_\beta v_\alpha^*\left(\begin{matrix}l\\ \kappa\end{matrix}\left|\begin{matrix}\boldsymbol{k}\\ j_\nu\end{matrix}\right.\right)(\varphi_{l_\lambda})_{\alpha\beta}\,\boldsymbol{u}_\beta\begin{pmatrix}l_{\bar\varphi_{l_\lambda}}\\ \kappa_{\bar\varphi_{l_\lambda}}\end{pmatrix} M_\kappa. \end{aligned} \tag{86.19}$$

Now use (86.1) to evaluate $\boldsymbol{u}_\beta\begin{pmatrix}l_{\bar\varphi_{l_\lambda}}\\ \kappa_{\bar\varphi_{l_\lambda}}\end{pmatrix}$:

$$\boldsymbol{u}_\beta\begin{pmatrix}l_{\bar\varphi_{l_\lambda}}\\ \kappa_{\bar\varphi_{l_\lambda}}\end{pmatrix} = \frac{1}{\sqrt{M_{\bar\varphi_{l_\lambda}} N}} \sum_{j'\nu'\boldsymbol{k}'} \exp(i\,\boldsymbol{k}'\cdot \boldsymbol{R}_{L_{\bar\varphi_{l_\lambda}}})\, e_\beta\left(\kappa_{\bar\varphi_{l_\lambda}}\left|\begin{matrix}\boldsymbol{k}'\\ j'_{\nu'}\end{matrix}\right.\right)\left(Q\left|\begin{matrix}\boldsymbol{k}'\\ j'_{\nu'}\end{matrix}\right.\right). \tag{86.20}$$

But

$$\exp i\,\boldsymbol{k}'\cdot \boldsymbol{R}_{L_{\bar\varphi_{l_\lambda}}} = \exp i(\boldsymbol{\varphi}_{l_\lambda}\cdot \boldsymbol{k}')\cdot \boldsymbol{R}_L. \tag{86.21}$$

Then inserting (86.20), (86.21) into (86.19), using also (86.3), we find

$$\begin{aligned} &\sum_{l\kappa\alpha}\sum_\beta\sum_{j'\nu'\boldsymbol{k}'} \frac{M_\kappa}{N\sqrt{M_\kappa\cdot M_{\kappa_{\bar\varphi_{l_\lambda}}}}} \exp\left(i\boldsymbol{R}_L\cdot(\boldsymbol{\varphi}_{l_\lambda}\cdot \boldsymbol{k}'-\boldsymbol{k})\right) \\ &\quad\times(\boldsymbol{\varphi}_{l_\lambda})_{\alpha\beta}\, e_\alpha^*\left(\kappa\left|\begin{matrix}\boldsymbol{k}\\ j_\nu\end{matrix}\right.\right) e_\beta\left(\kappa_{\bar\varphi_{l_\lambda}}\left|\begin{matrix}\boldsymbol{k}'\\ j'_{\nu'}\end{matrix}\right.\right) Q\begin{pmatrix}\boldsymbol{k}'\\ j'_{\nu'}\end{pmatrix}. \end{aligned} \tag{86.22}$$

But

$$\sum_l \frac{1}{N}\exp\left(i\boldsymbol{R}_L\cdot(\boldsymbol{\varphi}_{l_\lambda}\cdot \boldsymbol{k}'-\boldsymbol{k})\right) = \delta(\boldsymbol{\varphi}_{l_\lambda}\cdot \boldsymbol{k}'-\boldsymbol{k}). \tag{86.23}$$

But as $\boldsymbol{\varphi}_{l_\lambda}$ is a rotation in $\mathfrak{G}(\boldsymbol{k})$, the delta in (86.23) requires $\boldsymbol{k}=\boldsymbol{k}'$. Hence the sum on $\boldsymbol{k}'$ reduces to a single term and (86.22) simplifies when we use also $M_\kappa = M_{\kappa_{\bar\varphi_{l_\lambda}}}$ to

$$\sum_{\kappa\alpha}\sum_\beta\sum_{j'\nu'}(\boldsymbol{\varphi}_{l_\lambda})_{\alpha\beta}\, e_\alpha^*\left(\kappa\left|\begin{matrix}\boldsymbol{k}\\ j_\nu\end{matrix}\right.\right) e_\beta\left(\kappa_{\bar\varphi_{l_\lambda}}\left|\begin{matrix}\boldsymbol{k}\\ j'_{\nu'}\end{matrix}\right.\right) Q\begin{pmatrix}\boldsymbol{k}\\ j'_{\nu'}\end{pmatrix}. \tag{86.24}$$

Then using (82.12)

$$e_\beta\left(\kappa_{\bar{\varphi}l_\lambda}\middle|\begin{matrix}\boldsymbol{k}\\ j'_{\nu'}\end{matrix}\right)=D^{(\boldsymbol{k})}\left(\{\varepsilon|-\boldsymbol{R}_N(\kappa_{\bar{\varphi}l_\lambda},\kappa'')\}\right)e_\beta\left(\kappa''\middle|\begin{matrix}\boldsymbol{k}\\ j'_{\nu'}\end{matrix}\right)\delta_{\kappa_{\bar{\varphi}l_\lambda},\kappa''}. \tag{86.25}$$

From (86.25) and (82.16) we find for (86.24)

$$\sum_{\kappa\alpha}\sum_{\kappa''\beta}\sum_{j'\nu'}D^{(\boldsymbol{k})(\boldsymbol{e})}(\{\varphi_{l_\lambda}|\tau_{l_\lambda}\})_{\alpha\kappa,\beta\kappa''}e^*_\alpha\left(\kappa\middle|\begin{matrix}\boldsymbol{k}\\ j_\nu\end{matrix}\right)e_\beta\left(\kappa''\middle|\begin{matrix}\boldsymbol{k}\\ j'_{\nu'}\end{matrix}\right)Q\begin{pmatrix}\boldsymbol{k}\\ j'_{\nu'}\end{pmatrix}. \tag{86.26}$$

Then from (85.18), (85.19) we have for (86.26):

$$\sum_{j'\nu'}(D^{(\boldsymbol{k})(j)}(\{\varphi_{l_\lambda}|\tau_{l_\lambda}\}))_{\nu'\nu}\,\delta_{jj'}\,Q\begin{pmatrix}\boldsymbol{k}\\ j'_{\nu'}\end{pmatrix}. \tag{86.27}$$

Finally assembling (86.18) and (86.27) we have

$$Q_{\{\varphi_{l_\lambda}\}}\begin{pmatrix}\boldsymbol{k}\\ j_\nu\end{pmatrix}=\sum_{\nu'}D^{(\boldsymbol{k})(j)}(\{\varphi_{l_\lambda}|\tau_{l_\lambda}\})_{\nu'\nu}\,Q\begin{pmatrix}\boldsymbol{k}\\ j_{\nu'}\end{pmatrix}. \tag{86.28}$$

This is the penultimate result. Now defining

$$P_{\{\varphi_{l_\lambda}|\tau_{l_\lambda}\}}Q\begin{pmatrix}\boldsymbol{k}\\ j_\nu\end{pmatrix}\equiv Q_{\{\varphi_{l_\lambda}\}}\begin{pmatrix}\boldsymbol{k}\\ j_\nu\end{pmatrix} \tag{86.29}$$

we achieve in (86.28) the desired rule by which the normal coordinate $Q\begin{pmatrix}\boldsymbol{k}\\ j_\nu\end{pmatrix}$ can be taken to transform under an element in $\mathfrak{G}(\boldsymbol{k})$. As could be expected from (86.1) and (86.17), this is just the rule by which the eigenvector $\boldsymbol{e}\left(\middle|\begin{matrix}\boldsymbol{k}\\ j_\nu\end{matrix}\right)$ transforms. But it is most useful to use (86.17) and (86.29) to describe the effect of symmetry upon the lattice via the "rotated" (transformed) normal coordinates, keeping the "axes" $\boldsymbol{v}\left(\middle|\begin{matrix}\boldsymbol{k}\\ j_\nu\end{matrix}\right)$ unchanged. Recall (85.20), (85.21) so that the $D^{(\boldsymbol{k})(j)}$ is a physically irreducible allowable representation of $\mathfrak{G}(\boldsymbol{k})$.

We have established that $Q\begin{pmatrix}\boldsymbol{k}\\ j_\nu\end{pmatrix}$ has the transformation properties of a Bloch vector. Thus we may combine (86.14) and (86.28), (86.29) into a single rule. Let $P_{\{\varphi_p|\boldsymbol{t}(\varphi_p)\}}$ be any operation in $\mathfrak{G}$, then from (36.17), (36.18)

$$\begin{aligned}P_{\{\varphi_p|\boldsymbol{t}(\varphi_p)\}}Q\begin{pmatrix}\boldsymbol{k}_\tau\\ j_\nu\end{pmatrix}&\equiv P_{\{\varphi_p\}}Q\begin{pmatrix}\boldsymbol{k}_\tau\\ j_\nu\end{pmatrix}\equiv Q_{\{\varphi_p\}}\begin{pmatrix}\boldsymbol{k}_\tau\\ j_\nu\end{pmatrix}\\ &=\sum_{\sigma=1}^{s}\sum_{\mu=1}^{l_m}D^{(\star\boldsymbol{k})(j)}(\{\varphi_p|\boldsymbol{t}(\varphi_p)\})_{(\sigma\mu)(\tau\nu)}\,Q\begin{pmatrix}\boldsymbol{k}_\sigma\\ j_\mu\end{pmatrix}\end{aligned} \tag{86.30}$$

with an obvious, and minor change in notation (subscripts). Recall that the indices (σ,τ) refer to the block sub-matrices corresponding to matrix blocks going with wave vectors $\boldsymbol{k}_\sigma$ and $\boldsymbol{k}_\tau$, while (μ,ν) refer to the row and column indices of a single matrix.

I. Space-time symmetry and classical lattice dynamics

87. Introduction. In this part of the article we shall consider the effect of including time-reversal symmetry on an equal footing to the spatial or geometric symmetry, in classical lattice dynamics. The object here is to develop the modern approach, known as corepresentation theory following WIGNER [*1*], with specific objective of connecting that theory to lattice dynamics. Time reversal as a symmetry element enters the problem dynamically, in the sense of "reversal of the motion" rather than statically as the spatial symmetry does. In addition, time reversal is an anti-unitary symmetry element, which preserves the absolute modulus of the scalar product but does not simply preserve the actual scalar product.

This part begins with a treatment of time reversal which is somewhat more traditional in Sects. 88–94, being based on the identification of the time reversal operator as complex conjugation. Thus time reversal, in this manifestation, operates upon different variables than spatial transformation. Complex conjugation is a change (mapping) of the complex field (in which the eigenfunction exists) onto itself, while spatial transformations maps points in configuration space onto themselves). Since the basic lattice dynamical variables are real displacements the "physically irreducible" representations must also be real. The test for reality of a space group irreducible representation given by HERRING is presented in Sect. 93. In Sect. 94 we present an extension of a more useful test for reality due to FREI. Using this one can determine not only whether an irreducible representation is real, complex, or pseudo real, but more specifically, the actual symmetry of its complex conjugate, in case the representation is not real.

In Sects. 95–100 corepresentation theory is developed. This is intended as a general treatment, adopted for solid state physics. Particularly the close connections with different types of costars should facilitate future applications.

Having developed the general theory, specific applications are given in Sects. 101–104 to the problem of determining symmetry of eigenvectors and factorization of the dynamical matrix in any particular crystal.

A final remark: the work of this part is intended to be simultaneously of general import (reality tests and classification of the space group irreducible representations) and related to the lattice dynamical Hamiltonian. In that sense the space-time symmetry group is a dynamical symmetry group.

88. The antilinear, antiunitary transformation operator K and time reversal[1]**.** It is a matter of importance to our work to be able to classify the representations based on the normal modes $Q\begin{pmatrix}\boldsymbol{k}\\ j_\mu\end{pmatrix}$ according to their reality properties. This derives from the fact.that a transformation operator K which transforms a function into its complex conjugate plays an important role as one of the symmetry operators of the dynamical matrix $[\boldsymbol{D}(\boldsymbol{k})]$. This will be discussed in greater detail below. In this section we initiate the discussion of the transformation operator K.

[1] Originally discussed by E. P. WIGNER, Nachr. Akad. Wiss. Göttingen, Math.-Physik. Kl. 546 (1932).

We define the operator K by

$$K\,e = e^*, \tag{88.1}$$

where $\boldsymbol{e}$ is any of the eigenvectors with which we have been concerned. For a real eigenvector

$$K\,e^{(R)} = e^{(R)} \tag{88.2}$$

for an imaginary eigenvector

$$K\,e^{(I)} = -e^{(I)}. \tag{88.3}$$

Let a and b be constants, then

$$K(a\,e + b\,e') = a^*\,K\,e + b^*\,K\,e'. \tag{88.4}$$

If $\boldsymbol{e}$ is a complex eigenvector such as $\boldsymbol{e}\left(\left|\begin{matrix}\boldsymbol{k}\\ j_\lambda\end{matrix}\right.\right)$, and obeys the Hermitian scalar product rule (79.8), then

$$\left(K\,\boldsymbol{e}\left(\left|\begin{matrix}\boldsymbol{k}\\ j_\lambda\end{matrix}\right.\right)\right)^* \cdot \left(K\,\boldsymbol{e}\left(\left|\begin{matrix}\boldsymbol{k}\\ j'_\lambda\end{matrix}\right.\right)\right) = \left(\boldsymbol{e}\left(\left|\begin{matrix}\boldsymbol{k}\\ j_\lambda\end{matrix}\right.\right)^* \cdot \boldsymbol{e}\left(\left|\begin{matrix}\boldsymbol{k}\\ j'_\lambda\end{matrix}\right.\right)\right)^*. \tag{88.5}$$

Then, in regard to the Hermitian scalar product (79.8), K is an antiunitary transformation operator. K is also an antiunitary transformation operator in regard to the Hermitian scalar product (79.9) or (79.11) also, so that

$$\left(\left(K\,\boldsymbol{e}_\alpha\left(\kappa\left|\boldsymbol{k}\right.\right)\right)^* : K\,\boldsymbol{e}_\beta\left(\kappa'\left|\boldsymbol{k}\right.\right)\right)^* = \boldsymbol{e}_\alpha\left(\kappa\left|\boldsymbol{k}\right.\right)^* : \boldsymbol{e}_\beta\left(\kappa'\left|\boldsymbol{k}\right.\right). \tag{88.6}$$

It needs to be emphasized that the properties of antiunitarity, or antilinearity epitomized in (88.4)–(88.6) mark the operator K as qualitatively different from the transformation operators $P_{\{\varphi|\boldsymbol{t}\}}$ with which we previously worked. This will again be observed as we develop the calculus, or algebra, related to these operators, and as we develop the corepresentation theory (matrix homomorphism theory) based on functions.

As an abstract operator, K operates upon different variables than the coordinate transformation operators $P_{\{\varphi|\boldsymbol{t}\}}$. Hence as abstract operators K and $P_{\{\varphi|\boldsymbol{t}\}}$ commute:

$$K P_{\{\varphi|\boldsymbol{t}\}} = P_{\{\varphi|\boldsymbol{t}\}} K. \tag{88.7}$$

The set of transformation operators consisting of the coordinate transformation operators alone, form the spatial symmetry group $\mathfrak{G}$ of the crystal; the abstract group $\mathcal{G}$, consisting of

$$\mathcal{G} = \mathfrak{G} + K\mathfrak{G} \tag{88.8}$$

will be referred to as the space-time symmetry group $\mathcal{G}$, for reasons to be examined in the following.

The complete, time dependent solutions to the dynamical equation (79.4), or (85.1) are obtained by restoring the full time dependence of the displacements removed in (72.2). Then it follows that the time dependent eigenvector $\boldsymbol{\varepsilon}$ at wave vector $\boldsymbol{k}$

$$\boldsymbol{\varepsilon}\left(\left|\begin{matrix}\boldsymbol{k}\\ j_\mu\end{matrix}\right|t\right) \equiv \boldsymbol{e}\left(\left|\begin{matrix}\boldsymbol{k}\\ j_\mu\end{matrix}\right.\right) \exp -i\omega(\boldsymbol{k}|j_\mu)\,t \tag{88.9}$$

satisfies the equation

$$[\boldsymbol{D}(\boldsymbol{k})]\cdot\varepsilon\left(\begin{vmatrix}\boldsymbol{k}\\ j_\mu\end{vmatrix}t\right)+\frac{\partial^2\varepsilon}{\partial t^2}\left(\begin{vmatrix}\boldsymbol{k}\\ j_\mu\end{vmatrix}t\right)=0. \tag{88.10}$$

Next consider applying the operator K to (88.10)

$$K[\boldsymbol{D}(\boldsymbol{k})]\,K^{-1}\cdot K\,\varepsilon\left(\begin{vmatrix}\boldsymbol{k}\\ j_\mu\end{vmatrix}t\right)+K\frac{\partial^2}{\partial t^2}K^{-1}\cdot K\,\varepsilon\left(\begin{vmatrix}\boldsymbol{k}\\ j_\mu\end{vmatrix}t\right)=0. \tag{88.11}$$

But the operator $\partial^2/\partial t^2$ is real so that

$$K(\partial^2/\partial t^2)\,K^{-1}=(\partial^2/\partial t^2). \tag{88.12}$$

Then taking

$$K[\boldsymbol{D}(\boldsymbol{k})]\,K^{-1}=[\boldsymbol{D}(\boldsymbol{k})]^* \tag{88.13}$$

we obtain, using also the definition (78.5), and the reality of the elementary force constants $[\boldsymbol{\Phi}]$,

$$[\boldsymbol{D}(\boldsymbol{k})]^*=[\boldsymbol{D}(-\boldsymbol{k})]. \tag{88.14}$$

Compare (88.14) with the hermiticity (78.6) of $[\boldsymbol{D}(\boldsymbol{k})]$. Then, (78.6) states that the dynamical matrix $[\boldsymbol{D}(\boldsymbol{k})]$ at one fixed $\boldsymbol{k}$ is an Hermitian matrix; (88.14) states the relationship between the dynamical matrix $[\boldsymbol{D}(\boldsymbol{k})]$ at two distinct values of wave vector $\boldsymbol{k}$. Then using (88.12), (88.14) in (88.11) we obtain

$$[\boldsymbol{D}(-\boldsymbol{k})]\cdot K\,\varepsilon\left(\begin{vmatrix}\boldsymbol{k}\\ j_\mu\end{vmatrix}t\right)=(\partial^2/\partial t^2)\cdot K\,\varepsilon\left(\begin{vmatrix}\boldsymbol{k}\\ j_\mu\end{vmatrix}t\right). \tag{88.15}$$

Consider next the complete set of time dependent eigenvectors at wave vector $-\boldsymbol{k}$:

$$\varepsilon\left(\begin{vmatrix}-\boldsymbol{k}\\ j_\lambda\end{vmatrix}t\right)=\boldsymbol{e}\left(\begin{vmatrix}-\boldsymbol{k}\\ j_\lambda\end{matrix}\right)\exp-i\omega(-\boldsymbol{k}\,|\,j_\lambda)\,t. \tag{88.16}$$

These satisfy an equation

$$[\boldsymbol{D}(-\boldsymbol{k})]\cdot\varepsilon\left(\begin{vmatrix}-\boldsymbol{k}\\ j_\lambda\end{vmatrix}t\right)+(\partial^2/\partial t^2)\,\varepsilon\left(\begin{vmatrix}-\boldsymbol{k}\\ j_\lambda\end{vmatrix}t\right)=0. \tag{88.17}$$

Comparing (88.15) and (88.17) it is clear that $K\,\varepsilon\left(\begin{vmatrix}\boldsymbol{k}\\ j_\mu\end{vmatrix}t\right)$ is intimately related to $\varepsilon\left(\begin{vmatrix}-\boldsymbol{k}\\ j_\lambda\end{vmatrix}t\right)$. To examine the relationship we return to the time independent cases. The equation at wave vector $\boldsymbol{k}$ is

$$[\boldsymbol{D}(\boldsymbol{k})]\cdot\boldsymbol{e}\left(\begin{vmatrix}\boldsymbol{k}\\ j_\mu\end{matrix}\right)=\omega^2(\boldsymbol{k}\,|\,j)\,\boldsymbol{e}\left(\begin{vmatrix}\boldsymbol{k}\\ j_\mu\end{matrix}\right),\qquad \mu=1,\ldots,l_j \tag{88.18}$$

and the equation at wave vector $-\boldsymbol{k}$ is

$$[\boldsymbol{D}(-\boldsymbol{k})]\cdot\boldsymbol{e}\left(\begin{vmatrix}-\boldsymbol{k}\\ \bar{j}_\lambda\end{matrix}\right)=\omega^2(-\boldsymbol{k}\,|\,\bar{j})\,\boldsymbol{e}\left(\begin{vmatrix}-\boldsymbol{k}\\ \bar{j}_\lambda\end{matrix}\right),\qquad \lambda=1,\ldots,l_{\bar{j}}. \tag{88.19}$$

Applying the operator K to both sides of (88.18) we have, using also (88.14)

$$[\boldsymbol{D}(-\boldsymbol{k})]\cdot K\,\boldsymbol{e}\left(\left|\begin{matrix}\boldsymbol{k}\\ j_\mu\end{matrix}\right.\right)=\omega^2(\boldsymbol{k}|j)\,K\,\boldsymbol{e}\left(\left|\begin{matrix}\boldsymbol{k}\\ j_\mu\end{matrix}\right.\right),\qquad \mu=1,\ldots,l_j. \tag{88.20}$$

Hence the time independent eigenvector $K\,\boldsymbol{e}\left(\left|\begin{matrix}\boldsymbol{k}\\ j_\mu\end{matrix}\right.\right)$ satisfies the same eigenvalue equation as $\boldsymbol{e}\left(\left|\begin{matrix}-\boldsymbol{k}\\ j_\lambda\end{matrix}\right.\right)$ with the dynamical matrix $[\boldsymbol{D}(-\boldsymbol{k})]$.

Further, from (78.6), $[\boldsymbol{D}(\boldsymbol{k})]$ and $[\boldsymbol{D}(\boldsymbol{k})]^*=[\boldsymbol{D}(-\boldsymbol{k})]$ have the same set of $3r$ real eigenvalues. In fact, $[\boldsymbol{D}(\boldsymbol{k})]$ and $[\boldsymbol{D}(\boldsymbol{k})]^*$ are equivalent and can be made real. That is, a unitary transformation exists which will transform $[\boldsymbol{D}(\boldsymbol{k})]$ to $[\boldsymbol{D}(\boldsymbol{k})]^*$. Also, a unitary transformation exists which will transform $[\boldsymbol{D}(\boldsymbol{k})]$ to a real matrix. Now let us examine these statements further. From (79.16)

$$\boldsymbol{E}(\boldsymbol{k})^{-1}\cdot[\boldsymbol{D}(\boldsymbol{k})]\cdot\boldsymbol{E}(\boldsymbol{k})=\boldsymbol{\Delta}(\boldsymbol{k}) \tag{88.21}$$

and similarity transforming (88.21) by the operator K, or equivalently taking the complex conjugate of (88.21)

$$\boldsymbol{E}(\boldsymbol{k})^{*-1}\cdot[\boldsymbol{D}(\boldsymbol{k})]^*\cdot\boldsymbol{E}(\boldsymbol{k})^*=\boldsymbol{\Delta}^*(\boldsymbol{k})=\boldsymbol{\Delta}(\boldsymbol{k}), \tag{88.22}$$

where the last step follows owing to reality of $\boldsymbol{\Delta}(\boldsymbol{k})$. Then from (79.15), (88.13), (88.14)

$$\widetilde{\boldsymbol{E}(\boldsymbol{k})}\cdot[\boldsymbol{D}(-\boldsymbol{k})]\cdot\widetilde{\boldsymbol{E}(\boldsymbol{k})^{-1}}=\boldsymbol{\Delta}(\boldsymbol{k}) \tag{88.23}$$

or equating (88.21) to (88.23)

$$\boldsymbol{E}(\boldsymbol{k})^{-1}\cdot[\boldsymbol{D}(\boldsymbol{k})]\cdot\boldsymbol{E}(\boldsymbol{k})=\widetilde{\boldsymbol{E}(\boldsymbol{k})}\cdot[\boldsymbol{D}(-\boldsymbol{k})]\cdot\widetilde{\boldsymbol{E}(\boldsymbol{k})^{-1}}. \tag{88.24}$$

Then

$$\big(\boldsymbol{E}(\boldsymbol{k})\cdot\widetilde{\boldsymbol{E}(\boldsymbol{k})}\big)^{-1}\cdot[\boldsymbol{D}(\boldsymbol{k})]\cdot\big(\boldsymbol{E}(\boldsymbol{k})\cdot\widetilde{\boldsymbol{E}(\boldsymbol{k})}\big)=[\boldsymbol{D}(-\boldsymbol{k})]. \tag{88.25}$$

Hence the symmetric matrix

$$\boldsymbol{S}(\boldsymbol{k})\equiv\boldsymbol{E}(\boldsymbol{k})\cdot\widetilde{\boldsymbol{E}(\boldsymbol{k})} \tag{88.26}$$

effects a similarity transformation of $[\boldsymbol{D}(\boldsymbol{k})]$ to $[\boldsymbol{D}(-\boldsymbol{k})]$:

$$\boldsymbol{S}(\boldsymbol{k})^{-1}\cdot[\boldsymbol{D}(\boldsymbol{k})]\cdot\boldsymbol{S}(\boldsymbol{k})=[\boldsymbol{D}(-\boldsymbol{k})]. \tag{88.27}$$

From (88.26), and (79.14) we can obtain the elements of $\boldsymbol{S}(\boldsymbol{k})$; they are:

$$(\boldsymbol{S}(\boldsymbol{k}))_{\beta\kappa',\bar{\beta}\bar{\kappa}'}=\sum_j e_\beta\left(\kappa'\left|\begin{matrix}\boldsymbol{k}\\ j\end{matrix}\right.\right)e_{\bar{\beta}}\left(\bar{\kappa}'\left|\begin{matrix}\boldsymbol{k}\\ j\end{matrix}\right.\right). \tag{88.28}$$

Notice the difference between (88.28), and the orthonormality rules (79.6) or (79.9), (79.11). If the individual eigenvectors are real (88.28) will be identical to (79.9) or (79.11) so that

$$\text{if } \boldsymbol{e}\left(\left|\begin{matrix}\boldsymbol{k}\\ j\end{matrix}\right.\right)\text{ is real then } \boldsymbol{S}(\boldsymbol{k})=\boldsymbol{\Pi}_{3r} \tag{88.29}$$

where $\boldsymbol{\Pi}_{3r}$ is the $3r$ dimensional unit matrix. Clearly then, in the case of real eigenvectors, (88.27), and (88.14) then show that $[\boldsymbol{D}(\boldsymbol{k})]$ is real, and so equal to $[\boldsymbol{D}(-\boldsymbol{k})]$. But in general this will not be so.

Using (88.22) and (88.14) we find

$$[\boldsymbol{D}(-\boldsymbol{k})] = \boldsymbol{E}(\boldsymbol{k})^* \boldsymbol{\Delta}(\boldsymbol{k}) \widetilde{\boldsymbol{E}(\boldsymbol{k})}. \tag{88.30}$$

Taking the canonical form of $\boldsymbol{\Delta}(\boldsymbol{k})$ as in (79.17) we can obtain the components of both sides of (88.30) as

$$\sum_j \sum_\lambda e_\alpha^* \left(\kappa \left| \begin{matrix} \boldsymbol{k} \\ j_\lambda \end{matrix} \right.\right) \omega^2(\boldsymbol{k}|j)\, e_\beta \left(\kappa' \left| \begin{matrix} \boldsymbol{k} \\ j_\lambda \end{matrix} \right.\right) = D\begin{pmatrix} -\boldsymbol{k} \\ \kappa\ \kappa' \end{pmatrix}_{\alpha\beta}. \tag{88.31}$$

Next let us use the eigenvector of $\boldsymbol{D}(-\boldsymbol{k})$ to take the scalar product, and so to obtain, from (88.19) and (88.31):

$$\begin{aligned}
&\sum_j \sum_\lambda \sum_{\beta\kappa'} e_\alpha^* \left(\kappa \left| \begin{matrix} \boldsymbol{k} \\ j_\lambda \end{matrix} \right.\right) \omega^2(\boldsymbol{k}|j)\, e_\beta \left(\kappa' \left| \begin{matrix} \boldsymbol{k} \\ j_\lambda \end{matrix} \right.\right) e_\beta \left(\kappa' \left| \begin{matrix} -\boldsymbol{k} \\ j_{\bar{\mu}} \end{matrix} \right.\right) \\
&\quad = \sum_{\beta\kappa'} D\begin{pmatrix} -\boldsymbol{k} \\ \kappa\ \kappa' \end{pmatrix}_{\alpha\beta} e_\beta \left(\kappa' \left| \begin{matrix} -\boldsymbol{k} \\ \bar{j}_{\bar{\mu}} \end{matrix} \right.\right) = \omega^2(-\boldsymbol{k}\,\bar{j})\, e_\alpha \left(\kappa \left| \begin{matrix} -\boldsymbol{k} \\ \bar{j}_{\bar{\mu}} \end{matrix} \right.\right).
\end{aligned} \tag{88.32}$$

Anticipating a later result we take as the orthogonality rule for eigenvectors (Bloch vectors) at inequivalent $\boldsymbol{k}$, using a suitable scalar product (79.10)

$$\sum_{\beta\kappa'} e_\beta \left(\kappa' \left| \begin{matrix} \boldsymbol{k} \\ j_\lambda \end{matrix} \right.\right) e_\beta \left(\kappa' \left| \begin{matrix} -\boldsymbol{k} \\ \bar{j}_{\bar{\mu}} \end{matrix} \right.\right) = \delta(\boldsymbol{k}+\boldsymbol{k}')\, \delta_{j\bar{j}}\, \delta_{\lambda\bar{\mu}}. \tag{88.33}$$

and also we used (88.37), *vide infra*. Then the left hand side of (88.32) becomes

$$e_\alpha^* \left(\kappa \left| \begin{matrix} \boldsymbol{k} \\ j_\lambda \end{matrix} \right.\right) \omega^2(\boldsymbol{k}|j) = \omega^2(-\boldsymbol{k}|\bar{j})\, e_\alpha \left(\kappa \left| \begin{matrix} -\boldsymbol{k} \\ \bar{j}_{\bar{\mu}} \end{matrix} \right.\right). \tag{88.34}$$

But, as the real eigenvalues of $[\boldsymbol{D}(\boldsymbol{k})]$ and $[\boldsymbol{D}(-\boldsymbol{k})] = [\boldsymbol{D}(\boldsymbol{k})]^*$ are identical, apart from order of appearance, we may take

$$\omega^2(\boldsymbol{k}|j) = \omega^2(-\boldsymbol{k}|\bar{j}) \tag{88.35}$$

and then

$$e_\alpha^* \left(\kappa \left| \begin{matrix} \boldsymbol{k} \\ j_\lambda \end{matrix} \right.\right) = e_\alpha \left(\kappa \left| \begin{matrix} -\boldsymbol{k} \\ \bar{j}_{\bar{\mu}} \end{matrix} \right.\right) \tag{88.36}$$

or

$$K\boldsymbol{e}\left(\left| \begin{matrix} \boldsymbol{k} \\ j_\lambda \end{matrix} \right.\right) = \boldsymbol{e}^* \left(\left| \begin{matrix} \boldsymbol{k} \\ j_\lambda \end{matrix} \right.\right) = \boldsymbol{e}\left(\left| \begin{matrix} -\boldsymbol{k} \\ \bar{j}_{\bar{\mu}} \end{matrix} \right.\right). \tag{88.37}$$

It can be observed that elsewhere in this Encyclopaedia LIEBFRIED ([*4*], p. 174, Eq. (49.4)) has used (88.37) in the special case of no degeneracy ($l_j = 1$; or $l_{\bar{j}} = 1$) taking

$$\boldsymbol{e}^* \left(\left| \begin{matrix} \boldsymbol{k} \\ j \end{matrix} \right.\right) = -\boldsymbol{e}\left(\left| \begin{matrix} -\boldsymbol{k} \\ j \end{matrix} \right.\right). \tag{Liebfried}$$

LIEBFRIED's choice is not in contradiction with ours owing to the fashion by which we have chosen to define the orthogonality (88.33). In our work we adhere to (88.37) including for cases of non-degeneracy. For emphasis we need to point out that in general $j_{\bar{\mu}}$ and j_λ in (88.36) differ; their relationship will be discussed below, in (94.22) *et seq.* and in Sects. 95–100.

Coming back to the time dependent eigenvectors, (88.16), we have

$$K\,\varepsilon\left(\left|\begin{matrix}\boldsymbol{k}\\ j_\mu\end{matrix}\right| t\right)=\varepsilon\left(\left|\begin{matrix}-\boldsymbol{k}\\ \bar{j}_{\bar{\mu}}\end{matrix}\right| -t\right). \tag{88.38}$$

Hence, the transformation operator K will change an eigenvector giving displacements of wave vector $\boldsymbol{k}$ at time t into an eigenvector giving displacements of wave vector $-\boldsymbol{k}$ at time $-t$. Then it is justified to refer to the eigenvector $K\,\varepsilon\left(\left|\begin{matrix}\boldsymbol{k}\\ j_\mu\end{matrix}\right| t\right)$ as the time-reversed, or motion reversed displacement. From (88.37), (88.35), (88.20) we find that the time-independent eigenvector $\boldsymbol{e}\left(\left|\begin{matrix}\boldsymbol{k}\\ j_\mu\end{matrix}\right.\right)$ is degenerate with $\boldsymbol{e}^*\left(\left|\begin{matrix}\boldsymbol{k}\\ \bar{j}_{\bar{\mu}}\end{matrix}\right.\right)$. Consequently, even in the time independent situation, it is appropriate to refer to $\boldsymbol{e}^*\left(\left|\begin{matrix}\boldsymbol{k}\\ \bar{j}_\mu\end{matrix}\right.\right)$ as the time reversed eigenvector.

As in our previous work on the spatial transformation operators $P_{\{\varphi|\boldsymbol{t}\}}$ we may define the effect of K upon the complex normal coordinate $Q\begin{pmatrix}\boldsymbol{k}\\ j_\nu\end{pmatrix}$ by analogy to (88.37)

$$KQ\begin{pmatrix}\boldsymbol{k}\\ j_\nu\end{pmatrix}=Q\begin{pmatrix}\boldsymbol{k}\\ j_\nu\end{pmatrix}^*=Q\begin{pmatrix}-\boldsymbol{k}\\ \bar{j}_{\bar{\nu}}\end{pmatrix}. \tag{88.39}$$

Observe also that (88.39) is consistent with direct application of K to the expression (86.2) and use of (88.37). Again note that the precise relationship between unbarred indices j_ν and barred $\bar{j}_{\bar{\nu}}$ needs to be determined and will be discussed below.

89. The complete space-time symmetry group $\mathscr{G}$. Coming back to the force constant matrix $[\boldsymbol{\Phi}]$ of Sect. 71, we recall the invariance of $[\boldsymbol{\Phi}]$ under general spatial symmetry transformation

$$P_{\{\varphi|\boldsymbol{t}\}}[\boldsymbol{\Phi}]\,P_{\{\varphi|\boldsymbol{t}\}}^{-1}=[\boldsymbol{\Phi}]. \tag{89.1}$$

From the work of the last section we have also the invariance under time-reversal

$$K[\boldsymbol{\Phi}]\,K^{-1}=[\boldsymbol{\Phi}]. \tag{89.2}$$

Consequently the complete symmetry group of the potential energy V, the kinetic energy T, the Hamiltonian H and hence of the crystal, is the crystal space group $\mathfrak{G}$ plus the coset $K\mathfrak{G}$, or

$$\mathscr{G}=\mathfrak{G}+K\mathfrak{G}. \tag{89.3}$$

We refer to $\mathscr{G}$ as the full space-time symmetry group of the crystal.

Now turn to the dynamical matrix $[\boldsymbol{D}(\boldsymbol{k})]$. Recalling (81.26) we have for the transformation operators corresponding to spatial symmetry

$$P_{\{\varphi|\boldsymbol{\tau}(\varphi)\}}[\boldsymbol{D}(\boldsymbol{k})]\,P_{\{\varphi|\boldsymbol{\tau}|\varphi\}}^{-1}=[\boldsymbol{D}(\varphi\cdot\boldsymbol{k})]. \tag{89.4}$$

Now to establish (88.13), (88.14) in an alternate manner we may use (81.23), and the definitions of K in (88.1)–(88.4), along with property (88.7).

Thus

$$K[\boldsymbol{D}(\boldsymbol{k})]\,K^{-1} = K\cdot P^{(\boldsymbol{k})}\,[\boldsymbol{M}]^{-\frac{1}{2}}\,[\boldsymbol{\Phi}(0)]\,[\boldsymbol{M}]^{-\frac{1}{2}}\,K^{-1}. \tag{89.5}$$

But

$$\begin{aligned} KP^{(\boldsymbol{k})} &= K\sum_{\lambda} \exp i\,\boldsymbol{k}\cdot\boldsymbol{R}_{\lambda}\,P_{\{\varepsilon|-\boldsymbol{R}_{\lambda}\}} \\ &= \sum_{\lambda} \exp - i\,\boldsymbol{k}\cdot\boldsymbol{R}_{\lambda}\,P_{\{\varepsilon|-\boldsymbol{R}_{\lambda}\}}\,K = P^{(-\boldsymbol{k})}\,K. \end{aligned} \tag{89.6}$$

Furthermore $[\boldsymbol{\Phi}(0)]$ and $[\boldsymbol{M}]^{-\frac{1}{2}}$ are real, so that finally we have

$$K\cdot[\boldsymbol{D}(\boldsymbol{k})]\,K^{-1} = [\boldsymbol{D}(-\boldsymbol{k})] \tag{89.7}$$

owing to the antilinearity, and antiunitarity of the operator K it is usual to treat the transformation operator K on a different footing than the spatial symmetry operators. This possibility is also suggested by (89.7) and (88.37) since K inverts the wave vector $\boldsymbol{k}$. However, if $\boldsymbol{k}$ is a wave vector in the Brillouin zone then also $-\boldsymbol{k}$ must be. Thus in a sense K connects wave vectors which may or may not already be connected by the operators $P_{\{\varphi|\boldsymbol{t}\}}$. In this fashion of treating the problem, K is like an added symmetry, beyond the spatial symmetry already epitomized in $\mathfrak{G}$. In a unified treatment all operators would be treated on essentially the same footing: such a treatment will be given below in Sects. 95–102.

90. Eigenvectors $e\left(\begin{vmatrix}\boldsymbol{k}\\ j_{\lambda}\end{vmatrix}\right)$ and normal coordinates $Q\begin{pmatrix}\boldsymbol{k}\\ j_{\lambda}\end{pmatrix}$ as bases for representation of $\mathcal{G}$. In order to deal properly and completely with the effects of the transformation operator K, it is necessary for us to work with the entire space group irreducible representations $D^{(\star\boldsymbol{k})(m)}$. In the previous work of Sects. 77–86 we could restrict ourselves to working with the allowable irreducible representations $D^{(\boldsymbol{k})(m)}$ of $\mathfrak{G}(\boldsymbol{k})$ only, since the full $D^{(\star\boldsymbol{k})(m)}$ was obtained by straightforward induction from $D^{(\boldsymbol{k})(m)}$. But the transformation operator K is not a linear operator in the sense of the $P_{\{\varphi|\boldsymbol{t}\}}$ so that special care is required in analysing the effect of having K present among the transformation operators of the problem.

The complete set of eigenvectors which span a linear vector space upon which $D^{(\star\boldsymbol{k})(m)}$ is based may be taken as

$$\Sigma^{(\star\boldsymbol{k})(j)} \equiv \left\{ \boldsymbol{e}\left(\begin{vmatrix}\boldsymbol{k}\\ j_1\end{vmatrix}\right), \ldots, \boldsymbol{e}\left(\begin{vmatrix}\boldsymbol{k}\\ j_{l_j}\end{vmatrix}\right), \ldots, \boldsymbol{e}\left(\begin{vmatrix}\boldsymbol{k}_s\\ j_{l_j}\end{vmatrix}\right) \right\}. \tag{90.1}$$

But following (86.28), (86.29) the set of $(s\cdot l_j)$ complex normal coordinates

$$\Sigma^{(\star\boldsymbol{k})(j)} \equiv \left\{ Q\begin{pmatrix}\boldsymbol{k}\\ j_1\end{pmatrix}, \ldots, Q\begin{pmatrix}\boldsymbol{k}\\ j_{l_j}\end{pmatrix}, \ldots, Q\begin{pmatrix}\boldsymbol{k}_s\\ j_{l_j}\end{pmatrix} \right\} \tag{90.2}$$

can equally well be taken as a basis for the irreducible representation $D^{(\star\boldsymbol{k})(j)}$ since for (90.1) and (90.2)

$$P_{\{\varphi|\boldsymbol{t}\}}\,\Sigma^{(\star\boldsymbol{k})(j)} = \Sigma^{(\star\boldsymbol{k})(j)}\,D^{(\star\boldsymbol{k})(j)}(\{\varphi|\boldsymbol{t}\}). \tag{90.3}$$

Consider next the linear vector space

$$K\Sigma^{(\star k)(j)} = \Sigma^{(\star k)(j)*} = \left\{ e^* \left(\left| \begin{matrix} k \\ j_1 \end{matrix} \right. \right), \ldots, e^* \left(\left| \begin{matrix} k_s \\ j_{l_j} \end{matrix} \right. \right) \right\}. \tag{90.4}$$

Following (88.37)

$$K\Sigma^{(\star k)(j)} = \Sigma^{(\star -k)(\bar{\mu})}. \tag{90.5}$$

Then in terms of the complex normal coordinates we can equally well take

$$K\Sigma^{(\star k)(j)} = \left\{ Q \begin{pmatrix} k \\ j_1 \end{pmatrix}^*, \ldots, Q \begin{pmatrix} k_s \\ j_{l_j} \end{pmatrix}^* \right\} = \Sigma^{(\star -k)(\bar{\mu})}. \tag{90.6}$$

Using the space (90.6) as a basis for a representation we have

$$\begin{aligned} P_{\{\varphi|t\}} K\Sigma^{(\star k)(j)} &= P_{\{\varphi|t\}} \Sigma^{(\star -k)(\bar{\mu})} \\ &= \Sigma^{(\star -k)(\bar{\mu})} D^{(\star -k)(\bar{\mu})}(\{\varphi|t\}). \end{aligned} \tag{90.7}$$

In (90.7) we notice that the space $K\Sigma^{(\star k)(j)}$ is a basis for representation $D^{(\star -k)(\bar{\mu})}$ of $\mathfrak{G}$, where $\bar{\mu}$ is not yet related to j, but will be so related below.

Now apply the operator K to both sides of (90.3) and use (88.7):

$$\begin{aligned} KP_{\{\varphi|t\}} \Sigma^{(\star k)(j)} &= P_{\{\varphi|t\}} K\Sigma^{(\star k)(j)} \\ &= K\Sigma^{(\star k)(j)} \cdot D^{(\star k)(j)}(\{\varphi|t\})^*. \end{aligned} \tag{90.8}$$

$$D^{(\star k)(j)*} = D^{(\star -k)(\bar{\mu})}. \tag{90.9}$$

It thus follows that the complete set of eigenvectors (90.1) or complex normal coordinates (90.2) which are bases for irreducible representation $D^{(\star k)(j)}$ of $\mathfrak{G}$ are also bases for irreducible representation $D^{(\star k)(j)*}$ of $\mathfrak{G}$ when transformed by the time reversal operator K, as in (90.4) and (90.6). It should be immediately apparent that if $D^{(\star k)(j)}$ is irreducible in $\mathfrak{G}$ then so also is $D^{(\star k)(j)*}$.

91. Necessary degeneracy under the full space-time crystal symmetry group $\mathscr{G}$. The complete symmetry group of the lattice dynamic problem must be taken as $\mathscr{G}$, the full space-time symmetry group of (88.8). By necessary degeneracy we will mean the degeneracy associated with that set of vectors which span a linear vector space irreducible under $\mathscr{G}$. The vectors may be eigenvectors as in (90.1) and (90.4), or complex normal coordinates as in (90.2) and (90.6).

Clearly, owing to the operator K being a symmetry operator we have that the set of eigenvalues of $[\boldsymbol{D}(\boldsymbol{k})]$:

$$\{\omega^2(\boldsymbol{k}|j_1), \ldots, \omega^2(\boldsymbol{k}|j_{l_j}), \ldots, \omega^2(\boldsymbol{k}|j'_1), \ldots, \omega^2(\boldsymbol{k}|j'_{l_{j'}})\} \tag{91.1}$$

is identical to the set of eigenvalues of $[\boldsymbol{D}(\boldsymbol{k})]^* = [\boldsymbol{D}(-\boldsymbol{k})]$ which are

$$\{\omega^2(-\boldsymbol{k}|j_1), \ldots, \omega^2(-\boldsymbol{k}|j_{l_j}), \ldots, \omega^2(-\boldsymbol{k}|j'_1), \ldots, \omega^2(-\boldsymbol{k}|j'_{l_{j'}})\}. \tag{91.2}$$

Recall also (88.35). Hence the space spanned by the set of *all* eigenvectors, or normal coordinates at $\boldsymbol{k}$

$$\Sigma^{(\boldsymbol{k})(\{j\})} \equiv \left\{ \boldsymbol{e}\left(\Big|\begin{matrix}\boldsymbol{k}\\ j_1\end{matrix}\right), \ldots, \boldsymbol{e}\left(\Big|\begin{matrix}\boldsymbol{k}\\ j_{l_j}\end{matrix}\right), \ldots, \boldsymbol{e}\left(\Big|\begin{matrix}\boldsymbol{k}\\ j'_{l_{j'}}\end{matrix}\right) \right\}$$

or (91.3)

$$\Sigma^{(\boldsymbol{k})(\{j\})} = \left\{ Q\begin{pmatrix}\boldsymbol{k}\\ j_1\end{pmatrix}, \ldots, Q\begin{pmatrix}\boldsymbol{k}\\ j_{l_j}\end{pmatrix}, \ldots, Q\begin{pmatrix}\boldsymbol{k}\\ j'_{l_{j'}}\end{pmatrix} \right\}$$

is degenerate with the space spanned by the corresponding sets at $-\boldsymbol{k}$:

$$\Sigma^{(-\boldsymbol{k})(\{\mu\})} = \Sigma^{(\boldsymbol{k})(\{j\})*} = K\,\Sigma^{(\boldsymbol{k})(\{j\})} \tag{91.4}$$

$$\Sigma^{(\boldsymbol{k})(\{j\})*} \equiv \left\{ e^*\left(\Big|\begin{matrix}\boldsymbol{k}\\ j_1\end{matrix}\right), \ldots, e^*\left(\Big|\begin{matrix}\boldsymbol{k}\\ j'_{l_{j'}}\end{matrix}\right) \right\}$$

or (91.5)

$$\Sigma^{(\boldsymbol{k})(\{j\})*} = \left\{ Q^*\begin{pmatrix}\boldsymbol{k}\\ j_1\end{pmatrix}, \ldots, Q^*\begin{pmatrix}\boldsymbol{k}\\ j'_{l_{j'}}\end{pmatrix} \right\}.$$

The assertion of necessary degeneracy requires that the "physically irreducible representation" of $\mathscr{G}$ which can arise in the lattice dynamic problem shall be a real representation. Then if the space $\Sigma^{(*\boldsymbol{k})(j)}$ is real

$$\Sigma^{(*\boldsymbol{k})(j)} \equiv K\,\Sigma^{(*\boldsymbol{k})(j)} \tag{91.6}$$

it is a candidate for the basis for a physically irreducible representation in the lattice dynamic problem for space group $\mathfrak{G}$. If $\Sigma^{(*\boldsymbol{k})(j)}$ is not real, then the physically irreducible, real representation which may occur can be based upon the union or direct sum of the two spaces

$$\Sigma^{(*\boldsymbol{k})(j)} \oplus K\,\Sigma^{(*\boldsymbol{k})(j)}. \tag{91.7}$$

That is, the only physically irreducible representations which can arise must be based upon spaces Σ which are real. Recall that the underlying physical justification of this requirement resides in the reality of the physical displacement field $\boldsymbol{u}$ which describes the displacement of the crystal atoms from rest. Since this underlying physical field is real

$$K\,\boldsymbol{u} = \boldsymbol{u}^* = \boldsymbol{u} \tag{91.8}$$

it is clear that a restriction must exist upon the representation by which the eigenvectors and complex normal coordinates transform, in order that under all circumstances $\boldsymbol{u}$ shall be real. Observe that from (86.1)

$$K\,u_\alpha\begin{pmatrix}l\\ \kappa\end{pmatrix} = u_\alpha^*\begin{pmatrix}l\\ \kappa\end{pmatrix} = \frac{1}{\sqrt{M_\kappa N}} \sum_j \sum_\nu \sum_{\boldsymbol{k}} \exp i(-\boldsymbol{k}\cdot\boldsymbol{R}_L)\, e_\alpha^*\left(\kappa\Big|\begin{matrix}\boldsymbol{k}\\ j_\nu\end{matrix}\right) Q^*\begin{pmatrix}\boldsymbol{k}\\ j_\nu\end{pmatrix}. \tag{91.9}$$

Then the results (88.37) and (88.39) permit us to write (91.9) as

$$u_\alpha^*\begin{pmatrix}l\\ \kappa\end{pmatrix} = \frac{1}{\sqrt{M_\kappa N}} \sum_j \sum_\nu \sum_{\boldsymbol{k}} \exp i(-\boldsymbol{k}\cdot\boldsymbol{R}_L)\, e_\alpha\left(\kappa\Big|\begin{matrix}-\boldsymbol{k}\\ \bar{j}_{\bar{\nu}}\end{matrix}\right) Q\begin{pmatrix}-\boldsymbol{k}\\ \bar{j}_{\bar{\nu}}\end{pmatrix}. \tag{91.10}$$

But now taking the sums in (91.10) over $\bar{j}$, $\bar{\nu}$ and $(-\boldsymbol{k})$ and comparing with (86.1) we verify that (91.8) applies. The change of summation variable is permitted since these are dummy variables only.

Consequently we assert that any normal mode which will arise in the lattice vibration problem will transform as a basis for a real irreducible representation of the full space time group $\mathscr{G}$. Clearly, if

$$D^{(\star\boldsymbol{k})(j)} \equiv D^{(\star\boldsymbol{k})(j)*} \tag{91.11}$$

then the irreducible representation of $\mathfrak{G}$ is already real and it is thus acceptable as a "physically" irreducible representation of $\mathscr{G}$. If

$$D^{(\star\boldsymbol{k})(j)} \not\equiv D^{(\star\boldsymbol{k})(j)*} \tag{91.12}$$

then the "physically" irreducible representation of $\mathscr{G}$ is

$$D^{(\star\boldsymbol{k})(j)} \oplus D^{(\star\boldsymbol{k})(j)*} \tag{91.13}$$

which is a real, reducible representation of $\mathfrak{G}$.

In case (91.12) applies, time reversal symmetry does not change the results of the purely spatial symmetry analysis, carried out using $\mathfrak{G}$ above as the symmetry group. In case (91.13) applies the degeneracy is doubled owing to the effect of time reversal symmetry K.

It is thus a matter of importance to examine the reality of the irreducible representations of $\mathfrak{G}$ to determine whether or not including the full symmetry $\mathscr{G}$ causes an added degeneracy. We next turn to this problem.

A final important remark needs to be made. Clearly, the preceeding analysis merely establishes the *necessary* reality condition for a representation to arise as the representation by which the space of eigenvectors as normal modes transform. That is, *if* a representation of $\mathscr{G}$ arises for transformation of the space of eigenvectors or normal modes, it must be physically acceptable. The converse is not true: given $\mathscr{G}$ (full space-time symmetry group of a crystal), not every physically irreducible representation of $\mathscr{G}$ will occur in the lattice dynamic problem. Determination of the particular physically irreducible representations which arise for each crystal will be discussed later in Sect. 103.

92. Test for reality of $D^{(\star\boldsymbol{k})(j)}$ of $\mathfrak{G}$. As pointed out in the previous sections the question of necessary degeneracy under $\mathscr{G}$ is directly related to the reality of the representation $D^{(\star\boldsymbol{k})(j)}$ of the spatial symmetry group $\mathfrak{G}$ the crystal space group. In the present section we develop the theory by which the reality of representation $D^{(\star\boldsymbol{k})(j)}$ can be tested. It is to be assumed that all the irreducible representations $D^{(\star\boldsymbol{k})(j)}$ of the space group $\mathfrak{G}$ have been determined by methods discussed earlier. Now evidently if the allowable "small" irreducible representations $D^{(\boldsymbol{k})(m)}$ of $\mathfrak{G}(\boldsymbol{k})$ are all real, then the induced representation $D^{(\star\boldsymbol{k})(m)}$ of $\mathfrak{G}$ will be real. However, this is a sufficiency condition which is, in fact, too strong. That is, this condition would require

$$\chi^{(\boldsymbol{k})(m)}(\{\varphi_{l_\lambda}|\boldsymbol{t}\}) = \chi^{(\boldsymbol{k})(m)}(\{\varphi_{l_\lambda}|\boldsymbol{t}\})^* \tag{92.1}$$

for all elements $\{\varphi_{l_\lambda}|\boldsymbol{t}\}$ in $\mathfrak{G}(\boldsymbol{k})$. However, it does occur as will be seen later, that even if

$$\chi^{(k)(m)}(\{\varphi_{l_\lambda}|\boldsymbol{t}\}) \neq \chi^{(k)(m)}(\{\varphi_{l_\lambda}|\boldsymbol{t}\})^* \tag{92.2}$$

the representation $D^{(\star k)(m)}$ may be real.

In order to examine this question properly, it is necessary to study the full group irreducible representations *a priori.* It then develops that attention can indeed be restricted to the irreducible representation $D^{(k)(m)}$ of $\mathfrak{G}(\boldsymbol{k})$, but only after the analysis has been carried out.

In the lattice dynamic problem we have to deal only with representations of the single group $\mathfrak{G}$ so that complications which may arise owing to the presence of spin can be ignored. This has the particular consequence that

$$\begin{aligned} &\text{if} \quad D^{(\star k)(m)} \equiv D^{(\star k)(m)*} \\ &\text{then} \quad D^{(\star k)(m)} \text{ can be made real.} \end{aligned} \tag{92.3}$$

Now we give details of the reality tests which can be applied to the irreducible representations $D^{(\star k)(m)}$ of $\mathfrak{G}$.

In this work, a central role is played by the matrix

$$M^{(\star k\,(m)} \equiv \sum_{\mathfrak{G}} D^{(\star k)(m)}(\{\varphi|\boldsymbol{t}\}^2). \tag{92.4}$$

In (92.4), $M^{(\star k)(m)}$ is the sum over all elements $\{\varphi|\boldsymbol{t}\}$ in $\mathfrak{G}$ of the matrix representing the element

$$\{\varphi|\boldsymbol{t}\}^2 = \{\varphi^2|\varphi\cdot\boldsymbol{t}+\boldsymbol{t}\} \tag{92.5}$$

in the irreducible representation $D^{(\star k)(m)}$. First it is easily seen that $M^{(\star k)(m)}$ commutes with all the matrices $D^{(\star k)(m)}$ of the irreducible representation. Thus, let $\{\varphi_0|\boldsymbol{t}_0\}$ be an arbitrary element in $\mathfrak{G}$, and call the conjugate

$$\{\varphi_0|\boldsymbol{t}_0\}\cdot\{\varphi|\boldsymbol{t}\}\cdot\{\varphi_0|\boldsymbol{t}_0\}^{-1} \equiv \{\varphi|\boldsymbol{t}\}^{\bar{0}}. \tag{92.6}$$

Then

$$\begin{aligned} D^{(\star k)(m)}(\{\varphi_0|\boldsymbol{t}_0\})\cdot M^{(\star k)(m)} &= \sum_{\mathfrak{G}} D^{(\star k)(m)}(\{\varphi|\boldsymbol{t}\}^{\bar{0}}\cdot\{\varphi|\boldsymbol{t}\}^{\bar{0}}\cdot\{\varphi_0|\boldsymbol{t}_0\}) \\ &= \Big(\sum_{\mathfrak{G}} D^{(\star k)(m)}(\{\varphi|\boldsymbol{t}\}^{\bar{0}}\cdot\{\varphi|\boldsymbol{t}\}^{\bar{0}})\Big)\cdot D^{(\star k)(m)}(\{\varphi_0|\boldsymbol{t}_0\}) \\ &= M^{(\star k)(m)}\, D^{(\star k)(m)}(\{\varphi_0|\boldsymbol{t}_0\}). \end{aligned} \tag{92.7}$$

Note that the sum inside the large parenthesis can be identified as $M^{(\star k)(m)}$ owing to the invariance property of the sum over the group [*1, 18*]:

$$\sum_{\mathfrak{G}} = \sum_{\{\varphi_0|\boldsymbol{t}_0\}\cdot\mathfrak{G}\cdot\{\varphi_0|\boldsymbol{t}_0\}^{-1}}. \tag{92.8}$$

Eq. (92.8) expresses the invariance of the group sum under an inner automorphism: that is the sum is over all group elements, whether labelled by $\{\varphi_0|\boldsymbol{t}\}$ or $\{\varphi|\boldsymbol{t}\}^{\bar{0}}$.

But since (92.7) holds for all elements $\{\varphi_0|\boldsymbol{t}_0\}$ in G, we may apply SCHUR's lemma, and conclude that

$$M^{(\star k)(m)} = \mu^{(\star k)(m)}\, \boldsymbol{\Pi}_{m\cdot s}, \tag{92.9}$$

where $\boldsymbol{\Pi}_m$ is the $(s \cdot l_m)$ dimensional unit matrix and $\mu^{(\star\boldsymbol{k})(m)}$ is a constant. Taking the $\alpha\beta$ matrix element of (92.9) we have

$$(M^{(\star\boldsymbol{k})(m)})_{\alpha\beta} = \mu^{(\star\boldsymbol{k})(m)} \delta_{\alpha\beta}. \tag{92.10}$$

Setting $\alpha = \beta$ and summing we obtain

$$\operatorname{Tr} M^{(\star\boldsymbol{k})(m)} = \sum_{\mathfrak{G}} \chi^{(\star\boldsymbol{k})(m)}(\{\varphi|\boldsymbol{t}\}^2) = \mu^{(\star\boldsymbol{k})(m)}(s \cdot l_m). \tag{92.11}$$

The next task is to determine possible values of $\mu^{(\star\boldsymbol{k})(m)}$ which can arise in each of the cases of reality of the representation $D^{(\star\boldsymbol{k})(m)}$.

First take the case in which $D^{(\star\boldsymbol{k})(m)}$ and $D^{(\star\boldsymbol{k})(m)*}$ are inequivalent. The orthogonality relation for inequivalent irreducible representations then takes the form

$$\sum_{\mathfrak{G}} D^{(\star\boldsymbol{k})(m)}(\{\varphi|\boldsymbol{t}\})_{\mu\tau} (D^{(\star\boldsymbol{k})(m)}(\{\varphi|\boldsymbol{t}\})^*)^*_{\tau\mu} = 0 \tag{92.12}$$

or

$$\sum_{\mathfrak{G}} D^{(\star\boldsymbol{k})(m)}(\{\varphi|\boldsymbol{t}\})_{\mu\tau} D^{(\star\boldsymbol{k})(m)}(\{\varphi|\boldsymbol{t}\})_{\tau\mu} = 0. \tag{92.13}$$

Then summing (92.13) over τ and μ we obtain

$$\sum_{\mathfrak{G}} \chi^{(\star\boldsymbol{k})(m)}(\{\varphi|\boldsymbol{t}\})^2 = 0 \quad \text{if} \quad D^{(\star\boldsymbol{k})(m)} \not\doteq D^{(\star\boldsymbol{k})(m)*}. \tag{92.14}$$

Hence the case of inequivalence corresponds to

$$\mu^{(\star\boldsymbol{k})(m)} = 0 \tag{92.15}$$

in (92.11).

Now consider the case of equivalence (92.3). If $D^{(\star\boldsymbol{k})(m)}$ is equivalent to $D^{(\star\boldsymbol{k})(m)*}$, then there exists a unitary matrix U with the property that

$$U^{-1} D^{(\star\boldsymbol{k})(m)} U = D^{(\star\boldsymbol{k})(m)*} \tag{92.16}$$

or

$$D^{(\star\boldsymbol{k})(m)} U = U D^{(\star\boldsymbol{k})(m)*}. \tag{92.17}$$

Take the complex conjugate of (92.17), and multiply by U

$$U D^{(\star\boldsymbol{k})(m)*} U^* = U U^* D^{(\star\boldsymbol{k})(m)}. \tag{92.18a}$$

Then use (92.17)

$$D^{(\star\boldsymbol{k})(m)} U U^* = U U^* D^{(\star\boldsymbol{k})(m)}. \tag{92.18b}$$

Hence (UU^*) commutes with all matrices $D^{(\star\boldsymbol{k})(m)}$ of the irreducible representation. Hence by SCHUR's lemma:

$$U U^* = y \boldsymbol{\Pi}_{(s \cdot l_m)}, \tag{92.19}$$

where y is a constant, and $\boldsymbol{\Pi}_{(s \cdot l_m)}$ is the $(s \cdot l_m)$ dimensional unit matrix.

Now from (92.19)

$$U = y U^{-1*}. \tag{92.20}$$

But as U is unitary

$$U^{-1} = \tilde{U}^* = U^+. \tag{92.21}$$

Hence

$$U^{-1*} = \tilde{U}. \tag{92.22}$$

Then from (92.20)

$$U = y\tilde{U}. \tag{92.23}$$

Taking the transpose of (92.23)

$$\tilde{U} = yU \tag{92.24}$$

or from (92.23) and (92.24)

$$U = y^2 U. \tag{92.25}$$

Hence

$$y^2 = 1, \tag{92.26}$$

$$y = \pm 1. \tag{92.27}$$

Then from (92.23), we have either

$$U = \tilde{U} \tag{92.28}$$

in which case U is a symmetric matrix or

$$U = -\tilde{U} \tag{92.29}$$

in which case U is antisymmetric. The two cases (92.28) and (92.29) correspond to two distinct situations, both of which are compatible with the equivalence of $D^{(\star k)(m)}$ and $D^{(\star k)(m)*}$. First we deal with the situation (92.28).

To continue, consider the case that $D^{(\star k)(m)}$ can be made real. That is given $D^{(\star k)(m)}$ there exists a unitary matrix S such that

$$S^{-1} D^{(\star k)(m)} S = \bar{D}^{(\star k)(m)}, \tag{92.30}$$

where $\bar{D}^{(\star k)(m)}$ is real

$$\bar{D}^{(\star k)(m)*} = \bar{D}^{(\star k)(m)}. \tag{92.31}$$

Taking the complex conjugate of (92.30) we obtain:

$$\begin{aligned} S^{-1*} D^{(\star k)(m)*} S^* &= \bar{D}^{(\star k)(m)*} \\ &= \bar{D}^{(\star k)(m)} = S^{-1} D^{(\star k)(m)} S. \end{aligned} \tag{92.32}$$

Hence

$$S^* S^{-1} D^{(\star k)(m)} S S^{*-1} = D^{(\star k)(m)*}. \tag{92.33}$$

As S is unitary we may use (92.22) here too, so we find

$$(S\tilde{S})^{-1} D^{(\star k)(m)} (S\tilde{S}) = D^{(\star k)(m)*}. \tag{92.34}$$

But the matrix $(S\tilde{S})$ evidently transforms $D^{(\star k)(m)}$ into its complex conjugate and thus is to be identified with the U of (92.16).

Clearly

$$\widetilde{(S\tilde{S})} = S\tilde{S} \tag{92.35}$$

and consequently $(S\tilde{S})$ is a symmetric matrix as in (92.28). Evidently, the existence of a non-zero unitary matrix S, as in (92.30), corresponds to (92.28) so that if $D^{(\star k)(m)}$ is equivalent to $D^{(\star k)(m)*}$, *and* can be made real, then (92.28) applies, and U is symmetric; conversely if a symmetric U exists then $D^{(\star k)(m)}$ can be made real.

Now return again to (92.17), which we will transform into a form in which a connection with (92.14) can be demonstrated. Multiplying (92.17) with $D^{(\star k)(m)}$ and writing in explicitly the argument, or space group element, we obtain

$$D^{(\star k)(m)}(\{\varphi|t\}) D^{(\star k)(m)}(\{\varphi|t\}) U = D^{(\star k)(m)}(\{\varphi|t\}) U D^{(\star k)(m)}(\{\varphi|t\})^*. \tag{92.36}$$

Doing the matrix multiplication on the left hand side of (92.36), and taking the $\alpha\beta$ matrix element of the resulting matrices we have

$$\sum_{\gamma} D^{(\star k)(m)}(\{\varphi|t\}^2)_{\alpha\gamma} U_{\gamma\beta} = \sum_{\delta\varepsilon} D^{(\star k)(m)}(\{\varphi|t\})_{\alpha\delta} U_{\delta\varepsilon} D^{(\star k)(m)}(\{\varphi|t\})^*_{\varepsilon\beta}. \quad (92.37)$$

Now sum both sides of (92.37) over the group $\mathfrak{G}$. Then the left hand side becomes

$$\begin{aligned} \sum_{\gamma}\sum_{\mathfrak{G}} D^{(\star k)(m)}(\{\varphi|t\}^2)_{\alpha\gamma} U_{\gamma\beta} &= \sum_{\gamma} (M^{(\star k)(m)})_{\alpha\gamma} U_{\gamma\beta} \\ &= \sum_{\gamma} \mu^{(\star k)(m)} \delta_{\alpha\gamma} U_{\gamma\beta} \\ &= \mu^{(\star k)(m)} U_{\alpha\beta} \end{aligned} \quad (92.38)$$

where we used (92.10). Next consider the right hand side of (92.37), summed over $\mathfrak{G}$:

$$\begin{aligned} \sum_{\delta\varepsilon} U_{\delta\varepsilon} \sum_{\mathfrak{G}} D^{(\star k)(m)}(\{\varphi|t\})_{\alpha\delta} D^{(\star k)(m)}(\{\varphi|t\})^*_{\varepsilon\beta} &= \sum_{\delta\varepsilon} U_{\delta\varepsilon} \frac{g_p N}{s\cdot l_m} \delta_{\alpha\varepsilon}\delta_{\delta\beta} \\ &= U_{\beta\alpha}(g_p N/s\cdot l_m), \end{aligned} \quad (92.39)$$

where we used the orthogonality relations for the equivalent irreducible representations $D^{(\star k)(m)}$ and $D^{(\star k)(m)*}$ to obtain (92.39). Then from (92.38) and (92.39)

$$\mu^{(\star k)(m)} = (U_{\beta\alpha}/U_{\alpha\beta})(g_p N/s\cdot l_m). \quad (92.40)$$

But the ratio $(U_{\beta\alpha}/U_{\alpha\beta})$ can have only two values ± 1 according to (92.28), (92.29). Further in the case of interest, where $D^{(\star k)(m)}$ and $D^{(\star k)(m)*}$ are equivalent and can be made real (92.28) is applicable so

$$\mu^{(\star k)(m)} = (g_p N)/(s\cdot l_m). \quad (92.41)$$

For completeness, we write the possibility corresponding to the case (92.29) which is

$$\mu^{(\star k)(m)} = -(g_p N)/(s\cdot l_m). \quad (92.42)$$

Now we can assemble the three cases which have been discussed, results given in (92.15), (92.41), (92.42) and insert into (92.11). We shall also follow conventional nomenclature to distinguish these cases. When $D^{(\star k)(m)}$ can be made real as in (92.30) and (92.31) we call $D^{(\star k)(m)}$ potentially real. Then

$$\sum_{\mathfrak{G}} \chi^{(\star k)(m)}(\{\varphi|t\}^2) = (g_p N), \qquad \text{potentially real } D^{(\star k)(m)}. \quad (92.43)$$

When $D^{(\star k)(m)}$ is complex as in (92.14)

$$\sum_{\mathfrak{G}} \chi^{(\star k)(m)}(\{\varphi|t\}^2) = 0, \qquad \text{complex } D^{(\star k)(m)}. \quad (92.44)$$

When $D^{(\star k)(m)}$ is equivalent to $D^{(\star k)(m)*}$ as in (92.17), but cannot be made real, U must be antisymmetric and (92.42) applies; this case is called pseudoreal. Then

$$\sum_{\mathfrak{G}} \chi^{(\star k)(m)}(\{\varphi|t\}^2) = -(g_p N), \qquad \text{pseudoreal } D^{(\star k)(m)}. \quad (92.45)$$

In lattice dynamics only cases (92.43) and (92.44) can arise. The tests (92.43)–(92.45) are completely general and using the general results (49.3) we may test any irreducible representation for the applicable case. All the necessary information is at hand and it remains to work out the sums over the entire group in (92.43)–(92.45). While straightforward, this is evidently inconvenient. Hence in the next section we derive a simplification due to HERRING[1] which will permit us to work with the group $\mathfrak{G}(\boldsymbol{k})$ only, in testing $D^{(*\boldsymbol{k})(m)}$. Again it is to be emphasized that it is plausible that such a simplification should exist owing to the structure of $D^{(*\boldsymbol{k})(m)}$ of $\mathfrak{G}$ as an induced representation based on $D^{(*\boldsymbol{k})(m)}$ *of* $\mathfrak{G}(\boldsymbol{k})$, but as was emphasized earlier in this section we could not *a priori* make a limitation to $\mathfrak{G}(\boldsymbol{k})$ only. A careful examination of the criteria (92.43)–(92.45) shows that a stronger and more convenient test for reality of the representation can be given. These matters are discussed *seriatim* in Sects. 93 and 94 following.

93. Simplification of the reality test of $D^{(*k)(m)}$. A partial simplification of the reality tests (92.43)–(92.45) due to HERRING[1] can be developed, using (37.3) for the character which enters in the former expressions. In this fashion the character is expressed in terms of "dotted" characters, appertaining to $\mathfrak{G}(\boldsymbol{k})$. Now we write, for the element which is the argument in (92.43)–(92.45):

$$\{\varphi|\boldsymbol{t}\}=\{\varphi_p|\boldsymbol{\tau}_p+\boldsymbol{R}_L\}=\{\varepsilon|\boldsymbol{R}_L\}\cdot\{\varphi_p|\boldsymbol{\tau}_p\}. \tag{93.1}$$

Then

$$\begin{aligned}\{\varphi|\boldsymbol{t}\}^2&=\{\varepsilon|\boldsymbol{R}_L\}\cdot\{\varphi_p|\boldsymbol{\tau}_p\}\cdot\{\varepsilon|\boldsymbol{R}_L\}\{\varphi_p|\boldsymbol{\tau}_p\}\\&=\{\varepsilon|\boldsymbol{R}_L\}\cdot\{\varphi_p|\boldsymbol{\tau}_p\}\cdot\{\varepsilon|\boldsymbol{R}_L\}\cdot\{\varphi_p|\boldsymbol{\tau}_p\}^{-1}\cdot\{\varphi_p|\boldsymbol{\tau}_p\}^2\\&=\{\varepsilon|\boldsymbol{R}_L+\varphi_p\cdot\boldsymbol{R}_L\}\cdot\{\varphi_p|\boldsymbol{\tau}_p\}^2.\end{aligned} \tag{93.2}$$

Comparing with (37.3) we require the conjugate

$$\begin{aligned}&\{\varphi_\sigma|\boldsymbol{\tau}_\sigma\}^{-1}\cdot\{\varepsilon|\boldsymbol{R}_L+\varphi_p\cdot\boldsymbol{R}_L\}\cdot\{\varphi_p|\boldsymbol{\tau}_p\}^2\cdot\{\varphi_\sigma|\boldsymbol{\tau}_\sigma\}\\&\quad=\{\varphi_\sigma|\boldsymbol{\tau}_\sigma\}^{-1}\cdot\{\varepsilon|\boldsymbol{R}_L+\varphi_p\cdot\boldsymbol{R}_L\}\cdot\{\varphi_\sigma|\boldsymbol{\tau}_\sigma\}\cdot\{\varphi_\sigma|\boldsymbol{\tau}_\sigma\}^{-1}\cdot\{\varphi_p|\boldsymbol{\tau}_p\}^2\cdot\{\varphi_\sigma|\boldsymbol{\tau}_\sigma\}\\&\quad=\{\varepsilon|\varphi_\sigma^{-1}\cdot\boldsymbol{R}_L+\varphi_\sigma^{-1}\cdot\varphi_p\cdot\boldsymbol{R}_L\}\cdot\{\varphi_\sigma|\boldsymbol{\tau}_\sigma\}^{-1}\cdot\{\varphi_p|\boldsymbol{\tau}_p\}^2\cdot\{\varphi_\sigma|\boldsymbol{\tau}_\sigma\}.\end{aligned} \tag{93.3}$$

Then (37.3) becomes

$$\begin{aligned}&\chi^{(*\boldsymbol{k})(m)}(\{\varphi|\boldsymbol{t}\}^2)\\&\quad=\sum_{\sigma=1}^{s}\dot{\chi}^{(\boldsymbol{k})(m)}(\{\varepsilon|\varphi_\sigma^{-1}\cdot\boldsymbol{R}_L+\varphi_\sigma^{-1}\cdot\varphi_p\cdot\boldsymbol{R}_L\}\cdot\{\varphi_\sigma|\boldsymbol{\tau}_\sigma\}^{-1}\cdot\{\varphi_p|\boldsymbol{\tau}_p\}^2\{\varphi_\sigma|\boldsymbol{\tau}_\sigma\}).\end{aligned} \tag{93.4}$$

Now we notice the structure of the argument in (93.4) which is the product of a pure translation element with an element which is a rotation. But the dotted character in (93.4) will only be different from zero if condition (37.2) is satisfied. This a condition on the rotational part, that is, the second factor in (93.4). The condition is that

$$\{\varphi_\sigma|\boldsymbol{\tau}_\sigma\}^{-1}\cdot\{\varphi_p|\boldsymbol{\tau}_p\}^2\cdot\{\varphi_\sigma|\boldsymbol{\tau}_\sigma\}\quad\text{is in }\mathfrak{G}(\boldsymbol{k}) \tag{93.5}$$

[1] C. HERRING: Phys. Rev. **52**, 361 (1937).

or that

$$\varphi_\sigma^{-1} \cdot \varphi_p^2 \cdot \varphi_\sigma \cdot \boldsymbol{k} = \boldsymbol{k} + 2\pi \boldsymbol{B}_H. \tag{93.6}$$

It follows from (93.6) that

$$\varphi_p^2 \cdot \varphi_\sigma \cdot \boldsymbol{k} = \varphi_\sigma \cdot \boldsymbol{k} + 2\pi \boldsymbol{B}'_H \tag{93.7}$$

or

$$\varphi_p^2 \cdot \boldsymbol{k}_\sigma = \boldsymbol{k}_\sigma + 2\pi \boldsymbol{B}'_H. \tag{93.8}$$

Then if (93.5) holds for the fixed p and given σ involved, we may write the value of the character of the pure translational element factor in the dotted character as

$$\begin{aligned} \exp -i\boldsymbol{k} \cdot (\varphi_\sigma^{-1} \cdot \boldsymbol{R}_L + \varphi_\sigma^{-1} \cdot \varphi_p \cdot \boldsymbol{R}_L) &= \exp -i(\varphi_\sigma \cdot \boldsymbol{k} \cdot \boldsymbol{R}_L + \varphi_p^{-1} \cdot \varphi_\sigma \cdot \boldsymbol{k} \cdot \boldsymbol{R}_L) \\ &= \exp -i(\boldsymbol{k}_\sigma + \varphi_p^{-1} \cdot \boldsymbol{k}_\sigma) \cdot \boldsymbol{R}_L. \end{aligned} \tag{93.9}$$

Hence for (93.4) we have

$$\begin{aligned} &\chi^{(*\boldsymbol{k})(m)}(\{\varphi | \tau\}^2) \\ &= \sum_\sigma (\exp -i(\boldsymbol{k}_\sigma + \varphi_p^{-1} \cdot \boldsymbol{k}_\sigma) \cdot \boldsymbol{R}_L)\, \dot{\chi}^{(\boldsymbol{k})(m)} \cdot (\{\varphi_\sigma | \tau_\sigma\}^{-1} \cdot \{\varphi_p | \tau_p\}^2 \cdot \{\varphi_\sigma | \tau_\sigma\}). \end{aligned} \tag{93.10}$$

Now we may return to (92.43)–(92.45). For the sum over all elements in $\mathfrak{G}$ we write

$$\sum_{\mathfrak{G}} = \sum_p^{g_p} \sum_L^N. \tag{93.11}$$

Where the sum on p is over all coset representatives in $\mathfrak{G}/\mathfrak{T}$, and the sum on L is over all translations in $\mathfrak{T}$. First we carry out the sum over $\mathfrak{T}$ for fixed p to obtain

$$\sum_L (\exp -i(\boldsymbol{k}_\sigma + \varphi_p^{-1} \cdot \boldsymbol{k}_\sigma) \cdot \boldsymbol{R}_L) = \begin{cases} 0 & \text{if } \boldsymbol{k}_\sigma + \varphi_p^{-1} \cdot \boldsymbol{k}_\sigma \neq 2\pi \boldsymbol{B}_H \\ N & \text{if } \boldsymbol{k}_\sigma + \varphi_p^{-1} \cdot \boldsymbol{k}_\sigma = 2\pi \boldsymbol{B}_H. \end{cases} \tag{93.12}$$

Hence, the sum (93.12) is non-zero if

$$\varphi_p \cdot \boldsymbol{k}_\sigma = -\boldsymbol{k}_\sigma + 2\pi \boldsymbol{B}_H. \tag{93.13}$$

Evidently if (93.13) applies, then also will (93.7), (93.8), so we consider (93.13) the stronger condition.

Recapitulating the argument to this point we observe that we are examining the representation $D^{(*\boldsymbol{k})(m)}$ for which, as usual, one wave vector $\boldsymbol{k}$ is selected as canonical. The coset representatives labelled by σ are of course fixed as indicated in (36.2). The element labelled by p is arbitrary and may run through all coset representatives in $\mathfrak{G}$. Then, for the fixed $\boldsymbol{k}$ and the corresponding set of σ only those coset representatives p for which (93.13) applies make a contribution to the sum in (93.11). Then substituting into (93.11) we get

$$\begin{aligned} &\sum_{\mathfrak{G}} \chi^{(*\boldsymbol{k})(m)}(\{\varphi | \tau\}^2) \\ &= N \sum_{\sigma=1}^{s} \sum_{p=1}^{g_p} \dot{\chi}^{(\boldsymbol{k})(m)}(\{\varphi_\sigma | \tau_\sigma\}^{-1} \cdot \{\varphi_p | \tau_p\}^2 \cdot \{\varphi_\sigma | \tau_\sigma\}) \Delta_{p\sigma} = \pm N g_p, 0. \end{aligned} \tag{93.14}$$

In (93.14)

$$\Delta_{p\sigma}=\begin{cases}1 & \text{if } \varphi_p\cdot \boldsymbol{k}_\sigma=-\boldsymbol{k}_\sigma+2\pi \boldsymbol{B}_H\\ 0 & \text{otherwise}.\end{cases} \tag{93.15}$$

Hence

$$\sum_{\sigma=1}^{s}\sum_{p=1}^{g_p}\dot{\chi}^{(k)(m)}(\{\varphi_\sigma|\tau_\sigma\}^{-1}\cdot\{\varphi_p|\tau_p\}^2\{\varphi_\sigma|\tau_\sigma\})\Delta_{p\sigma}=(\pm g_p,0) \tag{93.16}$$

depending on the applicable case.

Now we demonstrate that each σ in the sum in (93.16) contributes an equal amount to the total. That is, each value of σ corresponds to one arm of $\star\boldsymbol{k}$ and each arm contributes the same amount to the total. Consider $\sigma=1$, and call all those particular coset representatives $\{\varphi_p|\tau_p\}$ such that $\Delta_{p1}=1$,

$$\{\varphi_{\bar{l}_\lambda}|\tau_{\bar{l}_\lambda}\} \quad \text{where } \varphi_{\bar{l}_\lambda}\cdot \boldsymbol{k}=-\boldsymbol{k}+2\pi\boldsymbol{B}_H. \tag{93.17}$$

Then

$$\{\varphi_{\bar{l}_\lambda}|\tau_{\bar{l}_\lambda}\}^2 \quad \text{is in } \mathfrak{G}(\boldsymbol{k}). \tag{93.18}$$

The value of any particular

$$\dot{\chi}^{(k)(m)}(\{\varphi_{\bar{l}_\lambda}|\tau_{\bar{l}_\lambda}\}^2) \tag{93.19}$$

may or may not be zero. However, if there is one such character (93.19) which does not vanish, then

$$\varphi_{\bar{l}_\lambda}\cdot\varphi_\sigma^{-1}\cdot\varphi_\sigma\cdot\boldsymbol{k}=-\varphi_\sigma^{-1}\cdot\varphi_\sigma\cdot\boldsymbol{k}+2\pi\boldsymbol{B}_H \tag{93.20}$$

and

$$(\varphi_\sigma\cdot\varphi_{\bar{l}_\lambda}\cdot\varphi_\sigma^{-1})\cdot\boldsymbol{k}_\sigma=-\boldsymbol{k}_\sigma+2\pi\boldsymbol{B}_H. \tag{93.21}$$

Calling

$$\varphi'_{\bar{l}_\lambda}\equiv\varphi_\sigma\cdot\varphi_{\bar{l}_\lambda}\cdot\varphi_\sigma^{-1} \tag{93.22}$$

then

$$(\varphi'_{\bar{l}_\lambda})^2=\varphi_\sigma\cdot\varphi_{\bar{l}_\lambda}^2\cdot\varphi_\sigma^{-1} \tag{93.23}$$

and the element corresponding to (93.23)

$$\{\varphi_\sigma|\tau_\sigma\}\cdot\{\varphi_{\bar{l}_\lambda}|\tau_{\bar{l}_\lambda}\}^2\{\varphi_\sigma|\tau_\sigma\}^{-1}\equiv\{\varphi'_{\bar{l}_\lambda}|\tau'_{\bar{l}_\lambda}\}^2 \tag{93.24}$$

is in $\mathfrak{G}(\boldsymbol{k}_\sigma)$, and

$$\{\varphi_\sigma|\tau_\sigma\}^{-1}\cdot\{\varphi'_{\bar{l}_\lambda}|\tau'_{\bar{l}_\lambda}\}^2\cdot\{\varphi_\sigma|\tau_\sigma\} \tag{93.25}$$

is in $\mathfrak{G}(\boldsymbol{k})$. Coming back to (93.16), we see that the value of

$$\dot{\chi}^{(k)(m)}(\{\varphi_\sigma|\tau_\sigma\}^{-1}\cdot\{\varphi'_{l_\lambda}|\tau'_{l_\lambda}\}^2\cdot\{\varphi_\sigma|\tau_\sigma\}) \tag{93.26}$$

will be identical to that of its counterpart with $\sigma=1$:

$$\dot{\chi}^{(k)(m)}(\{\varphi_{\bar{l}_\lambda}|\tau_{\bar{l}_\lambda}\}^2). \tag{93.27}$$

Hence, each nonvanishing element in the expression (93.6) with $\sigma=1$, will correspond to a nonvanishing element in the expression (93.16) for each $\sigma=2,\ldots,s$. Hence for fixed σ each sum is equal and so:

$$\sum_{p=1}^{g_p}\dot{\chi}^{(k)(m)}(\{\varphi_\sigma|\tau_\sigma\}^{-1}\cdot\{\varphi_p|\tau_p\}^2\cdot\{\varphi_\sigma|\tau_\sigma\})\Delta_{p\sigma}=\pm g_p/s,0. \tag{93.28}$$

Consequently, we can restrict ourselves conveniently to $\sigma=1$ and then we have the reality test due originally to HERRING for the three cases, as in (92.43)–(92.45):

$$\sum_{p=1}^{g_p} \dot{\chi}^{(k)(m)}(\{\varphi_p|\tau_p\}^2)\Delta_{p1}=+g_p/s, \quad \text{potentially real } D^{(*k)(m)}, \tag{93.29}$$

$$\sum_{p=1}^{g_p} \dot{\chi}^{(k)(m)}(\{\varphi_p|\tau_p\}^2)\Delta_{p1}=0, \quad \text{complex } D^{(*k)(m)}, \tag{93.30}$$

$$\sum_{p=1}^{g_p} \dot{\chi}^{(k)(m)}(\{\varphi_p|\tau_p\}^2)\Delta_{p1}=-g_p/s, \quad \text{pseudoreal } D^{(*k)(m)}, \tag{93.31}$$

where

$$\Delta_{p1}=\begin{cases}1 & \text{if } \varphi_p\cdot \boldsymbol{k}=-\boldsymbol{k}+2\pi \boldsymbol{B}_H\\ 0 & \text{otherwise.}\end{cases} \tag{93.32}$$

The usefulness of the formulae (93.29)–(93.31) is that all dotted characters refer to the one group $\mathfrak{G}(\boldsymbol{k})$. However, the Δ_{p1} is a troublesome restriction embodied in (93.32). The procedure to be followed is anyway straightforward in applying (93.29)–(93.31), once we are given the character table of the allowable irreducible representations of $\mathfrak{G}(\boldsymbol{k})$. However, while (93.29)–(93.32) test whether a representation is real, etc. they do not solve the problem of determining into which irreducible representation $(\overline{m})$ at $-\boldsymbol{k}$ a given allowable $D^{(*k)(m)}$ is transformed under time reversal, or under the operation K of complex conjugation. This is the work of Sect. 94.

94. Classification of $D^{(*k)(m)}$ according to reality by use of a new test. The criteria (93.29)–(93.31) may be directly used in lattice dynamic analyses of the reality of particular $D^{(*k)(m)}$ to test the possibility that they may occur among the representations by which phonons transform. Clearly, if a representation satisfies (93.29) it already obeys the reality criterion necessary for occurance among the phonon representations in the particular crystal with symmetry $\mathfrak{G}$. The case of a representation for which (93.30) applies needs particular examination. Then, when (93.30) applies the representation $D^{(*k)(m)}$ is complex and according to the analysis given in Sect. 87–93, the physically irreducible representations must be taken as $D^{(*k)(m)} \oplus D^{(*k)(m)*}$. In this case, there is an evident doubling of the necessary degeneracy of the particular lattice dynamic problem from that associated with the space group $\mathfrak{G}$, and characterized by $D^{(*k)(m)}$ to that associated with the space-time symmetry group $\mathcal{G}$, characterized by the direct sum $D^{(*k)(m)} \oplus D^{(*k)(m)*}$.

Now the representation $D^{(*k)(m)}$ is irreducible, thus $D^{(*k)(m)*}$ is irreducible. According to (90.9)

$$D^{(*k)(m)*}=D^{(*-k)(\overline{m})}, \tag{94.1}$$

where $D^{(*-k)(\overline{m})}$ is an irreducible representation with star $(*-\boldsymbol{k})$. However, according to our earlier work, we have determined all the irreducible representations of $\mathfrak{G}$. Hence, it is now possible to classify $D^{(*-k)(\overline{m})}$; that is to identify this representation with one of those already found.

Following FREI,[1] let us distinguish three classes of wave vectors $\boldsymbol{k}$, and stars $\star\boldsymbol{k}$:

$$\text{Class I:}\quad \boldsymbol{k}=-\boldsymbol{k}+2\pi\boldsymbol{B}_H;\quad \star\boldsymbol{k}=\star-\boldsymbol{k}, \tag{94.2}$$

$$\text{Class II:}\quad \boldsymbol{k}\neq-\boldsymbol{k}+2\pi\boldsymbol{B}_H;\quad \star\boldsymbol{k}=\star-\boldsymbol{k}, \tag{94.3}$$

$$\text{Class III:}\quad \boldsymbol{k}\neq-\boldsymbol{k}+2\pi\boldsymbol{B}_H;\quad \star\boldsymbol{k}\neq\star-\boldsymbol{k}. \tag{94.4}$$

Each of these classes I, II, III, requires separate examination in regard to reality behaviour. Our analysis in what follows differs from that of FREI as will be pointed out.

For class I wave vectors, we must have from (94.1)

$$D^{(\star k)(m)*}=D^{(\star k)(\overline{m})}. \tag{94.5}$$

Since in this case $-\boldsymbol{k}$ is equivalent to $\boldsymbol{k}$, we may consider the representation $D^{(\star k)(\overline{m})}$ to be induced from an allowable irreducible $D^{(k)(\overline{m})}$ of $\mathfrak{G}(\boldsymbol{k})$. Then, from (90.5), the basis for $D^{(k)(\overline{m})}$ can be taken as

$$\Sigma^{(k)(\overline{m})}=K\,\Sigma^{(k)(m)}. \tag{94.6}$$

Now let $P_{\{\varphi_{l_\lambda}|\boldsymbol{t}(\varphi)\}}$ be any operator in $\mathfrak{G}(\boldsymbol{k})$ then

$$P_{\{\varphi_{l_\lambda}|\boldsymbol{t}(\varphi)\}}\Sigma^{(k)(\overline{m})}=\Sigma^{(k)(\overline{m})}D^{(k)(\overline{m})}(\{\varphi_{l_\lambda}|\boldsymbol{t}(\varphi)\}). \tag{94.7}$$

From (94.6)

$$\chi^{(k)(\overline{m})}(\{\varphi_{l_\lambda}|\boldsymbol{t}(\varphi)\})=\chi^{(k)(m)}(\{\varphi_{l_\lambda}|\boldsymbol{t}(\varphi)\})^* \tag{94.8}$$

where all objects in (94.8) are available from the character table for allowable irreducible $D^{(k)(m)}$. Consequently, for any wave-vector of class I type a direct examination of the character table for the group $\mathfrak{G}(\boldsymbol{k})$ suffices to establish the representation $D^{(k)(\overline{m})}$. It is very important to realize that in principle every operator $P_{\{\varphi_{l_\lambda}|\boldsymbol{\tau}(\varphi)\}}$ must be examined and the characters accordingly, too. Consider the pure translation $\{\varepsilon|\boldsymbol{R}_L\}$ in $\mathfrak{T}$. From (94.2)

$$\text{Class I:}\quad \boldsymbol{k}=\pi\boldsymbol{B}_H \tag{94.9}$$

so

$$\chi^{(k)(m)}(\{\varepsilon|\boldsymbol{R}_L\})=\exp -i\pi\boldsymbol{B}_H\cdot\boldsymbol{R}_L=\pm 1. \tag{94.10}$$

Consequently, the choice among the three cases (93.29)–(93.31) can be obtained by direct inspection of the character tables of the elements in $\mathfrak{P}(\boldsymbol{k})$, or $\Pi(\boldsymbol{k})$ only. The consequences for added degeneracy due to time reversal are then evident for the class I wave vectors. If:

$$D^{(k)(m)}\ \text{is real},\qquad m=\overline{m}; \tag{94.11}$$

and no extra degeneracy.

If

$$D^{(k)(m)}\ \text{is complex},\qquad m\neq\overline{m};$$

and (94.12)

$$D^{(k)(m)}\quad\text{and}\quad D^{(k)(\overline{m})}\quad\text{stick}.$$

[1] V. FREI: Czech. J. Phys. 16, 207 (1966).

In case (94.12) applies, the physically irreducible representation which may arise for a phonon is $D^{(k)(m)} \oplus D^{(k)(\bar{m})}$ which has clearly twice the dimensionality of $D^{(k)(m)}$. This is an important case in practice as it corresponds to a *local* doubling of the degree of degeneracy compared to what would be permitted under purely spatial symmetry $\mathfrak{G}$ alone.

Next consider class II wave vectors. In this case $-\boldsymbol{k}$ is in the star ${}^{\star}\boldsymbol{k}$, but is not equivalent to $\boldsymbol{k}$. Consequently, unlike the previous analysis, it is not evident that we can use only the little group character tables of $\mathfrak{P}(\boldsymbol{k})$ and $\Pi(\boldsymbol{k})$ directly in order to determine the nature of these representations. We can now demonstrate that it is indeed possible to reach a conclusion in this case also by using only these tables, but we must first return to general principles to establish a relation between groups and representations at $\boldsymbol{k}$ and at $-\boldsymbol{k}$. Consider the coset representative $\{\varphi_{\bar{\sigma}}|\tau_{\bar{\sigma}}\}$ such that

$$\varphi_{\bar{\sigma}} \cdot \boldsymbol{k} = -\boldsymbol{k} + 2\pi \boldsymbol{B}_H. \tag{94.13}$$

Then we can decompose the entire space group $\mathfrak{G}$ into cosets with respect to $\mathfrak{G}(\boldsymbol{k})$, where the coset representatives are barred, or unbarred:

$$\mathfrak{G} = \mathfrak{G}(\boldsymbol{k}) + \{\varphi_{\bar{\sigma}}|\tau_{\bar{\sigma}}\}\, \mathfrak{G}(\boldsymbol{k}) + \cdots + \{\varphi_{\sigma}|\tau_{\sigma}\}\, \mathfrak{G}(\boldsymbol{k}) + \cdots. \tag{94.14}$$

Unbarred elements send $\boldsymbol{k}$ into some other vector in ${}^{\star}\boldsymbol{k}$, but not into $-\boldsymbol{k}$. The group $\mathfrak{G}(-\boldsymbol{k})$ is

$$\mathfrak{G}(-\boldsymbol{k}) = \{\varphi_{\bar{\sigma}}|\tau_{\bar{\sigma}}\}^{-1}\, \mathfrak{G}(\boldsymbol{k})\, \{\varphi_{\bar{\sigma}}|\tau_{\bar{\sigma}}\}. \tag{94.15}$$

By recapitulating the arguments of Sects. 91–93 it is evident that we may relate various characters. Thus, the character

$$\chi^{(k)(m)}(\{\varphi_l|\tau_l\}) \tag{94.16}$$

of $P_{\{\varphi_l|\tau_l\}}$ in irreducible representation $D^{(k)(m)}$ is based on the space $\Sigma^{(k)(m)}$. Now transform this space using the barred coset representative

$$\{\varphi_{\bar{\sigma}}|\tau_{\bar{\sigma}}\}\, \Sigma^{(k)(m)} = \Sigma^{(-k)(m)}, \tag{94.17}$$

where the right hand side of (94.17) is assured by the convention which was adopted in (35.3), that a purely spatial transformation will preserve the index m. But the element

$$P_{\{\varphi_{\bar{\sigma}}|\tau_{\bar{\sigma}}\}\{\varphi_l|\tau_l\}\{\varphi_{\bar{\sigma}}|\tau_{\bar{\sigma}}\}^{-1}} \equiv P_{\{\varphi_l|\tau_l\}^{\bar{\sigma}}} \tag{94.18}$$

has a character on (94.17) given as

$$\chi^{(-k)(m)}(\{\varphi_l|\tau_l\}^{\bar{\sigma}}). \tag{94.19}$$

Then also

$$\chi^{(-k)(m)}(\{\varphi_l|\tau_l\}^{\bar{\sigma}}) = \chi^{(k)(m)}(\{\varphi_l|\tau_l\}), \tag{94.20}$$

where (94.20) follows by applying (94.18) to (94.17). It is to be realized that both (94.16) and (94.19) arise as constituents in the expression (49.3), in which to be consistent with (94.14) we now must sum over σ and $\bar{\sigma}$

$$\chi^{(\star k)(m)}(\{\varphi_l|\tau_l\}) = \frac{1}{l_k} \sum_{\sigma, \bar{\sigma}} \dot{\chi}^{(k)(m)}(\{\varphi_\sigma|\tau_\sigma\}^{-1}\{\varphi_l|\tau_l\} \cdot \{\varphi_\sigma|\tau_\sigma\}) \tag{94.21}$$

for the full space group character of the element $\{\varphi_l|\tau_l\}$. Hence (94.16) and (94.19) are "pieces" of $\chi^{(\star k)(m)}$ for the same space group irreducible representation $D^{(\star k)(m)}$.

Now consider the space obtained by time-reversing $\Sigma^{(k)(m)}$:

$$K\,\Sigma^{(k)(m)} = \Sigma^{(k)(m)*} = \Sigma^{(-k)(\overline{m})} \tag{94.22}$$

and the space obtained by spatial transformation of (94.22):

$$\{\varphi_{\bar\sigma}|\tau_{\bar\sigma}\}\,K\,\Sigma^{(k)(m)} = \{\varphi_{\bar\sigma}|\tau_{\bar\sigma}\}\,\Sigma^{(-k)(\overline{m})} = \Sigma^{(k)(\overline{m})}. \tag{94.23}$$

The right hand side of (94.23) again follows owing to the convention used in (35.3) on the conservation of index, here $\overline{m}$. Again, it is worth emphasizing that we can always choose

$$P_{\{\varphi_\rho|\tau_\rho\}}\,\Sigma^{(k)(m)} = \Sigma^{(k_\rho)(m)} \tag{94.24}$$

as in (35.3), where m is the same on both sides of (94.24); however we must take for the antilinear operator K operating on the vector space:

$$K\,\Sigma^{(k)(m)} = \Sigma^{(-k)(\overline{m})} \tag{94.25}$$

with $\overline{m}$ to be determined as shown below.

Then the character of the element $P_{\{\varphi_l|\tau_l\}^{\bar\sigma}}$ applied to basis function (linear vector space) (94.20) will be

$$\chi^{(k)(m)}(\{\varphi_l|\tau_l\}^{\bar\sigma})^* \equiv \chi^{(k)(\overline{m})}(\{\varphi_l|\tau_l\}). \tag{94.26}$$

Eq. (94.26) completely determines the index $\overline{m}$ of the complex conjugate representation and space by direct inspection of the tables of $D^{(k)(j)}$ for $\mathfrak{G}(k)$ and solves the problem in this case. The formula (94.26) is stronger than HERRING's formula (93.29), (93.30) owing to the fact that it not only determines reality class, but also actually determines the specific identity $\overline{m}$ of the complex conjugate representations. Comparing (94.26) and (94.13) to determine whether or not $m=\overline{m}$ is a check, in this case, on (93.29) and (93.30). For wave vectors of this class II, we may again distinguish several consequences:

$$\overline{m}=m; \quad \text{no extra degeneracy}, \tag{94.27}$$

$$\overline{m}\neq m; \quad D^{(k)(m)} \text{ and } D^{(k)(\overline{m})} \text{ stick}. \tag{94.28}$$

In the case (94.28) we have again a local doubling compared to results for $\mathfrak{G}$ alone. Observe that (94.26) is particularly useful owing to the fact that all objects in it are defined in the one group, $\mathfrak{G}(k)$.

Finally for class III wave vectors k it is clear that

$$D^{(\star k)(m)} \not\doteq D^{(\star k)(\overline{m})}. \tag{94.29}$$

Hence the entire representations $D^{(\star k)(m)}$ and $D^{(\star -k)(\overline{m})}$ are to be taken together. But this his doubling is a "global" doubling and does not result in a local sticking at given wave vector k. The necessary degeneracy in this case is associated with the doubling of the full space group representations in the presence of $\mathscr{G}$; the space time group.

This concludes our survey of types of irreducible representation of $\mathfrak{G}$ as potential candidate for physically irreducible phonon representations under $\mathscr{G}$.

95. Physically irreducible representations of $\mathfrak{G}$ as corepresentations of $\mathscr{G}$. In the next few sections we shall indicate how to recast the previous analysis in a form in which the spatial operators P_R of $\mathfrak{G}$ are to be treated on an equal footing with the time-reversal operator K. This approach is by means of the theory of so-called corepresentations of the space-time group $\mathscr{G}$. We shall develop the relevent theory of corepresentations emphasizing in particular the important differences of this approach with that previously given in which the different operators were treated distinctly. The corepresentation approach gives a deeper insight into the theory. Although the basic corepresentation theory is given in [1], Chap. 26 we feel it is desirable in this article to give a development specifically directed toward the applications in lattice dynamics. Although there is some overlap with [1] it is kept to a minimum consistent with making this article self-contained.

Returning to the concept of the entire space-time group $\mathscr{G}$ of Sect. 89, we recognize that the structure of $\mathscr{G}$ is given in (89.3) as $\mathscr{G}=\mathfrak{G}+K\mathfrak{G}$ in terms of the spatial unitary, transformation operators $P_{\{\varphi|t\}}$ of $\mathfrak{G}$, and the antiunitary operators $KP_{\{\varphi|t\}}$ of the coset $K\mathfrak{G}$. To obtain the necessary degeneracy of a system for which $\mathscr{G}$ is the symmetry group, requires that we investigate the invariant and irreducible vector space i.e. the minimal invarient linear vector spaces spanned by eigenvectors or normal coordinates of the physical dynamical problem. Call the invariant, irreducible physical space $\boldsymbol{S}^{(j)}$. Then the operators in $\mathscr{G}$ will transform the basis vectors of $\boldsymbol{S}^{(j)}$ onto themselves. Then let

$$\boldsymbol{S}^{(j)}\equiv\{\psi_1^{(j)},\ldots,\psi_\alpha^{(j)},\ldots\}. \tag{95.1}$$

A transformation of $\boldsymbol{S}^{(j)}$ onto itself is specified by the matrix $D^{(j)}$ in the usual fashion. Attention must be paid however to the antiunitary nature of the operators in $K\mathfrak{G}$. Calling $D^{(\mathrm{co})(j)}$ the irreducible matrix system by which the irreducible invariant space $S^{(j)}$ transforms, we have

$$P_{\{\varphi|t\}}\boldsymbol{S}^{(j)}=\boldsymbol{S}^{(j)}D^{(\mathrm{co})(j)}(\{\varphi|\boldsymbol{t}\}), \tag{95.2}$$

$$KP_{\{\varphi|t\}}\boldsymbol{S}^{(j)}=\boldsymbol{S}^{(j)}D^{(\mathrm{co})(j)}(K\{\varphi|\boldsymbol{t}\}). \tag{95.3}$$

For the result of successive application of two such operators to the space $\boldsymbol{S}^{(j)}$ we have

$$\begin{aligned}P_{\{\varphi_2|t_2\}}P_{\{\varphi_1|t_1\}}\boldsymbol{S}^{(j)}&=\boldsymbol{S}^{(j)}D^{(\mathrm{co})(j)}(\{\varphi_2|\boldsymbol{t}_2\})\,D^{(\mathrm{co})(j)}(\varphi_1|\boldsymbol{t}_1\})\\&=\boldsymbol{S}^{(j)}D^{(\mathrm{co})(j)}(\{\varphi_2|\boldsymbol{t}_2\}\cdot\{\varphi_1|\boldsymbol{t}_1\})\end{aligned} \tag{95.4}$$

and

$$\begin{aligned}P_{\{\varphi_2|t_2\}}\cdot KP_{\{\varphi_1|t_1\}}\boldsymbol{S}^{(j)}&=\boldsymbol{S}^{(j)}D^{(\mathrm{co})(j)}(\{\varphi_2|\boldsymbol{t}_2\})\cdot D^{(\mathrm{co})(j)}(K\{\varphi_1|\boldsymbol{t}_1\})\\&=\boldsymbol{S}^{(j)}D^{(\mathrm{co})(j)}(\{\varphi_2|\boldsymbol{t}_2\}\cdot K\{\varphi_1|\boldsymbol{t}_1\}).\end{aligned} \tag{95.5}$$

However recalling (88.4)

$$\begin{aligned}(K P_{\{\varphi_2|t_2\}}) P_{\{\varphi_1|t_1\}} S^{(j)} &= S^{(j)} D^{(\mathrm{co})(j)}(K\{\varphi_2|t_2\}) \cdot D^{(\mathrm{co})(j)}(\{\varphi_1|t_1\})^* \\ &= S^{(j)} D^{(\mathrm{co})(j)}(K\{\varphi_2|t_2\} \cdot \{\varphi_1|t_1\})\end{aligned} \tag{95.6}$$

and

$$\begin{aligned}(K P_{\{\varphi_2|t_2\}}) \cdot (K P_{\{\varphi_1|t_1\}}) S^{(j)} &= S^{(j)} D^{(\mathrm{co})(j)}(K\{\varphi_2|t_2\}) \cdot D^{(\mathrm{co})(j)}(K\{\varphi_1|t_1\})^* \\ &= S^{(j)} D^{(\mathrm{co})(j)}((K\{\varphi_2|t_2\}) \cdot K\{\varphi_1|t_1\})).\end{aligned} \tag{95.7}$$

To emphasize the essential point of (95.4)–(95.7) we introduce some useful abbreviations. For the unitary elements in $\mathfrak{G}$ we write.

$$P_{\{\varphi_\mu|t_\mu\}} \equiv u_\mu \tag{95.8}$$

and for the antiunitary elements in $K\mathfrak{G}$

$$K P_{\{\varphi_\mu|t_\mu\}} \equiv a_\mu. \tag{95.9}$$

Then the matrix multiplications in (95.4)–(95.7) become

$$D(u_2) D(u_1) = D(u_2 u_1), \tag{95.10}$$

$$D(u_2) D(a_1) = D(u_2 a_1), \tag{95.11}$$

$$D(a_2) D(u_1)^* = D(a_2 u_1), \tag{95.12}$$

$$D(a_2) D(a_1)^* = D(a_2 a_1), \tag{95.13}$$

where D now stands for the corepresentation matrices.

The entire point of the corepresentation theory is that the determination of the inequivalent irreducible corepresentations of the space time group $\mathcal{G}$ amounts to a complete solution of the problem of necessary lattice dynamic degeneracy. It should be noted that from (95.13) there is a peculiarity regarding inverses as is seen if we allow $a_1 = a_2^{-1}$ so then

$$D(a_2) D(a_2^{-1})^* = D(a_2 a_2^{-1}) = D(\varepsilon) \tag{95.14}$$

from (95.14) we have

$$D(a_2)^{-1} = D(a_2^{-1})^*. \tag{95.15}$$

It should be recalled, however, that the corepresentation matrices which satisfy (95.10)–(95.15) are unitary, owing to the finiteness of the space-time group $\mathcal{G}$. Hence

$$D(a_2)^{-1} = D(a_2)^\dagger = \tilde{D}(a_2)^* \tag{95.16}$$

or, with (95.15)

$$D(a_2^{-1}) = \tilde{D}(a_2). \tag{95.17}$$

The set of matrices $D(u_\mu)$, $D(a_\mu)$ correspond to the set of elements in $\mathcal{G}$, but the correspondence is not a homomorphism between matrix group $D^{(\mathrm{co})}$ and operator group $\mathcal{G}$. Thus given that $\mathcal{G}$ consists of the set of unitary operators and antiunitary operators

$$\{u\} = \{P_R\}; \quad \{a\} = \{K P_R\} \tag{95.18}$$

a set of matrices $D^{(co)}$ is given, one for each operator, with the correspondence

$$D^{(co)}(u_\mu) \Rightarrow u_\mu, \tag{95.19}$$

$$D^{(co)}(a_\mu) \Rightarrow a_\mu \tag{95.20}$$

and the multiplication (95.10)–(95.13). Then this matrix set defines a *semi-linear representation* of $\mathcal{G}$. The term semi-linear representation is well established in the mathematical literature,[1] but WIGNER coined the term *corepresentation* [*1*] and it has become conventional in the physics literature to follow him in referring to a matrix set with the properties (95.10)–(95.13). Hence, these equations define a corepresentation of $\mathcal{G}$.

At this point, it will be useful to give the modified form of SCHUR's lemma appropriate to corepresentations. Let $\boldsymbol{S}^{(j)}$ be the irreducible space (95.1). Let $\boldsymbol{S}'^{(j)}$ be an equivalent irreducible space to $\boldsymbol{S}^{(j)}$. That is, let $\boldsymbol{S}'^{(j)}$ be linearly related to $\boldsymbol{S}^{(j)}$ via a unitary transformation matrix V. For the vectors in $\boldsymbol{S}'^{(j)}$:

$$\psi'^{(j)}_\alpha = \sum_\beta V_{\beta\alpha} \psi^{(j)}_\beta \tag{95.21}$$

or

$$\boldsymbol{S}'^{(j)} = V \boldsymbol{S}^{(j)}. \tag{95.22}$$

Then from (95.2), (95.3), (95.8), (95.9), (95.21)

$$\begin{aligned} u_\mu \psi'^{(j)}_\alpha &= \sum_\beta V_{\beta\alpha} u_\mu \psi^j_\beta \\ &= \sum_\beta V_{\beta\alpha} \sum_\gamma D_{\gamma\beta}(u_\mu) \psi^j_\gamma \\ &= \sum_\beta V_{\beta\alpha} \sum_\gamma D_{\gamma\beta}(u_\mu) \sum_\delta V^{-1}_{\delta\gamma} \psi'^{(j)}_\delta \\ &= \sum_\delta \Big(\sum_\beta \sum_\gamma V^{-1}_{\delta\gamma} D_{\gamma\beta}(u_\mu) V_{\beta\alpha}\Big) \psi'^{(j)}_\delta \end{aligned} \tag{95.23}$$

and

$$\begin{aligned} a_\mu \psi'^{(j)}_\alpha &= a_\mu \sum_\beta V_{\beta\alpha} \psi^{(j)}_\beta \\ &= \sum_\beta V^*_{\beta\alpha} a_\mu \psi^{(j)}_\beta \\ &= \sum_\beta V^*_{\beta\alpha} \sum_\gamma D(a_\mu)_{\gamma\beta} \sum_\delta V^{-1}_{\delta\gamma} \psi'^{(j)}_\delta \\ &= \sum_\delta \Big(\sum_\beta \sum_\gamma V^{-1}_{\delta\gamma} D_{\gamma\beta}(a_\mu) V^*_{\beta\alpha}\Big) \psi'^{(j)}_\delta \end{aligned} \tag{95.24}$$

or

$$u_\mu \psi'^{(j)}_\alpha = \sum_\delta D'(u_\mu)_{\delta\alpha} \psi'^{(j)}_\delta \tag{95.25}$$

$$a_\mu \psi'^{(j)}_\alpha = \sum_\delta D'(a_\mu)_{\delta\alpha} \psi'^{(j)}_\delta. \tag{95.26}$$

[1] A.W. CLIFFORD: Ann. Math. **38**, 533 (1937).

Clearly D and D' are equivalent corepresentations. They are related by

$$D'(u_\mu) = V^{-1} D(u_\mu) V, \tag{95.27}$$

$$D'(a_\mu) = V^{-1} D(a_\mu) V^*. \tag{95.28}$$

We shall take (95.27) and (95.28) to define equivalent corepresentations. Notice in (95.28) the appearance of the complex conjugate.

The modified Schur lemma then takes account of the occurrance of the complex conjugation in (95.28). Thus, let $D^{(i)}$ and $D^{(j)}$ be two irreducible corepresentations of a group. Let M be a Hermitian matrix such that

$$M D^{(i)}(u_\mu) = D^{(j)}(u_\mu) M, \tag{95.29}$$

$$M D^{(i)}(a_\mu) = D^{(j)}(a_\mu) M^*, \tag{95.30}$$

then either

$$M = 0 \text{ (null matrix) and } D^{(i)} \text{ is not equivalent to } D^{(j)} \tag{95.31}$$

or

$$M^{-1} \text{ exists and } D^{(i)} \text{ is equivalent to } D^{(j)}. \tag{95.32}$$

Also if $D^{(i)}$ is an irreducible corepresentation and there exists an Hermitian M such that

$$M D^{(i)}(u) = D^{(i)}(u) M, \tag{95.33}$$

$$M D^{(i)}(a) = D^{(i)}(a) M^*, \tag{95.34}$$

then

$$M \text{ is a constant matrix}. \tag{95.35}$$

The restriction of M in the previous argument to Hermitian matrices is of importance. It was emphasized by DIMMOCK,[2] although other authors [1] do not emphasize this point.

In the next sections, we shall continue with the examination of the structure of the corepresentation matrices of $\mathscr{G}$. Certain important differences from the ordinary representation matrices appear. We shall distinguish between cases where the time reversal produces "global" degeneracy from cases where a "local" added degeneracy arises. The latter can occur at points of sufficiently high symmetry especially where $-\boldsymbol{k}$ is already in the star of $\boldsymbol{k}$, or especially if $-\boldsymbol{k}$ is equivalent to $\boldsymbol{k}$.

First we introduce the idea of a costar: co $^\star\boldsymbol{k}$. This again leads to the classification of wave vectors into three types as in Sect. 94. The structure of the induced corepresentation matrices can then be blocked off as in Sect. 36. An examination of this shows that we have to deal, in the most interesting cases, with irreducible ray corepresentations of a certain point group, isomorphic to the factor group $\mathscr{G}/\mathfrak{T}$. This is the most important difference from previous cases, where the object which arose was the factor group group $\mathfrak{G}/\mathfrak{T}$. We must then characterize the factor system in this case and examine the differences which can occur from the usual ray representation case.

---

[2] J. DIMMOCK: J. Math. Phys. **4**, 1307 (1963).

96. Structure of corepresentations of $\mathscr{G}$: The costar, co $\star k$. The space time group $\mathscr{G}$ can be written as in (89.3) as

$$\mathscr{G} = \mathfrak{G} + K\,\mathfrak{G}. \tag{96.1}$$

Then, since $\mathfrak{G}$ possesses a normal subgroup $\mathfrak{T}$, it is clear that $\mathfrak{T}$ is a normal subgroup of $\mathscr{G}$. Hence, a space-time point group of the space-time group $\mathscr{G}$ could be given as

$$\mathscr{P} = \mathscr{G}/\mathfrak{T}. \tag{96.2}$$

Clearly we have for $\mathscr{P}$:

$$\mathscr{P} = \mathfrak{P} + K\,\mathfrak{P}. \tag{96.3}$$

Then $\mathscr{P}$ possesses both unitary and antiunitary elements. Another way of proceeding could utilize the structure of the normal subgroup

$$\mathscr{T} \equiv \mathfrak{T} + K\,\mathfrak{T}. \tag{96.4}$$

This is a group consisting of unitary plus antiunitary translations. It should be evident that both $\mathfrak{T}$ and $\mathscr{T}$ are normal in $\mathscr{G}$ owing to the fact that the operator K commutes with elements in $\mathfrak{T}$.

We shall start the study of irreducible representations of $\mathscr{G}$, by first reducing $\mathscr{T}$. As in Sect. 29, it is clear that Bloch vectors (functions) reduce $\mathfrak{T}$. So

$$\begin{aligned} P_{\{\varepsilon|\boldsymbol{R}_L\}}\,\psi^{(\boldsymbol{k})} &= D^{(\boldsymbol{k})}(\{\varepsilon|\boldsymbol{R}_L\})\,\psi^{(\boldsymbol{k})} \\ &= (\exp - i\,\boldsymbol{k}\cdot\boldsymbol{R}_L)\,\psi^{(\boldsymbol{k})}. \end{aligned} \tag{96.5}$$

Then also the Bloch vectors define an irreducible space under $\mathscr{T}$. That is since $K\psi^{(\boldsymbol{k})} = \psi^{(\boldsymbol{k})*}$ we have

$$\begin{aligned} K\cdot P_{\{\varepsilon|\boldsymbol{R}_L\}}\,\psi^{(\boldsymbol{k})} &= K P_{\{\varepsilon|\boldsymbol{R}_L\}}\cdot K\cdot K\psi^{(\boldsymbol{k})} \\ &= D^{(\boldsymbol{k})}(\{\varepsilon|\boldsymbol{R}_L\})^*\,K\psi^{(\boldsymbol{k})} \\ &= D^{(-\boldsymbol{k})}(\{\varepsilon|\boldsymbol{R}_L\})\,K\psi^{(\boldsymbol{k})} \end{aligned} \tag{96.6}$$

since

$$D^{(\boldsymbol{k})*} = D^{(-\boldsymbol{k})}. \tag{96.7}$$

Consequently $\psi^{(\boldsymbol{k})*} = \psi^{(-\boldsymbol{k})}$. Of course this only applies to the wave vector index $\boldsymbol{k}$. As we saw in Sects. 87–94, we must in general take

$$K\psi_{\alpha}^{(\boldsymbol{k})(m)} = \psi_{\bar{\alpha}}^{(-\boldsymbol{k})(\bar{m})} \tag{96.8}$$

with $(\bar{m}, \bar{\alpha})$ to be determined. Thus the antiunitary operator K changes $\boldsymbol{k}$ to $-\boldsymbol{k}$ but its effect upon the other indices specifying the behavior of the basis under operators of $\mathscr{G}$ remains to be discussed, in the corepresentation framework.

It proves useful as a simple introduction to corepresentation theory, to discuss the irreducible corepresentations of $\mathscr{T}$, using induction from $\mathfrak{T}$. The vector $\psi^{(\boldsymbol{k})}$ is an irreducible space under $\mathfrak{T}$. To obtain an irreducible space under $\mathscr{T}$ we need to adjoin the space $K\psi^{(\boldsymbol{k})} \equiv \psi^{(\boldsymbol{k})*}$. Thus consider the representation space consisting of the two functions $\psi^{(\boldsymbol{k})}$ and its time-reverse $K\psi^{(\boldsymbol{k})}$:

$$\Sigma^{(\mathrm{co}\,\boldsymbol{k})} \equiv \{\psi^{(\boldsymbol{k})}, K\psi^{(\boldsymbol{k})}\}. \tag{96.9}$$

In this space the unitary operators are represented as

$$D(\{\varepsilon|\boldsymbol{R}_L\}) = \begin{pmatrix} D^{(\boldsymbol{k})} & 0 \\ 0 & D^{(-\boldsymbol{k})} \end{pmatrix} \tag{96.10}$$

where we used (96.7). The antiunitary operators are represented as

$$D(K\{\varepsilon|\boldsymbol{R}_L\}) = \begin{pmatrix} 0 & D^{(\boldsymbol{k})} \\ D^{(-\boldsymbol{k})} & 0 \end{pmatrix}. \tag{96.11}$$

In (96.10) and (96.11) we do not give the (obvious) argument of the matrices. It is easy to verify that (96.10) and (96.11) satisfy (95.10)–(95.13). Hence (96.10) and (96.11) define a corepresentation, of the space-time translations group $\mathcal{T}$.

We examine the irreducibility of (96.10) and (96.11) by use of Schurs lemma, as in (95.33)–(95.35). Let

$$M = \begin{pmatrix} m_{11} & m_{12} \\ m_{21} & m_{22} \end{pmatrix} \tag{96.12}$$

be a Hermitian matrix, so

$$m_{11} = m_{11}^*; \quad m_{22} = m_{22}^*; \quad m_{12} = m_{21}^*. \tag{96.13}$$

From

$$MD(\{\varepsilon|\boldsymbol{R}_L\}) = D(\{\varepsilon|\boldsymbol{R}_L\})\, M \tag{96.14}$$

we find

$$m_{12}\, D^{(-\boldsymbol{k})} = m_{12}\, D^{(\boldsymbol{k})}. \tag{96.15}$$

Hence either

$$D^{(\boldsymbol{k})} \doteq D^{(-\boldsymbol{k})} \quad \text{and} \quad m_{12} \neq 0 \tag{96.16}$$

or

$$D^{(\boldsymbol{k})} \not\doteq D^{(-\boldsymbol{k})} \quad \text{and} \quad m_{12} = 0. \tag{96.17}$$

If (96.16) obtains

$$\boldsymbol{k} = -\boldsymbol{k} + 2\pi \boldsymbol{B}_H \quad \text{and} \quad m_{12} \neq 0. \tag{96.18}$$

If (96.17) obtains

$$\boldsymbol{k} \neq -\boldsymbol{k} + 2\pi \boldsymbol{B}_H \quad \text{and} \quad m_{12} = 0. \tag{96.19}$$

From

$$MD(K\{\varepsilon|\boldsymbol{R}_L\}) = D(K\{\varepsilon|\boldsymbol{R}_L\})\, M^* \tag{96.20}$$

we have

$$m_{11}\, D^{(\boldsymbol{k})} = m_{22}\, D^{(\boldsymbol{k})}, \tag{96.21}$$

$$m_{12}\, D^{(-\boldsymbol{k})} = m_{12}\, D^{(\boldsymbol{k})}. \tag{96.22}$$

Clearly (96.22) is trivially satisfied for either (96.18) or (96.19). From (96.21) and (96.13)

$$m_{11} = m_{22} = m, \tag{96.23}$$

where m is a real constant.

Then in case (96.19) applies, it follows that

$$M = m \begin{pmatrix} 1 & 0 \\ 0 & 1 \end{pmatrix}. \tag{96.24}$$

Consequently in this case D is irreducible. In case (96.18) applies we have

$$M=\begin{pmatrix} m & m_{12} \\ m_{12}^* & m \end{pmatrix} \tag{96.25}$$

as this is clearly not a constant matrix, D is not irreducible.

Summarizing to this point, if $\boldsymbol{k}$ is not equivalent to $-\boldsymbol{k}$ the irreducible corepresentation of $\mathscr{T}$ is

$$D(\{\varepsilon|\boldsymbol{R}_L\})=\begin{pmatrix} D^{(\boldsymbol{k})} & 0 \\ 0 & D^{(-\boldsymbol{k})} \end{pmatrix} \tag{96.26}$$

and

$$D(K\{\varepsilon|\boldsymbol{R}_L\})=\begin{pmatrix} 0 & D^{(\boldsymbol{k})} \\ D^{(-\boldsymbol{k})} & 0 \end{pmatrix}. \tag{96.27}$$

If $\boldsymbol{k}$ is equivalent to $-\boldsymbol{k}$, then (96.26) and (96.27) become respectively

$$D^{(\boldsymbol{k})}\begin{pmatrix} 1 & 0 \\ 0 & 1 \end{pmatrix} \quad \text{and} \quad D^{(\boldsymbol{k})}\begin{pmatrix} 0 & 1 \\ 1 & 0 \end{pmatrix}. \tag{96.28}$$

Now (96.28) are real, symmetric matrices. They can be diagonalized by similarity transform by the real, orthogonal matrix

$$S=\frac{1}{\sqrt{2}}\begin{pmatrix} 1 & 1 \\ -1 & 1 \end{pmatrix}. \tag{96.29}$$

In reduced form the corepresentation in this case is, for matrices representing unitary elements:

$$D(\{\varepsilon|\boldsymbol{R}_L\})=\begin{pmatrix} D^{(\boldsymbol{k})} & 0 \\ 0 & D^{(\boldsymbol{k})} \end{pmatrix} \tag{96.30}$$

and for matrices representing antiunitary elements:

$$D(K\{\varepsilon|\boldsymbol{R}_L\})=\begin{pmatrix} D^{(\boldsymbol{k})} & 0 \\ 0 & -D^{(\boldsymbol{k})} \end{pmatrix}. \tag{96.31}$$

The complete reduction of (96.30), (96.31) can be carried out by using the definition of equivalent corepresentations: (95.27) and (95.28). We can then see that (96.30) and (96.31) decomposes into the direct sum of two *equivalent* corepresentations.

To see this take as the matrix V of (95.27) and (95.28)

$$V=\begin{pmatrix} 1 & 0 \\ 0 & i \end{pmatrix}.$$

Then

$$V^{-1}D(\{\varepsilon|\boldsymbol{R}_L\})\,V=\begin{pmatrix} D^{(\boldsymbol{k})}(\{\varepsilon|\boldsymbol{R}_L\}) & 0 \\ 0 & D^{(\boldsymbol{k})}(\{\varepsilon|\boldsymbol{R}_L\}) \end{pmatrix} \tag{96.32}$$

and

$$V^{-1}D(K\{\varepsilon|\boldsymbol{R}_L\})\,V^*=\begin{pmatrix} D^{(\boldsymbol{k})}(K\{\varepsilon|\boldsymbol{R}_L\}) & 0 \\ 0 & D^{(\boldsymbol{k})}(K\{\varepsilon|\boldsymbol{R}_L\}) \end{pmatrix}. \tag{96.33}$$

Then in this case the irreducible corepresentation is one-dimensional and subduces $D^{(k)}$ of $\mathfrak{T}$, twice. Although this was a trivial example it may be useful as a specific illustration particularly in (98.52) *et seq.*

Consequently, we may proceed as in the analysis of the unitary group $\mathfrak{G}$. That is we assume that we are given an irreducible corepresentation of $\mathcal{G}$, which is in fully reduced form in $\mathcal{T}$. Recalling the analyses of Sects. 30–32 which carries through here owing to K commuting with the spatial unitary operators, we immediately can conclude that an irreducible corepresentation is specified by a set of wave vectors. But in this case, in general each $\boldsymbol{k}$ will be accompanied in the set by its negative: $-\boldsymbol{k}$. The resultant set is called a costar. We define

$$\text{co}\,{}^\star\boldsymbol{k} \equiv \{\boldsymbol{k}_1, -\boldsymbol{k}_1, \ldots, \boldsymbol{k}_s, -\boldsymbol{k}_s\}. \tag{96.34}$$

The costar consists of distinct, nonequivalent wave vectors.

It is evident that we must distinguish several cases or classes of wave vectors, and hence of costars as in Sect. 94:

$$\text{I:} \quad \boldsymbol{k} = -\boldsymbol{k} + 2\pi\boldsymbol{B}_H; \quad {}^\star\boldsymbol{k} = \text{co}\,{}^\star\boldsymbol{k}, \tag{96.35}$$

$$\text{II:} \quad \boldsymbol{k} \neq -\boldsymbol{k} + 2\pi\boldsymbol{B}_H; \quad {}^\star\boldsymbol{k} = \text{co}\,{}^\star\boldsymbol{k}, \tag{96.36}$$

$$\text{III:} \quad \boldsymbol{k} \neq -\boldsymbol{k} + 2\pi\boldsymbol{B}_H; \quad {}^\star\boldsymbol{k} \neq \text{co}\,{}^\star\boldsymbol{k}. \tag{96.37}$$

In (96.35), $\boldsymbol{k}$ and $-\boldsymbol{k}$ are equivalent wave vectors; in (96.36) $\boldsymbol{k}$ and $-\boldsymbol{k}$ are not equivalent but are in the same star; hence, the costar is identical to the star as it already contains $-\boldsymbol{k}$; in (96.37) neither of the previous conditions apply so that the costar contains twice as many wave vectors as the star since the negatives of all wave vectors now occur.

97. Corepresentations of $\mathcal{G}$: Class III costar. For the class III co ${}^\star\boldsymbol{k}$, of (96.37) no unitary rotation is present to take $\boldsymbol{k}_\sigma$ into $-\boldsymbol{k}_\sigma$, where $\boldsymbol{k}_\sigma$ is any one of the arms of ${}^\star\boldsymbol{k}$. Let us assume that the corepresentaion of $\mathcal{G}$ under discussion in this case is

$$D^{(\text{co}\,{}^\star\boldsymbol{k})(m)} \tag{97.1}$$

and that the representation subduced by it upon $\mathfrak{G}$ is diagonal. Then by the discussion in Sect. 96, we may take

$$D^{(\text{co}\,{}^\star\boldsymbol{k})(m)} \downarrow D^{({}^\star\boldsymbol{k})(m)} \oplus D^{({}^\star-\boldsymbol{k})(\overline{m})}. \tag{97.2}$$

We use the symbol $\downarrow$ for subduce; the left hand side of (97.2) refers to group $\mathcal{G}$; the right hand side to group $\mathfrak{G}$. That is we consider the representation $D^{(\text{co}\,{}^\star\boldsymbol{k})(m)}$ to be based upon the vector space $\Sigma^{(\text{co}\,{}^\star\boldsymbol{k})(m)}$ which is the union or direct sum

$$\begin{aligned}\Sigma^{(\text{co}\,{}^\star\boldsymbol{k})(m)} &= \Sigma^{({}^\star\boldsymbol{k})(m)} \oplus K\Sigma^{({}^\star\boldsymbol{k})(m)}\\ &= \Sigma^{({}^\star\boldsymbol{k})(m)} \oplus \Sigma^{({}^\star-\boldsymbol{k})(\overline{m})}.\end{aligned} \tag{97.3}$$

Observe that we have made the identification

$$K\Sigma^{({}^\star\boldsymbol{k})(m)} = \Sigma^{({}^\star\boldsymbol{k})(m)*} = \Sigma^{({}^\star-\boldsymbol{k})(\overline{m})}. \tag{97.4}$$

Evidently any unitary operator $\{\varphi_l|\tau_l\}$ in $\mathfrak{G}(\boldsymbol{k})$ is also in $\mathfrak{G}(-\boldsymbol{k})$, so the two groups are identical, hence have corresponding irreducible representations. We choose

$$D^{(\star\boldsymbol{k})(m)*} \doteqdot D^{(\star-\boldsymbol{k})(\bar{m})} \tag{97.5}$$

so that the index $\bar{m}$ occurs with $-\boldsymbol{k}$. Observe that $D^{(\star\boldsymbol{k})(m)*}$ must in general also be an irreducible representation of $\mathfrak{G}$ and *mutates mutandis* $D^{(\boldsymbol{k})(m)*}$ must also be an irreducible representation of $\mathfrak{G}(\boldsymbol{k})$ at $\boldsymbol{k}$. However, the property that $D^{(\boldsymbol{k})(m)*}$ transforms like an irreducible representation at $-\boldsymbol{k}$ e.g. $D^{(-\boldsymbol{k})(\bar{m})}$ immediately establishes the inequivalence of $D^{(\boldsymbol{k})(m)*}$ and $D^{(\boldsymbol{k})(m)}$. There is no contradiction. The irreducible representation of $\mathfrak{G}(\boldsymbol{k})$ produced by $\Sigma^{(\boldsymbol{k})(m)*}$ is inequivalent to that produced by $\Sigma^{(\boldsymbol{k})(m)}$; it is however equivalent to $D^{(-\boldsymbol{k})(\bar{m})}$, an irreducible representation of $\mathfrak{G}(-\boldsymbol{k})$. The reader should satisfy himself regarding these points.

We construct the corepresentation of $\mathscr{G}$ by induction from $\mathfrak{G}$. Again recalling

$$\mathscr{G} = \mathfrak{G} + K\mathfrak{G} \tag{97.6}$$

the corepresentation based upon the space (97.3) is for unitary elements:

$$D^{(\mathrm{co}\star\boldsymbol{k})(m)}(\{\varphi|\tau\}) = \begin{pmatrix} D^{(\star\boldsymbol{k})(m)}(\{\varphi|\tau\}) & 0 \\ 0 & D^{(\star-\boldsymbol{k})(\bar{m})}(\{\varphi|\tau\}) \end{pmatrix} \tag{97.7}$$

and for antiunitary elements:

$$D^{(\mathrm{co}\star\boldsymbol{k})(m)}(K\{\varphi|\tau\}) = \begin{pmatrix} 0 & D^{(\star\boldsymbol{k})(m)}(\{\varphi|\tau\}) \\ D^{(\star-\boldsymbol{k})(\bar{m})}(\{\varphi|\tau\}) & 0 \end{pmatrix}. \tag{97.8}$$

Owing to the inequivalence of $D^{(\star\boldsymbol{k})(m)}$ and $D^{(\star\boldsymbol{k})(m)*} = D^{(\star-\boldsymbol{k})(\bar{m})}$, an application of the Schur lemma immediately demonstrates the irreducibility of (97.7) and (97.8), with no more difficulty than the proof given in Sect. 96.

Consequently for class III wave vectors the time reversal operation has the effect of doubling the dimension of the necessary degeneracy from $(l_m \cdot s)$ to $(2 l_m \cdot s)$. The corepresentation matrices (97.7) and (97.8) epitomize the full effect of the space-time symmetry and their structure is of importance in the later work, in which selection rules for multiphonon processes will be obtained. Recapitulating this case, it is useful to observe the complete separation of the antiunitary time reversal operator K from the induction process involving $\mathfrak{G}(\boldsymbol{k})$. That is, $\mathfrak{G}(\boldsymbol{k})$ is a group of purely unitary operators, and the induction from $D^{(\boldsymbol{k})(m)}$ of $\mathfrak{G}(\boldsymbol{k})$ to $D^{(\star\boldsymbol{k})(m)}$ of $\mathfrak{G}$ precedes and is separate from the induction to $D^{(\mathrm{co}\star\boldsymbol{k})(m)}$ of $\mathscr{G}$.

98. Corepresentations of $\mathscr{G}$: Class II costar and general theory. Before analysing this case, which is one of the most important, let us review the star. For class II wave vectors of (96.36) the star $^\star\boldsymbol{k}$ already contains $-\boldsymbol{k}$, but $\boldsymbol{k}$ and $-\boldsymbol{k}$ are not equivalent. We may then decompose the unitary space group as

$$\mathfrak{G} = \mathfrak{G}(\boldsymbol{k}) + \{\varphi_{\bar{\sigma}}|\tau_{\bar{\sigma}}\}\,\mathfrak{G}(\boldsymbol{k}) + \cdots \tag{98.1}$$

where $\{\varphi_{\bar{\sigma}}|\tau_{\bar{\sigma}}\}$ is a coset representative whose rotational part has the property

$$\varphi_{\bar{\sigma}}\,\boldsymbol{k} = -\boldsymbol{k} + 2\pi \boldsymbol{B}_H. \tag{98.2}$$

Clearly in this case the number of arms in $\star\boldsymbol{k}$ is even, and the number of wave vectors which could be called "positive" is $s/2$. So $\star\boldsymbol{k}=\star-\boldsymbol{k}$. It is clear that for any spatial element $\{\varphi_{\bar{\sigma}'}|\tau_{\bar{\sigma}'}\}$ which takes $\boldsymbol{k}$ into a wave vector equivalent to $-\boldsymbol{k}$, the rotational part can be written

$$\varphi_{\bar{\sigma}'}=\varphi_{\bar{\sigma}}\,\varphi_l, \tag{98.3}$$

where φ_l is the rotational part of an element in $\mathfrak{G}(\boldsymbol{k})$. It follows that $(\varphi_{\bar{\sigma}}^{-1}\cdot\varphi_{\bar{\sigma}'})$ leaves $\boldsymbol{k}$ invariant and must be in $\mathfrak{G}(\boldsymbol{k})$. For the space-time group $\mathscr{G}$ we have

$$\mathscr{G}=\mathfrak{G}(\boldsymbol{k})+K\{\varphi_{\bar{\sigma}}|\tau_{\bar{\sigma}}\}\,\mathfrak{G}(\boldsymbol{k})+\cdots \tag{98.4}$$

or we may define the space-time group of the wave vector as

$$\mathscr{G}(\boldsymbol{k})\equiv\mathfrak{G}(\boldsymbol{k})+K\{\varphi_{\bar{\sigma}}|\tau_{\bar{\sigma}}\}\,\mathfrak{G}(\boldsymbol{k}). \tag{98.5}$$

Then (98.5) defines a group with antiunitary elements. It is convenient to determine the irreducible corepresentations of $\mathscr{G}(\boldsymbol{k})$ and then to induce the irreducible corepresentations of $\mathscr{G}$ from them. In passing it should be noted that not all space groups possess class II wave vectors: there must be both an operation $\{\varphi_{\bar{\sigma}}|\tau_{\bar{\sigma}}\}$ and a wave vector $\boldsymbol{k}$ such that (98.2) applies.

Observe that when $\mathscr{G}$ is decomposed with respect to cosets of $\mathscr{G}(\boldsymbol{k})$, only unitary operators occur as the coset representatives:

$$\mathscr{G}=\mathscr{G}(\boldsymbol{k})+\cdots+\{\varphi_\sigma|\tau_\sigma\}\,\mathscr{G}(\boldsymbol{k})+\cdots+\{\varphi_s|\tau_s\}\,\mathscr{G}(\boldsymbol{k}). \tag{98.6}$$

(98.6) is important and useful for the process of induction from $\mathscr{G}(\boldsymbol{k})$ to $\mathscr{G}$; *this* final step of induction can go just as for the purely unitary case discussed earlier.

Now consider the group $\mathscr{G}(\boldsymbol{k})$. Let us find its irreducible corepresentations. We shall do this by a generalized induction method. Observe the structure of (98.5) which (refer back to (95.8)–(95.35)) can be epitomized by

$$\mathscr{G}(\boldsymbol{k})=\mathfrak{G}(\boldsymbol{k})+a_0\,\mathfrak{G}(\boldsymbol{k}) \tag{98.7}$$

where

$$a_0\equiv K\{\varphi_{\bar{\sigma}}|\tau_{\bar{\sigma}}\} \tag{98.8}$$

and a_0^2 is an element in $\mathfrak{G}(\boldsymbol{k})$. Suppose we are given the space $\Sigma^{(\boldsymbol{k})(m)}$ producing the irreducible allowable representation $D^{(\boldsymbol{k})(m)}$ of $\mathfrak{G}(\boldsymbol{k})$. Consider the space $a_0\,\Sigma^{(\boldsymbol{k})(m)}$. Clearly, from the usual result for unitary operations

$$u\,\Sigma^{(\boldsymbol{k})(m)}=\Sigma^{(\boldsymbol{k})(m)}\,D^{(\boldsymbol{k})(m)}(u) \tag{98.9}$$

we have

$$u\cdot a_0\,\Sigma^{(\boldsymbol{k})(m)}=a_0\cdot a_0^{-1}\,u\,a_0\,\Sigma^{(\boldsymbol{k})(m)} \tag{98.10}$$

but the conjugate

$$a_0^{-1}\,u\,a_0\equiv u^{a_0} \tag{98.11}$$

is an element in $\mathfrak{G}(\boldsymbol{k})$, hence, unitary. Then

$$u\cdot a_0\,\Sigma^{(\boldsymbol{k})(m)}=a_0\,\Sigma^{(\boldsymbol{k})(m)}\cdot D^{(\boldsymbol{k})(m)}(u^{a_0})^*. \tag{98.12}$$

The matrix system based upon

$$\Sigma^{(\mathrm{co}\,\boldsymbol{k})(m)}\equiv\Sigma^{(\boldsymbol{k})(m)}+a_0\,\Sigma^{(\boldsymbol{k})(m)} \tag{98.13}$$

is then for unitary elements:

$$D(u)=\begin{pmatrix} D^{(\boldsymbol{k})(m)}(u) & 0 \\ 0 & D^{(\boldsymbol{k})(m)}(u^{a_0})^* \end{pmatrix}. \tag{98.14}$$

If a is any antiunitary element

$$a=u\,a_0 \tag{98.15}$$

then

$$\begin{aligned} a\,\Sigma^{(\boldsymbol{k})(m)} &= a_0\cdot a_0^{-1}\,u\,a_0\,\Sigma^{(\boldsymbol{k})(m)} \\ &= a_0\cdot\left(\Sigma^{(\boldsymbol{k})(m)}\,D^{(\boldsymbol{k})(m)}(u^{a_0})\right) \\ &= a_0\,\Sigma^{(\boldsymbol{k})(m)}\,D^{(\boldsymbol{k})(m)}(a_0^{-1}\,a)^* \end{aligned} \tag{98.16}$$

and

$$\begin{aligned} a\cdot a_0\,\Sigma^{(\boldsymbol{k})(m)} &= u\cdot a_0\,a_0\,\Sigma^{(\boldsymbol{k})(m)} \\ &= \Sigma^{(\boldsymbol{k})(m)}\,D^{(\boldsymbol{k})(m)}(u\,a_0\,a_0) \\ &= \Sigma^{(\boldsymbol{k})(m)}\,D^{(\boldsymbol{k})(m)}(a\,a_0). \end{aligned} \tag{98.17}$$

In obtaining (98.16) and (98.17) we emphasized the element a by substituting (98.15) as required. Combining (98.16) and (98.17) we find the typical matrix for antiunitary elements in $\mathscr{G}$:

$$D(a)=\begin{pmatrix} 0 & D^{(\boldsymbol{k})(m)}(a\,a_0) \\ D^{(\boldsymbol{k})(m)}(a_0^{-1}\,a)^* & 0 \end{pmatrix}. \tag{98.18}$$

It can be easily verified that (98.14) and (98.18) obey (95.10)–(95.13) and thus are a corepresentation. Now we address the problem of the irreducibility of (98.14), (98.18). There are some possibilities to be distinguished

$$\text{case A}':\quad D^{(\boldsymbol{k})(m)}(u) \doteqdot D^{(\boldsymbol{k})(m)}(u^{a_0})^* \tag{98.19}$$

and

$$\text{case B}:\quad D^{(\boldsymbol{k})(m)}(u) \not\doteqdot D^{(\boldsymbol{k})(m)}(u^{a_0})^*. \tag{98.20}$$

Keeping the possibilities (98.19) and (98.20) in mind, we test for irreducibility using the Schur lemma. Let M be a Hermitian matrix, of dimension $(2\,l_m\,s)$:

$$M=\begin{pmatrix} m_{11}\,\boldsymbol{\Pi} & m_{12}\,\boldsymbol{\Pi} \\ m_{12}^*\,\boldsymbol{\Pi} & m_{22}\,\boldsymbol{\Pi} \end{pmatrix} \tag{98.21}$$

where

$$m_{11} \text{ and } m_{22} \text{ are real numbers, } \boldsymbol{\Pi} \text{ is } (l_m\,s)\times(l_m\,s) \text{ unit matrix} \tag{98.22}$$

we evaluate

$$M\,D(u)=D(u)\,M \tag{98.23a}$$

and

$$M\,D(a)=D(a)\,M^* \tag{98.23b}$$

to determine what matrix M is consistent with these equations.

Immediately we find that in case (98.20) applies, D is irreducible. Thus in case B, where the subduced representations are inequivalent the induced representation is in fact irreducible.

Now consider case A′. Here $D^{(\boldsymbol{k})(m)}(u)$ and $D^{(\boldsymbol{k})(m)}(u^{a_0})^*$ are equivalent representations of $\mathfrak{G}(\boldsymbol{k})$. Then a unitary $(l_m \times l_m)$ matrix β exists such that

$$\beta^{-1} D^{(\boldsymbol{k})(m)}(u)\,\beta = D^{(\boldsymbol{k})(m)}(u^{a_0})^*. \tag{98.24}$$

Then let

$$u = a_0^{-1} u_1 a_0 \tag{98.25}$$

substitute into (98.24) and take the complex conjugate to obtain:

$$\beta^{-1*} D^{(\boldsymbol{k})(m)}(a_0^{-1} u_1 a_0)^* \beta^* = D^{(\boldsymbol{k})(m)}(a_0^{-2} u_1 a_0^2) \tag{98.26}$$

or substituting (98.24) as the left hand side of (98.26), and expanding out the right hand side since a_0^2 is unitary:

$$\beta^{-1*} \beta^{-1} D^{(\boldsymbol{k})(m)}(u_1)\,\beta\,\beta^* = D^{(\boldsymbol{k})(m)}(a_0^{-2})\, D^{(\boldsymbol{k})(m)}(u_1)\, D^{(\boldsymbol{k})(m)}(a_0^2). \tag{98.27}$$

This can be rearranged since

$$D^{(\boldsymbol{k})(m)}(a_0^{-2}) = D^{(\boldsymbol{k})(m)}(a_0^2)^{-1} \tag{98.28}$$

to give

$$D^{(\boldsymbol{k})(m)}(u_1)\,\beta\,\beta^* D^{(\boldsymbol{k})(m)}(a_0^2)^{-1} = \beta\,\beta^* D^{(\boldsymbol{k})(m)}(a_0^2)^{-1} D^{(\boldsymbol{k})(m)}(u_1). \tag{98.29}$$

But $D^{(\boldsymbol{k})(m)}$ is irreducible, so by the Schur lemma

$$\beta\,\beta^* D^{(\boldsymbol{k})(m)}(a_0^2)^{-1} = c\,\Pi \tag{98.30}$$

where Π is the unit matrix, and c is a number. Letting $u_1 = a_0^2$ and substituting into (98.26) we find

$$\beta^{-1*} D^{(\boldsymbol{k})(m)}(a_0^2)^* \beta^* = D^{(\boldsymbol{k})(m)}(a_0^2). \tag{98.31}$$

When (98.30) is put into (98.31) we obtain

$$\beta^{-1*}\left(\frac{\beta^*\beta}{c^*}\right)\beta^* = \frac{\beta\,\beta^*}{c}.$$

So

$$c = c^*. \tag{98.32}$$

From (98.30) we have, since all the matrices are unitary

$$|c| = 1. \tag{98.33}$$

Thus from (98.32) and (98.33)

$$c = \pm 1$$

and we have two alternatives from case A′

$$\text{case A:}\quad D^{(\boldsymbol{k})(m)}(a_0^2) = +\beta\,\beta^* \tag{98.34}$$

or

$$\text{case C:}\quad D^{(\boldsymbol{k})(m)}(a_0^2) = -\beta\,\beta^*. \tag{98.35}$$

These two alternatives permit a distinction regarding irreducibility to be made.

To obtain this we proceed by transforming (98.14) and (98.18). Define the $(2l_m \times 2l_m)$ matrix V by the direct sum

$$V \equiv \begin{pmatrix} 1 & 0 \\ 0 & \beta^{-1} \end{pmatrix}. \tag{98.36}$$

Then form the transform

$$\bar{D}(u) = V^{-1} D(u) V, \tag{98.37}$$

$$\bar{D}(a) = V^{-1} D(a) V^*. \tag{98.38}$$

Recall that β is unitary and then using (98.15), (98.24)

$$\bar{D}(u) = \begin{pmatrix} 1 & 0 \\ 0 & \beta \end{pmatrix} \begin{pmatrix} D^{(\boldsymbol{k})(m)}(u) & 0 \\ 0 & D^{(\boldsymbol{k})(m)}(u^{a_0})^* \end{pmatrix} \begin{pmatrix} 1 & 0 \\ 0 & \beta^{-1} \end{pmatrix}$$

$$= \begin{pmatrix} D^{(\boldsymbol{k})(m)}(u) & 0 \\ 0 & D^{(\boldsymbol{k})(m)}(u) \end{pmatrix} \tag{98.39}$$

and

$$\bar{D}(a) = \begin{pmatrix} 1 & 0 \\ 0 & \beta \end{pmatrix} \begin{pmatrix} 0 & D^{(\boldsymbol{k})(m)}(u)\, D^{(\boldsymbol{k})(m)}(a_0^2) \\ D^{(\boldsymbol{k})(m)}(u^{a_0})^* & 0 \end{pmatrix} \begin{pmatrix} 1 & 0 \\ 0 & \beta^{-1*} \end{pmatrix}$$

$$= \begin{pmatrix} 0 & D^{(\boldsymbol{k})(m)}(u)\, D^{(\boldsymbol{k})(m)}(a_0^2)\, \beta^{-1*} \\ D^{(\boldsymbol{k})(m)}(u)\, \beta & 0 \end{pmatrix} \tag{98.40}$$

Substituting from (98.34) and (98.35) into (98.40) we obtain for $\bar{D}(a)$:

$$\bar{D}(a) = \begin{pmatrix} 0 & \pm D^{(\boldsymbol{k})(m)}(u)\, \beta \\ D^{(\boldsymbol{k})(m)}(u)\, \beta & 0 \end{pmatrix} \tag{98.41}$$

where $a = u a_0$, as before. Thus (98.39) and (98.41) define the corepresentation, with two possibilities still open.

To determine irreducibility return to the Schur lemma, using the Hermitian matrix M, of (98.21), and evaluate (98.23) and (98.24) for $\bar{D}$. The only case of interest is

$$M \bar{D}(a) = \bar{D}(a) M^* \tag{98.42}$$

which gives

$$m_{12}\, D^{(\boldsymbol{k})(m)}(u)\, \beta = \pm m_{12}\, D^{(\boldsymbol{k})(m)}(u)\, \beta \tag{98.43}$$

and

$$m_{22}\, D^{(\boldsymbol{k})(m)}(u)\, \beta = m_{11}\, D^{(\boldsymbol{k})(m)}(u)\, \beta. \tag{98.44}$$

From (98.44)

$$m_{11} = m_{22} = m_1 \tag{98.45}$$

in any case. But (98.43) gives 2 possibilities. If the negative sign applies then, since $D^{(\boldsymbol{k})(m)}$ and β are non-null, we get

$$\text{case C:} \quad m_{12} = m_{21} = 0. \tag{98.46}$$

If the positive sign applies in (98.43) we get

$$\text{case A:} \quad m_{12} = m_{21} = m_2 \neq 0. \tag{98.47}$$

Then we obtain

$$\text{case A:} \quad M = \begin{pmatrix} m_1 \boldsymbol{\Pi} & m_2 \boldsymbol{\Pi} \\ m_2 \boldsymbol{\Pi} & m_1 \boldsymbol{\Pi} \end{pmatrix} \tag{98.48}$$

and

$$\bar{D}(a) = D^{(\boldsymbol{k})(m)}(u)\, \beta \otimes \begin{pmatrix} 0 & 1 \\ 1 & 0 \end{pmatrix} \tag{98.49}$$

while

$$\text{case C:} \quad M = m \begin{pmatrix} \boldsymbol{\Pi} & 0 \\ 0 & \boldsymbol{\Pi} \end{pmatrix} \tag{98.50}$$

and

$$\bar{D}(a) = D^{(\boldsymbol{k})(m)}(u)\,\beta \otimes \begin{pmatrix} 0 & -1 \\ 1 & 0 \end{pmatrix}. \tag{98.51}$$

Clearly in case C, $\bar{D}$ is an irreducible corepresentation since the only matrix which "commutes" (in the generalized sense of (98.23) and (98.24)) is a constant.

In case A, $\bar{D}$ is reducible. Generalizing the argument of (96.28)–(96.31), the reduction of $\bar{D}$ can be carried out in case A by a real orthogonal matrix of appropriate dimension.

$$\mathcal{S} = \frac{1}{\sqrt{2}} \begin{pmatrix} \boldsymbol{\Pi} & \boldsymbol{\Pi} \\ -\boldsymbol{\Pi} & \boldsymbol{\Pi} \end{pmatrix} \tag{98.52}$$

where $\boldsymbol{\Pi}$ are unit matrices of appropriate dimension, to the result

$$\bar{\bar{D}}(a) = \mathcal{S}^{-1}\,\bar{D}(a)\,\mathcal{S} = \begin{pmatrix} D^{(\boldsymbol{k})(m)}(u)\,\beta & 0 \\ 0 & -D^{(\boldsymbol{k})(m)}(u)\,\beta \end{pmatrix}. \tag{98.53}$$

As in reducing (96.30), (96.31), to (96.32), (96.33), a final reduction of (98.39), (98.53) can be achieved by transformation to an equivalent representation. Again use the matrix $\bar{V}$

$$\bar{V} = \begin{pmatrix} \boldsymbol{\Pi} & 0 \\ 0 & i\boldsymbol{\Pi} \end{pmatrix} \tag{98.54}$$

to find

$$\bar{V}^{-1}\,\bar{\bar{D}}(u)\,\bar{V}, \tag{98.55}$$

$$\bar{V}^{-1}\,\bar{\bar{D}}(a)\,\bar{V}^{*} \tag{98.56}$$

as in (96.32) and (96.33) the reduction in this case produces the irreducible (allowable) corepresentation of $\mathcal{G}(\boldsymbol{k})$:

$$D^{(\mathrm{co}\,\boldsymbol{k})(m)}(u) = D^{(\boldsymbol{k})(m)}(u), \tag{98.57}$$

$$D^{(\mathrm{co}\,\boldsymbol{k})(m)}(a) = D^{(\boldsymbol{k})(m)}(a\,a_0^{-1})\,\beta. \tag{98.58}$$

Now we may assume that we possess the allowable corepresentation D of the group $\mathcal{G}(\boldsymbol{k})$. This corepresentation is, according to the work just concluded, one of three kinds depending on:

$$\begin{aligned} \textit{case A:} \quad & D^{(\boldsymbol{k})(m)}(u) \doteqdot D^{(\boldsymbol{k})(m)}(a_0^{-1}\,u\,a_0)^{*} \\ \text{and} \quad & D^{(\boldsymbol{k})(m)}(a_0^2) = \beta\,\beta^{*} \\ \text{where} \quad & \beta^{-1}\,D^{(\boldsymbol{k})(m)}(u)\,\beta = D^{(\boldsymbol{k})(m)}(a_0^{-1}\,u\,a_0)^{*} \end{aligned} \tag{98.59}$$

and $D(u)$ given by (98.57), $D(a)$ given by (98.58);

$$\textit{case B:} \quad D^{(\boldsymbol{k})(m)}(u) \not\doteqdot D^{(\boldsymbol{k})(m)}(a_0^{-1}\,u\,a_0)^{*} \tag{98.60}$$

with $D(u)$ given by (98.14), and $D(a)$ by (98.18);

$$\begin{aligned} \textit{case } C\colon\quad & D^{(\boldsymbol{k})(m)}(u) \doteqdot D^{(\boldsymbol{k})(m)}(a_0^{-1}\,u\,a_0)^* \\ \text{and}\quad & D^{(\boldsymbol{k})(m)}(a_0^2) = -\beta\,\beta^* \\ \text{where}\quad & \beta^{-1}\,D^{(\boldsymbol{k})(m)}(u)\,\beta = D^{(\boldsymbol{k})(m)}(a_0^{-1}\,u\,a_0)^* \end{aligned} \tag{98.61}$$

and $D(u)$ given by (98.39), $D(a)$ given by (98.51).

As in dealing with the ordinary representations of $\mathfrak{G}$ induced from those of $\mathfrak{G}(\boldsymbol{k})$, we shall induce corepresentations of $\mathscr{G}$ from those of $\mathscr{G}(\boldsymbol{k})$. The decomposition of $\mathscr{G}$ with respect to $\mathscr{G}(\boldsymbol{k})$ was given in (98.6). Again notice that all coset representatives in (98.6) are unitary. In fact they have been chosen to correspond to the coset representatives which would arise for the *same* $\boldsymbol{k}$ in a decomposition of $\mathfrak{G}$ with respect to $\mathfrak{G}(\boldsymbol{k})$.

Consequently we may take over completely the argument leading to (36.7) and (36.8). Basic to this is the definition (36.6)

$$P_{\{\varphi_\sigma|\tau_\sigma\}}\,\psi_\alpha^{(\boldsymbol{k})(m)} = \psi_\alpha^{(\boldsymbol{k}_\sigma)(m)}. \tag{98.62}$$

In the presence of antiunitary operators also, a certain choice of basis is needed. Observe

$$P_{\{\varphi_\sigma|\tau_\sigma\}}\,a_0\,\psi_\alpha^{(\boldsymbol{k})(m)} = a_0^\sigma\cdot\psi_\alpha^{(\boldsymbol{k}_\sigma)(m)}, \tag{98.63}$$

where

$$a_0^\sigma \equiv P_{\{\varphi_\sigma|\tau_\sigma\}}\,a_0\,P_{\{\varphi_\sigma|\tau_\sigma\}}^{-1} \tag{98.64}$$

and $\psi_\alpha^{(\boldsymbol{k}_\sigma)(m)}$ is defined in (98.62). Now if we choose as the basis space in $\boldsymbol{k}_\sigma$:

$$\Sigma^{(\mathrm{co}\,\boldsymbol{k}_\sigma)(m)} \equiv \Sigma^{(\boldsymbol{k}_\sigma)(m)} \oplus a_0^\sigma\,\Sigma^{(\boldsymbol{k}_\sigma)(m)}, \tag{98.65}$$

where

$$\Sigma^{(\boldsymbol{k}_\sigma)(m)} = \{\psi_1^{(\boldsymbol{k}_\sigma)(m)},\ldots,\psi_{l_m}^{(\boldsymbol{k}_\sigma)(m)}\} \tag{98.66}$$

that we can proceed simply. That is we block off the entire corepresentation matrix $D^{(\mathrm{co}\,{}^\star\boldsymbol{k})(m)}$ into blocs corresponding to the coset decomposition (98.6), these are then labelled

$$D_{\sigma\tau}^{(\mathrm{co}\,{}^\star\boldsymbol{k})(m)} \tag{98.67}$$

where, as usual, $\sigma\,\tau$ must be taken from indices in the coset representatives in (98.6). As usual, the block matrix in (98.62) describes how the space $\Sigma^{(\mathrm{co}\,\boldsymbol{k}_\tau)(m)}$ is sent onto $\Sigma^{(\mathrm{co}\,\boldsymbol{k}_\sigma)(m)}$ by the (unspecified) operator which would be the argument of that matrix. Now we can proceed along lines similar to (36.7) and (36.8). The non-zero block matrices in the first row or column can immediately be given as:

$$D^{(\mathrm{co}\,{}^\star\boldsymbol{k})(m)}(\{\varphi_\sigma|\tau_\sigma\})_{\sigma 1} = E, \tag{98.68}$$

$$D^{(\mathrm{co}\,{}^\star\boldsymbol{k})(m)}(\{\varphi_\sigma|\tau_\sigma\}^{-1})_{1\sigma} = E, \tag{98.69}$$

where in (98.68) and (98.69), E is the unit l_m or $2\,l_m$ dimensional matrix. Observe that the complex conjugate of (98.68), and (98.69) is still the unit matrix.

Now we shall take the (1, 1) bloc matrix in $D^{(\mathrm{co}\,{}^\star\boldsymbol{k})(m)}$ to be

$$D_{11}^{(\mathrm{co}\,{}^\star\boldsymbol{k})(m)} \equiv D^{(\mathrm{co}\,\boldsymbol{k})(m)}. \tag{98.70}$$

Further, we shall define the dotted matrix

$$\dot{D}^{(\mathrm{co}\,\boldsymbol{k})(m)}(X)=\begin{cases}0 & \text{if } X \text{ is not in } \mathcal{G}(\boldsymbol{k})\\ D^{(\mathrm{co}\,\boldsymbol{k})(m)}(X) & \text{if } X \text{ is in } \mathcal{G}(\boldsymbol{k}).\end{cases} \tag{98.71}$$

Then let $\{\varphi_p|\tau_p)$ be an arbitrary unitary element in $\mathcal{G}$. Consider the (σ,τ) bloc matrix:

$$D^{(\mathrm{co}\,{}^\star\boldsymbol{k})(m)}(\{\varphi_\sigma|\tau_\sigma\}\cdot\{\varphi_p|\tau_p\}\cdot\{\varphi_\tau|\tau_\tau\}^{-1})_{\sigma\tau}=\dot{D}^{(\mathrm{co}\,\boldsymbol{k})(m)}(\{\varphi_p|\tau_p\}). \tag{98.72}$$

Consider the arbitrary antiunitary element $K\{\varphi_p|\tau_p\}$. Again the (σ,τ) bloc matrix in the full corepresentation is

$$D^{(\mathrm{co}\,{}^\star\boldsymbol{k})(m)}(\{\varphi_\sigma|\tau_\sigma\}\cdot K\{\varphi_p|\tau_p\}\cdot\{\varphi_\tau|\tau_\tau\}^{-1})_{\sigma\tau}=\dot{D}^{(\mathrm{co}\,\boldsymbol{k})(m)}(K\{\varphi_p|\tau_p\}). \tag{98.73}$$

Both (98.70), and (98.71) follow by taking the matrix elements of the relevant matrices, and using (98.68), (98.69).

Turning the argument around we find as in obtaining (36.10) for an arbitrary unitary element $\{\varphi_p|\tau_p\}$, and antiunitary element $K\{\varphi_p|\tau_p\}$:

$$D^{(\mathrm{co}\,{}^\star\boldsymbol{k})(m)}(\{\varphi_p|\tau_p\})_{\sigma\tau}=\dot{D}^{(\mathrm{co}\,\boldsymbol{k})(m)}(\{\varphi_\sigma|\tau_\sigma\}^{-1}\cdot\{\varphi_p|\tau_p\}\cdot\{\varphi_\tau|\tau_\tau\}) \tag{98.74}$$

and

$$D^{(\mathrm{co}\,{}^\star\boldsymbol{k})(m)}(K\{\varphi_p|\tau_p\})_{\sigma\tau}=\dot{D}^{(\mathrm{co}\,\boldsymbol{k})(m)}(\{\varphi_\sigma|\tau_\sigma\}^{-1}\cdot K\{\varphi_p|\tau_p\}\cdot\{\varphi_\tau|\tau_\tau\}). \tag{98.75}$$

The expressions (98.74) and (98.75) solve the problem. Their simple form, is a direct result of the coset decomposition (98.6). Had a different decomposition been used, an equivalent induced representation would have been obtained. In particular, instead of the s cosets in (98.6), a decomposition in cosets of $\mathfrak{G}(\boldsymbol{k})$ would produce $2s$ cosets, some with antiunitary coset representatives. A single induction from $\mathfrak{G}(\boldsymbol{k})$ to $\mathcal{G}$ would then be possible, instead of the two steps we used, leading from $\mathfrak{G}(\boldsymbol{k})$ to $\mathcal{G}(\boldsymbol{k})$ then to $\mathcal{G}$. It is easy to see that all roads lead to an equivalent result.

99. Corepresentations of $\mathcal{G}$: Class I costar. The class I wave vectors are, from (96.35)

$$\boldsymbol{k}=\pi\boldsymbol{B}_H. \tag{99.1}$$

For such a wave vector all spatial operators such as both $\{\varphi_l|\tau_l\}$ and $\{\varphi_{\bar\sigma}|\tau_{\bar\sigma}\}$ are contained in $\mathfrak{G}(\boldsymbol{k})$. The space-time group of the wave vector in this case is

$$\mathcal{G}(\boldsymbol{k})=\mathfrak{G}(\boldsymbol{k})+K\mathfrak{G}(\boldsymbol{k}). \tag{99.2}$$

Hence in this case the antiunitary coset representative a_0 is pure time reversal K. In this case the consequence of time reversal as discussed in (94.11) and (94.12) is relatively easily acertained.

Thus, going back to case A, if

$$D^{(\boldsymbol{k})(m)}(u)=D^{(\boldsymbol{k})(m)}(u)^* \tag{99.3}$$

then clearly $D^{(\boldsymbol{k})(m)}$ is a real representation and using (98.34) and $a_0=K$,

$$\beta\,\beta^*=D^{(\boldsymbol{k})(m)}(E) \tag{99.4}$$

and the irreducible corepresentation in this case is the set of matrices

$$D^{(\boldsymbol{k})(m)}(u),\quad D^{(\boldsymbol{k})(m)}(u)\,\beta \equiv D^{(\boldsymbol{k})(m)}(a). \tag{99.5}$$

From (99.4), and the property of β being unitary we obtain

$$\beta = \tilde{\beta}. \tag{99.6}$$

Thus β in this case is a symmetric, unitary matrix.

In case B we take over the discussion of Sect. 98 so that (98.14), and (98.18) are the corepresentation matrices.

In case C the equivalence

$$D^{(\boldsymbol{k})(m)}(u) \doteqdot D^{(\boldsymbol{k})(m)}(u)^* \tag{99.6}$$

does not mean that $D^{(\boldsymbol{k})(m)}$ is real, merely that it has a real character (trace). Then the irreducible corepresentation matrices are as in (98.39) and (98.51). In this case

$$\beta\,\beta^* = -D^{(\boldsymbol{k})(m)}(E) \tag{99.7}$$

and

$$\beta = -\tilde{\beta} \tag{99.8}$$

so that β is antisymmetric. Since β is also unitary (hence has determinant with unit modulus) the result (99.8) shows that in this case, β must be a matrix of even dimension. This is an important case since class I costars are on the Brillouin zone boundary and are usually selected for particular study.

The induction from the allowable irreducible corepresentation $D^{(\mathrm{co}\,\boldsymbol{k})(m)}$ of $\mathcal{G}(k)$ to $D^{(\mathrm{co}\,{}^*\boldsymbol{k})(m)}$ of $\mathcal{G}$ proceeds just as in (98.74) and (98.75).

100. Acceptable irreducible corepresentations of $\mathcal{G}(\boldsymbol{k})$ as irreducible ray corepresentations. A certain insight into the irreducible corepresentations of $\mathcal{G}(\boldsymbol{k})$ is obtained when they are considered as ray corepresentations of a certain point group with antiunitary operators.[1] Recall that in (41.11) the crystal point group of $\boldsymbol{k}$ was defined as the factor group

$$\mathfrak{G}(\boldsymbol{k})/\mathfrak{T} = \mathfrak{P}(\boldsymbol{k}) \tag{100.1}$$

which is isomorphic to some crystal point group

$$\mathfrak{P}(\boldsymbol{k}) \equiv \varepsilon, \ldots, \varphi_{l_\lambda}, \ldots, \varphi_{l_\kappa}. \tag{100.2}$$

In case of wave vectors of classes I and II for which $\boldsymbol{k}$ and $-\boldsymbol{k}$ are either equivalent or in the same star it is natural to augument $\mathfrak{P}(\boldsymbol{k})$ by adding certain antiunitary elements. Now consider the augmented group

$$\mathcal{P}(\boldsymbol{k}) \equiv \varepsilon, \varphi_{l_2}, \ldots, \varphi_{l_\kappa}, K\varphi_{\bar{\sigma}}, \ldots, K\varphi_{\bar{\sigma}}\varphi_{l_\lambda}, \ldots, K\varphi_{\bar{\sigma}}\varphi_{l_\kappa} \tag{100.3}$$

or the space-time point group:

$$\mathcal{P}(\boldsymbol{k}) = \mathfrak{P}(\boldsymbol{k}) + K\varphi_{\bar{\sigma}}\,\mathfrak{P}(\boldsymbol{k}). \tag{100.4}$$

---

[1] The use of ray corepresentations for electronic energy bands was discussed by N. V. Kudryavtseva, Soviet Phys. Solid State **7**, 803 (1965) [orig. FTT **7**, 998 (1965)], and references cited therein.

The group $\mathscr{P}(\boldsymbol{k})$ is an augmented group containing both unitary and antiunitary elements. It is clear that

$$\mathscr{P}(\boldsymbol{k}) = \mathcal{G}(\boldsymbol{k})/\mathfrak{T}. \tag{100.5}$$

As in Sect. 41, each element in the group $\mathscr{P}(\boldsymbol{k})$ stands for an entire coset in $\mathcal{G}(\boldsymbol{k})$:

$$\begin{aligned} &\varepsilon \leftrightarrow \mathfrak{T};\ \varphi_{l_\nu} \leftrightarrow \{\varphi_{l_\nu}|\tau_{l_\nu}\}\,\mathfrak{T}; \ldots; \\ &K\varphi_{\bar{\sigma}} \leftrightarrow K\{\varphi_{\bar{\sigma}}|\tau_{\bar{\sigma}}\}\,\mathfrak{T}; \ldots . \end{aligned} \tag{100.6}$$

The group $\mathscr{P}(\boldsymbol{k})$ contains all rotational elements which leave $\boldsymbol{k}$ invariant, plus those products of time-reversal K, and rotations taking $\boldsymbol{k}$ into $-\boldsymbol{k}+2\pi\boldsymbol{B}_H$.

Consider now the multiplication law for those corepresentation matrices in $\mathcal{G}(\boldsymbol{k})$, which "corepresent" the coset representatives in $\mathcal{G}(\boldsymbol{k})$ with respect to $\mathfrak{T}$. That is there are four types of rules to consider for binary products of unitary and antiunitary elements (recall 41.2):

$$\{\varphi_{l_\lambda}|\tau_{l_\lambda}\}\cdot\{\varphi_{l_\mu}|\tau_{l_\mu}\} = \{\varepsilon|\boldsymbol{R}_{L_{\lambda,\mu}}\}\cdot\{\varphi_{l_{\lambda,\mu}}|\tau_{l_{\lambda,\mu}}\}, \tag{100.7}$$

$$\{\varphi_{l_\lambda}|\tau_{l_\lambda}\}\cdot K\{\varphi_{\bar{\sigma}}|\tau_{\bar{\sigma}}\} = \{\varepsilon|\boldsymbol{R}_{L_{\lambda,\bar{\sigma}}}\}\cdot K\{\varphi_{l_{\lambda,\bar{\sigma}}}|\tau_{l_{\lambda,\bar{\sigma}}}\}, \tag{100.8}$$

$$K\{\varphi_{\bar{\sigma}}|\tau_{\bar{\sigma}}\}\cdot\{\varphi_{l_\lambda}|\tau_{l_\lambda}\} = \{\varepsilon|\boldsymbol{R}_{L_{\bar{\sigma},\lambda}}\}\cdot K\{\varphi_{\bar{\sigma},\lambda}|\tau_{\bar{\sigma},\lambda}\}, \tag{100.9}$$

$$K\{\varphi_{\bar{\sigma}}|\tau_{\bar{\sigma}}\}\cdot K\{\varphi_{\bar{\tau}}|\tau_{\bar{\tau}}\} = \{\varepsilon|\boldsymbol{R}_{L_{\bar{\sigma},\bar{\tau}}}\}\cdot\{\varphi_{\bar{\sigma},\bar{\tau}}|\tau_{\bar{\sigma},\bar{\tau}}\}. \tag{100.10}$$

In (100.10) we identified the second antiunitary coset representative as $K\{\varphi_{\bar{\tau}}|\tau_{\bar{\tau}}\}$, which may be e.g. (see (100.3)):

$$K\{\varphi_{\bar{\tau}}|\tau_{\bar{\tau}}\} \equiv K\{\varphi_{\bar{\sigma}}|\tau_{\bar{\sigma}}\}\cdot\{\varphi_{l_\lambda}|\tau_{l_\lambda}\}. \tag{100.11}$$

For present purposes this is not important. Now repeat the steps leading from (41.2) to (41.9) treating each of (100.7)–(100.11) with modification to take account of the fact that we are dealing with corepresentations of $\mathcal{G}(\boldsymbol{k})$, so that whenever the prefactor in a product is antiunitary there must be an appropriate complex conjugation.

Now in $D^{(\mathrm{co}\,\boldsymbol{k})(m)}$ we have for (100.7), as in (41.7):

$$\begin{aligned} D^{(\mathrm{co}\,\boldsymbol{k})(m)}(\{\varphi_{l_\lambda}|\tau_{l_\lambda}\}\cdot\{\varphi_{l_\mu}|\tau_{l_\mu}\}) &= \exp -i\,\boldsymbol{k}\cdot\boldsymbol{R}_{L_{\lambda,\mu}}\, D^{(\mathrm{co}\,\boldsymbol{k})(m)}(\{\varphi_{l_{\lambda,\mu}}|\tau_{l_{\lambda,\mu}}\}) \\ &= D^{(\mathrm{co}\,\boldsymbol{k})(m)}(\{\varphi_{l_\lambda}|\tau_{l_\lambda}\})\cdot D^{(\mathrm{co}\,\boldsymbol{k})(m)}(\{\varphi_{l_\mu}|\tau_{l_\mu}\}). \end{aligned} \tag{100.12}$$

For (100.8) we get

$$\begin{aligned} D^{(\mathrm{co}\,\boldsymbol{k})(m)}(\{\varphi_{l_\lambda}|\tau_{l_\lambda}\}\cdot K\{\varphi_{l_{\bar{\sigma}}}|\tau_{\bar{\sigma}}\}) &= \exp -i\,\boldsymbol{k}\cdot\boldsymbol{R}_{L_{\lambda,\bar{\sigma}}}\, D^{(\mathrm{co}\,\boldsymbol{k})(m)}(K\cdot\{\varphi_{l_{\lambda,\bar{\sigma}}}|\tau_{l_{\lambda,\bar{\sigma}}}\}) \\ &= D^{(\mathrm{co}\,\boldsymbol{k})(m)}(\{\varphi_{l_\lambda}|\tau_{l_\lambda}\}\cdot D^{(\mathrm{co}\,\boldsymbol{k})(m)}(K\cdot\{\varphi_{\bar{\sigma}}|\tau_{\bar{\sigma}}\}). \end{aligned} \tag{100.13}$$

For (100.9):

$$\begin{aligned} D^{(\mathrm{co}\,\boldsymbol{k})(m)}(K\{\varphi_{\bar{\sigma}}|\tau_{\bar{\sigma}}\}\cdot\{\varphi_{l_\lambda}|\tau_{l_\lambda}\}) &= \exp -i\,\boldsymbol{k}\cdot\boldsymbol{R}_{L_{\bar{\sigma},\lambda}}\, D^{(\mathrm{co}\,\boldsymbol{k})(m)}(K\{\varphi_{\bar{\sigma}\lambda}|\tau_{\bar{\sigma}\lambda}\}) \\ &= D^{(\mathrm{co}\,\boldsymbol{k})(m)}(K\{\varphi_{\bar{\sigma}}|\tau_{\bar{\sigma}}\})\cdot D^{(\mathrm{co}\,\boldsymbol{k})(m)}(\{\varphi_{l_\lambda}|\tau_{\bar{\sigma}\lambda}\})^* \end{aligned} \tag{100.14}$$

and for (100.10):

$$\begin{aligned} D^{(\mathrm{co}\,\boldsymbol{k})}(K\{\varphi_{\bar{\sigma}}|\tau_{\bar{\sigma}}\}\cdot K\{\varphi_{\bar{\tau}}|\tau_{\bar{\tau}}\}) &= \exp -i\,\boldsymbol{k}\cdot\boldsymbol{R}_{L_{\bar{\sigma},\bar{\tau}}}\, D^{(\mathrm{co}\,\boldsymbol{k})(m)}(\{\varphi_{\bar{\sigma}\bar{\tau}}|\tau_{\bar{\sigma}\bar{\tau}}\}) \\ &= D^{(\mathrm{co}\,\boldsymbol{k})(m)}(K\{\varphi_{\bar{\sigma}}|\tau_{\bar{\sigma}}\})\cdot D^{(\mathrm{co}\,\boldsymbol{k})(m)}(K\{\varphi_{\bar{\tau}}|\tau_{\bar{\tau}}\})^*. \end{aligned} \tag{100.15}$$

Evidently we have here to do again with a ray type imaging. Let us define the factor system for this situation as

$$r(\lambda, \mu) \equiv \exp - i\, \boldsymbol{k} \cdot \boldsymbol{R}_{L_{\lambda,\mu}}, \tag{100.16}$$

$$r(\lambda, \bar{\sigma}) \equiv \exp - i\, \boldsymbol{k} \cdot \boldsymbol{R}_{L_{\lambda,\bar{\sigma}}}, \tag{100.17}$$

$$r(\bar{\sigma}, \lambda) \equiv \exp - i\, \boldsymbol{k} \cdot \boldsymbol{R}_{L_{\bar{\sigma},\lambda}}, \tag{100.18}$$

$$r(\bar{\sigma}, \bar{\tau}) \equiv \exp - i\, \boldsymbol{k} \cdot \boldsymbol{R}_{L_{\bar{\sigma},\bar{\tau}}}. \tag{100.19}$$

Recall at this point the property of barred rotation operators: they send $\boldsymbol{k}$ into a vector equivalent to $-\boldsymbol{k}$. And K times a barred rotation is in the space-time point group of $\boldsymbol{k}$. Now repeating the associativity argument which lead to (41.18) for purely unitary operators, we find the requirement for the factor system to be, in case of a ray corepresentation:

$$r(\lambda, x_1)\, r(\lambda\, x_1, x_2) = r(\lambda, x_1\, x_2)\, r(x_1, x_2) \tag{100.20}$$

and

$$r(\bar{\sigma}, x_1)\, r(\bar{\sigma}\, x_1, x_2) = r(\bar{\sigma}, x_1\, x_2)\, r^*(x_1, x_2) \tag{100.21}$$

where in (100.20) and (100.21) x_1 and x_2 are any element (unitary or antiunitary), and λ is an element like $\boldsymbol{\varphi}_\lambda$ while $\bar{\sigma}$ refers to an *antiunitary* element like $K\boldsymbol{\varphi}_{\bar{\sigma}}$. We must verify that the choices (100.16)–(100.19) are consistent with (100.20) and (100.21), as we did in (41.19)–(41.23). The validity of (100.20) follows from (41.23); to demonstrate (100.21) choose the case where $x_1 \equiv \mu$; $x_2 = \nu$ are indices.

Now from (41.6) we write

$$\{\varepsilon | \boldsymbol{R}_{L_{\bar{\sigma},\mu}} + \boldsymbol{R}_{L_{\bar{\sigma}\mu,\nu}}\} = \{\varepsilon | \boldsymbol{\varphi}_{\bar{\sigma}} \cdot \boldsymbol{R}_{L_{\mu,\nu}} + \boldsymbol{R}_{L_{\bar{\sigma},\mu\nu}}\}. \tag{100.22}$$

Then

$$\begin{aligned} r(\bar{\sigma}, \mu)\, r(\bar{\sigma}\,\mu, \nu) &= \exp - i\, \boldsymbol{k} \cdot (\boldsymbol{R}_{L_{\bar{\sigma},\mu}} + \boldsymbol{R}_{L_{\bar{\sigma}\mu,\nu}}) \\ &= \exp - i\, \boldsymbol{k} \cdot (\boldsymbol{\varphi}_{\bar{\sigma}} \cdot \boldsymbol{R}_{L_{\mu,\nu}} + \boldsymbol{R}_{L_{\bar{\sigma},\mu\nu}}). \end{aligned} \tag{100.23}$$

But

$$\boldsymbol{k} \cdot \boldsymbol{\varphi}_{\bar{\sigma}} = \boldsymbol{\varphi}_{\bar{\sigma}}^{-1} \cdot \boldsymbol{k} = -\boldsymbol{k} + 2\pi\, \boldsymbol{B}_H. \tag{100.24}$$

Hence (100.23) becomes

$$= (\exp - i\, \boldsymbol{k} \cdot \boldsymbol{R}_{L_{\bar{\sigma},\mu\nu}})\,(\exp + i\, \boldsymbol{k} \cdot \boldsymbol{R}_{L_{\mu,\nu}}) \tag{100.25}$$

which we recognize as from (100.16)–(100.19) as

$$= r(\bar{\sigma}, \mu\,\nu)\, r^*(\mu, \nu) \tag{100.26}$$

which is required.

We may summarize the discussions of this section. The acceptable irreducible corepresentations of $\mathscr{G}(\boldsymbol{k})$ are irreducible ray corepresentations of the augmented space-time point group, with antiunitary elements, $\mathscr{P}(\boldsymbol{k})$. Given a set of matrices, one for each element in $\mathscr{P}(\boldsymbol{k})$:

$$D(\varepsilon), \ldots, D(\boldsymbol{\varphi}_{l_\lambda}), \ldots, D(K\boldsymbol{\varphi}_{\bar{\sigma}}), \ldots, D(K\boldsymbol{\varphi}_{\bar{\sigma}}\boldsymbol{\varphi}_{l_\kappa}) \tag{100.27}$$

and a factor system

$$r(1, 1), \ldots, r(\lambda, \mu), \ldots, r(\bar{\sigma}, \lambda), \ldots, r(\bar{\sigma}, \tau) \tag{100.28}$$

such that

$$D(\varphi_\lambda)\,D(\varphi_\mu)=r(\lambda,\mu)\,D(\varphi_{\lambda_\mu}) \tag{100.29}$$

and

$$D(K\,\varphi_{\bar\sigma})\,D(\varphi_\mu)^*=r(\bar\sigma,\mu)\,D(K\,\varphi_{\bar\sigma\,\mu}), \tag{100.30}$$

$$D(\varphi_\lambda)\,D(K\,\varphi_{\bar\sigma})=r(\lambda,\bar\sigma)\,D(K\,\varphi_{\lambda,\bar\sigma}), \tag{100.31}$$

$$D(K\,\varphi_{\bar\sigma})\,D(K\,\varphi_{\bar\tau})^*=r(\bar\sigma,\bar\tau)\,D(\varphi_{\bar\sigma\,\bar\tau}) \tag{100.32}$$

with the factor system given in (100.16)–(100.19) and satisfying also (100.20), (100.21), then the equivalence between the matrix set D and the corepresentation

$$D=D^{(\mathrm{co}\,\boldsymbol{k})\,(m)} \tag{100.33}$$

is established.

The analysis of all possible irreducible ray corepresentations of augmented crystal point groups has been carried forth by several authors.[2] Some of these authors use a different (p-equivalent) factor system. Further developments are to be expected along lines of compilation and condensation of the results.

101. Complex normal coordinates as bases for irreducible corepresentations of $\mathcal{G}$. The analysis given in Sects. 87–100 permits us to complete the work of Sect. 86. The physically irreducible representation is *per definitionum* an irreducible corepresentation. In (86.28) and (86.30) we obtained the transformation equation for complex (Fourier transformed) normal coordinates:

$$\begin{aligned}P_{\{\varphi_{l_\lambda}|\tau_{l_\lambda}\}}\,Q\begin{pmatrix}\boldsymbol{k}\\ j_\nu\end{pmatrix}&\equiv Q_{\{\varphi_{l_\lambda}\}}\begin{pmatrix}\boldsymbol{k}\\ j_\nu\end{pmatrix}\\ &=\sum_{\nu'} D^{(\boldsymbol{k})\,(j)}(\{\varphi_{l_\lambda}|\tau_{l_\lambda}\})_{\nu'\,\nu}\,Q\begin{pmatrix}\boldsymbol{k}\\ j_{\nu'}\end{pmatrix}.\end{aligned} \tag{101.1}$$

As $D^{(\boldsymbol{k})\,(j)}$ is irreducible, the space spanned by the set of complex normal coordinates

$$\Sigma^{(\boldsymbol{k})\,(j)}\equiv\left\{Q\begin{pmatrix}\boldsymbol{k}\\ j_1\end{pmatrix},\ldots,Q\begin{pmatrix}\boldsymbol{k}\\ j_\nu\end{pmatrix},\ldots,Q\begin{pmatrix}\boldsymbol{k}\\ j_{l_j}\end{pmatrix}\right\} \tag{101.2}$$

is invariant and irreducible under $\mathfrak{G}(\boldsymbol{k})$. Now, according to which class wave vector or case of irreducible corepresentation is under consideration, the space which is invariant and produces the irreducible corepresentation $D^{(\mathrm{co}\,\boldsymbol{k})\,(j)}$ will differ.

The simplest case is wave vectors of class III defined in (94.4). In this case time reversal doubles the "global" degeneracy. That is, proceeding first to construct

$$\Sigma^{(*\boldsymbol{k})\,(j)}\equiv\Sigma^{(\boldsymbol{k})\,(j)}\oplus\cdots\oplus P_{\{\varphi_\sigma|\tau_\sigma\}}\,\Sigma^{(\boldsymbol{k})\,(j)}\oplus\cdots\oplus P_{\{\varphi_s|\tau_s\}}\,\Sigma^{(\boldsymbol{k})\,(j)} \tag{101.3}$$

we then construct

$$\Sigma^{(*\boldsymbol{k})\,(j)*}=K\,\Sigma^{(*\boldsymbol{k})\,(j)}=\Sigma^{(*-\boldsymbol{k})\,(\bar{j})}. \tag{101.4}$$

---

[2] See reference 1, p. 196, of this section, and references to work of KARAVAEV, KUDRYARTSEVA, KOVALEV, and CHALDYSHEV cited therein.

The union of these spaces produces the basis of the irreducible corepresentation of $\mathscr{G}$:

$$\Sigma^{(\text{co}\star\boldsymbol{k})(j)} = \Sigma^{(\boldsymbol{k})(j)} \oplus K\Sigma^{(\boldsymbol{k})(j)} \oplus \cdots \oplus P_{\{\varphi_s|\tau_s\}}\Sigma^{(\boldsymbol{k})(j)} \oplus KP_{\{\varphi_s|\tau_s\}}\Sigma^{(\boldsymbol{k})(j)}. \quad (101.5)$$

In words, the complex conjugate of each of the members of (101.2) is degenerate with it, and linearly independent. In this case, of linear independence of $Q\begin{pmatrix}\boldsymbol{k}\\ j_\nu\end{pmatrix}$ and $Q^*\begin{pmatrix}\boldsymbol{k}\\ j_\nu\end{pmatrix} = Q\begin{pmatrix}-\boldsymbol{k}\\ j_\nu\end{pmatrix}$ real normal coordinates of the first kind are often introduced.[1] That is, a transformation to

$$q_1\begin{pmatrix}\boldsymbol{k}\\ j_\nu\end{pmatrix} \equiv \frac{1}{\sqrt{2}}\left(Q\begin{pmatrix}\boldsymbol{k}\\ j_\nu\end{pmatrix} + Q\begin{pmatrix}-\boldsymbol{k}\\ j_\nu\end{pmatrix}\right) \quad (101.6)$$

and

$$q_2\begin{pmatrix}\boldsymbol{k}\\ j_\nu\end{pmatrix} = \frac{1}{i\sqrt{2}}\left(Q\begin{pmatrix}\boldsymbol{k}\\ j_\nu\end{pmatrix} - Q\begin{pmatrix}-\boldsymbol{k}\\ j_\nu\end{pmatrix}\right) \quad (101.7)$$

produces two real objects. We also must divide the reciprocal space into "positive and negative" wave vectors in this situation by erecting a plane in the reciprocal space [8]. The transformation from $Q\begin{pmatrix}\boldsymbol{k}\\ j_\nu\end{pmatrix}$, $Q\begin{pmatrix}-\boldsymbol{k}\\ j_\nu\end{pmatrix}$ to the $q_1\begin{pmatrix}\boldsymbol{k}\\ j_\nu\end{pmatrix}$, $q_2\begin{pmatrix}\boldsymbol{k}\\ j_\nu\end{pmatrix}$ is unitary hence it produces an equivalent corepresentation. But this corepresentation is no longer in the simple, completely reduced form (97.7) when the subduced representation upon the unitary subgroup $\mathfrak{G}$ is obtained, because representations going with $\boldsymbol{k}$ and $-\boldsymbol{k}$ are mixed.

Another technique[2] often used in this case to obtain real normal coordinates of the second kind requires a canonical transformation. We will not discuss this transformation further here, since various subtle questions concerning the relation between the real normal coordinates and corepresentation theory have not yet been satisfactorily resolved, in the opinion of the author. Some discussion of this matter is given in [43], and also in Sect. 114.

Summarizing case III costars we come to the point that simply adjoining the two spaces as in (101.5) results in a basis which produces the physically irreducible corepresentation $D^{(\text{co}\star\boldsymbol{k})(j)}$ of $\mathscr{G}$. The degenerate eigenvalues, or squared frequencies associated with the vectors in (101.5) are

$$\begin{aligned}&\omega^2(\boldsymbol{k}|j_1), \ldots, \omega^2(\boldsymbol{k}|j_\lambda), \ldots, \omega^2(\boldsymbol{k}_s|j_{l_j}),\\ &\omega^2(-\boldsymbol{k}|\bar{j}_1), \ldots, \omega^2(-\boldsymbol{k}|\bar{j}_\lambda), \ldots, \omega^2(-\boldsymbol{k}_s|\bar{j}_{l_{\bar{j}}})\end{aligned} \quad (101.8)$$

a total of $s(l_j + l_{\bar{j}}) = 2l_j s$ degenerate eigenvalues. The index $\bar{j}$ must be determined by examining the character table (or the actual normal coordinate transformation) as discussed in (94.26). Forcing the eigenvectors, or normal coordinates, to be real, by imposing a transformation (101.6) and (101.7) may destroy the transformation properties under the spatial operators in $\mathscr{G}$.

For wave vectors of class II defined in (94.3) (and see Sect. 98), the situation is different. We must discuss the 3 possibilities, cases A, B, C as in (98.59), (98.60),

---

[1] See [8], Eq. (38.33).
[2] See [8], Eq. (38.38).

(98.61). For case A the basis

$$\Sigma^{(\mathrm{co}\,\boldsymbol{k})\,(j)} \equiv \left\{Q\begin{pmatrix}\boldsymbol{k}\\ j_1\end{pmatrix}, \ldots, Q\begin{pmatrix}\boldsymbol{k}\\ j_\lambda\end{pmatrix}, \ldots, Q\begin{pmatrix}\boldsymbol{k}\\ j_{l_j}\end{pmatrix}\right\} = \Sigma^{(\boldsymbol{k})\,(j)} \tag{101.9}$$

is identical to (101.2). There is no added degeneracy, local or global, due to time reversal symmetry.

For case B there is an added local degeneracy. The irreducible corepresentation of $\mathcal{G}(\boldsymbol{k})$ is based upon the union of two spaces, as in (98.13). From (101.1) and (95.62) we write

$$P_{\{\varphi_{\bar\sigma}|\boldsymbol{\tau}_{\bar\sigma}\}}\, Q\begin{pmatrix}\boldsymbol{k}\\ j_\nu\end{pmatrix} \equiv Q\begin{pmatrix}\boldsymbol{k}_{\bar\sigma}\\ j_\nu\end{pmatrix} = Q\begin{pmatrix}-\boldsymbol{k}\\ j_\nu\end{pmatrix} \tag{101.10}$$

and

$$K P_{\{\varphi_{\bar\sigma}|\boldsymbol{\tau}_{\bar\sigma}\}}\, Q\begin{pmatrix}\boldsymbol{k}\\ j_\nu\end{pmatrix} = Q\begin{pmatrix}-\boldsymbol{k}\\ j_\nu\end{pmatrix}^*. \tag{101.11}$$

Consequently for this case with $a_0 \equiv K P_{\{\varphi_{\bar\sigma}|\boldsymbol{\tau}_{\bar\sigma}\}}$:

$$\Sigma^{(\mathrm{co}\,\boldsymbol{k})\,(j)} = \left\{Q\begin{pmatrix}\boldsymbol{k}\\ j_1\end{pmatrix}, \ldots, Q\begin{pmatrix}\boldsymbol{k}\\ j_{l_j}\end{pmatrix}\right\} + \left\{Q\begin{pmatrix}-\boldsymbol{k}\\ j_1\end{pmatrix}^*, \ldots, Q\begin{pmatrix}-\boldsymbol{k}\\ j_{l_j}\end{pmatrix}^*\right\} \tag{101.12}$$

will produce the case II-B (98.60) irreducible corepresentation of $\mathcal{G}(\boldsymbol{k})$.

For case II-C (98.61) wave vectors we may again use the vector space (101.12).

Then in both cases II-B, and II-C, the representation of the unitary group $\mathfrak{G}(\boldsymbol{k})$ thereby produced will be as in (98.14) and (98.18) rather than (98.39) and (98.51). To obtain the representation of $\mathfrak{G}(\boldsymbol{k})$ in the usual form (98.39) and (98.51) for case II-C it is necessary to find the matrix β of (98.24). Then, with the unitary β determined by solving (98.24), the space

$$\Sigma^{(\mathrm{co}\,\boldsymbol{k})\,(j)} = \left\{Q\begin{pmatrix}\boldsymbol{k}\\ j_1\end{pmatrix}, \ldots, Q\begin{pmatrix}\boldsymbol{k}\\ j_{l_j}\end{pmatrix}\right\} + \left\{Q'\begin{pmatrix}-\boldsymbol{k}\\ j_1\end{pmatrix}^*, \ldots, Q'\begin{pmatrix}-\boldsymbol{k}\\ j_{l_j}\end{pmatrix}^*\right\} \tag{101.13}$$

with

$$Q'\begin{pmatrix}\boldsymbol{k}\\ j_\mu\end{pmatrix} = \sum_\nu \beta_{\nu\mu}\, Q\begin{pmatrix}\boldsymbol{k}\\ j_\nu\end{pmatrix} \tag{101.14}$$

gives the form (98.39) for these two cases.

From (101.9), (101.12), (101.13) the entire vector space $\Sigma^{(\mathrm{co}\,{}^\star\boldsymbol{k})\,(j)}$ may be found, using the process of induction. Then we obtain the space which is a basis for $D^{(\mathrm{co}\,{}^\star\boldsymbol{k})\,(j)}$.

For case I, A, B, C exactly the same argument applies, only now the antiunitary element $a_0 = K$. In this case then (101.12) becomes

$$\Sigma^{(\mathrm{co}\,\boldsymbol{k})\,(j)} = \left\{Q\begin{pmatrix}\boldsymbol{k}\\ j_1\end{pmatrix}, \ldots, Q\begin{pmatrix}\boldsymbol{k}\\ j_{l_j}\end{pmatrix}\right\} + \left\{Q\begin{pmatrix}\boldsymbol{k}\\ j_1\end{pmatrix}^*, \ldots, Q\begin{pmatrix}\boldsymbol{k}\\ j_{l_j}\end{pmatrix}^*\right\}. \tag{101.15}$$

Exactly the same remarks apply as in (101.13) and (101.14), only now the matrix β can be obtained, and its properties were given in (99.6), (99.8); it must be determined from (98.24) consistent with these equations.

102. Eigenvectors of $D(k)$ as bases for irreducible corepresentations of $\mathscr{G}$. Coming back to Sect. 83, after the discussion in Sects. 87–100, we shall discuss the transformation properties of the eigenvectors $e\left(\left|\begin{matrix}k\\ j_\mu\end{matrix}\right.\right)$ of (83.4) in terms of corepresentation theory. Recall that to establish (83.5) it was necessary to invoke the completeness of the set of degenerate eigenvectors of the dynamical matrix $[D(k)]$ at one wave vector k. But that completeness related only to the behavior under spatial symmetry; specifically under the unitary operators in $\mathfrak{G}(k)$.

The enlarged space-time symmetry group $\mathscr{G}$ which is the complete symmetry group of the problem admits the irreducible corepresentations. In particular an arbitrary eigenvector at wave vector k, eigenvalue $\omega^2(k|j_\mu)$ must be one of the degenerate set of eigenvectors $D^{(\mathrm{co}\,k)(j)}$: Consequently (83.5) must be replaced by

$$P_{\{\varphi_{l_\lambda}|\tau_{l_\lambda}\}}e^{\mathrm{co}}\left(\left|\begin{matrix}k\\ j_\mu\end{matrix}\right.\right)=\sum_\nu D^{(\mathrm{co}\,k)(j)}(\{\varphi_{l_\lambda}|\tau_{l_\lambda}\})_{\nu\mu}\,e^{\mathrm{co}}\left(\left|\begin{matrix}k\\ j_\nu\end{matrix}\right.\right). \tag{102.1}$$

In the sum (102.1) over ν we need to include all eigenvectors which span the irreducible corepresentation space $\Sigma^{(\mathrm{co}\,k)(j)}$. These eigenvectors exactly correspond to the various linear vector spaces discussed in Sect. 101, but now substitute $e\left(\left|\begin{matrix}k\\ j_\mu\end{matrix}\right.\right)$ for $Q\begin{pmatrix}k\\ j_\nu\end{pmatrix}$ as bases.

The rule (102.1) has a particular practical importance. Often, as a result of numerical calculation one obtains an eigenvector $e\left(\left|\begin{matrix}k\\ j_\mu\end{matrix}\right.\right)$ whose symmetry requires establishment. If the purely spatial operators are applied as in (102.1) it is important to know that the linear combination obtained is characterized by the corepresentation matrix $D^{(\mathrm{co}\,k)(j)}$ and *not* only $D^{(k)(j)}$. Thus to completely specify the symmetry of an eigenvector one should examine the transformation of the eigenvector under all operations of the full space-time group $\mathscr{G}(k)$, with the objective of determining the corepresentation type.

103. Determination of actual normal mode symmetry in a crystal. Now we are in a position to complete the task which was interrupted after Sects. 82 and 83. We can determine for a given crystal, with space time symmetry group $\mathscr{G}$, the symmetry of all those actual normal modes which will arise in lattice dynamics.

Let us review what we have available, to do this task. We have the complete set of irreducible corepresentations of $\mathscr{G}$. In particular we have the irreducible allowable corepresentation of $\mathscr{G}(k)$. For each of the latter, for each wave vector of class I, II, and III (94.2), (94.3), (94.4) we know whether the irreducible corepresentation is of type A, B, or C (98.59), (98.60), (98.61). Then we know whether the irreducible corepresentation subduces one irreducible representation of the unitary subgroup, or two. If the latter: that is, if the irreducible corepresentation is of types B or C and subduces the direct sum of two irreducible representations then it is known which two are joined in this way (e.g. the same $D^{(k)(m)}$ twice, as in C; or two inequivalent $D^{(k)(m)}$ as in B).

Now we are given also the representation $D^{(k)(e)}$ of $\mathfrak{G}(\boldsymbol{k})$ as in Sect. 82. The character system of $D^{(k)(e)}$ is given in (82.21) and (82.22) as:

$$\chi^{(k)(e)}(\{\varphi_{l_\lambda}|\tau_{l_\lambda}\}) \tag{103.1}$$

for every element in $\mathfrak{G}(\boldsymbol{k})$. But, as the set of eigenvectors are a basis for an allowable irreducible corepresentation of $\mathfrak{G}(\boldsymbol{k})$ we require the reduction

$$\chi^{(k)(e)} = \sum_m \chi^{(\text{co}\,k)(m)} c_m, \tag{103.2}$$

where the c_m give the number of times the m-th irreducible corepresentation $D^{(\text{co}\,k)(m)}$ of $\mathcal{G}(\boldsymbol{k})$ appears in the eigenvector representation $D^{(k)(e)}$. The character of $D^{(\text{co}\,k)(m)}$ is denoted $\chi^{(\text{co}\,k)(m)}$.

The simplest way to proceed is to carry out the reduction over the unitary subgroup only in order to determine the number of times the j-th irreducible representation of $\mathfrak{G}(\boldsymbol{k})$ appears:

$$c_j = \frac{1}{g} \sum_{\mathfrak{G}(k)} \chi^{(k)(e)}(\{\varphi_{l_\lambda}|\tau_{l_\lambda}\})\, \chi^{(k)(j)}(\{\varphi_{l_\lambda}|\tau_{l_\lambda}\})^* \tag{103.3}$$

the sum in (103.3) is over the unitary elements in $\mathfrak{G}(\boldsymbol{k})$. The character $\chi^{(k)(e)}$ is given in (82.21) which we repeat here:

$$\chi^{(k)(e)}(\{\varphi_{l_\lambda}|\tau_{l_\lambda}\}) = \pm(1+2\cos\varphi_{l_\lambda}) \sum_{\kappa''} D^{(k)}(\{\varepsilon| -\boldsymbol{R}_N(\kappa_{\bar{\varphi}_{l_\lambda}}, \kappa'')\})\, \delta_{\kappa\bar{\varphi}_{l_\lambda}, \kappa''} \tag{103.4}$$

where

$$\boldsymbol{R}_N(\kappa_{\bar{\varphi}_{l_\lambda}}, \kappa'') \tag{103.5}$$

is a certain lattice translation defined in (82.8) which depends on the two basis atoms connected by the rotation φ_{l_λ} and

$$\delta_{\kappa\bar{\varphi}_{l_\lambda}, \kappa''} \tag{103.6}$$

is determined by (82.16) and (82.3).

In this way we reduce the eigenvector representation $D^{(k)(e)}$ into irreducible representations of $D^{(k)(m)}$ of $\mathfrak{G}(\boldsymbol{k})$. Knowing whether the corresponding induced corepresentation is of type A, B, C we immediately can carry out the reduction of $D^{(k)(e)}$ into irreducible corepresentations of $\mathcal{G}$.

It is important to notice that besides the general knowledge of the structure of the space-time group $\mathcal{G}$, the specific atom locations, stoichiometry, and crystal basis structure are involved in the program leading to (103.3). This is because which basis atoms are present determines the factors $\boldsymbol{R}_N$, and the δ in (103.6), thereby determining the actual mode symmetries which arise.

This program will be illustrated in Sect. 138 below wherein phonon symmetries in rocksalt and diamond structures will be obtained.

104. Determination of eigenvectors $e\left(\left.\begin{matrix} \boldsymbol{k} \\ j \end{matrix}\right|\right)$ by symmetry: Factorization of the dynamical matrix. Following Sect. 103 it is known which irreducible corepresentations occur for the given crystal (material, and space-time group $\mathcal{G}$), at each $\boldsymbol{k}$. The reduction (103.2) and (103.3) gives us this.

Let us consider the determination of the actual eigenvectors $e\left(\left|{k \atop j}\right.\right)$ insofar as symmetry makes this possible. Of course it is evident that if we were given one of a degenerate set of eigenvectors: e.g. such as $e\left(\left|{k \atop j_\mu}\right.\right)$ belonging to $D^{(k)(j)}$ application of the symmetry operators of $\mathscr{G}$ would determine the partners. Such an $e\left(\left|{k \atop j_\mu}\right.\right)$, could be obtained perhaps with degenerate partners for example by numerical solution of the dynamical equation. Then in this case, application of the symmetry operators enables us to verify that the $e\left(\left|{k \atop j_\mu}\right.\right)$ spans the representation $D^{(k)(j)}$, of $\mathfrak{G}(k)$.

In general it is more useful to use an equivalent basis *a priori*, which will conveniently produce symmetrized eigenvectors or symmetry coordinates. The assumed basis is simply the set of unit cartesian displacements, one such component for each mechanical degree of freedom of the crystal: a total of $3rN$. The representation of $\mathscr{G}$ which is obtained on this basis is customarily called the "mechanical" or "total" representation [*17*]. We introduce the notation

$$\left\{\Delta X\begin{pmatrix} l \\ \kappa \end{pmatrix}, \Delta Y\begin{pmatrix} l \\ \kappa \end{pmatrix}, \Delta Z\begin{pmatrix} l \\ \kappa \end{pmatrix}\right\} \quad \text{or:} \quad \Delta X_\alpha\begin{pmatrix} l \\ \kappa \end{pmatrix}, \qquad \alpha=1,2,3 \tag{104.1}$$

to stand for the unit cartesian displacements. Thus $\Delta X_\alpha\begin{pmatrix} l \\ \kappa \end{pmatrix}$ is the α component of unit amplitude cartesian displacement of particle initially at $r^0\begin{pmatrix} l \\ \kappa \end{pmatrix}$, and the complete set of $3rN$ of these span all possible normal modes for the crystal.

Now let us construct the Bloch sums which are built upon these elementary displacements, and which are Bloch vectors with wave vectors k. These are obtained by application of the projection operator (25.4)

$$\Delta X_\alpha(\kappa|k)=\frac{1}{\sqrt{N}}\sum_L e^{ik\cdot R_L}\,\Delta X_\alpha\begin{pmatrix} l \\ \kappa \end{pmatrix}, \qquad \alpha=1,2,3;\ \kappa=1,\dots,r. \tag{104.2}$$

Clearly, for given k there are $3r$ of these Bloch sums. Using these as bases we may obtain a $3r$ dimensional representation of the little group $\Pi(k)=\mathfrak{G}(k)/\mathfrak{T}(k)$, defined in (39.9). Thus if $P_{\{\varphi_\lambda|t_\lambda\}}$ is an element in $\mathfrak{G}(k)/\mathfrak{T}(k)$ we may define the transformation of the Bloch sum (104.2) under this symmetry operation. Since the $3r$ Bloch sums (104.2) form a complete linear vector space for all normal modes with wave vector k, and also are complete in regard to all possible *unit* cartesian displacements, the effect of a symmetry operation $P_{\{\varphi_l|t_l\}}$ upon one such sum must be to produce a linear combination of all the $3r$ objects (104.2). In this fashion we obtain a representation which we shall call $D^{(k)(\Delta)}$ the unit cartesian displacement representation. This representation is also called the "total" representation. When $D^{(k)(\Delta)}$ is reduced into a direct sum of allowable irreducible representations of $\mathfrak{G}(k)/\mathfrak{T}(k)$ we can determine the specific representations which arise for the normal mode problem: that is, the symmetry of all normal modes which occur.

It may be useful to emphasize the essential point here. The space

$$\Sigma^{(\boldsymbol{k})(\Delta)} \equiv \{\Delta X(1|\boldsymbol{k}), \ldots, \Delta Z(r|\boldsymbol{k})\} \tag{104.3}$$

and the space

$$\Sigma^{(\boldsymbol{k})(e)} \equiv \left\{ \boldsymbol{e}\left(\begin{vmatrix}\boldsymbol{k}\\ j_1\end{vmatrix}\right), \ldots, \boldsymbol{e}\left(\begin{vmatrix}\boldsymbol{k}\\ j'_{l_{j'}}\end{vmatrix}\right) \right\} \tag{104.4}$$

are both complete linear vector spaces for the eigenvectors of the dynamical matrix at wave vector $\boldsymbol{k}$. The first: (104.3), is spanned by the $3r$ cartesian components of unit displacements of the atoms; the second (104.4) is spanned by the complete set of $3r$ eigenvectors of the dynamical matrix at $\boldsymbol{k}$. Because these are both complete, they must be equivalent. Hence some unitary matrix S exists such that

$$S^{-1}\,\Sigma^{(\boldsymbol{k})(\Delta)}\,S = \Sigma^{(\boldsymbol{k})(e)} \tag{104.5}$$

and the representations of the unitary subgroup $\mathfrak{G}(\boldsymbol{k})$ must be also equivalent:

$$S^{-1}\,D^{(\boldsymbol{k})(\Delta)}\,S = D^{(\boldsymbol{k})(e)}. \tag{104.6}$$

Let us verify (104.6) by obtaining the characters:

$$\operatorname{Tr} D^{(\boldsymbol{k})(\Delta)} \equiv \chi^{(\boldsymbol{k})(\Delta)}. \tag{104.7}$$

As usual, let us define the effect of $P_{\{\varphi_\lambda|t_\lambda\}}$ upon the unit Cartesian displacement $(\Delta X_\alpha)\begin{pmatrix} l\\ \kappa\end{pmatrix}$ by

$$(P_{\{\varphi_\lambda|\boldsymbol{t}_\lambda\}}\,\Delta X)_\alpha \begin{pmatrix} l\\ \kappa\end{pmatrix} \equiv (\Delta X_{\{\varphi\}_\lambda})_\alpha \begin{pmatrix} l\\ \kappa\end{pmatrix} = \sum_\beta \varphi_{\alpha\beta}(\Delta X)_\beta \begin{pmatrix} l_{\bar\varphi}\\ \kappa_{\bar\varphi}\end{pmatrix}. \tag{104.8}$$

Then, the rotated Bloch vector will be

$$\begin{aligned}(\Delta X_{\{\varphi\}_\lambda})_\alpha(\kappa|\boldsymbol{k}) &= \frac{1}{\sqrt{N}}\sum_L e^{i\boldsymbol{k}\cdot\boldsymbol{R}_L}(\Delta X_{\{\varphi\}_\lambda})_\alpha \begin{pmatrix} l\\ \kappa\end{pmatrix}\\ &= \frac{1}{\sqrt{N}}\sum_\beta(\varphi_\lambda)_{\alpha\beta}\sum_L e^{i\boldsymbol{k}\cdot\boldsymbol{R}_L}(\Delta X)_\beta \begin{pmatrix} l_{\bar\varphi}\\ \kappa_{\bar\varphi}\end{pmatrix}.\end{aligned} \tag{104.9}$$

Now let us change the lattice summation variable to be a sum over

$$\boldsymbol{R}_{L_{\bar\varphi_\lambda}} \equiv \boldsymbol{\varphi}_\lambda^{-1}\cdot\boldsymbol{R}_L - \boldsymbol{\varphi}_\lambda^{-1}\cdot\boldsymbol{t}_\lambda \tag{104.10}$$

so then

$$\boldsymbol{R}_L = \boldsymbol{\varphi}_\lambda\cdot\boldsymbol{R}_{L_{\bar\varphi_\lambda}} + \boldsymbol{t}_\lambda \tag{104.11}$$

$$\begin{aligned}(\Delta X_{\{\varphi_\lambda\}})_\alpha(\kappa|\boldsymbol{k}) &= \frac{1}{\sqrt{N}}\sum_\beta(\varphi_\lambda)_{\alpha\beta}\sum_{L_{\bar\varphi}} e^{i(\boldsymbol{k}\cdot\boldsymbol{\varphi}_\lambda\cdot\boldsymbol{R}_{L_{\bar\varphi_\lambda}}+\boldsymbol{k}\cdot\boldsymbol{t}_\lambda)}(\Delta X)_\beta \begin{pmatrix} l_{\bar\varphi_\lambda}\\ \kappa_{\bar\varphi_\lambda}\end{pmatrix}\\ &= e^{i\boldsymbol{k}\cdot\boldsymbol{t}_\lambda}\sum_\beta(\varphi_\lambda)_{\alpha\beta}(\Delta X)_\beta(\kappa_{\bar\varphi_\lambda}|\boldsymbol{k}).\end{aligned} \tag{104.12}$$

In obtaining (104.11) we use the property of $\{\boldsymbol{\varphi}_\lambda|\boldsymbol{t}_\lambda\}$ being in $\mathfrak{G}(\boldsymbol{k})/\mathfrak{T}(\boldsymbol{k})$. Owing to the property of $\{\boldsymbol{\varphi}_\lambda|\boldsymbol{t}_\lambda\}$ as a symmetry operation of the crystal, the basis index $\kappa_{\bar\varphi_\lambda}$ refers to one of the r basis particles.

If κ and $\kappa_{\bar{\varphi}_\lambda}$ are identical basis particles (i.e. belong to the same translational Bravais lattice in case the crystal consists of several equivalent interpenetrating Bravais lattices of the same chemical species) then $(\Delta X)_\beta(\kappa_{\bar{\varphi}_\lambda}|\boldsymbol{k})=(\Delta X)_\beta(\kappa|\boldsymbol{k})$. Hence in (104.12) there will be a non-vanishing contribution to the character of the representation based upon the $3r$ dimensional vector space (104.3). From (104.11) it is clear that we can immediately obtain the character (trace) of the representation $D^{(\boldsymbol{k})(\Delta)}$. It is, for the operation $\{\varphi_\lambda|\boldsymbol{t}_\lambda\}$:

$$\operatorname{Tr} D^{(\boldsymbol{k})(\Delta)}(\{\varphi_\lambda|\boldsymbol{t}_\lambda\})=\pm e^{i\boldsymbol{k}\cdot\boldsymbol{t}_\lambda}(1+2\cos\varphi_\lambda)\sum_{\kappa=1}^{r}\delta_{\kappa,\kappa_{\bar{\varphi}_\lambda}}. \tag{104.13}$$

Here

$$\delta_{\kappa,\kappa_{\bar{\varphi}_\lambda}}=\begin{cases}1 & \text{if } \kappa \text{ and } \kappa_{\bar{\varphi}_\lambda} \text{ are the same basis}\\ 0 & \text{otherwise}\end{cases} \tag{104.14}$$

The sum gives the number of basis particles left unchanged, or sent into particles of the same basis. The other factors in (104.13) arise from the translational phase (104.12), and the trace of the rotation matrix. Using (104.13) for each of the elements, or coset representatives in $\mathfrak{G}(\boldsymbol{k})/\mathfrak{T}(\boldsymbol{k})$ we obtain the complete character system $\chi^{(\boldsymbol{k})(\Delta)}$. This character system is equal to (103.4) (i.e. to (82.21)) which is not surprising since an entirely similar argument produced both. However the basis space $\Sigma^{(\boldsymbol{k})(\Delta)}$ is far more convenient to deal with at this stage in obtaining symmetry coordinates.

Recall that in (82.19) it was shown that $D^{(\boldsymbol{k})(e)}$ is a direct product; so therefore is $D^{(\boldsymbol{k})(\Delta)}$:

$$D^{(\boldsymbol{k})(\Delta)}=D^{(r)}\otimes D^{(\mathrm{perm})}. \tag{104.15}$$

It is to be noted also that $D^{(\mathrm{perm})}$ is a generalized permutation matrix with elements $D^{(\boldsymbol{k})}(\{\varepsilon|-\boldsymbol{R}_N\})$ in a row or column, thus establishing the connection with the wave vector $\boldsymbol{k}$. To obtain the correct linear combinations of the basic cartesian displacements in $\Sigma^{(\boldsymbol{k})(\Delta)}$ we need to reduce $D^{(\boldsymbol{k})(\Delta)}$ by finding the unitary matrix U so that:

$$U^{-1}D^{(\boldsymbol{k})(\Delta)}U=\bar{D}^{(\boldsymbol{k})(\Delta)}. \tag{104.16}$$

Here $\bar{D}^{(\boldsymbol{k})(\Delta)}$ is in fully reduced form:

$$\bar{D}^{(\boldsymbol{k})(\Delta)}=D^{(\boldsymbol{k})(j)}\oplus\cdots\oplus D^{(\boldsymbol{k})(j')}, \tag{104.17}$$

where (104.17) gives all the irreducible components of $\mathfrak{G}(\boldsymbol{k})$. Then the reduced spaces corresponding to (104.17) are given by

$$U^{-1}\Sigma^{(\boldsymbol{k})(\Delta)}=\Sigma^{(\boldsymbol{k})(j)}\oplus\cdots\oplus\Sigma^{(\boldsymbol{k})(j')}. \tag{104.18}$$

The essential point is the determination of U.

Now recall that from (103.2) and (103.3) we know *which* types of irreducible representations $D^{(\boldsymbol{k})(j)}$ appear in the reduction (104.17). For each type j of irreducible representation such that

$$c_j=1 \tag{104.19}$$

we may directly find the correct linear combination by applying the projection operator for the j-th allowable irreducible representation of $\mathfrak{G}(\boldsymbol{k})$. This can be (16.3), or (16.5) where the group elements are in $\mathfrak{G}(\boldsymbol{k})$. Calling this operator $P^{(\boldsymbol{k})(j)}$ we have:

$$P^{(\boldsymbol{k})(j)}\Sigma^{(\boldsymbol{k})(\Delta)}\to\Sigma^{(\boldsymbol{k})(j)}. \tag{104.20}$$

The projection operator $P^{(\boldsymbol{k})(j)}$ is of course known from the entire group theory analysis preceding.

If $c_j > 1$ then application of the projection operator $P^{(\boldsymbol{k})(j)}$ to various of the bases in $\Sigma^{(\boldsymbol{k})(\Delta)}$ (i.e. to various of the Cartesian components of unit atom displacements) will give *several* linearly independent spaces of identical symmetry $\Sigma^{(\boldsymbol{k})(j)}$. The correct linear combinations of these

$$\alpha_1 \Sigma^{(\boldsymbol{k})(j)} + \alpha_2 \Sigma^{(\boldsymbol{k})(j)} + \cdots \tag{104.21}$$

can only be determined by solution of the actual dynamical equation, except in certain simple, but important cases. As an example of the latter is the case at $\boldsymbol{k} = \boldsymbol{\Gamma} = (000)$ if more than one mode of same symmetry as the acoustic modes arises. Then the acoustic mode ($\omega^2 = 0$) is clearly determined as the three vectors with the following components

$$\begin{aligned} \Sigma_X^{(\boldsymbol{\Gamma})(\Delta)} &= \{\Delta X(1|\boldsymbol{\Gamma}), 0, 0, \Delta X(2|\boldsymbol{\Gamma}), \ldots\} \\ &\vdots \\ \Sigma_Z^{(\boldsymbol{\Gamma})(\Delta)} &= \{0, 0, \Delta Z(1|\boldsymbol{\Gamma}), \ldots, \Delta Z(r|\boldsymbol{\Gamma})\} \end{aligned} \tag{104.22}$$

Another $\boldsymbol{\Gamma}$ mode of this symmetry with $\omega^2(\boldsymbol{\Gamma}|j) \neq 0$ may be completely determined by the projection operator, *and* the requirement of orthogonality amongst the different modes (as in Sect. 47).

An alternative useful procedure[1] to determine U simply takes advantage of the fact that $D^{(\text{perm})}$ is an $r \times r$ matrix so that we first reduce $D^{(\text{perm})}$ by finding a matrix S', so that

$$S'^{-1} D^{(\text{perm})} S' = \bar{D}^{(\text{perm})} \tag{104.23}$$

and then finding a matrix S'' so that

$$S''^{-1} D^{(\boldsymbol{r})} S'' = \bar{D}^{(\boldsymbol{r})} \tag{104.24}$$

where the barred matrices are fully reduced in $\mathfrak{G}(\boldsymbol{k})$, and then continue to obtain (104.16). This sequential reduction is quite convenient. Specifically, we may write

$$\bar{D}^{(\text{perm})} = D^{(\boldsymbol{k})(\bar{j})} \oplus \cdots \oplus D^{(\boldsymbol{k})(\bar{j}')} \tag{104.25}$$

in terms of irreducible components of $\mathfrak{G}(\boldsymbol{k})$, (again using knowledge of the $c_{\bar{j}} \neq 0$), and straightforwardly find the S' which effects this reduction. Since the vector representation $D^{(\boldsymbol{r})}$ is also trivially reduced in $\mathfrak{G}(\boldsymbol{k})$ carrying out the final reduction to $\bar{D}^{(\boldsymbol{k})(\Delta)}$ may be simply a matter of an easy additional step or perhaps of using the Clebsch-Gordan coefficients, where available.[2,3] See also the discussion of Clebsch-Gordan coefficients in Sects. 18, 60, 134 of this article.

[1] This approach was apparently first used in crystals by K. Huang, Z. Physik **171**, 213 (1963). See also B. Klein: Ph. D. thesis, New York University (1969). Available from University microfilms, Ann Arbor, Mich.; and B. Klein in [*45*]; and in [*12*].

[2] D.B. Litvin and J. Zak: J. Math. Phys. **9**, 212 (1968).

[3] I. Itzkan: Clebsch-Gordon coefficients for the crystallographic space groups. Ph. D. thesis, Department of Physics, New York University, June 1969. — R. Berenson: Theory of Crystal Clebsch-Gordan Coefficients, Ph. D. Thesis, Physics Department, New York University, February 1974. Available from University Microfilms, Ann. Arbor, Mich. — R. Berenson, I. Itzkan and J.L. Birman: J. Math. Phys. (to be published). — J.F. Cornwall: phys. stat. solidi **37**, 225 (1970).

Direct inspection is also often a feasible method of finding U. That is, since U is a matrix of dimension $3r \times 3r$, and since the c_j are known, as is the explicit form of the constituent $D^{(\boldsymbol{k})(j)}$, the task of finding the elements of U can often be solved by inspection.

It may be helpful to notice the actual form of $D^{(\boldsymbol{k})(\Delta)}$ in carrying out the reduction. That is, by judicious choice of the order in which basis atoms are written down, the form of $D^{(\boldsymbol{k})(\Delta)}$ can be chosen in a most partitioned aspect i.e. in an off diagonal bloc form which may suggest U.

Summarizing: the actual work of finding the correct linear combinations for the $D^{(\boldsymbol{k})(j)}$ is easier done then described, using either the projection operator method, or by finding U directly. This will be illustrated later when we analyse diamond and rock salt space groups in Part M of the article, and particularly in Sect. 138.

To complete the task of obtaining the correct linear combinations it is necessary to account for time reversal. But here we again call upon our knowledge of the type of corepresentation which contains each $D^{(\boldsymbol{k})(j)}$ that occurs (i.e. for which $c_j \neq 0$ in the crystal at hand). For corepresentations *type A* where the subduced representation is $D^{(\boldsymbol{k})(j)}$ then the space $\Sigma^{(\boldsymbol{k})(j)}$ gives an irreducible corepresentation. For corepresentations of *types B and C*, the work of Sect. 98 shows that the space

$$\Sigma^{(\mathrm{co}\,\boldsymbol{k})(j)} \equiv \Sigma^{(\boldsymbol{k})(j)} \oplus P_{\boldsymbol{a}_0} \Sigma^{(\boldsymbol{k})(j)} \tag{104.26}$$

gives an irreducible corepresentation. It is simple to augment the eigenvectors comprising $\Sigma^{(\boldsymbol{k})(j)}$ by adjoining the transformed set obtained via application of $P_{\boldsymbol{a}_0}$. This will be illustrated later.

Concluding, we emphasize that the basic task is that of finding the set of symmetry coordinates for $\mathscr{G}(\boldsymbol{k})$. If each $c_j = 1$ this can be completely carried through. If $c_j > 1$ the eigenvectors are obtained, up to a linear combination. In either case to find numerical eigenfrequencies, or the remaining parameters, solution of the dynamical equation is required. If all $c_j = 1$, then a knowledge of all eigenvectors permits construction of the matrix $\boldsymbol{E}(\boldsymbol{k})$ of (79.14) and then complete factorization of $[\boldsymbol{D}(\boldsymbol{k})]$ as in (79.16) and (79.17). This is often useful in testing the effects of changes of force constants upon eigenvalues since one can obtain an explicit equation determining $\omega^2(\boldsymbol{k}|j)$ in this case from (79.16), (79.17), and (85.12), (85.13).

Often in other work (such as in the theory of electron–lattice interaction) it is necessary to have expressions in a given crystal for the "bare" eigenvectors of the lattice dynamic problem. We see that the use of projection operators suffices if $c_j = 1$, otherwise solution of a dynamical problem is needed to give them.

In cases of some remaining ambiguity of symmetry assignment to branches e.g. if an accurate force constant model is not known we shall later show that this may be resolved by analysis of multiphonon optical spectra due to the infrared or Raman effects. That is, using group theoretical methods certain alternatives can be posed, so that an observation of an optically allowed process in the phonon Raman spectrum may lead to determination of the energy of the single phonon of that symmetry and thus the resolution of question of assignment of phonons to particular branchs.

In this manner, using the unit cartesian displacements to generate the auxiliary vector space $\Sigma^{(\boldsymbol{k})(\Delta)}$ which is easily reduced, the determination of normal mode symmetries is accomplished. Clearly this determination is advisable whether or not actual numerical eigenvectors $\boldsymbol{e}\left(\left|\begin{matrix}\boldsymbol{k}\\ j_\mu\end{matrix}\right.\right)$ are obtained by solution of a dynamical problem. On the other hand the group theory analysis cannot give the relative ordering of the different branches in increasing energy, for that only a numerical analysis suffices. In addition, if for particular $\boldsymbol{k}$, $c_j > 1$ (i.e. 2 or more) there will be *a priori* ambiguities in assigning mode symmetry to branches. Again this is a reflection of the need to supplement the purely group theory analysis by solution of the dynamical equations.

J. Applications of results on symmetry adapted eigenvectors in classical lattice dynamics

105. Introduction. There are many ways in which the results of the last part can be used. Most obviously, the symmetry classification of eigenvectors is of use in itself. Besides this one can use the symmetry properties of the eigenvectors in developing a "tensor calculus". Just as in the more familiar quantum mechanical case, which will be illustrated below in Part L where one needs to evaluate matrix elements as integrals over products of functions, in classical lattice dynamics also an analogous situation exists, and quantities analogous to matrix elements arise, when an operator is contracted by means of eigenvectors. Such contraction is analogous to a scalar product and formulae analogous to the Wigner-Eckart formula are found. These permit the maximum use to be made of symmetry, especially if one possesses the relevant Clebsch-Gordan coefficients. As discussed in Sects. 18, 60, 134 the Clebsch-Gordan coefficients for space groups are only now beginning to become available but it can be expected that many will be calculated in the near future. The tensor calculus simplifies these type of calculations by demonstrating the occurrence of "reduced" matrix elements times factors entirely specified by symmetry as equal to the matrix element in question.

The tensor calculus is useful in determining critical points in the phonon distribution. This theory is a degenerate perturbation theory leading as usual in such cases to a secular equation. The basic matrix elements in that equation are determined by these symmetry considerations.

Many other applications of symmetry in the classical lattice dynamic problem exist and we touch on a few of them: calculation of coupling parameters (force constants) which is dealt with thoroughly in [*4*, *58*]; and calculation of other crystal invariants and covariants. Particularly the latter: electric moment and polarizibility are useful in our later treatment of infra-red absorption and Raman scattering in Sects. 123, 124.

106. Tensor calculus for lattice dynamics. We need to deal with objects analogous to the matrix elements of operators with eigenfunctions which arise in quantum mechanics. To deal with these quantities properly, we establish certain formulae analogous to the so-called Eckart-Wigner formulae [*1, 2, 24*].

In (79.8), (79.9), (79.10), (79.11) two scalar products involving eigenvectors $e\left(\begin{vmatrix}\boldsymbol{k}\\ j_\nu\end{vmatrix}\right)$ were defined. In the last few sections we have proved that these eigenvectors are bases for irreducible corepresentations of $\mathcal{G}$ which now permits further development of the theory. First we discuss the effect of unitary elements. So we first restrict attention to the fact that the eigenvectors span an irreducible representation of $\mathfrak{G}$.

α) Effect of unitary elements

In the following we shall examine the effect of unitary symmetry operators upon the matrix elements or the scalar products of objects whose transformation properties are specified under $\mathcal{G}$ or $\mathcal{G}(\boldsymbol{k})$ and thus certainly specified for $\mathfrak{G}$ and $\mathfrak{G}(\boldsymbol{k})$. The objective is to be able to determine whether the matrix element is non-zero; if it is non-zero then we shall determine the minimum number of independent basic or "reduced", matrix elements in terms of which the matrix element in question can be expressed [*25*].

We now consider any vector $\boldsymbol{X}$ in the $3r$ dimensional space (α, κ): such an object has $3r$ components labelled by letting $\alpha = 1, 2, 3$ and $\kappa = 1, \ldots, r$. Then if $\boldsymbol{X}$ and $\boldsymbol{Y}$ are two such vectors with components $X_\alpha(\kappa)$, $Y_\alpha(\kappa)$ we define the Hermitian scalar product

$$(\boldsymbol{X}, \boldsymbol{Y}) \equiv \sum_{\kappa, \alpha} X_\alpha(\kappa)^* \, Y_\alpha(\kappa) \equiv \boldsymbol{X} \cdot \boldsymbol{Y}. \tag{106.1}$$

In particular, consider

$$\left(e\left(\begin{vmatrix}\boldsymbol{k}\\ j_\nu\end{vmatrix}\right), e\left(\begin{vmatrix}\boldsymbol{k}'\\ j'_{\nu'}\end{vmatrix}\right)\right) \tag{106.2}$$

where

$$e\left(\begin{vmatrix}\boldsymbol{k}\\ j_\nu\end{vmatrix}\right) \text{ belongs to } D^{(\boldsymbol{k})(j)} \tag{106.3}$$

and is a basis function for the ν-th row of that irreducible representation. Consider the transformation of the scalar product (106.2) when the eigenvectors are transformed by a unitary operator $P_{\{\varphi_{l\lambda}|\boldsymbol{t}_{l\lambda}\}}$ in $\mathfrak{G}(\boldsymbol{k})$. Since a transformation by a unitary operator corresponds geometrically to a transformation to an equivalent coordinate system, the scalar product which is a number, must have the same value. First then, consider a pure translation:

$$\begin{aligned}&\left(P_{\{\varepsilon|\boldsymbol{R}_L\}}\, e\left(\begin{vmatrix}\boldsymbol{k}\\ j_\nu\end{vmatrix}\right), P_{\{\varepsilon|\boldsymbol{R}_L\}}\, e\left(\begin{vmatrix}\boldsymbol{k}'\\ j'_{\nu'}\end{vmatrix}\right)\right)\\ &\quad = D^{(\boldsymbol{k})}(\{\varepsilon|\boldsymbol{R}_L\})^* \, D^{(\boldsymbol{k}')}(\{\varepsilon|\boldsymbol{R}_L\}) \left(e\left(\begin{vmatrix}\boldsymbol{k}\\ j_\nu\end{vmatrix}\right), e\left(\begin{vmatrix}\boldsymbol{k}'\\ j'_{\nu'}\end{vmatrix}\right)\right).\end{aligned} \tag{106.4}$$

Now sum both sides of (106.4) over all elements in $\mathfrak{T}$. Then since the order of $\mathfrak{T}$ is N we have, using (24.8)

$$\begin{aligned} N \sum_L \left(e\left(\left|\begin{matrix}\boldsymbol{k}\\ j_\nu\end{matrix}\right.\right), e\left(\left|\begin{matrix}\boldsymbol{k}'\\ j'_{\nu'}\end{matrix}\right.\right)\right) &= \sum_L D^{(\boldsymbol{k})}(\{\varepsilon|\boldsymbol{R}_L\})^* D^{(\boldsymbol{k}')}(\{\varepsilon|\boldsymbol{R}_L\}) \left(e\left(\left|\begin{matrix}\boldsymbol{k}\\ j_\nu\end{matrix}\right.\right), e\left(\left|\begin{matrix}\boldsymbol{k}'\\ \eta'_{\nu'}\end{matrix}\right.\right)\right) \\ &= N\Delta(\boldsymbol{k}-\boldsymbol{k}') \left(e\left(\left|\begin{matrix}\boldsymbol{k}\\ j_\nu\end{matrix}\right.\right), e\left(\left|\begin{matrix}\boldsymbol{k}'\\ \eta'_{\nu'}\end{matrix}\right.\right)\right). \end{aligned} \tag{106.5}$$

Consequently only if:

$$\boldsymbol{k} \doteq \boldsymbol{k}' \quad \text{i.e.} \quad \boldsymbol{k} = \boldsymbol{k}' + 2\pi \boldsymbol{B}_H \tag{106.6}$$

will the scalar product (106.2) be non-zero. Let $\boldsymbol{k}=\boldsymbol{k}'$ then, and consider transformation of the eigenvectors by a rotational element in $\mathfrak{G}(\boldsymbol{k})$:

$$\left(e\left(\left|\begin{matrix}\boldsymbol{k}\\ j_\nu\end{matrix}\right.\right), e\left(\left|\begin{matrix}\boldsymbol{k}\\ j'_{\nu'}\end{matrix}\right.\right)\right) = \left(P_{\{\varphi_{l_\lambda}|\tau_{l_\lambda}\}}\, e\left(\left|\begin{matrix}\boldsymbol{k}\\ j_\nu\end{matrix}\right.\right), P_{\{\varphi_{l_\lambda}|\tau_{l_\lambda}\}}\, e\left(\left|\begin{matrix}\boldsymbol{k}\\ j'_{\nu'}\end{matrix}\right.\right)\right). \tag{106.7}$$

Use (83.6) and sum the result over all elements in $\mathfrak{G}(\boldsymbol{k})$ to get, after employing the orthonormality relations for irreducible representations of $\mathfrak{G}(\boldsymbol{k})$:

$$\begin{aligned} g(\boldsymbol{k}) \left(e\left(\left|\begin{matrix}\boldsymbol{k}\\ j_\nu\end{matrix}\right.\right), e\left(\left|\begin{matrix}\boldsymbol{k}\\ j'_{\nu'}\end{matrix}\right.\right)\right) &= \frac{g(\boldsymbol{k})}{l_j} \sum_\mu \sum_{\mu'} \delta_{jj'}\, \delta_{\mu\nu}\, \delta_{\mu'\nu'} \left(e\left(\left|\begin{matrix}\boldsymbol{k}\\ j_\mu\end{matrix}\right.\right), e\left(\left|\begin{matrix}\boldsymbol{k}\\ j'_{\mu'}\end{matrix}\right.\right)\right) \\ &= \frac{g(\boldsymbol{k})}{l_j}\, \delta_{jj'} \sum_\mu \left(e\left(\left|\begin{matrix}\boldsymbol{k}\\ j_\mu\end{matrix}\right.\right), e\left(\left|\begin{matrix}\boldsymbol{k}\\ j_\mu\end{matrix}\right.\right)\right) \delta_{\nu\nu'} \end{aligned}$$

or

$$l_j \left(e\left(\left|\begin{matrix}\boldsymbol{k}\\ j_\nu\end{matrix}\right.\right), e\left(\left|\begin{matrix}\boldsymbol{k}\\ j'_{\nu'}\end{matrix}\right.\right)\right) = \delta_{jj'}\, \delta_{\nu\nu'} \sum_\mu \left(e\left(\left|\begin{matrix}\boldsymbol{k}\\ j_\mu\end{matrix}\right.\right), e\left(\left|\begin{matrix}\boldsymbol{k}\\ j_\mu\end{matrix}\right.\right)\right). \tag{106.8}$$

From (106.8) it follows that the scalar product (106.7) vanishes unless $j=j'$, $\nu=\nu'$, and then it is independent of ν.

Assembling (106.8) and (106.5) we have in general

$$\left(e\left(\left|\begin{matrix}\boldsymbol{k}\\ j_\nu\end{matrix}\right.\right), e'\left(\left|\begin{matrix}\boldsymbol{k}'\\ j'_{\nu'}\end{matrix}\right.\right)\right) = \Delta(\boldsymbol{k}-\boldsymbol{k}')\, \delta_{jj'}\, \delta_{\nu\nu'}\, f(\boldsymbol{k}, j), \tag{106.9}$$

where

$$f(\boldsymbol{k}, j) \equiv \left(e\left(\left|\begin{matrix}\boldsymbol{k}\\ j_\mu\end{matrix}\right.\right), e'\left(\left|\begin{matrix}\boldsymbol{k}\\ j_\mu\end{matrix}\right.\right)\right) \tag{106.10}$$

is a reduced matrix element. Consistent with the usual normalization of the Hermitian product, or length of $e\left(\left|\begin{matrix}\boldsymbol{k}\\ j_\mu\end{matrix}\right.\right)$, as in (79.10) and (79.11), $f(\boldsymbol{k}, j)=1$ for the case discussed in (106.2)–(106.8). The general situation in which one takes a scalar product of two distinct eigenvectors such as in (106.9) is the analogue of the Wigner-Eckart matrix element formula. For such a case $f(\boldsymbol{k}, j)$ is the reduced matrix element which needs to be evaluated.

Consider the scalar product

$$\sum_{\alpha\kappa}\sum_{\beta\kappa'} e_\alpha\left(\kappa\left|\begin{matrix}\boldsymbol{k}\\ j_\nu\end{matrix}\right.\right)^* D_{\alpha\beta}\begin{pmatrix}\boldsymbol{k}\\ \kappa\,\kappa'\end{pmatrix} e_\beta\left(\kappa\left|\begin{matrix}\boldsymbol{k}\\ j'_{\nu'}\end{matrix}\right.\right) \tag{106.11}$$

which can be written as

$$\left(\boldsymbol{e}\left(\left|\begin{matrix}\boldsymbol{k}\\ j_\nu\end{matrix}\right.\right), \boldsymbol{D}(\boldsymbol{k})\,\boldsymbol{e}\left(\left|\begin{matrix}\boldsymbol{k}\\ j'_{\nu'}\end{matrix}\right.\right)\right) \tag{106.12}$$

since:

$$\left(\boldsymbol{D}(\boldsymbol{k})\,\boldsymbol{e}\left(\left|\begin{matrix}\boldsymbol{k}\\ j'_{\nu'}\end{matrix}\right.\right)\right)_{\alpha\kappa} \equiv \sum_{\beta\kappa'} D_{\alpha\beta}\begin{pmatrix}\boldsymbol{k}\\ \kappa\,\kappa'\end{pmatrix} e_\beta\left(\kappa'\left|\begin{matrix}\boldsymbol{k}\\ j'_{\nu'}\end{matrix}\right.\right). \tag{106.13}$$

When (106.12) is transformed by the operator $P_{\{\varphi_{l_\lambda}|\tau_{l_\lambda}\}}$ in $\mathfrak{G}(\boldsymbol{k})$, we use (81.33) and (106.9), (106.10) to obtain

$$\left(\boldsymbol{e}\left(\left|\begin{matrix}\boldsymbol{k}\\ j_\nu\end{matrix}\right.\right), D(\boldsymbol{k})\,\boldsymbol{e}\left(\left|\begin{matrix}\boldsymbol{k}\\ j'_{\nu'}\end{matrix}\right.\right)\right) = \delta_{jj'}\,\delta_{\nu\nu'}\,c(\boldsymbol{k};j). \tag{106.14}$$

But in this case, the reduced matrix element in (106.14) is, by the equation of motion (79.7):

$$c(\boldsymbol{k};j) = \omega^2(\boldsymbol{k}|j). \tag{106.15}$$

It is independent of ν, which is another way of stating (81.32), when $P_{\{\varphi_{l_\lambda}|\tau_{l_\lambda}\}}$ is in $\mathfrak{G}(\boldsymbol{k})$.

Clearly the type of result epitomized in (106.10), (106.14) only depends on taking the scalar product

$$\left(\boldsymbol{e}\left(\left|\begin{matrix}\boldsymbol{k}\\ j_\nu\end{matrix}\right.\right), \boldsymbol{O}\,\boldsymbol{e}\left(\left|\begin{matrix}\boldsymbol{k}\\ j'_{\nu'}\end{matrix}\right.\right)\right) \tag{106.16}$$

where $\boldsymbol{O}$ is an operator such as $[\boldsymbol{D}(\boldsymbol{k})]$ invariant under the elements of $\mathfrak{G}(\boldsymbol{k})$,

$$P_{\{\varphi_{l_\lambda}|\tau_{l_\lambda}\}}\,\boldsymbol{O}\,P^{-1}_{\{\varphi_{l_\lambda}|\tau_{l_\lambda}\}} = \boldsymbol{O}. \tag{106.17}$$

Expression (106.17) defines an invariant operator under the transformations in $\mathfrak{G}(\boldsymbol{k})$.

Generalizing (106.16), we may consider a covariant set of operators

$$\boldsymbol{O}\begin{pmatrix}\boldsymbol{k}''\\ m_\mu\end{pmatrix}, \qquad \mu = 1, \ldots, l_m \tag{106.18}$$

which belong to the μ-th row of irreducible representation $D^{(\boldsymbol{k}'')(m)}$ of $\mathfrak{G}(\boldsymbol{k})$:

$$P_{\{\varepsilon|\boldsymbol{R}_L\}}\,\boldsymbol{O}\begin{pmatrix}\boldsymbol{k}''\\ m_\mu\end{pmatrix} = D^{(\boldsymbol{k}'')}(\{\varepsilon|\boldsymbol{R}_L\})\,\boldsymbol{O}\begin{pmatrix}\boldsymbol{k}''\\ m_{\mu'}\end{pmatrix}$$

and

$$P_{\{\varphi_{l'_\lambda}|\tau_{l'_\lambda}\}}\,\boldsymbol{O}\begin{pmatrix}\boldsymbol{k}''\\ m_\mu\end{pmatrix} P^{-1}_{\{\varphi_{l'_\lambda}|\tau_{l'_\lambda}\}} = \sum_{\mu'} D^{(\boldsymbol{k}'')(m)}(\{\varphi_{l_\lambda}|\tau_{l_\lambda}\})_{\mu'\mu}\,\boldsymbol{O}\begin{pmatrix}\boldsymbol{k}''\\ m_{\mu'}\end{pmatrix}. \tag{106.19}$$

The set of operators (106.18) have been prescribed in respect to transformation under $\mathfrak{G}(\boldsymbol{k})$. As with any basis, the set (106.18) must in principle be considered a part of the entire set of operators in $\mathfrak{G}$:

$$O\begin{pmatrix}\boldsymbol{k}''\\ m_\mu\end{pmatrix},\ldots,O\begin{pmatrix}\boldsymbol{k}''_{\sigma''}\\ m_\mu\end{pmatrix},\ldots,O\begin{pmatrix}\boldsymbol{k}''_{s''}\\ m_\mu\end{pmatrix},\qquad \mu=1,\ldots,l_m$$

where the wave vectors range over *all* the inequivalent wave vectors of $^\star\boldsymbol{k}''$. In the immediately following analysis we shall only work in the subgroup framework, so that we restrict ourselves to the set (106.18). At this point, the reader may be advised to review the analysis of reduction coefficients given in Sect. 17, and Sects. 57–59. With this limitation to (106.18) the scalar product

$$\left(e\begin{pmatrix}\boldsymbol{k}\\ j_\nu\end{pmatrix}, O\begin{pmatrix}\boldsymbol{k}''\\ m_\mu\end{pmatrix} e\begin{pmatrix}\boldsymbol{k}'\\ j'_{\nu'}\end{pmatrix}\right) \tag{106.20}$$

can be analysed similarly to (106.2) and (106.12). First of all, it is clear that the matrix element will be non-zero, if, and only if, the corresponding little group reduction coefficient

$$(\boldsymbol{k}\,j\,|\,\boldsymbol{k}''\,m\,\boldsymbol{k}'j')\neq 0. \tag{106.21}$$

But two cases need to be distinguished for (106.21) depending on whether the non-zero coefficient equals 1 or an integer greater than 1.

When the reduction coefficient equals 1, *and* the wave vectors are such that $\mathfrak{G}(\boldsymbol{k}'')=\mathfrak{G}(\boldsymbol{k}')=\mathfrak{G}(\boldsymbol{k})$ a simple procedure suffices. This case means that all rotational elements φ_{l_λ} are common to the three wave vectors, and that the product of representations

$$D^{(\boldsymbol{k}'')(m)}\otimes D^{(\boldsymbol{k}')(j')} \tag{106.22}$$

contains the representation $D^{(\boldsymbol{k})(j)}$ only once. Then a single linear vector space $\Sigma^{(\boldsymbol{k})(j)}$ can be constructed, of dimension l_j with l_j independent basis vectors:

$$\Sigma^{(\boldsymbol{k})(j)}\equiv\{\theta_1^{(\boldsymbol{k})(j)},\ldots,\theta_{l_j}^{(\boldsymbol{k})(j)}\} \tag{106.23}$$

with $\theta_\nu^{(\boldsymbol{k})(j)}$ a symmetrized object going with ν-th row of representation $D^{(\boldsymbol{k})(j)}$

$$\theta_\nu^{(\boldsymbol{k})(j)}\equiv\sum_\mu\sum_{\nu'} U_{\boldsymbol{k}''m\mu,\,\boldsymbol{k}'j'\nu';\,\boldsymbol{k}j\nu}\, O\begin{pmatrix}\boldsymbol{k}''\\ m_\mu\end{pmatrix} e\left(\begin{vmatrix}\boldsymbol{k}'\\ j'_{\nu'}\end{vmatrix}\right). \tag{106.24}$$

The l_j objects $\theta_\nu^{(\boldsymbol{k})(j)}$ $(\nu=1,\ldots,l_j)$ may then be treated just as any basis. The matrix elements $U_{\boldsymbol{k}''m\mu,\,\boldsymbol{k}'j'\nu';\,\boldsymbol{k}j\nu}$ are the elements of the Clebsch-Gordon matrix reducing the direct product (106.22) into a direct sum, of which the representation of interest is $D^{(\boldsymbol{k})(j)}$. In the present case, by assumption only a single reduced matrix element is non-zero. Carrying over the analysis which gave us (106.9) we have

$$\left(e\left(\begin{vmatrix}\boldsymbol{k}\\ j_\nu\end{vmatrix}\right), \theta_\nu^{(\boldsymbol{k})(j)}\right)=f'(\boldsymbol{k},j), \tag{106.25}$$

where (106.25) gives an expression for the single non-zero matrix element. To evaluate the scalar product (106.25) we need an expression for $(\theta_\nu^{(\boldsymbol{k})(j)})_{\alpha\kappa}$. This can be found using an expression for the components

$$\left(O\begin{pmatrix}\boldsymbol{k}''\\ m_\mu\end{pmatrix} e\left(\begin{vmatrix}\boldsymbol{k}'\\ j'_{\nu'}\end{vmatrix}\right)\right)_{\alpha\kappa} = \sum_{\beta\kappa'} O_{\alpha\kappa,\beta\kappa'}\begin{pmatrix}\boldsymbol{k}''\\ m_\mu\end{pmatrix} e_\beta\left(\kappa'\begin{vmatrix}\boldsymbol{k}'\\ j'_{\nu'}\end{vmatrix}\right) \tag{106.26}$$

and then substituting (106.26) into (106.25) and using (79.10). This clearly requires that the operator $O\begin{pmatrix}\boldsymbol{k}\\ j_\nu\end{pmatrix}$ shall be expressible in the $3r$ dimensional space of the components of displacement vectors such as $e\left(\begin{vmatrix}\boldsymbol{k}'\\ j'_{\nu'}\end{vmatrix}\right)$. In the cases most often dealt with, such a decomposition into components labelled by (α, κ), $\alpha = 1, \ldots, 3$; $\kappa = 1, \ldots, r$ is possible. Otherwise it would be necessary to use the decomposition into the components labelled by m_μ of which there are also $3r$, and then employ the double dot scalar product of (79.9), (79.11).

For the case when the reduction coefficient (106.21) is equal to 2 or more there is more than one non-zero "reduced matrix element". Let us understand this as follows: Consider the *set* of $(l_\mu \cdot l_{\nu'})$ objects

$$O\begin{pmatrix}\boldsymbol{k}''\\ m_\mu\end{pmatrix} e\left(\begin{vmatrix}\boldsymbol{k}'\\ j'_{\nu'}\end{vmatrix}\right)_{\mu=1,\ldots,l_\mu;\,\nu'=1,\ldots,l_{\nu'}}. \tag{106.27}$$

According to hypothesis, the set of $(l_\mu \cdot l_{\nu'})$ objects contains functions which are a basis for as many linearly independent spaces $\Sigma^{(\boldsymbol{k})(j)}$ as the numerical value of the coefficient (106.21). For example, suppose the value of (106.21) were 2. Then there are two distinct linear combinations of the functions (106.27) which are bases for *each* row ν in the representation $D^{(\boldsymbol{k})(j)}$. To find these, an expeditious procedure is again to construct the projection operator

$$P_{\nu\nu}^{(\boldsymbol{k})(j)} \tag{106.28}$$

which will project out of any arbitrary function that part going with the ν-th row of $D^{(\boldsymbol{k})(j)}$. The construction of operators like (106.28) was discussed in Sect. 16, and explicitly given in (16.3). Recall now that we do indeed possess all the necessary ingredients to construct (106.21) by the two methods of Sect. 16: either the strictly algebraic method depending only on the group structure, or by the use of the full representation matrices $D^{(\boldsymbol{k})(j)}$ of $\mathfrak{G}(\boldsymbol{k})$. Applying (106.28) to basis (106.27) we obtain

$$P_{\nu\nu}^{(\boldsymbol{k})(j)} O\begin{pmatrix}\boldsymbol{k}''\\ m_\mu\end{pmatrix} e\left(\begin{vmatrix}\boldsymbol{k}'\\ j'_{\nu'}\end{vmatrix}\right). \tag{106.29}$$

Permitting μ and ν' to range through their permitted values we obtain the two linearly independent distinct functions

$$\theta_\nu^{(\boldsymbol{k})(j)},\ \theta_\nu'^{(\boldsymbol{k})(j)}. \tag{106.30}$$

Each of these will then give a linearly independent reduced matrix element when the scalar product is taken with $e\left(\begin{vmatrix}\boldsymbol{k}\\ j_\nu\end{vmatrix}\right)$: namely

$$\left(e\left(\begin{vmatrix}\boldsymbol{k}\\ j_\nu\end{vmatrix}\right), \theta_\nu^{(\boldsymbol{k})(j)}\right) \equiv c(\boldsymbol{k};j) \tag{106.31}$$

and

$$\left(e\left(\begin{vmatrix}\boldsymbol{k}\\ j_\nu\end{vmatrix}\right), \theta'^{(\boldsymbol{k})(j)}_\nu\right) \equiv c'(\boldsymbol{k}; j). \tag{106.32}$$

The essential point here is that the specific linearly independent basis functions are completely determined e.g. in (106.30). *Mutatis mutandis* cases where (106.21) is greater than 2 can be dealt with.

Now let us consider the case where all rotations φ_{l_λ} are *not* simultaneously in the three groups of the wave vector. To be specific, if the given $\boldsymbol{k}$ satisfies

$$\boldsymbol{k} = \boldsymbol{k}'' + \boldsymbol{k}' + 2\pi \boldsymbol{B}_H$$

then for operations in $\mathfrak{G}(\boldsymbol{k})$ but not in $\mathfrak{G}(\boldsymbol{k}'')$ and $\mathfrak{G}(\boldsymbol{k}')$ we have

$$\varphi_{l_\lambda} \cdot \boldsymbol{k} = \varphi_{l_\lambda} \cdot \boldsymbol{k}'' + \varphi_{l_\lambda} \cdot \boldsymbol{k}' + 2\pi \boldsymbol{B}'_H \tag{106.33}$$

or

$$\boldsymbol{k} = \boldsymbol{k}''_{l_\lambda} + \boldsymbol{k}'_{l_\lambda} + 2\pi \boldsymbol{B}'_H. \tag{106.34}$$

Then to obtain the *complete* basis function, we must apply the projection operator as follows, for given μ and ν':

$$P^{(\boldsymbol{k})(j)}_{\nu\nu}\, O\begin{pmatrix}\boldsymbol{k}''_{l_\lambda}\\ m_\mu\end{pmatrix} e\left(\begin{vmatrix}\boldsymbol{k}'_{l_\lambda}\\ j'_{\nu'}\end{vmatrix}\right), \qquad \lambda = 1 \ldots \tag{106.35}$$

for *all* λ such that φ_{l_λ} is in $\mathfrak{P}(\boldsymbol{k})$ but not in $\mathfrak{P}(\boldsymbol{k}'')$ and $\mathfrak{P}(\boldsymbol{k}')$. Then assembling together all functions from (106.35) with fixed μ and ν' we obtain the bases to be used in computing reduced matrix elements:

$$\theta^{(\boldsymbol{k})(j)}_\nu = \left\{P^{(\boldsymbol{k})(j)}_{\nu\nu}\, O\begin{pmatrix}\boldsymbol{k}''_{l_\lambda}\\ m_\mu\end{pmatrix} e\left(\begin{vmatrix}\boldsymbol{k}'_{l_\lambda}\\ j'_{\nu'}\end{vmatrix}\right)\right\} \tag{106.36}$$

$$\mu, \nu' \text{ fixed}; \ \lambda = 1 \ldots$$

$$\theta'^{(\boldsymbol{k})(j)}_\nu = \left\{P^{(\boldsymbol{k})(j)}_{\nu\nu}\, O\begin{pmatrix}\boldsymbol{k}''_{l_\lambda}\\ m_{\bar\mu}\end{pmatrix} e\left(\begin{vmatrix}\boldsymbol{k}'_{l_\lambda}\\ j'_{\bar\nu'}\end{vmatrix}\right)\right\} \tag{106.37}$$

$$\vdots$$

Clearly the problem we encounter here relates to the completeness of the subgroup reduction methods. The reader is advised to recall the discussion and comparison of fullgroup and subgroup methods and to verify the completeness of any basis function obtained in this fashion when attempting an actual computation. Verification may actually require redoing the entire problem in the fullgroup framework. This would require starting with the complete set of functions belonging to the complete space group representations:

$$O\begin{pmatrix}\boldsymbol{k}''\\ m_\mu\end{pmatrix} e\left(\begin{vmatrix}\boldsymbol{k}'\\ j'_{\nu'}\end{vmatrix}\right) \tag{106.38}$$

$$\boldsymbol{k}'' = \boldsymbol{k}''_1, \ldots, \boldsymbol{k}''_{s''}; \qquad \mu = 1, \ldots, l_m$$

$$\boldsymbol{k}' = \boldsymbol{k}'_1, \ldots, \boldsymbol{k}'_{s'}; \qquad \nu' = 1, \ldots, l_j,$$

and then applying projection operator to the set of all these, to obtain the set of functions transforming like $\Sigma^{(\star\boldsymbol{k})(m)}$. As discussed in Sect. 64 it is essential that one deal here with a vector space which is complete, and this must be carefully verified for (106.35)–(106.37).

Alternative, equivalent methods of treating the problem of "reduced matrix elements" have been given[1] based on the determination of the vector coupling coefficients: i.e. the matrix which reduces the direct product into a direct sum. At the time of writing, there seems no decisive practical advantage to doing the theory in one or another fashion; the subject is still evolving and one can anticipate further developments and clarification.

Summarizing: in this work we considered the effect of purely unitary operators from $\mathfrak{G}(\boldsymbol{k})$ upon the basis functions and thus the matrix elements of type (106.20). If we are to find value of the *specific* matrix element (106.20) with indices (ν, μ, ν') fixed, knowing that (106.21) is satisfied, we proceed by finding the correct number of independent linear combinations as in (106.24), or (106.30), or (106.36), (106.37) depending on the applicable case. Then, inverting, we shall obtain from (106.24)

$$O\begin{pmatrix}\boldsymbol{k}''\\ m_\mu\end{pmatrix} e\left(\left|\begin{matrix}\boldsymbol{k}'\\ j'_{\nu'}\end{matrix}\right.\right)=\theta_\nu^{(k)(j)}/U_{\boldsymbol{k}''m_\mu,\boldsymbol{k}'j'_{\nu'};\boldsymbol{k}j_\nu} \tag{106.39}$$

and then the scalar product (106.20) is obtained, in terms of the basic reduced matrix element (106.25) with correct constant factor as in (106.39). The same applies in case the projection operator is used to give the correct linear combination, thereby essentially determining the matrix elements of U.

In either case, the actual matrix element will be determined in terms of the number of non-vanishing reduced matrix elements, whose number is given by the value of the reduction coefficient (106.21).

β) *Effect of antiunitary elements*

Now we consider how to determine the effect of antiunitary symmetry operations upon the matrix elements. Again, the essential question is how to express a given matrix element in terms of the least number of independent or reduced matrix elements. In what follows it can already be assumed that the restrictions due to the presence of unitary elements have already been accounted for. Hence we require the *additional* restrictions due to the presence of both antiunitary and unitary elements.

We consider first the case where antiunitary elements are present in $\mathscr{G}(\boldsymbol{k})$: i.e. we take class I, and II type costars. For the antiunitary elements in $\mathscr{G}(\boldsymbol{k})$, the basic equation (106.7) on the invariance of the scalar product is no longer true. It is replaced by the result that the scalar product of eigenvectors transformed by antiunitary elements is the complex conjugate of the original scalar product. Thus

$$\left(e\left(\left|\begin{matrix}\boldsymbol{k}\\ j_\nu\end{matrix}\right.\right), e\left(\left|\begin{matrix}\boldsymbol{k}\\ j'_{\nu'}\end{matrix}\right.\right)\right)=\left(P_a\, e\left(\left|\begin{matrix}\boldsymbol{k}\\ j_\nu\end{matrix}\right.\right), P_a\, e\left(\left|\begin{matrix}\boldsymbol{k}\\ j'_{\nu'}\end{matrix}\right.\right)\right)^* \tag{106.40}$$

where P_a is an antiunitary operator in $\mathscr{G}(\boldsymbol{k})$. Then from the analysis in Sects. 101 and 102 we get for the second term in (106.40)

[1] D. LITVIN and J. ZAK: J. Math. Phys. **9**, 212 (1968).

$$\left(\sum_{\bar{\nu}} D^{(\mathrm{co}\,\boldsymbol{k})(j)}(a)_{\bar{\nu}\nu}\, \boldsymbol{e}\left(\left|\begin{matrix}\boldsymbol{k}\\ j_{\bar{\nu}}\end{matrix}\right.\right),\ \sum_{\bar{\nu}'} D^{(\mathrm{co}\,\boldsymbol{k})(j')}(a)_{\bar{\nu}'\nu'}\, \boldsymbol{e}\left(\left|\begin{matrix}\boldsymbol{k}\\ j'_{\bar{\nu}'}\end{matrix}\right.\right)\right)^{*}$$
$$=\sum_{\bar{\nu}}\sum_{\bar{\nu}'} D^{(\mathrm{co}\,\boldsymbol{k})(j)}(a)_{\bar{\nu}\nu}\, D^{(\mathrm{co}\,\boldsymbol{k})(j')}(a)^{*}_{\bar{\nu}'\nu'}\left(\boldsymbol{e}\left(\left|\begin{matrix}\boldsymbol{k}\\ j_{\bar{\nu}}\end{matrix}\right.\right),\boldsymbol{e}\left(\left|\begin{matrix}\boldsymbol{k}\\ j'_{\bar{\nu}'}\end{matrix}\right.\right)\right)^{*} \tag{106.41}$$

since the eigenvector $\boldsymbol{e}\left(\left|\begin{matrix}\boldsymbol{k}\\ j_{\nu}\end{matrix}\right.\right)$ is now to be considered belonging to the ν-th row of the corepresentation $D^{(\mathrm{co}\,\boldsymbol{k})(j)}$. At this point in the analysis one may simply retrace all steps leading to the results such as obtained above in the purely unitary case. This is an extremely tedious procedure and rather than do this we shall establish some results for simple scalar products, leaving the remainder of the analysis to be found in the literature.[2]

In considering even the simple scalar product (106.40) and (106.41) several cases need to be distinguished for class I and II wave vectors. Thus consider the case where the eigenvectors belong to the irreducible representation $D^{(\boldsymbol{k})(m)}$ of $\mathfrak{G}(\boldsymbol{k})$, and the corepresentation is of case A, with matrices given by (98.57) and (98.58). Then (106.41) can be simplified by use of

$$D^{(\mathrm{co}\,\boldsymbol{k})(j)}(a)_{\bar{\nu}\nu}=\sum_{\mu} D^{(\boldsymbol{k})(j)}(u)_{\bar{\nu}\mu}\,\beta_{\mu\nu} \tag{106.42}$$

and

$$D^{(\mathrm{co}\,\boldsymbol{k})(j)}(a)^{*}_{\bar{\nu}'\nu'}=\sum_{\mu'} D^{(\boldsymbol{k})(j)}(u)^{*}_{\bar{\nu}'\mu'}\,\beta^{*}_{\mu'\nu'} \tag{106.43}$$

we substitute (106.42) and (106.43) into (106.41) and sum over the *unitary* elements in $\mathscr{G}(\boldsymbol{k})$ using the orthonormality relations to obtain

$$g(\boldsymbol{k})\left(\boldsymbol{e}\left(\left|\begin{matrix}\boldsymbol{k}\\ j_{\nu}\end{matrix}\right.\right),\boldsymbol{e}\left(\left|\begin{matrix}\boldsymbol{k}\\ j'_{\nu'}\end{matrix}\right.\right)\right)=\frac{g(\boldsymbol{k})}{l_j}\sum_{\bar{\nu}}\sum_{\bar{\nu}'}\sum_{\mu}\sum_{\mu'}\delta_{\bar{\nu}\bar{\nu}'}\,\delta_{\mu\mu'}\,\delta_{jj'}\,\beta_{\mu\nu}\,\beta^{*}_{\mu'\nu'}$$
$$\left(\boldsymbol{e}\left(\left|\begin{matrix}\boldsymbol{k}\\ j_{\bar{\nu}}\end{matrix}\right.\right),\boldsymbol{e}\left(\left|\begin{matrix}\boldsymbol{k}\\ j'_{\bar{\nu}'}\end{matrix}\right.\right)\right)^{*}=\left(\frac{g(\boldsymbol{k})}{l_j}\right)\sum_{\bar{\nu}}\sum_{\mu}\delta_{jj'}\,\beta_{\mu\nu}\,\beta^{*}_{\mu\nu'}\left(\boldsymbol{e}\left(\left|\begin{matrix}\boldsymbol{k}\\ j_{\bar{\nu}}\end{matrix}\right.\right),\boldsymbol{e}\left(\left|\begin{matrix}\boldsymbol{k}\\ j'_{\bar{\nu}}\end{matrix}\right.\right)\right)^{*}. \tag{106.44}$$

But β is a unitary matrix so

$$\beta^{*}_{\mu\nu'}=(\beta^{-1})_{\nu'\mu} \tag{106.45}$$

and then (106.44) becomes

$$=\left(\frac{g(\boldsymbol{k})}{l_j}\right)\sum_{\bar{\nu}}\delta_{jj'}\,\delta_{\nu'\nu}\left(\boldsymbol{e}\left(\left|\begin{matrix}\boldsymbol{k}\\ j_{\bar{\nu}}\end{matrix}\right.\right),\boldsymbol{e}\left(\left|\begin{matrix}\boldsymbol{k}\\ j_{\bar{\nu}}\end{matrix}\right.\right)\right)^{*}, \tag{106.46}$$

$$=\frac{g(\boldsymbol{k})}{l_j}\,\delta_{jj'}\,\delta_{\nu'\nu'}\sum_{\bar{\nu}}\left(\boldsymbol{e}\left(\left|\begin{matrix}\boldsymbol{k}\\ j_{\bar{\nu}}\end{matrix}\right.\right),\boldsymbol{e}\left(\left|\begin{matrix}\boldsymbol{k}\\ j_{\bar{\nu}}\end{matrix}\right.\right)\right). \tag{106.47}$$

Consequently (106.47) produces the result

$$\left(\boldsymbol{e}\left(\left|\begin{matrix}\boldsymbol{k}\\ j_{\nu}\end{matrix}\right.\right),\boldsymbol{e}\left(\left|\begin{matrix}\boldsymbol{k}\\ j_{\nu}\end{matrix}\right.\right)\right)=\left(\boldsymbol{e}\left(\left|\begin{matrix}\boldsymbol{k}\\ j_{\nu}\end{matrix}\right.\right),\boldsymbol{e}\left(\left|\begin{matrix}\boldsymbol{k}\\ j_{\nu}\end{matrix}\right.\right)\right)^{*}\quad\{\text{independent of } \nu\}. \tag{106.48}$$

[2] A. Aviran and J. Zak: J. Math. Phys. 9, 2138 (1968).

Hence the scalar product is real, and non-vanishing, only for the functions from same row of the irreducible representation in this case. Thus in case A the presence of antiunitary symmetry restricts the scalar product to be real; this is an added restriction on $f(\boldsymbol{k},j)$ beyond (106.9). A certain amount of the work leading to (106.48) could have been obviated had we started with the two eigenvectors already belonging to the same row of $D^{(\mathrm{co}\,\boldsymbol{k})(j)}$.

For eigenvectors of case B let us subsume the results of the unitary group. The matrices for case B are given in (98.14) and (98.18). So we distinguish two cases in (106.40) and (106.41) which we write

$$\left(e\left(\begin{vmatrix}\boldsymbol{k}\\ j_\nu\end{vmatrix}\right), e\left(\begin{vmatrix}\boldsymbol{k}\\ j_\nu\end{vmatrix}\right)\right) \quad \nu=1,\ldots,l_j; \text{ or } \nu=l_j+1,\ldots,2l_j. \tag{106.49}$$

From (98.18) we have: for $\nu=1,\ldots,l_j$

$$D^{(\mathrm{co}\,\boldsymbol{k})(j)}(a)_{\bar\nu\nu}=D^{(\boldsymbol{k})(j)}(a_0^{-1}\,u\,a_0)^*_{\bar\nu\nu} \quad \text{with } \bar\nu=l_j+1,\ldots,2l_j \tag{106.50}$$

while for $\nu=(l_j+1),\ldots,2l_j$ we have

$$D^{(\mathrm{co}\,\boldsymbol{k})(j)}(a)_{\bar\nu\nu}=D^{(\boldsymbol{k})(j)}(u\,a_0^2)_{\bar\nu\nu} \quad \text{with } \bar\nu=1,\ldots,l_j. \tag{106.51}$$

Hence

$$\begin{aligned}&\left(e\left(\begin{vmatrix}\boldsymbol{k}\\ j_\nu\end{vmatrix}\right), e\left(\begin{vmatrix}\boldsymbol{k}\\ j_\nu\end{vmatrix}\right)\right), \quad \nu\leqq l_j\\ &\quad=\sum_{\bar\nu}\sum_{\bar\nu'} D^{(\boldsymbol{k})(j)}(a_0^{-1}\,u\,a_0)^*_{\bar\nu\nu}\,D^{(\boldsymbol{k})(j)}(a_0^{-1}\,u\,a_0)_{\bar\nu'\nu}\\ &\quad\quad\times\left(e\left(\begin{vmatrix}\boldsymbol{k}\\ j_{\bar\nu}\end{vmatrix}\right), e\left(\begin{vmatrix}\boldsymbol{k}\\ j_{\bar\nu'}\end{vmatrix}\right)\right)^*, \quad \bar\nu>l_j.\end{aligned} \tag{106.52}$$

Summing over the unitary elements in $\mathfrak{G}(\boldsymbol{k})$ we have

$$\begin{aligned}&g(\boldsymbol{k})\left(e\left(\begin{vmatrix}\boldsymbol{k}\\ j_\nu\end{vmatrix}\right), e\left(\begin{vmatrix}\boldsymbol{k}\\ j_\nu\end{vmatrix}\right)\right)\\ &\quad=\sum_{\bar\nu}(g(\boldsymbol{k})/l_j)\left(e\left(\begin{vmatrix}\boldsymbol{k}\\ j_{\bar\nu}\end{vmatrix}\right), e\left(\begin{vmatrix}\boldsymbol{k}\\ j_{\bar\nu}\end{vmatrix}\right)\right)^*, \quad \bar\nu>l_{j'},\end{aligned}$$

or

$$\begin{aligned}&l_j\left(e\left(\begin{vmatrix}\boldsymbol{k}\\ j_\nu\end{vmatrix}\right), e\left(\begin{vmatrix}\boldsymbol{k}\\ j_\nu\end{vmatrix}\right)\right), \quad \nu\leqq l_j\\ &\quad=\sum_{\bar\nu}\left(e\left(\begin{vmatrix}\boldsymbol{k}\\ j_{\bar\nu}\end{vmatrix}\right), e\left(\begin{vmatrix}\boldsymbol{k}\\ j_{\bar\nu}\end{vmatrix}\right)\right)^*, \quad \bar\nu>l_j.\end{aligned} \tag{106.53}$$

Likewise from the second alternative in (106.49) we obtain

$$\begin{aligned}&l_j\left(e\left(\begin{vmatrix}\boldsymbol{k}\\ j_\nu\end{vmatrix}\right), e\left(\begin{vmatrix}\boldsymbol{k}\\ j_\nu\end{vmatrix}\right)\right), \quad \nu>l_j\\ &\quad=\sum_{\bar\nu}\left(e\left(\begin{vmatrix}\boldsymbol{k}\\ j_{\bar\nu}\end{vmatrix}\right), e\left(\begin{vmatrix}\boldsymbol{k}\\ j_{\bar\nu}\end{vmatrix}\right)\right)^*, \quad \bar\nu<l_j.\end{aligned} \tag{106.54}$$

Now from (106.53), (106.54) we can see the relationship

$$\left.\begin{aligned}\left(e\left(\begin{vmatrix}\boldsymbol{k}\\ j_\nu\end{vmatrix}\right), e\left(\begin{vmatrix}\boldsymbol{k}\\ j_\nu\end{vmatrix}\right)\right) &= f(\boldsymbol{k},j)\\ \left(e\left(\begin{vmatrix}\boldsymbol{k}\\ j_{l_j+\nu}\end{vmatrix}\right), e\left(\begin{vmatrix}\boldsymbol{k}\\ j_{l_j+\nu}\end{vmatrix}\right)\right) &= f^*(\boldsymbol{k},j)\end{aligned}\right\} \quad \nu=1,\dots,l_j. \tag{106.55}$$

Note that (106.55) is independent of ν. Thus in case B, the scalar products are required to be complex conjugates as shown.

Turning to case C, we must use (98.51) in (106.40) and (106.41). Again we can restrict ourselves to the diagonal bases as a result of the analysis of unitary elements. Then we get in this case

$$D^{(\mathrm{co}\,\boldsymbol{k})(j)}(a)_{\bar{\nu}\nu} = (D^{(\boldsymbol{k})(j)}(u)\,\beta)_{\bar{\nu}\nu}, \qquad \nu=1,\dots,l_j;\; \bar{\nu}=l_j+1,\dots,2l_j \tag{106.56}$$

and

$$D^{(\mathrm{co}\,\boldsymbol{k})(j)}(a)_{\bar{\nu}\nu} = (-D^{(\boldsymbol{k})(j)}(u)\,\beta)_{\bar{\nu}\nu}, \qquad \nu=l_j+1,\dots,2l_j;\; \bar{\nu}=1,\dots,l_j. \tag{106.57}$$

Hence substituting into (106.40), (106.41) we get

$$\begin{aligned}&\left(e\left(\begin{vmatrix}\boldsymbol{k}\\ j_\nu\end{vmatrix}\right), e\left(\begin{vmatrix}\boldsymbol{k}\\ j_\nu\end{vmatrix}\right)\right), \qquad \nu<l_j\\ &\quad= \sum_{\bar{\nu}\bar{\nu}'}\sum_{\mu}\sum_{\mu'} D^{(\boldsymbol{k})(j)}(u)_{\bar{\nu}\mu}\,\beta_{\mu\nu}\,D^{(\boldsymbol{k})(j)}(u)^*_{\bar{\nu}'\mu'}\,\beta^*_{\mu'\nu}\left(e\left(\begin{vmatrix}\boldsymbol{k}\\ j_{\bar{\nu}}\end{vmatrix}\right), e\left(\begin{vmatrix}\boldsymbol{k}\\ j_{\bar{\nu}'}\end{vmatrix}\right)\right)^*.\end{aligned} \tag{106.58}$$

Carrying through the sum on the unitary elements u of $\mathfrak{G}(\boldsymbol{k})$ in exactly the same fashion as led to (106.55), we obtain the relation as in (106.55) between scalar products for $\nu<l_j$ and $\nu>l_j$.

Summarizing this part of the analysis we conclude that the existence of elements of antiunitary symmetry in the group, and the consequent result that eigenvectors belong to irreducible corepresentations of the group results in reality restrictions on matrix elements such as (106.48) and (106.55).

To continue, we must consider the more general matrix element such as (106.20). Then each of the factors in (106.20) should be considered as a basis of the appropriate corepresentation. So, the product:

$$D^{(\mathrm{co}\,\boldsymbol{k}'')(m)} \otimes D^{(\mathrm{co}\,\boldsymbol{k}')(j')} \tag{106.59}$$

needs to be reduced into a direct sum of corepresentations (generalized Maschke theorem). When the correct linear combinations are obtained, either by use of generalized Clebsch-Gordon coefficients, or by projection operator then the basic scalar products, which are the reduced matrix elements, can be found. Again, only the limited number of reduced matrix elements arise equal to the value of the reduction coefficient

$$(\mathrm{co}\,\boldsymbol{k}\,j\,|\,\mathrm{co}\,k''\,m,\ \mathrm{co}\,\boldsymbol{k}'\,j') \neq 0. \tag{106.60}$$

The principles of the analysis are just as in the purely unitary case, but we shall not give details as different cases are discussed in the literature.[3] Using the

[3] A. Aviran and J. Zak: J. Math. Phys. **9**, 2138 (1968).

properties of the corepresentation as developed in Sect. 98, all the cases can be determined.

Again it should be emphasized that in carrying through the complete analysis for matrix elements, the sub-group methods should only be used when taking great care, and this must be borne in mind when working the corepresentation theory. It is generally safer, if more tedious, to work with full group methods.

Some illustrations of additional selection rules due to antiunitary elements of symmetry will be given later.

107. Critical points

α) *Representation theory for the "symmetry set"*. In our previous work we have concerned ourselves with the eigenvalues and eigenfrequencies of the dynamical matrix at particular points or termini of wave vectors $\boldsymbol{k}$ in the Brillouin zone. Depending on whether the points associated with symmetry groups $\mathfrak{G}(\boldsymbol{k})$ or $\mathscr{G}(\boldsymbol{k})$ of high order, or of low order, various classifications and simplifications were possible in discussing necessary degeneracy, obtaining symmetrized eigenvectors, matrix elements of symmetric operators, etc.

In this section we turn to matters which are more related to the distribution of eigenvalues in the zone, or to the density of energy states. The density of energy states at energy E is the number of distinct eigenvectors of the dynamical matrix with energy between and E and $E+dE$. Evidently the density of states depends on $\boldsymbol{k}$. A critical point in the density of states is an energy E, or extremity of vector $\boldsymbol{k}$ at which there is a singularity in the density of energy states. In particular the slope (derivative of density of states with respect to energy) becomes infinite at a critical point. A succinct discussion of critical points is given in the companion article in this Encyclopedia by COCHRAN and COWLEY [*20*]. The subject is also discussed in [*23*, *47*].

The task of the present section of this article is to demonstrate the application of group theory methods to the determination of critical points. At the outset it needs to be emphasized that symmetry considerations will not give all the critical points of the energy distribution for a given crystal with specified crystal structure but merely a certain sub-set of all the critical points: these have been called[1] the "symmetry set". Additional critical points occur "dynamically". This is intended to convey the meaning that the "dynamical critical points" occur because the specific values of the force constants for a particular material result in a certain numerical energy distribution with certain critical points; and hence these dynamical critical points could *not* be predicted by symmetry. Over and above this, is a certain topological necessity[2] for the existence of critical points. That is, the "Morse Theory" (calculus of variations in the large) can be used to demonstrate that the existence of certain critical points in the energy distribution, classified according to their index, requires the existence of a specified number of other critical points with prescribed indices, and that the topological properties of the "space" on which the energy distribution is defined requires the existence of a minimal set [*23*, *47*].

[1] J.C. PHILLIPS: Phys. Rev. **104**, 1263 (1956).
[2] L. VAN HOVE: Phys. Rev. **89**, 1189 (1953).

The program of the present section is to derive the basic equations from which we can obtain the density of energy states in such a form that expeditious application of the symmetry analysis will determine critical points. In this section we restrict ourselves only to representation theory based on $\mathfrak{G}(\boldsymbol{k})$.

The eigenvalue problem for lattice frequencies is (79.7), (85.1).

$$[\boldsymbol{D}(\boldsymbol{k})]\cdot \boldsymbol{e}\begin{pmatrix}\boldsymbol{k}\\ j_\mu\end{pmatrix}=\omega^2(\boldsymbol{k}|j)\,\boldsymbol{e}\begin{pmatrix}\boldsymbol{k}\\ j_\mu\end{pmatrix}. \tag{107.1}$$

From (107.1) we obtain in principle the complete set of $3rN$ frequencies. Now, for each numerical value of squared frequency in (107.1) we could count the number of distinct eigenvectors irrespective of the indices $\boldsymbol{k}$, and j_μ. Since each eigenvector corresponds to a distinct state of oscillation, the number of such eigenvectors at numerical frequency ω^2 is the number of states at that energy. While this is a perfectly acceptable way of actually numerically calculating the density of states, and is so used, it is not convenient from the point of view of the present analysis. Recall [*4*] that it is convenient to consider the index j or j_μ if there is degeneracy to refer to the branch of the lattice vibration spectrum. Hence, on any branch (j fixed) there will be N states, one for each $\boldsymbol{k}$ in the zone. In the reduced zone scheme, which we are employing, the function $\omega^2(\boldsymbol{k}|j)$ will be multivalued: each $\boldsymbol{k}$ corresponds to $3r$ branch index values.

Consider that the frequencies of the lattice are arranged in such a way that without ambiguity we can define the j-th branch ($j=1, \ldots, 3r$). A labelling of the j-th branch is generally unique except for example, at points where branches cross. At a cross-over point it is not evident how to continue labelling a branch from one side of the crossing to the other. We adopt, following PHILIPS,[3] a convention of ordered labelling so that the j-th branch at wave vector $\boldsymbol{k}$ is labelled (in case of no degeneracy) by $\omega^2(\boldsymbol{k}|j)$ such that if $j'>j$ then

$$\omega^2(\boldsymbol{k}|j')>\omega^2(\boldsymbol{k}|j) \quad \text{if } j'>j. \tag{107.2}$$

This convention suffices to establish, away from points of degeneracy, a unique labelling in the reduced Brillouin zone. It may produce a generalized critical point owing to a discontinuous change in slope on one branch, at the cross over point. At such a singular critical point one or more components of the slope of ω^2 versus $\boldsymbol{k}$ change sign discontinuously, the other components vanish, but, $\omega^2(\boldsymbol{k}|j)$ is still continuous.

That is, as $\boldsymbol{k}$ varies continuously throughout the first Brillouin zone, a continuous variation in $\omega^2(\boldsymbol{k}|j)$ results with j fixed. Then if the total number of cells in the Born-Karman period is N, there are N such squared eigenfrequencies on each branch. The fraction of this total which lie, on branch j between $\omega^2(\boldsymbol{k}|j)$ and $\omega^2(\boldsymbol{k}|j)+\Delta\,\omega^2(\boldsymbol{k}|j)$ is

$$G_j(\omega^2)\,\Delta\,\omega^2=\frac{1}{N}\,[N_j(\omega^2+\Delta\,\omega^2)-N_j(\omega^2)], \tag{107.3}$$

where $N_j(\omega^2)$ is the total number of squared frequencies from 0 to the given value $\omega^2(\boldsymbol{k}|j)$ on the j-th branch. Then clearly we can write for the distribution function

[3] J.C. PHILLIPS: Phys. Rev. **104**, 1263 (1956).

for squared frequency $G_j(\omega^2)\Delta\omega^2$:

$$G_j(\omega^2)\Delta\omega^2 = \sum_{\boldsymbol{k}} \delta\left(\omega^2(\boldsymbol{k}|j) - \omega^2\right), \tag{107.4}$$

$$\omega^2(\boldsymbol{k}|j) \leqq \omega^2 \leqq \omega^2(\boldsymbol{k}|j) + \Delta\omega^2(\boldsymbol{k}|j). \tag{107.5}$$

If we prefer to consider $\boldsymbol{k}$ as a continuous variable we can write (107.4) as a volume integral (a factor of $(2\pi)^{-3}$ is not explicitly exhibited):

$$G_j(\omega^2)\Delta\omega^2 = \iiint d\boldsymbol{k}, \tag{107.6}$$

where volume is restricted by:

$$\omega^2(\boldsymbol{k}|j) \leqq \omega^2 \leqq \omega^2(\boldsymbol{k}|j) + \Delta\omega^2(\boldsymbol{k}|j). \tag{107.7}$$

In (107.6) and (107.7), it should be understood that the restrictions are actually restrictions upon the range of $\boldsymbol{k}$ to be considered: these are implicit equations for $\boldsymbol{k}(\omega^2)$, of course restricted to the j-th branch. Clearly (107.6) represents the volume in $\boldsymbol{k}$ space enclosed between the surfaces $\omega^2 = \omega^2(\boldsymbol{k}|j)$ and $\omega^2 = \omega^2(\boldsymbol{k}|j) + \Delta\omega^2(\boldsymbol{k}|j)$. We now convert (107.6) to a surface integral over one of these surfaces.

Let dS be an element of the surface $\omega^2(\boldsymbol{k}|j) = \omega^2$ where ω^2 is some prescribed value. Then a unit vector in the direction of increasing ω^2 is

$$\hat{\boldsymbol{n}} = \nabla_{\boldsymbol{k}}\,\omega^2(\boldsymbol{k}|j)/|\nabla_{\boldsymbol{k}}\,\omega^2(\boldsymbol{k}|j)|. \tag{107.8}$$

Then we can rewrite (107.6) for the volume in $\boldsymbol{k}$ space enclosed between the two relevant surfaces:

$$G_j(\omega^2)\Delta\omega^2 = \oint\!\!\oint_{\sigma} \frac{\nabla_{\boldsymbol{k}}\,\omega^2}{|\nabla_{\boldsymbol{k}}\,\omega^2|} \cdot \Delta\boldsymbol{k}\, dS, \tag{107.9}$$

where $\Delta\boldsymbol{k}$ is the increment in $\boldsymbol{k}$ which corresponds to the increment $\Delta\omega^2(\boldsymbol{k}|j)$, and σ is the surface $\omega^2 = \omega^2(\boldsymbol{k}|j)$. But if we make a vector Taylor expansion of $\omega^2(\boldsymbol{k}+\Delta\boldsymbol{k}|j)$ we get

$$\omega^2(\boldsymbol{k}+\Delta\boldsymbol{k}|j) = \omega^2(\boldsymbol{k}|j) + \nabla\omega^2(\boldsymbol{k}|j)\cdot\Delta\boldsymbol{k} + \cdots \tag{107.10}$$

or

$$\Delta\omega^2(\boldsymbol{k}|j) \cong \nabla\omega^2(\boldsymbol{k}|j)\cdot\Delta\boldsymbol{k}. \tag{107.11}$$

Hence

$$G_j(\omega^2)\Delta\omega^2 = \oint\!\!\oint_{\sigma} \frac{\Delta\omega^2(\boldsymbol{k}|j)}{|\nabla_{\boldsymbol{k}}\,\omega^2|}\, dS. \tag{107.12}$$

But $\Delta\omega^2(\boldsymbol{k}|j)$ is the constant increment, so

$$G_j(\omega^2)\Delta\omega^2 = \Delta\omega^2(\boldsymbol{k}|j) \iint_{\sigma} \frac{dS}{|\nabla_{\boldsymbol{k}}\,\omega^2|}. \tag{107.13}$$

This is the important result: the frequency distribution function or density of states $G_j(\omega^2)\Delta\omega^2$ is proportional to the surface integral

$$\iint_{\sigma} \frac{dS}{|\nabla_{\boldsymbol{k}}\,\omega^2|}, \tag{107.14}$$

where the integral is over a surface σ of constant ω^2. It seems clear that regions in the Brillouin zone for which

$$|\nabla_{\boldsymbol{k}}\, \omega^2(\boldsymbol{k}|j)|=0 \tag{107.15}$$

are potentially of great significance in giving large contributions to the integrand in (107.14), and hence to the slope $dG_j(\omega^2)/d\omega^2$. Different situations can arise depending on whether all components of $\nabla_{\boldsymbol{k}}\, \omega^2(\boldsymbol{k}|j)$ vanish, or only some [*20, 23*].

Since our principal interest in this article is in the group theoretical characterization of optical processes connected with the lattice vibrations we shall not explore the detailed analytical contribution to the density of states from various types of critical points. This type of analysis is contained in the contributions of Cochran and Cowley [*20*]. Here we shall discuss the more restricted problem: how to locate all the critical points necessitated by symmetry.

We procede by carrying out a perturbation theory analysis of Eq. (107.1) to determine the connection between the symmetry of the normal mode eigenvectors and the occurrence of critical points. Thus suppose we possess the solution of (107.1) at wave vector $\boldsymbol{k}_0$. Since necessary degeneracy, produced by transformation under symmetry, is an important element in the theory we write out the dynamical equations, eigenvectors, eigenvalues, etc. in full component form. Recall the discussion in Sects. 75, 85, 91.

Thus at $\boldsymbol{k}_0$ suppose the equation is

$$\sum_{\kappa'\beta} D_{\alpha\beta}\begin{pmatrix}\boldsymbol{k}_0\\ \kappa\ \kappa'\end{pmatrix} e_\beta\left(\kappa'\left|\begin{matrix}\boldsymbol{k}_0\\ j_\mu\end{matrix}\right.\right)=\omega^2(\boldsymbol{k}_0|j)\, e_\alpha\left(\kappa\left|\begin{matrix}\boldsymbol{k}_0\\ j_\mu\end{matrix}\right.\right), \qquad \mu=1,\ldots,l_j \tag{107.16}$$

where the degenerate eigenvectors with $\mu=1,\ldots,l_j$ all belong to eigenvalue $\omega^2(\boldsymbol{k}_0|j)$, which is l_j-fold degenerate. For the other eigenvectors we write the branch index using Latin index (e.g.: m):

$$\sum_{\kappa'\beta} D_{\alpha\beta}\begin{pmatrix}\boldsymbol{k}_0\\ \kappa\ \kappa'\end{pmatrix} e_\beta\left(\kappa'\left|\begin{matrix}\boldsymbol{k}_0\\ j_m\end{matrix}\right.\right)=\omega^2(\boldsymbol{k}_0|m)\, e_\alpha\left(\kappa\left|\begin{matrix}\boldsymbol{k}_0\\ j_m\end{matrix}\right.\right), \qquad m\neq\mu \tag{107.17}$$

with $m\neq\mu$ restricting these branches to lie outside the domain of degenerate states. Let the vector $\boldsymbol{k}$ be a wave vector near $\boldsymbol{k}_0$, so that

$$\boldsymbol{k}-\boldsymbol{k}_0=\xi \tag{107.18}$$

is a small vector. The problem to be solved, at $\boldsymbol{k}$, is then that of

$$\sum_{\kappa'\beta} D_{\alpha\beta}\begin{pmatrix}\boldsymbol{k}\\ \kappa\ \kappa'\end{pmatrix} e_\beta\left(\kappa'\left|\begin{matrix}\boldsymbol{k}\\ j_\mu\end{matrix}\right.\right)=\omega^2(\boldsymbol{k}|\mu)\, e_\alpha\left(\kappa\left|\begin{matrix}\boldsymbol{k}\\ j_\mu\end{matrix}\right.\right), \tag{107.19}$$

where we use the Greek index μ to refer to one of the branches originating from the degenerate manifold (107.16) as before. On this branch we shall use degenerate perturbation theory. First we make a Taylor expansion of the dynamical matrix and the eigenvalue as follows, keeping only linear terms:

$$D_{\alpha\beta}\begin{pmatrix}\boldsymbol{k}\\ \kappa\ \kappa'\end{pmatrix}=D_{\alpha\beta}\begin{pmatrix}\boldsymbol{k}_0\\ \kappa\ \kappa'\end{pmatrix}+(\xi\cdot\nabla_{\boldsymbol{k}})D_{\alpha\beta}\begin{pmatrix}\boldsymbol{k}_0\\ \kappa\ \kappa'\end{pmatrix}+\cdots \tag{107.20}$$

and

$$\omega^2(\boldsymbol{k}|\mu)=\omega^2(\boldsymbol{k}_0|j)+(\xi\cdot\nabla_{\boldsymbol{k}})\,\omega^2(\boldsymbol{k}_0|\mu)+\cdots \tag{107.21}$$

or

$$\omega^2(\boldsymbol{k}|\mu)=\omega^2(\boldsymbol{k}_0|j)^{(0)}+\omega^2(\boldsymbol{k}_0|\mu)^{(1)}+\cdots \tag{107.22}$$

Now let us take the perturbed eigenvectors as linear combinations of the unperturbed:

$$e_\alpha\left(\kappa\left|\begin{matrix}\boldsymbol{k}\\ j_\mu\end{matrix}\right.\right)=\sum_{\nu=1}^{l_j}B_{\nu\mu}\,e_\alpha\left(\kappa\left|\begin{matrix}\boldsymbol{k}_0\\ j_\nu\end{matrix}\right.\right). \tag{107.23}$$

In (107.23) we restrict consideration to the degenerate eigenvectors since we assume that these will make the largest contribution. The problem is to determine $B_{\nu\mu}$ and, simultaneously, the first order corrections to $\omega^2(\boldsymbol{k}_0|j)$. Substitute (107.23), (107.21), (107.20) into (107.19) and we obtain

$$\begin{aligned}\sum_{\kappa'\beta}\left\{D_{\alpha\beta}\begin{pmatrix}\boldsymbol{k}_0\\ \kappa\ \kappa'\end{pmatrix}+(\xi\cdot\nabla_{\boldsymbol{k}})D_{\alpha\beta}\begin{pmatrix}\boldsymbol{k}_0\\ \kappa\ \kappa'\end{pmatrix}\right\}\sum_\nu B_{\nu\mu}\,e_\beta\left(\kappa'\left|\begin{matrix}\boldsymbol{k}_0\\ j_\nu\end{matrix}\right.\right)\\ =\{\omega^2(\boldsymbol{k}_0|j)+(\xi\cdot\nabla_{\boldsymbol{k}})\,\omega^2(\boldsymbol{k}_0|\mu)\}\sum_\nu B_{\nu\mu}\,e_\alpha\left(\kappa\left|\begin{matrix}\boldsymbol{k}_0\\ j_\nu\end{matrix}\right.\right)\end{aligned} \tag{107.24}$$

or, using (107.16) to eliminate zero order terms we are left with a typical first order degenerate perturbation theory result (recall that we are only using the degenerate manifold of eigenvectors):

$$\sum_{\kappa'\beta}\sum_\nu B_{\nu\mu}(\xi\cdot\nabla_{\boldsymbol{k}})D_{\alpha\beta}\begin{pmatrix}\boldsymbol{k}_0\\ \kappa\ \kappa'\end{pmatrix}e_\beta\left(\kappa'\left|\begin{matrix}\boldsymbol{k}_0\\ j_\nu\end{matrix}\right.\right)=(\xi\cdot\nabla_{\boldsymbol{k}})\,\omega^2(\boldsymbol{k}_0|j_\mu)\sum_\nu B_{\nu\mu}\,e_\alpha\left(\kappa\left|\begin{matrix}\boldsymbol{k}_0\\ j_\nu\end{matrix}\right.\right). \tag{107.25}$$

Now we form the scalar product of (107.25) with $e_\alpha\left(\kappa\left|\begin{matrix}\boldsymbol{k}_0\\ j_\sigma\end{matrix}\right.\right)$ again restricting σ to be one of the degenerate branches, to obtain

$$\begin{aligned}&\sum_{\kappa\alpha}\sum_{\kappa'\beta}\sum_\nu B_{\nu\mu}\,e_\alpha^*\left(\kappa\left|\begin{matrix}\boldsymbol{k}_0\\ j_\sigma\end{matrix}\right.\right)(\xi\cdot\nabla_{\boldsymbol{k}})D_{\alpha\beta}\begin{pmatrix}\boldsymbol{k}_0\\ \kappa\ \kappa'\end{pmatrix}e_\beta\left(\kappa'\left|\begin{matrix}\boldsymbol{k}_0\\ j_\nu\end{matrix}\right.\right)\\ &\quad=\xi\cdot\nabla_{\boldsymbol{k}}\,\omega^2(\boldsymbol{k}_0|j_\mu)\sum_\nu B_{\nu\mu}\sum_{\kappa\alpha}e_\alpha^*\left(\kappa\left|\begin{matrix}\boldsymbol{k}_0\\ j_\sigma\end{matrix}\right.\right)e_\alpha\left(\kappa\left|\begin{matrix}\boldsymbol{k}_0\\ j_\nu\end{matrix}\right.\right)\\ &\quad=\xi\cdot\nabla_{\boldsymbol{k}}\,\omega^2(\boldsymbol{k}_0|j_\mu)\sum_\nu B_{\nu\mu}\,\delta_{\nu\sigma}.\end{aligned} \tag{107.26}$$

Hence the equation determining the coefficient $B_{\nu\mu}$ (the correct linear combinations) is

$$\begin{aligned}\sum_{\kappa\alpha}\sum_{\kappa'\beta}\sum_\nu B_{\nu\mu}\Bigg[e_\alpha^*\left(\kappa\left|\begin{matrix}\boldsymbol{k}_0\\ j_\sigma\end{matrix}\right.\right)(\xi\cdot\nabla_{\boldsymbol{k}})D_{\alpha\beta}\begin{pmatrix}\boldsymbol{k}_0\\ \kappa\ \kappa'\end{pmatrix}e_\beta\left(\kappa'\left|\begin{matrix}\boldsymbol{k}_0\\ j_\nu\end{matrix}\right.\right)\\ -\xi\cdot\nabla_{\boldsymbol{k}}\,\omega^2(\boldsymbol{k}_0|j_\mu)\,\delta_{\nu\sigma}\Bigg]=0,\qquad \sigma=1,\dots,l_j.\end{aligned} \tag{107.27}$$

This evidently leads to a determinant, the vanishing of which assures the existence of solutions. A sufficient condition for all roots of (107.27) $\xi\cdot\nabla_{\boldsymbol{k}}\,\omega^2(\boldsymbol{k}_0|\mu)$ to vanish is if all matrix elements vanish identically i.e. if

$$\sum_{\kappa\alpha}\sum_{\kappa'\beta} e^*_\alpha\left(\kappa\middle|\begin{matrix}\boldsymbol{k}_0\\ j_\sigma\end{matrix}\right)(\xi\cdot\nabla_{\boldsymbol{k}})D_{\alpha\beta}\begin{pmatrix}\boldsymbol{k}_0\\ \kappa\ \kappa'\end{pmatrix} e_\beta\left(\kappa'\middle|\begin{matrix}\boldsymbol{k}_0\\ j_\nu\end{matrix}\right)=0. \tag{107.28}$$

Thus we need to analyze the question of the vanishing of this quantity. Keeping in mind the analysis of Sect. 106, we rewrite (107.28) as scalar product of the kind dealt with there:

$$\left(\boldsymbol{e}\left(\middle|\begin{matrix}\boldsymbol{k}_0\\ j_\sigma\end{matrix}\right),\ \xi\cdot\nabla_{\boldsymbol{\kappa}}\boldsymbol{D}(\boldsymbol{k})_{\boldsymbol{k}_0}\boldsymbol{e}\left(\middle|\begin{matrix}\boldsymbol{k}_0\\ j_\nu\end{matrix}\right)\right). \tag{107.29}$$

Comparing with (106.20) *et seq.*, we have to deal with matrix elements of an operator

$$\boldsymbol{O}(\boldsymbol{k})\equiv(\xi\cdot\nabla_{\boldsymbol{k}}\boldsymbol{D}(\boldsymbol{k}))_{\boldsymbol{k}_0} \tag{107.30}$$

over the eigenvectors of a degenerate manifold belonging to eigenvalue $\omega^2(\boldsymbol{k}_0|j)$. In (107.29), (107.30) the subscript $\boldsymbol{k}_0$ on the gradient of the dynamical matrix indicates that the quantity is to be evaluated at $\boldsymbol{k}_0$.

We must first examine the transformation properties of the operator (107.30). Construct

$$P_{\{\varphi|t\}}(\xi\cdot\nabla_{\boldsymbol{k}}\boldsymbol{D}(\boldsymbol{k}))\,P^{-1}_{\{\varphi|t\}} \tag{107.31}$$

which is equal to:

$$P_{\{\varphi|t\}}(\xi\cdot\nabla_{\boldsymbol{k}})\,P^{-1}_{\{\varphi|t\}}\cdot P_{\{\varphi|t\}}\,D(\boldsymbol{k})\,P^{-1}_{\{\varphi|t\}}. \tag{107.32}$$

Now, we defined the $P_{\{\varphi|t\}}$ operators in (12.1)–(12.5) to transform functions of configuration space, then later obtained the effect of the operator on a Bloch function as in (30.1)–(30.10). But, the operator $(\xi\cdot\nabla_{\boldsymbol{k}})$ is independent of $\boldsymbol{r}$, and so we shall have

$$P_{\{\varphi|t\}}(\xi\cdot\nabla_{\boldsymbol{k}})\,P^{-1}_{\{\varphi|t\}}=(\xi\cdot\nabla_{\boldsymbol{k}}). \tag{107.33}$$

Then, using (107.33) and (81.26) we obtain

$$P_{\{\varphi|t\}}(\xi\cdot\nabla_k D(\boldsymbol{k}))\,P^{-1}_{\{\varphi|t\}}=(\xi\cdot\nabla_k)\,D(\varphi\cdot\boldsymbol{k}). \tag{107.34}$$

Before proceeding recall that in the Taylor series (107.20) the derivative is to be evaluated at $\boldsymbol{k}_0$ as indicated in the argument. Also we have

$$\begin{aligned}\nabla_k(\exp-i\varphi\cdot\boldsymbol{k}\cdot\boldsymbol{R}_L)&=\nabla_k(\exp-i\boldsymbol{k}\cdot\varphi^{-1}\cdot\boldsymbol{R}_L)\\ &=(-i\varphi^{-1}\cdot\boldsymbol{R}_L)\exp-i\boldsymbol{k}\cdot\varphi^{-1}\cdot\boldsymbol{R}_L\\ &=(-i\varphi^{-1}\cdot\boldsymbol{R}_L)\exp-i\varphi\cdot\boldsymbol{k}\cdot\boldsymbol{R}_L.\end{aligned} \tag{107.35}$$

Then since as in (81.24)–(81.26)

$$\boldsymbol{D}(\varphi\cdot\boldsymbol{k})=P^{(\varphi\cdot\boldsymbol{k})}[\boldsymbol{M}]^{-\frac{1}{2}}[\boldsymbol{\Phi}(0)]\,[\boldsymbol{M}]^{-\frac{1}{2}}$$

we get

$$(\xi\cdot\nabla_k)\,\boldsymbol{D}(\varphi\cdot\boldsymbol{k})=(\xi\cdot\nabla_k)\,P^{(\varphi\cdot\boldsymbol{k})}[\boldsymbol{M}]^{-\frac{1}{2}}[\boldsymbol{\Phi}(0)]\,[\boldsymbol{M}]^{-\frac{1}{2}}. \tag{107.36}$$

Then from (107.36), (107.35), and the definition of operator $P^{(\varphi\cdot k)}$ in (81.14) we get

$$(\xi\cdot\nabla_k)\,P^{(\varphi\cdot k)}=(k-k_0)\cdot\nabla_k\,P^{(\varphi\cdot k)}$$

$$=(k-k_0)\cdot\sum_{\{\varepsilon|-R_N\}}(-i\varphi^{-1}\cdot R_N)\,D^{(\varphi\cdot k)}(\{\varepsilon|-R_N\})^*\,P_{\{\varepsilon|-R_N\}} \tag{107.37}$$

$$=(k-k_0)\cdot\varphi^{-1}\sum_{\{\varepsilon|-R_N\}}(-i R_N)\,D^{(\varphi\cdot k)}(\{\varepsilon|-R_N\})^*\,P_{\{\varepsilon|-R_N\}}. \tag{107.38}$$

But the gradient in (107.36) is to be evaluated at k_0; hence under the summation in (107.38) we should take $D^{(\varphi\cdot k)}\to D^{(\varphi\cdot k_0)}$.

Now, let $P_{\{\varphi|t\}}$ be an element in $\mathfrak{G}(k_0)$. Then

$$D^{(\varphi\cdot k_0)}=D^{(k_0)}. \tag{107.39}$$

Hence (107.38) becomes

$$(\varphi\cdot(k-k_0))\cdot\sum_{\{\varepsilon|-R_N\}}(-i R_N)\,D^{(k_0)}(\{\varepsilon|-R_N\})^*\,P_{\{\varepsilon|-R_N\}}. \tag{107.40}$$

But the sum on the right hand side of (107.40) is evidently

$$(\nabla_k P^{(k)})_{k_0}. \tag{107.41}$$

Consequently, we can assemble (107.31)–(107.41) to obtain, for $P_{\{\varphi|t\}}$ an element in $\mathfrak{G}(k_0)$:

$$\begin{aligned}P_{\{\varphi|t\}}\big(\xi\cdot(\nabla_k D(k))_{k_0}\big)\,P_{\{\varphi|t\}}^{-1}&=\big(\xi\cdot\varphi^{-1}\cdot(\nabla_k D(k))_{k_0}\big)\\&=\big(\varphi\cdot\xi\cdot(\nabla_k D(k))_{k_0}\big).\end{aligned} \tag{107.42}$$

We conclude that the three components in the scalar product

$$\xi\cdot(\nabla_k D(k))_{k_0} \tag{107.43}$$

transform like a an ordinary polar vector under the operations of $\mathfrak{G}(k_0)$. Written in components (107.42) is

$$P_{\{\varphi|t\}}(\xi\cdot\nabla_k D(k))_{k_0}\,P_{\{\varphi|t\}}^{-1}=\sum_{\gamma\delta}\xi_\gamma(\varphi^{-1})_{\delta\gamma}\big((\partial/\partial k_\delta)\,D(k)\big)_{k_0}. \tag{107.44}$$

It is important to note that now the "rotation" φ occurs on the Cartesian components of the gradient ∇_k. As a result of the analysis we succeeded in shifting the operator to operate on the k part just as in (30.1)–(30.10) a similar argument was made for the Bloch vectors (functions). Thus the $(3s)^2$ components of $D(k)$ are unaffected, and in fact they are to be summed in taking the scalar product (107.28) with the eigenvectors $e\left(\left|\begin{matrix}k_0\\ j_\sigma\end{matrix}\right.\right)$ and $e\left(\left|\begin{matrix}k_0\\ j_\nu\end{matrix}\right.\right)$.

We can summarize this result with the following convenient symbolism denoting transformations of an operator like (106.18)

$$(\xi\cdot\nabla_k D(k))_{k_0}\sim D^{(r)}\begin{pmatrix}k_0\\ \alpha\end{pmatrix}, \tag{107.45}$$

where $D^{(r)}\begin{pmatrix}k_0\\ \alpha\end{pmatrix}$ is the representation by which the 3 components of a polar vector r transforms under the rotational operations in $\mathfrak{G}(k_0)$. That is, restricting the rotational elements φ to be those in $\mathfrak{G}(\boldsymbol{k}_0)$ or $\mathfrak{P}(\boldsymbol{k}_0)$, the operator on the left hand side of (107.45) transforms as an ordinary vector under these.

It follows that (106.20) and (106.21) can be applied. That is to determine whether the matrix element (107.28) vanishes it is sufficient to have the reduction coefficient corresponding to (107.28). This is in subgroup notation

$$(\boldsymbol{k}_0\, j_\sigma \,|\, \boldsymbol{k}_0\, \boldsymbol{r} \boldsymbol{k}_0\, j_\nu). \tag{107.46}$$

The usual warning about the precaution needed to ensure that the product basis is complete should be borne in mind.

A simple rubric describes this result. A sufficient condition for a critical point at $\boldsymbol{k}_0$ in the phonon frequency distribution is the vanishing of the reduction coefficient (107.46) and hence the absence of any linear term in the perturbation expansion (107.21) and (107.22). Alternatively put, we are actually reducing

$$D^{(\boldsymbol{k}_0)(j)*} \otimes D^{(\boldsymbol{k}_0)(\boldsymbol{r})} \otimes D^{(\boldsymbol{k}_0)(j)} \tag{107.47}$$

to determine whether the triple product contains the trivial representation. Since the slope is a function of only one branch (i.e. of its symmetry and topology) the same representation $D^{(\boldsymbol{k}_0)(j)}$ appears in the two factors in (107.47). In case $D^{(\boldsymbol{k}_0)(j)}$ is real, (107.47) can be written

$$[D^{(\boldsymbol{k}_0)(j)}]_{(2)} \otimes D^{(\boldsymbol{k}_0)(\boldsymbol{r})} \tag{107.48}$$

which is the product of the symmetrized square of $D^{(\boldsymbol{k}_0)(j)}$ with $D^{(\boldsymbol{k}_0)(\boldsymbol{r})}$.

The same argument applies to an examination of a possible vanishing of any one, or two, components of the gradient operator. So to test for a vanishing "δ" (for example k_x) component of $\nabla_k\, \omega^2(\boldsymbol{k}|j)$ at $\boldsymbol{k}_0$ the relevent x component of the operator (107.30) needs to be tested. Corresponding to (107.47) will be

$$D^{(\boldsymbol{k}_0)(\sigma)*} \otimes D^{(\boldsymbol{k}_0)(\boldsymbol{r}\delta)} \otimes D^{(\boldsymbol{k}_0)(\nu)} \tag{107.49}$$

which is the symmetrized square of $D^{(\boldsymbol{k}_0)(j)}$ product with component $(D^{(\boldsymbol{k}_0)(\boldsymbol{r})})_\delta$. Note that in groups $\mathfrak{P}(\boldsymbol{k}_0)$ of point symmetry lower than cubic O_h, different components of a polar vector $\boldsymbol{r}$ may transform according to different irreducible representations.

Examination of (107.47) and (107.49) for each of the specific branches of interest throughout the Brillouin zone suffices in all cases of interest to us to determine the candidates for critical points. Each such candidate can be classified according to whether it is an absolute minimum, maximum, or saddle point. This is determined, of course, by the number of vanishing components of the slope (gradient) at such a point. The type and index of the critical point is related to the type of singularity, or other extremum behavior, of the density of states of the phonon distribution in the vicinity of the critical point. This is discussed in the article of COCHRAN-COWLEY in this encyclopedia, vol. XXV/2a [*20*], as well as in [*23, 47*].

Finally we recall that the entire discussion given here relates to necessary critical points. Other, accidental, critical points occur due to the actual phonon dispersion $\omega^2(\boldsymbol{k}|j)$ caused by the actual values of force constants. Such dynamical critical points cannot be predicted by symmetry. They must however, in all cases, satisfy all the Morse relations when taken together with the necessary critical points due to symmetry.

β) Determination of potential critical points by point symmetry. In the previous subsection we obtained the important result that a sufficient condition for the wave vector $\boldsymbol{k}_0$ to be a point of zero slope, or critical point, for the surface of symmetry $D^{(\boldsymbol{k}_0)(j)}$ is that the subgroup reduction coefficient (107.46) shall vanish. The procedure is then straightforward for a given crystal. That is one should examine this reduction coefficient for all zone points $\boldsymbol{k}_0$ and for each possible allowable irreducible representation corresponding to the possible symmetry of a phonon for the crystal: thus the set of all such reduction coefficients merit examination.

In principle this is a tedious problem and to minimize the work involved, we may ask whether it is possible to restrict attention to particular points $\boldsymbol{k}_0$ in the Brillouin zone. To gain insight into this question we adopt a "global" approach[4] and rephrase the matter. Recall in (107.1) we had the result that the physical eigenfrequencies $\omega^2(\boldsymbol{k}_0|j)$ at wave vector $\boldsymbol{k}_0$ are the eigenvalues of the dynamical matrix

$$[\boldsymbol{D}(\boldsymbol{k}_0)]. \tag{107.50}$$

Further, from (79.16) and (79.17) we know that the unitary matrix $\boldsymbol{E}(\boldsymbol{k}_0)$ whose rows and columns are the eigenvectors of $[\boldsymbol{D}(\boldsymbol{k}_0)]$ will bring $[\boldsymbol{D}(\boldsymbol{k}_0)]$ to diagonal from (79.17). Consequently let us take Eq. (79.16)

$$\boldsymbol{E}(\boldsymbol{k}_0)^{-1}[\boldsymbol{D}(\boldsymbol{k}_0)]\cdot\boldsymbol{E}(\boldsymbol{k}_0)=\boldsymbol{\varDelta}(\boldsymbol{k}_0). \tag{107.51}$$

Taking the trace of both sides of the matrix equation (107.51) we have, since the trace of a matrix is invariant under unitary transformation

$$\begin{aligned}\operatorname{Tr}[\boldsymbol{D}(\boldsymbol{k}_0)]&=\operatorname{Tr}\boldsymbol{\varDelta}(\boldsymbol{k}_0)\\&=\omega^2(\boldsymbol{k}_0|j)+\cdots+\omega^2(\boldsymbol{k}_0|j_m)\\&=\sum_j c_j\,l_j\,\omega^2(\boldsymbol{k}_0|j)\\&\equiv f(\boldsymbol{k}_0),\end{aligned} \tag{107.52}$$

where c_j is the multiplicity of the eigenvalue $\omega^2(\boldsymbol{k}_0|j)$ and l_j is the degeneracy (if there is no accidental degeneracy then $c_j=1$). The essential point here is that the sum on the right hand side is some scalar function of the wave vector $\boldsymbol{k}_0$. We now take $c_j=1$ in the following.

Returning to (107.51) we write it

$$[\boldsymbol{D}(\boldsymbol{k}_0)]\cdot\boldsymbol{E}(\boldsymbol{k}_0)=E(\boldsymbol{k}_0)\,\varDelta(\boldsymbol{k}_0). \tag{107.53}$$

Now let us consider a transformation operator $P_{\{\varphi_0\}}$ from the group $\mathfrak{G}(k_0)$. Applying to (107.53) we have

$$P_{\{\varphi_0\}}\cdot[\boldsymbol{D}(\boldsymbol{k}_0)]\cdot P_{\{\varphi_0\}}^{-1}\cdot P_{\{\varphi_0\}}\boldsymbol{E}(\boldsymbol{k}_0)=P_{\{\varphi_0\}}\boldsymbol{E}(\boldsymbol{k}_0)\,\varDelta(\boldsymbol{k}_0). \tag{107.54}$$

But from (83.5) we see that each eigenvector (column) in $\boldsymbol{E}(\boldsymbol{k}_0)$ is a partner in a basis for an allowable irreducible representation of $\mathfrak{G}(k_0)$. Consequently the

[4] N.D. Kudryavtseva: Soviet Phys. Solid State **9**, 1850 (1968).

effect of the operator $P_{\{\boldsymbol{\varphi}_0\}}$ upon the $\boldsymbol{E}(\boldsymbol{k}_0)$ is to produce a $3r \times 3r$ dimensional matrix $D(\{\varphi_0\})$ which will be already in fully reduced form

$$P_{\{\boldsymbol{\varphi}_0\}} \boldsymbol{E}(\boldsymbol{k}_0) = D(\{\varphi_0\}) \boldsymbol{E}(\boldsymbol{k}_0) \tag{107.55}$$

with

$$D(\{\varphi_0\}) = \begin{pmatrix} D^{(\boldsymbol{k}_0)(1)} & 0 & 0 \\ 0 & D^{(\boldsymbol{k}_0)(j')} & 0 \\ 0 & 0 & D^{(\boldsymbol{k}_0)(j)} \end{pmatrix}. \tag{107.56}$$

But also from (81.26), neglecting for the moment that $\boldsymbol{\varphi}_0 \boldsymbol{k}_0 \mathrel{\ne} \boldsymbol{k}_0$:

$$P_{\{\boldsymbol{\varphi}_0\}} \cdot [\boldsymbol{D}(\boldsymbol{k}_0)] \cdot P_{\{\boldsymbol{\varphi}_0\}}^{-1} = [\boldsymbol{D}(\{\varphi_0\} \cdot \boldsymbol{k}_0)]. \tag{107.57}$$

Consequently we may write from (107.55)–(107.57)

$$\Delta(\boldsymbol{k}_0) = D^{-1}(\{\varphi_0\}) \cdot [\boldsymbol{D}(\{\varphi_0\} \cdot \boldsymbol{k}_0)] \cdot D(\{\varphi_0\}). \tag{107.58}$$

Taking the trace of both sides of (107.58) we have, using the fact that $D(\{\varphi_0\})$ is a unitary matrix (direct sum of unitary matrices)

$$\begin{aligned} \operatorname{Tr} \Delta(\boldsymbol{k}_0) &= \operatorname{Tr} \boldsymbol{D}(\{\varphi_0\} \cdot \boldsymbol{k}_0) \\ &= \operatorname{Tr} \Delta(\{\varphi_0\} \cdot \boldsymbol{k}_0) \end{aligned} \tag{107.59}$$

or in the notation defined in (107.52), for all $\{\varphi_0\}$

$$f(\boldsymbol{k}_0) = f(\{\varphi_0\} \cdot \boldsymbol{k}_0). \tag{107.60}$$

Now of course since $\{\varphi_0\}$ is in $\mathfrak{G}(\boldsymbol{k}_0)$ we expect (107.60) must be so on obvious grounds. But we can use (107.60) which is a scalar function of $\boldsymbol{k}_0$ to decide whether a point $\boldsymbol{k}_0$ is a possible critical point.

Consider that if $\boldsymbol{k}_0$ were *not* a potential critical point then some one of the squared eigenfrequencies (irrespective of which one in particular) would have an expansion around $\boldsymbol{k}_0$ with a linear term:

$$\omega^2(\boldsymbol{k}|j) = \omega^2(\boldsymbol{k}_0|j) + \left(\nabla_{\boldsymbol{k}} \omega^2(\boldsymbol{k}|j)\right)_{\boldsymbol{k}_0} \cdot (\boldsymbol{k} - \boldsymbol{k}_0). \tag{107.61}$$

Consequently in the vicinity of such a point the scalar function $f(\boldsymbol{k})$ would be

$$f(\boldsymbol{k}) = f(\boldsymbol{k}_0) + (\nabla f) \cdot (\boldsymbol{k} - \boldsymbol{k}_0). \tag{107.62}$$

However at $\boldsymbol{k}_0$ (107.60) must hold for all $\{\varphi_0\}$ in $\mathfrak{G}(\boldsymbol{k}_0)$.

Let us then turn the argument around. Tale $\boldsymbol{k}_0$ as the origin of coordinates. Then for a symmetry element $\{\varphi_0\}$ in $\mathfrak{G}(\boldsymbol{k}_0)$ to be consistent with the presence of a linear term in the expansion of the scalar invariant quantity $f(\boldsymbol{k})$ it must leave (107.62) invariant. But the wave vector $\boldsymbol{k}$ transforms like an ordinary polar vector (x, y, z) under rotation. Consequently only those rotations $\{\varphi_0\}$ which leave $\boldsymbol{k}$ invariant are consistent with (107.62).

Hence we have the rule that only those point groups are allowed in $\mathfrak{G}(\boldsymbol{k}_0)/\mathfrak{T}$ for which the rotations contained do *not* leave $\boldsymbol{k}_0$ invariant:

$$\{\varphi_0\} \cdot \boldsymbol{k}_0 = \boldsymbol{\varphi}_0 \cdot \boldsymbol{k}_0 \neq \boldsymbol{k}_0. \tag{107.63}$$

This is simultaneously a restriction on the possible wave vector $\boldsymbol{k}_0$ and on the possible $\{\varphi_0\}$: i.e. upon possible critical points.

For example, it is possible for φ_0 to be $\boldsymbol{i}$, the inversion. Hence the presence of inversion in the group $\mathfrak{P}(\boldsymbol{k}_0)$ is incompatible with a linear term in (107.62). For point groups not containing $\boldsymbol{i}$ it is necessary to inspect the transformation of the components (x, y, z) under the rotations in $\mathfrak{P}(\boldsymbol{k}_0)$. If x or y or z is invariant under all operations φ_0 of $\mathfrak{P}(\boldsymbol{k}_0)$ (with axes chosen for convenience) then that $\mathfrak{P}(\boldsymbol{k}_0)$ is consistent with a linear term in (107.62) and so cannot be a potential critical point. Of course this is simply a matter of inspection: the easiest is to examine whether any component of a vector transforms as the invariant (trivial) representation of the proposed group $\mathfrak{P}(\boldsymbol{k}_0)$, using the available tables of irreducible representations of the crystallographic point groups in three dimensions.

It is then simply seen that the possible point groups $\mathfrak{P}(\boldsymbol{k}_0)$ which *can* characterize a critical point are:

$$\text{any } \mathfrak{P}(\boldsymbol{k}_0) \text{ containing } \boldsymbol{i} \tag{107.64}$$

plus

$$\begin{aligned}&\text{groups } T_d,\ O,\ T,\ D_{2d},\ D_{3h},\ D_n,\ S_4,\ C_{3h}\\ &\text{and their subgroups } D_3,\ D_2,\ S_4,\ C_{3h}.\end{aligned} \tag{107.65}$$

All groups listed in (107.64) and (107.65) are incompatible with linear term in (107.62), and thus may be groups of point symmetry at a critical point.

It can easily be seen[4] that one can restrict attention to only the unitary operations: considering antiunitary operations (i.e. the full space-time point groups) does not change the results (107.64) and (107.65).

In any actual application of these arguments it is necessary to return to the analysis of the previous subsection in order to determine on just which branch at given $\boldsymbol{k}_0$, the critical point may lie. A little reflection will convince one that the considerations of the present subsection are completely consistent with those of the last one. But by working only with the scalar invariant $f(\boldsymbol{k}_0)$ defined in (107.62) we can achieve a preliminary identification of potential critical points: in crystals with space groups of high symmetry this may considerably reduce the work needed in order to find the set of symmetry critical points.

108. Compatibility or connectivity theory for representations. A topic which naturally should be considered along with the symmetry and critical point location discussed in previous sections is that of the compatibility of representations. Under this topic is meant the analysis of how representations are to be classified when we pass in a continuous fashion from a point, or line, or plane of high symmetry to one of lower symmetry.

Consider a point $\boldsymbol{k}$ of higher symmetry: it is characterized by space-time group $\mathcal{G}(\boldsymbol{k})$. The irreducible corepresentations $D^{(\mathrm{co}\,\boldsymbol{k})(j)}$, and then $D^{(\mathrm{co}\,{}^\star\boldsymbol{k})(j)}$ describe the admissable transformation of physical eigenvectors $\boldsymbol{e}\left(\begin{array}{c|}\boldsymbol{k}\\ j\end{array}\right)$.

Let k' be a point close to k. On passing from k to k' the space-time group $\mathcal{G}(k)$ may be lowered in symmetry to $\mathcal{G}(k')$. Evidently if this occurs

$$\mathcal{G}(k') \text{ is a subgroup of } \mathcal{G}(k). \tag{108.1}$$

Then for corepresentations we must have that

$$D^{(\mathrm{co}\,k)(j)} \quad \text{of } \mathcal{G}(k)$$

subduces a reducible representation upon $\mathcal{G}(k')$. That is the corepresentation $D^{(\mathrm{co}\,k)(j)}$ considered as corepresentation of $\mathcal{G}(k')$ is reducible into irreducible representations of that latter space-time group:

$$D^{(\mathrm{co}\,k)(j)} \text{ of } \mathcal{G}(k)\downarrow = D^{(\mathrm{co}\,k')(j_1')} + \cdots + D^{(\mathrm{co}\,k')(j_\nu')} \text{ of } \mathcal{G}(k'). \tag{108.2}$$

The representations $D^{(\mathrm{co}\,k_1')(j')}, \ldots, D^{(\mathrm{co}\,k')(j_\nu)}$ are said to be compatible with $D^{(\mathrm{co}\,k)(j)}$, and connected with it. That is as one passes from k to k' one goes from a higher symmetry situation to a lower and the corepresentations may break up into a sum of lower dimensional representations.

Conversely, on going from lower symmetry $\mathcal{G}(k')$ to higher symmetry $\mathcal{G}(k)$, the representations in the second line of (108.2) coalesce to give $D^{(\mathrm{co}\,k)(j)}$. The Froebenius reciprocity theorem actually covers this situation since the corepresentation $D^{(\mathrm{co}\,{}^\star k)(j')}$ may be considered as induced from any of the corepresentations in the second line of (108.2), when we are given the coset decomposition.

$$\mathcal{G}(k) = \mathcal{G}(k') + \{\varphi\,|\,\tau\}\,\mathcal{G}(k') + \cdots. \tag{108.3}$$

Summarizing: The connectivity, or compatibility, of representations, is a completely solved problem which requires knowledge of the groups $\mathcal{G}(k)$, $\mathcal{G}(k')$ at all points (especially neighboring points) in the zone. At each k, all the physical eigenvectors $e\left(\begin{array}{c|}k\\j\end{array}\right)$ which can occur are known, and the complete set of all corepresentations $D^{(\mathrm{co}\,k)(j)}, \ldots,$ also. Then using the subduction analysis (108.2) we determine the compatibility of lower group representations with those of the higher group. This will be illustrated in Sect. 136.

In practice there may be certain difficulties, still, in assignment using only the group theory, since the same subduced representation may occur more than once in (108.2). Clearly, using both the group theory analysis to test for the symmetry of the eigenvector, and the results of the solution of the dynamical equation to obtain relative ordering of eigenenergies the problem of labelling all branches correctly can be solved.

109. Construction of crystal invariants. In the quantitative treatment of crystal properties and in particular optical properties, a crucial role is played by various physical quantities which depend upon the displacements of the ions from their equilibrium positions. We can single out for emphasis here, the three typical quantities: V, the crystal potential energy; M, the crystal electric moment; and P the crystal polarizability. These are typical invariant and covariant quantities respectively, whose transformation properties we shall now examine. In the dynamical theory these objects, or the quantum mechanical operators related to them, are utilized directly in order to obtain quantitative expressions for the infra-red absorption coefficient or Raman scattering cross-section coefficient. Discussion of the use of these quantities in such theories is given briefly in this article in Sect. 124 and also elsewhere in this Encyclopedia [*20*] and in [*21*]. Here we concern ourselves with the utilization of the space symmetry group $\mathfrak{G}$ to specify the maximum information possible about these invariant and covariant crystal quantities. The present section is restricted to discussion of crystal invariants, using representation theory only, i.e. only the effect of the unitary group $\mathfrak{G}(\boldsymbol{k})$.

As a typical important invariant quantity, consider the crystal potential energy Φ. For completeness we briefly repeat some results given in Sect. 71. Now Φ is evidently a function of the set of instantaneous positions

$$\boldsymbol{\rho}\begin{pmatrix} l \\ \kappa \end{pmatrix} = \boldsymbol{r}\begin{pmatrix} l \\ \kappa \end{pmatrix} + \boldsymbol{u}\begin{pmatrix} l \\ \kappa \end{pmatrix}$$

of the constituent ions of the crystal. Consider an element $P_{\{\varphi\}}$ of $\mathfrak{G}$. By hypothesis, the elements of the group $\mathfrak{G}$ are those transformations of configuration space which produce an identical replica of the crystal, when the constituent atoms or ions of the crystal are at rest. Mathematically, the transformation defines an inner automorphism of the crystal constituents. Now let us call the value of the crystal potential energy Φ when the particles are in the instantaneous positions $\boldsymbol{\rho}\begin{pmatrix} l \\ \kappa \end{pmatrix}$:

$$\Phi\left(\left\{\boldsymbol{\rho}\begin{pmatrix} l \\ \kappa \end{pmatrix}\right\}\right). \tag{109.1}$$

Under the symmetry transformation $P_{\{\varphi\}}$ of $\mathfrak{G}$ each vector $\boldsymbol{\rho}\begin{pmatrix} l \\ \kappa \end{pmatrix}$ is transformed as

$$P_{\{\varphi|t\}}\,\rho_\alpha\begin{pmatrix} l_{\bar{\varphi}}+\boldsymbol{t} \\ \kappa_{\bar{\varphi}}+\boldsymbol{t} \end{pmatrix} = \sum_\beta \varphi_{\alpha\beta}\,\rho_\beta\begin{pmatrix} l \\ \kappa \end{pmatrix}. \tag{109.2}$$

Thus the atom initially at $\begin{pmatrix} l \\ \kappa \end{pmatrix}$ is transformed to the position $\begin{pmatrix} l_{\bar{\varphi}}+\boldsymbol{t} \\ \kappa_{\bar{\varphi}}+\boldsymbol{t} \end{pmatrix}$ i.e. rotated and translated by the symmetry operation. In addition, the displacement associated with this transformed point is the rotated displacement $\boldsymbol{\varphi}\cdot\boldsymbol{u}$. Thus the set of instantaneous positions $\boldsymbol{\rho}\begin{pmatrix} l \\ \kappa \end{pmatrix}$ is transformed into the set of instantaneous positions $P_{\{\varphi\}}\,\boldsymbol{\rho} \equiv \boldsymbol{\rho}_{\{\varphi\}}$. As usual,

$$\boldsymbol{\rho}_{\{\varphi\}\alpha}\begin{pmatrix} l \\ \kappa \end{pmatrix} = \sum_\beta \varphi_{\alpha\beta}\,\rho_\beta\begin{pmatrix} l_{\bar{\varphi}} \\ \kappa_{\bar{\varphi}} \end{pmatrix} \tag{109.3}$$

so that the instantaneous position of the atom at $\begin{pmatrix} l \\ \kappa \end{pmatrix}$ after the transformation involves a rotated displacement which was previously at $\begin{pmatrix} l_{\bar{\varphi}} \\ \kappa_{\bar{\varphi}} \end{pmatrix}$: On the other hand the transformation can be considered as one in which the crystal with its momentary pattern of instantaneous positions is viewed from a rotated coordinate system, related to the initial coordinate system by the operation $\{\varphi|\boldsymbol{t}\}^{-1}$. Clearly the potential energy is invariant to the description. Thus calling the transformed potential energy, i.e. the potential energy, after the change in instantaneous positions, $\Phi' = P_{\{\varphi\}}\Phi$ we have

$$P_{\{\varphi\}}\Phi(\{\boldsymbol{\rho}_{\{\varphi\}}\}) = \Phi(\{\boldsymbol{\rho}\}) \tag{109.4}$$

by definition. But, if $P_{\{\varphi\}}$ is a symmetry operation, by hypothesis the set of instantaneous positions $\{\boldsymbol{\rho}\}$ and the set $\{\boldsymbol{\rho}_{\{\varphi\}}\}$ are equivalent, so the value of the potential energy must be identical before and after transformation:

$$\Phi(\{\boldsymbol{\rho}\}) = \Phi(\{\boldsymbol{\rho}_{\{\varphi\}}\}), \tag{109.5a}$$

i.e. the form of Φ is invariant, since upon comparing (109.5a) and (109.4)

$$P_{\{\varphi\}}\Phi = \Phi. \tag{109.5b}$$

Clearly (109.5a) and (109.5b) is to apply for *any* operation in $\mathfrak{G}$. Thus as a generalized function of the spatial variables to which the configuration space operator $P_{\{\varphi\}}$ can be applied, Φ transforms as the identity representation: namely that representation for which all elements in $\mathfrak{G}$ are represented by the number $+1$. Simply as a matter of convenience, let us designate the identity representation as $(\boldsymbol{\Gamma})(1+)$. In fact later in the cubic groups the symbols $(\boldsymbol{\Gamma})(+1)$ stand for the identity representation. Conventionally $\boldsymbol{k} = \boldsymbol{\Gamma} = (0, 0, 0)$ is the null vector in the Brillouin zone, and $m = 1+$ is the appropriate symbol for the identity representation of $\mathfrak{G}(\boldsymbol{\Gamma})$. Thus we can rewrite (109.5a) and (109.5b) as

$$P_{\{\varphi\}}\Phi = D^{(\boldsymbol{\Gamma})(1+)}(\{\varphi\})\,\Phi \tag{109.6a}$$

or

$$\Phi \sim D^{(\boldsymbol{\Gamma})(1+)}. \tag{109.6b}$$

Now consider that we take in Φ the reference state $\left\{\boldsymbol{r}\begin{pmatrix} l \\ \kappa \end{pmatrix}\right\}$ and attempt to express Φ in power series of the set of displacements $\boldsymbol{u}\begin{pmatrix} l \\ \kappa \end{pmatrix}$. We have already carried out such a power series expansion in (67.8) up to terms bilinear in the $\boldsymbol{u}\begin{pmatrix} l \\ \kappa \end{pmatrix}$. To go further we consider the expansion

$$\Phi\left(\left\{\boldsymbol{\rho}\begin{pmatrix} l \\ \kappa \end{pmatrix}\right\}\right) = \Phi^{(0)} + \Phi^{(2)} + \Phi^{(3)} + \Phi^{(4)} + \cdots + \Phi^{(s)} + \cdots \tag{109.7}$$

where $\Phi^{(1)} = 0$ because of the assumed equilibrium of the crystal when the atoms are in their rest positions and $\boldsymbol{\rho}\begin{pmatrix} l \\ \kappa \end{pmatrix}_0 = \boldsymbol{r}\begin{pmatrix} l \\ \kappa \end{pmatrix}$. Again calling

$$V = \Phi\left(\left\{\boldsymbol{\rho}\begin{pmatrix} l \\ \kappa \end{pmatrix}\right\}\right) - \Phi^{(0)} \tag{109.8}$$

we may obtain expressions for V which are homogeneous of different degree in the components of the displacement $u_\alpha\begin{pmatrix} l \\ \kappa \end{pmatrix}$. Thus, e.g.

$$\Phi^{(3)} = \frac{1}{6} \sum_{l\kappa\alpha} \sum_{l'\kappa'\beta} \sum_{l''\kappa''\gamma} \Phi_{\alpha\beta\gamma}\begin{pmatrix} l & l' & l'' \\ \kappa & \kappa' & \kappa'' \end{pmatrix} u_\alpha\begin{pmatrix} l \\ \kappa \end{pmatrix} u_\beta\begin{pmatrix} l' \\ \kappa' \end{pmatrix} u_\gamma\begin{pmatrix} l'' \\ \kappa'' \end{pmatrix} \tag{109.9}$$

is of third degree in the components. By the argument just given, Φ is an invariant (109.6) and thus each term in the expansion (109.7) must be separately invariant, since no transformation can mix terms of various degree in (109.7). Now apply the transformation $P_{\{\varphi\}}$ of $\mathfrak{G}$. If we consider the third degree term in the potential $\Phi^{(3)}$ to be expressed in terms of the components of the transformed displacements $u_{\{\varphi\}}$ we can write for $\Phi^{(3)}$:

$$\begin{aligned}\Phi^{(3)} = \frac{1}{6} \sum_{l_\varphi\kappa_\varphi\bar{\alpha}} \sum_{l'_\varphi\kappa'_\varphi\bar{\beta}} \sum_{l''_\varphi\kappa''_\varphi\bar{\gamma}} \Phi'_{\bar{\alpha}\bar{\beta}\bar{\gamma}}\begin{pmatrix} l_\varphi & l'_\varphi & l''_\varphi \\ \kappa_\varphi & \kappa'_\varphi & \kappa''_\varphi \end{pmatrix} \\ \times (u_{\{\varphi\}})_{\bar{\alpha}}\begin{pmatrix} l_\varphi \\ \kappa_\varphi \end{pmatrix} (u_{\{\varphi\}})_{\bar{\beta}}\begin{pmatrix} l'_\varphi \\ \kappa'_\varphi \end{pmatrix} (u_{\{\varphi\}})_{\bar{\gamma}}\begin{pmatrix} l''_\varphi \\ \kappa''_\varphi \end{pmatrix}\end{aligned} \tag{109.10}$$

where, we introduce the new primed function Φ' to account for a possible different functional form of the expansion coefficients when expressed in terms of the rotated variables $u_{\{\varphi\}}$. By proceeding analogously to (71.1)–(71.10) we find

$$\Phi'_{\bar{\alpha}\bar{\beta}\bar{\gamma}}\begin{pmatrix} l_\varphi & l'_\varphi & l''_\varphi \\ \kappa_\varphi & \kappa'_\varphi & \kappa''_\varphi \end{pmatrix} = \sum_{\alpha\beta\gamma} \varphi_{\bar{\alpha}\alpha}\, \varphi_{\bar{\beta}\beta}\, \varphi_{\bar{\gamma}\gamma}\, \Phi_{\alpha\beta\gamma}\begin{pmatrix} l & l' & l'' \\ \kappa & \kappa' & \kappa'' \end{pmatrix} \tag{109.11}$$

so that the components $\Phi_{\alpha\beta\gamma}\begin{pmatrix} l & l' & l'' \\ \kappa & \kappa' & \kappa'' \end{pmatrix}$ transform as the components of a third rank tensor field under rotation. But by hypothesis $P_{\{\varphi\}}$ is a symmetry transformation hence, the form of the function is not changed by this transformation:

$$\Phi'_{\alpha\beta\gamma}\begin{pmatrix} l & l' & l'' \\ \kappa & \kappa' & \kappa'' \end{pmatrix} = \Phi_{\alpha\beta\gamma}\begin{pmatrix} l & l' & l'' \\ \kappa & \kappa' & \kappa'' \end{pmatrix}$$

or, since the argument is irrelevant:

$$\Phi'_{\bar{\alpha}\bar{\beta}\bar{\gamma}}\begin{pmatrix} l_\varphi & l'_\varphi & l''_\varphi \\ \kappa_\varphi & \kappa'_\varphi & \kappa''_\varphi \end{pmatrix} = \Phi_{\bar{\alpha}\bar{\beta}\bar{\gamma}}\begin{pmatrix} l_\varphi & l'_\varphi & l''_\varphi \\ \kappa_\varphi & \kappa'_\varphi & \kappa''_\varphi \end{pmatrix}. \tag{109.12}$$

Then (109.12) epitomizes the form-invariance of this tensor field (compare (109.6)). From (109.11) and (109.12)

$$\Phi_{\bar{\alpha}\bar{\beta}\bar{\gamma}}\begin{pmatrix} l_\varphi & l'_\varphi & l''_\varphi \\ \kappa_\varphi & \kappa_\varphi & \kappa_\varphi \end{pmatrix} = \sum_{\alpha\beta\gamma} \varphi_{\bar{\alpha}\alpha}\, \varphi_{\bar{\beta}\beta}\, \varphi_{\bar{\gamma}\gamma}\, \Phi_{\alpha\beta\gamma}\begin{pmatrix} l & l' & l'' \\ \kappa & \kappa' & \kappa'' \end{pmatrix} \tag{109.13}$$

which is an equation epitomizing the effects of rotational transformation upon the coupling coefficients in the third degree term.

Now writing

$$D^{(r)}(\{\varphi\})_{\bar{\alpha}\alpha} \equiv \varphi_{\bar{\alpha}\alpha} \tag{109.14}$$

as the matrix representing the transformation of a vector $\boldsymbol{r}$ in $\mathfrak{G}$, it is clear that the expression (109.13) can be written

$$\Phi_{\bar{\alpha}\bar{\beta}\bar{\gamma}}\begin{pmatrix} l_\varphi & l'_\varphi & l''_\varphi \\ \kappa_\varphi & \kappa'_\varphi & \kappa''_\varphi \end{pmatrix} = \sum_{\alpha\beta\gamma} D(\{\varphi\})_{\bar{\alpha}\bar{\beta}\bar{\gamma};\alpha\beta\gamma}\, \Phi_{\alpha\beta\gamma}\begin{pmatrix} l & l' & l'' \\ \kappa & \kappa' & \kappa'' \end{pmatrix} \tag{109.15}$$

where

$$D(\{\varphi\})_{\bar{\alpha}\bar{\beta}\bar{\gamma};\alpha\beta\gamma} \equiv \varphi_{\bar{\alpha}\alpha}\,\varphi_{\bar{\beta}\beta}\,\varphi_{\bar{\gamma}\gamma} \tag{109.16}$$

is evidently a matrix element of the general direct product matrix

$$D(\{\varphi\}) \equiv D^{(r)} \otimes D^{(r)} \otimes D^{(r)} = [D^{(r)}]_3 \tag{109.17}$$

which is the cube of the representations by which a polar vector transforms. Hence we can write

$$\Phi_{\bar{\alpha}\bar{\beta}\bar{\gamma}}\begin{pmatrix} l_\varphi & l'_\varphi & l''_\varphi \\ \kappa_\varphi & \kappa'_\varphi & \kappa''_\varphi \end{pmatrix} = \sum_{\alpha\beta\gamma} ([D^{(r)}]_3)_{\bar{\alpha}\bar{\beta}\bar{\gamma};\alpha\beta\gamma}\, \Phi_{\alpha\beta\gamma}\begin{pmatrix} l & l' & l'' \\ \kappa & \kappa' & \kappa'' \end{pmatrix}. \tag{109.18}$$

Of course if the symmetry element $P_{\{\varphi|t\}}$ does not change $\begin{pmatrix} l \\ \kappa \end{pmatrix}$ and $\begin{pmatrix} l' \\ \kappa' \end{pmatrix}$ and $\begin{pmatrix} l'' \\ \kappa'' \end{pmatrix}$ so that $\begin{pmatrix} l_\varphi \\ \kappa_\varphi \end{pmatrix} = \begin{pmatrix} l \\ \kappa \end{pmatrix}$ and $\begin{pmatrix} l'_\varphi \\ \kappa'_\varphi \end{pmatrix} = \begin{pmatrix} l''_\varphi \\ \kappa''_\varphi \end{pmatrix} = \begin{pmatrix} l'' \\ \kappa'' \end{pmatrix}$, then (109.18) will be a relation limiting the number of independent non vanishing coupling parameters of third degree. From (109.18) we immediately infer that the third degree force constants defined in (109.9) transform under rotations as the elements of a third rank Cartesian tensor, which, owing to symmetry is an invariant [*4*, *47*].

α) *The crystal Hamiltonian: Harmonic and anharmonic.* In order to quantize the theory of lattice vibrations it is necessary to put the theory in Hamiltonian form. Going back to the discussion in Sect. 67, we can proceed to do this.

The dynamic variables of the lattice dynamical problem are the $3rN$ Cartesian components of displacements $u_\alpha\begin{pmatrix} l \\ \kappa \end{pmatrix}$, $\alpha = 1, 2, 3$; $l = 1, \ldots, N$; $\kappa = 1, \ldots, r$. The kinetic energy of the lattice is

$$T = \tfrac{1}{2} \sum_{\kappa\alpha} M_\kappa \left(\dot{u}_\alpha \begin{pmatrix} l \\ \kappa \end{pmatrix} \right)^2. \tag{109.19}$$

In the harmonic approximation, the potential energy of the lattice, eliminating a constant, is

$$V = \tfrac{1}{2} \sum_{l\kappa\alpha} \sum_{l'\kappa'\beta} u_\alpha \begin{pmatrix} l \\ \kappa \end{pmatrix} \Phi_{\alpha\beta}\begin{pmatrix} l & l' \\ \kappa & \kappa' \end{pmatrix} u_\beta \begin{pmatrix} l' \\ \kappa' \end{pmatrix}. \tag{109.20}$$

Using the basic rule for transforming from the Cartesian displacements as variables to the complex normal coordinates, $Q\begin{pmatrix} \boldsymbol{k} \\ j \end{pmatrix}$, and the dynamical equation, and the orthogonality rules for eigenvectors $\boldsymbol{e}\left(\kappa \middle| \begin{matrix} \boldsymbol{k} \\ j \end{matrix}\right)$, we obtain for the kinetic energy

$$T = \tfrac{1}{2} \sum_{\boldsymbol{k}} \sum_{j} \dot{Q}\begin{pmatrix} \boldsymbol{k} \\ j \end{pmatrix}^* \dot{Q}\begin{pmatrix} \boldsymbol{k} \\ j \end{pmatrix}. \tag{109.21}$$

For the potential energy we have as in (80.3)

$$V=\tfrac{1}{2}\sum_{\boldsymbol{k}}\sum_{j}\omega^2(\boldsymbol{k}|j)\,Q\begin{pmatrix}\boldsymbol{k}\\ j\end{pmatrix}^{*} Q\begin{pmatrix}\boldsymbol{k}\\ j\end{pmatrix}. \tag{109.22}$$

Observe from (109.20) and (109.21) that the energy is not simply the sum of squares, since the coordinates $Q\begin{pmatrix}\boldsymbol{k}\\ j\end{pmatrix}$ are complex.

The crystal Hamiltonian is

$$\mathscr{H}=T+V \tag{109.23}$$

and evidently it is a bilinear hermitian form in the complex dynamical variables $Q\begin{pmatrix}\boldsymbol{k}\\ j\end{pmatrix}$. Writing (109.23) by use of (109.21), and (109.22) and adding the usual branch index to take account of degeneracy we obtain:

$$\mathscr{H}=\tfrac{1}{2}\sum_{\boldsymbol{k}}\sum_{\sigma=1}^{s}\sum_{\mu=1}^{l_m}\sum_{j}\left\{\dot{Q}\begin{pmatrix}\boldsymbol{k}_\sigma\\ j_\mu\end{pmatrix}^{*}\dot{Q}\begin{pmatrix}\boldsymbol{k}_\sigma\\ j_\mu\end{pmatrix}+\omega^2(\boldsymbol{k}_\sigma|j)\,Q\begin{pmatrix}\boldsymbol{k}_\sigma\\ j_\mu\end{pmatrix}^{*}Q\begin{pmatrix}\boldsymbol{k}_\sigma\\ j_\mu\end{pmatrix}\right\} \tag{109.24}$$

when the sum over j_μ is over those symmetries which actually do occur in the vibration problem.

It is useful to verify that (109.24) is invariant under the space-time symmetry group $\mathscr{G}$ of the crystal. Evidently $\mathscr{H}$ is manifestly invariant under the operation of time reversal-complex conjugation:

$$\mathscr{H}=\mathscr{H}^{*}=K\mathscr{H}K^{-1}, \tag{109.25}$$

where in the last step we emphasize that $\mathscr{H}$ is an operator. We can then examine the transformation of $\mathscr{H}$ under spatial, unitary operations, in $\mathfrak{G}$.

The normal modes $Q\begin{pmatrix}\boldsymbol{k}_\sigma\\ j_\alpha\end{pmatrix}$ transform irreducibly under operations in $\mathfrak{G}$, according to the work of Sect. 86. Consequently we can restrict attention to a single irreducible representation. Thus take

$$\mathscr{H}_j=\tfrac{1}{2}\sum_{\sigma}\sum_{\mu=1}^{l_j}\left\{\dot{Q}\begin{pmatrix}\boldsymbol{k}_\sigma\\ j_\mu\end{pmatrix}\dot{Q}\begin{pmatrix}\boldsymbol{k}_\sigma\\ j_\mu\end{pmatrix}^{*}+\omega^2(\boldsymbol{k}_\sigma|j)\,Q\begin{pmatrix}\boldsymbol{k}_\sigma\\ j_\mu\end{pmatrix}Q\begin{pmatrix}\boldsymbol{k}_\sigma\\ j_\mu\end{pmatrix}^{*}\right\} \tag{109.26}$$

where we also dispense with the sum over all wave vectors since we are focussing on a single irreducible representation. Now, under transformation by a symmetry operation we have by (86.30) with minor change in notation

$$P_{\{\varphi|\boldsymbol{t}\}}Q\begin{pmatrix}\boldsymbol{k}_\sigma\\ j_\mu\end{pmatrix}=\sum_{\nu,\sigma'}D^{(\star\boldsymbol{k})(j)}(\{\varphi|\boldsymbol{t}\})_{\nu\sigma',\mu\sigma}\,Q\begin{pmatrix}\boldsymbol{k}'_{\sigma'}\\ j_\nu\end{pmatrix}\equiv Q'\begin{pmatrix}\boldsymbol{k}_\sigma\\ j_\mu\end{pmatrix} \tag{109.27}$$

or when (109.27) is substituted into (109.26) we obtain the Hamiltonian in primed variables, which is

$$\mathscr{H}_j(\{\dot{Q}'\},\{Q'\})=\mathscr{H}_j(\{P_{\{\varphi|\boldsymbol{t}\}}\dot{Q}\},\{P_{\{\varphi|\boldsymbol{t}\}}Q\}). \tag{109.28}$$

In priming the vector $\boldsymbol{k}_\sigma$, on the right hand side of (109.27) we are, of course, allowing for the possibility that $P_{\{\varphi|\boldsymbol{t}\}}$ may not be in $\mathfrak{G}(\boldsymbol{k})$. Then the quadratic form

(109.26) becomes

$$\mathcal{H}_j = \tfrac{1}{2} \sum_{\sigma} \sum_{\mu} \sum_{\nu'} \sum_{\sigma'} \sum_{\nu''} \sum_{\sigma''} D^{(\star k)(j)}(\{\varphi|t\})_{\sigma''\nu''\sigma\mu}\, D^{(\star k)(j)}(\{\varphi|t\})^*_{\sigma'\nu'\sigma\mu} \times \left\{ \dot{Q}\begin{pmatrix} k_{\sigma'} \\ j_{\nu'} \end{pmatrix}^* \dot{Q}\begin{pmatrix} k_{\sigma''} \\ j_{\nu''} \end{pmatrix} + \omega^2(k_\sigma|j)\, Q\begin{pmatrix} k_{\sigma'} \\ j_{\nu'} \end{pmatrix}^* Q\begin{pmatrix} k_{\sigma''} \\ j_{\nu''} \end{pmatrix} \right\}. \tag{109.29}$$

But by hypothesis the representations $D^{(\star k)(j)}$ are unitary, hence

$$(D^{(\star k)(j)})^+ = (D^{(\star k)(j)})^{-1} \tag{109.30}$$

or

$$(D^{(\star k)(j)})^* = \widetilde{(D^{(\star k)(j)})^{-1}}. \tag{109.31}$$

In matrix elements then

$$(D^{(\star k)(j)})^*_{\mu\nu} = (D^{(\star k)(j)})^{-1}_{\nu\mu} \tag{109.32}$$

or

$$D^{(\star k)(j)}(\{\varphi|t\})_{\sigma'\nu'\sigma\mu} = D^{(\star k)(j)}(\{\varphi|t\})^{-1}_{\sigma\mu\sigma'\nu'}. \tag{109.33}$$

Hence we can write for the sum on μ and σ in (109.29)

$$\begin{aligned} &\sum_{\sigma}\sum_{\mu} D^{(\star k)(j)}(\{\varphi|t\})_{\sigma''\nu''\sigma\mu}\, D^{(\star k)(j)}(\{\varphi|t\}^{-1})_{\sigma\mu\sigma'\nu'} \\ &\quad = D^{(\star k)(j)}(\{\varphi|t\}\{\varphi|t\}^{-1})_{\sigma''\nu''\sigma'\nu'} \\ &\quad = D^{(\star k)(j)}(\{\varepsilon|\mathbf{0}\})_{\sigma''\nu''\sigma'\nu'} = \delta_{\nu''\nu'}\, \Delta(k_{\sigma'} - k_{\sigma''}). \end{aligned} \tag{109.34}$$

Then we obtain for (109.29) the result

$$\mathcal{H}_j = \tfrac{1}{2} \sum_{j\mu\tau} \left\{ \dot{Q}\begin{pmatrix} k_\tau \\ j_\mu \end{pmatrix}^* \dot{Q}\begin{pmatrix} k_\tau \\ j_\mu \end{pmatrix} + \omega^2(k_\tau|j)\, Q\begin{pmatrix} k_\tau \\ j_\mu \end{pmatrix}^* Q\begin{pmatrix} k_\tau \\ j_\mu \end{pmatrix} \right\}. \tag{109.35}$$

We may next sum over all the modes j which appear to regain (109.24). We thus verified that the Hamiltonian is a scalar invariant under transformation by symmetry operations.

The verification of the invariance of $\mathcal{H}$, while important, is not profound. It is to be expected on physical grounds as well as essentially by construction. Recall the theorem that for a finite group only a single Hermitian quadratic invariant form can be constructed from a given irreducible representation namely, the bilinear product of the basis vectors with themselves (Hermitian quadratic form) which requires (109.24) to be a scalar. This property will be used in Sect. 116, so it is important to have proved it explicitly here.

To go beyond the harmonic approximation requires adjoining the anharmonic potential of (109.9) to $\mathcal{H}$. Thus, the total classical lattice Hamiltonian is

$$\mathcal{H}_{\text{tot}} = \mathcal{H} + V_A, \tag{109.36}$$

where

$$V_A = \Phi^{(3)} + \cdots + \Phi^{(s)} + \cdots. \tag{109.37}$$

The series (109.37) consists of a sum of terms each of particular degree in normal modes, each suitably symmetrized, and represents the complete effect of symmetry in the lattice dynamic problem. We return to the series (109.37) after a brief digression.

β) Force constant coupling parameters. As a practical matter in carrying out lattice dynamic calculations of phonon frequencies one is often interested in obtaining an expression of the crystal potential energy in terms of a minimum set of unknown force constant parameters in the expressions for $\Phi^{(2)}$, as epitomized for example in the expression (67.8). That is suppose we desire to express the potential energy V in terms of a truncated expansion, retaining only a limited, finite, number of terms in the sum (67.8). We now make a short digression to discuss the use of symmetry in determining these coupling parameters.[1]

Let us consider the use of the transformation formula for second degree (harmonic) terms

$$\Phi_{\alpha\beta}\begin{pmatrix} l_{\varphi} & l'_{\varphi} \\ \kappa_{\varphi} & \kappa'_{\varphi} \end{pmatrix} = \sum_{\alpha'\beta'} \varphi_{\alpha\alpha'}\,\varphi_{\beta\beta'}\,\Phi_{\alpha'\beta'}\begin{pmatrix} l & l' \\ \kappa & \kappa' \end{pmatrix} \tag{109.38}$$

in which the $P_{\{\boldsymbol{\varphi}\}}$ is taken as the crystal symmetry operation whose rotational part is $\boldsymbol{\varphi}$. Now consider in particular the subset of operations in $\mathfrak{G}$ which leaves

$$\boldsymbol{r}\begin{pmatrix} l \\ \kappa \end{pmatrix} - \boldsymbol{r}\begin{pmatrix} l' \\ \kappa' \end{pmatrix} \tag{109.39}$$

invariant. This subset will determine a subgroup of $\mathfrak{G}$. In a symmorphic group the subgroup will be in fact a subgroup of $\mathfrak{P}$, the point group of the crystal. Call this subgroup $\mathfrak{G}\begin{pmatrix} l & l' \\ \kappa & \kappa' \end{pmatrix}$ defined so that for any element $\boldsymbol{\varphi}_{\lambda}$ in this group

$$\boldsymbol{\varphi}_{\lambda}\cdot\left(\boldsymbol{r}\begin{pmatrix} l \\ \kappa \end{pmatrix} - \boldsymbol{r}\begin{pmatrix} l' \\ \kappa' \end{pmatrix}\right) = \boldsymbol{r}\begin{pmatrix} l \\ \kappa \end{pmatrix} - \boldsymbol{r}\begin{pmatrix} l' \\ \kappa' \end{pmatrix}. \tag{109.40}$$

Then for any element $\boldsymbol{\varphi}_{\lambda}$ in this group:

$$\Phi_{\bar{\alpha}\bar{\beta}}\begin{pmatrix} l_{\varphi\lambda} & l'_{\varphi\lambda} \\ \kappa_{\varphi\lambda} & \kappa'_{\varphi\lambda} \end{pmatrix} = \Phi_{\bar{\alpha}\bar{\beta}}\begin{pmatrix} l & l' \\ \kappa & \kappa' \end{pmatrix}, \tag{109.41}$$

or

$$\Phi_{\bar{\alpha}\bar{\beta}}\begin{pmatrix} l & l' \\ \kappa & \kappa' \end{pmatrix} = \sum_{\alpha\beta} (\boldsymbol{\varphi}_{\lambda})_{\bar{\alpha}\alpha}(\boldsymbol{\varphi}_{\lambda})_{\bar{\beta}\beta}\,\Phi_{\alpha\beta}\begin{pmatrix} l & l' \\ \kappa & \kappa' \end{pmatrix} \tag{109.42}$$

when $\boldsymbol{\varphi}_{\lambda}$ is permitted to vary through all elements in $\mathfrak{G}\begin{pmatrix} l & l' \\ \kappa & \kappa' \end{pmatrix}$, various relations will be obtained among the coupling parameters appropriate to that particular pair of indices $\begin{pmatrix} l & l' \\ \kappa & \kappa' \end{pmatrix}$. In this fashion a complete set of independent coupling parameters appropriate to the given pair of indices will be obtained.

The group $\mathfrak{G}$ can be decomposed into a set of disjoint cosets with respect to $\mathfrak{G}\begin{pmatrix} l & l' \\ \kappa & \kappa' \end{pmatrix}$. Now let $\{\boldsymbol{\varphi}|\boldsymbol{t}\}$ be an element not in $\mathfrak{G}\begin{pmatrix} l & l' \\ \kappa & \kappa' \end{pmatrix}$. Then for this element, (109.38) will be a relationship between different coupling parameters of the same "shell", That is if $\begin{pmatrix} l & l' \\ \kappa & \kappa' \end{pmatrix}$ corresponds to a particular pair of second neighbors

[1] See, for example, the article by LEIBFRIED and LUDWIG [*4*], pp. 286–288 et seq.

then $\begin{pmatrix} l_\varphi & l'_\varphi \\ \kappa_\varphi & \kappa'_\varphi \end{pmatrix}$ will correspond to another equivalent pair and the coupling parameters will be related accordingly, via (109.42).

The principles of the determination of the set of independent coupling parameters corresponding to any "shell" of neighbors in $\Phi^{(2)}$ is thus clear. *Mutatis mutandis* the determination of the minimal set of coupling parameters for any "shell", for any order of term in the expansion (109.7). Since this is not our principal concern here, the reader is referred to the discussion by LEIBFRIED, LEIBFRIED and LUDWIG [4], and by LAX, who discusses the group of the bond [58].

γ) *Anharmonic terms in the potential.* To extract the restrictions which symmetry places upon the coefficients in the expansion of V, we shall transform (109.9) from explicit dependence upon the cartesian components of displacement to the complex normal coordinates $Q\begin{pmatrix} \boldsymbol{k} \\ j \end{pmatrix}$. From (80.9) we obtain

$$u_\alpha\begin{pmatrix} l \\ \kappa \end{pmatrix} = \frac{1}{\sqrt{M_\kappa}\sqrt{N}} \sum_{\boldsymbol{k}} \sum_{j} e_\alpha\left(\kappa \middle| \begin{matrix} \boldsymbol{k} \\ j \end{matrix}\right) Q\begin{pmatrix} \boldsymbol{k} \\ j \end{pmatrix} e^{i\boldsymbol{k}\cdot\boldsymbol{R}_L}. \tag{109.43}$$

Now we substitute (109.43) into the terms in (109.7) and simplify the resulting expressions.

Let us examine as a typical term, the cubic anharmonic term (109.9) when (109.43) is substituted. At this juncture there are two ways of proceeding. Let us first do the direct substitution. Recall the transformation property of the coupling parameters under a simple translation by a lattice vector $P_{\{\varepsilon|\boldsymbol{R}_{\bar{L}}\}}$. Thus under such a transformation the displacements $\boldsymbol{u}$ are shifted into displacements $\boldsymbol{u}_T$ where

$$(u_T)_\alpha \begin{pmatrix} l-\bar{l} \\ \kappa \end{pmatrix} = u_\alpha \begin{pmatrix} l \\ \kappa \end{pmatrix}. \tag{109.44}$$

The expression for the potential energy $\Phi^{(3)}$ is then

$$\begin{aligned} \Phi^{(3)} = \tfrac{1}{6} \sum_{\substack{l-\bar{l} \\ \kappa,\alpha}} \sum_{\substack{l'-\bar{l} \\ \kappa',\beta}} \sum_{\substack{l''-\bar{l} \\ \kappa'',\gamma}} (\Phi_T)_{\alpha\beta\gamma} \begin{pmatrix} l-\bar{l} & l'-\bar{l} & l''-\bar{l} \\ \kappa & \kappa' & \kappa'' \end{pmatrix} \\ \times (u_T)_\alpha \begin{pmatrix} l-\bar{l} \\ \kappa \end{pmatrix} (u_T)_\beta \begin{pmatrix} l'-\bar{l} \\ \kappa' \end{pmatrix} (u_T)_\gamma \begin{pmatrix} l''-\bar{l} \\ \kappa'' \end{pmatrix} \end{aligned} \tag{109.45}$$

where as usual

$$(\Phi_T)_{\alpha\beta\gamma} \begin{pmatrix} l-\bar{l} & l'-\bar{l} & l''-l' \\ \kappa & \kappa' & \kappa'' \end{pmatrix} = \Phi_{\alpha\beta\gamma} \begin{pmatrix} l & l' & l'' \\ \kappa & \kappa' & \kappa'' \end{pmatrix}. \tag{109.46}$$

But by translational invariance of the lattice, under translation by an arbitrary lattice vector $\boldsymbol{R}_{\bar{L}}$,

$$\Phi_{\alpha\beta\gamma} \begin{pmatrix} l & l' & l'' \\ \kappa & \kappa' & \kappa'' \end{pmatrix} = \Phi_{\alpha\beta\gamma} \begin{pmatrix} l-\bar{l} & l'-\bar{l} & l''-\bar{l} \\ \kappa & \kappa' & \kappa'' \end{pmatrix}. \tag{109.47}$$

Thus first of all we require that the cubic anharmonic term $\Phi^{(3)}$ is, as usual, invariant under lattice translation. Of greater moment from (109.47) we obtain, letting $\bar{l}=l$ (i.e. $\boldsymbol{R}_{\bar{L}}=\boldsymbol{R}_L$)

$$\Phi_{\alpha\beta\gamma}\begin{pmatrix} l & l' & l'' \\ \kappa & \kappa' & \kappa'' \end{pmatrix}=\Phi_{\alpha\beta\gamma}\begin{pmatrix} 0 & l'-l & l''-l \\ \kappa & \kappa' & \kappa'' \end{pmatrix}. \tag{109.48}$$

Of course we could have permitted the arbitrary $\bar{l}$ to equal in turn l' or l'' so that

$$\begin{aligned}\Phi_{\alpha\beta\gamma}\begin{pmatrix} l & l' & l'' \\ \kappa & \kappa' & \kappa'' \end{pmatrix}&=\Phi_{\alpha\beta\gamma}\begin{pmatrix} 0 & l'-l & l''-l \\ \kappa & \kappa' & \kappa'' \end{pmatrix}=\Phi_{\alpha\beta\gamma}\begin{pmatrix} l-l' & 0 & l''-l' \\ \kappa & \kappa' & \kappa'' \end{pmatrix}\\ &=\Phi_{\alpha\beta\gamma}\begin{pmatrix} l-l'' & l'-l'' & 0 \\ \kappa & \kappa' & \kappa'' \end{pmatrix}.\end{aligned} \tag{109.49}$$

To obtain further restrictions on the values of the coupling parameters for a given crystal, with space group $\mathfrak{G}$, we can apply the symmetry operations in $\mathfrak{G}$ using (109.17), (109.49) and so obtain relations among the coupling parameters. The independent coupling parameters are generally grouped into a restricted number of parameters of a given "shell". Since this is not our main objective, again we refer to the article of Leibfried and Ludwig [4].

We return to (109.9), and substitute (109.43), and (109.47), (109.49) to obtain

$$\begin{aligned}\Phi^{(3)}=&\frac{1}{6}\cdot\frac{1}{N^{\frac{3}{2}}}\sum_{\boldsymbol{k}j}\sum_{\boldsymbol{k}'j'}\sum_{\boldsymbol{k}''j''}Q\begin{pmatrix}\boldsymbol{k}\\ j\end{pmatrix}Q\begin{pmatrix}\boldsymbol{k}'\\ j'\end{pmatrix}Q\begin{pmatrix}\boldsymbol{k}''\\ j''\end{pmatrix}\\ &\times\sum_{l\kappa\alpha}\sum_{l'\kappa'\beta}\sum_{l''\kappa''\gamma}\frac{1}{(M_\kappa M_{\kappa'} M_{\kappa''})^{\frac{1}{2}}}\Phi_{\alpha\beta\gamma}\begin{pmatrix} 0 & l'-l & l''-l \\ \kappa & \kappa' & \kappa'' \end{pmatrix}\\ &\times e_\alpha\left(\kappa\middle|\begin{matrix}\boldsymbol{k}\\ j\end{matrix}\right)e_\beta\left(\kappa'\middle|\begin{matrix}\boldsymbol{k}'\\ j'\end{matrix}\right)e_\gamma\left(\kappa''\middle|\begin{matrix}\boldsymbol{k}''\\ j''\end{matrix}\right)\exp\{i(\boldsymbol{k}\cdot\boldsymbol{R}_L+\boldsymbol{k}'\cdot\boldsymbol{R}_{L'}+\boldsymbol{k}''\cdot\boldsymbol{R}_{L''})\}.\end{aligned} \tag{109.50}$$

Now in (109.50) we multiply the right hand side (under the summation) by unity, as:

$$\exp\{i(\boldsymbol{k}'\cdot\boldsymbol{R}_L-\boldsymbol{k}'\cdot\boldsymbol{R}_L+\boldsymbol{k}''\cdot\boldsymbol{R}_L-\boldsymbol{k}''\cdot\boldsymbol{R}_L)\}. \tag{109.51}$$

Then changing the lattice summation indices we have in the nine-fold sum in the right hand side of (109.50)

$$\begin{aligned}&\Big\{\sum_l \exp i(\boldsymbol{k}\cdot\boldsymbol{R}_L+\boldsymbol{k}'\cdot\boldsymbol{R}_L+\boldsymbol{k}''\cdot\boldsymbol{R}_L)\Big\}\\ &\quad\times\sum_{\kappa\alpha}\sum_{l'-l,\kappa'\beta}\sum_{l''-l,\gamma}\frac{1}{(M_\kappa M_{\kappa'} M_{\kappa''})^{\frac{1}{2}}}\Phi_{\alpha\beta\gamma}\begin{pmatrix} 0 & l'-l & l''-l \\ \kappa & \kappa' & \kappa'' \end{pmatrix}\\ &\quad\times e_\alpha\left(\kappa\middle|\begin{matrix}\boldsymbol{k}\\ j\end{matrix}\right)e_\beta\left(\kappa'\middle|\begin{matrix}\boldsymbol{k}'\\ j'\end{matrix}\right)e_\gamma\left(\kappa'\middle|\begin{matrix}\boldsymbol{k}''\\ j''\end{matrix}\right)\exp i\{\boldsymbol{k}'\cdot(\boldsymbol{R}_{L'}-\boldsymbol{R}_L)+\boldsymbol{k}''\cdot(\boldsymbol{R}_{L''}-\boldsymbol{R}_L)\}.\end{aligned} \tag{109.52}$$

The sum over l produces a lattice delta: $\Delta(\boldsymbol{k}+\boldsymbol{k}'+\boldsymbol{k}'')$ where:

$$\Delta(\boldsymbol{k}+\boldsymbol{k}'+\boldsymbol{k}'')=\begin{cases}1 & \text{if } \boldsymbol{k}+\boldsymbol{k}'+\boldsymbol{k}''=2\pi\boldsymbol{B}_H\\ 0 & \text{otherwise.}\end{cases} \tag{109.53}$$

Since the vectors $\boldsymbol{k}, \boldsymbol{k}', \boldsymbol{k}''$ are defined in the first Brillouin zone, the restriction (109.53) requires their sum to be zero, mod $2\pi\boldsymbol{B}_H$. We then define the Fourier transformed coupling parameters as

$$\Phi\begin{pmatrix}\boldsymbol{k} & \boldsymbol{k}' & \boldsymbol{k}''\\ j & j' & j''\end{pmatrix}=\sum_{\kappa\alpha}\sum_{\lambda'\kappa'\beta}\sum_{\lambda''\kappa''\gamma}\Phi_{\alpha\beta\gamma}\begin{pmatrix}0 & \lambda' & \lambda''\\ \kappa & \kappa' & \kappa''\end{pmatrix}\frac{1}{(M_\kappa M_{\kappa'} M_{\kappa''})^{\frac{1}{2}}}$$
$$\times e_\alpha\left(\kappa\middle|\begin{matrix}\boldsymbol{k}\\ j\end{matrix}\right) e_\beta\left(\kappa'\middle|\begin{matrix}\boldsymbol{k}'\\ j'\end{matrix}\right) e_\gamma\left(\kappa''\middle|\begin{matrix}\boldsymbol{k}''\\ j''\end{matrix}\right)\exp i\{\boldsymbol{k}'\cdot\boldsymbol{R}_{\lambda'}+\boldsymbol{k}''\cdot\boldsymbol{R}_{\lambda''}\}. \tag{109.54}$$

Hence we can write $\Phi^{(3)}$ as

$$\Phi^{(3)}=\frac{1}{6N^{\frac{1}{2}}}\sum_{\boldsymbol{k}j}\sum_{\boldsymbol{k}'j'}\sum_{\boldsymbol{k}''j''}\Phi\begin{pmatrix}\boldsymbol{k} & \boldsymbol{k}' & \boldsymbol{k}''\\ j & j' & j''\end{pmatrix} Q\begin{pmatrix}\boldsymbol{k}\\ j\end{pmatrix} Q\begin{pmatrix}\boldsymbol{k}'\\ j'\end{pmatrix} Q\begin{pmatrix}\boldsymbol{k}''\\ j''\end{pmatrix}\Delta(\boldsymbol{k}+\boldsymbol{k}'+\boldsymbol{k}''). \tag{109.55}$$

This is the standard form for the coupling coefficients and thus (109.55) is seen to be identical to a standard result.[2] Clearly in obtaining (109.55) use has been made of the translational symmetry of the crystal, and the Δ function restriction epitomizes this. However, the form (109.55) is not convenient for examining the complete (rotational plus translational) symmetry of the crystal.

Hence, we define coupling coefficients from the transformation of (109.11), using (109.10) to find

$$\Phi^{(3)}=\frac{1}{6N^{\frac{3}{2}}}\sum_{\boldsymbol{k}j}\sum_{\boldsymbol{k}'j'}\sum_{\boldsymbol{k}''j''}V\begin{pmatrix}\boldsymbol{k} & \boldsymbol{k}' & \boldsymbol{k}''\\ j & j' & j''\end{pmatrix} Q\begin{pmatrix}\boldsymbol{k}\\ j\end{pmatrix} Q\begin{pmatrix}\boldsymbol{k}'\\ j'\end{pmatrix} Q\begin{pmatrix}\boldsymbol{k}''\\ j''\end{pmatrix} \tag{109.56}$$

with

$$V\begin{pmatrix}\boldsymbol{k} & \boldsymbol{k}' & \boldsymbol{k}''\\ j & j' & j''\end{pmatrix}\equiv\sum_{l\kappa\alpha}\sum_{l'\kappa'\beta}\sum_{l''\kappa''\gamma}\Phi_{\alpha\beta\gamma}\begin{pmatrix}l & l' & l''\\ \kappa & \kappa' & \kappa''\end{pmatrix}$$
$$\times\frac{1}{(M_\kappa M_{\kappa'} M_{\kappa''})^{\frac{1}{2}}} e_\alpha\left(\kappa\middle|\begin{matrix}\boldsymbol{k}\\ j\end{matrix}\right) e_\beta\left(\kappa'\middle|\begin{matrix}\boldsymbol{k}'\\ j'\end{matrix}\right) e_\gamma\left(\kappa''\middle|\begin{matrix}\boldsymbol{k}''\\ j''\end{matrix}\right) \tag{109.57}$$
$$\times\exp\{i(\boldsymbol{k}\cdot\boldsymbol{R}_L+\boldsymbol{k}'\cdot\boldsymbol{R}_{L'}+\boldsymbol{k}''\cdot\boldsymbol{R}_{L''})\}.$$

Consider the expression for the anharmonic term $\Phi^{(3)}$ when we use as the basic set of dynamical variables the components of ion Cartesian displacement transformed by a general transformation $P_{\{\varphi_p\}}$ of $\mathfrak{G}$. That is we write instead of (109.11) the expression for $\Phi^{(3)}$ in terms of the variables $(u_{\varphi_p})_\alpha\begin{pmatrix}l\\ \kappa\end{pmatrix}$, etc. as:

$$\Phi'^{(3)}=\frac{1}{3!}\sum_{l\kappa\alpha}\sum_{l'\kappa'\beta}\sum_{l''\kappa''\gamma}\Phi'_{\alpha\beta\gamma}\begin{pmatrix}l & l' & l''\\ \kappa & \kappa' & \kappa''\end{pmatrix}(u_{\varphi_p})_\alpha\begin{pmatrix}l\\ \kappa\end{pmatrix}(u_{\varphi_p})_\beta\begin{pmatrix}l'\\ \kappa'\end{pmatrix}(u_{\varphi_p})_\gamma\begin{pmatrix}l''\\ \kappa''\end{pmatrix}. \tag{109.58}$$

The primed component of the effective force matrix is, by form invariance (109.5b), equal to the unprimed element, or we have

$$\Phi'^{(3)}=\tfrac{1}{6}\sum_{l\kappa\alpha}\sum_{l'\kappa'\beta}\sum_{l''\kappa''\gamma}\Phi_{\alpha\beta\gamma}\begin{pmatrix}l & l' & l''\\ \kappa & \kappa' & \kappa''\end{pmatrix}(u_{\varphi_p})_\alpha\begin{pmatrix}l\\ \kappa\end{pmatrix}(u_{\varphi_p})_\beta\begin{pmatrix}l'\\ \kappa'\end{pmatrix}(u_{\varphi_p})_\gamma\begin{pmatrix}l''\\ \kappa''\end{pmatrix}. \tag{109.59}$$

Using the rule for expressing the rotated ion displacements in terms of the initial plane waves and unrotated eigenvectors we have:

$$(u_{\varphi_p})_\alpha\begin{pmatrix}l\\ \kappa\end{pmatrix}=\frac{1}{\sqrt{M_\kappa N}}\sum_{\boldsymbol{k},\tau}\sum_{j,\nu}e^{i\boldsymbol{k}\cdot\boldsymbol{R}_L}e_\alpha\left(\kappa\middle|\begin{matrix}\boldsymbol{k}\\ j\end{matrix}\right)Q_{\{\varphi_p\}}\begin{pmatrix}\boldsymbol{k}_\tau\\ j_\nu\end{pmatrix}. \tag{109.60}$$

[2] See [13], Eq. (39.2).

Then using (86.30) we have:

$$Q_{\{\varphi_p\}}\begin{pmatrix}\boldsymbol{k}_\tau\\ j_\nu\end{pmatrix}=\sum_{\sigma=1}^{s}\sum_{\mu=1}^{l_m} D^{(\star\boldsymbol{k})(j)}(\{\varphi_p\})_{(\sigma\mu)(\tau\nu)}\,Q\begin{pmatrix}\boldsymbol{k}_\sigma\\ j_\mu\end{pmatrix}. \tag{109.61}$$

Substituting (109.60), and (109.61) into (109.59) we have:

$$\begin{aligned}\Phi'^{(3)}=\frac{1}{3!\,N^{\frac{3}{2}}}&\sum_{\boldsymbol{k}_\tau j_\nu}\sum_{\boldsymbol{k}'_{\tau'} j'_{\nu'}}\sum_{\boldsymbol{k}''_{\tau''} j''_{\nu''}} V\begin{pmatrix}\boldsymbol{k}_\tau & \boldsymbol{k}'_{\tau'} & \boldsymbol{k}''_{\tau''}\\ j_\nu & j'_{\nu'} & j''_{\nu''}\end{pmatrix}\\ &\times\sum_{\sigma\mu}\sum_{\sigma'\mu'}\sum_{\sigma''\mu''} D^{(\star\boldsymbol{k})(j)}(\{\varphi_p\})_{\sigma\mu\tau\nu}\,D^{(\star\boldsymbol{k}')(j')}(\{\varphi_p\})_{\sigma'\mu'\tau'\nu'}\\ &\times D^{(\star\boldsymbol{k}'')(j'')}(\{\varphi_p\})_{\sigma''\mu''\tau''\nu''}\,Q\begin{pmatrix}\boldsymbol{k}_\sigma\\ j_\mu\end{pmatrix}Q\begin{pmatrix}\boldsymbol{k}'_{\sigma'}\\ j'_{\mu'}\end{pmatrix}Q\begin{pmatrix}\boldsymbol{k}''_{\sigma''}\\ j''_{\mu''}\end{pmatrix}\end{aligned} \tag{109.62}$$

where V is defined in (109.57). The result of the transformation is to mix terms previously contained in $\Phi^{(3)}$, into those prescribed linear combinations given in (109.62). Then, the transformation has resulted in mixing of the $(s\cdot l_j)(s'\cdot l_{j'})(s''\cdot l_{j''})$ terms belonging to the full product space

$$\Sigma^{(\star\boldsymbol{k})(j)}\otimes\Sigma^{(\star\boldsymbol{k}')(j')}\otimes\Sigma^{(\star\boldsymbol{k}'')(j'')} \tag{109.63}$$

amongest one another: no terms from other spaces will be admixed. Clearly the product space (109.63) produces the representation

$$D^{(\star\boldsymbol{k})(j)}\otimes D^{(\star\boldsymbol{k}')(j')}\otimes D^{(\star\boldsymbol{k}'')(j'')}. \tag{109.64}$$

But by the previously given argument of invariance,

$$\Phi'^{(3)}=\Phi^{(3)}. \tag{109.65}$$

Recalling that we reserve the symbol $(\boldsymbol{\Gamma})(1+)$ for the invariant (identity) representation,

$$\Phi^{(3)}\sim D^{(\boldsymbol{\Gamma})(1+)}. \tag{109.66}$$

Consequently, the terms from the product space (109.63) will only appear if, in the reduction of the direct product, the identity representation $D^{(\boldsymbol{\Gamma})(1+)}$ appears. Thus, in the construction of the terms in (109.56) only those products of normal coordinates arise for which the reduction coefficient is nonvanishing i.e.

$$(\star\boldsymbol{k}\,m\,\star\boldsymbol{k}'\,m'\,\star\boldsymbol{k}''\,m''\,|\,\boldsymbol{\Gamma}(1+))\neq 0. \tag{109.67}$$

It can then easily be observed that this result encompasses the translational symmetry as well as the rotational: so for example the wave vector selection rule that

$$\boldsymbol{k}+\boldsymbol{k}'+\boldsymbol{k}''=2\pi\boldsymbol{B}_h \quad\text{or}\quad =0 \quad (\text{mod}\,2\pi\boldsymbol{B}_H) \tag{109.68}$$

which follows from translational symmetry is evidently subsumed under the requirement that in (109.67) $\boldsymbol{\Gamma}(1+)$ be contained. Of course the restriction (109.67) is still more severe and precise since the rotational transformation behavior is also included under (109.67).

We may generalize these results to any term in the series (109.7). Thus consider the s-th order terms in the expansion. These will be expressed as

$$\Phi^{(s)}=\frac{1}{s!}\cdot\frac{1}{N^{s/2}}\sum_{\boldsymbol{k}j}\sum_{\boldsymbol{k}'j'}\cdots\sum_{\boldsymbol{k}^{(s)}j^{(s)}} V\begin{pmatrix}\boldsymbol{k} & \boldsymbol{k}'\ldots\boldsymbol{k}^{(s)}\\ j & j'\ldots j^{(s)}\end{pmatrix} Q\begin{pmatrix}\boldsymbol{k}\\ j\end{pmatrix} Q\begin{pmatrix}\boldsymbol{k}'\\ j'\end{pmatrix}\ldots Q\begin{pmatrix}\boldsymbol{k}^{s}\\ j^{s}\end{pmatrix}, \tag{109.69}$$

where, by analogy to (109.57)

$$\begin{aligned} V\begin{pmatrix}\boldsymbol{k} & \boldsymbol{k}'\ldots\boldsymbol{k}^{(s)}\\ j & j'\ldots j^{(s)}\end{pmatrix} &\equiv \sum_{l\kappa\alpha}\sum_{l'\kappa'\beta}\cdots\sum_{l^{(s)}\kappa^{(s)}\gamma}\Phi_{\alpha\beta\ldots\gamma}\begin{pmatrix}l & l'\ldots l^{(s)}\\ \kappa & \kappa'\ldots\kappa^{(s)}\end{pmatrix}\\ &\times\frac{1}{(M_\kappa M_{\kappa'}\ldots M_{\kappa^s})^{\frac{1}{2}}}\, e_\alpha\left(\kappa\left|\begin{matrix}\boldsymbol{k}\\ j\end{matrix}\right.\right) e_\beta\left(\kappa'\left|\begin{matrix}\boldsymbol{k}'\\ j'\end{matrix}\right.\right)\ldots e_\gamma\left(\kappa^{(s)}\left|\begin{matrix}\boldsymbol{k}^{(s)}\\ j^{(s)}\end{matrix}\right.\right)\\ &\times\exp\{i\,\boldsymbol{k}\cdot\boldsymbol{R}_L+\boldsymbol{k}'\cdot\boldsymbol{R}_{L'}+\cdots+\boldsymbol{k}^{(s)}\cdot\boldsymbol{R}_{L^s}\}. \end{aligned} \tag{109.70}$$

The products (109.69) evidently transform according to the product representation

$$D^{(\star\boldsymbol{k})(j)}\otimes D^{(\star\boldsymbol{k}')(j')}\otimes\cdots\otimes D^{(\star\boldsymbol{k}^{(s)})(j^{(s)})}. \tag{109.71}$$

Equivalently, the products (109.69) may be grouped into spaces and the statement (109.71) is equivalent to the statement that the products

$$Q\begin{pmatrix}\boldsymbol{k}\\ j\end{pmatrix} Q\begin{pmatrix}\boldsymbol{k}'\\ j'\end{pmatrix}\ldots Q\begin{pmatrix}\boldsymbol{k}^{(s)}\\ j^{(s)}\end{pmatrix} \tag{109.72}$$

are bases for the product vector space

$$\Sigma^{(\star\boldsymbol{k})(j)}\otimes\Sigma^{(\star\boldsymbol{k}')(j')}\otimes\cdots\otimes\Sigma^{(\star\boldsymbol{k}^{(s)})(j^{(s)})}. \tag{109.73}$$

A necessary and sufficient condition that a particular product string of normal coordinates appear in $\Phi^{(s)}$ is that the corresponding character reduction coefficient be nonvanishing, i.e.

$$(\star\boldsymbol{k}j\,\star\boldsymbol{k}'j'\ldots\star\boldsymbol{k}^{(s)}j^{(s)}|\boldsymbol{\Gamma}\,1+)\neq 0. \tag{109.74}$$

As required, ordinary products should be replaced by symmetrized powers.

A trivial consequence of (109.67) or (109.74) is obtained from the corresponding wave vector selection rule, which in the case (109.67) is

$$(\star\boldsymbol{k}\,\star\boldsymbol{k}'\,\star\boldsymbol{k}''|\boldsymbol{\Gamma})\neq 0 \tag{109.75}$$

or, that for some set of three wave vectors (one from each of the stars):

$$\boldsymbol{k}_\sigma+\boldsymbol{k}'_{\sigma'}+\boldsymbol{k}''_{\sigma''}=0 \quad (\mathrm{mod}\ 2\pi\boldsymbol{B}_H). \tag{109.76}$$

For the most important applications, it is expansions such as (109.7) which are of importance in terms of series in the normal coordinates $Q\begin{pmatrix}\boldsymbol{k}\\ j_\nu\end{pmatrix}$ with each term in the series of some particular degree, as in (109.69). Thus one often writes down, *a priori*, an expansion like (109.7), for the Hamiltonian (e.g. the potential energy) or some other crystal invariant. The problem to be resolved, is then the inverse problem to that of the determination of the set of independent coordinate space coupling parameters as in (109.41) and (109.42) (second order coupling parameters), or (109.56) and (109.59) etc. (third or higher order coupling parameters). Namely in an expansion like (109.69) we can demand which combinations, or products, of the normal modes $Q\begin{pmatrix}\boldsymbol{k}\\ j_\nu\end{pmatrix}$ will have non-zero coefficients $V\begin{pmatrix}\boldsymbol{k}\ldots\\ j\ldots\end{pmatrix}$. The solution is given in (109.74). In case we are examining a term in (109.69) in which a given mode $Q\begin{pmatrix}\boldsymbol{k}\\ j\end{pmatrix}$ occurs p times, then the relevent product in (109.74) is *symmetrized* Kronecker power e.g. $[D^{(\star\boldsymbol{k})(j)}]_{(p)}$ etc. However, if two different (independent) modes which have the same symmetry occur, then the *ordinary* Kronecker power e.g. $[D^{(\star\boldsymbol{k})(j)}]_p$ occurs in (109.74).

The discussion of the construction of crystal invariants has been up to this point, entirely in the framework of the representation theory of the unitary spatial subgroup $\mathfrak{G}$. As we know the time reversal symmetry must also be incorporated into the theory. This can be accomplished in either of two ways, similarly to our discussion in Sects. 88–94 (or Sects. 95–102). Thus we can consider the operation K, of complex conjugation as introduced in Sect. 88: e.g. (88.1). Then the requirement of time reversal invariance is, in the case of the potential energy, a reality requirement:

$$K^{-1}\Phi K = \Phi^* = \Phi. \tag{109.77}$$

Now (109.77) must apply to each term in (109.7), since terms of different degrees are not mixed by the operation of complex conjugation. We can see that in detailed application (109.77) may result in the requirement of equality among certain of the coefficients in (109.69) in order that the reality (109.77) is satisfied. Each case should be examined in detail.

As an alternative approach, one could begin with the normal modes which are bases for irreducible corepresentations. Using these as bases, one would demand the construction of lattice invariants such as (109.7) and (109.69). But this approach requires ready availability of the corepresentation reduction coefficients analogous to (109.74) for the space groups. These have not yet been obtained. In principle however, the matter is straightforward—the entire preceeding analysis of this section carries over with the substitution of corepresentation theory. If for a space time group $\mathcal{G}$, the coefficient

$$(\text{co}\,{\star}\boldsymbol{k}j\;\text{co}\,{\star}\boldsymbol{k}'j'\ldots|\boldsymbol{\Gamma}1+)\neq 0 \tag{109.78}$$

then a term corresponding will appear in (109.74).

It is appropriate to note here that in addition to invariance requirements under the space-time group of the crystal, the coupling parameters (generalized force constants) may also be invariant under general homogeneous transformation (rotation and translation) of space. These questions are dealt with in [*4*] to which reference should be made for details.

110. Construction of crystal covariants: Electric moment and polarizability. Now we consider specifically two crystal covariants which are of importance in the general optic theory. These are the electric dipole moment of the crystal, and the electrical polarizability.[1] Each of these quantities may be considered to have a well defined value when the complete set of ion displacements $u_\alpha\binom{l}{\kappa}$ are specified. We write

$$M\left(\left\{u_\alpha\binom{l}{\kappa}\right\}\right) \quad \text{for electrical dipole moment,} \tag{110.1}$$

$$P\left(\left\{u_\alpha\binom{l}{\kappa}\right\}\right) \quad \text{for electric polarizability.} \tag{110.2}$$

The electric dipole moment $\boldsymbol{M}$ is a polar vector, while the electric polarizability $\boldsymbol{P}$ is a symmetric second rank tensor.[2] Thus the quantity $\boldsymbol{M}$ has three components on Cartesian axes

$$M_\alpha, M_\beta, M_\gamma \tag{110.3}$$

which transform under an arbitrary rotation (φ) as

$$M'_{\bar{\alpha}} = \sum_\alpha \varphi_{\bar{\alpha}\alpha} M_\alpha \tag{110.4}$$

where the $\varphi_{\bar{\alpha}\alpha}$ are, as before, the components of the rotation matrix which produces the coordinate transformation. The transformation (110.4) corresponds to the transformation of a polar vector, which would be epitomized as usual, by

$$P_{\{\varphi\}} \boldsymbol{M} = D^{(r)}(\{\varphi\}) \boldsymbol{M}, \tag{110.5}$$

where $D^{(r)}$ is the 3×3 matrix of the transformation of a polar vector. The representation $D^{(r)}$ may or may not be irreducible for the particular group $\mathfrak{G}$. The character, or trace of this representation is

$$\operatorname{Tr} D^{(r)}(\{\varphi\}) = \pm(1+2\cos\varphi), \tag{110.6}$$

where φ is the angle of the rotation $\{\varphi\}$. In the case of the cubic space groups, such as O_h^7 (diamond) or O_h^5 (rocksalt)

$$D^{(r)} = \Gamma^{(15-)} \tag{110.7}$$

which is in fact irreducible.

For the polarizability tensor $\boldsymbol{P}$, the rule which corresponds to (110.4) is obtained by considering the transformation of the components $P_{\alpha\beta}$. This is

$$P'_{\bar{\alpha}\bar{\beta}} = \sum_\alpha \sum_\beta \varphi_{\bar{\alpha}\alpha} \varphi_{\bar{\beta}\beta} P_{\alpha\beta} \tag{110.8}$$

and demonstrates that $\boldsymbol{P}$ is a second rank tensor. The polarizability tensor $\boldsymbol{P}$ is in addition generally taken symmetric [4][2]

$$P_{\alpha\beta} = P_{\beta\alpha}. \tag{110.9}$$

The transformation (110.8) may also be written

$$P_{\{\varphi\}} \boldsymbol{P} = D^{(rr)}(\{\varphi\}) \boldsymbol{P} = [D^{(r)}(\{\varphi\})]_{(2)} \boldsymbol{P}, \tag{110.10}$$

where $D^{(rr)}$ is the representation by which a symmetric second rank tensor transforms. Again, depending on the space group $\mathfrak{G}$, $D^{(rr)}$ may or may not be irreducible.

[1] See [*13*], Eqs. (39.11)–(39.18) for discussion of restrictions due to translational symmetry alone.

[2] Near resonance, antisymmetric components of $\boldsymbol{P}$ may become important in Raman scattering (see Sect. 124). Also morphic effects in certain crystal classes may involve those components.

In the case of space groups O_h^7 and O_h^5 the representation $[D^{(r)}]_{(2)}$ is reducible as

$$[D^{(r)}]_{(2)} = \Gamma^{(1+)} \oplus \Gamma^{(12+)} \oplus \Gamma^{(25+)}. \tag{110.11}$$

It may be well to emphasize at this point that the expressions (110.3) and (110.10) or (110.11) and (110.8) give the transformation of the entire objects $\boldsymbol{M}$ and $\boldsymbol{P}$; each of these is of course a multicomponent quantity and it is the components which transform among themselves under coordinate transformations. In particular, when each of the quantities $\boldsymbol{M}$, $\boldsymbol{P}$ is expressed in a Taylor series in the displacement components $u_\alpha \begin{pmatrix} l \\ \kappa \end{pmatrix}$ then each term, which is homogenous of particular degree in the components, must transform under rotation just as the entire objects (110.1) or (110.2).

Consider now $\boldsymbol{M}$, the electric dipole moment of the crystal. Let us write for $\boldsymbol{M}$ an expansion in Taylor series in the $u_\alpha \begin{pmatrix} l \\ \kappa \end{pmatrix}$. Thus for each component we have

$$M_\alpha = M_\alpha^{(1)} + M_\alpha^{(2)} + \cdots + M_\alpha^{(s)} + \cdots \tag{110.12}$$

where

$$M_\alpha^{(1)} = \sum_{l\kappa\beta} M_{\alpha,\beta} \begin{pmatrix} l \\ \kappa \end{pmatrix} u_\beta \begin{pmatrix} l \\ \kappa \end{pmatrix} \tag{110.13}$$

and

$$M_{\alpha,\beta} \begin{pmatrix} l \\ \kappa \end{pmatrix} = \left[\frac{\partial M_\alpha}{\partial u_\beta \begin{pmatrix} l \\ \kappa \end{pmatrix}} \right]_0 \tag{110.14}$$

and

$$M_\alpha^{(2)} = \sum_{l\kappa\beta} \sum_{l'\kappa'\gamma} M_{\alpha,\beta\gamma} \begin{pmatrix} l & l' \\ \kappa & \kappa' \end{pmatrix} u_\beta \begin{pmatrix} l \\ \kappa \end{pmatrix} u_\alpha \begin{pmatrix} l' \\ \kappa' \end{pmatrix} \tag{110.15}$$

and

$$M_{\alpha,\beta\gamma} \begin{pmatrix} l & l' \\ \kappa & \kappa' \end{pmatrix} = \left[\frac{\partial^2 M_\alpha}{\partial u_\beta \begin{pmatrix} l \\ \kappa \end{pmatrix} \partial u_\gamma \begin{pmatrix} l' \\ \kappa' \end{pmatrix}} \right]_0 \tag{110.16}$$

similarly for the terms of higher degree. Since the entire quantity $\boldsymbol{M}$ transforms as $D^{(r)}$, each of the series (110.13), (110.15) must. The terms in the series (110.12) can be expressed as a series in the normal coordinates as before. Thus we may write

$$M_\alpha^{(1)} = \sum_{\boldsymbol{k}j\nu} M_\alpha \begin{pmatrix} \boldsymbol{k} \\ j_\nu \end{pmatrix} Q \begin{pmatrix} \boldsymbol{k} \\ j_\nu \end{pmatrix} \tag{110.17}$$

with

$$M_\alpha \begin{pmatrix} \boldsymbol{k} \\ j_\nu \end{pmatrix} \equiv \sum_{l\kappa\beta} M_{\alpha,\beta} \begin{pmatrix} l \\ \kappa \end{pmatrix} \frac{1}{\sqrt{M_\kappa N}} e_\beta \left(\kappa \middle| \begin{matrix} \boldsymbol{k} \\ j_\nu \end{matrix} \right) \exp(i\boldsymbol{k} \cdot \boldsymbol{R}_L) \tag{110.18}$$

and

$$M_\alpha^{(2)} = \sum_{\boldsymbol{k}j\nu} \sum_{\boldsymbol{k}'j'\nu'} M_\alpha \begin{pmatrix} \boldsymbol{k} & \boldsymbol{k}' \\ j_\nu & j'_{\nu'} \end{pmatrix} Q \begin{pmatrix} \boldsymbol{k} \\ j_\nu \end{pmatrix} Q \begin{pmatrix} \boldsymbol{k}' \\ j'_{\nu'} \end{pmatrix} \tag{110.19}$$

with

$$M_\alpha\begin{pmatrix}\boldsymbol{k} & \boldsymbol{k}'\\ j_\nu & j'_{\nu'}\end{pmatrix} \equiv \sum_{l\kappa\beta}\sum_{l'\kappa'\gamma} M_{\alpha,\beta\gamma}\begin{pmatrix}l & l'\\ \kappa & \kappa'\end{pmatrix}\frac{e_\beta\left(\kappa\middle|\begin{smallmatrix}\boldsymbol{k}\\ j_\nu\end{smallmatrix}\right) e_\alpha\left(\kappa'\middle|\begin{smallmatrix}\boldsymbol{k}'\\ j'_{\nu'}\end{smallmatrix}\right)}{N\sqrt{M_\kappa M_{\kappa'}}} \times \exp i(\boldsymbol{k}\cdot\boldsymbol{R}_L + \boldsymbol{k}'\cdot\boldsymbol{R}_{L'}) \tag{110.20}$$

etc. As with the anharmonic terms in the potential, the complete space group symmetry requirement will restrict the sets of $Q\begin{pmatrix}\boldsymbol{k}\\ j_\nu\end{pmatrix}$ which may occur in the expansions (110.17), (110.18), etc.

Now consider the expression for the first degree term in $\boldsymbol{M}$ when the basic set of dynamical variables is taken as the rotated cartesian displacement components. From an analysis like (109.71)–(109.74)

$$M_{\bar{\alpha}}^{(1)} = \sum_\alpha \varphi_{\bar{\alpha}\alpha} M_\alpha^{(1)} \tag{110.21}$$

but also

$$M_{\bar{\alpha}}^{\prime(1)} = \sum_{l\kappa\beta} M_{\bar{\alpha},\beta}\begin{pmatrix}l\\ \kappa\end{pmatrix}(u_\varphi)_\beta\begin{pmatrix}l\\ \kappa\end{pmatrix}. \tag{110.22}$$

The transformation of the coefficients in (110.13) and (110.22) is seen to be

$$M'_{\bar{\alpha},\beta}\begin{pmatrix}l\\ \kappa\end{pmatrix} = \sum_{\bar{\beta}} \varphi_{\beta\bar{\beta}} M_{\bar{\alpha},\bar{\beta}}\begin{pmatrix}l_{\bar{\varphi}}\\ \kappa_{\bar{\varphi}}\end{pmatrix} \tag{110.23}$$

so that the transformation of these quantities is that of a polar vector. As with the corresponding coefficients in the expansion of the crystal potential energy, the invariance requirements for the "physical tensor" quantities $M_{\alpha,\beta}\begin{pmatrix}l\\ \kappa\end{pmatrix}$ give the form invariance of these quantities under a crystal symmetry transformation:

$$M'_{\alpha,\beta}\begin{pmatrix}l\\ \kappa\end{pmatrix} = M_{\alpha,\beta}\begin{pmatrix}l\\ \kappa\end{pmatrix}. \tag{110.24}$$

We next express the rotated displacements $(u_{\varphi_p})_\alpha\begin{pmatrix}l\\ \kappa\end{pmatrix}$ in terms of normal coordinates, and use form invariance (110.24) to obtain from (110.22)

$$M_{\bar{\alpha}}^{\prime(1)} = \sum_{l\kappa\beta} M'_{\bar{\alpha}\beta}\begin{pmatrix}l\\ \kappa\end{pmatrix}\frac{1}{\sqrt{M_\kappa N}}\sum_{\boldsymbol{k}_\tau}\sum_{j_\nu} e^{i\boldsymbol{k}_\tau\cdot\boldsymbol{R}_L} e_\beta\left(\kappa\middle|\begin{smallmatrix}\boldsymbol{k}_\tau\\ j_\nu\end{smallmatrix}\right) Q_{\{\varphi_p\}}\begin{pmatrix}\boldsymbol{k}_\tau\\ j_\nu\end{pmatrix} \tag{110.25}$$

or

$$M_{\bar{\alpha}}^{\prime(1)} = \sum_{\boldsymbol{k}_\tau}\sum_j\sum_\nu M_{\bar{\alpha}}\begin{pmatrix}\boldsymbol{k}_\tau\\ j_\nu\end{pmatrix} Q_{\{\varphi_p\}}\begin{pmatrix}\boldsymbol{k}_\tau\\ j_\nu\end{pmatrix}. \tag{110.26}$$

Then using (86.30) we obtain

$$M_{\bar{\alpha}}^{\prime(1)} = \sum_{\boldsymbol{k}_\tau}\sum_j\sum_\nu M_{\bar{\alpha}}\begin{pmatrix}\boldsymbol{k}_\tau\\ j_\nu\end{pmatrix}\sum_\sigma\sum_\mu D^{(*\boldsymbol{k})(j)}(\{\varphi_p\})_{\sigma\mu\tau\nu}\, Q\begin{pmatrix}\boldsymbol{k}_\sigma\\ j_\mu\end{pmatrix}. \tag{110.27}$$

This is the expression for the transformed $\bar{\alpha}$ component of $M^{(1)}$. But by the basic rule of transformation (110.21), and by (110.17)

$$M_{\bar{\alpha}}'^{(1)} = \sum_{\alpha} \varphi_{\bar{\alpha}\alpha} M_{\alpha}^{(1)} = \sum_{\alpha} \varphi_{\bar{\alpha}\alpha} \sum_{\boldsymbol{k}_\sigma} \sum_{j_\nu} M_{\alpha} \begin{pmatrix} \boldsymbol{k}_\sigma \\ j_\nu \end{pmatrix} Q \begin{pmatrix} \boldsymbol{k}_\sigma \\ j_\nu \end{pmatrix}. \tag{110.28}$$

But (110.27) must be identical to (110.28), hence we can equate them. First recall that

$$\varphi_{\bar{\alpha}\alpha} = D^{(r)}(\{\varphi_p\})_{\bar{\alpha}\alpha} \tag{110.29}$$

then

$$\begin{aligned} &\sum_{\boldsymbol{k}_\sigma} \sum_{j} \sum_{\nu} \sum_{\alpha} D^{(r)}(\{\varphi_p\})_{\bar{\alpha}\alpha} M_{\alpha} \begin{pmatrix} \boldsymbol{k}_\sigma \\ j_\nu \end{pmatrix} Q \begin{pmatrix} \boldsymbol{k}_\sigma \\ j_\nu \end{pmatrix} \\ &\quad = \sum_{\boldsymbol{k}_\tau} \sum_{j} \sum_{\nu} \sum_{\sigma} \sum_{\mu} D^{(\star \boldsymbol{k})(j)}(\{\varphi_p\})_{\sigma\mu\tau\nu} M_{\bar{\alpha}} \begin{pmatrix} \boldsymbol{k}_\tau \\ j_\nu \end{pmatrix} Q \begin{pmatrix} \boldsymbol{k}_\sigma \\ j_\nu \end{pmatrix}. \end{aligned} \tag{110.30}$$

A little reflection convinces one that, for the two sides to be equal, it suffices for the representation $D^{(\star \boldsymbol{k})(j)}$ by which the phonon $Q \begin{pmatrix} \boldsymbol{k}_\tau \\ j_\nu \end{pmatrix}$ transforms to be identical to $D^{(r)}$. But this is only possible if, for example in a cubic crystal

$$Q \begin{pmatrix} \boldsymbol{k}_\tau \\ j_\nu \end{pmatrix} \sim D^{(\Gamma)(15-)}. \tag{110.31}$$

Consequently the only modes for which a linear term exists in the expansion (110.17) are the three modes transforming like a polar vector: $Q \begin{pmatrix} \boldsymbol{\Gamma} \\ 15- \end{pmatrix}$. In a cubic crystal there are of course the three components of the three dimensional vector representation.

The argument for terms of second degree is carried out analogously. Then, a necessary and sufficient condition for a non-zero term of second degree (110.19) is that

$$D^{(\star \boldsymbol{k})(j)} \otimes D^{(\star \boldsymbol{k}')(j')} \quad \text{contain } D^{(r)}. \tag{110.32}$$

Evidently the necessary condition that particular pair of modes shall appear in the product series for $\boldsymbol{M}^{(2)}$ is that the reduction coefficient corresponding shall not vanish:

$$(\star \boldsymbol{k} j \, \star \boldsymbol{k} j' \,|\, \boldsymbol{\Gamma} r) \neq 0. \tag{110.33}$$

In (110.33), the symbol for the vector representation is taken as $\boldsymbol{\Gamma} r$, which in a cubic crystal is $\Gamma^{(15-)}$. The analysis can be carried out in subgroup, or full-group framework. It is apparent that in crystals containing inversion as a symmetry operation, modes of the same parity cannot in general combine to produce a term of second degree which will appear in $\boldsymbol{M}^{(2)}$. Other such simple and useful rubrics are easily obtained once one has reference to the space-group reduction coefficients for the particular space group. Some of these will be illustrated later for diamond and rocksalt space groups later on.

Terms of higher order can be treated in a similar fashion. Thus the term of s-th degree, $\boldsymbol{M}^{(s)}$ will involve a product of s factors, which are normal coordinates $Q \begin{pmatrix} \boldsymbol{k} \\ j \end{pmatrix}$. The necessary and sufficient condition for a prescribed product of particular terms to appear is that the direct product

$$D^{(\star k)(j)} \otimes D^{(\star k')(j')} \otimes \cdots \otimes D^{(\star k^{(s)})(j^{(s)})} \tag{110.34}$$

contain the representation $D^{(r)}$. If this representation is present then the particular linear combinations of the products which span the linear vector space $\Sigma^{(r)}$ can be determined by using projection operators. This is equivalent to the condition that the coupling term of s-th degree, which corresponds to (110.15) shall be non-zero. The entire argument for the higher order terms is the same as for those of lower order.

Again the matter of whether (110.34) does contain $D^{(r)}$ is equivalent to the non-vanishing of the reduction coefficient which corresponds. In the present case this is

$$(\star k j \, \star k' j' \ldots \star k^{(s)} j^{(s)} | \Gamma 15-) \tag{110.35}$$

for a cubic crystal. To determine the actual correct linear combinations requires the use of a projection operator, or equivalently having available the proper Clebsch-Gordan coefficients for the general tensors which arise here.

An exactly similar argument, *mutates mutandis*, applies to the construction of terms in the expansion of the polarizability $\boldsymbol{P}$ in powers of the displacements or normal coordinates

$$\boldsymbol{P} = \boldsymbol{P}^{(1)} + \boldsymbol{P}^{(2)} + \cdots + \boldsymbol{P}^{(s)} + \cdots. \tag{110.36}$$

Now, in (110.36) each term, of any degree, must separately be a symmetric tensor of second rank in order for the corresponding product of factors which are normal modes to occur in the series (110.36).

Returning to the notation of (110.10) we see that a necessary and sufficient condition for a term of some given degree in normal coordinates to appear in the expansion of polarizibility is if the reduction coefficient

$$(\star k j \, \star k' j' \ldots | [D^{(r)}]_{(2)}) \neq 0. \tag{110.37}$$

In case of a cubic crystal this implies that the coefficient for at least one component (110.11) shall be non-zero. Details of the proof follow as in the corresponding case for the dipole moment operator.

Finally then, we see the general scheme. For any crystal property which transforms covariantly under generalized rotation we must first find the representation of $\mathfrak{G}$, the space group by which the components of the covariant property transform. A necessary condition for the appearance, in an expansion of that property in products of normal coordinates is that the specified products contain a linear vector space which generates the same representation of $\mathfrak{G}$ as the covariant. In all cases to decide about particular products, we need to utilize the rules for reduction of ordinary and symmetrized Kronecker products and powers of space group irreducible representations, since the normal modes are by (86.30) bases for irreducible linear vector space.

The discussion in this section has been limited to the use of representation theory of the unitary subgroup but in principle of course it is the total space-time symmetry $\mathscr{G}$ which must be accommodated. As in the remarks at the end of Sect. 108, time reversal symmetry can be incorporated by using the operator K

to ensure reality of physical objects e.g.

$$K^{-1} M K = M^* = M .$$

A preferable approach, not yet carried through is to work throughout with the normal coordinates which are bases for corepresentations, and the corresponding reduction coefficients.

K. Space-time symmetry and quantum lattice dynamics

111. Introduction. The present part of this article consists of Sects. 111–118 which is concerned with the essentials of a quantum-mechanical treatment of the crystal or matter system. We are concerned in particular with obtaining a proper quantum-mechanical description of the relevant electronic and ionic (or nuclear) degrees of freedom of an insulating crystal in order to construct a quantum-mechanical theory of the infra-red absorption and Raman scattering by a crystal. The latter theory will be presented in Part L, Sects. 119–124. Thus these two parts of the article are closely related.

Let us first note that certain degrees of freedom, or quantum variables are ignored in our discussion since they are not of major relevance to our work on optical properties related to lattice vibrations or phonons. Thus for example we are not concerned with electron or nuclear spin, nor are we concerned with details of the electronic energy band structure of the solid. We do make use of the existence of an energy gap in the filled electron distribution as an important characteristic of an insulator in this part in Sects. 113, 114 and later in Sects. 120, 121, 124.

Even with this *caveat* any treatment of the relevant electronic and lattice degrees of freedom must necessarily be approximate since the electron-nuclear system is a many-body system for which a complete theory does not presently exist. One of the most useful treatments which will underlie our work in this and the next part is the adiabatic or Born-Oppenheimer[1] approximation. The immense utility of this approximation resides in the fashion by which it formally separates relevant electron variables from nuclear variables and permits each to be dealt with separately. The separation is not complete, since it is just the deformation of the electron states produced by motion of the nucleii which generates the potential (harmonic and anharmonic) in which the nucleii move. The nuclear potential function, which is simply assumed in the classical treatment as in Sect. 67 (harmonic), and Sect. 109 is shown to arise from the dependence of the total energy of the many-electron system upon nuclear displacements: it is in fact just the electronic energy with nuclear positions as parameters. The total matter system wave function is a product of a lattice part times a many-electron

[1] M. Born and J. R. Oppenheimer: Ann. d. Phys. **84**, 457 (1927).

part. It is just this factorization which will permit us to characterize the symmetry of the lattice eigenstate since the previous work on the space-time symmetry group $\mathcal{G}$, and its irreducible representations and corepresentations carries over to the quantum case as will be shown in Sects. 116–118. When we have characterized the symmetry of lattice eigenstates we shall be able to analyze the matrix elements relevant to infra-red absorption and Raman scattering using the Wigner-Eckart theorem and our previous work on space group reduction coefficients and Clebsch-Gordan coefficients. As will be demonstrated, any vibrational eigenstate can be exactly characterized, in harmonic approximation. In general excited states are composite states with overtones of one phonon, and combinations of others simultaneously present. Disentangling the symmetry of the state simply entails reducing the direct product of the relevant symmetrized powers, and Kronecker products. This type of analysis of overtone and combination states has already been very fruitful, as an application of the group theory in the analysis and prediction of spectra in many crystals: examples will be given in the applications cited in Part N.

The factorization given in the adiabatic theory has another important consequence for our later work in Part L. Because of the simple structure of the manifold of crystal states; namely, product of many-electron times vibrational eigenfunction, it will be possible within the adiabatic framework to define an electric moment operator $\mathcal{M}$, useful in the theory of infra-red absorption in Sect. 120, and a polarizability operator $\mathcal{P}$ useful in the theory of Raman scattering, in Sect. 121. Thus both aspects of the adiabatic framework are of considerable consequence to our work. As will be seen in Part N many predictions based on the adiabatic approximation are validated by comparison with experiment, giving support to using it.

Withall, the adiabatic approximation leaves out of consideration many very important effects of electron-lattice interaction, and besides it makes unproven assertion that the perturbation theory series upon which it is based is convergent. For these reasons *inter alia* there has recently been considerable work on the microscopic theory of crystal lattices from a modern many-body point of view. These more modern theories will be briefly discussed in Sect. 113 δ. The subject of the modern lattice dynamics will be authoritatively reviewed in another article in this Encyclopedia by Bilz and Wehner [*22*]. Also the theory of infra-red and Raman lattice optical properties of crystals can be largely freed from dependence upon the adiabatic approximation. Also in this connection modern many-body methods, using response function theory, have been applied. Some of this work will be briefly presented in Sect. 124 but this topic will also be given in greater depth in the article by Bilz and Wehner [*22*].

In Sects. 112–113 the traditional Born-Oppenheimer theory will be presented: despite its accessibility in standard works [*4*] we need to maintain notational consistency and also refer back to this treatment in later work. In Sect. 114 we discuss the passage from classical to quantum normal coordinates in a little detail. In Sects. 115–118 we discuss the symmetry of the general lattice eigenstate in harmonic approximation. The work of these sections is largely concerned with obtaining proper symmetry characterization of overtone and combination states by reducing symmetrized powers of space group irreducible representations.

Breakdown of adiabaticity in the case of insulating crystals may be associated with Jahn-Teller type electron-lattice coupling which may be invoked in certain phase transitions[2] or for insulators in which the electronic energy gap is of the same order of magnitude of the phonon energy but this very interesting and important subject is outside the domain of the present article so will not be further discussed here.

112. The many-body electron-ion Hamiltonian [*4*]. We denote the kinetic energy of an assembly of ions and electrons by T where

$$T = T_N + T_E \tag{112.1}$$

and

$$T_N = \frac{1}{2} \sum M \dot{\boldsymbol{X}}^2 = \sum \left(\frac{1}{2M}\right) \boldsymbol{P}^2 \tag{112.2}$$

is the nuclear kinetic energy, while

$$T_E = \frac{1}{2} \sum m \dot{\boldsymbol{x}}^2 = \sum \left(\frac{1}{2m}\right) \boldsymbol{p}^2 \tag{112.3}$$

represents the electronic kinetic energy. In this subsection we shall initially dispense with subscripts since we shall not be dealing with specific one-particle properties of the system. Let X and x refer to the nuclear and electronic coordinates respectively and the sums (112.2) and (112.3) are to be taken over all relevant nuclear and electronic degrees of freedom. Owing to the charge of these particles, the potential energy of the system is $U(X, x)$, where nucleii have charges Ze, electrons $-e$

$$U(X, x) = \sum \frac{ZZ'}{|\boldsymbol{X} - \boldsymbol{X}'|} + \sum \frac{e^2}{|\boldsymbol{x} - \boldsymbol{x}'|} - \sum \frac{Ze^2}{|\boldsymbol{X} - \boldsymbol{x}|} \tag{112.4}$$

represents the totality of all mutual Coulomb interactions.

The quantum mechanical Hamiltonian is obtained in the Schrödinger representation by converting (112.2) and (112.3) to operators by the replacements

$$\boldsymbol{P} \to \frac{\hbar}{i} \left(\frac{\partial}{\partial \boldsymbol{X}}\right); \quad \boldsymbol{p} \to \frac{\hbar}{i} \left(\frac{\partial}{\partial \boldsymbol{x}}\right). \tag{112.5}$$

Let us now consider that in T_N and T_E these replacements have been made. We must then solve the quantum mechanical many body Schrödinger equation

$$\mathscr{H}(\boldsymbol{P}, \boldsymbol{X}; \boldsymbol{p}, \boldsymbol{x})\, \Psi(\boldsymbol{X}, \boldsymbol{x}) = E\, \Psi(\boldsymbol{X}, \boldsymbol{x}), \tag{112.6}$$

where $\mathscr{H}$ is the quantum mechanical operator

$$\mathscr{H} = T_N + T_E + U \tag{112.7}$$

and Ψ is the many body eigenfunction. It needs to be remarked that at this stage no specification of existence of a lattice, or crystal, at equilibrium has been made.

[2] R.J. Elliott, R.T. Harley, W. Hayes, and S.R.P. Smith: Proc. Roy. Soc. (London), Ser. A **328**, 217 (1972).

One could, of course, *a priori* postulate the existence of an underlying space lattice $\mathfrak{G}$ such that when all sets of coordinates are simultaneously subjected to a suitable linear transformation from $\mathfrak{G}$, $\mathscr{H}$ is invariant, but at present the existence of such a group $\mathfrak{G}$ has not been rationalized from first principles.

113. Born-Oppenheimer adiabatic approximation [*4*]. We write the total system Hamiltonian as

$$\mathscr{H}=\mathscr{H}_0+T_N \tag{113.1}$$

where, as in (112.2) T_N is the nuclear kinetic energy. Then $\mathscr{H}_0$ is the Hamiltonian for all the electrons moving in the field of nucleii whose instantaneous positions are given as the set $\{X_i\}$. Then

$$\mathscr{H}_0=T_E+U. \tag{113.2}$$

α) *Born-Oppenheimer perturbation method.* Assume we possess a complete set of solutions to the "electronic" problem

$$\mathscr{H}_0\,\varphi_\nu(X,x)=(T_E+U)\,\varphi_\nu(X,x)=W_\nu(X)\,\varphi_\nu(X,x). \tag{113.3}$$

In (113.3), the $\varphi_\nu(X,x)$ is the electronic eigenfunction. This is considered as a function of the electronic coordinates $\{x\}$, and as parametrically depending upon the nuclear coordinates $\{X\}$. The eigenenergy $W_\nu(X)$ depends parametrically upon $\{X\}$; the quantum numbers $\{\nu\}$ specify the electronic state. We take the complete set of φ_ν to be orthonormalized for any prescribed $\{X_i\}$:

$$(\varphi_\nu,\varphi_{\nu'})_x\equiv\int\varphi_\nu^*(X,x)\,\varphi_\nu(X,x)\,d^3x=\delta_{\nu\nu'}. \tag{113.4}$$

Let

$$\kappa\equiv(m/M)^{\frac{1}{4}}, \tag{113.5}$$

where κ is a parameter of the theory, and m, M are electron mass and some typical ion mass respectively. If we assume there exists an equilibrium position X_i^0 for each ion then in this theory we take the deviation from equilibrium to be small. Then we write

$$X_i=X_i^0+\kappa u_i \tag{113.6}$$

where κ is as in (113.5) and u is a new dynamical variable, which is of course the displacement variable of the classical theory.

It follows that we can write for (113.2)

$$\mathscr{H}_0(X^0+\kappa u)\,\varphi_\nu(X^0+\kappa u,x)=W_\nu(X^0+\kappa u)\,\varphi_\nu(X^0+\kappa u,x). \tag{113.7}$$

Expanding all terms in (113.7) in series in κu we have:

$$\mathscr{H}_0(X^0+\kappa u,x)=\sum_\lambda\kappa^\lambda\,\mathscr{H}_0^\lambda(x), \tag{113.8}$$

$$\varphi_\nu(X^0+\kappa u,x)=\sum_\mu\kappa^\mu\,\varphi_\nu^\mu(X^0,x), \tag{113.9}$$

$$W_\nu(X^0+\kappa u)=\sum_\sigma\kappa^\sigma\,W_\nu^\sigma(X^0). \tag{113.10}$$

Substituting (113.8)–(113.10) into (113.7) we have for the term of the s-th degree in κ:

$$\kappa^s(\mathscr{H}_0^0 - W_\nu^0)\,\varphi_\nu^s = -\sum_{\lambda=0}^{s-1} (\mathscr{H}_0^{(s-\lambda)} - W_\nu^{(s-\lambda)})\,\varphi_\nu^\lambda\,\kappa^s. \tag{113.11}$$

Note that each equation remains homogeneous of degree s in the variable u. We shall assume that we know all the solutions of (113.11) to any required order.

Now return to T_N. From the definitions (112.2), and (113.6) it is clear that

$$\begin{aligned} T_N(X) &= \kappa^2 \sum \left(\frac{\hbar^2}{2m}\right)\left(\frac{\partial^2}{\partial u_j^2}\right) \\ &\equiv \kappa^2 H_1(u). \end{aligned} \tag{113.12}$$

Thus in terms of the new dynamical variable u the nuclear kinetic energy is of second degree. If we now return to the original many-body problem (112.6)

$$\mathscr{H}(X, x)\,\Psi(X, x) = E\Psi(X, x) \tag{113.13}$$

or

$$(\mathscr{H}_0 + T_N)\,\Psi(X, x) = E\Psi(X, x) \tag{113.14}$$

we can again expand all objects in a series using κ as the small parameter. Thus we write

$$\begin{aligned} \Psi(X, x) &= \Psi(X^0 + \kappa u, x) \\ &= \sum_\tau \kappa^\tau \Psi^{(\tau)}(u, x), \end{aligned} \tag{113.15}$$

$$E = \sum_\rho \kappa^\rho E^\rho. \tag{113.16}$$

Then the general equation for the term of s-th degree in κ is

$$\kappa^s(\mathscr{H}_0^0 - E^0)\,\Psi^{(s)} = -\kappa^s \sum_\lambda (\mathscr{H}_0^{(s-\lambda)} + H_1\,\delta_{s-\lambda,2} - E^{(s-\lambda)})\,\Psi^{(\lambda)}. \tag{113.17}$$

In (113.17) the term with H_1 is evidently to be taken only with other terms of second degree in κ. It is to be noted that $E^{(\rho)}$ is a constant, while $W_\nu^{(\rho)}$ is an object of ρ-th degree in u. Then comparing (113.7) with (113.11) term by term we obtain the well known conditions for solution of the general many-body equation (113.13), on the assumption that solutions of (113.11) are available:

$$E^{(0)} = W_\nu^0, \tag{113.18a}$$

$$E^{(1)} = W_\nu^{(1)} = \sum \left(\frac{\partial W_\nu}{\partial X}\right)\left(\frac{X - X^0}{\kappa}\right) = 0. \tag{113.18b}$$

Thus (113.18b) determines the equilibrium positions of the ions since

$$\left(\frac{\partial W_\nu}{\partial X}\right)_{X^0} = 0 \tag{113.19}$$

determines the set of equilibrium $\{X^0\}$. To this order the complete eigenfunction Ψ is given as

$$\Psi(x, u) = (\varphi_\nu^{(0)}(X^0, x) + \varphi_\nu^{(1)}(X^0, x))\,\chi_\nu^{(0)}(u). \tag{113.20}$$

To determine the unknown function $\chi_\nu^0(u)$ we require the second degree terms (in u^2). On comparing (113.11) and (113.17) to this order we find the equation determining $\chi_\nu^0(u)$ to be

$$[H_1+W_\nu^{(2)}(u)-E^{(2)}]\,\chi_\nu^0(u)=0 \tag{113.21}$$

or

$$\kappa^2\left(T_N+\sum\left(\frac{\partial^2 W_\nu}{\partial X_i\,\partial X_j}\right)_{\{X^0\}} u_i u_j - E^{(2)}\right)\chi_\nu^0(u)=0. \tag{113.22}$$

Clearly (113.22) is the usual Schrödinger equation for a system of coupled harmonic oscillators. At this level of approximation we can see that (113.22) can be put into the form of a diagonal quadratic form in the variables

$$\{\partial/\partial u_j\},\ \ \{u_j\} \tag{113.23}$$

so that the solution of (113.22) will be a product of simple harmonic oscillator eigenfunctions. We shall make use of this later.

Continuing it is easily shown by direct use of perturbation theory that accurate to fourth degree (κ^4) we can write the total wave function as

$$\Phi(X,x)=\varphi(X^0,x)\,\chi(u) \tag{113.24}$$

with $\chi(u)$ the solution of

$$[T_N(\partial/\partial u)+W_\nu(u)-E]\,\chi(u)=0, \tag{113.25}$$

where

$$\begin{aligned} W_\nu = W_\nu^{(0)} &+ \sum_{u,u'}\Phi^{(2)}(u,u')/2! + \sum_u\sum_{u'}\sum_{u''}\Phi^{(3)}(u,u',u'')/3! \\ &+ \sum_u\sum_{u'}\sum_{u''}\sum_{u'''}\Phi^{(4)}(u,u',u'',u''')/4!. \end{aligned} \tag{113.26}$$

The terms of third and fourth degree in (113.26) are anharmonic terms. The pattern of the argument is clear however, as regards obtaining the familiar adiabatic separation of the many-body system into an electronic and an ionic part indirectly coupled via introduction of (113.26) as the potential for nuclear motion.

If we now assume that the reference set of $X^{(0)}$ are the equilibrium values for which the ion positions are the lattice sites

$$X^{(0)}=\boldsymbol{r}^{(0)}\begin{pmatrix} l \\ \kappa \end{pmatrix} \tag{113.27}$$

then we may label the u variables as the familiar cartesian components of the ion displacements $u_\alpha\begin{pmatrix} l \\ \kappa \end{pmatrix}$, and (113.26) can then be rewritten in the familiar forms

$$W_\nu(u)=W_\nu^{(0)}+\sum_{l\kappa\alpha}\sum_{l'\kappa'\beta}\Phi_{\alpha\beta}\begin{pmatrix} l & l' \\ \kappa & \kappa' \end{pmatrix} u_\alpha\begin{pmatrix} l \\ \kappa \end{pmatrix} u_\beta\begin{pmatrix} l' \\ \kappa' \end{pmatrix}+\cdots. \tag{113.28}$$

The significance of $W_\nu(u)$ as the potential function for nuclear motion is thus made clear, as is also the intimate connection between the electronic state and this potential function. Central to the adiabatic argument is that the electronic state does not change during nuclear motion: i.e. the label representing the electronic quantum state: ν remains fixed. The argument just given justifies the separation of the total motion of the many-body system into a set of modes of motion characteristic of the lattice nuclear particles alone, yet emphasizes one aspect: the intimate connection of the electronic states with those modes.

β) *Born's method.* The analysis just given using perturbation theory only is valid to fourth degree[1,2] in κ. That is, no anharmonic terms in the potential function of fifth or higher degree can be formally justified if one uses the method of analysis just given.

As an alternate procedure,[3] one may again consider the electronic equation (113.3) to have been solved, and the eigenenergy $W_\nu(\boldsymbol{X})$ and eigenfunction $\varphi_\nu(\boldsymbol{X}, \boldsymbol{x})$ available. Now let us assume that the exact eigenfunction is expanded in the set $\varphi_\nu(\boldsymbol{X}, \boldsymbol{x})$:

$$\Psi(\boldsymbol{X}, \boldsymbol{x}) = \sum_\nu \varphi_\nu(\boldsymbol{X}, \boldsymbol{x})\, \chi_\nu(\boldsymbol{X}). \tag{113.29}$$

Owing to the completeness of the set of functions $\varphi_\nu(\boldsymbol{X}, \boldsymbol{x})$ this is a general expansion and the procedure now focusses upon the determination of the nuclear functions $\chi_\nu(\boldsymbol{X})$. Substituting (113.29) into the complete equation (112.6) and taking advantage of the orthonormality of the solutions of (113.1) we write:

$$\int \varphi_\nu^*(\boldsymbol{X}, \boldsymbol{x})\, \varphi_\mu(\boldsymbol{X}, \boldsymbol{x})\, d\boldsymbol{x} = \delta_{\nu\mu}, \tag{113.30}$$

where (113.30) is independent of nuclear coordinate $\boldsymbol{X}$. Upon carrying out this substitution we find as an equation for $\chi_\nu(\boldsymbol{X})$:

$$[T_N(\boldsymbol{X}) + W_\nu(\boldsymbol{X}) - E]\, \chi_\nu(\boldsymbol{X}) + \sum_{\nu\nu'} C_{\nu\nu'}(\boldsymbol{X}, \boldsymbol{P})\, \chi_{\nu'}(\boldsymbol{X}) = 0, \tag{113.31}$$

where

$$C_{\nu\nu'} \equiv \sum_\kappa \left(\frac{1}{M_\kappa}\right) (A_{\nu\nu'}^{(\kappa)} P_\kappa + B_{\nu\nu'}^{(\kappa)}) \tag{113.32}$$

with $\boldsymbol{P}_\kappa$ a typical nuclear momentum operator:

$$\boldsymbol{P}_\kappa = \frac{\hbar}{i} \frac{\partial}{\partial \boldsymbol{X}} \tag{113.33}$$

and

$$A_{\nu\nu'}^{(\kappa)}(\boldsymbol{X}) = \int \varphi_\nu^*(\boldsymbol{X}, \boldsymbol{x})\, P_\kappa\, \varphi_{\nu'}(\boldsymbol{X}, \boldsymbol{x})\, d\boldsymbol{x}^3, \tag{113.34}$$

$$B_{\nu\nu'}^{(\kappa)}(\boldsymbol{X}) = \tfrac{1}{2} \int \varphi_\nu^*(\boldsymbol{X}, \boldsymbol{x})\, P_\kappa^2\, \varphi_{\nu'}(\boldsymbol{X}, \boldsymbol{x})\, d\boldsymbol{x}^3. \tag{113.35}$$

Observe that in (113.34) and (113.35) the gradient operator with respect to nuclear coordinates operates upon the eigenfunction solution of the electronic problem which contains nuclear coordinates as parameters. If we assume that the different electronic states: $\nu \neq \nu'$ are not coupled, then $C_{\nu\nu'} = 0$ for $\nu \neq \nu'$. For $\nu = \nu'$, and for stationary states of our real Hamiltonian, $A_{\nu\nu}^{(\kappa)} = 0$. Hence (113.31) can be rewritten

$$(T_N(\boldsymbol{X}) + U_\nu(\boldsymbol{X}) - E)\, \chi_\nu(\boldsymbol{X}) = 0, \tag{113.36}$$

[1] M. Born and J. R. Oppenheimer: Ann. d. Phys. **84**, 457 (1927).
[2] Ref. [*13*], pp. 166–173.
[3] Ref. [*13*], Appendix VIII.

where a complete potential is

$$U_\nu(X) = W_\nu(X) + \sum_\kappa \frac{1}{M_\kappa} B_{\nu\nu}^{(\kappa)}. \tag{113.37}$$

The Eqs. (113.36) and (113.37) are the complete equation for the nuclear motion, under the supposition that interelectronic terms ($\nu \neq \nu'$) can be ignored, or vanish. In this equation anharmonic terms of *all* degrees may appear so that we may make the association of the complete nuclear potential energy function $U_\nu(X)$ of this section and the complete anharmonic potential energy function Φ, of Sect. 108.

Hence assuming some configuration of stable equilibrium exists, in which the set of all X^0 may be specified we may treat $T_N(\partial/\partial u)$ and $U_\nu(u)$ as functions of the cartesian displacement variables $u_\alpha\binom{l}{\kappa}$ and as before we write

$$U_\nu(u) \equiv \Phi_\nu(u) = \Phi_\nu^{(0)} + \Phi_\nu^{(2)} + \cdots + \Phi_\nu^{(s)}. \tag{113.38}$$

In our usual work the index ν which specifies the electronic state, plays no role and we neglect it. This is of course an assumption whose approximate validity for most insulating crystals seems justified.

Non-adiabaticity, or electron phonon coupling terms will arise in the theory in either of two equivalent ways. In the perturbation expansion method (113.8) *et seq.* terms in fifth order bring in the coupling. In the direct expansion method (113.29) *et seq.* the non-adiabatic terms come in via the $C_{\nu\nu'}$ off diagonal terms [*4*] with $\nu \neq \nu'$.

In the work which follows we shall take the adiabatic approximation for the matter system (electrons plus ions) to mean that the matter eigenfunction is a single term of (113.29)

$$\Psi_\nu(X\,\chi) = \Phi_\nu(X\,\chi)\, X_{\nu\{n\}}(X) \tag{113.39a}$$

or

$$|\Psi_{\text{adiabatic}}\rangle = |\Psi_{\text{electron}}\rangle\,|\Psi_{\text{lattice}}\rangle. \tag{113.39b}$$

It is of importance to appreciate that the electronic state above is actually a many-electron state, with all nuclear positions as parameters. A representation of such an electronic state as a Slater determinants of one-electron functions or a sum of such Slater determinants might be a working hypothesis under certain circumstance. Likewise the lattice eigenstate refers to all lattice particles (ions). If we now further consider the lattice eigenstate to be described in the harmonic approximation then the energy of the crystal corresponding to the state (113.29) is

$$E_\nu = \Phi_\nu^{(0)} + \kappa^2 E_{\nu\{n\}}^{(2)} \tag{113.40}$$

where the first term refers to the total electron energy in state ν with ions at equilibrium positions, while the second term is just the sum of the energies of all the nucleii. Again assuming the harmonic approximation then

$$\kappa^2 E_{\nu\{n\}}^{(2)} = \sum_{\{n\}} (n + \tfrac{1}{2})\,\hbar\omega. \tag{113.41}$$

That is the lattice energy is simply the sum of harmonic oscillator energies. The sum of $\{n\}$ is intended to signify a sum over all the sets of lattice quantum members.

For future reference (see Eq. (121.37)) note the expression for the difference in energy between two adiabatic eigenstates of the matter system. Let the two eigenstates be denoted (see Eq. (113.39 a)) $\Psi_\nu = \Phi_\nu \chi_{\nu\{n\}}$ and $\Psi_\mu = \Phi_\mu \chi_{\mu\{m\}}$ then the energy difference would be

$$E_\mu - E_\nu = (\Phi_\mu^{(0)} - \Phi_\nu^{(0)}) - \kappa^2 (E_{\mu\{m\}}^{(2)} - E_{\nu\{n\}}^{(2)}). \tag{113.42}$$

Now, anticipating our interest later in Sect. 121 observe that the energy difference $(E_{\mu\{m\}}^{(2)} - E_{\nu\{n\}}^{(2)})$ refers to the difference in energy between two different lattice eigenstates, one associated with electronic state μ, and the set of quantum numbers $\{m\}$, the other with ν and $\{n\}$. Usually, for small differences in numbers of phonons in the sets $\{m\}$ and $\{n\}$ this energy difference is much smaller than the energy difference $(\Phi_\mu^0 - \Phi_\nu^0)$ between electronic states. Hence it is a reasonable approximation, within the adiabatic framework to take the energy difference between adiabatic states

$$E_\mu - E_\nu \sim (\Phi_\mu^0 - \Phi_\nu^0) \tag{113.43}$$

in case of small phonon excitation. As will be apparent later on, the importance of this approximation, made within the adiabatic approximation, is that the right side is independent of vibrational quantum numbers.

γ) *Harmonic adiabatic approximation.* The harmonic adiabatic approximation is obtained, analogous to the classical harmonic case, when the series (113.20) is terminated at the term of second degree $\Phi^{(2)}$. In this case the Schrödinger equation for nuclear motion becomes:

$$[T_N(\partial/\partial u) + \Phi_\nu^{(0)} + \Phi_\nu^{(2)}(u) - E]\, \chi_\nu(u) = 0. \tag{113.44}$$

For the moment we add the index ν to emphasize that the existence and specific form of the nuclear potential energy depends on the electronic state of the system-specified by ν. As it stands, Eq. (113.44) defines the nuclear motion in terms of $3rN$ coupled harmonic oscillators. To decouple the oscillators we need to make the complex normal coordinate transformation as in (80.9) and (80.10). This unitary transformation will reduce the potential energy $\Phi_\nu^{(2)}$ to diagonal form as in (80.3) while leaving $T_N(\partial/\partial u)$ in its already diagonal form as in (80.5). In the classical case [*13*] the complex normal coordinate transformation is followed by the transformation to real normal coordinates in either of two ways. This will be discussed below in Sect. 114.

δ) *Recent theories.* There have been a number of recent attempts to examine the adiabatic approximation using modern Green function methods, and also different methods than those just presented above. We shall simply mention some of this work here and suggest the interested reader pursue the literature including the discussion in [*22*]. Diagrammatic methods for calculating energies and other physical properties related to correlation functions for the displacements (see in this context the brief discussion in Sect. 124 α–γ have been given by several workers,[4] based on a formalism of BAYM.[5] A generalized many-body dielectric

[4] J.W. GARLAND: Phys. Rev. **153**, 460 (1967). – P.N. KEATING: Phys. Rev. **175**, 1169 (1968); **187**, 1190 (1969).

[5] G. BAYM: Ann. Phys. **14**, 1 (1961).

response formalism which computes the response of the electron system to ion motion (this is, as the adiabatic theory instructs us, the source of the force constants) was presented by PICK, MARTIN, and COHEN[6] to discuss the origin of the force constants in lattice dynamics; this theory was applied by MARTIN[6] in several cases. A density matrix formalism was given by JOHNSON[7] which perhaps most closely approximates the type of perturbation theory familiar in Born-Oppenheimer approach. Calculations which attempt to make contact with the shell model of lattice vibrations were given by BILZ and GLISS.[8] Other attempts to go beyond the adiabatic theory were given using a series of canonical transformations on the complete Hamiltonian[10] and using a new many-body variational method.[11] Recent developments along these lines are reviewed briefly by RAJAGOPAL and COHEN[9] in several papers in the Proceedings of the International Conference on Phonons, Rennes (1971) [*45*], and in the Proceedings of the XII International Conference on Semiconductors, Stuttgart (1974) [*59*].[12]

For our present purposes the adiabatic theory seems adequate and permits confrontation with experiment. Consequently it forms the basis of our work, unless otherwise specifically indicated.

114. Normal coordinates and quantization. Let us now examine the steps involved in passing from the Schrödinger equation (113.44) to a separable equation in more useful and familiar form [*13*]. Observe that in (113.44) the kinetic energy term is a diagonal form in the derivatives with respect to the cartesian displacements as shown in (113.12): i.e. a sum of squares. But the potential energy term as shown in (113.28) is a general bilinear expression in the dynamical variables $\{u_j\}$. Now recall that the set of $\{u_j\}$ represent all cartesian displacements so that we could now restore all the indices missing in (113.6) by making the association:

$$u_j \rightarrow u_\alpha \begin{pmatrix} l \\ \kappa \end{pmatrix}. \tag{114.1}$$

Then the potential energy in the Schrödinger equation (113.44) is just the classical object (67.8). That is, the elementary force constants introduced in (67.7) are now identified: the force constants are second derivatives of the total electronic energy (in the adiabatic approximation) with respect to the nuclear displacements $u_\alpha \begin{pmatrix} l \\ \kappa \end{pmatrix}$. However, even in harmonic adiabatic approximation, (113.22) is not immediately solvable, since we have to deal with a system dynamically coupled by all the crossterms in (113.28).

The first step in solution requires that (113.28) be brought to diagonal (separable) form. This requires the introduction of normal coordinates. Of course

[6] R.M. PICK, M.H. COHEN, and R.M. MARTIN: Phys. Rev. **1B**, 910 (1970). – R.M. MARTIN: Phys. Rev. Lett. **21**, 536 (1968); Phys. Rev. **186**, 871 (1969).

[7] F.A. JOHNSON: Proc. Roy. Soc. (London), Ser. A **310**, 79, 89, 101, 111 (1969).

[8] G. GLISS and H. BILZ: Phys. Rev. Lett. **21**, 884 (1968).

[9] A.K. RAJAGOPAL and M.H. COHEN: Collective Phenomena **1**, 9 (1972).

[10] H. BILZ and E. ZYBELL: Unpublished.

[11] J. PRINCE: Ph. D. Thesis. New York: New York University 1967. Unpublished: available from University Microfilms Ann Arbor, Mich.

[12] See especially papers of H. BILZ, W. WEBER and F.A. JOHNSON in [*59*].

this is an exact repetition of the program followed in the classical case in Parts H and I. The reader may wish to refresh the recollection of the program followed for example in (73.1) et seq when real normal coordinates q_{j_ρ} were introduced by

$$u_\alpha\begin{pmatrix} l \\ \kappa \end{pmatrix} = \frac{1}{\sqrt{M_\kappa}} \sum_j \sum_\rho e_\alpha\left(\begin{matrix} l \\ \kappa \end{matrix}\middle| j_\rho\right) q_{j_\rho} \tag{114.2}$$

or in (80.9), (86.1) when complex normal coordinates $Q\begin{pmatrix} \boldsymbol{k} \\ j_\nu \end{pmatrix}$ were introduced by

$$u_\alpha\begin{pmatrix} l \\ \kappa \end{pmatrix} = \frac{1}{\sqrt{M_\kappa N}} \sum_j \sum_\nu \sum_{\boldsymbol{k}} \exp i\boldsymbol{k}\cdot\boldsymbol{R}_L\, e_\alpha\left(\kappa\middle|\begin{matrix} \boldsymbol{k} \\ j_\nu \end{matrix}\right) Q\begin{pmatrix} \boldsymbol{k} \\ j_\nu \end{pmatrix}. \tag{114.3}$$

Both of these transformations bring the classical potential energy to diagonal form, and thus decouple the terms in (113.22).

In the analysis in Sect. 101 it was established that the proper manner to take the full space-time crystal symmetry group $\mathscr{G}$ into account was to use the complex normal coordinates $Q\begin{pmatrix} \boldsymbol{k} \\ j_\nu \end{pmatrix}$ as the dynamical variables, since they are bases for irreducible corepresentations of $\mathscr{G}$. Thus from point of view of maximum utilization of symmetry these complex normal coordinates are the preferred set.

However in quantum theory it is preferable for the basic variables to be real, since complex classical coordinates or momenta introduce spurious gauge degrees of freedom when quantized as discussed in [*46*], p. 12 *et seq*. The conventional treatment [*13*] proceeds either by introducing real normal coordinates of the first kind $q_\lambda\begin{pmatrix} \boldsymbol{k} \\ j_\mu \end{pmatrix}$ with $\lambda = 1, 2$

$$q_{1,2}\begin{pmatrix} \boldsymbol{k} \\ j_\mu \end{pmatrix} = \begin{cases} \dfrac{1}{\sqrt{2}}\left\{Q\begin{pmatrix} \boldsymbol{k} \\ j_\mu \end{pmatrix} + Q\begin{pmatrix} -\boldsymbol{k} \\ j_\mu \end{pmatrix}\right\} \\ \dfrac{1}{i\sqrt{2}}\left\{Q\begin{pmatrix} \boldsymbol{k} \\ j_\mu \end{pmatrix} - Q\begin{pmatrix} -\boldsymbol{k} \\ j_\mu \end{pmatrix}\right\} \end{cases} \tag{114.4}$$

or canonical transformation to real normal coordinates of the second kind via

$$q\begin{pmatrix} \boldsymbol{k} \\ j_\mu \end{pmatrix} = \frac{1}{2}\left[Q\begin{pmatrix} \boldsymbol{k} \\ j_\mu \end{pmatrix} + Q\begin{pmatrix} -\boldsymbol{k} \\ j_\mu \end{pmatrix}\right] + \left(\frac{i}{2\omega(\boldsymbol{k}|j)}\right)\left[\dot{Q}\begin{pmatrix} \boldsymbol{k} \\ j_\mu \end{pmatrix} - \dot{Q}\begin{pmatrix} -\boldsymbol{k} \\ j_\mu \end{pmatrix}\right]. \tag{114.5}$$

The great advantage of the normal coordinates (114.5) is that they represent progressive waves [*13*] and when quantized are identified as phonons.

An examination of the three sets of real normal coordinates shows: the set $\{q_{j_\rho}\}$, with $j_\rho = 1, \ldots, s\cdot l_j$, defined in (114.2) are not bases for irreducible representation of $\mathfrak{G}$ or corepresentations of $\mathscr{G}$; the set of real normal coordinates of first kind $\left\{q_\lambda\begin{pmatrix} \boldsymbol{k} \\ j_\nu \end{pmatrix}\right\}$ with $\lambda = 1, 2$, and $j_\mu = 1, \ldots, l_j$ are not bases for an irreducible corepresentation unless $Q\begin{pmatrix} \boldsymbol{k} \\ j_\mu \end{pmatrix}$ are real; the set of real coordinates of the second kind $\left\{q\begin{pmatrix} \boldsymbol{k} \\ j_\mu \end{pmatrix}\right\}$ with $\mu = 1, \ldots, l_j$ defined in (114.5) are not bases for an irreducible representation unless $Q\begin{pmatrix}\boldsymbol{k}\\ j_\mu\end{pmatrix}$ are real. Since in general the reality condition is not satisfied it should be concluded that the conventional quantization procedure [*13, 47*], based on real normal coordinates is not consistent with maintaining the symmetry properties of the normal coordinates. At the time of writing this apparent inconsistency had not been resolved.

In what follows we shall adopt the conventional approach and simply make the *ansatz* that a set of normal coordinates $q\begin{pmatrix}\boldsymbol{k}\\ j_\mu\end{pmatrix}$ can be chosen to be bases for an irreducible representation of the unitary group $\mathfrak{G}$ and then extended to be bases for corepresentations of $\mathscr{G}$ and also can be quantized in the conventional manner, without introduction of spurious degrees of freedom. Then we shall assume that under transformation by P_u, where u is any unitary operator $\{\varphi|\boldsymbol{t}\}$ of $\mathfrak{G}$

$$P_u\, q\begin{pmatrix}\boldsymbol{k}_\sigma\\ j_\alpha\end{pmatrix}=\sum_{\tau\beta} D^{(\star\boldsymbol{k})(j)}(u)_{\tau\beta,\sigma\alpha}\, q\begin{pmatrix}\boldsymbol{k}_\tau\\ j_\beta\end{pmatrix}. \tag{114.6}$$

Then using the usual Schrödinger rule

$$p\begin{pmatrix}\boldsymbol{k}_\sigma\\ j_\mu\end{pmatrix}=\frac{\hbar}{i}\,\frac{\partial}{\partial q\begin{pmatrix}\boldsymbol{k}_\sigma\\ j_\mu\end{pmatrix}} \tag{114.7}$$

the quantum mechanical hamiltonian (113.22) is transformed to

$$\sum_{\boldsymbol{k}}\sum_{\sigma}\sum_{j}\sum_{\mu}\left\{-\hbar^2\frac{\partial^2}{\partial q\begin{pmatrix}\boldsymbol{k}_\sigma\\ j_\mu\end{pmatrix}^2}+\omega^2(\boldsymbol{k}_\sigma|j)\, q\begin{pmatrix}\boldsymbol{k}_\sigma\\ j_\mu\end{pmatrix}^2+\Phi_\nu^{(0)}-E\right\}\chi_\nu\left(\left\{q\begin{pmatrix}\boldsymbol{k}_\sigma\\ j_\mu\end{pmatrix}\right\}\right)=0. \tag{114.8}$$

Eq. (114.8) is the nuclear Schrödinger equation in the harmonic adiabatic approximation.

115. Lattice eigenfunctions in harmonic adiabatic approximation. The Eq. (114.8) is a sum of terms, hence separable, and ignoring for a moment the degeneracy among modes $q\begin{pmatrix}\boldsymbol{k}_\sigma\\ j_\mu\end{pmatrix}$ which is indicated since $\omega^2(\boldsymbol{k}_\sigma|j)$ is independent of μ, we treat each of the $3rN$ coordinates as independent. Thus we assume a product solution of (114.8)

$$\chi_\nu(\{q\})=\prod_{a=1}^{3rN}\chi_a\left(q\begin{pmatrix}\boldsymbol{k}_\sigma\\ j_\mu\end{pmatrix}\right), \tag{115.1}$$

where each function depends only upon one coordinate $q\begin{pmatrix}\boldsymbol{k}_\sigma\\ j_\mu\end{pmatrix}$. Then each function satisfies a single harmonic oscillator equation

$$\frac{1}{2}\left\{-\hbar^2\frac{\partial^2}{\partial q\begin{pmatrix}\boldsymbol{k}_\sigma\\ j_\mu\end{pmatrix}^2}+\omega^2(\boldsymbol{k}_\sigma|j)\, q\begin{pmatrix}\boldsymbol{k}_\sigma\\ j_\mu\end{pmatrix}^2-\varepsilon_a(n_{\sigma,j})\right\}\chi_a\left(q\begin{pmatrix}\boldsymbol{k}_\sigma\\ j_\mu\end{pmatrix}\right)=0. \tag{115.2}$$

Here

$$\varepsilon_a(n_{\sigma,j,\mu}) = [n_{\sigma,j,\mu} + \tfrac{1}{2}]\,\hbar\omega(\boldsymbol{k}_\sigma | j). \tag{115.3}$$

Since the l_j modes for given σ, j are degenerate, $\omega(\boldsymbol{k}_\sigma | j)$ is independent of μ. We may then denote the individual oscillator wave functions (dropping the index a) by:

$$\chi\left(n(\boldsymbol{k}_\sigma | j_\mu), q\begin{pmatrix}\boldsymbol{k}_\sigma\\ j_\mu\end{pmatrix}\right) = C\exp-\left[\frac{1}{2}\gamma_{\sigma,j}\left(q\begin{pmatrix}\boldsymbol{k}_\sigma\\ j_\mu\end{pmatrix}\right)^2\right] H_{n_{\sigma,j}}\left(\gamma_{\sigma,j}\, q\begin{pmatrix}\boldsymbol{k}_\sigma\\ j_\mu\end{pmatrix}\right), \tag{115.4}$$

where

$$\gamma_{\sigma,j} \equiv \frac{2\pi}{\hbar}\,\omega(\boldsymbol{k}_\sigma | j) \tag{115.5}$$

and $H_n(\gamma q)$ is the Hermite polynomial of order n, and C a normalizing constant

$$C = \left[\left(\frac{\gamma_{\sigma,j}}{\pi}\right)^{\frac{1}{2}} \frac{1}{2^{n_{\sigma,j,\mu}}(n_{\sigma,j,\mu})!}\right]^{\frac{1}{2}}. \tag{115.6}$$

The total vibrational energy is summed over all individual terms:

$$\varepsilon = \sum_{\boldsymbol{k},\sigma,j,\mu} (n_{\sigma,j,\mu} + \tfrac{1}{2})\,\hbar\omega(\boldsymbol{k}_\sigma | j) \tag{115.7}$$

in the harmonic-adiabatic approximation. The total energy of the system including electronic plus vibrational parts is

$$E = \Phi_\nu^0 + \sum_{\boldsymbol{k},\sigma,j,\mu} (n_{\sigma,j,\mu} + \tfrac{1}{2})\,\hbar\omega(\boldsymbol{k}_\sigma | j) \tag{115.8}$$

where Φ_ν^0 is the electronic energy and the index ν has been suppressed in the lattice frequencies.

The total vibrational eigenfunction of the system is given from (115.4) in the harmonic approximation as the product:

$$\begin{aligned}\chi_\nu(q) = \text{const} \times \exp&-\left[\tfrac{1}{2}\sum_{\boldsymbol{k},\sigma,j}\gamma_{\sigma,j}\sum_{\alpha} q\begin{pmatrix}\boldsymbol{k}_\sigma\\ j_\mu\end{pmatrix}^2\right]\\ &\times \prod_{\boldsymbol{k},\sigma,j}\prod_{\mu} H_{n_{\sigma,j,\mu}}\left(\gamma_{\sigma j}\, q\begin{pmatrix}\boldsymbol{k}_\sigma\\ j_\mu\end{pmatrix}\right).\end{aligned} \tag{115.9}$$

This harmonic eigenfunction will be the basis of some of our explicit analysis in what follows. It is perhaps worth emphasizing that the harmonic representation (115.9) is a consequence of the harmonic Hamiltonian. The eigenfunction (115.9) is still not complete since it is necessary to take account of the over-all symmetry due to the indistinguishability of the oscillators which are partners in a given irreducible representation. This essential symmetrization due to statistics will be discussed below, in Sect. 116.

The more general equation for the nuclear motion in the adiabatic theory is (113.25). While our general group theoretical analysis of selection rules and symmetry of eigenstates will apply whether or not the harmonic approximation is valid, a clear and concrete discussion of the various absorption and scattering processes is expedited by consideration of the product eigenfunction (115.9).

116. Symmetry of harmonic lattice eigenfunctions: Introduction. In (114.6) we took the normal coordinates $q\begin{pmatrix}\boldsymbol{k}\\ j_\mu\end{pmatrix}$ to transform as partners in an irreducible linear vector space, with transformation matrix characteristic of the irreducible representation of the unitary crystal space group $\mathfrak{G}$. We shall first consider the transformation of the harmonic lattice product eigenfunction (115.9) under such space group operations $\{\varphi|\boldsymbol{t}\}$. It turns out to be quite important that the transformation (114.6) is both linear and homogenous. Subsequent to this we shall invoke the Lemma of Necessary Degeneracy so that the classification of the individual states reverts to the familiar statement that states of different symmetry are to be considered distinct. We then make a connection between the previous discussion, and the theory of symmetry of the n-dimensional isotropic harmonic oscillator.[1] This group of three Sects. 116–118 is concluded with some discussion of the symmetry of the general adiabatic lattice eigenstate which is a solution of (113.18), and of the total crystal eigenstate (113.11) in the adiabatic approximation.

One alternate mode of representing the eigenfunction (115.9) is by giving an ordered set of integers corresponding to the ordered arrangement in which the factors are presented in (115.9). This representation of the harmonic eigenfunction is, of course, particularly appropriate when the discussion of lattice properties is to be put in the framework of the number, or "N" representation of second quantization. We may write this representation

$$|n(\star\boldsymbol{k}\, j), \ldots, n(\star\boldsymbol{k}'\, j'), \ldots, n(\star\boldsymbol{k}''\, j''), \ldots\rangle . \tag{116.1}$$

When the eigenfunction (115.9) is written in this manner, attention is called to the fact that what is important in the many-oscillator eigenfunction is not the occupancy of any particular individual partner state in a representation $D^{(\star\boldsymbol{k})(j)}$ but the total number of quanta $n(\star\boldsymbol{k}\, j)$. This number is to be distributed over all the partners.

According to the general principles of quantum mechanics [44] the eigenfunction of an assembly of equivalent phonons (bosons) must be symmetric under interchange of equivalent members of the degenerate set. Thus we need to symmetrize (115.9) under interchange of partners. Let $\mathscr{S}$ be the symmetrizer: it is evidently the product of individual symmetrizers for each representation

$$\mathscr{S} = \prod_{\boldsymbol{k}\, j} \mathscr{S}(\star\boldsymbol{k}\, j). \tag{116.2}$$

Then the property symmetrized vibrational eigenfunction is

$$\chi_\nu^{(\mathrm{sym})}(q) = \mathscr{S}\chi_\nu(q) \tag{116.3}$$

where $\chi_\nu(q)$ is given in (115.9).

Now consider a general unitary symmetry operation $\{\varphi|\boldsymbol{t}\}$ of $\mathfrak{G}$, under which, the normal coordinates transform as in (114.6) according to $D^{(\star\boldsymbol{k})(j)}$, an irreducible unitary representation of $\mathfrak{G}$. By the analysis of the invariance of the classical harmonic Hamiltonian given in Sect. 110 it is clear that the exponential factor

[1] G. BAKER: Phys. Rev. **103**, 1119 (1956).

in (115.9) only depends on the sum of squares of the normal coordinates:

$$-\tfrac{1}{2}\sum_{\sigma,j,\mu}\gamma_{\sigma,j}\sum\left[q\begin{pmatrix}\boldsymbol{k}_\sigma\\ j_\mu\end{pmatrix}\right]^2. \tag{116.4}$$

But as explicitly demonstrated in (110.10)–(110.15), the quadratic form (116.4) is a quadratic invariant under the transformation of q, by unitary transformations in $\mathfrak{G}$. Thus we need to examine the transformation of the second factor in (115.9) only, to determine how the entire lattice eigenfunction transforms under unitary spatial transformation.

117. Transformations of products of Hermite polynomials: Symmetrized Kronecker product.[1]

In order to determine the transformation of the symmetrized eigenfunction (116.3) under spatial symmetry we work in two stages. First, returning to (115.9) we examine the transformation of all terms in a given string (repeated product) referring to a given irreducible representation $D^{(\star\boldsymbol{k})(j)}$. That is select out of (115.9) the $s\cdot l_m$ terms with fixed $\boldsymbol{k}$, and all $\boldsymbol{k}_\sigma$ $(\sigma=1,\dots,s)$ in $\star\boldsymbol{k}$, and all associated l_μ $(\mu=1,\dots,l_j)$. After determing the transformation of such a basic string we then symmetrize.

The string belonging to $D^{(\star\boldsymbol{k})(j)}$ consists of two factors: the exponential prefactor independent of the occupancy numbers $n_{\sigma,j,\mu}$ as in (116.4) times a product of Hermite polynomials. The transformation of the exponential is trivial as pointed out below (116.4). Then our task is to determine the transformation of the "non-trivial" factor in such a string from (115.9). The string in question is

$$\prod_\sigma\prod_j\prod_\mu H_{n_{\sigma,j,\mu}}\left(\gamma_{\sigma j}\,q\begin{pmatrix}\boldsymbol{k}_\sigma\\ j_\mu\end{pmatrix}\right), \tag{117.1}$$

where $n_{\sigma,j,\mu}$ is the quantum state (number of phonons excited) of the oscillator $q\begin{pmatrix}\boldsymbol{k}_\sigma\\ j_\mu\end{pmatrix}$. To simplify notation slightly in the remainder of this subsection we occasionally supress index j which is assumed fixed, and write

$$n_{\sigma,j,\mu}\to n(\sigma,\mu).$$

Now recall that [27] a Hermite polynomial $H_n(\xi)$ is a polynomial of degree n in ξ. In the string (117.1) such polynomials multiply each other and the term of leading degree in the product is of form (again simplifying notation for a moment):

$$\xi_1^{n_1},\ \xi_2^{n_2},\ \dots,\ \xi_{sl_j}^{n_{sl_j}},$$

where

$$n_1+n_2+\cdots+n_{sl_j}=\sum_{\sigma\mu}n(\sigma,\mu)\equiv n(\star\boldsymbol{k}\,j)$$

which is the total number of quanta to be distributed over all the states of the manifold belonging to $D^{(\star\boldsymbol{k})(j)}$. Succeeding terms in the product (117.1) are sent into terms of the same degree by this linear transformation. Then it follows that

[1] See [16], p. 72 et seq.

we may restrict attention only to monomial terms of highest degree illustrated above as this will typify the transformation. This is the first step of the program, due in this context apparently to TISZA.[2] The same result follows if we consider that the Hermite polynomial $H_n(\xi)$ can be written in terms of a generating function

$$H_n(\xi)=(-1)^n \mathrm{e}^{\xi^2} \partial^n \mathrm{e}^{-\xi^2}/\partial \xi^n.$$

Then H_n transforms covariantly to ξ^n. This is true for each of the H_n occurring in the string, permitting us to restrict attention to the monomial of highest degree.[3]

Secondly we express the unitary spatial transformation in a transformed coordinate system in which it is diagonal. In that way under transformation the monomial is merely multiplied by some power of the diagonal matrix element. The next step is to symmetrize over all the states belonging to the same $D^{(*\boldsymbol{k})(j)}$. This step although not trivial, can be done straightforwardly. Here we illustrate it using the so-called "elementary" symmetric polynomials.

Finally the transformation of the entire eigenfunction (116.3) can be given: it is the product of individual factors, each of which refers to a given $D^{(*\boldsymbol{k})(j)}$ and each of which is individually symmetrized.

For the first step we work only with the representation $D^{(*\boldsymbol{k})(j)}$ whose basis is

$$\left\{q\begin{pmatrix}\boldsymbol{k}\\ j_1\end{pmatrix}, \ldots, q\begin{pmatrix}\boldsymbol{k}_\sigma\\ j_\mu\end{pmatrix}, \ldots, q\begin{pmatrix}\boldsymbol{k}_s\\ j_{l_j}\end{pmatrix}\right\}. \tag{117.2}$$

Consider an element $\{\boldsymbol{\varphi}|\boldsymbol{t}\}$ in $\mathfrak{G}$ represented by the matrix $D^{(*\boldsymbol{k})(j)}(\{\boldsymbol{\varphi}|\boldsymbol{t}\})$ as in (86.40). Let *this* unitary matrix be brought to diagonal form by some unitary transformation V:

$$V^{-1} D^{(*\boldsymbol{k})(j)}(\{\boldsymbol{\varphi}|\boldsymbol{t}\})\, V = D_d^{(*\boldsymbol{k})(j)}(\{\boldsymbol{\varphi}|\boldsymbol{t}\}), \tag{117.3}$$

where

$$D_d^{(*\boldsymbol{k})(j)}(\{\boldsymbol{\varphi}|\boldsymbol{t}\})_{\sigma\mu,\,\tau\nu} = d_\mu^{(\sigma)(j)}\, \delta_{\sigma\tau}\, \delta_{\mu\nu}. \tag{117.4}$$

Then V produces an equivalent basis

$$\left\{q'\begin{pmatrix}\boldsymbol{k}\\ j_1\end{pmatrix}, \ldots, q'\begin{pmatrix}\boldsymbol{k}_\sigma\\ j_\mu\end{pmatrix}, \ldots, q'\begin{pmatrix}\boldsymbol{k}_s\\ j_{l_j}\end{pmatrix}\right\} \tag{117.5}$$

such that for this element every (primed) normal coordinate is multiplied by a constant when transformed

$$P_{\{\boldsymbol{\varphi}|\boldsymbol{t}\}}\, q'\begin{pmatrix}\boldsymbol{k}_\sigma\\ j_\mu\end{pmatrix} = d_\mu^{(\sigma)(j)}\, q'\begin{pmatrix}\boldsymbol{k}_\sigma\\ j_\mu\end{pmatrix}. \tag{117.6}$$

Because of the invariance of the trace we have from (117.3):

$$\chi^{(*\boldsymbol{k})(j)}(\{\boldsymbol{\varphi}|\boldsymbol{t}\}) = \sum_{\sigma,\mu} d_\mu^{(\sigma)(j)}. \tag{117.7}$$

Because of (117.6) we have in general for the p-th power

$$\chi^{(*\boldsymbol{k})(j)}(\{\boldsymbol{\varphi}|\boldsymbol{t}\}^p) = \sum_{\sigma,\mu} (d_\mu^{(\sigma)(j)})^p. \tag{117.8}$$

---

[2] We follow, with modifications L. TISZA: Z. Physik **82**, 48 (1933).

[3] The above argument was pointed out by Dr. Y.C. CHO.

For future reference notice that (117.7), (117.8) could be considered as giving the values of the sums on the right side (individual $d_{\mu}^{(\sigma)(j)}$ are as yet undetermined) in terms of the presumably known characters on the left side of those equations which can be determined from the general rule (37.3).

Now return to (117.1). Consider the factors in (117.1) referring to the normal coordinates in (117.2). They are a product of factors for example:

$$H_{n(1,1)}\left(q\begin{pmatrix}\boldsymbol{k}\\ j_1\end{pmatrix}\right)\times\cdots\times H_{n(\sigma,\mu)}\left(q\begin{pmatrix}\boldsymbol{k}_\sigma\\ j_\mu\end{pmatrix}\right)\times\cdots\times H_{n(s,l_j)}\left(q\begin{pmatrix}\boldsymbol{k}_s\\ j_{l_j}\end{pmatrix}\right). \tag{117.9}$$

(For the moment suppress the label j and ignore symmetrization.) The string (117.9) corresponds to $n(1,1)$ phonons (the $n(1,1)$-st harmonic quantum state) of oscillator $q\begin{pmatrix}\boldsymbol{k}\\ j_1\end{pmatrix}$, plus ..., plus $n(\sigma,\mu)$ phonons of oscillator $q\begin{pmatrix}\boldsymbol{k}_\sigma\\ j_\mu\end{pmatrix}$, ..., plus $n(s_{1\,j})$ phonons of oscillator $q\begin{pmatrix}\boldsymbol{k}_\sigma\\ j_{l_j}\end{pmatrix}$. The total number of phonons in all states of the space (117.2) is

$$n(\star\boldsymbol{k}\,j)=\sum_{\sigma,\mu} n(\sigma,\mu). \tag{117.10}$$

Evidently symmetrization as in (116.2), (116.3) means physically that the fixed total number of phonons (117.10) is to be distributed in all possible ways over the states of the space (117.2). Now suppose we are dealing with a fixed string (117.9). Consider the factor

$$H_{n(\sigma,\mu)}\left(q\begin{pmatrix}\boldsymbol{k}_\sigma\\ j_\mu\end{pmatrix}\right). \tag{117.11}$$

The term of highest degree in (117.11) is a monomial of degree $n(\sigma,\mu)$ in the normal coordinate $q\begin{pmatrix}\boldsymbol{k}_\sigma\\ j_\mu\end{pmatrix}$. Now assume we work in the equivalent primed space (117.5). Then transformation by $\{\boldsymbol{\varphi}|\boldsymbol{\tau}\}$ merely multiplies the above mentioned monomial of highest degree by

$$(d_{\mu}^{(\sigma)(j)})^{n(\sigma,\mu)} \tag{117.12}$$

which is the $n(\sigma,m)$-th power of the corresponding diagonal element. Then the given string (117.9) is multiplied by the product

$$\prod_{\sigma,\mu}(d_{\mu}^{(\sigma)(j)})^{n(\sigma,\mu)} \tag{117.13}$$

one factor arising from each corresponding factor in (117.9).

Then if the fixed string (117.9) were the only term in the eigenstate, (117.13) would be the character of $\{\boldsymbol{\varphi}|\boldsymbol{\tau}\}$. However we are required to symmetrize the string (117.9) consistent only with the requirement that $n(\star\boldsymbol{k}, j)$ total phonons be present. We must distribute this number over all states symmetrically. For example all of them could be in one oscillator raising it to the $n(\star\boldsymbol{k}\,j)$ state, but then equivalence requires that such a state will equally be any of the oscillators in (117.2). In such a case the trace of the representation produced would be based upon the space:

$$\left\{H_{n(\star k\,j)}\left(q\begin{pmatrix}k\\ j_1\end{pmatrix}\right), \ldots, H_{n(\star k\,j)}\left(q\begin{pmatrix}k_\sigma\\ j_\mu\end{pmatrix}\right), \ldots, H_{n(\star k\,j)}\left(q\begin{pmatrix}k_s\\ j_{l_j}\end{pmatrix}\right)\right\} \tag{117.14}$$

giving

$$\sum_{(\mathrm{perm})}{}' \, (d_\mu^{(\sigma)\,(j)})^{n(\star k\,j)} \tag{117.15}$$

where the sum over permutations means summing over all values (permutations) of (σ, μ) keeping the form fixed as a single quantity raised to the power of the total quanta. In general suppose we distributed the fixed number $n(\star k, j)$ according to the specified distribution $\{n(\sigma, \mu)\}$. Then the corresponding eigenfunction is the string (117.9) and the corresponding symmetrized trace for that specified distribution is:

$$\sum_{(\mathrm{perm})}{}' \prod_{(\sigma,\mu)} (d_\mu^{(\sigma)\,(j)})^{n(\sigma,\mu)}, \tag{117.16}$$

where $\sum_{(\mathrm{perm})}'$ means summing over all terms in which the form of the product is preserved while indices are permuted through all permitted values.

But as *all* possible distributions may occur, so the trace must be the sum of all sums like (117.15) subject only to the condition that each distinct partition is consistent with the total number of phonons present being $n(\star k\,j)$.

Consider a particular partition; for simplicity take the partition for which (117.15) is the corresponding term. Then the sum (117.15) can be viewed as a completely symmetric sum of the $n(\star k, j)$-th power of each of $(s \cdot l_j)$ objects, namely the $d_\mu^{(\sigma)\,(j)}$. Another partition exists with $(n(\star k, j)-1)$ quanta in one oscillator, and one quantum in another. This gives rise to a symmetric polynomial of degree one in the $d_\mu^{(\sigma)\,(j)}$ for one oscillator, and degree $(n(\star k\,j)-1)$ in the $d_{\mu'}^{(\sigma')\,(j)}$ for one other oscillator in the set: the total degree of the polynomial is of course fixed at $n(\star k\,j)$. And so on for all partitions.

Then, the symmetrized product representation based upon

$$\mathscr{S}(\star k\,j) \prod_{\sigma,j,\mu} H_{n(\sigma,\mu)}\left(\gamma_{\sigma j}\, q\begin{pmatrix}k_\sigma\\ j_\mu\end{pmatrix}\right) \tag{117.17}$$

is evidently the symmetrized Kronecker $n(\star k\,j)$-th power of the representation $D^{(\star k)\,(j)}$ by which the set (117.2) transform. Thus we call the corresponding symmetrized character

$$\chi^{(\star k)\,(j)}(\{\varphi\,|\,t\})_{(n(\star k\,j))} = \mathsf{S} \sum_{(\mathrm{perm})}{}' \prod_{(\sigma,\mu)} (d_\mu^{(\sigma)\,(j)})^{n(\sigma,\mu)} \tag{117.18}$$

where S means sum over all partitions.

Now the problem is one of expressing the function on the right hand side of (117.18) in terms of the presumed known quantities (117.7), (117.8), which are expressed in terms of the known characters of powers of $\{\varphi\,|\,t\}$. To do this we have recourse to results from the theory of symmetric functions [*28*]. We need to express the terms in (117.18) in terms of the complete set of such elementary symmetric functions in given number of variables. This can be done in general straightforwardly [*28*], but we shall here only give the result and an illustration.

Consider the case of $n(\star\boldsymbol{k}\,j)=3$. Then we are dealing with second overtones (symmetrized Kronecker cube). There are three elementary symmetric polynomials of third degree. From (117.7), (117.8),

$$\chi^{(\star\boldsymbol{k})(j)}(\{\varphi|\boldsymbol{t}\})=\sum_{(\sigma,\mu)} d_\mu^{(\sigma)(j)} \tag{117.19}$$

then call

$$\Big(\sum_{\sigma,\mu} d_\mu^{(\sigma)(j)}\Big)^3 \equiv P_1 \tag{117.20}$$

it is evidently a symmetric polynomial of third degree. Another linearly independent one is the product

$$\Big(\sum_{\sigma,\mu} d_\mu^{(\sigma)(j)}\Big)\Big(\sum_{\sigma,\mu} (d_\mu^{(\sigma)(j)})^2\Big)\equiv P_2. \tag{117.21}$$

A final one is

$$\sum_{\sigma\{\mu} (d_\mu^{(\sigma)(j)})^3 \equiv P_3. \tag{117.22}$$

According to (117.18) the character of the symmetrized Kronecker cube is of form

$$\begin{aligned}\chi^{(\star\boldsymbol{k})(j)}(\{\varphi|\boldsymbol{t}\})_{(3)}= &\sum_{(\text{perm})}{}' (d_\mu^{(\sigma)(j)})^3+\sum_{(\text{perm})}{}' (d_\mu^{(\sigma)(j)})^2 (d_{\mu'}^{(\sigma')(j)})\\ &+\sum_{(\text{perm})}{}' (d_\mu^{(\sigma)(j)})(d_{\mu'}^{(\sigma')(j)})(d_{\mu''}^{(\sigma'')(j)}).\end{aligned} \tag{117.23}$$

The problem to be solved is that of finding the linear combination of (117.21)–(117.22) equal to the polynomial (117.23), i.e. the constants C_1, C_2, C_3 such that

$$C_1 P_1 + C_2 P_2 + C_3 P_3 = [\chi^{(\star\boldsymbol{k})(m)}(\{\varphi|\boldsymbol{t}\})]_{(3)}. \tag{117.24}$$

From the first term in (117.23) we see

$$C_1 + C_2 + C_3 = 1 \tag{117.25}$$

and by comparing terms in the expansions (117.20)–(117.22)

$$C_1=\tfrac{1}{6};\qquad C_2=\tfrac{1}{2};\qquad C_3=\tfrac{1}{3}. \tag{117.26}$$

Then:

$$\begin{aligned}&[\chi^{(\star\boldsymbol{k})(m)}(\{\varphi|\boldsymbol{t}\})]_{(3)}\\ &=\tfrac{1}{6}\{[\chi^{(\star\boldsymbol{k})(m)}(\{\varphi|\boldsymbol{t}\})]^3+3\,\chi^{(\star\boldsymbol{k})(m)}(\{\varphi|\boldsymbol{t}\})\,\chi^{(\star\boldsymbol{k})(m)}(\{\varphi|\boldsymbol{t}\}^2)+2\,\chi^{(\star\boldsymbol{k})(m)}(\{\varphi|\boldsymbol{t}\}^3)\}.\end{aligned} \tag{117.27}$$

The process which led to (117.27) may be continued to reduce the general case (117.18) to a sum of products of elementary characters, i.e. either products (powers) of the character of the specified element $\{\varphi|\boldsymbol{t}\}$ or of its powers. The basic theorem which can be used is that any symmetric polynomial can be written in one and only one way as a superposition of basic "elementary" polynomials [27]. The basic elementary polynomials are the P_j, of which $P_1, \ldots, P_3$ were enumerated above. The generalization of (117.27) to the symmetrized n-th power could be obtained in this manner but to the writer's knowledge it has not been done up to the time of writing. A related method was used by Tisza[2] and it gave rise to recursion relations for the symmetrized powers.

Another method to obtain the result makes use of the properties of the symmetric group [*17*]. Since we have not studied the symmetric group here we do not give the details of the proof but simply state the result. The character of $\{\varphi|\boldsymbol{t}\}$ in the symmetrized Kronecker n-th power is

$$[\chi^{(\star\boldsymbol{k})(m)}(\{\varphi|\boldsymbol{t}\})]_{(n)}=\sum\frac{(\chi^{(\star\boldsymbol{k})(m)}(\{\varphi|\boldsymbol{t}\}^{q_1}))^{r_1}\dots(\chi^{(\star\boldsymbol{k})(m)}(\{\varphi|\boldsymbol{t}\}^{q_\nu}))^{r_\nu}}{r_1!(q_1)^{r_1}\dots r_\nu!(q_2)^{r_\nu}}. \qquad (117.28)$$

The sum in (117.28) is over all possible partitions of the number n such that

$$n=r_1 q_1+\dots+r_\nu q_\nu \qquad (117.29)$$

where r_j and q_j are all integers.

This is then the solution of the problem posed at the beginning of this section. Evidently (117.28), (117.29) applies to each of the individual representations or subset of the symmetrized string (117.1) relating to a given irreducible representations $D^{(\star\boldsymbol{k})(j)}$. Direct expansion of (117.28) for $n=3$ shows that it gives agreement with the result (117.27) obtained by straightforward substitution.

The calculation of symmetrized Kronecker square, and Kronecker powers of a space group irreducible representation has been examined using a little group technique by BRADLEY and DAVIES.[4] In principle owing to the equivalence of subgroup, and full group methods of obtaining and reducing products of representations as discussed here in Sect. 64, the little group technique should be applicable to reduction of any ordinary or symmetrized power. Practical considerations may in the end favor one or another method. But, to date only the full group method seems to have been used to reduce symmetrized cubes.

118. Transformation of the lattice eigenfunction: Summary and some generalities. Let us now recapitulate. The work of the last section showed that the symmetrized product eigenfunction

$$\mathscr{S}(\star\boldsymbol{k}\,j)\exp-\left[\tfrac{1}{2}\sum_{\sigma,\mu}\gamma_{\sigma,j}\,q\begin{pmatrix}\boldsymbol{k}_\sigma\\ j_\mu\end{pmatrix}^2\right]\prod_{\sigma,\mu}H_{n_{\sigma,j,\mu}}\left(\gamma_{\sigma j}\,q\begin{pmatrix}\boldsymbol{k}_\sigma\\ j_\mu\end{pmatrix}\right) \qquad (118.1)$$

transforms as the symmetrized $n(\star\boldsymbol{k}\,j)$ power of $D^{(\star\boldsymbol{k})(j)}$ under transformation by space group operators. In (118.1) all coordinates belong to the space producing $D^{(\star\boldsymbol{k})(j)}$. Also $n(\star\boldsymbol{k}\,j)$ is the total number of quanta (phonons) in all states of this representation. The representation is denoted

$$[D^{(\star\boldsymbol{k})(j)}]_{(n(\star\boldsymbol{k},j))}. \qquad (118.2)$$

The complete eigenstate (116.3) transforms as the representation

$$[D^{(\star\boldsymbol{k})(j)}]_{(n)}\otimes[D^{(\star\boldsymbol{k}')(j')}]_{(n')}\otimes\dots\otimes[D^{(\star\boldsymbol{k}'')(j'')}]_{(n'')}, \qquad (118.3)$$

where

$$n\equiv n(\star\boldsymbol{k},j);\quad n'\equiv n(\star\boldsymbol{k}'\,j');\quad \text{etc.}$$

refers to the total number of quanta present of particular groups of degenerate oscillators. The representation (118.3) is generally reducible into a sum of irreduci-

[4] C.J. BRADLEY and B.L. DAVIES: J. Math. Phys. **11**, 1536 (1970).

ble representations. The reduction could be denoted

$$\begin{aligned}&[D^{(\star k)(j)}]_{(n)} \otimes [D^{(\star k')(j')}]_{(n')} \otimes \cdots \otimes [D^{(\star k'')(j'')}]_{(n'')} \\ &\quad = \sum_{k''' m'''} ([\star k\, j]_{(n)} \ldots [\star k''\, j'']_{(n'')} | \star k'''\, j''')\, D^{(\star k''')(j''')},\end{aligned} \tag{118.4}$$

where $([\star k\, j]_{(n)} \ldots [\star k''\, j'']_{(n'')} | \star k'''\, j''')$ is the relevant reduction coefficient, which we may obtain by methods given in Sects. 52–60.

According to the lemma of necessary degeneracy each state occurring in the reduced representation on the right of (118.4) should be considered distinct and then (118.4) gives rise to selection rules to physical processes of infra-red absorption or Raman scattering.

In the discussion of multiphonon processes in Sect. N, we shall return to this relationship.

A remark may now be addressed to the added symmetry which arises because the harmonic Hamiltonian (114.8) is separable into groups of harmonic Hamiltonians referring to those real normal coordinates in the same $(s \cdot l_j)$ dimensional irreducible vector space $\Sigma^{(\star k)(j)}$. We have to deal then with an $s \cdot l_j$ dimensional isotropic harmonic oscillator for each underlying irreducible vector space. The symmetry group of such an $(s \cdot l_j)$ dimensional isotropic oscillator, apart from the physical spatial symmetry with which we have been dealing, is the group $SU_{(s \cdot l_j)}$ – the special unitary group on $(s \cdot l_j)$ objects – these are the degenerate oscillator coordinates $q\begin{pmatrix} k_\sigma \\ j_\mu \end{pmatrix}$ in any group. It is clear that *any* $(s \cdot l_j)$ dimensional unitary transformation

$$q\begin{pmatrix} k_\sigma \\ l_\alpha \end{pmatrix} = \sum_{\sigma' \beta} U_{\sigma' \beta, \sigma \alpha}\, q\begin{pmatrix} k_{\sigma'} \\ l_\beta \end{pmatrix} \tag{118.5}$$

with

$$U^{-1}_{\sigma' \beta, \sigma \alpha} = U^{*}_{\sigma \alpha, \sigma' \beta}$$

is a symmetry operation of the $(s \cdot l_j)$ dimensional isotropic harmonic oscillator Hamiltonian. Then for each such group of coordinates in the same irreducible representation $D^{(\star k)(j)}$ there must be the added symmetry of the unitary group. Each eigenstate must simultaneously transform irreducibly under the unitary group, and the symmetrized Kronecker power representation of the spatial symmetry group $\mathfrak{G}$. Elements of the unitary group (118.5) do not appear to have any simple physical counterpart, although active work on questions related to the n dimensional isotropic harmonic oscillator are now in progress.[1, 2] It could be noted that irreducible vector spaces of the group $SU_{(s l j)}$ are based [*28*] upon symmetric tensors which give rise to irreducible representations of dimension

$$\begin{pmatrix} n(\star k\, j) + s\, l_j - 1 \\ s\, l_j - 1 \end{pmatrix}. \tag{118.7}$$

This is just the dimension of the representation of the symmetrized $n(\star k\, j)$-th power, whose character was given in (117.28).

---

[1] L. BIEDENHARN: J. Math. Phys. **4**, 436 (1963). – L. WEBER: Z. Physik **190**, 25 (1966).

[2] J. M. JAUCH and R. HILL: Phys. Rev. **57**, 641 (1940). – G. BAKER: Phys. Rev. **103**, 1119 (1956). See also [*61*] and references cited therein.

We conclude by remarking that the harmonic approximation gives an eigenfunction (116.3) which in the sense of many-body theory is correct to zeroth order. Higher corrections would include linear combinations of the basic set (116.3). But whatever the order of approximation adopted, as long as the Lemma of Necessary Degeneracy applies, any many-body state must belong to some distinct irreducible representation $D^{(\star k)(j)}$ of $\mathfrak{G}$ or corepresentation of G. As remarked in several places in this article a synthesis of group theory and many-body theory remains for future work.

In our later work we shall often proceed on the basis of the harmonic approximation because it permits explicit analysis and calculations of the symmetry of eigenstates and thus of transition processes. These predictions can be compared to experiment. Very frequently predictions of the harmonic theory are validated: examples given below are the critical point analyses of two-phonon Raman spectra in silicon and germanium (Sect. 142) in diamond structure; and NaCl, NaF in rocksalt structure; and various predictions regarding polarization of Raman scattering. When the harmonic approximation breaks down, it is often a signal that some interesting new effect is occurring, for example the proposed two-phonon bound state in diamond discussed in Sects. 124δ and 146β which is presumed due to anharmonic effects. All in all, with due regard for the possible need for inclusion of anharmonicity, the harmonic approximation is a very useful adjunct to analysis of lattice optical properties.

L. Interaction of radiation and matter: Infra-red absorption and Raman scattering by phonons

119. Introduction. In the present part of our article we shall develop the theory necessary to calculate the intensity of infra-red absorption or Raman scattering by a crystal. Evidently, only when the complete quantum-mechanical theory is available will we be able to carry out our program of utilizing the symmetry or group theory arguments to the maximum in the interpretation or prediction of the optical properties of crystals.

We give the theory of the interaction of an electromagnetic field with the ions and electrons comprising the crystal, starting from the general non-relativistic Hamiltonian for matter plus radiation. In Sect. 120 the theory of infra-red absorption by phonons is given. It is adequate for this work to reduce the most general form of the Hamiltonian to the semiclassical level. Then the intensity of infra-red absorption can be related to the square of matrix element of an electric moment operator taken between two different vibrational states of the crystal. In Sect. 121 a generalized Placzek theory of Raman scattering by phonons is given. This theory requires full quantum treatment of the radiation plus matter fields and produces the result that the intensity of Raman scattering by phonons is proportional to the square of the matrix element of a polarizability operator taken between two different vibrational states of the crystal. From the results so derived the symmetry restrictions as infra-red and Raman transition processes can be obtained by the use of group theory. The general principles of this analysis are given in Sects. 120, 121 where the transformation properties of the dipole moment and polarizibility operators are obtained. The results are obtained in Sects. 120, 121 are based upon the Born-Oppenheimer adiabatic approximation for the matter subsystem.

In Sects. 122, 123 more detailed examination is given of the restrictions imposed by symmetry on infra-red and Raman processes in crystals. The results of these two sections are widely used in interpretation of optical spectra of crystals. Particularly noteworthy is the rule of mutual exclusion for infra-red and Raman processes given in Sect. 122, which generalizes a similar rule for molecular vibrational spectra. The discussion of polarization effects in crystal vibrational spectra is also of great importance owing to the crucial part which crystal symmetry plays in determining polarization. It is precisely here that crystalline effects differ from molecular, or isotropic (powdered system) and taking maximum advantage of such polarization effects permits one to gain the maximum information regarding phonon symmetry, and interactions, which determine the scattering. Few analyses of Raman spectra have availed themselves of the full power of the polarization studies, and this remains an important avenue for the future.

Finally in Sect. 124 we give a review of some modern work on the quantum theory of Raman and infra-red processes. Necessarily the report given here will be brief but the objective is to make contact with other treatments in this Encyclopedia, as well as with recent literature. We discuss in Sect. 124 aspects of the many-body theory of these absorption and scattering processes, as well as the modern microscopic basis, especially of Raman scattering by phonons. A topic which may be particularly interesting concerns resonance Raman scattering and symmetry breaking, presented in Sect. 124η.

120. Infra-red absorption by phonons. The semiclassical radiation theory is adequate for our discussion of the infra-red absorption processes in crystals. The theory has been given in well known texts for the case of a single charged species such as electrons, or ions of given charge and mass, interacting with an electromagnetic field, as in Chap. X of [*29*]. However, our interest is in the case of an assembly of electrons plus ions interacting with the electromagnetic field, so it is desirable to give a brief outline of the relevant theory. Our goal is to write down the total Hamiltonian for electrons plus ions in semi-classical approximation. Then we shall identify a perturbation part of the Hamiltonian. This part will cause transitions between states of the unperturbed Hamiltonian corresponding to photon destruction and phonon creation Because these processes generally occur in the infra-red part of the electromagnetic spectrum we describe this as infra-red absorption by phonons.

α) *Semi-classical radiation theory for ions and electrons.* We assume the ions and electrons of the crystal are in an electromagnetic field produced by a 4-vector potential $(\boldsymbol{A}, \varphi)$. We choose to work in Lorentz gauge, and choose the scalar potential of the external field to vanish:

$$\nabla\cdot \boldsymbol{A}=0; \qquad \varphi=0. \tag{120.1}$$

Then the electric and magnetic field strengths are given as

$$\mathscr{H}=\nabla\times \boldsymbol{A}; \qquad \mathscr{E}=-(1/c)\,\partial \boldsymbol{A}/\partial t. \tag{120.2}$$

The Hamiltonian for the electrons plus ions in presence of the field is given by making the substitutions

$$\boldsymbol{p}\rightarrow\left(\boldsymbol{p}-\frac{e\boldsymbol{A}}{c}\right), \tag{120.3}$$

$$\boldsymbol{p}\rightarrow\left(\boldsymbol{p}+\frac{eZ\boldsymbol{A}}{c}\right) \tag{120.4}$$

for electron and ion momenta respectively. In (120.3) and (120.4) the vector potential must be evaluated at electron sites $\boldsymbol{x}$ and ion sites $\boldsymbol{X}$ but we will simplify the notation for the next few lines.

Then the total Hamiltonian (112.7) becomes, in symbolic terms

$$\begin{aligned}\mathscr{H}(\boldsymbol{x};\underline{\boldsymbol{X}};t)&=\sum\left(\tfrac{1}{2}m\right)(\boldsymbol{p}-e\boldsymbol{A}/c)^2+\sum\left(\tfrac{1}{2}M\right)\left(\boldsymbol{P}+\frac{eZ\boldsymbol{A}}{c}\right)^2\\&\quad+\sum\frac{ZZ'e^2}{|\underline{\boldsymbol{X}}-\underline{\boldsymbol{X}}'|}+\sum\frac{e^2}{|\boldsymbol{x}-\boldsymbol{x}'|}-\sum\frac{Ze^2}{|\underline{\boldsymbol{X}}-\boldsymbol{x}|}.\end{aligned} \tag{120.5}$$

In (120.5) the electron mass is m, the ion mass M, electron charge is e, ion charge is $-Ze$, the sums are over all electron and ion positions $\boldsymbol{x}$ and $\underline{\boldsymbol{X}}$ as appropriate.

If we expand the quadratic terms in (120.5) we obtain

$$\mathscr{H}(\boldsymbol{P},\boldsymbol{X};\boldsymbol{p},\boldsymbol{x};t)=\mathscr{H}(\boldsymbol{P},\boldsymbol{X};\boldsymbol{p},\boldsymbol{x})+\mathscr{H}'(\boldsymbol{P},\boldsymbol{X};\boldsymbol{p},\boldsymbol{x};t), \tag{120.6}$$

where we used the notation of (112.6). Let us now neglect terms quadratic in the vector potential. Then we obtain explicitly

$$\begin{aligned}\mathscr{H}&=\sum\left(\tfrac{1}{2}m\right)\boldsymbol{p}^2+\sum\left(\tfrac{1}{2}M\right)\boldsymbol{P}^2+\sum\frac{ZZ'e^2}{|\underline{\boldsymbol{X}}-\underline{\boldsymbol{X}}'|}+\sum\frac{e^3}{|\boldsymbol{x}-\boldsymbol{x}'|}\\&\quad-\sum\frac{Ze^2}{|\boldsymbol{x}-\underline{\boldsymbol{X}}|}-\sum\left(\frac{e}{mc}\right)\boldsymbol{A}\cdot\boldsymbol{p}+\sum\left(\frac{Ze}{Mc}\right)\boldsymbol{A}\cdot\boldsymbol{P}.\end{aligned} \tag{120.7}$$

From (120.7) we can identify the perturbing term as

$$\mathscr{H}'\equiv-\sum\left(\frac{e}{mc}\right)\boldsymbol{A}\cdot\boldsymbol{p}+\sum\left(\frac{Ze}{Mc}\right)\boldsymbol{A}\cdot\boldsymbol{P}, \tag{120.8}$$

This is the usual approximation in the semi-classical theory, with neglect of the $\boldsymbol{A}\cdot\boldsymbol{A}$ terms. This approximation is general valid for insulators (*vide infra* Sect. 121 ζ) but not for metals, or where a high density of free carriers is present.

Let us now prescribe that the vector potential $\boldsymbol{A}$ gives rise to electromagnetic plane waves:

$$\boldsymbol{A}=\boldsymbol{A}_0\,\mathrm{e}^{i(\boldsymbol{k}\cdot\boldsymbol{r}-\omega t)} \tag{120.9}$$

where A_0 in (120.9) is a constant vector amplitude, and from (120.2):

$$k \cdot A_0 = 0 \tag{120.10}$$

which is the requirement of transversality.

Writing out the terms in (120.8) explicitly, we have

$$\mathscr{H}' = -\sum_j \left(\frac{e}{mc}\right) A_0 e^{i(k \cdot r_j - \omega t)} \cdot p_j + \sum_\alpha \left(\frac{Ze}{Mc}\right) A_0 e^{i(k \cdot R_\alpha - \omega t)} \cdot P_\alpha, \tag{120.11}$$

It is of importance to realize that in (120.11) the coordinates of both electrons and of ions enters (r_j and R_α respectively), because the vector potential appearing in (120.8) must be evaluated at the electron and ion sites.

At this point in a usual treatment of interaction of radiation with a molecule or a system whose dimension is smaller than a wavelength of the light, one assumes that $k \cdot r \approx 0$ i.e. the radiation wavelength is much greater than any dimension of the system. But for a (formally) infinite crystal this cannot be assumed. Let us assume the electrons are strictly localized at the ion sites, or cells, following a discussion given by LAX and BURSTEIN.[1] Thus we replace r_j by R_α and the index j on electron variables by α. The perturbing Hamiltonian then becomes;

$$\mathscr{H}' = -\sum_\alpha \left(\frac{1}{c}\right) e^{i(k \cdot R_\alpha - \omega t)} A_0 \cdot \left(\frac{e}{m} p_\alpha - \frac{Ze}{M} P_\alpha\right), \tag{120.12}$$

In (120.12) the ions and electrons are now labelled by the common index α.

The replacement of the electron coordinate r_j as a dynamical variable, by the ion coordinate R_α is a very drastic assumption. It has, to the date of writing, not been fully justified by a more rigorous treatment. For example, even in insulators where there is not the assumed extreme localization which is implied by the substitution i.e. insulators with mixed, ionic-covalent binding one could expect sizable corrections. This approximation is very much akin to a rigid ion approximation,

β) Transition rate. According to our model, the perturbation (120.12) is to produce transitions between stationery states of the unperturbed Hamiltonian, $\mathscr{H}(r, R)$ in (112.7) where now we use r, and R for electron and ion position variables respectively. But the unperturbed Hamiltonian is, in this approximation, that of ions and electrons. In the adiabatic approximation the eigenfunctions are given in (113.24) as

$$\Psi^{(\text{adiabatic})} = \varphi(r, R)\, \chi(R). \tag{120.13}$$

The energy of this state is, as discussed in Sect. 113, Eq. (113.40)

$$E = \Phi^{(0)} + E^{(2)} \tag{120.14}$$

where $\Phi^{(0)}$ is the eigenenergy of the electronic problem with nuclei fixed, and $E^{(2)}$ is the energy of the nuclear problem. In the harmonic approximation, $E^{(2)}$ is the sum of harmonic oscillator energies as in (115.8). The particular transitions in which we are interested are those in which only the vibrational part of Ψ changes, while the electronic state of the system is unchanged, before and after the tran-

[1] M. LAX and E. BURSTEIN: Phys. Rev. **97**, 42 (1955).

sition. Hence, if we include electronic and vibrational (nuclear) quantum numbers in the specification of the eigenfunctions we take, as initial state

$$\Psi_i(\boldsymbol{r}, \boldsymbol{R}) = \varphi_\nu(\boldsymbol{r}, \boldsymbol{R})\, \chi_{\nu n}(\boldsymbol{R}) \tag{120.15}$$

where ν is the electronic and n the nuclear quantum index, and as final state

$$\Psi_f(\boldsymbol{r}, \boldsymbol{R}) = \varphi_{\bar\nu}(\boldsymbol{r}, \boldsymbol{R})\, \chi_{\bar\nu \bar n}(\boldsymbol{R}). \tag{120.16}$$

Of course, the nuclear eigenfunction and eigenenergy depend upon the electronic state index. The energy of states (120.15), (120.16) will be denoted $E_i = E_{\nu n}$ and $E_f = E_{\bar\nu \bar n}$ respectively.

Using ordinary first order time-dependent perturbation theory [29] the rate of transitions from Ψ_i to Ψ_f produced by the perturbation (120.12) is proportional to

$$\begin{aligned} \omega_{i\to f} = \sum_f \left| \left\langle \Psi_f \left| \sum_\alpha \left(\frac{i\hbar}{c}\right) \mathrm{e}^{i(\boldsymbol{k}\cdot \boldsymbol{R}_\alpha)} \boldsymbol{A}_0 \cdot \left(\frac{e}{m}\, \boldsymbol{p}_\alpha - \frac{Ze}{M}\, \boldsymbol{P}_\alpha\right) \right| \Psi_i \right\rangle \right|^2 \\ \times \delta(E_f - E_i - \hbar\omega). \end{aligned} \tag{120.17}$$

Before proceeding with the conventional analysis observe that in (120.17) the two terms inside the matrix element are of different order of magnitude in the spirit of the adiabatic approximation. That is, the operator in the matrix element (120.17) can be rewritten

$$\sum_\alpha \left(\frac{e}{mc}\right) \mathrm{e}^{i\boldsymbol{k}\cdot \boldsymbol{R}_\alpha} \boldsymbol{A}_0 \cdot \left(\boldsymbol{p}_\alpha - \left(\frac{Z}{e}\right)\left(\frac{m}{M}\right) \boldsymbol{P}_\alpha\right). \tag{120.18}$$

Since, the parameter of smallness in the adiabatic theory is $(m/M) \equiv \kappa^4$, and using (113.6) to define reduced ion displacements $\boldsymbol{u}_\alpha$:

$$\boldsymbol{P}_\alpha \equiv \frac{\hbar}{i}\, \partial/\partial \boldsymbol{X}_\alpha = \frac{\hbar}{i}\left(\frac{1}{\kappa}\right)(\partial/\partial \boldsymbol{u}_\alpha) \tag{120.19}$$

also, we have for (120.18)

$$-\sum_\alpha \left(\frac{e i \hbar}{m c}\right) \mathrm{e}^{i\boldsymbol{k}\cdot \boldsymbol{R}_\alpha} \boldsymbol{A}_0 \cdot (\partial/\partial \boldsymbol{r}_\alpha - Z\kappa^3\, \partial/\partial \boldsymbol{u}_\alpha). \tag{120.20}$$

Thus if the conventional argument is applied, by which the term

$$-(e i \hbar/m c)\, \boldsymbol{A}_0 \cdot \partial/\partial \boldsymbol{r}_\alpha \tag{120.21}$$

can be taken as a first order perturbation on the unperturbed electronic Schrödinger equation, then clearly the term (which originated from the ionic momentum):

$$+Z\kappa^3 (e i \hbar/m c)\, \boldsymbol{A}_0 \cdot \partial/\partial \boldsymbol{u}_\alpha \tag{120.22}$$

must be treated in a still higher order of perturbation theory since it is $(\kappa^3 Z)$ times the previous term in (120.21). Perhaps the most direct approach would be to do first order perturbation theory on the largest term (120.21) and do further (higher order) corrections in the smaller term (120.22). As of the time of writing this problem seems not to have been discussed in this way.

γ) *Analysis of the transition matrix element for infra-red lattice absorption.* In infra-red lattice absorption we are concerned with an actual photon absorption and the corresponding change in the state of excitation of the lattice oscillators. One or more phonons are created via the absorption; the "order" of the infra-red process is the number of such phonons produced. The electronic state of the system is assumed unchanged in the transition. Consequently we choose the index ν of the electron part of Ψ_i and Ψ_f to be the same: $\varphi_\nu(\boldsymbol{r}, \boldsymbol{R}) = \varphi_{\bar{\nu}}(\boldsymbol{r}, \boldsymbol{R})$.

In order to obtain the infra-red lattice vibration absorption cross sections, we treat the entire term (120.20) as a single perturbing term. Hence we evaluate (120.17) using the adiabatic states (120.15), (120.16). Denoting variables over which integration is performed by a subscript on the scalar product, we adopt the notational devices

$$\langle\ \rangle_{\boldsymbol{r}} \equiv \int \Pi\, d^3\boldsymbol{r} \tag{120.23a}$$

and

$$\langle\ \rangle_{\boldsymbol{R}} \equiv \int \Pi\, d^3\boldsymbol{R} \tag{120.23b}$$

which implies integration over *all* electron $\{\boldsymbol{r}\}$ and all *all* ion coordinates $\{\boldsymbol{R}\}$

$$\begin{aligned} &\langle \Psi_f | \sum_\alpha (1/c)\, e^{i\boldsymbol{k}\cdot\boldsymbol{R}_\alpha} \boldsymbol{A}_0 \cdot \left(\frac{e}{m}\boldsymbol{p}_\alpha - \frac{eZ}{M}\boldsymbol{P}_\alpha\right) |\Psi_i\rangle \\ &= \sum_\alpha (1/c)\, e^{i\boldsymbol{k}\cdot\boldsymbol{R}_\alpha} \boldsymbol{A}_0 \cdot \langle \varphi_\nu(\boldsymbol{r},\boldsymbol{R})\, \chi_{\nu\bar{n}}(\boldsymbol{R})| \frac{e}{m}\boldsymbol{p}_\alpha - \frac{Ze}{M}\boldsymbol{P}_\alpha |\varphi_\nu(\boldsymbol{r},\boldsymbol{R})\, \chi_{\nu n}(\boldsymbol{R})\rangle_{\boldsymbol{r},\boldsymbol{R}} \qquad (120.24)\\ &= \sum_\alpha (1/c)\, e^{i\boldsymbol{k}\cdot\boldsymbol{R}_\alpha} \boldsymbol{A}_0 \cdot \langle \chi_{\nu\bar{n}}(\boldsymbol{R})|\, \langle \varphi_\nu(\boldsymbol{r},\boldsymbol{R})| \frac{e}{m}\boldsymbol{p}_\alpha - \frac{Ze}{M}\boldsymbol{P}_\alpha |\varphi_\nu(\boldsymbol{r},\boldsymbol{R})\rangle_{\boldsymbol{r}}\, |\chi_{\nu n}(\boldsymbol{R})\rangle_{\boldsymbol{R}}. \end{aligned}$$

In the last line we factorized the integrations over $\boldsymbol{r}$ and $\boldsymbol{R}$. Since for the electron momentum operator $\boldsymbol{p}_\alpha$

$$\boldsymbol{p}_\alpha = m\, d\boldsymbol{r}_\alpha/dt \tag{120.25}$$

and for the ion momentum operator $\boldsymbol{P}_\alpha$

$$\boldsymbol{P}_\alpha = M\, d\boldsymbol{R}_\alpha/dt \tag{120.26}$$

if then follows that

$$\frac{e}{m}\boldsymbol{p}_\alpha - \frac{eZ}{M}\boldsymbol{P}_\alpha = d/dt\,(e\,\boldsymbol{r}_\alpha - Z\,e\,\boldsymbol{R}_\alpha) \tag{120.27}$$

$$= (d/dt)\,\boldsymbol{\mu}_\alpha, \tag{120.28}$$

In (120.28), the dipole moment operator $\boldsymbol{\mu}_\alpha$ is

$$\boldsymbol{\mu}_\alpha = e\,\boldsymbol{r}_\alpha - e Z\,\boldsymbol{R}_\alpha. \tag{120.29}$$

Clearly if $\boldsymbol{r}_\alpha$ and $\boldsymbol{R}_\alpha$ are measured as displacements from some initial unperturbed positions $\boldsymbol{r}_\alpha^{(0)} \equiv \boldsymbol{R}_\alpha^{(0)}$ then (120.29) may be interpreted as the dipole moment operator of the ion-electron pair in cell α.

Hence in (120.24) we have a matrix element of the operator $d\boldsymbol{\mu}_\alpha/dt$, between adiabatic states. But this can be simplified using the Heisenberg equation of motion for an operator

$$d\boldsymbol{\mu}_\alpha/dt = \frac{i}{\hbar}\,[\mathscr{H}, \boldsymbol{\mu}_\alpha]. \tag{120.30}$$

In (120.30), $\mathscr{H}$ is the total Hamiltonian. Take the matrix element of (120.30) using the adiabatic basis, and assuming that the adiabatic eigenfunctions are exact as in (113.39)–(113.43)

$$\mathscr{H}\,|\varphi_\nu\,\chi_{\nu n}\rangle = E_{\nu n}\,|\varphi_\nu\,\chi_{\nu n}\rangle \tag{120.31}$$

and similarly for the excited state. Then we obtain for the matrix element involved

$$\begin{aligned}\langle\varphi_\nu\,\chi_{\nu\bar{n}}|\,d\boldsymbol{\mu}_\alpha/dt\,|\varphi_\nu\,\chi_{\nu n}\rangle_{\boldsymbol{r},\boldsymbol{R}} &= \frac{i}{\hbar}\langle\varphi_\nu\,\chi_{\nu\bar{n}}|\,[\mathscr{H},\boldsymbol{\mu}_\alpha]\,|\varphi_\nu\,\chi_{\nu n}\rangle_{\boldsymbol{r},\boldsymbol{R}}\\ &= \frac{i}{\hbar}(E_{\nu\bar{n}}-E_{\nu n})\langle\varphi_\nu\,\chi_{\nu\bar{n}}|\,\boldsymbol{\mu}_\alpha\,|\varphi_\nu\,\chi_{\nu n}\rangle_{\boldsymbol{r},\boldsymbol{R}}\\ &= i\,\omega_{\bar{n}n}\,\langle\chi_{\nu\bar{n}}|\,\langle\varphi_\nu|\,\boldsymbol{\mu}_\alpha\,|\varphi_\nu\rangle_{\boldsymbol{r}}\,|\chi_{\nu n}\rangle_{\boldsymbol{R}}.\end{aligned} \tag{120.32}$$

Let us now define a "phased" diagonal electric dipole moment operator which will arise when (120.32) is substituted into (120.24) by

$$\mathscr{M}^\nu(\boldsymbol{R},\boldsymbol{k}) \equiv \sum_\alpha e^{i\boldsymbol{k}\cdot\boldsymbol{R}_\alpha^{(0)}}\,\langle\varphi_\nu(\boldsymbol{r},\boldsymbol{R})|\,\boldsymbol{\mu}_\alpha(\boldsymbol{r},\boldsymbol{R})\,|\varphi_\nu(\boldsymbol{r},\boldsymbol{R})\rangle_{\boldsymbol{r}}. \tag{120.33}$$

This quantity is a phased sum of the individual ("cell") dipole moment expectation values in the electronic state φ_ν. It also depends on the radiation wave vector $\boldsymbol{k}$, and is a function of the set of all the nuclear coordinates. It may be useful to remark that $\mathscr{M}^\nu(\boldsymbol{R},\boldsymbol{k})$ depends on nuclear coordinates $\boldsymbol{R}$, both explicitly, via the definition (120.29) of $\boldsymbol{\mu}_\alpha$, as well as implicitly, via the dependence of the electronic eigenfunctions $\varphi_\nu(\boldsymbol{r},\boldsymbol{R})$ upon $\boldsymbol{R}$. In the latter context, recall the discussion in Sect. 113, and in particular Eq. (113.9). Clearly it is the "deformability" of the electronic eigenfunctions which introduces the physically important dependence of $\mathscr{M}^\nu(\boldsymbol{R},\boldsymbol{k})$ upon $\boldsymbol{R}$. Returning to (120.24) the term which determines the rate of transitions $i\to f$ can be written

$$\boldsymbol{A}_0\cdot\langle\chi_{\nu\bar{n}}(\boldsymbol{R})|\,\mathscr{M}^\nu(\boldsymbol{R},\boldsymbol{k})\,|\chi_{\nu n}(\boldsymbol{R})\rangle_{\boldsymbol{R}}. \tag{120.34}$$

A task of utmost importance in the application of group theory to the infra-red lattice absorption in crystals is the analysis of the matrix element in (120.34) for each specific case.

δ) *Symmetry of the matrix element for infra-red absorption.* First let us determine the transformation properties of $\mathscr{M}^\nu(\boldsymbol{R},\boldsymbol{k})$. Let $\{\boldsymbol{\varphi}\}$ be a unitary space group transformation in the space group $\mathfrak{G}$. Under transformation by $\{\boldsymbol{\varphi}\}$ an inner automorphism on the ion positions in the crystal is produced. Symbollically, taking $(\boldsymbol{R})$ to represent the ion positions (i.e. the set of $\boldsymbol{R}_\alpha$), then under $\{\boldsymbol{\varphi}\}$

$$(\boldsymbol{R})\to(\boldsymbol{R}')=(\boldsymbol{\varphi})\,\boldsymbol{R}. \tag{120.35}$$

The electronic function transforms as

$$\varphi_\nu(\boldsymbol{r},\boldsymbol{R})\to\varphi'_\nu(\boldsymbol{r},\boldsymbol{R})=\varphi_\nu(\boldsymbol{r},\{\boldsymbol{\varphi}\}^{-1}\boldsymbol{R}). \tag{120.36}$$

In general the electronic eigenfunctions φ_ν will also obey the lemma of necessary degeneracy and so (120.36) may be expressed in terms of the appropriate linear combination of the other members of its degenerate set via the irreducible matrices of the representation. But we do not require this here so we can have (120.36) as it stands. The individual cell dipole moment will transform as an ordinary (polar) vector. Thus let $(\mu_\alpha)_l$ be the l-th cartesian component of the moment $\boldsymbol{\mu}_\alpha$. Then under transformation by $\{\varphi\}$ we have directly from (120.29)

$$(\mu_\alpha)_l(\boldsymbol{r}, \boldsymbol{R}) \to (\mu'_\alpha)_l(\boldsymbol{r}, \boldsymbol{R}) = \sum_m D^{(\boldsymbol{v})}(\{\varphi\})_{ml}(\mu_\alpha)_m(\boldsymbol{r}, \{\varphi\}^{-1}\boldsymbol{R}) \tag{120.37}$$

where $D^{(\boldsymbol{v})}$ is the representation by which a polar vector transforms. These transformations are only needed symbollically here, but are given explicitely in (104.8)–(104.12). Below we shall also use the explicit form as required.

Then assembling (120.35)–(120.37) with the definition of the operator $\mathscr{M}^\nu(\boldsymbol{R}, \boldsymbol{k})$ in (120.33) we have that under $\{\varphi\}$

$$\mathscr{M}_l^\nu(\boldsymbol{R}, \boldsymbol{k}) \to \mathscr{M}_l^{\nu\prime}(\boldsymbol{R}, \boldsymbol{k}) = \sum_m D^{(\boldsymbol{v})}(\{\varphi\})_{ml}\, \mathscr{M}_m^\nu(\{\varphi\}^{-1}\boldsymbol{R}, \boldsymbol{k}). \tag{120.38}$$

This is the significant result: the operator $\mathscr{M}^\nu$ transforms like an polar vector field (first rank tensor field) operator. In brief

$$\mathscr{M}^\nu \sim D^{(\boldsymbol{v})} \sim D^{(\boldsymbol{r})} \tag{120.39}$$

observe that whether or not (120.33) vanishes is governed by an examination of the symmetry of electronic eigenstates φ_ν, which are many-electron states, and the relevant reduction coefficient.

Before proceeding let us clarify this result. From the definition (120.33) it can be seen that $\mathscr{M}^\nu(\boldsymbol{R}, \boldsymbol{k})$ transforms like a Bloch sum of localized objects (centered at $\boldsymbol{R}_\alpha$) with wave vector $\boldsymbol{k}$ is equal to the wave vector of the radiation. Despite the formal difficulty of taking $\boldsymbol{k}=0$, one often does just this in analysis of transition processes. We shall keep the $\boldsymbol{k}$ dependence in mind when constructing the selection rules: as could be expected it is of greatest importance in question of momentum conservation.

The transformation property (120.39) produces restrictions on the matrix element in (120.34), according to the familiar dictates of the Wigner-Eckart theorem. Recall now that (120.34) now refers to a purely vibrational transition $n \to \bar{n}$. Let the symmetry of the nuclear eigenfunctions be given as

$$|\chi_{\nu n}\rangle \sim D^{(n)}, \tag{120.40}$$

$$|\chi_{\nu \bar{n}}\rangle \sim D^{(\bar{n})}, \tag{120.41}$$

then the matrix element

$$\langle \chi_{\nu\bar{n}} | \mathscr{M}^\nu(\boldsymbol{R}, \boldsymbol{k}) | \chi_{\nu n}\rangle \neq 0$$

if

$$D^{(n)} \otimes D^{(\bar{n})} \sim D^{(\boldsymbol{v})}. \tag{120.42}$$

That is the direct product of symmetries of initial times final states should contain the symmetry of a polar vector. The transition rate is then given as

$$\omega_{i\to f} = \sum_f |\boldsymbol{A}_0 \cdot \langle \chi_{\nu\bar{n}} | \mathscr{M}^\nu(\boldsymbol{R}, \boldsymbol{k}) | \chi_{\nu n}\rangle|^2\, \delta(E_f - E_i - \hbar\omega). \tag{120.43}$$

The basic selection rule (120.42) must be applied in every order, and will permit or restrict transitions depending on circumstances, as will be discussed below.

ε) *One phonon and multiphonon processes.*

i) *One phonon.* Physically important results can be obtained by expanding the operator $\mathcal{M}^{\nu}(\boldsymbol{R}, \boldsymbol{k})$ in configuration space in a Taylor series of ion displacements form equilibrium. Given the equilibrium ion positions at $\boldsymbol{R}_{\alpha}^{(0)}$, the instantaneous ion positions can be taken as small deviations from equilibrium. Returning to the usual notation let $\boldsymbol{R}_{\alpha} \to \boldsymbol{R}\begin{pmatrix} l \\ \kappa \end{pmatrix}$ and $\boldsymbol{\mu}_{\alpha} \to \boldsymbol{\mu}\begin{pmatrix} l \\ \kappa \end{pmatrix}$ we have:

$$\boldsymbol{R}\begin{pmatrix} l \\ \kappa \end{pmatrix} - \boldsymbol{R}^{(0)}\begin{pmatrix} l \\ \kappa \end{pmatrix} = \kappa\,\bar{\boldsymbol{u}}\begin{pmatrix} l \\ \kappa \end{pmatrix} \tag{120.44}$$

or using (80.9) we now identify the $\kappa\,\bar{\boldsymbol{u}}\begin{pmatrix} l \\ \kappa \end{pmatrix}$ of (120.44) with the $\boldsymbol{u}\begin{pmatrix} l \\ \kappa \end{pmatrix}$ of (80.9) so we obtain an expression in terms of normal coordinates:

$$\boldsymbol{R}\begin{pmatrix} l \\ \kappa \end{pmatrix} - \boldsymbol{R}^{(0)}\begin{pmatrix} l \\ \kappa \end{pmatrix} = \left(\frac{1}{\sqrt{M_{\kappa} N}}\right) \sum_{j\boldsymbol{k}'} \mathrm{e}^{i\boldsymbol{k}'\cdot\boldsymbol{R}_L}\, \boldsymbol{e}\left(\kappa \middle| \begin{matrix} \boldsymbol{k}' \\ j \end{matrix}\right) Q\begin{pmatrix} \boldsymbol{k}' \\ j \end{pmatrix}. \tag{120.45}$$

Depending on convenience we can use (120.44) or (120.45). Then let us expand the matrix elements in (120.33) in a Taylor series in displacements. Actually we could expand the eigenfunctions as in (113.9), and also the cell operator $\boldsymbol{\mu}\begin{pmatrix} l \\ \kappa \end{pmatrix}$, but for our purposes it is most convenient to proceed as follows:

$$\begin{aligned} \langle \varphi_{\nu} | \boldsymbol{\mu}\begin{pmatrix} l \\ \kappa \end{pmatrix} | \varphi_{\nu} \rangle_{\boldsymbol{r}} = {} & \boldsymbol{m}^{(0)}\begin{pmatrix} l \\ \kappa \end{pmatrix} + \sum_{l'\kappa'\alpha} \boldsymbol{m}^{(1)}_{,\alpha}\begin{pmatrix} l & l' \\ \kappa & \kappa' \end{pmatrix} u_{\alpha}\begin{pmatrix} l' \\ \kappa' \end{pmatrix} \\ & + \sum_{l'\kappa'\alpha} \sum_{l''\kappa''\beta} \boldsymbol{m}^{(2)}_{,\alpha\beta}\begin{pmatrix} l & l' & l'' \\ \kappa & \kappa' & \kappa'' \end{pmatrix} u_{\alpha}\begin{pmatrix} l' \\ \kappa' \end{pmatrix} u_{\beta}\begin{pmatrix} l'' \\ \kappa'' \end{pmatrix} + \cdots. \end{aligned} \tag{120.46}$$

We dispense with the index ν in the following, since it is assumed fixed. Since the left side of (120.46) can be interpreted as the dipole moment in cell $\begin{pmatrix} l \\ \kappa \end{pmatrix}$ then a term like $m^{(1)}\begin{pmatrix} l & l' \\ \kappa & \kappa' \end{pmatrix}$ could be interpreted as an effective charge on $\boldsymbol{R}\begin{pmatrix} l \\ \kappa \end{pmatrix}$. The sum of all such charges times displacement gives the total moment at $\boldsymbol{R}\begin{pmatrix} l \\ \kappa \end{pmatrix}$. Thus $m^{(1)}\begin{pmatrix} l & l' \\ \kappa & \kappa' \end{pmatrix}$ defines an effective charge matrix for the lattice. Obviously the space group symmetry $\mathfrak{G}$ produces restrictions on the number of independent terms in this effective charge matrix just as in the analogous case of the number of independent force constants for the lattice dynamic problem, as in Eqs. (109.38)–(109.43).

The next term in (120.46) is $m^{(2)}\begin{pmatrix} l & l' & l'' \\ \kappa & \kappa' & \kappa'' \end{pmatrix}$ which can be interpreted as an "induced" charge matrix. Similar remarks about restrictions due to symmetry can be made for the induced charge matrix.

Substituting (120.45) or (80.9) into (120.46) and then into (120.33) we find an expansion for $\mathcal{M}(\boldsymbol{R},\boldsymbol{k})$ in normal coordinates as:

$$\begin{aligned}\mathcal{M}(\boldsymbol{R},\boldsymbol{k})=\mathcal{M}^{(0)}(\boldsymbol{k})+\sum_{\boldsymbol{k}'j'}\mathcal{M}^{(1)}\left(\boldsymbol{k}\middle|\begin{matrix}\boldsymbol{k}'\\ j'\end{matrix}\right)Q\begin{pmatrix}\boldsymbol{k}'\\ j'\end{pmatrix}\\ +\sum_{\boldsymbol{k}''j''}\sum_{\boldsymbol{k}'j'}\mathcal{M}^{(2)}\left(\boldsymbol{k}\middle|\begin{matrix}\boldsymbol{k}' & \boldsymbol{k}''\\ j' & j''\end{matrix}\right)Q\begin{pmatrix}\boldsymbol{k}'\\ j'\end{pmatrix}Q\begin{pmatrix}\boldsymbol{k}''\\ j''\end{pmatrix}+\cdots.\end{aligned}\tag{120.47}$$

An explicit expression for the coefficients in (120.47) in terms of those in (120.46) can be given. For example

$$\begin{aligned}\mathcal{M}^{(1)}\left(\boldsymbol{k}\middle|\begin{matrix}\boldsymbol{k}'\\ j'\end{matrix}\right)=\sum_{l\kappa\alpha}\sum_{l'\kappa'}\left(\frac{1}{\sqrt{M_\kappa N}}\right)\boldsymbol{m}^{(1)}_{,\alpha}\begin{pmatrix}l & l'\\ \kappa & \kappa'\end{pmatrix}\\ \times e_\alpha\left(\kappa\middle|\begin{matrix}\boldsymbol{k}'\\ j'\end{matrix}\right)\exp\left(i\boldsymbol{k}\cdot\boldsymbol{R}\begin{pmatrix}l\\ \kappa\end{pmatrix}+i\boldsymbol{k}'\cdot\boldsymbol{R}\begin{pmatrix}l'\\ \kappa'\end{pmatrix}\right).\end{aligned}\tag{120.48}$$

Now, with (120.47) one can immediately draw some symmetry-related conclusions. Thus, every term in (120.47) must transform as $\mathcal{M}$. But, considering the radiation to be finite wave vector $\boldsymbol{k}\neq 0$, each term must transform with wave vector $\boldsymbol{k}$. Then in the term in (120.47) linear in $Q\begin{pmatrix}\boldsymbol{k}'\\ j'\end{pmatrix}$ we have

$$Q\begin{pmatrix}\boldsymbol{k}'\\ j'\end{pmatrix}\sim D^{(\boldsymbol{k})(v)}.\tag{120.49}$$

That is if (120.49) applies then the corresponding

$$\mathcal{M}^{(1)}\left(\boldsymbol{k}\middle|\begin{matrix}\boldsymbol{k}'\\ j'\end{matrix}\right)\neq 0.\tag{120.50}$$

Evidently the requirement that $\boldsymbol{k}'=\boldsymbol{k}$ simply is a momentum conservation condition. Now consider the term linear in $Q\begin{pmatrix}\boldsymbol{k}'\\ j'\end{pmatrix}$ substituted into (120.43) as:

$$\boldsymbol{A}_0\langle\chi_{0\bar{n}}|\,\mathcal{M}^{(1)}\left(\boldsymbol{k}\middle|\begin{matrix}\boldsymbol{k}'\\ j'\end{matrix}\right)Q\begin{pmatrix}\boldsymbol{k}'\\ j'\end{pmatrix}|\chi_{0n}\rangle_{\boldsymbol{R}}\tag{120.51}$$

where now choose $v=0$ signifying some fixed electronic state. But only if $|\chi_{0\bar{n}}\rangle$ differs from $|\chi_{0n}\rangle$ by having the oscillator $Q\begin{pmatrix}\boldsymbol{k}'\\ j'\end{pmatrix}$ singly excited will the matrix element (120.51) be non-zero. Again the Wigner-Eckart theorem applies: the direct product $D^{(n)}\otimes D^{(\bar{n})}$ must contain $D^{(\boldsymbol{k}')(j')}$ which in turn must be $D^{(\boldsymbol{k})(v)}$ for the matrix element (120.51) to be non-zero. This is a one-phonon processes. For simplicity we only consider photon absorption processes although the same type of analysis will apply to emission. The value of (120.51) is then easily obtained, using the expression for the value of coordinate in oscillator matrix elements [*31*]:

$$\boldsymbol{A}_0\cdot\mathcal{M}^{(1)}\left(\boldsymbol{k}\middle|\begin{matrix}\boldsymbol{k}'\\ j'\end{matrix}\right)\left(\frac{\hbar}{2\omega(\boldsymbol{k}'|j')}\right)^{\frac{1}{2}}(n+1)^{\frac{1}{2}}\,\delta_{\bar{n},n+1}\tag{120.52}$$

where n is generally zero (crystal initially in ground vibrational state).

From (120.52), and (120.43), we can obtain the intensity of infra-red absorption producing a single phonon. For incident radiation in the spectral range ω to $\omega + d\omega$ this is proportional to

$$\mathscr{I}(\omega)\, d\omega \sim \sum_{\boldsymbol{k}' j'} \left| \left(\frac{\hbar(n+1)}{2\omega(\boldsymbol{k}'|j')} \right)^{\frac{1}{2}} \delta_{\bar{n},\, n+1} \boldsymbol{A}_0 \cdot \mathscr{M}^{(1)} \left(\boldsymbol{k} \middle| \begin{matrix} \boldsymbol{k}' \\ j' \end{matrix} \right) \right|^2 \delta(\omega - \omega(\boldsymbol{k}'|j')). \quad (120.53)$$

In (120.53) the delta function results from energy conservation requirements. The spectrum of absorption then consists of a sharp delta peak at those incident frequencies ω corresponding to the phonon energy $\omega(\boldsymbol{k}|j')$ whose symmetry permits a non-zero coupling coefficient $\mathscr{M}^{(1)} \left(\boldsymbol{k} \middle| \begin{matrix} \boldsymbol{k}' \\ j' \end{matrix} \right)$. The "infra-red active" phonons are those whose symmetry is $D^{(\boldsymbol{v})}$ or in a cubic crystal, making the infinite wavelength assumption ($\boldsymbol{k}=0$) their symmetry must be $D^{(\boldsymbol{\Gamma})(15-)} \sim D^{(\boldsymbol{v})}$. In a real crystal the absorption is not a delta function but is broadened owing to final state phonon lifetime, caused by anharmonic coupling of the infra-red phonon to other phonons in the crystal, etc. These effects are generally treated by many-body methods [2–7] and are reviewed in other articles in this Encyclopedia [*20*, *22*], and in Sect. 124b of the present article, a very brief review will be given.

ii) *Two phonon.* Next in order of importance and intensity are two phonon processes. These arise from bilinear terms in (120.47) which contain products like $Q \begin{pmatrix} \boldsymbol{k}' \\ j' \end{pmatrix} Q \begin{pmatrix} \boldsymbol{k}'' \\ j'' \end{pmatrix}$. The matrix element analogous to (120.51) is:

$$\boldsymbol{A}_0 \cdot \langle \chi_{0\{\bar{n}\}} | \mathscr{M}^{(2)} \left(\boldsymbol{k} \middle| \begin{matrix} \boldsymbol{k}' & \boldsymbol{k}'' \\ j' & j'' \end{matrix} \right) Q \begin{pmatrix} \boldsymbol{k}' \\ j' \end{pmatrix} Q \begin{pmatrix} \boldsymbol{k}'' \\ j'' \end{pmatrix} | \chi_{0\{n\}} \rangle_{\boldsymbol{R}} \quad (120.54)$$

where $\{n\}$ and $\{\bar{n}\}$ are relevant set of vibrational quantum numbers for the states. This particular matrix element will be non-zero, and the corresponding process allowed in infra-red absorption if

$$D^{(n)} \otimes D^{(\bar{n})} \text{ contains } D^{(\boldsymbol{k}')(j')} \otimes D^{(\boldsymbol{k}'')(j'')} \quad (120.55)$$

which in turn must contain

$$D^{(\boldsymbol{k})(v)} \quad \text{or} \quad D^{(\boldsymbol{\Gamma})(v)}. \quad (120.56)$$

But this is the requirement that

$$(\boldsymbol{k}' j' \boldsymbol{k}'' j'' | \boldsymbol{k}\, v) \neq 0 \quad (120.57)$$

and

$$(\{n\}\, \{\bar{n}\} | \boldsymbol{k}' j' \boldsymbol{k}'' j'') \neq 0. \quad (120.58)$$

Eq. (120.57) prescribes the requirement that the coefficient $\mathscr{M}^{(2)} \begin{pmatrix} \boldsymbol{k}' & \boldsymbol{k}'' \\ j' & j'' \end{pmatrix}$ is non zero and (120.58) gives the usual limitation upon thc vibrational eigenstates

---

[2] Ref. [*32*], p. 265.

[3] See H. Bilz: Infra-red lattice vibration spectra of perfect crystals, in [*19*].

[4] I. P. Ipatova, A. A. Maradudin, and R. F. Wallis: Soviet Phys. Solid State **8**, 850 (1966) and Phys. Rev. **155**, 882 (1967).

[5] B. Szigeti: Proc. Roy. Soc. (London), Ser. A **258**, 377 (1960).

[6] V. S. Vinogradov: Soviet Phys. Solid State **4**, 519 (1962).

[7] R. Wehner: phys. stat. solidi **15**, 725 (1966).

between which transitions can occur, following the Wigner-Eckart theorem. For such a case, where (120.57), (120.58) are obeyed, the matrix element (120.54) is proportional to

$$A_0 \cdot \mathscr{M}^{(2)} \left(\boldsymbol{k} \middle| \begin{matrix} \boldsymbol{k}' & \boldsymbol{k}'' \\ j' & j'' \end{matrix} \right) C_0^{(2)} \delta_{\bar{n}, n+1} \delta_{\bar{n}', n'+1} \tag{120.59}$$

where $C_0^{(2)}$ is a constant depending upon which particular oscillators are increased by one. The total intensity of absorption, producing two phonons, is then

$$\begin{aligned} \mathscr{I}_2(\omega)\, d\omega \sim \sum_{\boldsymbol{k}' \boldsymbol{k}'', j' j''} & \left| C_0^{(2)} A_0 \cdot \mathscr{M}^{(2)} \left(\boldsymbol{k} \middle| \begin{matrix} \boldsymbol{k}' & \boldsymbol{k}'' \\ j' & j'' \end{matrix} \right) \right|^2 \\ & \times \delta\left[\left(\omega(\boldsymbol{k}'|j') + \omega(\boldsymbol{k}''|j'') \right) - \omega \right] \delta_{\bar{n}, n+1} \delta_{\bar{n}', n'+1} . \end{aligned} \tag{120.60}$$

Subject to (120.57), (120.58). As before, anharmonicity and final state interaction (lifetime effects) will broaden the absorption in (120.60). These many-body effects are discussed elsewhere in this Encyclopedia [*20*, *22*] and also briefly in Sect. 124b, as well as in recent literature.[1, 2]

If the various matrix elements and other factors in (120.60) can be assumed independent of $\boldsymbol{k}' j'$, $\boldsymbol{k}'' j''$ then the sum on these indices will produce the two phonon density of states

$$\sum_{\boldsymbol{k}' j' \boldsymbol{k}'' j''} \delta\left(\omega(\boldsymbol{k}'|j') + \omega(\boldsymbol{k}''|j'') - \omega \right) = \rho_2\left(\omega(\boldsymbol{k}'|j') + \omega(\boldsymbol{k}''|j'') \right). \tag{120.61}$$

iii) *Multiphonon processes.* The generalization to multiphonon processes of arbitrary order is immediate. In general, each term of given degree in (120.47) corresponds to a multiphonon process of the same degree. The vanishing or non-vanishing of a process of given order is entirely controlled by the appropriate reduction coefficients. Illustrations of this will be given below when we analyze specific crystals with diamond and rocksalt structure.

Attention has recently been given to the theory of high-order multiphonon summation infra-red absorption in cases where the number of phonons involved can range up to $n = 10$. Experimental work on this problem apparently indicates that there is an experimental frequency dependence of the absorption coefficient $\beta \sim \exp(-A\omega)$ with increasing absorption frequency ω. The first theories of this effect assume that some non-zero coupling function exists and then calculate the absorption coefficient by a Green function theory.[8–11] The analysis of symmetry on the n-phonon coupling coefficients via the n-phonon generalization of the requirements (120.57), (120.58) on the Clebsch-Gordan series, or reduction coefficients, has not been given as of the time of writing.

121. Raman scattering by phonons: Generalized Placzek theory. In order to develop a useful theory of Raman scattering by phonons we will quantize the radiation field, and so we shall treat the field variables as dynamical objects of the theory rather than as externally imposed as we did in the semi-classical

[8] M. Sparks and L. J. Sham: Solid State Comm. **11**, 1451 (1972); Phys. Rev. **B8**, 3037 (1973); Phys. Rev. Lett. **31**, 718 (1973).

[9] D. L. Mills and A. A. Maradudin: Phys. Rev. **B8**, 1617 (1973).

[10] B. Bendow, S. C. Ying, and S. P. Yukon: Phys. Rev. **B8**, 1679 (1973).

[11] H. Rosenstock: Phys. Rev. **B9**, 1963 (1974).

treatment of infra-red absorption given in the previous sections. This adds a little complexity to the theory. Actually a semi-classical approach to the Raman scattering by phonons can also be given. The basic object in such a theory is an assumed off-diagonal transition moment which is a mathematically simple object, albeit a little difficult to grasp conceptually. So we prefer to work in a "generalized Placzek" approach,[1] deriving the off-diagonal transition moment operator from a more first principles method.

The program is very similar to that of the previous section. After isolating the perturbation part of the Hamiltonian (coupling of the radiation field to ions and electrons) we use perturbation theory to calculate the appropriate transition rate. For scattering processes in this generalized Placzek theory, this requires the use of second order perturbation theory.

α) *Hamiltonian.* As in the previous section we consider that the radiation field is specified by giving the field variables at every point: these are the three components of the vector potential $\boldsymbol{A}(\boldsymbol{r})$, and the scalar potential of the radiation field $\varphi(\boldsymbol{r})$ which we take as zero; $\varphi=0$. Taking the components of $\boldsymbol{A}$ as field "coordinate variables", the Lagrangian density of the electromagnetic field can be taken as

$$\mathscr{L} \equiv \frac{1}{8\pi}(1/c\,\partial \boldsymbol{A}/\partial t)^2 - \frac{1}{8\pi}(\nabla \times \boldsymbol{A})^2. \tag{121.1}$$

Hence the momentum $\boldsymbol{M}$ conjugate to $\boldsymbol{A}$ is found as

$$\boldsymbol{M} = \partial \mathscr{L}/\partial \dot{\boldsymbol{A}} = \frac{1}{4\pi c^2}\,\partial \boldsymbol{A}/\partial t \tag{121.2}$$

and the Hamiltonian density of the field

$$\mathscr{H}_R = 2\pi c^2 \boldsymbol{M}^2 + \frac{1}{8\pi}(\nabla \times \boldsymbol{A})^2 \tag{121.3}$$

which agrees with the usual electromagnetic field energy density:

$$\mathscr{H}_R = \frac{1}{8\pi}(\mathscr{E}^2 + \mathscr{H}^2) \tag{121.4}$$

if we use the expression (120.2). Now we again choose as a basic set of functions in which to represent $\boldsymbol{A}$, the normalized vector plane waves;

$$V^{-\frac{1}{2}}\,\varepsilon_{\boldsymbol{k}\lambda} \exp i\boldsymbol{k}\cdot\boldsymbol{r} \equiv \boldsymbol{\mu}_{\boldsymbol{k}\lambda}(\boldsymbol{r}), \tag{121.5}$$

with wave vector $\boldsymbol{k}$ polarization λ. Then we expand

$$\boldsymbol{A}(\boldsymbol{r},t) = \sum_{\boldsymbol{k}\lambda} \left(q_{\boldsymbol{k}\lambda}(t)\,\boldsymbol{u}_{\boldsymbol{k}\lambda}(\boldsymbol{r}) + q^*_{\boldsymbol{k}\lambda}(t)\,\boldsymbol{u}^*_{\boldsymbol{k}\lambda}(\boldsymbol{r})\right) \tag{121.6}$$

$$\boldsymbol{M}(\boldsymbol{r},t) = \sum_{\boldsymbol{k}\lambda} \left(m_{\boldsymbol{k}\lambda}(t)\,\boldsymbol{u}_{\boldsymbol{k}\lambda}(\boldsymbol{r}) + m^*_{\boldsymbol{k}\lambda}(t)\,\boldsymbol{u}^*_{\boldsymbol{k}\lambda}(\boldsymbol{r})\right). \tag{121.7}$$

---

[1] In this section we shall follow the general approach originally due to G. PLACZEK, Handbuch der Radiologie (ed. E. MARX), Vol. VI, **2**, pp. 209–374 (1934). Leipzig: Akademische Verlagsgesellschaft 1934. See also Ref. [*4*], BORN-HUANG, Sects. 19–21, and also Chap. VII.

The canonical variables of the radiation field are $q_{\boldsymbol{k}\lambda}$, $m_{\boldsymbol{k}\lambda}$ and we quantize them by the equal-time commutation rules:

$$[q_{\boldsymbol{k}\lambda}(t), m^*_{\boldsymbol{k}'\lambda'}(t)] = [q^*_{\boldsymbol{k}\lambda}(t), m_{\boldsymbol{k}'\lambda'}(t)] = i\hbar\, \delta_{\boldsymbol{k}\boldsymbol{k}'}\, \delta_{\lambda'\lambda}. \tag{121.8}$$

The total radiation Hamiltonian is the space integral of (121.3) or:

$$H_R = \int \left\{ 2\pi c^2 \boldsymbol{M}^2 + \frac{1}{8\pi} (\boldsymbol{\nabla} \times \boldsymbol{A})^2 \right\} d\tau, \tag{121.9}$$

Now substituting (121.6), (121.7) in (121.9) we obtain

$$H_R = \sum_{\boldsymbol{k}\lambda} \left(4\pi c^2\, m_{\boldsymbol{k}\lambda}\, m^*_{\boldsymbol{k}\lambda} + \frac{\boldsymbol{k}^2}{4\pi}\, q_{\boldsymbol{k}\lambda}\, q^*_{\boldsymbol{k}\lambda} \right). \tag{121.10}$$

As well known, this is the Hamiltonian of a system of uncoupled simple Harmonic oscillators.[2] The Hamiltonian (121.10) can also be treated in the Schrödinger representation as we treated the electron-ion Hamiltonian (i.e. replacing the momentum $m_{\boldsymbol{k}\lambda} \to (\hbar/i)\, \partial/\partial q_{\boldsymbol{k}\lambda}$) but this is not necessary for our present purposes. With the radiation Hamiltonian given as (121.10), the eigenfunction of the unperturbed radiation field is a product of harmonic oscillator eigenfunctions

$$\Psi_{\text{rad}} = \prod_{\boldsymbol{k}} \chi(n_{\boldsymbol{k}}(q_{\boldsymbol{k}})) \tag{121.11}$$

just analogous to (115.1) for the lattice oscillators. As in (116.1) where an "N" representation of the phonon (vibrational) eigenfunction was given, an equally good representation of the photon (radiation) eigenfunction is in terms of the number N of quanta of the $\boldsymbol{k}$-th eigenstate excited, or:

$$|\Psi_{\text{rad}}\rangle = |n_{\boldsymbol{k}_1} \ldots n_{\boldsymbol{k}_l} \ldots\rangle. \tag{121.12}$$

Thus the eigenstates of the free radiation Hamiltonian can be prescribed in form (121.12).

The Hamiltonian of ions and electrons interacting has been previously discussed in Sect. 113. We can take the adiabatic approximation to be valid for the ion plus electron, or matter, part of the Hamiltonian. Consequently the Hamiltonian of the matter (ion plus electron) system plus the radiation field system is a sum of the two parts. The eigenfunction to this level of approximation is a product:

$$\Psi = |\Psi_{\text{adiabatic}}\rangle\, |\Psi_{\text{rad}}\rangle. \tag{121.13}$$

The interaction term (120.8) must now also be put in terms of the electromagnetic field variables. Substituting (121.6) into (120.8) we obtain

$$\begin{aligned} \mathscr{H}' = &- \sum_{i,\boldsymbol{k},\lambda} \left(\frac{e}{mc}\right) \left(q_{\boldsymbol{k}\lambda}(t)\, \boldsymbol{u}_{\boldsymbol{k}\lambda}(\boldsymbol{r}) \cdot \boldsymbol{p}_i + q^*_{\boldsymbol{k}\lambda}(t)\, \boldsymbol{u}^*_{\boldsymbol{k}\lambda}(\boldsymbol{r}) \cdot \boldsymbol{p}_i\right) \\ &+ \sum_{\alpha,\boldsymbol{k},\lambda} \left(\frac{Ze}{MC}\right) \left(q_{\boldsymbol{k}\lambda}(t)\, \boldsymbol{u}_{\boldsymbol{k}\lambda}(\boldsymbol{R}) \cdot \boldsymbol{P}_\alpha + q^*_{\boldsymbol{k}\lambda}(t)\, \boldsymbol{u}^*_{\boldsymbol{k}\lambda}(\boldsymbol{R}) \cdot \boldsymbol{P}_\alpha\right). \end{aligned} \tag{121.14}$$

[2] W. Heitler: Quantum Theory of Radiation. Oxford: University Press (1954).

Thus the total Hamiltonian here is the sum

$$\mathcal{H} = \mathcal{H}_{\text{rad}} + T_E + T_I + U + \mathcal{H}' \tag{121.15}$$

and the perturbation has been isolated from the other terms.

β) Transition rate for scattering. Now for the analysis of spontaneous Raman scattering in this framework we must consider transitions produced by the perturbation (121.14) between an initial state with (incident) number of photons n_1 present but no phonons, and a final state with one photon from state $\boldsymbol{k}_1$ destroyed, one photon created in state $\boldsymbol{k}_2$, and phonons present.

Consistent with the previous discussion we write for the initial state eigenfunction

$$\Psi_i = |\Psi_{\text{adiabatic}}\rangle_i \, |\Psi_{\text{rad}}\rangle_i, \tag{121.16}$$

$$\Psi_i = \varphi_\nu(\boldsymbol{r}\,\boldsymbol{R})\, \chi_{\nu n}(\boldsymbol{R})\, |n_{\boldsymbol{k}_1}, \ldots, n_{\boldsymbol{k}_2}, \ldots, n_{\boldsymbol{k}_l}, \ldots\rangle \tag{121.17}$$

the use of a mixed notation in (121.17) should cause no confusion. For the final state eigenfunction we write

$$\Psi_f = \varphi_\nu(\boldsymbol{r}\,\boldsymbol{R})\, \chi_{\nu \bar{n}}(\boldsymbol{R})\, |(n_{\boldsymbol{k}_1} - 1), \ldots, (n_{\boldsymbol{k}_2} + 1), \ldots, n_{\boldsymbol{k}_l}, \ldots\rangle. \tag{121.18}$$

Now we observe that the perturbation (121.14) cannot cause a transition between the states Ψ_i and Ψ_f in first order perturbation theory. Clearly owing to the linearity of $\mathcal{H}'$ in $q_{\boldsymbol{k}\lambda}$ the perturbation $\mathcal{H}'$ can only cause transitions in which a single photon changes its state. Thus $q_{\boldsymbol{k}\lambda}$ has matrix elements only between states $|\ldots n_{\boldsymbol{k}\lambda} \ldots\rangle$ and $|\ldots(n_{\boldsymbol{k}\lambda} \pm 1)\ldots\rangle$. In order to calculate the transition probability for scattering, which is a two photon process, we require the use of second order perturbation theory. Thus we must consider matrix elements of $\mathcal{H}'$ connecting the initial state (121.17) to intermediate states e.g. to

$$\Psi_{\text{int}} = \varphi_\mu(\boldsymbol{r}\,\boldsymbol{R})\, \chi_{\mu m}(\boldsymbol{R})\, |(n_{\boldsymbol{k}_1} - 1), \ldots, n_{\boldsymbol{k}_2}, \ldots, n_{\boldsymbol{k}_l}, \ldots\rangle \tag{121.19}$$

and then connecting Ψ_{int} to Ψ_f.

The total transition probability per unit time will be given formally as:

$$\omega_{i\to f} = \sum_f \left| \sum_{(\text{int})} \frac{\langle \Psi_f | \mathcal{H}' | \Psi_{\text{int}} \rangle \langle \Psi_{\text{int}} | \mathcal{H}' | \Psi_i \rangle}{(E_i - E_{\text{int}})} \right|^2 \delta(E_f - E_i) \tag{121.20a}$$

or equivalently as

$$\omega_{i\to f} = \left| \sum_{(\text{int})} \frac{\langle \Psi_f | \mathcal{H}' | \Psi_{\text{int}} \rangle \langle \Psi_{\text{int}} | \mathcal{H}' | \Psi_i \rangle}{(E_i - E_{\text{int}})} \right|^2 \rho(E_f). \tag{121.20b}$$

where $\rho(E_f)$ is the density of energy conserving final states.

A typical term of the kind which enters in the scattering amplitude is

$$\{\langle \Psi_f | \mathcal{H}' | \Psi_{\text{int}} \rangle \langle \Psi_{\text{int}} | \mathcal{H}' | \Psi_i \rangle / (E_i - E_{\text{int}})\} \tag{121.21}$$

or, if we use (121.14)–(121.19), to simplify the numerator of (121.21):

$$\sum_{\mu m}\sum_{\boldsymbol{k}\lambda}\sum_{\boldsymbol{k}'\lambda'}\sum_{ii'}\sum_{\alpha\alpha'}\langle\ldots(n_{\boldsymbol{k}_2}+1)\ldots(n_{\boldsymbol{k}_1}-1)\,\varphi_\nu\,\chi_{\nu\bar{n}}|$$

$$\times q_{\boldsymbol{k}\lambda}\,\boldsymbol{u}_{\boldsymbol{k}\lambda}\cdot\left\{-\left(\frac{e}{mc}\right)\boldsymbol{p}_i+\left(\frac{Ze}{mc}\right)\boldsymbol{P}_\alpha\right\}|\varphi_\mu\,\chi_{\mu m}\ldots n_{\boldsymbol{k}_2}\ldots(n_{\boldsymbol{k}_1}-1)\rangle \tag{121.22}$$

$$\times\langle\ldots n_{\boldsymbol{k}_2}\ldots(n_1-1)\ldots\varphi_\mu\,\chi_{\mu m}|\,q_{\boldsymbol{k}'\lambda'}\,\boldsymbol{u}_{\boldsymbol{k}'\lambda'}\cdot\left\{\left(-\frac{e}{mc}\,\boldsymbol{p}_{i'}+\frac{Ze}{mc}\right)\boldsymbol{P}_{\alpha'}\right\}|\varphi_\nu\,\chi_{\nu n}\ldots n_{\boldsymbol{k}_1}\ldots n_{\boldsymbol{k}_2}\ldots\rangle.$$

Owing to the assumptions made, the initial and final state eigenfunctions are fixed, which places restrictions on the $q_{\boldsymbol{k}\lambda}$ and $q_{\boldsymbol{k}'\lambda'}$ in the sum. The matrix element of "radiation oscillator" coordinate $q_{\boldsymbol{k}\lambda}$ has the value [*31, 60*]:

$$\langle\ldots(n_{\boldsymbol{k}\lambda}+1)\ldots|\,q_{\boldsymbol{k}\lambda}\,|\ldots n_{\boldsymbol{k}\lambda}\ldots\rangle=\left(\frac{(n_{\boldsymbol{k}\lambda}+1)\,2\pi\,\hbar\,c^2}{\omega_{\boldsymbol{k}\lambda}}\right)^{\frac{1}{2}} \tag{121.23}$$

where $\hbar\,\omega_{\boldsymbol{k}\lambda}$ is the energy of the quantum of the oscillator labelled by index $\boldsymbol{k}\lambda$ (i.e. with coordinate $q_{\boldsymbol{k}\lambda}$) and the other necessary matrix element is:

$$\langle\ldots(n_{\boldsymbol{k}'\lambda'}-1)\ldots|\,q_{\boldsymbol{k}'\lambda'}\,|\ldots n_{\boldsymbol{k}'\lambda'}\rangle=\left(\frac{n_{\boldsymbol{k}'\lambda'}\,2\pi\,c^2}{4\pi\,\omega_{\boldsymbol{k}'\lambda'}}\right)^{\frac{1}{2}}. \tag{121.24}$$

Hence for (121.22) we obtain, by eliminating the sums on $\boldsymbol{k}\lambda$, and $\boldsymbol{k}'\lambda'$, and assuming that the polarization indices λ and λ' are fixed:

$$\sum_{\mu\nu}\sum_{ii'}\sum_{\alpha\alpha'}\left(\frac{n_{\boldsymbol{k}'\lambda'}(n_{\boldsymbol{k}\lambda}+1)\,\hbar^2}{\omega_{\boldsymbol{k}\lambda}\,\omega_{\boldsymbol{k}'\lambda'}}\right)^{\frac{1}{2}}(2\pi\,c^2)\,\langle\varphi_\nu\,\chi_{\nu\bar{n}}|\,\boldsymbol{u}_{\boldsymbol{k}\lambda}\cdot\left\{-\left(\frac{e}{mc}\right)\boldsymbol{p}_i+\left(\frac{Ze}{mc}\right)\boldsymbol{P}_\alpha\right\}$$
$$\times|\varphi_\mu\,\chi_{\mu m}\rangle\,\langle\varphi_\mu\,\chi_{\mu m}|\,\boldsymbol{u}_{\boldsymbol{k}'\lambda'}\cdot\left\{-\left(\frac{e}{mc}\right)\boldsymbol{p}_i+\left(\frac{Ze}{mc}\right)\boldsymbol{P}_\alpha\right\}|\varphi_\nu\,\chi_{\nu n}\rangle. \tag{121.25}$$

Now in dealing with (121.25) we shall again utilize the two assertions previously employed in dealing with the infra-red absorption processes. First we use the assumption which effectively permits us to replace the electron index by its localized ion counterpart. So we write $\boldsymbol{p}_i\to\boldsymbol{p}_\alpha$ and take

$$\begin{aligned}\boldsymbol{u}_{\boldsymbol{k}_\lambda}(\boldsymbol{r}_i)&=\boldsymbol{\varepsilon}_{1\lambda}\exp i\,\boldsymbol{k}_1\cdot\boldsymbol{r}_1\to\boldsymbol{\varepsilon}_{1\lambda}\exp i\,\boldsymbol{k}_1\cdot\boldsymbol{R}_{\alpha'}\\ \boldsymbol{u}_{\boldsymbol{k}'_{\lambda'}}(\boldsymbol{r}_i)&=\boldsymbol{\varepsilon}_{2\lambda}\exp -i\,\boldsymbol{k}_2\cdot\boldsymbol{r}_2\to\boldsymbol{\varepsilon}_{2\lambda}\exp -i\,\boldsymbol{k}_2\cdot\boldsymbol{R}_\alpha\end{aligned} \tag{121.26}$$

where $\boldsymbol{\varepsilon}_{2\lambda}$ is a real polarization vector and the remainder is as in Sect. 120. Then (121.25) becomes the phased object:

$$\sum_{\alpha\alpha'}\left(\frac{n_{\boldsymbol{k}'\lambda'}(n_{\boldsymbol{k}\lambda}+1)\,\hbar^2}{\omega_{\boldsymbol{k}\lambda}\,\omega_{\boldsymbol{k}'\lambda'}}\right)^{\frac{1}{2}}(2\pi\,c^2)\exp i(-\boldsymbol{k}_2\cdot\boldsymbol{R}_\alpha+\boldsymbol{k}_1\cdot\boldsymbol{R}_{\alpha'})$$

$$\times\langle\chi_{\nu\bar{n}}|\,\langle\varphi_\nu|\,\boldsymbol{\varepsilon}_{1\lambda}\cdot\left\{-\left(\frac{e}{mc}\right)\boldsymbol{p}_\alpha+\left(\frac{Ze}{mc}\right)\boldsymbol{P}_{\alpha'}\right\}|\varphi_\mu\rangle_{\boldsymbol{r}}\,|\chi_{\mu m}\rangle_{\boldsymbol{R}} \tag{121.27}$$

$$\times\langle\chi_{\mu m}|\,\langle\varphi_\mu|\,\boldsymbol{\varepsilon}_{2\lambda}\cdot\left\{-\left(\frac{e}{mc}\right)\boldsymbol{p}_\alpha+\left(\frac{Ze}{mc}\right)\boldsymbol{P}_\alpha\right\}|\varphi_\nu\rangle_{\boldsymbol{r}'}\,|\chi_{\nu n}\rangle_{\boldsymbol{R}'}.$$

In (121.27) the subscripts on matrix elements indicate variables being integrated, as is conventional.

Next we use the results (120.25)–(120.32) in order to rewrite the product of matrix elements in (121.27) as:

$$\left\{\sum_{\alpha} \boldsymbol{\varepsilon}_{2\lambda} \cdot \langle \chi_{\nu\bar{n}}| \langle \varphi_\nu| \frac{i}{\hbar}[H, \boldsymbol{\mu}_\alpha] |\varphi_\mu\rangle_{\boldsymbol{r}} |\chi_{\mu m}\rangle_{\boldsymbol{R}} \exp - i\boldsymbol{k}_2 \cdot \boldsymbol{R}_\alpha\right\}$$
$$\times \left\{\sum_{\alpha'} \boldsymbol{\varepsilon}_{1\lambda'} \cdot \langle \chi_{\mu m}| \langle \varphi_\mu| \frac{i}{\hbar}[H, \boldsymbol{\mu}_{\alpha'}] |\varphi_\nu\rangle_{\boldsymbol{r}'} |\chi_{\nu n}\rangle_{\boldsymbol{R}'} \exp i\boldsymbol{k}_1 \cdot \boldsymbol{R}_{\alpha'}\right\} \tag{121.28}$$

or,

$$\left\{\sum_{\alpha} \frac{i}{\hbar}(E_{\nu\bar{n}} - E_{\mu m}) \langle \chi_{\nu\bar{n}}| \langle \varphi_\nu| \boldsymbol{\varepsilon}_{2\lambda} \cdot \boldsymbol{\mu}_\alpha |\varphi_\mu\rangle_{\boldsymbol{r}} |\chi_{\mu m}\rangle_{\boldsymbol{R}} \exp - i\boldsymbol{k}_2 \cdot \boldsymbol{R}_\alpha\right\}$$
$$\times \left\{\sum_{\alpha'} \frac{i}{\hbar}(E_{\mu m} - E_{\nu n}) \langle \chi_{\mu m}| \langle \varphi_\mu| \boldsymbol{\varepsilon}_{1\lambda'} \cdot \boldsymbol{\mu}_{\alpha'} |\varphi_\nu\rangle_{\boldsymbol{r}'} |\chi_{\nu n}\rangle_{\boldsymbol{R}'} \exp i\boldsymbol{k}_1 \cdot \boldsymbol{R}_{\alpha'}\right\}. \tag{121.29}$$

Let us now define a "phased" off-diagonal electric dipole operator in analogy with (120.33). Then call

$$\mathscr{M}_\lambda^{\nu\mu}(\boldsymbol{R}, -\boldsymbol{k}_2) \equiv \boldsymbol{\varepsilon}_{2\lambda} \cdot \mathscr{M}^{\nu\mu}(\boldsymbol{R}, -\boldsymbol{k}_2) \equiv \sum_{\alpha} \mathrm{e}^{-i\boldsymbol{k}_2 \cdot \boldsymbol{R}_\alpha} \langle \varphi_\nu| \boldsymbol{\varepsilon}_{2\lambda} \cdot \boldsymbol{\mu}_\alpha |\varphi_\mu\rangle_{\boldsymbol{r}}. \tag{121.30}$$

This is a phased sum of off-diagonal dipole moment operators.

Then (121.28) becomes

$$-\boldsymbol{\varepsilon}_\lambda \cdot \langle \chi_{\nu\bar{n}}| \mathscr{M}^{\nu\mu}(\boldsymbol{R}, -\boldsymbol{k}_2) |\chi_{\mu m}\rangle_{\boldsymbol{R}} \cdot \langle \chi_{\mu m}| \mathscr{M}^{\mu\nu}(\boldsymbol{R}', \boldsymbol{k}_1) |\chi_{\nu n}\rangle_{\boldsymbol{R}'} \cdot \boldsymbol{\varepsilon}_{\lambda'}$$
$$\times (\omega_{\nu\bar{n},\mu m})(\omega_{\mu m,\nu n}). \tag{121.31}$$

In principle, (121.31) can now be substituted into (121.20) and the sum over intermediate states carried out. This will produce only a formal result, not useful in case of insulating crystals. We therefore proceed to a further level of simplification by taking into account some qualitative aspects of the electronic spectrum of an insulator. The most important aspect of the spectrum in the present context is the existence of an energy gap.

γ) *Simplification of the scattering matrix elements for an insulator.* Returning to (121.20) observe that the sum over intermediate states can be taken as

$$\sum_{(\mathrm{int})} \to \sum_{\mu} \sum_{m} \tag{121.32}$$

where μ is the electronic and m the vibrational quantum number.

For the intermediate adiabatic states of major importance we choose states which belong to a different electronic manifold from the initial or final states: i.e. $\mu \neq \nu$. Then in an insulator the separation between states with $\mu \neq \nu$ is much greater than a phonon energy. Or we make the approximation (c.f. Eq. (120.39)–(120.43))

$$\omega_{\nu\bar{n},\mu m} \equiv \omega_{\nu\bar{n}} - \omega_{\mu m} = 1/\hbar(\Phi_\nu^0 + E_{\nu\bar{n}}^{(2)} - \Phi_\mu^0 - E_{\mu m}^{(2)})$$
$$\approx 1/\hbar(\Phi_\nu^0 - \Phi_\mu^0) = \omega_{\nu\mu} \tag{121.33}$$

and

$$\omega_{\mu m,\nu n} \approx \omega_{\mu\nu} = -\omega_{\nu\mu}. \tag{121.34}$$

Further we may take the system to be initially and finally in the ground electronic state $v=0$. Now consider the denominator in (121.20). For initial state energy we can take

$$E_i = \hbar\omega_i + \Phi^0 + E_n^{(2)} \tag{121.35}$$

where Φ^0 is the electronic energy of the ground state, and $\hbar\omega_i$ is the initial energy of the photon. The intermediate state energy is

$$E_{\text{int}} = \Phi_\mu^0 + E_{\mu m}^{(2)} \tag{121.36}$$

since the photon is absorbed, so that

$$(E_i - E_{\text{int}}) = \hbar\omega_i + \Phi^0 + E_n^{(2)} - \Phi_\mu^0 - E_{\mu m}^{(2)} \approx \hbar\omega_i + \Phi^0 - \Phi_\mu^0 \tag{121.37}$$

$$= \hbar\omega_i + \hbar\omega_{0\mu} \tag{121.38}$$

where $\hbar\omega_{0\mu}$ is the (purely electronic) excitation energy. The factor $\delta(E_f - E_i)$ in (121.20) can be written as

$$\delta(\hbar\omega_f - \hbar\omega_i + E_{\bar{n}}^{(2)} - E_n^{(2)}). \tag{121.39}$$

In writing (121.39) we have assumed that the system returns to its original (ground) electronic state after the scattering event so that the energy difference $(\hbar\omega_f - \hbar\omega_i)$ is simply that vibrational energy difference. Also we take

$$E_f = \hbar\omega_f + \Phi^{(0)} + E_{\bar{n}}^{(2)}. \tag{121.40}$$

At this point let us pause to remark that in this section we are taking the various virtual processes to occur in a particular time order. In fact, in a proper theory all time orders must occur, but the essential results will be found by our methods. A somewhat more complete discussion will be given in Sect. 124.

Upon comparing (121.31)–(121.36) with (121.20) we can observe now that the sum on the vibrational quantum number m of the intermediate state can indeed be carried out. In fact the only term depending on m is

$$\sum_m |\chi_{\mu m}\rangle_{\boldsymbol{R}} \langle\chi_{\mu m}|_{\boldsymbol{R}'} = \sum_m \chi_{\mu m}(\boldsymbol{R})\, \chi_{\mu m}^*(\boldsymbol{R}') = \delta(\boldsymbol{R} - \boldsymbol{R}') \tag{121.41}$$

which follows because of the completeness of the set $\chi_{\mu m}(\boldsymbol{R})$. This result permits us to write the expression for the total transition (scattering) rate in a useful form.

We have, for (121.20) for the total transition rate $w_{0n\to 0\bar{n}}$:

$$w_{0n\to 0\bar{n}} = \left| \sum_\mu \frac{(\omega_{0\mu})^2}{(\hbar\omega_i - \hbar\omega_{\mu 0})} \times \boldsymbol{\varepsilon}_{2\lambda} \cdot \langle\chi_{0\bar{n}}|\, \mathscr{M}^{0\mu}(\boldsymbol{R}, -\boldsymbol{k}_2)\, \mathscr{M}^{\mu 0}(\boldsymbol{R}, \boldsymbol{k}_1)\, |\chi_{0n}\rangle_{\boldsymbol{R}} \cdot \boldsymbol{\varepsilon}_{1\lambda'} \right|^2 \rho(E_f). \tag{121.42}$$

Observe in (121.42) that the expression for the transition rate involves a density of final states (energy conserving) and a matrix element squared. Let us define a tensor field function of $\boldsymbol{R}$, or polarizability operator, by

$$\boldsymbol{P}(\boldsymbol{R}, \boldsymbol{k}_1, \boldsymbol{k}_2) \equiv \sum_\mu C_{\mu 0}(\omega_i)\, \mathscr{M}^{0\mu}(\boldsymbol{R}, -\boldsymbol{k}_2)\, \mathscr{M}^{\mu 0}(\boldsymbol{R}, \boldsymbol{k}_1) \tag{121.43}$$

with

$$C_{\mu 0} \equiv (\omega_{0\mu})^2 (\hbar\omega_i - \hbar\omega_{\mu 0})^{-1}. \tag{121.44}$$

Now in the application of the theory presented here one usually assumes that incident and scattered frequencies are far from resonance so $\omega_i < \omega_{\mu 0} = (\Phi_\mu^0 - \Phi^0)/\hbar$. Then $C_{\mu 0}(\omega_i)$ is essentially independent of incident frequency ω_i.

Then (121.42) can be written

$$\omega_{0\,n \to 0\,\bar{n}} = |\langle \chi_{0\,\bar{n}} | \boldsymbol{\varepsilon}_{2\,\lambda} \cdot \boldsymbol{P}(\boldsymbol{R}, \boldsymbol{k}_1 - \boldsymbol{k}_2) \cdot \boldsymbol{\varepsilon}_{1\,\lambda'} | \chi_{0\,n} \rangle_{\boldsymbol{R}}|^2 \rho(E_f). \tag{121.45}$$

Evidently we have to do in (121.45) with matrix elements of the $\lambda\lambda'$ cartesian components of the tensor operator $\boldsymbol{P}$ corresponding to the polarization of incident $\boldsymbol{\varepsilon}_{1\,\lambda'}$ and scattered $\boldsymbol{\varepsilon}_{2\,\lambda}$ radiation.

δ) *Symmetry of the Raman scattering matrix element.* In (120.34)–(120.39) we analysed the symmetry of the operator for infra-red dipole absorption. Comparing the definition (120.32) with (121.30) observe that the major difference is that for Raman scattering the "off diagonal" matrix element occurs rather than the "diagonal" matrix element.

The operator $\boldsymbol{P}(\boldsymbol{R}, -\boldsymbol{k}_2, \boldsymbol{k}_1)$ transforms like the direct product of a Bloch sum at wave vector $-\boldsymbol{k}_2$ times a Bloch sum at wave vector $\boldsymbol{k}_1$. Thus under transformation by translational symmetry operation we have

$$\boldsymbol{P}(\boldsymbol{R}, -\boldsymbol{k}_2, \boldsymbol{k}_1) \sim D^{(\boldsymbol{k}_1 - \boldsymbol{k}_2)}. \tag{121.46}$$

This follows from the rule for the product of two Bloch functions. As in the case of infra-red absorption one often takes $\boldsymbol{k}_1$ and $\boldsymbol{k}_2$ to be zero (infinite wave length assumption) but (121.46) is rigorously true.

Proceeding as in the case of the previous section, Eqs. (120.34)–(120.37) we obtain for the transformation of the components

$$P_{\lambda\lambda'}(\boldsymbol{R}) \equiv \boldsymbol{\varepsilon}_{2\,\lambda} \cdot \boldsymbol{P}(\boldsymbol{R}, -\boldsymbol{k}_2, \boldsymbol{k}_1) \cdot \boldsymbol{\varepsilon}_{1\,\lambda'}, \tag{121.47}$$

$$\begin{aligned} P_{\lambda\lambda'}(\boldsymbol{R}) &\to P'_{\lambda\lambda'}(\boldsymbol{R}) \\ &= \sum_{\sigma} \sum_{\sigma'} D^{(v)}(\{\varphi\})_{\lambda\sigma} D^{(v)}(\{\varphi\})_{\lambda'\sigma'} P_{\sigma\sigma'}(\{\varphi\})^{-1}(\boldsymbol{R}, -\boldsymbol{k}_2, \boldsymbol{k}_1). \end{aligned} \tag{121.48}$$

Consequently from (121.46) and (121.48) we conclude that the quantity $\boldsymbol{P}_{\lambda\lambda'}(\boldsymbol{R}, -\boldsymbol{k}_2, \boldsymbol{k}_1)$ transforms as a second rank tensor field, of wave vector $(\boldsymbol{k}_1 - \boldsymbol{k}_2)$. As in the case of the infra-red absorption operator $\mathscr{M}^v$, there is a slight imprecision about our result: if we take $\boldsymbol{k}_2 = \boldsymbol{k}_1 = 0$, then the operator $\boldsymbol{P}(\boldsymbol{R})$ transforms actually as a second rank tensor field under rotation. Otherwise if $\boldsymbol{k}_{2'} = \boldsymbol{k}_1 \neq 0$ the full space group theory machinery is required in order to characterize this operator at finite wave vector. This imprecision is rather commonly found. The usual procedure is to treat the wave vector dependence exactly insofar as required in order to satisfy kinematic momentum conservation, and then to treat the operator $\boldsymbol{P}(\boldsymbol{R})$ as an object at wave vector zero. Then from (121.48) and by inspection of (121.31) we see that under rotation the operator $\boldsymbol{P}(\boldsymbol{R})$ transforms as a symmetric second rank tensor field

$$P_{\lambda\lambda'}(\boldsymbol{R}) = P_{\lambda'\lambda}(\boldsymbol{R})$$

so

$$\boldsymbol{P}(\boldsymbol{R}) \sim [D^{(v)}]_{(2)} \tag{121.49}$$

when the ion positions are subjected to the transformation

$$(\boldsymbol{R}) \rightarrow (\boldsymbol{R}') = \{\varphi\} \boldsymbol{R} \tag{121.50}$$

as in (120.34). That is, in the usual case, away from resonance, the polarizability operator $\boldsymbol{P}(\boldsymbol{R})$ transforms as the symmetrized Kronecker square of the representation by which a polar vector transforms.

The consequences of (121.48), and (121.49) for the selection rules for Raman scattering are immediate. In (121.45) we found that the Raman lattice scattering from state $|\chi_{0\,n}\rangle$ to $|\chi_{0\,\bar{n}}\rangle$ was controlled by

$$\langle \chi_{0\,\bar{n}} | \, \boldsymbol{P}(\boldsymbol{R}) \, | \chi_{0\,n} \rangle_{\boldsymbol{R}} . \tag{121.51}$$

The usual application of Wigner-Eckart tensor calculus reveals that the matrix element (121.45) will vanish unless

$$D^{(\bar{n})*} \otimes D^{(n)} \sim D^{(\boldsymbol{k}_1 - \boldsymbol{k}_2)(\boldsymbol{v})} \tag{121.52}$$

$$\sim [D^{(\boldsymbol{v})}]_{(2)} . \tag{121.53}$$

That is, the total momentum conservation condition (121.46) must be obeyed, as well as the requirement that the remaining symmetry restrictions be obeyed (121.49). If $\boldsymbol{k}_2$ is the scattered photon wave vector, $\boldsymbol{k}_1$ the initial, and $\boldsymbol{\eta}$ the wave vector corresponding to the change of vibrational state of the crystal

$$\boldsymbol{k}_2 - \boldsymbol{k}_1 = \boldsymbol{\eta} . \tag{121.54}$$

In many applications of symmetry to the scattering processes finite wave vector effects are ignored. Then the requirement for non-zero matrix element (121.45) is that the product of initial times final state eigenfunction shall contain the representation by which a symmetric second rank tensor transforms: (121.53). Again remark that a proper treatment of finite wave vector effects requires use of $\mathfrak{G}(\boldsymbol{\eta})$, but this will not be given here.

ε) *One phonon and multiphonon processes.* In strict analogy to (120.44)–(120.60) we can develop the single and multiphonon scattering theory. The operator $\boldsymbol{P}(\boldsymbol{R}, \boldsymbol{k}_1 - \boldsymbol{k}_2)$ can be expanded about the equilibrium reference ion positions $\boldsymbol{R}^0$. Again one could proceed, as in e.g. (120.46), or (120.47). To expand in configuration space requires that we expand the off-diagonal matrix elements of (121.30), again reverting to our usual notation $\boldsymbol{R}_\alpha \rightarrow \boldsymbol{R}\begin{pmatrix} l \\ \kappa \end{pmatrix}$:

$$\langle \varphi_\nu | \, \boldsymbol{\mu}\begin{pmatrix} l \\ \kappa \end{pmatrix} | \varphi_\mu \rangle = \mathcal{M}^{(0)}_{\nu\mu}\begin{pmatrix} l \\ \kappa \end{pmatrix} + \sum_{l'\kappa'\alpha} \mathcal{M}^{(1)}_{\nu\mu,\alpha}\begin{pmatrix} l & l' \\ \kappa & \kappa' \end{pmatrix} u_\alpha\begin{pmatrix} l \\ \kappa \end{pmatrix} + \cdots . \tag{121.55}$$

Now if (121.55) is substituted into (121.30), and a similar substitution made in the expression for $\mathcal{M}^{\mu\nu}_{\lambda'}(\boldsymbol{R}, \boldsymbol{k}_1)$ and then both are substituted into (121.43) we have an expression like

$$\boldsymbol{P}(\boldsymbol{R}) = \boldsymbol{p}^0 + \sum_{l\kappa\alpha} \boldsymbol{p}^{(1)}_\alpha \begin{pmatrix} l \\ \kappa \end{pmatrix} u_\alpha \begin{pmatrix} l \\ \kappa \end{pmatrix} + \cdots \tag{121.56}$$

with, e.g.

$$p^{(0)} \equiv \sum_{\mu} \sum_{l\kappa} \sum_{l'\kappa'} e^{i k_1 \cdot R\binom{l}{\kappa} - i k_2 \cdot R\binom{l'}{\kappa'}} C_{\mu 0}(\omega_i) \mathscr{M}^{(0)}_{\nu\mu}\binom{l}{\kappa} \mathscr{M}^{(0)}_{\mu\nu}\binom{l'}{\kappa'} \tag{121.57}$$

and algebraically more complicated expressions in higher order. A more useful expansion for our immediate needs corresponds to (120.47). It is

$$\begin{aligned} P(R, k_2 - k_1) = p^{(0)} &+ \sum_{k, j, \mu} P^{(1)}\left(k_1 - k_2 \middle| \begin{matrix} k \\ j_\mu \end{matrix}\right) Q\binom{k}{j_\mu} \\ &+ \sum_{k j \mu} \sum_{k' j' \mu'} P^{(2)}\left(k_1 - k_2 \middle| \begin{matrix} k & k' \\ j_\mu & j'_{\mu'} \end{matrix}\right) Q\binom{k}{j_\mu} Q\binom{k'}{j'_{\mu'}} + \cdots. \end{aligned} \tag{121.58}$$

The coefficients in (121.58) could be expressed in terms of the basic objects in the off-diagonal expansion (121.55), but we shall not pursue that here.

Before considering the result of using (121.58) in (121.51) we examine the non-zero terms in (121.58). Here the discussion of Sect. 109 now applies. The entire term (left hand side) of (121.58) transforms as a symmetric second rank tensor field. Thus each term in the series (121.58) must transform in the same way. Hence, if we take $k_2 = k_1 = 0$, then

$$Q\binom{k}{j_\mu} \sim [D^{(v)}]_{(2)}, \tag{121.59}$$

and then

$$P^{(1)}\left(0 \middle| \begin{matrix} k \\ j_\mu \end{matrix}\right) \neq 0. \tag{121.60}$$

Likewise in higher order,

$$Q\binom{k}{j_\mu} Q\binom{k'}{j'_{\mu'}} \sim D^{(k)(j)} \otimes D^{(k')(j')}$$

and if this direct product contains $[D^{(v)}]_{(2)}$ then

$$P^{(2)}\left(0 \middle| \begin{matrix} k & k' \\ j_\mu & j'_{\mu'} \end{matrix}\right) \neq 0. \tag{121.61}$$

The general condition that a non-zero term shall arise in (121.58) is that for the corresponding term, the reduction coefficient

$$({}^\star k\, j\, {}^\star k' j' | (k_2 - k_1)[v]_{(2)}) \neq 0 \tag{121.62}$$

or the corresponding condition on the subgroup reduction coefficients. To emphase the point and relate it to Sect. 109: each term in the series (121.58) must transform covariantly as does the entire term. To be exact again one requires a determination of the transformation of a second rank symmetric tensor field, with wave vector $(k_2 - k_1)$ under $\mathfrak{G}$, or $\mathscr{G}$.

Each term in (121.58), corresponds to a Raman scattering process of different order. Linear terms in $Q\binom{k}{j_\mu}$ are one phonon processes, bilinear are two phonon processes, etc. Thus upon substituting (121.58) into (121.47), and (121.51) we distinguish terms in each order. That is, we decompose the total transition rate $\omega_{0n\to 0\bar{n}}$ into partial transition rates of one, two, ... phonon processes, depending on the states $|\chi_{0n}\rangle$ and $|\chi_{0\bar{n}}\rangle$ involved.

Consider two phonon Stokes Raman scattering for example as produced by the bilinear term in (121.58). A term like

$$\sum_{\boldsymbol{k} j \mu} \sum_{\boldsymbol{k}' j' \mu'} \boldsymbol{P}^{(2)} \left(\boldsymbol{k}_1 - \boldsymbol{k}_2 \middle| \begin{matrix} \boldsymbol{k} & \boldsymbol{k}' \\ j_\mu & j'_{\mu'} \end{matrix} \right) Q \begin{pmatrix} \boldsymbol{k} \\ j_\mu \end{pmatrix} Q \begin{pmatrix} \boldsymbol{k}' \\ j'_{\mu'} \end{pmatrix} \tag{121.63}$$

will produce a transition from an initial state $|\chi_{0n}\rangle$ to a final state $|\chi_{0\bar{n}}\rangle$, with two additional quanta $\omega(\boldsymbol{k}|j)+\omega(\boldsymbol{k}'|j')$ present. Because in the final state the photon and crystal states are decoupled we write the final density of states as the product

$$\rho_f(E_f) = \rho_R(\hbar\,\omega_f)\, \rho_L\big(\omega(\boldsymbol{k}|j)+\omega(\boldsymbol{k}'|j')\big) \tag{121.64}$$

where R and L refer to radiation and lattice. The matrix element of (121.63) will give terms like

$$\sum_{\boldsymbol{k} j \mu} \sum_{\boldsymbol{k}' j' \mu'} \boldsymbol{P}^{(2)} \left(\boldsymbol{k}_1 - \boldsymbol{k}_2 \middle| \begin{matrix} \boldsymbol{k} & \boldsymbol{k}' \\ j_\mu & j'_{\mu'} \end{matrix} \right) \langle \chi_{0\bar{n}}|\, Q \begin{pmatrix} \boldsymbol{k} \\ j_\mu \end{pmatrix} Q \begin{pmatrix} \boldsymbol{k}' \\ j'_{\mu'} \end{pmatrix} |\chi_{0n}\rangle \tag{121.65}$$

or since

$$\langle \chi_{0\bar{n}}|\, Q \begin{pmatrix} \boldsymbol{k} \\ j_\mu \end{pmatrix} Q \begin{pmatrix} \boldsymbol{k}' \\ j'_{\mu'} \end{pmatrix} |\chi_{0n}\rangle \cong C \begin{pmatrix} \boldsymbol{k} & \boldsymbol{k}' \\ j_\mu & j'_{\mu'} \end{pmatrix} \neq 0, \tag{121.66}$$

if

$$\bar{n} = n+2 \tag{121.67}$$

where n, $\bar{n}$ refer to the *total* number of phonons present. The value of the constant in (121.66) depends on the occupancy of the phonon states ($n_{\boldsymbol{k}j}$, $n_{\boldsymbol{k}'j'}$, etc.), and also has a dependence upon the frequency of the modes.

Then, the total transition rate for all two phonon Stokes Raman scattering from (121.58), (121.67) for incident photon polarized $\boldsymbol{\varepsilon}_\lambda$, scattered photon polarized $\boldsymbol{\varepsilon}_{\lambda'}$ is proportional to

$$\omega_{0n\to 0(n+2)} \sim \left\{ \left| \sum_{\boldsymbol{k} j \mu} \sum_{\boldsymbol{k}' j' \mu'} P^{(2)}_{\lambda\lambda'} \left(\boldsymbol{k}_2 - \boldsymbol{k}_1 \middle| \begin{matrix} \boldsymbol{k} & \boldsymbol{k}' \\ j_\mu & j'_{\mu'} \end{matrix} \right) \right|^2 \rho_R(\hbar\,\omega_f)\, \rho_L\big(\omega(\boldsymbol{k}|j)+\omega(\boldsymbol{k}'|j')\big) \right\}. \tag{121.68}$$

If we select out a particular energy shift $\Delta\omega_2$, in the two phonon region,

$$\Delta\omega_2 = \omega(\boldsymbol{k}|j)+\omega(\boldsymbol{k}'|j'). \tag{121.69}$$

Now the transition rate (121.68) is proportional to the intensity of radiation scattered at a frequency shift $\Delta\omega_2$, into solid angle $d\Omega$ around the direction $\boldsymbol{\varepsilon}_{2\lambda'}$ of the scattered radiation. Calling this intensity $\mathscr{I}_{\Delta\omega_2}$ we can make the following argument which in many cases is extremely useful and may be quantitatively correct. Let us assume that all factors in (121.68) are slowly varying functions of the frequency shift $\Delta\omega_2$ (or what is the same, of the incident and scattered frequencies) and in particular that the coefficients $P^{(2)}_{\lambda\lambda'}$ have no sharp variations. Then the major frequency dependence of $\mathscr{I}_{\Delta\omega_2}$ with respect to $\Delta\omega_2$ is due to the variation of the joint density of states $\rho_L\big(\omega(\boldsymbol{k}|j)+\omega(\boldsymbol{k}'|j')\big)$ with $\Delta\omega_2$.

Consequently

$$\frac{d\mathcal{I}_{\Delta\omega_2}}{d\Delta\omega_2} \sim (\text{const}) \frac{d\rho_L(\Delta\omega_2)}{d\Delta\omega_2}. \tag{121.70}$$

But the derivative on the r.h.s. is the derivative of the joint density of states with respect to frequency. The latter quantity may have critical points determined by symmetry as in the discussion in Sect. 107. We shall return to the discussion of the critical points in the joint density of states for multiphonon absorption and scattering in Sects. 141–146.

For the Raman scattering processes, in any order similar considerations obtain. The term of appropriate degree in the normal coordinates is responsible for the multiphonon scattering process. First one needs to determine whether such a term is non-zero in the series (121.58), and then evaluate the transition rate, or scattering intensity.

Group theory plays a very basic role in determining the structure of the series (121.58): namely which coefficients are non-zero. As a useful approximation, in certain systems, one may investigate specific multiphonon scattering processes by studying the quantity $d\mathcal{I}_{\Delta\omega_n}/d\Delta\omega_n$ and relating it to critical points in the multiphonon density of states. To go beyond the this approximation within the generalized Placzek theory requires calculation of the matrix elements, or coupling coefficients and evaluation of the density of accessible final states as in (121.45). Work on this polarizability theory has been recently carried out using the many-body Green function formation and some of this is reported elsewhere in this Encyclopedia [*20*, *22*], and very briefly in Sect. 124.

More microscopic approaches going beyond the generalized Placzek theory or polarizability into a true microscopic scattering theory are presently being developed in view of the rapid expansion of interest and of experimental work on Raman scattering. Some of these approaches will be discussed below in Sect. 124.

ζ) *The* $\boldsymbol{A}\cdot\boldsymbol{A}$ *term in scattering. Hamiltonian.* Returning to (120.5) for the moment, we observe that there are terms quadratic in the vector potential, which we have previously neglected. For example there are the terms

$$\mathcal{H}' = \sum_j \left(\frac{1}{2m}\right)\left(\frac{e^2}{c^2}\right) \boldsymbol{A}^2(\boldsymbol{r}_j) + \sum_\alpha \left(\frac{1}{2M}\right)\left(\frac{eZ}{c}\right)^2 \boldsymbol{A}^2(\boldsymbol{R}_\alpha). \tag{121.71}$$

Let us go over to second quantized representation for the electromagnetic field using (121.5)–(121.10). Typical terms in the Hamiltonian operator will be of the type for the electrons

$$\left(\frac{e^2}{2mc^2}\right)\frac{1}{V}\,\boldsymbol{\varepsilon}^*_{\boldsymbol{k}_2\lambda_2}\cdot\boldsymbol{\varepsilon}_{\boldsymbol{k}_1\lambda_1}\,q^*_{\boldsymbol{k}_2\lambda_2}\,q_{\boldsymbol{k}_1\lambda_1}\,e^{i(\boldsymbol{k}_1-\boldsymbol{k}_2)\cdot\boldsymbol{r}_j} \tag{121.72}$$

and a similar term for the ions

$$\left(\frac{e^2}{2Mc^2}\right)\frac{1}{V}\,\boldsymbol{\varepsilon}^*_{\boldsymbol{k}_2\lambda_2}\cdot\boldsymbol{\varepsilon}_{\boldsymbol{k}_1\lambda_1}\,q^*_{\boldsymbol{k}_2\lambda_2}\,q_{\boldsymbol{k}_1\lambda_1}\,e^{i(\boldsymbol{k}_1-\boldsymbol{k}_2)\cdot\boldsymbol{R}_\alpha}. \tag{121.73}$$

Because the terms (121.72), (121.73) involve a photon creation and a photon destruction operator a transition can be produced in the radiation field, with photon $\boldsymbol{k}_1$ polarization λ_1 being destroyed while $\boldsymbol{k}_2$ with polarization λ_2 is created; at the same time a transition must occur in the electron, or ion subsystem. Consider (121.72) taken between two eigenstates (121.17). The result must be non-zero only if the two occupancy numbers of the radiation eigenfunction corresponding to $\boldsymbol{k}_2$ and $\boldsymbol{k}_1$ increase and decrease by one respectively; the lattice eigenstate is unchanged. The remaining integral over system variables is of form

$$\sum_j \langle \varphi_\mu(\boldsymbol{r}, \boldsymbol{R})| \, \mathrm{e}^{i(\boldsymbol{k}_2-\boldsymbol{k}_1)\cdot \boldsymbol{r}_j} \, |\varphi_\nu(\boldsymbol{r}, \boldsymbol{R})\rangle_{\boldsymbol{r}}. \tag{121.74}$$

It is immediately evident that for an integral like (121.74) not to vanish

$$\varphi_\mu^*(\boldsymbol{r}, \boldsymbol{R})\, \varphi_\nu(\boldsymbol{r}, \boldsymbol{R}) \sim D^{(\boldsymbol{k}_2-\boldsymbol{k}_1)}. \tag{121.75}$$

Thus if we take the (many-body) electron eigenfunction φ_μ to be specified by wave vector $\boldsymbol{k}_\mu$ and the φ_ν to have $\boldsymbol{k}_\nu$, then momentum conservation (translation invariance) requires that

$$\boldsymbol{k}_\mu - \boldsymbol{k}_\nu = \boldsymbol{k}_2 - \boldsymbol{k}_1. \tag{121.76}$$

We can take for granted the momentum conservation, which requires that the difference of photon momentum be absorbed by the electron system.

To go further let the many-body electron eigenfunctions φ_μ and φ_ν be taken in Hartree, or Hartree-Fock approximation as an antisymmetrized product of one particle eigenfunction:

$$\varphi_\nu(\boldsymbol{r}, \boldsymbol{R}) = \mathscr{A} \prod_m \psi^{(\boldsymbol{k}_1)(j_1)}(\boldsymbol{r}_j) \ldots \psi^{(\boldsymbol{k}_m)(j_m)}(\boldsymbol{r}_j) \tag{121.77}$$

where $\mathscr{A}$ is the antisymmetrizer. Then (121.74) becomes a sum of integrals each one like

$$\langle \psi^{(\boldsymbol{k}_{j'})(j')}| \, \mathrm{e}^{i(\boldsymbol{k}_2-\boldsymbol{k}_1)\cdot \boldsymbol{r}_j} \, |\psi^{(\boldsymbol{k}_{j''})(j'')}\rangle_{\boldsymbol{r}_j}. \tag{121.78}$$

Since $\boldsymbol{k}_1$ and $\boldsymbol{k}_2$ are small, expand

$$\mathrm{e}^{i(\boldsymbol{k}_2-\boldsymbol{k}_1)\cdot \boldsymbol{r}_j} = 1 + i(\boldsymbol{k}_2 - \boldsymbol{k}_1)\cdot \boldsymbol{r}_j + \cdots. \tag{121.79}$$

The first term, after (121.79) is substituted in (121.78), is the overlap integral

$$\langle \psi^{(\boldsymbol{k}_{j'})(j')} | \psi^{(\boldsymbol{k}_{j''})(j'')}\rangle_{\boldsymbol{r}_j} = \delta_{j' j''}\, \delta_{\boldsymbol{k}_{j'}, \boldsymbol{k}_{j''}}. \tag{121.80}$$

Let us assume that the $\boldsymbol{k}$ selection rule is satisfied. Then (121.80) is the overlap, or non-orthogonality, integral for two Bloch functions. It is well known that in general, if $\langle \psi^{(\boldsymbol{k}_{j'})(j')}$ and $\psi^{(\boldsymbol{k}_{j''})(j'')}\rangle_{\boldsymbol{r}_j}$ belong to different electronic energy bands, the integral (121.80) will vanish.

The result is a general one. The term $\boldsymbol{A}^2$ in the Hamiltonian for the system plus radiation will usually have non-zero matrix element only for two states in the same band. Generally this restricts the interest in such terms to processes in metals, or in highly doped insulators with plasma-type effects and Raman (inelastic) scattering from plasma type collective modes.

The case of the $\boldsymbol{A}^2$ term in (121.73) involving ionic coordinates can be treated by similar arguments. First of all, as we remarked in (120.21), (120.22) *passim* the magnitude of the direct photon-ion coupling term is smaller in the Hamiltonian by a factor $Z\kappa^3$ compared to the photon-electron coupling term. Then the matrix elements of (121.73) involve terms like

$$\langle \chi_{\mu \bar{n}}(R)|\, e^{i(\boldsymbol{k}_2-\boldsymbol{k}_1)\cdot \boldsymbol{R}_\alpha}\, |\chi_{\nu n}(R)\rangle_{\boldsymbol{R}}. \tag{121.81}$$

But in the long wave limit with $\boldsymbol{k}_2 = \boldsymbol{k}_1 = 0$, this term is an overlap integral

$$\langle \chi_{\mu \bar{n}}(R)|\chi_{\nu n}(R)\rangle_{\boldsymbol{R}}. \tag{121.82}$$

In general such an integral will not vanish, but if the electronic state is the same $\mu = \nu$ then $\chi_{\nu \bar{n}}$ and $\chi_{\nu n}$ belong to vibrational states of the same electronic manifold and

$$\langle \chi_{\nu \bar{n}}|\chi_{\nu n}\rangle_{\boldsymbol{R}} = \delta_{\bar{n} n}. \tag{121.83}$$

But for vibrational Raman scattering $\bar{n} \neq n$ so that the term vanishes.

To summarize, we see that the conventional arguments for neglect of both the $\boldsymbol{A}^2$ terms in (121.71) in phonon Raman scattering are certainly plausible and even rigorous in certain cases. We accept these arguments here, while recognizing that further work should be done to examine this matter particularly in regard to detailed calculation of the magnitude of Raman scattering tensor. Some recent discussion of this problem has been given.[3]

In any event, in the symmetry analysis of the matrix elements to obtain selection rules for infra-red and Raman scattering in insulators we shall proceed on the assumption that the $\boldsymbol{A}^2$ term vanishes.

122. A mutual exclusion selection rule for certain two phonon overtones in infra-red and Raman processes in crystals with an inversion. Before proceeding to discuss the infra-red and Raman processes further we can establish an important, if restricted, general result for certain of special two phonon processes known as overtones. In (120.52)–(120.58) we derived general formulae for two phonon infra-red absorption, and in (121.56)–(121.63) we gave the formulae for the two phonon Raman scattering. The essential point for the present discussion is (120.55) and (121.53).

To be specific consider these rules for two phonon overtones in a crystal whose space group $\mathfrak{G}$ contains inversion symmetry. For an overtone we consider the creation of two identical phonons in infra-red or Raman processes. Then if an overtone of $Q\begin{pmatrix}\boldsymbol{k}\\ j_\mu\end{pmatrix}$ is to be infra-red active

$$([{}^\star\boldsymbol{k}j]_{(2)}|\boldsymbol{\Gamma}\,\boldsymbol{v}) \neq 0 \tag{122.1}$$

while if the same overtone is to be Raman active

$$([{}^\star\boldsymbol{k}j]_{(2)}|\boldsymbol{\Gamma}\,[\boldsymbol{v}]_{(2)}) \neq 0. \tag{122.2}$$

We assume that the radiation wavelength in the moment operator $\mathscr{M}(\boldsymbol{R},\boldsymbol{\Gamma})$, and in the polarization operator $\mathfrak{P}(\boldsymbol{R},\boldsymbol{\Gamma})$ is infinite: $\boldsymbol{k} = \boldsymbol{k}_1 - \boldsymbol{k}_2 = \boldsymbol{\Gamma} = (0, 0, 0)$.

Now we need to determine the symmetrized square of the representation by which the phonon $Q\begin{pmatrix}\boldsymbol{k}\\ j_\mu\end{pmatrix}$ transforms, in case the space group contains inversion

---

[3] P. P. PLATZMAN and N. TZOAR: Phys. Rev. **182**, 510 (1969).

symmetry. We consider overtones corresponding to a general wave vector $\boldsymbol{k}$, and general star $\star\boldsymbol{k}$. The phonon $Q\begin{pmatrix}\boldsymbol{k}\\ j_\mu\end{pmatrix}$ and its g_p degenerate partners span the representation $D^{(\star\boldsymbol{k})(j)}$. Owing to assumption that $\boldsymbol{k}$ is a general wave vector, the star $\star\boldsymbol{k}$ is general, then (38.1), (45.4), and (45.7) applies in regard to the structure of the representation. Now if the star $\star\boldsymbol{k}$ is general, and the crystal contains inversion symmetry then the set of wave vectors in $\star\boldsymbol{k}$ can be partitioned, so that to each wave vector $\boldsymbol{k}_i$, there corresponds a wave vector $-\boldsymbol{k}_i$:

$$\star\boldsymbol{k}=(\boldsymbol{k}_1,\boldsymbol{k}_2,\dots,\boldsymbol{k}_{g_{p/2}},-\boldsymbol{k}_1,\dots,-\boldsymbol{k}_{g_{p/2}}). \tag{122.3}$$

Then in the symmetrized square

$$[D^{(\star\boldsymbol{k})(j)}]_{(2)} \tag{122.4}$$

the only product functions which belong to total wave vector $\boldsymbol{\Gamma}$ are evidently

$$\psi^{(\boldsymbol{k}_i)(j)}\cdot\psi^{(-\boldsymbol{k}_i)(j)},\qquad i=1,\dots,g_{p/2}. \tag{122.5}$$

(Of course in (122.4) all other product stars also arise but we are not concerned with them.)

Consider the space spanned by the set of product functions (122.5):

$$\Sigma^{(\boldsymbol{\Gamma})}\equiv\{\psi^{(\boldsymbol{k}_1)(j)}\cdot\psi^{(-\boldsymbol{k}_1)(j)},\dots\}. \tag{122.6}$$

The dimension of the representation is

$$\chi^{(\boldsymbol{\Gamma})}(\{\varepsilon|0\})=g_{p/2}. \tag{122.7}$$

For any pure rotational element $\{\varphi|\tau(\varphi)\}$

$$\varphi\cdot\boldsymbol{k}_m\neq\boldsymbol{k}_m \tag{122.8}$$

by hypothesis. However if $\varphi=\boldsymbol{i}$, the inversion then

$$\boldsymbol{i}\cdot\boldsymbol{k}_m=-\boldsymbol{k}_m \tag{122.9}$$

and

$$P_{\{\boldsymbol{i}|\tau(\boldsymbol{i})\}}\,\psi^{(\boldsymbol{k}_m)(j)}\cdot\psi^{(-\boldsymbol{k}_m)(j)}=\psi^{(\boldsymbol{k}_m)(j)}\cdot\psi^{(-\boldsymbol{k}_m)(j)}. \tag{122.10}$$

Thus the remainder of the character system based on the space (122.6) is

$$\chi^{(\boldsymbol{\Gamma})}(\{\boldsymbol{i}|\tau(\boldsymbol{i})\})=g_{p/2} \tag{122.11}$$

and for any other element

$$\chi^{(\boldsymbol{\Gamma})}(\{\varphi|\tau(\varphi)\})=0. \tag{122.12}$$

Now for any space group $\mathfrak{G}$ the entire point group

$$\mathfrak{P}=\mathfrak{G}/\mathfrak{T} \tag{122.13}$$

is sufficient to classify the representations at $\boldsymbol{\Gamma}$ since $\mathfrak{P}(\boldsymbol{\Gamma})=\mathfrak{P}$. For the point group of a space group with inversion we have

$$\mathfrak{P}=\varepsilon,\varphi_2,\dots,\varphi_{g_{p/2}},\boldsymbol{i},\boldsymbol{i}\cdot\varphi_2,\dots,\boldsymbol{i}\,\varphi_{g_{p/2}}. \tag{122.14}$$

Thus the point group of a space group with inversion is also a direct product of a lower order point group times the parity group. Consequently the irreducible representations of divide into even and odd:

$$\chi^{(j\pm)}(\boldsymbol{i}\,\varphi_p) = \pm\chi^{(j\pm)}(\varphi_p). \tag{122.15}$$

Then from (122.7) and (122.11) we conclude that the representation $\chi^{(\boldsymbol{\Gamma})}$ of the symmetric square can be reduced as

$$\chi^{(\boldsymbol{\Gamma})} = \sum_{j+} n_{j+}\,\chi^{(j+)} \tag{122.16}$$

only even parity representations occur. To determine the reduction coefficients

$$n_{j+} \equiv (\boldsymbol{\Gamma}|j+)$$

we use a well known argument like the reduction of the regular representation of any finite group. We find

$$(\boldsymbol{\Gamma}|j+) = \frac{1}{g_p}\sum_R \chi^{(\boldsymbol{\Gamma})}(R)\,\chi^{(j+)}(R)^*, \tag{122.17}$$

and the sum is over all elements R in $\mathfrak{P}$. Then

$$(\boldsymbol{\Gamma}|j+) = l_{j+}. \tag{122.18}$$

Consequently *all* even parity representations of $\mathfrak{P}$ occur, each one as often as its dimension, which is l_{j+}.

Finally consider the representations by which a vector and a symmetric second rank tensor transform under transformation by $\mathfrak{P}$. Call the relevent character systems $\chi^{(\boldsymbol{v})}$ and $\chi^{([\boldsymbol{v}])(2)}$. For a vector $\boldsymbol{v} \equiv (x, y, z)$ and

$$\chi^{(\boldsymbol{v})}(\boldsymbol{i}\,\varphi) = -\chi^{(\boldsymbol{v})}(\varphi). \tag{122.19}$$

Consequently

$$\chi^{(\boldsymbol{v})} = \sum_{j-} (\boldsymbol{v}|j-)\,\chi^{(j-)}. \tag{122.20}$$

Thus for a group with inversion only *odd* representations occur in the decomposition of $\chi^{(\boldsymbol{v})}$. For a symmetric second rank tensor $[\boldsymbol{v}]_{(2)}$ we have

$$\begin{aligned}\chi^{([\boldsymbol{v}]_{(2)})}(\varphi) &= \tfrac{1}{2}\left(\chi^{(\boldsymbol{v})}(\varphi^2) + (\chi^{(\boldsymbol{v})}(\varphi))^2\right)\\ &= \chi^{([\boldsymbol{v}]_{(2)})}(\boldsymbol{i}\,\varphi).\end{aligned} \tag{122.21}$$

It thus follows that

$$\chi^{([\boldsymbol{v}]_{(2)})} = \sum_{j+} ([\boldsymbol{v}]_{(2)}|j+)\,\chi^{(j+)}. \tag{122.22}$$

Consequently the representations of a symmetric second rank tensor decompose into a direct sum of *even* representations of $\mathfrak{P}$. The particular non-zero $(\boldsymbol{v}|j-)$ and $([\boldsymbol{v}]_{(2)}|j+)$ depend upon which group $\mathfrak{P}$ is involved in each case.

Now we can assemble these results to obtain the mutual exclusion rule for crystals whose space group $\mathfrak{G}$ contains inversion $\boldsymbol{i}$. Clearly the overtones transform as a sum of even representations as shown in (122.22). *The general overtone is active in Raman scattering since* (122.2) *is satisfied, and forbidden in infra-red since* (122.1) *is not satisfied.*[1-3]

Remark that the proof given applied only to the general wave vector. In case of crystals with inversion, overtones at high symmetry points may contain both even and odd representations of $\mathfrak{G}(\Gamma)$. An important example is in the diamond space group for overtones of phonons at $\star X$. This will be discussed in detail below (*vide infra* Sect. 133). Hence, each case of overtones of high symmetry phonons needs to be specially analysed.

The utility of this rule arises from the analysis of the two phonon part of the spectrum in Raman scattering and infra-red absorption using critical point theory. Of course in an actual application the possible contribution from high symmetry phonons requires separate examination.

A final important remark concerns the falsity of the converse of this rule. That is: if a crystal lacks inversion then it is not necessarily true in general that overtones are infra-red active. As an example refer to a comparison between the diamond and the zincblende space groups. One finds that certain phonon species in zincblende can be related to a "parent" diamond phonon species by considering inversion symmetry to be lost. But in both cases overtones can be infra-red forbidden. The purpose of this remark is cautionary. In interpreting two phonon spectra the rule proved above is useful, and exact, but it should not be extended beyond its domain of validity without proof in any particular case. Also resonance effects may break the rule (see Sect. 124η).[4]

123. Polarization effects in infra-red and Raman lattice processes. There are three major sources of polarization effects which may be observed in Raman scattering and infra-red absorption. Perhaps the most important both in principle and in applications is the tensor character of the scattering. Study of the polarization of scattered light is a powerful tool in the understanding of microscopic dynamics. The polarization of the scattered light is inextricably connected with the properties of the scattering tensor in the crystal which is a special form of second rank tensor. The use of the scattering tensor is very common in all Raman scattering studies. The tensor can be easily computed once it is realized that the components of the tensor are simply related to Clebsch-Gordan coefficients. This general relationship will be shown here, and then the use of the Raman tensor will be illustrated by working a typical case in which the intensity depends on polarization of incident and scattered photons. A second source producing polarization effects is related to finite wave vector of the moment operator $\boldsymbol{M}(\boldsymbol{R}_1\, \boldsymbol{k})$ and the polarizability operator $\boldsymbol{P}(\boldsymbol{R}, \boldsymbol{k}_1 - \boldsymbol{k}_2)$. A particularly well-known manifestation of this is related to macroscopic electric field effects which occur for crystals without inversion center, for polar optic phonons simultaneously Raman and infra-red active. Finally, and most simply is geometrical anisotropy as a source of polarization effects caused by inequivalence of crystal directions

---

[1] J.L. Birman: Phys. Rev. **127**, 1093 (1962); **131**, 1489 (1963).
[2] M. Lax: Comment on paper by J.L. Birman, in: [*32*], p. 712.
[3] R. Loudon: Phys. Rev. **137**, 1784 (1965).
[4] R. Berenson and J.L. Birman: Phys. Rev. **B9**, 4512 (1974); J.L. Birman: Phys. Rev. **B9**, 4518 (1974).

(axes). These matters will be discussed in this section. Also mention will be made of the use of polarization as a tool in a tentative identification, by Raman scattering, of a proposed novel two phonon bound state.

α) *Raman tensor and Clebsch-Gordan coefficients.* The rate of vibrational lattice Raman scattering by phonons was given in (121.45). That expression can be rewritten in a form more easily used, from which the essential polarization effects can be derived. In this subsection we shall demonstrate that the elements of the first order scattering tensor are actually simple Clebsch-Gordan coefficients. This demonstration should make these elements more comprehensible and aid in the calculation of scattering tensors for new situations, such as arise in symmetry-breaking resonance Raman scattering which will be discussed in Subsect. 124ε.

Let us rewrite (121.45) to give the rate of scattering as

$$\mathscr{I}_{n\to\bar{n}} = C \Big| \sum_{\rho,\sigma} (\varepsilon_\lambda)_\sigma R^{n\bar{n}}_{\sigma\rho} (\varepsilon_{\lambda'})_\rho \Big|^2 \tag{123.1}$$

where $I_{n\to\bar{n}}$ is the Raman scattering intensity from vibrational state $|\chi_{0n}\rangle$ to $|\chi_{0\bar{n}}\rangle$, and $(\varepsilon_\lambda)_\sigma (\varepsilon_{\lambda'})_\rho$ are the σ and ρ Cartesian components of the incident and scattered photon polarization vectors respectively. The Raman tensor is defined by

$$R^{n\bar{n}}_{\sigma\rho} \equiv \sum_f \langle \chi_{0\bar{n}} | P_{\sigma\rho} | \chi_{0n} \rangle. \tag{123.2}$$

As shown in Sect. 121 the polarizability operator $\boldsymbol{P}(\boldsymbol{R}, \boldsymbol{k}_2 - \boldsymbol{k}_1)$ is a second rank symmetric tensor field in $\boldsymbol{R}$ in the limit $\boldsymbol{k}_1 - \boldsymbol{k}_2 = \boldsymbol{\Gamma}$. It is worthwhile to recall that the requirement that the tensor be symmetric follows from the lack of any structure in the matrix elements $\langle \varphi_\nu | \boldsymbol{\varepsilon} \cdot \boldsymbol{\mu}_\alpha | \varphi_\mu \rangle$ of (121.31) and consequently the way this carries through to (121.48). Later in the microscopic theory of Sect. 124 this requirement will be relieved. Then we have, for transformation of the Cartesian indices (σ, ρ):

$$\boldsymbol{P}(R) \sim [D^{(v)}]_{(2)} \tag{123.3}$$

which is the symmetrized Kronecker square of the vector representation. In any space group the representation (123.3) can be decomposed into irreducible representations by

$$[D^{(v)}]_{(2)} = \sum_{\oplus j} ([v]_{(2)} | \boldsymbol{\Gamma} j)\, D^{(\boldsymbol{\Gamma})(j)}.$$

From (123.1) one immediately surmises that polarization, and anisotropy effects are to be expected, depending upon polarization of the incident and scattered radiation.

We now will examine this matter more carefully, but still retaining the restriction that $\boldsymbol{k}_2 = \boldsymbol{k}_1 = 0$. First introduce the normal coordinates $Q\begin{pmatrix}\boldsymbol{k}\\ j_\mu\end{pmatrix}$ which are assumed to be bases for $D^{(\boldsymbol{k})(j)}$. Then in (123.1) the sets of quantum numbers $\{n\}$ and $\{\bar{n}\}$ which label the lattice vibrational states $|\chi_n\rangle, |\chi_{\bar{n}}\rangle$ refer to these normal coordinates. We may write

$$\left\langle \left\{ Q\begin{pmatrix}\boldsymbol{k}\\ j_\mu\end{pmatrix} \right\} \middle| \chi_n \right\rangle \equiv \chi\left(\{n_{\boldsymbol{k}, j_\mu}\}, Q\begin{pmatrix}\boldsymbol{k}\\ j_\mu\end{pmatrix} \right). \tag{123.4}$$

In order to analyse the transition matrix element (123.2) we introduce the expansion of $P_{\sigma\rho}(\boldsymbol{R})$ into normal coordinates as in (121.58), which now appears as

$$P_{\sigma\rho}(\boldsymbol{R})=P_{\sigma\rho}^{(0)}+\sum_{\boldsymbol{k}j\mu}P_{\sigma\rho}^{(1)}\left(\boldsymbol{\Gamma}\middle|\begin{matrix}\boldsymbol{k}\\ j_\mu\end{matrix}\right)Q\begin{pmatrix}\boldsymbol{k}\\ j_\mu\end{pmatrix}+\cdots. \qquad (123.5)$$

Substituting (123.5) into (123.2) the matrix element is seen to decompose into a sum of individual matrix elements, each one referring to a single normal coordinate, which is only non-zero if the occupation number $n_{\boldsymbol{k}j_\mu}$ for the corresponding oscillator changes (in our case increases) by one. Then for a *specified* $Q\begin{pmatrix}\boldsymbol{k}\\ j_\mu\end{pmatrix}$ the intensity of one phonon Raman scattering can be written

$$\mathscr{I}(\boldsymbol{k},j_\mu)=C\sum_\mu\left|\sum_{\sigma\rho}\left(\varepsilon_{2\rho}P_{\sigma\rho}^{(1)}\left(\boldsymbol{\Gamma}\middle|\begin{matrix}\boldsymbol{k}\\ j_\mu\end{matrix}\right)\varepsilon_{1\sigma}\right)\right|^2. \qquad (123.6)$$

The object $P_{\sigma\rho}^{(1)}\left(\boldsymbol{\Gamma}\middle|\begin{matrix}\boldsymbol{k}\\ j_\mu\end{matrix}\right)$ $(\rho,\sigma=1,2,3)$ is the symmetric Raman scattering tensor. If we suppress $\boldsymbol{\Gamma}$ and also the superscript it is understood that the tensor refers to infinite photon wavelength: we call it $P_{\sigma\rho}(\boldsymbol{k},j_\mu)$. Evidently just this coefficient controls the amplitude of the one-phonon scattering.

Now consider a symmetry element $\{\boldsymbol{\varphi}|\boldsymbol{\tau}\}$ in $\mathfrak{G}$. Transform (123.5) by this element to obtain

$$P_{\sigma\rho}(\boldsymbol{R})\to P'_{\sigma\rho}(\{\boldsymbol{\varphi}|\boldsymbol{\tau}\}^{-1}\cdot\boldsymbol{R})=\sum_{\alpha\beta}[D^{(\boldsymbol{v})}(\boldsymbol{\varphi})_{(2)}]_{\sigma\rho,\alpha\beta}P_{\alpha\beta}(\{\boldsymbol{\varphi}|\boldsymbol{\tau}\}^{-1}\cdot\boldsymbol{R})=P_{\sigma\rho}(\boldsymbol{R}). \quad (123.7)$$

Here $D^{(\boldsymbol{v})}(\boldsymbol{\varphi})_{(2)}$ is the symmetrized square representation of $D^{(\boldsymbol{v})}$ which is assumed for the Cartesian indices (σ,ρ); and the last equality follows because $\{\boldsymbol{\varphi}|\boldsymbol{\tau}\}$ is a symmetry element. We could now expand the last two terms in the equation into normal coordinates and equate term by term, using (86.40) for the transformation of normal coordinates. An equivalent but more useful procedure is to assume that for fixed $\sigma\rho$ the set of quantities $P_{\sigma\rho}(\boldsymbol{k}j_\mu)$ transform as bases for the appropriate representation:

$$P_{\{\boldsymbol{\varphi}_\lambda|\boldsymbol{\tau}_\lambda\}}P_{\alpha\beta}(\boldsymbol{k},j_\mu)=\sum_\nu D^{(\boldsymbol{k})(j)}(\{\boldsymbol{\varphi}_\lambda|\boldsymbol{\tau}_\lambda\})_{\nu\mu}P_{\alpha\beta}(\boldsymbol{k}j_\nu) \qquad (123.8)$$

where $\{\boldsymbol{\varphi}_\lambda|\boldsymbol{\tau}_\lambda\}$ is in $\mathfrak{G}(\boldsymbol{k})$. Substitute (123.8) into (123.7) to obtain

$$P_{\sigma\rho}(\boldsymbol{k}j_\mu)=\sum_{\alpha\beta}[D^{(\boldsymbol{v})}(\boldsymbol{\varphi}_\lambda)_{(2)}]_{\sigma\rho,\alpha\beta}D^{(\boldsymbol{k})(j)}(\{\boldsymbol{\varphi}_\lambda|\boldsymbol{\tau}_\lambda\})^*_{\mu\nu}P_{\alpha\beta}(\boldsymbol{k}j_\nu). \qquad (123.9)$$

In the conventional treatment[1] this equation is used, for several different $\{\boldsymbol{\varphi}_\lambda|\boldsymbol{\tau}_\lambda\}$ in $\mathfrak{G}(\boldsymbol{k})$ one at a time in order to determine the independent non-zero elements $P_{\sigma\rho}(\boldsymbol{k}j_\mu)$.

Let U be the unitary matrix which brings $D_{(2)}^{(\boldsymbol{v})}$ to fully reduced form:

$$U^{-1}D_{(2)}^{(\boldsymbol{v})}U=\bar{\Delta} \qquad (123.10\text{a})$$

where

$$\bar{\Delta}_{lnl'n'}=\delta_{ll}D_{nn'}^{(l)}, \qquad (123.10\text{b})$$

[1] H. Poulet: Ann. Phys. (Paris) **10**, 908 (1955).

and we now assume $D^{(v)}$ is irreducible and that each $D^{(l)}$ occurs only once in the reduced $\bar{\Delta}$. Then (123.10a) becomes, letting S be an element in $\mathfrak{G}(\boldsymbol{k})$:

$$[D^{(v)}(S)_{(2)}]_{\sigma\rho,\alpha\beta} = \sum_{l n n'} U_{\sigma\rho l n} D^{(l)}(S)_{n n'} U^{-1}_{l n' \alpha\beta}. \tag{123.11}$$

Now substitute (123.11) into (123.9) and sum over all elements S and use the orthonormality relations for irreducible representations to obtain

$$l_j P_{\sigma\rho}(\boldsymbol{k} j_\mu) = \sum_{\alpha\beta v} U_{\sigma\rho \boldsymbol{k} j_\mu} U^{-1}_{\boldsymbol{k} j_v \alpha\beta} P_{\alpha\beta}(\boldsymbol{k} j_v) \tag{123.12}$$

where l_j is the dimension of $D^{(\boldsymbol{k})(j)}$.

This equation is easily solved by taking

$$P_{\sigma\rho}(\boldsymbol{k} j_\mu) = c(\boldsymbol{k} j)\, U_{\sigma\rho \boldsymbol{k} j_\mu}. \tag{123.13}$$

This is the essential result. The elements of the Raman scattering tensor are Clebsch-Gordan coefficients. The constant $c(\boldsymbol{k}, j)$ depends only on the irreducible representation. This result should clarify the physical and mathematical significance of the elements of the scattering tensor.

The result can be generalized to the second order scattering matrix also, to other scattering processes (magnon, etc.), as well as to relieve the restriction that $D^{(v)}$ is irreducible and that there is no multiplicity (each $D^{(l)}$ occurs only once). These matters are discussed in the literature along with various applications.[2]

β) Polarization effects in Raman scattering: The Raman tensor in a cubic crystal with inversion. Now we illustrate the material of the previous subsection. To be specific consider the case of a cubic space and point group $O_h = \mathfrak{G}(\boldsymbol{\Gamma})/T(\boldsymbol{\Gamma})$. In this case

$$[D^{(v)}]_{(2)} = \boldsymbol{\Gamma}^{(1+)} \oplus \boldsymbol{\Gamma}^{(12+)} \oplus \boldsymbol{\Gamma}^{(25+)}. \tag{123.14}$$

The irreducible representations appearing in (123.14) are respectively the identity, a two dimensional and a three dimensional irreducible representation.

In order to apply the previous result (123.13) in the present context it sufficies for us to use the available table of the Clebsch-Gordan coefficients for the crystallographic point groups [55] in particular for O_h. The required elements $P_{\alpha\beta}(\boldsymbol{\Gamma} j_\mu)$ are then simply read off from the tables. In the present situation each representation listed in (123.14) corresponds to an allowed phonon symmetry in Raman scattering.

It might be pointed out at this point that the notation for these matrices is not uniform. LOUDON, in well known articles[3] has used for fixed index j

$$R_{\mu,\alpha\beta}(\text{Loudon}) = P_{\alpha\beta}(\boldsymbol{\Gamma} j_\mu)\,(\text{ours})$$

while POULET and MATHIEU [52] use

$$P_{\alpha\beta}((i), m)\,(\text{Poulet-Mathieu})$$

---

[2] J.L. BIRMAN and R. BERENSON: Phys. Rev. **B9**, 4512 (1974). – J.L. BIRMAN: Phys. Rev. **B9** 4518 (1974).

[3] R. LOUDON: Proc. Roy. Soc. A **275**, 218 (1963). – R. LOUDON: Adv. Phys. **13**, 423 (1964).

which is practically identical to our notation. In the next few lines we give the matrices using both notations. The matrices are given on the usual Cartesian crystal axes, here denoted $\hat{\boldsymbol{x}}$, $\hat{\boldsymbol{y}}$, $\hat{\boldsymbol{z}}$ in terms of unit vectors.

$$\boldsymbol{P}(\boldsymbol{\Gamma}(1+)) = \begin{pmatrix} a & \cdot & \cdot \\ \cdot & a & \cdot \\ \cdot & \cdot & a \end{pmatrix} = \boldsymbol{R}, \tag{123.15}$$

$$\boldsymbol{P}(\boldsymbol{\Gamma}(12+), 1) = \begin{pmatrix} b & \cdot & \cdot \\ \cdot & b & \cdot \\ \cdot & \cdot & -2b \end{pmatrix} = \boldsymbol{R}_1, \tag{123.16}$$

$$\boldsymbol{P}(\boldsymbol{\Gamma}(12+), 2) = \begin{pmatrix} -\sqrt{3}b & \cdot & \cdot \\ \cdot & \sqrt{3}b & \cdot \\ \cdot & \cdot & \cdot \end{pmatrix} = \boldsymbol{R}_2, \tag{123.17}$$

$$\boldsymbol{P}(\boldsymbol{\Gamma}(25+), 1) = \begin{pmatrix} \cdot & \cdot & \cdot \\ \cdot & \cdot & d \\ \cdot & d & \cdot \end{pmatrix} = \boldsymbol{R}_1, \tag{123.18}$$

$$\boldsymbol{P}(\boldsymbol{\Gamma}(25+), 2) = \begin{pmatrix} \cdot & \cdot & d \\ \cdot & \cdot & \cdot \\ d & \cdot & \cdot \end{pmatrix} = \boldsymbol{R}_2, \tag{123.19}$$

$$\boldsymbol{P}(\boldsymbol{\Gamma}(25+), 3) = \begin{pmatrix} \cdot & d & \cdot \\ d & \cdot & \cdot \\ \cdot & \cdot & \cdot \end{pmatrix} = \boldsymbol{R}_3. \tag{123.20}$$

For any one phonon transition involving phonon $Q\begin{pmatrix}\boldsymbol{\Gamma}\\ j_\nu\end{pmatrix}$ where j is among the species in (123.14) we can immediately write down an expression for the scattering intensity in various polarizations. Consider the one phonon production of a phonon of representation $\boldsymbol{\Gamma}^{(25+)}$. The matrices are (123.18)–(123.20) for this representation and from them we have for the non-zero components of the Raman tensor corresponding to each partner in the irreducible representation:

$$R_{1,yz} = R_{1,zy} = R_{2,xz} = R_{2,zx} = R_{3,xy} = R_{3,yx} = d \tag{123.21}$$

where the first index corresponds to the number in (123.18)–(123.20).

Let us now work a typical case, keeping in mind this is for the $\boldsymbol{\Gamma}^{(25+)}$ phonon. Consider a photon incident in the x direction so

$$\boldsymbol{k}_1 = k_1 \hat{\boldsymbol{x}}. \tag{123.22}$$

The corresponding polarization vector is of course transverse, so take

$$\boldsymbol{\varepsilon}_\lambda = \varepsilon_{\lambda y} \hat{\boldsymbol{y}} + \varepsilon_{\lambda z} \hat{\boldsymbol{z}} \tag{123.23}$$

with

$$(\varepsilon_{\lambda y})^2 + (\varepsilon_{\lambda z})^2 = 1. \tag{123.24}$$

Let the scattered photon be propagating in the direction

$$\boldsymbol{k}_2 = (\cos\theta\,\hat{\boldsymbol{x}} + \sin\theta\,\hat{\boldsymbol{y}})\,k_2 \tag{123.25}$$

in the xy plane. The scattered radiation can be analysed into perpendicular or parallel components, defined respectively for polarization vector perpendicular to, or lying in, the xy plane. Thus:

$$\boldsymbol{\varepsilon}_{\lambda'\perp} = \varepsilon_{\lambda' z}\,\hat{\boldsymbol{z}} \tag{123.26}$$

and

$$\boldsymbol{\varepsilon}_{\lambda'\parallel} = -\sin\theta\,\hat{\boldsymbol{x}} + \cos\theta\,\hat{\boldsymbol{y}}. \tag{123.27}$$

By conservation of total wave vector we get for the phonon

$$\boldsymbol{k} = \boldsymbol{k}_1 - \boldsymbol{k}_2. \tag{123.28}$$

Neglecting the wave length difference between incident and scattered photon, we may take $k_1 = k_2$. Then simple geometry gives for the scattering phonon

$$\boldsymbol{k} = k(\sin\theta/2\,\hat{\boldsymbol{x}} - \cos\theta/2\,\hat{\boldsymbol{y}}). \tag{123.29}$$

The phonon polarization can be specified with respect to the direction $\boldsymbol{k}$. Thus for the three-fold degenerate mode under discussion we can define a longitudinal polarization, given by

$$\boldsymbol{\varepsilon}^L = (\sin\theta/2\,\hat{\boldsymbol{x}} - \cos\theta/2\,\hat{\boldsymbol{y}}) \tag{123.30}$$

and two transverse polarization modes, given by

$$\boldsymbol{\varepsilon}_1^T = \hat{\boldsymbol{z}} \tag{123.31}$$

and

$$\boldsymbol{\varepsilon}_2^T = (\cos\theta/2\,\hat{\boldsymbol{x}} + \sin\theta/2\,\hat{\boldsymbol{y}}). \tag{123.32}$$

Here, all three phonons (123.30)–(123.32) are degenerate, of course, and the distinction between them is artificial here, although see below (123.64)–(123.67).

Now we can calculate the intensity of scattered radiation in each polarization, produced by the Raman scattering from the triply degenerate $\boldsymbol{\Gamma}^{(25+)}$ optic (O) phonon using (123.21)–(123.32). In "perpendicular" polarization $\boldsymbol{\varepsilon}_{\lambda' z}$ we have:

$$\mathscr{I}_\perp^0 = C\,|\varepsilon_{\lambda y} R_{1,yz}\varepsilon_{\lambda' z}|^2 = C\,d^2\,(\varepsilon_{\lambda y})^2 \tag{123.33}$$

and in the "parallel" polarization: $\boldsymbol{\varepsilon}_{\lambda'\parallel}$;

$$\mathscr{I}_\parallel^0 = C\{|\varepsilon_{\lambda y} R_{3,yx}(-\sin\theta)|^2 + |\varepsilon_{\lambda z} R_{1,zy}\cos\theta|^2 + |\varepsilon_{\lambda z} R_{3,zx}(-\sin\theta)|^2\}$$

or

$$\mathscr{I}_\parallel^0 = C\,d^2\,(\varepsilon_{\lambda y}^2 \sin^2\theta + \varepsilon_{\lambda z}^2). \tag{123.34}$$

Clearly, a comparison of (123.33), and (123.34) shows a polarization effect, depending on the polarization of the incident beam. The basic anisotropy illustrated here relates to the fact that Raman scattering is a tensor process and the intensity of scattered radiation of a given polarization depends tensorially upon the incident polarization.

It is also evident that this type of polarization effect will produce depolarization effects in general. In molecular vibrational Raman scattering the term "depolarization ratio" is often used to describe the change in polarization of an initially polarized beam as a result of the scattering. Since the relative intensities of scattered radiation in each polarization are explicitly calculable, using arguments such as the above, for each incident polarization for crystal Raman scattering, the depolarization ratios in case of a solid do not appear quantitatively meaningful in spite of some efforts to use them.

For a quantitative description of the scattering of partially polarized light by a crystal one should use the Stokes parameter description of the incident and of the scattered radiation. The relationship between the Stokes parameters of initial and scattered beam can be directly related to the symmetry of the Raman scattering tensor.

As a final point, we remark upon polarization effects in two phonon scattering and by extension in multiphon scattering. The basic theory of the multiphonon scattering is contained in (121.42) and (121.51). Again we restrict attention to a cubic crystal. For a pair of phonons to be active in Raman scattering we have the condition (121.62) equivalent to the discussion preceeding it (121.59)–(121.61). Suppose two normal modes $Q\begin{pmatrix}\boldsymbol{k}\\ j_\mu\end{pmatrix}$, $Q\begin{pmatrix}\boldsymbol{k}'\\ j'_{\mu'}\end{pmatrix}$ satisfy the condition (121.62). They will then be active if the direct product contains one of the representations (123.14). But we can go further in principal. Let $Q\begin{pmatrix}\boldsymbol{k}\\ j_\mu\end{pmatrix} Q\begin{pmatrix}\boldsymbol{k}'\\ j'_{\mu'}\end{pmatrix}$ contain a sum of two of the active representations (123.14). Then by use of the appropriate projection operator or by use of the proper Clebsch-Gordan coefficients one can determine the correct linear combinations of the products which transform as a particular row of the allowed representation. Explicitly we may have for modes $\boldsymbol{\Gamma}^{(12+)}$

$$\sum_{\boldsymbol{k}\in{}^{\star}\boldsymbol{k}} \sum_{\mu} \sum_{\mu'} (\boldsymbol{\Gamma}(12+)\,1\,|\,\boldsymbol{k}\,\mu\,\mu')\, Q\begin{pmatrix}\boldsymbol{k}\\ j_\mu\end{pmatrix} Q\begin{pmatrix}-\boldsymbol{k}\\ j'_{\mu'}\end{pmatrix} \sim \boldsymbol{\Gamma}_1^{(12+)}. \tag{123.35}$$

Then the particular active phonons as in (123.35) will produce one partner of the $\boldsymbol{\Gamma}^{(12+)}$ representation and, after substituting in (121.51) we obtain a properly symmetrized, or "correct linear combination", two phonon state, which acts as a part of a "pure" representation.

If an experiment can be designed by appropriate choice of polarizations $\boldsymbol{\varepsilon}_\lambda$ and $\boldsymbol{\varepsilon}_{\lambda'}$ to separate out the distinct contributions of each of the pure symmetries of (123.14), then one could separately discuss the contributions from e.g. two phonon scattering to each type of polarization. We shall illustrate this below when we discuss comparison between experiment and theory for the rocksalt crystals. In those systems some significant progress has been made, as will be discussed in Sect. 146.

We conclude by presenting a useful condensation of the polarization effects in Raman scattering in cubic crystals. We can assemble (123.10), and the matrices (123.15)–(123.20). Thus, suppose we prepare a matrix I whose $(\alpha\beta)$ entries are the intensities observed of *any* symmetry species for incident photon polarization $(\boldsymbol{\varepsilon}_\lambda)_\alpha$ and scattered photon polarization $(\boldsymbol{\varepsilon}_{\lambda'})_\beta$. The matrices (123.15)–(123.20) it should be recalled, are expressed upon the usual Cartesian $(x\,y\,z)$ axes. Then the matrix I reads

$$I = C \begin{pmatrix} a^2+4b^2 & d^2 & d^2 \\ d^2 & a^2+4b^2 & d^2 \\ d^2 & d^2 & a^2+b^2 \end{pmatrix}. \tag{123.36}$$

Clearly in this setting it is not possible to resolve separately the individual symmetry species.

However, consider now a rotation of axes from the orthogonal set $x\,y\,z$ parallel to the crystal $(\boldsymbol{x}, \boldsymbol{y}, \boldsymbol{z})$ axes, to a new set $(x'\,y'\,z')$ parallel respectively to $[1,1,0]$, $[1,\bar{1},0]$, $[0,0,1]$ or: $1/\sqrt{2}\,(\boldsymbol{x}+\boldsymbol{y})$, $1/\sqrt{2}\,(-\boldsymbol{x}+\boldsymbol{y})$, $\boldsymbol{z}$. The matrix I then takes the form I' where

$$I' = C \begin{pmatrix} a^2+b^2+d^2 & 3b^2 & d^2 \\ 3b^2 & a^2+b^2+d^2 & d^2 \\ d^2 & d^2 & a^2+4b^2 \end{pmatrix}. \tag{123.37}$$

Using these axes, three measurements suffice to determine the specific contributions from individual species. Let incident and scattered radiation be polarized in the z' direction. Then one obtains a^2+4b^2. Let incident radiation be x' polarized, scattered radiation y' polarized. Then one obtains $3b^2$. Let incident be x' polarized, scattered radiation z' polarized. Then one obtains d^2. In this manner the constants a, b, d characterizing each species can be determined. The use of these matrices I and I' will be illustrated in Sect. 146 wherein two phonon Raman spectra will be discussed.

γ) Polarization effects due to macroscopic electric field

i) *Cubic crystals with inversion symmetry.* Consider a long wave length phonon, say in a cubic ionic crystal, which is active in infra-red absorption. If the wave length of this phonon were actually infinite then

$$Q(\text{I. R.}) \sim Q\begin{pmatrix} \boldsymbol{\Gamma} \\ 15- \end{pmatrix} \tag{123.38}$$

since the $\boldsymbol{\Gamma}^{(15-)}$ phonon corresponds in a cubic crystal to the transformation of a polar vector $\boldsymbol{v}$. Actually of course the phonon active in infra-red absorption will have finite $\boldsymbol{k} \neq \boldsymbol{\Gamma}$. Such a phonon produces ionic displacements in a crystal corresponding to an electric moment $\mathcal{M}$, or electric polarization: moment per unit volume. This electric moment is due to the long range effects of ion displacements from equilibrium. It can be visualized as if the actual configuration of the crystal, with the ions in their displaced configurations corresponding to the mode (123.38) were replaced by the sum of: the crystal with ions in their rest positions plus a fictitious dipole lattice centered at the undisplaced ion sites. The dipole lattice produces a field at each ion site which is in addition to the short range field produced by the close-by valence forces: the latter can be represented by a set of classical mechanical springs. This added electric field will produce an added term in the classical equations of motion for the ions. For our present purposes, the simplest way to regard this added term is as if it were an *external* electric field coupled to the crystal. This electric field will be coupled to the ion system by means of an added term in the ion Hamiltonian

$$\mathscr{H}' \equiv \mathscr{P} \cdot \mathscr{E} \tag{123.39}$$

where $\mathscr{P}$ is the ion lattice polarization produced by the displacements (123.38), and $\mathscr{E}$ is the resultant macroscopic electric field. The term (123.39) is a term linear in the $Q\begin{pmatrix}\Gamma\\15-\end{pmatrix}$.

Consider now that part of the lattice system Hamiltonian relating to the three degenerate modes $Q\begin{pmatrix}\Gamma\\15-\end{pmatrix}$. In the harmonic approximation it is simply

$$\mathscr{H} = \sum_{\nu=1}^{3} \left\{ \left(\dot{Q}\begin{pmatrix}\Gamma\\15-_\nu\end{pmatrix}\right)^2 + \omega^2(\Gamma|15-)\, Q\begin{pmatrix}\Gamma\\15-_\nu\end{pmatrix}^2 \right\}. \tag{123.40}$$

The notation in (123.40) indicates that the three modes ($\nu = 1, 2, 3$) are degenerate. Now in presence of the "external" macroscopic electric field $\mathscr{E}$, the perturbing term (123.38) must be added. It will be of form (sum on Cartesian components):

$$\mathscr{H}' = \mathscr{P}_0 \sum_{\nu=1}^{3} \alpha_\nu\, Q\begin{pmatrix}\Gamma\\15-_\nu\end{pmatrix} \mathscr{E}_\nu. \tag{123.41}$$

But in presence of the perturbation (123.41) the three-fold degeneracy ($\nu = 1, 2, 3$) will be broken. This is immediately evident physically. The "external" electric field $\mathscr{E}$ selects a preferred crystal direction (parallel to its axis) and so lowers the crystal point symmetry from cubic to uniaxial. Then the symmetry will change as:

$$\mathfrak{G} = O_h \rightarrow O_h \cap C_{2\infty}(\mathscr{E}) \equiv \mathfrak{G}'. \tag{123.42}$$

That is it changes from the complete cubic point group to that point group $\mathfrak{G}'$ which is the intersection of the cubic group O_h with the symmetry group $C_{2\infty}$ of the rotations about the electric field.

This lowering of symmetry will break the degeneracy of the modes $Q\begin{pmatrix}\Gamma\\15-_\nu\end{pmatrix}$:

$$Q\begin{pmatrix}\Gamma\\15-_\nu\end{pmatrix} \rightarrow Q\begin{pmatrix}\Gamma\\\parallel\end{pmatrix} \oplus Q\begin{pmatrix}\Gamma\\\perp\end{pmatrix} \tag{123.43}$$

into a longitudinal plus a transverse polarized optic normal mode:

$$\omega(\Gamma|15-) \rightarrow \omega(\Gamma|LO) \oplus \omega(\Gamma|TO). \tag{123.44}$$

This splitting of any polar optic mode in the presence of self-coupling to its own electric field is a universal phenomenon in all crystals with polar phonons. It is also basically a macroscopic phenomenon. Since the dynamics of this splitting has been thoroughly discussed in standard references on lattice dynamics [*4*] we shall not give derivations of the well known result that for a single polar optic mode in a cubic crystal split by self-coupling the frequencies (123.44) are given as

$$\frac{\omega^2(\Gamma|LO)}{\omega(\Gamma|TO)} = \frac{\varepsilon(0)}{\varepsilon(\infty)} \tag{123.45}$$

where $\varepsilon(0)$ is the static (low frequency, or "D.C.") dielectric coefficient, $\varepsilon(\infty)$ is the high frequency (optic) dielectric constant of the medium

$$\varepsilon(\infty)=n^2 \tag{123.46}$$

where n is the refractive index.

The contrast between polar optic modes which split by self-coupling and non-polar optic modes which do not split, will be illustrated below when we examine the lattice dynamics of rocksalt structures and diamond structure respectively. For a crystal with several polar optic branches, (123.44) must be generalized. But since our concern will be only with rocksalt and diamond structures we shall not discuss this generalization here, but refer the reader to recent literature [47].

For a cubic crystal with inversion symmetry the rule of mutual exclusion evidently applies, so that a polar optic infra-red active mode of symmetry $\boldsymbol{\Gamma}^{(15-)}$ will not be active in Raman scattering. According to the basic formula (120.42) along with the selection rule (120.42) infra-red absorption requires the non-vanishing of

$$\boldsymbol{A}_0\cdot\langle\chi_{\nu\bar{n}}|\,\mathcal{M}^{\nu}(\boldsymbol{R},\boldsymbol{k})\,|\chi_{\nu n}\rangle \tag{123.47}$$

where $\boldsymbol{A}_0$ is parallel to the polarization vector of the electromagnetic infra-red radiation, and perpendicular to the photon propagation direction:

$$\boldsymbol{A}_0\cdot\boldsymbol{k}=0. \tag{123.48}$$

Let us now specialize to a consideration of one-phonon processes as in the discussion given in the paragraph containing (120.51) *et seq.* But for a non-zero matrix element we require that the wave vector of the phonon created: $Q\begin{pmatrix}\boldsymbol{k}'\\ j'_{\nu'}\end{pmatrix}$ shall also equal $\boldsymbol{k}$:

$$\boldsymbol{k}'=\boldsymbol{k}. \tag{123.49}$$

Then the condition (123.48) becomes

$$\boldsymbol{A}_0\cdot\boldsymbol{k}'=0. \tag{123.50}$$

Thus $\boldsymbol{A}_0$ must be polarized perpendicular to the direction of propagation of the phonon. Consequently, when the macroscopic electric field effect is included and the phonon splitting into longitudinal and transverse is considered we can write (compare (120.48))

$$\mathcal{M}^{(1)}\left(\boldsymbol{k}\middle|\begin{matrix}\boldsymbol{k}'\\ j'\end{matrix}\right)=\mathcal{M}_T^{(1)}\left(\boldsymbol{k}\middle|\begin{matrix}\boldsymbol{k}'\\ j'\end{matrix}\right)+\mathcal{M}_L^{(1)}\left(\boldsymbol{k}\middle|\begin{matrix}\boldsymbol{k}'\\ j'\end{matrix}\right) \tag{123.51}$$

decomposing into transverse component (T) for which

$$\boldsymbol{k}'\cdot\mathcal{M}_T^{(1)}=0 \tag{123.52}$$

and longitudinal component (L) for which

$$\boldsymbol{k}'\cdot\mathcal{M}_L^{(1)}\neq 0. \tag{123.53}$$

Note this decomposition is essentially a decomposition of the eigenvector $\boldsymbol{e}\left(\kappa'\middle|\begin{matrix}\boldsymbol{k}\\ j\end{matrix}\right)$ into T and L components.

Now since $\boldsymbol{A}_0$ and $\mathcal{M}_T^{(1)}$ are both transverse to $\boldsymbol{k}'$, the direction of phonon propagation we shall have in general

$$\boldsymbol{A}_0 \cdot \langle \chi_{0\bar{n}} | \mathcal{M}_T^{(1)} \left(\boldsymbol{k} \middle| \begin{matrix} \boldsymbol{k}' \\ j \end{matrix}\right) Q \begin{pmatrix} \boldsymbol{k}' \\ \perp \end{pmatrix} |\chi_{0n}\rangle \neq 0 \tag{123.54}$$

while

$$\boldsymbol{A}_0 \cdot \langle \chi_{0\bar{n}} | \mathcal{M}_L^{(1)} \left(\boldsymbol{k} \middle| \begin{matrix} \boldsymbol{k}' \\ j \end{matrix}\right) Q \begin{pmatrix} \boldsymbol{k}' \\ \| \end{pmatrix} |\chi_{0n}\rangle = 0. \tag{123.55}$$

Hence only the *transverse* mode will be infra-red active and only the frequency $\omega(\boldsymbol{k}'|TO)$ will be observed. In the limit $\boldsymbol{k}' \to \boldsymbol{\Gamma}$ this will be

$$\omega(\boldsymbol{\Gamma}|TO) \text{ is I.R. active}. \tag{123.56}$$

For higher order infra-red processes similar results will be found involving combinations and overtones. The analysis must be carried out in the framework in which the macroscopic electric field is assumed present: thus the appropriate group to use is the intersection group (123.42). Then by analysis of the transformation properties of the operators and of the combination and overtone states, under this group one can determine whether or not a given process such as $(LO+TO)$ is active in infra-red absorption.

ii) *Cubic crystals without inversion.* Now we turn to Raman scattering effects related to the presence of the macroscopic electric field, if the crystal point group lacks inversion symmetry and there exists an optic phonon branch simultaneously infra-red and Raman active.[1] A typically important situation is that of a cubic, tetrahedral, class T_d. This case could be considered to arise from that of the class O_h by removing the inversion as a symmetry element. Then the representations $\boldsymbol{\Gamma}^{(15-)}$ *and* $\boldsymbol{\Gamma}^{(25+)}$ of O_h, both subduce $\boldsymbol{\Gamma}^{(15)}$ of T_d, and the mode $\boldsymbol{\Gamma}^{(15)}$ of T_d is simultaneously infra-red and Raman allowed. Thus one can write down the correlation table (at $\boldsymbol{k}=\boldsymbol{\Gamma}$) assembling these results

$$\begin{array}{lcl} O_h & & T_d \\ \boldsymbol{\Gamma}^{(15-)} \to LO+TO \ \text{(I.R.)} & \searrow & \\ & & \boldsymbol{\Gamma}^{(15)} \to LO \ \text{(RA)} + TO \ \text{(I.R. and RA)}. \\ \boldsymbol{\Gamma}^{(25+)} \ \text{(RA)} & \nearrow & \end{array} \tag{123.57}$$

In the first instance, one predicts that instead of a triply degenerate optic branch $\omega(\boldsymbol{\Gamma}|15-/25+)$ the Raman spectrum in a crystal with T_d point symmetry should consist of two distinct lines (in Stokes Raman scattering, for example) with frequencies $\omega(\boldsymbol{\Gamma}|LO)$, $\omega(\boldsymbol{\Gamma}|TO)$.

The coexistence of infra-red and Raman activity, i.e. absence of a center of inversion means that the crystal is piezoelectric, and this is also equivalent to the crystal showing the linear electrooptic effect (Pockels effect). Although we shall not give a comprehensive discussion of Raman scattering theory in piezoelectric crystals the basic new effect can be rather easily discussed in terms of the theory already given. We specialize to a cubic piezoelectric crystal belonging to point group T_d. For a crystal in this crystal class the electrooptic effect requires only a single coefficient for a complete description. Recall [*41*] that the electrooptic effect is a modulation of the crystal optical polarizability by an applied external electric field. But in the present context, the role of the "applied, or external" electric field is taken by the macroscopic electric field $\mathscr{E}$ accompanying the self-coupled long-wave optical polar phonon. Then, the linear electrooptic effect is the tensor coupling of the macroscopic electric field to the crystal polarizability operator. We can write this linear coupling as

$$\boldsymbol{P}(\boldsymbol{R}, \boldsymbol{k}_2 - \boldsymbol{k}_1) = [\boldsymbol{P}'(\boldsymbol{k}_2 - \boldsymbol{k}_1, \boldsymbol{k}_3)] \cdot \mathscr{E}(\boldsymbol{R}, \boldsymbol{k}_3). \tag{123.58}$$

In (123.58) we use $\mathscr{E}(\boldsymbol{R}, \boldsymbol{k}_3)$ to denote the macroscopic electric field accompanying the polar optic vibration, of wave vector $\boldsymbol{k}_3$, and

$$[\boldsymbol{P}'(\boldsymbol{k}_2 - \boldsymbol{k}_1, \boldsymbol{k}_3)] \tag{123.59}$$

denotes the tensor of third rank. But in a cubic crystal, with point symmetry T_d the third rank tensor (123.59) has only a single independent non-zero component, which we call d'

$$P'_{x,yz} = P'_{y,zx} = P'_{z,xy} = P'_{x,zy} = P'_{y,xz} = P'_{z,yx} = d'. \tag{123.60}$$

Then, from (123.58), and (123.60) we can find the appropriate polarizability component for any $\mathscr{E}$ by

$$P_{kl} = \sum P'_{m,kl} \mathscr{E}_m. \tag{123.61}$$

Comparing (123.60) and (123.21) note the similarity. In the present case however, the phonon polarization e.g. (123.30)–(123.32) will play an important role in determining the angular dependence, or polarization, of the scattered radiation. In general, the coefficient d' will differ from d. Anticipating some of the later discussion in Sect. 124 we can remark that according to present theory d represents the contribution to the scattering tensor due to deformation potential while d' represents the contribution to the scattering tensor due to electro-optic coupling, or to the Fröhlich mechanism.

It is commonly assumed that for Raman scattering by the longitudinal mode (LO), the sum of both contributions

$$d_L \equiv d + d' \tag{123.62}$$

is the appropriate coefficient while for the transverse mode TO only the deformation potential contribution applies

$$d_T = d. \tag{123.63}$$

With this assumption we can find the parallel and perpendicular polarized components of the Raman scattering for LO and TO phonons.

Consider first the LO mode, whose polarization is given in (123.30). In perpendicular scattering we find

$$\mathscr{I}_\perp^{LO} = C d_L^2 \varepsilon_{\lambda y}^2 \sin^2 \theta/2 \tag{123.64}$$

while for parallel scattering we find

$$\mathscr{I}_\parallel^{LO} = C d_L^2 \varepsilon_{\lambda z}^2 (\sin 3\theta/2)^2. \tag{123.65}$$

For the TO phonon with polarizations given by (123.31) and (123.32) we find for perpendicular scattering

$$\mathscr{I}_\perp^{TO} = C d_T^2 \varepsilon_{\lambda y}^2 \cos^2 \theta/2 \tag{123.66}$$

and for parallel scattering

$$\mathscr{I}_{\parallel}^{TO} = C\,d_T^2\,(\varepsilon_{\lambda y}^2 \sin\theta + \varepsilon_{\lambda z}^2 \cos^2 3\theta/2). \tag{123.67}$$

Comparison of (123.66) and (123.67) with (123.33), (123.34) shows the qualitative differences which arise when the macroscopic field effect is included in the calculation. The initial proposal that a macroscopic field effect should be considered in the theory of Raman scattering was due to POULET[1] who invoked this as explanation of the polarization "anomalies": i.e. the departures of polarization dependence from the predictions of (123.33), (123.34).

To summarize then, we observe that the major effects of the macroscopic electric field are to remove the three-fold degeneracy of the polar optic vibrations at $\boldsymbol{k}=\boldsymbol{\Gamma}$. In case of a cubic crystal the splitting is into a longitudinal non-degenerate single and a transverse doubly degenerate pair of modes. Consequently more distinct modes are present at $\boldsymbol{k}=\boldsymbol{\Gamma}$ than would be predicted by a group theory analyses neglecting the macroscopic field. The TO mode only will be observed in the infra-red absorption spectrum in the analysis given here. In case of an optic mode in a piezoelectric crystal which will be also Raman active one observes more frequencies than those found in the infra-red spectrum. Further, the polarization dependence of the Raman scattered light is changed from what would apply in a corresponding non-piezoelectric case.

Let us now remark that in principle there is an alternate way of doing the theory with macroscopic field included. This would consider the coupled problem of ionic lattice motion plus the electromagnetic field *a priori*. That is by considering the dynamical ionic lattice motion as being the source of the charges and currents which appear in Maxwell equations one could avoid the artifice of treating the macroscopic field as an "external" field. The entire system of ion-motion and coupled electromagnetic fields would then appear in a unified form. In such a treatment the proper symmetry group would be a group which was the union of the crystal space-time group $\mathscr{G}_{\text{cry}}$ and the space time group of the Maxwell equations $\mathscr{G}_{\text{max}}$. Then, to study the optical properties of such a system the *external* electromagnetic field would be introduced as a weak, coupled, probe. Such a complete theory appears not yet in existence, although it would be well worth the effort of constructing it. We can anticipate that such a theory would agree, in lowest order at least, with the results given in this subsection.

δ) Polarization effects and two phonon bound states. Very recently it has been proposed that a bound state of two interacting phonons can exist owing to the "many-body" nature of the interaction between phonons.[4] The basic physics in this proposal is that two phonons $Q\begin{pmatrix}\boldsymbol{k}\\ j_\nu\end{pmatrix}$ and $Q\begin{pmatrix}\boldsymbol{k}'\\ j'_{\nu'}\end{pmatrix}$ interact with each other via third order and fourth order anharmonic couplings. The total Hamiltonian is written

$$\mathscr{H} = \mathscr{H}_{\text{harm}}\left(Q\begin{pmatrix}\boldsymbol{k}\\ j_\nu\end{pmatrix}, Q\begin{pmatrix}\boldsymbol{k}'\\ j'_{\nu'}\end{pmatrix}\right) + \Phi^{(3)} + \Phi^{(4)} \tag{123.68}$$

---

[4] M. H. COHEN and J. RUVALDS: Phys. Rev. Letters **23**, 1378 (1969); — J. RUVALDS and A. ZAWADEWSKI: Phys. Rev. B**2**, 1172 (1970).

where in (123.68) the harmonic part is the usual quadratic form in $\dot{Q}\binom{\boldsymbol{k}}{j_v}$, $Q\binom{\boldsymbol{k}}{j_v}$, $\dot{Q}\binom{\boldsymbol{k}'}{j'_{v'}}$, $Q\binom{\boldsymbol{k}'}{j'_{v'}}$ and the terms $\Phi^{(3)}$ and $\Phi^{(4)}$ are anharmonic terms, cubic and quartic in normal modes, including of course the modes which will couple. It has been found that for particular pairs of modes, magnitude of coupling coefficients in $\Phi^{(3)}$ and $\Phi^{(4)}$, and dispersion of the uncoupled (harmonic) frequencies $\omega(\boldsymbol{k}|j)$, $\omega(\boldsymbol{k}'|j')$ that a two phonon bound state may be formed. We shall not go into any general details of this subject here in regard to method of proof but simply make some remarks.

The bound two phonon state is identified as an eigenstate of the special Hamiltonian (123.68) with eigenvalue

$$\omega(\underline{\bar{K}})_B > 2\omega(\boldsymbol{\Gamma}|0) \tag{123.69}$$

where $\omega(\boldsymbol{\Gamma}|0)$ is the uppermost optic phonon energy at $\boldsymbol{k}=\boldsymbol{\Gamma}$. In the normal situation the manifold of two phonon states spans a continuum (perhaps with gaps) from

$$0 < \omega < 2\omega(\boldsymbol{\Gamma}|0). \tag{123.70}$$

Hence the energy of the conjectured two phonon bound state lies above the two phonon continuum. It was found to be a sharp state. The methods of calculation used in the theoretical proof of existence of this bound state are to date only approximate so the theory should be considered provisional.

The form of the two phonon bound eigenstate which was found can be crudely written as

$$\chi(\boldsymbol{k}, \boldsymbol{k}') \sim \sum_{\boldsymbol{k}, v} \varphi(2\boldsymbol{k})\, Q\binom{\boldsymbol{k}}{j_v}\, Q\binom{-\boldsymbol{k}}{j_v}. \tag{123.71}$$

This state has total crystal momentum (wave vector) $\underline{\bar{K}} = \boldsymbol{k} - \boldsymbol{k} = 0$. The function of relative momentum $\varphi(2\boldsymbol{k})$ was taken spherically symmetric therefore transforming like $(\boldsymbol{\Gamma}1+)$ in the cubic group O_h. That is a spherically symmetric "envelope function" to use the language of effective mass theory of electrons in solids. Hence, the two phonon state (123.71) transforms as the reducible representation

$$\chi(2\boldsymbol{k}) \sim D^{(\boldsymbol{\Gamma})(1+)} \otimes [D^{(\boldsymbol{k})(j)}]_{(2)}. \tag{123.72}$$

In case of diamond, the major contribution to the proposed bound state was found to be from the overtone of the Raman active phonon $Q\binom{\boldsymbol{\Gamma}}{25+}$. Hence, this bound state in diamond transforms as

$$\chi \sim D^{(\boldsymbol{\Gamma})(1+)} \otimes [D^{(\boldsymbol{\Gamma})(25+)}]_{(2)} = D^{(\boldsymbol{\Gamma})(1+)} \oplus D^{(\boldsymbol{\Gamma})(12+)} \oplus D^{(\boldsymbol{\Gamma})(25+)}. \tag{123.73}$$

All the components in (123.73) are allowed as Raman active.

A sharp line of polarization $D^{(\boldsymbol{\Gamma})(1+)}$ and frequency $2666.9 \pm 0.5\,\mathrm{cm}^{-1}$ was observed in Raman scattering in diamond and has been identified as the $D^{(\boldsymbol{\Gamma})(1+)}$ component of the two phonon bound state since it occurs above the expected maximum in the two phonon continuum which is twice the Raman optic frequency, or $2665.0 \pm 1.0\,\mathrm{cm}^{-1}$. Other symmetry components of (123.73) have not yet been identified experimentally. Some discussion of this is given in Sect. 142.

The two phonon bound state has not yet at the time of writing been directly identified by neutron scattering or other methods. Thus it seems that the identification of a two phonon bound state should be accepted provisionally. But it is of note that the support for this proposed general mechanism was found by Raman scattering and in particular by polarization measurement, deriving directly from symmetry. We can anticipate further work on this subject to attempt to find other evidence for the bound state. See the discussion in Sect. 146 relating to some recent work in silicon and germanium where no evidence for the two phonon bound state has been found, as of the time of writing.

ε) *Polarization in infra-red absorption due to anisotropy.* In a sense the most conceptually simple source of polarization is the geometrical anisotropy which we can illustrate based upon the expressions (120.43), for infra-red absorption due to phonons.

Consider first, geometrical anisotropy present at infinite wave length. As simplest case suppose we consider a uniaxial crystal: a typical example would be a crystal with point group symmetry $C_{3v}=3m$. For such a crystal we can distinguish between parallel and perpendicular ($\|$ and $\perp$) components of a polar vector in respect to the crystalline axis. Then the vector amplitude $\boldsymbol{A}_0$ of the vector potential can be taken as $\boldsymbol{A}_0=\mathscr{A}_0\boldsymbol{\varepsilon}$, where $\mathscr{A}_0$ is a number and $\boldsymbol{\varepsilon}$ is a polarization vector. Owing to the transversality condition (120.10) we have

$$\boldsymbol{k}\cdot\boldsymbol{\varepsilon}=0 \tag{123.74}$$

where $\boldsymbol{k}$ is the propagation vector.

Now the basic selection rule for infra-red absorption is (120.43) and in it, the controlling factor is

$$\boldsymbol{A}_0\cdot\langle\chi_{\nu\bar{n}}|\,\mathscr{M}^{\nu}(\boldsymbol{R},\boldsymbol{k})\,|\chi_{\nu n}\rangle_{\boldsymbol{R}}. \tag{123.75}$$

We have proved that the tensor operator $\mathscr{M}^{\nu}(\boldsymbol{R},\boldsymbol{k})$ transforms as a polar vector $D^{(\boldsymbol{v})}$ (taking $\boldsymbol{k}=\boldsymbol{\Gamma}=(0,0,0)$). Let the unique axis of the crystal be z. Then a polar vector can be decomposed into components parallel and perpendicular to z:

$$\boldsymbol{v}=\boldsymbol{v}_{\|}+\boldsymbol{v}_{\perp} \tag{123.76}$$

and correspondingly the representation $D^{(\boldsymbol{v})}$ decomposes

$$D^{(\boldsymbol{v})}=D^{(\boldsymbol{v}_{\|})}\oplus D^{(\boldsymbol{v}_{\perp})}. \tag{123.77}$$

Consequently the operator $\mathscr{M}^{\nu}(\boldsymbol{R},\boldsymbol{k})$ also decomposes

$$\mathscr{M}^{(\nu)}=\mathscr{M}_{\|}^{(\nu)}+\mathscr{M}_{\perp}^{(\nu)}. \tag{123.78}$$

Note: the symbols ($\|$) and ($\perp$) in (123.78) refer to polarization with respect to the *crystal* axis z.

Likewise the phonon representation for single phonon infra-red processes as in (120.51) will decompose as is seen by comparing (123.77), and (120.51).

We immediately note then that geometrical anisotropy is included in the theory by writing the basic matrix element (123.75) as the sum of two terms

$$\mathscr{A}_0\boldsymbol{\varepsilon}\cdot\langle\chi_{\nu\bar{n}}|\,\mathscr{M}_{\|}^{(\nu)}(\boldsymbol{R},\boldsymbol{k})\,|\chi_{\nu n}\rangle_{\boldsymbol{R}} \tag{123.79}$$

plus

$$\mathscr{A}_0\boldsymbol{\varepsilon}\cdot\langle\chi_{\nu\bar{n}}|\,\mathscr{M}_{\perp}^{(\nu)}(\boldsymbol{R},\boldsymbol{k})\,|\chi_{\nu n}\rangle_{\boldsymbol{R}}. \tag{123.80}$$

Clearly the two basic "pure" situations are for light polarized parallel or perpendicular to the crystal z axis:

$$\varepsilon=\varepsilon_{\|} \quad \text{or} \quad \varepsilon=\varepsilon_{\perp}$$

in which the transition matrix element is either (123.79) or (123.80) alone.

Thus in case of a uniaxial crystal we can define two basic transition processes with matrix elements (123.79), (123.80). In general the transition matrix elements (123.79) and (123.80) will be unequal. Consequently, the transition rate for parallel polarized radiation: $\omega_{i\to f}(\|)$ will in general be unequal to $\omega_{i\to f}(\perp)$. This polarization will also carry over in regard to the actual phonons allowed when we make an expansion such as (120.47), and finally then to the frequency difference to be expected associated with $\|$, and $\perp$ polarized component.

For a one-phonon process, then, the intensity of absorption for radiation polarized parallel to z is given by making appropriate changes in (120.53).

$$\mathscr{I}_1(\omega)_{\|}\, d\omega \sim \sum_{k'j'} \left| \left(\frac{\hbar(n+1)}{2\omega(\boldsymbol{k}'|j')} \right)^{\frac{1}{2}} \mathscr{A}_0 \boldsymbol{\varepsilon} \cdot \mathscr{M}_{\|}^{(1)} \left(\boldsymbol{k} \middle| \begin{matrix} \boldsymbol{k}' \\ j' \end{matrix} \right) \right|^2 \delta\big(\omega-\omega(\boldsymbol{k}'|j')\big) \quad (123.81)$$

and for perpendicular polarization

$$\mathscr{I}_{\perp}(\omega)\, d\omega \sim \sum_{k'j'} \left| \left(\frac{\hbar(n+1)}{2\omega(\boldsymbol{k}'|j')} \right)^{\frac{1}{2}} \mathscr{A}_0 \boldsymbol{\varepsilon} \cdot \mathscr{M}_{\perp}^{(1)} \left(\boldsymbol{k} \middle| \begin{matrix} \boldsymbol{k}' \\ j' \end{matrix} \right) \right|^2 \delta\big(\omega-\omega(\boldsymbol{k}'|j')\big). \quad (123.82)$$

The anisotropy (polarization) of infra-red processes of higher order (multiphonon) can be obtained in just the analogous manner.

124. Aspects of modern quantum theories of lattice Raman scattering and infra-red absorption. This section is intended as an introduction to several aspects of the modern quantum theories of Raman and infra-red processes; in various places indication will be given of the role of symmetry.

In the first three Subsects. 124α–γ attention is given to analyses which are in the general many-body framework. In order to give a detailed quantitative theory of the optical properties of an anharmonic crystal many-body techniques are essential. These subsections make contact with the fuller and more thorough treatments given in the articles by COCHRAN and COWLEY [*20*] and BILZ and WEHNER [*22*] in the present Encyclopedia. A brief discussion is given in Subsect. 124γ on the connection of group theory (symmetry) analysis and the structure of the thermal Green function. The treatment of infra-red absorption and Raman scattering in this framework can be viewed as extensions of the treatments given in Sects. 120 and 121 respectively.

We shall first discuss the many-body polarizability theory of Raman scattering. This theory can be thought of as based upon the generalized Placzek theory which we presented in Sect. 121. The essential point of the many-body polarizability theory is to compute the scattering cross section of the crystal in the framework of modern response theory including all the interactions. As a physical picture one can consider as usual that the incident electromagnetic field induces a fluctuating dipole moment in the medium which reradiates the scattered field. The moment arises due to the electronic polarizability of the medium, which is itself modified by the interactions amongst the phonons due to anharmonicity. In particular there is a component of the moment at shifted frequencies corresponding to single phonon and multiphonon creation, and annihilation processes. But, owing to the interactions the individual phonons are no longer sharp (infinite lifetime) but decay into other phonons, hence the spectrum of scattered light is not merely a superposition of lines.

Similar remarks can be made regarding the many-body theory of infra-red absorption by phonons. Phonon line widths are no longer infinitely sharp, corresponding to sharpely defined energies. Absorption of radiation at any frequency ω actually involves contributions from single and multiphonon processes.

The remaining three Subsects. 124δ–ζ, are devoted to three closely related approaches to the microscopic theory of Raman scattering by phonons. The attempt is made to construct a fully microscopic theory including all the basic interactions amongst particles, or quasiparticles, as well as a quantitative evaluation of all the dynamical processes which occur involving these interactions. In all three approaches the essential microscopics is similar. Simply put, the incident external electromagnetic field creates a virtual excitation in the electronic system, the virtual excitation interacts with the lattice system producing one or more real phonons, and the virtual excitation again interacts with the radiation field to produce the scattered photon. The three pictures differ in the fashion that each of these processes is described.

In the first picture, of this group, the electronic excitation is prescribed as being due to the creation of a virtual electron-hole pair by the photon, via electron-radiation coupling. Each of these is taken in its own Bloch state. This is called the Bloch picture: electron and hole in the intermediate state move independently of one another.

In the second picture the electronic excitation is taken as creating a virtual exciton. By contrast to the first, this is called the "exciton picture". In this "exciton picture" the electron and hole in the intermediate state move in each others coulomb field, and can exist in bound or continuum exciton states.

In the final picture, the entire discussion is cast into a form involving the exact normal modes of the coupled photon-exciton system so called "polaritons", first introduced (although not named) by HUANG.[1] These are the exciton polaritons, or coulomb excitons. In this polariton picture the entire discussion involves polariton scattering processes and their ultimate relationship to the Raman cross-section.

In a certain sense the sequence of pictures: Bloch → Exciton → Polariton can be considered as a sequence in which an additional interaction is "turned on" in each stage: namely the coulomb electron-hole interaction in going from Bloch → Exciton and then the exciton-photon interaction going from Exciton → Polariton. We are referring to intermediate, virtual state character.

The point made here should be somewhat clearer after we discuss these approaches in more detail. Again we remark that we shall not give a comprehensive

[1] K. HUANG: E.R.A. Report L/T 239 (1950); – Proc. Roy. Soc. (London), Ser. A **208**, 352 (1951); and private communication to the author (May 1973).

discussion but refer the reader to the on-going and cited literature for additional details. Subsect. 124η touches on Resonance Raman scattering and symmetry breaking.

α) *Many-body polarizability theory of Raman scattering.* This theory has been discussed in other articles in this Encyclopedia,[2, 3] to which the reader is referred for details. Our objectives in this and the next subsection is to give enough detail to make clear where symmetry analysis may be applied to the calculation of the Green functions which arise, and also to indicate a point of apparent contact between this theory and the critical point analysis of Sects. 107, 120 ε, 121 ε, 140, 141, 143.

The basic formula for the Raman scattering rate is (121.45). We can rewrite this expression by calling I the intensity of Raman scattered light per unit solid angle which is proportional to the transition rate of (121.45):

$$I \sim \sum_{\lambda\lambda'\bar{\lambda}\bar{\lambda}'} \varepsilon_\lambda \varepsilon_{\lambda'} I_{\lambda\lambda'\bar{\lambda}\bar{\lambda}'} \varepsilon_{\bar{\lambda}} \varepsilon_{\bar{\lambda}'} \tag{124.1}$$

where we adopt a shortened notation and write $|\chi_{0n}\rangle \equiv |n\rangle$, etc., so that:

$$I_{\lambda\lambda'\bar{\lambda}\bar{\lambda}'} = \sum_{\bar{n}} \langle n| P_{\lambda\lambda'} |\bar{n}\rangle \langle \bar{n}| P_{\bar{\lambda}\bar{\lambda}'} |n\rangle \, \delta(\Delta\omega - \omega_{n\bar{n}}). \tag{124.2}$$

Recall that $\chi_{0\bar{n}}$ is the final vibrational state, χ_{0n} the initial vibrational state. The theory given in Sects. 120, 121 was at temperature equal to zero. To generalize to $T \neq 0$ we should include in (124.2) the appropriate thermal averaging over initial states. Then considering the energy change in scattering $\Delta\omega$ as a fixed quantity we can write the generalization of (124.2) as the thermal average of the scattering tensor

$$I_{\lambda\lambda'\bar{\lambda}\bar{\lambda}'}(\Delta\omega, T) = (1/Z) \sum_{n\bar{n}} \langle n| P_{\lambda\lambda'} |\bar{n}\rangle \langle \bar{n}| P_{\bar{\lambda}\bar{\lambda}'} |n\rangle \, e^{-\beta E_n} \delta(\Delta\omega - \omega_{n\bar{n}}) \tag{124.3}$$

where $\beta = (kT)^{-1}$ and

$$Z = \sum_n e^{-\beta E_n}. \tag{124.4}$$

The particular form (124.3) can be related to the spectral density decomposition of the thermodynamic Green function of the two operators $\hat{P}_{\lambda\lambda'}(\tau)\, \hat{P}_{\bar{\lambda}\bar{\lambda}'}$. In our previous discussion (Sect. 121) the quantities $P_{\lambda\lambda'}$ and $P_{\bar{\lambda}\bar{\lambda}'}$ were defined in (121.43) as Schrödinger (or "c") functions of ion coordinate, or displacement $\boldsymbol{R}$. At present we require that these quantities be expressed in the second quantized form.

Let the total lattice Hamiltonian of the crystal, including anharmonic effects be called $\mathscr{H}$, in second quantized form, then the required second quantized operators which correspond to $P_{\lambda\lambda'}$ and $P_{\bar{\lambda}\bar{\lambda}'}$ are given by

$$P_{\lambda\lambda'} \to \hat{P}_{\lambda\lambda'}(\tau) \equiv e^{\mathscr{H}\tau} \hat{P}_{\lambda\lambda'} e^{-\mathscr{H}\tau}, \tag{124.5}$$

$$P_{\bar{\lambda}\bar{\lambda}'} \to \hat{P}_{\bar{\lambda}\bar{\lambda}'}(0) \tag{124.6}$$

where τ is a real variable, co-responding to "imaginary time" in the usual Heisenberg picture. The second quantized operators $\hat{P}$ of (124.5), (124.6) operate on the

[2] W. Cochran and R. Cowley, in [*20*]. – H. Bilz and R.K. Wehner, in [*22*].

[3] R.A. Cowley, in [*19*]. See also A.D. Bruce and R.A. Cowley: J. Phys. C. (Solid State Physics) **5**, 595 (1972).

state vectors in a Hilbert-Fock space. The thermal Green function of these operators in the canonical ensemble[4] is defined as [20]

$$\mathscr{G}(P_{\lambda\lambda'}(\tau)\,P_{\bar\lambda\bar\lambda'}(0)) \equiv \mathscr{G}(P_{\lambda\lambda'}, P_{\bar\lambda\bar\lambda'}, \tau) \equiv \mathrm{Tr}\left(\mathrm{e}^{-\beta\mathscr{H}}\,T_\tau P_{\lambda\lambda'}(\tau)\,P_{\bar\lambda\bar\lambda'}(0)\right)/Z. \tag{124.7}$$

In (124.7) T_τ is the chronological ordering operator for τ. The thermal Green function (124.7) can be frequency Fourier analysed to give

$$\mathscr{G}(P_{\lambda\lambda'}(\tau)\,P_{\bar\lambda\bar\lambda'}(0)) = \sum_{n=-\infty}^{+\infty} \mathscr{G}(P_{\lambda\lambda'}, P_{\bar\lambda\bar\lambda'}, i\omega_n)\exp i\omega_n\tau \tag{124.8}$$

where the variable ω_n is given as:

$$\omega_n = 2\pi n/\beta; \qquad n = 0, 1, 2, \ldots. \tag{124.9}$$

The inverse of (124.8) is

$$\mathscr{G}(P_{\lambda\lambda'}, P_{\bar\lambda\bar\lambda'}, i\omega_n) = \tfrac{1}{2}\beta \int_{-\infty}^{+\infty} \mathscr{G}(P_{\lambda\lambda'}(\tau)\,P_{\bar\lambda\bar\lambda'}(0))\,\mathrm{e}^{-i\omega_n\tau}\,d\tau. \tag{124.10}$$

The thermal Green function (124.10) is given at an infinite sequence of discrete points $i\omega_n$ on the imaginary frequency axis. Continuing this function to the real axis (plus or minus an infinitesimal imaginary) we obtain by letting

$$i\omega_n \to \Omega \pm i\eta \qquad (\eta = 0^+) \tag{124.11}$$

the functions

$$\mathscr{G}(P_{\lambda\lambda'}, P_{\bar\lambda\bar\lambda'}, \Omega \pm i\eta). \tag{124.12}$$

Now, the function $\mathscr{G}(P_{\lambda\lambda'}, P_{\bar\lambda\bar\lambda'}, \Omega)$ can also be expressed in terms of the spectral density function $\rho(P_{\lambda\lambda'}, P_{\bar\lambda\bar\lambda'}, \omega)$ in the continuous variable ω. When this is done it is found that

$$\mathscr{G}(P_{\lambda\lambda'}, P_{\bar\lambda\bar\lambda'}, \Omega) = \frac{1}{\beta}\int_{-\infty}^{+\infty} (1-\mathrm{e}^{-\beta\omega})\,\rho(P_{\lambda\lambda'}, P_{\bar\lambda\bar\lambda'}, \omega)\left(\frac{d\omega}{\omega+\Omega}\right). \tag{124.13}$$

This is a certain integral transform relating the thermal Green function and its spectral density. The inverse of (124.13), from which the spectral density is obtained is:

$$\begin{aligned}\rho(P_{\lambda\lambda'}, P_{\bar\lambda\bar\lambda'}, \Omega) = {} & \frac{i\beta}{2\pi}(1-\mathrm{e}^{\beta\Omega})^{-1} \\ & \times \lim_{\eta\to 0}\,[\mathscr{G}\{(P_{\lambda\lambda'}, P_{\bar\lambda\bar\lambda'}, \Omega+i\eta) - \mathscr{G}(P_{\lambda\lambda'}, P_{\bar\lambda\bar\lambda'}, \Omega-i\eta)\}].\end{aligned} \tag{124.14}$$

Thus the spectral density is determined as the discontinuity of a thermal Green function across the real axis.

At this point it is necessary to make two remarks [20]. First is that the intensity of Raman scattering (124.3) is exactly the spectral density (taking $\Delta\omega = \Omega$):

$$I_{\lambda\lambda', \bar\lambda\bar\lambda'}(\Delta\omega, T) = \rho(P_{\lambda\lambda'}, P_{\bar\lambda\bar\lambda'}, \Delta\omega). \tag{124.15}$$

[4] A. A. Maradudin and A. E. Fein: Phys. Rev. **128**, 2589 (1962).

Secondly, the many-body graphical techniques described in [*20*] permit the calculation of just the thermal Green functions on the right hand side of (124.14). Note that other methods such as equation of motion, and functional variation techniques also may be used to obtain the thermal Green functions [*22*].

We turn to the specific question of the use of symmetry in these calculations. To answer this it is required that we examine carefully the structure of the calculation, in particular the way the frequency Fourier transformed thermal Green function $\mathscr{G}(P_{\lambda\lambda'}, P_{\bar{\lambda}\bar{\lambda}'}, i\omega_n)$ is evaluated. For definiteness we have in mind, only the diagram methods for the evaluation of the thermal Green function, but similar remarks can be made regarding the calculation of the thermal Green function by equation of motion, or functional integral methods [*22*].

As a first step we need to obtain the exact Hamiltonian $\mathscr{H}$ which appears in (124.7); for the moment consider this as a "*c*" operator. This is the total lattice Hamiltonian—all electronic effects having been incorporated into the dynamic calculation of the polarizibility tensor operator *P.* In Sect. 110 we discussed the total lattice Hamiltonian, which can be written for the present purposes as

$$\mathscr{H} = \mathscr{H}_{\text{harm}} + \mathscr{H}_A \tag{124.16}$$

where, comparing with (110.17),

$$\mathscr{H}_A \equiv V_A \tag{124.17}$$

the anharmonic part of the potential energy. One usually writes the Harmonic part of the Hamiltonian as

$$\mathscr{H}_{\text{harm}} = \mathscr{H}_0 . \tag{124.18}$$

Then as in (109.24), the Harmonic part is bilinear in the complex normal coordinates $Q\begin{pmatrix}\boldsymbol{k}\\ j_\mu\end{pmatrix}$ and $Q\begin{pmatrix}\boldsymbol{k}\\ j_\mu\end{pmatrix}^*$. The anharmonic part of the potential is the remainder: terms of higher than second degree in the complex normal coordinates.

Clearly, in principle group theory plays an essential role in determining the structure of $\mathscr{H}_A$, since only those particular combinations (products) of given degree for which all symmetry requirements are obeyed, will appear in the series for $\mathscr{H}_A$. We remind the reader particularly of Eqs. (109.69)–(109.75) *et seq.* which inform us that in each degree only these terms will appear in the series for $\mathscr{H}_A$ for which the relevent reduction coefficient is non-zero: i.e. such that a crystal invariant can be composed from the product of normal coordinates. Thus the first application of symmetry is the one familiar to us already, namely limitation of the terms in $\mathscr{H}_A$.

The next step is to obtain the polarizability tensor operator $P_{\lambda\lambda'}$. As in the paragraphs above, we restrict ourselves for the moment to considering $P_{\lambda\lambda'}$ as a "*c*" operator. Then, return to the basic definition, and the expansion of the polarizibility operator in terms of the complex normal coordinates Eq. (121.58). Each term in the expansion must transform as one of the irreducible representations contained in the representation by which a symmetric second rank tensor transforms. Again symmetry restricts the terms in (121.58), to those with non-zero reduction coefficient (121.62); and the obvious generalization for the terms of higher degree.

The remainder of the work now involves straightforward application of the techniques of many-body perturbation theory. We shall outline the steps and give some results. For details the reader is referred to the other articles in this Encyclopedia [20, 22]. The first step now, is to express the Hamiltonian (124.16) in terms of creation and annihilation operators. This is done by the substitution [5] of normal coordinate of the second kind:

$$Q\begin{pmatrix}\boldsymbol{k}\\ j_\nu\end{pmatrix}=\left(\frac{\hbar}{2\omega(\boldsymbol{k}\,|\,j)}\right)^{\frac{1}{2}}\left(a^+(-\boldsymbol{k}\,|\,j_\nu)+a(\boldsymbol{k}\,|\,j_\nu)\right) \qquad (124.19)$$
$$\equiv\left(\frac{\hbar}{2\omega(\boldsymbol{k}\,|\,j)}\right)^{\frac{1}{2}}A(\boldsymbol{k}\,|\,j)$$

where $A(\boldsymbol{k}\,|\,j)$ is defined as

$$A(\boldsymbol{k}\,|\,j_\nu)=a^+(-\boldsymbol{k}\,|\,j_\nu)+a(\boldsymbol{k}\,|\,j_\nu) \qquad (124.20)$$

(note $A^+(\boldsymbol{k}\,|\,j_\nu)=A(-\boldsymbol{k}\,|\,j_\nu)$). It turns out to be most useful to work with the second quantized $A(\boldsymbol{k}\,|\,j)$ operators rather than $a(\boldsymbol{k}\,|\,j)$. When (124.19) and (124.20) is introduced into the expression of $\mathscr{H}$ we obtain

$$\mathscr{H}_0=\sum_{\boldsymbol{k}j}\hbar\,\omega(\boldsymbol{k}\,|\,j)\,A^+(\boldsymbol{k}\,|\,j)\,A(\boldsymbol{k}\,|\,j), \qquad (124.21)$$

$$\mathscr{H}_A=\sum_{\boldsymbol{k}j\boldsymbol{k}'j'\boldsymbol{k}''j''}V^{(3)}\begin{pmatrix}\boldsymbol{k} & \boldsymbol{k}' & \boldsymbol{k}''\\ j & j' & j''\end{pmatrix}A(\boldsymbol{k}\,|\,j)\,A(\boldsymbol{k}'\,|\,j')\,A(\boldsymbol{k}''\,|\,j'') \qquad (124.22)$$
$$+\sum V^{(4)}\begin{pmatrix}\boldsymbol{k} & \boldsymbol{k}' & \boldsymbol{k}'' & \boldsymbol{k}'''\\ j & j' & j'' & j'''\end{pmatrix}A(\boldsymbol{k}\,|\,j)\,A(\boldsymbol{k}'\,|\,j')\,A(\boldsymbol{k}''\,|\,j'')\,A(\boldsymbol{k}'''\,|\,j''').$$

In (124.22) we retain only cubic plus quartic anharmonicity, as is conventional. When (124.19) and (123.20) is introduced into the expression for the polarizability operator (121.58) we have (suppressing the photon wave-vector indices $\boldsymbol{k}_1\,\boldsymbol{k}_2$)

$$\boldsymbol{P}-P^0=\sum_{\boldsymbol{k}j_\mu}P^{(1)}(\boldsymbol{k}\,|\,j_\mu)\left(\frac{\hbar}{2\omega(\boldsymbol{k}\,|\,j)}\right)^{\frac{1}{2}}A(\boldsymbol{k}\,|\,j) \qquad (124.23)$$
$$+\sum_{\boldsymbol{k}j_\mu}\sum_{\boldsymbol{k}'j'_{\mu'}}P^{(2)}(\boldsymbol{k}j\,|\,\boldsymbol{k}'j')\,A(\boldsymbol{k}\,|\,j)\,A(\boldsymbol{k}'\,|\,j')\left(\frac{\hbar^2}{4\omega(\boldsymbol{k}\,|\,j)\,\omega(\boldsymbol{k}'\,|\,j')}\right)^{\frac{1}{2}}+\cdots.$$

In order to carry out the calculation it is necessary to convert the expression (124.7) into a form suitable for diagrammatic analysis. This is accomplished by transforming all operators into the "imaginary time", or thermal interaction picture. For example the operator $\boldsymbol{A}(\boldsymbol{k}\,|\,j)$ in the interaction picture is defined as

$$\boldsymbol{A}(\boldsymbol{k}\,|\,j)(\beta)_{\mathrm{I.R.}}\equiv e^{\beta\mathscr{H}_0}A(\boldsymbol{k}\,|\,j)\,e^{-\beta\mathscr{H}_0}. \qquad (124.24)$$

This transformation permits the evaluation of (124.14). The basic technique utilized is reduction of (124.14) to a sum of thermal averages over the non-interacting canonical ensemble: i.e. in the canonical ensemble of a crystal characterized by Hamiltonian $\mathscr{H}_0$. For notational purposes, let $\hat{O}(\tau)$ be an operator which may

[5] M. BORN and K. HUANG, in [4], Eq. (38.38).

be a string of operators then define

$$\begin{aligned}\mathscr{G}_0(\hat{O},\tau) &\equiv (1/Z_0)\,\mathrm{Tr}\left(\mathrm{e}^{-\beta\mathscr{H}_0}\, T_\tau \hat{O}(\tau)\right)\\ &= (1/Z_0)\,\langle T_\tau \hat{O}(\tau)\rangle_0\end{aligned} \tag{124.25}$$

with

$$Z_0 \equiv \mathrm{Tr}\left(\mathrm{e}^{-\beta\mathscr{H}_0}\right). \tag{124.26}$$

Then, the thermal Green function (124.7) can be shown to be given by

$$\begin{aligned}&\mathscr{G}(P_{\lambda\lambda'}, P_{\bar{\lambda}\bar{\lambda}'}, \tau)\\ &= \sum_{n=0}^{\infty} \frac{(-1)^n}{n!} \left\langle T_\tau P_{\lambda\lambda'}(\tau)\, P_{\bar{\lambda}\bar{\lambda}}(0) \int_0^\beta d\tau_1 \ldots \int_0^\beta d\tau_n\, \mathscr{H}_A(\tau_1)\ldots\mathscr{H}_A(\tau_n)\right\rangle_{0c}\end{aligned} \tag{124.27}$$

where all operators in (124.27) are in the imaginary time picture. The symbol $\langle\ldots\rangle_{0c}$ means that one is to evaluate the operators within the bracket in the canonical ensemble of the harmonic Hamiltonian $\mathscr{H}_0$ and one is to retain only the connected diagrams.[4] Disconnected diagrams cancel the partition function Z. As an alternate and equivalent condensed mode of writing (124.27) we may put

$$\mathscr{G}(P_{\lambda\lambda'}, P_{\bar{\lambda}\bar{\lambda}'}, \tau) = \langle T_\tau\, \mathrm{e}^{-\int_0^\beta \mathscr{H}_A(\tau')\,d\tau'}\, P_{\lambda\lambda'}(\tau)\, P_{\bar{\lambda}\bar{\lambda}'}(0)\rangle_{0c}. \tag{124.28}$$

Now the structure of the theory should be evident from (124.28) and (124.23). Using (124.23) in (124.28), the thermal Green function can be expressed as a sum of terms in which each term involves one factor from $P_{\lambda\lambda'}(\tau)$ and one factor from $P_{\bar{\lambda}\bar{\lambda}'}(0)$. Thus

$$\begin{aligned}\mathscr{G}(P_{\lambda\lambda'}, P_{\bar{\lambda}\bar{\lambda}'}, \tau) &= \sum_{\boldsymbol{k} j\mu}\sum_{\boldsymbol{k}' j'\mu'} P^{(1)}(\boldsymbol{k}|j_\mu)\, P^{(1)}(\boldsymbol{k}'|j'_{\mu'}) \left(\frac{\hbar^2}{4\omega(\boldsymbol{k}\}\,\omega(\boldsymbol{k}'|j')}\right)^{\frac{1}{2}}\\ &\quad\times \langle T_\tau\, \mathrm{e}^{-\int_0^\beta \mathscr{H}_A(\tau')\,d\tau'}\, A(\boldsymbol{k}|j)(\tau)\, A^+(\boldsymbol{k}'|j')(0)\rangle_{0c} + \cdots.\end{aligned} \tag{124.29}$$

Consequently the thermal Green function (124.28) is expressible in a sum of: thermal one-phonon Green function, plus thermal two-phonon Green function, etc. The basic quantities are then the one and multiple phonon Green functions.[6-8]

Consider for the moment the one-phonon thermal Green function:

$$\mathscr{G}\left(A(\boldsymbol{k}|j), A^+(\boldsymbol{k}'|j'), \tau\right) = \left\langle T_\tau \left(\exp - \int_0^\beta \mathscr{H}_A(\tau')\,d\tau'\right) A(\boldsymbol{k}|j)(\tau)\, A^+(\boldsymbol{k}'|j')(0)\right\rangle_{0c}. \tag{124.30}$$

The lowest order approximation to (124.30) is the unperturbed Green function[4] which is diagonal in $\boldsymbol{k}, \boldsymbol{k}'$ and $j j'$,

$$\langle T_\tau A(\boldsymbol{k}|j)(\tau)\, A^+(\boldsymbol{k}|j)(0)\rangle_{0c} = n(\boldsymbol{k}|j)\, \mathrm{e}^{|\tau|\,\omega(\boldsymbol{k}|j)} + \left(n(\boldsymbol{k}|j)+1\right) \mathrm{e}^{-|\tau|\,\omega(\boldsymbol{k}|j)}. \tag{124.31}$$

The next approximation to (124.30) involves including the term linear in $\mathscr{H}_A(\tau_1)$, and higher approximations involve still higher powers of $\mathscr{H}_A(\tau')$ as in (124.27) with $n > 1$.

---

[6] W. Cochran and R. Cowley, in [20]. – H. Bilz and R. K. Wehner, in [22].
[7] R. A. Cowley: Proc. Phys. Soc. (London) **84**, 281 (1964); – J. Phys. Radium **26**, 659 (1965). – In [20].
[8] R. A. Cowley: J. Phys. Radium **26**, 659 (1965), Eq. (3.9).

In general, to determine the one-phonon thermal Green function requires solving a Dyson equation. This equation has been discussed elsewhere[9-11] and it has been remarked that owing to the anharmonicity modes labels j and j' need not be the same in (124.30) so that mode coupling occurs in general. However, because the crystal is translationally invariant, $\mathscr{G}(A(\boldsymbol{k}|j), A^{+}(\boldsymbol{k}'|j'), \tau)$ must be diagonal in $\boldsymbol{k}$. Consequently we can omit the prime in the wave vector of the second operator. The question of mode coupling is sufficiently important that we discuss it briefly from point of view of symmetry below in Subsect. 124γ.

For present purposes it suffices to cite the expressions[7] for the scattering intensity (124.2) and (124.15), in the approximation in which we consider only the one-phonon contributions, as in the first term in (124.29), and also we only retain the linear correction in (124.30). Then neglecting mode coupling ($j=j'$), and working at infinite wave-length ($\boldsymbol{k}=0$)

$$\mathscr{I}_{\lambda\lambda',\lambda\lambda'}(\Delta\omega)=\left(\frac{\exp\beta\Delta\omega-1}{2\pi}\right)^{-1} \times\sum_j \frac{4P^*(0j)\,P(0j)\,\omega(0j)^2\,\Gamma(0jj,\Delta\omega)}{\{\omega(0j)^2-(\Delta\omega)^2+2\omega(0j)\,\Delta(0jj,\Delta\omega)\}^2+4\omega(0j)^2\,\Gamma(0jj,\Delta\omega)^2} \tag{124.32}$$

where $\Gamma(0jj,\Delta\omega)$ is a lifetime (including a factor accounting for energy conservation) and $\Delta(0jj,\Delta\omega)$ is a self-energy shift. These quantities are

$$\begin{aligned}\Delta(0jj_1x)=12\sum_{\boldsymbol{k}_1 j_1} V\begin{pmatrix}0 & 0 & \boldsymbol{k}_1 & -\boldsymbol{k}_1\\ j & j & j_1 & j_1\end{pmatrix}(2\bar{n}_1+1)+2\sum_{\alpha\beta} V_{\alpha\beta}\begin{pmatrix}0 & 0\\ j & j\end{pmatrix} n^T_{\alpha\beta}\\ -18\sum_{\boldsymbol{k} j_1 j_2} V\begin{pmatrix}0 & \boldsymbol{k} & -\boldsymbol{k}\\ j & j_1 & j_2\end{pmatrix} V\begin{pmatrix}0 & -\boldsymbol{k} & \boldsymbol{k}\\ 0 & j_1 & j_2\end{pmatrix} R(X)\end{aligned} \tag{124.33}$$

where $U^T_{\alpha\beta}$ arises from thermal strain, and

$$\Gamma(0jj,x)=18\sum_{\boldsymbol{k} j_1 j_2} V\begin{pmatrix}0 & \boldsymbol{k} & -\boldsymbol{k}\\ j & j_1 & j_2\end{pmatrix} V\begin{pmatrix}0 & \boldsymbol{k} & -\boldsymbol{k}\\ j & j_2 & j_1\end{pmatrix} S(X). \tag{124.34}$$

In (124.33), (124.34)

$$\bar{n}_1\equiv[\exp\beta\,\omega(\boldsymbol{k}_1|j_1)-1]^{-1} \tag{124.35}$$

is the occupation number for normal mode $(\boldsymbol{k}_1|j_1)$ and $R(X)$ and $S(X)$ are real and imaginary parts of the function $\dfrac{A}{(X\pm i\eta)}$. They are explicitly given as

$$\begin{aligned}R(X)=&\frac{\bar{n}_1+\bar{n}_2+1}{(\omega_1+\omega_2+X)_P}+\frac{\bar{n}_1+\bar{n}_2+1}{(\omega_1+\omega_2-X)_P}\\ &+\frac{\bar{n}_2-\bar{n}_1}{(\omega_1-\omega_2-X)_P}+\frac{\bar{n}_1-\bar{n}_2}{(\omega_2-\omega_1-X)_P}\end{aligned} \tag{124.36}$$

---

[9] R.F. Wallis, I.P. Ipatova, and A.A. Maradudin: Soviet Phys. Solid State **8**, 850 (1966). – I.P. Ipatova, A.A. Maradudin, and R.F. Wallis: Phys. Rev. **155**, 882 (1967). – L.E. Gurevich and I.P. Ipatova: Soviet Phys. Solid State **4**, 1513 (1963); – Soviet Phys. JETP **18**, 162 (1964).

[10] A.A. Abrikosov: Soviet Phys. JETP **16**, 765 (1963).

[11] A.A. Maradudin and I.P. Ipatova: Tech. Report U. of California, Irvine, March 1966; – J. Math. Phys. **9**, 525 (1968).

and

$$S(X) = -\pi(\bar{n}_1 + \bar{n}_2 + 1)\big(\delta(\omega_1 + \omega_2 + X) - \delta(X - \omega_1 - \omega_2)\big) + \pi(\bar{n}_1 - \bar{n}_2)\big(\delta(\omega_1 - \omega_2 + X) - \delta(X - \omega_1 + \omega_2)\big). \tag{124.37}$$

In (124.32) the expression under the summation (apart from the expansion coefficients $P^{(1)}(0\,j)$) is the spectral density of the one-phonon thermal Green function.

In crystals such as rocksalt structure (Sect. 138) symmetry requires all first order coefficients $P^{(1)}(0\,|\,j)$ to vanish, since the $\boldsymbol{k} = \boldsymbol{\Gamma}$ phonons do not have symmetry of a second rank tensor. Consequently the first non-zero contribution to the thermal Green function, and hence to the scattering must arise from the second term (on the right hand side of (124.23), bilinear in the operators $A(\boldsymbol{k}\,|\,j)$). But this requires a two phonon Green function, since each factor gives a bilinear term. A great variety of terms then contribute to the scattering. A contribution to the scattering can be isolated of form

$$\mathscr{I}_{\lambda\lambda'\bar{\lambda}\bar{\lambda}'}(\Delta\omega) \sim (\exp\beta\Delta\omega - 1)^{-1} \sum_{\boldsymbol{k}jj'} P^{(2)}(\boldsymbol{k}\,j\,j')^* \, P^{(2)}(\boldsymbol{k}\,j\,j')\, S(\Delta\omega). \tag{124.38}$$

This term corresponds to second order, or two phonon, harmonic Raman scattering processes, such as given in (121.68). In general, at $T \neq 0$ of course $S(\Delta\omega)$ is a complicated function of $\Delta\omega$, as well as the mode indices, but if $T=0$ the term $S(\Delta\omega)$ is approximately

$$S(\Delta\omega) \sim \rho_L(\Delta\omega) \tag{124.39}$$

that is, it reproduces the joint density of states in (121.68). However the actual point of contact between the many-body theory given here and the critical point theory earlier presented is evidently via (124.38).

Regarding the two phonon region, COWLEY[12] has examined certain diagrams in detail and he observes that there can be, in certain cases a contradiction between conditions needed to observe a two phonon process, and to produce a sharp two phonon critical point. He concludes that the two phonon contribution will be large when the modes have large lifetimes; it will then be peaked about frequency transfer

$$\Delta\omega = \omega(\boldsymbol{k}\,|\,j) + 2\omega(\boldsymbol{k}\,|\,j)\,\Delta\big(\boldsymbol{k}\,j, \omega(\boldsymbol{k}\,j)\big) + \omega(-\boldsymbol{k}\,|\,j') + 2\omega(-\boldsymbol{k}\,|\,j')\,\Delta\big(-\boldsymbol{k}, j', \omega(-\boldsymbol{k}\,j')\big) \tag{124.40}$$

i.e. around the harmonic two phonon peak, but shifted by self-energy effects. The peak width will be given by

$$\Gamma\big(\boldsymbol{k}\,j, \omega(\boldsymbol{k}\,j)\big) + \Gamma\big(-\boldsymbol{k}\,j', \omega(-\boldsymbol{k}\,j')\big). \tag{124.41}$$

According to this, large anharmonicity is required in order to observe the two phonon peak, which will, as a consequence, be broad. But if the harmonic frequency is broadened, the critical point behavior may be lost. These arguments appear not to have been yet been properly and completely resolved. The quantitative comparisons needed have not been carried out to really determine, for given

[12] R.A. COWLEY: J. di Physique (Paris) **26**, 659 (1965).

materials, when the phonon lifetime effects will eliminate the sharp critical point behavior.

To date the major experimental evidence as will be seen in Sects. 142, 145 seems to indicate that critical points can be observed in two phonon Raman scattering in many materials with sufficiently great experimental resolution. Thus, in the cases where both inelastic neutron scattering data, and optical Raman scattering are available, very good agreement exists. The weight of evidence so far favors use of the critical point method but the very important questions raised by the many-body theory evidently call for resolution in order to determine in detail the quantitative limits of validity of the critical point method.

From the structure of (124.28) and the resulting transform to obtain the spectral density, we can conclude that in general the contributions to $\mathscr{I}_{\lambda\lambda',\bar{\lambda}\bar{\lambda}'}(\Delta\omega)$ for fixed $(\Delta\omega)$ occur from a sum of terms including linear, bilinear and higher order terms. Thus for fixed energy shift $\Delta\omega=\omega_1-\omega_2$, all orders contribute. Physically speaking, the anharmonicity couples all phonons so even a nominally "single phonon" event in Raman scattering must include contributions from many-phonon interactions.

Unfortunately, direct comparison of the results of the many-body polarizability theory of Raman scattering with experiment have been made difficult owing to the need to introduce actual lattice dynamic models for computation of the coefficients in the expansion of V_A and of $P^{(1)}, \ldots$ into normal coordinates. Thus an added unknown enters namely the model parameters. Quantitative comparison of theory and experiment for line shapes, and various temperature dependences are continuing at the time of writing.

β) *Many-body polarizability theory of infra-red absorption.* The theory of infra-red absorption by phonons has also been discussed in the many-body framework in other articles in this Encyclopedia.[13–15]

The basic formula for the infra-red absorption rate is (120.43) which gives the transition rate for the transition from an initial to a final state. Again adopting the shortened notation of (124.2) we can write (120.43) as

$$\omega_{i\to f}\sim I_{\mathrm{I.R.}}(\omega)\sim\sum_{\lambda\lambda'} A_0^2\,\varepsilon_\lambda\,\mathscr{I}_{\lambda\lambda'}\,\varepsilon_{\lambda'} \tag{124.42}$$

where we write

$$\mathscr{I}_{\lambda\lambda'}\equiv\sum_{\bar{n}}\langle n|\mathscr{M}_\lambda|\bar{n}\rangle\langle\bar{n}|\mathscr{M}_{\lambda'}|n\rangle\,\delta(\omega-\omega_{\bar{n}n}). \tag{124.43}$$

Here again we drop all reference to the electronic state index v. Since the initial and final states belong to the same electronic manifold, only the difference in vibrational energies appears in (124.43). The formula (124.43) is only applicable at $T=0$. To generalize to $T\neq 0$ we require a thermal average over the initial states. Then the expression (124.43) becomes replaced by the object

$$\mathscr{I}_{\lambda\lambda'}(\omega,T)=\frac{1}{Z}\sum_{n\bar{n}}\left(\langle n|\mathscr{M}_\lambda|\bar{n}\rangle\langle\bar{n}|\mathscr{M}_{\lambda'}|n\rangle\,\mathrm{e}^{-\beta E_n}\,\delta(\omega-\omega_{n\bar{n}})\right) \tag{124.44}$$

---

[13] W. COCHRAN and R. COWLEY, in [*20*], see also A.D. BRUCE: J. Phys. C. (Solid State Physics) **6**, 174 (1973).

[14] H. BILZ and R.K. WEHNER, in [*22*].

[15] H. BILZ, in [*19*].

where

$$\beta=(kT)^{-1} \tag{124.45}$$

and

$$Z=\sum_n \mathrm{e}^{-\beta E_n}. \tag{124.46}$$

The quantity defined in (124.44) is proportional to the transition rate, or absorption coefficient for the absorption of radiation of frequency ω at temperature T, producing a change in vibrational state from n to $\bar{n}$. At this point it is clear, comparing (124.44) with (124.3) that the quantities: $\mathscr{I}_{\lambda\lambda'}(\omega, T)$ for infra-red absorption at frequency ω, and $\mathscr{I}_{\lambda\lambda'\bar{\lambda}\bar{\lambda}'}(\Delta\omega, T)$ for Raman scattering (Stokes shift $\Delta\omega$) have the same formal structure. Consequently $\mathscr{I}_{\lambda\lambda'}(\omega, T)$ is also a spectral density of a thermodynamic Green function.

We define the thermal Green function of the electric dipole moment operator as (c.f. (124.7))

$$\mathscr{G}(\hat{M}_\lambda, \hat{M}_{\lambda'}, \tau)=\frac{1}{Z}\operatorname{Tr}\left(\mathrm{e}^{-\beta\mathscr{H}} T_\tau \hat{M}_\lambda(\tau)\hat{M}_{\lambda'}(0)\right). \tag{124.47}$$

In view of the formal similarity of (124.47) to (124.6) we can then take over the rest of the analysis, merely making the formal operator substitution of $\mathscr{M}_\lambda$ for $P_{\lambda\lambda'}$ throughout. In particular then the spectral density $\mathscr{I}_{\lambda\lambda'}(\omega, T)$ is given as

$$\begin{aligned}\mathscr{I}_{\lambda\lambda'}(\omega, T)&\sim\rho(M_\lambda, M_{\lambda'}, \omega)\\ &=\left(\frac{i\beta}{2\pi}\right)(1-\mathrm{e}^{\beta\omega})^{-1}\lim_{\eta\to 0^+}\{\mathscr{G}(M_\lambda, M_{\lambda'}, \omega+i\eta)-\mathscr{G}(M_\lambda, M_{\lambda'}, \omega-i\eta)\}.\end{aligned} \tag{124.48}$$

In order to evaluate (124.48) it is necessary to put the expressions into form suitable for diagrammatic analysis. This is done, as indicated in the previous subsection, obtaining the explicit expression for the anharmonic Hamiltonian $\mathscr{H}_A$, determined by crystal structure and symmetry as in (124.21) and (124.22). Then, the electric moment operator needs expansion as in Sect. 110. The expansion which corresponds to (124.23) is then as in (110.12) *et seq.*:

$$\mathscr{M}=\sum_{kj\mu}\mathscr{M}^{(1)}\begin{pmatrix}\boldsymbol{k}\\ j_\mu\end{pmatrix}A(\boldsymbol{k}\,|\,j_\mu)+\sum_{kj\mu}\sum_{k'j'\mu'}\mathscr{M}^{(2)}\begin{pmatrix}\boldsymbol{k} & \boldsymbol{k}'\\ j_\mu & j'_{\mu'}\end{pmatrix}A(\boldsymbol{k}\,|\,j_\mu)\,A(\boldsymbol{k}'\,|\,j'_{\mu'})+\cdots. \tag{124.49}$$

It is to be emphasized that the entire term (124.49) is symmetry determined in a given crystal structure. Since $\mathscr{M}$ is a polar vector only those terms with the symmetry of a polar vector enter. This type of restriction is of course familiar to us from previous analyses.

We can repeat, almost verbatim the work of (124.23)–(124.32). Clearly in a crystal with non-vanishing first order electric moment $\mathscr{M}^{(1)}(\boldsymbol{k}\,|\,j_\mu)$ the lowest order contributions to the Green function (124.47) will occur from the one-phonon thermal Green function. Again it needs to be emphasized that for fixed ω (i.e. fixed external frequency of the electromagnetic field) there will be contributions to $\mathscr{I}_{\lambda\lambda'}(\omega)$ from all terms in the series obtained by substituting (124.49) into (124.47). Thus, in principle there is no strict separation of contributions into one-phonon, two-phonon, ... processes since anharmonic interaction mixes them all.

In dealing with infra-red absorption from a many-body point of view it is more conventional to calculate the dielectric susceptibility rather than the absorption rate. Thus rather than evaluate the analogous expression to (124.32) but with $M^{(1)}(0\,j)$ replacing $P^{(1)}(0\,j)$, it is more usual to calculate the susceptibility, and then the absorption coefficient as the imaginary part of the susceptibility. This calculation is discussed in the cited articles [*20, 22*]. The susceptibility is given as

$$\chi_{\lambda\lambda'}(\omega)=\beta \underset{\eta\to 0^+}{\mathcal{G}}(M_\lambda, M_{\lambda'}, \omega+i\eta) \tag{124.50}$$

(compare (124.48)), and when evaluated in the same approximation as gave rise to (124.32) one obtains for the "one phonon" contribution:

$$\chi_{\lambda\lambda'}(\omega)=\sum_j \frac{2\omega(0\,j)\,M_\lambda(0\,j)^*\,M_{\lambda'}(0\,j)}{\left[\omega^2(0\,j)-\omega^2+2\omega(0\,j)\big(\Delta(0\,j\,j|\omega)-i\Gamma(0\,j\,j|\omega)\big)\right]}. \tag{124.51}$$

Again $\Delta(0\,j\,j|\omega)$ is the self-energy shift due to anharmonicity, and $\Gamma(0\,j\,j|\omega)$ is the lifetime.

A second-order (two phonon) contribution to the susceptibility which can be correlated with the critical point analysis can be obtained from the various diagrams contributing. It is of the form

$$\chi_{\lambda\lambda'}(\omega)\sim\sum_{\boldsymbol{k}jj'} M_\lambda^{(2)}\begin{pmatrix}\boldsymbol{k} & \boldsymbol{k}'\\ j & j'\end{pmatrix} M_{\lambda'}^{(2)}\begin{pmatrix}\boldsymbol{k} & -\boldsymbol{k}\\ j & j'\end{pmatrix} S(\omega). \tag{124.52}$$

As for Raman scattering, it can be shown that at low temperatures, and small anharmonicity, there is a close connection between this term (124.52) and the critical point analyses of multiphonon processes. The absorption coefficient (or transition rate) is proportional to $\operatorname{Im} S(\omega)$, and

$$\operatorname{Im} S(\omega)\sim\rho_2\big(\omega(\boldsymbol{k}\,|\,j)+\omega(-\boldsymbol{k}\,|\,j')\big). \tag{124.53}$$

Comparing with (120.60) one observes that the critical point theory arises from the terms which produced (124.53).

Again it should be remarked that a unified theory demonstrating the limit of validity of critical point analysis in presence of anharmonicity seems lacking. But in Part M comparison with experiment will be given, which shows good agreement in many cases with critical point theory.

γ) *Group theory and the thermal phonon Green functions.* From the discussion given in previous subsections, it is evident that the thermal phonon Green functions play an essential role in determining the response of a crystal to external influences. In the present situation let us consider the one phonon thermal Green function for a crystal which admits a space-symmetry group $\mathfrak{G}$. We ignore the antiunitary operations. Although we consider here only the one-phonon thermal Green function, the arguments used can be generalized to the multiphonon case, although this does not seem yet to have been done.

In the canonical ensemble of the interacting (anharmonic) system, the one phonon thermal Green function is defined as

$$\mathcal{G}\big(A(\boldsymbol{k}\,|\,j), A^+(\boldsymbol{k}'\,|\,j'), \tau\big)=\operatorname{Tr}\big(e^{-\beta\mathcal{H}}\,T_\tau\,A(\boldsymbol{k}\,|\,j)(\tau)\,A^+(\boldsymbol{k}'\,|\,j')(0)\big)/Z \tag{124.54}$$

with

$$Z = \operatorname{Tr} e^{-\beta \hat{\mathscr{H}}}. \tag{124.55}$$

Owing to translational invariance, all terms in $\hat{\mathscr{H}}$ conserve total momentum. Consequently if $\hat{P}$ is the total momentum operator

$$[\hat{\mathscr{H}}, \hat{P}]_- = 0. \tag{124.56}$$

Hence it follows that (124.54) must be diagonal in $\boldsymbol{k}$ or $\boldsymbol{k} = \boldsymbol{k}'$.

We then address ourselves to the matter of symmetry restrictions imposed upon the mode indices j and j', for (124.54) to be a non-vanishing. Recall that for each value of $\boldsymbol{k}$, the index j refers to an allowable irreducible representation of $\mathfrak{G}(\boldsymbol{k})$. It should also be recalled that the classical normal coordinates $Q\begin{pmatrix}\boldsymbol{k}\\ j_\mu\end{pmatrix}$ transform under operations in $\mathfrak{G}(\boldsymbol{k})$ like a partner of the μ-th row in allowable irreducible representation $D^{(\boldsymbol{k})(j)}$ of $\mathfrak{G}(\boldsymbol{k})$. In going over to the second quantized representation we assume that the transformation behavior of the operators $A(\boldsymbol{k}\,|\,j_\mu)$ is just the same as that of the classical $Q\begin{pmatrix}\boldsymbol{k}\\ j_\mu\end{pmatrix}$ from which they were derived. Recall the discussion of this matter and certain unresolved questions related to it, in Sect. 102. Consequently we can use here (86.28), (86.30), (101.1), (109.24)–(109.35), for $A(\boldsymbol{k}\,|\,j_\mu)$ and $\mathscr{H}$ respectively. Making the necessary changes to account for the fact that we are transforming operators, we can discuss the effect of symmetry transformation by a space group operator on the factors in (124.54).

Let $P_{\{\boldsymbol{\varphi}|\boldsymbol{t}\}}$ be an operator in the space group $\mathfrak{G}$. Then under transformation of $A(\boldsymbol{k}_\sigma\,|\,j_\mu)$ by this operator we have

$$P_{\{\boldsymbol{\varphi}|\boldsymbol{t}\}}^{-1}\, A(\boldsymbol{k}_\sigma\,|\,j_\mu)\, P_{\{\boldsymbol{\varphi}|\boldsymbol{t}\}} = \sum_{\nu\sigma'} D^{(\star\boldsymbol{k})(j)}(\{\boldsymbol{\varphi}\,|\,\boldsymbol{t}\})_{\nu\sigma',\mu\sigma}\, A(\boldsymbol{k}_{\sigma'}\,|\,j_\nu). \tag{124.57}$$

This just carries over to the operator $A(\boldsymbol{k}_\sigma\,|\,j_\mu)$ the rule for transforming $Q\begin{pmatrix}\boldsymbol{k}_\sigma\\ j_\mu\end{pmatrix}$.

Since $P_{\{\boldsymbol{\varphi}|\boldsymbol{t}\}}$ is defined to be an operation of symmetry of the total Hamiltonian we have

$$P_{\{\boldsymbol{\varphi}|\boldsymbol{t}\}}^{-1}\, \hat{\mathscr{H}}\, P_{\{\boldsymbol{\varphi}|\boldsymbol{t}\}} = \hat{\mathscr{H}}. \tag{124.58}$$

Then consider transformation of the thermal Green function given in (124.54) by the crystal symmetry operator $P_{\{\boldsymbol{\varphi}|\boldsymbol{t}\}}$. Since $P_{\{\boldsymbol{\varphi}|\boldsymbol{t}\}}$ is a linear unitary operator we have, ignoring the T_τ operator:

$$\begin{aligned}
&\operatorname{Tr}\big(e^{-\beta\hat{\mathscr{H}}} A(\boldsymbol{k}_\sigma\,|\,j_\mu)(\tau)\, A^+(\boldsymbol{k}_\sigma\,|\,j'_{\mu'})(0)\big)\\
&\quad= \operatorname{Tr}\big(P_{\{\boldsymbol{\varphi}|\boldsymbol{t}\}}^{-1}\, e^{-\beta\hat{\mathscr{H}}} A(\boldsymbol{k}_\sigma\,|\,j_\mu)(\tau)\, A^+(\boldsymbol{k}_\sigma\,|\,j'_{\mu'})(0)\, P_{\{\boldsymbol{\varphi}|\boldsymbol{t}\}}\big) \qquad (124.59)\\
&\quad= \operatorname{Tr}\big(P_{\{\boldsymbol{\varphi}|\boldsymbol{t}\}}^{-1}\, e^{-\beta\hat{\mathscr{H}}} P_{\{\boldsymbol{\varphi}|\boldsymbol{t}\}}\, P_{\{\boldsymbol{\varphi}|\boldsymbol{t}\}}^{-1}\, A(\boldsymbol{k}_\sigma\,|\,j_\mu)(\tau)\, P_{\{\boldsymbol{\varphi}|\boldsymbol{t}\}}\, P_{\{\boldsymbol{\varphi}|\boldsymbol{t}\}}^{-1}\, A^+(\boldsymbol{k}_\sigma\,|\,j'_{\mu'})(0)\, P_{\{\boldsymbol{\varphi}|\boldsymbol{t}\}}\big)\\
&\quad= \sum_{\nu\sigma'}\sum_{\bar{\nu}\bar{\sigma}'} D^{(\star\boldsymbol{k})(j)}(\{\boldsymbol{\varphi}\,|\,\boldsymbol{t}\})_{\nu\sigma',\mu\sigma}\, D^{(\star\boldsymbol{k})(j')}(\{\boldsymbol{\varphi}\,|\,\boldsymbol{t}\})^*_{\bar{\nu}\bar{\sigma}',\mu'\sigma}\\
&\qquad\times \operatorname{Tr}\big(e^{-\beta\hat{\mathscr{H}}} A(\boldsymbol{k}_{\sigma'}\,|\,j_\nu)(\tau)\, A^+(\boldsymbol{k}_{\sigma'}\,|\,j_{\bar{\nu}})(0)\big). \qquad (124.60)
\end{aligned}$$

It thus follows that under transformation by operator $P_{\{\varphi|\tau\}}$ the one phonon thermal Green function transforms like a basis for the direct product of the representations[10,11] by which the operator factors transform. Symbolically

$$\mathcal{G}\big(A(\boldsymbol{k}\,|\,j), A^{+}(\boldsymbol{k}'\,|\,j'), \tau\big) \sim D^{(\star\boldsymbol{k})(j)} \otimes D^{(\star\boldsymbol{k}')(j')*}. \tag{124.61}$$

Of course since $A(\boldsymbol{k}|j)$ and $A(\boldsymbol{k}'|j')$ are basically normal coordinates they must transform like the representation from which they were derived.

In (124.59) the trace operators is a form of scalar product, and $P_{\{\varphi|t\}}$ is a "rotation" so the Green function must be an invariant.

$$\mathcal{G}\big(A(\boldsymbol{k}\,|\,j), A^{+}(\boldsymbol{k}'\,|\,j'), \tau\big) \sim D^{(\Gamma)(1+)}. \tag{124.62}$$

From (124.61) and (124.62) we see that if the reduction coefficient

$$(\star\boldsymbol{k}\,j\,\star\boldsymbol{k}'\,j'\,|\,\Gamma(1+)) = 1$$

then

$$\mathcal{G}\big(A(\boldsymbol{k}\,|\,j), A^{+}(\boldsymbol{k}'\,|\,j'), \tau\big) \neq 0. \tag{124.63}$$

As in the discussion of selection rules for two phonon processes we see that (124.62) can only be satisfied for $\boldsymbol{k}=\boldsymbol{k}'$ as previously assumed from translation invariance. Further we recall from the discussion in Sect. 109 that a bilinear invariant Hermitian quadratic form can only be formed from partners which transform according to the same row of the same irreducible representation. Thus, also $j=j'$ and the only non-zero terms will be diagonal of form

$$\mathcal{G}\big(A(\boldsymbol{k}\,|\,j_\mu), A^{+}(\boldsymbol{k}\,|\,j_\mu), \tau\big). \tag{124.64}$$

The case of "mode mixing" occurs if the reduction coefficient in (124.63) is greater than 1. That is for example if

$$(\star\boldsymbol{k}\,j\,\star\boldsymbol{k}\,j'\,|\,\Gamma(1+)) = 2 \tag{124.65}$$

then there will be *two* non-vanishing thermal one phonon Green functions. This can only occur if the same irreducible representation $D^{(\star\boldsymbol{k})(j)}$, or $D^{(\boldsymbol{k})(j)}$ occurs twice among the phonons at $\star\boldsymbol{k}$ or $\boldsymbol{k}$. Then, just as for the analogous case in the two-phonon selection rules there will be two linearly independent linear combinations of bilinear expressions in the operators $A(\boldsymbol{k}|j_\mu)$ which will produce an object transforming as $D^{(\Gamma)(1+)}$. Then, to find the correct linear combinations requires solution of a 2×2 equation, in this case.

To summarize, the thermal one-phonon Green function can involve mode mixing for any operators which belong to the same row of the same irreducible representation. There will be as many distinct non-zero thermal one-phonon Green functions at given $\boldsymbol{k}$ as the multiplicity of a given acceptable irreducible representation of $\mathfrak{G}(\boldsymbol{k})$. If the multiplicity is two or more then the thermal one-phonon Green function satisfies a matrix Dyson equation, given in terms of frequency Fourier transform as

$$\begin{aligned}\mathcal{G}\big(A(\boldsymbol{k}\,|\,j_\mu), A^{+}(\boldsymbol{k}\,|\,j'_{\mu'}), i\,\omega_n\big) &= \mathcal{G}_0\big(A(\boldsymbol{k}\,|\,j'_\mu), A^{+}(\boldsymbol{k}\,|\,j_\mu), i\,\omega_n\big)\,\delta_{jj'}\,\delta_{\mu\mu'}\\ &-\beta\sum_{j''_{\mu''}} \mathcal{G}_0\big(A(\boldsymbol{k}\,|\,j_\mu), A^{+}(\boldsymbol{k}\,|\,j_\mu), i\omega_n\big)\,\Sigma(\boldsymbol{k}\,j j''\,|\,i\omega_n)\,\mathcal{G}\big(A(\boldsymbol{k}\,|\,j''_{\mu''}), A^{+}(\boldsymbol{k}\,|\,j'_{\mu'}), i\,\omega_n\big).\end{aligned} \tag{124.66}$$

Now (124.66) could be symbolized as follows if we suppress all unimportant indices

$$\mathcal{G}(\boldsymbol{k})_{j_\mu, j'_{\mu'}} = \mathcal{G}_0(\boldsymbol{k})_{j_\mu j'_{\mu'}} \delta_{jj'} \delta_{\mu\mu'} - \beta\, \mathcal{G}_0(\boldsymbol{k})_{j_\mu j_\mu} \sum_{j''_{\mu''}} \left(\Sigma(\boldsymbol{k})_{j_\mu j''_{\mu''}} \mathcal{G}(\boldsymbol{k})_{j''_{\mu''}, j'_{\mu'}}\right). \tag{124.67}$$

According to the arguments just given the matrix

$$\mathcal{G}(\boldsymbol{k})_{j_\mu j'_{\mu'}} \equiv \mathcal{G}\left(A(\boldsymbol{k}\,|\,j_\mu), A^+(\boldsymbol{k}\,|\,j'_{\mu'}), i\omega_n\right) \tag{124.68}$$

is of dimension equal to the multiplicity of the acceptable irreducible representation $D^{(\boldsymbol{k})(j)}$ of $\mathfrak{G}(\boldsymbol{k})$. If a given $D^{(\boldsymbol{k})(j)}$ only occurs once then the Green function $\mathcal{G}(\boldsymbol{k})_{j_\mu, j'_{\mu'}}$ is diagonal. For a general point in the Brillouin Zone, where the group $\mathfrak{G}(\boldsymbol{k})$ is simply $\mathfrak{T}$, there must be complete mode coupling and $\mathcal{G}(\boldsymbol{k})_{j_\mu, j'_{\mu'}}$ is an l_s dimensional matrix, as is also the self-energy $\Sigma(\boldsymbol{k})$ a matrix of the same dimension.

An application of these results to the calculation of the thermal Green function $\mathcal{G}(0)_{j_\mu, j_\mu}$ in the NaCl structure has been given by WALLIS, IPATOVA, and MARADUDIN.[9] In this structure, as will be discussed in detail below in Sect. 138, the $k=0$ acoustic modes transform according to the triply degenerate polar vector representation. In the absence of the macroscopic electric field, one should expect coupling between acoustic ($\boldsymbol{k}=0$) and optic ($\boldsymbol{k}=0$) modes. However owing to symmetry of the crystal, and the potential energy, under infinitesimal rotation it can be shown[9] that the coupling coefficients vanish (refer to (124.26))

$$V^{(3)}\begin{pmatrix} 0 & \boldsymbol{k}' & \boldsymbol{k}'' \\ j & j' & j'' \end{pmatrix} = V^{(4)}\begin{pmatrix} 0 & \boldsymbol{k}' & \boldsymbol{k}'' & \boldsymbol{k}''' \\ j & j' & j'' & j''' \end{pmatrix} = 0 \tag{124.69}$$

if j refers to an acoustic branch. Hence the acoustic branch at $\boldsymbol{k}$ exactly zero does not couple to the optic branches of the same symmetry at $\boldsymbol{k}=0$. Consequently the diagonal and off diagonal terms in $\mathcal{G}(0)_{j_\mu, j_\mu}$ where j_μ refers to acoustic branches vanish. For the optic branch in the absence of considering the long wave macroscopic field, there are three, equal, quantities $\mathcal{G}(0)_{0,0}$ where 0 refers to the optic mode. But when the macroscopic electric field is included, the same arguments used in Sect. 123γ (Eqs. (123.38), (123.44)) reveal the result that there will be a lifting of this degeneracy so that one obtains the three quantities

$$\mathcal{G}(0)_{TO, TO} \quad \text{(twice)}, \tag{124.70}$$

$$\mathcal{G}(0)_{LO, LO} \quad \text{(once)} \tag{124.71}$$

for the thermal Green functions. We shall not go further into this type of calculation here but refer the reader to the literature where the predictions of the theory are compared with experiment. In considering the comparison with experiment (which was thoroughly discussed by the cited authors) it is desirable to keep in mind that for temperatures $T \ll \Theta_D$ (where Θ_D is the Debye temperature) the crystal may be effectively considered to be at $T=0$. Then in presence of weak anharmonicity the many-body theory for both the infra-red, and the Raman effects goes over into the "critical point" theory. That is one may approximate many of the dynamical factors as giving simply density of states factors so the absorption, or scattering, processes then are proportional to the density of states.

We conclude then, that the symmetry analyses relevant to the simplification of the classical dynamical matrix, Sect. 85, and to the construction of two-phonon selection rules can be taken over to the analysis of the non-zero elements of the thermal one-phonon Green function.[16,17]

Higher order correlation functions (for example, two phonon thermal Green functions) can also be simplified by study of the transformation properties under the elements of $\mathfrak{G}$. But it seems that little work on this topic has yet been reported. It may be a fruitful subject for future study.

δ) Microscopic theory of Raman scattering: Bloch picture. Now we turn to a first variant of the microscopic theory of Raman scattering. We follow generally the development given by LOUDON[18] in an important paper which marks one of the milestones of the microscopic theory. Let us outline the structure of this theory. We shall make particular contact with the question of symmetry of the Raman scattering tensor.

We consider the crystal plus radiation field to be a large system contained within a large, ultimately infinite, box. The material system plus the radiation field are all considered as a single system. We treat the material system of electrons plus ions in the adiabatic Born-Oppenheimer approximation and the radiation field in second quantization as in Sect. 121.

The Hamiltonian of the system can now be written as in Sect. 121:

$$\mathscr{H} = \mathscr{H}_M + \mathscr{H}_R + \mathscr{H}_{EL} + \mathscr{H}_{ER} \tag{124.72}$$

where $\mathscr{H}_M$ is the matter Hamiltonian (electrons plus ions) which will be taken to have Born-Oppenheimer eigenstates, $\mathscr{H}_R$ is the radiation Hamiltonian, and the coupling terms are $\mathscr{H}_{EL}$ and $\mathscr{H}_{ER}$. The uncoupled Hamiltonian will be defined as

$$\mathscr{H}_0 \equiv \mathscr{H}_M + \mathscr{H}_R. \tag{124.73}$$

The eigenstates of (124.73) will be taken as products, as in (121.16):

$$\Psi = |\Psi_{\text{adiabatic}}\rangle\, |\Psi_{\text{rad}}\rangle \tag{124.74}$$

of the matter eigenstate times the radiation eigenstate. For the present purposes it is sufficient to indicate the constituents of the adiabatic state by giving a general quantum number for the electronic and for the vibrational states. We also describe the radiation field in the N (number) representation by giving the occupancy of the two states involved in inelastic scattering: respectively n_1 and n_2. Hence the initial eigenstate is

$$\Psi_i = |0;\, n_0\rangle_M\, |n_1, n_2 = 0\rangle_R \tag{124.75}$$

the final eigenstate is

$$\Psi_f = |0;\, n_0 + 1\rangle_M\, |n_1 - 1, n_2 = 1\rangle_R. \tag{124.76}$$

In initial and final matter eigenstates the index 0 refers to ground electronic state, while n_0 is the number of phonons of energy $\hbar\omega_0$. We consider only one phonon spontaneous Stokes Raman scattering.

[16] R. F. WALLIS *et al.* in Ref. [*12*].
[17] R. K. WEHNER: phys. stat. solidi **15**, 725 (1966).
[18] R. LOUDON: Proc. Roy. Soc. (London), Ser. A **275**, 218 (1963).

The electronic state manifold in this treatment is taken as that of the Bloch states of the crystal. Hence the index 0 refers to the situation for a typical insulator with all states occupied in the lowest available bands. In particular, we consider an insulator with simple spherical bands, with band extrema in conduction and valence bands at the zone center $\boldsymbol{k}=0$. This picture of Raman scattering is based upon Bloch states for the electrons. Denote the intermediate states

$$\Psi_{\text{int}}=|\alpha, n_0'\rangle_M\,|n_1, n_2\rangle_R \tag{124.77}$$

where α refers to some excited electronic state of the many electron system. In particular, for an insulator, an excited electronic state refers to a state with an electron in a conduction band Bloch state and a hole in a valence band Bloch state. The energy of such a state is denoted $\hbar\omega_\alpha$. It will prove useful to express (124.72) in second quantized form.

Let the electron Fermion operators be denoted

$$c_{\boldsymbol{k}p};\; c^+_{\boldsymbol{k}p} \tag{124.78}$$

where $\boldsymbol{k}$ is the wave vector and $p=v$ or c is the band index corresponding to valence and conduction band respectively; they satisfy the anticommutation rules

$$\{c_{\boldsymbol{k}p}, c^+_{\boldsymbol{k}'p'}\}_+=\delta_{\boldsymbol{k}\boldsymbol{k}'}\,\delta_{pp'}. \tag{124.79}$$

Let the phonon Bose operators be denoted

$$b_{\boldsymbol{k}j};\; b^+_{\boldsymbol{k}j} \tag{124.80}$$

where $\boldsymbol{k}$ is the phonon wave vector and j its branch index. Let the photon Bose operators be denoted

$$a_{\boldsymbol{\eta}\sigma};\; a^+_{\boldsymbol{\eta}\sigma} \tag{124.81}$$

where $\boldsymbol{\eta}$ is the photon wave vector, and σ the polarization index. They satisfy commutation rules

$$[b_{\boldsymbol{k}j}, b^+_{\boldsymbol{k}'j'}]_-=\delta_{\boldsymbol{k},\boldsymbol{k}'}\,\delta_{jj'}, \tag{124.82}$$

$$[a_{\boldsymbol{\eta}\sigma}, a^+_{\boldsymbol{\eta}'\sigma'}]_-=\delta_{\boldsymbol{\eta},\boldsymbol{\eta}'}\,\delta_{\sigma\sigma'}. \tag{124.83}$$

Then the uncoupled Hamiltonian becomes

$$\mathscr{H}_0=\sum_{\boldsymbol{k}p}\varepsilon(\boldsymbol{k},p)\,c^+_{\boldsymbol{k}p}c_{\boldsymbol{k}p}+\sum\hbar\omega_{\boldsymbol{k}}(b^+_{\boldsymbol{k}}\,b_{\boldsymbol{k}}+\tfrac{1}{2})+\sum_{\boldsymbol{\eta},\sigma}\hbar\omega(\boldsymbol{\eta},\sigma)\,a^+_{\boldsymbol{\eta}\sigma}a_{\boldsymbol{\eta}\sigma} \tag{124.84}$$

where the first term refers to the energy of the electrons, with $\varepsilon(\boldsymbol{k},p)$ equal to the one-particle Bloch energy of electron in band p; the second term is the phonon energy $\hbar\omega_{\boldsymbol{k}}$ in harmonic approximation; the last term is the radiation field energy. In terms of the one particle Bloch energies $\varepsilon(\boldsymbol{k},p)$ the energy of a pair state (e.g. α or β in (124.77)) in an insulator is the difference between excited electron and remaining hole or

$$\hbar\omega_\alpha\equiv\varepsilon(\boldsymbol{k}-\boldsymbol{k}',p)-\varepsilon(\boldsymbol{k},p') \tag{124.85}$$

is the energy of an excited electronic state: for example a state with one electron in the conduction band, one hole in the valence band $p=c$; $p'=v$.

The interaction terms are then

$$\mathscr{H}_{ER}=\sum_{k\eta}\sum_{vc} f(\boldsymbol{k},\boldsymbol{\eta},v,c)\,a_{-\eta\sigma}c^{+}_{\boldsymbol{k}-\boldsymbol{\eta},c}c_{\boldsymbol{k},v}+\text{c.c.} \tag{124.86}$$

where $f(\boldsymbol{k},\boldsymbol{\eta},v,c)$ is a coupling function, which can be written

$$f(\boldsymbol{k},\boldsymbol{\eta},v,c)=(2\pi/\omega_{\eta}\varepsilon_{\infty})\langle c|\,\boldsymbol{\varepsilon}_{\eta}\mathrm{e}^{i\boldsymbol{\eta}\cdot\boldsymbol{r}}\cdot\boldsymbol{p}\,|v\rangle. \tag{124.87}$$

In (124.87) the matrix element is an interband $(v-c)$ element. Generally we may write

$$\mathrm{e}^{i\boldsymbol{\eta}\cdot\boldsymbol{r}}=(1+i\boldsymbol{\eta}\cdot\boldsymbol{r}+\cdots) \tag{124.88 a}$$

and if the first term only is used in (124.88a), then (124.87) becomes, in dipole approximation

$$f(\boldsymbol{k},\boldsymbol{\eta},v,c)=(2\pi/v_{\eta}\varepsilon_{\infty})\,\boldsymbol{\varepsilon}_{\eta}\cdot(\boldsymbol{p})_{cv} \tag{124.88 b}$$

using an obvious shorthand for the momentum matrix element. Higher order terms in (124.88a) correspond to magnetic dipole, electric quadrupole, ... transitions and are very important in certain resonant scattering situations but we neglect them here. The electron-lattice interaction term is more complicated since knowledge of the coupling mechanism is required. Taking deformation potential coupling to the band electrons we obtain

$$\mathscr{H}_{EL}=\sum g(\boldsymbol{k},\boldsymbol{k}',p,p')\,b^{+}_{\boldsymbol{k}'}c^{+}_{\boldsymbol{k}-\boldsymbol{k}'}c_{\boldsymbol{k}p}+\text{c.c.} \tag{124.89}$$

where

$$\begin{aligned} g(\boldsymbol{k}\boldsymbol{k}'pp')&=(\hbar/\omega_{\boldsymbol{k}'})^{\frac{1}{2}}\langle p'|\,\Xi\,|p\rangle\\ &=(\hbar/\omega_{\boldsymbol{k}'})^{\frac{1}{2}}\,\Xi_{p'p}. \end{aligned} \tag{124.90}$$

Depending on details $p=p'$ or $p\neq p'$. Higher order terms are discussed in Sect. 124η *infra*.

Then, the second quantized Hamiltonian which we require for single-phonon Raman scattering is (124.84) plus (124.86), plus (124.89) and (124.87). Clearly the process which we wish to describe is one in which a transition occurs from Ψ_i to Ψ_f through intermediate states Ψ_{int}. A quick examination of the strings of operators, which occur in (124.86), (124.89) shows that the type of process which we want to calculate would require the string $\mathscr{H}_{ER}\mathscr{H}_{EL}\mathscr{H}_{ER}$ to operature upon Ψ_i in order to give a transition to Ψ_f. But this is certainly a third order process. We can do the third order perturbation theory in a compact fashion by performing a canonical transformation.[19] Thus let us consider the time independent Schrödinger equation for the total system radiation plus matter:

$$\mathscr{H}\Psi=E\Psi \tag{124.91}$$

where $\mathscr{H}$ is the total Hamiltonian. We seek to make a canonical transformation to a new Hamiltonian $\mathscr{H}'$ such that

$$\mathrm{e}^{-iS}\mathscr{H}\mathrm{e}^{iS}\mathrm{e}^{-iS}\Psi=E\,\mathrm{e}^{-iS}\Psi \tag{124.92}$$

and

$$\mathscr{H}'=\mathrm{e}^{-iS}\mathscr{H}\,\mathrm{e}^{iS} \tag{124.93}$$

and

$$\Psi'=\mathrm{e}^{-iS}\Psi. \tag{124.94}$$

---

[19] A.K. Ganguly and J.L. Birman: Phys. Rev. **162**, 806 (1967).

Now using a well known operator identity

$$e^{-iS}\mathcal{H}\,e^{iS}=\mathcal{H}-i[S,\mathcal{H}]_{-}-\tfrac{1}{2}[S,[S,\mathcal{H}]]_{-}+\cdots. \tag{124.95}$$

Let us choose the generating function S, of the transformation such that

$$i[S,\mathcal{H}_M+\mathcal{H}_R]_{-}=\mathcal{H}_{ER}+\mathcal{H}_{EL}. \tag{124.96}$$

The reason for selecting (124.96) will become apparent in a few lines. To obtain the operator S is also simple. We are guided by the requirement that (124.96) should be fulfilled. As a trial select

$$\begin{aligned}S=&\sum_{\boldsymbol{k}\boldsymbol{\eta}vc}\varphi_1(\boldsymbol{k}\boldsymbol{\eta}vc)\,g(\boldsymbol{k}\boldsymbol{\eta}vc)\,a^{+}_{\boldsymbol{\eta}\sigma}c^{+}_{\boldsymbol{k}-\boldsymbol{\eta}c}c_{\boldsymbol{k},v}\\&+\sum_{\boldsymbol{k}\boldsymbol{k}'pp'}\varphi_2(\boldsymbol{k}\boldsymbol{k}'pp')\,f(\boldsymbol{k}\boldsymbol{k}'pp')\,b^{+}_{\boldsymbol{k}'}a^{+}_{\boldsymbol{k}-\boldsymbol{k}'p'}a_{\boldsymbol{k}p}\end{aligned} \tag{124.97}$$

where φ_1 and φ_2 are unknown "c" functions. Using (124.97) in (124.96) we immediately find

$$\begin{aligned}S=&\left(\frac{1}{i}\right)\sum_{\boldsymbol{k}\boldsymbol{k}'pp'}\left\{\frac{f(\boldsymbol{k}\boldsymbol{k}'pp')\,b^{+}_{\boldsymbol{k}'}c^{+}_{\boldsymbol{k}-\boldsymbol{k}'p'}c_{\boldsymbol{k}p}}{\varepsilon(\boldsymbol{k},p)-\varepsilon(\boldsymbol{k}-\boldsymbol{k}',p')-\hbar\omega_{\boldsymbol{k}'}}\right\}\\&+\left(\frac{1}{i}\right)\sum_{\boldsymbol{k}\boldsymbol{\eta}vc}\left\{\frac{g(\boldsymbol{k},\boldsymbol{\eta},c,v)\,a_{-\boldsymbol{\eta}}c^{+}_{\boldsymbol{k}-\boldsymbol{\eta}c}c_{\boldsymbol{k}v}}{\hbar\omega_{-\boldsymbol{\eta}}-(\varepsilon(\boldsymbol{k}-\boldsymbol{\eta},c)-\varepsilon(\boldsymbol{k},v))}\right\}+\text{c.c.}\end{aligned} \tag{124.98}$$

When (124.98) is substituted into (124.96) we find for the structure of the transformed Hamiltonian

$$\mathcal{H}'=\mathcal{H}_M+\mathcal{H}_R-i[S,\mathcal{H}_{ER}+\mathcal{H}_{EL}]_{-}-\tfrac{1}{2}[S,[S,\mathcal{H}]]_{-}+\cdots \tag{124.99}$$

we later return to (124.93). Next consider the transformed eigenfunction (124.94)

$$\Psi'=\left(1-iS-\frac{S^2}{2}+\cdots\right)\Psi. \tag{124.100}$$

Since the canonical transformation has preserved $\mathcal{H}_M+\mathcal{H}_R$ as lowest order terms in (124.99) we assume that we may use the lowest order eigenfunction in (124.99) i.e. take

$$\Psi'\simeq\Psi. \tag{124.101}$$

Then, inspection of the operator strings in (124.99) shows that the term

$$-\tfrac{1}{2}[S,[S,\mathcal{H}_{ER}+\mathcal{H}_{EL}]]_{-} \tag{124.102}$$

which is contained in (124.93) has the desired operator string. Then, considering (124.102) as the perturbation, the rate of transitions proportional to the Raman scattering cross-section, is

$$w_{i\to f}=\sum_f|\langle\Psi_f|\,(-\tfrac{1}{2})\,[S,[S,\mathcal{H}_{ER}+\mathcal{H}_{EL}]]_{-}\,|\Psi_i\rangle|^2\,\delta(E_f-E_i) \tag{124.103}$$

where (E_f-E_i) are final and initial state energies respectively. For one phonon Stokes Raman scattering

$$E_f-E_i=\hbar(\omega_2+\omega_0-\omega_1). \tag{124.104}$$

Now substitute (124.102), (124.98) into (124.103) to obtain for the intensity of Raman scattering:

$$\mathscr{I} \sim c(\omega) \sum_{k\eta} \frac{n_1(n_0+1)}{\omega_0 \omega_1 \omega_2} |\xi^i_{0\eta} R^i_{12}(-\omega_1, \omega_2, \omega_0)|^2 \times \left(\frac{(2\pi)^3}{V}\right) \delta(\mathbf{k}_1 - (\mathbf{k}_2 + \boldsymbol{\eta}))\, \delta(\omega_2 + \omega_0 - \omega_1) \tag{124.105}$$

where in (124.105) $\hat{\xi}_{0\eta}$ is a phonon polarization vector and $R^{(i)}_{12}$ is the Raman tensor. The Raman tensor can be given explicitly in a conventional form, due to LOUDON:

$$\begin{aligned} R^i_{12}(-\omega_1, \omega_2, \omega_0) &= \frac{1}{V} \sum_{\alpha\beta} \left\{ \frac{p^2_{\alpha\beta} \Xi^i_{\beta\alpha} p^1_{\alpha 0}}{(\omega_\beta + \omega_0 - \omega_1)(\omega_\alpha - \omega_1)} + \frac{p^1_{0\beta} \Xi^i_{\beta\alpha} p^2_{\alpha 0}}{(\omega_\beta + \omega_0 + \omega_2)(\omega_\alpha + \omega_2)} \right. \\ &+ \frac{p^2_{0\beta} p^1_{\beta\alpha} \Xi^i_{\alpha 0}}{(\omega_\beta + \omega_0 - \omega_1)(\omega_\alpha + \omega_0)} + \frac{p^1_{0\beta} p^2_{\beta\alpha} \Xi^i_{\alpha 0}}{(\omega_\beta + \omega_0 + \omega_2)(\omega_\alpha + \omega_0)} \\ &+ \left. \frac{\Xi^i_{0\beta} p^2_{\beta\alpha} p^1_{\alpha 0}}{(\omega_\beta + \omega_2 - \omega_1)(\omega_\alpha - \omega_1)} + \frac{\Xi^i_{0\beta} p^1_{\beta\alpha} p^2_{\alpha 0}}{(\omega_\beta + \omega_2 - \omega_1)(\omega_\alpha + \omega_2)} \right\}. \end{aligned} \tag{124.106}$$

In (124.106) the sum on α, β is over all intermediate excited electronic states. (Recall the discussion in Sect. 121, where such sums on intermediate states also arose in the generalized Placzek theory.) Six terms arise in (124.106) owing to the possibility of various time orders which arise in (124.102) when the operator strings are worked out. Here 1, 2, 3 are $(x y z)$ components, $\mathbf{p}_{\alpha\beta}$ is the dipole matrix element between states $\alpha\beta$, and $\Xi^i_{\alpha\beta}$ is the $\alpha\beta$ matrix element of the deformation potential.

At this point the symmetry of the microscopic Raman scattering tensor (124.106) can be examined. This has been done by LOUDON.[18] Note that $\mathbf{p} \equiv i\hbar\nabla$ is a pure imaginary operator and the deformation potential operator Ξ is pure real. Then, if the electron wave functions, including spin, can be taken real (*real* corepresentations as in Sects. 94–100) the matrix elements satisfy

$$\mathbf{p}_{\alpha\beta} = -\mathbf{p}_{\beta\alpha}; \quad \text{and} \quad \Xi_{\alpha\beta} = \Xi_{\beta\alpha} \tag{124.107}$$

and of course

$$\omega_2 = \omega_1 - \omega_0. \tag{124.108}$$

When (124.107), (124.108) are put into (124.106) one finds the symmetry

$$R_{12}(-\omega_1, \omega_2, \omega_0) = R_{21}(-(\omega_1 - \omega_0), \omega_2 + \omega_0, -\omega_0). \tag{124.109}$$

This very interesting result actually relates the time reversed scattering process (Anti-Stokes) to the original (Stokes) process, via their scattering matrices. In general this microscopic theory shows that the Raman scattering tensor is not symmetric in the Cartesian indices 12, or $x y$. However if ω_0 is small compared to $(\omega_\alpha - \omega_1)$ and $(\omega_\alpha - \omega_2)$ i.e. *far* from resonance then

$$R_{12}(-\omega_1, \omega_2, \omega_0) \simeq R_{21}(-\omega_1, \omega_2, \omega_0). \tag{124.110}$$

Thus in this case the microscopic theory in Bloch picture provides a proof that the Raman tensor is a symmetric second rank tensor. Recall in this context the discussion in Sect. 121, especially Eq. (121.49).

To the writer's knowledge, the exact symmetry result in (124.109) or its analog (*vide infra*) in the exciton picture has not yet been tested experimentally. Presumably this should be done by some resonance scattering experiment in a properly selected material, in which it would be seen whether $R_{xy}(-\omega_1, \omega_2, \omega_0)$ equals $R_{yx}(-\omega_1, \omega_2, \omega_0)$. An equivalent way of stating this is whether an antisymmetric component (e.g. ($\boldsymbol{\Gamma}$ 15+) in a cubic crystal) is present in phonon scattering.[20]

ε) *Microscopic theory of Raman scattering: Exciton-Picture.*[19] The discussion given in Subsect. δ suffers from the limitation that the basic electronic states which were used are the Bloch states. Bloch states are one-particle states which, in the language of many-body theory, are non-interacting. That is to be specific the electron and hole in the virtual, intermediate, Bloch states do not interact. But it is well known that an electron and hole do interact via a screened or shielded coulomb interaction. The eigenstates resulting from this interaction are the exciton states of the system. Equivalently stated, the interaction between the electron and hole needs to be taken into account in constructing a proper, or complete set of states for the electron system. Thus the theory given in (δ) needs to be reformulated in the exciton basis. It turns out that the formal changes in the theory are relatively minimal. However some very important qualitative changes in the results of the theory do occur when we transform to exciton bases and we now discuss these, at least insofar as they relate to the predictions for one-phonon spontaneous Stokes Raman scattering.

We take as the basic Hamiltonian for Raman scattering the Hamiltonian for free radiation, phonon, and exciton fields plus the lowest order couplings between exciton and photon, and exciton phonon fields. We pause at this point to remark that by our choice of Hamiltonian, we already build into the theory, *à priori*, certain of the important physical interactions. We could, for example, have returned to the starting point of the previous subsection, namely the one-body Bloch electron Hamiltonian which is, from (124.84)

$$\mathscr{H}_{\text{Bloch}} = \sum_{\boldsymbol{k}\,p} \varepsilon(\boldsymbol{k}, p)\, c^+(\boldsymbol{k}, p)\, c(\boldsymbol{k}, p).$$

Then adding some two body coulomb interaction such as:

$$\mathscr{H}_{EE} = \sum_{\boldsymbol{k}_1, \ldots, \boldsymbol{k}_4} \sum_{v, c} V(\boldsymbol{k}_1 \boldsymbol{k}_2 \boldsymbol{k}_3 \boldsymbol{k}_4)\, c^+(\boldsymbol{k}, v)\, c^+(\boldsymbol{k}_2, c)\, c(\boldsymbol{k}_3, c)\, c(\boldsymbol{k}_4, v)\, \delta(\boldsymbol{k}_1 + \boldsymbol{k}_2 + \boldsymbol{k}_3 + \boldsymbol{k}_4)$$

can result in the theoretical prediction that excitons will be formed. This approach has been utilized by some authors[21] but seems to offer no advantage to the direct introduction of excitons at an early stage in the theory.

Consequently we write *à priori* for the electronic part of the Hamiltonian in exciton representation:

$$\mathscr{H}_{EX} = \sum E_{\lambda \underline{\boldsymbol{K}}}(c, v)\, \alpha^+_{\lambda \underline{\boldsymbol{K}}}(c, v)\, \alpha_{\lambda \underline{\boldsymbol{K}}}(c, v) \tag{124.111}$$

[20] Suggested by Prof. H. Bilz. At the time of writing (Sept. 1974) such tests are apparently under way in several laboratories.

[21] D.L. Mills and E. Burstein: Phys. Rev. **188**, 1465 (1969).

where $\alpha^+_{\lambda\underline{\bar{K}}}(c, v)$ is the exciton creation operator, which we take to obey Bose commutation rules:

$$[\alpha_{\lambda\underline{\bar{K}}}(c, v), \alpha^+_{\lambda'\underline{\bar{K}}'}(c', v')]_- = \delta_{cc'}\delta_{vv'}\delta_{\lambda\lambda'}\delta_{\underline{\bar{K}},\underline{\bar{K}}'}. \tag{124.112}$$

The exciton energy $E_{\lambda\underline{\bar{K}}}(c, v)$ for the exciton formed from parabolic bands c and v, with "inner" quantum number λ, center of mass wave vector $\underline{\bar{K}}$, is for bound states given by

$$E_{n,\underline{\bar{K}}} = E_g + \hbar^2 \underline{\bar{K}}^2/2(m_e^* + m_h^*) - R'/n^2 \tag{124.113}$$

where $n = 1, \ldots$ or for continuum states

$$E_{k,\underline{\bar{K}}} = E_g + \hbar^2 \underline{\bar{K}}^2/2(m_e^* + m_h^*) + \hbar^2 h^2/2\mu \tag{124.114}$$

where $R' = \mu e^4/2\hbar^2\kappa^2$ is the exciton Rydberg with κ the dielectric constant, m_e^* and m_h^* are electron and hole effective masses, E_g the direct energy gap, and $\mu^{-1} = m_e^{*-1} = m_h^{*-1}$ is the reduced mass.

The exciton wave function can be taken in an adequate approximation for our work as

$$\Psi_{\lambda\underline{\bar{K}}}(\boldsymbol{r}_e, \boldsymbol{r}_h) = C e^{i\underline{\bar{K}}\cdot\boldsymbol{R}} U_\lambda(\boldsymbol{\beta})\, \psi_c(\boldsymbol{r}_e)\, \psi_v(\boldsymbol{r}_h) \tag{124.115}$$

where

$$\boldsymbol{R} = \tfrac{1}{2}(\boldsymbol{r}_e + \boldsymbol{r}_h) \tag{124.116}$$

is the average electron-hole coordinate

$$\boldsymbol{\beta} = \boldsymbol{r}_e - \boldsymbol{r}_h \tag{124.117}$$

is the relative electron-hole coordinate $U_\lambda(\boldsymbol{\beta})$ is the effective mass envelope function with set of inner quantum numbers λ, $\psi_c(\boldsymbol{r}_e)$ and $\psi_v(\boldsymbol{r}_h)$ are the Bloch state eigenfunctions for electrons and holes respectively. Depending on whether λ corresponds to the discrete ($n = 1, 2, \ldots$) or continuous ($\boldsymbol{k}$) part of the spectrum the envelope function will be accordingly chosen. If $\underline{\bar{K}} \neq 0$, *spatial dispersion* is present.

For the uncoupled radiation plus phonon fields we choose the same terms as in (124.84). The interaction Hamiltonian will now be the sum of exciton-radiation plus exciton-phonon terms where here we keep only the lowest order term which contributes to the first order Raman scattering. These are

$$\mathscr{H}_{E\times R} = \sum_{\boldsymbol{k}\boldsymbol{\eta}vc} \bar{f}(\boldsymbol{k}, \boldsymbol{\eta}, v, c)\, \alpha^+_{\lambda\underline{\bar{K}}}(c, v)\, a_{\boldsymbol{\eta},\sigma}\, \delta_{\underline{\bar{K}},\boldsymbol{\eta}} \tag{124.118}$$

where $\bar{f}$ is the coupling function appropriate to the exciton-photon coupling and is related to the "bare" electron-photon coupling function f of the previous subsection by:

$$\bar{f}(\boldsymbol{k}, \boldsymbol{\eta}, v, c) = -N^{\frac{1}{2}}(e/m)(2\pi\hbar/V\kappa_\infty\omega_\eta)^{\frac{1}{2}}\, U^*_{c,v,\lambda,\underline{\bar{K}}}(0)\, f(\boldsymbol{k}, \boldsymbol{\eta}, v, c). \tag{124.119}$$

The exciton lattice interaction term must also now be reformulated in terms of the exciton variables. This interaction term can now be written

$$\begin{aligned}\mathscr{H}_{E\times L} = \sum \bar{G}(c, v, \lambda, \underline{\bar{K}}, c', v', \lambda', \underline{\bar{K}}')\, \alpha^+_{\lambda\underline{\bar{K}}}(c, v)\, \alpha_{\lambda'\underline{\bar{K}}'}(c', v')\\ \times b^+_{\boldsymbol{k}'}\, \delta(\underline{\bar{K}} - \underline{\bar{K}}' + \boldsymbol{k}') + \text{c.c.}\end{aligned} \tag{124.120}$$

with the coupling term given as

$$\bar{G}(c, v, \lambda, \underline{\bar{K}}, c', v', \lambda', \underline{\bar{K}}') = (\hbar/2MN\omega)^{\frac{1}{2}} \times (1/a) \sum_i [\langle c| \Xi |c'\rangle \, q_e(c, v, \lambda, \underline{\bar{K}}, c', v', \lambda', \underline{\bar{K}}') \, \delta_{vv'} - \langle v'| \Xi |v\rangle \, q_h(c, v, \lambda, \underline{\bar{K}}, c', v', \lambda', \underline{\bar{K}}') \, \delta_{cc'}] \tag{124.121}$$

where

$$q_e(c, v, \lambda, \underline{\bar{K}}, c', v', \lambda', \underline{\bar{K}}') = \sum_{\boldsymbol{\beta}} U^*_{cv\lambda\underline{\bar{K}}}(\boldsymbol{\beta}) \, U_{c'v'\lambda'\underline{\bar{K}}'}(\boldsymbol{\beta}) \, e^{i(\underline{\bar{K}} - \underline{\bar{K}}')\cdot\boldsymbol{\beta}} \tag{124.122}$$

and

$$q_h(c, v, \lambda, \underline{\bar{K}}, c', v', \lambda', \underline{\bar{K}}') = \sum_{\boldsymbol{\beta}} U^*_{cv\lambda\underline{\bar{K}}}(\boldsymbol{\beta}) \, U_{c'v'\lambda'\underline{\bar{K}}'}(\boldsymbol{\beta}). \tag{124.123}$$

The total Hamiltonian is then, in the exciton representation

$$\mathcal{H} = \mathcal{H}_{EX} + \mathcal{H}_L + \mathcal{H}_R + \mathcal{H}_{E\times L} + \mathcal{H}_{E\times R} \tag{124.124}$$

where we assemble the parts of (124.124) from (124.84), (124.118), and (124.120). The structure of the exciton-Raman Hamiltonian is then very similar to that of the Hamiltonian in Bloch representation. It is, neglecting all indices and subscripts, of the form

$$\mathcal{H} = \sum E\alpha^+\alpha + \sum \hbar\omega(b^+b + \tfrac{1}{2}) + \sum \hbar\omega \, a^+ a + \sum \bar{G}\alpha^+\alpha b^+ + \sum \bar{f} a \alpha^+ + \text{c.c.} \tag{124.125}$$

To extract the Raman scattering cross section from this Hamiltonian we need to proceed just as in case of the Bloch Hamiltonian. Note that now the eigenstates of the matter part of the Hamiltonian differ. Particularly the difference is in the electronic manifold. Thus here all electronic states (initial, final, and intermediate) refer to exciton states for an insulator. Below we shall be more specific. Now, the Hamiltonian (124.125) can be treated just as in the previous subsection by canonical transformation after breaking up the Hamiltonian into an unperturbed part $\mathcal{H}_0$ and a perturbing part $\mathcal{H}_1$:

$$\mathcal{H}_0 = \mathcal{H}_{EX} + \mathcal{H}_R + \mathcal{H}_L = \sum E\alpha^+\alpha + \sum \hbar\omega \, a^+ a + \sum \hbar\omega(b^+b + \tfrac{1}{2}) \tag{124.126}$$

and

$$\mathcal{H}_1 = \mathcal{H}_{E\times L} + \mathcal{H}_{E\times R} = \sum \bar{G}\alpha^+\alpha b^+ + \sum \bar{f}\alpha^+ a. \tag{124.127}$$

The generating function S in the present context is given by making the substitution of exciton operators $\boldsymbol{\alpha}$ for electron operators $\boldsymbol{c}$, and also substituting exciton energies E for one electron energies ε, and likewise for the coupling functions f, g replaced by $\bar{f}$ and $\bar{G}$ respectively.

The expression for one-phonon spontaneous Raman scattering is then (124.65) as before, but now the Raman tensor $R^{(i)}_{12}$ is given by [19]

$$R_{12}^{(i)}(-\omega_1,\omega_2,\omega_0)$$

$$\begin{aligned}
&= \frac{N h^2}{V} \sum_{\substack{\lambda c v\\ \lambda' c' v'}} \frac{U_{cv\lambda\chi_2}(0)\, U^*_{c'v'\lambda'\chi_2+\eta}(0)\,\langle v|\,\varepsilon_2\cdot \boldsymbol{p}\,|c\rangle \times \{q_e\langle c|\,\mathscr{D}^{(i)}\,|c'\rangle\,\delta_{vv'} - q_h\langle v'|\,\mathscr{D}^{(i)}\,|v\rangle\,\delta_{cc'}\}\,\langle c'|\,\varepsilon_1\cdot \boldsymbol{p}\,|v'\rangle}{[E_{\lambda\chi_2}(c v) - \hbar\omega_1 + \hbar\omega_0]\,[E_{\lambda'\chi_2+\eta}(c' v') - \hbar\omega_1]}\\
&+ \frac{U_{cv\lambda,-\chi_1}(0)\, U^*_{c'v'\lambda',\eta-\chi_1}(0)\,\langle v|\,\varepsilon_1\cdot \boldsymbol{p}\,|c\rangle \times \{q_e\langle c|\,\mathscr{D}^{(i)}\,|c'\rangle\,\delta_{vv'} - q_h\langle v'|\,\mathscr{D}^{(i)}\,|v\rangle\,\delta_{cc'}\}\,\langle c'|\,\varepsilon_2\cdot \boldsymbol{p}\,|v'\rangle}{[E_{\lambda,-\chi_1}(c v) + \hbar\omega_2 + \hbar\omega_0]\,[E_{\lambda',\eta-\chi_1}(c' v') + \hbar\omega_2]}\\
&+ \frac{U_{cv\lambda,-\chi_1}(0)\, U^*_{c'v'\lambda'\chi_2-\chi_1}(0)\,\langle v|\,\varepsilon_1\cdot \boldsymbol{p}\,|c\rangle \times \{q_e\langle c|\,\varepsilon_2\cdot \boldsymbol{p}\,|c'\rangle\,\delta_{vv'} - q_h\langle v'|\,\varepsilon_2\cdot \boldsymbol{p}\,|v\rangle\,\delta_{cc'}\}\,\langle c'|\,\mathscr{D}^{(i)}\,|v'\rangle}{[E_{\lambda,-\chi_1}(c v) + \hbar\omega_2 + \hbar\omega_0]\,[E_{\lambda',\chi_2-\chi_1}(c' v') + \hbar\omega_0]}\\
&+ \frac{U_{cv\lambda\chi_2}(0)\, U^*_{c'v'\lambda',\chi_2-\chi_1}(0)\,\langle v|\,\varepsilon_2\cdot \boldsymbol{p}\,|c\rangle \times \{q_e\langle c|\,\varepsilon_1\cdot \boldsymbol{p}\,|c'\rangle\,\delta_{vv'} - q_h\langle v'|\,\varepsilon_1\cdot \boldsymbol{p}\,|v\rangle\,\delta_{cc'}\}\,\langle c'|\,\mathscr{D}^{(i)}\,|v'\rangle}{[E_{\lambda\chi_2}(c v) - \hbar\omega_1 + \hbar\omega_0]\,[E_{\lambda',\chi_2-\chi_1}(c' v') + \hbar\omega_0]}\\
&+ \frac{U_{cv\lambda\eta}(0)\, U^*_{c'v'\lambda',\eta-\chi_1}(0)\,\langle v|\,\mathscr{D}^{(i)}\,|c\rangle \times \{q_e\langle c|\,\varepsilon_1\cdot \boldsymbol{p}\,|c'\rangle\,\delta_{vv'} - q_h\langle v'|\,\varepsilon_1\cdot \boldsymbol{p}\,|v\rangle\,\delta_{cc'}\}\,\langle c'|\,\varepsilon_2\cdot \boldsymbol{p}\,|v'\rangle}{[E_{\lambda\eta}(c v) - \hbar\omega_1 + \hbar\omega_2]\,[E_{\lambda',\eta-\chi_1}(c' v') + \hbar\omega_2]}\\
&+ \frac{U_{cv\lambda\eta}(0)\, U^*_{c'v'\lambda'\chi_2+\eta}(0)\,\langle v|\,\mathscr{D}^{(i)}\,|c\rangle \times \{q_e\langle c|\,\varepsilon_2\cdot \boldsymbol{p}\,|c'\rangle\,\delta_{vv'} - q_h\langle v'|\,\varepsilon_2\cdot \boldsymbol{p}\,|v\rangle\,\delta_{cc'}\}\,\langle c'|\,\varepsilon_1\cdot \boldsymbol{p}\,|v'\rangle}{[E_{\lambda\eta}(c v) - \hbar\omega_1 + \hbar\omega_2]\,[E_{\lambda',\chi_2+\eta}(c' v') - \hbar\omega_1]}.
\end{aligned} \tag{124.128}$$

To be noted is that the sum over intermediate states refers to exciton intermediate states, of which we need to sum over discrete plus continuum manifolds. Examination of (124.128) indicates the same symmetry as in the case of the Raman tensor derived on the Bloch picture, although certain important dynamical consequences differ (vide infra Subsect. η). In (124.128) we used $\mathscr{D}^{(i)}$ instead of $\Xi^{(i)}$ for the deformation potential, to be consistent with reference.[19]

ζ) *Microscopic theory of Raman scattering: Polariton picture.* Returning to (124.126), (124.127) we now remark that the division of the total Hamiltonian into a free part $\mathscr{H}_0$ plus an interaction part $\mathscr{H}_1$ is an arbitrary feature of the exciton treatment.

Consider the unperturbed Hamiltonian which is the sum of free exciton plus photon fields plus the linear coupling $\mathscr{H}_{ER}$. Then again suppressing all indices, and calling

$$\mathscr{H}_0' = \sum E\,\alpha^+\alpha + \sum \hbar\omega(a^+ a + \tfrac{1}{2}) + \sum \bar{f}\alpha^+ a \tag{124.129}$$

$$\mathscr{H}_1' = \mathscr{H} - \mathscr{H}_0' \tag{124.130}$$

so we have a different division of the Hamiltonian, which corresponds to a different picture of the system, and of Raman scattering.

The importance of the new division (124.129)–(124.130) is that it defines non-interacting Hamiltonian (124.129) which can be exactly diagonalized to produce new mixed exciton-photon modes now called polaritons.[21] Thus for a system characterized by the Hamiltonian (124.129), we can find a canonical transformation, (in fact, a linear transformation) which brings (124.129) to the form

$$\mathcal{H}_0' = \sum_{k\lambda\eta} W(\boldsymbol{k}, \lambda, \boldsymbol{\eta})\, A_{\lambda\eta}^{+}(\boldsymbol{k})\, A_{\lambda\eta}(\boldsymbol{k}) \tag{124.131}$$

with the $A_{\lambda,\eta}(\boldsymbol{k})$ new Bose operators

$$[A_{\lambda,\boldsymbol{\eta}}(\boldsymbol{k}), A_{\lambda',\boldsymbol{\eta}'}^{+}(\boldsymbol{k}')]_{-} = \delta_{\lambda\lambda'}\, \delta_{\boldsymbol{\eta},\boldsymbol{\eta}'}\, \delta_{\boldsymbol{k},\boldsymbol{k}'}. \tag{124.132}$$

The polariton operators $A_{\lambda,\boldsymbol{\eta}}(\boldsymbol{k})$ are given as

$$\begin{aligned} A_{\lambda,\eta}^{+}(\boldsymbol{k}) = {} & \chi(\boldsymbol{k})\, a^{+}(\boldsymbol{k}) + \sum_{\lambda,\eta} \chi_{\lambda\boldsymbol{\eta}}(\boldsymbol{k})\, \alpha_{\lambda,\eta}^{+}(\boldsymbol{k}) \\ & + \varphi(\boldsymbol{k})\, a(-\boldsymbol{k}) + \sum_{\lambda,\eta} \varphi_{\lambda,\eta}(\boldsymbol{k})\, \alpha_{\lambda,\eta}(-\boldsymbol{k}) \end{aligned} \tag{124.133}$$

with the "c" functions $\varphi(\boldsymbol{k})$ and $\chi(\boldsymbol{k})$ to be determined from the eigenvalue equation

$$[H_0', A_{\lambda,\boldsymbol{\eta}}^{+}]_{-} = W A_{\lambda,\boldsymbol{\eta}}^{+} \tag{124.134}$$

which of course also determines the eigenfrequencies $W_{\lambda,\boldsymbol{\eta}}(\boldsymbol{k})$. The simplest case to deal with both conceptually and for calculation is the case of a single, spatially dispersive exciton level. That is in (124.129), let us only take one exciton term $E_1(\boldsymbol{k})$. Then there are only four functions to determine. These can be found explicitly and are simply given as:

$$\chi_1 = (k/W)^{\frac{1}{2}}\, \bar{f}_1^{*}\, (W + E_1)\, \alpha(\boldsymbol{k}, \omega), \tag{124.135}$$

$$\varphi_1 = -(k/W)^{\frac{1}{2}}\, \bar{f}_1\, (W - E_1)\, \alpha(\boldsymbol{k}, \omega), \tag{124.136}$$

$$\chi_0 = 2(c\,k + W)(k\,W)^{-\frac{1}{2}}(W^2 - E_1^2)\, \alpha(\boldsymbol{k}, \omega), \tag{124.137}$$

$$\varphi_0 = 2(c\,k - W)(k\,W)^{-\frac{1}{2}}(W^2 - E_1^2)\, \alpha(\boldsymbol{k}, \omega) \tag{124.138}$$

with

$$\alpha(\boldsymbol{k}, \omega) = [(W^2 - E_1^2)^2 + 4E_1 |\bar{f}|^2 k^2]^{\frac{1}{2}}. \tag{124.139}$$

The new eigenvalues of the system are given by the implicit equation for W:

$$\frac{c^2 k^2}{W^2} = 1 + \frac{4k^2 \bar{f}^2}{E_1(\boldsymbol{k})(E_1(\boldsymbol{k})^2 - W^2)}. \tag{124.140}$$

The essential point of the transformation to polariton variables is that these are in principle the correct normal modes of the coupled exciton-photon system. Thus, a photon incident upon the crystal surface must be considered to create a polariton within the crystal. This polariton propagates, or is scattered in the

[21] Polaritons were apparently first discussed in the context of phonons by HUANG [*13*], Sect. 7, and references therein. HUANG (private communication to the author MAY (1973)), also discussed exciton polaritons about that time (1950) but they were not named by him at that time. Their popularity seems to derive from the work of U. FANO, Phys. Rev. **103**, 1202 (1956) and J.J. HOPFIELD, Phys. Rev. **112**, 1555 (1958).

crystal. It is the scattering of a polariton which will be identified with the process of Raman scattering.[22] To be specific when a polariton of frequency ω_1 (corresponding to the external incident photon) scatters (or decomposes) to produce a polariton of frequency ω_2 (corresponding to the scattered photon which will emerge from the crystal) plus a phonon of frequency $\omega_0=\omega_1-\omega_2$ a process of Stokes Raman scattering has occured. Note that, especially in case of the transverse polar phonon the resulting phonon must be considered also as a coupled, polariton mode, owing to coupling with the radiation field. To distinguish these, we can use the terminology "exciton polariton" for the one, "phonon-polariton" for the other. Raman scattering is then the decay process:

$$\{\text{exciton-polariton } (\omega_1) \rightarrow \text{exciton polariton } (\omega_2) + \text{phonon polariton } (\omega_0)\}. \tag{124.141}$$

The simplest way to calculate the cross section for this process is to consider the various polaritons in (124.131) as "quasi particles" and then to use the Golden rule[22] to compute the transition rate for particle (quasi-particle) scattering from initial to final state. The perturbation causing the transition may be taken as of the form

$$V^{(3)}=\sum \bar{f}^{(3)}\,\alpha^{+}_{\boldsymbol{k}-\boldsymbol{\eta}}\,\alpha_{\boldsymbol{k}}\,b^{+}_{\boldsymbol{\eta}}+\text{c.c.} \tag{124.142}$$

with appropriate indices on the coupling function $\bar{f}^{(3)}$. In writing (124.142) we write the coupling term in terms of a bilinear interaction in exciton operators, and linear in phonon operators. In principle it is best, as explained above to convert phonon states also to polariton states which corresponds in (124.142) to replacing the phonon operator $b^{+}_{\boldsymbol{\eta}}$ by a polariton operator $\beta^{+}_{\boldsymbol{\eta}}$. We shall not go into the details of this theory here.

The theory of Raman scattering in the polariton picture has been given by many authors, beginning with work of OVANDER,[22] which embodied this qualitative picture. Recent treatments of phonon Raman scattering in the polariton picture[23-26] have clarified many essential points of this approach. For example, in the simplest approach to the theory the transition rate in first Born approximation for the Raman process is proportional to

$$I\sim\frac{(k_2)^2}{\pi h^2\, v_{1g}\, v_{2g}}\,|\langle \boldsymbol{k}_1|\, V^{(3)}\,|\boldsymbol{k}_2,\boldsymbol{k}_0\rangle|^2. \tag{124.143}$$

The rate (124.143) includes the matrix element for the transition as well as the pre-multiplying phase-space factors, where $v_{1g}(v_{2g})$ are the group velocities of the initial (scattered) polariton defined by

$$v_{1g}\equiv v_g(\boldsymbol{k}_1)=(\partial W/\partial k)_{\boldsymbol{n}}. \tag{124.144}$$

---

[22] L.N. OVANDER: Soviet Phys. Usp. **8**, 337 (1965). – J.J. HOPFIELD: Phys. Rev. **182**, 945 (1969).

[23] B. BENDOW and J.L. BIRMAN: Phys. Rev. B, **1**, 1678 (1970) et seq.

[24] D.L. MILLS and E. BURSTEIN: Phys. Rev. **188**, 1465 (1969).

[25] See the recent review by A.S. BARKER, JR., and R. LOUDON: Rev. Mod. Phys. **44**, 18 (1972) and citations therein, especially to the work of MILLS, MILLS and BENTON, LOUDON and BARKER, TAIT, MAVROYANNIS and MARTIN. Another review is J.A. DEVERIN and C. MAVROYANNIS: Helv. Phys. Acta. **45**, 1005 (1972).

[26] R. ZEYHER, C.S. TING, and J.L. BIRMAN: Phys. Rev. **B10**, 1725 (1974).

The group velocities must be found from the eigenenergy dispersion $W(\boldsymbol{k})$, which is itself the solution of (124.140) in the simplest case. Again it is to be expected that the symmetry of (124.143) will be consonant with general tensor symmetry as given earlier. This question seems not to have been discussed up to the present time, at least to the present writer's knowledge. For a discussion of polariton theory in a bounded crystal see ref. [26].

η) *Resonance Raman scattering and symmetry breaking.* The availability of modern tunable lasers has opened a new era in Raman scattering and we shall now very briefly touch upon some matters related to resonance Raman scattering, which has benefited in particular from this. First it should be noted that each of the three pictures presented above (Sects. 124 δ, ε, ζ) gives, already at the level discussed different predictions for the dependence of the intensity of one-phonon Raman scattering upon incident frequency ω_1 when ω_1 approaches an actual absorption level or band. We can sketch the arguments. In Bloch picture (124.106) is the relevant tensor. In the simplest case let $\omega_1 \to \omega_\alpha$, where (recall (124.85)) ω_α is an energy separation (energy gap ω_g) between upper valence and lowest conduction bands. Then one term in (124.106) is large. Take the bands as simple spherical bands and take $\omega_\alpha = \omega_g + \hbar^2 \underline{\boldsymbol{K}}^2/2M$ where $\underline{\boldsymbol{K}}$ is a total wave vector M the total electron plus hole mass. Let all matrix elements be taken as constant in the resonance region, and convert the sum over states (α, β) to an integral on $\underline{\boldsymbol{K}}$. Then for $\omega_1 \sim \omega_g$

$$R_{12}(\omega_1) \sim (p^2 \Xi)\left[(\omega_g + \omega_0 - \omega_1)^{\frac{1}{2}} - (\omega_g - \omega_1)^{\frac{1}{2}}\right]. \tag{124.145}$$

Using exciton picture the scattering tensor is (124.128). A resonant behavior occurs in this case when the incident frequency $\hbar\omega_1 \to E_1(\boldsymbol{k})$ where $E_1(\boldsymbol{k}) = \hbar\omega_{1S}$ is the energy of an exciton, that is the $n=1$ discrete level of an exciton series. Again taking all matrix elements constant one term is most divergent in (124.128), and we obtain

$$R_{12}(\omega_1) \sim (p^2 \Xi)\left[(\omega_1 - \omega_{1S})(\omega_1 - (\omega_{1S} - \omega_0))\right]^{-1}. \tag{124.146}$$

It is a little more intricate to obtain resonance behavior in the polariton picture since assumptions regarding the frequency dependence of various group velocity factors (which follow from the dispersion) are needed, as well as the frequency dependence of the transformation coefficients which take one from bare exciton and photon variables to polariton variables. An immediate qualitative result from (124.143) is that the scattering intensity does not diverge at any frequency.[27] At present most experiments performed on resonant Raman scattering from insulators appear to agree with the result (124.146) based upon the exciton picture.[28]

Now we turn to matters more related to the work of this article namely: symmetry breaking effects at resonance. Again only brief discussion will be given. Near resonance in the exciton picture the dominant channel gives rise to a contribution to the one-phonon scattering which is like

$$\frac{\langle f|\,\mathscr{H}_{EXR}\,|b\rangle\,\langle b|\,\mathscr{H}_{EXL}\,|a\rangle\,\langle a|\,\mathscr{H}_{EXR}\,|i\rangle}{(\omega_1 - \omega_a)(\omega_1 - \omega_b)} \tag{124.147}$$

[27] J.J. HOPFIELD: Phys. Rev. **182**, 945 (1969). But see also reference [26].

[28] B. BENDOW, J.L. BIRMAN, A.K. GANGULY, T.C. DAMEN, R.C.C. LEITE, and J.F. SCOTT: Optics Comm. **1**, 267 (1970). – R. CALLENDER, S.S. SUSSMAN, M. SELDERS, and R.K. CHANG: Phys. Rev. **B7**, 388 (1973) and references cited in the latter.

where $|i\rangle$, $|f\rangle$, are initial and final states and $|a\rangle$, $|b\rangle$ are intermediate states. The usual symmetry effects arise upon considering the matrix elements in (124.147). Thus if the dipole approximation is valid as in (124.88 b) then (roughly speaking)

$$\langle f|\,\mathscr{H}_{EXR}\,|b\rangle \sim \langle f|\,\boldsymbol{p}\,|b\rangle\cdot\boldsymbol{\varepsilon}\sim D^{(\boldsymbol{v})} \tag{124.148}$$

where the last follows since $\boldsymbol{\varepsilon}$, the polarization vector transforms like an ordinary polar vector. The product of two factors of this sort in (124.147) transform as $D^{(\boldsymbol{v})}\otimes D^{(\boldsymbol{v})}$. In a cubic crystal

$$[D^{(\boldsymbol{v})}]_2 = D^{(\boldsymbol{\Gamma})(1+)}\oplus D^{(\boldsymbol{\Gamma})(12+)}\oplus D^{(\boldsymbol{\Gamma})(25+)}\oplus D^{(\boldsymbol{\Gamma})(15+)}. \tag{124.149}$$

Then the right side of (124.149) gives the symmetry of the phonons which can arise. Note that if we symmetrize (as is justified away from resonance) then representation $D^{(\boldsymbol{\Gamma})(15+)}$ falls away, and the only Raman allowed phonons must have symmetry $(\boldsymbol{\Gamma}\,1+)$; $(\boldsymbol{\Gamma}\,12+)$; $(\boldsymbol{\Gamma}\,25+)$. This observation is justified by examination of the structure of $\langle b|\,\mathscr{H}_{EXL}\,|a\rangle$.

Symmetry breaking can occur at resonance if: the dipole transition is forbidden so the dipole matrix element $\langle f|\,\boldsymbol{p}\,|b\rangle$ vanishes but a quadrupole matrix element is non-zero: $\langle f|\,\boldsymbol{r}\,\boldsymbol{p}\,|b\rangle \neq 0$. Then e.g. at the frequency corresponding to the quadrupole transition $\omega_1 \simeq \omega_{1Q}$ a small denominator in (124.147) can produce a ratio $\boldsymbol{k}\cdot\langle a|\,\boldsymbol{r}\,\boldsymbol{p}\,|i\rangle\cdot\boldsymbol{\varepsilon}/(\omega_1-\omega_{1Q})$ which is comparable in magnitude to the usual "dipole" ratio $\boldsymbol{\varepsilon}\cdot\langle a|\,\boldsymbol{p}\,|i\rangle/(\omega_1-\omega_{1S})$. To distinguish the usual situation from that now under discussion it is helpful to refer to the case where *both* $\langle f|\,\mathscr{H}_{EXR}\,|b\rangle$ *and* $\langle a|\,\mathscr{H}_{EXR}\,|i\rangle$ in (124.147) are non-zero in *dipole* order as Dipole-Dipole Raman scattering

$$DD \sim \sum \boldsymbol{\varepsilon}'\cdot\langle\boldsymbol{p}\rangle\,\langle\Xi\rangle\,\langle\boldsymbol{p}\rangle\cdot\boldsymbol{\varepsilon}/(\Delta\Omega)(\Delta\Omega') \tag{124.150}$$

while in the case $\langle f|\,\mathscr{H}_{EXR}\,|b\rangle$ is non-zero in dipole approximation, and $\langle a|\,\mathscr{H}_{EXR}\,|i\rangle$ is non-zero in quadrupole approximation we can denote this Quadrupole-Dipole Raman scattering

$$QD \sim \sum \boldsymbol{\varepsilon}'\,\langle\boldsymbol{p}\rangle\,\langle\Xi\rangle\,\langle Q\rangle\,\boldsymbol{\varepsilon}\,\boldsymbol{k}/(\Delta\Omega)(\Delta\Omega'). \tag{124.151}$$

Symmetry breaking could occur via any non-zero multipole exciton-radiation matrix element giving rise to Multipole-Dipole, Multipole-Multipole scattering.

Consider the symmetry breaking which accompanies the Quadrupole-Dipole mechanism. For the case of one dipole allowed matrix element, one quadrupole, the same argument as above (124.149) gives for the phonons $D^{(\boldsymbol{v})}\otimes D^{(\boldsymbol{Q})}$. In a cubic crystal $D^{(\boldsymbol{Q})} \doteq D^{(\boldsymbol{\Gamma})(1+)}\oplus D^{(\boldsymbol{\Gamma})(12+)}\oplus D^{(\boldsymbol{\Gamma})(25+)}$. Take the electric quadrupole component $D^{(\boldsymbol{\Gamma})(25+)}$. Then

$$D^{(\boldsymbol{v})}\otimes D^{(\boldsymbol{Q})} = D^{(\boldsymbol{\Gamma})(2-)}\oplus D^{(\boldsymbol{\Gamma})(12-)}\oplus D^{(\boldsymbol{\Gamma})(15-)}\oplus D^{(\boldsymbol{\Gamma})(25-)}. \tag{124.152}$$

Then every representation in (124.152) corresponds now to a permitted phonon for the $Q(\boldsymbol{\Gamma}\,25+)$-$D$ mechanism. Note the dramatic fashion in which for a cubic crystal with inversion $D^{(\boldsymbol{\Gamma})(15-)}$ is now permitted in Raman scattering via this new mechanism. This phonon is also permitted in infra-red absorption. This breaks the "rule of mutual exclusion in one phonon processes" in a cubic crystal with inversion: of course normally a phonon of $D^{(\boldsymbol{\Gamma})(15-)}$ is observable in the infra-red absorption but not in Raman scattering. Compare here the discussion of Sect. 122.

Dramatic illustration of these effects is provided by resonance Raman scattering in Cu_2O for incident laser frequency at the energy of the $1S$ Yellow exciton.[29] Then *all* phonons listed in (124.150) are seen in the Raman scattering.

The Raman scattered light has quite different polarization properties from those of the usual *D-D* situation discussed in Sect. 123. The origin of different polarization properties is easily seen from (124.151). The scattering intensity or cross-section for "in" resonance can be written in this case as proportional to

$$\sum_{j\sigma} \Big|\sum_{\alpha\beta\gamma} \varepsilon_{2\alpha} P_{\alpha\beta\gamma}(j\sigma)\, \varepsilon_{1\beta} k_{1\gamma}\Big|^2 \tag{124.153}$$

where $P_{\alpha\beta\gamma}(j\sigma)$ is a scattering tensor which is a crystal tensor of third rank. The scattering from mode $(j\sigma)$ which must be one of those given in (124.152) depends upon the wave vector of the incident radiation $\boldsymbol{k}_1$. Tables of these matrices are now available.[30] It is clear that the tensor $P_{\alpha\beta\gamma}(j\sigma)$ will differ from the usual scattering tensor $P_{\alpha\beta}(j\sigma)$ which arises in allowed Raman scattering as discussed in Sect. 123. At the most immediate level, it may be remarked that the physical origin of added anisotropy in *Q-D* Raman scattering is due to the anisotropy associated with quadrupole radiation matrix element.

Another example of symmetry breaking[31] in resonance scattering arises when the lowest non-zero term in the exciton-lattice matrix element $\langle b|\,\mathscr{H}_{EXL}\,|a\rangle$ is a term linear in phonon wave vector $\vec{q}$. Then, in this case

$$\langle b|\,\mathscr{H}_{EXL}\,|a\rangle \sim \boldsymbol{q}\,\Xi^{(1)}. \tag{124.154}$$

Returning to the usual *D-D* Raman scattering theory consider the situation where a single channel approximation is valid. Then instead of (124.150) one would obtain

$$\boldsymbol{\varepsilon}'\,\langle\boldsymbol{p}\rangle\,\boldsymbol{q}\,\langle\Xi^{(1)}\rangle\,\langle\boldsymbol{p}\rangle\cdot\boldsymbol{\varepsilon}/(\Delta\Omega)\,(\Delta\Omega'). \tag{124.155}$$

To discuss the symmetry breaking in this case note that in order for $\Xi^{(1)}$ to be non-zero the corresponding phonon needs to have symmetry $D^{(\boldsymbol{q})} \sim D^{(\boldsymbol{v})}$ since $\boldsymbol{q}$ is a polar vector. In fact $\Xi^{(1)}$ arises as the matrix element of a bilinear term in expansion of exciton-lattice interaction such as

$$\boldsymbol{q}\,\Xi^{(1)} = \langle a|\left(\partial^2 \mathscr{H}_{EXL}/\partial q\,\partial Q\begin{pmatrix}\boldsymbol{k}\\ j\end{pmatrix}\right)_0 Q\begin{pmatrix}\boldsymbol{k}\\ j\end{pmatrix}|b\rangle\,\boldsymbol{q}. \tag{124.156}$$

Then this symmetry breaking mechanism is restricted to the "polar optic" phonons $Q\begin{pmatrix}\boldsymbol{k}\\ j\end{pmatrix} \sim D^{(\boldsymbol{v})}$, since the total Hamiltonian $\mathscr{H}_{EXL}$ is an invariant and $\left\{\boldsymbol{q}\,Q\begin{pmatrix}\boldsymbol{k}\\ j\end{pmatrix}\right\}$ can be bilinear invariant only if this condition is met. In a cubic crystal the required phonon symmetry is $D^{(\Gamma)(15-)}$; thus also breaking the rule of mutual exclusion.

It is also evident that in this case the macroscopic scattering can be written

$$\sum_{\sigma} \Big|\sum_{\alpha\beta\gamma} \varepsilon_{\alpha\alpha} Q_{\alpha\beta\gamma}(j\sigma)\, \varepsilon_{1\beta} q_{\gamma}\Big|^2 \tag{124.157}$$

---

[29] A. Compaan and H. Z. Cummins: Phys. Rev. Letters **31**, 41 (1973).

[30] J. L. Birman and R. Berenson: Phys. Rev. **B9**, 4512 (1974). — J. L. Birman: Phys. Rev. **B9**, 4518 (1974).

[31] R. M. Martin in [*48*] and also: Phys. Rev. **B4**, 3676 (1971).

where $j=\mathfrak{v}$ is fixed, the sum is over partners and $Q_{\alpha\beta\gamma}(j\sigma)$ is the appropriate scattering tensor, closely related to the $P_{\alpha\beta\gamma}(j\sigma)$ of (124.153).

A close look shows that both of the above mentioned symmetry-breaking mechanisms can be simultaneously operative, and one would expect them to be observed.

The subject of symmetry breaking at resonance is, in particular, undergoing rapid development and the ongoing literature should be consulted for recent work.

M. Group theory of diamond and rocksalt space groups

125. Introduction. In this part we shall give specific applications of the general space-group theory to diamond and rocksalt space groups. These are two of the most important space groups of current theoretical and experimental interest. Not only are they of interest in and for themselves, but also as prototypes of the analysis of any non-symmorphic and symmorphic space groups.

In the next sections of this part we will discuss the geometry of these space groups, the irreducible representations, and the selection rules, especially the procedure by which reduction coefficients are obtained. Obviously, *all* such reduction coefficients cannot be exhibited but by giving specific illustrations of many different coefficients obtained by different techniques, it is hoped that the entire methodology will be clarified. Then, other tables in the literature can be consulted if available; if no tables are available, they can be constructed using the techniques given in this article.

The emphasis of the work is on the selection rules useful in lattice dynamic applications such as related to optical absorption and scattering. Thus we do not discuss the "double valued" or spin representations. These can be obtained, however by straightforward extension and applications of the methods given here. Some applications of time reversal symmetry restrictions are illustrated.

In addition to applications in lattice dynamic related processes the work presented here is applicable to scattering processes such as electron (or hole) intervalley and intravalley scatterings which enter into the determination of resistivity, and a mention of this application will be given.

126. Geometry of the rocksalt and diamond space groups. It proves convenient to discuss the geometrical aspects of these two space groups together. To specify the space group $\mathfrak{G}$ we require the generators of the translation group $\mathfrak{T}$, and the coset representatives $\{\varphi|\tau(\varphi)\}$ in the factor group decomposition $\mathfrak{G}/\mathfrak{T}$. The two groups [9]

$$O_h^5 - Fm3m \quad \text{(rocksalt)} \tag{126.1}$$

and

$$O_h^7 - Fd3m \quad \text{(diamond)} \tag{126.2}$$

have the identical translation subgroup $\mathfrak{T}$: i.e. that of the face-centered cubic translations which can be denoted $\mathfrak{T}$ or F. The translation group $\mathfrak{T}$ is generated by the three primitive vectors

$$\boldsymbol{t}_{xy}=\boldsymbol{t}_1;\ \boldsymbol{t}_{xz}=\boldsymbol{t}_2;\ \boldsymbol{t}_{yz}=\boldsymbol{t}_3, \quad \text{where, for example} \quad \boldsymbol{t}_{xy}=\tfrac{1}{2}\boldsymbol{a}_1+\tfrac{1}{2}\boldsymbol{a}_2 \tag{126.3}$$

and $\boldsymbol{a}_1, \boldsymbol{a}_2, \boldsymbol{a}_3$ are three orthonormal base vectors of the cubic cell, $|\boldsymbol{a}_i|=a$, $i=1, 2, 3$. In Table 1 this information is assembled to be used for both groups. Also in Table 1 the Fourier vector set $2\pi\boldsymbol{B}_j$ is given, where $\boldsymbol{B}_j$ is a reciprocal lattice vector. As discussed in Sect. 20 the irreducible representations of the translation group $\mathfrak{T}$ are based upon the Fourier vector set $2\pi\boldsymbol{B}_j$.

Table 1. The face centered cubic translation group

Cubic vector set:	$\boldsymbol{a}_1=(1,0,0)\,a$
	$\boldsymbol{a}_2=(0,1,0)\,a$
	$\boldsymbol{a}_3=(0,0,1)\,a$
Primitive vector set:	$\boldsymbol{t}_{xy}=(\frac{1}{2},\frac{1}{2},0)\,a=\boldsymbol{t}_1$
	$\boldsymbol{t}_{xz}=(\frac{1}{2},0,\frac{1}{2})\,a=\boldsymbol{t}_2$
	$\boldsymbol{t}_{yz}=(0,\frac{1}{2},\frac{1}{2})\,a=\boldsymbol{t}_3$
Fourier vector set:	$2\pi\boldsymbol{B}_1=(1,1,-1)(2\pi/a)$
	$2\pi\boldsymbol{B}_2=(1,-1,1)(2\pi/a)$
	$2\pi\boldsymbol{B}_3=(-1,1,1)(2\pi/a)$
	$2\pi\boldsymbol{B}_j\cdot\boldsymbol{t}_i=2\pi\delta_{ij}$

The rotational operators φ which appear in the coset representatives for both space groups are identical, since the factor groups $\mathfrak{G}/\mathfrak{T}$ are both isomorphic to O_h for both cases. In Table 2 the rotational symmetry operations φ of the point group O_h are given.

To complete the specification of these space groups, we need to enumerate the actual coset representatives in the two cases. For rocksalt this is simple owing to the fact that the group is symmorphic. Thus the set of coset representatives in this case is

$$\{\varepsilon|0\}, \ldots, \{\varphi|0\}, \ldots, \{\varphi_{48}|0\}, \tag{126.4}$$

where every rotational element appears without a fractional in the coset decomposition

$$\{\varepsilon|0\}\,\mathfrak{T}, \ldots, \{\varphi|0\}\,\mathfrak{T}, \ldots, \{\varphi_{48}|0\}\,\mathfrak{T}=O_h^5. \tag{126.5}$$

Hence the set of coset representatives alone themselves form the point group O_h. Otherwise stated the group $O_h^5=Fm3m$ is a split extension of $\mathfrak{T}$ by O_h (see Sect. 8).

For the diamond space group O_h^7, the rotational elements in the left column of Table 2 which are isomorphic to point group T_d appear as coset representatives without any fractional translation while the remaining 24 rotations occur with the identical fractional translation $\boldsymbol{\tau}_1=(1,1,1)(a/4)$. Thus the coset decomposition here is

$$\begin{gathered}\{\varepsilon|0\}\,\mathfrak{T}, \ldots, \{\varphi_{24}|0\}\,\mathfrak{T},\\ \{\boldsymbol{i}|\boldsymbol{\tau}_1\}\,\mathfrak{T}, \ldots, \{\varphi_{48}|\boldsymbol{\tau}_1\}\,\mathfrak{T}.\end{gathered} \tag{126.6}$$

Table 2. Rotational symmetry operations[a] for diamond and rocksalt; in diamond $\tau_1=(1,1,1)a/4$

	Type $\{\varphi\|0\}$ in diamond and rocksalt		Type $\{\varphi\|\tau_1\}$ in diamond Type $\{\varphi\|0\}$ in rocksalt
$\varphi_1=\varepsilon$	$x\,y\,z$	$\varphi_{25}=\boldsymbol{i}$	$\bar{x}\,\bar{y}\,\bar{z}$
$\varphi_2=\boldsymbol{\delta}_{2x}$	$x\,\bar{y}\,\bar{z}$	$\varphi_{26}=\boldsymbol{\rho}_x$	$\bar{x}\,y\,z$
$\varphi_3=\boldsymbol{\delta}_{2y}$	$\bar{x}\,y\,\bar{z}$	$\varphi_{27}=\boldsymbol{\rho}_y$	$x\,\bar{y}\,z$
$\varphi_4=\boldsymbol{\delta}_{2z}$	$\bar{x}\,\bar{y}\,z$	$\varphi_{28}=\boldsymbol{\rho}_z$	$x\,y\,\bar{z}$
$\varphi_5=\boldsymbol{\sigma}_{4x}$	$\bar{x}\,z\,\bar{y}$	$\varphi_{29}=\boldsymbol{\delta}_{4x}$	$x\,\bar{z}\,y$
$\varphi_6=(\boldsymbol{\sigma}_{4x})^{-1}$	$\bar{x}\,\bar{z}\,y$	$\varphi_{30}=(\boldsymbol{\delta}_{4x})^{-1}$	$x\,z\,\bar{y}$
$\varphi_7=\boldsymbol{\sigma}_{4y}$	$\bar{z}\,\bar{y}\,x$	$\varphi_{31}=(\boldsymbol{\delta}_{4y})$	$z\,y\,\bar{x}$
$\varphi_8=(\boldsymbol{\sigma}_{4y})^{-1}$	$z\,\bar{y}\,\bar{x}$	$\varphi_{32}=(\boldsymbol{\delta}_{4y})^{-1}$	$\bar{z}\,y\,x$
$\varphi_9=\boldsymbol{\sigma}_{4z}$	$y\,\bar{x}\,\bar{z}$	$\varphi_{33}=\boldsymbol{\delta}_{4z}$	$\bar{y}\,x\,z$
$\varphi_{10}=(\boldsymbol{\sigma}_{4z})^{-1}$	$\bar{y}\,x\,\bar{z}$	$\varphi_{34}=(\boldsymbol{\delta}_{4z})^{-1}$	$y\,\bar{x}\,z$
$\varphi_{11}=\boldsymbol{\rho}_{xy}$	$\bar{y}\,\bar{x}\,z$	$\varphi_{35}=\boldsymbol{\delta}_{2xy}$	$y\,x\,\bar{z}$
$\varphi_{12}=\boldsymbol{\rho}_{x\bar{y}}$	$y\,x\,z$	$\varphi_{36}=\boldsymbol{\delta}_{2x\bar{y}}$	$\bar{y}\,\bar{x}\,\bar{z}$
$\varphi_{13}=\boldsymbol{\rho}_{xz}$	$\bar{z}\,y\,\bar{x}$	$\varphi_{37}=\boldsymbol{\delta}_{2xz}$	$z\,\bar{y}\,x$
$\varphi_{14}=\boldsymbol{\rho}_{x\bar{z}}$	$z\,y\,x$	$\varphi_{38}=\boldsymbol{\delta}_{2x\bar{z}}$	$\bar{z}\,\bar{y}\,\bar{x}$
$\varphi_{15}=\boldsymbol{\rho}_{yz}$	$x\,\bar{z}\,\bar{y}$	$\varphi_{39}=\boldsymbol{\delta}_{2yz}$	$\bar{x}\,z\,y$
$\varphi_{16}=\boldsymbol{\rho}_{y\bar{z}}$	$x\,z\,y$	$\varphi_{40}=\boldsymbol{\delta}_{2y\bar{z}}$	$\bar{x}\,\bar{z}\,\bar{y}$
$\varphi_{17}=\boldsymbol{\delta}_{3xyz}$	$y\,z\,x$	$\varphi_{41}=\boldsymbol{\sigma}_{6xyz}$	$\bar{y}\,\bar{z}\,\bar{x}$
$\varphi_{18}=(\boldsymbol{\delta}_{3xyz})^{-1}$	$z\,x\,y$	$\varphi_{42}=(\boldsymbol{\sigma}_{6xyz})^{-1}$	$\bar{z}\,\bar{x}\,\bar{y}$
$\varphi_{19}=\boldsymbol{\delta}_{3\bar{x}\bar{y}z}$	$\bar{z}\,x\,\bar{y}$	$\varphi_{43}=\boldsymbol{\sigma}_{6\bar{x}\bar{y}z}$	$z\,\bar{x}\,y$
$\varphi_{20}=(\boldsymbol{\delta}_{3\bar{x}\bar{y}z})^{-1}$	$y\,\bar{z}\,\bar{x}$	$\varphi_{44}=(\boldsymbol{\sigma}_{6\bar{x}\bar{y}z})^{-1}$	$\bar{y}\,z\,x$
$\varphi_{21}=\boldsymbol{\delta}_{3\bar{x}y\bar{z}}$	$z\,\bar{x}\,\bar{y}$	$\varphi_{45}=\boldsymbol{\sigma}_{6\bar{x}y\bar{z}}$	$\bar{z}\,x\,y$
$\varphi_{22}=(\boldsymbol{\delta}_{3\bar{x}y\bar{z}})^{-1}$	$\bar{y}\,\bar{z}\,x$	$\varphi_{46}=(\boldsymbol{\sigma}_{6\bar{x}y\bar{z}})^{-1}$	$y\,z\,\bar{x}$
$\varphi_{23}=\boldsymbol{\delta}_{3x\bar{y}\bar{z}}$	$\bar{z}\,\bar{x}\,y$	$\varphi_{47}=\boldsymbol{\sigma}_{6x\bar{y}\bar{z}}$	$z\,x\,\bar{y}$
$\varphi_{24}=(\boldsymbol{\delta}_{3x\bar{y}\bar{z}})^{-1}$	$\bar{y}\,z\,\bar{x}$	$\varphi_{48}=(\boldsymbol{\sigma}_{6x\bar{y}\bar{z}})^{-1}$	$y\,\bar{z}\,x$

[a] In both diamond O_h^7 and rocksalt O_h^5 the origin is chosen at the site occupied by an atom [9].

It is clear that these coset representatives are not closed under multiplication. Hence $O_h^7=Fd3m$ is a more general central extension of $\mathfrak{T}$ be O_h (see Sect. 9). However, the subset of the first 24 cosets from (126.6) forms space group $T_d^2=F\bar{4}3m$.

The unit cells of rocksalt and diamond structure are illustrated in Figs. 1 and 2 respectively.

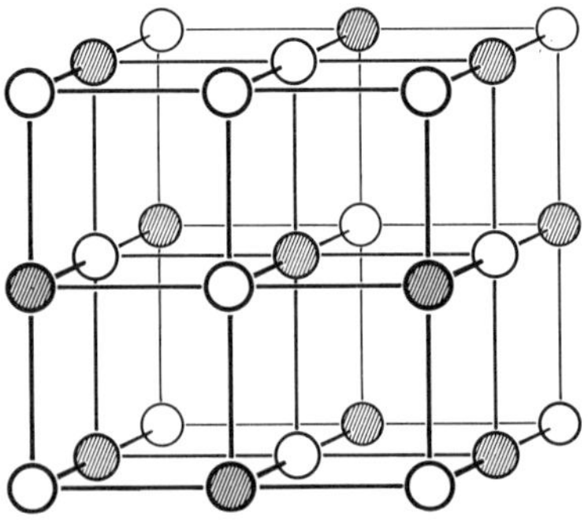

Fig. 1. The rocksalt-NaCl structure, unit cell. (From H. JAGODZINSKI, in: this Encyclopedia, Vol. VII/1, p. 84, Fig. 54)

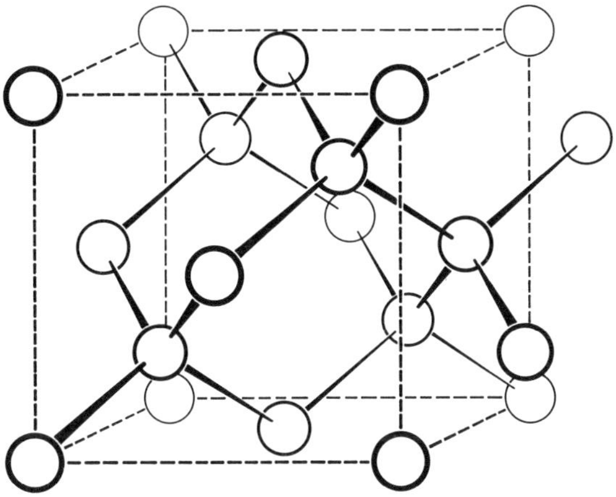

Fig. 2. The unit cell of diamond structure. (From H. JAGODZINSKI, in: this Encyclopedia, Vol. VII/1, p. 80, Fig. 48)

127. Irreducible representations in rocksalt. Our first task here is to select the set of all wave vectors from within or upon the surface of the first Brillouin zone to be used in constructing first the stars and, then complete irreducible representations. The first Brillouin zone for $\mathfrak{T}$ is shown on Fig. 3. Evidently there are many special wave vectors which lie at points, on lines or planes of symmetry in the zone. In Table 3 a list of the wave vectors of interest is given. In this table one wave vector from each star is singled out, and its coordinates are given explicitly. This wave vector is the canonical wave vector of its star, as defined in (32.10). It appears that by considering phonons whose symmetries are based upon these wave vectors (and then stars) corresponding to critical points the

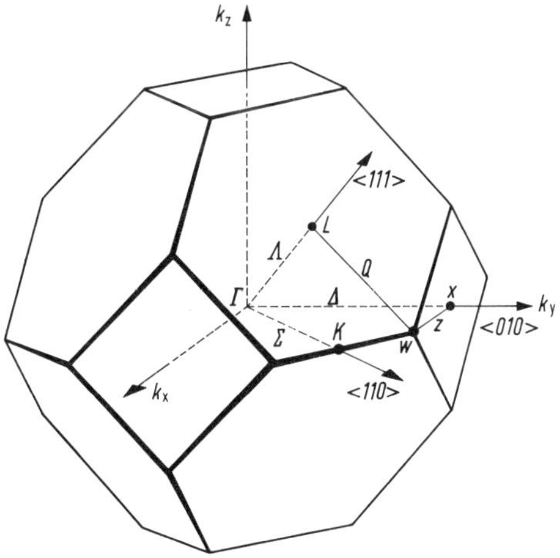

Fig. 3. Brillouin zone for the face centered cubic translation group. [From L. BOUCKAERT, R. SMOLUCHOWSKI, and E. WIGNER: Phys. Rev. **50**, 65 (1936), Fig. 4]

major available experimental results can be explained, for infra-red and Raman processes in perfect and imperfect crystals. This will be illustrated in Parts N and O.

Table 3. Coordinates of zone points for diamond and rocksalt

Points	Coordinates[a]	s = Multiplicity
$\boldsymbol{\Gamma}$	$(0, 0, 0)(1/a)$	1
$\boldsymbol{X}$	$(2\pi, 0, 0)(1/a)$	3
$\boldsymbol{L}$	$(\pi, \pi, \pi)(1/a)$	4
$\boldsymbol{W}$	$(2\pi, 0, -\pi)(1/a)$	6
$\boldsymbol{\Delta}$	$(\kappa, 0, 0)(1/a)$	6
$\boldsymbol{\Lambda}$	$(\kappa, \kappa, \kappa)(1/a)$	8
$\boldsymbol{Z}$	$(2\pi, 0, \kappa)(1/a)$	12
$\boldsymbol{\Sigma}$	$(\kappa, \kappa, 0)(1/a)$	12
$\boldsymbol{G}$	$(2\pi, \kappa, \kappa)(1/a)$	12
$\boldsymbol{M}$	$(\kappa_1, \kappa_1, \kappa_2)(1/a)$	24
$\boldsymbol{Q}$	$(\kappa, \pi, 2\pi-\kappa)(1/a)$	24
$\boldsymbol{N}$	$(\kappa_1, \kappa_2, 0)(1/a)$	24
$\boldsymbol{P}$	$(\kappa_1, \kappa_2, 2\pi)(1/a)$	24
$\boldsymbol{k}$	$(\kappa_1, \kappa_2, \kappa_3)(1/a)$	48

[a] Coordinates are given in terms of components on cubic axes.

The next step in our program is to determine the space group of the canonical wave vector $\boldsymbol{k}$ in each case. Owing to the property that O_h^5 is symmorphic, it is clear that the sub-space groups of O_h^5, which are the particular groups of the wave vector $\mathfrak{G}(\boldsymbol{k})$, are also symmorphic. Hence, in each case the factor group for any $\boldsymbol{k}$:

$$\mathfrak{G}(\boldsymbol{k})/\mathfrak{T} = \mathfrak{P}(\boldsymbol{k}) \tag{127.1}$$

is a point group which is a subgroup of O_h. It suffices then to enumerate the point groups and their irreducible representations which are relevant, for each canonical wave vector $\boldsymbol{k}$.

When the irreducible representations of all the point groups $\mathfrak{P}(\boldsymbol{k})$ have been assembled we possess the matrix in the (11) block matrix of the full matrix irreducible representation $D^{(\star\boldsymbol{k})(m)}$. It is then a simple matter of using Eq. (36.14) or the corresponding character Eq. (37.3) to obtain the character system in the full-group irreducible representation for any element in $\mathfrak{G}$.

The calculation will be illustrated for the rocksalt structure[1, 2] by giving the character system of the full group irreducible representation corresponding to stars $\boldsymbol{\Gamma}$, $\star\boldsymbol{L}$, $\star\boldsymbol{X}$. Selection rules for a rather large variety of processes can be given using only those character systems, owing to the special properties of the wave vector selection rules among these stars. In addition in order to analyze dynamics of processes involving other stars whose canonical vectors are listed in Table 3, the full matrices for these irreducible representations are needed to get various matrix elements. Aside from special cases of interest, we shall in these other cases, give the results in the form of tables of selection rules for the rocksalt structure.

[1] L. BOUCKAERT, R. SMOLUCHOWSKI, and E. WIGNER: Phys. Rev. **50**, 58 (1936).
[2] L.C. CHEN, R. BERENSON, and J.L. BIRMAN: Phys. Rev. **170**, 639 (1968).

We start with wave vector $\boldsymbol{k}=\boldsymbol{\Gamma}=(0,0,0)$. Clearly the space group of $\boldsymbol{\Gamma}$ is $\mathfrak{G}(\boldsymbol{\Gamma})=\mathfrak{G}$, the entire space group. The translation group $\mathfrak{T}(\boldsymbol{\Gamma})\equiv\mathfrak{T}$ is the entire translation group. Hence the factor group

$$\mathfrak{G}(\boldsymbol{\Gamma})/\mathfrak{T}(\boldsymbol{\Gamma})=\mathfrak{P}(\boldsymbol{\Gamma})=O_h \tag{127.2}$$

is the entire point group of the space group O_h: The irreducible representations of point group O_h are well known and are listed in Table 4. Observe that since every translation operation is represented in this irreducible representation by the unit matrix, we have in Table 4 actually the entire space group irreducible representations, including translations: each rotational element listed in the table, $\boldsymbol{\varphi}$, actually stands for an entire coset $\{\boldsymbol{\varphi}|0\}\,\mathfrak{T}$ in $\mathfrak{G}$.

Table 4. Irreducible representations of O_h. $\boldsymbol{\Gamma}^{(m)}=\mathfrak{C}(\boldsymbol{\Gamma})/\mathfrak{T}$

m	ε	$8\delta_{3xyz}$	$3\delta_{2x}$	$6\delta_{4x}$	$6\delta_{2xy}$	i	$8\sigma_{6xyz}$	$3\rho_x$	$6\sigma_{4x}$	$6\rho_{xy}$
1$\pm$	1	1	1	1	1	± 1	± 1	± 1	± 1	± 1
2$\pm$	1	1	1	-1	-1	± 1	± 1	± 1	∓ 1	∓ 1
12$\pm$	2	-1	2	0	0	± 2	∓ 1	± 2	0	0
15$\pm$	3	0	-1	1	-1	± 3	0	∓ 1	± 1	∓ 1
25$\pm$	3	0	-1	-1	1	± 3	0	∓ 1	∓ 1	± 1

Note to Table 4: Correspondence between our symbols (Table 4) and BOUCKAERT, SMOLUCHOWSKI, WIGNER: Phys. Rev. **50**, 58 (1936).

$$\begin{array}{ll} \boldsymbol{\Gamma}^{1+}=\boldsymbol{\Gamma}_1 & \boldsymbol{\Gamma}^{12-}=\boldsymbol{\Gamma}'_{12} \\ \boldsymbol{\Gamma}^{1-}=\boldsymbol{\Gamma}'_1 & \boldsymbol{\Gamma}^{15+}=\boldsymbol{\Gamma}'_{15} \\ \boldsymbol{\Gamma}^{2+}=\boldsymbol{\Gamma}_2 & \boldsymbol{\Gamma}^{15-}=\boldsymbol{\Gamma}_{15} \\ \boldsymbol{\Gamma}^{2-}=\boldsymbol{\Gamma}'_2 & \boldsymbol{\Gamma}^{25+}=\boldsymbol{\Gamma}'_{25} \\ \boldsymbol{\Gamma}^{12+}=\boldsymbol{\Gamma}_{12} & \boldsymbol{\Gamma}^{25-}=\boldsymbol{\Gamma}_{25} \end{array}$$

Consider $^\star\boldsymbol{L}$. The canonical arm listed in Table 3 is taken as $\boldsymbol{L}_1$. Then the arms of $^\star\boldsymbol{L}$ are

$$\begin{aligned} \boldsymbol{L}_1&=(1,1,1)(\pi/a); & \boldsymbol{L}_2&=(1,-1,-1)(\pi/a) \\ \boldsymbol{L}_3&=(-1,1,-1)(\pi/a); & \boldsymbol{L}_4&=(-1,-1,1)(\pi/a) \end{aligned} \tag{127.3}$$

and

$$^\star\boldsymbol{L}=\{\boldsymbol{L}_1,\boldsymbol{L}_2,\boldsymbol{L}_3,\boldsymbol{L}_4\}. \tag{127.4}$$

From Table 2 we note that the following proper and improper rotations leave $\boldsymbol{L}_1$ invariant (mod $2\pi\,\boldsymbol{B}_j$):

$$\varepsilon,\boldsymbol{\rho}_{x\bar{y}},\boldsymbol{\rho}_{x\bar{z}},\boldsymbol{\rho}_{y\bar{z}},\boldsymbol{\delta}_{3xyz},\boldsymbol{\delta}_{3xyz}^{-1},\boldsymbol{i},\boldsymbol{\delta}_{2x\bar{y}},\boldsymbol{\delta}_{2x\bar{z}},\boldsymbol{\delta}_{2y\bar{z}},\boldsymbol{\sigma}_{6xyz},\boldsymbol{\sigma}_{6xyz}^{-1}\equiv\mathfrak{P}(\boldsymbol{L}_1). \tag{127.5}$$

These rotations taken as coset representatives $\{\boldsymbol{\varphi}|0\}$, and then combined with $\mathfrak{T}$, form the space group $\mathfrak{G}(\boldsymbol{L}_1)$. Then we may write for the entire space group $\mathfrak{G}$, decomposed into cosets of $\mathfrak{G}(\boldsymbol{L}_1)$:

$$\mathfrak{G}=\mathfrak{G}(\boldsymbol{L}_1)+\{\boldsymbol{\delta}_{2x}|0\}\,\mathfrak{G}(\boldsymbol{L}_1)+\{\boldsymbol{\delta}_{2y}|0\}\,\mathfrak{G}(\boldsymbol{L}_1)+\{\boldsymbol{\delta}_{2z}|0\}\,\mathfrak{G}(\boldsymbol{L}_1). \tag{127.6}$$

We proceed by finding the irreducible representations of the space group $\mathfrak{G}(\boldsymbol{L}_1)$. First observe that the translation group $\mathfrak{T}(\boldsymbol{L}_1)$, is defined as the set of all $\boldsymbol{R}_L$ such that

$$\boldsymbol{k}\cdot\boldsymbol{R}_L=\boldsymbol{L}_1\cdot\boldsymbol{R}_L=2\pi n, \tag{127.7}$$

where n is an integer. From the definition of $\boldsymbol{L}_1$ we must have for any translation $\boldsymbol{R}_L$ in $\mathfrak{T}$

$$(l_1+l_2+l_3)=2n; \qquad n=0,1,\ldots \tag{127.8a}$$

or

$$=(2n+1); \qquad n=0,1,\ldots. \tag{127.8b}$$

Thus all translations $\{\varepsilon|\boldsymbol{R}_L\}$ with the property (127.8a) are mapped onto the identy matrix in $\mathfrak{G}(\boldsymbol{L}_1)$ while those with (127.8b) are mapped onto (-1) times the identity. Hence we may decompose the full translation group $\mathfrak{T}$ into cosets with respect to $\mathfrak{T}(\boldsymbol{L}_1)$ as:

$$\mathfrak{T}/\mathfrak{T}(\boldsymbol{L}_1)=\mathfrak{T}(\boldsymbol{L}_1)+\{\varepsilon|\boldsymbol{R}_L'\}\,\mathfrak{T}(\boldsymbol{L}_1). \tag{127.9}$$

Any element in the coset whose representative is $\{\varepsilon|\boldsymbol{R}_L'\}$ is represented by a diagonal matrix whose diagonal elements are

$$\exp(i\,\boldsymbol{L}_1\cdot\boldsymbol{R}_L')=-1 \tag{127.10}$$

since we may take the representative in (127.9) as

$$\{\varepsilon|\boldsymbol{R}_L'\}=\{\varepsilon|\boldsymbol{t}_{xy}\} \tag{127.11}$$

and so

$$\boldsymbol{t}_{xy}\cdot\boldsymbol{L}_1=\pi. \tag{127.12}$$

We next decompose $\mathfrak{G}(\boldsymbol{L}_1)$ into cosets with respect to $\mathfrak{T}(\boldsymbol{L}_1)$. The coset representatives of $\mathfrak{G}(\boldsymbol{L}_1)$ in this decomposition are of types

$$\{\varphi_{L_1}|0\},\quad \{\varphi_{L_1}|\boldsymbol{t}_{xy}\} \tag{127.13}$$

where φ_{L_1} is one of the rotations in (127.5). Then the little group of the wave vector $\boldsymbol{L}_1$ is $\mathfrak{G}(\boldsymbol{L}_1)/\mathfrak{T}(\boldsymbol{L}_1)$, a group of order 24.

It is clear that the irreducible representations of $\mathfrak{G}(\boldsymbol{L}_1)/\mathfrak{T}(\boldsymbol{L}_1)$ are simply related to those of the point group $\mathfrak{P}(\boldsymbol{L}_1)=D_{3d}$ [2]. Thus, for an acceptable irreducible representation $D^{(\boldsymbol{L}_1)(m)}$ of $\mathfrak{G}(\boldsymbol{L}_1)$ we must have, from (127.10):

$$\begin{aligned} D^{(\boldsymbol{L}_1)(m)}(\{\varphi_{L_1}|\boldsymbol{t}_{xy}\}) &= D^{(\boldsymbol{L}_1)(m)}(\{\varepsilon|\boldsymbol{t}_{xy}\})\,D^{(\boldsymbol{L}_1)(m)}(\{\varphi_{L_1}|0\})\\ &=(-1)\,D^{(\boldsymbol{L}_1)(m)}(\{\varepsilon|0\})\,D^{(\boldsymbol{L}_1)(m)}(\{\varphi_{L_1}|0\})\\ &=-D^{(\boldsymbol{L}_1)(m)}(\{\varphi_{L_1}|0\}).\end{aligned} \tag{127.14}$$

Hence only the irreducible representations of the point group D_{3d} need be considered. An alternative way of stating the same result is that the *matrix* group $D^{(\boldsymbol{L}_1)(m)}$ consisting of the set of all matrices

$$D^{(\boldsymbol{L}_1)(m)}(\{\varphi_{L_1}|0\}) \quad\text{and}\quad D^{(\boldsymbol{L}_1)(m)}(\{\varphi_{L_1}|\boldsymbol{t}_{xy}\}) \tag{127.15}$$

is isomorphic to a direct product group $C_2\otimes D_{3d}$. Thus adding translations is in fact trivial, when one possesses all the irreducible representations of $\mathfrak{P}(\boldsymbol{L}_1)=D_{3d}$. In Table 5 the character table of the irreducible representations of the point group $\mathfrak{P}(\boldsymbol{L}_1)$ are given.

Table 5. Irreducible representations of $D_{3d}=\mathfrak{G}(L_1)/\mathfrak{T}(L_1)$ in rocksalt (see text (127.15)), and diamond (see text (132.22))

	$\{\varepsilon\|0\}$	$\delta_{3xyz}, \delta_{3xyz}^{-1}$	$\delta_{2x\bar{y}}, \delta_{2x\bar{z}}, \delta_{2y\bar{z}}$	i	$\sigma_{6xyz}, \sigma_{6xyz}^{-1}$	$\rho_{x\bar{y}}, \rho_{x\bar{z}}, \rho_{y\bar{z}}$
$D^{(L_1)(1+)}$	1	1	1	1	1	1
$D^{(L_1)(2+)}$	1	1	-1	1	1	-1
$D^{(L_1)(3+)}$	2	-1	0	2	-1	0
$D^{(L_1)(1-)}$	1	1	1	-1	-1	-1
$D^{(L_1)(2-)}$	1	1	-1	-1	-1	1
$D^{(L_1)(3-)}$	2	-1	0	-2	1	0

To obtain the character table of the full group irreducible representation $D^{(*L)(m)}$ based on $D^{(L_1)(m)}$ we proceed as discussed in Sect. 37. The relevant formula is (37.3).

$$\chi^{(*L)(m)}(\{\varphi_p|0\})=\sum_{\sigma}\dot{\chi}^{(L_1)(m)}(\{\varphi_\sigma|0\}^{-1}\{\varphi_p|0\}\{\varphi_\sigma|0\}), \qquad (127.16)$$

where the $\{\varphi_\sigma|0\}$ are the coset representatives in (127.6) and the φ_p are general elements in $\mathfrak{G}$. Then, what is required is the conjugate of each element in $\mathfrak{G}$ with the conjugation performed with respect to the coset representative $\{\varphi_\sigma|0\}$ in (127.6). The result is easily found and is given in Table 6. Then, with the table of conjugates Table 6 and the basic character Table 5 of the point group D_{3d}, we can assemble the character table for the full irreducible representation. In Table 7 the set of characters for the coset representatives $\{\varphi_p|0\}$ of $\mathfrak{G}$ is given.

Table 6. Conjugate elements in $\mathfrak{P}(L_1), \ldots, \mathfrak{P}(L_4)$

φ_p	$\delta_{2x}^{-1}\varphi_p\delta_{2x}$	$\delta_{2y}^{-1}\varphi_p\delta_{2y}$	$\delta_{2z}^{-1}\varphi_p\delta_{2z}$
ε	ε	ε	ε
δ_{2x}	δ_{2x}	δ_{2x}	δ_{2x}
δ_{2y}	δ_{2y}	δ_{2y}	δ_{2z}
δ_{2z}	δ_{2z}	δ_{2z}	δ_{2z}
σ_{4x}	σ_{4x}	σ_{4x}^{-1}	σ_{4x}^{-1}
σ_{4x}^{-1}	σ_{4x}^{-1}	σ_{4x}	σ_{4x}
σ_{4y}	σ_{4y}^{-1}	σ_{4y}	σ_{4z}^{-1}
σ_{4y}^{-1}	σ_{4y}	σ_{4y}^{-1}	σ_{4z}
σ_{4z}	σ_{4z}^{-1}	σ_{4z}^{-1}	σ_{4z}
σ_{4z}^{-1}	σ_{4z}	σ_{4z}	σ_{4z}^{-1}
ρ_{xy}	$\rho_{x\bar{y}}$	$\rho_{x\bar{y}}$	ρ_{xy}
$\rho_{x\bar{y}}$	ρ_{xy}	ρ_{xy}	$\rho_{x\bar{y}}$
ρ_{xz}	$\rho_{x\bar{z}}$	ρ_{xz}	$\rho_{x\bar{z}}$
$\rho_{x\bar{z}}$	ρ_{xz}	$\rho_{x\bar{z}}$	ρ_{xz}
ρ_{yz}	ρ_{yz}	$\rho_{y\bar{z}}$	$\rho_{y\bar{z}}$
$\rho_{y\bar{z}}$	$\rho_{y\bar{z}}$	ρ_{yz}	ρ_{yz}
δ_{3xyz}	$\delta_{3x\bar{y}\bar{z}}$	$\delta_{3\bar{x}y\bar{z}}$	$\delta_{3\bar{x}\bar{y}z}$
δ_{3xyz}^{-1}	$\delta_{3x\bar{y}\bar{z}}^{-1}$	$\delta_{3\bar{x}y\bar{z}}^{-1}$	$\delta_{3\bar{x}y\bar{z}}^{-1}$
$\delta_{3\bar{x}\bar{y}z}$	$\delta_{3\bar{x}y\bar{z}}$	$\delta_{3x\bar{y}\bar{z}}$	δ_{3xyz}
$\delta_{3\bar{x}\bar{y}z}^{-1}$	$\delta_{3\bar{x}y\bar{z}}^{-1}$	$\delta_{3x\bar{y}\bar{z}}^{-1}$	δ_{3xyz}^{-1}
$\delta_{3\bar{x}y\bar{z}}$	$\delta_{3\bar{x}\bar{y}z}$	δ_{3xyz}	$\delta_{3x\bar{y}\bar{z}}$
$\delta_{3\bar{x}y\bar{z}}^{-1}$	$\delta_{3\bar{x}\bar{y}z}^{-1}$	δ_{3xyz}^{-1}	$\delta_{3x\bar{y}\bar{z}}^{-1}$
$\delta_{3x\bar{y}\bar{z}}$	δ_{3xyz}	$\delta_{3\bar{x}\bar{y}z}$	$\delta_{3\bar{x}y\bar{z}}$
$\delta_{3x\bar{y}\bar{z}}^{-1}$	δ_{3xyz}^{-1}	$\delta_{3\bar{x}\bar{y}z}^{-1}$	$\delta_{3\bar{x}y\bar{z}}^{-1}$

Table 7. Irreducible representations $D^{(\star L)(m)}$ of O_h^5. Coset representatives are all $\{\varphi_p|0\}$

(m)	$\mathfrak{C}_1$ ε	$\mathfrak{C}_2$ $8\delta_{3xyz}$	$\mathfrak{C}_3$ $3\delta_{2x}$	$\mathfrak{C}_4$ $6\delta_{4x}$	$\mathfrak{C}_5$ $6\delta_{2xy}$	$\mathfrak{C}_6$ i	$\mathfrak{C}_7$ $8\sigma_{6xyz}$	$\mathfrak{C}_8$ $3\rho_x$	$\mathfrak{C}_9$ $6\sigma_{4x}$	$\mathfrak{C}_{10}$ $6\rho_{xy}$
1$\pm$	4	1	0	0	2	± 4	± 1	0	0	± 2
2$\pm$	4	1	0	0	-2	± 4	± 1	0	0	∓ 2
3$\pm$	8	-1	0	0	0	± 8	∓ 1	0	0	0

Table reprinted from L.C. Chen, R. Berenson, and J.L. Birman: Phys. Rev. **170**, 639 (1968).

In view of our later work with the non-symmorphic diamond structure, it may be useful to point out that in the case of the symmorphic space groups like rocksalt, the complete specification of the irreducible representation can most economically be given if only the characters of coset representatives (pure rotations) is given. To obtain the character for a general space group element including a translation, we modify (127.16) as:

$$\chi^{(\star L)(m)}(\{\varphi_p|\boldsymbol{R}_M\}) = \sum_\sigma \exp -(i\,\boldsymbol{k}_\sigma \cdot \boldsymbol{R}_M)\,\dot{\chi}^{(L_1)(m)}(\{\varphi_\sigma^{-1} \cdot \varphi_p \cdot \varphi_\sigma|0\}) \tag{127.17}$$

with

$$\boldsymbol{k}_\sigma \equiv \boldsymbol{k} \cdot \varphi_\sigma^{-1} = \varphi_\sigma \cdot \boldsymbol{k} \tag{127.18}$$

thus a phase factor enters when a translation $\boldsymbol{R}_M$, is to be considered. However, since the set of all coset representatives in $\mathfrak{G}/\mathfrak{T}$ are pure rotations, and hence closed as a group, no problems arise in augmenting (127.16) to obtain (127.17). For emphasis we remark that (127.17) should be used in cojunction with the definition of the set of wave vectors in $\star\boldsymbol{k}$ (e.g. (127.3) and (127.4)) and with a table of conjugates of coset representatives such as Table 5.

We next consider the irreducible representations which are associated with $\star\boldsymbol{X}$. The arms of this star are

$$X_1 = (1,0,0)(2\pi/a); \qquad X_2 = (0,1,0)(2\pi/a); \qquad X_3 = (0,0,1)(2\pi/a) \tag{127.19}$$

and

$$\star\boldsymbol{X} = \{X_1, X_2, X_3\}. \tag{127.20}$$

Calling the canonical wave vector in $\star\boldsymbol{X}$: $\boldsymbol{X}_1$ we have the following rotational elements which appear as $\{\varphi_{X_1}|0\}$, as coset representatives in $\mathfrak{G}(\boldsymbol{X}_1)$:

$$\varepsilon, \delta_{2x}, \delta_{2y}, \delta_{2z}, \sigma_{4x}, \sigma_{4x}^{-1}, \rho_{yz}, \rho_{y\bar{z}}; \qquad i, \rho_x, \rho_y, \rho_z, \delta_{4x}, \delta_{4x}^{-1}, \delta_{2yz}, \delta_{2y\bar{z}}. \tag{127.21}$$

Table 8. Irreducible representations $D^{(X_1)(m)}$ of $\mathfrak{G}(X_1)/\mathfrak{T} = D_{4h}$ in O_h^5

m	ε	δ_{2x}	δ_{2y}, δ_{2z}	$\delta_{4x}, \delta_{4x}^{-1}$	$\delta_{2yz}, \delta_{2y\bar{z}}$	i	ρ_x	ρ_y, ρ_z	$\sigma_{4x}, \sigma_{4x}^{-1}$	$\rho_{yz}, \rho_{y\bar{z}}$
1$\pm$	1	1	1	1	1	± 1	± 1	± 1	± 1	± 1
2$\pm$	1	1	1	-1	-1	± 1	± 1	± 1	∓ 1	∓ 1
3$\pm$	1	1	-1	-1	1	± 1	± 1	∓ 1	∓ 1	± 1
4$\pm$	1	1	-1	1	-1	± 1	± 1	∓ 1	± 1	∓ 1
5$\pm$	2	-2	0	0	0	± 2	∓ 2	0	0	0

Table 9. Conjugate elements of $\mathfrak{P}(X_1)$: Rotations only

φ_{X_1}	$\delta_{3xyz}^{-1}\,\varphi_{X_1}\,\delta_{3xyz}$	$\delta_{3xyz}\,\varphi_{X_1}\,\delta_{3xyz}^{-1}$
ε	ε	ε
δ_{2x}	δ_{2z}	δ_{2y}
δ_{2y}	δ_{2x}	δ_{2z}
δ_{2z}	δ_{2y}	δ_{2x}
σ_{4x}	σ_{4z}	σ_{4y}
σ_{4x}^{-1}	σ_{4z}^{-1}	σ_{4y}^{-1}
ρ_{yz}	ρ_{xy}	ρ_{xz}
$\rho_{y\bar{z}}$	$\rho_{x\bar{y}}$	$\rho_{x\bar{z}}$
i	i	i
ρ_x	ρ_z	ρ_y
ρ_y	ρ_x	ρ_z
ρ_z	ρ_y	ρ_x
δ_{4x}	δ_{4z}	δ_{4y}
δ_{4x}^{-1}	δ_{4z}^{-1}	δ_{4y}^{-1}
δ_{2yz}	δ_{2xy}	δ_{2xz}
$\delta_{2y\bar{z}}$	$\delta_{2x\bar{y}}$	$\delta_{2x\bar{z}}$

In $\mathfrak{G}(X_1)$ each of these is to be combined with the elements of $\mathfrak{T}$. The point group $\mathfrak{P}(X_1)$ is $D_{4h}=D_4 \otimes C_h$ [2]. The irreducible representations of D_{4h} are given in Table 8. The coset decomposition of $\mathfrak{G}$ with respect to $\mathfrak{G}(X_1)$ is

$$\mathfrak{G}=\mathfrak{G}(X_1)+\{\delta_{3xyz}|0\}\,\mathfrak{G}(X_1)+\{\delta_{3xyz}^{-1}|0\}\,\mathfrak{G}(X_1). \tag{127.22}$$

The conjugates of the rotational elements (127.21) by the coset representatives $\{\delta_{3xyz}|0\}$, $\{\delta_{3xyz}^{-1}|0\}$ are given in Table 9. As in Table 6, it is only necessary to give the conjugate of a single element from each class in the point group O_h. However, since some elements which are in the same class in O_h are in different classes in $\mathfrak{P}(X_1)$ Table 9 contains a little redundant information. In Table 10 characters of the full group irreducible representations $D^{(*X)(m)}$ are given. These are constructed in accordance with (37.3). It will be observed that elements in the same class in O_h do indeed have the same character in the full group representation.

A sample calculation should make this clear. The elements $\delta_{2x}, \delta_{2y}, \delta_{2z}$ are all elements in $\mathfrak{P}(X_1)$ but are in different classes in $\mathfrak{P}(X_1)$ as indicated in Table 8. Then the character of the elements δ_{2x}, δ_{2y} in the full group irreducible represen-

Table 10. Irreducible representations $D^{(*X)(m)}$ of O_h^5 for coset representatives $\{\varphi_p|0\}$

(m)	$\mathfrak{C}_1$ ε	$\mathfrak{C}_2$ $8\delta_{3xyz}$	$\mathfrak{C}_3$ $3\delta_{2x}$	$\mathfrak{C}_4$ $6\delta_{4x}$	$\mathfrak{C}_5$ $6\delta_{2xy}$	$\mathfrak{C}_6$ i	$\mathfrak{C}_7$ $8\sigma_{6xyz}$	$\mathfrak{C}_8$ $3\rho_x$	$\mathfrak{C}_9$ $6\sigma_{4x}$	$\mathfrak{C}_{10}$ $6\rho_{xy}$
1±	3	0	3	1	1	± 3	0	± 3	± 1	± 1
2±	3	0	3	-1	-1	± 3	0	± 3	∓ 1	∓ 1
3±	3	0	-1	-1	1	± 3	0	∓ 1	∓ 1	± 1
4±	3	0	-1	1	-1	± 3	0	∓ 1	± 1	∓ 1
5±	6	0	-2	0	0	± 6	0	∓ 2	0	0

Table reprinted from same reference as footnote to Tables 4 and 7.

tation is:

$$\begin{aligned}\chi^{(\star X)(m)}(\{\delta_{2x}|0\}) &= \chi^{(X_1)(m)}(\{\delta_{2x}|0\}) + \chi^{(X_1)(m)}(\{\delta_{2y}|0\}) + \chi^{(X_1)(m)}(\{\delta_{2z}|0\}) \\ &= \chi^{(X_1)(m)}(\{\delta_{2x}|0\}) + 2\chi^{(X_1)(m)}(\{\delta_{2y}|0\})\end{aligned} \quad (127.23)$$

and

$$\begin{aligned}\chi^{(\star X)(m)}(\{\delta_{2y}|0\}) &= \chi^{(X_1)(m)}(\{\delta_{2y}|0\}) + \chi^{(X_1)(m)}(\{\delta_{2x}|0\}) + \chi^{(X_1)(m)}(\{\delta_{2z}|0\}) \\ &= \chi^{(X_1)(m)}(\{\delta_{2x}|0\}) + 2\chi^{(X_1)(m)}(\{\delta_{2y}|0\}).\end{aligned} \quad (127.24)$$

Thus in the full group irreducible representation

$$\chi^{(\star X)(m)}(\{\delta_{2x}|0\}) = \chi^{(\star X)(m)}(\{\delta_{2y}|0\}) \quad (127.25)$$

which must be the case since these two elements are in the same class in the full group $\mathfrak{G}$.

128. Some wave vector selection rules in rocksalt. Now using the material in Tables 4, 7, 10 it will be possible to obtain selection rules for all processes involving transitions among and between the irreducible representation $D^{(\Gamma)(m)}$, $D^{(\star L)(m)}$, $D^{(\star X)(m)}$, since these form a closed set of representations as we shall now see. As first step in obtaining the general selection rules we obtain the wave vector selection rules as discussed in Sect. 56. In Table 11 we reproduce the wave vector selection rules for direct products among $D^{(\Gamma)(m)}$, $D^{(\star L)(m)}$ and $D^{(\star X)(m)}$. The general expressions to be used are given in Sect. 56. It is observed that the set of stars Γ, $\star L$, $\star X$ is indeed closed. That is, in the product $\star \boldsymbol{k} \otimes \star \boldsymbol{k}'$, taking any two from the set will produce again members of this set only.

Table 11. Wave vector selection rules in O_h^5 and O_h^7: For products with Γ, $\star X$, $\star L$

Ordinary:	$\Gamma \otimes \Gamma = \Gamma$	*Symmetrized:*	$[\Gamma]_{(2)} = \Gamma$
	$\Gamma \otimes \star X = \star X$		$[2\Gamma]_{(2)} = 3\Gamma$
	$\Gamma \otimes \star L = \star L$		$[3\Gamma]_{(2)} = 6\Gamma$
	$\star X \otimes \star X = 3\Gamma \oplus 2\star X$		$[\star L]_{(2)} = 4\Gamma \oplus 2\star X$
	$\star X \otimes \star L = 3\star L$		$[2\star L]_{(2)} = 12\Gamma \oplus 8\star X$
	$\star L \otimes \star L = 4\Gamma \oplus 4\star X$		$[\star X]_{(2)} = 3\Gamma \oplus \star X$
			$[2\star X]_{(2)} = 9\Gamma \oplus 4\star X$
			$[\Gamma]_{(3)} = \Gamma$
			$[2\Gamma]_{(3)} = 4\Gamma$
			$[3\Gamma]_{(3)} = 10\Gamma$
			$[\star L]_{(3)} = 5\star L$
			$[2\star L]_{(3)} = 30\star L$
			$[\star X]_{(3)} = \Gamma \oplus 3\star X$
			$[2\star X]_{(3)} = 8\Gamma \oplus 16\star X$

129. Reduction of $\star X^{(4-)} \otimes \star X^{(5-)}$ in rocksalt: Illustration of the use of linear algebraic equations. Now we shall obtain selection rules. The two cases which we shall now work out are both of practical interest in the analysis of second order infra-red and Raman processes in the rocksalt structure. Besides these

Table 12. Illustration of the reduction process: Example $\star X^{(4-)} \otimes \star X^{(5-)}$ in O_h^5

Initial character table (Table 10)

	$\{\varepsilon\|0\}$	$\{\delta_{3xyz}\|0\}$	$\{\delta_{2x}\|0\}$	$\{\delta_{4x}\|0\}$	$\{\delta_{2xy}\|0\}$
$(4-)\otimes(5-)$	18	0	2	0	0

Augmented character table for $D^{(\star X)(m)}(\{\varphi_p|t_p\})$ for some selected elements

m	$\{\delta_{2x}\|t_1\}$	$\{\delta_{2x}\|t_3\}$	$\{\delta_{2y}\|t_1\}$	$\{\delta_{4x}\|t_1\}$	$\{\delta_{2yz}\|t_1\}$
$1+$	-1	-1	-1	-1	-1
$2+$	-1	-1	-1	$+1$	$+1$
$3+$	-1	3	-1	$+1$	-1
$4+$	-1	3	-1	-1	$+1$
$5+$	2	-2	$+2$	0	0
$(4-)\otimes(5-)$	-2	-6	-2	0	0

cases, all results of possible use in these structures will be tabulated and given. If other rules are required the methods described in this article should permit them to be obtained.

The two cases are

$$\star X^{(4-)} \otimes \star X^{(5-)} \tag{129.1}$$

which will be analysed in this section and

$$\star L^{(3-)} \otimes \star L^{(3+)} \tag{129.2}$$

which will be analysed in the next section. To obtain the reduction coefficients for these products, any or all of the methods discussed in Sects. 58–60, are open to us. Here we use the method of linear algebraic equations, described in Sect. 57.

We first obtain the character system of the product representation (129.1) which is to be reduced. Observe that in principle we require one character for each space group element $\{\varphi_p|t_p\}$ in the entire unitary group $\mathfrak{G}$. Hence the character system to be given in Table 12 for the product must be considered as only a partial character system for the entire group $\mathfrak{G}$. To obtain a complete and successful reduction requires that for each element $\{\varphi_p|t_p\}$ we have

$$\begin{aligned} &\chi^{(\star X)(4-)}(\{\varphi_p|t_p\})\,\chi^{(\star X)(5-)}(\{\varphi_p|t_p\}) \\ &\quad = \sum_{\star k'' m''} (\star X^{(4-)} \star X^{(5-)}|\star k'' m'')\,\chi^{(\star k'')(m'')}(\{\varphi_p|t_p\}). \end{aligned} \tag{129.3}$$

The sum is over those $\star k''$ which occur, as given by the wave vector selection rules, and over the possible m''. Now the wave vector rule instructs us that only $\star k''=\Gamma$, and $\star k''=\star X$ can occur (see Table 11). But there are 10 irreducible representations (m'') for each of these stars. Hence a total of 20 reduction coefficients must be determined in order to complete the reduction (129.3). These are

$$(\star X^{(4-)} \star X^{(5-)}|\Gamma m''), \qquad m''=1\pm, 2\pm, 12\pm, 15\pm, 25\pm, \tag{129.4}$$

$$(\star X^{(4-)} \star X^{(5-)}|\star X m''), \qquad m''=1\pm, 2\pm, 3\pm, 4\pm, 5\pm. \tag{129.5}$$

Further, the coefficients (129.4) and (129.5) are restricted by dimensionality considerations which follow from the wave vector selection rules (Table 11). Thus if $l_{m''}$ is the dimensionality of the representation $D^{(\Gamma)(m'')}$, or $D^{(\star X)(m'')}$, then

$$\sum_{m''} (\star X^{(4-)} \star X^{(5-)} | \Gamma m'') \, l_{m''} = 6 \tag{129.6}$$

and

$$\sum_{m''} (\star X^{(4-)} \star X^{(5-)} | \star X m'') \, l_{m''} = 12. \tag{129.7}$$

Now in order to continue, we require a sufficient number of Eq. (129.3) to completely determine the unknown coefficients (129.4) and (129.5). Thus, as 20 coefficients are to be determined, we need at most 20 independent linear algebraic Eq. (129.3). (Actually as we soon see far fewer are needed owing to the restrictions (129.6) and (129.7), and the requirement that the coefficients $(\star X^{(4-)} \star X^{(5-)} | \star k'' m'')$ be integers.) In order to augment the basic set of characters for the coset representatives listed in e.g. Tables 4, 7, 10, and to obtain sufficient member of independent linear algebraic equations we need to add translations, i.e. consider elements $\{\varphi_p | t_p\}$ where $t_p = R_p$, a lattice vector. The full group characters which need to be added are obtained modifying (127.17) and (127.18). Now we observe that the parity operation i is a symmetry operation of all three of the representations involved here, so that

$$D^{(\star k)(m\pm)}(\{i\,\varphi_p | 0\}) = \pm D^{(\star k)(m\pm)}(\{\varphi_p | 0\}), \tag{129.8}$$

where

$$\star k = \Gamma, \star X, \star L. \tag{129.9}$$

Hence the product of two odd representations, must give only even resultants. Thus of the 20 potential coefficients 10, of odd parity, are eliminated. On the other hand, we also should eliminate for consistency, the 5 classes in which improper rotational group elements occur as giving redundant information. Hence from this argument we require the characters of at most 5 additional space group elements combined with translations. Clearly if (127.17) and (127.16) is to give any added information it is necessary to augment the rotations which are among the coset representatives in $\mathfrak{G}(X_1)$ with translations. Then (127.17) becomes, using (127.16):

$$\begin{aligned} \chi^{(\star X)(m)}(\{\varphi_{X_1} | R_p\}) = {} & \exp(-i X_1 \cdot R_p)\, \chi^{(X_1)(m)}(\{\varphi_{X_1}\}) \\ & + \exp(-i X_2 \cdot R_p)\, \dot{\chi}^{(X_1)(m)}(\{\varphi_{X_1}^{\sigma}\}) \\ & + \exp(-i X_3 \cdot R_p)\, \dot{\chi}^{(X_1)(m)}(\{\varphi_{X_1}^{\sigma^{-1}}\}), \end{aligned} \tag{129.10}$$

$$\{\varphi_{X_1}^{\sigma}\} \equiv \{\delta_{3xyz}^{-1}\, \varphi_{X_1}\, \delta_{3xyz} | 0\}, \tag{129.11}$$

$$\{\varphi_{X_1}^{\sigma^{-1}}\} \equiv \{\delta_{3xyz}\, \varphi_{X_1}\, \delta_{3xyz}^{-1} | 0\}. \tag{129.12}$$

The conjugates (129.11) and (129.12) are to be read from Table 9.

In Table 12 successive steps in the reduction are illustrated and will now be described. In the first line top labelled $(4-)\otimes(5-)$ the non-redundant part of the character product which is to be reduced is given. Let us denote the unknown reduction coefficients by an abbreviated notation:

$$\begin{aligned} &a_1=(|\boldsymbol{\Gamma}1+); \quad a_2=(|\boldsymbol{\Gamma}2+); \quad a_3=(|\boldsymbol{\Gamma}12+);\\ &a_4=(|\boldsymbol{\Gamma}15+); \quad a_5=(|\boldsymbol{\Gamma}25+) \end{aligned} \tag{129.13}$$

and

$$\begin{aligned} &b_1=(|\star \boldsymbol{X}1+); \quad b_2=(|\star \boldsymbol{X}2+); \quad b_3=(|\star \boldsymbol{X}3+);\\ &b_4=(|\star \boldsymbol{X}4+); \quad b_5=(|\star \boldsymbol{X}5+). \end{aligned} \tag{129.14}$$

Then from (129.6) and (129.7):

$$a_1+a_2+2a_3+3a_4+3a_5=6, \tag{129.15}$$

$$3b_1+3b_2+3b_3+3b_4+6b_5=12. \tag{129.16}$$

Now for each group element we obtain one linear equation involving the a_m, b_m. We use the first line of Table 12 and the relevant columns of Tables 4 and 10. For element $\{\boldsymbol{\delta}_{3xyz}|0\}$, we have from (129.3) and the cited tables:

$$a_1+a_2-a_3=0. \tag{129.17}$$

For $\{\boldsymbol{\delta}_{2x}|0\}$ we obtain

$$a_1+a_2+2a_3-a_4-a_5+3b_1+3b_2-b_3-b_4-2b_5=2. \tag{129.18}$$

For $\{\boldsymbol{\delta}_{4x}|0\}$

$$a_1-a_2+a_4-a_5+b_1-b_2-b_3+b_4=0. \tag{129.19}$$

For $\{\boldsymbol{\delta}_{2xy}|0\}$

$$a_1-a_2-a_4+a_5+b_1-b_2+b_3-b_4=0. \tag{129.20}$$

At this point we have exhausted the information contained in coset representatives $\{\varphi_p|0\}$. It is clear that we do not possess enough information to solve (129.15)–(129.20) uniquely, even adding the evident requirement that a_m and b_m be positive integers:

$$a_m \geqq 0, \quad b_m \geqq 0 \quad \text{integers}. \tag{129.21}$$

Using (129.17) in (129.15) we obtain

$$a_3+a_4+a_5=2 \tag{129.22}$$

so that at least one of the three a_m in (129.22) must vanish.

To go further we require more information: i.e. added characters and equations. Thus we need to augment the character Table 10 by computing added characters and then using the resulting linear equations. In Table 12 some added characters are given; they were computed using (129.10)–(129.12). From the element $\{\boldsymbol{\delta}_{2x}|\boldsymbol{t}_1\}$ we have using Tables 4 and 12:

$$a_1+a_2+2a_3-a_4-a_5-b_1-b_2-b_3-b_4+2b_5=-2 \tag{129.23}$$

and from $\{\boldsymbol{\delta}_{2x}|\boldsymbol{t}_3\}$

$$a_1+a_2+2a_3-a_4-a_5-b_1-b_2+3b_3+3b_4-2b_5=-6 \tag{129.24}$$

and $\{\boldsymbol{\delta}_{2y}|\boldsymbol{t}_1\}$

$$a_1+a_2+2a_3-a_4-a_5-b_1-b_2-b_3-b_4+2b_5=-2. \tag{129.25}$$

Using $\{\boldsymbol{\delta}_{4x}|\boldsymbol{t}_1\}$

$$a_1 - a_2 + a_4 - a_5 - b_1 + b_2 + b_3 - b_4 = 0 \tag{129.26}$$

and from $\{\boldsymbol{\delta}_{2yz}|\boldsymbol{t}_1\}$

$$a_1 - a_2 - a_4 + a_5 - b_1 + b_2 - b_3 + b_4 = 0. \tag{129.27}$$

Now we conclude the reduction. Since the terms from $\boldsymbol{\Gamma}^{(m)}$ are the same for all elements with the same rotational part irrespective of translation as in (129.18)–(129.25) we have the following, obtained by subtraction among these:

(129.18)–(129.23)

$$b_1 + b_2 - b_5 = 1, \tag{129.28}$$

(129.18)–(129.27)

$$-b_3 - b_4 + b_5 = 1, \tag{129.29}$$

(129.26)–(129.19)

$$b_1 - b_2 - b_3 + b_4 = 0, \tag{129.30}$$

(129.27)–(129.20)

$$b_1 - b_2 + b_3 - b_4 = 0, \tag{129.31}$$

and (129.16)

$$b_1 + b_2 + b_3 + b_4 + 2b_5 = 4. \tag{129.32}$$

Clearly then from (129.31) and (129.32):

$$b_1 = b_2; \quad b_3 = b_4 \tag{129.33}$$

leaving from (129.28)–(129.33):

$$2b_1 - b_5 = 1, \tag{129.34}$$

$$-2b_3 + b_5 = 1, \tag{129.35}$$

$$b_1 + b_3 + b_5 = 2. \tag{129.36}$$

Then from (129.34) and (129.35)

$$b_1 + b_3 - b_5 = 0. \tag{129.37}$$

Hence from (129.36) and (129.37)

$$b_5 = 1 \tag{129.38}$$

and

$$b_1 = 1, \tag{129.39}$$

$$b_3 = 0. \tag{129.40}$$

Thus the solution for the b_m is:

$$b_1 = b_2 = 1; \quad b_3 = b_4 = 0; \quad b_5 = 1. \tag{129.41}$$

The a_m may then be directly obtained by substituting the solution for the b_m into (129.15)–(129.20) and solving the resulting equations:

$$a_1 + a_2 + 2a_3 - a_4 - a_5 = -2, \tag{129.42}$$

$$a_1 - a_2 + a_4 - a_5 = 0, \tag{129.43}$$

$$a_1 - a_2 - a_4 + a_5 = 0. \tag{129.44}$$

From (129.43) and (129.44) we find,

$$a_1 = a_2; \qquad a_4 = a_5. \tag{129.45}$$

So, using (129.17) and (129.42) we obtain

$$3a_3 - 2a_4 = -2. \tag{129.46}$$

From (129.22)

$$a_3 + 2a_4 = 2. \tag{129.47}$$

Then from (129.47), (129.46), (129.45), (129.42) we find

$$a_1 = a_2 = a_3 = 0 \tag{129.48}$$

and

$$a_4 = a_5 = 1. \tag{129.49}$$

Summarizing and returning to proper notation we have:

$$\begin{aligned} &D^{(\star X)(4-)} \otimes D^{(\star X)(5-)} \\ &\quad = D^{(\Gamma)(15+)} \oplus D^{(\Gamma)(25+)} \oplus D^{(\star X)(1+)} \oplus D^{(\star X)(2+)} \oplus D^{(\star X)(5+)}. \end{aligned} \tag{129.50}$$

It is useful at this point to emphasize some aspects of the reduction epitomized by (129.50). First: the reduction is complete and unique. That is, consistent with the general group theoretical results, a representation such as the direct product representation $D^{(\star X)(4-)} \otimes D^{(\star X)(5-)}$ is uniquely reducible into a direct sum of irreducible components [*18*]. Of course, in obtaining (129.50) we only used a selected few of the totality of elements of $\mathfrak{G}$ to get the coefficients. But by the completeness and uniqueness of the reduction, (129.50) holds for *any* element of $\mathfrak{G}$ taken as argument. This always serves as a check on the reduction once it has been made, in using this method.

On the other hand, when it is possible to carry out the reduction using the reduction group method (see Sect. 58) one has available the row and column orthonormality relations which check *all* the elements in the reduction group for the correctness of the character. The advantage of the reduction group procedure is that a mathematically precise and complete problem is solved: the reduction group method will be used when we reduce direct products of representation $D^{(\Gamma)(m)}$, $D^{(\star X)(m)}$, $D^{(\star L)(m)}$ in the diamond structure (*vide infra*). The reader who has become acquainted with the reduction group method may wish to obtain (129.50) by that method.

Here we see that for any given reduction within the full group method, the method of linear algebraic equations provides an economical procedure of carrying out the reduction. That is, one only works out a minimal independent set of full group irreducible representation characters. The number of such characters needed is severely limited to the number of unknown reduction coefficients, and these are always a small, finite number. For example in the case at hand 20 coefficients needed to be found. These were completely determined by not more than 15 independent characters for each participating irreducible representation. Using the character tables for the same 15 elements but including also the improper rotations, i.e. those combined with $\boldsymbol{i}$ the reduction of any products $D^{(\star X)(m)} \otimes D^{(\star X)(m')}$ can be effected.

To obtain reduction coefficients for symmetrized powers of representations e.g. $[D^{(\star X)(m)}]_{(2)}$ and $[D^{(\star X)(m)}]_{(3)}$ we use Eqs. (54.16) and (54.21). The elements for which characters are needed, must then be enlarged to include $\{\varphi_p|\boldsymbol{t}_p\}$, $\{\varphi_p|\boldsymbol{t}_p\}^2$, $\{\varphi_p|\boldsymbol{t}_p\}^3$. But these are easily found from the characters already given, and (127.17).

A final remark about the reduction. When one realizes that a particular product such as the one now being reduced in (129.1) contains representations $D^{(\boldsymbol{\Gamma})(m)}$, one might reason that these representations could be removed by using only the characters of the coset representatives $\{\varphi_p|0\}$. That is one might be inclined to use the formula (17.8) to reduce $\chi(R)$

$$a_j = \frac{1}{g} \sum_R \chi^{(j)}(R)^* \chi(R) \tag{129.51}$$

to obtain the reduction coefficients a_j belonging to $D^{(\boldsymbol{\Gamma})(m)}$. This procedure can be employed of course if one carries out the sum over *all* group elements R in $\mathfrak{G}$. Equivalently the sum can be carried out if one sums over the elements R in the reduction group $\mathfrak{R}$ (see below illustration in diamond). However, *no* meaningful result at all can be obtained if one sums the product characters (line 1 of Table 12) over the coset representatives R in $\mathfrak{P}$ only. It is of considerable pedagogic importance to recognize the impropriety of attempting to do the calculation in this fashion, and the reader should stop here and test his understanding at this point.

130. Reduction of $\star L^{(3-)} \otimes \star L^{(3+)}$ in rocksalt. Now let us work the second illustration of the reduction process. Again we use the full group methods, and the technique of linear algebraic equations. We now reduce the product (129.2), which is our second case. Owing to the double degeneracy of $D^{(\boldsymbol{L}_1)(3+)}$ and $D^{(\boldsymbol{L}_1)(3-)}$ the wave vector rules become

$$2\star \boldsymbol{L} \otimes 2\star \boldsymbol{L} = 16\boldsymbol{\Gamma} \oplus 16\star \boldsymbol{X}. \tag{130.1}$$

Hence in the product to be reduced

$$D^{(\star \boldsymbol{L})(3+)} \otimes D^{(\star \boldsymbol{L})(3-)}$$

we have 20 unknown coefficients to determine. Owing to the presence of inversion in each of the $\mathfrak{G}(\boldsymbol{k})$ involved, only odd representations can occur in the reduction of (130.1). Then we abbreviate the coefficients which may arise as:

$$\begin{aligned} &c_1 = (|\boldsymbol{\Gamma}1-); \quad c_2 = (|\boldsymbol{\Gamma}2-); \quad c_3 = (|\boldsymbol{\Gamma}12-);\\ &c_4 = (|\boldsymbol{\Gamma}15-); \quad c_5 = (|\boldsymbol{\Gamma}25-), \end{aligned} \tag{130.2}$$

$$\begin{aligned} &d_1 = (|\star \boldsymbol{X}1-); \quad d_2 = (|\star \boldsymbol{X}2-); \quad d_3 = (|\star \boldsymbol{X}3-);\\ &d_4 = (|\star \boldsymbol{X}4-); \quad d_5 = (|\star \boldsymbol{X}5-). \end{aligned} \tag{130.3}$$

From (130.1) and Tables 4 and 10:

$$c_1 + c_2 + 2c_3 + 3c_4 + 3c_5 = 16, \tag{130.4}$$

$$3d_1 + 3d_2 + 3d_3 + 3d_4 + 6d_5 = 48. \tag{130.5}$$

Table 13. Illustration $D^{(\star L)(3-)} \otimes D^{(\star L)(3+)}$ from Table 7 and augmented terms in O_h^5

$D^{(\star L)(m)}$	$\{\varepsilon\|0\}$	$\{\delta_{3xyz}\|0\}$	$\{\delta_{2x}\|0\}$	$\{\delta_{4x}\|0\}$	$\{\delta_{2xy}\|0\}$	$\{\delta_{2x}\|t_1\}$	$\{\delta_{2x}\|t_3\}$	$\{\delta_{2y}\|t_1\}$	$\{\delta_{4x}\|t_1\}$	$\{\delta_{2yz}\|t_1\}$
1±	4	1	0	0	2	0	0	0	0	0
2±	4	1	0	0	−2	0	0	0	0	0
3±	8	−1	0	0	0	0	0	0	0	0
(3+)⊗(3−)	64	+1	0	0	0	0	0	0	0	0

Table 14. Augmented character table. Note column head for element represented. See text following Eqs. (129.22) and (130.2) for discussion for O_h^5

Irred. rep.	$\{\varepsilon\|0\}$	$\{\delta_{2xyz}\|0\}$	$\{\delta_{2x}\|0\}$	$\{\delta_{4x}\|0\}$	$\{\delta_{2xy}\|0\}$	$\{\delta_{2x}\|t_1\}$	$\{\delta_{2x}\|t_3\}$	$\{\delta_{2y}\|t_1\}$	$\{\delta_{4x}\|t_1\}$	$\{\delta_{2yz}\|t_1\}$	
$\Gamma^{1\pm}$	1	1	1	1	1	1	1	1	1	1	c_1
$\Gamma^{2\pm}$	1	1	1	−1	−1	1	1	1	−1	−1	c_2
$\Gamma^{12\pm}$	2	−1	2	0	0	2	2	2	0	0	c_3
$\Gamma^{15\pm}$	3	0	−1	1	−1	−1	−1	−1	1	−1	c_4
$\Gamma^{25\pm}$	3	0	−1	−1	1	−1	−1	−1	−1	1	c_5
$\star X^{1\pm}$	3	0	3	1	1	−1	−1	−1	−1	−1	d_1
$\star X^{2\pm}$	3	0	3	−1	−1	−1	−1	−1	1	1	d_2
$\star X^{3\pm}$	3	0	−1	−1	1	−1	3	−1	1	−1	d_3
$\star X^{4\pm}$	3	0	−1	1	−1	−1	3	−1	−1	1	d_4
$\star X^{5\pm}$	6	0	−2	0	0	2	−2	2	0	0	d_5
$\star L^{1\pm}$	4	1	0	0	2	0	0	0	0	0	
$\star L^{2\pm}$	4	1	0	0	−2	0	0	0	0	0	
$\star L^{3\pm}$	8	−1	0	0	0	0	0	0	0	0	
$\star L^{3+} \otimes \star L^{3-}$	64	1	0	0	0	0	0	0	0	0	

In Tables 13 and 14 the illustration is again given to help in the reduction of the required product. Only elements for proper rotations are given, as the improper ones simply change sign. From line 4 of Table 13 and Tables 4, 10, and 14 we obtain the following equations for c_m and d_m from consideration of elements without translation:

$$c_1 + c_2 - c_3 = 1, \tag{130.6}$$

$$c_1 + c_2 + 2c_3 - c_4 - c_5 + 3d_1 + 3d_2 - d_3 - d_4 - 2d_5 = 0, \tag{130.7}$$

$$c_1 - c_2 + c_4 - c_5 + d_1 - d_2 - d_3 + d_4 = 0, \tag{130.8}$$

$$c_1 - c_2 - c_4 + c_5 + d_1 - d_2 + d_3 - d_4 = 0 \tag{130.9}$$

from elements with translation:

$$c_1 + c_2 + 2c_3 - c_4 - c_5 - d_1 - d_2 - d_3 - d_4 + 2d_5 = 0, \tag{130.10}$$

$$c_1 + c_2 + 2c_3 - c_4 - c_5 - d_1 - d_2 + 3d_3 + 3d_4 - 2d_5 = 0, \tag{130.11}$$

$$c_1 - c_2 + c_4 - c_5 - d_1 + d_2 + d_3 - d_4 = 0, \tag{130.12}$$

$$c_1 - c_2 - c_4 + c_5 - d_1 + d_2 - d_3 + d_4 = 0. \tag{130.13}$$

From (130.4)–(130.13) we quickly find:

$$d_1 = d_2 = d_3 = d_4 = 2; \quad d_5 = 4 \tag{130.14}$$

and

$$c_1 = c_2 = c_3 = 1; \quad c_4 = c_5 = 2. \tag{130.15}$$

Hence

$$\begin{aligned} D^{(\star \boldsymbol{L})(3-)} \otimes D^{(\star \boldsymbol{L})(3+)} \\ &= D^{(\boldsymbol{\Gamma})(1-)} \oplus D^{(\boldsymbol{\Gamma})(2-)} \oplus D^{(\boldsymbol{\Gamma})(12-)} \oplus 2D^{(\boldsymbol{\Gamma})(15-)} \oplus 2D^{(\boldsymbol{\Gamma})(25-)} \\ &\oplus 2D^{(\star \boldsymbol{X})(1-)} \oplus 2D^{(\star \boldsymbol{X})(2-)} \oplus 2D^{(\star \boldsymbol{X})(3-)} \oplus 2D^{(\star \boldsymbol{X})(4-)} \oplus 4D^{(\star \boldsymbol{X})(5-)}. \end{aligned} \tag{130.16}$$

131. Additional reduction coefficients in rocksalt. It should now be clear that the task of obtaining all reduction coefficients using the technique of solving

Table 15. Wave vector selection rules in rocksalt O_h^5 and diamond O_h^7 coefficients $(\boldsymbol{k}_j \boldsymbol{k}_{j'} | \boldsymbol{k}_{j''})$

$(\boldsymbol{\Gamma k}_{j'} \mid \boldsymbol{k}_{j'}) = 1$	for all $\boldsymbol{k}_{j'}$
$(\boldsymbol{XX} \mid \boldsymbol{\Gamma}) = 3$	
$(\boldsymbol{XX} \mid \boldsymbol{X}) = 2$	
$(\boldsymbol{XL} \mid \boldsymbol{L}) = 3$	
$(\boldsymbol{XW} \mid \boldsymbol{\Delta}) = 2;$	$\Delta_1 = (\pi, 0, 0)(1/a)$
$(\boldsymbol{XW} \mid \boldsymbol{W}) = 1$	
$(\boldsymbol{XZ} \mid \boldsymbol{\Delta}) = 2;$	$\Delta_1 = (\kappa, 0, 0)(1/a)$
$(\boldsymbol{XZ} \mid \boldsymbol{\Delta}') = 2;$	$\Delta_1' = (2\pi - \kappa, 0, 0)(1/a)$
$(\boldsymbol{XZ} \mid \boldsymbol{Z}) = 2;$	$Z_1 = (2\pi, 0, \kappa)(1/a)$
$(\boldsymbol{X\Sigma} \mid \boldsymbol{\Sigma}') = 1;$	$\Sigma_1' = (\kappa - 2\pi, \kappa - 2\pi, 0)(1/a)$
$(\boldsymbol{X\Sigma} \mid \boldsymbol{N}) = 1;$	$N_1 = (\kappa - 2\pi, \kappa, 0)(1/a)$
$(\boldsymbol{X\Delta} \mid \boldsymbol{\Delta}') = 1;$	$\Delta_1' = (\kappa - 2\pi, 0, 0)(1/a)$
$(\boldsymbol{X\Delta} \mid \boldsymbol{Z}) = 1;$	$Z_1 = (2\pi, 0, \kappa)(1/a)$
$(\boldsymbol{X\Lambda} \mid \boldsymbol{M}) = 1;$	$M_1 = (\kappa, \kappa, \kappa - 2\pi)(1/a)$
$(\boldsymbol{LL} \mid \boldsymbol{\Gamma}) = 4$	
$(\boldsymbol{LL} \mid \boldsymbol{X}) = 4$	
$(\boldsymbol{LW} \mid \boldsymbol{\Sigma}) = 2;$	$\Sigma_1 = (\pi, \pi, 0)(1/a)$
$(\boldsymbol{LZ} \mid \boldsymbol{M}) = 2;$	$M_1 = (\pi, \pi, \pi + \kappa)(1/a)$
$(\boldsymbol{L\Sigma} \mid \boldsymbol{M}) = 2;$	$M_1 = (\kappa - \pi, \kappa - \pi, \pi)(1/a)$
$(\boldsymbol{L\Delta} \mid \boldsymbol{M}) = 1;$	$M_1 = (\pi, \pi, \pi + \kappa)(1/a)$
$(\boldsymbol{L\Lambda} \mid \boldsymbol{\Lambda}') = 1;$	$\Lambda_1' = (\pi + \kappa, \pi + \kappa, \pi + \kappa)(1/a)$
$(\boldsymbol{L\Lambda} \mid \boldsymbol{M}) = 1;$	$M_1 = (\pi + \kappa, \pi + \kappa, \kappa - \pi)(1/a)$
$(\boldsymbol{WW} \mid \boldsymbol{\Gamma}) = 6$	
$(\boldsymbol{WW} \mid \boldsymbol{X}) = 2$	
$(\boldsymbol{WW} \mid \boldsymbol{\Sigma}) = 2;$	$\Sigma_1 = (\pi, \pi, 0)(1/a)$
$(\boldsymbol{WZ} \mid \boldsymbol{\Delta}) = 4;$	$\Delta_1 = (\pi + \kappa, 0, 0)(1/a)$
$(\boldsymbol{WZ} \mid \boldsymbol{N}) = 1;$	$N_1 = (\pi, 2\pi - \kappa, 0)(1/a)$
$(\boldsymbol{WZ} \mid \boldsymbol{N}') = 1;$	$N_1' = (\pi, \kappa, 0)(1/a)$
$(\boldsymbol{W\Sigma} \mid \boldsymbol{N}) = 1;$	$N_1 = (2\pi + \kappa, \kappa - \pi, 0)(1/a)$
$(\boldsymbol{W\Sigma} \mid \boldsymbol{P}) = 1;$	$P_1 = (\kappa, \kappa + \pi, 2\pi)(1/a)$
$(\boldsymbol{W\Sigma} \mid \boldsymbol{H}) = 1;$	$H_1 = (\kappa, \pi, 2\pi + \kappa)(1/a)$
$(\boldsymbol{W\Delta} \mid \boldsymbol{Z}) = 1;$	$Z_1 = (2\pi, 0, \pi + \kappa)(1/a)$
$(\boldsymbol{W\Delta} \mid \boldsymbol{N}) = 1;$	$N_1 = (2\pi + \kappa, \pi, 0)(1/a)$
$(\boldsymbol{W\Lambda} \mid \boldsymbol{k}) = 1;$	$k_1 = (2\pi + \kappa, \pi + \kappa, \kappa)(1/a)$
$(\boldsymbol{ZZ}' \mid \boldsymbol{\Delta}) = 2;$	$\Delta_1 = (\kappa + \kappa', 0, 0)(1/a)$
$(\boldsymbol{ZZ}' \mid \boldsymbol{\Delta}') = 2;$	$\Delta_1' = (\kappa - \kappa', 0, 0)(1/a)$
$(\boldsymbol{ZZ}' \mid \boldsymbol{\Delta}'') = 2;$	$\Delta_1'' = (\kappa + \kappa' - 2\pi, 0, 0)(1/a)$
$(\boldsymbol{ZZ}' \mid \boldsymbol{\Delta}''') = 2;$	$\Delta_1''' = (\kappa - \kappa' - 2\pi, 0, 0)(1/a)$
$(\boldsymbol{ZZ}' \mid \boldsymbol{N}) = 3;$	$N_1 = (\kappa, \kappa', 0)(1/a)$
$(\boldsymbol{ZZ}' \mid \boldsymbol{N}') = 1;$	$N_1' = (2\pi - \kappa, \kappa', 0)(1/a)$
$(\boldsymbol{Z\Sigma} \mid \boldsymbol{N}) = 1;$	$N_1 = (2\pi + \kappa, \kappa + \kappa', 0)(1/a)$
$(\boldsymbol{Z\Sigma} \mid \boldsymbol{N}') = 1;$	$N_1' = (2\pi + \kappa, \kappa - \kappa', 0)(1/a)$
$(\boldsymbol{Z\Sigma} \mid \boldsymbol{P}) = 1;$	$P_1 = (\kappa + \kappa', \kappa, 2\pi)(1/a)$
$(\boldsymbol{Z\Sigma} \mid \boldsymbol{P}') = 1;$	$P_1' = (\kappa - \kappa', \kappa, 2\pi)(1/a)$
$(\boldsymbol{Z\Sigma} \mid \boldsymbol{k}) = 1;$	$k_1 = (2\pi + \kappa, \kappa, \kappa')(1/a)$
$(\boldsymbol{Z\Delta} \mid \boldsymbol{Z}') = 1;$	$Z_1' = (2\pi, 0, \kappa + \kappa')(1/a)$
$(\boldsymbol{Z\Delta} \mid \boldsymbol{Z}'') = 1;$	$Z_1'' = (2\pi, 0, \kappa - \kappa')(1/a)$
$(\boldsymbol{Z\Delta} \mid \boldsymbol{N}) = 1;$	$N_1 = (2\pi + \kappa', \kappa, 0)(1/a)$
$(\boldsymbol{Z\Delta} \mid \boldsymbol{P}) = 1;$	$P_1 = (\kappa, \kappa', 2\pi)(1/a)$
$(\boldsymbol{Z\Lambda} \mid \boldsymbol{k}) = 1;$	$k_1 = (2\pi + \kappa', \kappa', \kappa + \kappa')(1/a)$
$(\boldsymbol{Z\Lambda} \mid \boldsymbol{k}') = 1;$	$k_1' = (2\pi + \kappa', \kappa', \kappa - \kappa')(1/a)$
$(\boldsymbol{\Sigma\Sigma}' \mid \boldsymbol{\Sigma}'') = 1;$	$\Sigma_1'' = (\kappa + \kappa', \kappa + \kappa', 0)(1/a)$
$(\boldsymbol{\Sigma\Sigma}' \mid \boldsymbol{\Sigma}''') = 1;$	$\Sigma_1''' = (\kappa - \kappa', \kappa - \kappa', 0)(1/a)$
$(\boldsymbol{\Sigma\Sigma}' \mid \boldsymbol{N}) = 1;$	$N_1 = (\kappa - \kappa', \kappa + \kappa', 0)(1/a)$
$(\boldsymbol{\Sigma\Sigma}' \mid \boldsymbol{k}) = 1;$	$k_1 = (\kappa, \kappa', \kappa + \kappa')(1/a)$
$(\boldsymbol{\Sigma\Sigma}' \mid \boldsymbol{k}') = 1;$	$k_1' = (\kappa, \kappa', \kappa - \kappa')(1/a)$
$(\boldsymbol{\Sigma\Delta} \mid \boldsymbol{N}) = 1;$	$N_1 = (\kappa, \kappa + \kappa', 0)(1/a)$
$(\boldsymbol{\Sigma\Delta} \mid \boldsymbol{N}') = 1;$	$N_1' = (\kappa, \kappa - \kappa', 0)(1/a)$
$(\boldsymbol{\Sigma\Delta} \mid \boldsymbol{M}) = 1;$	$M_1 = (\kappa, \kappa, \kappa')(1/a)$
$(\boldsymbol{\Sigma\Lambda} \mid \boldsymbol{M}) = 1;$	$M_1 = (\kappa + \kappa', \kappa + \kappa', \kappa')(1/a)$
$(\boldsymbol{\Sigma\Lambda} \mid \boldsymbol{M}') = 1;$	$M_1' = (\kappa - \kappa', \kappa - \kappa', \kappa')(1/a)$
$(\boldsymbol{\Sigma\Lambda} \mid \boldsymbol{k}) = 1;$	$k_1 = (\kappa + \kappa', \kappa - \kappa', \kappa')(1/a)$
$(\boldsymbol{\Delta\Delta}' \mid \boldsymbol{\Delta}'') = 1;$	$\Delta_1'' = (\kappa + \kappa', 0, 0)(1/a)$
$(\boldsymbol{\Delta\Delta}' \mid \boldsymbol{\Delta}''') = 1;$	$\Delta_1''' = (\kappa - \kappa', 0, 0)(1/a)$
$(\boldsymbol{\Delta\Delta}' \mid \boldsymbol{N}) = 1;$	$N_1 = (\kappa, \kappa', 0)(1/a)$
$(\boldsymbol{\Delta\Lambda} \mid \boldsymbol{M}) = 1;$	$M_1 = (\kappa', \kappa', \kappa + \kappa')(1/a)$
$(\boldsymbol{\Delta\Lambda} \mid \boldsymbol{M}') = 1;$	$M_1' = (\kappa', \kappa', \kappa' - \kappa)(1/a)$
$(\boldsymbol{\Lambda\Lambda}' \mid \boldsymbol{\Lambda}'') = 1;$	$\Lambda_1'' = (\kappa + \kappa', \kappa + \kappa', \kappa + \kappa')(1/a)$
$(\boldsymbol{\Lambda\Lambda}' \mid \boldsymbol{\Lambda}''') = 1;$	$\Lambda_1''' = (\kappa - \kappa', \kappa - \kappa', \kappa - \kappa')(1/a)$
$(\boldsymbol{\Lambda\Lambda}' \mid \boldsymbol{M}) = 1;$	$M_1 = (\kappa + \kappa', \kappa + \kappa', \kappa - \kappa')(1/a)$
$(\boldsymbol{\Lambda\Lambda}' \mid \boldsymbol{M}') = 1;$	$M_1' = (\kappa - \kappa', \kappa - \kappa', \kappa + \kappa')(1/a)$

Table 16. Wave vector selection rules in rocksalt O_h^5 and diamond O_h^7 coefficients $([l_m \boldsymbol{k}_j]_{(2)}|\boldsymbol{k}_{j'})$

$([\boldsymbol{\Gamma}]_{(2)}\|\boldsymbol{\Gamma}) = 1$						
$([2\boldsymbol{\Gamma}]_{(2)}\|\boldsymbol{\Gamma}) = 3$						
$([3\boldsymbol{\Gamma}]_{(2)}\|\boldsymbol{\Gamma}) = 6$						
$([\boldsymbol{L}]_{(2)}\|\boldsymbol{\Gamma}) = 4;$	$(\|\boldsymbol{X}) = 2$ [a]					
$([2\boldsymbol{L}]_{(2)}\|\boldsymbol{\Gamma}) = 12;$	$(\|\boldsymbol{X}) = 8$ [a]					
$([2\boldsymbol{X}]_{(2)}\|\boldsymbol{\Gamma}) = 9;$	$(\|\boldsymbol{X}) = 4$ [a]					
$([\boldsymbol{\Delta}]_{(2)}\|\boldsymbol{\Gamma}) = 3;$	$(\|\boldsymbol{\Delta}') = 1;$	$(\|\boldsymbol{\Sigma}) = 1$				
$([2\boldsymbol{\Delta}]_{(2)}\|\boldsymbol{\Gamma}) = 12;$	$(\|\boldsymbol{\Delta}') = 3;$	$(\|\boldsymbol{\Sigma}) = 4$				
$([\boldsymbol{\Lambda}]_{(2)}\|\boldsymbol{\Gamma}) = 4;$	$(\|\boldsymbol{\Delta}') = 2;$	$(\|\boldsymbol{\Lambda}') = 1;$	$(\|\boldsymbol{\Sigma}') = 1$			
$([2\boldsymbol{\Lambda}]_{(2)}\|\boldsymbol{\Gamma}) = 16;$	$(\|\boldsymbol{\Delta}') = 8;$	$(\|\boldsymbol{\Lambda}') = 3;$	$(\|\boldsymbol{\Sigma}') = 4$			
$([2\boldsymbol{W}]_{(2)}\|\boldsymbol{\Gamma}) = 12;$	$(\|\boldsymbol{X}) = 6;$	$(\|\boldsymbol{\Sigma}'') = 4$ [b]				
$([\boldsymbol{\Sigma}'']_{(2)}\|\boldsymbol{\Gamma}) = 6;$	$(\|\boldsymbol{X}) = 8;$	$(\|\boldsymbol{\Sigma}'') = 4$ [b]				
$([\boldsymbol{K}]_{(2)}\|\boldsymbol{\Gamma}) = 6;$	$(\|\boldsymbol{K}) = 2;$	$(\|\boldsymbol{M}) = 1;$	$(\|\boldsymbol{\Delta}) = 2;$	$(\|\boldsymbol{\Sigma}'') = 1$ [b]		
$([2\boldsymbol{Z}]_{(2)}\|\boldsymbol{\Gamma}) = 24;$	$(\|\boldsymbol{\Delta}') = 6;$	$(\|\boldsymbol{\Delta}'') = 4;$	$(\|\boldsymbol{N}) = 4;$	$(\|\boldsymbol{X}) = 8;$	$(\|\boldsymbol{\Sigma}) = 4;$	$(\|\boldsymbol{\Sigma}'') = 4$
$([\boldsymbol{k}]_{(2)}\|\boldsymbol{\Gamma}) = 24;$	$(\|\boldsymbol{k}') = 24$ [c]					

[a] Since all $\star\boldsymbol{X}^{(m)}$ have $l_m = 2$, we must use up two stars whenever an $(|\boldsymbol{X}) \neq 0$.
[b] $\Sigma_1'' = (\pi, \pi, 0)(1/a)$.
[c] $\boldsymbol{k}$ here is a general vector, whose symmetrized square contains only general vectors except for $\boldsymbol{\Gamma}$.

linear algebraic equations is straightforward. One always requires sufficiently many independent equations to find all reduction coefficients. This in turn requires choosing elements of the group which give rise to linearly independent equations. One may find oneself calculating some full group characters which do not give new information since there does not seem to be any simple systematic way of easily prescribing *a priori* which elements should be chosen for the character calculations. In contrast to this, the method of the reduction group which will be reviewed and used below in the calculation of reduction coefficients for the diamond structure does have a well-defined prescription for calculating characters. It turns out that more characters must be evaluated in the reduction group method, but this is a price one is often willing to pay in order to have available complete methods of checking the calculation at any stage.

To reduce the remaining products in rocksalt requires that we obtain all the wave-vector selection rules. These are actually common to rocksalt and diamond space groups and are given in Tables 15–17 for ordinary and symmetrized products.

Table 17. Wave vector selection rules in rocksalt and diamond coefficients $([l_m \boldsymbol{k}_j]_{(3)}|\boldsymbol{k}_{j'})$

$([\boldsymbol{\Gamma}]_{(3)}\|\boldsymbol{\Gamma}) = 1$						
$([2\boldsymbol{\Gamma}]_{(3)}\|\boldsymbol{\Gamma}) = 4$						
$([3\boldsymbol{\Gamma}]_{(3)}\|\boldsymbol{\Gamma}) = 10$						
$([\boldsymbol{L}]_{(3)}\|\boldsymbol{L}) = 5$						
$([2\boldsymbol{L}]_{(3)}\|\boldsymbol{L}) = 30$						
$([2\boldsymbol{X}]_{(3)}\|\boldsymbol{\Gamma}) = 8;$	$(\|\boldsymbol{X}) = 16$					
$([2\boldsymbol{W}]_{(3)}\|\boldsymbol{W}) = 26;$	$(\|\boldsymbol{\Delta}) = 24;$	$(\|\boldsymbol{L}) = 16$				
$([\boldsymbol{K}]_{(3)}\|\boldsymbol{\Gamma}) = 8;$	$(\|\boldsymbol{K}) = 8;$	$(\|\boldsymbol{N}) = 2;$	$(\|\boldsymbol{\Sigma}) = 2;$	$(\|\boldsymbol{\Sigma}') = 1;$	$(\|\boldsymbol{M}) = 4;$	$(\|\boldsymbol{H}) = 2;$
	$(\|\boldsymbol{L}) = 2;$	$(\|\boldsymbol{\Delta}) = 4$			$(\|\boldsymbol{L}) = 2;$	$(\|\boldsymbol{\Delta}) = 4$

In Tables A 1–A 11* the complete space group reduction coefficients are tabulated for ordinary products and symmetrized Kronecker square and cube products of the irreducible representations for the rocksalt space group O_h^5. These were all obtained by use of the technique of linear algebraic equations.

Although these results have appeared in the literature they are given here for ease of reference and completeness. These rules will be referred to in the detailed analyses given below for particular NaCl structure materials.

132. Irreducible representations $D^{(\Gamma)(m)}$, $D^{(*X)(m)}$, $D^{(*L)(m)}$ in diamond. The diamond space group O_h^7 is non-symmorphic and the coset representatives are given in (126.6). Owing to the translation group $\mathfrak{T}$ being face centered cubic $\mathfrak{T}$, as in O_h^5, the complete set of wave vectors for which irreducible representations are desired is the same as in rocksalt; as is the Brillouin zone shown in Fig. 3. Thus Tables 1–3 can be directly taken over without change.

Of course, in the remaining work differences will generally arise as the set of coset representatives $\{\varphi_p|\tau_p\}$ is not closed. To make this point entirely clear, we shall carry out a detailed illustrative analysis of selection rules for diamond for the same points in the zone as in the case of rocksalt. The differences which arise are both in the structure of the irreducible representations, and in the types of processes which may be "optically allowed" (e.g. in the electric dipole approximation) as infra-red or Raman processes.

The method which will be illustrated here is the reduction group method. As a matter of ease and compactness of making the calculation, the method of linear algebraic equations used in rocksalt is probably preferable. However, from a formal and pedagogic viewpoint, as distinct from the trial and error type calculations required in the algebraic equation method, the method of the reduction group is in principle rigorous and intimately connected with the homomorphism of the full group irreducible representation of the space group with the full abstract space group. In this sense of rigor, the reduction group method has conceptual advantages which may more than compensate for the added work needed in computing characters. We shall follow the analysis indicated in Sect. 58, in analyzing products from within the "closed" set $D^{(\Gamma)(m)}$, $D^{(*X)(m)}$, $D^{(*L)(m)}$ of the diamond structure O_h^7. Of course the reduction group method can only be used conveniently when a few representations (stars) are involved in the closed set of factors and resultants.

The first step in the reduction group method is to define the kernel of the representations involved. That is for the "closed" set of products of $D^{(\Gamma)(m)}$, $D^{(*X)(m)}$, $D^{(*L)(m)}$ we need the kernels of each of these representations. The kernel of a representation is the set of all group elements in $\mathfrak{G}$ which are mapped onto the matrix representing the identity $\{\varepsilon|0\}$. This collection we call $\mathfrak{K}^{(\Gamma)(m)}$, $\mathfrak{K}^{(*X)(m)}$, $\mathfrak{K}^{(*L)(m)}$, in each case. Then clearly

$$\mathfrak{K}^{(\Gamma)(m)} = \mathfrak{T}. \tag{132.1}$$

That is for the point Γ the entire translation group is mapped into the identity.

* These numbers refer to the tables in Appendix A.

For the star $\star X$, the group $\mathfrak{K}^{(\star X)(m)}$ consists of all translations $\boldsymbol{R}_L$ such that

$$X_1 \cdot \boldsymbol{R}_L = 2\pi p, \tag{132.2}$$

$$X_2 \cdot \boldsymbol{R}_L = 2\pi q, \tag{132.3}$$

$$X_3 \cdot \boldsymbol{R}_L = 2\pi r, \tag{132.4}$$

where p, q, r are integers, and

$$\boldsymbol{R}_L = l_1 \boldsymbol{t}_1 + l_2 \boldsymbol{t}_2 + l_3 \boldsymbol{t}_3 \tag{132.5}$$

with l integers, so that from (132.2)–(132.5) we have

$$l_1 + l_2 = 2p, \tag{132.6}$$

$$l_1 + l_3 = 2q, \tag{132.7}$$

$$l_2 + l_3 = 2r. \tag{132.8}$$

The pure translation elements in $\mathfrak{K}^{(\star X)(m)}$ satisfy (132.5)–(132.8). Then the translation group $\mathfrak{T}$ may then be decomposed into cosets with respect to $\mathfrak{K}^{(\star X)(m)}$ as

$$\mathfrak{T} = \mathfrak{K}^{(\star X)(m)} \oplus \{\varepsilon|\boldsymbol{t}_1\}\, \mathfrak{K}^{(\star X)(m)} \oplus \{\varepsilon|\boldsymbol{t}_2\}\, \mathfrak{K}^{(\star X)(m)} \oplus \{\varepsilon|\boldsymbol{t}_3\}\, \mathfrak{K}^{(\star X)(m)}. \tag{132.9}$$

For the star $\star L$ the group $\mathfrak{K}^{(\star L)(m)}$ consists of all $\boldsymbol{R}_L$ such that

$$L_1 \cdot \boldsymbol{R}_L = 2\pi s, \tag{132.10}$$

$$L_2 \cdot \boldsymbol{R}_L = 2\pi t, \tag{132.11}$$

$$L_3 \cdot \boldsymbol{R}_L = 2\pi u, \tag{132.12}$$

$$L_4 \cdot \boldsymbol{R}_L = 2\pi v, \tag{132.13}$$

where s, t, u, v are integers, or with $\boldsymbol{R}_L$ of form (132.5) we have

$$l_1 + l_2 + l_3 = 2s, \tag{132.14}$$

$$-l_3 = 2t, \tag{132.15}$$

$$-l_2 = 2u, \tag{132.16}$$

$$-l_1 = 2v. \tag{132.17}$$

Then the coset decomposition of $\mathfrak{T}$ with respect to $\mathfrak{K}^{(\star L)(m)}$ is

$$\begin{aligned} \mathfrak{T} = {} & \mathfrak{K}^{(\star L)(m)} \oplus \{\varepsilon|\boldsymbol{t}_1\}\, \mathfrak{K}^{(\star L)(m)} \oplus \{\varepsilon|\boldsymbol{t}_2\}\, \mathfrak{K}^{(\star L)(m)} \oplus \{\varepsilon|\boldsymbol{t}_3\}\, \mathfrak{K}^{(\star L)(m)} \\ & \oplus \{\varepsilon|\boldsymbol{Q}\}\, \mathfrak{K}^{(\star L)(m)} \oplus \{\varepsilon|\boldsymbol{t}_1 + \boldsymbol{Q}\}\, \mathfrak{K}^{(\star L)(m)} \oplus \{\varepsilon|\boldsymbol{t}_2 + \boldsymbol{Q}\}\, \mathfrak{K}^{(\star L)(m)} \\ & \oplus \{\varepsilon|\boldsymbol{t}_3 + \boldsymbol{Q}\}\, \mathfrak{K}^{(\star L)(m)}, \end{aligned} \tag{132.18}$$

where

$$\boldsymbol{Q} = \boldsymbol{t}_1 + \boldsymbol{t}_2 + \boldsymbol{t}_3. \tag{132.19}$$

Clearly $\mathfrak{K}^{(\star L)(m)}$ is a common translational normal subgroup of $D^{(\Gamma)(m)}$, $D^{(\star L)(m)}$, $D^{(\star X)(m)}$ such that the elements in $\mathfrak{K}^{(\star L)(m)}$ are the minimal set of elements simultaneously in $\mathfrak{K}^{(\Gamma)(m)}$, $\mathfrak{K}^{(\star X)(m)}$ and $\mathfrak{K}^{(\star L)(m)}$.

Now call

$$\mathfrak{K}^{(R)} \equiv \mathfrak{K}^{(\star L)(m)}. \qquad (132.20)$$

We then define the reduction group $\mathfrak{R}$ for products of $D^{(\Gamma)(m)}$, $D^{(\star X)(m)}$, $D^{(\star L)(m)}$ as follows. Take the coset decomposition of $\mathfrak{G}$ with respect to $\mathfrak{K}^{(R)}$:

$$\begin{aligned}\mathfrak{G} = \{\varphi_p|\tau_p\}\,\mathfrak{K}^{(R)} \oplus \{\varphi_p|\tau_p+t_j\}\,\mathfrak{K}^{(R)} \oplus \{\varphi_p|\tau_p+Q\}\,\mathfrak{K}^{(R)}\\ \oplus \{\varphi_p|\tau_p+t_j+Q\}\,\mathfrak{K}^{(R)} \qquad p=1,\ldots,48;\ j=1,2,3.\end{aligned} \qquad (132.21)$$

The set of $8\times 48 = 384$ coset representatives in (132.21) define the reduction group $\mathfrak{R}$ for these products. Observe that the product of two such coset representatives is again a coset representative, modulus an element from $\mathfrak{K}^{(R)}$. But in $D^{(\Gamma)(m)}$, $D^{(\star X)(m)}$, $D^{(\star L)(m)}$ each element in $\mathfrak{K}^{(R)}$ is represented by the matrix identity. The matrices representing these 384 coset representatives are a closed set of matrices and form a group: the reduction matrix group. The irreducible representations $D^{(\Gamma)(m)}$, $D^{(\star X)(m)}$, $D^{(\star L)(m)}$ of $\mathfrak{G}$ are evidently also representations of $\mathfrak{R}$. *A priori* these are each simply the representation subduced upon $\mathfrak{R}$ by these irreducible representations of $\mathfrak{G}$, and hence there is no need for them to be irreducible in $\mathfrak{R}$. We shall in fact observe that the subduced representations are irreducible in $\mathfrak{R}$.

Now we proceed to construct the necessary character tables using the general theory. Again we start with $\star L$. The canonical arm is L_1, and the remainders are as in (127.3) and (127.4). Clearly the same rotations as in (127.5) leave L_1 invariant (mod $2\pi B_j$). However, in this case the last 6 elements in (127.5) occur with fractional translation τ_1, among coset representatives. That is, the coset representatives which appear in $\mathfrak{G}/\mathfrak{T}$, and whose rotational part leaves L_1 invariant are

$$\begin{aligned}&\mathfrak{G}(L_1)/\mathfrak{T} =\\ &\{\varepsilon|0\},\ \{\delta_{3xyz}|0\},\ \{\delta_{3xyz}^{-1}|0\},\ \{\delta_{2x\bar{y}}|\tau_1\},\ \{\delta_{2x\bar{z}}|\tau_1\},\ \{\delta_{2y\bar{z}}|\tau_1\};\\ &\{i|\tau_1\},\ \{\sigma_{6xyz}|\tau_1\},\ \{\sigma_{6xyz}^{-1}|\tau_1\},\ \{\rho_{x\bar{y}}|0\},\ \{\rho_{x\bar{z}}|0\},\ \{\rho_{y\bar{z}}|0\}.\end{aligned} \qquad (132.22)$$

Now observe the curious property of the set of coset representatives (132.22) namely that they are closed under multiplication. That is, in spite of the presence of fractional translations among (132.22), the set is a group, since every rotation either leaves τ, unchanged, or sends τ, into its negative. Then clearly we may use the identical argument as in (132.20) and (132.21) and instead of concerning ourselves with the full little group $\mathfrak{G}(L_1)/\mathfrak{T}(L_1)$ we only require the point group D_{3d}, as in the earlier case. Recall that in Table 5 we gave the character table for the allowable irreducible representations of $\mathfrak{G}(L_1)/\mathfrak{T}$ for O_h^5. This table can be taken over for O_h^7 making only a change of identification so that 6 coset representatives are now to be taken with the fractional τ_1: namely the 6 so enumerated in (132.22). With this change the appropriate character table is then available. Using (127.17) the complete character table for irreducible representations of O_h^7 of type $D^{(\star L)(m)}$ can be easily obtained. These will be presented below* in form to be used with $D^{(\star X)(m)}$.

Now consider $\star X$, taking X_1 as the canonical wave vector. The rotational elements which leave X_1 invariant are again listed in (127.21). Now there are essential differences between O_h^5 and O_h^7 caused by the presence of the fractional τ_1.

* In Appendix C a treatment by ray representation theory is given.

Perhaps the most convenient manner to proceed is to contrast the structure of $\mathfrak{G}(X_1)/\mathfrak{T}(X_1)$ for O_h^5 and O_h^7 by using the abstract method of generators and relations [10]. Recall that the group $\mathfrak{G}(X_1)/\mathfrak{T}(X_1)$ for O_h^5 was isomorphic to $D_{4h}=D_4 \otimes C_i$. Making the identifications

$$A \equiv \boldsymbol{\delta}_{4x}; \quad B=\boldsymbol{\delta}_{2y}; \quad C=\boldsymbol{i} \tag{132.23}$$

we take A, B, C as generators of D_{4h}, with relations

$$A^4=B^2=C^2=E; \quad AC=CA; \quad BC=CB; \quad AB=BA^3. \tag{132.24}$$

Clearly (132.23) and (132.24) define a group of order 16 with 10 classes and 10 irreducible representations, as given in Table 8. In diamond, however, the elements $\boldsymbol{\delta}_{4x}$ and $\boldsymbol{i}$ occur only with fractional translation $\boldsymbol{\tau}_1$ in the coset representatives; thus the coset representatives of $\mathfrak{G}/\mathfrak{T}$ whose rotational part leaves $\boldsymbol{X}_1$ invariant are:

$$\begin{aligned}\mathfrak{G}(X_1)/\mathfrak{T}= &\\ &\{\varepsilon|0\}, \quad \{\boldsymbol{\delta}_{2x}|0\}, \quad \{\boldsymbol{\delta}_{2y}|0\}, \quad \{\boldsymbol{\delta}_{2z}|0\}, \quad \{\boldsymbol{\sigma}_{4x}|0\}, \quad \{\boldsymbol{\sigma}_{4x}^{-1}|0\},\\ &\{\boldsymbol{\rho}_{yz}|0\}, \quad \{\boldsymbol{\rho}_{y\bar{z}}|0\}, \quad \{\boldsymbol{i}|\boldsymbol{\tau}_1\}, \quad \{\boldsymbol{\rho}_x|\boldsymbol{\tau}_1\}, \quad \{\boldsymbol{\rho}_y|\boldsymbol{\tau}_1\}, \quad \{\boldsymbol{\rho}_z|\boldsymbol{\tau}_1\},\\ &\{\boldsymbol{\delta}_{4x}|\boldsymbol{\tau}_1\}, \quad \{\boldsymbol{\delta}_{4x}^{-1}|\boldsymbol{\tau}_1\}, \quad \{\boldsymbol{\delta}_{2yz}|\boldsymbol{\tau}_1\}, \quad \{\boldsymbol{\delta}_{2y\bar{z}}|\boldsymbol{\tau}_1\}.\end{aligned} \tag{132.25}$$

However, the group $\mathfrak{T}(X_1)$ consists of all $\boldsymbol{R}_L$ such that

$$\boldsymbol{X}_1 \cdot \boldsymbol{R}_L = 2\pi p, \tag{132.26}$$

where p is an integer or

$$l_1 + l_2 = 2p. \tag{132.27}$$

Hence if we consider the decomposition of $\mathfrak{T}$ with respect to cosets of $\mathfrak{T}(X_1)$ we may take

$$\mathfrak{T}=\mathfrak{T}(X_1)+\{\varepsilon|\boldsymbol{t}_1\}\,\mathfrak{T}(X_1). \tag{132.28}$$

Consequently the cosets in $\mathfrak{G}(X_1)/\mathfrak{T}(X_1)$ are represented by

$$\{\boldsymbol{\varphi}_{X_1}|\boldsymbol{\tau}_1\}, \quad \{\boldsymbol{\varphi}_{X_1}|\boldsymbol{\tau}_1+\boldsymbol{t}_1\}. \tag{132.29}$$

Note that since

$$\boldsymbol{X}_1 \cdot \boldsymbol{t}_1 = \boldsymbol{X}_1 \cdot \boldsymbol{t}_2 = \pi \tag{132.30}$$

so

$$\exp - i\boldsymbol{X}_1 \cdot \boldsymbol{t}_1 = \exp - i\boldsymbol{X}_1 \cdot \boldsymbol{t}_2 = -1. \tag{132.31}$$

It is equally valid to take $\{\varepsilon|\boldsymbol{t}_2\}$ as coset representative in (132.28). The relevant point is that the coset representative is to be considered as an abstract (translation) element whose matrix representative is the diagonal matrix with elements (-1). Then we define an abstract group by means of the associations (generators)

$$A=\{\boldsymbol{\delta}_{4x}|\boldsymbol{\tau}_1\}; \quad B=\{\boldsymbol{\delta}_{2y}|0\}; \quad C=\{\boldsymbol{i}|\boldsymbol{\tau}_1\}; \quad D=\{\varepsilon|\boldsymbol{t}_1\}; \quad E=\{\varepsilon|0\} \tag{132.32}$$

and the relations

$$A^4=B^2=C^2=D^2=E, \tag{132.33}$$

$$AB=BA^3; \; AC=DCA; \; BC=DCB; \; DA=AD; \; DB=BD; \; DC=CD. \tag{132.34}$$

Table 18a. Character table of group $\mathfrak{G}(X_1)/\mathfrak{T}(X_1)$ in diamond O_h^7

	${}^1X_1$	${}^1X_2$	${}^1X_3$	${}^1X_4$	${}^2X_1^+$	${}^2X_2^+$	${}^2X_3^+$	${}^2X_4^+$	${}^2X_5^+$	${}^2X_1^-$	${}^2X_2^-$	${}^2X_3^-$	${}^2X_4^-$	${}^2X_5^-$
$\{\varepsilon\|0\}$	2	2	2	2	1	1	1	1	2	1	1	1	1	2
$\{\varepsilon\|\boldsymbol{t}_{xy}\}$	−2	−2	−2	−2	1	1	1	1	2	1	1	1	1	2
$\{\boldsymbol{\delta}_{2y}, \boldsymbol{\delta}_{2z}\|0, \boldsymbol{t}_{xy}\}$	0	0	0	0	1	−1	1	−1	0	1	−1	1	−1	0
$\{\boldsymbol{\delta}_{2x}\|0\}$	2	2	−2	−2	1	1	1	1	−2	1	1	1	1	−2
$\{\boldsymbol{\delta}_{2x}\|\boldsymbol{t}_{xy}\}$	−2	−2	2	2	1	1	1	1	−2	1	1	1	1	−2
$\{\boldsymbol{\delta}_{4x}, \boldsymbol{\delta}_{4x}^{-1}\|\boldsymbol{\tau}_1, \boldsymbol{\tau}_1+\boldsymbol{t}_{xy}\}$	0	0	0	0	1	1	−1	−1	0	1	1	−1	−1	0
$\{\boldsymbol{\delta}_{2yz}\|\boldsymbol{\tau}_1\}, \{\boldsymbol{\delta}_{2y\bar{z}}\|\boldsymbol{\tau}_1+\boldsymbol{t}_{xy}\}$	0	0	2	−2	1	−1	−1	1	0	1	−1	−1	1	0
$\{\boldsymbol{\delta}_{2y\bar{z}}\|\boldsymbol{\tau}_1\}, \{\boldsymbol{\delta}_{2yz}\|\boldsymbol{\tau}_1+\boldsymbol{t}_{xy}\}$	0	0	−2	2	1	−1	−1	1	0	1	−1	−1	1	0
$\{\boldsymbol{i}\|\boldsymbol{\tau}_1, \boldsymbol{\tau}_1+\boldsymbol{t}_{xy}\}$	0	0	0	0	1	1	1	1	2	−1	−1	−1	−1	−2
$\{\boldsymbol{\rho}_y, \boldsymbol{\rho}_z\|\boldsymbol{\tau}_1, \boldsymbol{\tau}_1+\boldsymbol{t}_{xy}\}$	0	0	0	0	1	−1	1	−1	0	−1	1	−1	1	0
$\{\boldsymbol{\rho}_x\|\boldsymbol{\tau}_1, \boldsymbol{\tau}_1+\boldsymbol{t}_{xy}\}$	0	0	0	0	1	1	1	1	−2	−1	−1	−1	−1	2
$\{\boldsymbol{\sigma}_{4x}, \boldsymbol{\sigma}_{4x}^{-1}\|0, \boldsymbol{t}_{xy}\}$	0	0	0	0	1	1	−1	−1	0	−1	−1	1	1	0
$\{\boldsymbol{\rho}_{yz}\|0\}, \{\boldsymbol{\rho}_{y\bar{z}}\|0\}$	2	−2	0	0	1	−1	−1	1	0	−1	1	1	−1	0
$\{\boldsymbol{\rho}_{yz}\|\boldsymbol{t}_{xy}\}, \{\boldsymbol{\rho}_{y\bar{z}}\|\boldsymbol{t}_{xy}\}$	−2	2	0	0	1	−1	−1	1	0	−1	1	1	−1	0

This group is an extension of the group D_{4h}. For an acceptable irreducible representation of this group $\mathfrak{G}(X_1)/\mathfrak{T}(X_1)$, which we call $D^{(X_1)(m)}$, we must have in the character system the relation among characters:

$$\chi^{(X_1)(m)}(\{\varepsilon|\boldsymbol{t}_1\}) = \exp - i(\boldsymbol{X}_1 \cdot \boldsymbol{t}_1)\, \chi^{(X_1)(m)}(\{\varepsilon|0\}) \\ = -\chi^{(X_1)(m)}(\{\varepsilon|0\}). \tag{132.35}$$

It should be noted that some of the relations (132.33) and (132.34) arise because we are defining the abstract little group $\mathfrak{G}(X_1)/\mathfrak{T}(X_1)$ in terms of group elements represented by matrics which have the correct property in respect to the underlying Bloch vector space $\psi^{(X_1)(m)}(\boldsymbol{r})$ so that any element of space group (i.e. any translation) with the property (132.30) and (132.31) can be called D in the defining relations (132.33) and (132.34).

The character table of the group defined in (132.33) and (132.34) can be found in a straightforward fashion by finding the classes and class multiplication coefficients, and then reducing in the usual fashion. In Table 18a the complete character table of the group is given.[1] It will be observed that only 4 of the 14 irreducible representations are acceptable, namely those for which (132.35) holds. These representations: $D^{(X_1)(1)}$, $D^{(X_1)(2)}$, $D^{(X_1)(3)}$, $D^{(X_1)(4)}$ are all two dimensional. Physically this implies a sticking together of energy surfaces at this point, so that a non-degenerate case cannot occur. Recall the discussion in Sect. 40 regarding the non-allowable irreducible representations which applies particularly in this situation for the 10 non-acceptable representations.

Now, with the acceptable irreducible representations of $\mathfrak{G}(\Gamma)/\mathfrak{T}(\Gamma)$, $\mathfrak{G}(X_1)/\mathfrak{T}(X_1)$, and $\mathfrak{G}(L_1)/\mathfrak{T}(L_1)$ we may construct the character table of the irreducible representation of the reduction group $\mathfrak{R}$. As a computational matter this simply means using (127.16) and (127.17) to obtain the character of all elements in $\mathfrak{R}$,

[1] Adapted from R. LOUDON and R.J. ELLIOTT: J. Phys. Chem. Solids **15**, 146 (1960).

i.e. in the set of representatives

$$\mathfrak{R}/\mathfrak{K} \equiv \{\varphi_p|\tau_p\},\ \{\varphi_p|\tau_p+t_j\},\ \{\varphi_p|\tau_p+Q\}_1,\ \{\varphi_p|\tau_p+t_j+Q\} \qquad p=1,\ldots,48;\ j=1,2,3. \tag{132.36}$$

In Table B 1 * all these characters are listed.

Now using this table, reduction coefficients are easily found, by the familiar character rule (58.13), as discussed below.

133. Reduction coefficients.

α) Products of $D^{(\Gamma)(m)}$, $D^{(\star X)(m)}$, $D^{(\star L)(m)}$ *in diamond.* To reduce the direct products of these representatives we proceed as with any finite group, we obtain first the product characters, e.g. $\chi^{(\star X)(3)}\,\chi^{(\star X)(4)}$ and now we reduce this product representation by using the rule (58.13) to determine the specific reduction coefficients

$$(\star X3\ \star X4|\star k''\ m'') = \frac{1}{g_R}\sum_T \chi^{(\star X)(3)}(T)\,\chi^{(\star X)(4)}(T)\,\chi^{(\star k'')(m)}(T)^* \qquad \star k''=\Gamma, \text{ or } \star X, \tag{133.1}$$

where $g_R=384$ is the order of $\mathfrak{R}$, and T is any element in $\mathfrak{R}$. We quickly find

$$\begin{aligned} D^{(\star X)(3)} \oplus D^{(\star X)(4)} = {} & D^{(\Gamma)(2+)} \oplus D^{(\Gamma)(2-)} \oplus D^{(\Gamma)(12+)} \\ & \oplus D^{(\Gamma)(12-)} \oplus D^{(\Gamma)(15+)} \oplus D^{(\Gamma)(15-)} \\ & \oplus D^{(\star X)(1)} \oplus D^{(\star X)(2)} \oplus D^{(\star X)(3)} \oplus D^{(\star X)(4)}. \end{aligned} \tag{133.2}$$

In this fashion all such products can be worked out, using the group $\mathfrak{R}$. All such products were reduced, with results given in Table B 2.

β) Additional reduction coefficients in diamond. The reduction group method was also used to reduce all products of type

$$D^{(\star W)(m)} \otimes D^{(\star W)(m')} \quad \text{and} \quad D^{(\star \Sigma)(m)} \otimes D^{(\star \Sigma)(m)} \qquad \text{for } \Sigma_1=(1,1,0)(\pi/a).$$

Hence using either the reduction group method, or the method of linear algebraic equations we have obtained all selection rules for reduction of ordinary products, or symmetrized powers of representations in O_h^7.

Complete tables of the reduction coefficients for all diamond structure irreducible representations are given in Tables B 2 to B 10.*

134**. Clebsch-Gordan coefficients in diamond structure for $D^{(\star X)(m)} \otimes D^{(\star X)(m')}$.

As discussed in the previous section it may be assumed now that all space group reduction coefficients are available in both diamond and rocksalt structures. The next step in the analysis of direct products is to obtain the Clebsch-Gordan coefficients. Recall that, as discussed in Sect. 18 and 60, the Clebsch-Gordan coefficients are the elements of the unitary matrix U which: similarity transforms *all* matrices of the specified direct product of two irreducible representations into

* These numbers refer to Appendix B.

** This section was largely written by Dr. RHODA BERENSON.

fully reduced (decomposed) form; or equivalently produces the "correct linear combinations".

Despite the evident importance of obtaining all Clebsch-Gordan coefficients only few have been calculated to date. Should the coefficients become readily available it is clear that many applications will be found for them since it is always desirable to make maximum use of symmetry in a calculation and the Clebsch-Gordan coefficients permit such maximum use of symmetry, owing to the way they can be used to obtain "correct linear combinations" of functions.

However in this section we shall confine attention to one of the very few cases so far available, namely the direct products of irreducible representations:

$$D^{(\star\mathbf{X})(m)} \otimes D^{(\star\mathbf{X})(m')} \tag{134.1}$$

in diamond. It will be recalled that the basic wave vector rule is

$$\star\boldsymbol{X} \otimes \star\boldsymbol{X} = 3\boldsymbol{\Gamma} \oplus 2\,\star\boldsymbol{X} \tag{134.2}$$

but in diamond owing to the irreducible representations at $\boldsymbol{X}_1$ all being two dimensional we must have

$$2\,\star\boldsymbol{X} \otimes 2\,\star\boldsymbol{X} = 6\boldsymbol{\Gamma} \oplus 4\,\star\boldsymbol{X}. \tag{134.3}$$

As a particular example utilizing the results of Sect. 60 we shall only consider that case (134.3). Consulting Table 31 we immediately see that for all products of type (134.1) in diamond, whatever are the values of m and m', each resultant representation in the reduction occurs only once, so that $\gamma = 1$ and we thus suppress it in the following. For example we have, using a shortened notation:

$$\begin{aligned}\star\boldsymbol{X}(1) \otimes \star\boldsymbol{X}(1) = {} & \boldsymbol{\Gamma}(1+) \oplus \boldsymbol{\Gamma}(2-) \oplus \boldsymbol{\Gamma}(12+) \oplus \boldsymbol{\Gamma}(12-) \oplus \boldsymbol{\Gamma}(15-) \oplus \boldsymbol{\Gamma}(25+) \\ & \oplus \star\boldsymbol{X}(1) \oplus \star\boldsymbol{X}(2) \oplus \star\boldsymbol{X}(3) \oplus \star\boldsymbol{X}(4).\end{aligned} \tag{134.4}$$

We will now use the procedure outlined in Sect. 60 to calculate the Clebsch-Gordan coefficients.

The canonical wave vectors are chosen to simplify the calculation of $\boldsymbol{X}_i \otimes \boldsymbol{X}_j = \boldsymbol{X}_k$ i.e.

$$\boldsymbol{k}_1 = \boldsymbol{k}_x = \frac{2\pi}{a}(1,0,0); \quad \boldsymbol{k}_1' = \boldsymbol{k}_y = \frac{2\pi}{a}(0,1,0); \quad \boldsymbol{k}_1'' = \boldsymbol{k}_z = \frac{2\pi}{a}(0,0,1). \tag{134.5}$$

The coset representatives, $\{\boldsymbol{\varphi}_\sigma | \boldsymbol{\tau}_\sigma\}$, are then

$$\{\varphi_1|\boldsymbol{\tau}_1\} = \{\varepsilon|\boldsymbol{0}\}; \quad \{\varphi_2|\boldsymbol{\tau}_2\} = \{\boldsymbol{\delta}_{3xyz}|\boldsymbol{0}\}; \quad \{\varphi_3|\boldsymbol{\tau}_3\} = \{\boldsymbol{\delta}_{3xyz}^{-1}|\boldsymbol{0}\} \tag{134.6}$$

for l, l', and l''.

In calculating the Clebsch-Gordan coefficients, the matrices used were those tabulated by KOVALEV [56]. These are ray representations of the point groups $\mathfrak{P}(\boldsymbol{k}) = \mathfrak{G}(\boldsymbol{k})/\mathfrak{T}$. They satisfy

$$\Gamma^{(j)}(\varphi_\alpha)\,\Gamma^{(j)}(\varphi_\beta) = \omega(\alpha,\beta)\,\Gamma^{(j)}(\varphi_{\alpha\beta}) \tag{134.7}$$

where for the particular factor system used by KOVALEV

$$\omega(\alpha,\beta) = \exp\{-i(\boldsymbol{\varphi}_\alpha^{-1}\cdot\boldsymbol{k}_1 - \boldsymbol{k}_1)\cdot\boldsymbol{\tau}_\beta\}. \tag{134.8}$$

These ray representations are then related to the space group representations by

$$D^{(\boldsymbol{k})(j)}(\{\varphi_\alpha|\tau_\alpha\}) = \mathrm{e}^{-i\boldsymbol{k}_1\cdot\boldsymbol{\tau}_\alpha} D^{(j)}(\varphi_\alpha). \tag{134.9}$$

For the point $\boldsymbol{X}$ in diamond Kovalev's table T 159 is needed where $D^{(x_1)} = \hat{\tau}^3$; $D^{(x_2)} = \hat{\tau}^4$, $D^{(x_3)} = \hat{\tau}^{(2)}$ and $D^{(x_4)} = \hat{\tau}^1$. The matrices for point Γ are found in table T 194. In all Kovalev tables the canonical wave vector $\boldsymbol{k}_1$ is taken as $\boldsymbol{k}_z = \frac{2\pi}{a}(0, 0, 1)$.

The Clebsch-Gordan coefficients for the (111) block are obtained from Eq. (60.10)

$$U_{1a\,1a',\,1a''}\, U^*_{1\bar{a}\,1\bar{a}',\,1\bar{a}''} = \frac{l''}{h} \sum_x D^{(\boldsymbol{k})(l)}(\{\varphi_x|\tau_x\})_{a\bar{a}}\, D^{(\boldsymbol{k}')(l')}(\{\varphi_x|\tau_x\})_{a'\bar{a}'}\, D^{(\boldsymbol{k}'')(l'')*}(\{\varphi_x|\tau_x\})_{a''\bar{a}''}. \tag{134.10}$$

The coefficients for the (111) block are given in Table B 5.

The $(\sigma\sigma'\sigma'')$ block is given by (60.21)

$$U(\sigma\sigma'\sigma'') = D^{(\boldsymbol{k})(l)}(\{\varphi_k|\boldsymbol{t}_k\}) \otimes D^{(\boldsymbol{k}')(l')}(\{\varphi_{k'}|\boldsymbol{t}_{k'}\})\, U(111)\, D^{(\boldsymbol{k}'')(l'')}(\{\varphi_{k''}|\boldsymbol{t}_{k''}\})^{-1} \tag{134.11}$$

where

$$\{\varphi_k|\boldsymbol{t}_k\} \equiv \{\varphi_\sigma|\tau_\sigma\}^{-1}\{\varphi_\Sigma|\tau_\Sigma\}; \qquad \{\varphi_{k'}|\boldsymbol{t}_{k'}\} = \{\varphi_{\sigma'}|\tau_{\sigma'}\}^{-1}\{\varphi_\Sigma|\tau_\Sigma\}$$

and

$$\{\varphi_{k''}|\boldsymbol{t}_{k''}\} = \{\varphi_{\sigma''}|\tau_{\sigma''}\}^{-1}\{\varphi_\Sigma|\tau_\Sigma\} \tag{134.12}$$

and

$$\{\varphi_\Sigma|\tau_\Sigma\}\,\boldsymbol{k} \equiv \boldsymbol{k}_\sigma; \qquad \{\varphi_\Sigma|\tau_\Sigma\}\,\boldsymbol{k}' \equiv \boldsymbol{k}'_{\sigma'}; \qquad \{\varphi_\Sigma|\tau_\Sigma\}\,\boldsymbol{k}'' = \boldsymbol{k}''_{\sigma''}. \tag{134.13}$$

Tables B6 and B7 list all the necessary information for calculating the non zero $(\sigma\sigma'\sigma'')$ blocks.

As Table B6 indicates, the (111), (222), and (333) blocks have identical coefficients, as do the (123), (231), and (312) blocks. From Eq. (134.11)

$$U(123) = D^{(\boldsymbol{k})(l)}(\{\rho_{y\bar{z}}|\boldsymbol{0}\}) \otimes D^{(\boldsymbol{k}')(l')}(\{\rho_{x\bar{z}}|\boldsymbol{0}\})\, U(111) \otimes D^{(\boldsymbol{k}'')(l'')}(\{\rho_{x\bar{y}}|\boldsymbol{0}\})^{-1}. \tag{134.14}$$

In order to calculate the coefficients for $\boldsymbol{X}\otimes\boldsymbol{X} = \Gamma$ we first notice that for the (111) block $\boldsymbol{k}_1 + \boldsymbol{k}'_1 - \boldsymbol{k}''_1 \neq 2\pi\boldsymbol{B}_L$ and, therefore, all coefficients are zero. We have, therefore, chosen the (321) block as the canonical or principle block and all other $(\sigma\sigma'\sigma'')$ blocks will be determined from the (321) block. Table B8 lists the coefficients for the (321) block and Tables B9 and B10 list the information necessary to obtain the remaining blocks.

An alternate procedure would have been to use the projection operator method. The projection operator

$$P^{(*\boldsymbol{k}'')(l'')}_{\sigma''\mu''\sigma''\mu''} = \frac{l''}{g} \sum_\lambda D^{(*\boldsymbol{k}'')(l'')}(\{\varphi_\lambda|\tau_\lambda\})^*_{\sigma''\mu''\sigma''\mu''}\, P_{\{\varphi_\lambda|\tau_\lambda\}}. \tag{134.15}$$

Since

$$P^{(\boldsymbol{k}'')(l'')}_{\sigma''\mu''\sigma''\mu''}\, \Psi^{(\boldsymbol{k}'')(l'')}_{\sigma''\mu''} = \Psi^{(\boldsymbol{k}'')(l'')}_{\sigma''\mu''} \tag{134.16}$$

we can use projection operators as a check on the coefficients given in Tables B5 to B10.

For example, consider $X^{(3)} \otimes X^{(4)} = X^{(1)}$ and $X^{(3)} \otimes X^{(4)} = \Gamma^{(12+)}$ in diamond. Using projection operators we get the following basis functions which are written as linear combinations of the six Bloch functions $\psi_\alpha(k)$, $\alpha = x\,y\,z$, $\boldsymbol{\kappa} = \mathbf{0}, \tau$.

$$\psi_{11}^{(k_x)(X^3)} = (\psi_z(\mathbf{0}) + \psi_y(\tau))/\sqrt{2}, \tag{134.17}$$
$$\psi_{12}^{(k_x)(X^3)} = i(\psi_y(\mathbf{0}) + \psi_z(\tau))/\sqrt{2}, \tag{134.18}$$
$$\psi_{21}^{(k_x)(X^3)} = (\psi_x(\mathbf{0}) + \psi_z(\tau))/\sqrt{2}, \tag{134.19}$$
$$\psi_{22}^{(k_x)(X^3)} = i(\psi_z(\mathbf{0}) + i\psi_x(\tau))/\sqrt{2}, \tag{134.20}$$
$$\psi_{31}^{(k_x)(X^3)} = (\psi_y(\mathbf{0}) + \psi_x(\tau))/\sqrt{2}, \tag{134.21}$$
$$\psi_{32}^{(k_x)(X^3)} = i(\psi_x(\mathbf{0}) + \psi_y(\tau))/\sqrt{2}, \tag{134.22}$$

$$\psi_{11}^{(k_y)(X^4)} = (\psi_z(\mathbf{0}) - \psi_x(\tau))/\sqrt{2}, \tag{134.23}$$
$$\psi_{12}^{(k_y)(X^4)} = i(-\psi_x(\mathbf{0}) + \psi_z(\tau))/\sqrt{2}, \tag{134.24}$$
$$\psi_{21}^{(k_y)(X^4)} = (\psi_x(\mathbf{0}) - \psi_y(\tau))/\sqrt{2}, \tag{134.25}$$
$$\psi_{22}^{(k_y)(X^4)} = i(\psi_x(\tau) - \psi_y(\mathbf{0}))/\sqrt{2}, \tag{134.26}$$
$$\psi_{31}^{(k_y)(X^4)} = (\psi_y(\mathbf{0}) - \psi_z(\tau))/\sqrt{2}, \tag{134.27}$$
$$\psi_{32}^{(k_y)(X^4)} = i(-\psi_z(\mathbf{0}) + \psi_y(\tau))/\sqrt{2}. \tag{134.28}$$

From Tables B5 to B10 the Clebsch-Gordan coefficients for $\star\boldsymbol{X}^{(3)} \otimes \star\boldsymbol{X}^{(4)} = \boldsymbol{X}^{(1)}$ and $\star\boldsymbol{X}^{(3)} \otimes \star\boldsymbol{X}^{(4)} = \boldsymbol{\Gamma}^{(12+)}$ are obtained.[1] First for $\star X$:

$$U_{111}^{(k_x)(X^3)(k_y)(X^4)(k_z)(X^1)} = \begin{pmatrix} \frac{1}{2} & \frac{1}{2} \\ 0 & 0 \\ 0 & 0 \\ -i/2 & i/2 \end{pmatrix} = U_{222} = U_{333}, \tag{134.29}$$

$$U_{123}^{(k_x)(X^3)(k_y)(X^4)(k_z)(X^1)} = \begin{pmatrix} -i/2 & i/2 \\ 0 & 0 \\ 0 & 0 \\ \frac{1}{2} & \frac{1}{2} \end{pmatrix} = U_{231} = U_{312}. \tag{134.30}$$

As a specific application we find:

$$\begin{aligned} \psi_{11}^{(k_z)(X^1)} = \tfrac{1}{2}\{&\psi_{11}^{(k_x)(X^3)}\,\psi_{11}^{(k_y)(X^4)} - i\,\psi_{12}^{(k_x)(X^3)}\,\psi_{12}^{(k_y)(X^4)} \\ &- i\,\psi_{21}^{(k_x)(X^3)}\,\psi_{31}^{(k_y)(X^4)} + \psi_{22}^{(k_x)(X^3)}\,\psi_{32}^{(k_y)(X^4)}\}. \end{aligned} \tag{134.31}$$

Now for $\boldsymbol{\Gamma}$ we obtain:

$$U_{321}^{(k_x)(X^3)(k_y)(X^4)(k_0)(\Gamma^{12+})} = \begin{pmatrix} 0 & 0 \\ 1/\sqrt{6} & -1/\sqrt{6} \\ 1/\sqrt{6} & -1/\sqrt{6} \\ 0 & 0 \end{pmatrix}, \tag{134.32}$$

[1] In Tables B.5 to B.10 the full representations are denoted $\boldsymbol{X}^{(j)}$ not $\star\boldsymbol{X}^{(j)}$.

$$U_{211}=\begin{pmatrix} 0 & 0 \\ \varepsilon^2/\sqrt{6} & -\varepsilon/\sqrt{6} \\ \varepsilon^2/\sqrt{6} & -\varepsilon/\sqrt{6} \\ 0 & 0 \end{pmatrix}, \tag{134.33}$$

$$U_{131}=\begin{pmatrix} 0 & 0 \\ \varepsilon/\sqrt{6} & -\varepsilon^2/\sqrt{6} \\ \varepsilon/\sqrt{6} & -\varepsilon^2/\sqrt{6} \\ 0 & 0 \end{pmatrix}, \tag{134.34}$$

where $\varepsilon \equiv e^{2\pi i/3}$. Then for example,

$$\begin{aligned} \psi_{11}^{\Gamma(12+)} = \frac{1}{\sqrt{6}} & \{\psi_{31}^{(k_x)(X^3)}\psi_{22}^{(k_y)(X^4)} + \psi_{32}^{(k_x)(X^3}\psi_{21}^{(k_y)(X^4)} \\ & + \varepsilon^2\psi_{21}^{(k_x)(X^3)}\psi_{12}^{(k_y)(X^4)} + \varepsilon^2\psi_{22}^{(k_x)(X^3)}\psi_{11}^{(k_y)(X^4)} \\ & + \varepsilon\psi_{11}^{(k_x)(X^3)}\psi_{32}^{(k_y)(X^4)} + \varepsilon\psi_{12}^{(k_x)(X^3)}\psi_{31}^{(k_y)(X^4)}\}. \end{aligned} \tag{134.35}$$

An application of the projection operators

$$P_{11}^{(k_z)(X^1)} = \frac{l_{X^1}}{g}\sum_{\lambda} D_{11}^{(k_z)(X^1)}(\{\varphi_\lambda|\tau_\lambda\})^* P_{\{\varphi_\lambda|\tau_\lambda\}} \tag{134.36}$$

and

$$P_{11}^{(\Gamma 12+)} = \frac{l_{\Gamma 12+}}{g}\sum_{\lambda} D_{11}^{(\Gamma 12+)}(\{\varphi_\lambda|\tau_\lambda\})^* P_{\{\varphi_\lambda|\tau_\lambda\}} \tag{134.37}$$

demonstrates that the Clebsch-Gordan coefficients described above indeed yield linear combinations of basis functions with the correct symmetry.[2]

135. Test of effect of time reversal symmetry in diamond and rocksalt structure. The complete exploitation of the effect of the antiunitary time reversal operator in the diamond and rocksalt structure requires that we examine each irreducible representation, and each phonon species and test for reality, so as to determine whether added degeneracy is produced by the time reversal operator K. Having done this each selection rule should also be examined to determine whether additional restrictions occur due to time reversal. This program as outlined in Sects. 87–94, has not yet been carried out, nor has the progress of reformulating the theory completely in terms of the modern, corepresentation approach given in Sects. 95–102.

Here we shall give results of a restricted study of effect of time reversal in the two groups using the approach of Sects. 87–94. We shall examine the same two typical wave vectors (see Table 3) in each space group: $\star\boldsymbol{X}$ which is of class I since $-\boldsymbol{X}_i \equiv \boldsymbol{X}_i$ for $i=1, 2, 3$, and $\star\boldsymbol{W}$ which is class II since $\star-\boldsymbol{W} \equiv \star\boldsymbol{W}$ although $-\boldsymbol{W}_i$ is not equivalent to $\boldsymbol{W}_i$ for any i.

[2] R. Berenson: Theory of Crystal Clebsch-Gordan Coefficients. Ph. D. Thesis, New York University, February 1974 (unpublished). Available from University Microfilms, Ann. Arbor, Mich. Also cf. references in Sect. 60.

For a class I wave vector, it will be recalled that (94.11) and (94.12) give an immediate test by inspection of

$$\chi^{(\boldsymbol{k})(m)} \quad \text{and} \quad \chi^{(\boldsymbol{k})(m)*}. \tag{135.1}$$

If these two numbers are equal there is no added degeneracy, if unequal there is added degeneracy.

For a class II wave vector from (94.26), we inspect

$$\chi^{(\boldsymbol{k})(m)}(\{\varphi_l|\tau_l\}), \tag{135.2}$$

and

$$\chi^{(\boldsymbol{k})(m)}(\{\varphi_l|\tau_l\}^{\bar{\sigma}})^* \equiv \chi^{(\boldsymbol{k})(\bar{m})}(\{\varphi_l|\tau_l\}), \tag{135.3}$$

where the conjugation in (135.3) is carried out with respect to element $\{\varphi_{\bar{\sigma}}|\tau_{\bar{\sigma}}\}$ defined in (94.11) as an element whose rotational part takes $\boldsymbol{k}$ to $-\boldsymbol{k}$. In this fashion we determine whether or not $D^{(\boldsymbol{k})(m)}$ and $D^{(\boldsymbol{k})(\bar{m})}$ are equivalent or inequivalent, and hence in the latter case stick together, thus doubling the degeneracy.

In spite of the lengthy discussion needed to establish these results in Sects. 91–94 they are quite simple in application.

In diamond, for $^\star\boldsymbol{X}$ we consider representation $D^{(^\star\boldsymbol{X})(m)}$. We use as one wave vector say $\boldsymbol{X}_1$. The character table $\chi^{(\boldsymbol{X}_1)(m)}$ is given in Table 18a. Observe that all characters are real. Consequently for

$$\text{diamond: all } D^{(\boldsymbol{X}_1)(m)} \text{ are real.} \tag{135.4}$$

Consequently there is no added degeneracy. For $^\star\boldsymbol{W}$ consider $D^{(^\star\boldsymbol{W})(m)}$. We refer to $\boldsymbol{W}_1=(2\pi, 0, -\pi)(1/a)$ as the canonical wave vector. The character table is given in Table 33 to which reference should be made.* Observe the complex characters for classes $C_3, C_4, C_9, \ldots, C_{12}$. In this case, however, we need to use the test (135.3). It is not necessary to use the entire table but for our purposes it is adequate to examine (135.3) for a single element. Choose $\varphi_l=\boldsymbol{\sigma}_{4z}$. This is an element chosen so that the behaviour of the complex character in $\chi^{(\boldsymbol{W}_1)(1)}(\{\boldsymbol{\sigma}_{4z}|0\})$ and in $\chi^{(\boldsymbol{W}_1)(2)}(\{\boldsymbol{\sigma}_{4z}|0\})$ can be examined. Now the spatial element $\{\boldsymbol{\rho}_{x\bar{y}}|0\}$ has the property that

$$\boldsymbol{\rho}_{x\bar{y}}\,\boldsymbol{W}_1 = -\boldsymbol{W}_1 \not\equiv \boldsymbol{W}_1. \tag{135.5}$$

Consequently, we may take $\boldsymbol{\rho}_{x\bar{y}}$ as $\varphi_{\bar{\sigma}}$ in (135.3). Then from (135.3) we find

$$\begin{aligned}\chi^{(\boldsymbol{W}_1)(1)}(\{\boldsymbol{\rho}_{x\bar{y}}|0\}\{\boldsymbol{\sigma}_{4z}|0\}\{\boldsymbol{\rho}_{x\bar{y}}|0\})^* &= \chi^{(\boldsymbol{W}_1)(1)}(\{\boldsymbol{\sigma}_{4z}^{-1}|0\})^* \\ &= \chi^{(\boldsymbol{W}_1)(1)}(\{\boldsymbol{\sigma}_{4z}|0\}).\end{aligned} \tag{135.6}$$

Consequently, in spite of the appearance of complex numbers in the character tables for these two representations, both $D^{(\boldsymbol{W}_1)(1)}$ and $D^{(\boldsymbol{W}_1)(2)}$ are real. No further degeneracy is caused by time reversal. Note that, while (135.6) was established only for $D^{(\boldsymbol{W}_1)(1)}$ the identical argument applies to $D^{(\boldsymbol{W}_1)(2)}$, in the case of diamond since the essential matter is the conjugation of the elements in (135.6).

In the case of rocksalt the character table for point $\boldsymbol{X}_1$ is given in Table 8. Clearly all representations are real, and since this is a class I wave vector we can immediately conclude that the representations are real. Hence, there is no effect of time reversal producing any extra degeneracy. For point $\boldsymbol{W}_1$, the representations all have real characters, in fact they are easily obtained from the character table

* See p. 395 *infra*.

Table 19. Character table for W_1 in rocksalt. Vector: $W_1 = (2\pi, 0, -\pi)(1/a)$

Element $\{\varphi\|0\}$	$W_1^{(1)}$	$W_1^{(2')}$	$W_1^{(1')}$	$W_1^{(2)}$	$W_1^{(3)}$
ε	1	1	1	1	2
δ_{2z}	1	1	1	1	-2
$\sigma_{4z}, \sigma_{4z}^{-1}$	1	1	-1	-1	0
$\delta_{2xy}, \delta_{2\bar{x}y}$	1	-1	1	-1	0
ρ_x, ρ_y	1	-1	-1	1	0

$\rho_{x\bar{y}}\, \sigma_{4z}\, \rho_{x\bar{y}} = \sigma_{4z}^{-1}; \quad \rho_{x\bar{y}}\, \rho_x\, \rho_{x\bar{y}} = \rho_y.$

of the point group D_{2d}. In Table 19 the characters of the representations of D_{2d} are given. Also, it can be seen that conjugation with respect to the element $\rho_{x\bar{y}}$ simply permutes the elements in each class. Consequently the criterion (135.3) tells us that no added degeneracy occurs.

Thus, continuing, it is found in general that the time reversal operation does not cause additional sticking in either rocksalt or diamond.

At present there has been little study of the detailed application of time reversal, antiunitary symmetry to matrix elements.[1-4]

136. Connectivity and labelling of irreducible representations in diamond and rocksalt structures: Consequences for selection rules. Owing to the presence of the fractional translation in the coset representatives of the symmetry group of diamond, one finds the consequent appearance of phase factors in the character tables of the irreducible representations. An equivalent way of stating the same result is that the factor set for the p-inequivalent irreducible ray representations of the relevant point groups are not trivial (i.e. not all equal to one).

There is a very important consequence of the presence of phase factors which follows directly by inspection of the character tables for the allowable irreducible representations. We may in fact work either in the subgroup formalism concentrating attention on the little group characters, or in the full group formalism, using the full-group irreducible representations. The question has to do with the connectivity[1] and labelling of space group irreducible representations as the wave vector $\boldsymbol{k}$ is considered to increase from within the first Brillouin zone, to the surface, then on further prolongation of the wave vector $\boldsymbol{k}$. From the very definition of the wave vector, and first Brillouin zone (as given in Sect. 23) we recall that if we are given two wave vectors $\boldsymbol{k}$ and $\boldsymbol{k}'$ such that (see (23.6))

$$\boldsymbol{k}' = \boldsymbol{k} + 2\pi\, \boldsymbol{B}_H \tag{136.1}$$

then $\boldsymbol{k}$ and $\boldsymbol{k}'$ are equivalent, and in particular as in (23.6), (23.9), then $\boldsymbol{k}$ and $\boldsymbol{k}'$ define the identical irreducible representation of $\mathfrak{T}$. Consequently it follows that each

[1] M. LAX: Proceedings of the Exeter Conference on Physics of Semiconductors (ed. A.C. STICKLAND), p. 395. Published by Institute of Physics and the Physical Society London SW. 7 (1962).
[2] M. HULIN: phys. stat. solidi **21**, 607 (1967).
[3] A. AVIRON and J. ZAK: J. Math. Phys. **9**, 2138 (1968).
[4] G.F. KARAVAEV: Fiz. Tverd. Tela **6**, 3676 (1964).

[1] C. HERRING: J. Franklin Inst. **233**, 525 (1942) discussed these questions for the non-symmorphic groups diamond O_h^7 and hexagonal close-packed D_{6h}^4.

Table 18b[a]. Allowable irreducible representations of $\mathfrak{G}(\Delta_1)/\mathfrak{T}(\Delta_1)$ in O_h^7, $\Delta_1=(\kappa,0,0)$

$\chi^{(\Delta_1)(j)}$	Coset				
	$\{\varepsilon\|0\}$	$\{\delta_{2x}\|0\}$	$\{\delta_{4x}\|\tau_1\}$	$\{\rho_y\|\tau_1\}$	$\{\rho_{yz}\|0\}$
$\Delta_1 1$	1	1	ω	ω	1
$\Delta_1 2$	1	1	$-\omega$	ω	-1
$\Delta_1 2'$	1	1	$-\omega$	$-\omega$	1
$\Delta_1 1'$	1	1	ω	$-\omega$	-1
$\Delta_1 5$	2	-2	0	0	0

$\omega \equiv \exp - i\kappa a/4$

[a] Notation follows HERRING ref. [1] *loc. cit.*

wave vector prolonged beyond the surface of the first Brillouin zone can be mapped back into a wave vector inside the first zone by (136.1). In some simple case the reentrant wave vectors comes in simply at a diametrically opposite face of the zone. This case will be illustrated below, for the prolongation of the wave vector beyond point $X_1=(2\pi,0,0)(1/a)$ in diamond. For other wave vectors, the reentrant prolongation as defined via (136.1) is to another face (non-diametrically opposed). In any event, the connectivity of wave vectors is simply obtained from (136.1).

The more interesting and important problem concerns the connectivity of irreducible representations, and then the implications of this connectivity for selection rules. We will illustrate these connectivity questions for two important classes of representations: based upon $^\star\Delta$ and $^\star\Sigma$. In Table 18b we give the little group allowable irreducible characters $\chi^{(\Delta_1)(j)}$ for the group $\mathfrak{G}(\Delta_1)/\mathfrak{T}(\Delta_1)$ in diamond, and in Table 18c we give the full group characters for $\chi^{(\star\Delta)(j)}$. Now attention should be given in particular to the two pairs of representations:

$$D^{(\star\Delta)(1)} \quad \text{and} \quad D^{(\star\Delta)(2')} \tag{136.2}$$

and

$$D^{(\star\Delta)(2)} \quad \text{and} \quad D^{(\star\Delta)(1')}. \tag{136.3}$$

Observe that the only difference between these inequivalent irreducible representations is the sign of the characters for elements (coset representatives)

$$\{\rho_x|\tau_1\} \quad \text{and} \quad \{\delta_{4x}|\tau_1\} \tag{136.4}$$

for each pair.

Now consider the sequence of increasing $\boldsymbol{k}$ vectors along the path of

$$\Gamma \to \Delta_1 \to X_1 \to -\Delta_1 \to \Gamma. \tag{136.5}$$

Table 18c. Full group irreducible representations ($\star\Delta j$) in O_h^7; see Table 18b

$\chi^{(\star\Delta)(j)}$	Coset				
	$\{\varepsilon\|0\}$	$\{\delta_{2x}\|0\}$	$\{\rho_{x\bar{y}}\|0\}$	$\{\rho_x\|\tau_1\}$	$\{\delta_{4x}\|\tau_1\}$
$\star\Delta 1$	6	2	2	$4\cos\kappa a/4$	$2\cos\kappa a/4$
$\star\Delta 2$	6	2	-2	$4\cos\kappa a/4$	$-2\cos\kappa a/4$
$\star\Delta 2'$	6	2	2	$-4\cos\kappa a/4$	$-2\cos\kappa a/4$
$\star\Delta 1'$	6	2	-2	$-4\cos\kappa a/4$	$2\cos\kappa a/4$
$\star\Delta 5$	12	-4	0	0	0

That is we consider $\boldsymbol{k}$ to be a vector in the [100] direction, increasing to the zone edge, and beyond. In components (136.5) is

$$\begin{aligned}(0,0,0)&\to(\kappa,0,0)\to(2\pi/a,0,0)\to(2\pi/a+\kappa,0,0)\\&=(-2\pi/a+\kappa,0,0)\to(0,0,0).\end{aligned}\qquad(136.6)$$

Recall that 2π times a reciprocal lattice vector is given as $2\pi\boldsymbol{B}_H=(4\pi/a,0,0)$, as in Table 3. Now consider two wave vectors related by

$$\boldsymbol{k}=(\kappa,0,0)=(\kappa',0,0)+(4\pi/a,0,0)=\boldsymbol{k}'+2\pi\boldsymbol{B}_H.\qquad(136.7)$$

But

$$\exp-i\kappa a/4=e^{-i\pi}\exp-i\kappa' a/4$$

or the phase changes sign:

$$\omega(\kappa)=-\omega(\kappa').\qquad(136.8)$$

Thus if we consider the prolongation of a vector $\Delta_1\equiv(\kappa,0,0)$ as in the sequence (136.5) we have the important conclusion that upon passing beyond $\boldsymbol{X}_1$ the sign of ω changes, and the representation pairs become interchanged. Let us emphasize the significance of this result. The basis $\psi^{(\Delta_1)(j)}$ vectors (functions) of the representation e.g. of $\Delta_1\,1$, are prescribed but, considered as a function of Δ_1, when $\boldsymbol{k}$ moves beyond the first zone, the function $\psi^{(\Delta_1)(1)}$ which (for $\Delta_1<\boldsymbol{X}_1$) behaves as a basis for $\Delta_1\,1$, becomes (for $\Delta_1>\boldsymbol{X}_1$) a basis for representation $\Delta_1\,2'$. The physics contained in the function $\psi^{(\Delta_1)(1)}$ does not change, but the name of the representation is changed. Consequently a continuous cycle of a representation corresponding to (136.5) for wave vectors is

$$\boldsymbol{\Gamma}15-\to\Delta_1\,1\to\boldsymbol{X}_1\,1\to(-\Delta_1)\,2'\to\boldsymbol{\Gamma}25+\to(\Delta_1)\,2'\to\boldsymbol{X}_1\,1\to\Delta_1\,1\to\boldsymbol{\Gamma}15-.\qquad(136.9)$$

In exactly the same fashion one can derive a continuous cycle for the other pair of representations:

$$\boldsymbol{\Gamma}25-\to\Delta_1\,2\to\boldsymbol{X}_1\,2\to(-\Delta_1)\,1'\to\boldsymbol{\Gamma}15+\to\Delta_1\,1'\to\boldsymbol{X}_1\,2\to\Delta_1\,2\to\boldsymbol{\Gamma}25-.\qquad(136.10)$$

But since the doubly degenerate $\Delta_1\,5$ has no phase its cycles are:

$$\boldsymbol{\Gamma}15-\to\Delta_1\,5\to\boldsymbol{X}_1\,3\to(-\Delta_1)\,5\to\boldsymbol{\Gamma}15-\qquad(136.11)$$

and

$$\boldsymbol{\Gamma}25+\to\Delta_1\,5\to\boldsymbol{X}_1\,4\to(-\Delta_1)\,5\to\boldsymbol{\Gamma}25+.\qquad(136.12)$$

As could be anticipated, exactly the same result is obtained by explicit examination of the transformation of the basis functions

$$\psi^{(\Delta_1)(1)},\ \psi^{(\Delta_1)(2')}\quad\text{and}\quad\psi^{(\Delta_1)(2)},\ \psi^{(\Delta_1)(1')}$$

with particular reference to transformation under the test operator which can distinguish these functions, e.g. $P_{\{\boldsymbol{\rho}_y|\boldsymbol{\tau}_1\}}$.

One way of understanding these results is that strictly speaking the restriction of wave vectors to the first zone (e.g. that $|\Delta_1|<|\boldsymbol{X}_1|$) serves to keep the irreducible representations of $\mathfrak{T}$ distinct (recall Sect. 23) so that two equivalent wave vectors should not be double-counted if they are related by (136.1). But, on the other hand, the representation must be a continuous function of $\boldsymbol{k}$. The continuity requirement can result in a "double periodicity" for certain representations in a non-symmorphic crystal, such as diamond O_h^7.

Consider now the consequence of this result, first for the classical phonon dynamical problem. In the notation of Sect. 79 we have (see (79.7))

$$[\boldsymbol{D}(\boldsymbol{k})] \cdot \boldsymbol{e}\left(\left|\begin{matrix}\boldsymbol{k}\\ j_\lambda\end{matrix}\right.\right) = \omega^2(\boldsymbol{k}|j)\, \boldsymbol{e}\left(\left|\begin{matrix}\boldsymbol{k}\\ j_\lambda\end{matrix}\right.\right). \tag{136.13}$$

It will be recalled that the $\boldsymbol{k}$ dependence of the dynamical matrix $[\boldsymbol{D}(\boldsymbol{k})]$ enters via the projection operator $P^{(\boldsymbol{k})}$ as in (81.20) and (81.21). From (81.21) it immediately follows that

$$[\boldsymbol{D}(\boldsymbol{k})] = [\boldsymbol{D}(\boldsymbol{k} + 2\pi \boldsymbol{B}_H)]. \tag{136.14}$$

Consequently it follows that the set of eigenvectors of the dynamical matrices $[\boldsymbol{D}(\boldsymbol{k})]$ and $[\boldsymbol{D}(\boldsymbol{k}+2\pi\boldsymbol{B}_H)]$ must be identical (except perhaps for a sign, or in general, a phase factor of unit modulus). Thus the set

$$\left\{\boldsymbol{e}\left(\left|\begin{matrix}\boldsymbol{k}\\ 1\end{matrix}\right.\right), \ldots, \boldsymbol{e}\left(\left|\begin{matrix}\boldsymbol{k}\\ l_m\end{matrix}\right.\right)\right\} \tag{136.15}$$

can be taken identical to the set

$$\left\{\boldsymbol{e}\left(\left|\begin{matrix}\boldsymbol{k}+2\pi \boldsymbol{B}_H\\ 1\end{matrix}\right.\right), \ldots, \boldsymbol{e}\left(\left|\begin{matrix}\boldsymbol{k}+2\pi \boldsymbol{B}_H\\ l_m\end{matrix}\right.\right)\right\}. \tag{136.16}$$

Likewise for the eigenvalues, the set

$$\{\omega^2(\boldsymbol{k}|1), \ldots, \omega^2(\boldsymbol{k}|l_m)\} \tag{136.17}$$

corresponds to

$$\{\omega^2(\boldsymbol{k}+2\pi \boldsymbol{B}_H|1), \ldots, \omega^2(\boldsymbol{k}+2\pi \boldsymbol{B}_H|l_m)\}. \tag{136.18}$$

But the identity of the sets (136.15), and (136.16) and the sets (136.17) and (136.18) must be understood as in general

$$\omega^2(\boldsymbol{k}|j) = \omega^2(\boldsymbol{k}+2\pi \boldsymbol{B}_H|j') \tag{136.19}$$

where j' may or may not be identical to j. Each case must be separately studied. The situation here is reminiscent of the examination of effect of time reversal upon an eigenvector as in Sect. 94. Explicitly then, we have for diamond, from (136.19) and (136.9):

$$\omega^2(\Delta_1|1) = \omega^2(\Delta_1 + 2\pi \boldsymbol{B}_H|2') \tag{136.20}$$

and

$$\omega^2(\Delta_1|2) = \omega^2(\Delta_1 + 2\pi \boldsymbol{B}_H|1'). \tag{136.21}$$

The physical importance of (136.20) and (136.21) is seen in case of selection rules for processes in which a phonon of energy $\omega(\boldsymbol{k}|j)$ participates: i.e. an electron scattering. An example of practical, recent importance occurs when the selection rule for electron scattering between equivalent distinct conduction band minima (intravalley scattering) in Si is analysed.[2] This "g" process is governed by the (subgroup) reduction coefficient

$$([\Delta_1\, 1]_{(2)}|2\Delta_1\, j), \tag{136.22}$$

[2] D.L. Rode: phys. stat. solidi (b), **53**, 245 (1972).

where Δ_1 is the wave vector at the conduction band minimum in Silicon:

$$\Delta_1 = 0.83\,(2\pi/a, 0, 0). \tag{136.23}$$

Now consulting Table B.3 we find (line 8):

$$([\Delta_1\, 1]_{(2)} | 2\Delta_1\, 1) = 1. \tag{136.24}$$

This requires the phonon to be of wave vector $2\Delta_1$ and species $j=1$. But

$$\begin{aligned} 2\Delta_1 &= 1.66\,(2\pi/a, 0, 0) \\ &= -0.34\,(2\pi/a, 0, 0) + (4\pi/a, 0, 0). \end{aligned} \tag{136.25}$$

Then $2\Delta_1$ is greater than X_1 so that we can immediately use (136.20) to write

$$\omega^2(2\Delta_1|1) = \omega^2(-0.34\,X_1|2'). \tag{136.26}$$

Consequently we see that the phonon involved in the process governed by (136.24) is one which *in the first Brillouin zone* has symmetry $\Delta_1\, 2'$. Consulting Table 23 we note that this is in fact a phonon of type "LO". The selection rules (e.g. (136.22)) require correct interpretation in the light of the proper compatibility and continuity requirements. The rules themselves are self consistent with no ambiguity. In using the rules one must pay attention to the continuity especially when momentum transfers occur putting one outside the zone, as in the case illustrated.[3]

We can anticipate similar results for other non-symmorphic crystals, although at the time of writing no other examples have come to our attention. In diamond, representations of type $\Sigma_1 j$ are simply connected. Likewise for representations $\Lambda_1 j$.

The problem does not arise in case of symmorphic space groups.

N. Phonon symmetry, infra-red absorption and Raman scattering in diamond and rocksalt space groups

137. Introduction. Having determined reduction coefficients for the irreducible representations of both space groups, it now remains for us to continue the program and connect these results with physical processes involving phonons in both structures. We must first determine the symmetry species of the phonons; that is, of what space group irreducible representations they are bases. This will be done for both structures.

An enumeration of the critical points in both structures can then be given, insofar as these are predicted by symmetry. Determination of the symmetry set

[3] H.W. STREITWOLF: phys. stat. solidi **37**, K 47 (1970), has discussed this case. He reached the correct conclusion although he mistakenly claimed that earlier work by M. LAX and J.J. HOPFIELD: Phys. Rev. **124**, 115 (1961) and J.L. BIRMAN: Phys. Rev. **127**, 1093 (1962) was in disagreement. See M. LAX and J.L. BIRMAN: phys. stat. solidi **49**, K 153 (1972).

of critical points and classifying them according to the Morse theory is straightforward, in each structure. In addition, very comprehensive critical point analysis has been carried out for several diamond structure crystals: germanium, silicon, and diamond; on the basis of the additional information about phonon dispersion provided by a combination of detailed calculation and by inelastic neutron scattering. These will be reviewed. Following examination of the critical points in the phonon dispersion (i.e. single phonon dispersion) it is useful to give results of critical point analyses of the joint, or two phonon, density of states in various diamond structure, for comparison with optical studies in the two phonon energy region, insofar as now available.

The preceding information can then be used in an analysis, or interpretation of the infra-red absorption and Raman scattering spectra in diamond structure crystals. Results will be presented which are current as of the time of writing; it is to be expected that improvements in quantitative interpretation will continue to be made even though a considerable understanding has been already achieved for the diamond structure.

The same information is then presented for the rocksalt structure for specific materials. It will be seen that detailed interpretation in rocksalt has proceeded along slightly different lines than in diamond and is at present perhaps not as far advanced.

It will be recalled that in Sect. 123γ a discussion was given of the effect of the macroscopic electric field upon the splitting of the degenerate optic mode in a cubic crystal with inversion symmetry: this long wave (finite wave vector) effect also produced a differentiation in infra-red and Raman spectra. The entire discussion given in that subsection is directly appropriate to the case of rocksalt symmetry. The TO mode (component of the split optic mode) is infra-red active (see (123.56)), while the LO mode is inactive. In Raman scattering neither is allowed. Although we shall not give explicit discussion of the application of these macroscopic field rules and results to analyses of the rocksalt spectra, in fact the results to be given in Sects. 140–144 have been obtained by taking macroscopic field effects into account in determining TO-LO phonon energy splittings. The most striking of the results such as the anomalous angular dependences of Raman scattering discussed in (123.57)–(123.67) occur only for cubic crystals without inversion (e.g. zincblende structure) and do not apply to either rocksalt or diamond crystals. However symmetry breaking effects, such as mentioned in Sect. 124η can produce dramatic modification of the selection rules, and anisotropic scattering, at resonance, even in a crystal with cubic symmetry O_h.

138. Phonon symmetry in rocksalt and diamond. As discussed in Sect. 104 we proceed to determine phonon symmetry by considering that each of the atoms in the crystal is displaced in accord with an independent cartesian component, to produce the mechanical or Δ representation. Then this representation will be reduced to determine the symmetry species or irreducible (physically irreducible) representation contained in it. We illustrate the calculation for both rocksalt and diamond by obtaining symmetry species of the normal modes at the stars Γ, $\star X$, $\star L$. The method of calculation is to focus upon the group of the canonical wave vector of the star e.g. $\mathfrak{G}(\boldsymbol{\Gamma})$, $\mathfrak{G}(\boldsymbol{X}_1)$, $\mathfrak{G}(\boldsymbol{L}_1)$ and then induce the full representation.

First consider the rocksalt structure. Owing to the symmorphic structure of the group O_h^5, each lattice point occupied by an atom possesses full symmetry O_h. At $\boldsymbol{\Gamma}$ the displacements in each unit cell in the crystal are in phase. It is then evident that the character of element $\{\varphi_p|0\}$ in the representation produced by the displacement basis at $\boldsymbol{\Gamma}$, is simply

$$\chi^{(\boldsymbol{\Gamma})(\Delta)}(\{\varphi_p|0\}) = \pm n(\varphi_p)(1+2\cos\varphi_p), \tag{138.1}$$

where as usual $n(\varphi_p)$ is the number of atoms left unchanged (in place) by element $\{\varphi_p|0\}$, and φ_p is the angle of rotation of the operation $\{\varphi_p|0\}$, the $\pm$ is for proper/improper elements respectively. In Table 20, the column labelled $\boldsymbol{\Gamma}$ gives the character system of the representation (138.1) and its reduction in the last row.

Now consider the situation at $\boldsymbol{X}_1$. We need the characters $\chi^{(\boldsymbol{X}_1)(\Delta)}$ for the elements of $\mathfrak{G}(\boldsymbol{X}_1)$ listed in Table 8. The entire procedure is quite straightforward also at $\boldsymbol{k} \neq \boldsymbol{\Gamma}$ when one appreciates that one is applying a symmetry operation to the entire, phased sum of Cartesian displacements as in (104.2) *et seq*. Hence, when a symmetry operation $\{\varphi_p|0\}$ or $\{\varphi_p|\tau_p\}$ shifts an atom into the position of an equivalent atom, two cases may arise: the atoms may belong to different Bravais sub-lattices or the two atom sites may belong to the same Bravais sub-lattice. The first case does not occur in rocksalt, but does occur in diamond *(vide infra)*. In the second case we must consider the translation $\{\varepsilon|\boldsymbol{R}_L\}$ required to shift back the atoms in the same Bravais lattice to their original positions. The phase factor $\exp - i\,\boldsymbol{k}\cdot\boldsymbol{R}_L$ is then required in order to compute the character. An example will make this clear, at $\boldsymbol{X}_1$ in rocksalt. Consider the character of element $\{\delta_{2y}|0\}$. The character involves a sum of the contributions from the displacement vectors on the atoms at (0, 0, 0) and the atom at $(\frac{1}{2}, \frac{1}{2}, \frac{1}{2})$. As the element $\{\delta_{2y}|0\}$

Table 20. Reduction of displacement representation in rocksalt O_h^5: NaCl type crystal. Wave vectors $\boldsymbol{\Gamma}$, $\boldsymbol{X}_1$, $\boldsymbol{L}_1$. Each column gives the rotational symmetry elements φ_l in $\mathfrak{G}(\boldsymbol{k})$, and the character of the displacement representation at that $\boldsymbol{k}$

$\boldsymbol{\Gamma}$		$\boldsymbol{X}_1$		$\boldsymbol{L}_1$	
$\{\varepsilon\|0\}$	6	$\{\varepsilon\|0\}$	6	$\{\varepsilon\|0\}$	6
$\{\delta_{3xyz}\|0\}$	0	δ_{2x}	−2	$\{\delta_{3xy2}\|0\}$, $\{\delta_{3xyz}^{-1}\|0\}$	0
$\{\delta_{2x}\|0\}$	−2	δ_{2y}, δ_{2z}	−2	$\{\delta_{2x\bar{y}}\|0\}$, $\{\delta_{2x\bar{z}}\|0\}$, $\{\delta_{2y\bar{z}}\|0\}$	0
$\{\delta_{4x}\|0\}$	+2	$\delta_{4x}, \delta_{4x}^{-1}$	+2	$\boldsymbol{i}$	0
$\{\delta_{2xy}\|0\}$	−2	$\delta_{2yz}, \delta_{2y\bar{z}}$	−2	$\sigma_{6xyz}, \sigma_{6\ xyz}^{-\ 1}$	0
$\{\boldsymbol{i}\|0\}$	−6	$\boldsymbol{i}$	−6	$\rho_{x\bar{y}}, \rho_{x\bar{z}}, \rho_{y\bar{z}}$	2
$\{\sigma_{6xyz}\|0\}$	−0	ρ_x	2		
$\{\rho_x\|0\}$	+2	ρ_y, ρ_z	2		
$\{\sigma_{4x}\|0\}$	−2	$\sigma_{4x}, \sigma_{4x}^{-1}$	−2		
$\{\rho_{xy}\|0\}$	−2	$\rho_{yz}, \rho_{y\bar{z}}$	2		
Reduced representation:					
$2D^{(\boldsymbol{\Gamma})(15-)}$		$2D^{(\boldsymbol{X}_1)(4-)} \oplus 2D^{(\boldsymbol{X}_1)(5-)}$		$D^{(\boldsymbol{L}_1)(1+)} \oplus D^{(\boldsymbol{L}_1)(3+)} \oplus D^{(\boldsymbol{L}_1)(2-)} \oplus D^{(\boldsymbol{L}_1)(3-)}$ [a]	

[a] Note GANESAN *et al.* incorrectly give all parities at $\star\boldsymbol{L}$ as even in J. Phys. **26**, 640 (1966), in contrast to BURSTEIN, JOHNSON, and LOUDON, Phys. Rev. **139** A, 1240 (1965).

leaves the origin atom unchanged, and is a proper rotation, a contribution

$$(1+2\cos\pi) = -1 \quad \text{from } (0,0,0) \tag{138.2}$$

is found. The atom $(\frac{1}{2},\frac{1}{2},\frac{1}{2})a$ is shifted into $(-\frac{1}{2},\frac{1}{2},-\frac{1}{2})a$, so that $\boldsymbol{R}_L=(1,0,1)a=2\boldsymbol{t}_2$, is required to shift back. The phase factor here is $\exp -i\,\boldsymbol{X}_1\cdot\boldsymbol{R}_L=1$ so the contribution is

$$(1+2\cos\pi) = -1 \quad \text{from } (\tfrac{1}{2},\tfrac{1}{2},\tfrac{1}{2})a, \tag{138.3}$$

Hence

$$\chi^{(\boldsymbol{X}_1)(\Delta)}(\{\boldsymbol{\delta}_{2y}|0\}) = -2. \tag{138.4}$$

In this fashion characters of all relevant elements in $\mathfrak{G}(\boldsymbol{X}_1)$ have been found. These are given in Table 20, and the reduction in the last row.

Similarly the displacement representation can be found and reduced for $\boldsymbol{L}_1$. Observe the calculation of characters for $\{\boldsymbol{\delta}_{2x\bar{y}}|0\}$. The contribution from the basis atom at (0, 0, 0) which is left in place is -1. The atom at $(\frac{1}{2},\frac{1}{2},\frac{1}{2})a$ is shifted to $(-\frac{1}{2},-\frac{1}{2},-\frac{1}{2})a$. Hence a translation $\boldsymbol{R}_L=\boldsymbol{t}_1+\boldsymbol{t}_2+\boldsymbol{t}_3$ produces a phase

$$\exp -i\,\boldsymbol{L}_1\cdot\boldsymbol{R}_L = -1.$$

Hence the contribution from this atom is $(-1)\cdot(-1)=+1$. Then

$$\chi^{(\boldsymbol{L}_1)(\Delta)}(\{\boldsymbol{\delta}_{2x\bar{y}}|0\}) = 0. \tag{138.5}$$

In like manner the remainder of Table 20 has been obtained, and the reduction is given in the last row.

Results of such a calculation of the reduction of the Cartesian displacement representation for all points, lines and planes of high symmetry are given in Table 21.

Also in Table 21 we present phonon assignments e.g. of polarizations and branch (transverse optic, etc.). In some cases these assignments are manifestly non-group theoretical information, i.e. cannot be obtained from the symmetry analysis alone. The assignments of polarization and branch require that one have available the actual eigenvectors of the dynamical matrix, and thus determine the transformation properties of the eigenvectors. Actual phonon calculation[1,2] for certain rocksalt structures have been performed with results given in Fig. 4. Use of these results enables the assignments shown in the last column Table 21 to be made. Although these assignments are not absolute, insofar as they must depend upon the force matrix and assumed force constants to some degree, we shall use them for the alkali halide rocksalt structures with which we shall be concerned. Refer also to the article by Cochran and Cowley in this Encyclopedia, for further discussion [*20*].

Now we consider the assignment of phonon symmetry in the diamond O_h^7 structure. Again we use the method of Cartesian displacements to obtain and reduce the displacement representation and to determine the symmetry species of the normal modes at $\boldsymbol{\Gamma}$, $\star\boldsymbol{X}$, $\star\boldsymbol{L}$. The presence of fractional translations makes for some added difficulty in practice, not in principle. We repeat the type of analysis given for rocksalt. The character table for $\boldsymbol{\Gamma}$ is given in Table 22. Observe

[1] A. Karo and J. R. Hardy: Phys. Rev. **141**, 701 (1966).
[2] R. A. Cowley: Phys. Rev. **131**, 1030 (1963).

Table 21. Symmetry species of phonons rocksalt O_h^5

Star	Reduced Representation	Assignment[a]
Γ	$D^{(\Gamma)(15-)}$	Acoustic
	$D^{(\Gamma)(15-)}$	Optic
$\star X$	$D^{(\star X)(4-)}$	LA
	$D^{(\star X)(5-)}$	TA
	$D^{(\star X)(4-)}$	LO
	$D^{(\star X)(5-)}$	TO
$\star L$	$D^{(\star L)(1+)}$	LA
	$D^{(\star L)(3+)}$	TA
	$D^{(\star L)(2-)}$	LO
	$D^{(\star L)(3-)}$	TO
$\star W$	$2D^{(\star W)(3)}$	
	$D^{(\star W)(1)}$	
	$D^{(\star W)(2')}$	
$\star \Delta$	$D^{(\star \Delta)(1)}$	LA
	$D^{(\star \Delta)(5)}$	TA
	$D^{(\star \Delta)(1)}$	LO
	$D^{(\star \Delta)(5)}$	TO
$\star \Lambda$	$D^{(\star \Lambda)(1)}$	LA
	$D^{(\star \Lambda)(3)}$	TA
	$D^{(\star \Lambda)(1)}$	LO
	$D^{(\star \Lambda)(3)}$	TO
$\star Z$	$2D^{(\star Z)(1)}$	
	$2D^{(\star Z)(3)}$	
	$2D^{(\star Z)(4)}$	
$\star S$, $\star U$	(same as $\star Z$)	
$\star \Sigma$, $\star K$	(same as $\star Z$)	
$\star \Pi$	$4D^{(\star \Pi)(1)}$	
	$2D^{(\star \Pi)(2)}$	
$\star Q$	$3(\star Q)(1)$	
	$3(\star Q)(2)$	

[a] Assignments follow BURSTEIN, JOHNSON, and LOUDON: Phys. Rev. **139** A, 1240 (1965).

that at $\boldsymbol{\Gamma}$ any space group element with fractional τ_1 has vanishing character. Further, on comparing with Table 20 for rocksalt observe that the triply degenerate optic modes have opposite parity from the acoustic and in fact belong to altogether different symmetry species. Also notice that in rocksalt optic and acoustic modes have the same transformation properties and both sets belong to the same representation $D^{(\Gamma)(15-)}$, while this is not true in diamond.

For the point $\boldsymbol{X}_1$ we obtain the character system whose relevant part is shown in Table 22. Again, in obtaining the indicated characters account must be taken of the phase change introduced when atoms are shifted within one Bravais sublattice. Reduction of the representation is also given. Note that the phonon symmetry does not permit parity designation at this point even though $\{\boldsymbol{i}|\tau_1\}$ is in $\mathfrak{G}(\boldsymbol{X}_1)/\mathfrak{T}(\boldsymbol{X}_1)$: the modes are *not* simply even or odd.

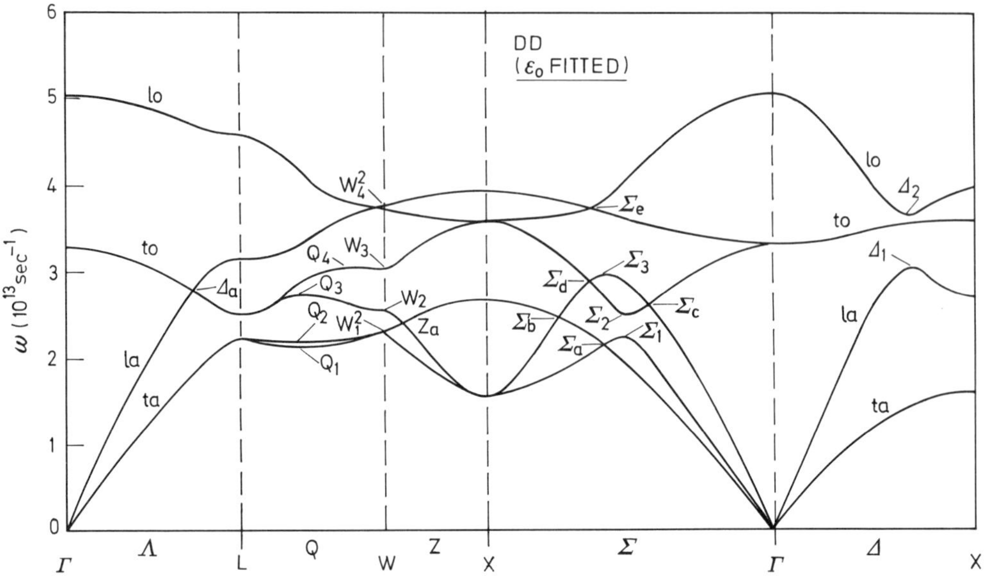

Fig. 4. Phonon dispersion in NaCl. (Calculated by A. M. KARO and J. R. HARDY: Phys. Rev. **141**, 701 (1966), Fig. 2d])

Finally we obtain at $\boldsymbol{L}_1$ the character system for the displacement representation shown, and reduce it in Table 22. Interestingly enough, the reduced representation is identical in rocksalt and diamond, as can be seen on comparing Tables 20 and 22.

As in the case of rocksalt, unambiguous assignment of polarization and branch indices is not possible on the basis of group theory considerations alone. One must supplement the straightforward group theory with an examination of

Table 22. Reduction of displacement representation in diamond O_h^7. Wave vectors $\boldsymbol{\Gamma}$, $\boldsymbol{X}_1$, $\boldsymbol{L}_1$

	$\boldsymbol{\Gamma}$		$\boldsymbol{X}_1$		$\boldsymbol{L}_1$
$\{\varepsilon\|0\}$	6	$\{\varepsilon\|0\}$	6	$\{\varepsilon\|0\}$	6
$\{\boldsymbol{\delta}_{3xyz}\|0\}$	0	$\{\boldsymbol{\delta}_{2x}\|0\}$	-2	$\{\boldsymbol{\delta}_{3xyz}\|0\}$	0
$\{\boldsymbol{\delta}_{2x}\|0\}$	-2	$\{\boldsymbol{\delta}_{2x}\|\boldsymbol{t}_{xy}\}$	2	$\{\boldsymbol{\delta}_{2x\bar{y}}\|\boldsymbol{\tau}_1\}$	0
$\{\boldsymbol{\delta}_{4x}\|\boldsymbol{\tau}_1\}$	0	$\{\boldsymbol{\delta}_{2yz}\|\boldsymbol{\tau}_1\}$	0	$\{\boldsymbol{i}\|\boldsymbol{\tau}_1\}$	0
$\{\boldsymbol{\delta}_{2xy}\|\boldsymbol{\tau}_1\}$	0	$\{\boldsymbol{\delta}_{2y\bar{z}}\|\boldsymbol{\tau}_1\}$	0	$\{\boldsymbol{\sigma}_{6xyz}\|\boldsymbol{\tau}_1\}$	0
$\{\boldsymbol{i}\|\boldsymbol{\tau}_1\}$	0	$\{\boldsymbol{\rho}_{yz}\|0\}$	2	$\{\boldsymbol{\rho}_{x\bar{y}}\|0\}$	2
$\{\boldsymbol{\sigma}_{6xyz}\|\boldsymbol{\tau}_1\}$	0	$\{\boldsymbol{\rho}_{y\bar{z}}\|\boldsymbol{t}_{xy}\}$	-2		
$\{\boldsymbol{\rho}_x\|\boldsymbol{\tau}_1\}$	0				
$\{\boldsymbol{\sigma}_{4x}\|0\}$	-2				
$\{\boldsymbol{\rho}_{xy}\|0\}$	2				
Reduction					
$D^{(\boldsymbol{\Gamma})(15-)} \oplus D^{(\boldsymbol{\Gamma})(25+)}$		$D^{(\boldsymbol{X}_1)(1)} \oplus D^{(\boldsymbol{X}_1)(3)} \oplus D^{(\boldsymbol{X}_1)(4)}$		$D^{(\boldsymbol{L}_1)(1+)} \oplus D^{(\boldsymbol{L}_1)(3+)} \oplus D^{(\boldsymbol{L}_1)(2-)} \oplus D^{(\boldsymbol{L}_1)(3-)}$	

Table 23. Symmetry species of phonons in diamond O_h^7

Star	Reduced representation	Assignment[a]
Γ	$D^{(\Gamma)(15-)}$	Acoustic
	$D^{(\Gamma)(25+)}$	Optic
$\star X$	$\star D^{(\star X)(3)}$	TA
	$D^{(\star X)(4)}$	TO
	$D^{(\star X)(1)}$	$LA=LO$
$\star L$	$D^{(\star L)(3+)}$	TA
	$D^{(\star L)(1+)}$	LA
	$D^{(\star L)(3-)}$	TO
	$D^{(\star L)(2-)}$	LO
$\star \Lambda$	$D^{(\star \Lambda)(3)}$	TA
	$D^{(\star \Lambda)(1)}$	LA
	$D^{(\star \Lambda)(3)}$	TO
	$D^{(\star \Lambda)(1)}$	LO
$\star \Sigma$	$D^{(\star \Sigma)(4)}$	A
	$D^{(\star \Sigma)(3)}$	A
	$D^{(\star \Sigma)(1)}$	A
	$D^{(\star \Sigma)(3)}$	
	$D^{(\star \Sigma)(2)}$	
	$D^{(\star \Sigma)(1)}$	
$\star W$	$D^{(\star W)(2)}$	
	$D^{(\star W)(1)}$	
	$D^{(\star W)(2)}$	
$\star \Delta$	$D^{(\star \Delta)(5)}$	TA
	$D^{(\star \Delta)(1)}$	LA
	$D^{(\star \Delta)(5)}$	TO
	$D^{(\star \Delta)(2')}$	LO

[a] After ref. [3] this page.

the transformation properties of the eigenvectors of the dynamical matrix. This in turn necessitates the solution of the dynamical matrix equations using some prescribed, and assumed known, force field. By now several such calculations have been made with a variety of force fields and in Table 23 we present assignments which have been made using a combination of the group theory results obtained as above, and also solutions to the dynamical matrix.[3] Assignments now known with reasonable certainty are given.

Perhaps it should be emphasized that irrespective of the order in energy of assignments such as LO, TO, made for particular modes of certain frequencies, which may depend on specific models, the nature and number of distinct type of symmetry species at each star is completely fixed and determined. Thus in diamond at $\star X$, the phonons do, without question, belong to types (irreducible representations) $D^{(\star X)(1)}$, $D^{(\star X)(3)}$, $D^{(\star X)(4)}$. Which of these is to be assigned LO+LA is precisely fixed by examining the pattern of cartesian displacement vectors which are a basis for $D^{(\star X)(1)}$, etc. However the transverse modes TO and TA cannot be

[3] J. L. WARREN, J. L. YARNELL, G. DOLLING and R. A. COWLEY: Phys. Rev. **158**, 805 (1967).

so easily assigned; it is this assignment which may perhaps be model-dependent. It is also of importance to realize that the polarization index, e.g. (longitudinal or transverse) does *not* apply in general. Only at points of high symmetry, or on lines of high symmetry is a mode polarized. The only proper designation for any mode is its symmetry designation (irreducible representation) in general.

Certain of the assignments may be clarified when one considers the compatibility relations between high symmetry points and adjacent lines, and planes. We may consider the task of enumerating the phonon symmetry species in rocksalt and diamond to be provisionally given subject to the *caveat* indicated in the preceding paragraphs regarding certain aspects of assignments of polarizations, etc.

139. Compatibility and phonon symmetry in diamond and rocksalt. The theory of determination of compatible symmetries of phonons when one moves in $\boldsymbol{k}$ space from a point, line, or plane of higher symmetry to one of lower symmetry was discussed in Sects. 51 and 108. It is a straightforward matter in principle of relating the group $\mathfrak{G}(\boldsymbol{k})$ of the higher symmetry point and the group $\mathfrak{G}(\boldsymbol{k}')$ of the lower symmetry point and determining the representations $D^{(\boldsymbol{k}')(m)}$ subduced by the allowable phonon species of symmetry $D^{(\boldsymbol{k})(m)}$. Since determination of the subduced representation is a simple matter we merely list in Tables 24–29 the relevant results [*30*]. In the figures which will illustrate phonon dispersion in diamond and rocksalt crystals, use will be made of the compatibility relations: they are given on the figures in the labelling.

Table 24. Compatibility table for diamond and rocksalt: $\Gamma \to {}^\star\Delta, {}^\star\Sigma, {}^\star\Lambda$

Γ	${}^\star\Delta$	${}^\star\Sigma$	${}^\star\Lambda$
$D^{(\Gamma)(1+)}$	$D^{({}^\star\Delta)(1)}$	$D^{({}^\star\Sigma)(1)}$	$D^{({}^\star\Lambda)(1)}$
$D^{(\Gamma)(1-)}$	$D^{({}^\star\Delta)(1')}$	$D^{({}^\star\Sigma)(2)}$	$D^{({}^\star\Lambda)(2)}$
$D^{(\Gamma)(2+)}$	$D^{({}^\star\Delta)(2)}$	$D^{({}^\star\Sigma)(4)}$	$D^{({}^\star\Lambda)(2)}$
$D^{(\Gamma)(2-)}$	$D^{({}^\star\Delta)(2')}$	$D^{({}^\star\Sigma)(3)}$	$D^{({}^\star\Lambda)(1)}$
$D^{(\Gamma)(12+)}$	$\{D^{({}^\star\Delta)(1)}$ $D^{({}^\star\Delta)(2)}$	$D^{({}^\star\Sigma)(1)}$ $D^{({}^\star\Sigma)(4)}$	$D^{({}^\star\Lambda)(3)}\}$
$D^{(\Gamma)(12-)}$	$\{D^{({}^\star\Delta)(1')}$ $D^{({}^\star\Delta)(2')}$	$D^{({}^\star\Sigma)(2)}$ $D^{({}^\star\Sigma)(3)}$	$D^{({}^\star\Lambda)(3)}\}$
$D^{(\Gamma)(15+)}$	$\{D^{({}^\star\Delta)(1')}$ $D^{({}^\star\Delta)(5)}$	$D^{({}^\star\Sigma)(2)}$ $D^{({}^\star\Sigma)(3)}$ $D^{({}^\star\Sigma)(4)}$	$D^{({}^\star\Lambda)(2)}$ $D^{({}^\star\Lambda)(3)}\}$
$D^{(\Gamma)(15-)}$	$\{D^{({}^\star\Delta)(1)}$ $D^{({}^\star\Delta)(5)}$	$D^{({}^\star\Sigma)(1)}$ $D^{({}^\star\Sigma)(3)}$ $D^{({}^\star\Sigma)(4)}$	$D^{({}^\star\Lambda)(1)}$ $D^{({}^\star\Lambda)(3)}\}$
$D^{(\Gamma)(25+)}$	$\{D^{({}^\star\Delta)(1')}$ $D^{({}^\star\Delta)(5)}$	$D^{({}^\star\Sigma)(1)}$ $D^{({}^\star\Sigma)(2)}$ $D^{({}^\star\Sigma)(3)}$	$D^{({}^\star\Lambda)(1)}$ $D^{({}^\star\Lambda)(3)}\}$
$D^{(\Gamma)(25-)}$	$\{D^{({}^\star\Delta)(2)}$ $D^{({}^\star\Delta)(5)}$	$D^{({}^\star\Sigma)(1)}$ $D^{({}^\star\Sigma)(2)}$ $D^{({}^\star\Sigma)(4)}$	$D^{({}^\star\Lambda)(2)}$ $D^{({}^\star\Lambda)(3)}\}$

Table 25. Compatibility for $\star X \to \star \Delta$, $\star Z$, $\star S$ in rocksalt

$\star X$	Δ	$\star Z$	$\star S$
$D^{(\star X)(1)}$	$D^{(\star \Delta)(1)}$	$D^{(\star Z)(1)}$	$D^{(\star S)(1)}$
$D^{(\star X)(2)}$	$D^{(\star \Delta)(2)}$	$D^{(\star Z)(1)}$	$D^{(\star S)(4)}$
$D^{(\star X)(3)}$	$D^{(\star \Delta)(2')}$	$D^{(\star Z)(4)}$	$D^{(\star S)(1)}$
$D^{(\star X)(4)}$	$D^{(\star \Delta)(1')}$	$D^{(\star Z)(4)}$	$D^{(\star S)(4)}$
$D^{(\star X)(5)}$	$D^{(\star \Delta)(5)}$	$D^{(\star Z)(1)}$	$D^{(\star S)(1)}$
		$D^{(\star Z)(4)}$	$D^{(\star S)(4)}$
$D^{(\star X)(m-)}$	Change $(\star \Delta)(n')$ to $(\star \Delta)(n)$ and v. versa; $n=1,2$	Change $(\star Z)(1) \rightleftarrows (2)$, $(3) \rightleftarrows (4)$	Change $1 \rightleftarrows 2$, $4 \rightleftarrows 3$

Table 26. Compatibility for $\star X$ and $\star \Delta$, $\star Z$, $\star S$ in diamond

$\star X$	$\star \Delta$	$\star Z$	$\star S$
$D^{(\star X)(1)}$	$D^{(\star \Delta)(1)}$	$D^{(\star Z)(1)}$	$D^{(\star S)(1)}$
	$D^{(\star \Delta)(2')}$		$D^{(\star S)(3)}$
$D^{(\star X)(2)}$	$D^{(\star \Delta)(1')}$	$D^{(\star Z)(1)}$	$D^{(\star S)(2)}$
	$D^{(\star \Delta)(2)}$		$D^{(\star S)(4)}$
$D^{(\star X)(3)}$	$D^{(\star \Delta)(5)}$	$D^{(\star Z)(1)}$	$D^{(\star S)(3)}$
			$D^{(\star S)(4)}$
$D^{(\star X)(4)}$	$D^{(\star \Delta)(5)}$	$D^{(\star Z)(1)}$	$D^{(\star S)(1)}$
			$D^{(\star S)(2)}$

Table 27. Compatibility $\star L \to \star \Lambda$, $\star Q$, for diamond and rocksalt

$\star L$	$\star \Lambda$	$\star Q$
$D^{(\star L)(1\pm)}$	$D^{(\star \Lambda)(1)}$	$D^{(\star Q)(1)}$
$D^{(\star L)(2\pm)}$	$D^{(\star \Lambda)(2)}$	$D^{(\star Q)(2)}$
$D^{(\star L)(3\pm)}$	$D^{(\star \Lambda)(3)}$	$D^{(\star Q)(3)}$

Table 28. Compatibility $\star W \to \star Z$, $\star Q$ for rocksalt

$\star W$	$\star Z$	$\star Q$
$D^{(\star W)(1)}$	$D^{(\star Z)(1)}$	$D^{(\star Q)(1)}$
$D^{(\star W)(2)}$	$D^{(\star Z)(2)}$	$D^{(\star Q)(2)}$
$D^{(\star W)(3)}$	$\left\{ D^{(\star Z)(3)} \right.$	$D^{(\star Q)(1)}$
	$\left. D^{(\star Z)(4)} \right.$	$D^{(\star Q)(2)}$
$D^{(\star W)(1')}$	$D^{(\star Z)(2)}$	$D^{(\star Q)(1)}$
$D^{(\star W)(2')}$	$D^{(\star Z)(1)}$	$D^{(\star Q)(2)}$

Table 29. Compatibility $\star W \to \star Z$, $\star Q$ in diamond

$\star W$	$\star Z$	$\star Q$
$D^{(\star W)(1)}$	$D^{(\star Z)(1)}$	$D^{(\star Q)(1)}$
		$D^{(\star Q)(2)}$
$D^{(\star W)(2)}$	$D^{(\star Z)(1)}$	$D^{(\star Q)(1)}$
		$D^{(\star Q)(2)}$

140. Critical points for phonons in diamond structure: Germanium, silicon and diamond. Now we turn to apply the theory of critical points in lattice vibration spectra discussed in Sect. 107 to the crystals of diamond space groups. We shall proceed systematically by first locating and classifying the symmetry set of critical points which are determined from group theory considerations alone. Having located these required critical points, we may proceed in either of several fashions. If precise inelastic neutron results are available, additional critical points may be located from these experimentally determined results. These dynamical critical points need to be classified in accord with the general theory. If no inelastic neutron experiments have been performed, but calculated phonon dispersions are available, the added dynamical critical points can be obtained from the calculation. Finally, the Morse relations can be used in order to determine whether the topological requirements, connecting the number and type of critical points for each branch, have been fulfilled. If not, then added critical points must exist on the branch. The location of these added dynamical points cannot be precisely specified by the theory. One would use the Morse theory as a guide to aid in searching then for such points rather than to determine the exact location of the point, i.e. one would search on the branch by interpolation, or extrapolation of known results. So far as is known to the present author, aside from illustrative calculations[1] on model problems with various arbitrary force constants the Morse theory has not so far had extensive application in practical cases to the prediction of new physical critical points, but has merely been used to verify that one has a set of critical points consistent with the topological requirements, and thus complete in that limited sense.

Singular critical points have been identified in both observed and calculated phonon dispersion in various crystals with diamond structure. From earlier work we saw that singular critical points, with discontinuous first derivatives in one principal direction only, contribute a discontinuity to the derivative of the density of states, $dg/d\omega$, if the critical point is a maximum or minimum.[2,3] Other singular cases, such as discontinuity in more than a single first derivative for maxima P_0 or minima P_3, or cases of saddle points ($P_1 P_2$ or $F_1 F_2$) with one or more discontinuous derivatives produce singular behavior in derivatives higher than the first in the density of states function $g(\omega)$. The notation is defined more precisely below.

Thus we shall restrict ourselves in the optical analysis to the following kinds of critical points: ordinary maxima and minima, denoted P_0 and P_3, singular maxima and minima with one discontinuous derivative denoted $P_0(1)$, $P_3(1)$ and ordinary saddle points denoted P_1 and P_2 and fluted saddle points denoted F_1 and F_2 with no discontinuous first derivatives. For completeness, we shall enumerate all the critical points which have been identified to date in diamond and rocksalt structures, but the actual critical point optical analysis will only be given for the above-mentioned types.

It is important to emphasize again that our object is the analysis of experimental data on infra-red optical absorption and Raman scattering. Hence for

---

[1] H. B. Rosenstock: J. Phys. Chem. Solids **2**, 44 (1957). – J. C. Phillips and H. B. Rosenstock: J. Phys. Chem. Solids **5**, 288 (1958).

[2] F. A. Johnson and R. Loudon: Proc. Roy. Soc. (London), Ser. A **281**, 274 (1964), especially Fig. 2.

[3] J. C. Phillips: Phys. Rev. **113**, 147 (1959), Table VI.

example in infra-red absorption the data available is not an exact analytical function $\alpha(\omega)$, but an experimental tracing whose major features which are to be studied are changes in slope, and maxima and minima. At the present stage of the theory there are two ways to proceed. We can attempt to compute multi-phonon density of states such as two phonon additive (combination and overtone) density of states and compare it directly with the observed infra-red absorption and Raman scattering. This assumes exact constancy of the matrix element which determines the proportionality between density of states, and intensity of absorption. In this fashion we attempt to determine the detailed frequency dependence of the density of multiphonon states – in particular the singular derivatives and related features – experimentally, and correlate this with the predicted features attempting to assign detailed correspondance, on the assumption that the known critical points are responsible. Alternatively one can assume crudely that, any critical point behavior will appear as a simple step or even a "delta" function feature and one then attempts to correlate peaks, slope changes and other features (including sharp dips in intensity of absorption or scattering) with the location of the critical points, without being overly concerned with detailed quantitative agreement or lack of same in relative intensities.

The first task in locating critical points is to find the set determined by symmetry. That is we need to find the set of wave vectors $\boldsymbol{k}_0$, such that (recall (107.47)–(107.49))

$$\nabla\omega(\boldsymbol{k}_0|j)=0. \tag{140.1}$$

Now owing to the result, [*20*, *47*] that we should also count as critical points, those points at which n first derivatives change sign discontinuously and the remaining $(3-n)$ derivatives vanish, we should examine all components of $\nabla\omega(\boldsymbol{k}_0|j)$ in all directions. Thus if $\hat{\boldsymbol{e}}$ represents a unit polarization vector, we need to examine

$$\hat{\boldsymbol{e}}\cdot\nabla\omega(\boldsymbol{k}_0|j) \tag{140.2}$$

in all significant directions $\hat{\boldsymbol{e}}$, around $\boldsymbol{k}_0$. As in the cases of the selection rules, we shall work out the results at points $\boldsymbol{\Gamma}$, $\boldsymbol{X}$, $\boldsymbol{L}$ in detail, for directions of principal interest.

We proceed to examine the product representation $D^{(\boldsymbol{k}_0)(j)*}\otimes D^{(\nabla)}\otimes D^{(\boldsymbol{k}_0)(j)}$ as called for by the discussion of Sect. 107. We are asking a local question concerning the vanishing of the slope of $\omega(\boldsymbol{k}|j)$ at one point in the Brillouin zone. To answer such a local question it is useful that we work in the small, that is, we use the subgroup method to investigate the relevant matrix elements. But, as $D^{(\boldsymbol{k}_0)(j)*}=D^{(-\boldsymbol{k}_0)(\bar{j})}$ (recall Sect. 92 *et seq.*) we have two options. We may take advantage of the fact that $D^{(\nabla)}$ belongs to $D^{(\boldsymbol{\Gamma})}$ so that

$$D^{(\nabla)}\otimes D^{(\boldsymbol{k}_0)(j)}=\sum c_{j'}\,D^{(\boldsymbol{k}_0)(j')} \tag{140.3}$$

can be reduced into a sum of representations $D^{(\boldsymbol{k}_0)(j')}$ at wave vector $\boldsymbol{k}_0$. Then we reduce

$$D^{(-\boldsymbol{k}_0)(\bar{j})}\otimes D^{(\boldsymbol{k}_0)(j')}. \tag{140.4}$$

If (140.4) contains $D^{(\boldsymbol{\Gamma})(1+)}$, then the slope will in general not vanish. Since we are asking a completely local question the relevant group operations and characters to be used in (140.3) and (140.4) are that of point group $\mathfrak{G}(\boldsymbol{k}_0)/\mathfrak{T}(\boldsymbol{k}_0)$. This is clear since the operator ∇ transforms like $D^{(\Gamma)}$ and any operation not in that point group will simply change the vector $\boldsymbol{k}_0$ to another in the same star.

Alternatively, it sometimes proves more convenient to reduce

$$D^{(-\boldsymbol{k}_0)(\bar{j})} \otimes D^{(\boldsymbol{k}_0)(j)} = \sum_{j'} c_{j'} D^{(\boldsymbol{\Gamma})(j')} \tag{140.5}$$

into a sum of irreducible representations at Γ. Then if among the $D^{(\boldsymbol{\Gamma})(j)}$ there are representations by which $D^{(\nabla)}$ transforms, the slope is non-zero. Again it is to be emphasized that in working in the small, in this manner only the operations in

$$\mathfrak{G}(\boldsymbol{k}_0)/\mathfrak{T}(\boldsymbol{k}_0) \equiv \mathfrak{P}(\boldsymbol{k}_0) \tag{140.6}$$

are to be considered, not the operations in $\mathfrak{G}/\mathfrak{T} = \mathfrak{P}$. Recall that $\mathfrak{P}$ is the *entire* point group and $\mathfrak{P}(\boldsymbol{k}_0)$ is a subgroup of it. As always it is the unitarity of the operator $\mathfrak{P}_{\{\boldsymbol{\varphi}\}}$ which permits us to conclude that if $\{\boldsymbol{\varphi}\}$ is in $\mathfrak{P}(\boldsymbol{k}_0)$:

$$[\boldsymbol{e}]^+ [\boldsymbol{\xi} \cdot \nabla D] [\boldsymbol{e}] = [\boldsymbol{e}_{\{\boldsymbol{\varphi}\}}]^+ [\boldsymbol{\xi} \cdot \nabla_{\{\boldsymbol{\varphi}\}} D] [\boldsymbol{e}_{\{\boldsymbol{\varphi}\}}] \tag{140.7}$$

as in (81.33), (107.42) *et seq.*, so that only if (140.7) is unchanged under all operations in $P(\boldsymbol{k}_0)$ will the "matrix element" be non-zero, and therefore the slope $\nabla\omega(\boldsymbol{k}_0|j) \neq 0$.

Diamond structure: Point $\boldsymbol{\Gamma}$

Now we consider the set of critical points for the diamond structure. Even in this structure, about which a very great deal is known, there remain uncertainties in detail. These will be indicated as we go along. At the point $\boldsymbol{\Gamma}$ there are the two three-fold degenerate modes $D^{(\boldsymbol{\Gamma})(15-)}$ acoustic, and $D^{(\boldsymbol{\Gamma})(25+)}$ optic. For the acoustic branch $D^{(\boldsymbol{\Gamma})(15-)}$ we are concerned with a non-analytic minimum for which the perturbation theory of Sect. 107 breaks down. The gradiant $\nabla\omega$ is not defined at $\boldsymbol{k}=0$. Thus there is a singular point since $\omega(\boldsymbol{\Gamma}|A)=0$ and the dynamical matrix is singular, which can be taken as P_0 (i.e. P_j with $j=0$). Now we adopt the convention following JOHNSON and LOUDON[4] that $P_j(\mu)$ refers to a critical point of index j with μ principal directions for which the branch has a discontinuous first derivative. The generalized acoustic minimum is then a point $P_0(3)$ and since the density of states is $g(\omega) \sim \omega^2$ no discontinuity or singularity appears in $d g(\omega)/d\omega|_{\boldsymbol{k}=\boldsymbol{\Gamma}}$.

The optic mode $D^{(\boldsymbol{\Gamma})(25+)}$ permits the application of the perturbation theory. Then since

$$D^{(\boldsymbol{\Gamma})(25+)} \otimes D^{(\boldsymbol{\Gamma})(25+)} \tag{140.8}$$

contains only even representations, and $D^{(\nabla)} = D^{(\boldsymbol{\Gamma})(15-)}$ it is clear that the slope vanishes in the 3 optic branches at this point. As far as now known, the highest phonon frequency in all diamond structure materials is at $\boldsymbol{\Gamma}$ in the optic branch. Hence this point is a P_3 in each of the three optic branches.

Diamond structure: Point $\boldsymbol{X}_1$

Next consider the point $\boldsymbol{X}_1$ in the zone. Here, all representations are doubly degenerate and the relevant phonon representations given in Table 22 are $D^{(\boldsymbol{X}_1)(4)}$;

---

[4] *op. cit.*

Table 30. Character table for critical points near $\boldsymbol{X}_1$ in diamond

Class[a]		$X^{(1)}$	$X^{(2)}$	$X^{(3)}$	$X^{(4)}$	$1\pm$	$2\pm$	$3\pm$	$4\pm$	$5\pm$	x	y, z
C_1	$\{\varepsilon\|0\}$	2	2	2	2	1	1	1	1	2	1	2
C_2	$\{\varepsilon\|\boldsymbol{t}_{xy}\}$	-2	-2	-2	-2	1	1	1	1	2	1	2
C_3	$\{\boldsymbol{\delta}_{2y}, \boldsymbol{\delta}_{2z}\|0, \boldsymbol{t}_{xy}\}$	0	0	0	0	1	1	-1	-1	0	-1	0
C_4	$\{\boldsymbol{\delta}_{2x}\|0\}$	2	2	-2	-2	1	1	1	1	-2	1	-2
C_5	$\{\boldsymbol{\delta}_{2x}\|\boldsymbol{t}_{xy}\}$	-2	-2	2	2	1	1	1	1	-2	1	-2
C_6	$\{\boldsymbol{\delta}_{4x}, \boldsymbol{\delta}_{4x}^{-1}\|\boldsymbol{\tau}_1, \boldsymbol{\tau}_1+\boldsymbol{t}_{xy}\}$	0	0	0	0	1	-1	-1	1	0	1	0
C_7	$\{\boldsymbol{\delta}_{2yt}\|\boldsymbol{\tau}_1\}, \{\boldsymbol{\delta}_{2y\bar{z}}\|\boldsymbol{\tau}_1+\boldsymbol{t}_{xy}\}$	0	0	2	-2	1	-1	1	-1	0	-1	0
C_8	$\{\boldsymbol{\delta}_{2y\bar{z}}\|\boldsymbol{\tau}_1\}, \{\boldsymbol{\delta}_{2yz}\|\boldsymbol{\tau}_1+\boldsymbol{t}_{xy}\}$	0	0	-2	2	1	-1	1	-1	0	-1	0
C_9	$\{\boldsymbol{i}\|\boldsymbol{\tau}_1, \boldsymbol{\tau}_1+\boldsymbol{t}_{xy}\}$	0	0	0	0	± 1	± 1	± 1	± 1	± 2	-1	-2
C_{10}	$\{\boldsymbol{\rho}_y, \boldsymbol{\rho}_z\|\boldsymbol{\tau}_1, \boldsymbol{\tau}_1+\boldsymbol{t}_{xy}\}$	0	0	0	0	± 1	± 1	∓ 1	∓ 1	0	1	0
C_{11}	$\{\boldsymbol{\rho}_x\|\boldsymbol{\tau}_1, \boldsymbol{\tau}_1+\boldsymbol{t}_{xy}\}$	0	0	0	0	± 1	± 1	± 1	± 1	∓ 2	-1	2
C_{12}	$\{\boldsymbol{\sigma}_{4x}, \boldsymbol{\sigma}_{4x}^{-1}\|0, \boldsymbol{t}_{xy}\}$	0	0	0	0	± 1	∓ 1	∓ 1	± 1	0	-1	0
C_{13}	$\{\boldsymbol{\rho}_{yz}\|0\}, \{\boldsymbol{\rho}_{y\bar{z}}\|0\}$	2	-2	0	0	± 1	∓ 1	± 1	∓ 1	0	1	0
C_{14}	$\{\boldsymbol{\rho}_{yz}\|\boldsymbol{t}_{xy}\}, \{\boldsymbol{\rho}_{y\bar{z}}\|\boldsymbol{t}_{xy}\}$	-2	$+2$	0	0	± 1	∓ 1	± 1	∓ 1	0	1	0
											$4-$	$5-$

[a] Classes refer to Table 18a (p. 366) in same order.

$D^{(\boldsymbol{X}_1)(1)}$; $D^{(\boldsymbol{X}_1)(3)}$. As discussed earlier, in all cases the relevant group is an extension of the point group D_{4h}; or otherwise put, the acceptable irreducible representations are ray representations of the point group D_{4h}. Now consider the mode $D^{(\boldsymbol{X}_1)(4)}$. In Table 30 we reproduce the relevant character information about this mode and the other two, for use in the reduction. Then we observe that $D^{(\boldsymbol{V})}$ transforms as $D^{(\bar{\boldsymbol{\Gamma}})(4-)} \oplus D^{(\bar{\boldsymbol{\Gamma}})(5-)}$ where we use $\bar{\boldsymbol{\Gamma}}$ to indicate that we are dealing here not with the entire group $\mathfrak{P}_0$ but with the subgroup $\mathfrak{P}(\boldsymbol{k}) = \mathfrak{P}(D_{4h})$. Observe too that we took advantage of the natural factoring of $D^{(\boldsymbol{V})}$ in D_{4h} into $D^{(\boldsymbol{V}_x)}$ and $D^{(\boldsymbol{V}_{z,y})}$. Clearly $D^{(\boldsymbol{V}_x)}$ refers to the x component of the gradient parallel to the line Δ. So we find from Table 30:

$$D^{(\boldsymbol{V}_x)} \otimes D^{(\boldsymbol{X}_1)(4)} = D^{(\boldsymbol{X}_1)(3)} \tag{140.9}$$

and

$$D^{(\boldsymbol{V}_{zy})} \otimes D^{(\boldsymbol{X}_1)(4)} = D^{(\boldsymbol{X}_1)(1)} \oplus D^{(\boldsymbol{X}_1)(2)}. \tag{140.10}$$

where $\boldsymbol{V}_{zy}$ refers to the perpendicular component. But then

$$D^{(\boldsymbol{X}_1)(4)} \otimes D^{(\boldsymbol{V}_x)} \otimes D^{(\boldsymbol{X}_1)(4)} = 2\left[D^{(\bar{\boldsymbol{\Gamma}})(2+)} \oplus D^{(\bar{\boldsymbol{\Gamma}})(2-)} \oplus D^{(\bar{\boldsymbol{\Gamma}})(4+)} \oplus D^{(\bar{\boldsymbol{\Gamma}})(4-)}\right] \tag{140.11}$$

which does not contain $D^{(\bar{\boldsymbol{\Gamma}})(1+)}$; and also

$$\begin{aligned} D^{(\boldsymbol{X}_1)(4)} \otimes D^{(\boldsymbol{V}_{zy})} \otimes D^{(\boldsymbol{X}_1)(4)} &= 2\, D^{(\boldsymbol{X}_1)(4)} \otimes D^{(\boldsymbol{X}_1)(1)} \\ &= 2\, D^{(\bar{\boldsymbol{\Gamma}})(5+)} \oplus 2\, D^{(\bar{\boldsymbol{\Gamma}})(5-)} \end{aligned} \tag{140.12}$$

which does not contain $D^{(\bar{\boldsymbol{\Gamma}})(1+)}$.

Consequently all three components of the gradient vanish, and the branch of symmetry $D^{(\boldsymbol{X}_1)(4)}$ has zero slope in all directions. Thus this is clearly a critical point P_j. To classify the index of this point we require some detailed analytic knowledge of the behavior of the frequency surfaces in in the neighborhood of $\boldsymbol{X}_1$.

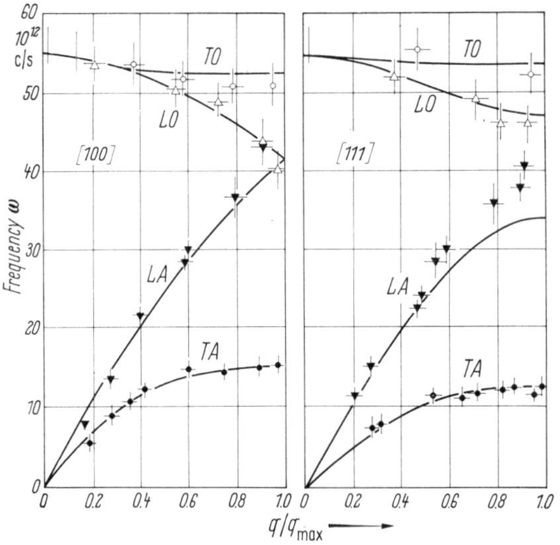

Fig. 5. Phonon dispersion in germanium. Calculated (solid curves) and measured values (points). (Cited in W. COCHRAN and R. A. COWLEY, in: this Encyclopdia, Vol. XXV/2a, p. 97, Fig. 16)

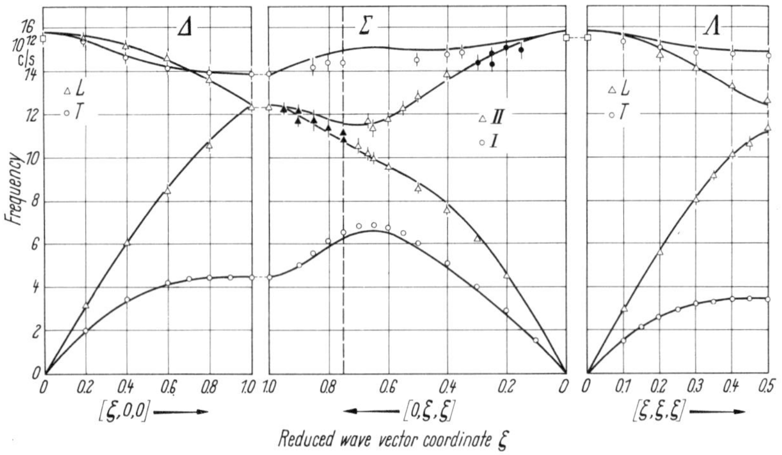

Fig. 6. Phonon dispersion in silicon. Calculated (solid curves) and measured values. (Cited in W. COCHRAN and R. A. COWLEY, in: this Encyclopedia, Vol. XXV/2a, p. 98, Fig. 17)

In Figs. 5–8 we give recent results, as of the time of writing for phonon dispersion in diamond-structure materials. Fig. 5 is taken from shell model calculations and experiment for Ge,[5] Fig. 6 from shell model and experiment for Si,[6] Fig. 7a and b from shell model and experiment for diamond,[7] and Fig. 8 from a

[5] See W. COCHRAN and R. A. COWLEY, in [*20*], p. 97, Fig. 16.

[6] *ibid.*, p. 98, Fig. 17.

[7] See G. DOLLING and R. A. COWLEY: Proc. Phys. Soc. (London) **88**, 463 (1966) Fig. 4. – J. L. WARREN, J. L. YARNELL, G. DOLLING, and R. A. COWLEY: Phys. Rev. **158**, 805 (1967).

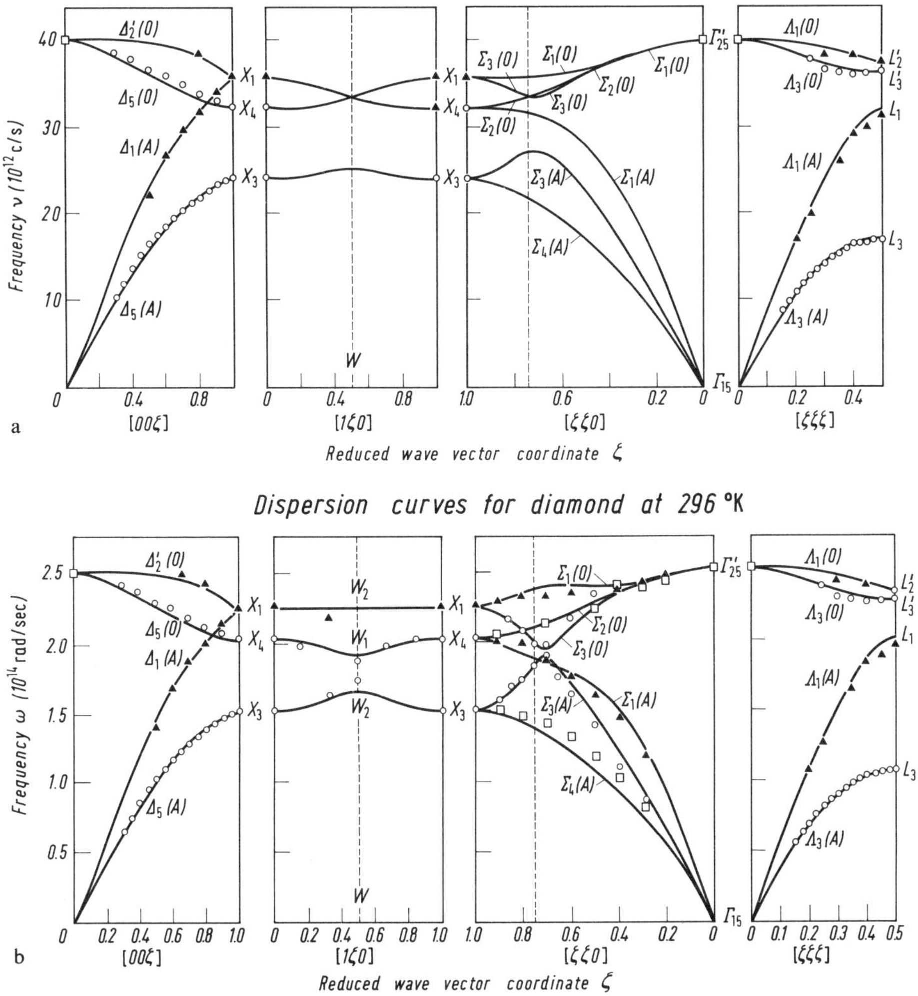

Fig. 7a and b. Phonon dispersion in diamond. Calculated (solid curves) and measured values (points). [After J. L. WARREN, J. L. YARNELL, G. DOLLING, and R. A. COWLEY: Phys. Rev. **158**, 805 (1967), Fig. 1]

shell model calculation for Ge.[8] By examining these figures we observe first that the diamond dispersion curves of Fig. 7a and b are quite different from the rest particularly in regard to ordering of states at $\boldsymbol{X}$. Note Ge and Si (Figs. 5, 6, 8); for the latter two cases the dispersion curves are quite similar.

From Figs. 5, 6, 8 we see that the mode $D^{(X_1)(4)}$, the TO mode, is an analytic minimum in Si and Ge and thus has index $j=0$. It may be observed that the three principal axes around $\boldsymbol{X}$ can be taken in the direction Δ, and along the two orthogonal $\boldsymbol{Z}$ directions in the plane square face of the zone and by inspection the branch rises in all these direction. However, it will be necessary in the sequel to study carefully the possibility of fluted critical points arising at $\boldsymbol{X}$. To do this we

[8] F. A. JOHNSON and R. LOUDON: Proc. Roy. Soc. (London), Ser. A **281**, 274 (1964), Fig. 1.

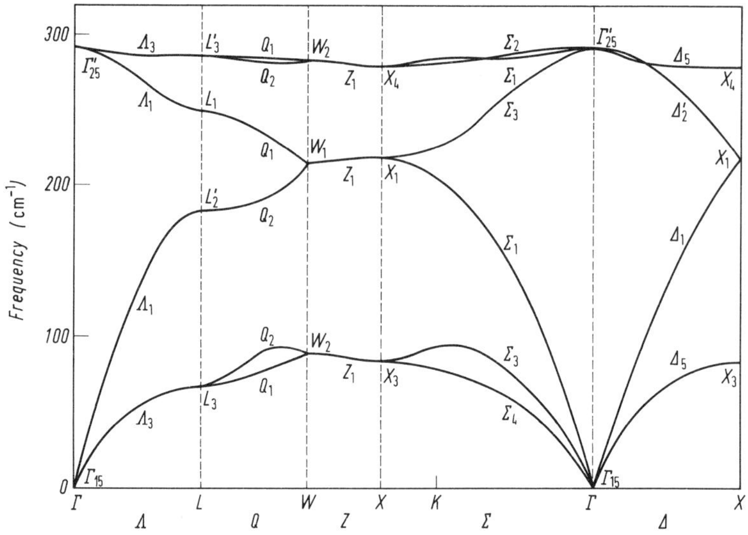

Fig. 8. Phonon dispersion calculation for germanium. [After F. A. JOHNSON and R. LOUDON: Proc. Roy. Soc. A **281**, 274 (1964), Fig. 1]

need to study the behavior over the entire square face of the Brillouin zone around $\boldsymbol{X}$; that is in the directions $\boldsymbol{X}-\boldsymbol{\Sigma}-\boldsymbol{K}'(U)$. Then we are dealing with a fluted point if the sign of $\omega(\boldsymbol{k}|j)-\omega(\boldsymbol{X}|j)$, on the branch j near $\boldsymbol{X}$, has the following behavior:

$$\operatorname{sgn}[\omega(\boldsymbol{Z}|j)-\omega(\boldsymbol{X}|j)]=-\operatorname{sgn}[\omega(\boldsymbol{\Sigma}|j)-\omega(\boldsymbol{X}|j)]. \tag{140.13}$$

In this case, there are 4 regions on the square face with $[\omega-\omega(\boldsymbol{X}|j)]>0$ and 4 regions with $[\omega-\omega(\boldsymbol{X}|j)]<0$.

If

$$\omega(\boldsymbol{\Delta}|j)-\omega(\boldsymbol{X}|j)>0 \quad \text{then } D^{(\boldsymbol{X}_1)(j)} \text{ is } F_1 \tag{140.14}$$

while if

$$\omega(\boldsymbol{\Delta}|j)-\omega(\boldsymbol{X}|j)<0 \quad \text{then } D^{(\boldsymbol{X}_1)(j)} \text{ is } F_2. \tag{140.15}$$

The F_j points are to be counted as P_j points[9] of weight 3. Now the mode $D^{(\boldsymbol{X}_1)(4)}$ in diamond appears fluted in regard to the branch which in $\boldsymbol{\Sigma}$ direction is $\Sigma_2(0)$. We classify $D^{(\boldsymbol{X}_1)(4)}$ for the upper branch as F_1 in diamond. The lower of the two branches degenerate at $\boldsymbol{X}_1$, we index as P_2. These results are summarized in Table 31.

Next consider the mode of symmetry $D^{(\boldsymbol{X}_1)(1)}$: it is the longitudinal mode: degenerate LO and LA. Repeating the analysis as for the mode $D^{(\boldsymbol{X}_1)(4)}$ we find

$$\begin{aligned} D^{(\boldsymbol{X}_1)(1)}\otimes D^{(\boldsymbol{V}_x)}\otimes D^{(\boldsymbol{X}_1)(1)} &= D^{(\boldsymbol{X}_1)(1)}\otimes D^{(\boldsymbol{X}_1)(1)} \\ &= D^{(\boldsymbol{\Gamma})(1+)}\oplus D^{(\boldsymbol{\Gamma})(2-)}\oplus D^{(\boldsymbol{\Gamma})(3+)}\oplus D^{(\boldsymbol{\Gamma})(4-)} \end{aligned} \tag{140.16}$$

and

$$\begin{aligned} D^{(\boldsymbol{X}_1)(1)}\otimes D^{(\boldsymbol{V}_{zy})}\otimes D^{(\boldsymbol{X}_1)(1)} &= D^{(\boldsymbol{X}_1)(1)}\otimes[D^{(\boldsymbol{X}_1)(3)}\oplus D^{(\boldsymbol{X}_1)(4)}] \\ &= 2[D^{(\boldsymbol{\Gamma})(2+)}\oplus D^{(\boldsymbol{\Gamma})(2-)}\oplus D^{(\boldsymbol{\Gamma})(4+)}\oplus D^{(\boldsymbol{\Gamma})(4-)}]. \end{aligned} \tag{140.17}$$

[9] F. A. JOHNSON and R. LOUDON: *op. cit.* — J. C. PHILLIPS: *op. cit.*

Table 31. One-phonon critical points and indices in diamond[a]

		$\boldsymbol{\Gamma}^{(1)}$	$\boldsymbol{X}^{(3)}$	$\boldsymbol{L}^{(4)}$	$\boldsymbol{W}^{(6)}$	$\boldsymbol{\Sigma}^{(12)}$	$\boldsymbol{Q}^{(24)}$	$\boldsymbol{S}^{(24)}$
TO 1	C	P_3	F_1	P_2?	$P_1(2)$?	P_j		
	Si	P_3	P_0	P_2	$P_1(2)$	$P_2\ P_1(1)$		
	Ge	P_3	P_0	P_2	$P_1(2)$	$P_2\ P_1(1)$		
TO 2	C	P_3	P_2	P_2?	$P_3(2)$?	$P_j(1)$	P_1?	P_2?
	Si	P_3	P_0	P_2	$P_3(2)$	$P_2\ P_1(1)$	P_1	P_2
	Ge	P_3	P_0	P_2	$P_3(2)$	$P_2\ P_1(1)$	P_1	P_2
LO	C	P_3	$F_1(1)/P_0(1)$	P_2	$P_0(2)$?	$P_j,\ P_{j'}(1)$		
	Si	P_3	$P_2(1)$	P_2	$P_0(2)$	P_1		
	Ge	P_3	$F_1(1)$	P_2	$P_0(2)$	—		
LA	C	$P_0(3)$	$P_3(1)/F_2(1)$	P_1?	$P_2(2)$?	$P_j(1),\ P_{j'}(1),\ P_{j''}$		
	Si	$P_0(3)$	$P_3(1)$	P_1	$P_2(2)$	—		
	Ge	$P_0(3)$	$P_3(1)$	P_1	$P_2(2)$	—		
TA 1	C	$P_0(3)$	P_1	P_1	$P_1(2)$?	$P_j(1),\ P_{j'}(1)$	P_2?	
	Si	$P_0(3)$	P_1	P_1	$P_1(2)$	P_3	P_2	
	Ge	$P_0(3)$	P_1	P_1	$P_1(2)$	P_3	P_2	
TA 2	C	$P_0(3)$	F_2	P_1	$P_3(2)$?	—		
	Si	$P_0(3)$	F_2	P_1	$P_3(2)$	—		
	Ge	$P_0(3)$	F_2	P_1	$P_3(2)$	—		

[a] Question marks refer to uncertainty in assignment as in text. Number in parenthesis is the number of arms in the star. Assignments of indices follow discussion in text.

In (140.16) and (140.17) all representations $D^{(\boldsymbol{\Gamma})(j)}$ refer to the point group D_{4h}. As (140.16) contains $D^{(\boldsymbol{\Gamma})(1+)}$, at least one component of $\nabla_{\boldsymbol{k}}\,\omega(\boldsymbol{k})$ is non-zero by symmetry. But, as $D^{(\boldsymbol{X}_1)(1)}$ is a double degenerate mode, this implies that in a secular equation for this mode around the point $\boldsymbol{X}_1$, in the direction Δ, for which the compatibility relation requires a splitting $D^{(\boldsymbol{X}_1)(1)} \to D^{(\Delta)(2')} \oplus D^{(\Delta)(1)}$ that both branches enter $\boldsymbol{X}_1$ with finite slope. However, from (140.17) we see that the slope $\nabla\omega$ of both branches $D^{(\Delta)(2')}$ and $D^{(\Delta)(1)}$ vanishes at $\boldsymbol{X}_1$ in the transverse plane perpendicular to Δ (i.e. in $k_z\,k_y$ plane). Hence the branches which meet at X_1 have two vanishing first derivatives, and one discontinuous first derivative. Thus for these two modes $n=1$, in $P_j(n)$ or $F_j(n)$.

Now we turn to the dispersion curves to discuss these branches. In Si the LO branch has a minimum in the Σ direction (branch $\Sigma_3(0)$, Fig. 6). Hence $\boldsymbol{X}_1$ point is index 2 for this branch since it decreases in $\boldsymbol{Z}$ and $\boldsymbol{\Sigma}$ directions. Further, since

$$\operatorname{sgn}[\omega(\boldsymbol{Z}\,|\,\mathrm{LO})-\omega(\boldsymbol{X}_1\,|\,\mathrm{LO})]=\operatorname{sgn}[\omega(\Sigma\,|\,\mathrm{LO})-\omega(\boldsymbol{X}_1\,|\,\mathrm{LO})] \tag{140.18}$$

the LO branch has an analytic saddle point $P_2(1)$ at $\boldsymbol{X}_1$. In Ge the LO branch may not have a $\boldsymbol{\Sigma}$ minimum so an alternate assignment may be needed. Assuming the LO branch *rises* in the $\boldsymbol{\Sigma}$ direction, we have again the situation of (140.13)–(140.15) shown in Figs. 5 and 8. The LO branch in Ge should then be indexed $F_1(1)$. For diamond we do not have precise information about the dispersion in the $\boldsymbol{Z}$ directions compatible with $D^{(\boldsymbol{X}_1)(1)}$, hence we cannot determine relative signs as in (140.18). Assuming negative $\boldsymbol{Z}$ curvature would give this mode index $F_1(1)$ and if positive $\boldsymbol{Z}$ curvature, a $P_0(1)$. All the foregoing results are incorporated in Table 31.

For the LA mode in Si we have behavior $P_3(1)$ as the branch energy decreases in all directions. For LA in Ge the same is true. For diamond, negative $\boldsymbol{Z}$ curvature gives $P_3(1)$; while positive $\boldsymbol{Z}$ curvature gives $F_2(1)$ (see Table 31).

Now consider mode of symmetry $D^{(X_1)(3)}$. Repeating the analysis as in (140.11) and (140.12) we find a vanishing gradient in all directions. Again we consider separately the TA 1 and TA 2 branches. In Si the TA 1 branch is P_1, likewise in Ge, and diamond (Figs. 5–8). The TA 2 branch is fluted and classed F_2 in all compounds (see Table 31).

Point $\boldsymbol{L}_1$

Now turn to the point $\boldsymbol{L}_1$. The relevant point group $\mathfrak{P}(\bar{\Gamma})$, in D_{3h}. In Table 32 we present the character table needed for the analysis. It is now straightforward to see from point group theory and the matrix element that all branches at $\boldsymbol{L}_1$ have zero slope in all directions. To analyze the branches we can use a set of orthogonal axes based on the directions $\boldsymbol{\Lambda}$, $\boldsymbol{Q}(\boldsymbol{L}-\boldsymbol{W})$, and $\boldsymbol{M}(\boldsymbol{L}-\boldsymbol{K}')$ or $(\boldsymbol{L}-\boldsymbol{U})$. The dispersion curves in the directions $\boldsymbol{M}$ from $\boldsymbol{L}_1$ are generally not available but one may estimate them crudely.

Table 32. Critical points near $\boldsymbol{L}_1$ in diamond

Class[a]	$L_1^{(1\pm)}$	$L_1^{(2\pm)}$	$L_1^{(3\pm)}$	Z'	(X', Y')[b]
$\{\varepsilon\|0\}$	1	1	2	1	2
$\{\delta_{3xyz}\|0\}$	1	1	-1	1	-1
$\{\delta_{2x\bar{y}}\|\tau_1\}$	1	-1	0	-1	0
$\{i\|\tau_1\}$	± 1	± 1	± 2	-1	-2
$\{\sigma_{6xyz}\|\tau_1\}$	± 1	± 1	∓ 1	-1	1
$\{\rho_{x\bar{y}}\|\tau_1\}$	± 1	∓ 1	0	1	0
				$2-$	$3-$

[a] See Tables B 1 and 22.
[b] Components $\boldsymbol{Z}'$ and $(\boldsymbol{X}', \boldsymbol{Y}')$ are respectively $\parallel$ and $\perp$ to [111] direction.

Consider first silicon, and in decreasing order of frequency, consider the TO 1 branch $D^{(L_1)(3')}$. This branch is very flat (Fig. 8) but appears to be P_2. The same appears to be the case in Ge (see Fig. 8). In diamond, Fig. 7, where the TO modes at $\boldsymbol{L}_1$ are below the LO, we have insufficient information for a decision. It appears to the writer that fluted behavior can well occur at $\boldsymbol{L}$, consistent with PHILIPS' criteria, but failing an exact theory or experiment we shall simply follow existing assignments as for Si and Ge, and call the TO 1 mode here a P_2(?). The TO 2 mode is also taken as P_2 for all three materials.

Next take the mode of symmetry $D^{(L_1)(1)}$. In Si and Ge this is the LO mode and is evidently P_2. In diamond this is the LA mode and we assign it provisionally P_1. For mode $D^{(L_1)(2')}$ which is LA in Si and Ge, the assignment of P_1 is obvious. In diamond this is the LO mode and we assign it P_2. For mode $D^{(L_1)(3)}$ which is TA in all materials, we assign P_1 as the index. The possibility of fluted behavior in the hexagonal plane on $\boldsymbol{L}$ seems not settled, at the time of writing.

Table 33. Critical point near $\boldsymbol{W}_1$ in diamond. $\boldsymbol{W}_1=(2\pi, 0, -\pi)(1/a)$

Element	Class[a]	$\boldsymbol{W}_1^{(1)}$	$\boldsymbol{W}_1^{(2)}$	$(\hat{\boldsymbol{\varepsilon}}_z\cdot\boldsymbol{V}_k\equiv\boldsymbol{V}_z)$[b]	$(\hat{\boldsymbol{\varepsilon}}_{Q_1}\cdot\boldsymbol{V}_k, \hat{\boldsymbol{\varepsilon}}_{Q_2}\cdot\boldsymbol{V}_k)$[b]
$\{\varepsilon\|\boldsymbol{0}\}$	C_1	2	2	1	2
$\{\varepsilon\|\boldsymbol{t}_{xy}\}$	C_2	-2	-2	1	2
$\{\varepsilon\|\boldsymbol{t}_{yz}\}$	C_3	$-2i$	$-2i$	1	2
$\{\varepsilon\|\boldsymbol{t}_{zx}\}$	C_4	$2i$	$2i$	1	2
$\{\delta_{2z}\|\boldsymbol{0}\}\ \{\delta_{2z}\|\boldsymbol{t}_{xy}\}$	C_5	0	0	1	-2
$\{\delta_{2z}\|\boldsymbol{t}_{yz}\}\ \{\delta_{2z}\|\boldsymbol{t}_{zx}\}$	C_6	0	0	1	-2
$\{\delta_{2xy}\|\tau, \tau+\boldsymbol{t}_{xy}\}\ \{\delta_{2\bar{x}y}\|\tau+\boldsymbol{t}_{yz}, \tau+\boldsymbol{t}_{zx}\}$	C_7	0	0	-1	2
$\{\delta_{2\bar{x}y}\|\tau, \tau+\boldsymbol{t}_{xy}\}\ \{\delta_{2xy}\|\tau+\boldsymbol{t}_{yz}, \tau+\boldsymbol{t}_{zx}\}$	C_8	0	0	-1	2
$\{\sigma_{4z}\|\boldsymbol{0}\}\ \{\sigma_{4z}^{-1}\|\boldsymbol{t}_{yz}\}$	C_9	$(1-i)$	$-(1-i)$	-1	0
$\{\sigma_{4z}\|\boldsymbol{t}_{xy}\}\ \{\sigma_{4z}^{-1}\|\boldsymbol{t}_{zx}\}$	C_{10}	$-(1-i)$	$(1-i)$	-1	0
$\{\sigma_{4z}\|\boldsymbol{t}_{yz}\}\ \{\sigma_{4z}^{-1}\|\boldsymbol{t}_{xy}\}$	C_{11}	$-(1+i)$	$(1+i)$	-1	0
$\{\sigma_{4z}\|\boldsymbol{t}_{zx}\}\ \{\sigma_{4z}^{-1}\|\boldsymbol{0}\}$	C_{12}	$(1+i)$	$-(1+i)$	-1	0
$\{\rho_x\|\tau, \tau+\boldsymbol{t}_{xy}\}\ \{\rho_y\|\tau+\boldsymbol{t}_{yz}, \tau+\boldsymbol{t}_{zx}\}$	C_{13}	0	0	1	0
$\{\rho_y\|\tau, \tau+\boldsymbol{t}_{xy}\}\ \{\rho_x\|\tau+\boldsymbol{t}_{yz}, \tau+\boldsymbol{t}_{zx}\}$	C_{14}	0	0	1	0
				$\bar{\boldsymbol{\Gamma}}^{(B_2)}$	$\bar{\boldsymbol{\Gamma}}^{(E)}$

[a] R. LOUDON and R.J. ELLIOTT: J. Phys. Chem. Solids **15**, 146 (1960).
[b] $\hat{\boldsymbol{\varepsilon}}_z=(0, 0, 1)$; $\hat{\boldsymbol{\varepsilon}}_{Q_1}=(1, 1, 0)$; $\hat{\boldsymbol{\varepsilon}}_{Q_2}=(1, -1, 0)$. These are an orthogonal triad.

Point $\boldsymbol{W}_1$

At the point $\boldsymbol{W}_1$ in the zone we construct Table 33 to examine the slope behavior along the three orthogonal directions namely $\boldsymbol{Z}$, and the two $\boldsymbol{Q}$ directions in the plane perpendicular to Z, passing through $\boldsymbol{W}$. The analysis is indicated in the table. The allowable irreducible representations of $\boldsymbol{W}_1$, which are listed in Table 33, for diamond, are ray representations, with particular factor system of the underlying point group $\mathfrak{P}(\boldsymbol{W}_1)$: it is isomorphic to point group D_{2d}. We choose as three unit polarization vectors: $\hat{\boldsymbol{\varepsilon}}_z$, $\hat{\boldsymbol{\varepsilon}}_{Q_1}$, and $\hat{\boldsymbol{\varepsilon}}_{Q_2}$, which point along the three orthogonal directions. Then as in the table, under the rotations in D_{2d}, the operators $\hat{\boldsymbol{\varepsilon}}_z\cdot\boldsymbol{V}_k$ transforms as a one dimensional representation B_2 in a standard listing of irreducible representations for D_{2d} given in [2] p. 126, while $(\hat{\boldsymbol{\varepsilon}}_{Q_1}\cdot\boldsymbol{V}_k, \hat{\boldsymbol{\varepsilon}}_{Q_2}\cdot\boldsymbol{V}_k)$ transform as a basis for a two dimensional E representation. Both of these may, as in the preceding discussion of the $\boldsymbol{X}$, point be given pseudo-$\bar{\boldsymbol{\Gamma}}$ labels which is done in the last line of the table in the corresponding columns, where these are called $\bar{\boldsymbol{\Gamma}}^{(B_2)}$ and $\bar{\boldsymbol{\Gamma}}^{(E)}$. Multiplying out the representations and identifying the result we find that for *both* $\boldsymbol{W}_1^{(1)}$ and $\boldsymbol{W}_1^{(2)}$ representations, the slope vanishes for direction parallel to Z, while for both $\boldsymbol{W}_1^{(1)}$ and $\boldsymbol{W}_1^{(2)}$ the slope is non-zero in the two orthogonal $\boldsymbol{Q}$ directions passing through $\boldsymbol{W}_1$. Then, owing to the double degeneracy of modes $D^{(\boldsymbol{W})(1)}$ and $D^{(\boldsymbol{W})(2)}$ at $\boldsymbol{W}_1$ and along $\boldsymbol{Z}$, but the splitting in directions $\boldsymbol{Q}$, all branches have 2 directions of discontinuous first derivatives, $n=2$. Assuming all branches have analytic behavior (unlike $\boldsymbol{X}$) we read off the assignments from the figures, and insert in Table 31.

Line $\boldsymbol{\Sigma}$

Consider next line $\boldsymbol{\Sigma}$. From Table 34 we establish that perpendicular to $\boldsymbol{\Sigma}$ all branches have zero slope. Although the case $\boldsymbol{\Sigma}=(\pi, \pi, 0)(1/a)$ is given the results are general on all $\boldsymbol{\Sigma}$. Thus by inspecting the lines $\boldsymbol{\Sigma}$ we can determine existence of added critical points in each branch. Assigning indices to these $\boldsymbol{\Sigma}$ critical points is another matter, and to undertake this we require detailed topological information only available by numerical calculation. For Si and Ge (Figs. 6 and 8) we follow JOHNSON and LOUDON [10] and we take over their results. The TO 1 branch in Si, has one critical point near $\boldsymbol{X}$, and another (singular?) critical point due to crossing (accidental degeneracy). These are indexed P_2 and $P_1(1)$ respectively following these authors, although it one could perhaps use $F_1(1)$ for the latter. The next, TO 2 branch, has also a critical point due to the crossing and a minimum besides. These are $P_2(1)$ and P_1 respectively, again it appears that one might take it singular for the crossing (?). In Si the LO optic (Σ_3) branch has a strong minimum (Figs. 5, 6, 8) classed as P_1. In Ge the branch has no critical point. The LA branch (Σ_1) in Si and Ge has no critical point. The TA 1 branch (Σ_3) in Si has a maximum classed P_3 which we assume also is true for Ge, while TA 2 (Σ_4) has no critical point on it.

Table 34. Critical point near $\boldsymbol{\Sigma}_1$ in diamond. $\boldsymbol{\Sigma}_1=(\pi, \pi, 0)(1/a)$[a]

Class		$\boldsymbol{\Sigma}_1^{(1)}$	$\boldsymbol{\Sigma}_1^{(2)}$	$\boldsymbol{\Sigma}_1^{(3)}$	$\boldsymbol{\Sigma}_1^{(4)}$	$\hat{\varepsilon}_\Sigma \cdot \nabla$	$\hat{\varepsilon}_\Sigma \cdot \nabla$	$\hat{\varepsilon}_2 \cdot \nabla$
C_1	$\{\varepsilon\|0\}$	1	1	1	1	1	1	1
C_2	$\{\rho_{x\bar{y}}\|0\}$	1	−1	1	−1	11	−1	1
C_3	$\{\rho_z\|\tau_1\}$	1	−1	−1	1	1	1	−1
C_4	$\{\delta_{2xy}\|\tau_1\}$	1	1	−1	−1	1	−1	−1

All branches have zero slope in directions perpendicular to the line $\boldsymbol{\Sigma}$. Take orthogonal triad: $\hat{\varepsilon}_{\Sigma_1}, \hat{\varepsilon}_{\Sigma'}, \hat{\varepsilon}_Z$ where $\boldsymbol{\Sigma}=[\pi, \pi, 0]$; $\boldsymbol{\Sigma}'=[\pi, -\pi, 0]$; $\boldsymbol{Z}=[0, 0, 1]$.

[a] Labelling of representations agrees with J.L. BIRMAN: Phys. Rev. **131**, 1492 (1963) and G.F. KOSTER [49].

Along $\boldsymbol{\Sigma}$ in diamond the situation is quite different, owing principally to the different order of states at the $\boldsymbol{X}$ points. Here, inspection of Fig. 7 is not enough to determine all indices j of the points. We simply list the distinct critical points which can be observed in each branch. The TO 1 $(\Sigma_1(0))$ branch has a maximum P_j. The TO 2 $(\Sigma_2(0))$ has a singular cross-over $P_j(1)$. The LO $(\Sigma_3(0))$ branch has a pronounced minimum plus a cross-over P_j, $P_{j_1}(1)$. The next branch (ordered labelling) has 2 cross-overs and a maximum thus LA $(\Sigma_1(A))$ has $P_j(1)$, $P_{j'}(1)$, $P_{j''}$. The TA 1 $(\Sigma_3(A))$ branch has $P_j(1)$, $P_j(1)$. The lowest branch TA 2 has no critical points.

Line Q

Along $\boldsymbol{Q}$ it is found (Table 35) that slopes vanish perpendicular to it but inspection of figures is needed to find added critical points. Failing accidental degeneracy all along $\boldsymbol{Q}$, for branches $\boldsymbol{Q}_1$ and $\boldsymbol{Q}_2$, there must be at least some critical points along Q in branches connecting TO $(\boldsymbol{L})$ and TO $(\boldsymbol{W})$, and TA $(\boldsymbol{L})$ and TA $(\boldsymbol{W})$. Again we follow JOHNSON and LOUDON here for Si and Ge (see Table 31). In diamond there is additional uncertainty owing to lack of computed

[10] F.A. JOHNSON and R. LOUDON: *ibid.*, Table I.

Table 35. Critical point near Q in diamond. $Q_1 = (\kappa, \pi, 2\pi - \kappa)(1/a)$

Class	$Q_1^{(1)}$	$Q_1^{(2)}$	$\hat{\varepsilon}_Q \cdot V_k^{(a)}$	$\hat{\varepsilon}_1 \cdot V_k^{(a)}$
$\{\varepsilon \vert 0\}$	1	1	1	1
$\{\delta_{2x\bar{z}} \vert \tau_1\}$	1	−1	1	−1

The only non-zero slope is for longitudinal component, in the line containing Q_1. All other components vanish.

or observed frequencies on Q. We simply repeat the (provisional) assignments here as already used for Si and Ge, subject to check via the Morse relations.

Finally, critical point assignment on planes S for TO2 mode was made as before.

It may be remarked that, earlier, other one phonon critical point assignments have been proposed for the diamond structure, but less was known both experimentally and theoretically at the time when they were proposed.[11] The assignments of BILZ and coworkers are fairly close to those more recent ones of JOHNSON and LOUDON. We shall later refer to this work in comparing experiment and theory. At the time of writing the most accurate proposal would appear to be that of JOHNSON and LOUDON.

It will be recalled that each type of critical point produces a different type contribution to the density of states (and different) discontinuities. These are illustrated in Fig. 9. See the discussion in [*20*, *47*].

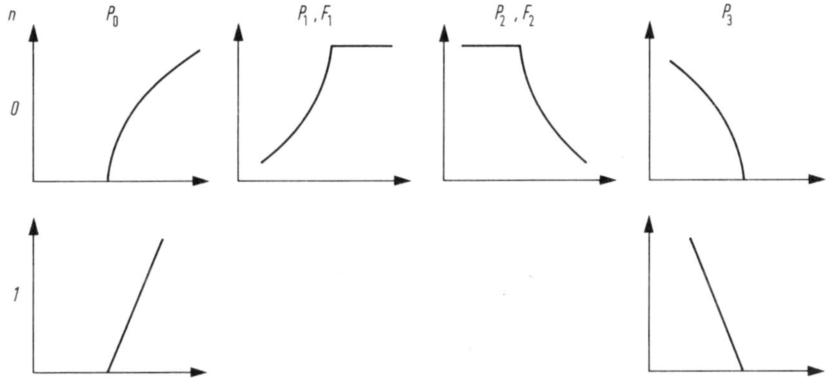

Fig. 9. Shapes of critical point contribution to the density of states for $P_j(n)$; $j=1, 2, 3$; $n=0, 1$. [After F.A. JOHNSON and R. LOUDON: Proc. Roy. Soc. A **281**, 274 (1964), Fig. 1]

141. Two phonon density of states and critical points in diamond structure. Now we can construct the analogous table of critical points for the two phonon dispersion. Again, part of the work is purely analytical and can be carried through using the group theory (symmetry) analysis only, and part requires that we have available detailed dispersion curves. In principal however, the procedure is the

[11] H. BILZ, R. GEICK, and K.F. RENK, in: Proceedings of the Conference on Lattice Dynamics, Copenhagen 1963 (ed. R.F. WALLIS). New York: Pergamon Press 1965.

same as in the case of the one-phonon density of states. We have to distinguish in this work the cases where the two phonons arise from the identical branch, and are thus degenerate – the case we call *overtones*, and the case in which the phonons arise from two different branches – these are *combinations* [*31*]. The discussion in Sects. 117 and 118 should be recalled at this point. In all cases the selection rules or reduction coefficients must be in accord with the process which we intend to analyse: infra-red or Raman.

Consider first the symmetry analysis. It is clear that the dispersion surfaces of the two-phonon overtone branches simply mimic those of the single phonon prototype, aside from a frequency scale change (doubling). Hence the location of the critical points on overtone branches is identical with that of the corresponding single phonon branches of Table 31. Then Table 31 should be taken over to apply also to the overtone case, for each of the materials with this structure e.g. C, Si, Ge.

For the combination branches our concern is with the dispersion surfaces $\omega(\boldsymbol{k}|j)+\omega(-\boldsymbol{k}|j')$, and the critical points of the combination branches are those $\boldsymbol{k}_0$ for which

$$\nabla_{\boldsymbol{k}}[\omega(\boldsymbol{k}|j)+\omega(-\boldsymbol{k}|j')]|_{\boldsymbol{k}_0=\boldsymbol{k}}=0. \tag{141.1}$$

Clearly we can have a combination branch critical point if both branches j and j' have a critical point at $\boldsymbol{k}_0$ and $-\boldsymbol{k}_0$ respectively and/or also if

$$\nabla_{\boldsymbol{k}}\,\omega(\boldsymbol{k}|j)_{\boldsymbol{k}_0}=-\nabla_{\boldsymbol{k}}\,\omega(-\boldsymbol{k}|j')_{\boldsymbol{k}_0}. \tag{141.2}$$

In the first case

$$\nabla_{\boldsymbol{k}}\,\omega(\boldsymbol{k}|j)_{\boldsymbol{k}_0}=\nabla_{\boldsymbol{k}}\,\omega(-\boldsymbol{k}|j')_{\boldsymbol{k}_0}=0 \tag{141.3}$$

but as the set of points for which the individual branches have critical points is already known (Table 31) the set of critical points arising from case (141.3) is easily determined. The problem of indexing the critical points on the two-phonon branches is again one requiring detailed knowledge of the dispersion characteristics of the branch in the vicinity of the point in question. To determine the points $\boldsymbol{k}_0$ for which (141.2) applies we certainly require detailed knowledge of the dispersion surfaces since critical points may arise at general points in the zone, where slopes of the constituent summand branches are equal and opposite. In Table 36 a listing of two-phonon critical points is given for Si and Ge: as obtained by JOHNSON and LOUDON.[1] As of time of writing a similar table for diamond is not yet available, but see the discussion on diamond in Sects. 142 and 146.

142. Interpretation of lattice Raman and infra-red spectra in crystals of the diamond structure. From the work of the previous sections we have now available the information needed to interpret the multiphonon infra-red and Raman lattice spectra of diamond structure crystals within the approximation that these spectra are governed by the density of states and the selection rules. That is, we do not consider any variation of matrix element with details of the transitions: "constant matrix element" assumption. Hence the intensity of an infra-red absorption, or Raman scattering event at frequency ω is proportional to the product of the

[1] F.A. JOHNSON and R. LOUDON: Proc. Roy. Soc. (London), Ser. A **281**, 274 (1964).

Table 36. Two phonon processes in diamond structure, silicon and germanium

Point	Two phonons	Type	Index[b]	Activity[a]
		Overtones		
Γ	$[\Gamma^{(25+)}]_{(2)}$	$2O$	P_3	R
$\star X$	$[\star X^{(4)}]_{(2)}$	$2TO$	P_0	R
	$[\star X^{(1)}]_{(2)}$	$2L$	$(P_2(1)+P_3(1))$	R
	$[\star X^{(3)}]_{(2)}$	$2TA$	P_1+F_2	R
$\star L$	$[\star L^{(3-)}]_{(2)}$	$2TO$	P_2	R
	$[\star L^{(1+)}]_{(2)}$	$2LO$	P_2	R
	$[\star L^{(2-)}]_{(2)}$	$2LA$	P_1	R
	$[\star L^{(3+)}]_{(2)}$	$2TA$	P_1	R
$\star\Sigma$	$[\star\Sigma^{(2)}]_{(2)}$	$2TO1$	$P_2+P_1(1)$	R
	$[\star\Sigma^{(1)}]_{(2)}$	$2TO2$	$P_2(1)+P_1$	R
	$[\star\Sigma^{(3)}]_{(2)}$	$2LO$	P_1	R
	$[\star\Sigma^{(3)}]_{(2)}$	$2TA1$	P_3	R
$\star Q$	$[\star Q^{(2)}]_{(2)}$	$2TO2$	P_1	R
	$[\star Q^{(2)}]_{(2)}$	$2TA1$	P_2	R
$\star S$	$[\star S^{(2)}]_{(2)}$	$2TO2$	P_2	R
		Combinations		
$\star X$	$\star X^{(4)}\otimes\star X^{(1)}$	$TO1,2+LO, LA$	$P_3(1)$ $+(P_2(1)$ or $F_1(1))$	$R; IR$
	$\star X^{(4)}\otimes\star X^{(3)}$	$TO1,2+TA1,2$	P_1+F_2	$R; IR$
	$\star X^{(1)}\otimes\star X^{(3)}$	$L+TA1,2$	$P_0(1)+F_2(1)$	$R; IR$
$\star L$	$\star L^{(3-)}\otimes\star L^{(1+)}$	$TO+LO$	P_2	IR
	$\star L^{(3-)}\otimes\star L^{(2-)}$	$TO+LA$	P_1	R
	$\star L^{(3-)}\otimes\star L^{(3+)}$	$TO+TA$	P_1	IR
	$\star L^{(1+)}\otimes\star L^{(2-)}$	$LO+LA$	P_3	IR
	$\star L^{(1+)}\otimes\star L^{(3+)}$	$LO+TA$	P_0+P_2	R
	$\star L^{(2-)}\otimes\star L^{(3+)}$	$LA+TA$	P_1	IR
$\star W$	$\star W^{(2)}\otimes\star W^{(2)}$	$TO(1)+TO(2)$	P_1	R
	$\star W^{(1)}\otimes\star W^{(1)}$	$LO+LA$	P_3	R
	$\star W^{(2)}\otimes\star W^{(2)}$	$TA(1)+TA(2)$	P_2	R
$\star\Delta$	$\star\Delta^{(1)}\otimes\star\Delta^{(5)}$	$LO+TA1,2$	P_3	$R; IR$
$\star\Sigma$	$\star\Sigma^{(1)}\otimes\star\Sigma^{(3)}$	$TO1,2+LO$	P_1	$R; IR$
	$\star\Sigma^{(1)}\otimes\star\Sigma^{(3)}$	$TO1,2+TA1$	P_3	$R; IR$
	$\star\Sigma^{(3)}\otimes\star\Sigma^{(3)}$	$LO+TA1$	P_2	$R; IR$
$\star Q$		$TO1,2+TA2$	P_2	$R; IR$
		$LO+LA$	P_1	$R; IR$
		$LO+TA1$	P_2	$R; IR(15R)$
$\star S_I$		$TA1+TA2$	P_1	$R; IR$
$\star S_{II}$		$LO+LA$	P_2	$R; IR$
		$LO+TA1$	P_1	$R; IR$
		$LO+TA2$	P_2	$R; IR$
		$TA1+TA2$	P_2	$R; IR$

[a] R means Raman active (at least one allowed representation present; IR means infra-red active, $\Gamma^{(15-)}$ present). From J.L. BIRMAN: Phys. Rev. **131**, 1492 (1963). [b] See discussion in text.

frequency dependent density of states $\rho_L(\omega)$ and square of the constant matrix element $\mathscr{M}$. The formula is given in (121.68)–(121.70).

Now there are two ways of proceeding. First possibility is to consider that for allowed transitions the intensity $I(\omega)$ is actually proportional to $\rho_L(\omega)$, in all detail. In this case we expect the experimental variation of $I(\omega)$ with ω to mimic exactly the variation of $\rho_L(\omega)$. In particular we look for slope discontinuities corresponding to the density of states associated with each kind of critical point for each branch considered as illustrated in Fig. 9. Thus in studying the two phonon infra-red absorption region we should find features corresponding to all the critical points for allowed overtone and combination branches. For the overtones the type and location of critical points is identical to the one-phonon counterparts; for the indexing of the critical points on combination branches we see Table 36. In Table 37 a listing of allowed 2 phonon processes is given for the diamond structure.

Now if we attempt to carry out such a detailed critical point analysis of the spectrum we require a fairly good knowledge of the approximate expected location of slope discontinuities. This can be provided by good neutron scattering data and accurate phonon dispersion calculation to locate all the major critical point phonons. Then, using the selection rules and the critical point analysis the exact energy location of the slope discontinuities can be determined, and from the energy-dependence of the absorption coefficient, or scattering cross-section in the vicinity of the slope discontinuity, the actual index of the critical point responsible for the feature can be verified. In Table 36 we enumerate the combined results, indicating the two phonon process involved, the activity (infra-red or Raman) and the nature of the singularity expected. In this table we give a double labelling: the symmetry species label, and also the labelling according to the convention of JOHNSON and LOUDON (see Table 31) in which designations such as TO 1, TA 1, etc., are presented. Only those points are listed which can be expected to produce slope discontinuities according to the criterion discussed earlier, i.e. $P_j(\mu)$ points with $\kappa<2$ for $j=0, 3$ and $\mu<1$ for $j=1, 2$.

It is then seen from Table 36 that in the second order, two phonon infra-red absorption spectrum, we should see 19 slope discontinuities; in the two phonon Raman scattering intensity, there should be 37 slope discontinuities.

It should be noted that not all of the two phonon features predicted in Table 36 will be seen, but the table will enable interpretations of these features which are observed to be given. Lack of observation is due to experimental uncertainties as well as to the influence of omitted theoretical factors such as anharmonicity which casue broadening of the features of the spectrum and may obscure the slope discontinuities. In Table 37 we recapitulate selection rules for two phonon processes in diamond structure, and in Table 38 we recapitulate selection rules for three phonon processes in the diamond structure.

A critical point analysis of the three-phonon branches has not been carried out owing to practical difficulties because of the large number of possibilities, and also because anharmonic effects become larger, diminishing the validity of critical point assumptions. Again, selected ones of all the permitted three phonon processes are observed, insofar as they produce slope discontinuities in the three phonon spectrum. These will be identified below.

An alternate possibility for interpreting the spectra is to utilize the critical point hypothesis in a coarser form. That is to assume the intensity of *allowed* processes is governed only by a joint density of states of the two branches, without regard to detailed slope discontinuities. This case would call for a plot of the joint density of states and a direct comparison of the relative intensity of observed features with the density of modes in a given frequency interval; account is taken

Table 37. Reduction coefficients corresponding to Raman- and infrared-active representations of overtones and combinations in diamond structure O_h^7

Combinations or overtones	Raman-active			Infrared-active
	$\Gamma^{(1+)}$	$\Gamma^{(12+)}$	$\Gamma^{(25+)}$	$\Gamma^{(15-)}$
$[\Gamma^{(25+)}]_{(2)}$	1	1	1	0
$[{}^\star X^{(1)}]_{(2)}$	1	1	1	0
$[{}^\star X^{(2)}]_{(2)}$	1	1	1	0
$[{}^\star X^{(3)}]_{(2)}$	1	1	1	0
$[{}^\star X^{(4)}]_{(2)}$	1	1	1	0
${}^\star X^{(1)} \otimes {}^\star X^{(2)}$	0	1	0	1
${}^\star X^{(1)} \otimes {}^\star X^{(3)}$	0	0	1	1
${}^\star X^{(1)} \otimes {}^\star X^{(4)}$	0	0	1	1
${}^\star X^{(2)} \otimes {}^\star X^{(3)}$	0	0	1	1
${}^\star X^{(2)} \otimes {}^\star X^{(4)}$	0	0	1	1
${}^\star X^{(3)} \otimes {}^\star X^{(4)}$	0	1	0	1
$[{}^\star L^{(1+)}]_{(2)}$	1	0	1	0
$[{}^\star L^{(2-)}]_{(2)}$	1	0	1	0
$[{}^\star L^{(3+)}]_{(2)}$	1	1	2	0
$[{}^\star L^{(3-)}]_{(2)}$	1	1	2	0
${}^\star L^{(1+)} \otimes {}^\star L^{(2-)}$	0	0	0	1
${}^\star L^{(1+)} \otimes {}^\star L^{(3+)}$	0	1	1	0
${}^\star L^{(1+)} \otimes {}^\star L^{(3-)}$	0	0	0	1
${}^\star L^{(2-)} \otimes {}^\star L^{(3+)}$	0	0	0	1
${}^\star L^{(2-)} \otimes {}^\star L^{(3-)}$	0	1	1	0
${}^\star L^{(3+)} \otimes {}^\star L^{(3-)}$	0	0	0	2
$[{}^\star W^{(1)}]_{(2)}$	1	1	1	0
$[{}^\star W^{(2)}]_{(2)}$	1	1	1	0
${}^\star W^{(1)} \otimes {}^\star W^{(2)}$	0	1	2	2
${}^\star W^{(2)} \otimes {}^\star W^{(2)}$	1	1	1	1
$[{}^\star \Sigma^{(1)}]_{(2)}$	1	1	1	0
$[{}^\star \Sigma^{(2)}]_{(2)}$	1	1	1	0
$[{}^\star \Sigma^{(3)}]_{(2)}$	1	1	1	0
${}^\star \Sigma^{(1)} \otimes {}^\star \Sigma^{(2)}$	0	0	1	0
${}^\star \Sigma^{(1)} \otimes {}^\star \Sigma^{(3)}$	0	0	1	1
${}^\star \Sigma^{(1)} \otimes {}^\star \Sigma^{(4)}$	0	1	0	1
${}^\star \Sigma^{(2)} \otimes {}^\star \Sigma^{(3)}$	0	1	0	1
${}^\star \Sigma^{(1)} \otimes {}^\star \Sigma^{(1)}$	1	1	1	1
${}^\star \Sigma^{(2)} \otimes {}^\star \Sigma^{(2)}$	1	1	1	1
${}^\star \Sigma^{(3)} \otimes {}^\star \Sigma^{(3)}$	1	1	1	1

This is a somewhat expanded version of Table 36 above; these entries can be read off from Tables B2 to B4; they are presented here for ease of comparison with experiment (see text).

Table 38. Three-phonon processes in diamond structure[f]

Species	Activity[e]	Type[a]
	Overtones	
$[\Gamma^{(25+)}]_{(3)}$	R	$3\,O(\Gamma)$
$[{}^{\star}X^{(4)}]_{(3)}$	R	$3\,TO(X)$
$[{}^{\star}X^{(1)}]_{(3)}$	$D;\,R$	$3\,L(X)$
$[{}^{\star}X^{(3)}]_{(3)}$	D	$3\,TA(X)$
$[{}^{\star}L^{(m)}]_{(3)}$	No activity	
$[{}^{\star}W^{(m)}]_{(3)}$	No activity	
	Simple combinations	
${}^{\star}X^{(4)}\otimes{}^{\star}X^{(1)}\otimes\Gamma^{(25+)}$	$D(2);\,R$	$TO(X)+L(X)+O(\Gamma)$
${}^{\star}X^{(4)}\otimes{}^{\star}X^{(3)}\otimes\Gamma^{(25+)}$	$D(3);\,R$	$TO(X)+TA(X)+O(\Gamma)$
${}^{\star}X^{(1)}\otimes{}^{\star}X^{(3)}\otimes\Gamma^{(25+)}$	$D(2);\,R$	$L(X)+TA(X)+O(\Gamma)$
${}^{\star}X^{(4)}\otimes{}^{\star}X^{(1)}\otimes{}^{\star}X^{(3)}$	$D(3);\,R$	$TO(X)+L(X)+TA(X)$
${}^{\star}L^{(3-)}\otimes{}^{\star}L^{(1+)}\otimes\Gamma^{(25+)}$	$D(3)$	$TO(L)+LO(L)+O(\Gamma)$
${}^{\star}L^{(3-)}\otimes{}^{\star}L^{(2-)}\otimes\Gamma^{(25+)}$	R	$TO(L)+LA(L)+O(\Gamma)$
${}^{\star}L^{(3-)}\otimes{}^{\star}L^{(3+)}\otimes\Gamma^{(25+)}$	$D(6)$	$TO(L)+TA(L)+O(\Gamma)$
${}^{\star}L^{(1+)}\otimes{}^{\star}L^{(2-)}\otimes\Gamma^{(25+)}$	$D(2)$	$LO(L)+LA(L)+O(\Gamma)$
${}^{\star}L^{(1+)}\otimes{}^{\star}L^{(3+)}\otimes\Gamma^{(25+)}$	R	$LO(L)+TA(L)+O(\Gamma)$
${}^{\star}L^{(2-)}\otimes{}^{\star}L^{(3+)}\otimes\Gamma^{(25+)}$	$D(3)$	$LA(L)+TA(L)+O(\Gamma)$
${}^{\star}L^{(3-)}\otimes{}^{\star}L^{(1+)}\otimes{}^{\star}X^{(4)}$	$D(3);\,R$	$TO(L)+LO(L)+TO(X)$
$\otimes{}^{\star}X^{(1)}$	$D(3);\,R$	$+L(X)$
$\otimes{}^{\star}X^{(3)}$	$D(3);\,R$	$+TA(X)$
${}^{\star}L^{(3-)}\otimes{}^{\star}L^{(2-)}\otimes{}^{\star}X^{(4)}$	$D(3);\,R$	$TO(L)+LA(L)+TO(X)$
$\otimes{}^{\star}X^{(1)}$	$D(3);\,R$	$+L(X)$
$\otimes{}^{\star}X^{(3)}$	$D(3);\,R$	$+TA(X)$
${}^{\star}L^{(3-)}\otimes{}^{\star}L^{(3+)}\otimes{}^{\star}X^{(4)}$	$D(6);\,R$	$TO(L)+TA(L)+TO(X)$
$\otimes{}^{\star}X^{(1)}$	$D(6);\,R$	$+L(X)$
$\otimes{}^{\star}X^{(3)}$	$D(6);\,R$	$+TA(X)$
${}^{\star}L^{(1+)}\otimes{}^{\star}L^{(2-)}\otimes{}^{\star}X^{(4)}$	$D;\,R$	$LO(L)+LA(L)+TO(X)$
$\otimes{}^{\star}X^{(1)}$	$D(2);\,R$	$+L(X)$
$\otimes{}^{\star}X^{(3)}$	$D(2);\,R$	$+TA(X)$
${}^{\star}L^{(1+)}\otimes{}^{\star}L^{(3+)}\otimes{}^{\star}X^{(4)}$	$D(3);\,R$	$LO(L)+TA(L)+TO(X)$
$\otimes{}^{\star}X^{(1)}$	$D(3);\,R$	$+L(X)$
$\otimes{}^{\star}X^{(3)}$	$D(3);\,R$	$+TA(X)$
${}^{\star}L^{(2-)}\otimes{}^{\star}L^{(3+)}\otimes{}^{\star}X^{(4)}$	$D(3);\,R$	$LA(L)+TA(L)+TO(X)$
$\otimes{}^{\star}X^{(1)}$	$D(3);\,R$	$+L(X)$
$\otimes{}^{\star}X^{(3)}$	$D(3);\,R$	$+TA(X)$
${}^{\star}W^{(1)}\otimes{}^{\star}W^{(1)}\otimes\Gamma^{(25+)}$	$D(5);\,R$	$O(\Gamma)+\begin{cases}O_1(W)+O_2(W)\,[b]\\ O_1(W)+A_1(W)\,[b]\\ A_1(W)+A_2(W)\,[b]\end{cases}$
${}^{\star}W^{(1)}\otimes{}^{\star}W^{(2)}\otimes\Gamma^{(25+)}$	$D(5);\,R$	
${}^{\star}W^{(1)}\otimes{}^{\star}W^{(1)}\otimes{}^{\star}X^{(4)}$	$D(3);\,R$	$TO(X)+$ } c
$\otimes{}^{\star}X^{(1)}$	$D(3);\,R$	$L(X)+$ } c
$\otimes{}^{\star}X^{(3)}$	$D(3);\,R$	$TA(X)+$ } c
${}^{\star}W^{(1)}\otimes{}^{\star}W^{(2)}\otimes{}^{\star}X^{(4)}$	$D(3);\,R$	$TO(X)+$ } b
$\otimes{}^{\star}X^{(1)}$	$D(3);\,R$	$L(X)+$ } b
$\otimes{}^{\star}X^{(3)}$	$D(3);\,R$	$TA(X)+$ } b
	General combinations	
$[{}^{\star}X^{(4)}]_{(2)}\otimes\Gamma^{(25+)}$	$D;\,R$	$2\,TO(X)+O(\Gamma)$
$[{}^{\star}X^{(1)}]_{(2)}\otimes\Gamma^{(25+)}$	$D(2);\,R$	$2\,L(X)+O(\Gamma)$
$[{}^{\star}X^{(3)}]_{(2)}\otimes\Gamma^{(25+)}$	$D(2);\,R$	$2\,TA(X)+O(\Gamma)$

Table 38 (continued)

Species	Activity	Type [a]
	General combinations	
$[{}^{\star}X^{(4)}]_{(2)} \otimes {}^{\star}X^{(1)}$	$D(2)$; R	$2\,TO(X)+L(X)$
$\otimes\, {}^{\star}X^{(3)}$	$D(2)$; R	$+TA(X)$
$[{}^{\star}X^{(1)}]_{(2)} \otimes {}^{\star}X^{(4)}$	D; R	$2\,L(X)+TO(X)$
$\otimes\, {}^{\star}X^{(3)}$	$D(2)$; R	$+TA(X)$
$[{}^{\star}X^{(3)}]_{(2)} \otimes {}^{\star}X^{(4)}$	D; R	$2\,TA(X)+TO(X)$
$\otimes\, {}^{\star}X^{(1)}$	$D(2)$; R	$+L(X)$
$[{}^{\star}L^{(3-)}]_{(2)} \otimes \Gamma^{(25+)}$	R	$2\,TO(L)+O(\Gamma)$
$[{}^{\star}L^{(1+)}]_{(2)} \otimes \Gamma^{(25+)}$	R	$2\,LO(L)+O(\Gamma)$
$[{}^{\star}L^{(2-)}]_{(2)} \otimes \Gamma^{(25+)}$	R	$2\,LA(L)+O(\Gamma)$
$[{}^{\star}L^{(3+)}]_{(2)} \otimes \Gamma^{(25+)}$	R	$2\,TA(L)+O(\Gamma)$
$[{}^{\star}L^{(3-)}]_{(2)} \otimes {}^{\star}X^{(4)}$	$D(3)$; R	$2\,TO(L)+TO(X)$
$\otimes\, {}^{\star}X^{(1)}$	$D(3)$; R	$+L(X)$
$\otimes\, {}^{\star}X^{(3)}$	$D(3)$; R	$+TA(X)$
$[{}^{\star}L^{(1+)}]_{(2)} \otimes {}^{\star}X^{(4)}$	D; R	$2\,LO(L)+TO(X)$
$\otimes\, {}^{\star}X^{(1)}$	D; R	$+L(X)$
$\otimes\, {}^{\star}X^{(3)}$	D; R	$+TA(X)$
$[{}^{\star}L^{(2-)}]_{(2)} \otimes {}^{\star}X^{(4)}$	D; R	$2\,LA(L)+TO(X)$
$\otimes\, {}^{\star}X^{(1)}$	D; R	$+L(X)$
$\otimes\, {}^{\star}X^{(3)}$	D; R	$+TA(X)$
$[{}^{\star}L^{(3+)}]_{(2)} \otimes {}^{\star}X^{(4)}$	$D(3)$; R	$2\,TA(L)+TO(X)$
$\otimes\, {}^{\star}X^{(1)}$	$D(3)$; R	$+L(X)$
$\otimes\, {}^{\star}X^{(3)}$	$D(3)$; R	$+TA(X)$
$[{}^{\star}W^{(1)}]_{(2)} \otimes \Gamma^{(25+)}$	$D(2)$; R	$O(\Gamma)+$ } b
$[{}^{\star}W^{(2)}]_{(2)} \otimes \Gamma^{(25+)}$	$D(2)$; R	$O(\Gamma)+$ } b
$[{}^{\star}W^{(1)}]_{(2)} \otimes {}^{\star}X^{(4)}$	$D(2)$; R	$TO(X)+$ } d
$\otimes\, {}^{\star}X^{(1)}$	$D(2)$; R	$L(X)+$ } d
$\otimes\, {}^{\star}X^{(3)}$	$D(3)$; R	$LA(X)+$ } d

[a] Where possible, we follow Lax-Hopfield assignments; see Phys. Rev. **124**, 115 (1961).
[b] See Table 36 regarding assigning species to branches at W.
[c] The two phonons of same symmetry must come from different branches. The same rule obtains if the different branches of the same symmetry are ${}^{\star}W^{(1)}$.
[d] Refer to footnote b. Also note that the same selection rule holds for ${}^{\star}W^{(2)}$. Thus, each of the three ${}^{\star}X^{(m)}$ phonons may combine with the first overtone of each of the three ${}^{\star}W^{(m)}$ phonons.
[e] R means Raman active (see footnote a Table 36; $D(n)$ means that Γ^{15-} occurs n times in the reduction so there is "n-fold" infra-red dipole activity.
[f] Three phonon selection rules. J. L. Birman: Phys. Rev. **131**, 1492 (1963), Table III.

of the selection rules by omitting those branch contributions which are forbidden. For example, since overtones cannot be active in the infra-red the contribution from overtone branches cannot contribute to any infra-red process,* so these are omitted from the calculated density of states for comparison with experiment. This possibility of analysis will also be used in the analysis of the infra-red spectra of both diamond, and rocksalt structures.

Now we turn to some detailed analysis of diamond, silicon and germanium spectra.

* Recall Sect. 122.

Diamond

For diamond the measured phonon dispersion has recently become available,[1] using results of inelastic neutron scattering; the results are shown in Fig. 7b. So far, these results have not been used in a complete critical point analysis of the two-phonon spectrum in diamond. It will be observed below that the order in which various branches occur, along with their symmetry designation, differs markedly in diamond from the two other crystals with this structure which we discuss: germanium and silicon. Note particularly the order of states at $^{\star}X$ and the crossings of the Σ line.

Using the neutron data, a shell model calculation was carried out, as shown in Fig. 7 from which single phonon and joint density of states for two-phonon processes were calculated.[2] The details of these calculations are not available, so we shall only quote their published results.

In Fig. 10a we give the measured (experimental) infra-red absorption of diamond.[3] Features labelled $a-m$ in the figure are various slope discontinuities. It

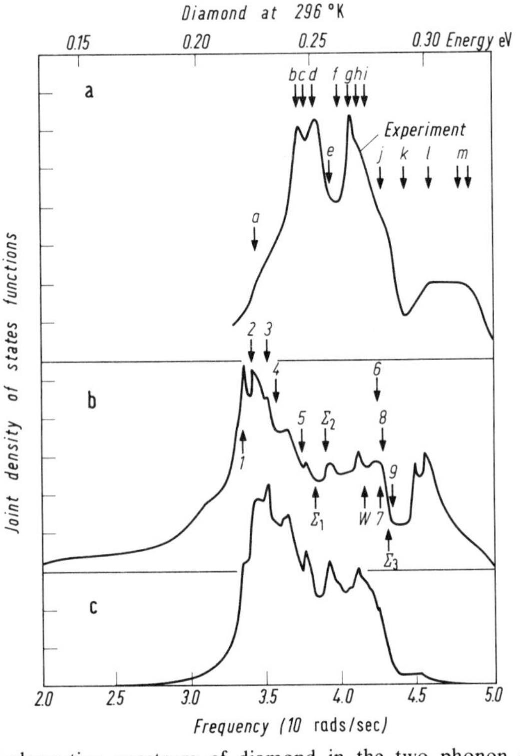

Fig. 10a–c. Infra-red absorption spectrum of diamond in the two phonon energy region. Joint density-of-states functions for diamond at 296 °K. Curve a is the experimental infra-red absorption, curve b is a joint density of states, and curve c the infra-red spectrum calculated with second-nearest-neighbour interactions. [After J.R. HARDY and D. SMITH: Phil. Mag. **6**, 1165 (1961)]

[1] J.L. WARREN, J.L. YARNELL, G. DOLLING, and R.A. COWLEY: Phys. Rev. **158**, 805 (1967).

[2] J.L. WARREN: Reference 1 this section.

[3] J.R. HARDY and S.D. SMITH: Phil. Mag. **6**, 1163 (1961).

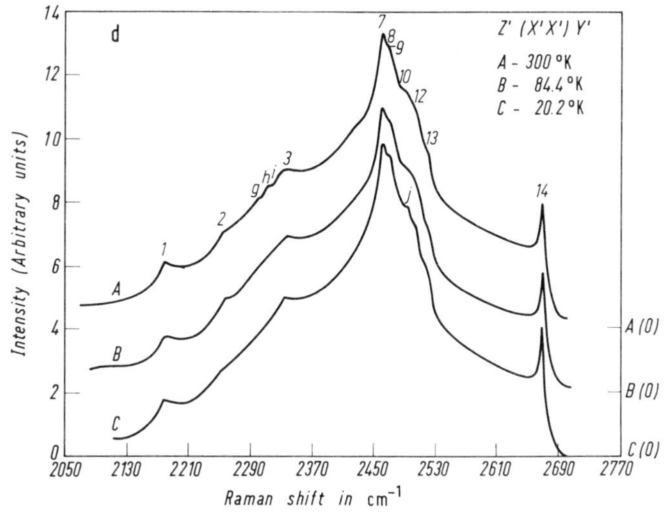

Fig. 10d. Raman scattering in diamond. Temperature dependence of the second-order Raman spectrum of diamond for $Z'(X'X')Y'$ scattering in the 2050–2770-cm^{-1} region. The spectra are displaced vertically with respect to one another for clarity. The relative intensity scales of curves *A*, *B* and *C* are identical and *A*(0), *B*(0), *C*(0) denote the base lines of *A*, *B*, and *C*, respectively. The spectra were excited with 4880 Å radiation of the Ar^+ laser. [From S.A. SOLIN and A.K. RAMDAS: Phys. Rev. *B*, **1**, 1687 (1970)]

will be noticed that features *b*, *d*, *g* seem to be maxima but may in fact be critical points of various sorts (P_0 or P_1). In Fig. 10b and c we give a calculated density of states, and comparison with experiment. It will be observed that in the region of 2-phonon absorption quite acceptable agreement is reached between calculated and observed curve shapes. In Table 39 the reported critical point analysis, is reproduced. On comparing with Table 36 it is observed that all the observed features of the two-phonon infra-red spectrum correspond to allowed two-phonon processes; however, the reverse is not true: some permitted processes cannot be located in the spectrum.

Recently, a comprehensive study of Raman scattering in diamond was carried out by SOLIN and RAMDAS[4] including one and two phonon processes and their temperature and polarization properties. An example of the results obtained for two-phonon Raman scattering is shown in Fig. 10d taken from these authors work. It will be noted that the spectrum shows considerable structural detail (many of the features are identified by circled numbers and letters). In this particular polarization all the Raman active representations contribute so that no separation is possible into particular active representations. The features listed in that Fig. 10d have been interpreted in terms of various two-phonon Stokes processes (i.e. the photon creates two phonons). Temperature dependence of the scattering rules out any anti-Stokes (or mixed Stokes – anti-Stokes) processes as contributing.

In Table 40a we list the phonon energies in diamond identified by these authors, and in Table 40b a complete listing of all assignments for second order Raman

[4] S.A. SOLIN and A.K. RAMDAS: Phys. Rev. **B1**, 1687 (1970), earlier work was reported by R.S. KRISHNAN: Proc. Indian Acad. Sci. Sect. A **26**, 399 (1947).

Table 39. Critical-point analysis of infrared absorption in diamond. The kink positions (a to m) are those listed by HARDY and SMITH[a]. Frequencies are given in 10^{14} rad/sec. The code letters refer to Fig. 2, curves *A* and *B*

Kink positions, curve-*A* code	Frequency	Assignments, curve-*B* code	Modes	Model calculation	Neutron data
a	3.42	2	$LOTA(\boldsymbol{L})$	3.40	3.38
b	3.71	4	$ML(\boldsymbol{W})$	3.55	3.60
c	3.75	5	$LTA(\boldsymbol{X})$	3.74	3.75
d	3.81	Σ_1	$\Sigma_3 O \Sigma_3 A$	3.81	3.83
e	3.92	Σ_2	$UL(\boldsymbol{W})$	3.88	(3.92)[b]
f	3.98	...	(unknown)		
g	4.06	W	$UM(\boldsymbol{W})$	4.14	(4.07)
h	4.10	...	(unknown)	4.11[c]	
i	4.16	6	$\Sigma_1 O \Sigma_3 A$	4.24	4.19
j	4.27	7	$LTO(\boldsymbol{X})$	4.26	4.26
		or 8	$TOLA(\boldsymbol{L})$	4.27	4.23
k	4.42	9	$LOLA(\boldsymbol{L})$	4.34	4.29
		or Σ_3	$\Sigma_1 O \Sigma_3 O$	4.31	4.29
l	4.59	...	(unknown)		
m	4.81[d]	...		4.83	4.84
Not observed		1	$TOTA(\boldsymbol{L})$	3.34	3.32
Not observed		3	$TOTA(\boldsymbol{X})$	3.51	3.54

[a] J.R. HARDY and S.D. SMITH: Phil. Mag. **6**, 1163 (1961).
[b] A bracketed figure in the final column indicates an estimated "experimental" value.
[c] Combination presumably arising from non-symmetry-point modes. (The assignment and footnote for kinks g and h could clearly be interchanged with out significantly worsening the agreement.)
[d] Probably a 3-phonon combination $TA(\boldsymbol{X})+TA(\boldsymbol{L})+TO(\boldsymbol{L})$; some of the other features may also arise from 3-phonon contributions.
J.L. WARREN *et al.*: Phys. Rev. **158**, 808 (1967), Table III.

and infra-red optical processes in diamond is given. In Table 40b the symbol + (−) in Theory (Th) column means the process (in that polarization) is theoretically allowed (forbidden). The symbol + (−) under the Experimental (Exp) column means the process is (is not) observed. The values of energies listed are for room temperature. The data of Fig. 10d was not interpreted according to the critical point theory in which actual slope discontinuities were identified (as predicted in Table 36 for Si and Ge).

Leaving aside the verification of critical indices in diamond, inspection of Table 40a and b indicates excellent agreement of theory and experiment. In particular, the major experimentally observed Raman and infra-red features can be consistently fit with a set of phonon energies which are also in agreement with the neutron results. No violations of selection rules are observed (i.e. no forbidden transitions are seen) although not all the predicted processes are observed. The latter may be due to breakdown of the "critical point" approximation owing to anomalously weak intensity, anharmonic effects, etc.

In summary, the typical error of phonon energies deduced from optical data is reported as 0.5%; that from the neutron spectroscopy has errors of order of ±3%. This is a remarkable illustration of the power of optical techniques.

Table 40a. Phonon energies at critical points in diamond[c]

Symmetry point	Phonon	Energy (cm^{-1}) Neutron	Optical[a]	"Brout sum" (from optical data) ($10^6\ cm^{-2}$)
Γ	0		1332 ± 0.5	5.322
X	TO	1072 ± 26	1069	6.396
	L	1184 ± 21	1185	
	TA	807 ± 32	807	
L	TO	1210 ± 37	1206	6.122
	LO	1242 ± 37	1252	
	TA	552 ± 16	563	
	LA	1035 ± 32	1006	
W	TO	993 ± 53	999	6.426
	L	1168 ± 53	1179	
	TA	918 ± 11	908	
$\Sigma(q/q_{max} \sim 0.7)$[b]	$\Sigma^{(1)}(0)$	1231 ± 32	1230	6.331 ± 0.024
	$\Sigma^{(2)}(0)$	1120 ± 21	1109	
	$\Sigma^{(3)}(0)$	1046 ± 21	1045	
	$\Sigma^{(1)}(A)$	982 ± 11	988	
	$\Sigma^{(3)}(A)$	993 ± 16	980	
	$\Sigma^{(4)}(A)$	748 ± 16		

[a] Error in these numbers is $\pm 5\ cm^{-1}$ except as noted.
[b] Only $\Sigma_1(0)$, $\Sigma_3(0)$, and $\Sigma_3(A)$ show critical points in the one-phonon dispersion curves. Note: The value of $\Sigma^{(4)}(A)$ was not determined from optical data.
[c] S.A. Solin and A.K. Ramdas: Phys. Rev. B, **1**, 1687 (1970), Table VII.

In Sect. 146 we shall discuss some results of the polarization studies of Raman scattering by diamond.

Silicon

The lattice absorption bands of silicon have been measured[5] and in Fig. 11 we show the results. A considerable amount of structure exists, as is shown in the figure by slope changes, and a few strong and sharp peaks.

This data has been interpreted by Johnson and Loudon[6] in terms of the two phonon critical point assignments of Table 36.

In Table 41a we give the list of two phonon features as interpreted by them. A striking feature of this comparison is the verification of the *type* of critical point in the 11 cases cited. That is, predicted critical points from the lattice dynamics agree with the observations based on optical spectra. As an example of the type of work possible, we also reproduce in Fig. 12 a more detailed re-measurement of the transmission, in the region 580–640 cm^{-1} reported by these authors. Each of the indicated discontinuities shown by an arrow corresponds to a different type of critical point as in Table 40 and it will be observed that predicted and observed types of slopes agree well. There is some additional structures in the figure whose interpretation is not reported.

[5] F.A. Johnson: Proc. Phys. Soc. (London) **73**, 265 (1959).
[6] F.A. Johnson and R. Loudon: *loc. cit.*

Table 40b. Assignments for the second-order Raman and infrared spectrum of diamond[f]

Serial[a] no.	Raman (cm^{-1})	Infrared[b] (cm^{-1})	Assignment[c]	Calculated (cm^{-1}) Neutron[e]	Calculated (cm^{-1}) Optical[e]	Raman activity $Z'(X'X')Y'$ Th	$Z'(X'X')Y'$ Exp	$Y'(Z'Z')X'$ Th	$Y'(Z'Z')X'$ Exp	$Z'(X'Z')Y'$ Th	$Z'(X'Z')Y'$ Exp	$Z'(Y'X')Y'$ Th	$Z'(Y'X')Y'$ Exp	Infrared activity Th	Infrared activity Exp
		1815	$LO(L^{(2-)})+TA(L^{(3+)})$	1799 ± 56	1815	—	—	—	—	—	?	—	?	+	+
I	1817		$2TA(W^{(2)})$	1836 ± 22	1816	+	+	+	+	+	?	+	?	—	—
II	1864	1871	$TA(X^{(3)})+TO(X^{(4)})$	1879 ± 58	1876	+	+	+	+	+	?	+	?	+	+
		1968	$\Sigma^{(1)}(A)+\Sigma^{(3)}(A)$	1975 ± 40	1976	+	—	—	—	+	?	—	?	+	+
		1992	$L(X^{(1)})+TA(X^{(3)})$	1991 ± 53	1992	+	—	—	—	+	?	—	?	+	+
III	1998		$2TO(W^{(1)})$	1986 ± 106	1998	+	+	+	+	+	?	+	?	—	—
IV	2011		$2LA(L^{(1+)})$	2070 ± 64	2012	+	+	+	+	+	?	—	?	—	—
V	2025	2025	$\Sigma^{(3)}(0)+\Sigma^{(3)}(A)$	2039 ± 37	2025	+	+	+	+	+	?	+	?	+	+
		2041	$\Sigma^{(3)}(0)+\Sigma^{(1)}(A)$	2028 ± 32	2041	+	—	—	—	+	?	—	?	+	+
		2081	$\Sigma^{(2)}(0)+\Sigma^{(3)}(A)$	2113 ± 56	2089	+	—	+	—	—	—	+	—	+	+
		2113	$\Sigma^{(3)}(0)+\Sigma^{(1)}(A)$[d]	2086 ± 88		+	—	—	—	+	—	—	—	+	+
		2154	$\Sigma^{(2)}(0)+\Sigma^{(3)}(0)$	2166 ± 48	2154	+	—	+	—	—	—	+	—	+	+
1	2177	2178	$L(W^{(2)})+TO(W^{(1)})$	2161 ± 106	2178	+	+	+	+	+	+	+	+	+	+
		2210	$\Sigma^{(1)}(0)+\Sigma^{(3)}(A)$	2224 ± 48	2210	+	—	—	—	+	—	—	—	+	+
2	2254		$L(X^{(1)})+TO(X^{(4)})$	2256 ± 47	2254	+	+	—	—	+	+	—	—	+	—
		2267	$LA(L^{(1+)})+LO(L^{(2-)})$	2275 ± 72	2258	—	—	—	—	—	—	—	—	+	+
3	2333	2331					+		+		+		+		+
		2355					—		—		—		—		+
15	2370		$2L(X^{(1)})$	2371 ± 40	2370	+	—	+	—	+	+	+	—	—	—
5	2422		$2TO(L^{(3-)})$	2420 ± 74	2412	+	—	+	+	+	—	+	—	—	—
		2436					—		—		—		—		+
7	2458		$TO(L^{(3-)})+LO(L^{(2-)})$	2452 ± 72	2458	+	+	+	+	+	+	+	+	—	—
8	2461		$2\Sigma^{(1)}(0)$	2461 ± 64	2460	+	+	+	+	+	+	+	+	—	—
9	2467						+		+		?		?		—
10	2485						+		+		+		+		—
12	2054		$2LO(L^{(2-)})$	2484 ± 72	2504	+	+	+	+	+	+	—	—	—	—
13	2519						+		+		+		+		—
14	2667		$2O(\Gamma^{(25+)})$	2665 ± 1	2665	+	+	+	+	+	+	+	+	—	—

[a] The serial numbers refer to Figs. 30a–30c and Fig. 10d.

[b] With the exception of the kinks at 1871, 2331, and 2355 cm^{-1}, all the values in this column are from J. R. HARDY and S. SMITH: Phil. Mag. **6**, 1163 (1961); the former are from R. WEHNER *et al.*: Solid State Comm. **5**, 307 (1967).

[c] The phonons are labeled here by the type and the irreducible representation as shown in Fig. 7; the notation for the latter follows as in text.

[d] This corresponds to $(q/q_{max}) \sim 0.5$, where q is the wave vector of the phonon.

[e] Calculated from the phonon energies chosen from neutron data and optical data as given in Table 55a.

[f] Table from S. A. SOLIN and A. K. RAMDAS: Phys. Rev. B. **1**, 1687 (1970), Table V.

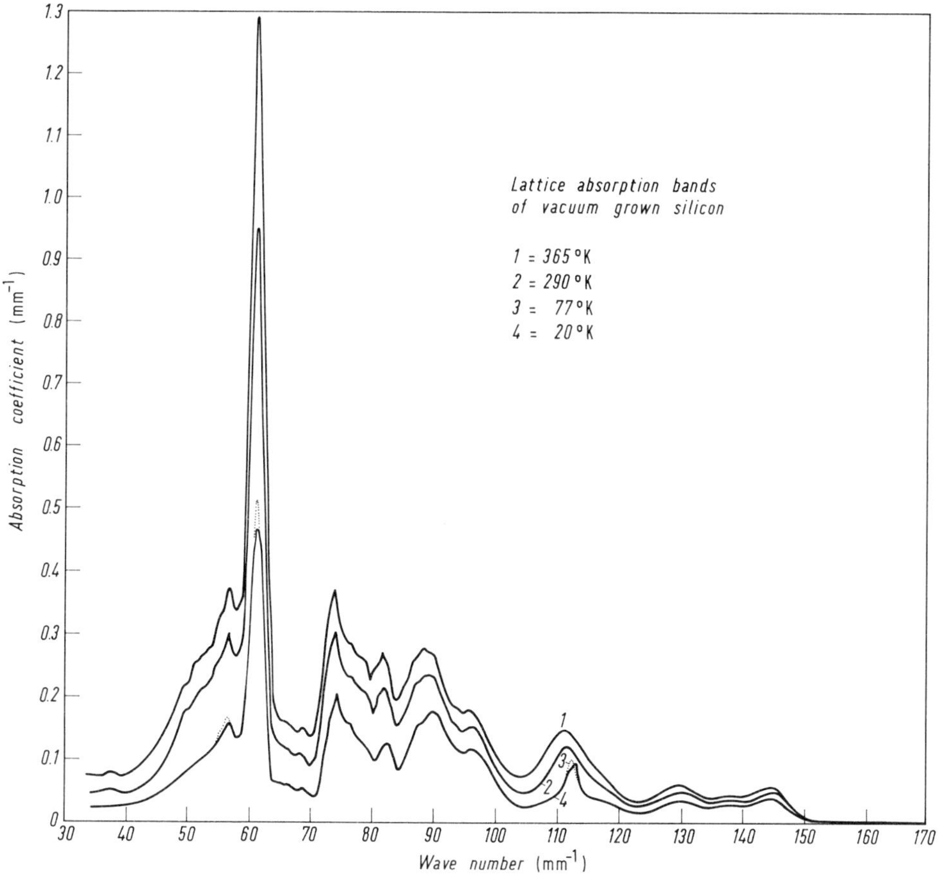

Fig. 11. Infra-red absorption spectrum of silicon in the two phonon energy region. [After F. A. JOHNSON: Proc. Phys. Soc. **73**, 265 (1959)]

Table 41a. Comparison of theory with experiment for silicon-two-phonon infra-red absorption

Point	Combination	Frequency (cm^{-1})		Type	
		Neutron	Infra-red	Calc.	Expt.
X	$TO+L$	874	875	$P_3(1)$	$P_3(1)$
	$TO+TA$	613	612	P_1+F_2	P_1+F_2
	$L+TA$	561	—	$P_0(1)$	—
L	$TO+LO$	909	917	P_2	P_2
	$TO+TA$	603	605	P_1	P_1
	$LO+LA$	798	800	P_3	P_3
	$LA+TA$	492	488	P_1	P_1
Δ	$LO+TA$	630	636	P_3	P_3
Σ	$TO1+TA1$	705	700	P_3	P_3
	$TO1+LO$	850	860	P_1	P_1
	$LO+TA1$	630	624	P_2	P_2

Table 41 b. Three-phonon combinations in silicon

Combination	Frequency (cm^{-1})	
	Neutron	Infra-red
$3\,TA(\boldsymbol{X})$	450	447
$2\,L(\boldsymbol{X})+TA(\boldsymbol{X})$	972	972
$2\,TA(\boldsymbol{X})+TO(\boldsymbol{X})$	763	763
$2\,TA(\boldsymbol{X})+O(\boldsymbol{\Gamma})$	818	820
$LA(\boldsymbol{L})+TA(\boldsymbol{L})+O(\boldsymbol{\Gamma})$	1010	1011
$TO(\boldsymbol{L})+TA(\boldsymbol{L})+TA(\boldsymbol{X})$	753	752
$2\,LO(\boldsymbol{L})+TA(\boldsymbol{X})$	990	1002?
$LO(\boldsymbol{L})+TA(\boldsymbol{L})+TO(\boldsymbol{X})$	997	1002?
$LO(\boldsymbol{L})+LA(\boldsymbol{L})+TA(\boldsymbol{X})$	948	948
$LO(\boldsymbol{L})+TA(\boldsymbol{L})+TA(\boldsymbol{X})$	684	690?
$2\,TA(\boldsymbol{L})+TO(\boldsymbol{X})$	691	690?

Tables 41 a, 41 b from F. A. JOHNSON and R. LOUDON: Proc. Roy. Soc. (London), Ser. A, **281**, 285 (1964), Table 5.

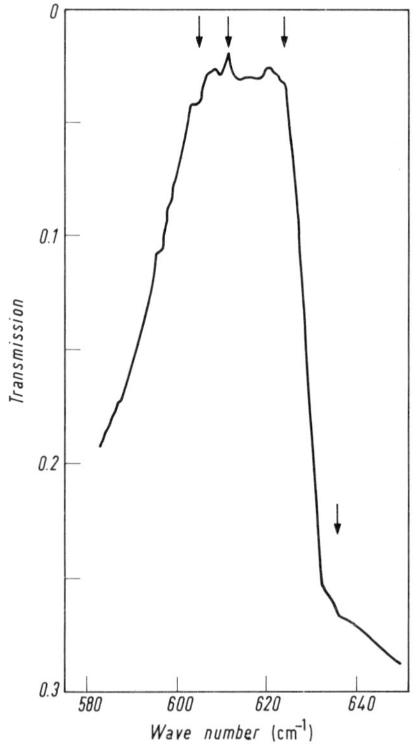

Fig. 12. Details of the infra-red absorption (two phonon region) in silicon. [After F. A. JOHNSON and R. LOUDON: Proc. Roy. Soc. A **281**, 274 (1964), Fig. 1]

Table 41c. Phonon frequencies in silicon (Raman scattering)

Star	Γ	$\star X$		$\star L$		$\star W$	
		TO	TA	TO	TA		
Frequency	519±1	460±2	151±2	490±2	113±2	470± 2	From [a]
(cm^{-1})	517±2	449±3	155±5	493±2	113±2	478±13	From [b]

[a] From P. TEMPLE and C. HATHAWAY: Phys. Rev. **B7**, 3685 (1973).
[b] From M. BALKANSKI and M. NUSIMOVICI: Phys. stat. solidi **5**, 635 (1964).

Table 42. Comparison of phonon frequencies in silicon

Point	Branch	Neutron	Infra-red
Γ	O	518± 8	522
X	TO	463±10	463
	L	411± 7	412
	TA	150± 2	149
L	TO	489±10	491
	LO	420±11	426
	LA	378±10	374
	TA	114± 2	114

From F.A. JOHNSON and R. LOUDON: Proc. Roy. Soc. (London), Ser. A, **281**, 285 (1964), Table 4.

Those three phonon infra-red transitions in silicon which have been interpreted are listed in Table 41b. Again it will be noted that all the assigned transitions correspond to allowed processes, but the converse is not true as can be seen by comparison with Table 38. A summary of results is given in Table 42.

The multiphonon Raman spectrum in silicon was recently extensively measured by TEMPLE and HATHAWAY[7] including polarization characteristics. These authors observed two phonon Raman peaks which they ascribe to a variety of overtones and combinations originating from Γ; $\star X$; $\star L$; $\star W$ and $\star \Sigma$. From their data they were able to determine the one phonon frequencies at a variety of critical points and a precis of their results is given in Table 41c. It can be seen by comparing Tables 41a, b, c that generally excellent agreement was obtained between the Raman scattering data interpreted using critical points analysis plus the group theory and neutron data; the infra-red results of Johnson and Loudon on perfect crystals are given and also the results due to BALKANSKI and NUSIMOVICI taken on imperfect crystals where symmetry breaking activates the normally forbidden one phonon processes at zone edge critical points. An interesting observation reported by TEMPLE and HATHAWAY is that for the two phonon overtones the $(\Gamma 1+)$ component of the Raman scattered light is much stronger than either the $(\Gamma 12+)$ or $(\Gamma 25+)$. It will be recalled (see Table 37 of selection rules) that

[7] P.A. TEMPLE and C.E. HATHAWAY: Phys. Rev. **B7**, 3685 (1973).

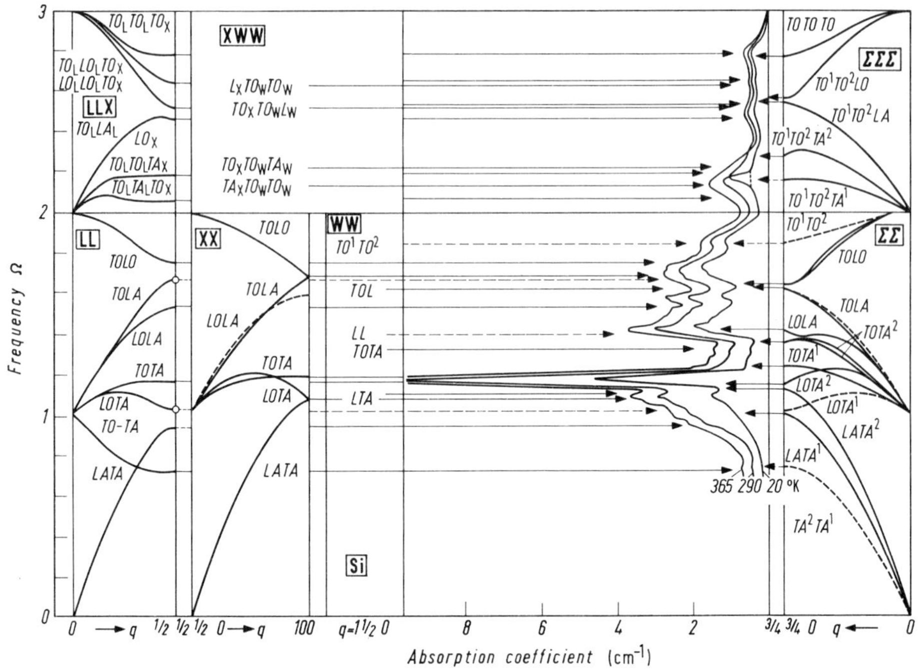

Fig. 13. Comparison between measured infra-red absorption curves in silicon, and calculated two and three phonon combination branches. Full lines allowed combinations, dashed lines forbidden combinations; circle is forbidden at the end point of the direction under consideration. [After H. BILZ, R. GEICK, and K. F. RENK, in: Proceedings of the International Conference on Lattice Dynamics, Copenhagen 1963 (ed. R. F. WALLIS), Fig. 6. New York: Pergamon Press 1965]

for overtones all three representations may occur. The two phonon Raman scattering from silicon was also measured by WEINSTEIN and CARDONA,[8] with similar results.

An earlier analysis[9] of the infra-red absorption in Si due to BILZ and coworkers is shown in Fig. 13. It will be observed that the features of the measured two phonon absorption can be correlated with critical points of the two and even three-phonon dispersion calculation corresponding to symmetry allowed processes. It seems reasonable to conclude from this agreement that the major spectral features are due to symmetry and density of states.

Germanium

Infra-red lattice absorption bands in germanium have been measured by several authors.[10, 11] The results are given in Fig. 14a, b.

[8] B. A. WEINSTEIN and M. CARDONA: Solid State Comm. **10**, 961 (1972).

[9] H. BILZ, R. GEICK, and K. RENK, in: Proceedings of the International Conference on Lattice Dynamics, Copenhagen 1963 (ed. R. F. WALLIS), p. 355. New York: Pergamon Press 1965.

[10] R. J. COLLINS and H. Y. FAN: Phys. Rev. **93**, 674 (1954).

[11] S. J. FRAY, F. A. JOHNSON, and J. E. QUARRINGTON: Proc. Phys. Soc. (London) **85**, 153 (1965).

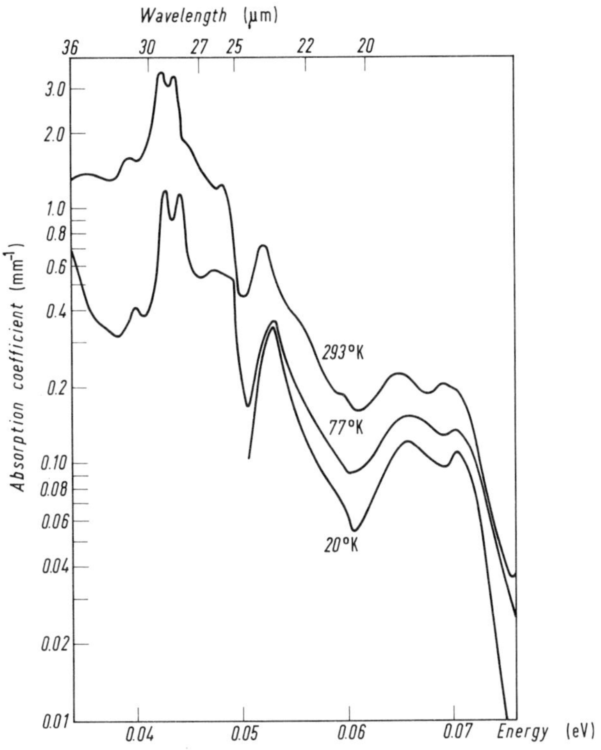

a

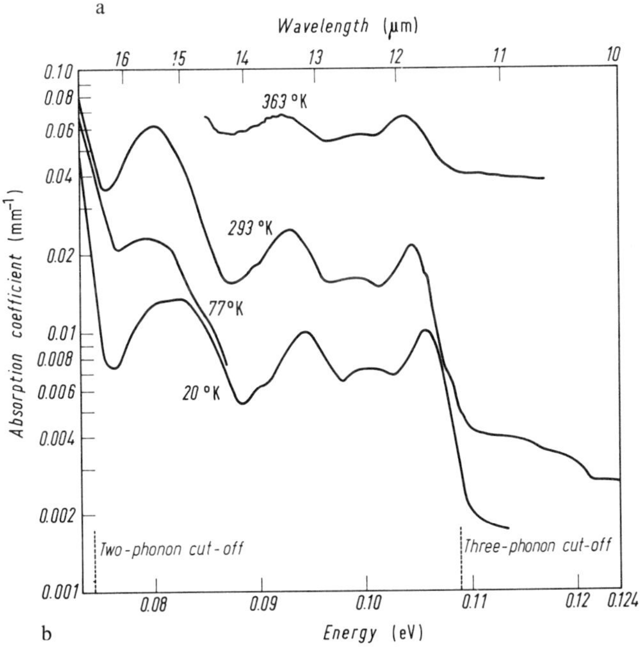

b

Fig. 14a and b. Lattice bands in germanium. [After S.J. FRAY, F.A. JOHNSON and J.E. QUARRINGTON: Proc. Phys. Soc. (London) **85**, 153 (1965)]

Table 43a. Comparison of theory with experiment for germanium[a]

Point	Combination	Frequency (cm^{-1})		Type	
		Neutron	Infra-red	Calc.	Expt.
X	$TO+L$	505	496	$P_3(1)$	$P_3(1)$
	$TO+TA$	357	352	P_1+F_2	P_1+F_2
	$L+TA$	312	310?	$P_0(1)$	$P_0(1)$
L	$TO+LO$	527	527	P_2	P_2
	$TO+TA$	345	343	P_1	P_1
	$LO+LA$	462	460	P_3	P_3
	$LA+TA$	280	—	P_1	—
Δ	$LO+TA$	350	350	P_3	P_3

Table 43b. Three-phonon combinations in germanium[a]

Combination	Frequency (cm^{-1})	
	Neutron	Infra-red
$2L(X)+TA(X)$	542	536
$2TA(X)+TO(X)$	439	435
$2TA(X)+O(\Gamma)$	464	464
$TO(L)+TA(L)+TA(X)$	427	425
$LO(L)+TA(L)+TA(X)$	394	394
$2LA(L)+TA(X)$	512	508
$LA(L)+TA(L)+TO(X)$	555	545
$LA(L)+TA(L)+TA(X)$	362	358

[a] From F.A. JOHNSON and R. LOUDON: Proc. Roy. Soc. (London), Ser. A, **281**, 285 (1964), Tables 3 and 4.

These data have been interpreted by JOHNSON and LOUDON and in Table 43a we reproduce their table giving correlation between predicted and observed energy and type of the slope discontinuity. It will be noted that seven such discontinuities are identified, with exact agreement regarding prediction and observation. The remaining two phonon features are not reported. These authors also report on the identification of several allowed three-phonon features which we reproduce in Table 43b. As a summary we give the phonon frequencies in Ge at the high symmetry critical points which fit the infra-red data compared to the values determined from inelastic neutron scattering: this is reproduced in Table 44.

Recently WEINSTEIN and CARDONA measured second order Raman spectra of germanium.[12] Their reported spectra appear in very good agreement with critical point analysis based upon available neutron scattering data.[13] They also observed (see remark on silicon in the previous paragraph) that the ($\Gamma 1+$) portion of the two phonon overtone spectrum was most intense, while the other permitted components ($\Gamma 12+$) and ($\Gamma 25+$) appear weakly. The neutron scattering ex-

[12] B.A. WEINSTEIN and M. CARDONA: Phys. Rev. **B7**, 2545 (1973).

[13] G. NILSSON and G. NELIN: Phys. Rev. **B3**, 364 (1971). — G. NELIN and G. NILSSON: Phys. Rev. **B5**, 3151 (1972).

Table 44. Comparison of phonon frequencies in germanium

Point	Branch	Neutron	Infra-red
Γ	O	300 ± 10	298
X	TO	275 ± 10	269
	L	230 ± 13	227
	TA	82 ± 5	83
L	TO	280 ± 10	279
	LO	247 ± 10	248
	LA	215 ± 10	212
	TA	65 ± 3	64

From F.A. Johnson and R. Loudon: Proc. Roy. Soc. (London), Ser. A, **281**, 285 (1964), Table 9.

periment is a very interesting one since Nelin and Nilsson were able to obtain the one phonon density of states directly from the neutron scattering cross-sections. It appears as if this technique will be a very powerful one when it can be applied and thus provide ample opportunity to check details of lattice dynamic calculations, which heretofore have been checked against available phonon dispersion in principle directions. The new work will apparently permit both a comparison in principle directions and a comparison of the entire density of states on a given branch.

C, Si, Ge

A useful comparison of all three compounds with diamond structure has been presented by Bilz, Geick, and Renk[14] [*32*] showing the normalized absorption coefficients superimposed on a common scale. This is reproduced in Fig. 15. It is evident that the infra-red absorption spectra germanium and silicon is quite similar while that of diamond solid curve differs radically. This is evidence for the close similarity in lattice phonon dispersion of Ge and Si. This similarity has been used to do a homologous scaling of phonons in these two materials.[15] In turn this reflects close similarity in bonding which, for example, permits a similar shell model calculation to be made in both cases. For diamond, on the contrary the binding forces are quite different being typical of covalent bonds,[16] the phonon dispersion is consequently different, and so the qualitative and quantitative features of the optical spectrum differ.

In summary, critical point theory has been used in the detailed interpretation of the optical spectra of these materials, but more remains to be done. For example, one of the most important quantitative features yet to be studied is the variation of matrix element $\mathcal{M}$ with frequency. For example in the two-phonon infra-red spectrum, the relative intensity of the various parts of the spectrum is not always in agreement with simply the numberical value of the joint density of states, even if all transitions are allowed. Observe for example

[14] H. Bilz, R. Geick, and K. Renk: *loc. cit.* [*32*].

[15] M. Lax: in [*32*].

[16] M.J.M. Musgrave and J.A. Pople: Proc. Roy. Soc. (London), Ser. A **268**, 474 (1962); a pioneer paper is M. Born: Ann. Phys. **4**, 44 (1914). Recently this type of model was used by R. Tubino, L. Piseri, and G. Zerbi: J. Chem. Phys. **56**, 1022 (1972).

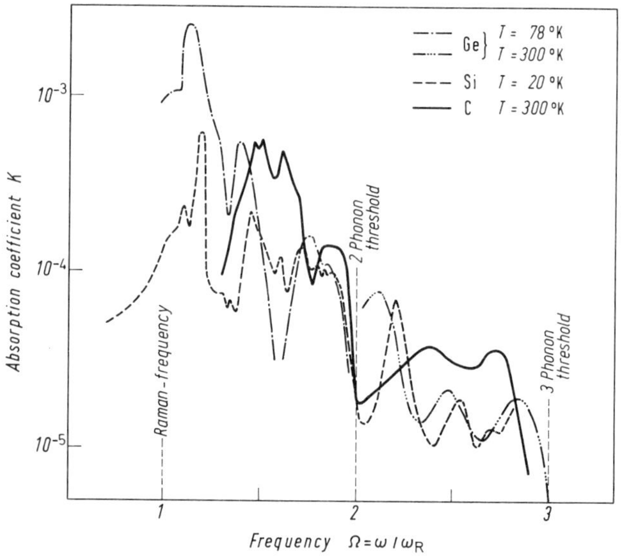

Fig. 15. Infra-red absorption bands in the two and three phonon regions for germanium, silicon and diamond. Notice that while germanium and silicon scale, diamond does not appear homologous. This of course reflects the quite different phonon dispersions of diamond compared to the others. [After H. BILZ, R. GEICK, and K. F. RENK, in: Proceedings of the International Conference on Lattice Dynamics, Copenhagen 1963 (ed. R. F. WALLIS), Fig. 6. New York: Pergamon Press 1965]

in Fig. 10, the multiphonon experimental curve (*a*) compared to joint density of states (*b*). There would appear to be a correction needed for quantitative agreement, which would lower the intensity at higher frequency with respect to that at lower frequency. A broad, structureless renormalization of the spectrum via matrix element $\mathcal{M}$ would provide this, if $\mathcal{M}$ were generally smaller in the multiphonon region toward higher energy.

Work along these lines has been reported, within what could be called polarizibility theory.[17,18] The subject is still developing and still further activity is to be expected.

143. Symmetry set of critical points in rocksalt structure. The critical point theory has apparently not been applied in as fine detail to the interpretation of spectra of rocksalt crystals as in case of diamond structure although work along these lines in under way. Hence we shall briefly review the relevant work which has been done.

Since the location of critical points is a local matter, we can proceed as in the case of the diamond structure using only factor group information. Further, rocksalt is a symmorphic space group so the actual computation is only con-

[17] G. DOLLING and R. A. COWLEY: Proc. Phys. Soc. (London) **88**, 463 (1964). — G. DOLLING, R. A. COWLEY, and A. D. B. WOODS: Can. J. Phys. **43**, 1397 (1965).

[18] R. A. COWLEY: J. Phys. (Paris) **26**, 659 (1965); — Advan. Phys. **12**, 421 (1963); — Proc. Phys. Soc. (London) **84**, 281 (1964).

Table 45. Symmetry set of critical points for phonons in rocksalt structure O_h^5 (see Table 21)

k	$\mathfrak{P}(k)$	Representation	P_n	Component of $\nabla\omega$ vanishing
Γ	O_h	$O(\Gamma 15-)$	P_3	All vanish
		or $(LO+TO)$[a]	P_3	All vanish (max.)
		$A(\Gamma 15-)$	P_0	Non-analytic c.p., ∴ none vanish
X_1	D_{4h}	$LA(X_1\,4-)$		All vanish
		$TA(X_1\,5-)$		All vanish
		$LO(X_1\,4-)$		All vanish
		$TO(X_1\,5-)$		All vanish
L_1	D_{3H}	$LA(L_1\,1+)$		All vanish
		$TA(L_1\,3+)$		$\nabla\omega \parallel L_1$ vanishes
		$LO(L_1\,2-)$		All vanish
		$TO(L_1\,3-)$		$\nabla\omega \parallel L_1$ vanishes
W_1	D_{2d}	$(W_1\,3)$		$\nabla\omega$ vanishes $\perp W_1$
		$(W_1\,1)$		All vanish
		$(W_1\,2')$		All vanish
Δ_1	C_{4v}	$LA(\Delta_1\,1)$		$\nabla\omega$ vanishes $\perp \Delta_1$
		$TA(\Delta_1\,5)$		$\nabla\omega$ vanishes $\perp \Delta_1$
		$LO(\Delta_1\,1)$		$\nabla\omega$ vanishes $\perp \Delta_1$
		$TO(\Delta_1\,5)$		$\nabla\omega$ vanishes $\perp \Delta_1$
Λ_1	C_{3v}	$LA(\Lambda_1\,1)$		$\nabla\omega$ vanishes $\perp \Lambda_1$
		$TA(\Lambda_{1'}\,3)$		None vanish
		$LO(\Lambda_1\,1)$		$\nabla\omega$ vanishes $\perp \Lambda_1$
		$TO(\Lambda_{1'}\,3)$		None vanish
Z	C_{2v}	$Z_{1'}\,1$		$\nabla\omega$ vanishes $\perp Z_1$
		$Z_{1'}\,3$		$\nabla\omega$ vanishes $\perp Z_1$
		$Z_{1'}\,4$		$\nabla\omega$ vanishes $\perp Z_1$
$S_1\,U_1\,\Sigma_1\,K$		Same as Z		
Q	C_2	$Q_1\,1$		$\nabla\omega$ vanishes $\perp Q_1$
		$Q_1\,2$		$\nabla\omega$ vanishes $\perp Q_1$

[a] Including the macroscopic electric field effect. Note no change in the vanishing components.

cerned with point group manipulations since all factor groups are either point groups or the direct product of a point group and a simple Abelian group. Carrying through the same manipulations as before it is found that the zone points Γ, X_1, L_1, W_1, are critical points for all branches, and the lines Σ and Q have vanishing transverse components of $\nabla\omega$. Then wherever branches have a maximum or minimum along Σ and Q lines, there will be an added critical point. A summary of these results is given in Table 45.

In Fig. 16 the results of a detailed model calculation of phonon dispersion in NaCl are given.[1] The locations of the critical points, and their labelling by Karo and Hardy is given on Fig. 16. While there is almost enough information given in the figure in regard to phonon dispersion of branches in different directions, to classify the critical points according to index as $P_j(n)$ such a classification has

[1] A.M. Karo and J.R. Hardy: Phys. Rev. **141**, 701 (1966).

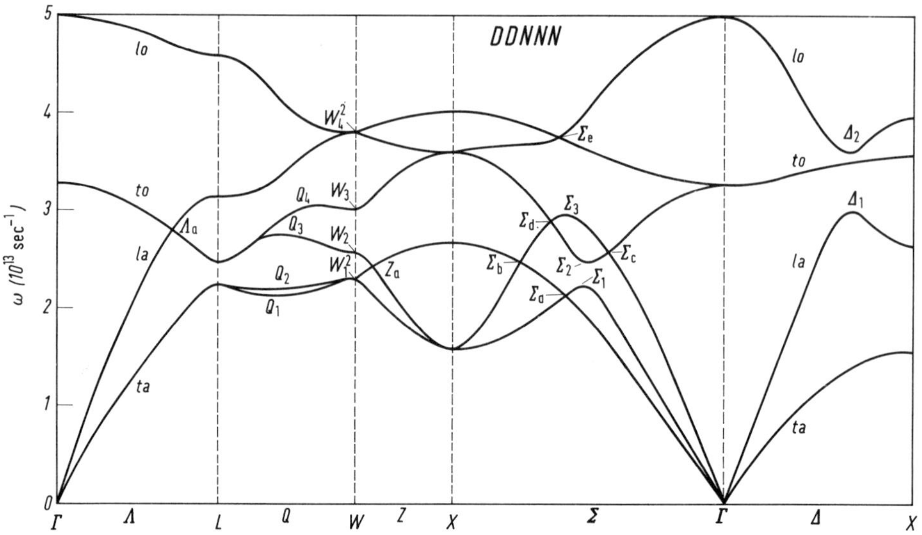

Fig. 16. A model calculation of phonon dispersion in NaCl. [After A. M. KARO and J. R. HARDY: Phys. Rev. **141**, 701 (1966), Fig. 2c]

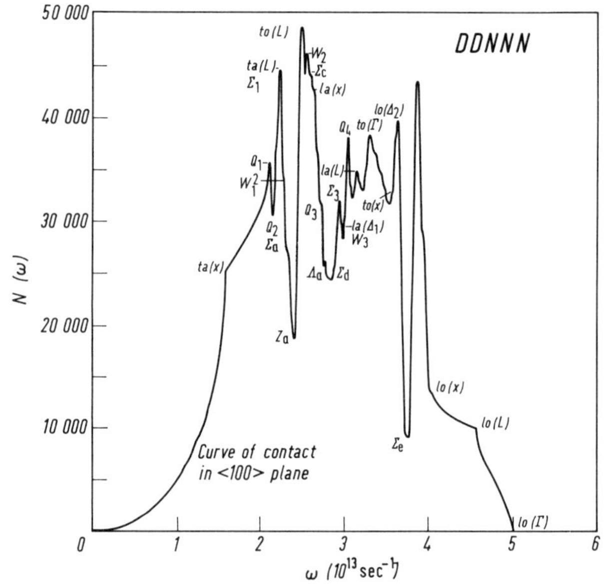

Fig. 17. A calculation of the density of states (one phonon) in NaCl. The model corresponds to that used in Fig. 16. [After A. M. KARO and J. R. HARDY: Phys. Rev. **141**, 696 (1966), Fig. 3]

not been given so we will not attempt it. Rather, we reproduce in Fig. 17 the calculated density of states in NaCl using a refined grid adequate to bring out details of the structure. Observe the labelling of the single phonon critical points in Figs. 16 and 17.

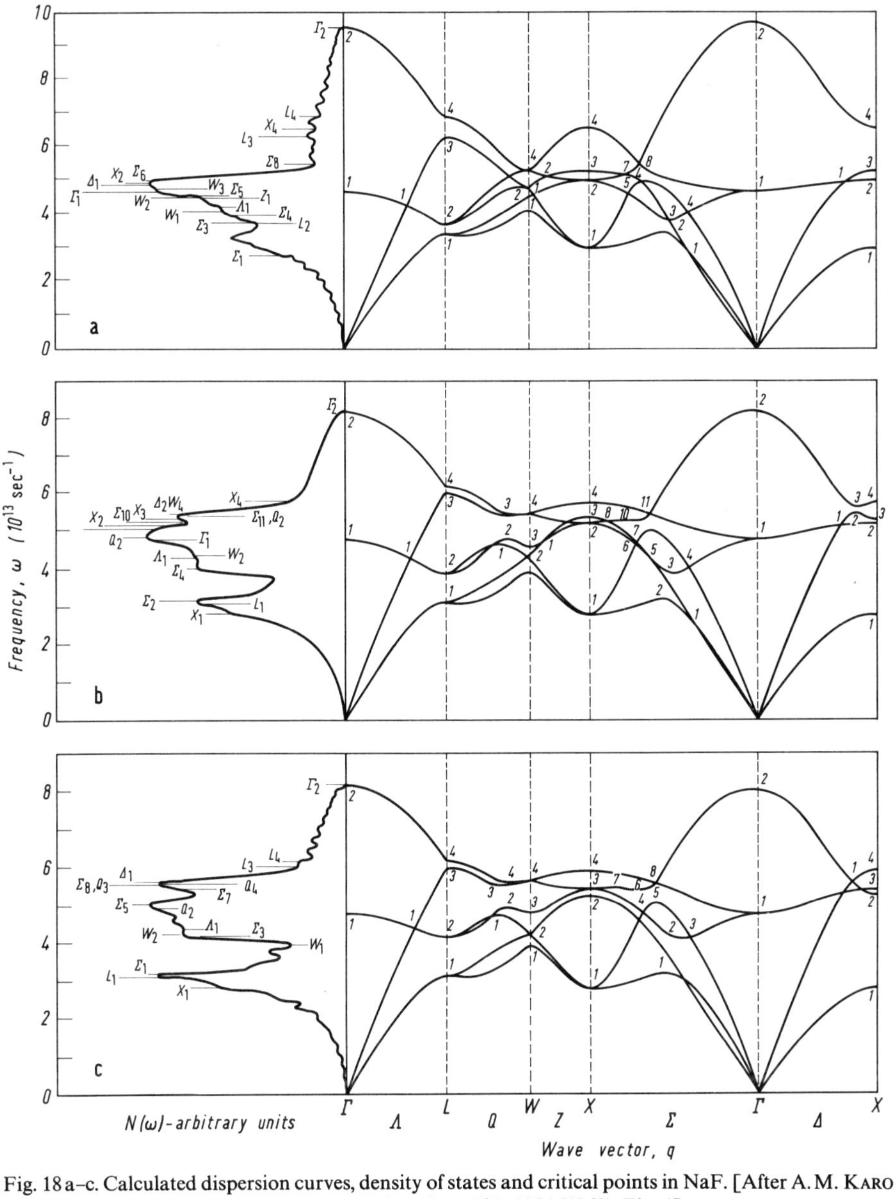

Fig. 18 a–c. Calculated dispersion curves, density of states and critical points in NaF. [After A. M. KARO and J. R. HARDY: Phys. Rev. **181**, 1274 (1969), Fig. 1]

These authors have also used a variety of models including the rigid ion and certain deformable ion models, to compute phonon dispersion[2] [*33*] in sequence of alkali fluoride crystals with rocksalt crystal structure: NaF, KF, RbF, and CsF. At the time of writing, particular attention has been given to NaF. In Fig. 18 we

[2] J. R. HARDY and A. M. KARO, p. 99 in [*31*].

give examples of this work, showing also the identified critical points obtained from the detailed phonon dispersion. Actual indices of the critical points have not yet been reported: compare Sect. 142 on diamond.

144. Two phonon density of states and critical points in rocksalt – NaCl. Using the same models as for the single phonon dispersion, the two-phonon dispersion

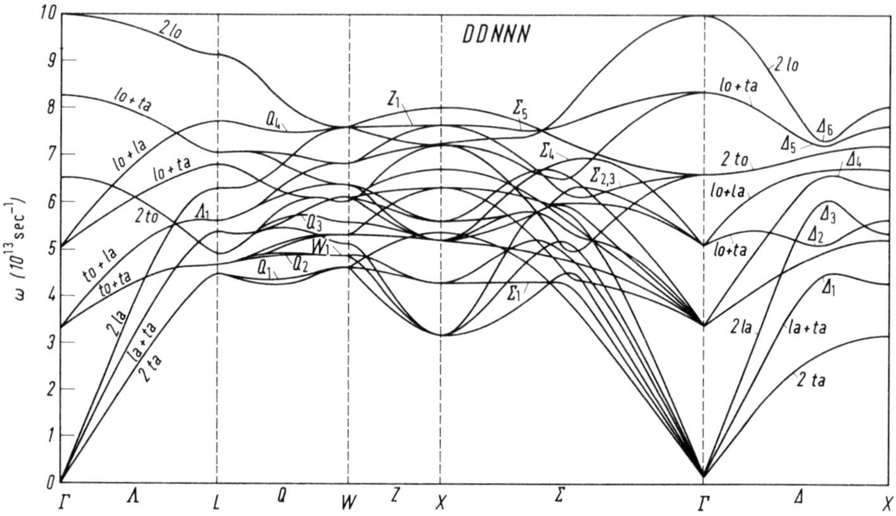

Fig. 19. Calculated dispersion curves for two phonons in NaCl. [After A. M. KARO and J. R. HARDY: Phys. Rev. **141**, 701 (1966), Fig. 5c]

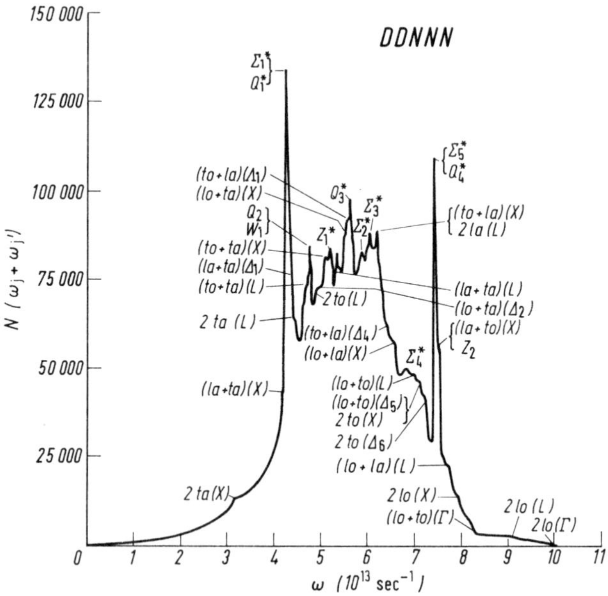

Fig. 20. Calculated two phonon density of states and critical points in rocksalt. See model of Fig. 19. [After A. M. KARO and J. R. HARDY: Phys. Rev. **141**, 696 (1966), Fig. 4]

Table 46a. Phonon symmetries in the rocksalt structure

Symmetry point	Phonon species
Γ	$\Gamma_{15}(O)+\Gamma_{15}(A)$
X	$X_4'(LO)+X_5'(TO)+X_4'(LA)+X_5'(TA)$
L	$L_2'(LO)+L_3'(TO)+L_1(LA)+L_3(TA)$[a]
W	$W_1+W_2'+2W_3$
Δ	$\Delta_1(LO)+\Delta_5(TO)+\Delta_1(LA)+\Delta_5(TA)$
Λ	$\Lambda_1(LO)+\Lambda_3(TO)+\Lambda_1(LA)+\Lambda_3(TA)$
Σ	$2\Sigma_1+2\Sigma_3+2\Sigma_4$
Z	$2Z_1+2Z_3+2Z_4$
Q	$3Q_1+3Q_2$

[a] Note: $L_2' \equiv L^{(2-)}$; $L_3' \equiv L^{(3-)}$; $L_1 \equiv L^{(1+)}$; $L_3 \equiv L^{(3+)}$ the notation of Table 46a follows BURSTEIN, JOHNSON, LOUDON, *loc. cit.*, which in turn follows BOUCKAERT, SMOLUCHOWSKI, WIGNER, *loc. cit.*

Table 46b. Infrared-active two-phonon combinations[a]

Symmetry point	Active combinations
Γ	None
X	None
L	$TO+LA$, $TO+TA$, $LO+LA$, $LO+TA$
W	W_1+W_2', W_1+W_3, $W_2'+W_3$
Δ, Λ, Q	All combinations
Σ	All combinations except $\Sigma_3+\Sigma_4$
Z	All combinations except Z_3+Z_4
No overtones are infra-red active	

[a] From E. BURSTEIN, F. A. JOHNSON, and R. LOUDON: Phys. Rev. **139**, A 1240–1241 (1965), Tables I and II.

curves for several alkalie halides have been computed.[1] For NaCl this is given in Fig. 19 and the corresponding two-phonon density of states in Fig. 20. Again these authors have labelled the critical point features to correspond.

Using the tables of reduction coefficients given above in Tables A 1 to A 11 for rocksalt structure, and the listed assignment of phonon symmetries at points and lines of symmetry as in Table 21 we can prepare a table of active two-phonon processes at known critical points in the halides for both Raman scattering and infra-red absorption in this structure.[2] This is presented in Table 46, which gives these rules. Comparing with Table 37 for diamond note that in both cases all overtones are infra-red forbidden, or inactive. This result is a direct consequence of the space group selection rules; a concise statement of these which deals particularly with space groups containing the inversion has been given in Sect. 122.

[1] J. R. HARDY and A. M. KARO: Phys. Rev. **141**, 703 (1966).
[2] E. BURSTEIN, F. A. JOHNSON, and R. LOUDON: Phys. Rev. **139**, A 1240 (1965).

Also note that at $\star X$ no combinations are infra-red active, contrary to the case of diamond; other similarities and differences can be observed from the two tables.

145. Interpretation of lattice Raman and infra-red spectra in some rocksalt structure crystals.

NaCl

The crystal which has apparently been most carefully studied to date in regard to comparison of experiment and theory is NaCl. The two phonon region of Raman scattering was earlier measured by WELSH and co-workers[1] using mercury lamp radiation and results are shown in lower part of Fig. 21. A cal-

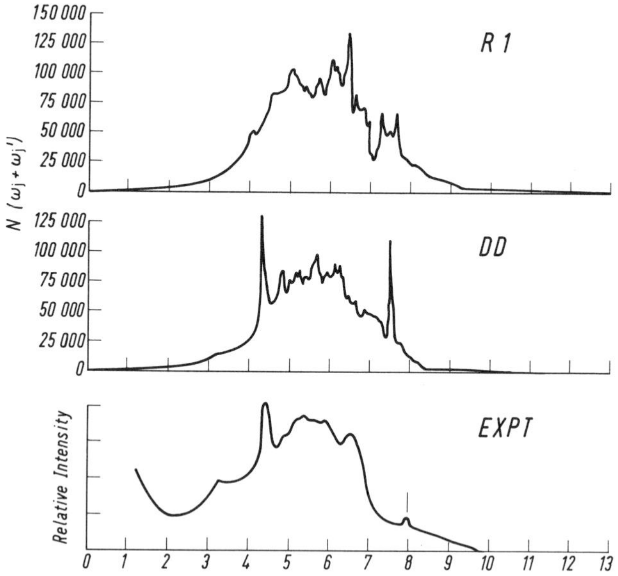

Fig. 21. Second-order Raman scattering in NaCl. Comparison of experiments and a model calculation as in model of A. M. KARO and J. M. HARDY: Phys. Rev. **141**, 701 (1966), Fig. 2c

culated two phonon density of states is shown directly above it for comparison, as given by KARO and HARDY.[2] It is seen that a good fit is obtained. The calculated two phonon density of states has considerably more structure: slope changes and maxima and minima than the observed Raman scattering profile: but the major features and a good many of the detailed features do superimpose. Notice that the Raman data was taken at 300 °K, while the theory should be compared to low temperature (0 °K) data. Also recall the density of states approximation ignores variation of matrix elements and departure from harmonicity.

[1] H. L. WELSH, M. F. CRAWFORD, and W. J. STAPLE: Nature **164**, 737 (1949).
[2] J. R. HARDY and A. M. KARO: Phys. Rev. **141**, 703 (1966).

Table 47. Second-order Raman peaks in NaCl. (Peak positions in cm^{-1})[b]

Experimental	Interpretation	Calculated for X point
31	$TO-LA(X)$	30
55	$LA-TA(X)$; $LO-LA(\Delta)$; $LA-TA(L)$	56
174	$2TA(X)$	174
234[a]	$LA+TA(X)$; $LA+TA(\Delta)$; $2TA(L)$	230
256	$TO+TA(X)$; $2TO(L)$	260
275	$LA+TA(L)$	
285	$2LA(X)$	286
299	$LO+TA(X)$; $2LA(\Delta)$	299
314	$TO+LA(X)$; $2LA(L)$	316
346	$2TO(X)$; $LO+LA(\Delta)$; $LO+TO(L)$	346
415 (broad)	$2LO(X)$ (among others)	414

[a] Strongest peak.
[b] From E. BURSTEIN, F.A. JOHNSON, and R. LOUDON: Phys. Rev. **139**, A 1243 (1965).

Table 48. Room-temperature phonon frequencies in NaCl. (Calculated by HARDY and KARO)[b]

Branch	L	Δ	X	Deduced at X from Raman effect (cm^{-1})
	(cm^{-1})	(cm^{-1})	(cm^{-1})	
LO	236	196[a]	208	212
TO	123	183	183	173
LA	160	151[a]	140	143
TA	112	75	83	87

[a] The Δ_{LO} and Δ_{LA} branches have extrema at a q/q_{max} of about 0.7.
[b] From E. BURSTEIN, F.A. JOHNSON, and R. LOUDON: Phys. Rev. **139**, A 1243 (1965).

An interpretation of the same data has been given[3] using phonons only at the $^{\star}X$, $^{\star}L$, $^{\star}\Delta$ points, which accounts for the location of major features of the spectrum: of course this is not as accurate as the density of states calculation, but emphasizes the importance of the critical point phonons. This interpretation is given in Table 47. In Table 48 some values of phonon frequencies at various energies are listed for NaCl.

The room temperature two phonon infra-red absorption spectrum of NaCl[4] is reproduced in Fig. 22. From Table 46 it is seen that the infra-red selection rules in this structure are much more restrictive than in the diamond structure. The figure apparently reflects this insofar as only a limited amount of structure is exhibited, and only four subsidiary peaks can be defined. (Perhaps higher resolution results would have more structure.) At the time of writing a complete analysis of these results has not been given, taking account of both selection rules and critical point phonons. In Sect. 146 some discussion of the Raman scattering

[3] E. BURSTEIN, F.A. JOHNSON, and R. LOUDON: Phys. Rev. **139**, A, 1240 (1965).
[4] C. SMART, G.R. WILKINSON, A.M. KARO, and J.R. HARDY, p. 387 in [*32*].

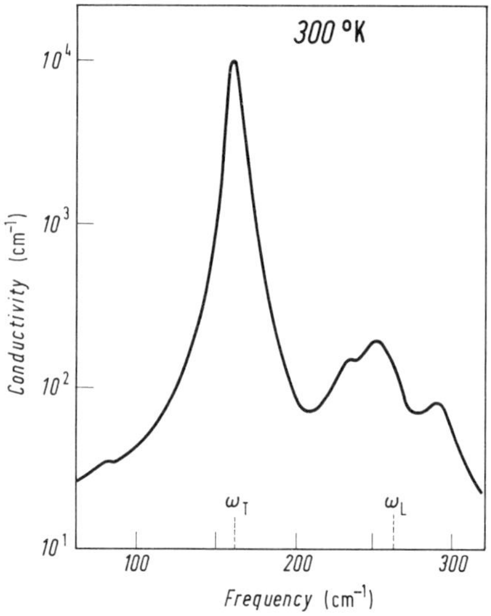

Fig. 22. Infra-red absorption spectrum of NaCl. Note the absence of structure by comparison to the Raman scattering spectrum, in the two-phonon region. [After C. SMART, G. R. WILKINSON, A. M. KARO, and J. R. HARDY, in: Proceedings of the International Conference on Lattice Dynamics, Copenhagen 1963 (ed. by R. F. WALLIS), p. 383. New York: Pergamon Press 1965]

is given. In the earlier work,[5] partial account of the selection rules was taken and the density of states available at that time was less accurate than needed for a definitive analysis, although general agreement with experiments was found.

Recently, the second order Raman spectrum of NaCl has been reinvestigated by a number of workers. For example, KRAUZMAN[6] has published the spectrum for NaCl which is given in Fig. 23, along with the interpretation indicated on the figure. Additional interpretation of various detailed features of the spectrum has also been given by KRAUZMAN and this is given in Table 49. Noteworthy is the great detail of interpretation given, by use of the selection rules. Also noteworthy is the fact that it has not yet been possible to identify the indices $P_j(n)$ and make the detailed correlation of spectral shapes as was done for Si and Ge.

A report[7] that a laser measurement of the two-phonon Raman spectrum of NaCl was in sufficient disagreement with the results just given as to necessitate reinterpretation of the spectrum, was apparently due to a misunderstanding.[8] Hence it appears that the interpretation of the Raman and infra-red two-phonon spectrum of NaCl given above should be provisionally accepted as correct.[9]

[5] H. BILZ, L. GENZEL, and H. HAPP: Z. Physik **160**, 535 (1960). — H. BILZ and L. GENZEL: Z. Physik **169**, 53 (1962).

[6] M. KRAUZMAN, p. 109 in [*32*]. Also M. KRAUZMAN: Thèse de Doctorat D'État ès Sciences Physiques, Fac. des Sciences, Paris, 10 Mars, 1969.

[7] J. R. HARDY, A. M. KARO, I. MORRISON, C. T. SENNETT, and J. P. RUSSELL: Phys. Rev. **179**, 837 (1969). See esp. p. 838 and the footnote therein.

[8] J. M. WORLOCK: Private communication.

[9] Also the reported work by A. I. STEKHANOV and A. P. COROLKOV, p. 119 in [*32*] for which only an abstract exists, seems to agree with the other work reported above.

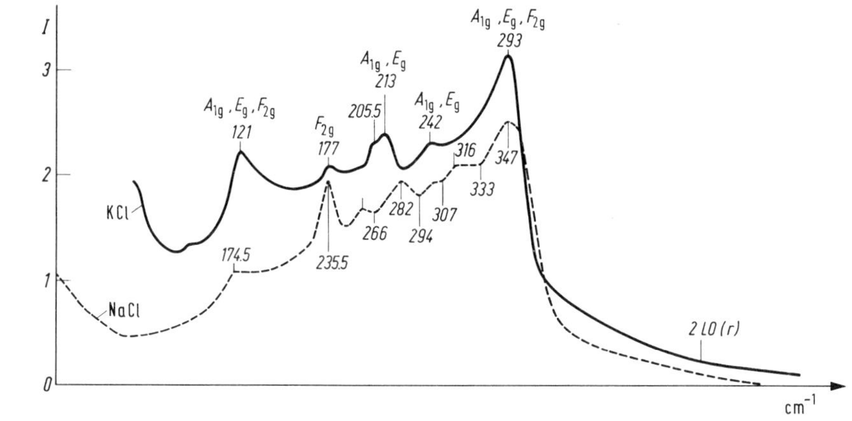

Fig. 23. Raman scattering in NaCl and KCl: Scattering intensity (arb. units) vs. wave length. [After M. KRAUZMAN, in: Light Scattering Spectra of Solids (ed. by G. WRIGHT), p. 116. Berlin-Heidelberg-New York: Springer 1969]

Table 49. Interpretation of NaCl spectra

Experimental		Calculated		
ν cm^{-1}	Modes	Attributions	Selection rules	ν cm^{-1}
		$TO\{z\} - TA\{x\bar{y}\}(\Sigma)$	F	53
55	F	$LA - TA(X)$	F	55
		$LA - TA(L)$	E F	55
		$Q_1 - Q_1$	A E F	57
60.5	F	$W_3^0 - W_3^A$	A E F	59
		$Q_1 - Q_2$	E F	59
87	E	$LO - TO(L)$	E F	86
		$TO - TA(X)$	A E F	86.5
104	A E F	$LO - TO(\Gamma)$	A E F	102
174.5	A E F	$2\,TA(X)$	A E F	175
		$2\,W_3^A$	A E F	230
		$2\,Q_2$	A E F	231
231	A	$Q_1 + Q_2$	E F	233
233	E	$LA + TA(X)$	F	230
235.5	F	$2\,Q_1$	A E F	236
239	A E	$Z_3 + Z_4$	F	236.5
		$2\,TA(L)$	A E F	236
		$TA\{z\} + TA\{x\bar{y}\}(\Sigma)$	F	238
248	A E	$2\,TA\{x\bar{y}\}(\Sigma)$	A E F	248
251.5	F	$LA + TA(\Delta)$	F	250
		$W_1 + W_3$	F	253
258	A	$TO + TA(X)$	A E F	261.5
259	F	$LO + TA(\Delta)$	F	260
266	A	$2\,TO\{x\bar{y}\}(\Sigma)$	A E F	266
		$LO + TA(X)$	F	270
273.5	F	$TA\{x\bar{y}\} + TO\{x\bar{y}\}(\Sigma)$	A E F	271
		$W_{2'} + W_3^A$	F	273.5
		$Z_1 + Z_3$	F	270

Table 49 (continued)

Experimental		Calculated		
$\nu\,cm^{-1}$	Modes	Attributions	Selection rules	$\nu\,cm^{-1}$
276	A E	$2W_1$	A E	276
280	F	$2TO(L)$	A E F	280
282	F	$TA\{z\}+LA(\Sigma)$	F	282
286	A	$2LA(X)$	A E	285
288	F	$W_3^A+W_3^0$	A E F	289
		Q_1+Q_2	E F	289
294	A	Q_1+Q_1	A E F	293
		$2Q_1$	A E F	294
300	F	Q_1+Q_2	E F	300
307	A	?		
314	E	$LO+TA\{x\bar{y}\}(\Sigma)$	E	314
		$LA+TO\{x\bar{y}\}(\Sigma)$	E	314
316	F	$TO+LA(X)$	F	316.5
317	A	$2W_{2'}$	A E	317
332	A	$2LA(\Delta)$	A E	331
333	F	$W_{2'}+W_3^0$	F	332.5
		$TO+LA(\Delta)$	F	336
343	E	$LO+LA(\Delta)$	A E	343
347–355	A	$2LA(L)$	A F	346
		$2TO(X)$	A E F	348
		$2W_3^0$	A E F	348
		$2Q_1$	A E F	349
		$LO+LA(\Sigma)$	A E F	350
		$2LO(\Delta)$	A E	356
354	F	$LO+TO(\Delta)$	F	350
		$LO+TO\{z\}(\Sigma)$	F	352
		$LO+TO(X)$	F	356.5
360	E	$2LO(X)$	A E	365
381	F	?		
394	A	?		
524–543	A	$2LO(\Gamma)$	A E F	528

From M. KRAUZMAN: in [*33*].

NaF

A careful analysis of the lattice dynamics and the second order Raman spectrum of NaF has been put forth recently.[10] The observed second order Raman spectrum of NaF is shown on Fig. 24 (note non-linear scale). In the figure the principal features of the spectrum are indicated by arrows leading to various letters. In Fig. 25 the infra-red absorption of NaF is given the high energy part evidently reflects the two-phonon absorption processes. This seems to be the most recently available data: it would evidently be most useful to have more refined studies made.

[10] J. R. HARDY, A. M. KARO, I. MORRISON, C. T. SENNETT, and J. P. RUSSELL: Phys. Rev. **179**, 837 (1969).

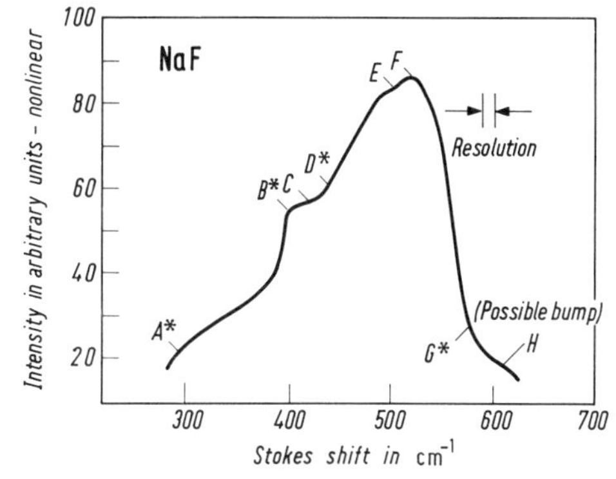

Fig. 24

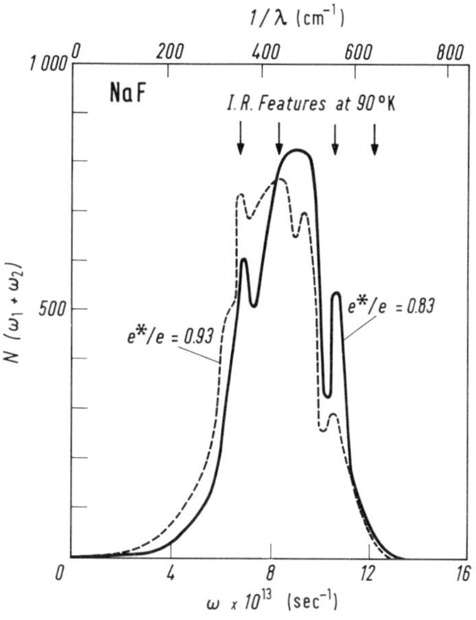

Fig. 25

Fig. 24. Second-order Raman spectrum of NaF. [After J.R. HARDY, A.M. KARO, I.W. MORRISON, C.T. SENNETT, and J.P. RUSSELL: Phys. Rev. **179**, 838 (1969), Fig. 1]

Fig. 25. Two phonon (combined) density of states in NaF. [After C. SMART, G.R. WILKINSON, A.M. KARO, and J.R. HARDY, in: Proceedings of the International Conference on Lattice Dynamics, Copenhagen 1963 (ed. by R.F. WALLIS), p. 383, Fig. 3a]

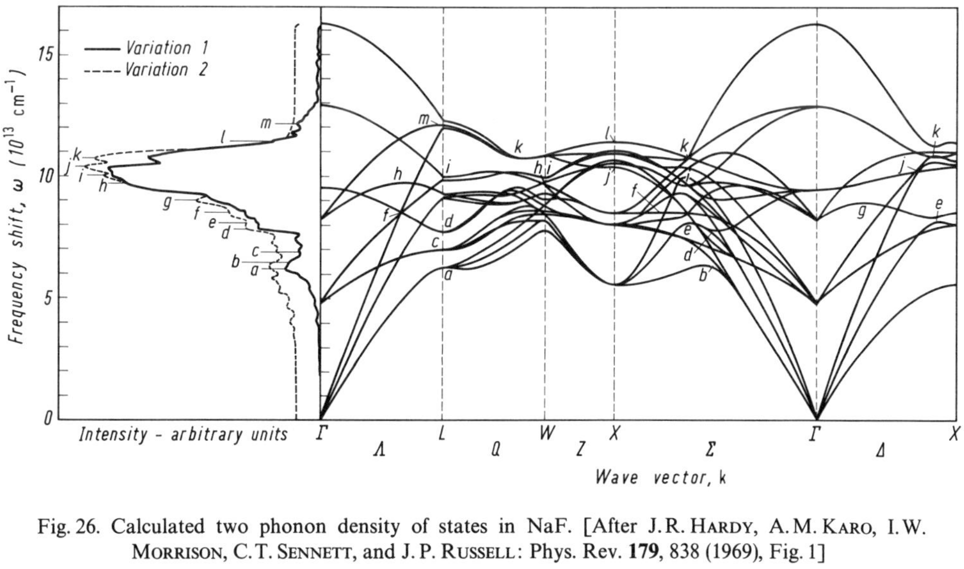

Fig. 26. Calculated two phonon density of states in NaF. [After J.R. HARDY, A.M. KARO, I.W. MORRISON, C.T. SENNETT, and J.P. RUSSELL: Phys. Rev. **179**, 838 (1969), Fig. 1]

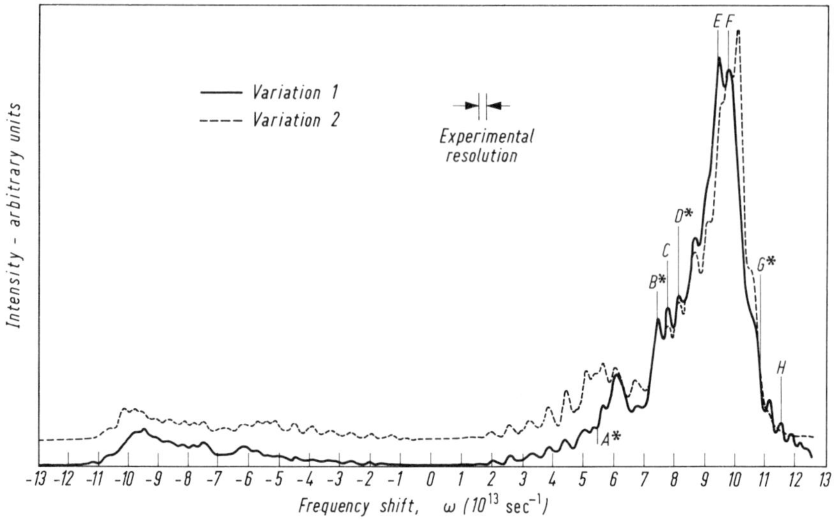

Fig. 27. Comparison between experimental and theoretical two phonon density of states in NaF. [After J.R. HARDY, A.M. KARO, I.W. MORRISON, C.T. SENNETT, and J.P. RUSSELL: Phys. Rev. **179**, 838 (1969), Fig. 1]

To interpret these data, HARDY, KARO and collaborators[11] have calculated the lattice phonon dispersion in NaF for single and two phonons using several models of varying complexity including ionic deformation effects. As in the NaCl case, a one and two-phonon density of states can be evaluated for each model, and critical points identified. In Fig. 26 the results of calculation of the two phonon density of states in NaF, exhibiting the critical points (but not classifying them according to index) are given for the best model devised by these authors. In Fig. 26 the model parameters were taken at a value appropriate for 300 °K and compared with the prominent features of measurements as given in Fig. 24. From the agreement in location of predicted and measured features, it is clear that the model agrees with the major features of the experiment. No further detailed analysis is possible at time of writing.

Other alkali halides

Some other alkali halides for which the two phonon Raman spectra have been analysed include KBr, KI and NaI. This work was carried out using the critical point theory and the selection rules.[12,13] In Tables 50 and 51 some of these results are presented.

It is clear that a considerable quantitative understanding of these alkali halide crystals spectra is achieved by use of the critical point theory, and the entire group theory (selection rule) apparatus. For comparison with a treatment of 2-phonon

[11] A.M. KARO and J.R. HARDY: Phys. Rev. **181**, 1272 (1969).

[12] E. BURSTEIN, F.A. JOHNSON, and R. LOUDON: Phys. Rev. **139** A, 1240 (1965).

[13] M. KRAUZMAN: Thèse (footnote 6, p 424).

Table 50. Interpretation of KBr spectra

Experimental		Calculated		
$\nu\,\mathrm{cm}^{-1}$	Modes	Attributions	Selection rules	$\nu\,\mathrm{cm}^{-1}$
46.5	A E F	$TO-LA(X)$	F	45
		$LO-TO(\Delta)$	E F	46.5
		$LO-TO(L)$	E F	47.5
		$LA-TA(\Delta)$	F	47.5
61	A E	$LO-LA(X)$	A E	60.5
76	A E F	$TO-TA(X)$	A 2E F	76
86	A E (F)?	$2\,TA(X)$	A 2E F	84
116	F	$LA+TA(X)$	F	115
125	F	$TA\{z\}+TA\{x\bar{y}\}(\Sigma)$	F	124
		$LA+TA(\Delta)$	F	125
135	E F	$2\,TA\{x\bar{y}\}(\Sigma)$	A E F	135
138	A	$2\,TA(L)$	A E 2F	140
145	A	$2\,LA(X)$	A E	146
150	F	$W_1+W_3^A$	F	150
157	F	$LO+TA(\Delta)$	F	157
158	A	$TO\{x\bar{y}\}+TA\{x\bar{y}\}(\Sigma)$	A E F	157
		$TO+TA(X)$	A 2E F	160
165	E F	$LA+TA(L)$	E F	162
178	A (E)?	$2\,LA(\Sigma)$	A E F	177
		$2\,W_1$	A E	178
182	F	$2\,TO\{x\bar{y}\}(\Sigma)$	A E F	182
184.5	A	$2\,LA(L)$	A F	184
193	A (E)? F	$TO+LA(X)$	F	191
		$W_3^0+W_3^A$	A 2E F	191.5
		$2\,TO(L)$	A E 2F	193
200	E F	$W_{2'}+W_1$	E	200
		$TO+LA(\Delta)$	F	200
207	F	$TO\{z\}+LA(\Sigma)$	F	208
208	A E	$LO+LA(X)$	A E	206.5
		$LO+LA(\Delta)$	A E	208
214	A E F	$LO+LA(\Sigma)$	A E F	215
221	E F	$W_3^0+W_1$	F	219.5
		$2\,W_{2'}$	A E	222
231	A E	$2\,TO(\Delta)$	A 2E F	232
235.5	A F	$2\,TO(X)$	A 2E F	236
		$LO+TO(\Delta)$	F	236
242	A	$2\,LO(\Delta)$	A E	242
246	E (F)?	$LO+TO\{z\}(\Sigma)$	F	245
251.5	F	$LO+TO(X)$	F	251.5
259	A	$2\,W_3^0$	A 2E F	261
290	A F	$2\,LO(L)$	A F	288
316–336 end	A E F	$2\,LO(\Gamma)$	A E F	326

From M. KRAUZMAN: in [33].

Raman scattering and infra-red absorption using the many-body polarizability and response theory (Sect. 124α, β). See the work of BRUCE and COWLEY.[14]

[14] A. D. BRUCE and R. A. COWLEY: J. Phys. C. Solid state Phys. **5**, 595 (1972); — A. D. BRUCE, *ibid*, **5**, 2909 (1972); — *ibid*, **6**, 174 (1973).

Table 51. Interpretation of KI spectra

Experimental		Calculated		
$\nu\,\mathrm{cm}^{-1}$	Modes	Attributions	Selection rules	$\nu\,\mathrm{cm}^{-1}$
62	$A\ E$	$2\,TA(X)$	$A\ 2E\ F$	62
70	$A\ E$	$TO-TA(X)$	$A\ 2E\ F$	71.5
77	F	$LO-TA(\Delta)$	F	75
		$LO-TA(X)$	F	77
		$W_3^0-W_3^A$	$A\ 2E\ F$	77
		$2\,W_3^A$	$A\ 2E\ F$	78
85	$A\ E$	$2\,TA\{x\bar{y}\}(\Sigma)$	$A\ E\ F$	85
90	F	$W_1+W_3^A$	F	90
		$LA+TA(\Delta)$	F	91
97.5	F	$LA+TA\{z\}(\Sigma)$	F	97
		$TA\{z\}+TO\{x\bar{y}\}(\Sigma)$	F	98
102	$A\ E$	$2\,W_1$	$A\ E$	102
		$2\,LA(X)$	$A\ E$	102
		$2\,TA(L)$	$A\ E\ 2F$	103
		$LA+TA\{x\bar{y}\}(\Sigma)$	E	103
116?	F?	$LA+TA(L)$	$E\ F$	116
122.5	E	$2\,LA(\Sigma)$	$A\ E\ F$	122
126?	E?	?		
129	$A\ F$	$2\,LA(L)$	$A\ F$	129
130	E	$2\,LA(\Delta)$	$A\ E$	132
134.5	A	$TA+TO(X)$	$A\ 2E\ F$	133.5
136.5	E	$W_1+W_{2'}$	E	137
139.5	F	$TA+LO(X)$	F	139
144	E	$TO\{x\bar{y}\}+LA(\Sigma)$	E	144
150	$A\ E$	$2\,TO\{x\bar{y}\}(\Sigma)$	$A\ E\ F$	150
153	F	$TO+LA(X)$	F	153.5
158	E	$LO+TA\{x\bar{y}\}(\Sigma)$	E	158
		$LO+LA(X)$	$A\ E$	159
162	$A\ F$	$LA+TO\{z\}(\Sigma)$	F	163
166.5	$A\ (F)$?	$LA+TO(\Delta)$	F	166
		$LA+LO(\Delta)$	$A\ E$	167
		$W_1+W_3^0$	F	167
173.5	A	$2\,W_{2'}$	$A\ E$	172
178	E	$LO+LA(\Sigma)$	$A\ E\ F$	176
182	E	?		
186	$A\ F$	$2\,TO(L)$	$A\ E\ 2F$	186
196	$A\ E$	$2\,TO(\Delta)$	$A\ 2E\ F$	198
199.5	A	$2\,LO(\Delta)$	$A\ E$	200
205	$A\ (E)?\ F$	$2\,TO(X)$	$A\ 2E\ F$	205
213	E	$2\,LO(X)$	$A\ E$	216
216	A			
218	$E\ F$	$TO+LO(L)$	$E\ F$	218
229–232	$A\ E\ F$	$2\,W_3^0$	$A\ 2E\ F$	232
250	$A\ F$	$2\,LO(L)$	$A\ F$	250
270–305		$2\,LO(\Gamma)$		278
end				

From M. Krauzman: in [33].

146. Polarization effects in two phonon Raman scattering in rocksalt and diamond structures. The analysis of the spectra given in previous sections has neglected the use of polarization effects for both incident and scattered radiation. The theory of polarization effects in Raman scattering has been discussed in Sect. 123 β.

α) *Rocksalt.* Polarized second order Raman scattering spectra have been reported for NaCl, KI, KCl, RbI, TlBr, ZnS and CuCl recently.[1] The spectra were analysed in terms of the selection rules, and the relevant calculated two phonon density of states from phonon dispersion calculations. Results for NaCl will be cited in the present section. A complete critical point analysis, for example, along the lines of the work on Ge, and Si has not yet been reported, although the data certainly appears of sufficiently high precision so this could be attempted.

According the theory discussed in Sect. 123 the non-resonant Raman scattering process in dipole approximation is characterized by matrix elements of a tensor operator $[\boldsymbol{P}]$ which is a symmetric second rank tensor. In a cubic crystal this operator transforms as a reducible representation as discussed in Sects. 110 and 123 and so

$$[\boldsymbol{P}] \simeq \boldsymbol{\Gamma}^{(1+)} \oplus \boldsymbol{\Gamma}^{(12+)} \oplus \boldsymbol{\Gamma}^{(25+)}. \tag{146.1}$$

Now each of the three constitutent irreducible representations in (146.1) can separately be active in the scattering. As shown in Sect. 124 Raman-Stokes scattering will occur if the crystal which is initially in the vibrational ground state can make a transition to an excited state with phonons present of symmetry of one or more of the species given in (146.1). Let us consider each such irreducible representation in (146.1) separately, as was done in (123.14)–(123.37). Also, the most convenient geometry seems to be that leading to (123.37): primed axes are defined just above that matrix.

Polarized Raman spectra for NaCl were obtained by KRAUZMAN and these were shown on Fig. 23. In Fig. 28 some further examples of this work is illustrated. Different curves correspond to different settings of polarizer-analyser combinations. The group theory notation used by KRAUZMAN involves the correspondence

$$A \leftrightarrow \Gamma^{(1+)}; \quad E \leftrightarrow \Gamma^{(12*)}; \quad F \leftrightarrow \Gamma^{(25+)} \tag{146.2}$$

and also (see (123.15)–(123.20)):

$$a = \lambda_A; \quad b = \lambda_E; \quad d = \lambda_F. \tag{146.3}$$

In Fig. 29 a comparison of these experiments and the two phonon density of states from the Karo-Hardy calculation is given. It will be noted that there is a good agreement between various major features such as peaks. This excellent data seems capable of careful analysis in the spirit of locating van Hove singularities, as in the Johnson-Loudon work as Ge and Si, but this has not yet been reported. Note for example what appear to be square root singularities in the data near 170 cm^{-1}, etc. In Table 49 the interpretation of the spectral features by KRAUZMAN was given.

[1] M. KRAUZMAN: Thèse (footnote 6, p. 424).

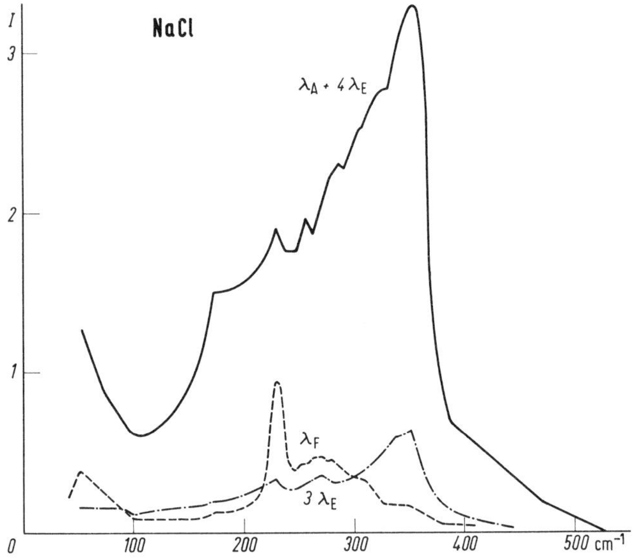

Fig. 28. Experimental spectra of Raman scattering in NaCl. Different attributions refer to different polarization contributions. [After M. KRAUZMAN: Thèse, Fig. 8-5]

The attributions in Table 49 permit an interpretation of the two phonon infra-red spectrum of NaCl to be given. In Table 52 such an interpretation is put forth of the data shown on Fig. 25. The agreement is excellent.

Excellent data for other alkalie halide crystals is also given by KRAUZMAN.

β) Diamond. In (123.37) we gave the form of the Raman scattering matrix corresponding to one of the conventional choice of axes for Raman scattering in a cubic crystal: this corresponds to

$$X'=[1,1,0]; \quad Y'=[1,\bar{1},0]; \quad Z'=[0,0,1]. \tag{146.4}$$

Table 52. Interpretation of NaCl infra-red absorption at 300 °K (see Fig. 25)[a]

ν_{exp} cm^{-1}	Process	ν_{calc} cm^{-1}
80	$LA-TA(\Delta)$	82
162	$TO(\Gamma)$	
236	$\begin{cases} Q_1+Q_2 \\ Z_3+Z_4 \end{cases}$	233 236.5
254	$\begin{cases} LA+TA(\Delta) \\ W_1+W_3^A \\ TA+TO(L) \end{cases}$	250 253 258
292	$\begin{cases} W_3^A+W_3^0 \\ Q_1+Q_2 \\ Q_1+Q_1 \end{cases}$	289 289 293

[a] M. KRAUZMAN: Thèse, Table 8–4. Paris 1969.

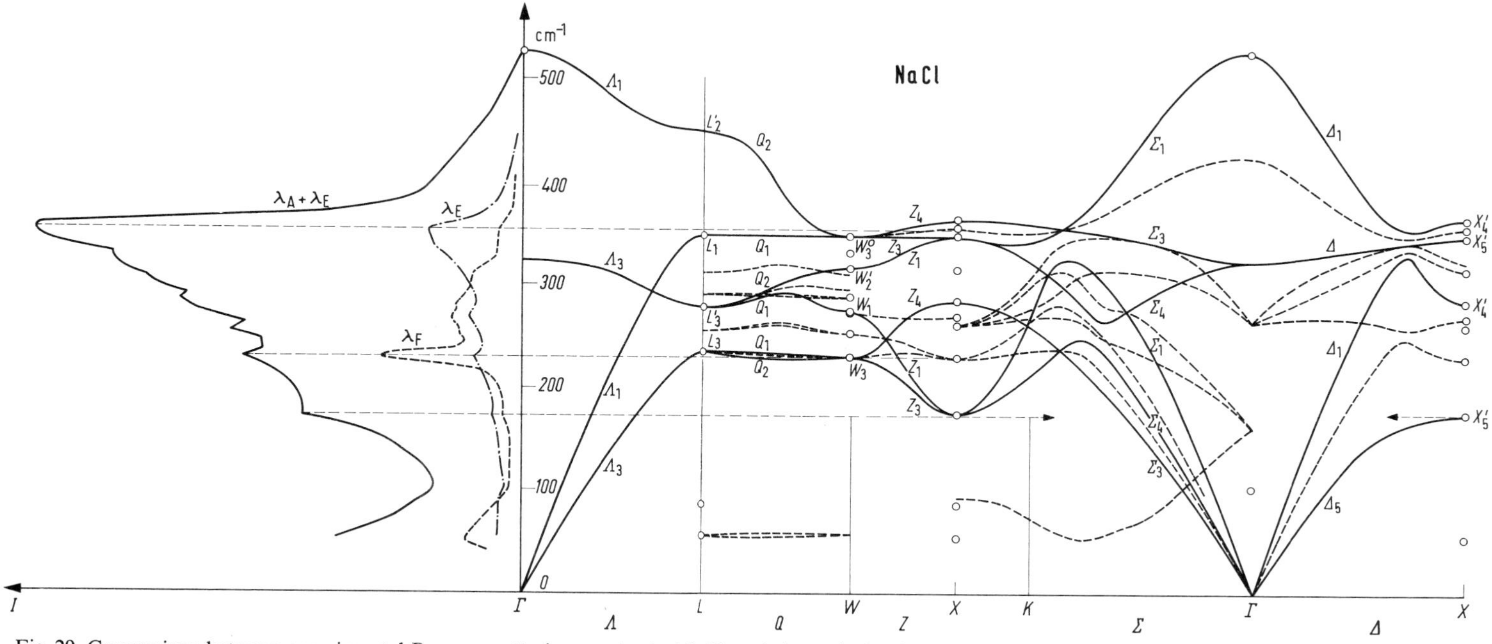

Fig. 29. Comparison between experimental Raman scattering spectra in NaCl, and theoretical calculations, based on density of states computed from phonon dispersion. Note correlation between critical features. [After M. KRAUZMAN: Thèse, Fig. 8-6]. In this figure the experimental curve is drawn on the left — see Fig. 28

The recent work on Raman scattering by diamond was carried out using polarized incident light, and an analyser to detect polarization of the scattered light. To describe these experiments, it is customary to use the notation

$$i(j\,k)\,l, \tag{146.5}$$

where the incident wave is propagating in the direction i, with $\mathscr{E}$ field polarized parallel to j; the scattered wave is propagating in direction l, with $\mathscr{E}$ field polarized parallel to k. As discussed just below Eq. (123.37), three measurements suffice to determine contribution of individual representations to the scattering.[2]

In Fig. 10a we illustrated the $Z'(X'X')\,Y'$ scattered spectrum. From (123.37) this spectrum is a superposition of three contributions from $\boldsymbol{\Gamma}^{(1+)} \oplus \boldsymbol{\Gamma}^{(12+)} \oplus \boldsymbol{\Gamma}^{(25+)}$. In Fig. 30a we illustrate the $Z'(Y'X')\,Y'$ spectrum. It is due to the $\boldsymbol{\Gamma}^{(12+)}$ representation. In Fig. 30b the $Z'(X'Z')\,Y'$ spectrum is shown: it is a pure $\boldsymbol{\Gamma}^{(25+)}$ spectrum. In Fig. 30c the $Y'(Z'Z')\,X'$ spectrum is given: it is $\boldsymbol{\Gamma}^{(1+)} \oplus \boldsymbol{\Gamma}^{(12+)}$.

The analyses given of major features of these polarized spectra were presented in Table 40b, and we refer to that table and the remarks made in the discussion of diamond in Sect. 142 regarding the import of the agreement between theory and experiment.

We conclude the discussion of this work by noting the sharp feature labelled (14) in Figs. 10d and 30a, b, c. Recall the discussion given in Sect. 123d. There it was noted that a "two-phonon bound state" had been proposed to account for a sharp feature observed at a frequency greater than twice the Raman frequency in diamond. Twice the Raman frequency is $2\omega(0) = 2664\ \mathrm{cm}^{-1}$ and the feature now under discussion occurs at $2667\ \mathrm{cm}^{-1}$, which is clearly visible in Fig. 10d, and in Fig. 30a, c. From the polarization studies this feature is of $\boldsymbol{\Gamma}^{(1+)}$ symmetry. More work is needed to decide on the validity of the proposal that this is a two phonon bound state.

The recent work on two phonon Raman scattering in silicon[3,4] and germanium[5] referred to in Sect. 146 did not show an analogous sharp peak above the "two-phonon cut off" $2\omega(0)$ in these materials. There are some anomalies in the case of germanium near $2\omega(0)$ not yet explained i.e. perhaps indicating deviation from two phonon density of states. In view of the differences between diamond, silicon and germanium considerable activity concerning the interpretation of the two phonon spectra is to be expected.

---

[2] L. KLEINMAN: Solid State Communications **3**, 47 (1967) has given some examples of calculation of depolarization of Raman scattering. KLEINMAN's results are based upon obtaining partial reduction coefficients by what amounts to the method based on use of a partial linear vector space. But in certain cases his results contradict those obtainable using the full basis (full group method). It seems that KLEINMAN did not verify the completeness of his partial vector space, and so his work is in error. Recall here the caution discussed on p. 358, Eq. (129.51), *et seq.*

[3] P.A. TEMPLE and C.E. HATHAWAY: Phys. Rev. **B7**, 3685 (1973).

[4] B.A. WEINSTEIN and M. CARDONA: Solid State Comm. **10**, 961 (1972).

[5] B.A. WEINSTEIN and M. CARDONA: Phys. Rev. **B7**, 2545 (1973).

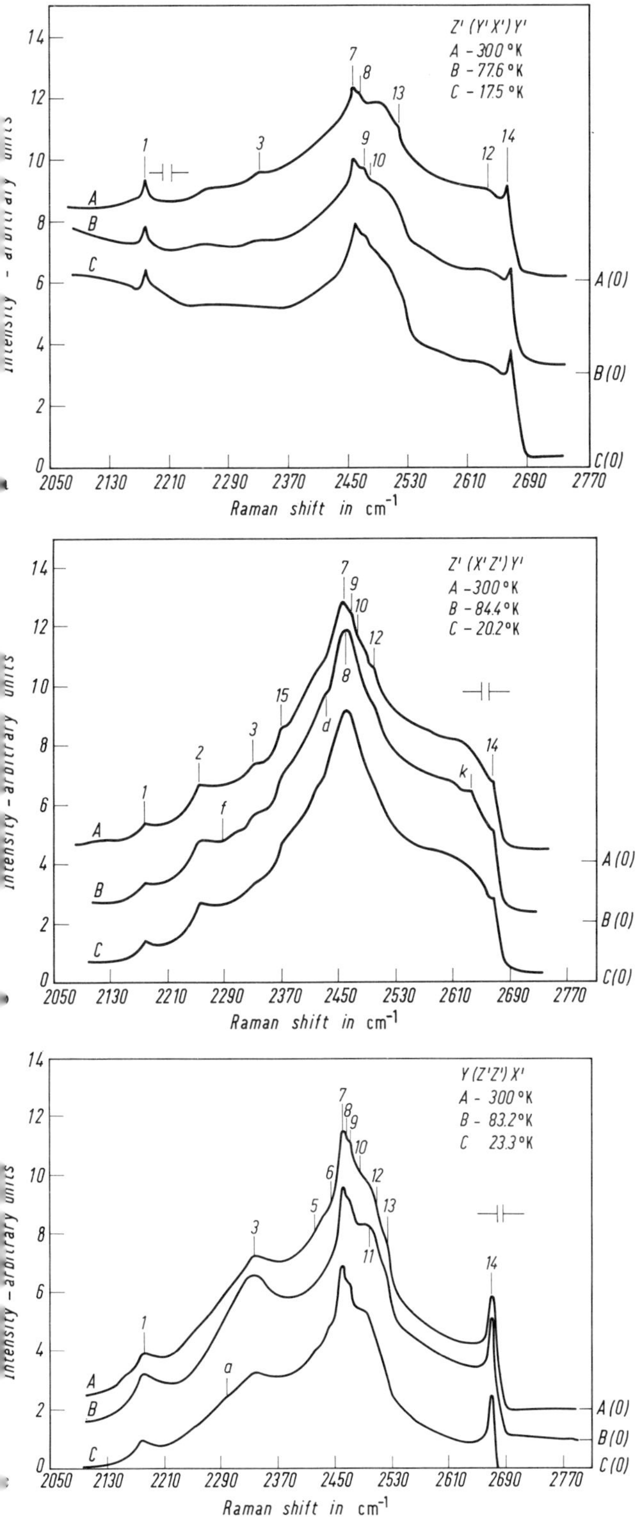

Fig. 30a–c. Temperature dependence of the second-order Raman spectrum of diamond; a) for $Z'(Y'X')Y'$ scattering in the range 2050–2770 cm^{-1}; b) in the range 2050–2770 cm^{-1} for $Z'(X'Z')Y'$ scattering; c) for $Y'(Z'Z')X'$ scattering in the range 2050–2770 cm^{-1}. The spectra were recorded with the 4880 Å radiation of the Ar^+ laser. Also see Table 40b (p. 408)

O. Some aspects of the optical properties of crystals with broken symmetry: Point imperfections and external stresses

147. Introduction. We shall conclude the work of this article with some very brief discussion of a topic of increasing interest: broken symmetry. The entire work of the article to this point has been concerned with the analysis of maximal crystal symmetry, that is: the crystal space-time group $\mathscr{G}$ containing the space group $\mathfrak{G}$ as a subgroup, and the consequences of such symmetry for the lattice dynamics and related optical effects of crystals. Symmetry breaking can come about in a variety of fashions. For example impurities or imperfections of varying complexity can be incorporated into the crystal and external generalized stresses can be applied. The total system can now be taken to possess a lowered symmetry which is defined as the resultant symmetry present. In favorable cases sufficient symmetry remains in the composite system to enable interesting and symmetry-related effects to be analysed. Then, the symmetry group of the composite or imperfect system will be a non-trivial subgroup $\mathscr{G}^s$ or $\mathfrak{G}^s$ of $\mathscr{G}$ and $\mathfrak{G}$. A key theme to be examined in such favorable cases is the relationship between those properties originally classified under the groups $\mathscr{G}$ or $\mathfrak{G}$ of the perfect crystal, and the same properties which can now be classified under the subgroups $\mathscr{G}^s$ and $\mathfrak{G}^s$.

A thorough treatment of systems with broken symmetry would be at least as extensive as that given in Sect. 1–146 in the case of maximal space-time crystal symmetry. In particular in such a treatment, every aspect of the group theory, lattice dynamics, and optical properties of the imperfect crystal would need to be examined, and if possible correlated with its counterpart in the perfect crystal, and it would be found that in favorable circumstances (for example isolated point defects; uniform homogeneous stress) that the imperfect crystal mimics the perfect counterpart. But such a complete treatment is beyond the scope of the present article.

We shall confine ourselves here to giving a brief review of certain selected aspects of the theory of the imperfect crystal, with particular attention to those aspects which can be directly related to our previous work on the perfect crystal. The reader wishing to go deeper into the theory of the imperfect crystal is advised to consult the ongoing literature for recent work in this active area. In particular there have been a number of recent conferences devoted in part [*33*, *45*, *48*] or in a whole [*34*, *50*] to the subject of the physics of imperfect crystals, and in particular their optical properties.

In Sects. 148–153 we will confine ourselves to the simplest situation of a crystal with an isolated point defect substitutionally located at a host lattice site. In most of that work it is assumed that the only distinguishing characteristic of the defect is its mass which differs from the host atom it replaces. Then in Sect. 148 the symmetry group of the system with defect is defined: it is a point site group given in Sect. 60. In Sect. 149, 150 correlations are established between the perfect crystal phonons and the band phonons of the imperfect crystal; local-type modes are also introduced. In Sect. 151 a precis of the lattice dynamics of a crystal with mass defect (e.g. an isotope) is given and it is pointed out very briefly how symmetry may permit simplification (factorization) of the dynamical matrix as in the case of the perfect crystal. Sect. 152–153 review elements of the theory of infra-red absorption and Raman scattering, respectively again with emphasis on selection rules and symmetry relationship. Sect. 154 closes the article by discussing how symmetry can be broken by external agencies: e.g. generalized stress. Of perhaps greatest interest is the fashion by which symmetry breaking, via defects or external stresses, permits the observation of processes which are normally forbidden, or silent, in the perfect crystal and in this way the broken symmetry can be a powerful lever to gain information even about the perfect crystal.

On this theme of broken symmetry the article ends, save for a recapitulation in Part P.

148. Symmetry group of the imperfect crystal with a point defect. When an imperfection is present in a crystal, the symmetry of the resultant system is lowered. In particular if the imperfection is a mass different from the atom it replaces, we have a substitutional mass defect point imperfection.

We consider this case of a single, isolated, substitutional point imperfection in an otherwise perfect crystal. For simplicity we idealize the situation so that the isolated imperfection is a structureless mass (say an isotope) replacing the host atom without alteration of any force constant in the lattice dynamical problem. In the following sections we shall study some aspects of the dynamical problem of the mass defect: e.g. the perturbed eigenfrequencies, the infra-red absorption and Raman scattering from the perturbed crystal, and related optical properties. But here we examine the symmetry of the system.

Introduction of the impurity clearly destroys the complete translational symmetry of the crystal. No longer is the appropriate group to describe the symmetry the pure space group $\mathfrak{G}$ or space-time group $\mathscr{G}$. Rather, the group of symmetry transformations is that group of operations performed about the defect site, which leave the site invariant and replicates the remainder of the crystal. Thus, the invariance group of a crystal with imperfection at site $\boldsymbol{r}\begin{pmatrix} l \\ \kappa \end{pmatrix}$ is the site group

$$\mathfrak{P}\begin{pmatrix} l \\ \kappa \end{pmatrix}. \tag{148.1}$$

Here $\mathfrak{P}\begin{pmatrix} l \\ \kappa \end{pmatrix}$ is that set of point group operations (pure rotations)

$$\{\varepsilon|0\}, \ldots, \{\boldsymbol{\varphi}_{h_s}|0\} \equiv \mathfrak{P}\begin{pmatrix} l \\ \kappa \end{pmatrix} \tag{148.2}$$

which when performed about $\boldsymbol{r}\begin{pmatrix} l \\ \kappa \end{pmatrix}$ as origin, leave the crystal invariant. Of course the set (148.2) also occurs as a subgroup in $\mathfrak{G}$, and $\mathscr{G}$.

It follows then that all representations now entering the theory are to be representations of the group $\mathfrak{P}\begin{pmatrix} l \\ \kappa \end{pmatrix}$. The site group $\mathfrak{P}\begin{pmatrix} l \\ \kappa \end{pmatrix}$ is generally distinct from the factor group $\mathfrak{P} \equiv \mathfrak{G}/\mathfrak{T}$ since the factor group contains all rotations in $\mathfrak{G}$, while the site group contains only those rotations which do in fact leave the site invariant and replicate the crystal.[1]

As an important example, the site group of either atom in the basis in diamond space group is the tetrahedral point group T_d, while the factor group of diamond is O_h the full cubic point group. In rocksalt, both atoms in the basis have site group O_h. It would be trivial for the reader to verify these assertions by examining the effect of each coset representative in Table 2 applied at the origin. More instructive would be to determine the site group of other points in the unit cell of each group which may then be compared to the results given in [9].

149. "Band" phonons in imperfect diamond and rocksalt crystal. In this and the following section we examine aspects of the symmetry of phonons in the perturbed lattice with point imperfections. The idealized situation which we treat is one in which the perfect crystal symmetry is broken by the introduction of a single, isolated mass defect substituting for a host atom, or ion. In these two sections we neglect dynamical questions involved in the calculation of actual eigenfrequencies and eigenvectors, which will be discussed in Sect. 151. We shall anticipate one result from that analysis however: that the perturbed crystal with mass defect possesses two "types" of phonons: "band" phonons which are only slightly perturbed from their properties in the perfect crystal (small frequency shifts); and "local" phonons or "resonance" modes which have no counterpart in the perfect crystal since their displacement pattern is generally localized around the defect.

The phonons in the perfect lattice are characterized by their transformation under the space group $\mathfrak{G}$, neglecting time-reversal symmetry. They are bases for irreducible representations $D^{(\star \boldsymbol{k})(j)}$ of $\mathfrak{G}$. Now let the crystal symmetry group be reduced by placing an imperfection at $\boldsymbol{r}\begin{pmatrix} l \\ \kappa \end{pmatrix}$. Assume that "essentially" the band phonons are unchanged: i.e. eigenvectors and eigenfrequencies are just as in the unperturbed crystal. Then the symmetry group of the crystal has been changed from

$$\mathfrak{G} \to \mathfrak{P}\begin{pmatrix} l \\ \kappa \end{pmatrix}. \tag{149.1}$$

The same phonons which previously spanned $D^{(\star \boldsymbol{k})(j)}$ of $\mathfrak{G}$ now need to be classified under the irreducible representations $D^{(m)}$ of the site group $\mathfrak{P}\begin{pmatrix} l \\ \kappa \end{pmatrix}$. But this is merely the subduction problem adapted to the present case. That is

$$D^{(\star \boldsymbol{k})(j)} \text{ of } \mathfrak{G} \downarrow D^{(m)} \text{ of } \mathfrak{P}\begin{pmatrix} l \\ \kappa \end{pmatrix}. \tag{149.2}$$

To carry out the subduction (149.2) we need to employ the character tables of full group irreducible representations of $\mathfrak{G}$, and those of $\mathfrak{P}\begin{pmatrix} l \\ \kappa \end{pmatrix}$. Then using the

[1] H. WINSTON and R. H. HALFORD: J. Chem. Phys. **17**, 607 (1950) (Site group).

Table 53. Space-group reduction coefficients for the diamond lattice[a]

	P_1	P_2	P_3	P_4	P_5
$\Gamma^{(1+)}$	1				
$\Gamma^{(2+)}$		1			
$\Gamma^{(12+)}$			1		
$\Gamma^{(15+)}$					1
$\Gamma^{(25+)}$				1	
$\Gamma^{(1-)}$		1			
$\Gamma^{(2-)}$	1				
$\Gamma^{(12-)}$			1		
$\Gamma^{(15-)}$				1	
$\Gamma^{(25-)}$					1
$\star X^{(1)}$	1		1	1	
$\star X^{(2)}$		1	1		1
$\star X^{(3)}$				1	1
$\star X^{(4)}$				1	1
$\star L^{(1+)}$	1			1	
$\star L^{(2+)}$		1			1
$\star L^{(3+)}$			1	1	1
$\star L^{(1-)}$		1			1
$\star L^{(2-)}$	1			1	
$\star L^{(3-)}$			1	1	1
$\star W^{(1)}$	1		1	1	2
$\star W^{(2)}$		1	1	2	1
$\star \Delta^{(1)}$	1		1	1	
$\star \Delta^{(2)}$		1	1		1
$\star \Delta^{(2')}$	1		1	1	
$\star \Delta^{(1')}$		1	1		1
$\star \Delta^{(5)}$				2	2
$\star \Lambda^{(1)}$	2			2	
$\star \Lambda^{(2)}$		2			2
$\star \Lambda^{(3)}$			2	2	2
$\star \Sigma^{(1)}$	1		1	2	1
$\star \Sigma^{(2)}$		1	1	1	2
$\star \Sigma^{(3)}$	1		1	2	1
$\star \Sigma^{(4)}$		1	1	1	2
$\star Z$	1	1	2	3	3
$\star Q^{(1)}$	1	1	2	3	3
$\star Q^{(2)}$	1	1	2	3	3
$\star \boldsymbol{k}$	2	2	4	6	6

[a] From R. LOUDON: Proc. Phys. Soc. (London) **84**, 381 (1964).

familiar formula for reduction of a reducible representation (17.9) adapted to the present situation we have:

$$a_m = 1/g_p \sum_{R \in \mathfrak{P}} \chi^{(\star\boldsymbol{k})(j)}(R)\, \chi^{(m)}(R)^* \tag{149.3}$$

which gives the number of times the m-th irreducible representation of $\mathfrak{P}\binom{l}{\kappa}$ appears in the subduced representation. In (149.3) the sum on the right hand side

Table 54. Space-group reduction coefficients for the face-centred cubic lattice[a]

	$\Gamma^{(1+)}$	$\Gamma^{(2+)}$	$\Gamma^{(12+)}$	$\Gamma^{(15+)}$	$\Gamma^{(25+)}$	$\Gamma^{(1-)}$	$\Gamma^{(2-)}$	$\Gamma^{(12-)}$	$\Gamma^{(15-)}$	$\Gamma^{(25-)}$
$\star X^{(1+)}$	1		1							
$\star X^{(2+)}$		1	1							
$\star X^{(3+)}$					1					
$\star X^{(4+)}$				1						
$\star X^{(5+)}$				1	1					
$\star X^{(1-)}$						1		1		
$\star X^{(2-)}$							1	1		
$\star X^{(3-)}$										1
$\star X^{(4-)}$									1	
$\star X^{(5-)}$									1	1
$\star L^{(1+)}$	1				1					
$\star L^{(2+)}$		1		1						
$\star L^{(3+)}$			1	1	1					
$\star L^{(1-)}$						1				1
$\star L^{(2-)}$							1		1	
$\star L^{(3-)}$								1	1	1
$\star W^{(1)}$	1		1							1
$\star W^{(1')}$					1	1		1		
$\star W^{(2)}$				1			1	1		
$\star W^{(2')}$		1	1						1	
$\star W^{(3)}$				1	1				1	1
$\star \Delta^{(1)}$	1		1						1	
$\star \Delta^{(2)}$		1	1							1
$\star \Delta^{(2')}$					1		1	1		
$\star \Delta^{(1')}$				1		1		1		
$\star \Delta^{(5)}$				1	1				1	1
$\star \Lambda^{(1)}$	1				1		1		1	
$\star \Lambda^{(2)}$		1		1		1				1
$\star \Lambda^{(3)}$			1	1	1			1	1	1
$\star \Sigma^{(1)}$	1		1		1				1	1
$\star \Sigma^{(2)}$				1	1	1		1		1
$\star \Sigma^{(3)}$				1	1		1	1	1	
$\star \Sigma^{(4)}$		1	1	1					1	1
$\star Z^{(1)}$	1	1	2						1	1
$\star Z^{(2)}$				1	1	1	1	2		
$\star Z^{(3)}$				1	1				1	1
$\star Z^{(4)}$				1	1				1	1
$\star Q^{(1)}$	1		1	1	2	1		1	1	2
$\star Q^{(2)}$		1	1	2	1		1	1	2	1
$\star \boldsymbol{k}$	1	1	2	3	3	1	1	2	3	3

[a] From R. LOUDON: Proc. Phys. Soc. (London) **84**, 381 (1964).

should be restricted to elements R in $\mathfrak{P}\binom{l}{\kappa}$, the characters $\chi^{(\star\boldsymbol{k})(j)}$ and $\chi^{(m)}$ are appropriate to the groups $\mathfrak{G}$ and $\mathfrak{P}\binom{l}{\kappa}$ respectively.

Now we can straightforwardly illustrate this procedure on the two space groups of our interest: diamond and rocksalt. The site groups $\mathfrak{P}\binom{l}{\kappa}$ are known

Table 55. Subduction of $D^{(*X)(1)}$ of O_h^7 onto T_d

Class	E	$8C_3$	$3C_2$	$6\sigma_d$	$6S_4$
$\chi^{(*X)(1)}$	6	0	2	2	0
	$=P_1+P_3+P_4$				

Table 56. Local modes in imperfect diamond site group T_d

Represent.	E	$8C_3$	$3C_2$	$6\sigma_d$	$6S_4$
$P_1=A_1$	1	1	1	1	1
$P_2=A_2$	1	1	1	-1	-1
$P_3=E$	2	-1	2	0	0
$P_5=F_1$	3	0	-1	-1	1
$P_4=F_2(x_1\, y_1\, z)$	3	0	-1	1	-1
$\chi^{(\Delta)}_{\text{n.n.}}$	15	0	-1	3	$-1=A_1+E+F_1+3F_2$
$\chi^{(\Delta)}_{\text{n.n.n.}}$	33	0	-3	7	$-3=2A_1+2E+F_1+7F_2$

for diamond and rocksalt space groups to be T_d and O_h respectively. In Tables 53 and 54 we give the results of the subduction analysis following LOUDON.[1]

As an example of how to obtain the subduced representation consider the case of the longitudinal mode at the [*100*] zone edge in diamond structure. This mode, in full group notation is $D^{(*X)(1)}$. For the full group characters of $D^{(*X)(1)}$ see Table B 1. In order to carry out the subduction as given in Table 53, and also in Eq. (149.3) we require the characters $\chi^{(*X)(1)}(R)$ of the elements R common to both $\mathfrak{G}$ and $\mathfrak{P}\begin{pmatrix} l \\ \kappa \end{pmatrix}$. In the present case, these are groups O_h^7 and T_d respectively. For T_d, the elements are simply the point group rotations. Thus we need to read off from Table B 1 the appropriate characters. In Table 55 this is given. Note the definition of the labelling of the irreducible representations of T_d as given in Table 53 and Table 56. It then follows immediately that the subduction is

$$D^{(*X)(1)} \quad \text{of} \quad O_h^7 \downarrow P_1 \oplus P_3 \oplus P_4 \quad \text{of} \quad T_d. \tag{149.4}$$

The notation given in (149.4) is explained in Table 56 where the usual molecular notation is also given. It is then clear in any case how to carry out the subduction and the results of Table 53 are easily obtained. In general, for any space group one always requires the full group irreducible representation in order to perform the subduction.

150. Local phonons in imperfect diamond and rocksalt crystals. In the case where a local mode exists[1,2] in a crystal with an isolated structureless mass

[1] R. LOUDON: Proc. Phys. Soc. (London) **84**, 379 (1964).

[1] A.A. MARADUDIN: Theoretical and experimental aspects of the effects of point defects and disorder as the vibrations of crystals, in: Solid State Physics (eds. F. SEITZ and D. TURNBULL), Part 1, Vol. 18, Part 2, Vol. 19. New York: Academic Press 1966.

[2] I.M. LIFSCHITZ: Soviet Phys. Usp. **7**, 549 (1965).

defect we have to do with spatially localized displacements attached principally to the impurity which coexist with the band modes; the latter are slightly perturbed perfect crystal modes. The total number of modes of the system is of course conserved at $3rN$. Attaching the perturbed modes to the impurity means treating a subset of γ modes separately from the (essentially) unperturbed $(3rN-\gamma)$ band modes. If the imperfection is a simple mass defect without any change in its force constant to be other atoms or ions in the crystal, we may idealize the matter as if we deal with a molecule consisting of the impurity plus its immediate neighbors. In the present section we wish to examine the symmetry of such localized modes.

To be specific consider an isolated substitutional mass defect in diamond structure. We may define an imperfection cluster, or molecule, as the substitutional defects and its 4 first neighbors. If we consider each atom to be given a virtual cartesian displacement, we can obtain the mechanical or displacement representation in the usual fashion. In Table 56 this is shown in terms of the character system for the site group of diamond, i.e. the group T_d with operations performed around the impurity as origin. In obtaining the characters $\chi^{(\Delta)}_{\text{n.n.}}$ only the 5 atoms mentioned were considered.

Observe that in $\chi^{(\Delta)}_{\text{n.n.}}$, the representation P_4 or F_2 occurs. Since the origin is taken at the defect, *any* motion involving the defect must be F_2, since that is the representation by which the cartesian displacements $(\Delta x, \Delta y, \Delta z)$ of the defect transform i.e. a polar vector. Thus the local mode in which the imperfection moves must always be of symmetry F_2 when indexed using the site group. As another approximation consider that the imperfection interacts with 4 nearest, plus 6 nearest atoms. Now the cluster consists of 11 atoms or 33 degrees of freedom.

Table 57[a]. The (un-normalized) eigenvectors $\{\psi^{(\sigma i)}(l\kappa)\}$ for a substitutional impurity possessing T_d symmetry in a crystal possessing the same symmetry about the impurity site*

$\psi(l\kappa)$; $(l\kappa)=$	A_1	$E^{(1)}$	$E^{(2)}$
0 0 0	0 0 0	0 0 0	0 0 0
1 1 1	α α α	α $-\alpha$ 0	α α -2α
$\bar{1}$ $\bar{1}$ 1	$-\alpha$ $-\alpha$ α	$-\alpha$ α 0	$-\alpha$ $-\alpha$ -2α
1 $\bar{1}$ $\bar{1}$	α $-\alpha$ $-\alpha$	α α 0	α α 2α
$\bar{1}$ 1 $\bar{1}$	$-\alpha$ α $-\alpha$	$-\alpha$ $-\alpha$ 0	$-\alpha$ α 2α

$\psi(l\kappa)$; $(l\kappa)=$	$F_1^{(1)}$	$F_1^{(2)}$	$F_1^{(3)}$	$F_2^{(1)}$	$F_2^{(2)}$	$F_2^{(3)}$
0 0 0	0 0 0	0 0 0	0 0 0	x_0 0 0	0 x_0 0	0 0 x_0
1 1 1	α $-\alpha$ 0	α 0 $-\alpha$	0 α $-\alpha$	α β β	β α β	β β α
$\bar{1}$ $\bar{1}$ 1	$-\alpha$ α 0	α 0 α	0 α α	α β $-\beta$	β α $-\beta$	$-\beta$ $-\beta$ α
1 $\bar{1}$ $\bar{1}$	$-\alpha$ $-\alpha$ 0	$-\alpha$ 0 $-\alpha$	0 $-\alpha$ α	α $-\beta$ $-\beta$	$-\beta$ α β	$-\beta$ β α
$\bar{1}$ 1 $\bar{1}$	α α 0	$-\alpha$ 0 α	0 $-\alpha$ $-\alpha$	α $-\beta$ β	$-\beta$ α $-\beta$	β $-\beta$ α

[a] From A. A. MARADUDIN: Reports. Prog. Phys. **28**, 378 (1969), Table 2.

* It is assumed that the coordination number of the impurity site is 4, and that the impurity interacts with no more than its four nearest neighbours. The components of the position vectors $\{\boldsymbol{x}(l\kappa)\}$ of the nearest neighbours to the impurity are given in units of one quarter of the lattice parameter for the diamond structure.

The character system is shown in Table 56 as $\chi^{(\Delta)}_{\text{n.n.n.}}$. In this case F_2 occurs seven times. Clearly, each of the F_2 modes involves coupled motion of both the impurity with F_2 symmetry and the neighbors displacements also with the symmetry F_2. We expect that in one of these F_2 modes the largest amplitude of motion will be the impurity, the smallest will be the motion of neighbors and conversely for the other F_2 modes, but this would require solution of the dynamical problem.

It is easy to understand the significance of the remaining modes in this cluster picture. In those (non-F_2) modes the impurity is at rest, and they must be band modes classified according to the site group T_d of $\mathfrak{P}\begin{pmatrix} l \\ \kappa \end{pmatrix}$. If the dynamical problem included more and more neighbors the "molecule" would converge more closely toward the actual imperfect crystal and one of the F_2 modes would correspond to the impurity vibration eigenfunction and energy (local or resonance) while all other modes would asymptotically go over to the band modes essentially of the perfect crystal.

An example of the displacements in different symmetry modes is shown in Table 57 taken from the work of Dettmann and Ludwig[3] who determined eigenvectors for a "near-neighbor" model of imperfect diamond. Note that in the F_2 mode the imperfection is displaced while in the other it is stationary.

Table 58. Local modes in imperfect rocksalt site group O_h

	E	$8C_3$	$3C_4^2$	$6C_2$	$6C_4$	i	$8S_6$	$3\sigma_h$	$6\sigma_d$	$6S_4$
$\chi^{(\Delta)}_{\text{n.n.}}$	21	0	-3	-1	3	-3	0	5	3	-1
	$=\Gamma^{(1+)}\oplus\Gamma^{(12+)}\oplus\Gamma^{(25+)}\oplus 3\Gamma^{(15-)}\oplus\Gamma^{(25-)}$									

For rocksalt the imperfection plus its first neighbors are a 7 atomic "octahedral" molecule. These generate a 21 dimensional mechanical representation. The character system is given in Table 58 and reduced there. As in the case of diamond a first neighbor-force-constant calculation shows that only in the modes $\Gamma^{(15-)}$ is there any impurity atom motion. Again it follows that all other modes are part of the band continuum and correspond to the band mode subduced (or reclassified) under the site group $\mathfrak{P}\begin{pmatrix} l \\ \kappa \end{pmatrix}$.

151. Dynamical aspects of perturbed crystal vibrations. The lattice dynamics of perfect crystals was analysed in Sects. 67–86. A brief review of the needed equations will be given in (151.1)–(151.9). Recall the equations of motion (67.19):

$$M_\kappa(l)\,\ddot{u}_\alpha\begin{pmatrix} l \\ \kappa \end{pmatrix} + \sum_{l'\kappa'\beta} \Phi_{\alpha\beta}\begin{pmatrix} l & l' \\ \kappa & \kappa' \end{pmatrix} u_\beta\begin{pmatrix} l' \\ \kappa' \end{pmatrix} = 0 \qquad (151.1)$$

[3] W. Ludwig, in: Ergeb. der exakten Naturwissenschaften (ed. S. Flügge and F. Trendelenburg), Bd. 35. Berlin-Heidelberg-New York: Springer 1964.

where $M_\kappa(l)$ is the mass at site κ in each cell, and is independent of l, the $u_\alpha\begin{pmatrix} l \\ \kappa \end{pmatrix}$ are the elementary cartesian components of displacements, and the $\Phi_{\alpha\beta}\begin{pmatrix} l & l' \\ \kappa & \kappa' \end{pmatrix}$ the force matrix of the entire crystal. The eigenvectors of (151.1) are obtained by diagonalizing the matrix (72.6)

$$\left(D_{\alpha\beta}\begin{pmatrix} l & l' \\ \kappa & \kappa' \end{pmatrix} - \omega^2\, \delta_{\alpha\beta}\, \delta_{ll'}\, \delta_{\kappa\kappa'}\right) \tag{151.2}$$

where

$$D_{\alpha\beta}\begin{pmatrix} l & l' \\ \kappa & \kappa' \end{pmatrix} \equiv [M_\kappa\, M_{\kappa'}]^{-\frac{1}{2}}\, \Phi_{\alpha\beta}\begin{pmatrix} l & l' \\ \kappa & \kappa' \end{pmatrix}. \tag{151.3}$$

Owing to the complete spatial symmetry of the force matrix $\boldsymbol{\Phi}$ or $\boldsymbol{D}$, under all elements in the space group $\mathfrak{G}$, it proved useful to Fourier transform (151.3) to obtain the dynamical matrix (78.5)

$$D_{\alpha\beta}\begin{pmatrix} \boldsymbol{k} \\ \kappa\;\; \kappa' \end{pmatrix} \equiv \sum_{ll'} \Phi_{\alpha\beta}\begin{pmatrix} l & l' \\ \kappa & \kappa' \end{pmatrix} \frac{\exp -i\,\boldsymbol{k}\cdot(\boldsymbol{R}_L - \boldsymbol{R}_{L'})}{[M_\kappa\, M_{\kappa'}]^{\frac{1}{2}}}. \tag{151.4}$$

The eigenvectors of the dynamical matrix (151.4) are obtained as in (79.4) by diagonalizing the matrix

$$\left(D_{\alpha\beta}\begin{pmatrix} \boldsymbol{k} \\ \kappa\;\; \kappa' \end{pmatrix} - \omega^2(\boldsymbol{k}\,|\,j_\mu)\, \delta_{\alpha\beta}\, \delta_{\kappa\kappa'}\right). \tag{151.5}$$

These eigenvectors were the vectors $\boldsymbol{e}_\alpha\left(\kappa\,\middle|\,\begin{matrix} \boldsymbol{k} \\ j_\mu \end{matrix}\right)$, chosen to be orthonormalized as in (79.8)–(79.11):

$$\sum_{\kappa\alpha} e_\alpha^*\left(\kappa\,\middle|\,\begin{matrix} \boldsymbol{k} \\ j_\mu \end{matrix}\right) e_\alpha\left(\kappa\,\middle|\,\begin{matrix} \boldsymbol{k}' \\ j'_{\nu'} \end{matrix}\right) = \delta_{jj'}\, \delta_{\mu\nu}\, \delta_{\boldsymbol{k}\boldsymbol{k}'} \tag{151.6}$$

and

$$\sum_{j\mu} e_\beta^*\left(\kappa'\,\middle|\,\begin{matrix} \boldsymbol{k} \\ j_\mu \end{matrix}\right) e_\alpha\left(\kappa\,\middle|\,\begin{matrix} \boldsymbol{k} \\ j_\mu \end{matrix}\right) = \delta_{\alpha\beta}\, \delta_{\kappa\kappa'}. \tag{151.7}$$

The normal coordinates for this problem are given in terms of these eigenvectors by (80.9), (80.10) as:

$$Q\begin{pmatrix} \boldsymbol{k} \\ j_\mu \end{pmatrix} = \frac{1}{\sqrt{N}} \sum_l \sum_{\kappa\alpha} \mathrm{e}^{-i\boldsymbol{k}\cdot\boldsymbol{R}_L}\, e_\alpha^*\left(\kappa\,\middle|\,\begin{matrix} \boldsymbol{k} \\ j_\mu \end{matrix}\right) u_\alpha\begin{pmatrix} l \\ \kappa \end{pmatrix} \sqrt{M_\kappa}. \tag{151.8}$$

The harmonic time dependence is implicit in (151.8). Neglecting the time reversal operation, the $Q\begin{pmatrix} \boldsymbol{k} \\ j_\mu \end{pmatrix}$ are bases for irreducible representation of $\mathfrak{G}$. The inverse of (151.8) is as in (80.9)

$$u_\alpha\begin{pmatrix} l \\ \kappa \end{pmatrix} = \frac{1}{\sqrt{M_\kappa N}} \sum_{j\mu} \sum_{\boldsymbol{k}} \mathrm{e}^{i\boldsymbol{k}\cdot\boldsymbol{R}_L}\, e_\alpha\left(\kappa\,\middle|\,\begin{matrix} \boldsymbol{k} \\ j_\mu \end{matrix}\right) Q\begin{pmatrix} \boldsymbol{k} \\ j_\mu \end{pmatrix}. \tag{151.9}$$

In the imperfect lattice both masses and force constants may be changed from their perfect lattice counterparts. In the harmonic approximation the forms of (151.1) will be retained. However, let us take the imperfection mass as $M'_\kappa(l)$ and the changed force constants as $\Theta_{\alpha\beta}\begin{pmatrix} l & l' \\ \kappa & \kappa' \end{pmatrix}$. The equations of motion are now instead of (151.1):

$$M'_\kappa(l)\,\ddot{u}_\alpha\begin{pmatrix} l \\ \kappa \end{pmatrix} + \sum_{l'\kappa'\beta} \Theta_{\alpha\beta}\begin{pmatrix} l & l' \\ \kappa & \kappa' \end{pmatrix} u_\beta\begin{pmatrix} l' \\ \kappa' \end{pmatrix} = 0. \tag{151.10}$$

By hypothesis the problem (151.10) has now the lowered symmetry $\mathfrak{P}\begin{pmatrix} l \\ \kappa \end{pmatrix}$ of the site group. From the discussion in Sects. 72–76 the eigenvectors of (151.10) can be taken to transform as bases for irreducible representations now of the point site group $\mathfrak{P}\begin{pmatrix} l \\ \kappa \end{pmatrix}$. The translational symmetry is lost, as well as any rotational symmetry not in the site group. A direct general solution of (151.10) could be attempted but in general this is not convenient.

A connection can now be made with the work of Sect. 150. We seek the solution of (151.10) only for the displacements $\boldsymbol{u}\begin{pmatrix} l \\ \kappa \end{pmatrix}$, of the imperfection at site $\boldsymbol{r}\begin{pmatrix} l \\ \kappa \end{pmatrix}$. As a radical assumption we first suppose that the imperfection is only coupled to nearest neighbors all other force constants vanishing. Then (151.10) can be truncated and becomes

$$M'_\kappa(l)\,\ddot{u}_\alpha\begin{pmatrix} l \\ \kappa \end{pmatrix} + \sum_{\substack{l'\kappa'\beta \\ \text{n.n.}}} \Theta_{\alpha\beta}\begin{pmatrix} l & l' \\ \kappa & \kappa' \end{pmatrix} u_\beta\begin{pmatrix} l & l' \\ \kappa & \kappa' \end{pmatrix}$$

and

$$(l, \kappa) = (\text{set of imperfection} + \text{n.n.}). \tag{151.11}$$

The set of Eq. (151.11) then can be solved in the "molecular" approximation for the displacements of the imperfection plus coupled near neighbors. Recall from Sect. 150 that the local mode solution which is sought has overall symmetry

$$Q_j \sim \boldsymbol{\Gamma}^{(v)} \tag{151.12}$$

i.e. it transforms as a vector under transformations in $\mathfrak{P}\begin{pmatrix} l \\ \kappa \end{pmatrix}$. All modes except the Q_j of (151.12) in which the imperfection amplitude of displacement is largest are to be assigned as band modes, following the previous discussion.

Aside from the molecular approximation, solutions to (151.11) can be obtained by a Green function method[1] which lends itself to perturbation theory.[2] Thus, define a matrix

$$C_{\alpha\beta}\begin{pmatrix} l & l' \\ \kappa & \kappa' \end{pmatrix} \equiv \varepsilon_\kappa(l)\,\omega^2\,\delta_{\alpha\beta}\,\delta_{ll'}\,\delta_{\kappa\kappa'} + \Theta_{\alpha\beta}\begin{pmatrix} l & l' \\ \kappa & \kappa' \end{pmatrix} - \Phi_{\alpha\beta}\begin{pmatrix} l & l' \\ \kappa & \kappa' \end{pmatrix} \tag{151.13}$$

---

[1] I.M. Lifschitz: Soviet Phys. Usp. **7**, 549 (1965).

[2] A.G. Dawber and R.J. Elliott: Proc. Roy. Soc. (London), Ser. A **273**, 222 (1963).

with an auxiliary variable $\varepsilon_\kappa(l)$:

$$\varepsilon_\kappa(l) \equiv (M_\kappa(l) - M'_\kappa(l))/M_\kappa(l). \tag{151.14}$$

Then the equation of motion of the perturbed lattice (151.10) becomes

$$M_\kappa(l)\,\ddot{u}_\alpha\begin{pmatrix} l \\ \kappa \end{pmatrix} + \sum_{l'\kappa'\beta} \Phi_{\alpha\beta}\begin{pmatrix} l & l' \\ \kappa & \kappa' \end{pmatrix} u_\beta\begin{pmatrix} l' \\ \kappa' \end{pmatrix} = \sum_{l'\kappa'\beta} C_{\alpha\beta}\begin{pmatrix} l & l' \\ \kappa & \kappa' \end{pmatrix} u_\beta\begin{pmatrix} l' \\ \kappa' \end{pmatrix}. \tag{151.15}$$

The system (151.15) is an inhomogeneous system of equations, the homogeneous counterpart of which (for the perfect lattice) is (151.1). Since (151.15) still describes a harmonic quadratic problem, a transformation to normal coordinates and normal modes exists. Then we write

$$u_\alpha\begin{pmatrix} l \\ \kappa \end{pmatrix} = \sum_j w_\alpha\left(j\middle|\begin{matrix} l \\ \kappa \end{matrix}\right) q_j \tag{151.16}$$

or

$$q_j = \sum_{l\kappa\alpha} u_\alpha\begin{pmatrix} l \\ \kappa \end{pmatrix} w_\alpha\left(j\middle|\begin{matrix} l \\ \kappa \end{matrix}\right). \tag{151.17}$$

To find the q_j we must solve (151.15) or the equivalent secular equation:

$$\begin{aligned} -\omega^2 M_\kappa\, w_\alpha\left(j\middle|\begin{matrix} l \\ \kappa \end{matrix}\right) + \sum_{l'\kappa'\beta} \Phi_{\alpha\beta}\begin{pmatrix} l & l' \\ \kappa & \kappa' \end{pmatrix} w_\beta\left(j\middle|\begin{matrix} l' \\ \kappa' \end{matrix}\right) \\ = \sum_{l'\kappa'\beta} C_{\alpha\beta}\begin{pmatrix} l & l' \\ \kappa & \kappa' \end{pmatrix} w_\beta\left(j\middle|\begin{matrix} l' \\ \kappa' \end{matrix}\right). \end{aligned} \tag{151.18}$$

Define the Green function, or inverse matrix to the homogeneous problem (151.2) as the object $G_{\alpha\beta}\left(\begin{matrix} l & l' \\ \kappa & \kappa' \end{matrix}\middle|\omega\right)$ where

$$-\omega^2 M_\kappa G_{\alpha\gamma}\left(\begin{matrix} l & l'' \\ \kappa & \kappa'' \end{matrix}\middle|\omega\right) + \sum_{l'\kappa'\beta} \Phi_{\alpha\beta}\begin{pmatrix} l & l' \\ \kappa & \kappa' \end{pmatrix} G_{\beta\gamma}\left(\begin{matrix} l' & l'' \\ \kappa' & \kappa'' \end{matrix}\middle|\omega\right) = \delta_{\alpha\gamma}\delta_{ll''}\delta_{\kappa\kappa''}. \tag{151.19}$$

The Green matrix in (151.19) can be found from the eigensolutions of the unperturbed problem. It is:

$$G_{\alpha\beta}\left(\begin{matrix} l & l'' \\ \kappa & \kappa'' \end{matrix}\middle|\omega\right) = \sum_{j\mu\boldsymbol{k}} \left(\frac{e_\alpha^*\left(\kappa\middle|\begin{matrix} \boldsymbol{k} \\ j_\mu \end{matrix}\right) e_\beta\left(\kappa''\middle|\begin{matrix} \boldsymbol{k} \\ j_\mu \end{matrix}\right) e^{i\boldsymbol{k}\cdot(\boldsymbol{R}_L - \boldsymbol{R}_{L''})}}{N M_\kappa^{\frac{1}{2}} M_{\kappa''}^{\frac{1}{2}} [\omega^2(k|j) - \omega^2]} \right). \tag{151.20}$$

If the Green matrix satisfies (151.19) then we can write an "integral equation" for (151.18) as

$$w_\alpha\left(j\middle|\begin{matrix} l \\ \kappa \end{matrix}\right) = \sum_{l'\kappa'\beta} \sum_{l''\kappa''\gamma} G_{\alpha\beta}\left(\begin{matrix} l & l' \\ \kappa & \kappa' \end{matrix}\middle|\omega\right) C_{\beta\gamma}\begin{pmatrix} l' & l'' \\ \kappa' & \kappa'' \end{pmatrix} w_\gamma\left(j\middle|\begin{matrix} l' \\ \kappa' \end{matrix}\right). \tag{151.21}$$

Again this is verified by direct substitution into (151.18) as in (151.19). A perturbation theory applies to (151.21) if we take the number of cells and sites affected by the imperfection to be small (i.e. the only terms considered are the imperfection and its nearest neighbors). Clearly (151.21) can be written as

$$\sum_{\substack{l'\kappa'\beta\\ l''\kappa''\gamma}} \left(G_{\alpha\beta}\left(\begin{matrix} l & l' \\ \kappa & \kappa' \end{matrix}\middle|\,\omega\right) C_{\beta\gamma}\left(\begin{matrix} l' & l'' \\ \kappa' & \kappa'' \end{matrix}\right) - \delta_{\alpha\gamma}\,\delta_{ll'}\,\delta_{\kappa\kappa'} \right) w_\gamma\left(j\,\middle|\begin{matrix} l' \\ \kappa' \end{matrix}\right) = 0 \qquad (151.22)$$

and if the effect of the perturbation can be restricted to a small number of cells then the condition for a solution of (151.22) is

$$\left\| \sum_{l''\kappa''\beta} \left(G_{\alpha\beta}\left(\begin{matrix} l & l' \\ \kappa & \kappa' \end{matrix}\middle|\,\omega\right) C_{\beta\gamma}\left(\begin{matrix} l' & l'' \\ \kappa' & \kappa'' \end{matrix}\right) - \delta_{\alpha\gamma}\,\delta_{ll'}\,\delta_{\kappa\kappa'} \right) \right\| = 0. \qquad (151.23)$$

This is an implicit equation for ω which for the given perturbation matrix $\boldsymbol{C}$ has solutions only at the eigenfrequencies of the perturbed problem, and at those frequencies the amplitude $\boldsymbol{w}$ can be found.

The case of the mass defect without change of force constants gives for the perturbation matrix $\boldsymbol{C}$:

$$C_{\alpha\beta}\left(\begin{matrix} l & l' \\ \kappa & \kappa' \end{matrix}\right) = -M_\kappa\,\varepsilon_\kappa(l)\,\omega^2\,\delta_{ll'}\,\delta_{\kappa\kappa'}\,\delta_{\alpha\beta}. \qquad (151.24)$$

So the secular equation is then a 3×3 equation involving only x, y, z components:

$$\| M_\kappa\,\varepsilon_\kappa(l)\,\omega^2\,G_{\alpha\beta}(0|\omega) + \delta_{\alpha\beta} \| = 0. \qquad (151.25)$$

Here we took the imperfection at $\left(\begin{matrix} l \\ \kappa \end{matrix}\right) = 0$. In the case of a cubic crystal we have the possibility of an exact evaluation of the Green function (151.20)

$$G_{\alpha\beta}(0|\omega) = \frac{1}{N\sqrt{M_\kappa M_{\kappa'}}} \sum_{j\mu\boldsymbol{k}} \left(\frac{e^*_\alpha\left(0\,\middle|\begin{matrix}\boldsymbol{k}\\ j_\mu\end{matrix}\right) e_\beta\left(0\,\middle|\begin{matrix}\boldsymbol{k}\\ j_\mu\end{matrix}\right)}{\omega^2(\boldsymbol{k}|j) - \omega^2} \right). \qquad (151.26)$$

But for a cubic crystal symmetry can be applied to (151.26). Thus $G_{\alpha\beta}(0|\omega)$ is an object which transforms under rotation as a second rank symmetric tensor and has only one independent nonzero component:

$$G_{xx}(0|\omega) = G_{yy}(0|\omega) = G_{zz}(0|\omega) \qquad (151.27)$$

a result which also follows from the second orthogonality rule (151.7) upon summing over μ. Then (151.26) becomes

$$G_{\alpha\alpha}(0|\omega) = \frac{1}{N\sqrt{M_\kappa M_{\kappa'}}} \sum_{j\boldsymbol{k}} \frac{\left| e_\alpha\left(0\,\middle|\begin{matrix}\boldsymbol{k}\\ j_\mu\end{matrix}\right)\right|^2}{\omega^2(\boldsymbol{k}|j) - \omega^2}. \qquad (151.28)$$

For a monoatomic cubic crystal with equal masses (for example diamond) and where the sites are equivalent, then (151.6) gives $\left| \boldsymbol{e}_\alpha\left(0\,\middle|\begin{matrix}\boldsymbol{k}\\ j_\mu\end{matrix}\right)\right|^2 = (1/3r)$ so the Green function is immediately simplified to

$$G_{\alpha\alpha}(0|\omega) = \frac{1}{3sNM_\kappa} \sum_{\boldsymbol{k}j} \frac{1}{\omega^2(\boldsymbol{k}|j) - \omega^2}. \qquad (151.29)$$

Then the secular equation for perturbed eigenfrequencies (151.25) becomes the product of three identical diagonal terms (each of which must vanish) giving a three-fold degenerate root:

$$1+\frac{\varepsilon\omega^2}{3sN}\sum_{\boldsymbol{k}j}\frac{1}{\omega^2(\boldsymbol{k}|j)-\omega^2}=0. \tag{151.30}$$

This Eq. (151.30) gives the eigenvalues in the perturbed lattice.

Of importance also are the eigenvectors which are to be obtained from (151.21) as:

$$\begin{aligned} w_\beta\left(j\left|\begin{matrix} l' \\ \kappa' \end{matrix}\right.\right) &= \frac{1}{N}\sum_{\alpha\boldsymbol{k}j\kappa}\frac{e_\beta^*\left(\kappa'\left|\begin{matrix}\boldsymbol{k}\\ j\end{matrix}\right.\right)e_\alpha^*\left(\kappa\left|\begin{matrix}\boldsymbol{k}\\ j\end{matrix}\right.\right)}{\omega^2(\boldsymbol{k}|j)-\omega^2(j)} \\ &\quad\times\left(\varepsilon_\kappa\sqrt{\frac{M_\kappa}{M_{\kappa'}}}\right)\omega^2(j)\,\mathrm{e}^{-i\boldsymbol{k}\cdot\boldsymbol{R}_{L'}}\,w_\alpha(j|0). \end{aligned} \tag{151.31}$$

This gives the wave amplitude at site $\binom{l'}{\kappa'}$ in terms of that at the origin.

Eqs. (151.30) and (151.31) were first discussed in some detail and numerical results given when applied to the theory of the mass defect in silicon, by DAWBER and ELLIOTT. Since that early work many cases were studied, including isolated defects in crystals of lower symmetry, pairs of defects in cubic crystals, etc.[3] This important area of work can be followed by reference to the proceedings of several recent conferences [*33, 34, 45, 48, 50*].

For our purposes it suffices that this approach provides the dynamical justification for the symmetry related statements made in Sects. 149, 150 concerning identification of local modes as symmetry $\Gamma^{(v)}$ and the correlation between band modes in the imperfect crystal and host modes via the subduction argument. Generally for a light impurity with $\varepsilon<0$ in (151.14) among the eigenfrequency and eigenvector of the solution of (151.30) are localized modes while for the heavy impurity resonance modes may occur. In both cases the broken symmetry permits a quasi-continuum of modes correlated with the perfect crystal modes.

152. Infra-red-absorption in the perturbed system. The theory given in Sect. 120 for infra-red absorption by phonons in the perfect crystal can be carried over to the case of crystal with imperfections. We consider only the case of an isolated point mass defect as was discussed in Sect. 151. In that case the major work of Sect. 120 we require is the use of semi-classical radiation theory which will permit us to define a moment operator like that of (120.33), although of course, the translational symmetry (periodicity) is broken. We may then expect that the rate of infra-red absorption will be given proportional to a matrix element like (120.34). We also anticipate that local modes (if they exist) and band modes will be infra-red active if they have proper symmetry, $\Gamma^{(v)}$ to couple with the radiation field. Let us give a précis of a theory.

[3] A. A. MARADUDIN: Footnote 1, Sect. 150

We are concerned with an optical absorption process by which the crystal makes a transition from the ground vibrational state $|g\rangle$ to some excited vibrational state $|e\rangle$. The perturbation causing this transition in semi-classical approximation is

$$\mathscr{H}' = \frac{e\hbar}{mc}\,\boldsymbol{\mu}\cdot\boldsymbol{A} \simeq \frac{e\hbar}{mc}\,\mathscr{E}\,\boldsymbol{\mu}\cdot\hat{\boldsymbol{\varepsilon}}.$$

$\hat{\boldsymbol{\varepsilon}}$ is the polarization of the electromagnetic wave and $\boldsymbol{\mu}$ an appropriate dipole moment. The transition matrix element is

$$\mathscr{M} \sim \langle e|\boldsymbol{\mu}\cdot\hat{\boldsymbol{\varepsilon}}|g\rangle. \tag{152.1}$$

The operator $\boldsymbol{\mu}$ transforms as a polar vector

$$\boldsymbol{\mu} \sim \boldsymbol{\Gamma}^{(v)} \tag{152.2}$$

so that an elementary application of the Wigner-Eckart theorem leads to the prediction that *any* excited state $|e\rangle$ with symmetry $\boldsymbol{\Gamma}^{(v)}$ is accessible by a single quantum (one-phonon) absorption process from the initial (ground/state with symmetry $\boldsymbol{\Gamma}^{(1+)}$. For example in perturbed diamond and rocksalt space groups the mode of symmetry $\Gamma^{(15-)}$ is optically active in infra-red absorption. From Sects. 149 and 150 we deduce that the local mode of a substitutional mass defect is active. Also those band modes which subduce $\Gamma^{(15-)}$ are active. The band modes are part of what is effectively a continuum so that some weighting is to be expected, for example by projecting out of the continuum all states with symmetry $\boldsymbol{\Gamma}^{(v)}$.

The quantitative theory substantiates this picture. Let the mass defect be charged, while all other host atoms are unchanged. In the perturbed crystal let $\boldsymbol{u}(0)$ be the displacement vector of the displaced imperfection centered at the origin 0. Then the induced dipole moment during vibration is

$$\boldsymbol{\mu} = e\,\boldsymbol{u}(0) \tag{152.3}$$

and the transition matrix element (152.1) is

$$\mathscr{M} \sim \langle e|\hat{\boldsymbol{\varepsilon}}\cdot\boldsymbol{u}(0)|g\rangle. \tag{152.4}$$

The f value or oscillator strength for single-phonon absorption can be defined as

$$f = \frac{2M\omega}{\hbar}\,|\mathscr{M}|^2 \tag{152.5}$$

where M is the impurity mass. Now from (151.16) the displacement of the atom at the origin is

$$\boldsymbol{u}(0) = \sum_j \boldsymbol{w}(j|0)\,q_j. \tag{152.6}$$

Then it can be shown that use of the normalization condition with (152.6) gives

$$f \sim 1/3\,M \sum_\alpha (w_\alpha(j|0))^2. \tag{152.7}$$

For the local mode

$$\sum_{\alpha} |w_\alpha|^2 \sim 1/M \tag{152.8}$$

so

$$f \sim 1.$$

For the band mode each atom displacement is weighted by $|MN|^{-\frac{1}{2}}$, owing to the wave-like character of these. Then in this case

$$\sum_{\alpha} |w_\alpha|^2 \sim 1/rMN. \tag{152.9}$$

But also the band modes have a density of states

$$g(\omega) \tag{152.10}$$

as defined in Sect. 106. We may take the band mode density of states in the perturbed crystal equal to that in the unperturbed crystal. The f value in this case for absorption in an interval $\Delta\omega$ about ω would be

$$f \sim \frac{M}{3} \cdot \frac{1}{rNM} \cdot g(\omega)\,\Delta\omega, \quad \text{or} \quad \frac{f}{\Delta\omega} \sim \frac{1}{3rN}\, g(\omega). \tag{152.11}$$

An absorption coefficient can be described for both cases:

$$\textit{Local mode:} \quad K(\omega) = \frac{2\pi^2}{nc}\,\delta\, e^2\, \frac{\gamma(\omega)}{M} \tag{152.12}$$

where $\gamma(\omega)$ is some shape function for the local mode line width and

$$\textit{Band:} \quad K(\omega) \simeq \frac{2\pi^2}{3nc}\,\delta\, e^2\, \frac{1}{MN}\, g(\omega). \tag{152.13}$$

In (152.12), (152.13) n is the refractive index, δ the concentration of imperfections.

A very important point concerning local and band absorption now relates to the symmetry argument. The local mode for the case of the pure mass defect will always be one-phonon infra-red active owing to its symmetry being $\boldsymbol{\Gamma}^{(v)}$. Band modes will be infra-red active if they subduce $\boldsymbol{\Gamma}^{(v)}$. If active, the absorption coefficient is proportional to the "active" density of states $g(\omega)$. It would be more appropriate then to replace

$$g(\omega) \quad \text{by} \quad g'_v(\omega) \tag{152.14}$$

where $g'_v(\omega)$ is the density of band states with subduced vector representation $\boldsymbol{\Gamma}^{(v)}$ around the impurity site in the perturbed crystal. Hence the one-phonon absorption coefficient in the imperfect crystal with mass defect, should consist essentially of a sharp, intense local mode peak superimposed upon a continuum whose structure mirrors that of $g'_v(\omega)$, for a lighter atom imperfection; for a heavier atom a broad "in-band" peak corresponding to the resonance mode should be seen: the resonance mode will encompass band modes which subduce $\boldsymbol{\Gamma}^{(v)}$.

To illustrate some of the general features discussed above, we show on Fig. 31 the infra-red spectrum of silicon doped with boron and phosphorous.[1,2] In the

[1] M. Balkanski and W. Nazariewicz: J. Phys. Chem. Solids **25**, 437 (1964); **27**, 671 (1966).

[2] J.F. Angress, S.D. Smith, and K.F. Renk, in [32], p. 467. — J.F. Angress, A.R. Goodwin, and S.D. Smith: Proc. Roy. Soc. (London), Ser. A **287**, 64 (1965).

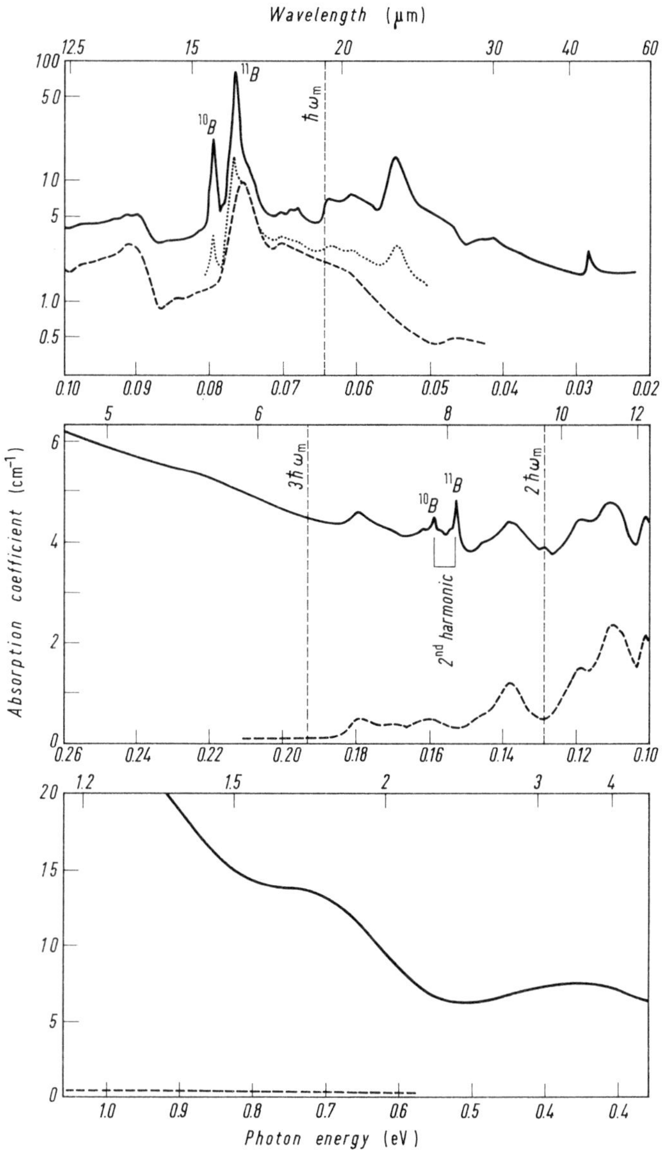

Fig. 31. The infra-red spectrum of silicon doped with both boron and phosphorus in concentrations of $5\times10^{19}\,\mathrm{cm}^{-3}$ (full curve) and $5\times10^{18}\,\mathrm{cm}^{-3}$ (dotted curve). The spectrum of pure silicon (dashed curve) is shown for comparison. [After J. F. ANGRESS, A. R. GOODWIN, and S. D. SMITH: Proc. Roy. Soc. A **287**, 64 (1965), Fig. 1a]

region below the energy labelled $\hbar\omega_m$ one sees absorption greater than that in the pure crystal. The absorption in that region ($\hbar\omega < \hbar\omega_m$) is due to defect-activated lattice, or band modes. In Fig. 32 the analysis is shown and the correlation between features in the absorption spectrum and the critical points in the calculated phonon

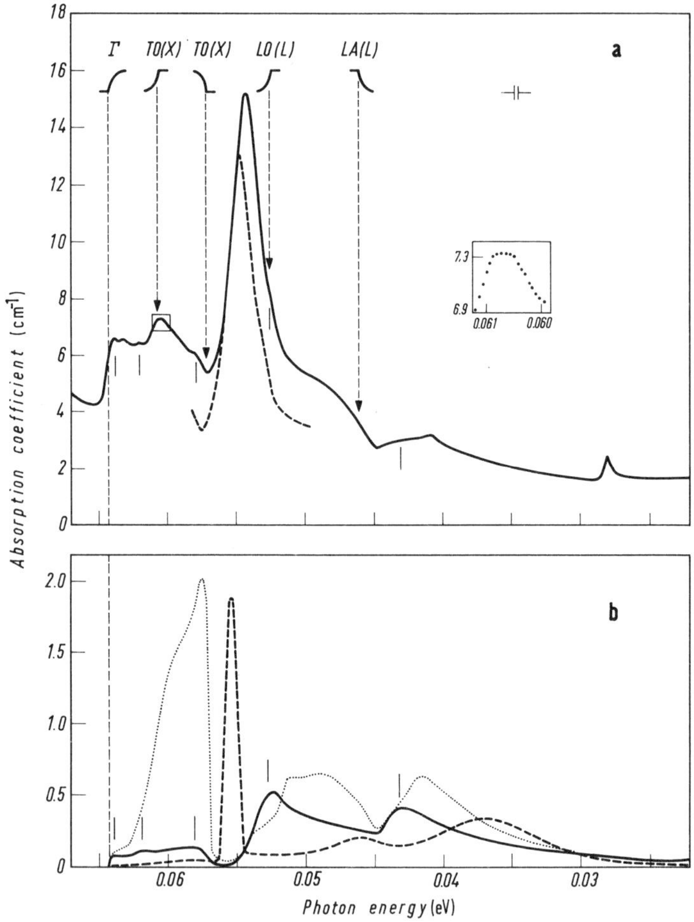

Fig. 32. a) Band mode absorption of silicon doped with both boron and phosphorus in concentrations of $5 \times 10^{19}\,cm^{-3}$. Full curve, 290 °K; dashed curve, 80 °K. Inset are the experimental points over the *TO* region. Critical point energies and shapes (JOHNSON and LOUDON, 1964) are shown. b) Comparison of theoretical and experimental calculation for infra red absorption of doped Si, with *P* and *B*. [After J.F. ANGRESS, A.R. GOODWIN, and S.D. SMITH: Proc. Roy. Soc. A **287**, 64 (1965), Fig. 5]

spectrum of silicon is illustrated. Noteworthy also is the agreement between one phonon density of states and the general shape of the curve.

The identification of local mode absorption due to the boron impurity is facilitated by the presence of two isotopes of boron ^{10}B and ^{11}B. The attribution is shown on Fig. 31.

The calculation of band mode density of states might have given improved agreement if it had been made including selection rules i.e. by only calculating $g'_v(\omega)$ instead of $g(\omega)$.

Work on these types of problems has recently moved ahead in several directions of which we mention only a few. According to the discussion in Sects. 149 and 150 all the imperfect crystal modes band and local need to be classified under the site group $\mathfrak{P}\binom{l}{\kappa}$. Then as with the perfect lattice theory or a molecular vibration problem, we can expect combinations and overtones of all such modes in higher order processes. A necessary condition for infra-red absorption activity is that the product representation $D^{(j)} \otimes D^{(j')}$ of the excited state (combination or overtone) should contain $\Gamma^{(\boldsymbol{v})}$. A detailed dynamical theory has been undertaken[3] for the Boron-Silicon and Lithuum-Silicon systems with results for actual integrated absorption and frequency dependence of the absorption coefficient, in good agreement with experiment.

Pairing effects have also been studied in the case that the pairs are close neighbors in the diamond structure,[4] for example a theory of the local modes and infrared absorption in silicon containing B-Li pairs was developed. The site symmetry group in this case is dependent upon the atomistic model taken for the defect. For B-Li both substitutional in near-neighbor sites the imperfection group in C_{3v}, the active representations are then

$$\Gamma^{(\boldsymbol{v})}(C_{3v}) = \Gamma^{(1)} \oplus \Gamma^{(3)}$$

while for an interstitial-substitutional pair a symmetry of T_d may be achieved. For details of this work particularly the important use of symmetry to factorize the Green matrix for the pair problem, the original papers should be consulted. The actual symmetry C_{3v} or T_d of the complex evidently plays a decisive role.

Not as much work on mass defects in rocksalt structure seems to have been reported. In the rocksalt it appears that to date the U center is the most accessible defect for study.[5,6] Also recently much work on molecular inclusions in rocksalt structure (e.g. OH^- ions in KCl) was reported. But as this would take us too far afield we refer to the literature [*33*, *34*, *48*, *50*].

153. Raman scattering in the perturbed system. As in the case of the perfect crystal, the theory of Raman scattering in the imperfect system of an isolated point impurity in otherwise perfect crystal can be developed using a variety of techniques including the many-body methods (response theory) ordinary perturbation theory, etc. Again it is outside the scope of this article to do more than make a few remarks concerning symmetry-related aspects of Raman scattering.

Recall that according to the polarizability theory or generalized Placzek theory given in Sect. 121 it is possible to define a polarizability operator for the matter system in the presence of the electromagnetic field. In particular, the entire analysis leading to Eq. (121.45) will go through as in the perfect crystal, apart from any results which depend upon the translational periodicity and which give rise to dependence of the polarizability operator $\boldsymbol{P}(\boldsymbol{R})$ upon the wave vectors. However, in the work which led to (121.45), the adiabatic approximation and related assumptions were used which it seems reasonable to take over to the perturbed

---

[3] L. Bellomante and M. H. L. Pryce: Proc. Phys. Soc. (London) **89**, 967, 973 (1966); *et seq.*

[4] R. J. Elliott and P. Pfeuty, in [*34*], p. 193.

[5] Work on the U center is actively going on. See H. Bilz, R Zeyher, and R. K. Wehner: phys. stat. solidi **20**, K 167 (1967). Also see [*50*].

[6] See also M. V. Klein, in [*34*].

case. It then follows that in the perturbed crystal the structure of the theory described in Sect. 121 carries over so that phonon Raman scattering will be described by a theory in which the operator $\boldsymbol{P}(\boldsymbol{R})$ will be expanded in Taylor series of normal coordinates and put into (121.45) with successive terms representing one, two, phonon processes.

First recall that the polarizability operator $\boldsymbol{P}(\boldsymbol{R})$ has the symmetry of a second-rank symmetric tensor, for processes away from resonance. This follows simply from the generalized Placzek theory and presumably may be carried over from a more microscopic theory. Thus

$$\boldsymbol{P}(\boldsymbol{R}) \cong \Gamma^{(\boldsymbol{v})}_{(2)}. \tag{153.1}$$

The essential selection rule for Raman scattering in imperfect crystals derives from (153.1). At this level the identification, or prediction, of spectra in the imperfect crystal is no more complicated than for an isolated molecule. In a cubic crystal

$$\boldsymbol{P}(\boldsymbol{R}) \lesssim \Gamma^{(1+)} \oplus \Gamma^{(12+)} \oplus \Gamma^{(25+)}. \tag{153.2}$$

Consequently we anticipate Raman scattering to occur in the perturbed crystal from the band states which subduce any of the active modes (153.2). Evidently this will be true in first order (one phonon) as well as succeeding orders.

Of course, the basic matrix element for Raman scattering will be as in (121.45)

$$\langle \chi_{\bar{n}} | \boldsymbol{P}(\boldsymbol{R}) | \chi \rangle \tag{153.3}$$

and the symmetry analysis concerns the vanishing or not of this matrix element for specified vibrational transition from state $|\chi_n\rangle$ to $|\chi_{\bar{n}}\rangle$ in the perturbed crystal.

In a cubic crystal with point symmetry O_h at the defect site there will be a rule of mutual exclusion regarding the local mode: since (as discussed in Sects. 149–150) it has symmetry of a polar vector $\Gamma^{(\boldsymbol{v})}$, the local mode can be active in infra-red but not Raman scattering.

A quantitative theory of Raman scattering by diamond-type crystals with isolated mass defect was given by XINH[1] in Ref. [*34*]. In the case of alkali halide crystals (rocksalt structure) the approximation of taking the defect as a simple mass defect seems unrealistic, since very common defects in those structures are molecular inclusions: example $KCl:OH^-$, $KCl:NO_2^-$ etc., and their analysis requires study of the molecule within the crystal. Reference to [*33*, *34*, *48*, *50*] is advisable for those interested in pursuing this subject. A microscopic theory of Raman scattering in imperfect crystals generalizing the exciton theory presented in Sect. 124ε was given by GANGULY and BIRMAN[2] in [*33*] and by MULLAZZI[3] but seems not yet to have been tested experimentally in crystals of diamond or rocksalt structure. LEIGH and SZIGETTI[4] discussed a polarizability theory of impurity Raman scattering.

---

[1] N.X. XINH (thesis unpublished); referred to in A.A. MARADUDIN, footnote 1, Sect. 150, and also Westinghouse Research Laboratories paper 65-9 F 5-442, p. 8.

[2] A. GANGULY and J.L. BIRMAN, in [*33*].

[3] E. MULAZZI: Phys. Rev. Letters **25**, 228 (1970).

[4] R.S. LEIGH and B. SZIGETTI, in [*33*], p. 477.

154. Symmetry breaking and induced lattice absorption and scattering. An extremely interesting and powerful technique for studying optical properties of crystals is the application of generalized external stresses to the crystal, such as electric or magnetic fields, or mechanical stress, and examination of the perturbed infra-red absorption or Raman scattering. For low intensity stresses the major effect will be a symmetry breaking.

That is let the spatial symmetry of the perfect crystal with Hamiltonian $\mathcal{H}$ be $\mathfrak{G}(\mathcal{H})$ or the space-time symmetry $\mathcal{G}(\mathcal{H})$. Let the generalized stress $\boldsymbol{S}$ have spatial symmetry or invariance group, $\mathfrak{G}(\boldsymbol{S})$, and space-time symmetry group $\mathcal{G}(\boldsymbol{S})$. Then, in the presence of the stress the total spatial symmetry group of the system is

$$\mathfrak{G}_{\text{tot}} = \mathfrak{G}(\mathcal{H}) \cap \mathfrak{G}(\boldsymbol{S}) \tag{154.1}$$

and the space time symmetry group

$$\mathcal{G}_{\text{tot}} = \mathcal{G}(\mathcal{H}) \cap \mathcal{G}(\boldsymbol{S}). \tag{154.2}$$

That is the net symmetry is the intersection of the two symmetry groups—those elements in common. In point of fact, the group $\mathfrak{G}(\boldsymbol{S})$ is generally a lower symmetry group than $\mathfrak{G}(\mathcal{H})$ and often a subgroup. In that case the resultant symmetry has been lowered. By the Lemma of Necessary Degeneracy, all properties of the resultant system should be classified under $\mathfrak{G}_{\text{tot}}$, and its irreducible representations. Generally, two types of properties will be of interest. Eicher tensor properties which can define macroscopic responses, and terms in an expansion of covariant quantities, or properties of the nature of selection rules for transitions between different states of the new symmetry $\mathfrak{G}_{\text{tot}}$.

α) *Symmetry breaking.* As one specific example, we shall consider a case which has been discussed relating to the application of a static electric field to diamond.[1] Let the electric field $\mathcal{E}$ be applied parallel to a cube axis [0 0 1]. The $\mathcal{E}$ field is a polar vector object. In isotropic space it transforms under O_3 like the three components of the spherical harmonic $Y_{lm}(\theta, \varphi)$ as basis for the $D^{(l=1)}$ irreducible representation of the spatial group O_3 (or SU_2). The invariance group of the electric field $\mathcal{E}$ is the group of all operations consisting of rotations about the $\mathcal{E}$ field and reflections in plane passing through $\mathcal{E}$. The group so defined is the two dimensional rotation-reflection group O_2, with z axis chosen to pass through the electric field vector $\mathcal{E}$. The group O_2 is a group of infinite order with a denumerable infinity of one and two dimensional representations [*1*].

When the electric field $\mathcal{E}$ is applied parallel to z axis in the diamond crystal the total symmetry group is the intersection group

$$\mathfrak{G}_{\text{tot}} = O_h^7 \cap O_2 = C_{4v}. \tag{154.3}$$

When the $\mathcal{E}$ field is taken parallel to the crystal [0 0 1] direction the resultant symmetry group is the set of all rotations leaving z invariant, or the point group C_{4v} as shown in Table 1. Consequently we have a reduction in symmetry, so to say, on both sides, since C_{4v} is a subgroup of both O_2 and O_h^7. In diamond

---

[1] E. Anastassakis, A. Filler, and E. Burstein, in: Light Scattering Spectra of Solids (ed. G. Wright), p. 421. Berlin-Heidelberg-New York: Springer 1969.

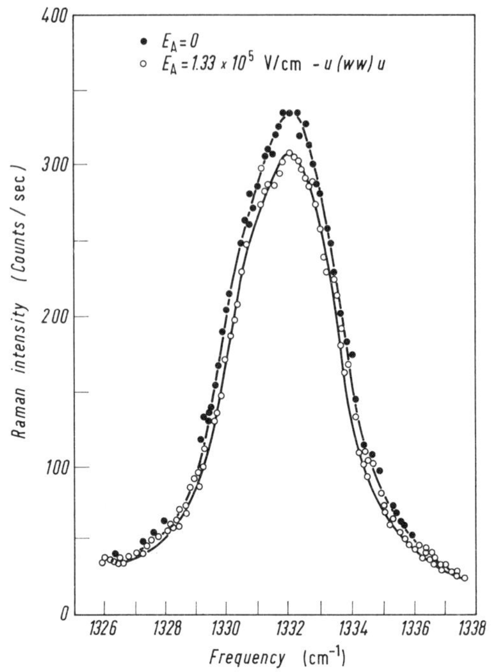

Fig. 33. Electric field effect upon first order Raman scattering in diamond. The broadening may be an indication of symmetry-breaking. [After E. ANASTASSAKIS, A. FILLER, and E. BURSTEIN, in: Light Scattering Spectra of Solids (ed. by G. WRIGHT), Fig. 1. Berlin-Heidelberg-New York: Springer 1969]

structure with an electric field present we must reclassify all modes according to the new group: C_{4v}. Using the subduction tables it is easy to show that the Raman mode $\Gamma^{(25+)}$ of diamond becomes

$$\Gamma^{(25+)} \quad \text{of} \quad O_h^7 \downarrow B_2 \oplus E \quad \text{of} \quad C_{4v}. \tag{154.4}$$

Thus the applied electric field splits the triply degenerate Raman mode into two components, separately polarized parallel and perpendicular to the $\mathscr{E}$ field. No direct observation of this splitting seems to have been reported, but what may be an indirect indication of splitting was observed by ANASTASSAKIS, FILLER and BURSTEIN.[1] They found a broadening of the Raman shifted line when an electric field was applied to a diamond crystal. This is shown in Fig. 33. The line broadening mechanisms have not been quantitatively analysed in this case although the electric field effect on infra-red absorption has been studied[2] theoretically so it is not certain that the observed broading should be attributed to the splitting due to symmetry breaking. Other mechanisms such as anharmonic couplings and/or higher order dipole moments coupled to the external electric field, may also produce line broadening.

β) *Morphic effects.* In a macroscopic or phenomenological treatment, Raman scattering is due to the modulated polarizability tensor of the medium. Recall

[2] H. HARTMANN: Phys. Rev. **147**, 663 (1966).

the discussion in Sect. 121 especially Eq. (121.58). These type of symmetry breaking effects have been called "morphic" effects by BURSTEIN and collaborators.[3a] The lowest order terms in an expansion of $\boldsymbol{P}$ in the present case can be obtained by considering the polarizability as a function of displacements and in particular we single out the Raman active normal mode $Q^{(j)}$ (normal mode) *and* of the electric field $\mathscr{E}$:

$$\boldsymbol{P}(\mathscr{E}, Q) = \underline{\boldsymbol{P}}^0 + \boldsymbol{P}^{(1,0)}:\mathscr{E} + \boldsymbol{P}^{(0,1)}:\boldsymbol{Q}^{(j)} + \boldsymbol{P}^{(1,1)}:\mathscr{E}\boldsymbol{Q}^{(j)} + \boldsymbol{P}^{(2,0)}:\mathscr{E}\mathscr{E} + \cdots. \tag{154.5}$$

It is important to emphasize that (154.5) is an expression of the polarizability $\boldsymbol{P}$ of the diamond crystal. Thus (154.5) must be constant with the crystalline symmetry of the diamond group, so that each term of each degree must transform like $\boldsymbol{P}$ in O_h, when $Q^{(j)}$ and $\mathscr{E}$ are transformed. Consider the terms in (154.5). A term like

$$\boldsymbol{P}^{(1,1)}:\mathscr{E}Q^{(j)} \tag{154.6}$$

could be considered to produce a field-modulated Raman scattering. But the structure of this term must be consistent with the covariance of $\underline{\boldsymbol{P}}$. That is: $\underline{\boldsymbol{P}}$ is a second rank, tensor field, and in diamond transforms as

$$\underline{\boldsymbol{P}} \simeq \Gamma^{(1+)} \oplus \Gamma^{(12+)} \oplus \Gamma^{(25+)}. \tag{154.7}$$

Now in diamond $\mathscr{E}$ transforms like a polar vector

$$\mathscr{E} \approx \Gamma^{(15-)} \tag{154.8}$$

and the Raman normal mode

$$Q^{(R)} \sim \Gamma^{(25+)}. \tag{154.9}$$

So the term (154.6) transforms as

$$\Gamma^{(15-)} \times \Gamma^{(25+)} \tag{154.10}$$

which obviously does not contain any of the components of $\underline{\boldsymbol{P}}$ since these are all even parity by (154.7). Hence in diamond we cannot make a covariant like $[\boldsymbol{P}]$ from the components of (154.6), and so

$$P^{(1,1)} = 0. \tag{154.11}$$

To produce a field modulation of the one-phonon Raman scattering requires the term

$$\underline{\boldsymbol{P}}^{(2,1)}\,\mathscr{E}\mathscr{E}\,Q^{(j)}. \tag{154.12}$$

This term transforms as

$$[\Gamma^{(15-)}]_{(2)} \otimes \Gamma^{(25+)} \tag{154.13}$$

and so components of $P^{(2,1)}$ may be non-zero since (154.13) contains (154.7). In fact, reducing (154.13), we find

$$\Gamma^{(1+)} \oplus \Gamma^{(12+)} \oplus 2\,\Gamma^{(15+)} \oplus 3\,\Gamma^{(25+)}. \tag{154.14}$$

Consequently, the components of (154.12) can be used to construct covariant objects properly. It is then this term which (*inter alia*) can produce a field modulated Raman scattering. Some further illustrations are given in [1]. An important

[3a] E. ANASTASSAKIS and E. BURSTEIN: J. Phys. Chem. Solids **32**, 313, 563; **33**, 519 (1972).

qualitative prediction of the theory is the appearance of certain terms in the Raman scattering tensor owing only to the non-vanishing field. Thus these authors predict a non-zero scattering in certain polarization owing to second order effect (154.12). This would be a test of this particular mechanism. No experimental verification of this prediction seems to have been yet obtained.

Another symmetry breaking effect in diamond which is induced by the static electric field is: induced first order infra-red absorption. In the perfect crystal the $\boldsymbol{k}=\boldsymbol{\Gamma}$ mode in diamond is of symmetry $\Gamma^{(25+)}$ and thus infra-red inactive. Then returning to Sect. 122, (122.10) *et passim*, we need to expand the dipole moment operator (122.10) in a mixed series in the normal coordinates $Q^{(j)}$ and the applied electric field $\mathscr{E}$. Writing symbollically

$$\boldsymbol{\mu}=\mu_v^0+\mu^{(1,0)}\,Q=\mu^{(0,1)}\,\mathscr{E}+\mu^{(1,1)}\,\mathscr{E}Q+\cdots \tag{154.15}$$

we have the possibility of a matrix element of form

$$\mu^{(1,1)}\,\mathscr{E}\,\langle\chi_{v\bar{\kappa}}|\,Q\,|\chi_{v\kappa}\rangle \tag{154.16}$$

arising in (122.11). Now in diamond the operator $\boldsymbol{\mu}$ transforms as $\Gamma^{(v)}$, i.e. $\Gamma^{(15-)}$ so that each non zero term in (154.15) must transform as $\Gamma^{(15-)}$. Clearly the term

$$\mu^{(1,1)}\,\mathscr{E}Q^{(j)} \tag{154.17}$$

meets this requirement since

$$\mathscr{E}Q\sim\Gamma^{(15-)}\oplus\Gamma^{(25+)}\sim\Gamma^{(15-)} \tag{154.18}$$

for $Q^{(j)}$ the Raman mode. Hence a matrix element (154.16) will be non-zero due to the electric field coupling the mode $\boldsymbol{\Gamma}^{(25+)}$, via the term $\mu^{(1,1)}$. Such an electric field induced infra-red absorption at the one phonon Raman frequency has apparently been observed recently[3b] when there was a report of an apparent shift of peak position of the Raman mode in addition to the absorption induced by the electric field. Further work on such field and stress induced absorption will be surely profitable in the future.

In the rocksalt structure, the $\boldsymbol{\Gamma}$ phonon is odd parity symmetry $\boldsymbol{\Gamma}^{(15-)}$ and thus infra-red active but inactive in Raman scattering. Again, the symmetry argument produces insight into the phenomenon of symmetry-breaking via application of external perturbation. As with diamond, an electric field along a cube axis will lower the symmetry of the system to C_{4v}. Then the $\boldsymbol{\Gamma}^{(15-)}$ phonon reduces in symmetry to

$$\boldsymbol{\Gamma}^{(15-)}\quad\text{of}\quad O_h^7\downarrow A_1\oplus E\quad\text{of}\quad C_{4v}. \tag{154.19}$$

But in the group C_{4v}, the components of polarizability tensor $\boldsymbol{P}$ transform as

$$\boldsymbol{P}\sim 2A\oplus B_1\oplus B_2\oplus E. \tag{154.20}$$

Consequently symmetry permits the field induced Raman scattering activity of the infra-red mode in rocksalt. That is, the components A_1 and E of (154.19) now are observable in Raman scattering. A dynamical theory of this effect does not yet exist adequate to predict magnitudes, line shapes etc. While field induced Raman scattering has not yet been observed in rocksalt, clear experimental

[3b] E. Anastassakis, S. Iwasa, and E. Burstein: Phys. Rev. Letters **17**, 1051 (1966).

evidence for it in the perovskite structure has been given.[4] A theory of this effect also has been given and is in good agreement with experiments.[5]

So far no theory or experiments in either diamond or rocksalt structure have been reported in which magnetic field dependence of Raman or infra-red lattice effects were studied.[6]

Likewise, and perhaps more important are stress effects. Actually photoelastic studies in diamond and rocksalt are very old but those studies mostly concerned themselves with measurements of macroscopic constants such as the photoelastic coefficient. This potentially very rewarding area has recently begun to be studied.[7,8]

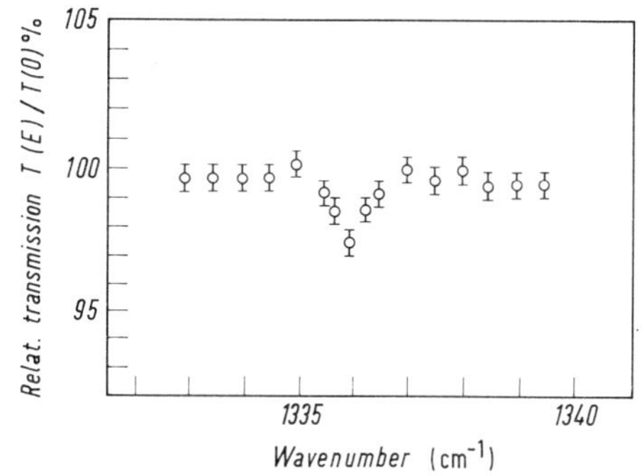

Fig. 34. Electric field induced symmetry breaking and optical absorption in diamond. [After E. BURSTEIN, in: Dynamical Processes in Solid State Optics (eds. R. KUBO and N. KAMIMURA), p. 34. New York: McGraw Hill Book Co (1969)]

Still other, more refined aspects of symmetry breaking can be envisaged related to application of external force gradients to a crystal. Instead of expanding the dipole moment, or polarizability in the stress, one could consider expansions in spatial derivatives are tensors of higher order. For example the derivative of the stress tensor with respect to spatial coordinates is a third rank tensor, symmetric in the first two indices. These intriguing possibilities were put forth in work by HUMPHREYS and MARADUDIN,[9] and BERENSON[10] showed the close relationship of the scattering tensors which arise in these processes to Clebsch-Gordan coefficients.

On this note of broken symmetry the article ends, having come full circle. As a rule symmetry is not completely broken but the new symmetry of the system with inner plus outer stresses present must stand by itself as the new symmetry group to be analysed.

---

[4] J.M. WORLOCK and P. FLEURY: Phys. Rev. Letters **19**, 1176 (1967); **18**, 665 (1967). Also J.M. WORLOCK, in [*33*], p. 415.

[5] V. DVORAK: Phys. Rev. **159**, 652 (1967).

[6] Such effects were reported in CdS on the magnetic field dependence of the Raman scattering by T.C. DAMAN and J. SHAH: Bull. Am. Phys. Soc., Ser. 2, **16**, 29 (1971), Abstract AK 10.

[7] F. CERDEIRI, C.J. BUCHENAUER, F.H. POLLAK, and M. CARDONA: Bull. Am. Phys. Soc., Ser. 2, **16**, 29 (1971), Abstract AK 9.

[8] G.S. HOBSON and E.G.S. PAIGE: Proc. Phys. Soc. (London) **88**, 437 (1966).

[9] L.B. HUMPHREYS and A.A. MARADUDIN: Solid State Comm. **11**, 1003 (1972).

[10] See ref. [2], p. 371 and references to Sect. 60.

P. Respice, adspice, prospice

α) *Respice.* When this article was conceived as a contribution to this Encyclopedia the author was mindful of the state of development and knowledge of the several subjects which would be presented here. The intention was to develop these topics taking account of their varied levels of elaboration and so provide at one and the same time a review from a unified viewpoint and a useful basis for further advances.

It was quite clear that the theory of crystal space groups was already the most mature and even somewhat classical branch which would be presented. Already long behind us was the work of SCHÖNFLIESS-FEDEROV [7] on the enumeration of crystal space groups in three dimensions and the work of BOUCKAERT, SMOLUCHOWSKI and WIGNER,[1] and SEITZ[2] on the classification of irreducible representations of space groups. Work on the general methods of obtaining space group selection rules was considerably more recent but also represented satisfactory solutions of well posed problems in theoretical physics. Despite the existence of a mature and coherent literature related to this branch of the application of group theory in physics, and the availability of several excellent texts exposing group theory from the viewpoint of anticipating its use by physicists there was surprisingly little day-to-day use of group theory, by physicists who might be expected to understand and use this tool as comfortably as perturbation theory was used in either its old fashioned or modern many-body versions. Consequently the author set out to write a clear and explicit exposition of group theory methods with the objective of assisting the reader to overcome inhibitions or barriers to understanding and use. This meant in particular proving those important statements so often left to the reader to "supply proofs", when the reader lacks the technical sophistication or confidence to so do. But of course we had to assume some introductory level of understanding such as provided in the usual graduate level course in theoretical physics as an example the Landau and Lifschitz course [*51*].

We thus first set out to develop the theory of the structure of crystal space groups and their irreducible representations. The approach taken was to develop the subject presuming that the reader had sufficient motivation to follow straightforward elaboration of this theory — rather closely connected to the usual development of the theory of finite groups and applications. So, in the early part of the work (Sects. 1–65) we concentrated on step by step procedure: we construct the space group irreducible representations by what is essentially a process of inducing from a subgroup, although we do not *a priori* give an abstract definition of induction rather we allow induction to appear as a necessary consequence of a straightforward and systematic characterization of the irreducible representations. Then we formally define the induction procedure for irreducible representations $D^{(\star\boldsymbol{k})(j)}$ and characters $\chi^{(\star\boldsymbol{k})(j)}$. Having developed the subject to a level where the reader should be comfortable with the structure of the representations and characters, the terminology of ray representations is introduced. This is a more compact and

[1] L. P. BOUCKAERT, R. SMOLUCHOWSKI, and E. P. WIGNER: Phys. Rev. **50**, 58 (1936).

[2] F. SEITZ: Ann. Math. **37**, 17 (1936).

economical description of the allowable $D^{(\boldsymbol{k})(m)}$ which can be used to induce irreducible $D^{(\star\boldsymbol{k})(j)}$.

Then we turn to the calculation of reduction coefficients in Sects. 52–65. These are the members of the Clebsch-Gordan series. More importantly as the reader is reminded, they give directly the selection rules for physical transition processes. We contrast various methods useful in obtaining reduction coefficients and show their equivalence. Hence, as long as any method is properly used it gives the correct result – but the reader must be certain he or she knows completely the proper use since otherwise errors may mar the result. This is explained in Sect. 64.

Also in this part, the calculation of space group Clebsch-Gordan coefficients is discussed. These quantities represent the most complete and sophisticated information which can be obtained for any group and enable one to maximize the symmetrization of the relevant quantity be it an eigenfunction or an eigenvector or another quantity. Thus these coefficients may permit (in favorable cases) a complete factorization of a calculation of a matrix element or analogous quantity into a symmetry related part which is generally a Clebsch-Gordan coefficient, and a reduced matrix element which realizes the result of specific interactions being present.

In all the work to this point we assume the readers motivation and interest is maintained because he or she can transfer over knowledge obtained in familiar work [*1*, *51*] which instructs us regarding the importance of symmetry in determining the correct linear vector spaces which are bases for a calculation. This should sustain an interest and curiosity regarding the structure of the crystal space group and its representations. Since the work of these sections is largely mathematical, albeit developed with a physicist's interests in view, the material is general and stands by itself irrespective of application. It can find application to problems of phase transitions and symmetry change, electrons in solids via energy band theory and optical properties related to electronic transitions, electron-phonon interaction and scattering processes including transport theory and by straightforward extension to problems of magnetic crystals, spin waves and related problems. With the basis given here it is hoped the reader can easily comprehend those extensions.

Now we turn to the physical problem of major interest here: namely the connection with lattice dynamics. The usual treatments [*4*] of lattice dynamics are given with crystal symmetry added post-facto. We develop the theory here in Sects. 66–86 to demonstrate the fashion by which symmetry is inextricably intertwined with the physics. The eigenvectors of the dynamical equation span just the irreducible linear vector spaces or representation modules of representation theory. The reader who appreciates this simply stated result and the consequences which flow from it understands the major importance of the study of group theory in physics. This result was shown originally by WIGNER[3] for the simpler problem of molecular vibrations but was extended soon after to solids by SEITZ.[4]

---

[3] E. P. WIGNER: Göttinger Nachrichten, Math.-Phys. Klasse, 546 (1932).

[4] R. B. BARNES, R. BRATTEIN, and F. SEITZ: Phys. Rev. **48**, 582 (1935).

In order to make this result meaningful we examine it in detail, actually exhibiting the transformations required to obtain symmetry adapted normal coordinates and showing how this reduces or factorizes the dynamical matrix.

By introducing the dynamical matrix and therefore crystal dynamics we must now focus on dynamical symmetry, which requires that the symmetry group be enlarged to include time-reversal. This subject is discussed first from the viewpoint of extending the spatial symmetry group by an operation of complex conjugation and then from the more modern corepresentation viewpoint of WIGNER [1]. Particular features of our treatment which are intended to improve the utility of the analysis of the effect of time-reversal are the discussion of the classification of the representations $D^{(*k)(j)}$ according to a new test due originally to FREI, but modified here to make it more convenient in application. The remainder of the discussion of corepresentation theory is intended to be general, in its mathematical aspects, and thus useful to any problem with space-time symmetry group $\mathcal{G}$, as well as specifically aimed toward concrete application in calculating eigenvectors in lattice dynamics.

The part of the article which follows knits together many of the formal results obtained earlier. In this work selected applications are given which show how symmetry permits simplification and rationalization of the calculations of invariant and covariant quantities related to important physical questions in a crystal: the Hamiltonian, the symmetry-related critical points in density of states, electric moment and polarizability operators. Enough detail is given so it is hoped the reader can confidently apply these symmetry methods to novel cases of his interest. Many of the results given here are found piecemeal in various places in the literature: it is hoped that a more unified view will bring out the underlying similarities involved.

The next major topics are the development of the quantum theories of lattice dynamics and of infra-red absorption and Raman scattering by phonons. In this work symmetry plays a familiar, if somewhat elaborate role. Once the structure of the crystal space group and its representations and reduction coefficients are known, the rest is application and elaboration in fashion familiar from analogous problems in atomic, molecular or nuclear physics. But to know how and where to apply, and to elaborate, requires that the relevant quantum theory be at hand. Many levels of sophistication are possible, but the major explicit developments of the theory of quantum lattice dynamics are presented in harmonic approximation in order to make clear how to obtain the symmetry of a many-phonon eigenstate in harmonic approximation. But generalization to include anharmonic terms permitted by symmetry and needed for a complete many-body theory is immediate and follows the methods set down in the discussion of the classical tensor calculus. The work on infra-red absorption and Raman scattering is developed in a semi-classical radiation theory framework and also, to varying degrees of depth in more modern microscopic formalisms. In all cases the effect of symmetry is explicitly exhibited in the locale appropriate to it. This again illustrates our strategy of developing the intertwined dynamical theory and symmetry theory side by side rather than separately or *seriatem* as is often the case.

The reason for selection of diamond and rocksalt space groups for illustration was given earlier: they are interesting in themselves as example of crystal symmetry types respectively non-symmorphic and symmorphic, and also many crystals with these symmetries are presently in the forefront of research. The reader who takes the trouble to work through the calculations of selection rules, tests of time reversal and connectivity should consolidate his understanding of the general material presented earlier. The physically interesting processes analysed in Sects. 137–154 are in part the most topical of all the material presented in the article. Work is still actively underway examining the phonon dispersion in many crystals of these structures, and the optical properties such as infra-red absorption and Raman scattering are continuing to be actively studied. Consequently some of the material given such as particular experimental results cited, along with the experimental interpretations appertaining may be dated. But as illustrations of the best information or work available at time of writing they should be useful it is hoped in illuminating the methods and the results of analyses even if some numerical values or identifications may later be modified. Again it is suggested that a reader who wishes to consolidate his understanding work through the illustrative material and then update as needed, after referring to ongoing literature. All in all, the material in these sections should be a forceful illustration of the utility of group theory analysis in terms of optical processes identified via application of selection rules, polarization of scattered light predicted and observed, critical point features in phonon density of states identified and the effect of symmetry breaking in producing new active degrees of freedom as well as in activating formally silent modes. A *caveat* is needed: group theory is not the entire story and throughout the actual work of illustration it should be appreciated just where dynamical information such as calculations of phonon dispersion; relative energies of different modes; exciton-photon or exciton-phonon coupling coefficients, are needed.

It has been my hope throughout that the reader who reached the end of the journey with me will feel no qualms about attacking a new problem with these group theory methods, having seen that they are worth the effort.

β) Adspice. At the present time there has been a veritable explosion in the level of activity of work relating to the several topics which were developed in this article. Perhaps the most obvious illustration is in the area of light scattering, including Raman scattering by phonons. International conferences on this topic succeed one another with dizzying regularity [*33*, *45*, *48*] and even at the time of writing smaller international meetings [*62*] such as colloquia, or seminars are in planning stage. This makes any attempt to write a definitive treatise on the theory or experimental results rather unrealistic. The data changes as experimental technique improves and interpretation likewise changes. But the general theoretical tools needed in the analysis change much less rapidly. Quantum theory including many-body perturbation theory and group theory remain vital to the prediction, understanding, and interpretation of phenomena in solids including lattice dynamics and optical phenomena.

Perhaps more surprising in view of our earlier remarks that the group theory side of our subject has a long and honorable history has been the upsurge in work sharpening and extending the group theory techniques and enlarging their domain of application. Already mentioned in the text in Sects. 60 and 134 is the development of reliable calculational methods to obtain space group Clebsch-Gordan coefficients. Also coming to the fore now are techniques making more use of ray representation and ray corepresentation theory. The basis of these methods in space group theory were given in Sects. 41–44 and 100 but it is only recently that significant progress has been made in actual calculations. The very economy of these methods has seemed in the past to prevent their wide acceptance but with experience this unfamiliarity should be overcome. It seems unlikely that new physical advances will emerge from such methods but perhaps by clearing away some notational complexity new features may arise. Another recent development in application of group theory has been the development of digital computer routines to facilitate the construction of tables of irreducible representations and corepresentations of space groups [54] and also computer analyses of the structure of space groups, in 1, 2, 3, 4, dimensions.[5] Development of computer routines and subroutines to assist in carrying through the eigenvalue-eigenvector program discussed in Sects. 79–85 and 101–103 has been reported[6] and seems to be a rapidly developing technique which will be useful in the future.

Another indication of the continuing high level of activity in this field is the appearance of several recent texts whose subject matter has some overlap with the present article. The work by MARADUDIN, MONTROLL, WEISS, and IPATOVA [47] covers very many dynamical aspects of the theory of lattice dynamics in harmonic approximation, and also gives excellent coverage of the theory of the imperfect crystal as well as of many types of experimental measurements not considered in this article. The work by POULET and MATHIEU [52] covers the general results of the group theory analysis and the semiclassical radiation theory of infra-red absorption and Raman scattering; this work gives a particularly thorough treatment of polarization effects in Raman scattering, and of macroscopic field effects, which we presented in Sect. 123. A comprehensive treatise by BRADLEY and CRACKNELL on the mathematical theory of crystal symmetry [53] gives tables which can be used to construct irreducible representations and corepresentations of all 230 "gray" crystal space group as well as of all 1651 magnetic of Shubnikov groups.

The reader of this article is advised to make good use of these three recent additions to the literature, as well as to the other standard works [1–4] to which earlier reference was made.

γ) *Prospice.* The extraordinary activity now typifying the various topics presented here make it extremely unlikely that any specific predictions will have much validity beyond the time they appear in print.

Certain themes may be worthy of mention however. It does seem evident that resonance light scattering processes will play a central role in providing new experimental data for interpretation. Such work already has had the effect of changing light scattering from a branch of spectroscopy wholely concerned with identifying phonons and their symmetry to a greater emphasis on microscopic dynamics and the basic radiation-matter interaction using light scattering as the

[5] J. NEUBUSER: Proc. IInd Colloquium on Group Theory in Physics, Nijmegen, June 1973 (ed. A. JANNER). Available from Prof. JANNER, University of Nijmegen, Netherlands.

[6] T.G. WORLTON and J.L. WARREN: Computer Phys. Comm. **3**, 88 (1972). – J.L. WARREN: Rev. Mod. Phys. **40**, 38 (1970).

probe by which this is examined. Also the probing of collective degrees of freedom such as the proposed two-phonon bound state may be expected to grow. Many different materials with different crystal structure and different coupling strengths will undoubtedly give rise to novel effects presently unexpected.

On the theoretical side we can expect work to move ahead vigorously, partly under the stimulus of experiments, partly in the attempt to formulate better and more harmonious, unified, and general theories encompassing symmetry theory and dynamical theory. As with any living branch of science the future can only be guessed at, or seen through a glass darkly.

If the foundations already built are insufficient to sustain the future work the history of science teaches us to press, ahead, and simultaneously to rebuild that which is inadequate.

Q. Acknowledgements

During the course of preparation of this article I have been helped by many present and former colleagues. I want to thank in particular Dr. A. K. Ganguly, Professor M. Nusimovici; and Dr. Y. C. Cho each of whom read the article in a penultimate version and made important contributions, enabling me to eliminate errors. Dr. Rhoda Berenson contributed a large part of Sects. 18, 60, and 134: the latter representing much of her own work. In various ways my understanding of the theory presented here benefited from association with Dr. B. Bendow, Dr. C. S. Ting, and Dr. R. Zeyher.

Parts of this article were completed during my stay as Professeur Associé, Groupe de Physique des Solides de l'École Normale Superieure, et Laboratoire de Physique des Solides, Université de Paris in 1969–70. I thank Professor P. Nozières, Professor M. Balkanski, and Professor M. Hulin for making my stay in Paris possible.

During the course of writing this article the research program of my group has been partially supported by the United States Army Research Office, Durham, the United States Aerospace Research Laboratories, Dayton, and in the very final stages by the National Science Foundation of the United States.

It is a pleasure to thank Mrs. E. De Crescenzo for her patient, tireless, accurate, and cheerful work of typing and retyping the largest part of this article.

I must also thank Chancellor Sidney Borowitz for his encouragement at various stages in the work.

It is a pleasure to acknowledge the splendid cooperation and help I received from the Springer-Verlag, especially Mr. K. Koch, in meeting my requests for printing this article, including many difficult and complicated typographical requirements.

If errors and shortcomings remain in this work, as undoubtedly they do, despite the generous help I received, they are my own responsibility and I would appreciate the assistance of readers in pointing them out for future correction.

Appendix A

Complete tables of reduction coefficients-selection rules for rocksalt structure O_h^5

Tables A.1 to A.11 see following pages

Table A.1. Reduction coefficients of type $({}^{\star}\boldsymbol{k}\, m\, \Gamma\, m'|)$ for rocksalt O_h^5. In Tables A.1–A.11, the entire full-group irreducible representation symbol $D^{({}^{\star}\boldsymbol{k})(m'')}$ is written $\boldsymbol{k}^{(m'')}$ for compactness. All symbols in the tables refer to the entire full-group irreducible representations. The + sign is used instead of ⊕ to indicate direct sum in the tables; likewise × instead of ⊗. The factors in the product of full-group irreducible representations are given as row and column headings; representations which arise are listed in the tables, using the abbreviated notation for compactness. The tables also contain wave vector rules where helpful, as well as indicating equalities among various reduction coefficients

$\boldsymbol{k}_{j'}^{(m')}$	$\boldsymbol{k}_j^{(m)}$			
	$\Gamma^{(2+)}$	$\Gamma^{(12+)}$	$\Gamma^{(15+)}$	$\Gamma^{(25+)}$
	$\Gamma \times \boldsymbol{k} = \boldsymbol{k}$		$(\Gamma^{(1+)}\boldsymbol{k}_{j'}^{(m')} \mid \boldsymbol{k}_{j''}^{(m'')}) = \delta_{j'j''}\,\delta_{m'm''}$	
Γ^{2+}	Γ^{1+}			
Γ^{12+}	Γ^{12+}	$\Gamma^{1+}+\Gamma^{2+}+\Gamma^{12+}$		
Γ^{15+}	Γ^{25+}	$\Gamma^{15+}+\Gamma^{25+}$	$\Gamma^{1+}+\Gamma^{12+}+\Gamma^{15+}+\Gamma^{25+}$	
Γ^{25+}	Γ^{15+}	$\Gamma^{15+}+\Gamma^{25+}$	$\Gamma^{2+}+\Gamma^{12+}+\Gamma^{15+}+\Gamma^{25+}$	$\Gamma^{15+}\times\Gamma^{15+}$
L^{1+}	L^{2+}	L^{3+}	$L^{2+}+L^{3+}$	$L^{1+}+L^{3+}$
L^{2+}	L^{1+}	L^{3+}	$L^{1+}+L^{3+}$	$L^{2+}+L^{3+}$
L^{3+}	L^{3+}	$L^{1+}+L^{2+}+L^{3+}$	$L^{1+}+L^{2+}+2L^{3+}$	$L^{1+}+L^{2+}+2L^{3+}$
X^{1+}	X^{2+}	$X^{1+}+X^{2+}$	$X^{4+}+X^{5+}$	$\Gamma^{15+}\times X^{2+}$
X^{2+}	X^{1+}	$X^{1+}+X^{2+}$	$X^{3+}+X^{5+}$	$\Gamma^{15+}\times X^{1+}$
X^{3+}	X^{4+}	$X^{3+}+X^{4+}$	$X^{2+}+X^{5+}$	$\Gamma^{15+}\times X^{4+}$
X^{4+}	X^{3+}	$X^{3+}+X^{4+}$	$X^{1+}+X^{5+}$	$\Gamma^{15+}\times X^{3+}$
X^{5+}	X^{5+}	$2X^{5+}$	$X^{1+}+X^{2+}+X^{3+}+X^{4+}+X^{5+}$	$\Gamma^{15+}\times X^{5+}$
For Γ, L, and X $(\boldsymbol{k}_j^{(m+)}\boldsymbol{k}_{j'}^{(m'+)} \mid \boldsymbol{k}_{j''}^{(m''+)}) = (\boldsymbol{k}_j^{(m-)}\boldsymbol{k}_{j'}^{(m'-)} \mid \boldsymbol{k}_{j''}^{(m''+)}) = (\boldsymbol{k}_j^{(m-)}\boldsymbol{k}_{j'}^{(m'+)} \mid \boldsymbol{k}_{j''}^{(m''-)})$.				
W^{1}	$W^{2'}$	$W^{1}+W^{2'}$	$W^{2}+W^{3}$	$\Gamma^{15+}\times W^{2'}$
$W^{1'}$	W^{2}	$W^{1'}+W^{2}$	$W^{2'}+W^{3}$	$\Gamma^{15+}\times W^{2}$
W^{2}	$W^{1'}$	$W^{1'}+W^{2}$	$W^{1}+W^{3}$	$\Gamma^{15+}\times W^{1'}$
$W^{2'}$	W^{1}	$W^{1}+W^{2'}$	$W^{1'}+W^{3}$	$\Gamma^{15+}\times W^{1}$
W^{3}	W^{3}	$2W^{3}$	$W^{1}+W^{1'}+W^{2}+W^{2'}+W^{3}$	$\Gamma^{15+}\times W^{3}$
$(\Gamma^{(i-)}W^{(m)}\mid) = (\Gamma^{(i+)}W^{(m')}\mid)$, $(\Gamma^{(i-)}W^{(m')}\mid) = (\Gamma^{(i+)}W^{(m)}\mid)$, $m=1,2$; $(\Gamma^{(i-)}W^{(3)}\mid) = (\Gamma^{(i+)}W^{(3)}\mid)$.				
Δ^{1}	Δ^{2}	$\Delta^{1}+\Delta^{2}$	$\Delta^{1'}+\Delta^{5}$	$\Delta^{2'}+\Delta^{5}$
$\Delta^{1'}$	$\Delta^{2'}$	$\Delta^{1'}+\Delta^{2'}$	$\Delta^{1}+\Delta^{5}$	$\Delta^{2}+\Delta^{5}$
Δ^{2}	Δ^{1}	$\Delta^{1}+\Delta^{2}$	$\Delta^{2'}+\Delta^{5}$	$\Delta^{1'}+\Delta^{5}$
$\Delta^{2'}$	$\Delta^{1'}$	$\Delta^{1'}+\Delta^{2'}$	$\Delta^{2}+\Delta^{5}$	$\Delta^{1}+\Delta^{5}$
Δ^{5}	Δ^{5}	$2\Delta^{5}$	$\Delta^{1}+\Delta^{1'}+\Delta^{2}+\Delta^{2'}+\Delta^{5}$	$\Delta^{1}+\Delta^{1'}+\Delta^{2}+\Delta^{2'}+\Delta^{5}$
$(\Gamma^{(i-)}\Delta^{(m)}\mid) = (\Gamma^{(i+)}\Delta^{(m')}\mid)$, $(\Gamma^{(i-)}\Delta^{(m')}\mid) = (\Gamma^{(i+)}\Delta^{(m)}\mid)$, $m=1,2$; $(\Gamma^{(i-)}\Delta^{(5)}\mid) = (\Gamma^{(i+)}\Delta^{5}\mid)$.				
Z^{1}	Z^{1}	$2Z^{1}$	$Z^{2}+Z^{3}+Z^{4}$	$\Gamma^{15+}\times Z^{1}$
Z^{2}	Z^{2}	$2Z^{2}$	$Z^{1}+Z^{3}+Z^{4}$	$\Gamma^{15+}\times Z^{2}$
Z^{3}	Z^{3}	$2Z^{3}$	$Z^{1}+Z^{2}+Z^{4}$	$\Gamma^{15+}\times Z^{3}$
Z^{4}	Z^{4}	$2Z^{4}$	$Z^{1}+Z^{2}+Z^{3}$	$\Gamma^{15+}\times Z^{4}$
$(\Gamma^{(i-)}Z^{1}\mid) = (\Gamma^{(i+)}Z^{2}\mid)$, $(\Gamma^{(i-)}Z^{2}\mid) = (\Gamma^{(i+)}Z^{1}\mid)$, $(\Gamma^{(i-)}Z^{3}\mid) = (\Gamma^{(i+)}Z^{4}\mid)$, $(\Gamma^{(i-)}Z^{4}\mid) = (\Gamma^{(i+)}Z^{3}\mid)$.				
Λ^{1}	Λ^{2}	Λ^{3}	$\Lambda^{2}+\Lambda^{3}$	$\Gamma^{15+}\times\Lambda^{2}$
Λ^{2}	Λ^{1}	Λ^{3}	$\Lambda^{1}+\Lambda^{3}$	$\Gamma^{15+}\times\Lambda^{1}$
Λ^{3}	Λ^{3}	$\Lambda^{1}+\Lambda^{2}+\Lambda^{3}$	$\Lambda^{1}+\Lambda^{2}+2\Lambda^{3}$	$\Gamma^{15+}\times\Lambda^{3}$
$(\Gamma^{(1-)}\Lambda^{i}\mid) = (\Gamma^{(2+)}\Lambda^{i}\mid)$, $(\Gamma^{(2-)}\Lambda^{i}\mid) = (\Gamma^{(1+)}\Lambda^{i}\mid)$, $(\Gamma^{(12-)}\Lambda^{i}\mid) = (\Gamma^{(12+)}\Lambda^{i}\mid)$, $(\Gamma^{(15-)}\Lambda^{i}\mid) = (\Gamma^{(25+)}\Lambda^{i}\mid)$, $(\Gamma^{(25-)}\Lambda^{i}\mid) = (\Gamma^{(15+)}\Lambda^{i}\mid)$.				
Σ^{1}	Σ^{4}	$\Sigma^{1}+\Sigma^{4}$	$\Sigma^{2}+\Sigma^{3}+\Sigma^{4}$	$\Gamma^{15+}\times\Sigma^{4}$
Σ^{2}	Σ^{3}	$\Sigma^{2}+\Sigma^{3}$	$\Sigma^{1}+\Sigma^{3}+\Sigma^{4}$	$\Gamma^{15+}\times\Sigma^{3}$
Σ^{3}	Σ^{2}	$\Sigma^{2}+\Sigma^{3}$	$\Sigma^{1}+\Sigma^{2}+\Sigma^{4}$	$\Gamma^{15+}\times\Sigma^{2}$
Σ^{4}	Σ^{1}	$\Sigma^{1}+\Sigma^{4}$	$\Sigma^{1}+\Sigma^{2}+\Sigma^{3}$	$\Gamma^{15+}\times\Sigma^{1}$
$(\Gamma^{(1-)}\Sigma^{1}\mid) = (\Gamma^{(2-)}\Sigma^{4}\mid) = (\Gamma^{(2+)}\Sigma^{3}\mid)$, $(\Gamma^{(1-)}\Sigma^{2}\mid) = (\Gamma^{(2-)}\Sigma^{3}\mid) = (\Gamma^{(2+)}\Sigma^{4}\mid)$, $(\Gamma^{(1-)}\Sigma^{3}\mid) = (\Gamma^{(2-)}\Sigma^{2}\mid) = (\Gamma^{(2+)}\Sigma^{1}\mid)$, $(\Gamma^{(1-)}\Sigma^{4}\mid) = (\Gamma^{(2-)}\Sigma^{1}\mid) = (\Gamma^{(2+)}\Sigma^{2}\mid)$, $(\Gamma^{(12-)}\Sigma^{1}\mid) = (\Gamma^{(12-)}\Sigma^{4}\mid) = (\Gamma^{(12+)}\Sigma^{3}\mid)$, $(\Gamma^{(12-)}\Sigma^{2}\mid) = (\Gamma^{(12-)}\Sigma^{3}\mid) = (\Gamma^{(12+)}\Sigma^{1}\mid)$, $(\Gamma^{(15-)}\Sigma^{1}\mid) = (\Gamma^{(25-)}\Sigma^{4}\mid) = (\Gamma^{(15+)}\Sigma^{2}\mid)$, $(\Gamma^{(15-)}\Sigma^{2}\mid) = (\Gamma^{(25-)}\Sigma^{3}\mid) = (\Gamma^{(15+)}\Sigma^{1}\mid)$, $(\Gamma^{(15-)}\Sigma^{3}\mid) = (\Gamma^{(25-)}\Sigma^{2}\mid) = (\Gamma^{(15+)}\Sigma^{4}\mid)$, $(\Gamma^{(15-)}\Sigma^{4}\mid) = (\Gamma^{(25-)}\Sigma^{1}\mid) = (\Gamma^{(15+)}\Sigma^{3}\mid)$.				
Q^{1}	Q^{2}	$Q^{1}+Q^{2}$	$Q^{1}+2Q^{2}$	$2Q^{1}+Q^{2}$
Q^{2}	Q^{1}	$Q^{1}+Q^{2}$	$2Q^{1}+Q^{2}$	$Q^{1}+2Q^{2}$
$(\Gamma^{(i-)}Q^{j}\mid) = (\Gamma^{(i+)}Q^{j}\mid)$.				

[a] Tables A.1–A.11 are taken in large part from L. C. Chen, R. Berenson, and J. L. Birman: Phys. Rev. **170**, 639 (1968).

Table A.2. Reductions of type $(\star\boldsymbol{k}\, m\, \star\boldsymbol{L} m'|)$ for O_h^5. See also caption to Table A.1

$k_{j'}^{(m')}$	$k_j^{(m)}$		
	L^{1+}	L^{2+}	L^{3+}
	$L\times L=4\Gamma+4X$		
L^{1+}	$\Gamma^{1+}+\Gamma^{25+}+X^{1+}+X^{3+}+X^{5+}$	$\Gamma^{2+}+\Gamma^{15+}+X^{2+}+X^{4+}+X^{5+}$	$\Gamma^{12+}+\Gamma^{15+}+\Gamma^{25+}+X^{1+}+X^{2+}+X^{3+}+X^{4+}+2X^{5+}$
L^{2+}		$L^{1+}\times L^{1+}$	$L^{1+}\times L^{3+}$
L^{3+}			$\Gamma^{1+}+\Gamma^{2+}+\Gamma^{12+}+2\Gamma^{15+}+2\Gamma^{25+}+2X^{1+}+2X^{2+}+2X^{3+}+2X^{4+}+4X^{5+}$
	$L\times X=3L$		
X^{1+}	$L^{1+}+L^{3+}$	$L^{1+}\times X^{2+}$	$L^{1+}+L^{2+}+2L^{3+}$
X^{2+}	$L^{2+}+L^{3+}$	$L^{1+}\times X^{1+}$	$L^{1+}+L^{2+}+2L^{3+}$
X^{3+}	$L^{1+}+L^{3+}$	$L^{1+}\times X^{2+}$	$L^{1+}+L^{2+}+2L^{3+}$
X^{4+}	$L^{2+}+L^{3+}$	$L^{1+}\times X^{1+}$	$L^{1+}+L^{2+}+2L^{3+}$
X^{5+}	$L^{1+}+L^{2+}+2L^{3+}$	$L^{1+}\times X^{5+}$	$2L^{1+}+2L^{2+}+4L^{3+}$
	$L\times W=2\Sigma(\pi,\pi,0)(1/a)$		
W^1	$\Sigma^1+\Sigma^2$	$\Sigma^3+\Sigma^4$	$\Sigma^1+\Sigma^2+\Sigma^3+\Sigma^4$
$W^{1'}$	$\Sigma^1+\Sigma^2$	$\Sigma^3+\Sigma^4$	$\Sigma^1+\Sigma^2+\Sigma^3+\Sigma^4$
W^2	$\Sigma^3+\Sigma^4$	$\Sigma^1+\Sigma^2$	$\Sigma^1+\Sigma^2+\Sigma^3+\Sigma^4$
$W^{2'}$	$\Sigma^3+\Sigma^4$	$\Sigma^1+\Sigma^2$	$\Sigma^1+\Sigma^2+\Sigma^3+\Sigma^4$
W^3	$\Sigma^1+\Sigma^2+\Sigma^3+\Sigma^4$	$\Sigma^1+\Sigma^2+\Sigma^3+\Sigma^4$	$2\Sigma^1+2\Sigma^2+2\Sigma^3+2\Sigma^4$
$(L^{(m-)}W^{(m')}|)=(L^{(m+)}W^{(m')}|)$.			
	$L\times\Delta(\pi,0,0)(1/a)=2\Sigma(\pi,\pi,0)(1/a)$		
Δ^1	$\Sigma^1+\Sigma^3$	$\Sigma^2+\Sigma^4$	$\Sigma^1+\Sigma^2+\Sigma^3+\Sigma^4$
$\Delta^{1'}$	$\Sigma^2+\Sigma^4$	$\Sigma^1+\Sigma^3$	$\Sigma^1+\Sigma^2+\Sigma^3+\Sigma^4$
Δ^2	$\Sigma^2+\Sigma^4$	$\Sigma^1+\Sigma^3$	$\Sigma^1+\Sigma^2+\Sigma^3+\Sigma^4$
$\Delta^{2'}$	$\Sigma^1+\Sigma^3$	$\Sigma^2+\Sigma^4$	$\Sigma^1+\Sigma^2+\Sigma^3+\Sigma^4$
Δ^5	$\Sigma^1+\Sigma^2+\Sigma^3+\Sigma^4$	$\Sigma^1+\Sigma^2+\Sigma^3+\Sigma^4$	$2\Sigma^1+2\Sigma^2+2\Sigma^3+2\Sigma^4$
$(L^{(n-)}\Delta^{(m)}|)=(L^{(n)}\Delta^{(m')}|)$, $(L^{(n-)}\Delta^{(m')}|)=(L^{(n)}\Delta^{(m)}|)$, $(L^{(n-)}\Delta^{(5)}|)=(L^{(n)}\Delta^{(5)}|)$.			
	$L\times\Sigma(\pi,\pi,0)(1/a)=8\Delta$		
Σ^1	$\Delta^1+\Delta^{2'}+2\Delta^5$	$L^{1+}\times\Sigma^2$	$2\Delta^1+2\Delta^{1'}+2\Delta^2+2\Delta^{2'}+4\Delta^5$
Σ^2	$\Delta^{1'}+\Delta^2+2\Delta^5$	$L^{1+}\times\Sigma^1$	$2\Delta^1+2\Delta^{1'}+2\Delta^2+2\Delta^{2'}+4\Delta^5$
Σ^3	$\Delta^{1'}+\Delta^2+2\Delta^5$	$L^{1+}\times\Sigma^1$	$2\Delta^1+2\Delta^{1'}+2\Delta^2+2\Delta^{2'}+4\Delta^5$
Σ^4	$\Delta^1+\Delta^{2'}+2\Delta^5$	$L^{1+}\times\Sigma^2$	$2\Delta^1+2\Delta^{1'}+2\Delta^2+2\Delta^{2'}+4\Delta^5$
$(L^{(1-)}\Sigma^1|)=(L^{(1-)}\Sigma^4|)=(L^{(2-)}\Sigma^2|)=(L^{(2-)}\Sigma^3|)=(L^{(1+)}\Sigma^2|)$, $(L^{(1-)}\Sigma^2|)=(L^{(1-)}\Sigma^3|)=(L^{(2-)}\Sigma^1|)=(L^{(2-)}\Sigma^4|)=(L^{(1+)}\Sigma^1|)$, $(L^{(3-)}\Sigma^{(m)}|)=(L^{(3+)}\Sigma^{(m)}|)$.			
	$L\times Z(2\pi,\kappa,0)(1/a)=2M(\pi,\pi,\pi+\kappa)(1/a)$		
Z^1	M^1+M^2	M^1+M^2	$2M^1+2M^2$
Z^2	M^1+M^2	M^1+M^2	$2M^1+2M^2$
Z^3	M^1+M^2	M^1+M^2	$2M^1+2M^2$
Z^4	M^1+M^2	M^1+M^2	$2M^1+2M^2$
$(L^{(n-)}Z^{(m)}|)=(L^{(n)}Z^{(m)}|)$.			
	$L\times\Lambda(\kappa,\kappa,\kappa)(1/a)=\bar{\Lambda}(\pi+\kappa,\pi+\kappa,\pi+\kappa)(1/a)+M(\pi+\kappa,\pi+\kappa,\pi-\kappa)(1/a)$		
Λ^1	Λ^1+M^1	Λ^2+M^2	$\Lambda^3+M^1+M^2$
Λ^2	Λ^2+M^2	Λ^1+M^1	$\Lambda^3+M^1+M^2$
Λ^3	$\Lambda^3+M^1+M^2$	$\Lambda^3+M^1+M^2$	$\Lambda^1+\Lambda^2+\Lambda^3+M^1+M^2$
$(L^{(n-)}\Lambda^1|)=(L^{(n)}\Lambda^2|)$, $(L^{(n-)}\Lambda^2|)=(L^{(n)}\Lambda^1|)$, $(L^{(n-)}\Lambda^3|)=(L^{(n)}\Lambda^3|)$.			
	$L\times Q=2N(\kappa+\pi,\kappa-\pi,0)(1/a)+2\Sigma(\kappa-\pi,\kappa-\pi,0)(1/a)+2\bar{\Sigma}(\pi+\kappa,\pi+\kappa,0)(1/a)$		
Q^1	$N^1+N^2+\Sigma^1+\Sigma^2+\bar{\Sigma}^1+\bar{\Sigma}^2$	$N^1+N^2+\Sigma^3+\Sigma^4+\bar{\Sigma}^3+\bar{\Sigma}^4$	$2N^1+2N^2+\Sigma^1+\Sigma^2+\Sigma^3+\Sigma^4+\bar{\Sigma}^1+\bar{\Sigma}^2+\bar{\Sigma}^3+\bar{\Sigma}^4$
Q^2	$Q^1\times L^{2+}$	$Q^1\times L^{1+}$	$Q^1\times L^{3+}$
$(L^{(n-)}Q^m|)=(L^{(n+)}Q^m|)$.			

Table A.3. Reductions of type $(\star \boldsymbol{k}\, m\, \star \boldsymbol{X}\, m'|)$ for O_h^5. See also caption to Table A.1

$k_{j'}^{(m')}$	$k_j^{(m)}$ X^{1+}	X^{2+}	X^{3+}	X^{4+}	X^{5+}
	$X \times X = 3\Gamma + 2X$				
X^{1+}	$\Gamma^{1+} + \Gamma^{12+} + X^{1+} + X^{2+}$	$X^1 \times X^2$			
X^{2+}	$\Gamma^{2+} + \Gamma^{12+} + X^{1+} + X^{2+}$	$X^1 \times X^1$			
X^{3+}	$\Gamma^{25+} + X^{5+}$	$X^1 \times X^4$	$\Gamma^{1+} + \Gamma^{12+} + X^{3+} + X^{4+}$		
X^{4+}	$\Gamma^{15+} + X^{5+}$	$X^1 \times X^3$	$\Gamma^{2+} + \Gamma^{12+} + X^{3+} + X^{4+}$	$X^3 \times X^3$	
X^{5+}	$\Gamma^{15+} + \Gamma^{25+} + X^{3+} + X^{4+} + X^{5+}$	$X^1 \times X^5$	$\Gamma^{15+} + \Gamma^{25+} + X^{1+} + X^{2+} + X^{5+}$	$X^3 \times X^5$	$\Gamma^{1+} + \Gamma^{2+} + 2\Gamma^{12+} + \Gamma^{15+} + \Gamma^{25+} + X^{1+} + X^{2+} + X^{3+} + X^{4+} + 2X^{5+}$
	$X \times W = W + 2\Delta(\pi, 0, 0)(1/a)$				
W^1	$W^1 + \Delta^1 + \Delta^2$	$X^{1+} \times W^{2'}$	$W^{1'} + \Delta^5$	$X^{3+} \times W^{2'}$	$W^3 + \Delta^{1'} + \Delta^{2'} + \Delta^5$
$W^{1'}$	$W^{1'} + \Delta^{1'} + \Delta^{2'}$	$X^{1+} \times W^2$	$W^1 + \Delta^5$	$X^{3+} \times W^2$	$W^3 + \Delta^1 + \Delta^2 + \Delta^5$
W^2	$W^2 + \Delta^{1'} + \Delta^{2'}$	$X^{1+} \times W^{1'}$	$W^{2'} + \Delta^5$	$X^{3+} \times W^{1'}$	$X^{5+} \times W^{1'}$
$W^{2'}$	$W^{2'} + \Delta^1 + \Delta^2$	$X^{1+} \times W^1$	$W^2 + \Delta^5$	$X^{3+} \times W^1$	$X^{5+} \times W^1$
W^3	$W^3 + 2\Delta^5$	$X^{1+} \times W^3$	$W^3 + \Delta^1 + \Delta^{1'} + \Delta^2 + \Delta^{2'}$	$X^{3+} \times W^3$	$W^1 + W^{1'} + W^2 + W^{2'} + \Delta^1 + \Delta^{1'} + \Delta^2 + \Delta^{2'} + 2\Delta^5$
$(X^{(n-)} W^m\|) = (X^{(n)} W^{m'}\|)$, $(X^{(n-)} W^{(m')}\|) = (X^{(n)} W^{(m)}\|)$, $(X^{(n-)} W^3\|) = (X^{(n)} W^3\|)$.					
	$X \times \Delta(\pi, 0, 0)(1/a) = \Delta(\pi, 0, 0)(1/a) + 2W(2\pi, \pi, 0)(1/a)$				
Δ^1	$\Delta^1 + W^1 + W^{2'}$	$X^{1+} \times \Delta^2$	$\Delta^{2'} + W^3$	$X^{3+} \times \Delta^2$	$\Delta^5 + W^{1'} + W^2 + W^3$
$\Delta^{1'}$	$\Delta^{1'} + W^{1'} + W^2$	$X^{1+} \times \Delta^{2'}$	$\Delta^2 + W^3$	$X^{3+} \times \Delta^{2'}$	$\Delta^5 + W^{1'} + W^{2'} + W^3$
Δ^2	$\Delta^2 + W^1 + W^{2'}$	$X^{1+} \times \Delta^1$	$\Delta^{1'} + W^3$	$X^{3+} \times \Delta^1$	$X^{5+} \times \Delta^1$
$\Delta^{2'}$	$\Delta^{2'} + W^{1'} + W^2$	$X^{1+} \times \Delta^{1'}$	$\Delta^1 + W^3$	$X^{3+} \times \Delta^{1'}$	$X^{5+} \times \Delta^{1'}$
Δ^5	$\Delta^5 + 2W^3$	$X^{1+} \times \Delta^5$	$\Delta^5 + W^1 + W^{1'} + W^2 + W^{2'}$	$X^{3+} \times \Delta^5$	$\Delta^1 + \Delta^{1'} + \Delta^2 + \Delta^{2'} + W^1 + W^{1'} + W^2 + W^{2'} + 2W^3$

$(X^{(n-)}\Delta^{(m)}|) = (X^{(n)}\Delta^{(m')}|)$, $(X^{(n-)}\Delta^{(m')}|) = (X^{(n)}\Delta^{(m)})$, $(X^{(n-)}\Delta^5|) = (X^{(n)}\Delta^5|)$.
In general $X \times \Delta(\kappa, 0, 0)(1/a) = \Delta(\kappa - 2\pi, 0, 0)(1/a) + Z(2\pi, \kappa, 0)(1/a)$.
The compatibility relations imply $Z^1 \leftrightarrow W^1 + W^{2'}$, $Z^2 \leftrightarrow W^{1'} + W^2$, Z^3 and $Z^4 \leftrightarrow W^3$.

Table A.3 (continued)

$k_{j'}^{(m')}$	$k_j^{(m)}$				
	X^{1+}	X^{2+}	X^{3+}	X^{4+}	X^{5+}
	$X \times \Sigma(\pi, \pi, 0)(1/a) = 3\Sigma(\pi, \pi, 0)(1/a)$				
Σ^1	$2\Sigma^1 + \Sigma^4$	$X^{1+} \times \Sigma^4$	$\Sigma^1 + \Sigma^2 + \Sigma^3$	$X^{3+} \times \Sigma^4$	$\Sigma^1 + 2\Sigma^2 + 2\Sigma^3 + \Sigma^4$
Σ^2	$2\Sigma^2 + \Sigma^3$	$X^{1+} \times \Sigma^3$	$\Sigma^1 + \Sigma^2 + \Sigma^4$	$X^{3+} \times \Sigma^3$	$2\Sigma^1 + \Sigma^2 + \Sigma^3 + 2\Sigma^4$
Σ^3	$2\Sigma^3 + \Sigma^2$	$X^{1+} \times \Sigma^2$	$\Sigma^1 + \Sigma^3 + \Sigma^4$	$X^{3+} \times \Sigma^2$	$X^{5+} \times \Sigma^2$
Σ^4	$2\Sigma^4 + \Sigma^1$	$X^{1+} \times \Sigma^1$	$\Sigma^2 + \Sigma^3 + \Sigma^4$	$X^{3+} \times \Sigma^1$	$X^{5+} \times \Sigma^1$
$(X^{(m-)}\Sigma^1\|) = (X^{(m)}\Sigma^2\|)$, $(X^{(m-)}\Sigma^2\|) = (X^{(m)}\Sigma^1\|)$, $(X^{(m-)}\Sigma^3\|) = (X^{(m)}\Sigma^4\|)$, $(X^{(m-)}\Sigma^4\|) = (X^{(m)}\Sigma^3\|)$.					
	$X \times Z(2\pi, \kappa, 0)(1/a) = 2\Delta(\kappa, 0, 0)(1/a) + 2\bar{\Delta}(2\pi - \kappa, 0, 0)(1/a) + Z(2\pi, \kappa, 0)(1/a)$				
Z^1	$\Delta^1 + \Delta^2 + \Delta^1 + \Delta^2 + Z^1$	$X^{1+} \times Z^1$	$\Delta^5 + \Delta^5 + Z^2$	$X^{3+} \times Z^1$	$\Delta^{1'} + \Delta^{2'} + \Delta^5 + \Delta^{1'} + \Delta^{2'} + \bar{\Delta}^5 + Z^3 + Z^4$
Z^2	$\Delta^{1'} + \Delta^{2'} + \Delta^{1'} + \Delta^{2'} + Z^2$	$X^{1+} \times Z^2$	$\Delta^5 + \Delta^5 + Z^1$	$X^{3+} \times Z^2$	$\Delta^1 + \Delta^2 + \Delta^5 + \bar{\Delta}^1 + \Delta^2 + \Delta^5 + Z^3 + Z^4$
Z^3	$\Delta^5 + \Delta^5 + Z^4$	$X^{1+} \times Z^3$	$\Delta^1 + \Delta^2 + \Delta^{1'} + \Delta^{2'} + Z^3$	$X^{3+} \times Z^3$	$\Delta^{1'} + \Delta^{2'} + \Delta^5 + \Delta^1 + \Delta^2 + \bar{\Delta}^5 + Z^1 + Z^2$
Z^4	$\Delta^5 + \Delta^5 + Z^3$	$X^{1+} \times Z^4$	$\Delta^{1'} + \Delta^{2'} + \Delta^1 + \Delta^2 + Z^4$	$X^{3+} \times Z^4$	$\Delta^1 + \Delta^2 + \Delta^5 + \bar{\Delta}^{1'} + \Delta^{2'} + \bar{\Delta}^5 + Z^1 + Z^2$
$(X^{(m-)}Z^1\|) = (X^{(m)}Z^2\|)$, $(X^{(m-)}Z^2\|) = (X^{(m)}Z^1\|)$, $(X^{(m-)}Z^3\|) = (X^{(m)}Z^4\|)$, $(X^{(m-)}Z^4\|) = (X^{(m)}Z^3\|)$.					
	$X \times \Lambda(\kappa, \kappa, \kappa)(1/a) = M(\kappa, \kappa, \kappa - 2\pi)(1/a)$				
Λ^1	M^1	M^2	M^1	M^2	$M^1 + M^2$
Λ^2	M^2	M^1	M^2	M^1	$M^1 + M^2$
Λ^3	$M^1 + M^2$	$M^1 + M^2$	$M^1 + M^2$	$M^1 + M^2$	$2M^1 + 2M^2$
$(X^{(m-)}\Lambda^1\|) = (X^{(m)}\Lambda^2\|)$, $(X^{(m-)}\Lambda^2\|) = (X^{(m)}\Lambda^1\|)$, $(X^{(m-)}\Lambda^3\|) = (X^{(m)}\Lambda^3\|)$.					
	$Q \times X = 2M(\kappa, \kappa, \pi)(1/a) + Q$				
Q^1	$Q^1 + M^1 + M^2$	$Q^2 + M^1 + M^2$	$Q^1 \times X^{1+}$	$Q^1 \times X^{2+}$	$Q^1 + Q^2 + 2M^1 + 2M^2$
Q^2	$Q^1 \times X^{2+}$	$Q^1 \times X^{1+}$	$Q^1 \times X^{2+}$	$Q^1 \times X^{1+}$	$Q^1 \times X^{5+}$
$(X^{(m-)}Q^{(n)}\|) = (X^{(m+)}Q^{(n)}\|)$.					

Table A.4. Reductions of type $(\star k\, m \star W m'|)$ for O_h^5. See also caption to Table A.1

$k_{j'}^{(m')}$	$k_j^{(m)}$ W^1	$W^{1'}$	W^2	$W^{2'}$	W^3
	$W \times W = 6\Gamma + 2X + 2\Sigma$				
W^1	$\Gamma^{1+} + \Gamma^{12+} + \Gamma^{25-} + X^{1+} + X^{3-} + \Sigma^1 + \Sigma^4$				
$W^{1'}$	$\Gamma^{1-} + \Gamma^{12-} + \Gamma^{25+} + X^{1-} + X^{3+} + \Sigma^2 + \Sigma^3$	$W^1 \times W^1$			
W^2	$\Gamma^{2-} + \Gamma^{12-} + \Gamma^{15+} + X^{2-} + X^{4+} + \Sigma^2 + \Sigma^3$	$W^1 \times W^{2'}$	$W^1 \times W^1$		
$W^{2'}$	$\Gamma^{2+} + \Gamma^{12+} + \Gamma^{15-} + X^{2+} + X^{4-} + \Sigma^1 + \Sigma^4$	$W^1 \times W^2$	$W^1 \times W^{1'}$	$W^1 \times W^1$	
W^3	$\Gamma^{15\pm} + \Gamma^{25\pm} + X^{5\pm} + \Sigma^1 + \Sigma^2 + \Sigma^3 + \Sigma^4$	$W^1 \times W^3$	$W^1 \times W^3$	$W^1 \times W^3$	$\Gamma^{1\pm} + \Gamma^{2\pm} + \Gamma^{15\pm} + \Gamma^{25\pm} + 2\Gamma^{12\pm} + 2\Sigma^1 + 2\Sigma^2 + 2\Sigma^3 + 2\Sigma^4 + X^{1\pm} + X^{2\pm} + X^{3\pm} + X^{4\pm}$
	$W \times \Delta(\pi, 0, 0)(1/a) = 4X + 2\Sigma(\pi, \pi, 0)(1/a)$				
Δ^1	$X^{1+} + X^{2+} + X^{5-} + \Sigma^1 + \Sigma^4$	$W^1 \times \Delta^{1'}$	$W^1 \times \Delta^{1'}$	$W^1 \times \Delta^1$	$W^1 \times \Delta^5$
$\Delta^{1'}$	$X^{1-} + X^{2-} + X^{5+} + \Sigma^2 + \Sigma^3$	$W^1 \times \Delta^1$	$W^1 \times \Delta^1$	$W^1 \times \Delta^{1'}$	$W^1 \times \Delta^5$
Δ^2	$W^1 \times \Delta^1$	$W^1 \times \Delta^{1'}$	$W^1 \times \Delta^{1'}$	$W^1 \times \Delta^1$	$W^1 \times \Delta^5$
$\Delta^{2'}$	$W^1 \times \Delta^{1'}$	$W^1 \times \Delta^1$	$W^1 \times \Delta^1$	$W^1 \times \Delta^{1'}$	$W^1 \times \Delta^5$
Δ^5	$X^{3\pm} + X^{4\pm} + X^{5\pm} + \Sigma^1 + \Sigma^2 + \Sigma^3 + \Sigma^4$	$W^1 \times \Delta^5$	$W^1 \times \Delta^5$	$W^1 \times \Delta^5$	$2X^{1\pm} + 2X^{2\pm} + 2X^{5\pm} + 2\Sigma^1 + 2\Sigma^2 + 2\Sigma^3 + 2\Sigma^4$
	$W \times \Sigma(\pi, \pi, 0)(1/a) = 6L + 4\Delta(\pi\ 0, 0)(1/a) + 4W$				
Σ^1	$L^{1\pm} + L^{3\pm} + \Delta^1 + \Delta^2 + \Delta^5 + W^1 + W^{2'} + W^3$	$W^1 \times \Sigma^2$	$W^1 \times \Sigma^3$	$W^1 \times \Sigma^4$	$L^{1\pm} + L^{2\pm} + 2L^{3\pm} + \Delta^1 + \Delta^{1'} + \Delta^2 + \Delta^{2'} + 2\Delta^5 + W^1 + W^{1'} + W^2 + W^{2'} + 2W^3$
Σ^2	$L^{1\pm} + L^{3\pm} + \Delta^{1'} + \Delta^{2'} + \Delta^5 + W^{1'} + W^2 + W^3$	$W^1 \times \Sigma^1$	$W^1 \times \Sigma^4$	$W^1 \times \Sigma^3$	$L^{1\pm} + L^{2\pm} + 2L^{3\pm} + \Delta^1 + \Delta^{1'} + \Delta^2 + \Delta^{2'} + 2\Delta^5 + W^1 + W^{1'} + W^2 + W^{2'} + 2W^3$
Σ^3	$L^{2\pm} + L^{3\pm} + \Delta^{1'} + \Delta^{2'} + \Delta^5 + W^{1'} + W^2 + W^3$	$W^1 \times \Sigma^4$	$W^1 \times \Sigma^1$	$W^1 \times \Sigma^2$	$L^{1\pm} + L^{2\pm} + 2L^{3\pm} + \Delta^1 + \Delta^{1'} + \Delta^2 + \Delta^{2'} + 2\Delta^5 + W^1 + W^{1'} + W^2 + W^{2'} + 2W^3$
Σ^4	$L^{2\pm} + L^{3\pm} + \Delta^1 + \Delta^2 + \Delta^5 + W^1 + W^{2'} + W^3$	$W^1 \times \Sigma^3$	$W^1 \times \Sigma^2$	$W^1 \times \Sigma^1$	$L^{1\pm} + L^{2\pm} + 2L^{3\pm} + \Delta^1 + \Delta^{1'} + \Delta^2 + \Delta^{2'} + 2\Delta^5 + W^1 + W^{1'} + W^2 + W^{2'} + 2W^3$
	$W \times Q = k(\kappa, \pi-\kappa, \pi)(1/a) + \bar{k}(\kappa, \pi+\kappa, \pi)(1/a) + 2\Sigma(2\pi+\kappa, 2\pi+\kappa, 0)(1/a) + 2\bar{\Sigma}(\kappa, \kappa, 0)(1/a)$				
Q^1	$k + \bar{k} + \Sigma^1 + \Sigma^2 + \bar{\Sigma}^1 + \bar{\Sigma}^2$	$Q^1 \times W^1$	$k + \bar{k} + \Sigma^3 + \Sigma^4 + \bar{\Sigma}^3 + \bar{\Sigma}^4$	$Q^1 \times W^1$	$2k + 2\bar{k} + \Sigma^1 + \Sigma^2 + \Sigma^3 + \Sigma^4 + \bar{\Sigma}^1 + \bar{\Sigma}^2 + \bar{\Sigma}^3 + \bar{\Sigma}^4$
Q^2	$Q^1 \times W^2$	$Q^1 \times W^2$	$Q^1 \times W^1$	$Q^1 \times W^1$	$Q^1 \times W^3$

Table A.5. Reductions of type ($\star k\, m \star \Delta\, m'|$) for O_h^5. See also caption to Table A.1

$k_{j'}^{(m')}$	$k_j^{(m)}$				
	Δ^1	$\Delta^{1'}$	Δ^2	$\Delta^{2'}$	Δ^5
	$\Delta \times \Delta = 6\Gamma + 2X + 2\Sigma(\pi,\pi,0)(1/a)$				
Δ	$\Gamma^{1+} + \Gamma^{12+} + \Gamma^{15-} + X^{1+} + X^{4-} + \Sigma^1 + \Sigma^4$				
$\Delta^{1'}$	$\Gamma^{1-} + \Gamma^{12-} + \Gamma^{15+} + X^{1-} + X^{4+} + \Sigma^2 + \Sigma^3$	$\Delta^1 \times \Delta^1$			
Δ^2	$\Gamma^{2+} + \Gamma^{12+} + \Gamma^{25-} + X^{2+} + X^{3-} + \Sigma^1 + Z^4$	$\Delta^1 \times \Delta^{2'}$	$\Delta^1 \times \Delta^1$		
$\Delta^{2'}$	$\Gamma^{2-} + \Gamma^{12-} + \Gamma^{25+} + X^{2-} + X^{3+} + \Sigma^2 + \Sigma^3$	$\Delta^1 \times \Delta^2$	$\Delta^1 \times \Delta^{1'}$	$\Delta^1 \times \Delta^1$	
Δ^5	$\Gamma^{15\pm} + \Gamma^{25\pm} + X^{5\pm} + \Sigma^1 + \Sigma^2 + \Sigma^3 + \Sigma^4$	$\Delta^1 \times \Delta^5$	$\Delta^1 \times \Delta^5$	$\Delta^1 \times \Delta^5$	$\Gamma^{1\pm} + \Gamma^{2\pm} + \Gamma^{12\pm} + \Gamma^{15\pm} + \Gamma^{25\pm} + X^{1\pm} + X^{2\pm}$ $+ X^{3\pm} + X^{4\pm} + 2\Sigma^1 + 2\Sigma^2 + 2\Sigma^3 + 2\Sigma^4$
	$\Delta(\pi,0,0)(1/a) \times \Sigma(\pi,\pi,0)(1/a) = 4W + 4\Delta + 6L$				
Σ^1	$L^{1+} + L^{2-} + L^{3\pm} + W^1 + W^{2'} + W^3 + \Delta^1 + \Delta^2 + \Delta^5$	$\Delta^1 \times \Sigma^2$	$\Delta^1 \times \Sigma^4$	$\Delta^1 \times \Sigma^3$	$L^{1\pm} + L^{2\pm} + 2L^{3\pm} + W^1 + W^{1'} + W^2 + W^{2'}$ $+ 2W^3 + \Delta^1 + \Delta^{1'} + \Delta^2 + \Delta^{2'} + 2\Delta^5$
Σ^2	$L^{1-} + L^{2+} + L^{3\pm} + W^{1'} + W^2 + W^3 + \Delta^{1'} + \Delta^{2'} + \Delta^5$	$\Delta^1 \times \Sigma^1$	$\Delta^1 \times \Sigma^3$	$\Delta^1 \times \Sigma^4$	$L^{1\pm} + L^{2\pm} + 2L^{3\pm} + W^1 + W^{1'} + W^2 + W^{2'}$ $+ 2W^3 + \Delta^1 + \Delta^{1'} + \Delta^2 + \Delta^{2'} + 2\Delta^5$
Σ^3	$L^{1+} + L^{2-} + L^{3\pm} + W^{1'} + W^2 + W^3 + \Delta^{1'} + \Delta^{2'} + \Delta^5$	$\Delta^1 \times \Sigma^4$	$\Delta^1 \times \Sigma^2$	$\Delta^1 \times Z^1$	$L^{1\pm} + L^{2\pm} + 2L^{3\pm} + W^1 + W^{1'} + W^2 + W^{2'}$ $+ 2W^3 + \Delta^1 + \Delta^{1'} + \Delta^2 + \Delta^{2'} + 2\Delta^5$
Σ^4	$L^{1-} + L^{2+} + L^{3\pm} + W^1 + W^{2'} + W^3 + \Delta^1 + \Delta^2 + \Delta^5$	$\Delta^1 \times \Sigma^3$	$\Delta^1 \times \Sigma^1$	$\Delta^1 \times \Sigma^2$	$L^{1\pm} + L^{2\pm} + 2L^{3\pm} + W^1 + W^{1'} + W^2 + W^{2'}$ $+ 2W^3 + \Delta^1 + \Delta^{1'} + \Delta^2 + \Delta^{2'} + 2\Delta^5$
	$\Delta(\kappa,0,0)(1/a) \times \Lambda(\kappa',\kappa',\kappa')(1/a) = M(\kappa'+\kappa,\kappa',\kappa')(1/a) + \overline{M}(\kappa'-\kappa,\kappa',\kappa')(1/a)$				
Λ^1	$M^1 + \overline{M}^1$	$\Delta^1 \times \Lambda^2$	$\Delta^1 \times \Lambda^2$	$\Delta^1 \times \Lambda^1$	$\Delta^1 \times \Lambda^1$
Λ^2	$M^2 + \overline{M}^2$	$\Delta^1 \times \Lambda^1$	$\Delta^1 \times \Lambda^1$	$\Delta^1 \times \Lambda^2$	$\Delta^1 \times \Lambda^1$
Λ^3	$M^1 + M^2 + \overline{M}^1 + \overline{M}^2$	$\Delta^1 \times \Lambda^3$	$\Delta^1 \times \Lambda^3$	$\Delta^1 \times \Lambda^3$	$2M^1 + 2M^2 + 2\overline{M}^1 + 2\overline{M}^2$
	$\Delta(k,0,0)(1/a) \times Q = k(2\kappa,\pi,2\pi-\kappa)(1/a) + \bar{k}(\kappa,\pi+\kappa,2\pi-\kappa)(1/a) + 2N(2\pi+\kappa,\pi,0)(1/a)$				
Q^1	$k + \bar{k} + N^1 + N^2$	$Q^1 \times \Delta^1$	$Q^1 \times \Delta^1$	$Q^1 \times \Delta^1$	$2k + 2\bar{k} + 2N^1 + 2N^2$
Q^2	$Q^1 \times \Delta^1$	$Q^1 \times \Delta^1$	$Q^1 \times Q^1$	$Q^1 \times \Delta^1$	$Q^1 \times \Delta^5$

Table A.6. Reductions of type $(\star\boldsymbol{k}\,m\,\star\boldsymbol{\Sigma}\,m'|)$ for O_h^5. See also caption to Table A.1

$\boldsymbol{k}_{j'}^{(m')}$	$\boldsymbol{k}_j^{(m)}$			
	$\boldsymbol{\Sigma}^1$	$\boldsymbol{\Sigma}^2$	$\boldsymbol{\Sigma}^3$	$\boldsymbol{\Sigma}^4$
	$\boldsymbol{\Sigma}(\pi,\pi,0)(1/a)\times\boldsymbol{\Sigma}(\pi,\pi,0)(1/a)=12\boldsymbol{\Gamma}+12\boldsymbol{X}+8\boldsymbol{\Sigma}(\pi,\pi,0)(1/a)$			
$\boldsymbol{\Sigma}^1$	$\boldsymbol{\Gamma}^{1+}+\boldsymbol{\Gamma}^{12+}+\boldsymbol{\Gamma}^{15-}+\boldsymbol{\Gamma}^{25\pm}$ $+2\boldsymbol{X}^{1+}+\boldsymbol{X}^{2+}+\boldsymbol{X}^{3\pm}+\boldsymbol{X}^{4-}$ $+\boldsymbol{X}^{5+}+2\boldsymbol{X}^{5-}+2\boldsymbol{\Sigma}^1+2\boldsymbol{\Sigma}^2$ $+2\boldsymbol{\Sigma}^3+2\boldsymbol{\Sigma}^4$	$\boldsymbol{\Gamma}^{1-}+\boldsymbol{\Gamma}^{12-}+\boldsymbol{\Gamma}^{15+}+\boldsymbol{\Gamma}^{25\pm}$ $+2\boldsymbol{X}^{1-}+\boldsymbol{X}^{2-}+\boldsymbol{X}^{3\pm}+\boldsymbol{X}^{4+}$ $+\boldsymbol{X}^{5-}+2\boldsymbol{X}^{5+}+2\boldsymbol{\Sigma}^1+2\boldsymbol{\Sigma}^2$ $+2\boldsymbol{\Sigma}^3+2\boldsymbol{\Sigma}^4$	$\boldsymbol{\Gamma}^{2-}+\boldsymbol{\Gamma}^{12-}+\boldsymbol{\Gamma}^{15\pm}+\boldsymbol{\Gamma}^{25+}$ $+\boldsymbol{X}^{1-}+2\boldsymbol{X}^{2-}+\boldsymbol{X}^{3+}+\boldsymbol{X}^{4\pm}$ $+2\boldsymbol{X}^{5+}+\boldsymbol{X}^{5-}+2\boldsymbol{\Sigma}^1+2\boldsymbol{\Sigma}^2$ $+2\boldsymbol{\Sigma}^3+2\boldsymbol{\Sigma}^4$	$\boldsymbol{\Gamma}^{2+}+\boldsymbol{\Gamma}^{12+}+\boldsymbol{\Gamma}^{15\pm}+\boldsymbol{\Gamma}^{25-}$ $+\boldsymbol{X}^{1+}+2\boldsymbol{X}^{2+}+\boldsymbol{X}^{3-}+\boldsymbol{X}^{4\pm}$ $+\boldsymbol{X}^{5+}+2\boldsymbol{X}^{5-}+2\boldsymbol{\Sigma}^1+2\boldsymbol{\Sigma}^2$ $+2\boldsymbol{\Sigma}^3+2\boldsymbol{\Sigma}^4$
$\boldsymbol{\Sigma}^2$		$\boldsymbol{\Sigma}^1\times\boldsymbol{\Sigma}^1$	$\boldsymbol{\Sigma}^1\times\boldsymbol{\Sigma}^4$	$\boldsymbol{\Sigma}^1\times\boldsymbol{\Sigma}^3$
$\boldsymbol{\Sigma}^3$			$\boldsymbol{\Sigma}^1\times\boldsymbol{\Sigma}^1$	$\boldsymbol{\Sigma}^1\times\boldsymbol{\Sigma}^2$
$\boldsymbol{\Sigma}^4$				$\boldsymbol{\Sigma}^1\times\boldsymbol{\Sigma}^1$
	$\boldsymbol{\Sigma}(\kappa,\kappa,0)(1/a)\times\boldsymbol{\Lambda}(\kappa',\kappa',\kappa')(1/a)=\overline{\boldsymbol{M}}(\kappa'-\kappa,\kappa'-\kappa,\kappa')(1/a)+\boldsymbol{k}(\kappa'+\kappa,\kappa'-\kappa,\kappa')(1/a)+\boldsymbol{M}(\kappa'+\kappa,\kappa'+\kappa,\kappa')$			
$\boldsymbol{\Lambda}^1$	$\boldsymbol{M}^1+\overline{\boldsymbol{M}}^1+k$	$\boldsymbol{\Sigma}^1\times\boldsymbol{\Lambda}^2$	$\boldsymbol{\Sigma}^1\times\boldsymbol{\Lambda}^1$	$\boldsymbol{\Sigma}^1\times\boldsymbol{\Lambda}^2$
$\boldsymbol{\Lambda}^2$	$\boldsymbol{M}^2+\overline{\boldsymbol{M}}^2+k$	$\boldsymbol{\Sigma}^1\times\boldsymbol{\Lambda}^1$	$\boldsymbol{\Sigma}^1\times\boldsymbol{\Lambda}^2$	$\boldsymbol{\Sigma}^1\times\boldsymbol{\Lambda}^1$
$\boldsymbol{\Lambda}^3$	$\boldsymbol{M}^1+\boldsymbol{M}^2+\overline{\boldsymbol{M}}^1+\overline{\boldsymbol{M}}^2+2k$	$\boldsymbol{\Sigma}^1\times\boldsymbol{\Lambda}^3$	$\boldsymbol{\Sigma}^1\times\boldsymbol{\Lambda}^3$	$\boldsymbol{\Sigma}^1\times\boldsymbol{\Lambda}^3$
	$\boldsymbol{\Sigma}(\kappa,\kappa,0)(1/a)\times\boldsymbol{Z}(2\pi,\kappa',0)(1/a)=\boldsymbol{N}(2\pi+\kappa,\kappa+\kappa',0)(1/a)+\overline{\boldsymbol{N}}(2\pi+\kappa,\kappa-\kappa',0)(1/a)+\boldsymbol{N}'(2\pi+\kappa+\kappa',2\pi+\kappa,0)(1/a)$ $+\boldsymbol{N}''(2\pi+\kappa-\kappa',2\pi+\kappa,0)(1/a)+\boldsymbol{k}(2\pi+\kappa,\kappa,\kappa')(1/a)$			
$\boldsymbol{Z}^1$	$\boldsymbol{N}^1+\overline{\boldsymbol{N}}^1+\boldsymbol{N}'^1+\boldsymbol{N}''^1+k$	$\boldsymbol{\Sigma}^1\times\boldsymbol{Z}^2$	$\boldsymbol{\Sigma}^1\times\boldsymbol{Z}^2$	$\boldsymbol{\Sigma}^1\times\boldsymbol{Z}^1$
$\boldsymbol{Z}^2$	$\boldsymbol{N}^2+\overline{\boldsymbol{N}}^2+\boldsymbol{N}'^2+\boldsymbol{N}''^2+k$	$\boldsymbol{\Sigma}^1\times\boldsymbol{Z}^1$	$\boldsymbol{\Sigma}^1\times\boldsymbol{Z}^1$	$\boldsymbol{\Sigma}^1\times\boldsymbol{Z}^2$
$\boldsymbol{Z}^3$	$\boldsymbol{N}^2+\overline{\boldsymbol{N}}^2+\boldsymbol{N}'^1+\boldsymbol{N}''^1+k$	$\boldsymbol{\Sigma}^1\times\boldsymbol{Z}^4$	$\boldsymbol{\Sigma}^1\times\boldsymbol{Z}^4$	$\boldsymbol{\Sigma}^1\times\boldsymbol{Z}^3$
$\boldsymbol{Z}^4$	$\boldsymbol{N}^1+\overline{\boldsymbol{N}}^1+\boldsymbol{N}'^2+\boldsymbol{N}''^2+k$	$\boldsymbol{\Sigma}^1\times\boldsymbol{Z}^3$	$\boldsymbol{\Sigma}^1\times\boldsymbol{Z}^3$	$\boldsymbol{\Sigma}^1\times\boldsymbol{Z}^4$

Table A.7. Reductions of type $(^{\star}\Lambda\, m\, ^{\star}\Lambda\, m'|)$ for O_h^5. See also caption to Table A.1

$\boldsymbol{k}_{j'}^{(m')}$	$\boldsymbol{k}_j^{(m)}$		
	Λ^1	Λ^2	Λ^3
	$\Lambda(\kappa,\kappa,\kappa)(1/a)\times\Lambda(\kappa,\kappa,\kappa)(1/a)=8\Gamma+4\Delta(2\kappa,0,0)(1/a)$ $+2\Sigma(2\kappa,2\kappa,0)(1/a)+\bar{\Lambda}(2\kappa,2\kappa,2\kappa)(1/a)$		
Λ^1	$\Gamma^{1+}+\Gamma^{2-}+\Gamma^{15-}+\Gamma^{25+}$ $+\Lambda^1+\Delta^1+\Delta^{2'}+\Delta^5$ $+\Sigma^1+\Sigma^3$	$\Gamma^{1-}+\Gamma^2+\Gamma^{15+}+\Gamma^{25-}$ $+\Lambda^2+\Delta^{1'}+\Delta^2+\Delta^5$ $+\Sigma^2+\Sigma^4$	$\Gamma^{12\pm}+\Gamma^{15\pm}+\Gamma^{25\pm}$ $+\Delta^1+\Delta^{1'}+\Delta^2+\Delta^{2'}+2\Delta^5$ $+\Sigma^1+\Sigma^2+\Sigma^3+\Sigma^4+\Lambda^3$
Λ^2		$\Lambda^1\times\Lambda^1$	$\Lambda^1\times\Lambda^3$
Λ^3			$\Gamma^{1\pm}+\Gamma^{2\pm}+\Gamma^{12\pm}+2\Gamma^{15\pm}$ $+2\Gamma^{25\pm}+2\Delta^1+2\Delta^{1'}+2\Delta^2$ $+2\Delta^{2'}+4\Delta^5+2\Sigma^1+2\Sigma^2$ $+2\Sigma^3+2\Sigma^4+\Lambda^1+\Lambda^2+\Lambda^3$

When $\boldsymbol{k}=\frac{1}{2}\pi$: $\Lambda^1\to L^{1+}+L^{2-}$, $\Lambda^2\to L^{2+}+L^{1-}$, $\Lambda^3\to L^{3\pm}$.

Table A.8. Reductions of type $(\star\boldsymbol{k}\, m \star\boldsymbol{Z}\, m'|)$ for O_h^5. See also caption to Table A.1

$k_{j'}^{(m')}$	$k_j^{(m)}$ Z^1	Z^2	Z^3	Z^4
	$Z(2\pi,\kappa,0)(1/a)\times Z=12\Gamma+4X+2\Delta(2\kappa,0,0)(1/a)+2\bar{\Delta}(2\kappa-2\pi,0,0)(1/a)+2\Sigma(\kappa,\kappa,0)(1/a)+2\bar{\Sigma}(2\pi+\kappa,2\pi+\kappa,0)(1/a)+2N(2\pi-\kappa,\kappa,0)(1/a)$			
Z^1	$\Gamma^{1+}+\Gamma^{2+}+2\Gamma^{12+}+\Gamma^{15-}+\Gamma^{25-}+X^{1+}+X^{2+}+X^{3-}+X^{4-}+\Delta^1+\Delta^2+\bar{\Delta}^1+\bar{\Delta}^2+\Sigma^1+\bar{\Sigma}^4+\bar{\Sigma}^1+\Sigma^4+2N^1$	$\Gamma^{1-}+\Gamma^{2-}+2\Gamma^{12+}+\Gamma^{15+}+\Gamma^{25+}+X^{1-}+X^{2-}+X^{3+}+X^{4+}+\Delta^{1'}+\Delta^{2'}+\bar{\Delta}^{1'}+\bar{\Delta}^{2'}+\Sigma^2+\Sigma^3+\bar{\Sigma}^2+\bar{\Sigma}^3+2N^2$	$\Gamma^{15\pm}+\Gamma^{25\pm}+X^{5\pm}+\Delta^5+\bar{\Delta}^5+\Sigma^1+\Sigma^4+\bar{\Sigma}^2+\bar{\Sigma}^3+N^1+N^2$	$\Gamma^{15\pm}+\Gamma^{25\pm}+X^{5\pm}+\Delta^5+\bar{\Delta}^5+\Sigma^2+\Sigma^3+\bar{\Sigma}^1+\bar{\Sigma}^4+N^1+N^2$
Z^2		$Z^1\times Z^1$	$Z^1\times Z^4$	$Z^1\times Z^3$
Z^3			$\Gamma^{1+}+\Gamma^{2+}+2\Gamma^{12}+\Gamma^{15-}+\Gamma^{25-}+X^{1-}+X^{2-}+X^{3+}+X^{4+}+\Delta^1+\Delta^2+\bar{\Delta}^{1'}+\bar{\Delta}^{2'}+\Sigma^1+\Sigma^4+\bar{\Sigma}^1+\bar{\Sigma}^4+2N^2$	$\Gamma^{1-}+\Gamma^{2-}+2\Gamma^{12-}+\Gamma^{15+}+\Gamma^{25+}+X^{1+}+X^{2+}+X^{3-}+X^{4-}+\Delta^{1'}+\Delta^{2'}+\bar{\Delta}^1+\bar{\Delta}^2+\Sigma^2+\Sigma^3+\bar{\Sigma}^2+\bar{\Sigma}^3+2N^1$
Z^4				$Z^3\times Z^3$
	$Z\times W=2\Delta(\pi+\kappa,0,0)(1/a)+2\bar{\Delta}(\pi-k,0,0)(1/a)+N(\pi,2\pi-\kappa,0)(1/a)+\bar{N}(\pi,\kappa,0)(1/a)$			
W^1	$\Delta^1+\Delta^2+\bar{\Delta}^1+\bar{\Delta}^2+N^1+\bar{N}^1$	$Z^1\times W^{1'}$	$\Delta^5+\bar{\Delta}^5+N^2+\bar{N}^1$	$Z^3\times W^{1'}$
$W^{1'}$	$\Delta^{1'}+\Delta^{2'}+\bar{\Delta}^{1'}+\bar{\Delta}^{2'}+N^2+\bar{N}^2$	$Z^1\times W^1$	$\Delta^5+\bar{\Delta}^5+N^1+\bar{N}^2$	$Z^3\times W^1$
W^2	$Z^1\times W^{1'}$	$Z^1\times W^1$	$Z^3\times W^{1'}$	$Z^3\times W^1$
$W^{2'}$	$Z^1\times W^1$	$Z^1\times W^{1'}$	$Z^3\times W^1$	$Z^3\times W^{1'}$
W^3	$2\Delta^5+2\Delta^5+N^1+N^2+\bar{N}^1+\bar{N}^2$	$Z^1\times W^3$	$\Delta^1+\Delta^{1'}+\Delta^2+\Delta^{2'}+\bar{\Delta}^1+\Delta^{1'}+\Delta^2+\bar{\Delta}^{2'}+N^1+N^2+\bar{N}^1+\bar{N}^2$	$Z^3\times W^3$
	$Z(2\pi,\kappa,0)(1/a)\times\Delta(\kappa,0,0)(1/a)=4X+Z(2\pi,2\kappa,0)(1/a)+2\Sigma(2\pi+\kappa,2\pi+\kappa,0)(1/a)+N(2\pi+\kappa,\kappa,0)(1/a)$			
Δ^1	$X^{1+}+X^{2+}+X^{5-}+Z^1+\Sigma^1+\Sigma^4+N^1$	$Z^1\times\Delta^{1'}$	$X^{3+}+X^{4+}+X^{5-}+Z^3+\Sigma^1+\Sigma^4+N^2$	$Z^3\times\Delta^{1'}$
$\Delta^{1'}$	$X^{1-}+X^{2-}+X^{5+}+Z^2+\Sigma^2+\Sigma^3+N^2$	$Z^1\times\Delta^1$	$X^{3-}+X^{4-}+X^{5+}+Z^4+\Sigma^2+\Sigma^3+N^1$	$Z^3\times\Delta^1$
Δ^2	$Z^1\times\Delta^1$	$Z^1\times\Delta^{1'}$	$Z^3\times\Delta^1$	$Z^3\times\Delta^1$
$\Delta^{2'}$	$Z^1\times\Delta^{1'}$	$Z^1\times\Delta^1$	$Z^3\times\Delta^{1'}$	$Z^3\times\Delta^{1'}$
Δ^5	$X^{3\pm}+X^{4\pm}+X^{5\pm}+Z^3+Z^4+\Sigma^1+\Sigma^2+\Sigma^3+\Sigma^4+N^1+N^2$	$Z^1\times\Delta^5$	$X^{1\pm}+X^{2\pm}+X^{5\pm}+Z^1+Z^2+\Sigma^1+\Sigma^2+\Sigma^3+\Sigma^4+N^1+N^2$	$Z^3\times\Delta^5$
	$Q\times Z=2N(\kappa,\pi,0)(1/a)+2\bar{N}(2\pi-\kappa,\pi,0)(1/a)+2M(\kappa,\kappa,\pi+\kappa)(1/a)+2\bar{M}(\kappa,\kappa,\pi-\kappa)(1/a)+k(\kappa,\pi,-2\kappa)(1/a)+\bar{k}(2\kappa,\pi,2\pi-\kappa)(1/a)$			
Q^1	$N^1+N^2+\bar{N}^1+\bar{N}^2+M^1+M^2+\bar{M}^1+\bar{M}^2+k+\bar{k}$	$Q^1\times Z^1$	$Q^1\times Z^1$	$Q^1\times Z^1$
Q^2	$Q^1\times Z^1$	$Q^1\times Z^1$	$Q^1\times Z^1$	$Q^1\times Z^1$

Table A.9. Reductions of type ($^\star Q\, m\, ^\star Q\, m'|$) for O_h^5. See also caption to Table A.1

$k_{j'}^{(m')}$	$k_j^{(m)}$ Q^1	Q^2
	$Q\times Q = 24\Gamma + 8X + 4Z(2\kappa, 2\pi, 0)(1/a) + 2\Sigma(2\pi+2\kappa, 2\pi+2\kappa, 0)(1/a) + 2\bar{\Sigma}(2\kappa, 2\kappa, 0)(1/a)$ $+4\Sigma'(\pi+\kappa, \pi+\kappa, 0)(1/a) + 4\Sigma''(\pi-\kappa, \pi-\kappa, 0)(1/a) + 2M(\pi+\kappa, \pi+\kappa, 2\kappa)(1/a)$ $+2\bar{M}(\kappa-\pi, \kappa-\pi, 2\kappa)(1/a) + 4N(\pi+\kappa, \pi-\kappa, 0)(1/a)$ $+4k(\pi+\kappa, \pi-\kappa, 2\kappa)(1/a) + 8\Delta(2\kappa, 0, 0)(1/a)$	
Q^1	$\Gamma^{1+}+\Gamma^{12+}+\Gamma^{15+}+2\Gamma^{25+}$ $+\Gamma^{1-}+\Gamma^{12-}+\Gamma^{15-}+2\Gamma^{25-}$ $+X^{1+}+X^{3+}+X^{5+}+X^{1-}+X^{3-}+X^{5-}$ $+\Delta^1+\Delta^{1'}+\Delta^2+\Delta^{2'}+2\Delta^5$ $+Z^1+Z^2+Z^3+Z^4$ $+M^1+M^2+\bar{M}^1+\bar{M}^2$ $+2N^1+2N^2+4k$ $+\Sigma^1+\Sigma^2+\bar{\Sigma}^1+\bar{\Sigma}^2$ $+\Sigma'^1+\Sigma'^2+\Sigma'^3+\Sigma'^4$ $+\Sigma''^1+\Sigma''^2+\Sigma''^3+\Sigma''^4$	$\Gamma^{2+}+\Gamma^{12+}+2\Gamma^{15+}+\Gamma^{25+}$ $+\Gamma^{2-}+\Gamma^{12-}+2\Gamma^{15-}+\Gamma^{25-}$ $+X^{2+}+X^{4+}+X^{5+}+X^{2-}+X^{4-}+X^{5-}$ $+\Delta^1+\Delta^{1'}+\Delta^2+\Delta^{2'}+2\Delta^5$ $+Z^1+Z^2+Z^3+Z^4$ $+M^1+M^2+\bar{M}^1+\bar{M}^2$ $+2N^1+2N^2+4k$ $+\Sigma^3+\Sigma^4+\bar{\Sigma}^3+\bar{\Sigma}^4$ $+\Sigma'^1+\Sigma'^2+\Sigma'^3+\Sigma'^4$ $+\Sigma''^1+\Sigma''^2+\Sigma''^3+\Sigma''^4$
Q^2	$Q^1\times Q^2$	$Q^1\times Q^1$

Table A.10. Reduction coefficients for symmetrized Kronecker squares for O_h^5. See also caption to Table A.1

$([k_j^{(m)}]_{(2)}$	$k_{j''}^{(m'')})$
$[\Gamma^{(m)}]_{(2)}$	Γ^{1+}; $m=1^\pm, 2^\pm$
$[\Gamma^{(12\pm)}]_{(2)}$	$\Gamma^{1+}+\Gamma^{12+}$
$[\Gamma^{(m)}]_{(2)}$	$\Gamma^{12+}+\Gamma^{25+}$; $m=15^\pm, 25^\pm$
$[L^{(m)}]_{(2)}$	$\Gamma^{1+}+\Gamma^{25+}+X^{1+}+X^{3+}$; $m=1^\pm, 2^\pm$
$[L^{3\pm}]_{(2)}$	$\Gamma^{1+}+\Gamma^{12+}+\Gamma^{15+}+2\Gamma^{25+}+2X^{1+}+X^{2+}+2X^{3+}+X^{4+}+X^{5+}$
$[X^{(m)}]_{(2)}$	$\Gamma^{1+}+\Gamma^{12+}+X^{1+}$; $m=1^\pm, 2^\pm$
$[X^{(n)}]_{(2)}$	$\Gamma^{1+}+\Gamma^{12+}+X^{3+}$; $n=3^\pm, 4^\pm$
$[X^{5\pm}]_{(2)}$	$\Gamma^{1+}+\Gamma^{2+}+2\Gamma^{12+}+\Gamma^{25+}+X^{1+}+X^{3+}+X^{5+}$
$[W^{(m)}]_{(2)}$	$\Gamma^{1+}+\Gamma^{12+}+X^{1+}+X^{3-}+\Sigma^1$; $m=1, 1', 2, 2'$
$[W^3]_{(2)}$	$\Gamma^1+\Gamma^{2\pm}+2\Gamma^{12+}+\Gamma^{12-}+\Gamma^{25+}+X^{1\pm}+X^{2+}+X^{3\pm}+X^{4-}+2\Sigma^1+\Sigma^2+\Sigma^3$
$[\Delta^{(m)}]_{(2)}$	$\Gamma^{1+}+\Gamma^{12+}+X^{1-}+X^{4-}+\Sigma^1$; $m=1, 1', 2, 2'$
$[\Delta^5]_{(2)}$	$\Gamma^{1\pm}+\Gamma^{2+}+2\Gamma^{12+}+\Gamma^{12-}+\Gamma^{25+}+X^{1+}+X^{2\pm}+X^{3\pm}+X^{4-}+2\Sigma^1+\Sigma^2+\Sigma^3$
$[\Sigma^{(m)}]_{(2)}$	$\Gamma^{1+}+\Gamma^{12+}+\Gamma^{25+}+2X^{1+}+X^{2+}+X^{3\pm}+X^{4-}+X^{5-}+2\Sigma^1+\Sigma^2+\Sigma^3$; $m=1, 2, 3, 4$
$[\Lambda^{(m)}]_{(2)}$	$\Gamma^{1+}+\Gamma^{25+}+\Delta^1+\Delta^{2'}+\Sigma^1+\Lambda^1$; $m=1, 2$
$[\Lambda^3]_{(2)}$	$\Gamma^{1\pm}+\Gamma^{12+}+\Gamma^{15+}+\Gamma^{25-}+2\Gamma^{25+}+2\Delta^1+\Delta^{1'}+2\Delta^{2'}+\Delta^2+\Delta^5+\Sigma^1+\Sigma^2+\Sigma^4+\Lambda^1+\Lambda^3$
$[Z^{(m)}]_{(2)}$	$\Gamma^{1+}+\Gamma^{2+}+\Gamma^{12+}+X^{1+}+X^{3-}+\Delta^1+\Delta^2+\bar{\Delta}^1+\Sigma^1+\bar{\Sigma}^1+N^1$; $m=1, 2$
$[Z^{(n)}]_{(2)}$	$\Gamma^{1+}+\Gamma^{2+}+2\Gamma^{12+}+X^{1-}+X^{3+}+\Delta^1+\Delta^2+\bar{\Delta}^{1'}+\Sigma^1+\bar{\Sigma}^1+N^2$; $n=3, 4$
$[Q^{(m)}]_{(2)}$	$\Gamma^{1+}+\Gamma^{12+}+\Gamma^{15+}+2\Gamma^{25+}+X^{1+}+X^{3+}+X^{1-}+X^{3-}+\Delta^{1+}+\Delta^{1'}+\Delta^{2+}+\Delta^{2'}$ $+Z^1+Z^3+N^1+N^2+M^1+\bar{M}^1+2k$ $+\Sigma^1+\Sigma^2+\bar{\Sigma}^1+\Sigma'^3+\Sigma'^4+\Sigma''^3+\Sigma''^4$; $m=1, 2$

Table A.11. Reduction coefficients for symmetrized Kronecker cubes for O_h^5. See also caption to Table A.1

$([\boldsymbol{k}_j^{(m)}]_{(3)}$	$\boldsymbol{k}_{j''}^{(m'')})$
$[\Gamma^{(m)}]_{(3)}$	$\Gamma^m;\quad m=1^\pm, 2^\pm$
$[\Gamma^{12\pm}]_{(3)}$	$\Gamma^{1\pm}+\Gamma^{2\pm}+\Gamma^{12\pm}$
$[\Gamma^{15\pm}]_{(3)}$	$\Gamma^{2\pm}+2\Gamma^{15\pm}+\Gamma^{25\pm}$
$[\Gamma^{25\pm}]_{(3)}$	$\Gamma^{1\pm}+\Gamma^{15\pm}+2\Gamma^{25\pm}$
$[L^{(1\pm)}]_{(3)}$	$3L^{1\pm}+L^{3\pm}$
$[L^{(2\pm)}]_{(3)}$	$3L^{2\pm}+L^{3\pm}$
$[L^{(3\pm)}]_{(3)}$	$6L^{1\pm}+6L^{2\pm}+9L^{3\pm}$
$[X^{1\pm}]_{(3)}$	$\Gamma^{1\pm}+2X^{1\pm}+X^{2\pm}$
$[X^{2\pm}]_{(3)}$	$\Gamma^{2\pm}+X^{1\pm}+2X^{2\pm}$
$[X^{3\pm}]_{(3)}$	$\Gamma^{1\pm}+2X^{3\pm}+X^{4\pm}$
$[X^{4\pm}]_{(3)}$	$\Gamma^{2\pm}+X^{3\pm}+2X^{4\pm}$
$[X^{5\pm}]_{(3)}$	$\Gamma^{1\pm}+\Gamma^{2\pm}+\Gamma^{15\pm}+\Gamma^{25\pm}+X^{1\pm}+X^{2\pm}+X^{3\pm}+X^{4\pm}+6X^{5\pm}$
$[W^1]_{(3)}$	$3W^1+W^{2'}+\Delta^1+\Delta^2+\Delta^5+L^{1+}+L^{1-}$
$[W^{1'}]_{(3)}$	$3W^{1'}+W^{2'}+\Delta^{1'}+\Delta^{2'}+\Delta^5+L^{1+}+L^{1-}$
$[W^2]_{(3)}$	$3W^2+W^{1'}+\Delta^{1'}+\Delta^{2'}+\Delta^5+L^{2+}+L^{2-}$
$[W^{2'}]_{(3)}$	$3W^{2'}+W^1+\Delta^1+\Delta^2+\Delta^5+L^{2+}+L^{2-}$
$[W^3]_{(3)}$	$W^1+W^{1'}+W^2+W^{2'}+11W^3+3\Delta^1+3\Delta^{1'}+3\Delta^2+3\Delta^{2'}+6\Delta^5$ $+2L^{1+}+2L^{1-}+2L^{2+}+2L^{2-}+2L^{3+}+2L^{3-}$
$[\Delta^1]_{(3)}$	$3\Delta^1+\Delta^2+W^1+W^{2'}+W^3+L^{1+}+L^{2-}$
$[\Delta^{1'}]_{(3)}$	$3\Delta^{1'}+\Delta^{2'}+W^{1'}+W^2+W^3+L^{1-}+L^{2+}$
$[\Delta^2]_{(3)}$	$\Delta^1+3\Delta^2+W^1+W^{2'}+W^3+L^{1-}+L^{2+}$
$[\Delta^{2'}]_{(3)}$	$\Delta^{1'}+3\Delta^{2'}+W^{1'}+W^2+W^3+L^{1+}+L^{2-}$
$[\Delta^3]_{(3)}$	$\Delta^1+\Delta^{1'}+\Delta^2+\Delta^{2'}+11\Delta^3+3W^1+3W^{1'}+3W^2+3W^{2'}+6W^3$ $+2L^{1+}+2L^{2+}+2L^{1-}+2L^{2-}+2L^{3+}+2L^{3-}$
$[\Sigma^1]_{(3)}$	$2\Gamma^{1+}+\Gamma^{1-}+\Gamma^{2-}+\Gamma^{15-}+2\Gamma^{25+}+\Gamma^{25-}+2X^{1+}+X^{1-}+X^{2-}$ $+2X^{3+}+X^{3-}+X^{4-}+2X^{5\pm}+10\Sigma^1+4\Sigma^2+4\Sigma^3+7\Sigma^4$
$[\Sigma^2]_{(3)}$	$\Gamma^{1+}+2\Gamma^{1-}+\Gamma^{2+}+\Gamma^{15+}+\Gamma^{25+}+2\Gamma^{25-}+X^{1+}+2X^{1-}$ $+X^{2+}+2X^{3-}+X^{4+}+2X^{5\pm}+4\Sigma^1+10\Sigma^2+7\Sigma^3+4\Sigma^4+X^{3+}$
$[\Sigma^3]_{(3)}$	$\Gamma^{1+}+\Gamma^{2+}+2\Gamma^{2-}+\Gamma^{15+}+2\Gamma^{15-}+\Gamma^{25+}+X^{1+}+X^{2+}+2X^{2-}+X^{3+}$ $+X^{4+}+2X^{4-}+2X^{5\pm}+4\Sigma^1+7\Sigma^2+10\Sigma^3+4\Sigma^4+X^{3+}$
$[\Sigma^4]_{(3)}$	$\Gamma^{1-}+2\Gamma^{2+}+\Gamma^{2-}+2\Gamma^{15+}+\Gamma^{15-}+\Gamma^{25-}+X^{1-}+2X^{2+}+X^{2-}+X^{3-}$ $+2X^{4+}+X^{4-}+2X^{5\pm}+7\Sigma^1+4\Sigma^2+4\Sigma^3+10\Sigma^4$

Appendix B

Complete tables of reduction coefficients-selection rules for the diamond space group O_h^7

Tables B.1 to B.10 see following pages

Table B.1. Complete character table for reduction group for Γ, $\star X$, $\star L$ in diamond O_h^7. Reduction group characters for $\Gamma^{(m)}$, $\star L^{(m)}$, $\star X^{(m)}$. $\star L^{(m)}(\{|\varphi|t+Q\}) = -\star L^{(m)}(\{\varphi|t\})$

Element $\{\varphi\|\tau+t\}$	$\Gamma^{(1\pm)}$	$\Gamma^{(2\pm)}$	$\Gamma^{(12\pm)}$	$\Gamma^{(15\pm)}$	$\Gamma^{(25\pm)}$	$\star L^{(1\pm)}$	$\star L^{(2\pm)}$	$\star L^{(3\pm)}$	$\star X^{(1)}$	$\star X^{(2)}$	$\star X^{(3)}$	$\star X^{(4)}$
$\{\varepsilon\|0\}$	1	1	2	3	3	4	4	8	6	6	6	6
$\|t_{xy}\}$	1	1	2	3	3	0	0	0	−2	−2	−2	−2
$\|t_{xz}\}$	1	1	2	3	3	0	0	0	−2	−2	−2	−2
$\|t_{yz}\}$	1	1	2	3	3	0	0	0	−2	−2	−2	−2
$\{\delta_{2x}\|0\}$	1	1	2	−1	−1	0	0	0	2	2	−2	−2
	1	1	2	−1	−1	0	0	0	−2	−2	2	2
	1	1	2	−1	−1	0	0	0	−2	−2	2	2
	1	1	2	−1	−1	0	0	0	2	2	−2	−2
$\{\delta_{2y}\|0\}$	1	1	2	−1	−1	0	0	0	2	2	−2	−2
	1	1	2	−1	−1	0	0	0	−2	−2	2	2
	1	1	2	−1	−1	0	0	0	2	2	−2	−2
	1	1	2	−1	−1	0	0	0	−2	−2	2	2
$\{\delta_{2z}\|0\}$	1	1	2	−1	−1	0	0	0	2	2	−2	−2
	1	1	2	−1	−1	0	0	0	2	2	−2	−2
	1	1	2	−1	−1	0	0	0	−2	−2	2	2
	1	1	2	−1	−1	0	0	0	−2	−2	2	2
$\{\rho_{x\bar{y}}\|0\}$	±1	∓1	0	∓1	±1	±2	∓2	0	2	−2	0	0
	±1	∓1	0	∓1	±1	∓2	±2	0	2	−2	0	0
	±1	∓1	0	∓1	±1	0	0	0	−2	2	0	0
	±1	∓1	0	∓1	±1	0	0	0	−2	2	0	0
$\{\rho_{xy}\|0\}$	±1	∓1	0	∓1	±1	±2	∓2	0	2	−2	0	0
	±1	∓1	0	∓1	±1	±2	∓2	0	2	−2	0	0
	±1	∓1	0	∓1	±1	0	0	0	−2	2	0	0
	±1	∓1	0	∓1	±1	0	0	0	−2	2	0	0
$\{\rho_{x\bar{z}}\|0\}$	±1	∓1	0	∓1	±1	±2	∓2	0	2	−2	0	0
	±1	∓1	0	∓1	±1	0	0	0	−2	2	0	0
	±1	∓1	0	∓1	±1	∓2	±2	0	2	−2	0	0
	±1	∓1	0	∓1	±1	0	0	0	−2	2	0	0
$\{\rho_{xz}\|0\}$	±1	∓1	0	∓1	±1	±2	∓2	0	2	−2	0	0
	±1	∓1	0	∓1	±1	0	0	0	−2	2	0	0
	±1	∓1	0	∓1	±1	±2	∓2	0	2	−2	0	0
	±1	∓1	0	∓1	±1	0	0	0	−2	2	0	0

$\{\boldsymbol{\rho}_{y\bar{z}}\|0\}$	± 1	∓ 1	0	∓ 1	± 1	± 2	∓ 2	0	2	-2	0	0
	± 1	∓ 1	0	∓ 1	± 1	0	0	0	-2	2	0	0
	± 1	∓ 1	0	∓ 1	± 1	0	0	0	-2	2	0	0
	± 1	∓ 1	0	∓ 1	± 1	∓ 2	± 2	0	2	-2	0	0
$\{\boldsymbol{\rho}_{yz}\|0\}$	± 1	∓ 1	0	∓ 1	± 1	± 2	∓ 2	0	2	-2	0	0
	± 1	∓ 1	0	∓ 1	± 1	0	0	0	-2	2	0	0
	± 1	∓ 1	0	∓ 1	± 1	0	0	0	-2	2	0	0
	± 1	∓ 1	0	∓ 1	± 1	± 2	∓ 2	0	2	-2	0	0
$\{\boldsymbol{\sigma}_{4x}\|0\}$	± 1	∓ 1	0	± 1	∓ 1	0	0	0	0	0	0	0
	± 1	∓ 1	0	± 1	∓ 1	0	0	0	0	0	0	0
	± 1	∓ 1	0	± 1	∓ 1	0	0	0	0	0	0	0
	± 1	∓ 1	0	± 1	∓ 1	0	0	0	0	0	0	0
$\{\boldsymbol{\sigma}_{4x}^{-1}\|0\}$	± 1	∓ 1	0	± 1	∓ 1	0	0	0	0	0	0	0
	± 1	∓ 1	0	± 1	∓ 1	0	0	0	0	0	0	0
	± 1	∓ 1	0	± 1	∓ 1	0	0	0	0	0	0	0
	± 1	∓ 1	0	± 1	∓ 1	0	0	0	0	0	0	0
$\{\boldsymbol{\sigma}_{4y}\|0\}$	± 1	∓ 1	0	± 1	∓ 1	0	0	0	0	0	0	0
	± 1	∓ 1	0	± 1	∓ 1	0	0	0	0	0	0	0
	± 1	∓ 1	0	± 1	∓ 1	0	0	0	0	0	0	0
	± 1	∓ 1	0	± 1	∓ 1	0	0	0	0	0	0	0
$\{\boldsymbol{\sigma}_{4y}^{-1}\|0\}$	± 1	∓ 1	0	± 1	∓ 1	0	0	0	0	0	0	0
	± 1	∓ 1	0	± 1	∓ 1	0	0	0	0	0	0	0
	± 1	∓ 1	0	± 1	∓ 1	0	0	0	0	0	0	0
	± 1	∓ 1	0	± 1	∓ 1	0	0	0	0	0	0	0
$\{\boldsymbol{\sigma}_{4z}\|0\}$	± 1	∓ 1	0	± 1	∓ 1	0	0	0	0	0	0	0
	± 1	∓ 1	0	± 1	∓ 1	0	0	0	0	0	0	0
	± 1	∓ 1	0	± 1	∓ 1	0	0	0	0	0	0	0
	± 1	∓ 1	0	± 1	∓ 1	0	0	0	0	0	0	0
$\{\boldsymbol{\sigma}_{4z}^{-1}\|0\}$	± 1	∓ 1	0	± 1	∓ 1	0	0	0	0	0	0	0
	± 1	∓ 1	0	± 1	∓ 1	0	0	0	0	0	0	0
	± 1	∓ 1	0	± 1	∓ 1	0	0	0	0	0	0	0
	± 1	∓ 1	0	± 1	∓ 1	0	0	0	0	0	0	0

Table B.1 (continued)

Element $\{\varphi\|\tau+t\}$	$\Gamma^{(1\pm)}$	$\Gamma^{(2\pm)}$	$\Gamma^{(12\pm)}$	$\Gamma^{(15\pm)}$	$\Gamma^{(25\pm)}$	$\star L^{(1\pm)}$	$\star L^{(2\pm)}$	$\star L^{(3\pm)}$	$\star X^{(1)}$	$\star X^{(2)}$	$\star X^{(3)}$	$\star X^{(4)}$
$\{\delta_{3xyz}\|0\}$	1	1	−1	0	0	1	1	−1	0	0	0	0
	1	1	−1	0	0	−1	−1	+1	0	0	0	0
	1	1	−1	0	0	−1	−1	+1	0	0	0	0
	1	1	−1	0	0	−1	−1	+1	0	0	0	0
$\{\delta_{3xyz}^{-1}\|0\}$	1	1	−1	0	0	1	1	−1	0	0	0	0
	1	1	−1	0	0	−1	−1	1	0	0	0	0
	1	1	−1	0	0	−1	−1	1	0	0	0	0
	1	1	−1	0	0	−1	−1	1	0	0	0	0
$\{\delta_{3x\bar{y}\bar{z}}\|0\}$	1	1	−1	0	0	+1	+1	−1	0	0	0	0
	1	1	−1	0	0	+1	+1	−1	0	0	0	0
	1	1	−1	0	0	+1	+1	−1	0	0	0	0
	1	1	−1	0	0	−1	−1	+1	0	0	0	0
$\{\delta_{3x\bar{y}\bar{z}}^{-1}\|0\}$	1	1	−1	0	0	+1	+1	−1	0	0	0	0
	1	1	−1	0	0	+1	+1	−1	0	0	0	0
	1	1	−1	0	0	+1	+1	−1	0	0	0	0
	1	1	−1	0	0	−1	−1	+1	0	0	0	0
$\{\delta_{3\bar{x}y\bar{z}}\|0\}$	1	1	−1	0	0	+1	+1	−1	0	0	0	0
	1	1	−1	0	0	+1	+1	−1	0	0	0	0
	1	1	−1	0	0	−1	−1	+1	0	0	0	0
	1	1	−1	0	0	+1	+1	−1	0	0	0	0
$\{\delta_{3\bar{x}y\bar{z}}^{-1}\|0\}$	1	1	−1	0	0	+1	+1	−1	0	0	0	0
	1	1	−1	0	0	+1	+1	−1	0	0	0	0
	1	1	−1	0	0	−1	−1	+1	0	0	0	0
	1	1	−1	0	0	+1	+1	−1	0	0	0	0
$\{\delta_{3\bar{x}\bar{y}z}\|0\}$	1	1	−1	0	0	+1	+1	−1	0	0	0	0
	1	1	−1	0	0	−1	−1	+1	0	0	0	0
	1	1	−1	0	0	+1	+1	−1	0	0	0	0
	1	1	−1	0	0	+1	+1	−1	0	0	0	0
$\{\delta_{3\bar{x}\bar{y}z}^{-1}\|0\}$	1	1	−1	0	0	+1	+1	−1	0	0	0	0
	1	1	−1	0	0	−1	−1	+1	0	0	0	0
	1	1	−1	0	0	+1	+1	−1	0	0	0	0
	1	1	−1	0	0	+1	+1	−1	0	0	0	0

$\{\boldsymbol{i}\|\tau_1\}$	±1	±1	±2	±3	±3	∓2	∓2	∓4	0	0	0	0
	±1	±1	±2	±3	±3	∓2	∓2	∓4	0	0	0	0
	±1	±1	±2	±3	±3	∓2	∓2	∓4	0	0	0	0
	±1	±1	±2	±3	±3	∓2	∓2	∓4	0	0	0	0
$\{\boldsymbol{\rho}_x\|\tau_1\}$	±1	±1	±2	∓1	∓1	0	0	0	0	0	0	0
	±1	±1	±2	∓1	∓1	0	0	0	0	0	0	0
	±1	±1	±2	∓1	∓1	0	0	0	0	0	0	0
	±1	±1	±2	∓1	∓1	0	0	0	0	0	0	0
$\{\boldsymbol{\rho}_y\|\tau_1\}$	±1	±1	±2	∓1	∓1	0	0	0	0	0	0	0
	±1	±1	±2	∓1	∓1	0	0	0	0	0	0	0
	±1	±1	±2	∓1	∓1	0	0	0	0	0	0	0
	±1	±1	±2	∓1	∓1	0	0	0	0	0	0	0
$\{\boldsymbol{\rho}_z\|\tau_1\}$	±1	±1	±2	∓1	∓1	0	0	0	0	0	0	0
	±1	±1	±2	∓1	∓1	0	0	0	0	0	0	0
	±1	±1	±2	∓1	∓1	0	0	0	0	0	0	0
	±1	±1	±2	∓1	∓1	0	0	0	0	0	0	0
$\{\boldsymbol{\delta}_{2x\bar{y}}\|\tau_1\}$	1	−1	0	−1	1	0	0	0	0	0	−2	2
	1	−1	0	−1	1	0	0	0	0	0	−2	2
	1	−1	0	−1	1	−2	+2	0	0	0	2	−2
	1	−1	0	−1	1	−2	+2	0	0	0	2	−2
$\{\boldsymbol{\delta}_{2xy}\|\tau_1\}$	1	−1	0	−1	1	−2	+2	0	0	0	2	−2
	1	−1	0	−1	1	−2	+2	0	0	0	2	−2
	1	−1	0	−1	1	0	0	0	0	0	−2	2
	1	−1	0	−1	1	0	0	0	0	0	−2	2
$\{\boldsymbol{\delta}_{2x\bar{z}}\|\tau_1\}$	1	−1	0	−1	1	0	0	0	0	0	−2	2
	1	−1	0	−1	1	−2	+2	0	0	0	2	−2
	1	−1	0	−1	1	0	0	0	0	0	−2	2
	1	−1	0	−1	1	−2	+2	0	0	0	2	−2
$\{\boldsymbol{\delta}_{2xy}\|\tau_1\}$	1	−1	0	−1	1	−2	+2	0	0	0	2	−2
	1	−1	0	−1	1	0	0	0	0	0	−2	2
	1	−1	0	−1	1	−2	+2	0	0	0	2	−2
	1	−1	0	−1	1	0	0	0	0	0	−2	2

Table B.1 (continued)

Element $\{\varphi \mid \tau + t\}$	$\Gamma^{(1\pm)}$	$\Gamma^{(2\pm)}$	$\Gamma^{(12\pm)}$	$\Gamma^{(15\pm)}$	$\Gamma^{(25\pm)}$	$\star L^{(1\pm)}$	$\star L^{(2\pm)}$	$\star L^{(3\pm)}$	$\star X^{(1)}$	$\star X^{(2)}$	$\star X^{(3)}$	$\star X^{(4)}$
$\{\delta_{2\bar{y}\bar{z}} \mid \tau_1\}$	1	−1	0	−1	1	0	0	0	0	0	−2	2
	1	−1	0	−1	1	−2	+2	0	0	0	2	−2
	1	−1	0	−1	1	−2	+2	0	0	0	2	−2
	1	−1	0	−1	1	0	0	0	0	0	−2	2
$\{\delta_{2yz} \mid \tau_1\}$	1	−1	0	−1	1	−2	+2	0	0	0	2	−2
	1	−1	0	−1	1	0	0	0	0	0	−2	2
	1	−1	0	−1	1	0	0	0	0	0	−2	2
	1	−1	0	−1	1	−2	+2	0	0	0	2	−2
$\{\delta_{4x} \mid \tau_1\}$	1	−1	0	1	−1	0	0	0	0	0	0	0
	1	−1	0	1	−1	0	0	0	0	0	0	0
	1	−1	0	1	−1	0	0	0	0	0	0	0
	1	−1	0	1	−1	0	0	0	0	0	0	0
$\{\delta_{4x}^{-1} \mid \tau_1\}$	1	−1	0	1	−1	0	0	0	0	0	0	0
	1	−1	0	1	−1	0	0	0	0	0	0	0
	1	−1	0	1	−1	0	0	0	0	0	0	0
	1	−1	0	1	−1	0	0	0	0	0	0	0
$\{\delta_{4y} \mid \tau_1\}$	1	−1	0	1	−1	0	0	0	0	0	0	0
	1	−1	0	1	−1	0	0	0	0	0	0	0
	1	−1	0	1	−1	0	0	0	0	0	0	0
	1	−1	0	1	−1	0	0	0	0	0	0	0
$\{\delta_{4y}^{-1} \mid \tau_1\}$	1	−1	0	1	−1	0	0	0	0	0	0	0
	1	−1	0	1	−1	0	0	0	0	0	0	0
	1	−1	0	1	−1	0	0	0	0	0	0	0
	1	−1	0	1	−1	0	0	0	0	0	0	0
$\{\delta_{4z} \mid \tau_1\}$	1	−1	0	1	−1	0	0	0	0	0	0	0
	1	−1	0	1	−1	0	0	0	0	0	0	0
	1	−1	0	1	−1	0	0	0	0	0	0	0
	1	−1	0	1	−1	0	0	0	0	0	0	0
$\{\delta_{4z}^{-1} \mid \tau_1\}$	1	−1	0	1	−1	0	0	0	0	0	0	0
	1	−1	0	1	−1	0	0	0	0	0	0	0
	1	−1	0	1	−1	0	0	0	0	0	0	0
	1	−1	0	1	−1	0	0	0	0	0	0	0

$\{\boldsymbol{\sigma}_{6xyz} \mid \boldsymbol{\tau}_1\}$	± 1	± 1	∓ 1	0	0	± 1	± 1	∓ 1	0	0	0	0
	± 1	± 1	∓ 1	0	0	∓ 1	∓ 1	± 1	0	0	0	0
	± 1	± 1	∓ 1	0	0	∓ 1	∓ 1	± 1	0	0	0	0
	± 1	± 1	∓ 1	0	0	∓ 1	∓ 1	± 1	0	0	0	0
$\{\boldsymbol{\sigma}_{6xyz}^{-1} \mid \boldsymbol{\tau}_1\}$	± 1	± 1	∓ 1	0	0	± 1	± 1	∓ 1	0	0	0	0
	± 1	± 1	∓ 1	0	0	∓ 1	∓ 1	± 1	0	0	0	0
	± 1	± 1	∓ 1	0	0	∓ 1	∓ 1	± 1	0	0	0	0
	± 1	± 1	± 1	0	0	∓ 1	± 1	± 1	0	0	0	0
$\{\boldsymbol{\sigma}_{6x\bar{y}\bar{z}} \mid \boldsymbol{\tau}_1\}$	± 1	± 1	∓ 1	0	0	∓ 1	∓ 1	± 1	0	0	0	0
	± 1	± 1	∓ 1	0	0	∓ 1	∓ 1	± 1	0	0	0	0
	± 1	± 1	∓ 1	0	0	∓ 1	∓ 1	± 1	0	0	0	0
	± 1	± 1	∓ 1	0	0	± 1	± 1	∓ 1	0	0	0	0
$\{\boldsymbol{\sigma}_{6x\bar{y}\bar{z}}^{-1} \mid \boldsymbol{\tau}_1\}$	± 1	± 1	∓ 1	0	0	∓ 1	∓ 1	± 1	0	0	0	0
	± 1	± 1	∓ 1	0	0	∓ 1	∓ 1	± 1	0	0	0	0
	± 1	± 1	∓ 1	0	0	∓ 1	∓ 1	± 1	0	0	0	0
	± 1	± 1	∓ 1	0	0	± 1	± 1	∓ 1	0	0	0	0
$\{\boldsymbol{\sigma}_{6\bar{x}y\bar{z}} \mid \boldsymbol{\tau}_1\}$	± 1	± 1	∓ 1	0	0	∓ 1	∓ 1	± 1	0	0	0	0
	± 1	± 1	∓ 1	0	0	∓ 1	∓ 1	± 1	0	0	0	0
	± 1	± 1	∓ 1	0	0	± 1	± 1	∓ 1	0	0	0	0
	± 1	± 1	∓ 1	0	0	∓ 1	∓ 1	± 1	0	0	0	0
$\{\boldsymbol{\sigma}_{6\bar{x}\bar{y}\bar{z}}^{-1} \mid \boldsymbol{\tau}_1\}$	± 1	± 1	∓ 1	0	0	∓ 1	∓ 1	± 1	0	0	0	0
	± 1	± 1	∓ 1	0	0	∓ 1	∓ 1	± 1	0	0	0	0
	± 1	± 1	∓ 1	0	0	± 1	± 1	∓ 1	0	0	0	0
	± 1	± 1	∓ 1	0	0	∓ 1	∓ 1	± 1	0	0	0	0
$\{\boldsymbol{\sigma}_{6\bar{x}\bar{y}z} \mid \boldsymbol{\tau}_1\}$	± 1	± 1	∓ 1	0	0	∓ 1	∓ 1	± 1	0	0	0	0
	± 1	± 1	∓ 1	0	0	± 1	± 1	∓ 1	0	0	0	0
	± 1	± 1	∓ 1	0	0	∓ 1	∓ 1	± 1	0	0	0	0
	± 1	± 1	∓ 1	0	0	∓ 1	∓ 1	± 1	0	0	0	0
$\{\boldsymbol{\sigma}_{6\bar{x}\bar{y}z}^{-1} \mid \boldsymbol{\tau}_1\}$	± 1	± 1	∓ 1	0	0	∓ 1	∓ 1	± 1	0	0	0	0
	± 1	± 1	∓ 1	0	0	± 1	± 1	∓ 1	0	0	0	0
	± 1	± 1	∓ 1	0	0	∓ 1	∓ 1	± 1	0	0	0	0
	± 1	± 1	∓ 1	0	0	∓ 1	∓ 1	± 1	0	0	0	0

Table B.2.[a] Reduction coefficients for diamond space group O_h^7

$(\Gamma^{(1+)} k_j^{(m)})=1$
$(\Gamma^{(2)}\Gamma^{(2)}|\Gamma^{(1)})=1$
$(\Gamma^{(2)}\Gamma^{(12)}|\Gamma^{(12)})=1$
$(\Gamma^{(2)}\Gamma^{(15)}|\Gamma^{(25)})=1$
$(\Gamma^{(2)}\Gamma^{(25)}|\Gamma^{(15)})=1$
$(\Gamma^{(12)}\Gamma^{(12)}|\Gamma^{(1)})=(|\Gamma^{(2)})=(|\Gamma^{(12)})=1$
$(\Gamma^{(12)}\Gamma^{(15)}|\Gamma^{(15)})=(|\Gamma^{(25)})=1$
$(\Gamma^{(12)}\Gamma^{(25)}|\Gamma^{(15)})=(|\Gamma^{(25)})=1$
$(\Gamma^{(15)}\Gamma^{(25)}|\Gamma^{(1)})=(|\Gamma^{(12)})=(|\Gamma^{(15)})=(|\Gamma^{(25)})=1$
$(\Gamma^{(15)}\Gamma^{(25)}|\Gamma^{(2)})=(|\Gamma^{(12)})=(|\Gamma^{(15)})=(|\Gamma^{(25)})=1$
$(\Gamma^{(25)}\Gamma^{(25)}|)=(\Gamma^{(15)}\Gamma^{(15)}|)$

(For the above, the product of even × odd yields odd representations, etc.)

$(\Gamma^{(1-)}L^{(1+)}|L^{(1-)})=1$
$(\Gamma^{(1-)}L^{(1-)}|L^{(1+)})=1$
$(\Gamma^{(1-)}L^{(2+)}|L^{(2-)})=1$
$(\Gamma^{(1-)}L^{(2-)}|L^{(2+)})=1$
$(\Gamma^{(1-)}L^{(3+)}|L^{(3-)})=1$
$(\Gamma^{(1-)}L^{(3-)}|L^{(3+)})=1$
$(\Gamma^{(1-)}X^{(1)}|X^{(2)})=1$
$(\Gamma^{(1-)}X^{(2)}|X^{(1)})=1$
$(\Gamma^{(1-)}X^{(3)}|X^{(3)})=1$
$(\Gamma^{(1-)}X^{(4)}|X^{(4)})=1$
$(\Gamma^{(2+)}L^{(1+)}|)=(\Gamma^{(2-)}L^{(1-)}|)=(\Gamma^{(1-)}L^{(2-)}|)$
$(\Gamma^{(2+)}L^{(1-)}|)=(\Gamma^{(2-)}L^{(1+)}|)=(\Gamma^{(1-)}L^{(2+)}|)$
$(\Gamma^{(2+)}L^{(2+)}|)=(\Gamma^{(2-)}L^{(2-)}|)=(\Gamma^{(1-)}L^{(1-)}|)$
$(\Gamma^{(2+)}L^{(2-)}|)=(\Gamma^{(2-)}L^{(2+)}|)=(\Gamma^{(1-)}L^{(1+)}|)$
$(\Gamma^{(2+)}L^{(3+)}|)=(\Gamma^{(2-)}L^{(3-)}|)=(\Gamma^{(1-)}L^{(3-)}|)$
$(\Gamma^{(2+)}L^{(3-)}|)=(\Gamma^{(2-)}L^{(3+)}|)=(\Gamma^{(1-)}L^{(3+)}|)$
$(\Gamma^{(2+)}X^{(1)}|)=(\Gamma^{(2-)}X^{(2)}|)=(\Gamma^{(1-)}X^{(1)}|)$
$(\Gamma^{(2+)}X^{(2)}|)=(\Gamma^{(2-)}X^{(1)}|)=(\Gamma^{(1-)}X^{(2)}|)$
$(\Gamma^{(2+)}X^{(3)}|)=(\Gamma^{(2-)}X^{(3)}|)=(\Gamma^{(1-)}X^{(4)}|)$
$(\Gamma^{(2+)}X^{(4)}|)=(\Gamma^{(2-)}X^{(4)}|)=(\Gamma^{(1-)}X^{(3)}|)$
$(\Gamma^{(12+)}L^{(1+)}|L^{(3+)})=1$
$(\Gamma^{(12+)}L^{(1-)}|L^{(3-)})=1$
$(\Gamma^{(12+)}L^{(2+)}|L^{(3+)})=1$
$(\Gamma^{(12+)}L^{(2-)}|L^{(3-)})=1$
$(\Gamma^{(12+)}L^{(3+)}|L^{(1+)})=(|L^{(2+)})=(|L^{(3+)})=1$
$(\Gamma^{(12+)}L^{(3-)}|L^{(1-)})=(|L^{(2-)})=(|L^{(3-)})=1$
$(\Gamma^{(12+)}X^{(1)}|X^{(1)})=(|X^{(2)})=1$
$(\Gamma^{(12+)}X^{(2)}|X^{(1)})=(|X^{(2)})=1$
$(\Gamma^{(12+)}X^{(3)}|X^{(3)})=(|X^{(4)})=1$
$(\Gamma^{(12+)}X^{(4)}|X^{(3)})=(|X^{(4)})=1$
$(\Gamma^{(12-)}L^{(m\pm)}|)=(\Gamma^{(12+)}L^{(m\mp)}|)$
$(\Gamma^{(12-)}X^{(m)}|)=(\Gamma^{(12+)}X^{(m)}|)$
$(\Gamma^{(15+)}L^{(1+)}|L^{(2+)})=(|L^{(3+)})=1$
$(\Gamma^{(15+)}L^{(1-)}|L^{(2-)})=(|L^{(3-)})=1$
$(\Gamma^{(15+)}L^{(2+)}|L^{(1+)})=(|L^{(3+)})=1$
$(\Gamma^{(15+)}L^{(2-)}|L^{(1-)})=(|L^{(3-)})=1$
$(\Gamma^{(15+)}L^{(3+)}|L^{(1+)})=(|L^{(2+)})=1;$
$\quad(|L^{(3+)})=2$
$(\Gamma^{(15+)}L^{(3-)}|L^{(1-)})=(|L^{(2-)})=1;$
$\quad(|L^{(3-)})=2$
$(\Gamma^{(15+)}X^{(1)}|X^{(2)})=(|X^{(3)})=(|X^{(4)})=1$
$(\Gamma^{(15+)}X^{(2)}|X^{(1)})=(|X^{(3)})=(|X^{(4)})=1$
$(\Gamma^{(15+)}X^{(3)}|X^{(1)})=(|X^{(2)})=(|X^{(4)})=1$
$(\Gamma^{(15+)}X^{(4)}|X^{(1)})=(|X^{(2)})=(|X^{(3)})=1$
$(\Gamma^{(15-)}L^{(m\pm)}|)=(\Gamma^{(15+)}L^{(m\mp)}|)$
$(\Gamma^{(15-)}X^{(1)}|)=(\Gamma^{(15+)}X^{(2)}|)$
$(\Gamma^{(15-)}X^{(2)}|)=(\Gamma^{(15+)}X^{(1)}|)$
$(\Gamma^{(15-)}X^{(3)}|)=(\Gamma^{(15+)}X^{(3)}|)$
$(\Gamma^{(15-)}X^{(4)}|)=(\Gamma^{(15+)}X^{(4)}|)$
$(\Gamma^{(25+)}L^{(1+)}|L^{(1+)})=(|L^{(3+)})=1$
$(\Gamma^{(25+)}L^{(1-)}|L^{(1-)})=(|L^{(3-)})=1$
$(\Gamma^{(25+)}L^{(2+)}|L^{(2+)})=(|L^{(3+)})=1$
$(\Gamma^{(25+)}L^{(2-)}|L^{(2-)})=(|L^{(3-)})=1$
$(\Gamma^{(25+)}L^{(3+)}|L^{(1+)})=(|L^{(2+)})=1;$
$\quad(|L^{(3+)})=2$
$(\Gamma^{(25+)}L^{(3-)}|L^{(1-)})=(|L^{(2-)})=1;$
$\quad(|L^{(3-)})=2$
$(\Gamma^{(25+)}X^{(1)}|X^{(1)})=(|X^{(3)})=(|X^{(4)})=1$
$(\Gamma^{(25+)}X^{(2)}|X^{(2)})=(|X^{(3)})=(|X^{(4)})=1$
$(\Gamma^{(25+)}X^{(3)}|X^{(1)})=(|X^{(2)})=(|X^{(3)})=1$
$(\Gamma^{(25+)}X^{(4)}|X^{(1)})=(|X^{(2)})=(|X^{(4)})=1$
$(\Gamma^{(25-)}L^{(m\pm)}|)=(\Gamma^{(25+)}L^{(m\mp)}|)$
$(\Gamma^{(25-)}X^{(1)}|)=(\Gamma^{(25+)}X^{(2)}|)$
$(\Gamma^{(25-)}X^{(2)}|)=(\Gamma^{(25+)}X^{(1)}|)$
$(\Gamma^{(25-)}X^{(3)}|)=(\Gamma^{(25+)}X^{(3)}|)$
$(\Gamma^{(25-)}X^{(4)}|)=(\Gamma^{(25+)}X^{(4)}|)$

$(L^{(1+)}L^{(1+)}|\Gamma^{(1+)})=(|\Gamma^{(25+)})=(|X^{(1)})=(|X^{(3)})=1$
$(L^{(1+)}L^{(1-)}|\Gamma^{(1-)})=(|\Gamma^{(25-)})=(|X^{(2)})=(|X^{(3)})=1$
$(L^{(1+)}L^{(2+)}|\Gamma^{(2+)})=(|\Gamma^{(15+)})=(|X^{(2)})=(|X^{(4)})=1$
$(L^{(1+)}L^{(2-)}|\Gamma^{(2-)})=(|\Gamma^{(15-)})=(|X^{(1)})=(|X^{(4)})=1$
$(L^{(1+)}L^{(3+)}|\Gamma^{(12+)})=(|\Gamma^{(15+)})=(|\Gamma^{(25+)})=(|X^{(1)})$
$\quad=(|X^{(2)})=(|X^{(3)})=(|X^{(4)})=1$
$(L^{(1+)}L^{(3-)}|\Gamma^{(12-)})=(|\Gamma^{(15-)})=(|\Gamma^{(25-)})=(|X^{(1)})$
$\quad=(|X^{(2)})=(|X^{(3)})=(|X^{(4)})=1$
$(L^{(1+)}X^{(1)}|L^{(1+)})=(|L^{(2-)})=(|L^{(3+)})=(|L^{(3-)})=1$
$(L^{(1+)}X^{(2)}|L^{(1-)})=(|L^{(2+)})=(|L^{(3+)})=(|L^{(3-)})=1$
$(L^{(1+)}X^{(3)}|L^{(1+)})=(|L^{(1-)})=(|L^{(3+)})=(|L^{(3-)})=1$
$(L^{(1+)}X^{(4)}|L^{(2+)})=(|L^{(2-)})=(|L^{(3+)})=(|L^{(3-)})=1$
$(L^{(1-)}L^{(1-)}|)=(L^{(1+)}L^{(1+)}|)$
$(L^{(1-)}L^{(2+)}|)=(L^{(1+)}L^{(2-)}|)$
$(L^{(1-)}L^{(2-)}|)=(L^{(1+)}L^{(2+)}|)$
$(L^{(1-)}L^{(3+)}|)=(L^{(1+)}L^{(3-)}|)$
$(L^{(1-)}L^{(3-)}|)=(L^{(1+)}L^{(3+)}|)$
$(L^{(1-)}X^{(1)}|)=(L^{(1+)}X^{(2)}|)$
$(L^{(1-)}X^{(2)}|)=(L^{(1+)}X^{(1)}|)$
$(L^{(1-)}X^{(3)}|)=(L^{(1+)}X^{(3)}|)$
$(L^{(1-)}X^{(4)}|)=(L^{(1+)}X^{(4)}|)$
$(L^{(2+)}L^{(2+)}|)=(L^{(1+)}L^{(1+)}|)$
$(L^{(2+)}L^{(2-)}|)=(L^{(1+)}L^{(1-)}|)$
$(L^{(2+)}L^{(3+)}|)=(L^{(1+)}L^{(3+)}|)$
$(L^{(2+)}L^{(3-)}|)=(L^{(1+)}L^{(3-)}|)$
$(L^{(2+)}X^{(1)}|)=(L^{(1+)}X^{(2)}|)$
$(L^{(2+)}X^{(2)}|)=(L^{(1+)}X^{(1)}|)$
$(L^{(2+)}X^{(3)}|)=(L^{(1+)}X^{(4)}|)$
$(L^{(2+)}X^{(4)}|)=(L^{(1+)}X^{(3)}|)$

[a] Tables B.2–B.4 are taken in large part from J. L. BIRMAN: Phys. Rev. **127**, 1093 (1962).

Table B.2 (continued)

$(L^{(2-)}L^{(2-)}|)=(L^{(1+)}L^{(1+)}|)$

$(L^{(2-)}L^{(3+)}|)=(L^{(1+)}L^{(3-)}|)$

$(L^{(2-)}L^{(3-)}|)=(L^{(1+)}L^{(3+)}|)$

$(L^{(2-)}X^{(1)}|)=(L^{(1+)}X^{(1)}|)$

$(L^{(2-)}X^{(2)}|)=(L^{(1+)}X^{(2)}|)$

$(L^{(2-)}X^{(3)}|)=(L^{(1+)}X^{(4)}|)$

$(L^{(2-)}X^{(4)}|)=(L^{(1+)}X^{(3)}|)$

$(L^{(3+)}L^{(3+)}|\Gamma^{(1+)})=(|\Gamma^{(2+)})=(|\Gamma^{(12+)})=1;$
$(|\Gamma^{(15+)})=(|\Gamma^{(25+)})=(|X^{(1)})$
$=(|X^{(2)})=(|X^{(3)})=(|X^{(4)})=2$

$(L^{(3+)}L^{(3-)}|\Gamma^{(1-)})=(|\Gamma^{(2-)})=(|\Gamma^{(12-)})=1;$
$(|\Gamma^{(15-)})=(|\Gamma^{(25-)})=(|X^{(1)})$
$=(|X^{(2)})=(|X^{(3)})=(|X^{(4)})=2$

$(L^{(3+)}X^{(1)}|L^{(1+)})=(|L^{(1-)})=(|L^{(2+)})=(|L^{(2-)})=1;$
$(|L^{(3+)})=(|L^{(3-)})=2$

$(L^{(3+)}X^{(2)}|)=(L^{(3+)}X^{(3)}|)=(L^{(3+)}X^{(4)}|)$
$=(L^{(3+)}X^{(1)}|)$

$(L^{(3-)}L^{(3-)}|)=(L^{(3+)}L^{(3+)}|)$

$(L^{(3-)}X^{(1)}|)=(L^{(3-)}X^{(2)}|)=(L^{(3-)}X^{(3)}|)$
$=(L^{(3-)}X^{(4)}|)=(L^{(3+)}X^{(1)}|)$

$(X^{(1)}X^{(1)}|\Gamma^{(1+)})=(|\Gamma^{(2-)})=(|\Gamma^{(12+)})=(|\Gamma^{(12-)})$
$=(|\Gamma^{(15-)})=(|\Gamma^{(25+)})=(|X^{(1)})$
$=(|X^{(2)})=(|X^{(3)})=(|X^{(4)})=1$

$(X^{(1)}X^{(2)}|\Gamma^{(1-)})=(|\Gamma^{(2+)})=(|\Gamma^{(12+)})=(|\Gamma^{(12-)})$
$=(|\Gamma^{(15+)})=(|\Gamma^{(25-)})=(|X^{(1)})$
$=(|X^{(2)})=(|X^{(3)})=(|X^{(4)})=1$

$(X^{(1)}X^{(3)}|\Gamma^{(15+)})=(|\Gamma^{(15-)})=(|\Gamma^{(25+)})=(|\Gamma^{(25-)})$
$=(|X^{(1)})=(|X^{(2)})=(|X^{(3)})$
$=(|X^{(4)})=1$

$(X^{(1)}X^{(4)}|)=(X^{(1)}X^{(3)}|)$

$(X^{(2)}X^{(2)}|)=(X^{(1)}X^{(1)}|)$

$(X^{(2)}X^{(3)}|)=(X^{(1)}X^{(3)}|)$

$(X^{(2)}X^{(4)}|)=(X^{(1)}X^{(3)}|)$

$(X^{(3)}X^{(3)}|\Gamma^{(1+)})=(|\Gamma^{(1-)})=(|\Gamma^{(12+)})=(|\Gamma^{(12-)})$
$=(|\Gamma^{(25+)})=(|\Gamma^{(25-)})=(|X^{(1)})$
$=(|X^{(2)})=(|X^{(3)})=(|X^{(4)})=1$

$(X^{(3)}X^{(4)}|\Gamma^{(2+)})=(|\Gamma^{(2-)})=(|\Gamma^{(12+)})=(|\Gamma^{(12-)})$
$=(|\Gamma^{(15+)})=(|\Gamma^{(15-)})=(|X^{(1)})$
$=(|X^{(2)})=(|X^{(3)})=(|X^{(4)})=1$

$(X^{(4)}X^{(4)}|)=(X^{(3)}X^{(3)}|)$

$(\Gamma^{(1-)}W^{(1)}|W^{(2)})=1$

$(\Gamma^{(1-)}W^{(2)}|W^{(1)})=1$

$(\Gamma^{(2+)}W^{(m)}|)=(\Gamma^{(1-)}W^{(m)}|)$

$(\Gamma^{(2-)}W^{(m)}|)=(\Gamma^{(1+)}W^{(m)}|)$

$(\Gamma^{(12+)}W^{(1)}|W^{(1)})=(|W^{(2)})=1$

$(\Gamma^{(12+)}W^{(2)}|)=(\Gamma^{(12+)}W^{(1)}|)$

$(\Gamma^{(12-)}W^{(m)}|)=(\Gamma^{(12+)}W^{(m)}|)$

$(\Gamma^{(15+)}W^{(1)}|W^{(2)})=1; \quad (|W^{(1)})=2$

$(\Gamma^{(15+)}W^{(2)}|W^{(1)})=1; \quad (|W^{(2)})=2$

$(\Gamma^{(15-)}W^{(1)}|)=(\Gamma^{(15+)}W^{(2)}|)=(\Gamma^{(25+)}W^{(1)}|)$
$=(\Gamma^{(25-)}W^{(2)}|)$

$(\Gamma^{(15-)}W^{(2)}|)=(\Gamma^{(15+)}W^{(1)}|)=(\Gamma^{(25+)}W^{(2)}|)$
$=(\Gamma^{(25-)}W^{(1)}|)$

$(W^{(1)}W^{(1)}|\Gamma^{(1+)})=(|\Gamma^{(2-)})=(|\Gamma^{(12+)})=(|\Gamma^{(12-)})$
$=(|\Gamma^{(15-)})=(|\Gamma^{(25+)})=(|X^{(1)})$
$=(|X^{(2)})=(|X^{(3)})=(|X^{(4)})=1;$
$(|\Sigma^{(1)})=(|\Sigma^{(2)})=(|\Sigma^{(3)})=(|\Sigma^{(4)})$
$=(|\Gamma^{(15+)})=(|\Gamma^{(25-)})=2$

$(W^{(1)}W^{(2)}|\Gamma^{(1-)})=(|\Gamma^{(2+)})=(|\Gamma^{(12+)})=(|\Gamma^{(12-)})$
$=(|\Gamma^{(15+)})=(|\Gamma^{(25-)})=(|X^{(1)})$
$=(|X^{(2)})=(|X^{(3)})=(|X^{(4)})=1;$
$(|\Sigma^{(1)})=(|\Sigma^{(2)})=(|\Sigma^{(3)})=(|\Sigma^{(4)})$
$=(|\Gamma^{(15-)})=(|\Gamma^{(25+)})=2$

$(W^{(2)}W^{(2)}|)=(W^{(1)}W^{(1)}|)$

$(\Sigma^{(1)}\Sigma^{(1)}|\Sigma^{(1)})=(|\Sigma^{(2)})=(|\Sigma^{(3)})=(|\Sigma^{(4)})$
$=(|X^{(1)})=(|X^{(4)})=2;$
$=(|X^{(2)})=(|X^{(3)})=(|\Gamma^{(1+)})$
$=(|\Gamma^{(12+)})=(|\Gamma^{(15-)})=(|\Gamma^{(25+)})$
$=(|\Gamma^{(25-)})=1$

$(\Sigma^{(1)}\Sigma^{(2)}|\Sigma^{(1)})=(|\Sigma^{(2)})=(|\Sigma^{(3)})=(|\Sigma^{(4)})$
$=(|X^{(2)})=(|X^{(4)})=2;$
$(|X^{(1)})=(|X^{(3)})=(|\Gamma^{(1-)})$
$=(|\Gamma^{(12-)})=(|\Gamma^{(15+)})=(|\Gamma^{(25+)})$
$=(|\Gamma^{(25-)})=1$

$(\Sigma^{(1)}\Sigma^{(3)}|\Sigma^{(1)})=(|\Sigma^{(2)})=(|\Sigma^{(3)})=(|\Sigma^{(4)})$
$=(|X^{(1)})=(|X^{(3)})=2;$
$(|X^{(2)})=(|X^{(4)})=(|\Gamma^{(2-)})$
$=(|\Gamma^{(12-)})=(|\Gamma^{(15+)})=(|\Gamma^{(15-)})$
$=(|\Gamma^{(25+)})=1$

$(\Sigma^{(1)}\Sigma^{(4)}|\Sigma^{(1)})=(|\Sigma^{(2)})=(|\Sigma^{(3)})=(|\Sigma^{(4)})$
$=(|X^{(2)})=(|X^{(3)})=2;$
$(|X^{(1)})=(|X^{(4)})=(|\Gamma^{(2+)})$
$=(|\Gamma^{(12+)})=(|\Gamma^{(15+)})=(|\Gamma^{(15-)})$
$=(|\Gamma^{(25-)})=1$

$(\Sigma^{(m)}\Sigma^{(m)}|)=(\Sigma^{(1)}\Sigma^{(1)}|); \quad m=2,3,4$

$(\Sigma^{(3)}\Sigma^{(4)}|)=(\Sigma^{(1)}\Sigma^{(2)}|)$

$(\Sigma^{(2)}\Sigma^{(4)}|)=(\Sigma^{(1)}\Sigma^{(3)}|)$

$(\Sigma^{(2)}\Sigma^{(3)}|)=(\Sigma^{(1)}\Sigma^{(4)}|)$

$(\Gamma^{(1+)}\Delta^{(m)}|\Delta^{(m)})=1$

$(\Gamma^{(1-)}\Delta^{(1)}|\Delta^{(4)})=1 \quad (\Gamma^{(2-)}\Delta^{(1)}|\Delta^{(3)})=1$

$(\Gamma^{(1-)}\Delta^{(2)}|\Delta^{(3)})=1 \quad (\Gamma^{(2-)}\Delta^{(2)}|\Delta^{(4)})=1$

$(\Gamma^{(1-)}\Delta^{(3)}|\Delta^{(2)})=1 \quad (\Gamma^{(2-)}\Delta^{(3)}|\Delta^{(1)})=1$

$(\Gamma^{(1-)}\Delta^{(4)}|\Delta^{(1)})=1 \quad (\Gamma^{(2-)}\Delta^{(4)}|\Delta^{(2)})=1$

$(\Gamma^{(1-)}\Delta^{(5)}|\Delta^{(5)})=1 \quad (\Gamma^{(2-)}\Delta^{(5)}|\Delta^{(5)})=1$

$(\Gamma^{(2+)}\Delta^{(1)}|\Delta^{(2)})=1 \quad (\Gamma^{(12+)}\Delta^{(1)}|\Delta^{(1)})=(|\Delta^{(2)})=1$

$(\Gamma^{(2+)}\Delta^{(2)}|\Delta^{(1)})=1 \quad (\Gamma^{(12+)}\Delta^{(2)}|)=(\Gamma^{(12+)}\Delta^{(1)}|)$

$(\Gamma^{(2+)}\Delta^{(3)}|\Delta^{(4)})=1 \quad (\Gamma^{(12+)}\Delta^{(3)}|\Delta^{(3)})=(|\Delta^{(4)})=1$

$(\Gamma^{(2+)}\Delta^{(4)}|\Delta^{(3)})=1 \quad (\Gamma^{(12+)}\Delta^{(4)}|)=(\Gamma^{(12+)}\Delta^{(3)}|)$

$(\Gamma^{(2+)}\Delta^{(5)}|\Delta^{(5)})=1 \quad (\Gamma^{(12+)}\Delta^{(5)}|\Delta^{(5)})=2$

$(\Gamma^{(12-)}\Delta^{(1)}|)=(\Gamma^{(12+)}\Delta^{(3)}|)=(\Gamma^{(12-)}\Delta^{(2)}|)$

$(\Gamma^{(12-)}\Delta^{(3)}|)=(\Gamma^{(12+)}\Delta^{(1)}|)=(\Gamma^{(12-)}\Delta^{(4)}|)$

$(\Gamma^{(12-)}\Delta^{(5)}|)=(\Gamma^{(12+)}\Delta^{(5)}|)$

$(\Gamma^{(15+)}\Delta^{(1)}|\Delta^{(4)})=(|\Delta^{(5)})=1$

$(\Gamma^{(15+)}\Delta^{(2)}|\Delta^{(3)})=(|\Delta^{(5)})=1$

$(\Gamma^{(15+)}\Delta^{(3)}|\Delta^{(2)})=(|\Delta^{(5)})=1$

$(\Gamma^{(15+)}\Delta^{(4)}|\Delta^{(1)})=(|\Delta^{(5)})=1$

Table B.2 (continued)[a, b]

$(\Gamma^{(15+)}\Delta^{(5)}|\Delta^{(1)})=(|\Delta^{(2)})=(|\Delta^{(3)})$
$=(|\Delta^{(4)})=(|\Delta^{(5)})=1$
$(\Gamma^{(15-)}\Delta^{(1)}|)=(\Gamma^{(15+)}\Delta^{(4)}|)$
$(\Gamma^{(15-)}\Delta^{(2)}|)=(\Gamma^{(15+)}\Delta^{(3)}|)$
$(\Gamma^{(15-)}\Delta^{(3)}|)=(\Gamma^{(15+)}\Delta^{(2)}|)$
$(\Gamma^{(15-)}\Delta^{(4)}|)=(\Gamma^{(15+)}\Delta^{(1)}|)$
$(\Gamma^{(15-)}\Delta^{(5)}|)=(\Gamma^{(15+)}\Delta^{(5)}|)$
$(\Gamma^{(25+)}\Delta^{(1)}|)=(\Gamma^{(15+)}\Delta^{(2)}|)$
$(\Gamma^{(25+)}\Delta^{(2)}|)=(\Gamma^{(15+)}\Delta^{(1)}|)$
$(\Gamma^{(25+)}\Delta^{(3)}|)=(\Gamma^{(15+)}\Delta^{(4)}|)$
$(\Gamma^{(25+)}\Delta^{(4)}|)=(\Gamma^{(15+)}\Delta^{(3)}|)$
$(\Gamma^{(25+)}\Delta^{(5)}|)=(\Gamma^{(15+)}\Delta^{(5)}|)$
$(\Gamma^{(25-)}\Delta^{(1)}|)=(\Gamma^{(25+)}\Delta^{(4)}|)$
$(\Gamma^{(25-)}\Delta^{(2)}|)=(\Gamma^{(25+)}\Delta^{(3)}|)$
$(\Gamma^{(25-)}\Delta^{(3)}|)=(\Gamma^{(25+)}\Delta^{(2)}|)$
$(\Gamma^{(25-)}\Delta^{(4)}|)=(\Gamma^{(25+)}\Delta^{(1)}|)$
$(\Gamma^{(25-)}\Delta^{(5)}|)=(\Gamma^{(25+)}\Delta^{(5)}|)$
$(\Gamma^{(1+)}\Lambda^{(m)}|\Lambda^{(m)})=1$
$(\Gamma^{(1-)}\Lambda^{(1)}|\Lambda^{(2)})=1$
$(\Gamma^{(1-)}\Lambda^{(2)}|\Lambda^{(1)})=1$
$(\Gamma^{(1-)}\Lambda^{(3)}|\Lambda^{(3)})=1$
$(\Gamma^{(2+)}\Lambda^{(m)}|)=(\Gamma^{(1-)}\Lambda^{(m)}|)$
$(\Gamma^{(2-)}\Lambda^{(m)}|)=(\Gamma^{(1+)}\Lambda^{(m)}|)$
$(\Gamma^{(12+)}\Lambda^{(1)}|\Lambda^{(3)})=1$
$(\Gamma^{(12+)}\Lambda^{(2)}|)=(\Gamma^{(12+)}\Lambda^{(1)}|)$
$(\Gamma^{(12+)}\Lambda^{(3)}|\Lambda^{(1)})=(|\Lambda^{(2)})=(|\Lambda^{(3)})=1$
$(\Gamma^{(12-)}\Lambda^{(m)}|)=(\Gamma^{(12+)}\Lambda^{(m)}|)$
$(\Gamma^{(15+)}\Lambda^{(1)}|\Lambda^{(2)})=(|\Lambda^{(3)})=1;$
$(\Gamma^{(15-)}\Lambda^{(2)}|)=(\Gamma^{(15+)}\Lambda^{(1)}|)$
$(\Gamma^{(15+)}\Lambda^{(2)}|\Lambda^{(1)})=(|\Lambda^{(3)})=1;$
$(\Gamma^{(15-)}\Lambda^{(1)}|)=(\Gamma^{(15+)}\Lambda^{(2)}|)$
$(\Gamma^{(15\pm)}\Lambda^{(3)}|\Lambda^{(1)})=(|\Lambda^{(2)})=1;$
$(|\Lambda^{(3)})=2; \quad (\Gamma^{(15-)}\Lambda^{(3)}|)$
$(\Gamma^{(25\pm)}\Lambda^{(m)}|)=(\Gamma^{(15\mp)}\Lambda^{(m)}|)$

$(\Delta^{(1)}\Delta^{(1)}|\Gamma^{(1+)})=(|\Gamma^{(12+)})=(|\Gamma^{(15-)})=(|\Sigma^{(1)})$
$=(|\Sigma^{(4)})=(|\Delta'^{(1)})=1$
$(\Delta^{(1)}\Delta^{(2)}|\Gamma^{(2+)})=(|\Gamma^{(12+)})=(|\Gamma^{(25-)})=(|\Sigma^{(1)})$
$=(|\Sigma^{(4)})=(|\Delta'^{(2)})=1$
$(\Delta^{(1)}\Delta^{(3)}|\Gamma^{(2-)})=(|\Gamma^{(12-)})=(|\Gamma^{(25+)})=(|\Sigma^{(2)})$
$=(|\Sigma^{(3)})=(|\Delta'^{(3)})=1$
$(\Delta^{(1)}\Delta^{(4)}|\Gamma^{(1-)})=(|\Gamma^{(12-)})=(|\Gamma^{(15+)})=(|\Sigma^{(2)})$
$=(|\Sigma^{(3)})=(|\Delta'^{(4)})=1$
$(\Delta^{(1)}\Delta^{(5)}|\Gamma^{(15+)})=(|\Gamma^{(15-)})=(|\Gamma^{(25+)})=(|\Gamma^{(25-)})$
$=(|\Sigma^{(1)})=(|\Sigma^{(2)})=(|\Sigma^{(3)})$
$=(|\Sigma^{(4)})=(|\Delta'^{(5)})=1$

$(\Delta^{(5)}\Delta^{(5)}|\Gamma^{(1+)})=(|\Gamma^{(1-)})=(|\Gamma^{(2+)})=(|\Gamma^{(2-)})$
$=(|\Gamma^{(15+)})=(|\Gamma^{(15-)})=(|\Gamma^{(25+)})$
$=(|\Gamma^{(25-)})=(|\Delta'^{(1)})=(|\Delta'^{(2)})$
$=(|\Delta'^{(3)})=(|\Delta'^{(4)})=1;$
$(|\Gamma^{(12+)})=(|\Gamma^{(12-)})=(|\Sigma^{(1)})$
$=(|\Sigma^{(2)})=(\Sigma^{(3)})=(|\Sigma^{(4)})=2$

$(\Lambda^{(1)}\Lambda^{(1)}|\Gamma^{(1+)})=(|\Gamma^{(2-)})=(|\Gamma^{(15-)})=(|\Gamma^{(25+)})$
$=(|\Lambda'^{(1)})=(|\Sigma'^{(1)})=(|\Sigma'^{(3)})$
$=(|\Delta'^{(1)})=(|\Delta'^{(3)})=(|\Delta'^{(5)})=1$
$(\Lambda^{(1)}\Lambda^{(2)}|\Gamma^{(1-)})=(|\Gamma^{(2+)})=(|\Gamma^{(15+)})=(|\Gamma^{(25-)})$
$=(|\Lambda'^{(2)})=(|\Sigma'^{(2)})=(|\Sigma'^{(4)})$
$=(|\Delta'^{(2)})=(|\Delta'^{(4)})=(|\Delta'^{(5)})=1$
$(\Lambda^{(1)}\Lambda^{(3)}|\Gamma^{(12+)})=(|\Gamma^{(12-)})=(|\Gamma^{(15+)})=(|\Gamma^{(15-)})$
$=(|\Gamma^{(25+)})=(|\Gamma^{(25-)})=(|\Lambda'^{(3)})$
$=(|\Sigma'^{(1)})=(|\Sigma'^{(2)})=(|\Sigma'^{(3)})$
$=(\Sigma'^{(4)})=(|\Delta'^{(1)})=(|\Delta'^{(2)})$
$=(|\Delta'^{(3)})=(|\Delta'^{(4)})=1;$
$(|\Delta'^{(5)})=2$
$(\Lambda^{(3)}\Lambda^{(3)}|\Gamma^{(1+)})=(|\Gamma^{(1-)})=(|\Gamma^{(2+)})=(|\Gamma^{(2-)})$
$=(|\Gamma^{(12+)})=(|\Gamma^{(12-)})=(|\Lambda'^{(1)})$
$=(|\Lambda'^{(2)})=(|\Lambda'^{(3)})=1;$
$(|\Gamma^{(15+)})=(|\Gamma^{(15-)})=(|\Gamma^{(25+)})=(|\Gamma^{(25-)})$
$=(|\Sigma'^{(1)})=(|\Sigma'^{(2)})=(|\Sigma'^{(3)})$
$=(|\Sigma'^{(4)})=(|\Delta'^{(1)})=(|\Delta'^{(2)})$
$=(|\Delta'^{(3)})=(|\Delta'^{(4)})=2;$
$(|\Delta'^{(5)})=4$

$(\Delta^{(1)}\Delta^{(1)}|)=(\Delta^{(2)}\Delta^{(2)}|)=(\Delta^{(3)}\Delta^{(3)}|)=(\Delta^{(4)}\Delta^{(4)}|)$
$(\Delta^{(1)}\Delta^{(2)}|)=(\Delta^{(3)}\Delta^{(4)}|)$
$(\Delta^{(1)}\Delta^{(3)}|)=(\Delta^{(2)}\Delta^{(4)}|)$
$(\Delta^{(1)}\Delta^{(4)}|)=(\Delta^{(2)}\Delta^{(3)}|)$
$(\Delta^{(1)}\Delta^{(5)}|)=(\Delta^{(2)}\Delta^{(5)}|)=(\Delta^{(3)}\Delta^{(5)}|)=(\Delta^{(4)}\Delta^{(5)}|)$
$(\Lambda^{(1)}\Lambda^{(1)}|)=(\Lambda^{(2)}\Lambda^{(2)}|)$
$(\Lambda^{(1)}\Lambda^{(3)}|)=(\Lambda^{(2)}\Lambda^{(3)}|)$
$(\Delta^{(1)}\Lambda^{(1)}|M^{(1)})=(|\Sigma^{(1)})=(|\Sigma^{(3)})=1;$
$(\Delta^{(3)}\Lambda^{(1)}|)=(\Delta^{(2)}\Lambda^{(2)}|)=(\Delta^{(4)}\Lambda^{(2)}|)=(\Delta^{(1)}\Lambda^{(1)}|);$
$(\Delta^{(1)}\Lambda^{(2)}|M^{(2)})=(|\Sigma^{(2)})=(|\Sigma^{(4)})=1;$
$(\Delta^{(3)}\Lambda^{(2)}|)=(\Delta^{(2)}\Lambda^{(1)}|)=(\Delta^{(4)}\Lambda^{(1)}|)=(\Delta^{(1)}\Lambda^{(2)}|);$
$(\Delta^{(1)}\Lambda^{(3)}|M^{(1)})=(|M^{(2)})=(|\Sigma^{(1)})=(|\Sigma^{(2)})$
$=(|\Sigma^{(3)})=(|\Sigma^{(4)})=1;$
$(\Delta^{(m)}\Lambda^{(3)}|)=(\Delta^{(1)}\Lambda^{(3)}|); \quad m=2,3,4;$
$(\Delta^{(5)}\Lambda^{(3)}|M^{(1)})=(|M^{(2)})=(|\Sigma^{(1)})=(|\Sigma^{(2)})$
$=(|\Sigma^{(3)})=(|\Sigma^{(4)})=2$

[a] In the last 11 lines of this table, $\Sigma_1 \equiv (\pi, \pi, 0)(1/a)$: see note b of Table 16.

[b] Our $\star\Delta^{(3)}$ corresponds to HERRING's Δ_2'; our $\star\Delta^{(4)}$ to his Δ_1'; we reserve primes to denote *different* wave vectors (stars) of a given type, e.g., on a line or plane.

Cf. C. HERRING: J. Franklin Inst. **233**, 525 (1942). — W. DÖRING and V. ZEHLER: Ann. Physik **13**, 214 (1953). We use essentially the notations of HERRING [see also M. LAX and J. J. HOPFIELD: Phys. Rev. **124**, 115 (1961), where some errors in HERRING's paper are corrected], exceptions are noted in the appropriate table. In labeling representations $\star L^{(m\pm)}$ we follow HERRING (*op. cit.*) and LAX and HOPFIELD, rather than DÖRING.

Table B.3. Reduction coefficients for symmetrized Kronecker square in diamond

$([\boldsymbol{\Gamma}^{(m)}]_{(2)}|\boldsymbol{\Gamma}^{(1+)}) = 1; \quad m = 1\pm, 2\pm$
$([\boldsymbol{\Gamma}^{(m)}]_{(2)}|\boldsymbol{\Gamma}^{(1+)}) = (|\boldsymbol{\Gamma}^{(12+)}) = 1; \quad m = 12\pm$
$([\boldsymbol{\Gamma}^{(m)}]_{(2)}|\boldsymbol{\Gamma}^{(1+)}) = (|\boldsymbol{\Gamma}^{(12+)}) = (|\boldsymbol{\Gamma}^{(25+)}) = 1; \quad m = 15\pm, 25\pm$

$([\boldsymbol{L}^{(m)}]_{(2)}|\boldsymbol{\Gamma}^{(1+)}) = (|\boldsymbol{\Gamma}^{(25+)}) = (|\boldsymbol{X}^{(1)}) = 1; \quad m = 1\pm, 2\pm$
$([\boldsymbol{L}^{(m)}]_{(2)}|\boldsymbol{\Gamma}^{(1+)}) = (|\boldsymbol{\Gamma}^{(12+)}) = (|\boldsymbol{\Gamma}^{(15+)}) = (|\boldsymbol{X}^{(2)}) = (|\boldsymbol{X}^{(4)}) = 1; \quad (|\boldsymbol{\Gamma}^{(25+)}) = (|\boldsymbol{X}^{(1)}) = 2; \quad m = 3\pm$

$([\boldsymbol{X}^{(m)}]_{(2)}|\boldsymbol{\Gamma}^{(1+)}) = (|\boldsymbol{\Gamma}^{(2-)}) = (|\boldsymbol{\Gamma}^{(12+)}) = (|\boldsymbol{\Gamma}^{(12-)}) = (|\boldsymbol{\Gamma}^{(25+)}) = (|\boldsymbol{X}^{(1)}) = (|\boldsymbol{X}^{(4)}) = 1; \quad m = 1, 2$
$([\boldsymbol{X}^{(m)}]_{(2)}|\boldsymbol{\Gamma}^{(1+)}) = (|\boldsymbol{\Gamma}^{(1-)}) = (|\boldsymbol{\Gamma}^{(12+)}) = (|\boldsymbol{\Gamma}^{(12-)}) = (|\boldsymbol{\Gamma}^{(25+)}) = (|\boldsymbol{X}^{(1)}) = (|\boldsymbol{X}^{(4)}) = 1; \quad m = 3, 4$

$([\boldsymbol{\Delta}^{(m)}]_{(2)}|\boldsymbol{\Gamma}^{(1+)}) = (|\boldsymbol{\Gamma}^{(12+)}) = (|\boldsymbol{\Sigma}^{(1)}) = (|\boldsymbol{\Delta}'^{(1)}) = 1; \quad m = 1, 2, 3, 4$ [a]
$([\boldsymbol{\Delta}^{(5)}]_{(2)}|\boldsymbol{\Gamma}^{(1+)}) = (|\boldsymbol{\Gamma}^{(1-)}) = (|\boldsymbol{\Gamma}^{(2+)}) = (|\boldsymbol{\Gamma}^{(12-)}) = (|\boldsymbol{\Gamma}^{(25+)}) = (|\boldsymbol{\Delta}'^{(1)}) = (|\boldsymbol{\Delta}'^{(2)}) = (|\boldsymbol{\Delta}'^{(3)})$
$= (|\boldsymbol{\Sigma}^{(2)}) = (|\boldsymbol{\Sigma}^{(3)}) = 1; \quad (|\boldsymbol{\Gamma}^{(12+)}) = (|\boldsymbol{\Sigma}^{(1)}) = 2$

$([\boldsymbol{\Lambda}^{(m)}]_{(2)}|\boldsymbol{\Gamma}^{(1+)}) = (|\boldsymbol{\Gamma}^{(25+)}) = (|\boldsymbol{\Delta}'^{(1)}) = (|\boldsymbol{\Delta}'^{(3)}) = (|\boldsymbol{\Lambda}'^{(1)}) = (|\boldsymbol{\Sigma}'^{(1)}) = 1; \quad m = 1, 2$
$([\boldsymbol{\Lambda}^{(3)}]_{(2)}|\boldsymbol{\Gamma}^{(1+)}) = (|\boldsymbol{\Gamma}^{(1-)}) = (|\boldsymbol{\Gamma}^{(12+)}) = (|\boldsymbol{\Gamma}^{(15+)}) = (|\boldsymbol{\Gamma}^{(25-)}) = (|\boldsymbol{\Delta}'^{(2)}) = (|\boldsymbol{\Delta}'^{(4)}) = (|\boldsymbol{\Delta}'^{(5)})$
$= (|\boldsymbol{\Sigma}'^{(2)}) = (|\boldsymbol{\Sigma}'^{(4)}) = (|\boldsymbol{\Lambda}'^{(1)}) = (|\boldsymbol{\Lambda}'^{(3)}) = 1; \quad (|\boldsymbol{\Gamma}^{(25+)}) = (|\boldsymbol{\Delta}'^{(1)}) = (|\boldsymbol{\Delta}'^{(3)}) = (|\boldsymbol{\Sigma}'^{(1)}) = 2$

$([\boldsymbol{W}^{(m)}]_{(2)}|\boldsymbol{\Gamma}^{(1+)}) = (|\boldsymbol{\Gamma}^{(2-)}) = (|\boldsymbol{\Gamma}^{(12+)}) = (|\boldsymbol{\Gamma}^{(12-)}) = (|\boldsymbol{\Gamma}^{(15+)}) = (|\boldsymbol{\Gamma}^{(25+)})$
$= (|\boldsymbol{X}^{(1)}) = (|\boldsymbol{X}^{(2)}) = (|\boldsymbol{X}^{(4)}) = (|\boldsymbol{\Sigma}^{(2)}) = (|\boldsymbol{\Sigma}^{(3)}) = 1; \quad (|\boldsymbol{\Sigma}^{(1)}) = 2; \quad m = 1, 2$ [b]

$([\boldsymbol{\Sigma}^{(m)}]_{(2)}|\boldsymbol{\Gamma}^{(1+)}) = (|\boldsymbol{\Gamma}^{(12+)}) = (|\boldsymbol{\Gamma}^{(25+)}) = (|\boldsymbol{X}^{(2)}) = (|\boldsymbol{X}^{(4)}) = (|\boldsymbol{\Sigma}^{(2)}) = (|\boldsymbol{\Sigma}^{(3)}) = 1;$
$(|\boldsymbol{X}^{(1)}) = (|\boldsymbol{\Sigma}^{(1)}) = 2; \quad m = 1, 2, 3, 4$ [b]

$([\boldsymbol{k}]_{(2)}|\boldsymbol{\Gamma}^{(1+)}) = (|\boldsymbol{\Gamma}^{(2+)}) = 1; \quad (|\boldsymbol{\Gamma}^{(12+)}) = 2; \quad (|\boldsymbol{\Gamma}^{(15+)}) = (|\boldsymbol{\Gamma}^{(25+)}) = 3$ [c]

[a] See footnote a of Table 31.
[b] See footnote b of Table 31.
[c] Only representations $\boldsymbol{\Gamma}^{(m)}$ were determined, using the basis function method.

Table B.4. Reduction coefficients for symmetrized Kronecker cubes in diamond

$([\boldsymbol{\Gamma}^{(m)}]_{(3)}|\boldsymbol{\Gamma}^{(m)}) = 1; \quad m = 1\pm, 2\pm$
$([\boldsymbol{\Gamma}^{(12+)}]_{(3)}|\boldsymbol{\Gamma}^{(1+)}) = (|\boldsymbol{\Gamma}^{(2+)}) = (|\boldsymbol{\Gamma}^{(12+)}) = 1$
$([\boldsymbol{\Gamma}^{(12-)}]_{(3)}|\boldsymbol{\Gamma}^{(1-)}) = (|\boldsymbol{\Gamma}^{(2-)}) = (|\boldsymbol{\Gamma}^{(12-)}) = 1$
$([\boldsymbol{\Gamma}^{(15+)}]_{(3)}|\boldsymbol{\Gamma}^{(2+)}) = (|\boldsymbol{\Gamma}^{(25+)}) = 1; \quad (|\boldsymbol{\Gamma}^{(15+)}) = 2$
$([\boldsymbol{\Gamma}^{(15-)}]_{(3)}|\boldsymbol{\Gamma}^{(2-)}) = (|\boldsymbol{\Gamma}^{(25-)}) = 1; \quad (|\boldsymbol{\Gamma}^{(15-)}) = 2$
$([\boldsymbol{\Gamma}^{(25\pm)}]_{(3)}|\boldsymbol{\Gamma}^{(1\pm)}) = (|\boldsymbol{\Gamma}^{(15\pm)}) = 1; \quad (|\boldsymbol{\Gamma}^{(25\pm)}) = 2$

$([\boldsymbol{X}^{(1)}]_{(3)}|\boldsymbol{\Gamma}^{(1+)}) = (|\boldsymbol{\Gamma}^{(2-)}) = (|\boldsymbol{\Gamma}^{(15-)}) = (|\boldsymbol{\Gamma}^{(25+)}) = (|\boldsymbol{X}^{(3)}) = (|\boldsymbol{X}^{(4)}) = 1; \quad (|\boldsymbol{X}^{(2)}) = 2; \quad (|\boldsymbol{X}^{(1)}) = 4$
$([\boldsymbol{X}^{(3)}]_{(3)}|\boldsymbol{\Gamma}^{(2+)}) = (|\boldsymbol{\Gamma}^{(2-)}) = (|\boldsymbol{\Gamma}^{(15+)}) = (|\boldsymbol{\Gamma}^{(15-)}) = (|\boldsymbol{X}^{(1)}) = (|\boldsymbol{X}^{(2)}) = 1; \quad (|\boldsymbol{X}^{(4)}) = 2; \quad (|\boldsymbol{X}^{(3)}) = 4$
$([\boldsymbol{X}^{(4)}]_{(3)}|\boldsymbol{\Gamma}^{(1+)}) = (|\boldsymbol{\Gamma}^{(1-)}) = (|\boldsymbol{\Gamma}^{(25+)}) = (|\boldsymbol{\Gamma}^{(25-)}) = (|\boldsymbol{X}^{(1)}) = (|\boldsymbol{X}^{(2)}) = 1; \quad (|\boldsymbol{X}^{(3)}) = 2; \quad (|\boldsymbol{X}^{(4)}) = 4$

$([\boldsymbol{L}^{(1+)}]_{(3)}|\boldsymbol{L}^{(2-)}) = (|\boldsymbol{L}^{(3+)}) = 1; \quad (|\boldsymbol{L}^{(1+)}) = 2$
$([\boldsymbol{L}^{(2-)}]_{(3)}|\boldsymbol{L}^{(1+)}) = (|\boldsymbol{L}^{(3-)}) = 1; \quad (|\boldsymbol{L}^{(2-)}) = 2$
$([\boldsymbol{L}^{(3+)}]_{(3)}|\boldsymbol{L}^{(1-)}) = (|\boldsymbol{L}^{(2-)}) = (|\boldsymbol{L}^{(3-)}) = 2; \quad (|\boldsymbol{L}^{(1+)}) = (|\boldsymbol{L}^{(2+)}) = 4; \quad (|\boldsymbol{L}^{(3+)}) = 7$
$([\boldsymbol{L}^{(3-)}]_{(3)}|\boldsymbol{L}^{(1+)}) = (|\boldsymbol{L}^{(2+)}) = (|\boldsymbol{L}^{(3+)}) = 2; \quad (|\boldsymbol{L}^{(1-)}) = (|\boldsymbol{L}^{(2-)}) = 4; \quad (|\boldsymbol{L}^{(3-)}) = 7$

$([\boldsymbol{K}^{(1)}]_{(3)}|\boldsymbol{\Gamma}^{(1+)}) = (|\boldsymbol{\Gamma}^{(1-)}) = (|\boldsymbol{\Gamma}^{(25+)}) = (|\boldsymbol{\Gamma}^{(25-)}) = 1$ [a]
$([\boldsymbol{K}^{(2)}]_{(3)}|\boldsymbol{\Gamma}^{(m)}) = ([\boldsymbol{K}^{(1)}]_{(3)}|\boldsymbol{\Gamma}^{(m)})$ [a]
$([\boldsymbol{K}^{(3)}]_{(3)}|\boldsymbol{\Gamma}^{(2+)}) = (|\boldsymbol{\Gamma}^{(2-)}) = (|\boldsymbol{\Gamma}^{(15+)}) = (|\boldsymbol{\Gamma}^{(15-)}) = 1$ [a]
$([\boldsymbol{K}^{(4)}]_{(3)}|\boldsymbol{\Gamma}^{(m)}) = ([\boldsymbol{K}^{(3)}]_{(3)}|\boldsymbol{\Gamma}^{(m)})$ [a]

[a] For these representations we only obtained coefficients of types $(|\boldsymbol{\Gamma}^{(m)})$, using the basis function method.

Table B.5. Principle block of Clebsch-Gordan coefficients for the product $D^{(\star X)(l)} \otimes D^{(\star X)(l')} \to D^{(\star X)(l'')}$ in diamond.

In all entries below canonical wave vectors are

$$\boldsymbol{k}_1 = \frac{2\pi}{a}(1,0,0) = \boldsymbol{X}_1; \quad \boldsymbol{k}_1' = \frac{2\pi}{a}(0,1,0) = \boldsymbol{X}_2'; \quad \boldsymbol{k}_1'' = \frac{2\pi}{a}(0,0,1) = \boldsymbol{X}_3''$$

The entries $U_{1a1a',1a''}$ are written in matrix form:

$$\begin{pmatrix} U_{1111,11} & U_{1111,12} \\ U_{1112,11} & U_{1112,12} \\ U_{1211,11} & U_{1211,12} \\ U_{1212,11} & U_{1212,12} \end{pmatrix}$$

Also, due to the symmetry of the representation matrices for $\boldsymbol{X}^{(1)}$ and $\boldsymbol{X}^{(2)}$, we have $\boldsymbol{X}^{(2)} \otimes \boldsymbol{X}^{(2)} = \boldsymbol{X}^{(1)} \otimes \boldsymbol{X}^{(1)}$

In Tables B.5–B.10 all full group representations are denoted $\boldsymbol{X}^{(j)}$ not ${}^\star\boldsymbol{X}^{(j)}$.

$D^{(\boldsymbol{k})(j)} \otimes D^{(\boldsymbol{k}')(j')}$	$\boldsymbol{X}^{(1)}$	$\boldsymbol{X}^{(2)}$	$\boldsymbol{X}^{(3)}$	$\boldsymbol{X}^{(4)}$
$\boldsymbol{X}^{(1)} \otimes \boldsymbol{X}^{(1)}$	$\begin{pmatrix} 1/\sqrt{8} & i/\sqrt{8} \\ i/\sqrt{8} & 1/\sqrt{8} \\ i/\sqrt{8} & 1/\sqrt{8} \\ 1/\sqrt{8} & i/\sqrt{8} \end{pmatrix}$	$\begin{pmatrix} 1/\sqrt{8} & -i/\sqrt{8} \\ -i/\sqrt{8} & 1/\sqrt{8} \\ -i/\sqrt{8} & 1/\sqrt{8} \\ 1/\sqrt{8} & -i/\sqrt{8} \end{pmatrix}$	$\begin{pmatrix} 1/\sqrt{8} & -i/\sqrt{8} \\ -1/\sqrt{8} & -i/\sqrt{8} \\ 1/\sqrt{8} & i/\sqrt{8} \\ -1/\sqrt{8} & i/\sqrt{8} \end{pmatrix}$	$\begin{pmatrix} 1/\sqrt{8} & -i/\sqrt{8} \\ 1/\sqrt{8} & i/\sqrt{8} \\ -1/\sqrt{8} & -i/\sqrt{8} \\ -1/\sqrt{8} & i/\sqrt{8} \end{pmatrix}$
$\boldsymbol{X}^{(1)} \otimes \boldsymbol{X}^{(2)}$	$\begin{pmatrix} 1/\sqrt{8} & -i/\sqrt{8} \\ -i/\sqrt{8} & 1/\sqrt{8} \\ -i/\sqrt{8} & 1/\sqrt{8} \\ 1/\sqrt{8} & -i/\sqrt{8} \end{pmatrix}$	$\begin{pmatrix} 1/\sqrt{8} & i/\sqrt{8} \\ i/\sqrt{8} & 1/\sqrt{8} \\ i/\sqrt{8} & 1/\sqrt{8} \\ 1/\sqrt{8} & i/\sqrt{8} \end{pmatrix}$	$\begin{pmatrix} 1/\sqrt{8} & i/\sqrt{8} \\ -1/\sqrt{8} & i/\sqrt{8} \\ 1/\sqrt{8} & -i/\sqrt{8} \\ -1/\sqrt{8} & -i/\sqrt{8} \end{pmatrix}$	$\begin{pmatrix} 1/\sqrt{8} & i/\sqrt{8} \\ 1/\sqrt{8} & -i/\sqrt{8} \\ -1/\sqrt{8} & i/\sqrt{8} \\ -1/\sqrt{8} & -i/\sqrt{8} \end{pmatrix}$
$\boldsymbol{X}^{(1)} \otimes \boldsymbol{X}^{(3)}$	$\begin{pmatrix} 1/\sqrt{8} & 1/\sqrt{8} \\ -i/\sqrt{8} & i/\sqrt{8} \\ -1/\sqrt{8} & -1/\sqrt{8} \\ -i/\sqrt{8} & i/\sqrt{8} \end{pmatrix}$	$\begin{pmatrix} 1/\sqrt{8} & 1/\sqrt{8} \\ i/\sqrt{8} & -i/\sqrt{8} \\ -1/\sqrt{8} & -1/\sqrt{8} \\ i/\sqrt{8} & -i/\sqrt{8} \end{pmatrix}$	$\begin{pmatrix} 0 & 1/2 \\ i/2 & 0 \\ 0 & 1/2 \\ -i/2 & 0 \end{pmatrix}$	$\begin{pmatrix} 1/2 & 0 \\ 0 & -i/2 \\ 1/2 & 0 \\ 0 & i/2 \end{pmatrix}$
$\boldsymbol{X}^{(1)} \otimes \boldsymbol{X}^{(4)}$	$\begin{pmatrix} 1/\sqrt{8} & -1/\sqrt{8} \\ -i/\sqrt{8} & -i/\sqrt{8} \\ 1/\sqrt{8} & -1/\sqrt{8} \\ i/\sqrt{8} & i/\sqrt{8} \end{pmatrix}$	$\begin{pmatrix} 1/\sqrt{8} & -1/\sqrt{8} \\ i/\sqrt{8} & i/\sqrt{8} \\ 1/\sqrt{8} & -1/\sqrt{8} \\ -i/\sqrt{8} & -i/\sqrt{8} \end{pmatrix}$	$\begin{pmatrix} 1/2 & 0 \\ 0 & i/2 \\ -1/2 & 0 \\ 0 & i/2 \end{pmatrix}$	$\begin{pmatrix} 0 & 1/2 \\ -i/2 & 0 \\ 0 & -1/2 \\ -i/2 & 0 \end{pmatrix}$
$\boldsymbol{X}^{(2)} \otimes \boldsymbol{X}^{(3)}$	$\begin{pmatrix} 1/\sqrt{8} & 1/\sqrt{8} \\ i/\sqrt{8} & -i/\sqrt{8} \\ -1/\sqrt{8} & -1/\sqrt{8} \\ i/\sqrt{8} & -i/\sqrt{8} \end{pmatrix}$	$\begin{pmatrix} 1/\sqrt{8} & 1/\sqrt{8} \\ -i/\sqrt{8} & i/\sqrt{8} \\ -1/\sqrt{8} & -1/\sqrt{8} \\ -i/\sqrt{8} & i/\sqrt{8} \end{pmatrix}$	$\begin{pmatrix} 0 & 1/2 \\ -i/2 & 0 \\ 0 & 1/2 \\ i/2 & 0 \end{pmatrix}$	$\begin{pmatrix} 1/2 & 0 \\ 0 & i/2 \\ 1/2 & 0 \\ 0 & -i/2 \end{pmatrix}$
$\boldsymbol{X}^{(2)} \otimes \boldsymbol{X}^{(4)}$	$\begin{pmatrix} 1/\sqrt{8} & -1/\sqrt{8} \\ i/\sqrt{8} & i/\sqrt{8} \\ 1/\sqrt{8} & -1/\sqrt{8} \\ -i/\sqrt{8} & -i/\sqrt{8} \end{pmatrix}$	$\begin{pmatrix} 1/\sqrt{8} & -1/\sqrt{8} \\ -i/\sqrt{8} & -i/\sqrt{8} \\ 1/\sqrt{8} & -1/\sqrt{8} \\ i/\sqrt{8} & i/\sqrt{8} \end{pmatrix}$	$\begin{pmatrix} 1/2 & 0 \\ 0 & -i/2 \\ -1/2 & 0 \\ 0 & -i/2 \end{pmatrix}$	$\begin{pmatrix} 0 & 1/2 \\ i/2 & 0 \\ 0 & -1/2 \\ i/2 & 0 \end{pmatrix}$

Table B.5 (continued)

$D^{(\boldsymbol{k})(j)} \otimes D^{(\boldsymbol{k}')(j')}$	$\boldsymbol{X}^{(1)}$	$\boldsymbol{X}^{(2)}$	$\boldsymbol{X}^{(3)}$	$\boldsymbol{X}^{(4)}$
$\boldsymbol{X}^{(3)} \otimes \boldsymbol{X}^{(3)}$	$\begin{pmatrix} 0 & 0 \\ 1/2 & 1/2 \\ i/2 & -i/2 \\ 0 & 0 \end{pmatrix}$	$\begin{pmatrix} 0 & 0 \\ 1/2 & 1/2 \\ -i/2 & i/2 \\ 0 & 0 \end{pmatrix}$	$\begin{pmatrix} 1/\sqrt{2} & 0 \\ 0 & 0 \\ 0 & 0 \\ 0 & i/\sqrt{2} \end{pmatrix}$	$\begin{pmatrix} 0 & 1/\sqrt{2} \\ 0 & 0 \\ 0 & 0 \\ -i/\sqrt{2} & 0 \end{pmatrix}$
$\boldsymbol{X}^{(3)} \otimes \boldsymbol{X}^{(4)}$	$\begin{pmatrix} 1/2 & 1/2 \\ 0 & 0 \\ 0 & 0 \\ -i/2 & i/2 \end{pmatrix}$	$\begin{pmatrix} 1/2 & 1/2 \\ 0 & 0 \\ 0 & 0 \\ i/2 & -i/2 \end{pmatrix}$	$\begin{pmatrix} 0 & 0 \\ 1/\sqrt{2} & 0 \\ 0 & -i/\sqrt{2} \\ 0 & 0 \end{pmatrix}$	$\begin{pmatrix} 0 & 0 \\ 0 & -i/\sqrt{2} \\ 1/\sqrt{2} & 0 \\ 0 & 0 \end{pmatrix}$
$\boldsymbol{X}^{(4)} \otimes \boldsymbol{X}^{(4)}$	$\begin{pmatrix} 0 & 0 \\ i/2 & -i/2 \\ 1/2 & 1/2 \\ 0 & 0 \end{pmatrix}$	$\begin{pmatrix} 0 & 0 \\ -i/2 & i/2 \\ 1/2 & 1/2 \\ 0 & 0 \end{pmatrix}$	$\begin{pmatrix} 0 & 1/\sqrt{2} \\ 0 & 0 \\ 0 & 0 \\ -i/\sqrt{2} & 0 \end{pmatrix}$	$\begin{pmatrix} 1/\sqrt{2} & 0 \\ 0 & 0 \\ 0 & 0 \\ 0 & i/\sqrt{2} \end{pmatrix}$

Table B.6. Symmetry elements for calculating the $(\sigma\,\sigma'\,\sigma'')$ blocks of Clebsch-Gordan coefficients for $D^{(^{*}\boldsymbol{X})(l)} \otimes D^{(^{*}\boldsymbol{X})(l')} \to D^{(^{*}\boldsymbol{X})(l'')}$ in diamond

σ	σ'	σ''	$\boldsymbol{\varphi}_{\Sigma}$	$\boldsymbol{\tau}_{\Sigma}$	$\boldsymbol{\varphi}_{k}$	$\boldsymbol{\varphi}_{k'}$	$\boldsymbol{\varphi}_{k''}$
1	1	1	$\boldsymbol{\varepsilon}$	$\mathbf{0}$	$\boldsymbol{\varepsilon}$	$\boldsymbol{\varepsilon}$	$\boldsymbol{\varepsilon}$
2	2	2	$\boldsymbol{\delta}_{3xyz}$	$\mathbf{0}$	$\boldsymbol{\varepsilon}$	$\boldsymbol{\varepsilon}$	$\boldsymbol{\varepsilon}$
3	3	3	$\boldsymbol{\delta}_{3xyz}^{-1}$	$\mathbf{0}$	$\boldsymbol{\varepsilon}$	$\boldsymbol{\varepsilon}$	$\boldsymbol{\varepsilon}$
1	2	3	$\boldsymbol{\rho}_{y\bar{z}}$	$\mathbf{0}$	$\boldsymbol{\rho}_{y\bar{z}}$	$\boldsymbol{\rho}_{x\bar{z}}$	$\boldsymbol{\rho}_{x\bar{y}}$
2	3	1	$\boldsymbol{\rho}_{x\bar{y}}$	$\mathbf{0}$	$\boldsymbol{\rho}_{y\bar{z}}$	$\boldsymbol{\rho}_{x\bar{z}}$	$\boldsymbol{\rho}_{x\bar{y}}$
3	1	2	$\boldsymbol{\rho}_{x\bar{z}}$	$\mathbf{0}$	$\boldsymbol{\rho}_{y\bar{z}}$	$\boldsymbol{\rho}_{x\bar{z}}$	$\boldsymbol{\rho}_{x\bar{y}}$

Table B.7. Matrices for calculating the $(\sigma\,\sigma'\,\sigma'')$ blocks of Clebsch-Gordan coefficients for $D^{(^{*}\boldsymbol{X})(l)} \otimes D^{(^{*}\boldsymbol{X})(l')} \to D^{(^{*}\boldsymbol{X})(l'')}$ in diamond

	$D^{(\boldsymbol{k})(l)}(\{\boldsymbol{\rho}_{y\bar{z}}\|\mathbf{0}\})$	$D^{(\boldsymbol{k}')(l')}(\{\boldsymbol{\rho}_{x\bar{z}}\|\mathbf{0}\})$	$D^{(\boldsymbol{k}'')(l'')}(\{\boldsymbol{\rho}_{x\bar{y}}\|\mathbf{0}\})$
$\boldsymbol{X}^{(1)}$	$\begin{pmatrix} 1 & 0 \\ 0 & 1 \end{pmatrix}$	$\begin{pmatrix} 1 & 0 \\ 0 & 1 \end{pmatrix}$	$\begin{pmatrix} 1 & 0 \\ 0 & 1 \end{pmatrix}$
$\boldsymbol{X}^{(2)}$	$\begin{pmatrix} -1 & 0 \\ 0 & -1 \end{pmatrix}$	$\begin{pmatrix} -1 & 0 \\ 0 & -1 \end{pmatrix}$	$\begin{pmatrix} -1 & 0 \\ 0 & -1 \end{pmatrix}$
$\boldsymbol{X}^{(3)}$	$\begin{pmatrix} 0 & i \\ -i & 0 \end{pmatrix}$	$\begin{pmatrix} 0 & i \\ -i & 0 \end{pmatrix}$	$\begin{pmatrix} 0 & i \\ -i & 0 \end{pmatrix}$
$\boldsymbol{X}^{(4)}$	$\begin{pmatrix} 0 & -i \\ i & 0 \end{pmatrix}$	$\begin{pmatrix} 0 & -i \\ i & 0 \end{pmatrix}$	$\begin{pmatrix} 0 & -i \\ i & 0 \end{pmatrix}$

Table B.8. Principle block of Clebsch-Gordan coefficients for $D^{(*\mathbf{X})(l)} \otimes D^{(*\mathbf{X})(l')} \to D^{(*\mathbf{X})(l'')}$ in diamond

	Γ^{1+}	Γ^{1-}	Γ^{2+}	Γ^{2-}	Γ^{12+}	Γ^{12-}
$\mathbf{X}^{(1)} \otimes \mathbf{X}^{(1)}$	$\begin{pmatrix} 0 \\ 1/\sqrt{6} \\ 1/\sqrt{6} \\ 0 \end{pmatrix}$			$\begin{pmatrix} 1/\sqrt{6} \\ 0 \\ 0 \\ 1/\sqrt{6} \end{pmatrix}$	$\begin{pmatrix} 0 & 0 \\ 1/\sqrt{6} & 1/\sqrt{6} \\ 1/\sqrt{6} & 1/\sqrt{6} \\ 0 & 0 \end{pmatrix}$	$\begin{pmatrix} 1/\sqrt{6} & -1/\sqrt{6} \\ 0 & 0 \\ 0 & 0 \\ 1/\sqrt{6} & -1/\sqrt{6} \end{pmatrix}$
$\mathbf{X}^{(1)} \otimes \mathbf{X}^{(2)}$		$\begin{pmatrix} 0 \\ 1/\sqrt{6} \\ 1/\sqrt{6} \\ 0 \end{pmatrix}$	$\begin{pmatrix} 1/\sqrt{6} \\ 0 \\ 0 \\ 1/\sqrt{6} \end{pmatrix}$		$\begin{pmatrix} 1/\sqrt{6} & -1/\sqrt{6} \\ 0 & 0 \\ 0 & 0 \\ 1/\sqrt{6} & -1/\sqrt{6} \end{pmatrix}$	$\begin{pmatrix} 0 & 0 \\ 1/\sqrt{6} & 1/\sqrt{6} \\ 1/\sqrt{6} & 1/\sqrt{6} \\ 0 & 0 \end{pmatrix}$
$\mathbf{X}^{(3)} \otimes \mathbf{X}^{(3)}$	$\begin{pmatrix} 1/\sqrt{6} \\ 0 \\ 0 \\ -1/\sqrt{6} \end{pmatrix}$	$\begin{pmatrix} 1/\sqrt{6} \\ 0 \\ 0 \\ 1/\sqrt{6} \end{pmatrix}$			$\begin{pmatrix} 1/\sqrt{6} & 1/\sqrt{6} \\ 0 & 0 \\ 0 & 0 \\ -1/\sqrt{6} & -1/\sqrt{6} \end{pmatrix}$	$\begin{pmatrix} 1/\sqrt{6} & 1/\sqrt{6} \\ 0 & 0 \\ 0 & 0 \\ 1/\sqrt{6} & 1/\sqrt{6} \end{pmatrix}$
$\mathbf{X}^{(3)} \otimes \mathbf{X}^{(4)}$			$\begin{pmatrix} 0 \\ 1/\sqrt{6} \\ 1/\sqrt{6} \\ 0 \end{pmatrix}$	$\begin{pmatrix} 0 \\ 1/\sqrt{6} \\ -1/\sqrt{6} \\ 0 \end{pmatrix}$	$\begin{pmatrix} 0 & 0 \\ 1/\sqrt{6} & -1/\sqrt{6} \\ 1/\sqrt{6} & -1/\sqrt{6} \\ 0 & 0 \end{pmatrix}$	$\begin{pmatrix} 0 & 0 \\ 1/\sqrt{6} & -1/\sqrt{6} \\ -1/\sqrt{6} & 1/\sqrt{6} \\ 0 & 0 \end{pmatrix}$

	Γ^{15+}	Γ^{15-}	Γ^{25+}	Γ^{25-}
$\mathbf{X}^{(1)} \otimes \mathbf{X}^{(1)}$		$\begin{pmatrix} 0 & 0 & 0 \\ 0 & 0 & 1/\sqrt{2} \\ 0 & 0 & -1/\sqrt{2} \\ 0 & 0 & 0 \end{pmatrix}$	$\begin{pmatrix} 0 & 0 & 1/\sqrt{2} \\ 0 & 0 & 0 \\ 0 & 0 & 0 \\ 0 & 0 & -1/\sqrt{2} \end{pmatrix}$	
$\mathbf{X}^{(1)} \otimes \mathbf{X}^{(2)}$	$\begin{pmatrix} 0 & 0 & 0 \\ 0 & 0 & 1/\sqrt{2} \\ 0 & 0 & -1/\sqrt{2} \\ 0 & 0 & 0 \end{pmatrix}$			$\begin{pmatrix} 0 & 0 & 1/\sqrt{2} \\ 0 & 0 & 0 \\ 0 & 0 & 0 \\ 0 & 0 & -1/\sqrt{2} \end{pmatrix}$
$\mathbf{X}^{(1)} \otimes \mathbf{X}^{(3)}$	$\begin{pmatrix} -i/\sqrt{8} & 1/\sqrt{8} & 0 \\ i/\sqrt{8} & 1/\sqrt{8} & 0 \\ i/\sqrt{8} & 1/\sqrt{8} & 0 \\ i/\sqrt{8} & -1/\sqrt{8} & 0 \end{pmatrix}$	$\begin{pmatrix} -i/\sqrt{8} & 1/\sqrt{8} & 0 \\ -i/\sqrt{8} & -1/\sqrt{8} & 0 \\ i/\sqrt{8} & 1/\sqrt{8} & 0 \\ -i/\sqrt{8} & 1/\sqrt{8} & 0 \end{pmatrix}$	$\begin{pmatrix} i/\sqrt{8} & 1/\sqrt{8} & 0 \\ -i/\sqrt{8} & 1/\sqrt{8} & 0 \\ -i/\sqrt{8} & 1/\sqrt{8} & 0 \\ -i/\sqrt{8} & -1/\sqrt{8} & 0 \end{pmatrix}$	$\begin{pmatrix} i/\sqrt{8} & 1/\sqrt{8} & 0 \\ i/\sqrt{8} & -1/\sqrt{8} & 0 \\ -i/\sqrt{8} & 1/\sqrt{8} & 0 \\ i/\sqrt{8} & 1/\sqrt{8} & 0 \end{pmatrix}$
$\mathbf{X}^{(1)} \otimes \mathbf{X}^{(4)}$	$\begin{pmatrix} -i/\sqrt{8} & 1/\sqrt{8} & 0 \\ -i/\sqrt{8} & -1/\sqrt{8} & 0 \\ -i/\sqrt{8} & -1/\sqrt{8} & 0 \\ i/\sqrt{8} & -1/\sqrt{8} & 0 \end{pmatrix}$	$\begin{pmatrix} -i/\sqrt{8} & 1/\sqrt{8} & 0 \\ i/\sqrt{8} & 1/\sqrt{8} & 0 \\ -i/\sqrt{8} & -1/\sqrt{8} & 0 \\ -i/\sqrt{8} & 1/\sqrt{8} & 0 \end{pmatrix}$	$\begin{pmatrix} i/\sqrt{8} & 1/\sqrt{8} & 0 \\ i/\sqrt{8} & -1/\sqrt{8} & 0 \\ i/\sqrt{8} & -1/\sqrt{8} & 0 \\ -i/\sqrt{8} & -1/\sqrt{8} & 0 \end{pmatrix}$	$\begin{pmatrix} i/\sqrt{8} & 1/\sqrt{8} & 0 \\ -i/\sqrt{8} & 1/\sqrt{8} & 0 \\ i/\sqrt{8} & -1/\sqrt{8} & 0 \\ i/\sqrt{8} & 1/\sqrt{8} & 0 \end{pmatrix}$
$\mathbf{X}^{(3)} \otimes \mathbf{X}^{(3)}$			$\begin{pmatrix} 0 & 0 & 0 \\ 0 & 0 & 1/\sqrt{2} \\ 0 & 0 & 1/\sqrt{2} \\ 0 & 0 & 0 \end{pmatrix}$	$\begin{pmatrix} 0 & 0 & 0 \\ 0 & 0 & 1/\sqrt{2} \\ 0 & 0 & -1/\sqrt{2} \\ 0 & 0 & 0 \end{pmatrix}$
$\mathbf{X}^{(3)} \otimes \mathbf{X}^{(4)}$	$\begin{pmatrix} 0 & 0 & 1/\sqrt{2} \\ 0 & 0 & 0 \\ 0 & 0 & 0 \\ 0 & 0 & -1/\sqrt{2} \end{pmatrix}$	$\begin{pmatrix} 0 & 0 & 1/\sqrt{2} \\ 0 & 0 & 0 \\ 0 & 0 & 0 \\ 0 & 0 & 1/\sqrt{2} \end{pmatrix}$		

Table B.9. Symmetry elements for calculation of the $(\sigma\sigma'\sigma'')$ blocks of Clebsch-Gordan coefficients for $D^{(\star X)(l)} \otimes D^{(\star X)(l')} \rightarrow D^{(\Gamma)(l'')}$ in diamond

σ	σ'	σ''	φ_Σ	τ_Σ	φ_k	$\varphi_{k'}$	$\varphi_{k''}$
3	2	1	ε	$\mathbf{0}$	ε	ε	ε
2	1	1	$\boldsymbol{\delta}_{3xyz}^{-1}$	$\mathbf{0}$	ε	ε	$\boldsymbol{\delta}_{3xyz}^{-1}$
1	3	1	$\boldsymbol{\delta}_{3xyz}$	$\mathbf{0}$	ε	ε	$\boldsymbol{\delta}_{3xyz}$

Table B.10. Matrices for calculation of the $(\sigma\sigma'\sigma'')$ blocks of Clebsch-Gordan coefficients for $D^{(\star X)(l)} \otimes D^{(\star X)(l')} \rightarrow D^{(\Gamma)(l'')}$ in diamond

$$\Gamma^{1\pm}(\{\boldsymbol{\delta}_{3xyz}|\mathbf{0}\})=1 \qquad \Gamma^{1\pm}(\{\boldsymbol{\delta}_{3xyz}^{-1}|\mathbf{0}\})=1$$

$$\Gamma^{2\pm}(\{\boldsymbol{\delta}_{3xyz}|\mathbf{0}\})=1 \qquad \Gamma^{2\pm}(\{\boldsymbol{\delta}_{3xyz}^{-1}|\mathbf{0}\})=1$$

$$\Gamma^{12\pm}(\{\boldsymbol{\delta}_{3xyz}|\mathbf{0}\})=\begin{pmatrix}\varepsilon^2 & 0\\ 0 & \varepsilon\end{pmatrix} \qquad \Gamma^{12\pm}(\{\boldsymbol{\delta}_{3xyz}^{-1}|\mathbf{0}\})=\begin{pmatrix}\varepsilon & 0\\ 0 & \varepsilon^2\end{pmatrix}$$

$$\Gamma^{15\pm}(\{\boldsymbol{\delta}_{3xyz}|\mathbf{0}\})=\begin{pmatrix}0 & 0 & 1\\ 1 & 0 & 0\\ 0 & 1 & 0\end{pmatrix} \qquad \Gamma^{15\pm}(\{\boldsymbol{\delta}_{3xyz}^{-1}|\mathbf{0}\})=\begin{pmatrix}0 & 1 & 0\\ 0 & 0 & 1\\ 1 & 0 & 0\end{pmatrix}$$

$$\Gamma^{25\pm}(\{\boldsymbol{\delta}_{3xyz}|\mathbf{0}\})=\begin{pmatrix}0 & 0 & 1\\ 1 & 0 & 0\\ 0 & 1 & 0\end{pmatrix} \qquad \Gamma^{25\pm}(\{\boldsymbol{\delta}_{3xyz}^{-1}|\mathbf{0}\})=\begin{pmatrix}0 & 1 & 0\\ 0 & 0 & 1\\ 1 & 0 & 0\end{pmatrix}$$

$$\varepsilon = e^{2\pi i/3}$$

Appendix C

Illustration of ray representation method: Point X in diamond

In Sects. 41–44 of this article we discussed the properties of projective or ray irreducible representations of point groups, especially of the point group of the wave vector $\mathfrak{P}(\boldsymbol{k})$, and in particular in Sect. 44 we showed the equivalence between little group and ray representation methods. Either of these methods can be used to obtain the irreducible representations of the space group of the wave vector $\mathfrak{G}(\boldsymbol{k})$. In Sect. 132 the little group method was used in order to generate acceptable irreducible representations for the stars $^\star X$, $^\star L$, Γ in the diamond structure. In this appendix we propose to illustrate the use of standard tables in order to construct the acceptable representations. Several standard works are now available including those by BRADLEY and CRACKNELL [*42*], KOVALEV [*56*], and ZAK, CASHER, GLUCK, GUR [*57*]. It appears that Kovalev's tables [*56*] are widely used so we shall draw our illustration from them. The reader is advised that the Kovalev tables contain some errors, therefore checking the results obtained internally as well as against the other sources is always desirable.

Let φ_{l_λ} be an element in point group $\mathfrak{P}$ isomorphic to the point group of the wave vector $\mathfrak{P}(\boldsymbol{k})$ as in (41.11). The set of irreducible matrices $D^{(\boldsymbol{k})(m)}$ which multiply as

$$D^{(\boldsymbol{k})(m)}(\varphi_{l_\lambda}) \cdot D^{(\boldsymbol{k})(m)}(\varphi_{l_\mu}) = r^{(\boldsymbol{k})}(\lambda, \mu)\, D^{(\boldsymbol{k})(m)}(\varphi_{l_{\lambda\mu}}) \tag{C.1}$$

with

$$r^{(\boldsymbol{k})}(\lambda, \mu) \equiv \exp - i\boldsymbol{k} \cdot \boldsymbol{R}_{L_{\lambda,\mu}} \tag{C.2}$$

are an acceptable irreducible ray representation, where $\boldsymbol{R}_{L_{\lambda,\mu}}$ is the lattice vector given in (41.4). The ray representation matrices for all point groups using this factor system have been tabulated,[1] consequently one can simply read off the desired matrices from Hurley's tables. The Kovalev [*56*] matrices are defined as

$$\hat{D}^{(\boldsymbol{k})(m)}(\varphi_{l_\lambda}) \equiv \exp i\boldsymbol{k} \cdot \boldsymbol{\tau}_{l_\lambda}\, D^{(\boldsymbol{k})(m)}(\varphi_{l_\lambda}) \tag{C.3}$$

where $\boldsymbol{\tau}_{l_\lambda}$ is the fractional associated with $\{\varphi_{l_\lambda} | \boldsymbol{\tau}_{l_\lambda}\}$. The factor set associated here is

$$\hat{r}^{(\boldsymbol{k})}(\lambda, \mu) \equiv \exp - i(\varphi_{l_\lambda} \cdot \boldsymbol{k} - \boldsymbol{k}) \cdot \boldsymbol{\tau}_{l_\mu}. \tag{C.4}$$

Clearly then we obtain the desired $D^{(\boldsymbol{k})(m)}$ for each coset representative from the Kovalev tabulated $\hat{D}^{(\boldsymbol{k})(m)}$ by using (C.3) i.e. multiplying by $\exp - i\boldsymbol{k} \cdot \boldsymbol{\tau}_{l_\lambda}$. This is a simple gauge transformation.

Now we turn to the case of $^\star X$ in diamond. In the discussion in Sect. 132 (see above Eq. (132.23)) we chose $\boldsymbol{X}_1$ as the canonical wave vector. But for com-

[1] A.C. HURLEY: Phil. Trans. Roy. Soc. (London) **260**, 1 (1966).

parison with Kovalev it is more convenient here to choose $\boldsymbol{X}_3 = \left(00, \frac{2\pi}{a}\right)$ as canonical. The work of this appendix can easily be translated back by constructing the relevant conjugate subgroups.[2] When constructing the full group representation it makes no difference which wave vector in the star is chosen as *a priori* canonical, as can be verified by using the results given *vide infra* to induce the full $D^{(\star \boldsymbol{X})(m)}$ and then comparing with Table B.1, which contains full character systems. The reader should at this point review the discussion given in (132.23)–(132.36).

The factor group $\mathfrak{G}(\boldsymbol{X}_3)/\mathfrak{T} \equiv \mathfrak{P}(\boldsymbol{X}_3)$ is isomorphic to the dihedral group D_{4h}, which can be defined by three generators A, B, C. We choose these here as

$$A \sim \delta_{4z}; \quad B \sim \rho_y; \quad C \sim \rho_z \tag{C.5}$$

by contrast to the choice in (132.23). However the relations (132.24) are unchanged. Each of the generators above is combined with fractional τ_1. The factor set for proper ray representations (C.2) is easily seen to be the following, in terms of the generators

$$r(A,A) = r(B,A) = r(B,B) = r(B,C) = r(A,C) = -1 \tag{C.6}$$

and

$$r(A^3,B) = r(C,C) = r(C,B) = r(C,A) = +1. \tag{C.7}$$

To convert over to the Kovalev factor system using (C.4) requires use of the multiplicative factor $\exp -i\boldsymbol{X}_3 \cdot \boldsymbol{\tau}_1 = -i$. Consequently we have

$$D^{(\boldsymbol{k})(j)}(\{\varphi_{x_3} | \tau_1\}) = -i\hat{D}^{(\boldsymbol{k})(j)}(\varphi_{x_3}) \tag{C.8}$$

where φ_{x_3} is a rotation in $\mathfrak{P}$ and $\hat{D}^{(\boldsymbol{k})(j)}$ is a Kovalev matrix. The factor $-i$ multiplies *only* those matrices whose coset representative has a fractional. A further relabelling is required to bring the Kovalev matrices $\hat{\tau}^{(j)}(h_\alpha)$ (where $j = 1, 2, 3, 4$, and h_α is a space group operator) into concordance with the usual labelling due to HERRING.[3] For the space group elements the correspondence is

$$\{\delta_{4z} | \tau_1\} \to h_{14} \to A; \quad \{\rho_y | \tau_1\} \to h_{27} \to B; \quad \{\rho_z | \tau_1\} \to h_{28} \to C. \tag{C.9}$$

For irreducible representations using our conventional labelling $D^{(\boldsymbol{k})(j)} \to D^{(\boldsymbol{X}_3)(j)}$ where we chose the wave vector $\boldsymbol{X}_3$ as canonical. Then using Table 159 of KOVALEV (*loc. cit.*) we have

$$\begin{aligned} D^{(\boldsymbol{X}_3)(1)} &\to \hat{\tau}^{(3)}; \quad & D^{(\boldsymbol{X}_3)(2)} &\to \hat{\tau}^{(4)}; \\ D^{(\boldsymbol{X}_3)(3)} &\to \hat{\tau}^{(2)}; \quad & D^{(\boldsymbol{X}_3)(4)} &\to \hat{\tau}^{(1)}. \end{aligned} \tag{C.10}$$

Then we give the matrices for each of the generators in each irreducible representation in Table C.1. These generating matrices can be used to obtain all the matrices for the representation when we respect the product rule (C.1). These matrices can now be used to produce the full group irreducible matrices, $D^{(\star \boldsymbol{X}_3)(j)}$.

This illustration is intended to bridge the gap between tables of ray (projective) representations in the literature and our work in order to make it easier to work new cases.

[2] J. L. BIRMAN: Table 2 in [*32*].

[3] C. HERRING: J. Franklin Inst. **233**, 525 (1942).

Table C.1. Generating matrices for irreducible representations at $\boldsymbol{X}_3$

	$\{\delta_{4z}\|\tau_1\}$	$\{\rho_y\|\tau_1\}$	$\{\rho_z\|\tau_1\}$
$D^{(\boldsymbol{X}_3)(1)}$	$\begin{pmatrix} -i & 0 \\ 0 & i \end{pmatrix}$	$\begin{pmatrix} -i & 0 \\ 0 & i \end{pmatrix}$	$\begin{pmatrix} 0 & -i \\ i & 0 \end{pmatrix}$
$D^{(\boldsymbol{X}_3)(2)}$	$\begin{pmatrix} -i & 0 \\ 0 & i \end{pmatrix}$	$\begin{pmatrix} i & 0 \\ 0 & -i \end{pmatrix}$	$\begin{pmatrix} 0 & i \\ -i & 0 \end{pmatrix}$
$D^{(\boldsymbol{X}_3)(3)}$	$\begin{pmatrix} -1 & 0 \\ 0 & 1 \end{pmatrix}$	$\begin{pmatrix} 0 & i \\ i & 0 \end{pmatrix}$	$\begin{pmatrix} 0 & i \\ -i & 0 \end{pmatrix}$
$D^{(\boldsymbol{X}_3)(4)}$	$\begin{pmatrix} -1 & 0 \\ 0 & 1 \end{pmatrix}$	$\begin{pmatrix} 0 & -i \\ -i & 0 \end{pmatrix}$	$\begin{pmatrix} 0 & i \\ -i & 0 \end{pmatrix}$

Appendix D

Tables for the zincblende structure: $F\bar{4}3m$; T_d^2

In this appendix we shall present tables which should be useful in dealing with selection rules and related matters in the zincblende structure. The space group symbols for this structure are $F\bar{4}3m$ or T_d^2. In order to avoid duplication of material given earlier when discussing the diamond structure, maximum use will be made of the tables previously given.

Table D.1. Symmetry of zincblende structure; T_d^2: $F\bar{4}3m$

a) Translation group:	Face centered cubic: see Table 1.
b) Rotational elements:	Factor set $\{\varphi_i\|\mathbf{0}\}$, $i=1, \ldots, 24$, isomorphic to T_d; see left column, Table 2.
c) Zone points, wave vectors:	As in Table 3 with following exceptions for multiplicity: Λ, $s=4$; $\boldsymbol{M}$, $s=12$; $\boldsymbol{k}$, $s=24$
d) Wave vector selection rules:[a]	As in Tables 15 and 16, with following exceptions:

Ordinary	$(\boldsymbol{LZ}\|\boldsymbol{M})=2$;	$\boldsymbol{M}_1=(\pi, \pi, -\kappa-\pi)(1/a)$
	$(\boldsymbol{LZ}\|\boldsymbol{M}')=2$;	$\boldsymbol{M}_1'=(\pi, \pi, \kappa-\pi)(1/a)$
	$(\boldsymbol{L\Sigma}\|\boldsymbol{M})=1$;	$\boldsymbol{M}_1=(\kappa+\pi, \kappa+\pi, \pi)(1/a)$
	$(\boldsymbol{L\Sigma}\|\boldsymbol{M}')=1$;	$\boldsymbol{M}_1'=(\kappa-\pi, \kappa-\pi, \pi)(1/a)$
	$(\boldsymbol{L\Sigma}\|\boldsymbol{k})=1$;	$\boldsymbol{k}_1=(\pi+\kappa, \kappa-\pi, -\pi)(1/a)$
	$(\boldsymbol{L\Delta}\|\boldsymbol{M})=1$;	$\boldsymbol{M}_1=(\pi, \pi, \pi+\kappa)(1/a)$
	$(\boldsymbol{L\Delta}\|\boldsymbol{M}')=1$;	$\boldsymbol{M}_1=(\pi, +\pi, \pi-\kappa)(1/a)$
	$(\boldsymbol{\Sigma\Lambda}\|\boldsymbol{M})=1$;	$\boldsymbol{M}_1=(\kappa+\kappa', \kappa+\kappa', \kappa')(1/a)$
	$(\boldsymbol{\Sigma\Lambda}\|\boldsymbol{M}')=1$;	$\boldsymbol{M}_1'=(\kappa-\kappa', \kappa-\kappa', \kappa')(1/a)$
	$(\boldsymbol{\Sigma\Lambda}\|\boldsymbol{k})=1$;	$\boldsymbol{k}_1=(\kappa+\kappa', \kappa-\kappa', -\kappa')(1/a)$
	$(\boldsymbol{\Delta\Lambda}\|\boldsymbol{M})=1$;	$\boldsymbol{M}_1=(\kappa, \kappa, \kappa'+\kappa)(1/a)$
	$(\boldsymbol{\Delta\Lambda}\|\boldsymbol{M}'=1$;	$\boldsymbol{M}_1'=(\kappa, \kappa, \kappa-\kappa')(1/a)$
	$(\boldsymbol{\Lambda\Lambda}'\|\boldsymbol{\Lambda}'')=1$;	$\boldsymbol{\Lambda}_1''=(\kappa+\kappa', \kappa+\kappa', \kappa+\kappa')(1/a)$
	$(\boldsymbol{\Lambda\Lambda}'\|\boldsymbol{M})=1$;	$\boldsymbol{M}_1=(\kappa-\kappa', \kappa-\kappa', \kappa+\kappa')(1/a)$

Symmetrized	$([\boldsymbol{X}]_{(2)}\|\boldsymbol{\Gamma})=3$;	$(\|\boldsymbol{X})=1$						
	$([\boldsymbol{\Lambda}]_{(2)}\|\boldsymbol{\Lambda}')=1$;	$(\|\boldsymbol{\Delta}')=1$						
	$([2\boldsymbol{\Lambda}]_{(2)}\|\boldsymbol{\Lambda}')=3$;	$(\|\boldsymbol{\Delta}')=4$						
	$([\boldsymbol{W}]_{(2)}\|\boldsymbol{\Gamma})=3$;	$(\|\boldsymbol{X})=2$;	$(\|\boldsymbol{\Sigma})=1$					
	$([\boldsymbol{Z}]_{(2)}\|\boldsymbol{\Gamma})=6$;	$(\|\boldsymbol{\Delta})=2$;	$(\|\boldsymbol{\Delta}')=1$;	$(\|\boldsymbol{k})=1$;	$(\|\boldsymbol{X})=2$;	$(\|\boldsymbol{\Sigma})=1$;	$(\|\boldsymbol{\Sigma}')=1$	
	$([\boldsymbol{k}]_{(2)}\|\boldsymbol{\Gamma})=12$;	$(\|\boldsymbol{k}')=12$						
	$([\boldsymbol{X}]_{(3)}\|\boldsymbol{\Gamma})=1$;	$(\|\boldsymbol{X})=3$						
	$([\boldsymbol{W}]_{(3)}\|\boldsymbol{W})=4$;	$(\|\boldsymbol{\Delta})=4$;	$(\|\boldsymbol{L})=2$					

[a] Rules not listed here are identical to those in diamond: see Tables 15, 16. When the latter are used for zinc blende, use the appropriate wave vectors (see Table 3, and above).

Table D.2. Irreducible representations in zincblende at symmetry points[a]

Character table for the representations of the single group of Γ

24	Γ	Γ_1	Γ_2	Γ_{12}	Γ_{15}	Γ_{25}
1	E	1	1	2	3	3
3	C_4^2	1	1	2	-1	-1
8	C_3	1	1	-1	0	0
6	JC_4	1	-1	0	-1	1
6	JC_2	1	-1	0	1	-1

Character table for the representations of the single group of Δ. (The two distinct operations JC_2 are those about the two twofold axes perpendicular to Δ.) Δ_3 and Δ_4 stick together

4	Δ	Δ_1	Δ_2	Δ_3	Δ_4
1	E	1	1	1	1
1	JC_2	1	-1	1	-1
1	JC_2	1	-1	-1	1
1	$C_{4\parallel}^2$	1	1	-1	-1

Character table for the representations of the single group of Λ

6	Λ	Λ_1	Λ_2	Λ_3
1	E	1	1	2
2	C_3	1	1	-1
3	JC_2	1	-1	0

Character table for the representations of the single groups of Σ and Z

2	Σ	Z	Σ_1	Σ_2
1	E	E	1	1
1	JC_2	$C_{4\perp}^2$	1	-1

Character table for the representations of the single group of X

8	X	X_1	X_2	X_3	X_4	X_5
1	E	1	1	1	1	2
2	$C_{4\perp}^2$	1	1	-1	-1	0
1	$C_{4\parallel}^2$	1	1	1	1	-2
2	$JC_{4\parallel}$	1	-1	-1	1	0
2	JC_2	1	-1	1	-1	0

Character table for the representations of the single group of W. (The two distinct operations JC_4 are those about the fourfold axis parallel to the face diagonal containing W.) W_3 and W_4 are degenerate by time-reversal symmetry

4	W	W_1	W_2	W_3	W_4
1	E	1	1	1	1
1	JC_4	1	-1	i	$-i$
1	JC_4	1	-1	$-i$	i
1	C_4^2	1	1	-1	-1

[a] From R. H. PARMENTER: Phys. Rev. **100**, 573 (1955).

Table D.3. Compatibility tables for the representations of the single groups connecting the zinc blende (T_d^2) with the rocksalt (O_h^5) and the diamond (O_h^7) structures[a]

T_d^2	O_h^5	T_d^2	O_h^7
Γ_1	Γ_1 or $\Gamma_{2'}$	Γ_1	Γ_1 or $\Gamma_{2'}$
Γ_2	Γ_2 or $\Gamma_{1'}$	Γ_2	Γ_2 or $\Gamma_{1'}$
Γ_{12}	Γ_{12} or $\Gamma_{12'}$	Γ_{12}	Γ_{12} or $\Gamma_{12'}$
Γ_{15}	Γ_{15} or $\Gamma_{25'}$	Γ_{15}	Γ_{15} or $\Gamma_{25'}$
Γ_{25}	Γ_{25} or $\Gamma_{15'}$	Γ_{25}	Γ_{25} or $\Gamma_{15'}$
Δ_1	Δ_1 or $\Delta_{2'}$	Δ_1	Δ_1 or $\Delta_{2'}$
Δ_2	Δ_2 or $\Delta_{1'}$	Δ_2	Δ_2 or $\Delta_{1'}$
Δ_3, Δ_4	Δ_5	Δ_3, Δ_4	Δ_5
Λ_1	Λ_1	Λ_1	Λ_1
Λ_2	Λ_2	Λ_2	Λ_2
Λ_3	Λ_3	Λ_3	Λ_3
Σ_1	Σ_1 or Σ_3	Σ_1	Σ_1 or Σ_3
Σ_2	Σ_2 or Σ_4	Σ_2	Σ_2 or Σ_4
Z_1	Z_1 or Z_2	Z_1, Z_2	Z_1
Z_2	Z_3 or Z_4		
X_1	X_1 or $X_{2'}$	X_1, X_3	X_1
X_2	X_2 or $X_{1'}$		
X_3	X_3 or $X_{4'}$	X_2, X_4	X_2
X_4	X_4 or $X_{3'}$		
X_5	X_5 or $X_{5'}$	X_5	X_3 or X_4
W_1	W_1 or W_2	W_1, W_3	W_1
W_2	$W_{1'}$ or $W_{2'}$		
W_3, W_4	W_3	W_2, W_4	W_2

[a] From R.H. PARMENTER: Phys. Rev. **100**, 573 (1955).

Zincblende is a symmorphic subgroup of diamond of index 2:

$$O_h^7 = T_d^2 + \{\boldsymbol{i}|\tau_1\}\ T_d^2 \tag{D.1}$$

where the coset representative $\{\boldsymbol{i}|\tau_1\}$ is the inversion plus fractional translation τ_1. The translation group and reciprocal space are then identical to that given in Table 1. The rotational symmetry operators are the set $\{\varphi|\mathbf{0}\}$ in the left column of Table 2. Space group operators are then elements $\{\varphi_i|\boldsymbol{R}_L\}$ where $i=1, \ldots, 24$, and $\boldsymbol{R}_L$ is an element in the face-centered translation group $\mathfrak{T}$. Wave vectors and their components are identical in zincblende and diamond but the multiplicity differs, for certain $\boldsymbol{k}$, owing to the factor group having only 24 rotational elements. All this information is summarized in Table D.1.

The irreducible representations for zincblende have all been given by PARMENTER,[1] and we follow his notations and designations in this appendix, with

[1] R.H. PARMENTER: Phys. Rev. **100**, 573 (1955).

Table D.4. Reduction coefficients for zincblende[a]

$(\boldsymbol{\Gamma}^{(1)}\,\mathbf{k}_j^{(m)}|\mathbf{k}_j^{(m)})=1$

$(\boldsymbol{\Gamma}^{(2)}\,\mathbf{L}^{(1)}|\mathbf{L}^{(2)})=1$

$(\boldsymbol{\Gamma}^{(2)}\,\mathbf{L}^{(2)}|\mathbf{L}^{(1)})=1$

$(\boldsymbol{\Gamma}^{(2)}\,\mathbf{L}^{(3)}|\mathbf{L}^{(3)})=1$

$(\boldsymbol{\Gamma}^{(2)}\,\mathbf{X}^{(1)}|\mathbf{X}^{(2)})=1$

$(\boldsymbol{\Gamma}^{(2)}\,\mathbf{X}^{(2)}|\mathbf{X}^{(1)})=1$

$(\boldsymbol{\Gamma}^{(2)}\,\mathbf{X}^{(3)}|\mathbf{X}^{(4)})=1$

$(\boldsymbol{\Gamma}^{(2)}\,\mathbf{X}^{(4)}|\mathbf{X}^{(3)})=1$

$(\boldsymbol{\Gamma}^{(2)}\,\mathbf{X}^{(5)}|\mathbf{X}^{(5)})=1$

$(\boldsymbol{\Gamma}^{(12)}\,\mathbf{L}^{(1)}|\mathbf{L}^{(3)})=1$

$(\boldsymbol{\Gamma}^{(12)}\,\mathbf{L}^{(2)}|\mathbf{L}^{(3)})=1$

$(\boldsymbol{\Gamma}^{(12)}\,\mathbf{L}^{(3)}|\mathbf{L}^{(1)})=(|\mathbf{L}^{(2)})=(|\mathbf{L}^{(3)})=1$

$(\boldsymbol{\Gamma}^{(12)}\,\mathbf{X}^{(1)}|\mathbf{X}^{(1)})=(|\mathbf{X}^{(2)})=1$

$(\boldsymbol{\Gamma}^{(12)}\,\mathbf{X}^{(2)}|\mathbf{X}^{(1)})=(|\mathbf{X}^{(2)})=1$

$(\boldsymbol{\Gamma}^{(12)}\,\mathbf{X}^{(3)}|\mathbf{X}^{(3)})=(|\mathbf{X}^{(4)})=1$

$(\boldsymbol{\Gamma}^{(12)}\,\mathbf{X}^{(4)}|\mathbf{X}^{(3)})=(|\mathbf{X}^{(4)})=1$

$(\boldsymbol{\Gamma}^{(12)}\,\mathbf{X}^{(5)}|\mathbf{X}^{(5)})=2$

$(\boldsymbol{\Gamma}^{(15)}\,\mathbf{L}^{(1)}|\mathbf{L}^{(1)})=(|\mathbf{L}^{(3)})=1$

$(\boldsymbol{\Gamma}^{(15)}\,\mathbf{L}^{(2)}|\mathbf{L}^{(2)})=(|\mathbf{L}^{(3)})=1$

$(\boldsymbol{\Gamma}^{(15)}\,\mathbf{L}^{(3)}|\mathbf{L}^{(1)})=(|\mathbf{L}^{(2)})=1; \quad (|\mathbf{L}^{(3)})=2$

$(\boldsymbol{\Gamma}^{(15)}\,\mathbf{X}^{(1)}|\mathbf{X}^{(3)})=(|\mathbf{X}^{(5)})=1$

$(\boldsymbol{\Gamma}^{(15)}\,\mathbf{X}^{(2)}|\mathbf{X}^{(4)})=(|\mathbf{X}^{(5)})=1$

$(\boldsymbol{\Gamma}^{(15)}\,\mathbf{X}^{(3)}|\mathbf{X}^{(1)})=(|\mathbf{X}^{(5)})=1$

$(\boldsymbol{\Gamma}^{(15)}\,\mathbf{X}^{(4)}|\mathbf{X}^{(2)})=(|\mathbf{X}^{(5)})=1$

$(\boldsymbol{\Gamma}^{(15)}\,\mathbf{X}^{(5)}|\mathbf{X}^{(1)})=(|\mathbf{X}^{(2)})=(|\mathbf{X}^{(3)})=(|\mathbf{X}^{(4)})$
$=(|\mathbf{X}^{(5)})=1$

$(\boldsymbol{\Gamma}^{(25)}\,\mathbf{L}^{(1)}|\mathbf{L}^{(2)})=(|\mathbf{L}^{(3)})=1$

$(\boldsymbol{\Gamma}^{(25)}\,\mathbf{L}^{(2)}|\mathbf{L}^{(1)})=(|\mathbf{L}^{(3)})=1$

$(\boldsymbol{\Gamma}^{(25)}\,\mathbf{L}^{(3)}|\mathbf{L}^{(1)})=(|\mathbf{L}^{(2)})=1; \quad (|\mathbf{L}^{(3)})=2$

$(\boldsymbol{\Gamma}^{(25)}\,\mathbf{X}^{(1)}|\mathbf{X}^{(4)})=(|\mathbf{X}^{(5)})=1$

$(\boldsymbol{\Gamma}^{(25)}\,\mathbf{X}^{(2)}|\mathbf{X}^{(3)})=(|\mathbf{X}^{(5)})=1$

$(\boldsymbol{\Gamma}^{(25)}\,\mathbf{X}^{(3)}|\mathbf{X}^{(2)})=(|\mathbf{X}^{(5)})=1$

$(\boldsymbol{\Gamma}^{(25)}\,\mathbf{X}^{(4)}|\mathbf{X}^{(1)})=(|\mathbf{X}^{(5)})=1$

$(\boldsymbol{\Gamma}^{(25)}\,\mathbf{X}^{(5)}|)=(\boldsymbol{\Gamma}^{(15)}\,\mathbf{X}^{(5)}|)$

$(\mathbf{L}^{(1)}\,\mathbf{L}^{(1)}|\boldsymbol{\Gamma}^{(1)})=(|\boldsymbol{\Gamma}^{(15)})=(|\mathbf{X}^{(1)})=(|\mathbf{X}^{(3)})$
$=(|\mathbf{X}^{(5)})=1$

$(\mathbf{L}^{(1)}\,\mathbf{L}^{(2)}|\boldsymbol{\Gamma}^{(2)})=(|\boldsymbol{\Gamma}^{(25)})=(|\mathbf{X}^{(2)})=(|\mathbf{X}^{(4)})$
$=(|\mathbf{X}^{(5)})=1$

$(\mathbf{L}^{(1)}\,\mathbf{L}^{(3)}|\boldsymbol{\Gamma}^{(12)})=(|\boldsymbol{\Gamma}^{(15)})=(|\boldsymbol{\Gamma}^{(25)})$
$=(|\mathbf{X}^{(1)})=(|\mathbf{X}^{(2)})$
$=(|\mathbf{X}^{(3)})=(|\mathbf{X}^{(4)})=1;$
$(|\mathbf{X}^{(5)})=2$

$(\mathbf{L}^{(1)}\,\mathbf{X}^{(1)}|\mathbf{L}^{(1)})=(|\mathbf{L}^{(3)})=1$

$(\mathbf{L}^{(1)}\,\mathbf{X}^{(2)}|\mathbf{L}^{(2)})=(|\mathbf{L}^{(3)})=1$

$(\mathbf{L}^{(1)}\,\mathbf{X}^{(3)}|)=(\mathbf{L}^{(1)}\,\mathbf{X}^{(1)}|)$

$(\mathbf{L}^{(1)}\,\mathbf{X}^{(4)}|)=(\mathbf{L}^{(1)}\,\mathbf{X}^{(2)}|)$

$(\mathbf{L}^{(1)}\,\mathbf{X}^{(5)}|\mathbf{L}^{(1)})=(|\mathbf{L}^{(2)})=1; \quad (|\mathbf{L}^{(3)})=2$

$(\mathbf{L}^{(2)}\,\mathbf{L}^{(2)}|)=(\mathbf{L}^{(1)}\,\mathbf{L}^{(1)}|)$

$(\mathbf{L}^{(2)}\,\mathbf{L}^{(3)}|)=(\mathbf{L}^{(1)}\,\mathbf{L}^{(3)}|)$

$(\mathbf{L}^{(2)}\,\mathbf{X}^{(1)}|)=(\mathbf{L}^{(1)}\,\mathbf{X}^{(2)}|)$

$(\mathbf{L}^{(2)}\,\mathbf{X}^{(2)}|)=(\mathbf{L}^{(1)}\,\mathbf{X}^{(1)}|)$

$(\mathbf{L}^{(2)}\,\mathbf{X}^{(3)}|)=(\mathbf{L}^{(1)}\,\mathbf{X}^{(2)}|)$

$(\mathbf{L}^{(2)}\,\mathbf{X}^{(4)}|)=(\mathbf{L}^{(1)}\,\mathbf{X}^{(1)}|)$

$(\mathbf{L}^{(2)}\,\mathbf{X}^{(5)}|)=(\mathbf{L}^{(1)}\,\mathbf{X}^{(5)}|)$

$(\mathbf{L}^{(3)}\,\mathbf{L}^{(3)}|\boldsymbol{\Gamma}^{(1)})=(|\boldsymbol{\Gamma}^{(2)})=(|\boldsymbol{\Gamma}^{(12)})=1;$
$(|\boldsymbol{\Gamma}^{(15)})=(|\boldsymbol{\Gamma}^{(25)})$
$=(|\mathbf{X}^{(1)})=(|\mathbf{X}^{(2)})=(|\mathbf{X}^{(3)})$
$=(|\mathbf{X}^{(4)})=2; \quad (|\mathbf{X}^{(5)})=4$

$(\mathbf{L}^{(3)}\,\mathbf{X}^{(1)}|\mathbf{L}^{(1)})=(|\mathbf{L}^{(2)})=1; \quad (|\mathbf{L}^{(3)})=2$

$(\mathbf{L}^{(3)}\,\mathbf{X}^{(2)}|)=(\mathbf{L}^{(3)}\,\mathbf{X}^{(1)}|)$

$(\mathbf{L}^{(3)}\,\mathbf{X}^{(3)}|)=(\mathbf{L}^{(3)}\,\mathbf{X}^{(1)}|)$

$(\mathbf{L}^{(3)}\,\mathbf{X}^{(4)}|)=(\mathbf{L}^{(3)}\,\mathbf{X}^{(1)}|)$

$(\mathbf{L}^{(3)}\,\mathbf{X}^{(5)}|\mathbf{L}^{(1)})=(|\mathbf{L}^{(2)})=2; \quad (|\mathbf{L}^{(3)})=4$

$(\mathbf{X}^{(1)}\,\mathbf{X}^{(1)}|\boldsymbol{\Gamma}^{(1)})=(|\boldsymbol{\Gamma}^{(12)})=(|\mathbf{X}^{(1)})=(|\mathbf{X}^{(2)})=1$

$(\mathbf{X}^{(1)}\,\mathbf{X}^{(2)}|\boldsymbol{\Gamma}^{(2)})=(|\boldsymbol{\Gamma}^{(12)})=(|\mathbf{X}^{(1)})=(|\mathbf{X}^{(2)})=1$

$(\mathbf{X}^{(1)}\,\mathbf{X}^{(3)}|\boldsymbol{\Gamma}^{(15)})=(|\mathbf{X}^{(5)})=1$

$(\mathbf{X}^{(1)}\,\mathbf{X}^{(4)}|\boldsymbol{\Gamma}^{(25)})=(|\mathbf{X}^{(5)})=1$

$(\mathbf{X}^{(1)}\,\mathbf{X}^{(5)}|\boldsymbol{\Gamma}^{(15)})=(|\boldsymbol{\Gamma}^{(25)})=(|\mathbf{X}^{(3)})=(|\mathbf{X}^{(4)})$
$=(|\mathbf{X}^{(5)})=1$

$(\mathbf{X}^{(2)}\,\mathbf{X}^{(2)}|)=(\mathbf{X}^{(1)}\,\mathbf{X}^{(1)}|)$

$(\mathbf{X}^{(2)}\,\mathbf{X}^{(3)}|)=(\mathbf{X}^{(1)}\,\mathbf{X}^{(4)}|)$

$(\mathbf{X}^{(2)}\,\mathbf{X}^{(4)}|)=(\mathbf{X}^{(1)}\,\mathbf{X}^{(3)}|)$

$(\mathbf{X}^{(2)}\,\mathbf{X}^{(5)}|)=(\mathbf{X}^{(1)}\,\mathbf{X}^{(5)}|)$

$(\mathbf{X}^{(3)}\,\mathbf{X}^{(3)}|\boldsymbol{\Gamma}^{(1)})=(|\boldsymbol{\Gamma}^{(12)})=(|\mathbf{X}^{(3)})=(|\mathbf{X}^{(4)})=1$

$(\mathbf{X}^{(3)}\,\mathbf{X}^{(4)}|\boldsymbol{\Gamma}^{(2)})=(|\boldsymbol{\Gamma}^{(12)})=(|\mathbf{X}^{(3)})=(|\mathbf{X}^{(4)})=1$

$(\mathbf{X}^{(3)}\,\mathbf{X}^{(5)}|\boldsymbol{\Gamma}^{(15)})=(|\boldsymbol{\Gamma}^{(25)})=(|\mathbf{X}^{(1)})=(|\mathbf{X}^{(2)})$
$=(|\mathbf{X}^{(5)})=1$

$(\mathbf{X}^{(4)}\,\mathbf{X}^{(4)}|)=(\mathbf{X}^{(3)}\,\mathbf{X}^{(3)}|)$

$(\mathbf{X}^{(4)}\,\mathbf{X}^{(5)}|)=(\mathbf{X}^{(3)}\,\mathbf{X}^{(5)}|)$

$(\mathbf{X}^{(5)}\,\mathbf{X}^{(5)}|\boldsymbol{\Gamma}^{(1)})=(|\boldsymbol{\Gamma}^{(2)})=(|\boldsymbol{\Gamma}^{(15)})=(|\boldsymbol{\Gamma}^{(25)})$
$=(|\mathbf{X}^{(1)})=(|\mathbf{X}^{(2)})=(|\mathbf{X}^{(3)})$
$=(|\mathbf{X}^{(4)})=1;$
$(|\boldsymbol{\Gamma}^{(12)})=(|\mathbf{X}^{(5)})=2$

$(\mathbf{W}^{(1)}\,\mathbf{W}^{(1)}|\boldsymbol{\Gamma}^{(1)})=(|\boldsymbol{\Gamma}^{(12)})=(|\boldsymbol{\Gamma}^{(25)})$
$=(|\mathbf{X}^{(1)})=(|\mathbf{X}^{(4)})$
$=(|\boldsymbol{\Sigma}^{(1)})=(|\boldsymbol{\Sigma}^{(2)})=1$

$(\mathbf{W}^{(1)}\,\mathbf{W}^{(2)}|\boldsymbol{\Gamma}^{(2)})=(|\boldsymbol{\Gamma}^{(12)})=(|\boldsymbol{\Gamma}^{(15)})$
$=(|\mathbf{X}^{(2)})=(|\mathbf{X}^{(3)})$
$=(|\boldsymbol{\Sigma}^{(1)})=(|\boldsymbol{\Sigma}^{(2)})=1$

$(\mathbf{W}^{(1)}\,\mathbf{W}^{(3)}|\boldsymbol{\Gamma}^{(15)})=(|\boldsymbol{\Gamma}^{(25)})=(|\mathbf{X}^{(5)})$
$=(|\boldsymbol{\Sigma}^{(1)})=(|\boldsymbol{\Sigma}^{(2)})=1$

$(\mathbf{W}^{(3)}\,\mathbf{W}^{(3)}|\boldsymbol{\Gamma}^{(1)})=(|\boldsymbol{\Gamma}^{(12)})=(|\boldsymbol{\Gamma}^{(25)})$
$=(|\mathbf{X}^{(2)})=(|\mathbf{X}^{(3)})$
$=(|\boldsymbol{\Sigma}^{(1)})=(|\boldsymbol{\Sigma}^{(2)})=1$

$(\mathbf{W}^{(3)}\,\mathbf{W}^{(4)}|\boldsymbol{\Gamma}^{(2)})=(|\boldsymbol{\Gamma}^{(12)})=(|\boldsymbol{\Gamma}^{(15)})$
$=(|\mathbf{X}^{(1)})=(|\mathbf{X}^{(4)})$
$=(|\boldsymbol{\Sigma}^{(1)})=(|\boldsymbol{\Sigma}^{(2)})=1$

$(\mathbf{W}^{(1)}\,\mathbf{W}^{(4)}|)=(\mathbf{W}^{(1)}\,\mathbf{W}^{(3)}|)$

$(\mathbf{W}^{(2)}\,\mathbf{W}^{(2)}|)=(\mathbf{W}^{(1)}\,\mathbf{W}^{(1)}|)$

$(\mathbf{W}^{(2)}\,\mathbf{W}^{(3)}|)=(\mathbf{W}^{(1)}\,\mathbf{W}^{(3)}|)$

$(\mathbf{W}^{(4)}\,\mathbf{W}^{(4)}|)=(\mathbf{W}^{(3)}\,\mathbf{W}^{(3)}|)$

[a] From J. L. Birman: Phys. Rev. **127**, 1093 (1962).

Table D.5. Additional reduction coefficients in zincblende*

$(\Gamma^{(2)}\,\Delta^{(1)}|\Delta^{(2)}) = (\Gamma^{(2)}\,\Delta^{(2)}|\Delta^{(1)}) = 1$
$(\Gamma^{(2)}\,\Delta^{(m)}|\Delta^{(m)}) = 1; \quad m = 3, 4$
$(\Gamma^{(12)}\,\Delta^{(1)}|\Delta^{(1)}) = (|\Delta^{(2)}) = 1$
$(\Gamma^{(12)}\,\Delta^{(2)}|) = (\Gamma^{(12)}\,\Delta^{(1)}|)$
$(\Gamma^{(12)}\,\Delta^{(3)}|\Delta^{(3)}) = (|\Delta^{(4)}) = 1$
$(\Gamma^{(12)}\,\Delta^{(4)}|) = (\Gamma^{(12)}\,\Delta^{(3)}|)$
$(\Gamma^{(15)}\,\Delta^{(1)}|\Delta^{(1)}) = (|\Delta^{(3)}) = (|\Delta^{(4)}) = 1$
$(\Gamma^{(15)}\,\Delta^{(2)}|\Delta^{(2)}) = (|\Delta^{(3)}) = (|\Delta^{(4)}) = 1$
$(\Gamma^{(15)}\,\Delta^{(3)}|\Delta^{(1)}) = (|\Delta^{(2)}) = (|\Delta^{(3)}) = 1$
$(\Gamma^{(25)}\,\Delta^{(1)}|) = (\Gamma^{(15)}\,\Delta^{(2)}|)$
$(\Gamma^{(25)}\,\Delta^{(2)}|) = (\Gamma^{(15)}\,\Delta^{(1)}|)$
$(\Gamma^{(25)}\,\Delta^{(3)}|) = (\Gamma^{(15)}\,\Delta^{(3)}|)$
$(\Gamma^{(2)}\,\Lambda^{(1)}|\Lambda^{(2)}) = 1$
$(\Gamma^{(2)}\,\Lambda^{(2)}|\Lambda^{(1)}) = 1$
$(\Gamma^{(2)}\,\Lambda^{(3)}|\Lambda^{(3)}) = 1$
$(\Gamma^{(12)}\,\Lambda^{(1)}|\Lambda^{(3)}) = 1$
$(\Gamma^{(12)}\,\Lambda^{(2)}|\Lambda^{(3)}) = 1$
$(\Gamma^{(12)}\,\Lambda^{(3)}|\Lambda^{(1)}) = (|\Lambda^{(2)}) = (|\Lambda^{(3)}) = 1$
$(\Gamma^{(15)}\,\Lambda^{(1)}|\Lambda^{(1)}) = (|\Lambda^{(3)}) = 1$
$(\Gamma^{(15)}\,\Lambda^{(2)}|\Lambda^{(2)}) = (|\Lambda^{(3)}) = 1$
$(\Gamma^{(15)}\,\Lambda^{(3)}|\Lambda^{(1)}) = (|\Lambda^{(2)}) = 1; \quad (|\Lambda^{(3)}) = 2$
$(\Gamma^{(25)}\,\Lambda^{(1)}|\Lambda^{(2)}) = (|\Lambda^{(3)}) = 1$
$(\Gamma^{(25)}\,\Lambda^{(2)}|\Lambda^{(1)}) = (|\Lambda^{(3)}) = 1$
$(\Gamma^{(25)}\,\Lambda^{(3)}|) = (\Gamma^{(15)}\,\Lambda^{(3)}|)$

$(\Delta^{(1)}\,\Delta^{(1)}|\Gamma^{(1)}) = (|\Gamma^{(12)}) = (|\Gamma^{(15)}) = (|\Sigma^{(1)}) = (|\Sigma^{(2)}) = (|\Delta'^{(1)}) = 1$
$(\Delta^{(1)}\,\Delta^{(2)}|\Gamma^{(2)}) = (|\Gamma^{(12)}) = (|\Gamma^{(25)}) = (|\Sigma^{(1)}) = (|\Sigma^{(2)}) = (|\Delta'^{(2)}) = 1$
$(\Delta^{(1)}\,\Delta^{(3)}|\Gamma^{(15)}) = (|\Gamma^{(25)}) = (|\Sigma^{(1)}) = (|\Sigma^{(2)}) = (|\Delta'^{(3)}) = 1$ [a]
$(\Delta^{(3)}\,\Delta^{(3)}|\Gamma^{(1)}) = (|\Gamma^{(12)}) = (|\Gamma^{(15)}) = (|\Sigma^{(1)}) = (|\Sigma^{(2)}) = (|\Delta'^{(2)}) = 1$ [b]
$(\Delta^{(3)}\,\Delta^{(3)}|\Gamma^{(2)}) = (|\Gamma^{(12)}) = (|\Gamma^{(25)}) = (|\Sigma^{(1)}) = (|\Sigma^{(2)}) = (|\Delta'^{(1)}) = 1$ [b]

$(\Lambda^{(1)}\,\Lambda^{(1)}|\Delta'^{(1)}) = (|\Delta'^{(3)}) = (|\Lambda'^{(1)}) = 1$ [a]
$(\Lambda^{(1)}\,\Lambda^{(2)}|\Delta'^{(2)}) = (|\Delta'^{(3)}) = (|\Lambda'^{(2)}) = 1$ [a]
$(\Lambda^{(1)}\,\Lambda^{(3)}|\Delta'^{(1)}) = (|\Delta'^{(2)}) = (|\Delta'^{(3)}) = (|\Delta'^{(4)}) = (|\Lambda'^{(3)}) = 1$
$(\Lambda^{(3)}\,\Lambda^{(3)}|\Lambda'^{(1)}) = (|\Lambda'^{(2)}) = (|\Lambda'^{(3)}) = 1; \quad (|\Delta'^{(1)}) = (|\Delta'^{(2)}) = (|\Delta'^{(3)}) = (|\Delta'^{(4)}) = 2$

$(\Lambda^{(1)}(-\Lambda)^{(1)}|\Gamma^{(1)}) = (|\Gamma^{(15)}) = (|\Sigma'^{(1)}) = 1$ [c]
$(\Lambda^{(1)}(-\Lambda)^{(2)}|\Gamma^{(2)}) = (|\Gamma^{(25)}) = (|\Sigma'^{(2)}) = 1$
$(\Lambda^{(1)}(-\Lambda)^{(3)}|\Gamma^{(12)}) = (|\Gamma^{(15)}) = (|\Gamma^{(25)}) = (|\Sigma'^{(1)}) = (|\Sigma'^{(2)}) = 1$
$(\Lambda^{(3)}(-\Lambda)^{(3)}|\Gamma^{(1)}) = (|\Gamma^{(2)}) = (|\Gamma^{(12)}) = 1; \quad (|\Gamma^{(15)}) = (|\Gamma^{(25)}) = (|\Sigma'^{(1)}) = (|\Sigma'^{(2)}) = 2$

* See footnote Table D.4.

[a] $\star\Delta^{(3)}$ and $\star\Delta^{(4)}$ are degenerate and interchangeable.

[b] The ambiguity in this reduction arises because of the degeneracy of $\star\Delta^{(3)}$ and $\star\Delta^{(4)}$.

[c] See text, Eqs. (7.1) and (7.2).

Table D.6. Reduction coefficients for symmetrized Kronecker square in zincblende*

$([\Gamma^{(m)}]_{(2)}|\Gamma^{(1)}) = 1; \quad m=1,2$
$([\Gamma^{(12)}]_{(2)}|\Gamma^{(1)}) = (|\Gamma^{(12)}) = 1$
$([\Gamma^{(m)}]_{(2)}|\Gamma^{(1)}) = (|\Gamma^{(12)}) = (|\Gamma^{(15)}) = 1; \quad m=15, 25$

$([L^{(m)}]_{(2)}|\Gamma^{(1)}) = (|\Gamma^{(15)}) = (|X^{(1)}) = (|X^{(3)}) = 1; \quad m=1,2$
$([L^{(3)}]_{(2)}|\Gamma^{(1)}) = (|\Gamma^{(12)}) = (|\Gamma^{(25)}) = (|X^{(2)}) = (|X^{(4)}) = (|X^{(5)}) = 1;$
$(|\Gamma^{(15)}) = (|X^{(1)}) = (|X^{(3)}) = 2$

$([X^{(m)}]_{(2)}|\Gamma^{(1)}) = (|\Gamma^{(12)}) = (|X^{(1)}) = 1; \quad m=1,2$
$([X^{(m)}]_{(2)}|\Gamma^{(1)}) = (|\Gamma^{(12)}) = (|X^{(3)}) = 1; \quad m=3,4$
$([X^{(5)}]_{(2)}|\Gamma^{(1)}) = (|\Gamma^{(2)}) = (|\Gamma^{(15)}) = (|X^{(1)}) = (|X^{(3)}) = (|X^{(5)}) = 1; \quad (|\Gamma^{(12)}) = 2$

$([\Delta^{(m)}]_{(2)}|\Gamma^{(1)}) = (|\Gamma^{(12)}) = (|\Delta'^{(1)}) = (|\Sigma^{(1)}) = 1; \quad m=1,2$
$([\Delta^{(m)}]_{(2)}|\Gamma^{(2)}) = (|\Gamma^{(12)}) = (|\Delta'^{(1)}) = (|\Sigma^{(1)}); \quad m=3,4$

$([\Lambda^{(m)}]_{(2)}|\Delta'^{(1)}) = (|\Lambda'^{(1)}) = 1; \quad m=1,2$
$([\Lambda^{(3)}]_{(2)}|\Lambda'^{(1)}) = (|\Lambda'^{(3)}) = (|\Delta'^{(2)}) = (|\Delta'^{(3)}) = 1$ [a]; $\quad (|\Delta'^{(1)}) = 2$

$([W^{(1)}]_{(2)}|\Gamma^{(1)}) = (|\Gamma^{(12)}) = (|X^{(1)}) = (|X^{(4)}) = (|\Sigma^{(1)}) = 1$
$([W^{(2)}]_{(2)}|) = ([W^{(1)}]_{(2)}|)$
$([W^{(3)}]_{(2)}|\Gamma^{(1)}) = (|\Gamma^{(12)}) = (|X^{(2)}) = (|X^{(3)}) = (|\Sigma^{(1)}) = 1$
$([W^{(4)}]_{(2)}|) = ([W^{(3)}]_{(2)}|)$

$([Z^{(1)}]_{(2)}|\Gamma^{(1)}) = (|\Gamma^{(2)}) = 1; \quad (|\Gamma^{(12)}) = 2$ [b]
$([Z^{(2)}]_{(2)}|\Gamma^{(m)}) = ([Z^{(1)}]_{(2)}|\Gamma^{(m)})$ [b]

$([k]_{(2)}|\Gamma^{(1)}) = (|\Gamma^{(2)}) = 1; \quad (|\Gamma^{(12)}) = 2$
$(|\Gamma^{(15)}) = (|\Gamma^{(25)}) = 3$ [b]

* See footnote Table D.4.
[a] $\star\Delta^{(3)}$ and $\star\Delta^{(4)}$ are degenerate and interchangeable.
[b] Only representations $\Gamma^{(m)}$ were obtained, using the basis function method.

Table D.7. Reduction coefficients for symmetrized Kronecker cube in zincblende*

$([\Gamma^{(m)}]_{(3)}|\Gamma^{(m)}) = 1; \quad m=1,2$
$([\Gamma^{(12)}]_{(3)}|\Gamma^{(1)}) = (|\Gamma^{(2)}) = (|\Gamma^{(12)}) = 1$
$([\Gamma^{(15)}]_{(3)}|\Gamma^{(1)}) = (|\Gamma^{(25)}) = 1; \quad (|\Gamma^{(15)}) = 2$
$([\Gamma^{(25)}]_{(3)}|\Gamma^{(2)}) = (|\Gamma^{(15)}) = 1; \quad (|\Gamma^{(25)}) = 2$

$([X^{(1)}]_{(3)}|\Gamma^{(1)}) = (|X^{(2)}) = 1; \quad (|X^{(1)}) = 2$
$([X^{(3)}]_{(3)}|\Gamma^{(1)}) = (|X^{(4)}) = 1; \quad (|X^{(3)}) = 2$
$([X^{(5)}]_{(3)}|\Gamma^{(1)}) = (|\Gamma^{(2)}) = (|\Gamma^{(15)}) = (|\Gamma^{(25)}) = (|X^{(1)})$
$= (|X^{(2)}) = (|X^{(3)}) = (|X^{(4)}) = 1; \quad (|X^{(5)}) = 6$

$([L^{(1)}]_{(3)}|L^{(3)}) = 1; \quad (|L^{(1)}) = 3;$
$([L^{(3)}]_{(3)}|L^{(1)}) = (|L^{(2)}) = 6; \quad (|L^{(3)}) = 9$

* See footnote Table D.4.

Table D.8.* Critical points and phonon species in T_d^2

Critical points	Phonon	Species
Γ	$O(\Gamma)$	$\Gamma^{(15)}$
$\star X$	$TO(X)$	$\star X^{(5)}$
	$LO(X)$	$\star X^{(1)}$ or $\star X^{(3)}$
	$LA(X)$	$\star X^{(1)}$ or $\star X^{(3)}$
	$TA(X)$	$\star X^{(5)}$
$\star L$	$TO(L)$	$\star L^{(3)}$
	$LO(L)$	$\star L^{(1)}$
	$LA(L)$	$\star L^{(1)}$
	$TA(L)$	$\star L^{(3)}$
$\star W$	$O_1(W), O_2(W), O_3(W)$ $A_1(W), A_2(W), A_3(W)$[a]	$\star W^{(m)}$; $m=1,2,3,4$

* From J. L. BIRMAN: Phys. Rev. **131**, 1489 (1963).

[a] Contrary to some implications in the literature there is no division of the modes at $\star W$ into transverse and longitudinal, either in zincblende or diamond. The only justification for this usage, in the case of diamond, resides in the fact that in the two parameter approximation (in which α and $\beta \neq 0$ in F. HERMAN, J. Phys. Chem. Solids **8**, 405 (1959)); the three branches at $\star W$ (each is 12-fold degenerate with $s=6$, $l_m=2$) are degenerate with the branches at X. Thus, we prefer the usage here $O_j(W)$ or $A_j(W)$ for optic or acoustic branches, respectively.

exceptions noted in the tables. Since zincblende is a symmorphic group only ordinary (vector) representations arise: all ray factors are equal to unity for the irreducible representations of the point groups $\mathfrak{P}(\boldsymbol{k})$, including for $\boldsymbol{k}$ on the zone boundary. The only matter to attend to then is one of labelling. For this reason we reproduce Parmenter's character tables here for the irreducible representations as Tables D.2. We also give his compatibility tables for ease of reference as Table D.3.

The wave vector selection rules which differ from those in diamond and rock-salt space groups involve the stars $\star\Lambda$, $\star H$, $\star M$, and these are given Table D.1, bottom, for the ordinary and the symmetrized powers.

In Table D.4 we give reduction coefficients for the reduction of the zincblende ordinary Kronecker products analogous to the diamond products reduced and presented in Table B.2, in both cases these were obtained using the reduction group.

In Table D.5 selection rules are given for zincblende analogous to those of diamond given Table B.2; these were obtained in both cases by direct inspection. Because of Eqs. (D.1) and (D.2), there are two ways of carrying out the direct inspection reduction for $\star\Lambda^{(m)} \otimes \star(-\Lambda)^{(m')}$ or $\star\Lambda^{(m)} \otimes \star\Lambda^{(m')}$; both of these products have identical characters in $\mathfrak{G}/\mathfrak{T}$. To eliminate ambiguity we merely need to use the correct wave vector selection rules as in Eqs. (D.1) or (D.2). Note also the degeneracy between $\star\Delta^{(3)}$ and $\star\Delta^{(4)}$ permitting them to be interchanged in the reductions. Tables D.6 and D.7 give reduction coefficients for symmetrized squares and cubes respectively.

Table D.9.* Two-phonon processes in zincblende

Species	Activity[b]	Type
	Overtones	
$[\Gamma^{(15)}]_{(2)}$	$D; R$	$2O(\Gamma)$
$[{}^{\star}X^{(5)}]_{(2)}$	$D; R$	$2\,TO(X)$ and $2\,TA(X)$
$[{}^{\star}X^{(1)}]_{(2)}$	R }	$2\,LO(X)$ and $2\,LA(X)$
$[{}^{\star}X^{(3)}]_{(2)}$	R }	
$[{}^{\star}L^{(3)}]_{(2)}$	$D(2); R$	$2\,TO(L)$ and $2\,TA(L)$
$[{}^{\star}L^{(1)}]_{(2)}$	$D; R$	$2\,LO(L)$ and $2\,LA(L)$
$[{}^{\star}W^{(m)}]_{(2)}$	R }	$2O_1(W)$; $2O_2(W)$; $2O_3(W)$;
$m=1, 2, 3, 4$		$2A_1(W)$; $2A_2(W)$; $2A_3(W)$
	Combinations	
${}^{\star}X^{(5)} \otimes {}^{\star}X^{(1)}$	$D; R$ }	$TO(X)+LO(X)$; $TO(X)+LA(X)$;
$\otimes {}^{\star}X^{(3)}$	$D; R$ }	$TA(X)+LO(X)$; $TA(X)+LA(X)$
$\otimes {}^{\star}X^{(5)}$	$D; R$	$TO(X)+TA(X)$
${}^{\star}X^{(3)} \otimes {}^{\star}X^{(1)}$	$D; R$	$LO(X)+LA(X)$
${}^{\star}L^{(3)} \otimes {}^{\star}L^{(1)}$	$D; R$	$TO(L)+LO(L)$; $TO(L)+LA(L)$;
		$TA(L)+LO(L)$; $TA(L)+LA(L)$
${}^{\star}L^{(3)} \otimes {}^{\star}L^{(3)}$	$D; R$	$TO(L)+TA(L)$
${}^{\star}L^{(1)} \otimes {}^{\star}L^{(1)}$	$D; R$	$LO(L)+LA(L)$
${}^{\star}W^{(1)} \otimes {}^{\star}W^{(1)}$	R	a
${}^{\star}W^{(2)} \otimes {}^{\star}W^{(2)}$	R	a
${}^{\star}W^{(3)} \otimes {}^{\star}W^{(3)}$	R	a
${}^{\star}W^{(4)} \otimes {}^{\star}W^{(4)}$	R	a
${}^{\star}W^{(1)} \otimes {}^{\star}W^{(2)}$	$D; R$	a
${}^{\star}W^{(1)} \otimes {}^{\star}W^{(3)}$	$D; R$	a
${}^{\star}W^{(1)} \otimes {}^{\star}W^{(4)}$	$D; R$	a
${}^{\star}W^{(2)} \otimes {}^{\star}W^{(3)}$	$D; R$	a
${}^{\star}W^{(2)} \otimes {}^{\star}W^{(4)}$	$D; R$	a
${}^{\star}W^{(3)} \otimes {}^{\star}W^{(4)}$	$D; R$	a

* From J. L. BIRMAN: Phys. Rev. **131**, 1489 (1963).

[a] In listing these combinations it is not possible to be more specific because of ambiguity of assignments at ${}^{\star}W$. However, in this list of combinations, it is to be noted that the two phonons participating, even if of the same symmetry, must arise from different branches.

[b] In this table as in Table D.10, we use the symbol $D(n)$ to signify that the representation $\Gamma^{(15)}$ occurs n times in the reduction and therefore there is an "n-fold" infrared dipole activity. The same convention was used for diamond (see Tables 38). For a discussion of this see the reference cited as footnote * above.

In Table D.8 we give a listing of phonon symmetry assignment in zincblende at the principal critical points. As in case of diamond certain ambiguities can only be resolved by experiment or by calculation of the symmetry of eigenvectors determined from a full lattice dynamic calculation. In Table D.9 and D.10 the activity of two and three phonon processes (infra-red dipole and/or Raman) is given. These tables are analogous, although less complete, than the ones for diamond given as Tables 36 and 38 respectively.

Table D.10. Three-phonon processes in zincblende*

Species	Activity[b]	Type
	Overtones	
$[\Gamma^{(15)}]_{(3)}$	D; R	$3\,O(\boldsymbol{\Gamma})$
$[{}^{\star}X^{(5)}]_{(3)}$	D; R	$3\,TO(\boldsymbol{X})$ and $3\,TA(\boldsymbol{X})$
$[{}^{\star}X^{(1)}]_{(3)}$	R	$3\,LO(\boldsymbol{X})$ and $3\,LA(\boldsymbol{X})$
$[{}^{\star}X^{(3)}]_{(3)}$	R	
$[{}^{\star}L^{(m)}]_{(3)}$; $m=1, 3$	No activity	
$[{}^{\star}W^{(m)}]_{(3)}$; $m=1, 2, 3, 4$	No activity	
	Simple combinations	
${}^{\star}X^{(5)} \otimes {}^{\star}X^{(1)} \otimes \Gamma^{(15)}$	$D(2)$; R	$TO(\boldsymbol{X})+O(\boldsymbol{\Gamma})$ $+\{LA(\boldsymbol{X})$ or $LO(\boldsymbol{X})\}$
$\otimes {}^{\star}X^{(3)} \otimes \Gamma^{(15)}$	$D(2)$; R	$TO(\boldsymbol{X})+O(\boldsymbol{\Gamma})$ $+\{LA(\boldsymbol{X})$ or $LO(\boldsymbol{X})\}$
$\otimes {}^{\star}X^{(5)} \otimes \Gamma^{(15)}$	$D(5)$; R	$TO(\boldsymbol{X})+TA(\boldsymbol{X})+O(\boldsymbol{\Gamma})$
${}^{\star}X^{(1)} \otimes {}^{\star}X^{(3)} \otimes \Gamma^{(15)}$	D; R	$LO(\boldsymbol{X})+LA(\boldsymbol{X})+O(\boldsymbol{\Gamma})$
$\otimes {}^{\star}X^{(5)} \otimes \Gamma^{(15)}$	$D(2)$; R	$TA(\boldsymbol{X})+O(\boldsymbol{\Gamma})$ $+\{LA(\boldsymbol{X})$ or $LO(\boldsymbol{X})\}$
${}^{\star}X^{(3)} \otimes {}^{\star}X^{(5)} \otimes \Gamma^{(15)}$	$D(2)$; R	$TA(\boldsymbol{X})+O(\boldsymbol{\Gamma})$ $+\{LA(\boldsymbol{X})$ or $LO(\boldsymbol{X})\}$
${}^{\star}X^{(5)} \otimes {}^{\star}X^{(3)} \otimes {}^{\star}X^{(1)}$	D; R	$TO(\boldsymbol{X})+LA(\boldsymbol{X})+LO(\boldsymbol{X})$; $TA(\boldsymbol{X})+LA(\boldsymbol{X})+LO(\boldsymbol{X})$
${}^{\star}X^{(5)} \otimes {}^{\star}X^{(1)} \otimes {}^{\star}X^{(5)}$	$D(3)$; R	$TO(\boldsymbol{X})+TA(\boldsymbol{X})+\{LA(\boldsymbol{X})$ or $LO(\boldsymbol{X})\}$[a]
${}^{\star}X^{(5)} \otimes {}^{\star}X^{(3)} \otimes {}^{\star}X^{(5)}$	$D(3)$; R	$TO(\boldsymbol{X})+TA(\boldsymbol{X})+\{LO(\boldsymbol{X})$ or $LA(\boldsymbol{X})\}$[a]
${}^{\star}L^{(3)} \otimes {}^{\star}L^{(1)} \otimes \Gamma^{(15)}$	$D(3)$; R	$TO(\boldsymbol{L})+LO(\boldsymbol{L})+O(\boldsymbol{\Gamma})$; $TO(\boldsymbol{L})+LA(\boldsymbol{L})+O(\boldsymbol{\Gamma})$; $TA(\boldsymbol{L})+LO(\boldsymbol{L})+O(\boldsymbol{\Gamma})$; $TA(\boldsymbol{L})+LA(\boldsymbol{L})+O(\boldsymbol{\Gamma})$
${}^{\star}L^{(1)} \otimes {}^{\star}L^{(1)} \otimes \Gamma^{(15)}$	$D(2)$; R	$LO(\boldsymbol{L})+LA(\boldsymbol{L})+O(\boldsymbol{\Gamma})$
${}^{\star}L^{(3)} \otimes {}^{\star}L^{(3)} \otimes \Gamma^{(15)}$	$D(6)$; R	$TO(\boldsymbol{L})+TA(\boldsymbol{L})+O(\boldsymbol{\Gamma})$
${}^{\star}L^{(3)} \otimes {}^{\star}L^{(1)} \otimes {}^{\star}X^{(5)}$	$D(6)$; R	$TO(\boldsymbol{L})+LO(\boldsymbol{L})+TO(\boldsymbol{X})$; $TO(\boldsymbol{L})+LA(\boldsymbol{L})+TO(\boldsymbol{X})$; $TA(\boldsymbol{L})+LO(\boldsymbol{L})+TO(\boldsymbol{X})$; $TA(\boldsymbol{L})+LA(\boldsymbol{L})+TO(\boldsymbol{X})$; $TO(\boldsymbol{L})+LO(\boldsymbol{L})+TA(\boldsymbol{X})$; $TO(\boldsymbol{L})+LA(\boldsymbol{L})+TA(\boldsymbol{X})$; $TA(\boldsymbol{L})+LO(\boldsymbol{L})+TA(\boldsymbol{X})$; $TA(\boldsymbol{L})+LA(\boldsymbol{L})+TA(\boldsymbol{X})$
$\otimes {}^{\star}X^{(1)}$	$D(3)$; R	$TO(\boldsymbol{L})+LO(\boldsymbol{L})+\{LA(\boldsymbol{X})$ or $LO(\boldsymbol{X})\}$; $TO(\boldsymbol{L})+LA(\boldsymbol{L})+\{LA(\boldsymbol{X})$ or $LO(\boldsymbol{X})\}$; $TA(\boldsymbol{L})+LO(\boldsymbol{L})+\{LA(\boldsymbol{X})$ or $LO(\boldsymbol{X})\}$; $TA(\boldsymbol{L})+LA(\boldsymbol{L})+\{LA(\boldsymbol{X})$ or $LO(\boldsymbol{X})\}$
$\otimes {}^{\star}X^{(3)}$	$D(3)$; R	
${}^{\star}L^{(1)} \otimes {}^{\star}L^{(1)} \otimes {}^{\star}X^{(5)}$	$D(3)$; R	$LO(\boldsymbol{L})+LA(\boldsymbol{L})+TO(\boldsymbol{X})$; $LO(\boldsymbol{L})+LA(\boldsymbol{L})+TA(\boldsymbol{X})$
$\otimes {}^{\star}X^{(1)}$	$D(2)$; R	$LO(\boldsymbol{L})+LA(\boldsymbol{L})+\{LA(\boldsymbol{X})$ or $LO(\boldsymbol{X})\}$
$\otimes {}^{\star}X^{(3)}$	$D(2)$; R	
${}^{\star}L^{(3)} \otimes {}^{\star}L^{(3)} \otimes {}^{\star}X^{(5)}$	$D(12)$; R	$TO(\boldsymbol{L})+TA(\boldsymbol{L})+TO(\boldsymbol{X})$; $TO(\boldsymbol{L})+TA(\boldsymbol{L})+TA(\boldsymbol{X})$;
$\otimes {}^{\star}X^{(1)}$	$D(6)$; R	$TO(\boldsymbol{L})+TA(\boldsymbol{L})+LA(\boldsymbol{X})$;
$\otimes {}^{\star}X^{(3)}$	$D(6)$; R	$TO(\boldsymbol{L})+TA(\boldsymbol{L})+LO(\boldsymbol{X})$
${}^{\star}W^{(m)} \otimes {}^{\star}W^{(m')} \otimes \Gamma^{(15)}$ $m=1, 2, 3, 4$; $m'=1, 2, 3, 4$	$D(2$, or $3)$; R	All combinations of $O(\boldsymbol{\Gamma})$ and phonons from two distinct branches of ${}^{\star}\boldsymbol{W}$ are allowed.
${}^{\star}W^{(m)} \otimes {}^{\star}W^{(m')} \otimes {}^{\star}X^{(m'')}$	D; R	All combinations of $TO(\boldsymbol{X})$, or $TA(\boldsymbol{X})$, or $LO(\boldsymbol{X})$, or $LA(\boldsymbol{X})$ and phonons from two distinct branches of ${}^{\star}\boldsymbol{W}$ are allowed.

Table D.10 (continued)

Species	Activity[b]	Type
	General combinations	
$[{}^\star\boldsymbol{X}^{(5)}]_{(2)} \otimes \boldsymbol{\Gamma}^{(15)}$	$D(4);\ R$	$2\,TO(\boldsymbol{X})+O(\boldsymbol{\Gamma})$; $2\,TA(\boldsymbol{X})+O(\boldsymbol{\Gamma})$
$[{}^\star\boldsymbol{X}^{(1)}]_{(2)} \otimes \boldsymbol{\Gamma}^{(15)}$	$D(2);\ R$	$O(\boldsymbol{\Gamma})+LO(\boldsymbol{X})$;
$[{}^\star\boldsymbol{X}^{(3)}]_{(2)} \otimes \boldsymbol{\Gamma}^{(15)}$	$D(2);\ R$	$O(\boldsymbol{\Gamma})+LA(\boldsymbol{X})$
$[{}^\star\boldsymbol{X}^{(5)}]_{(2)} \otimes {}^\star\boldsymbol{X}^{(5)}$	$D(3);\ R$	$2\,TO(\boldsymbol{X})+TA(\boldsymbol{X})$[c]; $2\,TA(\boldsymbol{X})+TO(\boldsymbol{X})$[c]
$\otimes {}^\star\boldsymbol{X}^{(1)}$	$D(2);\ R$	$2\,TO(\boldsymbol{X})+LO(\boldsymbol{X})$; $2\,TO(\boldsymbol{X})+LA(\boldsymbol{X})$;
$\otimes {}^\star\boldsymbol{X}^{(3)}$	$D(2);\ R$	$2\,TA(\boldsymbol{X})+LO(\boldsymbol{X})$; $2\,TA(\boldsymbol{X})+LA(\boldsymbol{X})$
$[{}^\star\boldsymbol{X}^{(1)}]_{(2)} \otimes {}^\star\boldsymbol{X}^{(5)}$	$D;\ R$	$2\,LO(\boldsymbol{X})+TO(\boldsymbol{X})$; $2\,LO(\boldsymbol{X})+TA(\boldsymbol{X})$;
$\otimes {}^\star\boldsymbol{X}^{(3)}$	$D;\ R$	$2\,LO(\boldsymbol{X})+LA(\boldsymbol{X})$
$[{}^\star\boldsymbol{X}^{(3)}]_{(2)} \otimes {}^\star\boldsymbol{X}^{(5)}$	$D;\ R$	$2\,LA(\boldsymbol{X})+TO(\boldsymbol{X})$; $2\,LA(\boldsymbol{X})+TA(\boldsymbol{X})$;
$\otimes {}^\star\boldsymbol{X}^{(1)}$	$D;\ R$	$2\,LA(\boldsymbol{X})+LO(\boldsymbol{X})$
$[{}^\star\boldsymbol{L}^{(3)}]_{(2)} \otimes \boldsymbol{\Gamma}^{(15)}$	$D(5);\ R$	$2\,TO(\boldsymbol{L})+O(\boldsymbol{\Gamma})$; $2\,TA(\boldsymbol{L})+O(\boldsymbol{\Gamma})$
$[{}^\star\boldsymbol{L}^{(1)}]_{(2)} \otimes \boldsymbol{\Gamma}^{(15)}$	$D(2);\ R$	$2\,LO(\boldsymbol{L})+O(\boldsymbol{\Gamma})$; $2\,LA(\boldsymbol{L})+O(\boldsymbol{\Gamma})$
$[{}^\star\boldsymbol{L}^{(3)}]_{(2)} \otimes {}^\star\boldsymbol{X}^{(5)}$	$D(7);\ R$	$2\,TO(\boldsymbol{L})+TO(\boldsymbol{X})$; $2\,TO(\boldsymbol{L})+TA(\boldsymbol{X})$; $2\,TA(\boldsymbol{L})+TO(\boldsymbol{X})$; $2\,TA(\boldsymbol{L})+TA(\boldsymbol{X})$
$\otimes {}^\star\boldsymbol{X}^{(1)}$	$D(3);\ R$	$2\,TO(\boldsymbol{L})+LO(\boldsymbol{X})$; $2\,TO(\boldsymbol{L})+LA(\boldsymbol{X})$;
$\otimes {}^\star\boldsymbol{X}^{(3)}$	$D(3);\ R$	$2\,TA(\boldsymbol{L})+LO(\boldsymbol{X})$; $2\,TA(\boldsymbol{L})+LA(\boldsymbol{X})$
$[{}^\star\boldsymbol{W}^{(m)}]_{(2)} \otimes \boldsymbol{\Gamma}^{(15)}$; $m=1, 2, 3, 4$	$D(2);\ R$	All combinations of $O(\boldsymbol{\Gamma})$ and phonon overtones of ${}^\star\boldsymbol{W}^{(m)}$ are allowed.
$[{}^\star\boldsymbol{W}^{(m)}]_{(2)} \otimes {}^\star\boldsymbol{X}^{(m')}$	$D;\ R$	All combinations of $TO(\boldsymbol{X})$, or $TA(\boldsymbol{X})$, or $LA(\boldsymbol{X})$, or $TA(\boldsymbol{X})$ and phonon overtones of ${}^\star\boldsymbol{W}^{(m)}$ are allowed.

* From J. L. Birman: Phys. Rev. **131**, 1489 (1963).

[a] The two ${}^\star\boldsymbol{X}^{(5)}$ modes are from different branches.

[b] See footnote to Table D.9.

[c] Here two ${}^\star\boldsymbol{X}^{(5)}$ modes are from the same branch, the third is from the other branch of symmetry ${}^\star\boldsymbol{X}^{(5)}$.

References

[1] WIGNER, E. P.: Group Theory and Application to Quantum Mechanics of Atomic Spectra. New York: Academic Press 1959.

[2] HAMERMESH, M.: Group Theory. Reading, Mass.: Addison-Wesley Publ. Inc. 1962.

[3] BIRMAN, J. L.: Group Theory Methods and Techniques with Applications to Physics. New York: J. Wiley & Sons Inc. (In preparation.)

[4] BORN, M., and K. HUANG: Dynamical Theory of Crystal Lattices. Oxford: University Press 1954. – LEIBFRIED, G., in: this Encyclopedia vol. VII, part 1, p. 104. Berlin-Göttingen-Heidelberg: Springer 1955. – LEIBFRIED, G., and W. LUDWIG, in: Solid State Phys. **12**, 275–444 (1961).

[5] JAGODZINSKI, H., in: this Encyclopedia, vol. VII, part 1, pp. 1–103. Berlin-Göttingen-Heidelberg: Springer 1955.

[6] ZACHARIASEN, W. H.: Theory of X-Ray Diffraction in Crystals. New York: J. Wiley & Sons Inc. 1945.

[7] SCHÖNFLISS, A.: Kristallsystem und Kristallstructur. Leipzig 1891. – Theorie der Kristallstructuren. Berlin 1923. – FEDEROV, E. VON: Verh. Kgl. Mineral. Ges., Petersburg **27**, 448 (1891).

[8] BORN, M., and M. G. MAYER: Handbuch der Physik, vol. XXIV/2 (ed. A. SMEKAL), 2. Auflage, pp. 623–790. Berlin: Julius Springer 1933.

[9] International Tables for X-Ray Crystallography, vol. 1: Symmetry Groups (eds. N. F. M. HENRY and K. LONSDALE). Birmingham: Kynoch Press 1965.

[10] HALL, M.: Theory of Groups. Macmillan 1950.

[11] BURROW, M.: Representation Theory of Finite Groups. New York: Academic Press 1965.

[12] LOMONT, J. S.: Applications of Finite Groups. New York: Academic Press 1959.

[13] BORN, M., and K. HUANG: Dynamical Theory of Crystal Lattices, p. 174, p. 300. Oxford: University Press 1954.

[14] CURTIS, C. W., and I. REINER: Representation Theory of Finite Groups and Associative Algebras. New York: Interscience Publishers 1962.

[15] WEYL, H.: Theory of Groups and Quantum Mechanics. Reprint. New York: Dover Publishing Co., Early discussion of the ray representation in quantum mechanics, pp. 180–184.

[16] MURNAGHAN, F. D.: The Theory of Group Representations. New York: Dover Publishing Co. 1963.

[17] LYUBARSKII, G. YA.: The Application of Group Theory in Physics. New York: Pergamon Press 1960.

[18] BOERNER, H.: Representations of Groups. 2nd edition. Amsterdam: North Holland Publishing Co. 1972.

[19] Phonons in Perfect Lattices and in Lattices with Point Imperfections (ed. R. W. H. STEVENSON). New York: Plenum Press 1966.

[20] COCHRAN, W., and R. A. COWLEY: Phonons in Perfect Crystals. In: this Encyclopedia: vol. XXV/2a, p. 59. Berlin-Heidelberg-New York: Springer 1967.

[21] KWOK, P. W.: Green Function Method in Lattice Dynamics. In: Solid State Physics (eds. F. SEITZ and D. TURNBULL), vol. 20. New York: Academic Press 1967.

[22] BILZ, H., D. STRAUCH, and R. K. WEHNER: Infrared Absorption and Raman Scattering in Pure and Perturbed Crystals. In: this Encyclopedia, vol. XXV/2d (in preparation).

[23] MARADUDIN, A. A., W. W. MONTROLL, and G. H. WEISS: Theory of Lattice Dynamics in the Harmonic Approximation, Supplement 3, in: Solid State Physics (eds. F. SEITZ and D. TURNBULL), 1st edition. New York: Academic Press 1963.

[24] FANO, U., and G. RACAH: Irreducible Tensorial Sets. New York: Academic Press 1959.

[25] SHALIT, A. DE, and I. TALMI: Nuclear Shell Theory. New York: Academic Press 1963.

[26] Hove, L. van, N. M. Hugenholz, and L. P. Howland: Problems in Quantum Theory of Many Particle Systems. New York: W. J. Benjamin Publ. 1963. – Choquard, P.: The Anharmonic Crystal. New York: W. J. Benjamin Publ. 1967.
[27] Courant, R., and D. Hilbert: Methods of Mathematical Physics. New York: Interscience Publishers 1953.
[28] Weyl, H.: Classical Groups, p. 30 and 37. New Jersey: Princeton University Press 1947. – van der Warden, B. H.: Algebra, vol. 1, p. 78 et seq. New York: F. Ungar Publ. Co. 1952.
[29] Schiff, L. I.: Quantum Mechanics. New York: McGraw-Hill Book Co. 1949.
[30] See Slater, J. C.: Quantum Theory of Molecules and Solids, vol. 2: Symmetry and Energy Bands in Crystals. New York: McGraw-Hill Book Co. 1965.
[31] Wilson, E. B., J. C. Decius, and P. C. Cross: Molecular Vibrations. New York: McGraw-Hill Book Co. 1955.
[32] Proceedings of the International Conference on Lattice Dynamics, Copenhagen 1963 (ed. R. F. Wallis). New York: Pergamon Press 1965.
[33] Light Scattering Spectra of Solids. Proceedings of the International Conference on Light Scattering by Solids at New York University 1968 (ed. G. F. Wright). Berlin-Heidelberg-New York: Springer 1969.
[34] Localized Excitations in Solids (ed. R. F. Wallis). Proceedings of the International Conference on Localized Excitations in Solids, Irvine California, 1968. New York: Plenum Press 1969. This conference contains many papers related to the subject of this part of the article.
[35] An early reference, still very valuable is: Theory of optical properties of imperfections in nonmetals, by D. L. Dexter, in: Solid State Phys. **6**, 353–411 (1958).
[36] Dettmann, K.: Diplomarbeit: Lokalisierte Schwingungszustände in kubischen Kristallen mit Punktdefekten. Fakultät der Rheinisch-Westfälischen Technischen Hochschule Aachen 1963.
[37] Ludwig, W.: Habilitationsschrift: Zur Dynamik von Kristallen mit Punktdefekten. Fakultät der Rheinisch-Westfälischen Technischen Hochschule Aachen 1962.
[38] LeCompte, J.: Spectroscopie dans l'infrarouge. In: this Encyclopedia, vol. XXVI, p. 244. Berlin-Göttingen-Heidelberg: Springer 1958.
[39] Mizushima, S.: Raman Effect. In: this Encyclopedia, vol. XXVI, p. 171. Berlin-Göttingen-Heidelberg: Springer 1959.
[40] Abrikosov, A. A., L. P. Gorkov, and I. Y. Dzyaloshinski: Quantum Field Theoretical Methods in Statistical Physics. New York: Pergamon Press 1965.
[41] Nye, J. F.: Physical Properties of Crystals. Oxford, N. Y.: Oxford University Press 1957.
[42] Bradley, C. S., and A. P. Cracknell: The Mathematical Theory of Symmetry in Solids. Oxford: Clarendon Press 1972.
[43] Birman, J. L.: A Quick Trip through Group Theory and Lattice Dynamics, Chap. 2 of "Lattice Dynamics" (eds. A. A. Maradudin and G. K. Horton). Amsterdam: North Holland Publishing Co. 1974
[44] Dirac, P. A. M.: Principles of Quantum Mechanics. 3rd ed. Oxford: Oxford University Press 1947.
[45] Nusimovici, M. A. (ed.): Phonons. Paris: Flammarion 1971.
[46] Wentzel, G.: Einführung in die Quantentheorie der Wellenfelder. Wien: F. Deuticke 1943.
[47] Maradudin, A. A., E. W. Montroll, G. H. Weiss, and I. P. Ipatova: Theory of Lattice Dynamics in the Harmonic Approximation, 2nd edition. New York: Academic Press 1971.
[48] Balkanski, M. (ed.): Light Scattering in Solids. Paris: Flammarion 1971.
[49] Koster, G. F.: In: Advances in Solid State Physics (eds. F. Seitz and D. Turnball), vol. 5. New York: Academic Press 1957.
[50] Zavt, G. S. (ed.): Physics of Impurity Centers in Crystals, Proceedings of International Seminar at Tallin, 1970. Published by Academy of Sciences Estonion SSR Tallin 1972.
[51] Landau, L., and E. Lifschitz: Course in Theoretical Physics Engl. Transl. Addison-Wesley, New York, 1957–1972. Volume 3 on: Quantum Mechanics – Non Relativistic Theory and Volume 5 on: Statistical Physics have Standard Text Material on Group Theory.
[52] Poulet, H., and J. P. Mathieu: Spectres de Vibration et Symétrie des Cristaux. Paris: Gordon-Breach 1971
[53] Bradley, D. J., and A. J. Cracknell: Mathematical Theory of Symmetry in Crystals. Oxford University Press 1972.
[54] Love, W. F., and S. C. Miller, Jr.: Tables of Irreducible Representations of Space Groups and Corepresentations of Magnetic Space Groups. Colorado: Pruett Publishing Co. 1967

[55] KOSTER, G.F., J.O. DIMMOCK, R.G. WHEELER, and H. STATZ: Properties of the Thirty-Two Point Groups. MIT Press 1963.
[56] KOVALEV, O.V.: Irreducible Representations of the Space Groups (transl. by A.M. GROSS). New York: Gordon and Breach 1965.
[57] ZAK, J., A. CASHER, M. GLUCK, and Y. GUR: The Irreducible Representations of Space Groups. New York: W.A. Benjamin, Inc. 1969.
[58] LAX, M.: Symmetry Principles in Solid State and Molecular Physics. New York: J. Wiley & Sons Inc. 1974.
[59] PILKUHN, M., ed.: Proc. 12th International Conference on Semiconductors, Stuttgart 1974. Stuttgart: Teubner 1974.
[60] HEITLER, W.: Quantum Theory of Radiation. 3rd ed. Oxford: Oxford University Press.
[61] GILMORE, R.: Lie Groups, Lie Algebras, and Some of Their Applications. New York: J. Wiley & Sons 1974.
[62] Atomic Structure and Properties of Solids (ed. E. BURSTEIN) (Proc. Course LII, International School of Physics, E. FERMI). New York: Academic Press 1972.

Index of key equations. Sects. 1–124

Multiplication of space group elements

$$\{\varphi_3|\boldsymbol{t}(\varphi_3)\}\cdot\{\varphi_2|\boldsymbol{t}(\varphi_2)\}=\{\varphi_3\cdot\varphi_2|\varphi_3\cdot\boldsymbol{t}(\varphi_2)+\boldsymbol{t}(\varphi_3)\}. \tag{6.8}$$

Multiplication of coset representatives

$$\{\varphi_\sigma|\boldsymbol{\tau}_\sigma\}\cdot\{\varphi_\rho|\boldsymbol{\tau}_\rho\}=\{\varepsilon|\boldsymbol{R}_{\sigma\rho}\}\cdot\{\varphi_{\sigma\rho}|\boldsymbol{\tau}_{\sigma\rho}\}, \tag{7.6}$$

where

$$\boldsymbol{R}_{\sigma\rho}\equiv\varphi_\sigma\cdot\boldsymbol{\tau}_\rho+\boldsymbol{\tau}_\sigma-\boldsymbol{\tau}_{\sigma\rho}.$$

Transformation operator on function

$$P_S\,\psi(S\,\boldsymbol{r})=\psi(\boldsymbol{r}). \tag{12.4}$$

Representation generated by functions

$$P_S\,\psi_a=\sum_{n=1}^{l}D(S)_{na}\,\psi_n,\qquad a=1,\ldots,l. \tag{14.5}$$

Schur's lemma

If there is an M:

$$M\,D(\{\varphi|\boldsymbol{t}(\varphi)\})=D(\{\varphi|\boldsymbol{t}(\varphi)\})\,M \tag{15.5}$$

and the *only* matrix M satisfying this equation is $M=m\,D(\{\varepsilon|\boldsymbol{0}\})$ where m is a constant, then D is irreducible.

Idempotent projection operator

$$P_{mn}^{(l)}=\frac{l}{N\,g_p}\sum_{\{\varphi|\boldsymbol{t}(\varphi)\}}D^{(l)}(\{\varphi|\boldsymbol{t}(\varphi)\})_{mn}^{*}\,P_{\{\varphi|\boldsymbol{t}(\varphi)\}}. \tag{16.3}$$

Reduction coefficients

$$(l\,l'|m)=\frac{1}{g_p\,N}\sum_{\mathfrak{G}}\chi^{(l\otimes l')}(\{\varphi|\boldsymbol{t}(\varphi)\})\cdot\chi^{(m)}(\{\varphi|\boldsymbol{t}(\varphi)\})^{*}. \tag{17.8}$$

Clebsch-Gordan coefficient

$$\psi_{\mu''}^{(l'')\gamma}=\sum_{\mu\mu'}\begin{pmatrix} l & l' & l'' & \gamma \\ \mu & \mu' & \mu'' & \end{pmatrix}\psi_\mu^{(l)}\,\psi_{\mu'}^{(l')}. \tag{18.4}$$

Irreducible representation of $\mathfrak{T}$

$$D^{(\boldsymbol{k})}(\{\varepsilon|\boldsymbol{R}_L\})=\exp-i\,\boldsymbol{k}\cdot\boldsymbol{R}_L. \tag{23.2}$$

Block diagonal structure of irreducible representation of $\mathfrak{G}$

$$D^{(\star \boldsymbol{k})}(\{\varepsilon|\boldsymbol{R}_L\})=\begin{pmatrix} \boldsymbol{\Pi}_m \exp -i\,\boldsymbol{k}_1\cdot\boldsymbol{R}_L & 0 & \dots & 0 \\ 0 & \boldsymbol{\Pi}_m \exp -i\,\boldsymbol{k}_\nu\cdot\boldsymbol{R}_L & & \\ \vdots & & \ddots & \\ & & & \boldsymbol{\Pi}_m \exp -i\,\boldsymbol{k}_s\cdot\boldsymbol{R}_L \end{pmatrix} \tag{33.10}$$

$$D^{(\star \boldsymbol{k})(m)}(\{\varphi_\lambda|\,\boldsymbol{t}(\varphi_\lambda)\})=\begin{pmatrix} 0 & \dots & 0 & 0 & \dots \\ 0 & & \vdots & \vdots & \\ & & & D^{(\boldsymbol{k}_\nu)(m)}(\{\varphi_\lambda|\boldsymbol{t}(\varphi_\lambda)\})_{\mu\nu} & \\ \vdots & & & \vdots & \\ D^{(\boldsymbol{k}_\lambda)(m)}(\{\varphi_\lambda|\boldsymbol{t}(\varphi_\lambda)\})_{\lambda 1} & & & 0 & \dots \\ \vdots & & & \vdots & \\ 0 & & & & \ddots \end{pmatrix} \tag{33.11}$$

Irreducible representations of $\mathfrak{G}$ in terms of dotted matrices

$$D^{(\star \boldsymbol{k})(m)}(\{\varphi_p|\tau_p\})_{\sigma\tau}=\dot{D}^{(\boldsymbol{k})(m)}(\{\varphi_\sigma|\tau_\sigma\}^{-1}\cdot\{\varphi_p|\tau_p\}\cdot\{\varphi_\tau|\tau_\tau\}) \tag{36.14}$$

or matrix elements

$$\begin{aligned} D^{(\star \boldsymbol{k})(m)}(\{\varphi_p\})_{(\sigma a)(\tau b)} &\equiv \big(D^{(\star \boldsymbol{k})(m)}(\{\varphi_p\})_{\sigma\tau}\big)_{ab} \\ &= \big(\dot{D}^{(\boldsymbol{k})(m)}(\{\varphi_\sigma\}^{-1}\cdot\{\varphi_p\}\cdot\{\varphi_\tau\})\big)_{ab}. \end{aligned} \tag{36.18}$$

with $\{\varphi_p\}\equiv\{\varphi_p|\tau_p\}$.

Character of irreducible representation of $\mathfrak{G}$

$$\chi^{(\star \boldsymbol{k})(m)}(\{\varphi_p|\tau_p\})=\sum_{\sigma=1}^{s}\dot{\chi}^{(\boldsymbol{k})(m)}(\{\varphi_\sigma|\tau_\sigma\}^{-1}\cdot\{\varphi_p|\tau_p\}\cdot\{\varphi_\sigma|\tau_\sigma\}). \tag{37.3}$$

Ray representation of $\mathfrak{P}$

$$D^{(\boldsymbol{k})(m)}(\{\varphi_{l_\lambda}|\tau_{l_\lambda}\})\cdot D^{(\boldsymbol{k})(m)}(\{\varphi_{l_\mu}|\tau_{l_\mu}\})=r^{(\boldsymbol{k})}(\lambda,\mu)\,D^{(\boldsymbol{k})(m)}(\{\varphi_{l_{\lambda\mu}}|\tau_{l_{\lambda\mu}}\}), \tag{41.17}$$

where $|r^{(\boldsymbol{k})}(\lambda,\mu)|=1$.

Direct product of two space group representations

$$\chi^{(\star \boldsymbol{k}\otimes\star \boldsymbol{k}')(m\otimes m')}(\{\varphi|\boldsymbol{t}(\varphi)\})=\chi^{(\star \boldsymbol{k})(m)}(\{\varphi|\boldsymbol{t}(\varphi)\})\cdot\chi^{(\star \boldsymbol{k}')(m')}(\{\varphi|\boldsymbol{t}(\varphi)\}). \tag{53.8}$$

Symmetrized square of representations

$$\mathrm{Tr}[D^{(\star \boldsymbol{k})(m)}]_{(2)}\equiv[\chi^{(\star \boldsymbol{k})(m)}]_{(2)}, \tag{54.15}$$

$$[\chi^{(\star \boldsymbol{k})(m)}(\varphi|\boldsymbol{t}(\varphi)\})]_{(2)}=\tfrac{1}{2}\{(\chi^{(\star \boldsymbol{k})(m)}(\{\varphi|\boldsymbol{t}(\varphi)\}))^2+\chi^{(\star \boldsymbol{k})(m)}(\{\varphi|\boldsymbol{t}(\varphi)\}^2)\}. \tag{54.16}$$

Symmetrized cube of representations

$$\begin{aligned} [\chi^{(\star \boldsymbol{k})(m)}(\{\varphi|\boldsymbol{t}(\varphi)\})]_{(3)} = \tfrac{1}{6}\{&(\chi^{(\star \boldsymbol{k})(m)}(\{\varphi|\boldsymbol{t}(\varphi)\}))^3 \\ &+3\chi^{(\star \boldsymbol{k})(m)}(\{\varphi|\boldsymbol{t}(\varphi)\})\cdot\chi^{(\star \boldsymbol{k})(m)}(\{\varphi|\boldsymbol{t}(\varphi)\}^2) \\ &+2\chi^{(\star \boldsymbol{k})(m)}(\{\varphi|\boldsymbol{t}(\varphi)\}^3)\}. \end{aligned} \tag{54.21}$$

Space group reduction coefficient

$(\star k\, m\, \star k'\, m' | \star k''\, m'')$

$$= \frac{1}{g_p N} \sum_{\{\varphi|t(\varphi)\}} \chi^{(\star k)(m)}(\{\varphi|t(\varphi)\}) \cdot \chi^{(\star k')(m')}(\{\varphi|t(\varphi)\})\, \chi^{(\star k'')(m'')}(\{\varphi|t(\varphi)\})^*. \qquad (55.5)$$

Wave vector reduction coefficient

$$\star k \otimes \star k' = \sum_{\star k''} \oplus (\star k\, \star k' | \star k'')\, \star k''. \qquad (56.3)$$

Subgroup reduction coefficient

$(\{k_\sigma + k'_{\sigma'}\}\, m\, m' | k''\, m'')$

$$= \left(\frac{1}{g(k'')}\right) \left(\frac{1}{n_{k''k_\sigma}}\right) \sum_p \sum_{\lambda''=1}^{k''} \dot{\chi}^{(k)(m)}(\{\varphi_p^{l_{\lambda''\sigma}}\})\, \dot{\chi}^{(k')(m')}(\{\varphi_p^{l_{\lambda''\sigma'}}\})\, \dot{\chi}^{(k'')(m'')}(\{\varphi_p | t_p\})^*. \qquad (63.2)$$

Lattice equation of motion (harmonic)

$$M_\kappa\, \ddot{u}_\alpha \begin{pmatrix} l \\ \kappa \end{pmatrix} + \sum_{l'\kappa'\beta} \Phi_{\alpha\beta} \begin{pmatrix} l & l' \\ \kappa & \kappa' \end{pmatrix} u_\beta \begin{pmatrix} l' \\ \kappa' \end{pmatrix} = 0. \qquad (67.19)$$

Symmetry of force constant matrix

$$\Phi'_{\alpha'\beta'} \begin{pmatrix} l_\varphi & l'_\varphi \\ \kappa_\varphi & \kappa'_\varphi \end{pmatrix} = \sum_\alpha \sum_\beta \varphi_{\alpha\alpha'}\, \Phi_{\alpha\beta} \begin{pmatrix} l & l' \\ \kappa & \kappa' \end{pmatrix} \varphi_{\beta\beta'} = \Phi_{\alpha'\beta'} \begin{pmatrix} l_\varphi & l'_\varphi \\ \kappa_\varphi & \kappa'_\varphi \end{pmatrix}. \qquad (71.24)$$

Orthonormality of the real eigenvectors

$$\sum_{\kappa l \alpha} e_\alpha \left(\begin{matrix} l \\ \kappa \end{matrix} \middle| j_\rho \right) e_\alpha \left(\begin{matrix} l \\ \kappa \end{matrix} \middle| j'_{\rho'} \right) = \delta_{jj'}\, \delta_{\rho\rho'}, \qquad (72.17)$$

$$\sum_j \sum_\rho e_\alpha \left(\begin{matrix} l \\ \kappa \end{matrix} \middle| j_\rho \right) e_\beta \left(\begin{matrix} l' \\ \kappa' \end{matrix} \middle| j_\rho \right) = \delta_{\alpha\beta}\, \delta_{ll'}\, \delta_{\kappa\kappa'}. \qquad (72.18)$$

Physical displacements in terms of complex normal coordinates

$$u_\alpha \begin{pmatrix} l \\ \kappa \end{pmatrix} = \frac{1}{\sqrt{M_\kappa N}} \sum_j \sum_k e^{i k \cdot R_L}\, e_\alpha \left(\kappa \middle| \begin{matrix} k \\ j \end{matrix}\right) Q \begin{pmatrix} k \\ j \end{pmatrix}. \qquad (80.9)$$

Complex normal coordinates and physical displacements

$$Q \begin{pmatrix} k \\ j \end{pmatrix} = \frac{1}{\sqrt{N}} \sum_L \sum_{\kappa\alpha} e^{-i k \cdot R_L}\, e^*_\alpha \left(\kappa \middle| \begin{matrix} k \\ j \end{matrix}\right) u_\alpha \begin{pmatrix} l \\ \kappa \end{pmatrix} \sqrt{M_\kappa}. \qquad (80.10)$$

Symmetry transformation of an eigenvector

$$P_{\{\varphi|\tau(\varphi)\}}\, e_\alpha \left(\kappa \middle| \begin{matrix} k \\ j \end{matrix}\right) = \sum_\beta \varphi_{\alpha\beta}\, e_\beta \left(\varphi^{-1} \cdot r_\kappa - \varphi^{-1} \cdot \tau(\varphi) \middle| \begin{matrix} \varphi \cdot k \\ j \end{matrix}\right). \qquad (81.18)$$

Symmetry transformation of the dynamical matrix

$$P_{\{\varphi|\tau(\varphi)\}} \cdot [D(k)] \cdot P^{-1}_{\{\varphi|\tau(\varphi)\}} = P^{(\varphi\cdot k)} [M]^{-\frac{1}{2}}\, [\Phi(0)]\, [M]^{-\frac{1}{2}} = [D(\varphi \cdot k)]. \qquad (81.26)$$

Representation $D^{(\boldsymbol{k})(e)}$ based on eigenvectors of dynamical matrix

$$P_{\{\varphi_{l_\lambda}|\tau_{l_\lambda}\}}\, e_\alpha\left(\kappa\left|\begin{matrix}\boldsymbol{k}\\ j\end{matrix}\right.\right)=\sum_\beta\sum_{\kappa''} D^{(\boldsymbol{k})(e)}(\{\varphi_{l_\lambda}|\tau_{l_\lambda}\})_{\alpha\kappa,\beta\kappa''}\, e_\beta\left(\kappa''\left|\begin{matrix}\boldsymbol{k}\\ j\end{matrix}\right.\right), \qquad \alpha=1,\dots,3;\ \kappa=1,\dots,r. \tag{82.15}$$

Trace of $D^{(\boldsymbol{k})(e)}$

$$\begin{aligned}\operatorname{Tr} D^{(\boldsymbol{k})(e)}(\{\varphi_{l_\lambda}|\tau_{l_\lambda}\})&\equiv\chi^{(\boldsymbol{k})(e)}(\{\varphi_{l_\lambda}|\tau_{l_\lambda}\})\\ &=\sum_{\kappa''}\left(\pm(1+2\cos\varphi_{l_\lambda})\right) D^{(\boldsymbol{k})}(\{\varepsilon|-\boldsymbol{R}_N(\kappa_{\bar\varphi_{l_\lambda}},\kappa'')\})\,\delta_{\kappa_{\bar\varphi_{l_\lambda}},\kappa''}.\end{aligned} \tag{82.21}$$

Representation $D^{(\boldsymbol{k})(j)}$ based on eigenvectors

$$\begin{aligned}D^{(\boldsymbol{k})(j)}(\{\varphi_{l_\lambda}|\tau_{l_\lambda}\})_{\nu\mu}&=\sum_{\alpha\kappa}\sum_{\beta\kappa''} e_\alpha^*\left(\kappa\left|\begin{matrix}\boldsymbol{k}\\ j_\nu\end{matrix}\right.\right) D^{(\boldsymbol{k})(e)}(\{\varphi_{l_\lambda}|\tau_{l_\lambda}\})_{\alpha\kappa,\beta\kappa''}\, e_\beta\left(\kappa''\left|\begin{matrix}\boldsymbol{k}\\ j_\mu\end{matrix}\right.\right)\\ &=\sum_{\alpha\kappa}\sum_{\beta\kappa''} e_\alpha^*\left(\kappa\left|\begin{matrix}\boldsymbol{k}\\ j_\nu\end{matrix}\right.\right)(\varphi_{l_\lambda})_{\alpha\beta}\, D^{(\boldsymbol{k})}(\{\varepsilon|-\boldsymbol{R}_N(\kappa_{\bar\varphi_{l_\lambda}},\kappa'')\})\,\delta_{\kappa_{\bar\varphi_{l_\lambda}},\kappa''}\, e_\beta\left(\kappa''\left|\begin{matrix}\boldsymbol{k}\\ j_\mu\end{matrix}\right.\right).\end{aligned} \tag{83.8}$$

Representation based on complex normal coordinates

$$\begin{aligned}P_{\{\varphi_p|\boldsymbol{t}(\varphi_p)\}}\, Q\begin{pmatrix}\boldsymbol{k}_\tau\\ j_\nu\end{pmatrix}&\equiv P_{\{\varphi_p\}}\, Q\begin{pmatrix}\boldsymbol{k}_\tau\\ j_\nu\end{pmatrix}\equiv Q_{\{\varphi_p\}}\begin{pmatrix}\boldsymbol{k}_\tau\\ j_\nu\end{pmatrix}\\ &=\sum_{\sigma=1}^{s}\sum_{\mu=1}^{l_m} D^{(\star\boldsymbol{k})(j)}(\{\varphi_p|\boldsymbol{t}(\varphi_p)\})_{(\sigma\mu)(\tau\nu)}\, Q\begin{pmatrix}\boldsymbol{k}_\sigma\\ j_\mu\end{pmatrix}.\end{aligned} \tag{86.30}$$

Reality tests for representation $D^{(\star\boldsymbol{k})(m)}$ (HERRING)

$$\sum_{p=1}^{g_p}\dot\chi^{(\boldsymbol{k})(m)}(\{\varphi_p|\tau_p\}^2)\,\varDelta_{p1}=+g_p/s,\qquad \text{potentially real } D^{(\star\boldsymbol{k})(m)}, \tag{93.29}$$

$$\sum_{p=1}^{g_p}\dot\chi^{(\boldsymbol{k})(m)}(\{\varphi_p|\tau_p\}^2)\,\varDelta_{p1}=0,\qquad \text{complex } D^{(\star\boldsymbol{k})(m)}, \tag{93.30}$$

$$\sum_{p=1}^{g_p}\dot\chi^{(\boldsymbol{k})(m)}(\{\varphi_p|\tau_p\}^2)\,\varDelta_{p1}=-g_p/s,\qquad \text{pseudoreal } D^{(\star\boldsymbol{k})(m)}, \tag{93.31}$$

where

$$\varDelta_{p1}=\begin{cases}1 & \text{if } \varphi_p\cdot\boldsymbol{k}=-\boldsymbol{k}+2\pi\,\boldsymbol{B}_H\\ 0 & \text{otherwise}.\end{cases} \tag{93.32}$$

Local reality tests for representation $D^{(\boldsymbol{k})(m)}$ (FREI-BIRMAN)

Class I

$$D^{(\boldsymbol{k})(m)} \text{ is real},\qquad m=\bar m; \tag{94.11}$$

and no extra degeneracy.

If

$$D^{(\boldsymbol{k})(m)} \text{ is complex},\qquad m\neq\bar m;$$

and

$$D^{(\boldsymbol{k})(m)} \text{ and } D^{(\boldsymbol{k})(\bar m)} \text{ stick}. \tag{94.12}$$

Class II

$$\chi^{(\boldsymbol{k})(m)}(\{\varphi_l|\tau_l\}^{\bar{\sigma}})^* \equiv \chi^{(\boldsymbol{k})(\bar{m})}(\{\varphi_l|\tau_l\}), \tag{94.26}$$

$$\bar{m}=m; \qquad \text{no extra degeneracy}, \tag{94.27}$$

$$\bar{m}\neq m; \qquad D^{(\boldsymbol{k})(m)} \text{ and } D^{(\boldsymbol{k})(\bar{m})} \text{ stick.} \tag{94.28}$$

Class III

$$D^{(\star\boldsymbol{k})(m)} \not\doteq D^{(\star\boldsymbol{k})(\bar{m})}. \tag{94.29}$$

Corepresentation properties

Class III costar

$$D^{(\mathrm{co}\,\star\boldsymbol{k})(m)}(\{\varphi|\tau\}) = \begin{pmatrix} D^{(\star\boldsymbol{k})(m)}(\{\varphi|\tau\}) & 0 \\ 0 & D^{(\star-\boldsymbol{k})(\bar{m})}(\{\varphi|\tau\}) \end{pmatrix} \tag{97.7}$$

and for antiunitary elements:

$$D^{(\mathrm{co}\,\star\boldsymbol{k})(m)}(K\{\varphi|\tau\}) = \begin{pmatrix} 0 & D^{(\star\boldsymbol{k})(m)}(\{\varphi|\tau\}) \\ D^{(\star-\boldsymbol{k})(\bar{m})}(\{\varphi|\tau\}) & 0 \end{pmatrix}. \tag{97.8}$$

Class II costar

Case A

$$D^{(\mathrm{co}\,\boldsymbol{k})(m)}(u) = D^{(\boldsymbol{k})(m)}(u), \tag{98.57}$$

$$D^{(\mathrm{co}\,\boldsymbol{k})(m)}(a) = D^{(\boldsymbol{k})(m)}(a a_0^{-1})\beta. \tag{98.58}$$

$$D^{(\boldsymbol{k})(m)}(u) \doteq D^{(\boldsymbol{k})(m)}(a_0^{-1} u a_0)^*$$
$$\text{and} \quad D^{(\boldsymbol{k})(m)}(a_0^2) = \beta\beta^* \tag{98.59}$$
$$\text{where} \quad \beta^{-1} D^{(\boldsymbol{k})(m)}(u)\beta = D^{(\boldsymbol{k})(m)}(a_0^{-1} u a_0)^*.$$

Case B

$$D^{(\mathrm{co}\boldsymbol{k})(m)}(u) = \begin{pmatrix} D^{(\boldsymbol{k})(m)}(u) & 0 \\ 0 & D^{(\boldsymbol{k})(m)}(u^{a_0})^* \end{pmatrix}, \tag{98.14}$$

$$D^{(\mathrm{co}\boldsymbol{k})(m)}(a) = \begin{pmatrix} 0 & D^{(\boldsymbol{k})(m)}(a a_0) \\ D^{(\boldsymbol{k})(m)}(a_0^{-1} a)^* & 0 \end{pmatrix} \tag{98.18}$$

$$\text{and} \qquad D^{(\boldsymbol{k})(m)}(u) \not\doteq D^{(\boldsymbol{k})(m)}(a_0^{-1} u a_0)^*. \tag{98.60}$$

Case C

$$\begin{aligned} D^{(\mathrm{co}\boldsymbol{k})(m)}(u) &= \begin{pmatrix} 1 & 0 \\ 0 & \beta \end{pmatrix} \begin{pmatrix} D^{(\boldsymbol{k})(m)}(u) & 0 \\ 0 & D^{(\boldsymbol{k})(m)}(u^{a_0})^* \end{pmatrix} \begin{pmatrix} 1 & 0 \\ 0 & \beta^{-1} \end{pmatrix} \\ &= \begin{pmatrix} D^{(\boldsymbol{k})(m)}(u) & 0 \\ 0 & D^{(\boldsymbol{k})(m)}(u) \end{pmatrix}, \end{aligned} \tag{98.39}$$

$$D^{(\mathrm{co}\boldsymbol{k})(m)}(a) = D^{(\boldsymbol{k})(m)}(u)\beta \otimes \begin{pmatrix} 0 & -1 \\ 1 & 0 \end{pmatrix}. \tag{98.51}$$

$$D^{(\boldsymbol{k})(m)}(u) \doteqdot D^{(\boldsymbol{k})(m)}(a_0^{-1}\, u\, a_0)^*$$

$$\text{and} \quad D^{(\boldsymbol{k})(m)}(a_0^2) = -\beta\,\beta^* \tag{98.61}$$

$$\text{where} \quad \beta^{-1}\, D^{(\boldsymbol{k})(m)}(u)\,\beta = D^{(\boldsymbol{k})(m)}(a_0^{-1}\, u\, a_0)^*.$$

Class I costar

Case A

The corepresentation is:

$$D^{(\boldsymbol{k})(m)}(u), \qquad D^{(\boldsymbol{k})(m)}(u)\,\beta \equiv D^{(\boldsymbol{k})(m)}(a) \tag{99.5}$$

with

$$\beta = \tilde{\beta}. \tag{99.6}$$

Case B

$$D^{(\mathrm{co}\boldsymbol{k})(m)}(u) = \begin{pmatrix} D^{(\boldsymbol{k})(m)}(u) & 0 \\ 0 & D^{(\boldsymbol{k})(m)}(u^{a_0})^* \end{pmatrix}, \tag{98.14}$$

$$D^{(\mathrm{co}\boldsymbol{k})(m)}(a) = \begin{pmatrix} 0 & D^{(\boldsymbol{k})(m)}(a\,a_0) \\ D^{(\boldsymbol{k})(m)}(a_0^{-1}\,a)^* & 0 \end{pmatrix}. \tag{98.18}$$

Case C

$$\begin{aligned} D^{(\mathrm{co}\boldsymbol{k})(m)}(u) &= \begin{pmatrix} 1 & 0 \\ 0 & \beta \end{pmatrix} \begin{pmatrix} D^{(\boldsymbol{k})(m)}(u) & 0 \\ 0 & D^{(\boldsymbol{k})(m)}(u^{a_0})^* \end{pmatrix} \begin{pmatrix} 1 & 0 \\ 0 & \beta^{-1} \end{pmatrix} \\ &= \begin{pmatrix} D^{(\boldsymbol{k})(m)}(u) & 0 \\ 0 & D^{(\boldsymbol{k})(m)}(u) \end{pmatrix}, \end{aligned} \tag{98.39}$$

$$D^{(\mathrm{co}\boldsymbol{k})(m)}(a) = D^{(\boldsymbol{k})(m)}(u)\,\beta \otimes \begin{pmatrix} 0 & -1 \\ 1 & 0 \end{pmatrix}, \tag{98.51}$$

$$\beta = -\tilde{\beta}. \tag{99.8}$$

Reduction of the representation generated by eigenvectors

$$c_j = \frac{1}{g} \sum_{\mathfrak{G}(\boldsymbol{k})} \chi^{(\boldsymbol{k})(e)}(\{\boldsymbol{\varphi}_{l_\lambda}|\boldsymbol{\tau}_{l_\lambda}\})\, \chi^{(\boldsymbol{k})(j)}(\{\boldsymbol{\varphi}_{l_\lambda}|\boldsymbol{\tau}_{l_\lambda}\})^*. \tag{103.3}$$

Trace of the "total" or cartesian representation

$$\operatorname{Tr} D^{(\boldsymbol{k})(\Delta)}(\{\boldsymbol{\varphi}_\lambda|\boldsymbol{t}_\lambda\}) = \pm e^{i\boldsymbol{k}\cdot\boldsymbol{t}_\lambda}(1+2\cos\varphi_\lambda) \sum_{\kappa=1}^{r} \delta_{\kappa,\kappa_{\bar{\varphi}\lambda}}. \tag{104.13}$$

Here

$$\delta_{\kappa,\kappa_{\bar{\varphi}\lambda}} = \begin{cases} 1 & \text{if } \kappa \text{ and } \kappa_{\bar{\varphi}_\lambda} \text{ are the same basis} \\ 0 & \text{otherwise.} \end{cases}$$

Reduced matrice element of lattice tensor operator

$$\left(\boldsymbol{e}\left(\left|\begin{matrix}\boldsymbol{k}\\ j_\nu\end{matrix}\right.\right), \boldsymbol{\theta}_\nu^{(\boldsymbol{k})(j)}\right) \equiv c(\boldsymbol{k}; j). \tag{106.31}$$

Possible point groups for critical points

$$\text{Any } \mathfrak{P}(\boldsymbol{k}_0) \text{ containing } \boldsymbol{i} \tag{107.64}$$

plus

$$\text{groups } T_d,\ O,\ T,\ D_{2d},\ D_{3h},\ D_n,\ S_4,\ C_{3h} \text{ and their subgroups } D_3,\ D_2,\ S_4,\ C_{3h}. \tag{107.65}$$

Connectivity of corepresentations

$$D^{(\mathrm{co}\boldsymbol{k})(j)} \text{ of } \mathscr{G}(\boldsymbol{k})\!\downarrow = D^{(\mathrm{co}\boldsymbol{k}')(j'_1)} + \cdots + D^{(\mathrm{co}\boldsymbol{k}')(j'_\nu)} \text{ of } \mathscr{G}(\boldsymbol{k}'). \tag{108.2}$$

Condition for a non-zero cubic anharmonic term

$$({}^\star\boldsymbol{k}\, m\, {}^\star\boldsymbol{k}'\, m'\, {}^\star\boldsymbol{k}''\, m'' | \boldsymbol{\Gamma}(1+)) \neq 0. \tag{109.67}$$

Condition for non-zero anharmonic term of s-th degree

$$({}^\star\boldsymbol{k}\, j\, {}^\star\boldsymbol{k}'\, j' \ldots {}^\star\boldsymbol{k}^{(s)}\, j^{(s)} | \boldsymbol{\Gamma} 1+) \neq 0. \tag{109.74}$$

Condition for non-zero second order electric moment

$$({}^\star\boldsymbol{k}\, j\, {}^\star\boldsymbol{k}\, j' | \boldsymbol{\Gamma}\, \boldsymbol{r}) \neq 0. \tag{110.33}$$

Condition for non-zero polarizability term of s-th degree

$$({}^\star\boldsymbol{k}\, j\, {}^\star\boldsymbol{k}'\, j' \ldots | [D^{(\boldsymbol{r})}]_{(2)}) \neq 0. \tag{110.37}$$

Adiabatic Born-Oppenheimer wave function

$$|\Psi_{\text{adiabatic}}\rangle = |\Psi_{\text{electron}}\rangle |\Psi_{\text{lattice}}\rangle. \tag{113.39b}$$

Adiabatic energy

$$E_\nu = \Phi_\nu^{(0)} + \kappa^2 E^{(2)}_{\nu\{n\}}. \tag{113.40}$$

Harmonic eigenfunction for ions

$$\chi_\nu(q) = \text{const} \times \exp - \left[\tfrac{1}{2} \sum_{\boldsymbol{k},\sigma,j} \gamma_{\sigma,j} \sum_{\alpha} q \begin{pmatrix} \boldsymbol{k}_\sigma \\ j_\mu \end{pmatrix}^2 \right] \times \prod_{\boldsymbol{k},\sigma,j} \prod_{\mu} H_{n_\sigma,j,\mu} \left(\gamma_{\sigma j}\, q \begin{pmatrix} \boldsymbol{k}_\sigma \\ j_\mu \end{pmatrix} \right). \tag{115.9}$$

Symmetry transformation of harmonic eigenfunctions

$$[D^{({}^\star\boldsymbol{k})(j)}]_{(n)} \otimes [D^{({}^\star\boldsymbol{k}')(j')}]_{(n')} \otimes \cdots \otimes [D^{({}^\star\boldsymbol{k}'')(j'')}]_{(n'')} = \sum_{\boldsymbol{k}'''m'''} ([{}^\star\boldsymbol{k}\, j]_{(n)} \ldots [{}^\star\boldsymbol{k}''\, j'']_{(n'')} | {}^\star\boldsymbol{k}'''\, j''')\, D^{({}^\star\boldsymbol{k}''')(j''')}. \tag{118.4}$$

Symmetry of matrix element for infra-red absorption

$$\langle \chi_{\nu\bar{n}} | \mathscr{M}^{\nu}(\boldsymbol{R}, \boldsymbol{k}) | \chi_{\nu n} \rangle \neq 0$$

if

$$D^{(n)} \otimes D^{(\bar{n})} \sim D^{(\boldsymbol{v})}. \tag{120.42}$$

Symmetry of matrix element for Raman scattering

$$\langle \chi_{0\bar{n}} | \boldsymbol{\varepsilon}_{2\lambda} \cdot \boldsymbol{P}(\boldsymbol{R}, \boldsymbol{k}_1 - \boldsymbol{k}_2) \cdot \boldsymbol{\varepsilon}_{1\lambda'} | \chi_{0n} \rangle_{\boldsymbol{R}} \tag{121.45}$$

$$D^{(\bar{n})*} \otimes D^{(n)} \sim D^{(\boldsymbol{k}_1 - \boldsymbol{k}_2)(\boldsymbol{v})}, \tag{121.52}$$

$$\sim [D^{(\boldsymbol{v})}]_{(2)}. \tag{121.53}$$

Raman scattering tensor elements as Clebsch-Gordan coefficients

$$P_{\sigma\rho}(\boldsymbol{k} j_\mu) = c(\boldsymbol{k} j)\, U_{\sigma\rho\boldsymbol{k} j_\mu}. \tag{123.13}$$

Symmetry of one-phonon thermal Green function

$$\mathscr{G}\big(A(\boldsymbol{k}|j), A^+(\boldsymbol{k}'|j'), \tau\big) \sim D^{(\star\boldsymbol{k})(j)} \otimes D^{(\star\boldsymbol{k}')(j')*}. \tag{124.61}$$

Time reverse symmetry of Raman scattering tensor

$$R_{12}(-\omega_1, \omega_2, \omega_0) = R_{21}\big(-(\omega_1 - \omega_0), \omega_2 + \omega_0, -\omega_0\big). \tag{124.109}$$

Symmetry breaking in Raman scattering (quadrupole-dipole)

$$QD \sim \sum \boldsymbol{\varepsilon}' \langle \boldsymbol{p} \rangle \langle \Xi \rangle \langle Q \rangle\, \boldsymbol{\varepsilon}\boldsymbol{k} \big/ (\Delta\Omega)(\Delta\Omega'), \tag{124.151}$$

in O_h:

$$D^{(\boldsymbol{v})} \otimes D^{(\boldsymbol{Q})} = D^{(\boldsymbol{\Gamma})(2-)} \oplus D^{(\boldsymbol{\Gamma})(12-)} \oplus D^{(\boldsymbol{\Gamma})(15-)} \oplus D^{(\boldsymbol{\Gamma})(25-)}, \tag{124.152}$$

$$\sum_{j\sigma} \Big| \sum_{\alpha\beta\gamma} \varepsilon_{2\alpha} P_{\alpha\beta\gamma}(j\sigma) \varepsilon_{1\beta} k_{1\gamma} \Big|^2. \tag{124.153}$$

Symmetry breaking in Raman scattering (phonon wave vector effects)

$$\boldsymbol{\varepsilon}' \langle \boldsymbol{p} \rangle q \langle \Xi^{(1)} \rangle \langle \boldsymbol{p} \rangle \cdot \boldsymbol{\varepsilon} \big/ (\Delta\Omega)(\Delta\Omega'), \tag{124.155}$$

$$\sum_{\sigma} \Big| \sum_{\alpha\beta\gamma} \varepsilon_{\alpha\alpha} Q_{\alpha\beta\gamma}(j\sigma) \varepsilon_{1\beta} q_\gamma \Big|^2. \tag{124.157}$$

Index of tables

Table 1. The face centered cubic translation group. 343
Table 2. Rotational symmetry operations for diamond and rocksalt. 344
Table 3. Coordinates of zone points for diamond and rocksalt. 346
Table 4. Irreducible representations of O_h. $\Gamma^{(m)} = \mathfrak{C}(\Gamma)/\mathfrak{T}$. 347
Table 5. Irreducible representations of $D_{3d} = \mathfrak{G}(L_1)/\mathfrak{T}(L_1)$ in rocksalt and diamond. 349
Table 6. Conjugate elements in $\mathfrak{P}(L_1), \ldots, \mathfrak{P}(L_4)$. 349
Table 7. Irreducible representations $D^{(\star L)(m)}$ of O_h^5. 350
Table 8. Irreducible representation $D^{(X_1)(m)}$ of $\mathfrak{G}(X_1)/\mathfrak{T} = D_{4h}$ in O_h^5. 350
Table 9. Conjugate elements of $\mathfrak{P}(X_1)$: Rotations only. 351
Table 10. Irreducible representations $D^{(\star X)(m)}$ of O_h^5 for coset representatives $\{\varphi_p|0\}$. 351
Table 11. Wave vector selection rules in O_h^5 and O_h^7: For products with Γ, $\star X$, $\star L$. 352
Table 12. Illustration of the reduction process: Example $\star X^{(4-)} \otimes \star X^{(5-)}$ in O_h^5. 353
Table 13. Illustration $D^{(\star L)(3-)} \otimes D^{(\star L)(3+)}$ in O_h^5. 359
Table 14. Augmented character table for $D^{(\star L)(3-)} \otimes D^{(\star L)(3+)}$ in O_h^5. 359
Table 15. Wave vector selection rules in rocksalt O_h^5 and diamond O_h^7 coefficients $(\boldsymbol{k}_j \boldsymbol{k}_{j'} | \boldsymbol{k}_{j''})$. 360
Table 16. Wave vector selection rules in rocksalt O_h^5 and diamond O_h^7 coefficients $([l_m \boldsymbol{k}_j]_{(2)} | \boldsymbol{k}_{j'})$. 361
Table 17. Wave vector selection rules in rocksalt and diamond coefficients $([l_m \boldsymbol{k}_j)_{(3)} | \boldsymbol{k}_{j'})$. 361
Table 18a. Character table of group $\mathfrak{G}(X_1)/\mathfrak{T}(X_1)$ in diamond O_h^7. 366
Table 18b. Allowable irreducible representations of $\mathfrak{G}(\Delta_1)/\mathfrak{T}(\Delta_1)$ in O_h^7, $\Delta_1 = (\kappa, 0, 0)$. 374
Table 18c. Full group irreducible representations $(\star \Delta j)$ in O_h^7. 374
Table 19. Character table for W_1 in rocksalt. Vector: $W_1 = (2\pi, 0, -\pi)(1/a)$. 373
Table 20. Reduction of displacement representation in rocksalt O_h^5: NaCl type crystal. Wave vectors Γ, X_1, L_1. Each column gives the rotational symmetry elements φ_l in $\mathfrak{G}(\boldsymbol{k})$, and the character of the displacement representation at that $\boldsymbol{k}$. 379
Table 21. Symmetry species of phonons rocksalt O_h^5. 381
Table 22. Reduction of displacement representation in diamond O_h^7. Wave vectors Γ, X_1, L_1. 382
Table 23. Symmetry species of phonons in diamond O_h^7. 383
Table 24. Compatibility table for diamond and rocksalt: $\Gamma \to \star\Delta$, $\star\Sigma$, $\star\Lambda$. 384
Table 25. Compatibility for $\star X \to \star\Delta$, $\star Z$, $\star S$ in rocksalt. 385
Table 26. Compatibility for $\star X$ and $\star\Delta$, $\star Z$, $\star S$ in diamond. 385
Table 27. Compatibility $\star L \to \star\Lambda$, $\star Q$, for diamond and rocksalt. 385
Table 28. Compatibility $\star W \to \star Z$, $\star Q$ for rocksalt. 385
Table 29. Compatibility $\star W \to \star Z$, $\star Q$ in diamond. 385
Table 30. Character table for critical points near X_1 in diamond. 389
Table 31. One-phonon critical points and indices in diamond. 393
Table 32. Critical points near L_1 in diamond. 394
Table 33. Critical point near W_1 in diamond. $W_1 = (2\pi, 0, -\pi)(1/a)$. 395
Table 34. Critical point near Σ_1 in diamond. $\Sigma_1 = (\pi, \pi, 0)(1/a)$. 396
Table 35. Critical point near Q in diamond. $Q_1 = (\kappa, \pi, 2\pi - \kappa)(1/a)$. 397
Table 36. Two phonon processes in diamond structure: silicon and germanium. 399
Table 37. Reduction coefficients corresponding to Raman- and infrared-active representations of overtones and combinations in diamond structure O_h^7. 401
Table 38. Three-phonon processes in diamond structure. 402, 403
Table 39. Critical-point analysis of infrared absorption in diamond. 406
Table 40a. Phonon energies at critical points in diamond. 407
Table 40b. Assignments for the second-order Raman and infrared spectrum of diamond. 408

Table 41 a. Comparison of theory with experiment for silicon-two-phonon infra-red absorption. 409
Table 41 b. Three-phonon combinations in silicon. 410
Table 41 c. Phonon frequencies in silicon (Raman scattering). 411
Table 42. Comparison of phonon frequencies in silicon. 411
Table 43 a. Comparison of theory with experiment for germanium. 414
Table 43 b. Three-phonon combinations in germanium. 414
Table 44. Comparison of phonon frequencies in germanium. 415
Table 45. Symmetry set of critical points for phonons in rocksalt structure O_h^5. 417
Table 46 a. Phonon symmetries in the rocksalt structure. 421
Table 46 b. Infrared-active two-phonon combinations. 421
Table 47. Second-order Raman peaks in NaCl. 423
Table 48. Room-temperature phonon frequencies in NaCl. 423
Table 49. Interpretation of NaCl spectra. 425, 426
Table 50. Interpretation of KBr spectra. 429
Table 51. Interpretation of KI spectra. 430
Table 52. Interpretation of NaCl infra-red absorption at 300 °K. 432
Table 53. Space-group reduction coefficients for the diamond lattice. 439
Table 54. Space-group reduction coefficients for the face-centred cubic lattice. 440
Table 55. Subduction of $D^{(\star X)(1)}$ of O_h^7 onto T_d. 441
Table 56. Local modes in imperfect diamond site group T_d. 441
Table 57. The (un-normalized) eigenvectors $\{\psi^{(\sigma l)}(l\kappa)\}$ for a substitutional impurity possessing T_d symmetry in a crystal possessing the same symmetry about the impurity site. 442
Table 58. Local modes in imperfect rocksalt site group O_h. 443

Table A.1. Reduction coefficients of type $(\star\boldsymbol{k}\, m\, \boldsymbol{\Gamma}\, m'|)$ for rocksalt O_h^5. 468
Table A.2. Reductions of type $(\star\boldsymbol{k}\, m\, \star\boldsymbol{L}\, m'|)$ for O_h^5. 469
Table A.3. Reductions of type $(\star\boldsymbol{k}\, m\, \star\boldsymbol{X}\, m'|)$ for O_h^5. 470, 471
Table A.4. Reductions of type $(\star\boldsymbol{k}\, m\, \star\boldsymbol{W} m'|)$ for O_h^5. 472
Table A.5. Reductions of type $(\star\boldsymbol{k}\, m\, \star\boldsymbol{\Delta}\, m'|)$ for O_h^5. 473
Table A.6. Reductions of type $(\star\boldsymbol{k}\, m\, \star\boldsymbol{\Sigma}\, m'|)$ for O_h^5. 474
Table A.7. Reductions of type $(\star\boldsymbol{\Lambda}\, m\, \star\boldsymbol{\Lambda}\, m'|)$ for O_h^5. 475
Table A.8. Reductions of type $(\star\boldsymbol{k}\, m\, \star\boldsymbol{Z}\, m'|)$ for O_h^5. 476
Table A.9. Reductions of type $(\star\boldsymbol{Q}\, m\, \star\boldsymbol{Q}\, m'|)$ for O_h^5. 477
Table A.10. Reduction coefficients for symmetrized Kronecker squares for O_h^5. 477
Table A.11. Reduction coefficients for symmetrized Kronecker cubes for O_h^5. 478

Table B.1. Complete character table for reduction group for $\boldsymbol{\Gamma}$, $\star\boldsymbol{X}$, $\star\boldsymbol{L}$ in diamond O_h^7. Reduction group characters for $\boldsymbol{\Gamma}^{(m)}$, $\star\boldsymbol{L}^{(m)}$, $\star\boldsymbol{X}^{(m)}$, $\star\boldsymbol{L}^{(m)}(\{|\boldsymbol{\varphi}|\boldsymbol{t}+\boldsymbol{Q}\})=-\star\boldsymbol{L}^{(m)}(\{\boldsymbol{\varphi}|\boldsymbol{t}\})$. 480–485
Table B.2. Reduction coefficients for diamond space group O_h^7. 486–488
Table B.3. Reduction coefficients for symmetrized Kronecker square in diamond. 489
Table B.4. Reduction coefficients for symmetrized Kronecker cubes in diamond. 489
Table B.5. Principle block of Clebsch-Gordan coefficients for the product $D^{(\star X)(l)}\otimes D^{(\star X)(l')}\to D^{(\star X)(l'')}$ in diamond. 490, 491
Table B.6. Symmetry elements for calculating the $(\sigma\,\sigma'\,\sigma'')$ blocks of Clebsch-Gordan coefficients for $D^{(\star X)(l)}\otimes D^{(\star X)(l')}\to D^{(\star X)(l'')}$ in diamond. 491
Table B.7. Matrices for calculating the $(\sigma\,\sigma'\,\sigma'')$ blocks of Clebsch-Gordan coefficients for $D^{(\star X)(l)}\otimes D^{(\star X)(l')}\to D^{(\star X)(l'')}$ in diamond. 491
Table B.8. Principle block of Clebsch-Gordan coefficients for $D^{(\star X)(l)}\otimes D^{(\star X)(l')}\to D^{(\star X)(l'')}$ in diamond. 492
Table B.9. Symmetry elements for calculation of the $(\sigma\,\sigma'\,\sigma'')$ blocks of Clebsch-Gordan coefficients for $D^{(\star X)(l)}\otimes D^{(\star X)(l')}\to D^{(\Gamma)(l'')}$ in diamond. 493
Table B.10. Matrices for calculation of the $(\sigma\,\sigma'\,\sigma'')$ blocks of Clebsch-Gordan coefficients for $D^{(\star X)(l)}\otimes D^{(\star X)(l')}\to D^{(\Gamma)(l'')}$ in diamond. 493

Table C.1. Matrices for irreducible representations at $\boldsymbol{X}_3$ in diamond O_h^7. 497

Table D.1. Symmetry of zincblende structure; T_d^2: $F\bar{4}3m$. 499
Table D.2. Irreducible representations in zincblende at symmetry points. 500
Table D.3. Compatibility tables for the representations of the single groups connecting the zincblende (T_d^2) with the rocksalt (O_h^5) and the diamond (O_h^7) structures. 501
Table D.4. Reduction coefficients for zincblende. 502
Table D.5. Additional reduction coefficients in zincblende. 503
Table D.6. Reduction coefficients for symmetrized Kronecker square in zincblende. 504
Table D.7. Reduction coefficients for symmetrized Kronecker cube in zincblende. 504
Table D.8. Critical points and phonon species in T_d^2. 505
Table D.9. Two-phonon processes in zincblende. 506
Table D.10. Three-phonon processes in zincblende. 507, 508

Index of figures

Fig. 1. The rocksalt-NaCl structure, unit cell. 344
Fig. 2. The unit cell of diamond structure. 345
Fig. 3. Brillouin zone for the face centered cubic translation group. 345
Fig. 4. Phonon dispersion in NaCl. 382
Fig. 5. Phonon dispersion in germanium. Calculated and measured values. 390
Fig. 6. Phonon dispersion in silicon. Calculated and measured values. 390
Fig. 7a and b. Phonon dispersion in diamond. Calculated and measured values. 391
Fig. 8. Phonon dispersion calculation for germanium. 392
Fig. 9. Shapes of critical point contribution to the density of states for $P_j(n)$; $j=1, 2, 3$; $n=0, 1$. 397
Fig. 10a–c. Infra-red absorption spectrum of diamond in the two phonon energy region. Joint density-of-states functions for diamond at 296 °K. 404
Fig. 10d. Raman scattering in diamond. Temperature dependence of the second-order Raman spectrum of diamond for $Z'(X'X')Y'$ scattering in the 2050–2770-cm^{-1} region. 405
Fig. 11. Infra-red absorption spectrum of silicon in the two phonon energy region. 409
Fig. 12. Details of the infra-red absorption (two phonon region) in silicon. 410
Fig. 13. Comparison between measured infra-red absorption curves in silicon, and calculated two and three phonon combination branches. 412
Fig. 14a and b. Lattice bands in germanium. 413
Fig. 15. Infra-red absorption bands in the two and three phonon regions for germanium, silicon and diamond. 416
Fig. 16. A model calculation of phonon dispersion in NaCl. 418
Fig. 17. A calculation of the density of states (one phonon) in NaCl. The model corresponds to that used in Fig. 16. 418
Fig. 18a–c. Calculated dispersion curves, density of states and critical points in NaF. 419
Fig. 19. Calculated dispersion curves for two phonons in NaCl. 420
Fig. 20. Calculated two phonon density of states and critical points in rocksalt. See model of Fig. 19. 420
Fig. 21. Second-order Raman scattering in NaCl. Comparison of experiments and a model calculation. 422
Fig. 22. Infra-red absorption spectrum of NaCl. 424
Fig. 23. Raman scattering in NaCl and KCl: Scattering intensity (arb. units) vs. wave length. 425
Fig. 24. Second-order Raman spectrum of NaF. 427
Fig. 25. Two phonon (combined) density of states in NaF. 427
Fig. 26. Calculated two phonon density of states in NaF. 427
Fig. 27. Comparison between experimental and theoretical two phonon density of states in NaF. 428
Fig. 28. Experimental spectra of Raman scattering in NaCl. Different attributions refer to different polarization contributions. 432
Fig. 29. Comparison between experimental Raman scattering spectra in NaCl, and theoretical calculations, based on density of states computed from phonon dispersion. 433
Fig. 30a–c. Temperature dependence of the second-order Raman spectrum of diamond; a) for $Z'(Y'X')Y'$ scattering in the range 2050–2770 cm^{-1}; b) in the range 2050–2770 cm^{-1} for $Z'(X'Z')Y'$ scattering; c) for $Y'(Z'Z')X'$ scattering in the range 2050–2770 cm^{-1}. 435
Fig. 31. The infra-red spectrum of silicon doped with both boron and phosphorus in concentrations of 5×10^{19} cm^{-3} and 5×10^{18} cm^{-3}. The spectrum of pure silicon is shown for comparison. 451
Fig. 32. a) Band mode absorption of silicon doped with both boron and phosphorus in concentrations of 5×10^{19} cm^{-3}. Critical point energies and shapes (Johnson and Loudon, 1964) are shown. b) Comparison of theoretical and experimental calculation for infra-red absorption of doped Si, with *P* and *B*. 452
Fig. 33. Electric field effect upon first order Raman scattering in diamond. The broadening may be an indication of symmetry-breaking. 456
Fig. 34. Electric field induced symmetry breaking and optical absorption in diamond. 459

Sachverzeichnis

(Deutsch-Englisch)

Bei gleicher Schreibweise in beiden Sprachen sind die Stichwörter nur einmal aufgeführt

Abelsche Gruppe, Translationen als, *Abelian group, translations as* 40

Adiabatische Näherung, allgemeine Formulierung, *adiabatic approximation, general statement* 257

— —, Born-Oppenheimer 253

Akustische Schwingungen, Symmetrie und gesamte Darstellung, *acoustic modes, symmetry and total representation* 207

Akzeptable Darstellung von $\mathfrak{G}(\boldsymbol{k})$, *acceptable representation of* $\mathfrak{G}(\boldsymbol{k})$ 56

Algebraische Gleichungen für Reduzierungskoeffizienten, *algebraic equations for reduction coefficients* 95

Anharmonisches Potential, kubische Terme, *anharmonic potential, cubic terms* 234, 239

— —, — — und Reduzierungskoeffizienten, *and reduction coefficients* 242

— —, — — und Transformationen, *and transformations* 234, 242

— —, Terme allgemeiner Ordnung, *terms of general degree* 243

— —, — — — und Reduzierungskoeffizienten, *and reduction coefficients* 243

— —, — — — und Transformation, *and transformation* 243

Anti-unitärer Operator, *K* als, *anti-unitary operator, K as an* 160

— — —, kommutiert mit unitärem Operator, *commutes with unitary operators* 160

Auslenkungen, Fourier-Transformation, *displacements, Fourier transformation* 136

— von Teilchen, *particle* 115

— — —, Matrix von, *matrix of* 116

— — — und allgemeine Symmetrie, *and general symmetry* 119, 232

— — — und komplexe Normalkoordinaten, *and complex normal coordinates* 142

— — — und Translationssymmetrie, *and translation symmetry* 117

Auswahlregel für Infrarot-Absorption, *selection rule for infra-red absorption* 278

— — Raman-Streuung, *Raman scattering* 290

Automorphismus, innerer, äußerer, einer Gruppe, *automorphisms, inner, outer, of group* 52

Band-Schwingungen im gestörten Kristall, *band modes in perturbed crystal* 440

Basis, Funktions-, *basis functions* 28

—, kristallographische, *crystallographic* 1

Bloch-Funktion, Definition von $\psi^{(\boldsymbol{k})}$, *Bloch function, definition of* $\psi^{(\boldsymbol{k})}$ 46

Bloch-Vektor, äquivalent, *Bloch vector, equivalent* 51

—, Definition von $\psi^{(\boldsymbol{k})}$, *definition of* $\psi^{(\boldsymbol{k})}$ 46

—, Transformierung bei Drehung, *transformed by rotation* 50

Born-Karman, (zyklische) Randbedingungen, *Born-Karman, (cyclic) boundary conditions* 12

Born-Oppenheimer, adiabatische Näherung, *Born-Oppenheimer, adiabatic approximation* 253

—, Parameter $\kappa \equiv (m/M)^{1/4}$ 253

—, Potentialfunktion für Kerne, *potential function for nucleii* 257

—, störungstheoretische Gleichungen, *perturbation theory equations* 253

Bornsche Methode für adiabatische Näherung, *Born method for adiabatic approximation* 256

Brillouin-Zone, erste, Definition, *Brillouin zone, first, defined* 43

—, —, Schönfleiss-Zelle, *Schönfleiss cell* 44

—, —, und Wigner-Seitz-Polyeder, *and Wigner-Seitz polyhedron* 44

Charaktere, induzierte, von $\mathfrak{G}$, *characters, induced, of* $\mathfrak{G}$ 60

—, punktiert, von $\mathfrak{G}$, *dotted, of* $\mathfrak{G}$ 61

Charakter-System, vollständige Untergruppe, *character system, complete subgroup* 105

Clebsch-Gordan-Koeffizienten, Berechnung von, *Clebsch-Gordan coefficients, calculation of* 37

—, Definitionen, *definitions* 34, 101

Clebsch-Gordan-Koeffizienten für Raumgruppen, *Clebsch-Gordan coefficients for space groups* 101
— — —, $(\sigma, \sigma, \sigma'')$-Block 103
— — —, (1, 1, 1)-Block 103
—, Matrix von, *matrix of* 36
— und Raman-Streutensor, *and Raman scattering tensor* 301
—, Unitarität, *unitarity* 37
Clebsch-Gordan-Serien, Analogie für Raumgruppen, *Clebsch-Gordan series, analogy for space groups* 94
—, Koeffizienten s. Reduzierungskoeffizienten, *coefficients see Reduction coefficients* 33

Darstellung einer Gruppe als Matrixgruppe, *representation of group as matrix group* 28
— — —, irreduzible, Definition, *irreducible, definition* 29
— — —, Unitarität, *unitarity* 30
Darstellungen einer Gruppe, direkte Produkte, *representations of group, direct products* 32
— — —, — —, Reduzierung, *reduction* 32
Darstellung von $\mathfrak{G}(\boldsymbol{k})$, basierend auf Eigenvektoren, *representation of* $\mathfrak{G}(\boldsymbol{k})$, *based on eigenvectors* 148, 150
— — —, basierend auf komplexen Normalkoordinaten, *based on complex normal coordinates* 155
— — —, Äquivalenz von $D^{(\boldsymbol{k})(e)}$ und $D^{(\boldsymbol{k})(j)}$, *equivalence of* $D^{(\boldsymbol{k})(e)}$ *and* $D^{(\boldsymbol{k})(j)}$ 154
Depolarisierungsquotient und Raman-Streuung, *depolarization ratio and Raman scattering* 304
Diamant, Clebsch-Gordan-Koeffizienten, *diamond, Clebsch-Gordan coefficients* 368
—, irreduzible Darstellungen, *irreducible representations* 362
—, — — für Γ, *at* Γ 362
—, — — für $\star L$, *at* $\star L$ 364
—, — — für $\star X$, *at* $\star X$ 363
—, — —, Fortsetzbarkeit längs Δ, *connectivity along* Δ 377
—, Phononen, Symmetrie, *phonons, symmetry* 382
—, —, kritische Punkte, *critical points* 386, 398
—, —, Zustandsdichte: 2-Phononen, *density of states: 2 phonon* 400
—, Raumgruppe $Fd3m$, O_h^7, *space group* $Fd3m$, O_h^7 21, 343
—, Untergruppe $F\bar{4}3m$, T_d^2, *subgroup* $F\bar{4}3m$, T_d^2 22
Direktes Produkt von Raumgruppen-Darstellungen, Charakter, *direct product of space group representations, character* 87
— — — —, Potenzen, *powers* 88
Dreh-Operatoren $\{\varphi|0\}$, *rotational operators* $\{\varphi|0\}$ 14
Drehung, dreidimensional, *rotation, 3 dimensions* 16
Drehungen, dyadische Form, nicht-kartesische Achsen, *rotations, dyadic form, non-cartesian axes* 17
Dynamische Matrix $[\boldsymbol{D}(\boldsymbol{k})]$, Definition, *dynamical matrix* $[\boldsymbol{D}(\boldsymbol{k})]$, *defined* 137
— — —, gestörter Kristall, *perturbed crystal* 443
— — —, Hermitezität, *hermitian property* 137
— — —, Kristallsymmetrie und, *crystal symmetry and* 145, 146

Ebene Wellen, Orthogonalität und Vollständigkeit für, *plane waves, orthonormality and completeness for* 136
— — und Fourier-Transformation, *and Fourier transformation* 136
Eigenenergie, Verschiebung, *self energy, shift* 321
Eigenfrequenzen der hermiteschen dynamischen Matrix $[\boldsymbol{D}(\boldsymbol{k})]$, *eigenfrequencies of the hermitian dynamical matrix* $[\boldsymbol{D}(\boldsymbol{k})]$ 138
— — reellen dynamischen Matrix, *real dynamical matrix* 126
Eigenvektoren $\boldsymbol{e}_{j_\rho}$ der dynamischen Matrix, *eigenvectors* $\boldsymbol{e}_{j_\rho}$ *of the dynamical matrix* 126
— — — —, Orthonormalität, *orthonormality* 127
— — — — und Symmetrie, *and symmetry* 131
— — — —, Vollständigkeit, *completeness* 129
— —, Darstellung basierend auf, *representation based on* 134
— —, Matrix von, *matrix of* 128
— —, notwendige Entartung, *necessary degeneracy* 132
Eigenvektoren von $[\boldsymbol{D}(\boldsymbol{k})]$, *eigenvectors of* $[\boldsymbol{D}(\boldsymbol{k})]$ 138
— — — als Basis für die Darstellung von $\mathcal{G}$, *as basis for representation of* $\mathcal{G}$ 165
— — — als Basis für die Darstellung von $\mathfrak{G}(\boldsymbol{k})$, *as basis for representation of* $\mathfrak{G}(\boldsymbol{k})$ 148, 150
— — — als Tensorfeld erster Stufe, *as first rank tensor field* 145
— — —, Beziehung zu reellen Eigenvektoren $\boldsymbol{e}_j$, *related to real eigenvectors* $\boldsymbol{e}_j$ 140
— — —, Einfluß von Symmetrietransformationen, *effect of symmetry transformation* 144, 147
— — —, Matrix von, *matrix of* 140
— — —, skalare Produkte, Definition, *scalar products, defined* 139
— — — und Ko-Darstellungen, *and corepresentations* 202
Einheitsoperator, *identity operator* 10

Erlaubte Darstellung von $\mathfrak{G}(\boldsymbol{k})$, *allowable representation of* $\mathfrak{G}(\boldsymbol{k})$ 56

Faktorsystem der Erweiterung von $\mathfrak{P}$ durch $\mathfrak{T}$, *factor set of extension of* $\mathfrak{P}$ *by* $\mathfrak{T}$ 19
— einer Strahldarstellung, Definition, *of ray representation, definition* 68
Fortsetzbarkeit von Darstellungen, *connectivity of representations* 230, 374
Fourier-Gitter, Definition, *Fourier lattice, defined* 44

Gebundene Zwei-Phonon-Zustände (im Diamant) und Raman-Streuung, *two-phonon bound states (in diamond) and Raman scattering* 310, 433
— —, Symmetrie von, *symmetry of* 311
Gestörter Kristall, Gruppe eines Gitterpunktes, *perturbed crystal, site group* 437
— —, Symmetrie-Gruppe, *symmetry group* 437
Gitter-Eigenfunktion, Transformation von, *lattice eigenfunction, transformation of* 269
Gitter-Hypothese, *lattice hypothesis* 9
Green-Funktion, thermische, *Green function, thermal* 319
—, mode mixing 326
—, Symmetrie von, *symmetry of* 326
Gruppe der Gitterpunkte, Definition, *site group, definition* 19, 437
— von $\boldsymbol{k}$, $\mathfrak{G}(\boldsymbol{k})$, Darstellung basierend auf Eigenvektoren, *group of* $\boldsymbol{k}$, $\mathfrak{G}(\boldsymbol{k})$, *representations based on eigenvectors* 148
Gruppen-Algebra für zweiseitige Ideale oder Idempotente, *group algebra, two-sided ideals or idempotents* 31
—, Projektionsoperator, *projection operators* 31, 33

Hamilton-Operator als Funktion komplexer Normalkoordinaten, *Hamiltonian as function of complex normal coordinates* 141
— und komplexe Normalkoordinaten, *and complex normal coordinates* 237
— — — —, Nachweis von Invarianz, *demonstration of invariance* 237
— — reelle Normalkoordinaten, *real normal coordinates* 130
—, Vielteilchensystem, *many electron—many ion* 252
Harmonische, adiabatische Näherung, Definition, *harmonic adiabatic approximation, defined* 258
—, — —, Gitter-Eigenfunktion als ein Produkt, *lattice eigenfunctions as a product* 262
—, — —, symmetrische Eigenfunktion, *symmetrized eigenfunction* 264
—, — — und Energie der Gitterschwingungen, *and vibrational lattice energy* 262

Induzierte Darstellung von $\mathfrak{G}$ aus $\mathfrak{G}(\boldsymbol{k})$, *induced representation of* $\mathfrak{G}$ *from* $\mathfrak{G}(\boldsymbol{k})$ 57, 60
— — von Raumgruppen aus Untergruppen, *of space group from subgroups* 82
Infrarot-Absorption, Ein-Phonon, *infrared absorption, one phonon* 279
—, —, Auswahlregeln, *selection rule* 280
—, Mehrphonon, *multiphonon* 282
—, Vielteilchen-Theorie, *many body theory* 322
—, —, Suszeptibilität, *susceptibility* 324
— durch Phononen, *by phonons* 272
— — —, Absorptionsrate, *rate* 277
— — —, Auswahlregeln, *selection rule* 278
— — —, Hamilton-Operator, *Hamiltonian* 272
— — —, Lax-Burstein-Näherung, *Lax-Burstein approximation* 274
—, Zwei-Phonon, *two phonon* 282
—, —, Auswahlregeln, *selection rule* 282
—, —, in Diamant, Ge, Si, *in diamond, Ge, Si* 414
Irreduzible Darstellung von $\mathfrak{G}$, Block-Matrix-Struktur, *irreducible representation of* $\mathfrak{G}$, *block matrix structure* 53
Irreduzible Darstellungen $D^{(\star k)(m)}$ der Raumgruppe $\mathfrak{G}$, *irreducible representations* $D^{(\star k)(m)}$ *of space group* $\mathfrak{G}$ 49
— —, Vollständigkeit aller, *completeness of all* 29
— — von Raumgruppen, Elemente, *of space group, elements* 60
— — von $\mathfrak{T}$, subduziert von $D^{(\star k)(m)}$ aus $\mathfrak{G}$, *of* $\mathfrak{T}$, *subduced by* $D^{(\star k)(m)}$ *of* $\mathfrak{G}$ 49

Kartesische Auslenkungs-Darstellung als direktes Produkt, *cartesian displacement representation as a direct product* 207
— —, Spur von, *trace of* 206
— — und mechanische Darstellung, *and mechanical representation* 204
Kinetische Gitterenergie, klassische, *kinetic energy of lattice, classical* 116
Klasse I-Ko-Stern und Ko-Darstellung von $\mathscr{G}$, *class I costar and corepresentation of* $\mathscr{G}$ 195
Klasse I-Wellenvektoren und reelle Eigenschaft von Darstellungen, *class I wave vectors and reality of representations* 177
Klasse II-Ko-Stern und Fälle A, B, C, *class II costar and cases A, B, C* 193, 194
— und $\mathscr{G}(\boldsymbol{k})$, *and* $\mathscr{G}(\boldsymbol{k})$ 189
— und Ko-Darstellungen von $\mathscr{G}$, *and corepresentations of* $\mathscr{G}$ 188
— und punktierte Ko-Darstellungsmatrizen, *and dotted corepresentation matrices* 195
Klasse II-Wellenvektoren und reelle Eigenschaft von Darstellungen, *class II wave vectors and reality of representation* 179

Klasse III-Ko-Star und Ko-Darstellungen, *class III costar and corepresentation* 187
Klasse III-Wellenvektoren und (fehlende) reelle Eigenschaft von Darstellungen, *class III wave vectors and (lack of) reality of representations* 179'
Klasse I,II,III-Wellenvektoren, Definition, *class I,II,III wave vectors, defined* 177
Kochsalzstruktur, Auswahlregeln für Wellenvektoren, *rocksalt, wave vector selection rules* 352
—, irreduzible Darstellungen, *irreducible representations* 345
—, — — Γ 347
—, — — $\star L$ 347
—, — — $\star X$ 351
—, NaCl und Raman, Infrarot, *NaCl and Raman, infra-red* 422
—, NaF 426
—, Phonon, kritische Punkte, *critical points* 416
—, —, Symmetrie, *symmetry* 379
—, Raumgruppe $Fm3m$, O_h^5, *space group* $Fm3m$, O_h^5 21, 343
—, Reduzierung von $\star L^{(3-)} \otimes \star L^{(3+)}$, *reduction of* $\star L^{(3-)} \otimes \star L^{(3+)}$ 359
—, — — $\star X^{(4)} \otimes \star X^{(5)}$ 352
Ko-Darstellungen von $\mathscr{G}$, Basisraum $S^{(j)}$, *corepresentations of* $\mathscr{G}$, *basis space* $S^{(j)}$ 180
— — —, Multiplikation für Matrizen in, *multiplication for matrices in* 181
— — — und Lemma von Schur, *and Schur Lemma* 183
— — — und komplexe Normalkoordinaten, *and complex normal coordinates* 199
— — — und notwendige Entartung, *and necessary degeneracy* 179
— von $\mathscr{T}$, induziert von $\mathfrak{T}$, *of* $\mathscr{T}$, *induced from* $\mathfrak{T}$ 184
— — —, — — —, irreduzibler Raum für, *irreducible space for* 184
Kompatibilität und subduzierte Darstellungen, *compatibility and subduced representations* 231
— von Darstellungen, *of representations* 230
— — — für benachbarte Sterne, *for adjacent stars* 86
Komplexe Darstellungen von $\mathfrak{G}$, Test (Frei), modifiziert, *complex representation of* $\mathfrak{G}$, *test (Frei), modified* 176
— — — $\mathfrak{G}$, Test (Herring) 176
Komplexe Normalkoordinaten, Basis für irreduzible Darstellung von $\mathscr{G}$, *complex normal coordinates, bases for irreducible representation of* $\mathscr{G}$ 165
— —, Definition, *defined* 141
— —, Hamiltonoperator als Funktion von, *Hamiltonian as function of* 141
— —, Symmetrietransformation von, *symmetry transformation of* 158
— — und Ko-Darstellungen, *and corepresentations* 201
— — und physikalische Auslenkungen, *and physical displacements* 142
Komplex-Konjugations-Operator K und Leibfried-Konvention, *complex conjugation operator* K *and Leibfried convention* 163
— — und Zeitumkehr, *and time reversal* 164
Konjugierte Darstellungen von $\mathfrak{T}$, *conjugate representation of* $\mathfrak{T}$ 51
Konjugierte Koordinate und Impuls, Massengewichtete Auslenkungen, *conjugate coordinate and momentum, mass weighted displacements* 137
Ko-Stern co $\star\boldsymbol{k}$, Definition, *costar co* $\star\boldsymbol{k}$, *defined* 187
— —, Klasse I, II, III, Definition, *class I, II, III, defined* 187
Kräfte-Matrix für dynamische Gleichungen, *force matrix for dynamical equations* 125
—, Symmetrie und reelle Eigenschaft, *symmetry and reality* 137
Kraftkonstante, Kopplungsparameter, Definition, *force constant, coupling parameters, defined* 238
—, —, Transformation 238
Kristall-Invariante, potentielle Energie, *crystal invariants, potential energy* 231
Kristall-Kovariante, elektrisches Moment, *crystal covariants, electric moment* 245
—, — —, Transformation 246
—, — —, und Reduzierungskoeffizient, *and reduction coefficient* 248
—, Polarisierbarkeit, *polarizability* 249
Kritische Punkte in der Zustandsdichte, Definition, *critical points in density of states, defined* 220
— — — — —, „dynamische", "*dynamical*" 220
— — — — —, hinreichende Bedingung für, und Reduzierungskoeffizient, *sufficient condition for, and reduction coefficient* 227
— — — — —, Index von, *index of* 227
— — — — —, mögliche Punktsymmetrie für, *possible point symmetry for* 230
— — — — — $P_j(\mu)$, Definition, $P_j(\mu)$, *defined* 388
— — — — —, „Symmetriemenge", "*symmetry set*" 220
— — — — — und „Morse-Theorie", *and "Morse theory"* 221

Kritischer-Punkt-Index, Definition, $P_j(\mu)$, *critical point index, defined*, $P_j(\mu)$ 388
Kronecker-Potenzen siehe Direktes Produkt, *Kronecker powers see Direct product*
—, kubische, *cube* 90
—, quadratische, *square* 89
—, symmetrische, *symmetrized* 89
Kronecker-Produkt siehe Direktes Produkt, *Kronecker product see Direct product*

Lagrange-Operator, Funktion komplexer Normalkoordinaten, *Lagrangian, function of complex normal coordinates* 137
—, klassische Gitter, *classical lattice* 116
Lemma über notwendige Entartung und Eigenvektoren, *lemma of necessary degeneracy and eigenvectors* 132
„Little group" von **k**, $\Pi(\mathbf{k})$, Definition, *"little group" of* **k**, $\Pi(\mathbf{k})$, *definition* 63
— — — —, —, irreduzible Darstellungen, erlaubte, *irreducible representations, allowable* 64
— — — —, —, irreduzible Darstellungen, nicht erlaubte, *irreducible representations, non-allowable* 65
Lokalisierte Schwingungen, *local modes* 440
Lösbarkeit von Raumgruppen, *solvability of space groups* 85

Makroskopisches, elektrisches Feld und Symmetrie-Erniedrigung, *macroscopic electric field and symmetry lowering* 306
—, — — und Lyddane-Sachs-Teller, *and Lyddane-Sachs-Teller* 306
Maschke's Theorem, Formulierung von, *Maschke's theorem, statement of* 30
Mechanische Darstellung als direktes Produkt, *mechanical representation as a direct product* 206
— — oder „volle" Darstellung, *or "total" representation* 204
— —, Spur von, *trace of* 206
Morphische Effekte, Definition von, *morphic effects, definition of* 456
— —, — —, Symmetrie-Effekt, *symmetry effect* 458
Multiplikation (Zusammensetzung) von Operatoren, *multiplication (composition) of operators* 10, 15

Nicht-ideales Kristall, Diamant und Kochsalz, lokalisierte Schwingungen, *imperfect crystal, diamond and rocksalt, local modes* 441
— —, — — —, lokalisierte Schwingungen, Symmetrie, *local modes, symmetry* 440
— —, Infrarot-Absorption, *infra-red absorption* 449
— —, Raman-Streuung, *Raman scattering* 453
Normalkoordinaten, reelle, *normal coordinates, real* 129
—, —, und Massen-gewichtete Auslenkungen, *and mass-weighted displacements* 130
—, —, und Quantisierung, *and quantization* 259
Normalschwingungen, Symmetrie in einem Kristall, *normal modes, symmetry in a crystal* 202
Notwendige Entartung und Raum-Zeit-Gruppe $\mathscr{G}$, *necessary degeneracy and space-time group* $\mathscr{G}$ 166
— — von Eigenvektoren, *of eigenvectors* 132, 152

Operator des elektrischen Dipolmoments mit Phase, diagonal, Definition, *electric dipole moment operator, phased, diagonal, definition* 277
— — — —, mit Phase, nicht diagonal, *phased, off diagonal* 287
— — — —, Symmetrie, *symmetry* 278
Orthogonale Transformation, *orthogonal transformation* 8

Physikalisch irreduzible Darstellung, Definition, *physically irreducible representation, defined* 132
— — — und Raum-Zeit-Gruppe $\mathscr{G}$, *and space-time group* $\mathscr{G}$ 167
Polare, optische Schwingung, Aufspaltung *LO-TO, polar optic mode, splitting LO-TO* 307
—, — —, Kristall ohne Inversion, *crystal without inversion* 308
—, — —, — — —, Aufspaltung und Aktivität, *splitting and activity* 308
—, — — und Regel gegenseitiger Ausschließung, *and rule of mutual exclusion* 307
Polarisierbarkeits-Operator als Tensorfeld 2. Stufe, *polarizability operator as second rank tensor field* 290
— für Raman-Streuung, *for Raman scattering* 288
—, Symmetrie, *symmetry* 289
Polarisierung bei Infrarot-Absorption, *polarization in infra-red absorption* 312
— — Raman-Streuung, *Raman scattering* 298, 301
— — — bei Zwei-Phononen, *two phonon* 304
Polarisierungseffekte, makroskopisches, elektrisches Feld (Poulet-Effekt), *polarization effects, macroscopic electric field (Poulet effect)* 309

Polariton, Operatoren, Definition, *polariton, operators, definition* 338
—, Streuung und Raman-Theorie, *scattering and Raman theory* 338
Potentiell reelle Darstellung von 𝔊, Test, *potentially real representation of 𝔊, test* 176, 179
Potentielle Gitterenergie, klassische, *potential energy of lattice, classical* 115
— —, positiv semidefinit, *positive semidefinite* 115
— — und allgemeine Symmetrie, *and general symmetry* 123
— — und Born-Oppenheimer-Theorie, *and Born-Oppenheimer theory* 254
— — und Translationssymmetrie, *and translation symmetry* 118
Projektionsoperator P^l_{mr}, Definition von, *projection operator P^l_{mr}, definition* 31
— — für Translationsgruppe 𝔗, *for translation group 𝔗* 46
Pseudo-reelle Darstellung von 𝔊, Test, *pseudo-real representation of 𝔊, test* 176, 179
Punktgruppe 𝔓, Definition, *point group 𝔓, definition* 13
— —, Untergruppe von 𝔊, *subgroup of 𝔊* 23
Punktierte Matrix, Definition, *dotted matrix, defined* 58

Raman-Streuung, anomale (Poulet), makroskopisches Feld, *Raman scattering anomalies (Poulet), macroscopic field* 310
—, Diamant (polarisiert), *diamond (polarized)* 434
—, Diamant-Kristall, *diamond crystal* 404
—, Diamant-Struktur und Zwei-Phonon-Zustandsdichte, *diamond structure and 2 phonon density of states* 400
—, Dipol-Dipol, *dipole-dipole* 340
— durch Phononen (Hamilton-Operator), *by phonons (Hamiltonian)* 285
—, Fröhlich-Intraband (verboten), *Fröhlich intraband, forbidden* 342
—, Germanium 412
—, Kochsalz (polarisiert), *rocksalt (polarized)* 431
—, Multipol-Dipol (verboten), *multipole-dipole, forbidden* 341
—, Placzek-Näherung, *Placzek approximation* 287
—, Polarisierbarkeitsoperator, *polarizibility operator* 289
—, Quantentheorie (Exziton), *quantum theory (Exciton)* 333
—, — —, Tensor 336
—, — (Polariton) 336
—, — (Bloch-Darstellung), *(Bloch picture)* 328
—, — —, Hamilton-Operator, *Hamiltonian* 330
—, — —, Kopplungsfunktion, Elektron-Strahlung, *coupling functions, electron radiation* 330
—, — —, Tensor 332
—, Resonanz, *resonance* 339
—, — und Symmetrie-Verletzung, *and symmetry breaking* 340
—, Silizium, *silicon* 407
—, Stokes, Ein-Phonon, *Stokes, one phonon* 292
—, —, Zwei-Phonon, *two phonon* 292
—, —, — und Ableitung der Zustandsdichte, *and derivative of density of states* 293
—, —, —, Wirkungsquerschnitt, *cross section* 292
— und der $A \cdot A$-Term, *and the $A \cdot A$ term* 293
—, Vielteilchen-Formulierung, *many body formalism* 315
—, —, Ein-Phonon, *one phonon* 321
—, —, spektrale Dichte, *spectral density* 316
—, —, Zwei-Phonon und Zustandsdichte, *two phonon and density of states* 321
Raman-Tensor, Definition, *Raman tensor, defined* 299
—, Loudon- und Poulet-Konvention, *Loudon and Poulet conventions* 301
—, Symmetrie von, *symmetry of* 332
— und Clebsch-Gordan-Koeffizienten, *and Clebsch-Gordan coefficients* 301
Raumgruppe 𝔊 als Erweiterung, *space group 𝔊 as extension* 19
— —, Definition 7, 16
— —, irreduzible Darstellung $D^{(*k)(m)}$, *irreducible representation $D^{(*k)(m)}$* 48
— —, — — —, Basisfunktion, *basis functions* 49
— —, — — —, Dimension 48
— —, konjugierte Untergruppe von, *conjugate subgroup of* 24
— —, nicht symmorphisch, *non symmorphic* 22
— —, Ordnung, *order* 18
— —, symmorphisch, *symmorphic* 21
Raum-Zeit-Gruppe, Definition von $\mathcal{G}$, *space-time group, $\mathcal{G}$ defined* 160, 165
Raum-Zeit-Punktgruppe $𝔓(\boldsymbol{k})$, Definition, *space-time point group $𝔓(\boldsymbol{k})$ defined* 196
— —, Strahl-Ko-Darstellung von, *ray corepresentation of* 199
Reduziertes Matrixelement, Definition, *reduced matrix element, defined* 212
— —, mehrere, *several* 214
— — und Wigner-Eckart-Theorem, *and Wigner-Eckart theorem* 216

Reduzierungsgruppe, Definition, *reduction group, definition* 97
Reduzierungskoeffizient, Basisfunktionen für, *reduction coefficient, basis functions for* 99
— der vollen Gruppe, Definition, *full-group, defined* 91
— für den Stern, *for stars* 94
—, „little group"-Methode, *little group method* 112
—, Untergruppe, *subgroup* 107
— siehe auch Untergruppe Reduzierungskoeffizient, *see also Subgroup reduction coefficient*
— siehe Stern, Reduzierungskoeffizient, *see Star, reduction coefficient*
Reduzierungskoeffizienten für Produkte von Darstellungen, *reduction coefficients for products of representations* 33
— für Raumgruppen, *for space groups* 87
— — —, normale, *ordinary* 91
— — —, symmetrisierte, *symmetrized* 92
Reelle Darstellung von $\mathscr{G}$, *real representation of* $\mathscr{G}$ 168
— — von $\mathfrak{G}$, *of* $\mathfrak{G}$ 169
— — — —, Frei-Test (modifiziert), *Frei test (modified)* 177
— — — —, Herring-Test 176
— — — —, Test 172
— — — — und Normalschwingungen, *and normal modes* 167
Reelle Normalkoordinaten und Ko-Darstellungen, *real normal coordinates and corepresentations* 199
— — — Kristallsymmetrie, *crystal symmetry* 133
— —, Transformation 134
Regel gegenseitiger Ausschließung in Kristallen mit Inversion, *mutual exclusion rule in crystals with inversion* 298
— — —, Verkehrtheit von, *falsity of converse* 298
— — —, verletzt in der Resonanz, *broken at resonance* 340
Replika, *Replica* 6
Reziproke Vektoren, Definition von $\boldsymbol{b}_i$, *reciprocal vectors, definition of* $\boldsymbol{b}_i$ 40
Reziprokes Gitter, Definition, *reciprocal lattice, definition* 41
— —, Vektor $\boldsymbol{B}_H$ (*rel*-Vektoren), *vectors* $\boldsymbol{B}_H$ *(rel vectors)* 41

Schur, Lemma von, Test für Irreduzibilität, *Schur lemma, test for irreducibility* 29
—, — —, und Ko-Darstellungen, *and corepresentations* 183
Semi-direkte Produktgruppe, Definition, *semi-direct product group, definition* 19
— —, symmorphische Raumgruppe als, *symmorphic space group as* 74
Skalarprodukt in Gitterdynamik, *scalar product in lattice dynamics* 210
—, „single dot" und „double dot", Definition, *single dot and double dot, defined* 127, 128
— von einem Operator mit Eigenvektoren, *of an operator with eigenvectors* 213
Stern, allgemein, *star, general* 61
—, direktes Produkt von, *direct product of* 93
—, Reduzierungskoeffizienten, normale, *reduction coefficients, ordinary* 91
—, —, symmetrisierte, *symmetrized* 92
—, speziell, *special* 62
Strahldarstellung von $\mathfrak{P}(\boldsymbol{k})$ als Multiplikator, *ray representations of* $\mathfrak{P}(\boldsymbol{k})$ *as multiplier* 69
— — —, assoziiert, *associated* 70
— — —, Definition 69
— — —, Eichtransformation, *gauge transformations* 70
— — — oder projektive, *or projective* 69
Strahl-Ko-Darstellung von $\mathfrak{P}(\boldsymbol{k})$, *ray corepresentation of* $\mathfrak{P}(\boldsymbol{k})$ 197
Streumatrix, zusammengesetzte Matrix in kubischen Kristallen, *scattering matrix, composite in cubic crystal* 305
Streu-Tensor, Raman, Definition, *scattering tensor, Raman, defined* 299
Streu-Tensoren in kubischen Kristallen, *scattering tensors in cubic crystals* 302
— — — — und Polarisierung, *and polarization* 303
Subduzierte Darstellung, Definition, *subduced representation, definition* 49
— — und Kompatibilität, *and compatibility* 86
— — von $\mathfrak{G}$ auf $\mathfrak{T}$, *of* $\mathfrak{G}$ *onto* $\mathfrak{T}$ 52
Symmetrie, Translations- $\{\varepsilon|\boldsymbol{t}\}$, *symmetry, translational* $\{\varepsilon|\boldsymbol{t}\}$ 7
Symmetrieoperation, Definition, Gleitebene, *symmetry operation, definition, glide plane* 7
—, —, Schraubenachse, *screw axis* 7
— $\{\varphi|\boldsymbol{t}\}$, Multiplikation, $\{\varphi|\boldsymbol{t}\}$, *multiplication* 16
—, Symmetrie-Transformation, „reelle", *symmetry transformation, "real"* 6
—, —, „annehmbare", *"imagined"* 6
Symmetrie-Operator auf Funktionen, *symmetry operator on functions* 26
— — —, Gruppe von, *group of* 27
— — —, idempotent 31
Symmetrie-Verletzung, externe Verzerrung, *symmetry breaking, external stresses* 455
Symmetrische Polynome, elementar, *symmetric polynomials, elementary* 267

Symmetrische Polynome und symmetrische Potenzen, *symmetric polynomials and symmetrized powers* 268
Symmetrisiertes Kronecker-Produkt und Symmetrie harmonischer Gitterfunktionen, *symmetrized Kronecker product and harmonic lattice eigenfunction symmetry* 264
— —, Charakter von, *character of* 269
— — von hermiteschen Polynomen, *of hermite polynomials* 264

Tensorrechnung und Gitterdynamik, *tensor calculus and lattice dynamics* 210
Transformations-Operator der komplexen Konjugation, *transformation operator of complex conjugation* 122
— für Auslenkungen, *for displacements* 119, 121
— für Koordinaten, *for coordinates* 120
Translationsgruppe $\mathfrak{T}$ als direktes Produkt, *translation group $\mathfrak{T}$ as direct product* 11
— —, irreduzible Darstellung, *irreducible representation* 40, 42
— —, normale Untergruppe von $\mathfrak{G}$, *normal subgroup of $\mathfrak{G}$* 17, 23

Überlagerungsgruppe $\mathfrak{P}^\star(\boldsymbol{k})$, Definition, *covering group $\mathfrak{P}^\star(\boldsymbol{k})$, defined* 68
— —, auch Darstellungsgruppe, *also representation group* 71
Unitäre Gruppe $SU_{(sl_j)}$ und Gitter-Eigenfunktionen, *unitary group $SU_{(sl_j)}$ and lattice eigenfunction* 270
Untergruppen-Reduzierungskoeffizient, Definition, *subgroup reduction coefficient, defined* 107
—, verglichen mit dem der vollen Gruppe, *compared to full group* 109, 111

Verletzte Symmetrie, Störstellen im Kristall, *broken symmetry, imperfection in crystals* 436
Vollständige Darstellung als ein direktes Produkt, *total representation as a direct product* 206
— —, Definition, *defined* 204
— —, Spur, *trace of* 206

Wellenvektor, äquivalent, *wave vector, equivalent* 43
—, Auswahlregeln, *selection rules* 93
—, Definition von $\boldsymbol{k}$, *definition of* $\boldsymbol{k}$ 42
—, kanonischer, $\boldsymbol{k}$ zu $^\star\boldsymbol{k}$, *canonical,* $\boldsymbol{k}$ *of* $^\star\boldsymbol{k}$ 59
—, nicht-äquivalent, *inequivalent* 54
—, Raumgruppe von $\boldsymbol{k}$, $\mathfrak{G}(\boldsymbol{k})$, *space group of* $\boldsymbol{k}$, $\mathfrak{G}(\boldsymbol{k})$ 53
—, Reduzierungskoeffizienten, *reduction coefficients* 93
—, Stern von, Definition, *star of, definition* 61
Wellenvektoren, Klasse I, II, III, Definition, *wave vectors, class I, II, III, defined* 177
Wigner-Eckart-Formeln, Analogie für Gitterdynamik, *Wigner-Eckart formula, analogy for lattice dynamics* 212
— und anti-unitäre Elemente, *and antiunitary elements* 216
— — — —, Wellenvektoren Fall A, *and case A wave vectors* 217
— — — —, Wellenvektoren Fall B, *and case B wave vectors* 218
— — — —, Wellenvektoren Fall C, *and case C wave vectors* 219

Zeitumkehr, Test am Diamant $^\star X$, $^\star W$, *time reversal, test in diamond* $^\star X$, $^\star W$ 372
—, — — Kochsalz, *rocksalt* 373
Zeitumkehr-Operator K, angewandt auf Eigenvektor, *time reversal operator K, applied to eigenvector* 161
— —, Definition, *defined* 160
Zusätzliche Entartung und Wellenvektoren I. Klasse, *extra degeneracy and class I wave vectors* 177
— — und Wellenvektoren II. Klasse, *and class II wave vectors* 179
— — und Wellenvektoren III. Klasse, *and class III wave vectors* 180
Zustandsdichte, Phonon (quadrierte Frequenz), *density of states, phonon (squared frequency)* 232
Zweige von Gitterfrequenzen, *branches of lattice frequencies* 155, 221

Subject Index

(English-German)

Where English and German spellings of a word are identical the *German* version is omitted

Abelian group, translations as, *Abelsche Gruppe, Translationen als* 40
Acceptable representation of $\mathfrak{G}(\boldsymbol{k})$, *akzeptable Darstellung von* $\mathfrak{G}(\boldsymbol{k})$ 56
Acoustic modes, symmetry and total representation, *akustische Schwingungen, Symmetrie und gesamte Darstellung* 207
Adiabatic approximation, Born-Oppenheimer, *adiabatische Näherung* 253
— —, general statement, *allgemeine Formulierung* 257
Algebraic equations for reduction coefficients, *algebraische Gleichungen für Reduzierungskoeffizienten* 95
Allowable representation of $\mathfrak{G}(\boldsymbol{k})$, *erlaubte Darstellung von* $\mathfrak{G}(\boldsymbol{k})$ 56
Anharmonic potential, cubic terms, *anharmonisches Potential, kubische Terme* 234, 239
— —, — — and reduction coefficients, *und Reduzierungskoeffizienten* 242
— —, — — and transformations, *und Transformationen* 234, 242
— —, terms of general degree, *Terme allgemeiner Ordnung* 243
— —, — — — — and reduction coefficients, *und Reduzierungskoeffizienten* 243
— —, — — — — and transformation, *und Transformation* 243
Antiunitary operator, K as an, *anti-unitärer Operator, K als* 160
— — — commutes with unitary operators, *kommutiert mit unitärem Operator* 160
Automorphisms, inner, outer, of group, *Automorphismus, innerer, äußerer, einer Gruppe* 52

Band modes in perturbed crystal, *Band-Schwingungen im gestörten Kristall* 440
Basis, crystallographic, *Basis, kristallographische* 1
— functions, *Funktions-* 28
Bloch function, definition of $\psi^{(\boldsymbol{k})}$, *Bloch-Funktion, Definition von* $\psi^{(\boldsymbol{k})}$ 46
Bloch vector, definition of $\psi^{(\boldsymbol{k})}$, *Bloch-Vektor, Definition von* $\psi^{(\boldsymbol{k})}$ 46
— —, equivalent, *äquivalent* 51
— —, transformed by rotation, *Transformierung bei Drehung* 50
Born method for adiabatic approximation, *Bornsche Methode für adiabatische Näherung* 256
Born-Karman, (cyclic) boundary conditions, *Born-Karman, (zyklische) Randbedingungen* 12
Born-Oppenheimer, adiabatic approximation, *adiabatische Näherung* 253
—, *Parameter* $\kappa \equiv (m/M)^{1/4}$ 253
—, perturbation theory equations, *störungstheoretische Gleichungen* 253
—, potential function for nucleii, *Potentialfunktion für Kerne* 257
Branches of lattice frequencies, *Zweige von Gitterfrequenzen* 155, 221
Brillouin zone, first, defined, *Brillouin-Zone, erste, Definition* 43
— —, —, Schönfleiss cell, *Schönfleiss-Zelle* 44
— —, —, and Wigner-Seitz polyhedron, *und Wigner-Seitz-Polyeder* 44
Broken symmetry, imperfection in crystals, *verletzte Symmetrie, Störstellen im Kristall* 436

Cartesian displacement representation and mechanical representation, *kartesische Auslenkungs-Darstellung und mechanische Darstellung* 204
— — — as a direct product, *als direktes Pro*dukt 207
— — —, trace of, *Spur von* 206
Characters, dotted, of $\mathfrak{G}$, *Charaktere, punktiert, von* $\mathfrak{G}$ 61
—, induced, of $\mathfrak{G}$, *induzierte, von* $\mathfrak{G}$ 60
Character system, complete subgroup, *Charakter-System, vollständige Untergruppe* 105
Class I costar and corepresentations of $\mathscr{G}$, *Klasse I-Ko-Stern und Ko-Darstellung von* $\mathscr{G}$ 195

Class I wave vectors and reality of representations, *Klasse I-Wellenvektoren und reelle Eigenschaft von Darstellungen* 177
Class II costar and cases A, B, C, *Klasse II-Ko-Stern und Fälle* A, B, C 193, 194
— — and corepresentations of $\mathscr{G}$, *und Ko-Darstellungen von* $\mathscr{G}$ 188
— — and dotted corepresentation matrices, *und punktierte Ko-Darstellungsmatrizen* 195
— — and $\mathscr{G}(\boldsymbol{k})$, *und* $\mathscr{G}(\boldsymbol{k})$ 189
Class II wave vectors and reality of representation, *Klasse II-Wellenvektoren und reelle Eigenschaft von Darstellungen* 179
Class III costar and corepresentations, *Klasse III-Ko-Star und Ko-Darstellungen* 187
Class III wave vectors and (lack of) reality of representations, *Klasse III-Wellenvektoren und (fehlende) reelle Eigenschaft von Darstellungen* 179
Class I, II, III wave vectors, defined, *Klasse I, II, III-Wellenvektoren, Definition* 177
Clebsch-Gordan coefficients, calculation of, *Clebsch-Gordan-Koeffizienten, Berechnung von* 37
— —, definitions, *Definitionen* 34, 101
— — for space groups, *für Raumgruppen* 101
— — — — —, $(\sigma, \sigma, \sigma'')$ block 103
— — — — —, (1, 1, 1) block 103
— —, matrix of, *Matrix von* 36
— — and Raman scattering tensor, *und Raman-Streutensor* 301
— —, unitarity, *Unitarität* 37
Clebsch-Gordan series, analogy for space groups, *Analogie für Raumgruppen* 94
— —, coefficients see reduction coefficients, *Koeffizienten siehe Reduzierungskoeffizienten* 33
Compatibility and subduced representations, *Kompatibilität und subduzierte Darstellungen* 231
— of representations, *von Darstellungen* 230
— — — for adjacent stars, *für benachbarte Sterne* 86
Complex conjugation operator K and Leibfried convention, *Komplex-Konjugations-Operator K und Leibfried-Konvention* 163
— — — and time reversal, *und Zeitumkehr* 164
Complex normal coordinates, bases for irreducible representation of $\mathscr{G}$, *komplexe Normalkoordinaten, Basis für irreduzible Darstellung von* $\mathscr{G}$ 165
— — — and corepresentations, *und Ko-Darstellungen* 201
— — — and physical displacements, *und physikalische Auslenkungen* 142
— — —, defined, *Definition* 141
— — —, Hamiltonian as function of, *Hamiltonoperator als Funktion von* 141
— — —, symmetry transformation of, *Symmetrietransformation von* 158
Complex representation of $\mathfrak{G}$, test (Frei), modified, *komplexe Darstellungen von* $\mathfrak{G}$, *Test (Frei), modifiziert* 176
— — — $\mathfrak{G}$, test (Herring) 176
Conjugate coordinate and momentum, mass weighted displacements, *konjugierte Koordinate und Impuls, Massen-gewichtete Auslenkungen* 137
— representations of $\mathfrak{T}$, *Darstellungen von* $\mathfrak{T}$ 51
Connectivity of representations, *Fortsetzbarkeit von Darstellungen* 230, 374
Corepresentations of $\mathscr{G}$, basis space $S^{(j)}$, *Ko-Darstellungen von* $\mathscr{G}$, *Basisraum* $S^{(j)}$ 180
— — — and complex normal coordinates, *und komplexe Normalkoordinaten* 199
— — — and necessary degeneracy, *und notwendige Entartung* 179
— — — and Schur Lemma, *und Lemma von Schur* 183
— — —, multiplication for matrices in, *Multiplikation für Matrizen in* 181
— of $\mathscr{T}$, induced from $\mathfrak{T}$, *von* $\mathscr{T}$, *induziert von* $\mathfrak{T}$ 184
— — —, — — —, irreducible space for, *irreduzibler Raum für* 184
Costar co$\star\boldsymbol{k}$, defined, *Ko-Stern co*$\star\boldsymbol{k}$, *Definition* 187
— —, class I, II, III, defined, *Klasse I, II, III, Definition* 187
Covering group $\mathfrak{P}^\star(\boldsymbol{k})$, defined, *Überlagerungsgruppe* $\mathfrak{P}^\star(\boldsymbol{k})$, *Definition* 68
— — — also representation group, *auch Darstellungsgruppe* 71
Critical point index, defined, $P_j(\mu)$, *Kritischer-Punkt-Index, Definition,* $P_j(\mu)$ 388
Critical points in density of states, defined, *Kritische Punkte in der Zustandsdichte, Definition* 220
— — — — — —, "dynamical", „*dynamische*" 220
— — — — — —, index of, *Index von* 227
— — — — — —, and "Morse theory", *und „Morse-Theorie*" 221
— — — — — —, $P_j(\mu)$, defined, $P_j(\mu)$, *Definition* 388
— — — — — —, possible point symmetry for, *mögliche Punktsymmetrie für* 230
— — — — — —, sufficient condition for, and reduction coefficient, *hinreichende Bedingung für, und Reduzierungskoeffizient* 227
— — — — — —, "symmetry set", „*Symmetriemenge*" 220

Crystal covariants, electric moment, *Kristall-Kovariante, elektrisches Moment* 245
— —, — —, and reduction coefficient, *und Reduzierungskoeffizient* 248
— —, — —, transformation 246
— —, polarizability, *Polarisierbarkeit* 249
Crystal invariants, potential energy, *Kristall-Invariante, potentielle Energie* 231

Density of states, phonon (squared frequency), *Zustandsdichte, Phonon (quadrierte Frequenz)* 232
Depolarization ratio and Raman scattering, *Depolarisierungsquotient und Raman-Streuung* 304
Diamond, Clebsch-Gordan coefficients, *Diamant, Clebsch-Gordan-Koeffizienten* 368
—, irreducible representations, *irreduzible Darstellungen* 362
—, — — at $\boldsymbol{\Gamma}$, *für* $\boldsymbol{\Gamma}$ 362
—, — — at $\star\boldsymbol{L}$, *für* $\star\boldsymbol{L}$ 364
—, — — at $\star\boldsymbol{X}$, *für* $\star\boldsymbol{X}$ 363
—, — —, connectivity along $\varDelta$, *Fortsetzbarkeit längs* $\varDelta$ 377
—, phonons, critical points, *Phononen, kritische Punkte* 386, 398
—, —, density of states: 2 phonon, *Zustandsdichte: 2-Phononen* 400
—, —, symmetry, *Symmetrie* 382
—, space group $Fd3m$, O_h^7, *Raumgruppe* $Fd3m$, O_h^7 21, 343
—, subgroup $F\bar{4}3m$, T_d^2, *Untergruppe* $F\bar{4}3m$, T_d^2 22
Direct product of space group representations, *direktes Produkt von Raumgruppen-Darstellungen* 87
— — — — — —, character, *Charakter* 87
— — — — — —, powers, *Potenzen* 88
Displacements, Fourier transformation, *Auslenkungen, Fourier-Transformation* 136
—, particle, *von Teilchen* 115
—, —, and complex normal coordinates, *und komplexe Normalkoordinaten* 142
—, —, and general symmetry, *und allgemeine Symmetrie* 119, 232
—, —, matrix of, *Matrix von* 116
—, —, and translation symmetry, *und Translationssymmetrie* 117
Dotted matrix, defined, *punktierte Matrix, Definition* 58
Dynamical matrix $[\boldsymbol{D}(\boldsymbol{k})]$, crystal symmetry and, *dynamische Matrix* $[\boldsymbol{D}(\boldsymbol{k})]$, *Kristallsymmetrie und* 145, 146
— — —, defined, *Definition* 137
— — —, hermitian property, *Hermitezität* 137
— —, perturbed crystal, *gestörter Kristall* 443

Eigenfrequencies of the hermitian dynamical matrix $[\boldsymbol{D}(\boldsymbol{k})]$, *Eigenfrequenzen der hermiteschen dynamischen Matrix* $[\boldsymbol{D}(\boldsymbol{k})]$ 138
— — — real dynamical matrix, *reelen dynamischen Matrix* 126
Eigenvectors $\boldsymbol{e}_{j_\rho}$ of the dynamical matrix, *Eigenvektoren* $\boldsymbol{e}_{j_\rho}$ *der dynamischen Matrix* 126
— — — — — —, completeness, *Vollständigkeit* 129
— — — — — —, orthonormality, *Orthonormalität* 127
— — — — — —, and symmetry, *und Symmetrie* 131
— —, matrix of, *Matrix von* 128
— —, necessary degeneracy, *notwendige Entartung* 132
— —, representation based on, *Darstellung basierend auf* 134
Eigenvectors of $[\boldsymbol{D}(\boldsymbol{k})]$, *Eigenvektoren von* $[\boldsymbol{D}(\boldsymbol{k})]$ 138
— — — and corepresentations of $\mathscr{G}$, *und Ko-Darstellungen von* $\mathscr{G}$ 202
— — —, as basis for representation of $\mathscr{G}$, *als Basis für die Darstellung von* $\mathscr{G}$ 165
— — —, as basis for representation of $\mathfrak{G}(\boldsymbol{k})$, *als Basis für die Darstellung von* $\mathfrak{G}(\boldsymbol{k})$ 148, 150
— — —, as first rank tensor field, *als Tensorfeld erster Stufe* 145
— — —, effect of symmetry transformation, *Einfluß von Symmetrietransformationen* 144, 147
— — —, matrix of, *Matrix von* 140
— — —, related to real eigenvectors $\boldsymbol{e}_j$, *Beziehung zu reellen Eigenvektoren* $\boldsymbol{e}_j$ 140
— — —, scalar products, defined, *skalare Produkte, Definition* 139
Electric dipole moment operator, phased, diagonal, definition, *Operator des elektrischen Dipolmoments mit Phase, diagonal, Definition* 277
— — — —, —, off diagonal, *nicht diagonal* 287
— — — —, symmetry, *Symmetrie* 278
Extra degeneracy and class I wave vectors, *zusätzliche Entartung und Wellenvektoren I. Klasse* 177
— — and class II wave vectors, *und Wellenvektoren II. Klasse* 179
— — and class III wave vectors, *und Wellenvektoren III. Klasse* 180

Factor set of extension of $\mathfrak{P}$ by $\mathfrak{T}$, *Faktorsystem der Erweiterung von* $\mathfrak{P}$ *durch* $\mathfrak{T}$ 19
— — of ray representation, definition, *einer Strahldarstellung, Definition* 68

Force constant, coupling parameters, defined, *Kraftkonstante, Kopplungsparameter, Definition* 238
— —, — —, *Transformation* 238
Force matrix for dynamical equations, *Kräfte-Matrix für dynamische Gleichungen* 125
— —, symmetry and reality, *Symmetrie und reelle Eigenschaft* 137
Fourier lattice, defined, *Fourier-Gitter, Definition* 44
Full-group reduction coefficients, defined, *Reduzierungskoeffizient der vollen Gruppe, Definition* 91

Green function, mode mixing, *Green-Funktion, mode mixing* 326
— —, symmetry of, *Symmetrie von* 326
— —, thermal, *thermische* 319
Group algebra, projection operators, *Gruppen-Algebra, Projektionsoperator* 31, 33
— —, two-sided ideals or idempotents, *für zweiseitige Ideale oder Idempotente* 31
Group of $\boldsymbol{k}$, $\mathfrak{G}(\boldsymbol{k})$, representations based on eigenvectors, *Gruppe von* $\boldsymbol{k}$, $\mathfrak{G}(\boldsymbol{k})$, *Darstellung basierend auf Eigenvektoren* 148

Hamiltonian, and complex normal coordinates, *Hamilton-Operator und komplexe Normakoordinaten* 237
—, — — — —, demonstration of invariance, *Nachweis von Invarianz* 237
—, — real normal coordinates, *reelle Normalkoordinaten* 130
—, as function of complex normal coordinates, *als Funktion komplexer Normalkoordinaten* 141
—, many electron – many ion, *Vielteilchensystem* 252
Harmonic adiabatic approximation, defined, *harmonische, adiabatische Näherung, Definition* 258
— — —, lattice eigenfunctions as a product, *Gitter-Eigenfunktion als ein Produkt* 262
— — —, symmetrized eigenfunction, *symmetrische Eigenfunktion* 264
— — — and vibrational lattice energy, *und Energie der Gitterschwingungen* 262

Identity operator, *Einheitsoperator* 10
Imperfect crystal, diamond and rocksalt, local modes, *nichtideales Kristall, Diamant und Kochsalz, lokalisierte Schwingungen* 441
— —, — — —, local modes, symmetry, *lokalisierte Schwingungen, Symmetrie* 440
— —, infra-red absorption, *Infrarot-Absorption* 449
— —, Raman scattering, *Raman-Streuung* 453

Induced representation of $\mathfrak{G}$ from $\mathfrak{G}(\boldsymbol{k})$, *Induzierte Darstellung von* $\mathfrak{G}$ *aus* $\mathfrak{G}(\boldsymbol{k})$ 57, 60
— — of space group from subgroups, *von Raumgruppen aus Untergruppen* 82
Infra-red absorption, many body theory, *Infrarot-Absorption, Vielteilchen-Theorie* 322
— —, — — —, susceptibility, *Suszeptibilität* 324
— —, multiphonon, *Mehrphonon* 282
— —, one phonon, *Ein-Phonon* 279
— —, — —, selection rule, *Auswahlregeln* 280
— — by phonons, *von Phononen* 272
— — — —, Hamiltonian, *Hamilton-Operator* 272
— — — —, Lax-Burstein approximation, *Lax-Burstein-Näherung* 274
— — — —, rate, *Absorptionsrate* 277
— — — —, selection rule, *Auswahlregeln* 278
— —, two phonon, *Zwei-Phonon* 282
— —, — — in diamond, Ge, Si, *in Diamant, Ge, Si* 414
— —, — —, selection rule, *Auswahlregeln* 282
Irreducible representations, completeness of all, *irreduzible Darstellungen, Vollständigkeit aller* 29
— — $D^{(*\boldsymbol{k})(m)}$ of space group $\mathfrak{G}$, $D^{(*\boldsymbol{k})(m)}$ *der Raumgruppe* $\mathfrak{G}$ 49
— — of $\mathfrak{G}$, block matrix structure, *von* $\mathfrak{G}$, *Block-Matrix-Struktur* 53
— — of space group, elements, *von Raumgruppen, Elemente* 60
— — of $\mathfrak{T}$, subduced by $D^{(*\boldsymbol{k})(m)}$ of $\mathfrak{G}$, *von* $\mathfrak{T}$, *subduziert von* $D^{(*\boldsymbol{k})(m)}$ *aus* $\mathfrak{G}$ 49

Kinetic energy of lattice, classical, *kinetische Gitterenergie, klassische* 116
Kronecker powers see Direct product, *Kronecker-Potenzen siehe Direktes Produkt*
— —, cube, *kubische* 90
— —, square, *quadratische* 89
— —, symmetrized, *symmetrische* 89
Kronecker product see Direct product, *Kronecker-Produkt siehe Direktes Produkt*

Lagrangian, classical lattice, *Lagrange-Operator, klassische Gitter* 116
—, function of complex normal coordinates, *Funktion komplexer Normalkoordinaten* 137
Lattice eigenfunction, transformation of, *Gitter-Eigenfunktion, Transformation von* 269
— hypothesis, *Gitter-Hypothese* 9

Lemma of necessary degeneracy and eigenvectors, *Lemma über notwendige Entartung und Eigenvektoren* 132
Little group of **k**, $\Pi(\mathbf{k})$, definition, "*Little group*" *von* **k**, $\Pi(\mathbf{k})$, *Definition* 63
— — — —, —, irreducible representation, allowable, *irreduzible Darstellung erlaubte* 64
— — — —, —, irreducible representation, non-allowable, *irreduzible Darstellung, nicht erlaubte* 65
Local modes, *lokalisierte Schwingungen* 440

Macroscopic electric field and Lyddane-Sachs-Teller, *Makroskopisches, elektrisches Feld und Lyddane-Sachs-Teller* 306
— — — and symmetry lowering, *und Symmetrie-Erniedrigung* 306
Maschke's theorem, statement of, *Maschke's Theorem, Formulierung von* 30
Mechanical representation as a direct product, *mechanische Darstellung als direktes Produkt* 206
— — or "total" representation, *oder „volle" Darstellung* 204
— —, trace of, *Spur von* 206
Morphic effects, definition of, *morphische Effekte, Definition von* 456
— —, — —, symmetry effect, *Symmetrie-Effekt* 458
Multiplication (composition) of operators, *Multiplikation (Zusammensetzung) von Operatoren* 10, 15
Mutual exclusion rule in crystals with inversion, *Regel gegenseitiger Ausschließung in Kristallen mit Inversion* 298
— — —, broken at resonance, *verletzt in der Resonanz* 340
— — —, falsity of converse, *Verkehrtheit der Umkehrung* 298

Necessary degeneracy and space-time group $\mathcal{G}$, *notwendige Entartung und Raum-Zeit-Gruppe* $\mathcal{G}$ 166
— — of eigenvectors, *von Eigenvektoren* 132, 152
Normal coordinates, real, *Normalkoordinaten, reelle* 129
— —, — and mass-weighted displacements, *und Massen-gewichtete Auslenkungen* 130
— —, — and quantization, *und Quantisierung* 259
Normal modes, symmetry in a crystal, *Normalschwingungen, Symmetrie in einem Kristall* 202

Orthogonal transformation, *orthogonale Transformation* 8

Perturbed crystal, site group, *gestörter Kristall, Gruppe eines Gitterpunktes* 437
— —, symmetry group, *Symmetrie-Gruppe* 437
Physically irreducible representation, defined, *physikalisch irreduzible Darstellung, Definition* 132
— — —, and space-time group $\mathcal{G}$, *und Raum-Zeit-Gruppe* $\mathcal{G}$ 167
Plane waves and Fourier transformation, *ebene Wellen und Fourier-Transformation* 136
— —, orthonormality and completeness for, *Orthonormalität und Vollständigkeit für* 136
Point group $\mathfrak{P}$, definition, *Punktgruppe* $\mathfrak{P}$, *Definition* 13
— — —, subgroup of $\mathfrak{G}$, *Untergruppe von* $\mathfrak{G}$ 23
Polar optic mode and rule of mutual exclusion, *polare, optische Schwingung und Regel gegenseitiger Ausschließung* 307
— — —, crystal without inversion, *Kristall ohne Inversion* 308
— — —, — — —, splitting and activity, *Aufspaltung und Aktivität* 308
— — —, splitting *LO-TO*, *Aufspaltung LO-TO* 307
Polariton, operators, definition, *Polariton, Operatoren, Definition* 338
—, scattering and Raman theory, *Streuung und Raman-Theorie* 338
Polarizability operator as second rank tensor field, *Polarisierbarkeits-Operator als Tensorfeld 2. Stufe* 290
— — for Raman scattering, *für Raman-Streuung* 288
— —, symmetry, *Symmetrie* 289
Polarization effects, macroscopic electric field (Poulet effect), *Polarisierungseffekte, makroskopisches, elektrisches Feld (Poulet-Effekt)* 309
Polarization in infra-red absorption, *Polarisierung bei Infrarot-Absorption* 312
— — Raman scattering, *Raman-Streuung* 298, 301
— — — —, two phonon, *Zwei-Phononen* 304
Potential energy of lattice and Born-Oppenheimer theory, *potentielle Gitterenergie und Born-Oppenheimer-Theorie* 254
— — — — and general symmetry, *und allgemeine Symmetrie* 123
— — — — and translation symmetry, *und Translationssymmetrie* 118
— — — —, classical, *klassische* 115
— — — —, positive semidefinite, *positiv semidefinit* 115

Potentially real representation of $\mathfrak{G}$, test, *potentiell reelle Darstellung von* $\mathfrak{G}$, *Test* 176, 179
Projection operator P^l_{mr}, definition, *Projektionsoperator* P^l_{mr}, *Definition von* 31
— — —, for translation group $\mathfrak{T}$, *für Translationsgruppe* $\mathfrak{T}$ 46
Pseudo-real representation of $\mathfrak{G}$, test, *pseudoreelle Darstellung von* $\mathfrak{G}$, *Test* 176, 179

Raman scattering and the $A \cdot A$ term, *Raman-Streuung und der* $A \cdot A$-*Term* 293
— — anomalies (Poulet), macroscopic field, *anomale (Poulet), makroskopisches Feld* 310
— — by phonons (Hamiltonian), *durch Phononen (Hamilton-Operator)* 285
— —, diamond (polarized), *Diamant (polarisiert)* 434
— —, diamond crystal, *Diamant-Kristall* 404
— —, diamond structure and 2 phonon density of states, *Diamant-Struktur und Zwei-Phonon-Zustandsdichte* 400
— —, dipole-dipole, *Dipol-Dipol* 340
— —, Fröhlich intraband, forbidden, *Fröhlich-Intraband (verboten)* 342
— —, *Germanium* 412
— —, many body formalism, *Vielteilchen-Formulierung* 315
— —, — — —, one phonon, *Ein-Phonon* 321
— —, — — —, spectral density, *spektrale Dichte* 316
— —, — — —, two phonon and density of states, *Zwei-Phonon und Zustandsdichte* 321
— —, multipole-dipole, forbidden, *Multipol-Dipol (verboten)* 341
— —, Placzek approximation, *Placzek-Näherung* 287
— —, polarizibility operator, *Polarisierbarkeitsoperator* 289
— —, quantum theory (Exciton), *Quantentheorie (Exziton)* 333
— —, — — —, tensor 336
— —, — — (Polariton) 336
— —, — — (Bloch picture), *(Bloch-Darstellung)* 328
— —, — — — —, coupling functions, electron radiation, *Kopplungsfunktion, Elektron-Strahlung* 330
— —, — — — —, Hamiltonian, *Hamilton-Operator* 330
— —, — — — —, tensor 332
— —, resonance, *Resonanz* 339
— —, — and symmetry breaking, *und Symmetrie-Verletzung* 340
— — rocksalt (polarized), *Kochsalz (polarisiert)* 431
— —, silicon, *Silizium* 407
— —, Stokes, one phonon, *Stokes, Ein-Phonon* 292
— —, —, two phonon, *Zwei-Phonon* 292
— —, —, — — and derivative of density of states, *und Ableitung der Zustandsdichte* 293
— —, —, — — cross section, *Wirkungsquerschnitt* 292
Raman tensor and Clebsch-Gordan coefficients, *Raman-Tensor und Clebsch-Gordan-Koeffizienten* 301
— —, defined, *Definition* 299
— —, Loudon and Poulet conventions, *Loudon- und Poulet-Konvention* 301
— —, symmetry of, *Symmetrie von* 332
Ray corepresentation of $\mathfrak{P}(\boldsymbol{k})$, *Strahl-Ko-Darstellung von* $\mathfrak{P}(\boldsymbol{k})$ 197
Ray representations of $\mathfrak{P}(\boldsymbol{k})$ as multiplier, *Strahldarstellung von* $\mathfrak{P}(\boldsymbol{k})$ *als Multiplikator* 69
— — — —, associated, *assoziiert* 70
— — — —, definition 69
— — — —, gauge transformations, *Eichtransformation* 70
— — — — or projective, *oder projektive* 69
Real normal coordinates and corepresentations, *reelle Normalkoordinaten und Ko-Darstellungen* 199
— — — — crystal symmetry, *Kristallsymmetrie* 133
— — —, transformation 134
Real representation of $\mathcal{G}$, *reelle Darstellung von* $\mathcal{G}$ 168
— — — — and normal modes, *und Normalschwingungen* 167
— — of $\mathfrak{G}$, *von* $\mathfrak{G}$ 169
— — — —, Frei test (modified), *Frei-Test (modifiziert)* 177
— — — —, Herring-test 176
— — — —, test 172
Reciprocal lattice, definition, *reziprokes Gitter, Definition* 41
— —, vectors $\boldsymbol{B}_H$ (*rel* vectors), *Vektor* $\boldsymbol{B}_H$ *(rel-Vektoren)* 41
Reciprocal vectors, definition of $\boldsymbol{b}_i$, *reziproke Vektoren, Definition von* $\boldsymbol{b}_i$ 40
Reduced matrix element and Wigner-Eckart theorem, *reduziertes Matrixelement und Wigner-Eckart-Theorem* 216
— — —, defined, *Definition* 212
— — —, several, *mehrere* 214
Reduction coefficient see also Subgroup reduction coefficient, *Reduzierungskoeffizient siehe auch Untergruppe Reduzierungskoeffizient* 107

— — see Star, reduction coefficient, *siehe Stern, Reduzierungskoeffizient*
— —, basis functions for, *Basisfunktionen für* 99
— — for stars, *für den Stern* 94
— —, full-group, defined, *der vollen Gruppe, Definition* 91
— —, little group method, "*little group*"-*Methode* 112
— —, subgroup, *Untergruppe* 107
Reduction coefficients for products of representations, *Reduzierungskoeffizienten für Produkte von Darstellungen* 33
— — for space groups, *für Raumgruppen* 87
— — — — —, ordinary, *normale* 91
— — — — —, symmetrized, *symmetrisierte* 92
Reduction group, definition, *Reduzierungsgruppe, Definition* 97
Replica, *Replika* 6
Representation of group as matrix group, *Darstellung einer Gruppe als Matrixgruppe* 28
— of $\mathfrak{G}(\boldsymbol{k})$ based on complex normal coordinates, *von* $\mathfrak{G}(\boldsymbol{k})$, *basierend auf komplexen Normalkoordinaten* 155
— — — — — eigenvectors, *Eigenvektoren* 148, 150
— — —, equivalence of $D^{(\boldsymbol{k})(e)}$ and $D^{(\boldsymbol{k})(j)}$, *Äquivalenz von* $D^{(\boldsymbol{k})(e)}$ *und* $D^{(\boldsymbol{k})(j)}$ 154
Representations of group, direct products, *Darstellungen einer Gruppe, direkte Produkte* 32
— — —, — —, reduction, *Reduzierung* 32
— — —, irreducible, definition, *irreduzible, Definition* 29
— — —, unitarity, *Unitarität* 30
Rocksalt, irreducible representations, *Kochsalzstruktur, irreduzible Darstellungen* 345
—, — — Γ 347
—, — — $\star L$ 347
—, — — $\star X$ 351
—, NaCl and Raman, infra-red, *NaCl und Raman, Infrarot* 422
—, NaF 426
—, phonon, critical points, *kritische Punkte* 416
—, —, symmetry, *Symmetrie* 379
—, reduction of $\star L^{(3-)} \otimes \star L^{(3+)}$, *Reduzierung von* $\star L^{(3-)} \otimes \star L^{(3+)}$ 359
—, — — $\star X^{(4)} \otimes \star X^{(5)}$ 352
—, space group $Fm3m$, O_h^5, *Raumgruppe* $Fm3m$, O_h^5 21, 343
—, wave vector selection rules, *Auswahlregeln für Wellenvektoren* 352
Rotation, 3 dimensions, *Drehung, dreidimensional* 16
Rotational Operators $\{\varphi|0\}$, *Dreh-Operatoren* $\{\varphi|0\}$ 14
Rotations, dyadic form, non-cartesian axes, *Drehungen, dyadische Form, nicht-kartesische Achsen* 17

Scalar product in lattice dynamics, *Skalarprodukt in Gitterdynamik* 210
— — of an operator with eigenvectors, *von einem Operator mit Eigenvektoren* 213
— —, single dot and double dot, defined, „*single dot*" *und* „*double dot*", *Definition* 127, 128
Scattering matrix, composite in cubic crystal, *Streumatrix, zusammengesetzte Matrix in kubischen Kristallen* 305
Scattering tensor, Raman, defined, *Streu-Tensor, Raman, Definition* 299
Scattering tensors in cubic crystals, *Streu-Tensoren in kubischen Kristallen* 302
— — — — — and polarization, *und Polarisierung* 303
Schur lemma and corepresentations, *Schur-Lemma und Ko-Darstellungen* 183
— —, test for irreducibility, *Test für Irreduzibilität* 29
Selection rule for infra-red absorption, *Auswahlregel für Infrarot-Absorption* 278
— — — Raman scattering, *Raman-Streuung* 290
Self energy, shift, *Eigenenergie, Verschiebung* 321
Semi-direct product group, definition, *semidirekte Produktgruppe, Definition* 19
— — —, symmorphic space group as, *symmorphische Raumgruppe als* 74
Site group, definition, *Gruppe der Gitterpunkte, Definition* 19, 437
Solvability of space groups, *Lösbarkeit von Raumgruppen* 85
Space group $\mathfrak{G}$ as extension, *Raumgruppe* $\mathfrak{G}$ *als Erweiterung* 19
— — —, conjugate subgroup of, *konjugierte Untergruppe von* 24
— — —, Definition 7, 16
— — —, irreducible representation $D^{(\star\boldsymbol{k})(m)}$, *irreduzible Darstellung* $D^{(\star\boldsymbol{k})(m)}$ 48
— — —, — — —, basis functions, *Basisfunktion* 49
— — —, — — —, dimension 48
— — —, non symmorphic, *nicht symmorphisch* 22
— — —, order, *Ordnung* 18
— — —, symmorphic, *symmorphisch* 21
Space-time group, $\mathcal{G}$ defined, *Raum-Zeit-Gruppe, Definition von* $\mathcal{G}$ 160, 165

Space-time point group $\mathfrak{P}(\boldsymbol{k})$, defined, *Raum-Zeit-Punktgruppe* $\mathfrak{P}(\boldsymbol{k})$, *Definition* 196
— — — —, ray corepresentation of, *Strahl-Ko-Darstellung von* 199
Star, direct product of, *Stern, direktes Produkt von* 93
—, general, *allgemein* 61
—, reduction coefficients, ordinary, *Reduzierungskoeffizienten, normale* 91
—, — —, symmetrized, *symmetrisierte* 92
Star, special, *Stern, speziell* 62
Subduces representation and compatibility, *subduzierte Darstellung und Kompatibilität* 86
— —, definition 49
— — of $\mathfrak{G}$ onto $\mathfrak{T}$, *von* $\mathfrak{G}$ *auf* $\mathfrak{T}$ 52
Subgroup reduction coefficient, defined, *Untergruppen-Reduzierungskoeffizient, Definition* 107
— — —, compared to full group, *verglichen mit dem der vollen Gruppe* 109, 111
Symmetric polynomials and symmetrized powers, *symmetrische Polynome und symmetrische Potenzen* 268
— —, elementary, *elementar* 267
Symmetrized Kronecker product and harmonic lattice eigenfunction symmetry, *symmetrisiertes Kronecker-Produkt und Symmetrie harmonischer Gitterfunktionen* 264
— — —, character of, *Charakter von* 269
— — — of hermite polynomials, *von hermiteschen Polynomen* 264
Symmetry, translational $\{\boldsymbol{\varepsilon}|\boldsymbol{t}\}$, *Symmetrie, Translations-* $\{\boldsymbol{\varepsilon}|\boldsymbol{t}\}$ 7
Symmetry breaking, external stresses, *Symmetrie-Verletzung, externe Verzerrung* 455
Symmetry operation, definition, glide plane, *Symmetrieoperation, Definition, Gleitebene* 7
— —, —, screw axis, *Schraubenachse* 7
— — $\{\boldsymbol{\varphi}|\boldsymbol{t}\}$ multiplication, $\{\boldsymbol{\varphi}|\boldsymbol{t}\}$, *Multiplikation* 16
— —, symmetry transformation, "imagined", *Symmetrie-Transformation, "annehmbare"* 6
— —, — —, "real", *„reelle"* 6
Symmetry operator on functions, *Symmetrieoperator auf Funktionen* 26
— — — —, group of, *Gruppe von* 27
— — — —, idempotent 31

Tensor calculus and lattice dynamics, *Tensorrechnung und Gitterdynamik* 210
Time reversal, test in diamond $\star X$, $\star W$, *Zeitumkehr, Test am Diamant* $\star X$, $\star W$ 372
— —, — — rocksalt, *Kochsalz* 373
Time reversal operator K, applied to eigenvector, *Zeitumkehr-Operator* K, *angewandt auf Eigenvektor* 161
— — — —, defined, *Definition* 160
Total representation as a direct product, *vollständige Darstellung als ein direktes Produkt* 206
— —, defined, *Definition* 204
— —, trace of, *Spur* 206
Transformation operators for coordinates, *Transformations-Operator für Koordinaten* 120
— — for displacements, *für Auslenkungen* 119, 121
— — of complex conjugation, *der komplexen Konjugation* 122
Translation group $\mathfrak{T}$ as direct product, *Translationsgruppe* $\mathfrak{T}$ *als direktes Produkt* 11
— — —, irreducible representation, *irreduzible Darstellung* 40, 42
— — —, normal subgroup of $\mathfrak{G}$, *normale Untergruppe von* $\mathfrak{G}$ 17, 23
Two-phonon bound states (in diamond) and Raman scattering, *gebundene Zwei-Phonon-Zustände (in Diamant) und Raman-Streuung* 310, 433
— — — — —, symmetry of, *Symmetrie von* 311

Unitary group $SU_{(sl_j)}$ and lattice eigenfunction, *unitäre Gruppe* $SU_{(sl_j)}$ *und Gitter-Eigenfunktionen* 270

Wave vector, canonical, $\boldsymbol{k}$ of $\star\boldsymbol{k}$, *Wellenvektor, kanonischer,* $\boldsymbol{k}$ *zu* $\star\boldsymbol{k}$ 59
— —, definition of $\boldsymbol{k}$, *Definition von* $\boldsymbol{k}$ 42
— —, equivalent, *äquivalent* 43
— —, inequivalent, *nicht-äquivalent* 54
— —, reduction coefficients, *Reduzierungskoeffizienten* 93
— —, selection rules, *Auswahlregeln* 93
— —, space group of $\boldsymbol{k}$, $\mathfrak{G}(\boldsymbol{k})$, *Raumgruppe von* $\boldsymbol{k}$, $\mathfrak{G}(\boldsymbol{k})$ 53
— —, star of, definition, *Stern von, Definition* 61
Wave vectors, class I, II, III, defined, *Wellenvektoren, Klasse I, II, III, Definition* 177
Wigner-Eckart formula, analogy for lattice dynamics, *Wigner-Eckart-Formeln, Analogie für Gitterdynamik* 212
— — and antiunitary elements, *und anti-unitäre Elemente* 216
— — — — —, case A wave vectors, *und Wellenvektoren Fall A* 217
— — — — —, case B wave vectors, *und Wellenvektoren Fall B* 218
— — — — —, case C wave vectors, *und Wellenvektoren Fall C* 219

K. R. Lang

Astrophysical Formulae

A Compendium for the Physicist and Astrophysicist

46 figures
Approx. 800 pages
1974
Cloth DM 192,–
US $78.40
ISBN 3-540-06605-5

Prices are subject to change without notice

This volume is a reference source for the fundamental formulae of astrophysics. Over 2,100 formulae are included and the original papers for the formulae are referenced together with papers on modern applications in a bibliography of over 1,900 entries. Included in the 69 tables are positions and basic physical data on astronomical objects such as planets, the sun, planetary nebulae, ionized (H II) regions, magnetic stars, white dwarfs, galactic clusters, globular clusters, galaxies, clusters of galaxies, pulsars, radio galaxies, and quasi-stellar objects. Also included are supplementary data such as Gaunt factors, interstellar molecular lines, intense solar emission and absorption lines, transition probabilities and collision strengths for forbidden transitions, mass excess and solar system abundance of the elements, reaction rate constants, properties of stars, etc. The 46 figures include: the frequency spectrum of black-body, synchrotron, free-free, free-bound, and Cerenkov radiation; line intensity ratios as a function of electron density and electron temperature; Grotrian diagrams for the abundant elements, neutrino emission regimes; element abundances synthesizes in a "big bang"; spectrum of cosmic ray electrons, X-rays and γ rays; the redshift-magnitude relation; galactic rotation curves; luminosity functions of stars, galaxies, quasars, and radio galaxies; scale factors for the universe, and the spectrum of the isotropic background radiation.

Contents: Continuum Radiation. – Monochromatic (Line) Radiation. – Gas Processes. – Nuclear Astrophysics and High Energy Particles. – Astrometry and Cosmology.

Springer-Verlag
Berlin · Heidelberg · New York

Topics in Applied Physics

Founded by H. K. V. Lotsch

Springer-Verlag has recently introduced this new book series, devoted to research achievements of current interest. Each volume will deal with a different topic under the editorship of a recognized authority in the field. It will cover application-oriented aspects of the topic under consideration, the basic physical principles being summarized in a comprehensive introduction. The contributors to each volume are internationally known experts. The publication periods are comparable with those of scientific journals to keep pace with the rapidly accumulating results.

Volume 1:

Dye Lasers

Editor: F. P. Schäfer
With contributions by
K. H. Drexhage, T. W. Hänsch,
E. P. Ippen, F. P. Schäfer,
C. V. Shank, B. B. Snavely
114 figures. XI, 285 pages. 1973
Cloth DM 65,–; US $26.60
ISBN 3-540-06438-9

Contents
F. P. Schäfer: Principles of Dye Laser Operation.
B. B. Snavely: Continuous-Wave Dye Lasers.
C. V. Shank, E. P. Ippen: Mode-Locking of Dye Lasers.
K. H. Drexhage: Structure and Properties of Laser Dyes.
T. W. Hänsch: Applications of Dye Lasers.

Volume 3 will be about

Raman Scattering in Solids

Editor: M. Cardona
In preparation

Contents
A. Pniczuk: Phenomenology.
L. Falicov: Resonant Raman Scattering: Deformation Potential.
R. M. Martin: Resonant Raman Scattering: Fröhlich Interaction.
M. V. Klein: Electronic Raman Scattering.
A. S. Pine: Brillouin Scattering in Semiconductors.
M. H. Brodsky: Scattering in Disordered Systems.
Y. R. Shen: Stimulated Raman Scattering.

Prices are subject to change without notice

Springer-Verlag
Berlin · Heidelberg · New York
München Johannesburg London Madrid New Delhi Paris
Rio de Janeiro Sydney Tokyo Utrecht Wien